Stripline-Like Transmission Lines for **Microwave Integrated Circuits**

Bharathi Bhat
Professor
Centre for Applied Research Electronics
Indian Institute of Technology
New Delhi

Shiban K. Koul
Associate Professor
Centre for Applied Research in Electronics
Indian Institute of Technology
New Delhi

PUBLISHING FOR ONE WORLD

NEW AGE INTERNATIONAL (P) LIMITED, PUBLISHERS
(formerly Wiley Eastern Ltd.)
New Delhi · Bangalore · Chennai · Cochin · Guwahati · Hyderabad
Jalandhar · Kolkata · Lucknow · Mumbai · Ranchi
Visit us at www.newagepublishers.com

Published by New Age International (P) Ltd., Publishers
First Edition: 2007

ISBN : 978-81-224-2124-8

Rs. 595.00

C-07-04-1428

1 2 3 4 5 6 7 8

Printed in India at Sarasgraphics, Rai.

PUBLISHING FOR ONE WORLD

NEW AGE INTERNATIONAL (P) LIMITED, PUBLISHERS
(formerly Wiley Eastern Limited)
4835/24, Ansari Road, Daryaganj, New Delhi - 110002
Visit us at **www.newagepublishers.com**

To

Jaya and (Late) Gopal Bhat

and

Jaykishori, Bansi Lal and Veena Koul

Preface

The use of stripline and microstripline as viable transmission media for planar microwave circuitry was conceived as early as in 1950s. However, their implementation for building practical components could not gain momentum until around 1965, mainly due to the non-availability of low-loss dielectric substrates and microwave semiconductor devices. With technological advances in these areas since then, there has been a tremendous growth of activity in planar transmission line techniques for microwave integrated circuits (MICs). Over the years, several variants of the above basic lines such as the suspended stripline, microstrip-like stripline and sandwiched stripline have emerged, each offering certain desirable characteristics. The introduction of anisotropic dielectric substrates, in recent years, has added further flexibility in circuit design. Considering the advantages and importance of a number of modified versions in practical MICs, a real need was felt to consolidate the information available, and bring out a unified treatment at one place. This book is designed to fulfill this need.

The book begins with an introductory chapter summarizing the basic elements of MICs and their fabrication technologies. Chapter 2 describes the various quasi-static analytical methods used for solving planar transmission lines. The most important of these from the point of view of generality is the 'variational method in combination with the transverse transmission line technique'. A unified analytical approach based on this technique is developed for a general multilayer structure using isotropic/anisotropic substrates. This method is systematically applied to a variety of single- and coupled-conductor stripline-like transmission lines in Chapters 3 and 4. The unified formulas summarized in these chapters can be readily programmed to generate basic design data on effective dielectric constant, guide wavelength, and characteristic impedance of a class of such structures.

Chapters 5 to 8 present the propagation characteristics of a wide variety of configurations. These are classified under single conductor, edge-coupled, broadside-coupled and broadside edge-coupled structures. The characteristic features of some special structures are also included in Chapter 8. Extensive tables of data on various structures using commercially available isotropic as well as anisotropic substrates are appended towards the end of Chapters 5 to 8.

While the unified analytical formulas derived under quasi-static approximation and the tables of data are adequate for most practical applications, readers would be interested to know the frequency range upto which such formulas could be utilized. Chapter 9 is therefore exclusively devoted to the hybrid-mode analysis of stripline-like transmission lines and their frequency dependent characteristics. Suspended striplines which are popularly used even beyond the microwave frequency range are specially focussed.

Chapter 10 deals with the analyses and characterization of discontinuities. The unified analytical technique developed in Chapter 2 is extended to the analysis of rectangular patches, open circuit end-effects, and gap capacitances of strip conductors in stripline-like transmission lines. In Chapter 11 are included the analysis of losses and details of power handling capability. The final chapter is on circuit applications. It includes the salient constructional features, transitions to coaxial lines and waveguides, and design aspects of a number of MIC components. The relative advantages of different types of stripline-like configurations in component design are highlighted.

Bulk of the material presented in Chapters 2 to 10 is the outcome of the research carried out by the authors during the last eight years at the Centre for Applied Research in Electronics, I.I.T., Delhi. With Chapters 1, 11, 12, the book provides an exhaustive coverage of the analysis, design, and applications of a class of stripline-like transmission lines for microwave integrated circuits. Starting from the fundamental principles, the book builds up the analytical techniques towards the solution of various structures in a lucid and systematic manner so as to be of direct utility for class room teaching. Selected portions of the book already form part of a master's level course at I.I.T. Delhi in the area of MICs. To an intelligent research student, this is a self teaching text for acquiring a basic foundation of theoretical and practical techniques in the area. The book is also intended to be a reference data handbook for R & D engineers involved in MIC design. Engineers who are not interested in the details of the analysis can skip most of the sections in Chapters 2 to 4, 9, and 10 without loosing continuity and use the rest of the book. Unified analytical formulas, graphical illustrations, and extensive tables of data are included (Chapters 5 to 8) to serve as ready design aid.

The book contains a large number of figures and design graphs all of which were drawn by Mr. R.P. Kapoor. The authors wish to thank him for his patience and precision drafting work.

BHARATHI BHAT
SHIBAN K KOUL

All the typographical errors found in the first edition have been rectified in this reprint volume.

Contents

CHAPTER 1

Elements of Microwave Integrated Circuits

1.1 Evolution of Microwave Integrated Circuits

Low frequency circuits are generally built up from lumped elements, active devices in the form of vacuum tubes or transistors, and conducting connections between the various passive elements and active devices. Typical lumped elements include capacitors, inductors, and resistors whose values are assumed to be concentrated in the respective elements. This is a valid assumption at low frequencies at which the dimensions of these lumped elements are small in comparison with the wavelength of operation. However, at higher frequencies, where the wavelengths are comparable to the dimensions of these components, stray elements begin to appear. Typical strays include the inductance between the plates of a capacitor, capacitance between the turns of an inductor coil, the inductance of the interconnecting conducting wires or strips, and the capacitance between them. Yet another undesirable effect at higher frequencies is the emergence of radiation from the circuit. Radiation occurs when the transit time of the signal through the circuit becomes comparable to or greater than the time period of the RF signal. These effects, therefore, become significant at microwave frequencies.

The above limitations of the conventional lumped circuits prompted the development of a new approach, called the distributed circuit approach for use at microwave frequencies. Instead of lumped elements, circuits that are made up of transmission lines in which the electromagnetic fields are bound in the transverse direction came into existence in the thirties. Of the various types of transmission lines, the coaxial line and the rectangular waveguide (Fig. 1.1) became the most popular. The emphasis was shifted from the currents flowing inside the conductors to the electromagnetic waves propagating in the space inside these transmission lines. In a coaxial line, the electromagnetic fields are confined to the dielectric region between the inner and outer conductors. The lowest order mode is a transverse electromagnetic (TEM) mode, which means that the electric and magnetic field lines of force lie in planes transverse to the longitudinal direction along which the wave propagates. In this case, it is possible to define a voltage and a current uniquely for any given cross-section, and the transmission line can be characterized by capacitances and inductances distributed along the length. The circuit parameters are, therefore, referred to as distributed along the length of the line. In a rectangular waveguide, the electromagnetic fields may be visualized in terms of plane waves bouncing back and forth between the waveguide walls, giving propagation only in the longitudinal direction. The field configuration can be described in terms of transverse electric (TE) and transverse magnetic (TM) modes. Since the fields are non-TEM in nature, it is not possible to define a unique voltage and/or current. However, there exists an useful analogy between the electric and magnetic field strength of TE and TM waves, and the voltages and currents in suitably loaded transmission lines, thus enabling the guide to be represented as a distributed network.

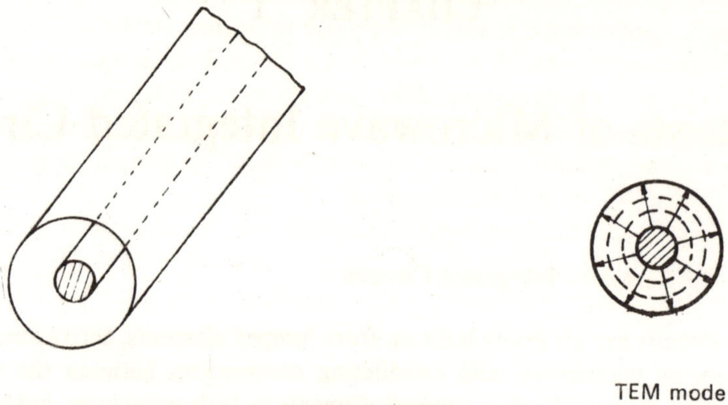

(a) Coaxial line

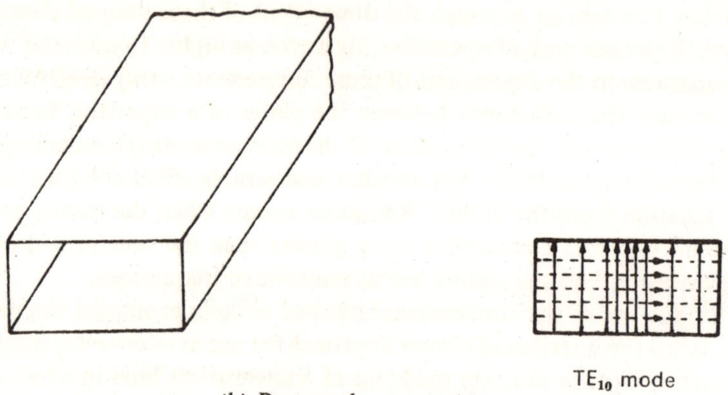

(b) Rectangular waveguide

FIG. 1.1 Coaxial line and rectangular waveguide with electromagnetic field configurations for the dominant modes. (——— E (electric) field lines, — — — H (magnetic) field lines.)

Short sections of transmission lines terminated at one end in either a short circuit or an open circuit behave as reactive elements when viewed from the second terminal. The reactive element may be inductive or capacitive, depending on the wavelength of operation. Circuit elements also take the form of obstacles placed within the transmission line. The resonant combination of a coil and a condenser at lower frequencies gets replaced by a resonant cavity in the microwave region.

Microwave circuits and technology utilizing coaxial lines and waveguides went through a silent but rapid development during the Second World War. Substantial progress was also achieved in the development of techniques for microwave power generation and amplification leading to the emergence of klystrons, magnetrons, travelling wave tubes, etc. It was in the late 1940s and early 1950s, when the microwave technology was beginning to emerge from the cocoon of war time secrecy that its applications started encompassing the non-military fields as well. Since then, the entire field of microwaves has undergone tremendous revolution both in terms of techniques and technologies. The 1950s saw the entry of a new era of microwave semiconductor device technology. Solid state devices were introduced promising an encouraging future for microwave applications. It was in the years following 1965 that major breakthrough was achieved leading to a rapid progress in the microwave semiconductor device technology. A number of devices were developed for performing a variety of microwave functions such as power generation, mixing, switching, amplification, etc. These include the IMPATT diode, Gunn diode, Schottky barrier diode, PIN diode, gallium arsenide field effect transistor (GaAs FET), and silicon bipolar transistor.

The availability of small size semiconductor devices made it necessary to look for transmission media compatible with these devices. The answer was realized in the form of planar transmission lines. The first planar transmission line, called the 'strip transmission line', was proposed by Barrett and Barnes [1] as early as in 1951. The structure consists of a thin strip conductor sandwiched between two dielectric sheets with conductive plating on the two outer sides (Fig. 1.2a). In 1952, another basic transmission line, called the 'microstrip line', was conceived [2]. It consists of a dielectric substrate with a strip conductor on one side and a complete conducting layer on the opposite side (Fig. 1.2b). Unlike the stripline, the microstrip is basically an open structure, and requires a high dielectric constant substrate to confine the electromagnetic fields near the strip conductor. Although various aspects of these basic transmission lines such as the impedance, radiation, discontinuity effects, etc., were studied in the 1950s [3], microwave components using these circuit forms did not become popular, mainly because of the non-availability of low loss dielectric materials with reproducible characteristics at microwave frequencies. With the advancement in material technology and also the fabrication processes, namely, the deposition of metallic films using the thin- and thick-film technologies, the interest in planar transmission lines was revived during the 1960s. The integration of miniature solid state microwave devices with the planar transmission lines resulted in the beginning of the evolution of a new circuit form called the 'microwave integrated circuit' (MIC) in the late 1960s [4, 5]. In the year 1968, a third type of basic transmission line named the 'slotline' was proposed by Cohn [6]. The slotline consists of a slot etched from the conducting layer on one side of the dielectric substrate, the other side being bare (Fig. 1.2c). Subsequently, a variety of other planar transmission lines, which form variants of the stripline, microstrip line, and slotline were evolved [7–18], ever widening the scope of applicability of MICs in practical systems. Figure 1.3 shows some of the commonly used configurations such as the suspended stripline, suspended microstrip, inverted microstrip, coplanar waveguide, and coplanar strips.

Developments in vacuum deposition techniques and photolithography made it possible to realize lumped-circuit elements also in planar version [19–24]. Some typical forms such as the single turn inductor, spiral inductor, and interdigital capacitor are shown in Fig. 1.4a-c. The technology includes the deposition of thin films of metals such as gold, copper, and nichrome on a dielectric substrate. The same technology also facilitates the fabrication of chip capacitors (Fig. 1.4d), and resistors, which can be connected to the circuit by means of wire bonding. These lumped elements offer significant size reduction as compared with the distributed circuit elements at lower microwave frequencies.

The process of fabricating MICs by soldering or bonding semiconductor devices on the passive circuitry made of planar transmission lines and lumped elements has been termed

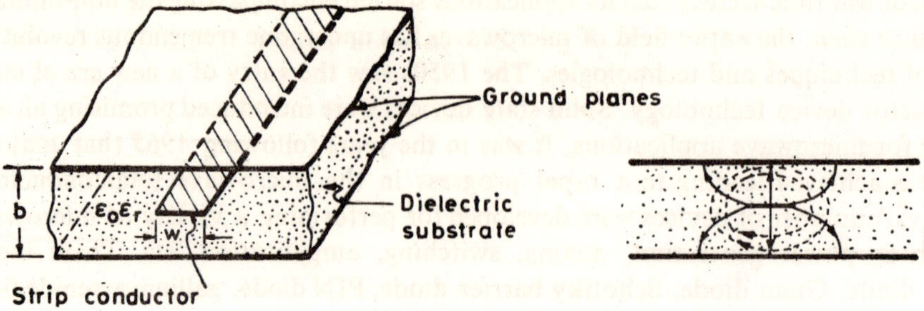

(a) Stripline (TEM mode)

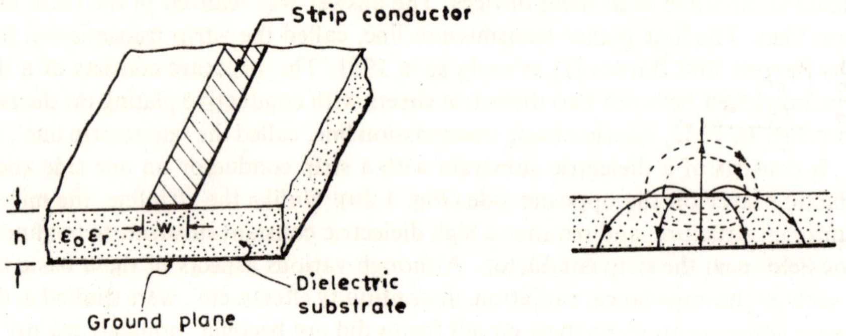

(b) Microstrip line (quasi-TEM mode)

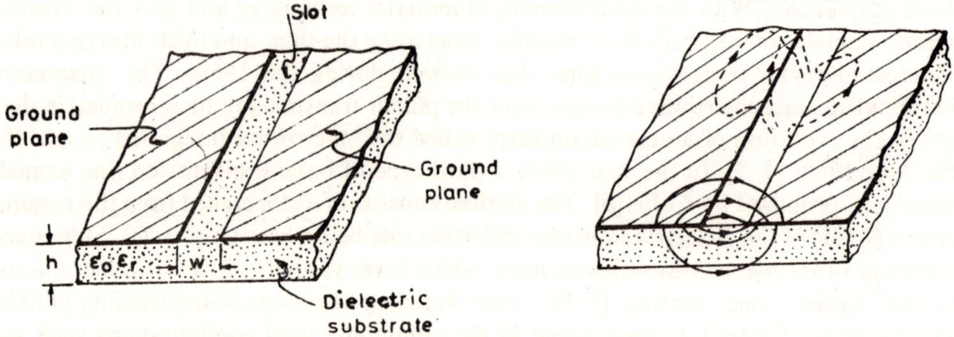

(c) Slotline (non-TEM mode)

FIG. 1.2 Basic planar transmission lines. (——— *E* field lines, — — — *H* field lines.)

'hybrid technology', and the circuits realized are called 'hybrid MICs'. Since 1970, there has been a phenomenal progress in the development of planar transmission line techniques [9–18, 23–30], solid state devices [31–40], and also the hybrid technology [41–48], leading to a variety of practical hybrid MIC components [49–54]. The advent of this technology has resulted in considerable reduction in size and weight, ease of mass production with improved reproducibility, and potential low cost of functional circuits. The technology offers the advantage of combining multicircuit functions without any interconnecting wires, thus permitting compact integrated modules with highly reliable performance.

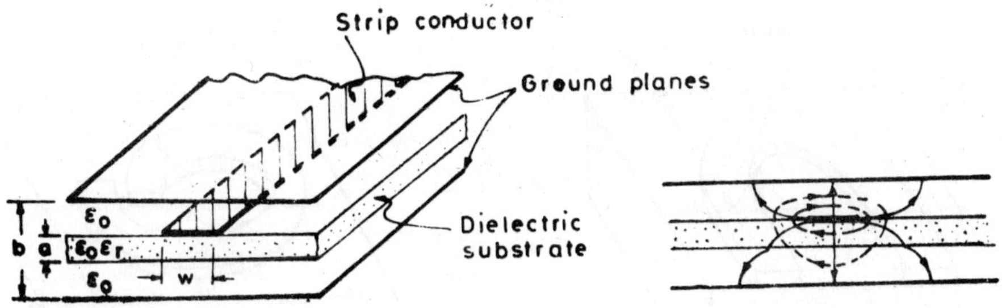

(a) Suspended stripline

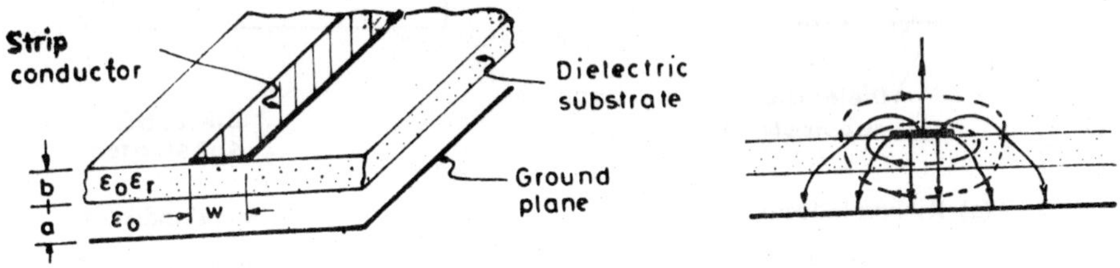

(b) Suspended microstrip

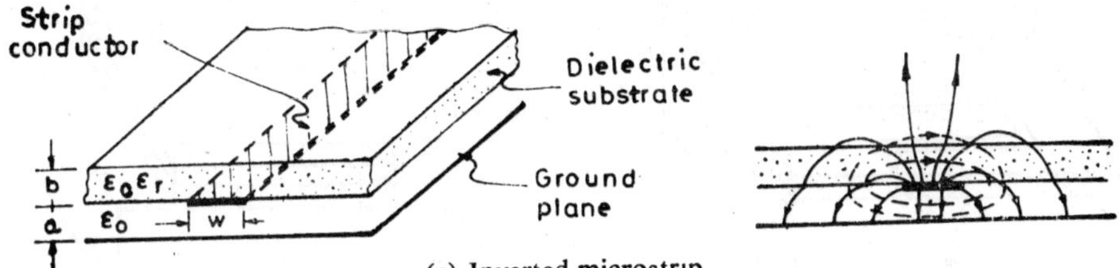

(c) Inverted microstrip

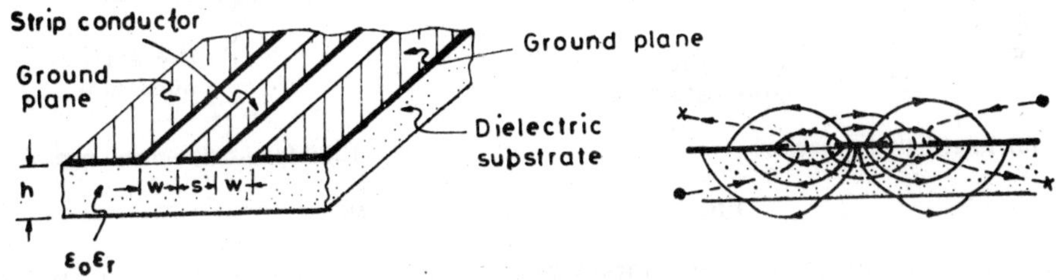

(d) Coplanar waveguide

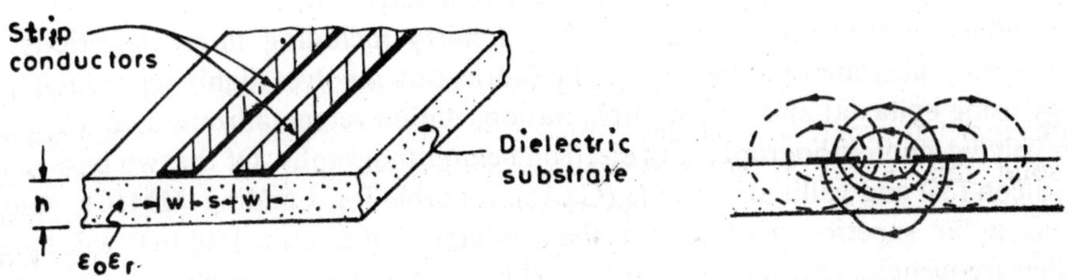

(e) Coplanar strips

FIG. 1.3 Some variants of stripline, microstrip line, and slotline. (—— E field lines, - - - H field lines.)

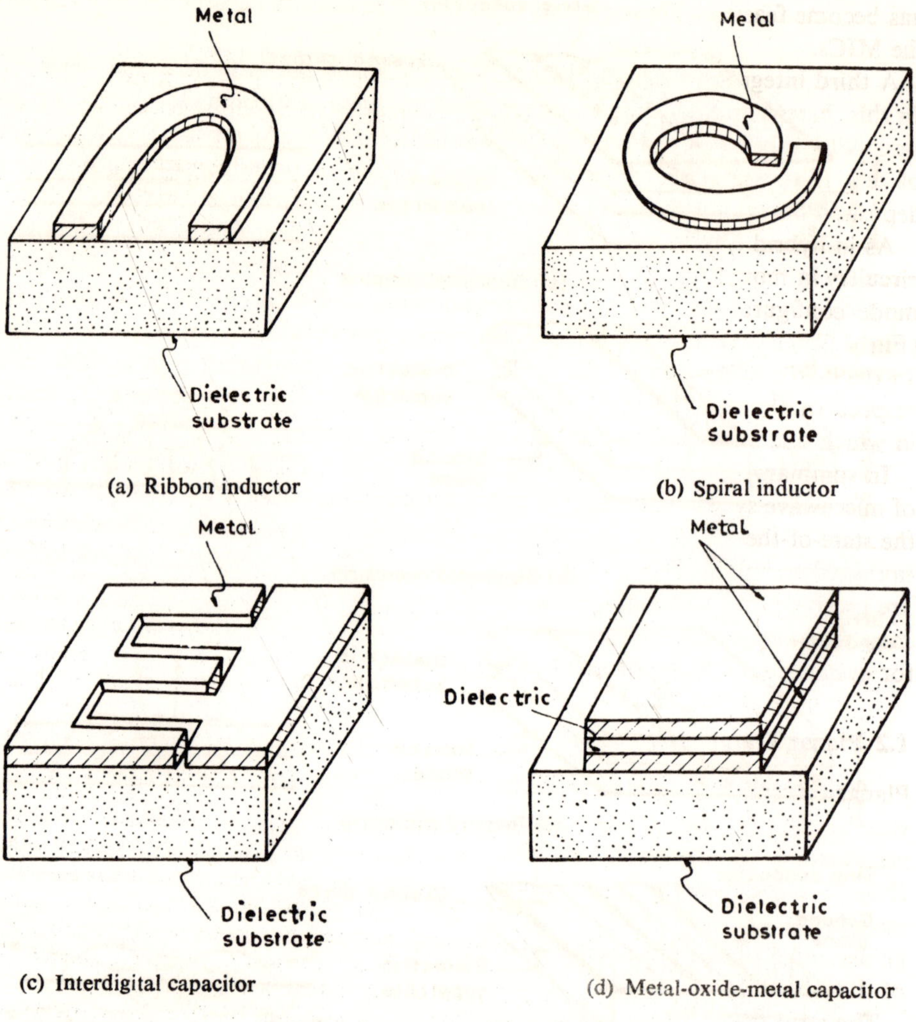

(a) Ribbon inductor (b) Spiral inductor

(c) Interdigital capacitor (d) Metal-oxide-metal capacitor

FIG. 1.4 Lumped elements for MICs.

A new generation of microwave components having a much higher degree of miniaturization and integration is presently being conceived through 'monolithic technology' [55–58]. In the microwave monolithic integrated circuit (MMIC), active devices are grown *in-situ* on or in a semiconducting substrate. Planar passive circuitry including interconnections is then deposited on the substrate or grown in it. The fabrication involves highly reproducible device technology using epitaxial and/or ion implantation, multilevel metallization, and a composite process involving photolithography and electron beam lithography. Of the two base materials, namely, silicon (Si) and gallium arsenide (GaAs), suitable for building MMICs, GaAs has emerged as a far superior material for the production of circuits [59] that will operate at much higher frequencies and speeds than Si is able to. With the development of semi-insulating GaAs as a low-loss substrate and the Schottky barrier field-effect transistor, monolithic integration of active devices, bias networks, and passive microwave circuitry on a single chip has become feasible. These circuits offer broad bandwidth capabilities many times wider than the MICs.

A third integration scheme suitable for MICs is in the form of a quasi-monolithic circuit. In this, hetero-epitaxy techniques are used to grow semiconductor layers on passive high resistivity substrates. Silicon-on-sapphire (SOS), and GaAs on spinel circuits form examples of this. In these, active devices are grown in the semiconductor, and passive circuitry is deposited on the substrates in areas where the semiconductor is removed.

As in hybrid MICs, a basic consideration in the implementation of a particular microwave circuitry in monolithic or quasi-monolithic form involves the choice of suitable transmission-mode configuration. The various transmission media for MMICs include the microstrip (Fig. 1.2b), slotline (Fig. 1.2c), coplanar waveguide (Fig. 1.3d) and coplanar strips (Fig. 1.3e). In the coplanar waveguide, the center conductor is maintained at a positive potential with respect to the two adjacent ground planes. The coplanar strip structure forms its 'inverse', in which one of the two conductors is a ground plane.

In summary, the last two decades have seen a revolution in techniques and technologies of microwave systems through the application of microwave integrated circuits. It is presently the state-of-the-art to use hybrid MICs for essentially all advanced microwave radar, communication, navigation, and sensing systems. With the emerging GaAs technology, MMICs are receiving increasing attention for the next generation of microwave components. The monolithic approach offers promising future also for millimeter wave integrated circuits and for systems desiring extremely wide band capabilities.

1.2 Planar Transmission Lines

Planar transmission lines form the basic transmission media for microwave integrated circuits, whether the circuits are fabricated in hybrid or monolithic form. There are three basic versions of the planar transmission line—the stripline, the microstrip line, and the slotline (Fig. 1.2). Many other forms which are variants of these basic lines are also available. These include the suspended stripline, suspended microstrip, inverted microstrip, coplanar waveguide, and coplanar strips (Fig. 1.3). In this section, the fundamental electrical characteristics of these commonly used planar structures for hybrid and monolithic MICs are outlined.

The strip transmission line consists of a flat strip conductor situated symmetrically between two ground planes with the intervening space filled with a homogeneous dielectric. The dominant mode of propagation is the transverse electromagnetic (TEM) mode, having electric and magnetic field components lying entirely in the transverse plane. Figure 1.2a shows the cross-section of a symmetric stripline along with the field configuration. The guide wavelength λ for the TEM mode is given by $\lambda^a/\sqrt{\epsilon_r}$, where λ^a is the free space wavelength and ϵ_r is the relative permittivity of the dielectric. In order for the dominant mode to exist at the exclusion of all the higher order modes, the ground plane spacing must be less than $\lambda/2$. Since every point on the surface of the strip conductor is at the same potential with respect to the ground plane, and the current flows only in the longitudinal direction, the ratio of the voltage to the current in the line, uniquely gives the characteristic impedance Z of the line. It is given by $Z = Z^a/\sqrt{\epsilon_r}$, where Z^a is the characteristic impedance of the line with air as the dielectric medium ($\epsilon_r = 1$). Z^a is determined by the geometry of the line, in particular, by the ratio w of the strip to the spacing b between the ground planes. Practical circuits make use of dielectric substrates with ϵ_r greater than 1. Hence, by choosing an appropriate dielectric constant, the circuit dimensions can be kept to conveniently small values.

Unlike the stripline, the microstrip line is an inhomogeneous transmission line structure. It consists of a strip conductor on a flat dielectric substrate, the reverse side of which is metallized to provide a ground plane. The cross-sectional view of the microstrip line along with the electric and magnetic field lines is shown in Fig. 1.2b. In this structure, a pure TEM mode cannot exist because it is not possible to satisfy the boundary conditions for this mode at the surface of the dielectric. However, at low frequencies, the mode of propagation closely resembles the TEM mode, and hence, is termed the quasi-TEM mode. If the dielectric constant ϵ_r of the substrate is sufficiently high, then, most of the electromagnetic energy gets confined to the dielectric in the vicinity of the strip conductor, and radiation is minimized. The wavelength λ is related to the free space wavelength λ^a by the expression $\lambda = \lambda^a/\sqrt{\epsilon_f}$. Here, ϵ_f is the effective dielectric constant of the medium, and its value lies in the range $1 < \epsilon_f < \epsilon_r$. The characteristic impedance Z of the microstrip line is given by $Z = Z^a/\sqrt{\epsilon_f}$, where Z^a is the characteristic impedance of the microstrip line with the dielectric replaced by air.

The slotline consists of a slot etched from the top conducting layer of a dielectric substrate, the bottom side being bare (Fig. 1.2c). The electric field is essentially transverse to the slot because of the voltage difference across the gap. The magnetic field in a slotline has both a longitudinal and a transverse component. This implies that a rotating magnetic field vector is present at some points, which can be used for coupling to the gyromagnetic effects in ferrites. The dominant mode of propagation is essentially non-TEM, and hence, the characteristic impedance and guide wavelength are dependent on frequency, although at a slow rate. This feature is in contrast with the TEM nature of the stripline whose propagation parameters are independent of frequency, and also with the quasi-TEM nature of the microstrip line whose propagation parameters are, to a first-order, independent of frequency. The slotline differs also from the waveguide in that it has no cut-off frequency.

Among the three basic transmission lines described above, the homogeneous stripline is ideal for passive integrated circuits such as filters, couplers, and other hybrids. For active components such as oscillators and amplifiers, only packaged and sealed semiconductor devices can be mounted by making through holes in the substrate. The integration of semiconductor devices into the stripline package is, therefore, not as convenient as with the microstrip. The microstrip, being an open structure, is easily amenable for series mounting of discrete devices. Although the mode of propagation is quasi-TEM, that is, not strictly TEM as in the case of homogeneous stripline, the dispersion associated with it is negligible for most practical applications. The slotline, on the other hand, offers the advantages of an open structure, especially for shunt mounting of discrete chips, and the medium is basically dispersive. The microstrip, being more versatile, is the widely accepted transmission line for most applications at microwave frequencies. However, as the frequency is increased above the microwave range, it becomes necessary to use progressively thinner substrates in order to ensure that the propagation is confined to the quasi-TEM mode without the presence of any higher order mode. The width of the strip conductor also becomes progressively narrower, posing fabricational problems. Further, the losses in microstrip increase considerably. In order to circumvent these problems, some modified versions of the stripline and the microstrip line have been evolved, of which, the suspended stripline, the suspended microstrip, and the inverted microstrip are the more common ones. In a suspended stripline, a thin dielectric substrate containing the strip conductor is suspended midway between the top and bottom ground planes as shown in Fig. 1.3a. In a practical version, the structure is provided with side walls with grooves to support the substrate. The suspended microstrip (Fig. 1.3b) and the inverted microstrip (Fig. 1.3c) incorporate an air gap between the dielectric substrate carrying the strip conductor and the ground plane. In the suspended microstrip, the strip conductor is on the upper surface of the dielectric substrate, and in the inverted microstrip, it is situated on the lower surface of the substrate facing the ground plane. The presence of the air gap in these structures reduces the concentration of electromagnetic energy near the ground plane. Hence, the conductor loss, which is the dominant contributing factor to the total loss in the structure, is far less compared with that in the microstrip. Another effect of the air gap is to reduce the effective dielectric constant of the medium which results in an increased width of the strip conductor. The dimensional tolerances in these structures can therefore be relaxed, resulting in considerable fabricational ease at higher microwave and even millimeter wave frequencies.

Coplanar lines are a class of transmission lines in which all the conductors are situated on one side of the dielectric substrate with no metallization on the reverse side. Examples are the slotline (Fig. 1.2c), coplanar waveguide (Fig. 1.3d) and coplanar strips (Fig. 1.3e). Since the electric field in a slotline is essentially across the slot, it enables shunt mounting of components and active devices. The propagating wave has elliptically polarized regions that can be utilized for the fabrication of non-reciprocal ferrite components. Slotline can be easily coupled to a microstrip line by depositing the microstrip line on the opposite side of the same substrate. Coupling between a slot and a microstrip offers interesting circuit applications such as in broadband transitions, couplers, and filters. The coplanar waveguide consists of a strip conductor separated by a slot on either side from the two adjacent ground planes. The configuration of coplanar strips is complementary to the coplanar waveguide, with one of the two strips forming a ground plane. Unlike the slotline which is non-TEM in nature, the coplanar waveguide and coplanar strips propagate quasi-TEM modes. At higher microwave frequencies, however, the contribution of the longitudinal magnetic field is sufficiently large, resulting in non-TEM mode of propagation. One distinct advantage of both these lines is that they permit simple integration of hybrid components both for series and parallel connections. In the fabrication of monolithic integrated circuits, these coplanar lines offer convenient transmission media in addition to the microstrip.

1.3 Lumped Elements for Microwave Integrated Circuits

With the planar transmission lines as the basic transmission media, a variety of microwave circuit functions can be performed by using combinations of individual components built from these transmission lines. These components are generally of the order of a wavelength or more in length, in the medium of propagation. Such circuits built from sections of transmission lines are described as 'distributed circuits'. At lower microwave frequencies, these distributed circuits consume considerable space. The use of lumped elements, on the other hand, can result in a substantial reduction in size [19–24]. Minute lumped elements in planar version can be realized by depositing high conductivity patterns on dielectric substrates following the same vacuum deposition and photoetching techniques as used in the distributed circuit fabrication. The purpose of the substrate here, however, is mainly to provide physical support and isolation between the various elements. The microwave energy within the substrate is small. Consequently, unlike in the case of distributed circuits, the electrical and mechanical properties of the dielectric substrate do not play much importance. With the advancement in technological processes, the size of the lumped elements has been reduced so as to extend the operational frequencies up to about 15 GHz. The circuit techniques employed for the fabrication of lumped elements are also applicable in the construction of monolithic and quasi-monolithic microwave integrated circuits. An integrated combination of lumped and distributed elements affords considerable flexibility in the design of circuits.

Figure 1.4 illustrates typical lumped-element configurations used in MICs. A single turn inductor consists of a strip conductor deposited on a dielectric substrate (Fig. 1.4a). By varying the width of the strip and the diameter of the loop, inductance values in the range of a few nanohenries can be obtained. Higher values can be achieved by spiralling the conductor pattern as shown in Fig. 1.4b. Figure 1.4c shows an interdigital capacitor in which the capacitance is formed by the fringing field between the interdigital gap. By increasing the number of interdigital sections, capacitance values ranging from a fraction of a picofarad to a few picofarads can be realized. Larger capacitance values can be obtained using a metal-oxide-metal (MOM) sandwich capacitor as shown in Fig. 1.4d. The oxide layer is generally silicon dioxide, which serves as a dielectric between the two metallic layers. Unlike the single layer interdigital capacitor, the fabrication of the MOM capacitor involves deposition of three layers. The values of inductances and capacitances obtainable with these lumped elements are convenient for use with practical microwave unencapsulated device chips, since they combine with parasitic reactances of the device chips to enable resonances in the microwave frequency range. Lumped resistors in the form of minute rectangular patches can be realized by evaporating nickel-chromium or by sputtering tantalum directly on the circuit pattern.

Besides the planar lumped elements described above, discrete chip resistors and multilayer ceramic capacitors in chip and beam lead form are available at lower microwave frequencies. The chip resistors are generally used for biasing purposes and line terminations. The chip capacitors offer a wide range of capacitance values starting from about 0.5 pF to 1000 pF. These are used in special hybrid circuits when large capacitance values are required.

Figure 1.5 shows some typical lumped and distributed circuits. Figure 1.5a shows a series resonant circuit, and Fig. 1.5b shows a parallel resonant circuit in lumped form. A microstrip low pass filter is shown in Fig. 1.5c which essentially consists of series inductors represented by thin lines, and shunt capacitors represented by broad rectangular patches. Each of these elements is realized from a short section of transmission line having a length smaller than quarter wavelength, and hence, the circuit may be termed semi-lumped. The microstrip band pass filter shown in Fig. 1.5d is an example of a typical distributed circuit. The circuit consists of parallel-coupled resonators, where, each resonator is a section of transmission line approximately half-wavelength long, and open circuited at either ends.

Besides consideration in terms of size, an important aspect to be taken into account while making a choice between a lumped and a distributed circuit element is the quality factor Q of the element. The Q of the distributed circuit varies from 100 to 1000 over the entire microwave band, depending on the type of transmission line used. For example, Q of the microstrip lies in the range 100 to 250, and that of the suspended stripline is nearly four times the Q of the microstrip; the homogeneous stripline falling in between these two. In the case of lumped elements, at lower microwave frequencies up to C-band, the Q of single turn- and

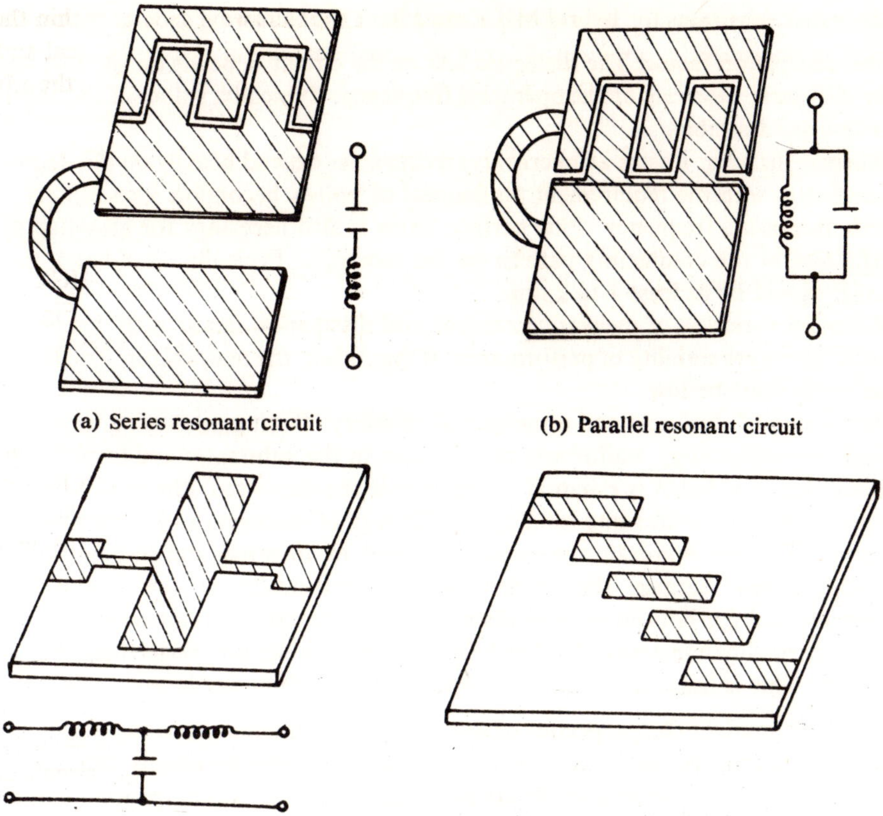

(a) Series resonant circuit (b) Parallel resonant circuit

(c) Low pass filter (d) Band pass filter

FIG. 1.5 Examples of lumped and distributed MICs.

spiral-inductors ranges from 50 to 150, while that of interdigital and MOM capacitors starts from a high value of the order of 1000 at *L*-band, with their values steadily decreasing with increasing frequency. Beyond *C*-band, the *Q* of these lumped elements reduces to values lower than those of the distributed circuits. Further, as the frequency is increased, the size of the lumped elements does not shrink drastically, while the parasitics associated with these elements begin to effect the performance. In view of these features, the use of lumped elements has been essentially confined to the lower microwave frequency range up to *C*-band, although, their performance allows integration up to about 15 GHz.

1.4 Substrates for Microwave Integrated Circuits

The first step in the design of any MIC is the choice of the substrate. Selection of the substrate material depends on three major factors—the type of fabrication technology, i.e., hybrid or monolithic, the type of transmission line used, and the frequency band of operation. Other factors include the desired electrical and mechanical properties and cost. Monolithic MICs use essentially semiconductor substrates. Hybrid MICs use dielectric substrates which are generally isotropic although anisotropic dielectrics are also used for certain special applications.

The dielectric substrates for hybrid MICs must have the following general properties:

1. Low dissipation factor: The dielectric loss in the circuit depends on the loss tangent of the dielectric substrate at the operating frequency. Typically, values of the order of 10^{-4} or less are desirable.
2. Polished surfaces: Losses at microwave frequencies depend heavily on the degree of surface finish, with the requirement on fineness of polish becoming increasingly stringent with increasing frequency. Fine surface finish is also necessary for attaining good line definition of the conducting pattern on the substrate. Typically, at *X*-band, the surface finish should be at least 1 to 2 μm.
3. Minimum variation of dielectric constant and dissipation factor with temperature: In order to ensure stability of performance of the circuit, the temperature coefficient of the dielectric must be low.
4. Uniformity of thickness and dimensional stability: During processing, especially using thick-film technology, uniformity of thickness in the substrates and from one substrate to another in a batch is essential in order that the deposited film may be uniform. A typical degree of uniformity is 0.5%. Uniformity of thickness and its dimensional stability are of prime importance in circuits realized in inverted- and suspended-microstrip configurations because they reflect on the uniformity of the air gap between the substrate and the bottom ground plane which is critical in circuit design.
5. High thermal conductivity: Higher heat conductivity permits better heat dissipation in circuits. This is important in circuits operating at high power levels.

Table 1.1 summarizes the properties of some substrate materials used in hybrid MICs. For stripline applications, the dielectric substrate should be flexible and easily machinable. This is necessary for proper clamping of the stripline package and to ensure that the dielectric medium is homogeneous, with no air gap at the edge of the strip conductor. A wide range of stripline dielectrics, such as the irradiated polyolefin, cross-linked polystyrene, polyphenylene, etc., have been reported over the years. Of more recent origin, are the flexible, low-loss copper-clad laminates trade named as RT-duroid and Epsilam-10. RT-duroid is a glass microfiber-reinforced polytetrafluoroethylene (PTFE), and is available in large sheets of various standard thicknesses for low as well as high dielectric constants. RT-duroid 5880 with a lower dielectric constant ($\epsilon_r = 2.22$) is commonly used for most stripline applications over the entire microwave frequency range. RT-duroid 6010 and Epsilam-10 having higher dielectric constants are useful for stripline applications at *L*-band. They are also alternatives to alumina in the fabrication of microstrip circuits requiring forming to a radius and clamping. Some of the other Cu-clad laminates which have been recently introduced include Kapton, Cu-flon and Rutile.

For microstrip and slotline applications, high purity alumina substrates have won widespread acceptance at microwave frequencies. These are ceramics possessing a good combination of dielectric, conductive, mechanical, and thermal characteristics. With an alumina content of about 99.5% and good surface smoothness, low-loss properties are achieved permitting their use at these frequencies. The frequency range of operation can be extended by using these substrates in special configurations such as the suspended stripline, the suspended microstrip, and the inverted microstrip. Fused quartz has a lower dielectric constant and is available in substrates with thickness as low as 0.002 inch and with a high degree of surface polish ($\simeq 0.006\ \mu$m). It is therefore an excellent medium for use at millimeter wave frequencies.

Table 1.1 Properties of substrates used in hybrid MICs

Material	ϵ_r (Approx.)	tan $\delta \times 10^4$ at 10 GHz	Surface roughness (μm)	Thermal conductivity, K (W/cm/°C)	Remarks/MIC applications
RT-duroid 5880	2.16–2.24	5–15	0.75–1.0* 4.25–8.75**	0.0026	Cu-plating, flexible/stripline, microstrip.
RT-duroid 6010	10.2–10.7	10-60	-do-	0.0041	Cu-plating, flexible/ stripline at L-band, microstrip.
Epsilam-10	10–13	20	–	0.0037	Cu-plating, flexible/ stripline at L-band, microstrip.
Alumina 99.5%	9.6–10.4	0.5-3	0.05–0.25	0.37	Cr-Au layer/microstrip, microstrip-like lines, slotline, coplanar lines.
Fused quartz 99.9%	3.75–3.8	1	0.006-0.025	0.01	Cr-Au layer, optical finish/microstrip-like lines at millimeter wave frequencies.
Beryllia (BeO)	6.6	1	0.05-1.25	2.5	High conductivity/compound substrate.
Glass	5	20	0.025	0.01	Lossy/lumped element.
Kapton	3–3.5	–	–	–	Flexible substrate/stripline.
Cu-flon	2.1	4.5	–	–	Flexible substrate/stripline, microstrip.
Rutile (TiO_2)	100	4	0.25–2.5	0.02	High dielectric constant/ microstrip, slotline, coplanar lines.
Ferrite/garnet	13–16	2	0.25	0.03	Porous/non-reciprocal devices in slotline, coplanar lines.
Sapphire (single crystal)	$\epsilon_{r\perp}=9.4$ $\epsilon_{r\parallel}=11.6$	0.4–0.7	0.005–0.025	0.4	Well-defined and repeatable electrical properties. anisotropic/microstrip.
Pyrolytic boron nitride	$\epsilon_{r\perp}=3.4$ $\epsilon_{r\parallel}=5.12$	–	–	–	Anisotropic/microstrip, suspended stripline.

*Average peak to valley difference in height (rolled copper).
**Average peak to valley difference in height (electro-deposited copper).

The undesirable features of quartz are its high cost, ease of breakage, and non-drillability. In devices such as power amplifiers where high dissipation rates are encountered, materials with a high coefficient of thermal conductivity are required. Polycrystalline BeO which has a thermal conductivity of the order of 2.5 W/cm/°C is well suited for such applications. It can be used as an insert in a combined substrate where metal cannot be used for transfer of heat. It is particularly suited as a base substrate for GaAs, since the thermal coefficients of expansion of the two materials match well. By flip-flop mounting GaAs pellets, directly onto

BeO matching substrates, high power devices with excellent thermal performance can be realized. Glass is a lossy material and can be used only at lower frequencies in the fabrication of lumped elements.

Ferrites and garnets are used essentially in the fabrication of non-reciprocal components. They are used either in the form of substrates or as inserts in dielectric substrates such as alumina. Sapphire and pyrolytic boron nitride are anisotropic materials with equivalent dielectric constants close to those of alumina and quartz, respectively. Sapphire substrates have some advantages over alumina, which include optical grade surface finish, well-defined and repeatable electrical properties, and higher thermal conductivity. Since the anisotropy of the substrate can be easily taken into account in the design, sapphire has become an attractive medium for MICs requiring a high level of reproducibility. The anisotropic property of sapphire and pyrolytic boron nitride can be used for achieving certain desirable characteristics such as phase velocity equalization in inhomogeneous coupled structures.

Materials suitable for monolithic MICs are summarized in Table 1.2. Microwave monolithi circuits essentially make use of silicon (Si) and gallium arsenide (GaAs) as substrate materials. Silicon technology for monolithic circuits is well established and high power devices such as IMPATTs and PIN diodes are within the state-of-the-art. One major disadvantage of Si is that its intrinsic insulating qualities get degraded due to temperature above 800°C. GaAs is a far superior insulator material as compared with Si. Further, GaAs could operate at higher temperatures and also has an advantage in terms of radiation hardness over Si. Other advantages of GaAs over Si are its much higher electron mobility, wider band gap, and its high peak (threshold) velocity. By introducing suitable dopant impurities into the material, a wide range of values of electrical conductivity can be realized. The electrical properties of the substrate can be selected by doping so that it can act either as a semiconducting layer or as an insulating mechanical support. The high resistivity properties of the semi-insulating GaAs substrates are utilized for realizing passive components whereas the high mobility properties of the semiconducting layer that is either epitaxially grown onto the semi-insulating substrate or implanted into it are utilized for active devices. In monolithic circuits based on silicon-on-sapphire (SOS) approach, sapphire is used as a base substrate to eliminate the losses associated with semi-insulating silicon.

Table 1.2 Properties of substrates used in MMICs

Material	ϵ_r (approx.)	tan $\delta \times 10^4$ at 10 GHz	Surface roughness (μm)	Thermal conductivity K (W/cm/°C)	Resistivity (Ω-cm)	Electron mobility (cm²/V-sec)	Density (gm/cc)
Si	11.7-12	40-150	< 0.025	0.9	—	700*	2.33
Semi-insulating Si	11.7-12	10-100	< 0.025	0.9	10^3-10^5	—	2.33
GaAs	12-13	10-60	< 0.025	0.3	—	4300*	5.32
Semi-insulating GaAs	12-13	16	≃ 0.025	0.3	10^7-10^9	—	5.32
Sapphire	11.6 (*c*-axis)	0.4-0.7	≃ 0.005	0.4	> 10^{14}	—	3.98

*At 10^{17}/cm³ doping.

1.5 Hybrid Technology

The fabrication of hybrid MICs involves forming the passive circuit pattern on an inert dielectric substrate and then attaching active devices to the passive circuitry. The quality of MICs, in addition to depending on substrate properties, is critically influenced by factors such as the adhesive strength of the conducting pattern, homogeneity of the conductor lines and their geometry definition. These depend essentially on the metal system and the process used to delineate the circuit. The fabrication technology used in hybrid circuit can be divided into two basic categories, namely, thin-film technology and thick-film technology [41–47]. A composite technique, which combines some of the process steps involved in the thin- and thick-film technologies is called the mid-film technology [48]. The degree to which the aforementioned circuit parameters are controlled using these processing techniques is discussed in this section.

1.5.1 Thin-Film Technology

Thin-film technology utilizes vacuum evaporation or RF/D.C. cathode sputtering techniques for the metal deposition. The deposited metal film should adhere well to the substrate while having good conductivity at the operating frequencies. These are contradictory requirements, because materials with good conductivity such as silver (Ag), copper (Cu), gold (Au), and aluminium (Al) have poor adherence property, and materials such as chrome (Cr), titanium (Ti) and tantalum (Ta) which adhere well to the dielectrics are poor electrical conductors. Thin-film technology, therefore, employs a composite metal system. Systems commonly used are Cr-Au, Cr-Cu, Cr-Cu-Au, Ti-Au and Ti-Pd-Au. The process of deposition, for example, in the case of Cr-Au, involves providing a thin flash of chrome of approximately 100-200 Å thick, followed by a deposited layer of Au of similar thickness in the same vacuum run. The gold layer is then increased in thickness to a few 'skin depths' by electroplating. In the Cr-Cu system, copper offers an inexpensive alternative to gold. Copper also provides very low sheet resistance and is amenable to soldering of devices. A disadvantage of copper is that it corrodes easily unless protected by a layer of passivating material. This problem can be overcome by treating the copper pattern with a gold electroless plating solution. The protective gold layer also facilitates easy bonding of semiconductor and other passive chips, which is rather difficult on a copper surface. The Ti-Pd-Au system uses palladium as a barrier layer between the titanium seed layer and the top gold layer, to make it metallurgically stable for high temperature operation. The circuit pattern in all these cases is formed by photolithographic techniques. A photoresist mask of the desired circuit pattern is produced at the metallized surface and the exposed undesired metallization is etched by means of a wet chemical process. Circuits demanding more stringent dimensional tolerances and sharp edge definitions can be etched by the ion-etching technique.

Lumped inductors and interdigital capacitors form part of the conducting pattern, and are provided by the same process in the same thickness of metallization as the rest of the conducting lines. These components usually have much smaller line widths and gap widths (of the order of $10-20\ \mu$m), and hence, demand finer edge definition and minimal chemical-etch undercut. Special precaution needs to be taken in the design of the photomasks and during etching. Lumped inductors take the form of spirals or meander lines, and have values in the range 1 to 30 nH. The interdigital geometry of the capacitors gives low values of capacitances in the range 0.1 to 2 pF. Larger capacitance values up to about 100 pF are realized using MOM (metal-oxide-metal) sandwich capacitors with SiO and SiO_2 dielectric films. These are available in chip form and can be directly bonded to the circuit.

Resistive films are useful in MICs for bias networks, terminations, and attenuators. These films should have a low temperature coefficient of resistivity and good stability. The commonly used resistive materials are nichrome and tantalum. By depositing these films, resistance values in the range 50 Ω to 5 KΩ, generally required in MICs, can be realized. Using abrasive trimming or laser trimming, resistance tolerances to an accuracy of better than 1% can be achieved. The formation of resistive films to realize resistors requires additional processes of deposition and etching. In many circuits, this complexity is obviated by directly bonding chip resistors on the conducting pattern.

1.5.2 Thick-Film Technology

Thick-film technology is essentially based on the screen-printing process. It involves printing of certain kinds of inks or pastes on to ceramic substrates, and melting, sintering, or reacting them into continuous layers to form conductors, resistors, and capacitors. The inks/pastes for conducting patterns consist of a suspension in organic vehicle of finely powdered metals (typically Ag, Au, Pd-Au, Pt-Au and Ag-Pd) with a glass frit or binder. Insulator materials are similarly constituted, but have no metallic component. Resistor pastes generally have complex compositions of metals and compounds to meet the requirements of a wide range of resistivities combined with low temperature coefficients of resistance.

Thick-film circuits are generally formed on high purity alumina substrates because of their higher thermal conductivity, strength, stability, and good electrical properties. The screen used during printing usually consists of a piece of finely woven mesh, of either stainless steel or nylon, mounted under tension on a metal frame. The circuit pattern is first photolithographically formed on the screen mesh using an ultra-violet sensitive filler emulsion. The finished screen is mounted in the screen printer having a narrow gap ($\simeq$ 0.5 mm) above the substrate. The appropriate paste is then placed on the printing screen, and a squeegee is made to traverse the screen under pressure, bringing it into contact with the substrate and at the same time forcing the paste through the open meshes defining the circuit pattern. The substrate containing the deposited pattern is then fired in a firing furnace. The firing process includes burning off of the organic vehicle at 200-500°C, sintering of metal particles and melting of glass particles at a temperature of the order of 850°C. This process establishes a continuous bond between the metal and the substrate. When different types of pastes are used, more than one cycle of screen printing, drying, and firing is required. The patterns can be etched if further improvement in pattern delineation is necessary.

Compared with the thin-film technology, the screen-printing process is simpler and inexpensive. However, the pattern definition that can be achieved is inferior to that of thin films. This is mainly because the material used is not pure but interdispersed with glass. Another disadvantage is that the d.c. resistance of the fired conductor is higher than that of pure evaporated films, and the RF loss is higher. Because of these limitations, thick-film technology is used for producing MICs at lower operating frequencies extending up to about 9 GHz.

At higher frequencies, MICs require increasingly thinner line dimensions and narrow gaps between the lines. Even in the lower microwave frequency range, circuits incorporating spiral inductors and interdigital capacitors demand fine line definitions for these lumped elements. For improved line definition, it becomes necessary to adopt etching, thereby reducing the simplicity of the manufacture process.

1.5.3 Mid-Film Technology

An intermediate process which strikes a compromise between the high resolution capability of thin-film technology and the low cost and large area capability of thick-film technology is called the mid-film technology [48]. As in thin- and thick-film technologies, high purity alumina substrate is the most suitable dielectric base for mid-film circuits, although a variety of other substrate materials including fused silica and glass can also be used. In the mid-film process, the substrate is first coated with a thin layer of an organic photosensitive resin and then exposed to ultra-violet light through a positive photomask. The exposed portion of the mid-film resin cross-links and becomes particle non-receptive. A solid powder of metallic conductor (generally Au or Ag) and glass frit are then applied to the surface. These get embedded in unexposed portions of the light-sensitive film. This is followed by firing the substrate in a belt conveyer furnace at a temperature of about 920°C. This process removes the organic resin by oxidation and bonds the powder particles to the substrate.

Compared with the thick-film technology which is limited to line widths of greater than about 100 μm, the mid-film process enables fabrication of narrower line widths of the order of 50 μm. The mid-film conductors are compatible with the thick-film resistors and dielectric materials and are also amenable for attachment of discrete chip components. Considering the cost-effectiveness of the process as well as the RF performance of the resulting circuits, the mid-film process appears to offer its best capability in the frequency range 3-12 GHz.

1.6 Monolithic Technology

An essential feature of monolithic technology which differentiates it from hybrid technology is that, the active devices, passive components, and interconnections are all formed into the bulk or onto the surface of the same semiconductor material [55-59]. Substrates must have high resistivities of the order of 2000 Ω-cm or greater to keep transmission losses within the acceptable limits and to provide good isolation. High resistivity Si which offers resistivity in the range 2×10^3 to 10^4 Ω-cm and semi-insulating GaAs possessing even higher resistivities in the range 10^7 to 10^9 Ω-cm are widely accepted as the suitable substrate materials for monolithic MICs.

The processes generally used in the realization of active devices are epitaxial growth and ion implantation techniques. In the epitaxial technique, a doped semiconductor layer is deposited on a high resistivity Si or semi-insulating GaAs substrate. Generally, an intervening high resistivity epitaxial buffer layer is also deposited to prevent diffusion of impurities from the substrate during the active layer growth. In ion implantation, dopant atoms are implanted directly into the surface of the substrate material by impinging beams of ion of the desired element. Ion implantation is preferred over the epitaxial technique, as the former permits precise control of active region doping, thus offering a highly reproducible device technology. It also facilitates selective doping and control of multiple doping densities for optimization of active devices. The substrates, however, require a high temperature anneal cycle to activate the implanted dopants and remove radiation damage. Quality substrates should therefore retain their intrinsic properties during the annealing cycle. The chrome-compensated semi-insulating GaAs offers stable resistivities over the temperature range (850-900°C) required in the fabrication of microwave devices. Silicon has a disadvantage in this respect because the intrinsic insulating qualities get degraded at temperatures above 800°C. However, with the use of ion implantation and selective laser annealing, it is possible to process Si wafers without increasing the temperature of the whole substrate. Active devices can therefore be fabricated without degradation of the high resistivity substrate required for the passive circuits.

From the point of view of production of devices operating in the higher microwave frequency range, GaAs offers a distinct advantage over Si. The electron mobility of GaAs is nearly six times that of Si and hence the device can operate at much higher frequencies and speeds than a Si device. The most commonly used active device on GaAs is the metal-semiconductor field effect transistor (MESFET).

Resistors can be fabricated using either the intrinsic resistivity of the GaAs active layer or employing thin metal or cermet films giving uniform and low temperature coefficients of resistance. Capacitors are realized in the form of interdigital planar structures for low value capacitances up to about 2 pF and metal-insulator-metal (MIM) structures for higher values up to about 200 pF. The MIM capacitor consists of two overlapping metal layers separated by a dielectric layer such as polyimide or silicon nitride. The metallic layers are produced using either plated gold or sputtered gold in conjunction with ion beam milling. Planar inductors are formed using gold films, in the shapes of a single turn loop for low values up to about 1 nH and square or circular spirals for values up to about 30 nH.

Delineation of device geometries and circuit pattern is accomplished through the lithographic process. In the case of photolithography, diffraction effects limit the resolution of lines to dimensions of the order of a wavelength of light. Electron beam lithography offers a considerably higher degree of resolution and is used especially for realizing submicron gates of devices at higher microwave frequencies. A composite process using photolithography and electron beam lithography is desirable in the fabrication of MMICs. The circuit patterns of MMICs are defined using the same lithographic techniques as are used to define the one-micron FET gates. The dimensional tolerances in the patterns are controlled far more precisely than in standard MIC technology.

The important benefits of MMICs over the hybrid MICs are their wider bandwidth capability, multifunction integration onto a single chip, and higher system reliability. However, in view of the complicated technological processes involved and the high cost investment and development, MMICs are far more expensive to produce compared with their hybrid counterparts. The production of MMICs is therefore restricted to meeting the military requirements demanding stringent specifications or to components having a large market potential. Such applications include monolithic power and low noise amplifiers for phased array modules, broadband amplifiers and oscillators for electronic countermeasures and home receivers for a direct broadcast satellite TV. For low volume production, the hybrid MICs are much more cost-effective, and hence, will continue to dominate in most commercial applications.

1.7 Concluding Remarks

The past two decades have seen significant advances in the area of microwave integrated circuits. The hybrid technology of fabricating MICs is now well established. Using advanced hybrid techniques, circuits operating up to frequencies as high as 140 GHz have been developed. Implementation of the monolithic approach in MICs, especially using GaAs, has started more recently. MMICs operating up to a frequency of about 110 GHz have been reported. Although the potential capabilities of MMICs are well known, in view of the present high cost development, batch production of MMICs is not cost-effective unless there is a large volume requirement. With the advances that are taking place, a cost-effective approach which makes use of the combination of hybrid and monolithic MICs is emerging. In this approach monolithic chips are incorporated in hybrid MICs wherever appropriate. This reduces the number of interconnections considerably, resulting in substantial reduction in cost and enhanced circuit reliability.

1.8 Prelude to the Book

Both hybrid and monolithic MICs make use of planar transmission lines as the basic transmission media. Hybrid MICs offer considerable device flexibility permitting the use of a variety of planar transmission line configurations. On the other hand, for MMICs, the configurations are restricted to microstrip and coplanar lines by virtue of the process technology involved. While the process technologies of MICs, and MIC techniques using the basic transmission lines (striplines, microstrip lines and slotlines) are documented in several books, a number of useful variants of these basic lines which have emerged over the years have not been treated in consolidated form. The present book covers in detail, the analysis, design, and circuit applications of a variety of stripline-like transmission lines. Unified analytical techniques are presented which are applicable to multilayer structures using isotropic as well as anisotropic substrates. The book includes extensive design data on a number of practical single and coupled conductor stripline-like transmission lines.

REFERENCES

1. R.M. Barrett and M.H. Barnes, Microwave printed circuits, *Radio TV News*, Vol. 46, 16, Sept. 1951.
2. D.D. Grieg and H.F. Englemann, Microstrip—A new transmission technique for the kilomegacycle range, *Proc IRE.*, Vol. 40, pp. 1644-1650, Dec. 1952.
3. *IRE Trans. Microwave Theory Tech.*, Vol. MTT-3, March 1955.
4. Special issue on microwave integrated circuits, *IEEE Trans. Electron Devices*, Vol. ED-15, July 1968.
5. M.V. Schneider, Microstrip lines for microwave integrated circuits, *Bell Syst. Tech. J.*, Vol. 48, pp. 1421-1444, May-June 1969.
6. S.B. Cohn, Slot line as an alternative transmission medium for integrated circuits, *IEEE G-MTT Int. Microwave Symp., Digest* pp. 104-109, 1968.
7. S. Yamamoto, T. Azakami and K. Itakura, Slit-coupled strip transmission lines, *IEEE Trans. Microwave Theory Tech.*, Vol. MTT-14, pp. 542-553, Nov. 1966.
8. C.P. Wen, Coplanar waveguide—A surface strip transmission line suitable for nonreciprocal gyromagnetic device applications, *IEEE Trans. Microwave Theory Tech.*, Vol. MTT-17, pp. 1087-1090, Dec. 1969.
9. E. Yamashita and K. Atsuki, Stripline with rectangular outer conductor and three dielectric layers, *IEEE Trans. Microwave Theory Tech.*, Vol. MTT-18, pp. 238-244, May 1970.
10. J.B. Knorr and K.D. Kuchler, Analysis of coupled slots and coplanar strips on dielectric substrate, *IEEE Trans. Microwave Theory Tech.*, Vol. MTT-23, pp. 541-548, July 1975.
11. J.L. Allen and M.F. Estes, Broadside coupled strips in a layered dielectric medium, *IEEE Trans. Microwave Theory Tech.*, Vol. MTT-20, pp. 662-669, Oct. 1972. Also correction, *IEEE Trans. Microwave Theory Tech.*, Vol. MTT-23, p. 779, Sept. 1975.
12. B.E. Spielman, Dissipation loss effects in isolated and coupled transmission lines, *IEEE Trans. Microwave Theory Tech.*, Vol. MTT-25, pp. 648-656, Aug. 1977.
13. I.J. Bahl and P. Bhartia, Characteristics of inhomogeneous coupled striplines, *IEEE Trans. Microwave Theory Tech.*, Vol. MTT-28, pp. 529-535, June 1980.

14. S.K. Koul and B. Bhat, Characteristic impedance of microstrip-like transmission lines for millimeter wave applications, *Arch. Elek. Ubertragung*, Vol. 35, pp. 253-258, June 1981.
15. B. Bhat and S.K. Koul, Unified approach to solve a class of strip and microstrip-like transmission lines, *IEEE Trans. Microwave Theory Tech.*, Vol. MTT-30, pp. 679-686, May 1982. Also correction, *IEEE Trans. Microwave Theory Tech.*, Vol. MTT-30, p. 2067, Nov. 1982.
16. S.K. Koul and B. Bhat, Broadside, edge-coupled symmetric strip transmission lines, *IEEE Trans. Microwave Theory Tech.*, Vol. MTT-30, pp. 1874-1880, Nov. 1982.
17. S.K. Koul and B. Bhat, Generalized analysis of microstrip-like transmission lines and coplanar strips with anisotropic substrates for MIC, electrooptic modulator and SAW applications, *IEEE Trans. Microwave Theory Tech.*, Vol. MTT-31, pp. 1051-1058, Dec. 1983.
18. J.H Cloete, Rectangular bars coupled through a finite-thickness slot, *IEEE Trans. Microwave Theory Tech.*, Vol. MTT-32, pp. 39-46, Jan. 1984.
19. M. Caulton, The lumped element approach to microwave integrated circuits, Microwave J., Vol. 13, pp. 51-57, Feb. 1970.
20. M. Caulton, B. Hershenov, S.P. Knight and R.E. DeBrecht, Status of lumped elements in microwave integrated circuits—present and future, *IEEE Trans. Microwave Theory Tech.*, Vol. MTT-19, pp. 588-599, July 1971.
21. C.S. Aitchison, R. Davies, I.D. Higgins, S.R. Longley, B.H. Newton, J.F. Wells and J.C. Williams, Lumped circuit elements at microwave frequencies, *IEEE Trans. Microwave Theory Tech.*, Vol. MTT-19, pp. 928-937, Dec. 1971.
22. R.L. Camisa, J.B. Klatskin and A. Mikelsons, Lumped element GaAs FET power amplifiers, *RCA Review*, Vol. 42, pp. 557-575, Dec. 1981.
23. L. Young and H. Sobol (eds.), *Advances in Microwaves*, Vol. 8, Academic Press, New York, 1974.
24. J. Frey and K. Bhasin (eds.), *Microwave Integrated Circuits*, 2nd Edn., Artech House, Dedham, Mass., 1985. (A reprint volume.)
25. M.A.R. Gunston, *Microwave Transmission Line Impedance Data*, Van Nostrand Reinhold, London, 1972.
26. H. Howe, *Stripline Circuit Design*, Artech House, Dedham, Mass., 1974.
27. K.C. Gupta and A. Singh (eds.), *Microwave Integrated Circuits*, Halsted Press, John Wiley & Sons, New York, 1974.
28. K.C. Gupta, R. Garg and, I.J. Bahl, *Microstrip Lines and Slotlines*, Artech House, Dedham, Mass., 1979.
29. T.C. Edwards, *Foundations for Microstrip Circuit Design*, John Wiley & Sons, New York, 1981.
30. K.C. Gupta, R. Garg and R. Chadha, *Computer-aided Design of Microwave Circuits*, Artech House, Dedham, Mass., 1981.
31. R.V. Garver, *Microwave Diode Control Devices*, Artech House, Dedham, Mass., 1976.
32. Special issue on microwave field-effect transistors, *IEEE Trans. Microwave Theory Tech.*, Vol. MTT-24, June 1976.
33. Y. Harada and H. Fukuda, A novel beam lead GaAs Schottky-barrier diode fabricated by using thick polyimide film, *IEEE Trans. Electron Devices*, Vol. ED-26, pp. 1799-1804, Nov. 1979.
34. C.T. Rucker, J.W. Amoss, G.N. Hill, and N.W. Cox, Multichip IMPATT power combining, a summary with new analytical and experimental results, *IEEE Trans. Microwave Theory Tech.*, Vol. MTT-27, pp. 951-957, Dec. 1979.
35. Special issue on gigabit logic for microwave systems, *IEEE Trans. Microwave Theory Tech.*, Vol. MTT-28, May 1980.
36. C.T. Rucker, J.W. Amoss and G.N. Hill, Chip level IMPATT combining at 40 GHz, *IEEE Trans. Microwave Theory Tech.*, Vol. MTT-29, pp. 1266-1271, Dec. 1981.
37. J.W. Archer, B.B. Cregger, R.J. Mattauch, and J.D. Oliver, Jr., Harmonic generators have high efficiency, *Microwaves*, Vol. 21, pp. 84-87, May 1982.
38. M. Caulton, A. Rosen, P.J. Stabile, and A. Gombar, PIN diodes for low-frequency high-power switching applications, *IEEE Trans. Microwave Theory Tech.*, Vol. MTT-30, pp. 875-882, June 1982.
39. Special issue on GaAs IC's, *IEEE Trans. Microwave Theory Tech.*, Vol. MTT-30, July 1982.
40. Special issue on power and low-noise GaAs FET circuits and applications, *IEEE Trans. Microwave Theory Tech.*, Vol. MTT-32, March 1984.

41. M. Caulton and H. Sobol, Microwave integrated circuit technology—A survey, *IEEE J. Solid State Circuits*, Vol. SC-5, pp. 292-303, Dec. 1970.
42. C.A. Harper, *Handbook of Materials and Processes for Electronics*, McGraw-Hill, Maidenhead, England, 1970.
43. M.L. Topfer, *Thick-film Microelectronics. Fabrication, Design and Applications*, Van Nostrand Reinhold, London, 1971.
44. P.J. Holmes and R.J. Loasby, *Handbook of Thick-film Technology*, Electrochemical Publications, Great Britain, 1976.
45. J.E. Curran, R.V. Jeanes and H. Sewell, A technology of thin-film hybrid microwave circuits, *IEEE Trans. Parts, Hybrids, Packag.*, Vol. PHP-12, No. 4, pp. 304-309, Dec. 1976.
46. J P. Ramy, M.T. Cotte, J.P. Bolloch, R. Schnitzler, J.J. Guena and C. Thebault, Optimization of the thick- and thin-film technologies for microwave circuits on alumina and fused silica substrates, *IEEE Trans. Microwave Theory Tech.*, Vol. MTT-26, pp. 814-820, Oct. 1978.
47. C. Newport, A review of thick-film microwave integrated circuit technology, *Microelectronics J.*, Vol. 15, pp. 44-52, Feb. 1984.
48. D.R. Conn, H.M. Naguib and C.M. Anderson, Mid-film for microwave integrated circuits, *IEEE Trans. Components, Hybrids, Manuf. Technol.*, Vol. CHMT-5, No. 1, pp. 185-191, March 1982.
49. J.F. White, *Microwave Semiconductor Engineering*, Van Nostrand Reinhold, New York, 1982.
50. T.T. Ha, *Solid-State Amplifier Design*, John Wiley & Sons, New York, 1981.
51. T.H. Oxley, Microwave integrated circuit techniques, *GEC. J. Sci. and Tech.*, Vol. 43, pp. 21-31, Jan. 1976.
52. J.G. Galli, B.E. Gilchrist, S.T. Rempel, D.E. Peck and R.J. Forse, Integration shrinks microwave front ends, *Microwave Syst. News*, Vol. 10, pp. 119-130, Sept./Oct. 1980.
53. E. Hagihara, H. Ogawa, N. Imai and M. Akaike, A 26 GHz miniaturized MIC transmitter/receiver, *IEEE Trans. Microwave Theory Tech.*, Vol. MTT-30, pp. 235-242, March 1982.
54. Y. Tokumitsu, M. Ishizaki, M. Iwakuni and T. Saito, 50 GHz IC components using alumina substrates, *IEEE Trans. Microwave Theory Tech.*, Vol., MTT-31, pp. 121-127, Feb. 1983.
55. D. Ch'en and D.R. Decker, Monolithic microwave integrated circuits (MMICs)— The next generation of microwave components, *Microwave J.*, Vol. 23, pp. 67-78, May 1980.
56. R.S. Pengelly, Monolithic microwave circuits, *Electronics and power*, Vol. 20, pp. 379-384, May 1981.
57. R.A. Pucel, Design considerations for monolithic microwave circuits, *IEEE Trans. Microwave Theory Tech.*, Vol. MTT-29, pp. 513-534, June 1981.
58. R.S. Pengelly, Hybrid vs. monolithic microwave circuits—A matter of cost, *Microwave Syst. News*, Vol. 13, pp. 77-114, Jan. 1983.
59. R.S. Pengelly, *Microwave Field-Effect Transistors*, Research Studies Press, A division of John Wiley & Sons, UK, 1982.

CHAPTER 2

Planar Transmission Lines—Methods of Analysis

2.1 Introduction

The fundamental mode of propagation in conventional transmission lines such as the two-wire line, the coaxial line, and the balanced stripline is essentially a transverse-electromagnetic (TEM) mode. In the ideal case, when the conductors are considered to have infinite conductivity and the medium in which the lines are embedded is homogeneous, lossless, and isotropic, the basic mode is a pure TEM mode. Practical lines, in general, have such small losses that we can determine their propagation parameters, namely, the characteristic impedance, guide wavelength, and phase velocity by considering them as ideal lines. The conductor and dielectric losses can then be determined by making a perturbation calculation.

A conventional strip transmission line is a balanced line in which a flat conducting strip is placed symmetrically between two large ground planes, with the space between the ground planes filled with a homogeneous dielectric. In this book, we consider, in addition to this basic configuration, several of its useful variants. These include structures having off-set strip conductors and multilayer iso/anisotropic dielectrics. While the fundamental mode of propagation in lines embedded in homogeneous dielectric is a TEM mode, lines in inhomogeneous media cannot support a pure TEM mode. The field distribution for the dominant mode in the latter case, however, closely resembles the TEM mode and is usually referred to as the quasi-TEM mode. This means that the longitudinal components of the electric (E) and magnetic (H) fields are so small that they can be neglected. If the ground plane spacing and the width of the strip conductor are sufficiently less than $\lambda/2$ (where λ is the wavelength in the structure at the highest frequency of operation), then the structure can support the dominant mode propagation at the exclusion of all the higher order modes. With this quasi-TEM mode approximation, the calculation of propagation parameters of the line reduces essentially to the solution of the two-dimensional Laplace's equation subject to the boundary conditions determined by the geometry of the line. Bulk of the book is devoted to the study of strip-line-like transmission lines which satisfy this criterion. This is adequate for most of the practical applications. One separate chapter, however, has been set aside for discussing the frequency-dependent solutions of such structures operating beyond the quasi-static limit.

There are many methods which exist for the solution of Laplace's equation in two dimensions. The commonly reported ones are, the conformal transformation method [1, 2], the variational method [2], and the finite-difference method [3, 4]. The conformal transformation technique is exact, and has been applied to obtain closed form expression for the characteristic impedance of the homogeneous stripline [2]. For structures with inhomogeneous medium, however, the application of this method becomes prohibitively complicated. The variational method, on the other hand, is approximate, but it offers a simpler way of determining the propagation parameters of stripline-like transmission lines. When combined with the 'transverse transmission line' technique [5] of determining the potential function or the Green's function, the derivation of capacitance of the structure, and hence, its propagation parameters becomes simple even when multilayer dielectrics are involved [6]. This method also enables determination of propagation parameters of stripline-like structures with multilayer anisotropic substrates [7]. The use of the finite-difference method essentially involves a numerical evaluation and is suitable for simple structures which can be solved within a reasonable computer time.

After a brief review of the general TEM analysis of a two-conductor transmission line, the basic principle and approach underlying the aforementioned four methods are presented in this chapter. The variational method and the 'transverse transmission line' technique are described in detail. The generality and simplicity of this approach are demonstrated by deriving the expression for the capacitance of a multilayer structure. In the subsequent chapters, these results are applied to study a wide variety of single and coupled iso/anisotropic stripline-like transmission lines. The finite-difference method, as reported in the literature for the homogeneous stripline and for structures with a limited amount of inhomogeneity [3, 4] is summarized in this chapter. Finally, a technique involving the combination of a variational series and conformal mapping method as reported by Smith [8] specifically for the suspended stripline, is outlined.

2.2 TEM Analysis and Definitions

2.2.1 Plane TEM Waves

Consider a plane TEM wave propagating along the z-direction of a $x-y-z$ rectangular coordinate system. The medium is assumed to be lossless, isotropic, and homogeneous, having a magnetic permeability μ and electric permittivity ϵ. For a time dependence of $e^{j\omega t}$, Maxwell's equations are

$$\nabla \times \mathbf{E}_t = -j\omega\mu\mathbf{H}_t \tag{2.1a}$$

$$\nabla \times \mathbf{H}_t = j\omega\epsilon\mathbf{E}_t \tag{2.1b}$$

$$\nabla \cdot \mathbf{E}_t = 0 \tag{2.1c}$$

$$\nabla \cdot \mathbf{H}_t = 0 \tag{2.1d}$$

The permeability (absolute) of the medium may be written as $\mu = \mu_0\mu_r$, where μ_0 is the permeability of free space and has a value $4\pi \times 10^{-7}$ henry/m, and μ_r is the relative permeability of the medium. Similarly, the permittivity (absolute) of the medium may be written as $\epsilon = \epsilon_0\epsilon_r$, where ϵ_0 is the permittivity or dielectric constant of free space and has a value $10^{-9}/36\pi$ farads/m, and ϵ_r is the relative dielectric constant of the medium. For a perfect dielectric medium having no magnetic property $\mu_r = 1$.

The subscript t in (2.1) indicates that the field vectors have transverse, that is, x and y, components only. Splitting the vector operator ∇ into a transverse part $\nabla_t = \hat{x}\,\partial/\partial x + \hat{y}\,\partial/\partial y$

and a longitudinal part $\nabla_z = \hat{z}\,(\partial/\partial z)$, we can reduce (2.1) to the following:

$$\hat{z} \times \frac{\partial \mathbf{E}_t}{\partial z} = -j\omega\mu \mathbf{H}_t \tag{2.2a}$$

$$\hat{z} \times \frac{\partial \mathbf{H}_t}{\partial z} = j\omega\epsilon \mathbf{E}_t \tag{2.2b}$$

$$\nabla_t \times \mathbf{E}_t = 0 \tag{2.2c}$$

$$\nabla_t \times \mathbf{H}_t = 0 \tag{2.2d}$$

$$\nabla_t \cdot \mathbf{E}_t = 0 \tag{2.2e}$$

$$\nabla_t \cdot \mathbf{H}_t = 0 \tag{2.2f}$$

From the theory of vector fields, it is known that, when the curl of a vector is identically zero, the vector may be derived from the gradient of a suitable scalar function. Thus, from (2.2c) and (2.2d) $\mathbf{E}_t$ and $\mathbf{H}_t$ can be assumed to be of the form

$$\mathbf{E}_t = g_1(z)\,\nabla_t\phi \tag{2.3a}$$

$$\mathbf{H}_t = g_2(z)\,\nabla_t\psi \tag{2.3b}$$

where g_1, g_2, ϕ and ψ are all scalar functions. It may be noted that ϕ and ψ are functions of the transverse coordinates x and y only. Combining (2.3a) with (2.2e), and (2.3b) with (2.2f), it can be seen that both ϕ and ψ satisfy Laplace's equation in the transverse plane.

$$\nabla_t^2\phi = 0 \tag{2.4a}$$

$$\nabla_t^2\psi = 0 \tag{2.4b}$$

The scalar functions ϕ and ψ are not independent because $\mathbf{E}_t$ and $\mathbf{H}_t$ are related by (2.1a) and (2.1b). Differentiating (2.2a) with respect to z and using (2.2b), we get

$$\hat{z} \times \left(\hat{z} \times \frac{\partial^2 \mathbf{E}_t}{\partial z^2}\right) = \omega^2\mu\epsilon \mathbf{E}_t \tag{2.5}$$

which simplifies to

$$\frac{\partial^2 \mathbf{E}_t}{\partial z^2} + \beta_z^2 \mathbf{E}_t = 0 \tag{2.6}$$

where $\beta_z^2 = \omega^2\mu\epsilon$. Similarly, it can be shown that H_t also satisfies the same differential equation. Assuming propagation in the positive z direction, the solution of (2.6) can be written in the form

$$\mathbf{E}_t = \mathbf{A}(x, y)\exp(-j\beta_z z) \tag{2.7}$$

where $\mathbf{A}(x, y)$ is a function of only the transverse coordinates. Comparing with (2.3a), $\mathbf{E}_t$ can be written as

$$\mathbf{E}_t = B\nabla_t\phi(x, y)\exp(-j\beta_z z) \tag{2.8}$$

where B is a constant. Substituting for $\mathbf{E}_t$ from (2.8) in (2.2a), we obtain

$$-j\beta_z \hat{z} \times B\nabla_t\phi(x, y)\exp(-j\beta_z z) = -j\omega\mu \mathbf{H}_t$$

which simplifies to

$$\mathbf{H}_t = \frac{\hat{z} \times \mathbf{E}_t}{\eta} \tag{2.9}$$

where $\eta = \sqrt{\mu/\epsilon}$ is the intrinsic impedance of the medium.

2.2.2 *Ideal Two-Conductor Transmission Line*

Consider an infinitely long two-conductor transmission line oriented parallel to the z-axis as shown in Fig. 2.1. The conductors are assumed to be perfectly conducting and have uniform cross-sections in the z-direction. The medium surrounding the conductors is homogeneous, lossless, and isotropic with electrical parameters μ and ϵ. The dominant mode of propagation on this line is the TEM mode. It has been shown in the preceding section that the potential function ϕ satisfies Laplace's equation in the x-y plane, and the transverse electric and magnetic fields are given by

$$\mathbf{E}_t = -\nabla_t \phi \exp(-j\beta_z z) \tag{2.10}$$

$$\mathbf{H}_t = \frac{\hat{z} \times \mathbf{E}_t}{\eta} \tag{2.11}$$

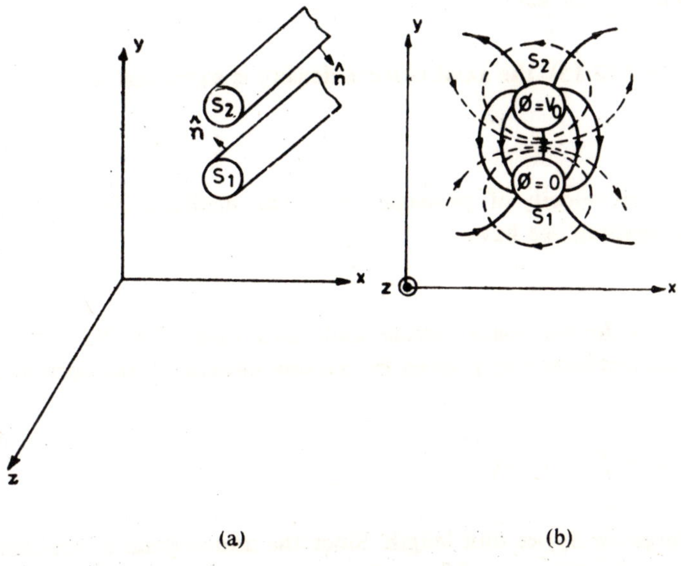

FIG. 2.1 (a) Two-conductor transmission line and (b) E and H field lines in the transverse plane. (——— E_t, - - - H_t)

The negative sign in (2.10) is chosen only to conform with the convention used in electrostatics. A solution for ϕ exists only if one conductor is at a potential which is different from the other one. Without any loss of generality, we may assume conductor S_1 to be at zero potential and conductor S_2 at potential V_0. In the transverse plane, $\nabla_t \times \mathbf{E}_t = 0$, and hence, the line integral of $\mathbf{E}_t$ from S_1 to S_2 is unique and does not depend on the path of integration. Consequently, there is no induced potential around any closed curve in the x-y plane:

$$V_0 = -\int_{S_1}^{S_2} \mathbf{E}_t \cdot \mathbf{dl} = \int_{S_1}^{S_2} \nabla_t \phi \cdot \mathbf{dl} = \int_{S_1}^{S_2} d\phi \tag{2.12}$$

In the above equation, the factor $\exp(-j\beta_z z)$ has been suppressed. Associated with the transverse electric field expressed by (2.10), there is a unique voltage wave $V = V_0 \exp(-j\beta_z z)$.

Since the conductors are assumed to be perfectly conducting, the tangential component of the electric field is zero at the conductor surfaces and the normal component of the displacement vector $\mathbf{D}_t = \epsilon \mathbf{E}_t$ is discontinuous by the surface charge density ρ_s. Thus, we have

$$\rho_s = \epsilon \hat{n} \cdot \mathbf{E}_t \tag{2.13}$$

where $\hat{n}$ is the unit outward normal to the conductors.

The surface current density $\mathbf{J}_s$ is given by

$$\mathbf{J}_s = \hat{n} \times \mathbf{H}_t = \hat{z} J \tag{2.14}$$

Substituting for $\mathbf{H}_t$ from (2.11) in (2.14) and simplifying, we get

$$\hat{z} J = \frac{\hat{n} \times (\hat{z} \times \mathbf{E}_t)}{\eta} = \hat{z} \frac{(\hat{n} \cdot \mathbf{E}_t)}{\eta} \tag{2.15}$$

Combining (2.13) and (2.15), the axial current density is expressed as

$$|J| = \frac{\rho_s}{\epsilon \eta} = v \rho_s \tag{2.16}$$

where $v = 1/\sqrt{\mu \epsilon}$ is the velocity of propagation in the medium. For lines embedded in an uniform dielectric medium, we have

$$v = v^a / \sqrt{\epsilon_r} \tag{2.17}$$

where $v^a = 1/\sqrt{\mu_0 \epsilon_0}$ is the free space velocity and has a value 2.99793×10^8 m/s. The total axial current I_0 on conductor S_2 is given by the line integral of the current density around S_2, and hence,

$$I_0 = \oint_{S_2} |J| \, dl = \oint_{S_2} v \rho_s \, dl = Qv \tag{2.18}$$

where Q is the charge on S_2 per unit length. Since the unit normal is oppositely directed on the two conductors, the direction of flow of current on conductor S_1 is opposite to that on S_2. The total current on each conductor is the same and the current wave associated with the magnetic field wave H_t is $I = I_0 \exp(-j\beta_z z)$. The characteristic impedance Z of the transmission line is given by the ratio of the potential difference between the conductors to the total current flowing on the conductor for a wave travelling in either the positive or negative z-direction, i.e.,

$$Z = \frac{V_0}{I_0} = \frac{V_0}{Qv} = \frac{1}{Cv} \tag{2.19}$$

where C is the electrostatic capacitance between the two conductors per unit length of the conductors. It may be noted that the capacitance is associated with the static field distribution in the transverse plane and is independent of the frequency of operation.

2.2.3 Propagation Parameters of TEM Transmission Lines

The most important propagation parameters useful from the point of view of circuit design are the guide wavelength and the characteristic impedance. It has been shown in the preceding section that if the capacitance C of the transmission line can be determined, then we can obtain the characteristic impedance from $Z = 1/Cv$. The phase velocity v of the electromagnetic wave in the line embedded in a dielectric medium is determined from (2.17) from the knowledge of the relative dielectric constant of the homogeneous dielectric medium. If the dielectric is replaced by air ($\epsilon_r = 1$), the characteristic impedance of the line can be obtained from

$$Z^a = \frac{1}{C^a v^a} \tag{2.20}$$

where C^a is the capacitance per unit length of the transmission line with dielectric replaced by air. From (2.17), (2.19), and (2.20) and noting that $C/C^a = \epsilon_r$, we obtain the following alternate expression for the characteristic impedance Z of the transmission line:

$$Z = \frac{Z^a}{\sqrt{\epsilon_r}} \tag{2.21}$$

Because $v^a = f\lambda^a$ and $v = f\lambda$, using (2.17) the guide wavelength λ in the TEM line can be expressed in terms of the free space wavelength λ^a as

$$\lambda = \frac{\lambda^a}{\sqrt{\epsilon_r}} \tag{2.22}$$

2.2.4 Propagation Parameters of Quasi-TEM Transmission Lines

Transmission lines in mixed dielectric media cannot support a TEM mode. Microstrip line (Fig. 1.2b) and suspended stripline (Fig. 1.3a) are examples of such lines. Under the quasi-static approximation, the propagation parameters of such lines can be defined by replacing the mixed dielectric by an equivalent uniform dielectric medium. The relative dielectric constant of this equivalent medium is termed as the 'effective dielectric constant' and is defined as

$$\epsilon_f = \frac{C}{C^a} \tag{2.23}$$

where C is the capacitance per unit length of the transmission line, and C^a is the capacitance per unit length of the same line with all dielectrics replaced by air. The phase velocity v, the guide wavelength λ, and the characteristic impedance Z are then given by

$$v = v^a/\sqrt{\epsilon_f} \tag{2.24}$$

$$\lambda = \lambda^a/\sqrt{\epsilon_f} \tag{2.25}$$

$$Z = Z^a/\sqrt{\epsilon_f} \tag{2.26}$$

Propagation parameters calculated using (2.23)-(2.26) are known to give quite accurate results adequate for most practical applications.

2.3 Method of Conformal Transformation

2.3.1 Principle of Conformal Transformation [1, 9]

It is well known that an analytic function provides a conformal transformation at any point where the function is regular and has a non-zero derivative. This conformal property of analytic functions is useful in arriving at the solutions of two-dimensional Laplace's equation subject to certain types of boundary conditions.

In order to illustrate the principle of conformal transformation, consider the functional relationship

$$W = F(Z) \tag{2.27}$$

where $Z = x + jy$ and $W = u + jv$. Here, u and v are clearly functions of x and y. We represent the complex number Z on one Argand diagram called the Z-plane (x-y plane), and the complex number W on another Argand diagram called the W-plane (u-v plane) as shown in Fig. 2.2. If W is an analytic function of the complex variable Z, then corresponding to each point Z in the x-y plane, $dW/dZ = F'(Z)$ must have an unique value in the u-v plane. Further, the real and imaginary parts of an analytic function must satisfy the Cauchy-Riemann equations [9]:

$$\frac{\partial u}{\partial x} = \frac{\partial v}{\partial y}, \qquad \frac{\partial u}{\partial y} = -\frac{\partial v}{\partial x} \tag{2.28}$$

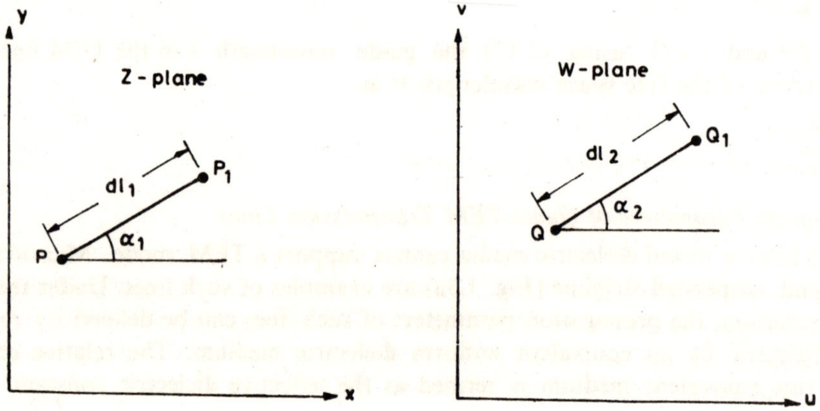

FIG. 2.2 Conformal transformation of a straight line from Z-plane to W-plane.

Differentiating these equations with respect to x and y, it can be easily shown that u and v satisfy Laplace's differential equation in two dimensions:

$$\frac{\partial^2 u}{\partial x^2} + \frac{\partial^2 u}{\partial y^2} = 0 \tag{2.29a}$$

$$\frac{\partial^2 v}{\partial x^2} + \frac{\partial^2 v}{\partial y^2} = 0 \tag{2.29b}$$

Consider now two points P and P_1 very close to each other in the Z-plane. Let dl_1 be the distance between the points and α_1 be the angle between PP_1 and the x-axis. Corresponding to each of the two points P and P_1, there will be an unique value in the W-plane. These values are represented by points Q and Q_1 as shown in Fig. 2.2. Let dl_2 be the distance between the points Q and Q_1, and α_2 be the angle between QQ_1 and the u-axis. If we represent Z in complex form as

$$Z = r\, e^{j\theta} \tag{2.30}$$

then we have,

$$dZ = dl_1\, e^{j\alpha_1} \tag{2.31a}$$

$$dW = dl_2\, e^{j\alpha_2} \tag{2.31b}$$

Differentiating the transforming function $W = F(Z)$, we have

$$dW = F'(Z)\, dZ \tag{2.32}$$

Since $F'(Z)$ must be a complex number, we can write

$$F'(Z) = A\, e^{j\phi} \tag{2.33}$$

where A is the modulus and ϕ is the argument of $F'(Z)$. Substituting for $F'(Z)$ and dZ from (2.33) and (2.31a), respectively, in (2.32) and equating with (2.31b), we have

$$dl_2\, e^{j\alpha_2} = A\, dl_1\, e^{j(\alpha_1 + \phi)} \tag{2.34}$$

so that

$$dl_2 = A\, dl_1 \tag{2.35a}$$

and

$$\alpha_2 = \alpha_1 + \phi \tag{2.35b}$$

Further, from (2.32), (2.33), and (2.35a), we have

$$\frac{dl_2}{dl_1} = A = |F'(Z)| = \left|\frac{dW}{dZ}\right| \tag{2.36}$$

Thus, each small displacement from a point P in the Z-plane, when transformed to the W-plane, is increased by a factor $A = |F'(Z)|$ and is rotated through an angle $\phi = \alpha_2 - \alpha_1 = \angle F'(Z)$. Such a process of mapping from one plane to the other is known as conformal mapping. The functions u and v are called conjugate functions. Since

$$\nabla_t u = \hat{x}\frac{\partial u}{\partial x} + \hat{y}\frac{\partial u}{\partial y} \quad \text{and} \quad \nabla_t v = \hat{x}\frac{\partial v}{\partial x} + \hat{y}\frac{\partial v}{\partial y},$$

it can be shown by using (2.28) that

$$\nabla_t u \cdot \nabla_t v = 0 \tag{2.37}$$

This relation states that the families of curves corresponding to the conjugate functions u and v form an orthogonal system.

The method of conjugate functions is very useful in the solution of two-dimensional electrostatic problems. Consider, for example, a two-conductor TEM transmission line extending uniformly in the z-direction. Figure 2.3a shows the cross-sectional view of the two conductors in the x-y plane. The potential function in the x-y plane is the solution of Laplace's equation $\nabla_t^2 \phi = 0$, subject to the boundary conditions, namely, $\phi = V_1$ on conductor S_1 and $\phi = V_2$ on conductor S_2. The solution of this problem is complicated, since S_1 and S_2 do not coincide with the constant coordinate axes. The analysis, however, can be simplified by choosing, instead of a cartesian coordinate system, another kind of system in which S_1 and S_2 coincide with the constant coordinate curves in that system.

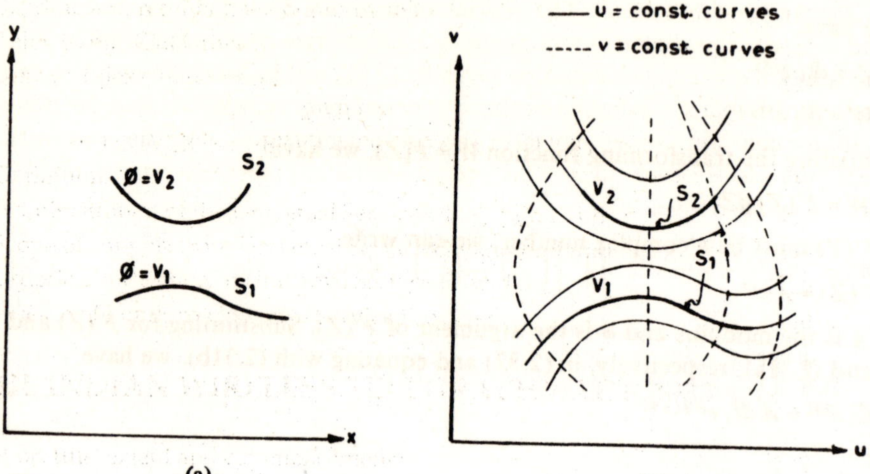

(a) (b)

FIG. 2.3 (a) Cross section of a two-conductor transmission line and (b) transformation in the u-v coordinate system.

Let u, v, w form a right-handed curvilinear coordinate frame with unit vectors $\hat{u}$, $\hat{v}$, $\hat{w}$ along the coordinate curves and scale factors h_1, h_2, and h_3. Since we are considering only the TEM mode solution, we shall confine ourselves to the two dimensional u-v plane (no component of **E** or **H** along the w-coordinate). The curvilinear coordinates u, v are functions of the cartesian coordinates x, y; that is, $u=u(x, y)$ and $v=v(x, y)$. The vector operator in the u-v plane is

$$\nabla_t=\frac{\hat{u}}{h_1}\frac{\partial}{\partial u}+\frac{\hat{v}}{h_2}\frac{\partial}{\partial v} \tag{2.38}$$

so that

$$\nabla_t u\cdot\nabla_t u=\frac{1}{h_1^2}=\left(\frac{\partial u}{\partial x}\right)^2+\left(\frac{\partial u}{\partial y}\right)^2 \tag{2.39a}$$

and

$$\nabla_t v\cdot\nabla_t v=\frac{1}{h_2^2}=\left(\frac{\partial v}{\partial x}\right)^2+\left(\frac{\partial v}{\partial y}\right)^2 \tag{2.39b}$$

By virtue of the Cauchy-Riemann conditions (2.28), we have

$$\frac{1}{h_1^2}=\frac{1}{h_2^2}=\frac{1}{h^2}=\left|\frac{dW}{dZ}\right|^2 \tag{2.40}$$

The Laplace's equation in the u-v coordinate system becomes

$$\frac{\partial}{\partial u}\left(\frac{h_2}{h_1}\frac{\partial \phi}{\partial u}\right)+\frac{\partial}{\partial v}\left(\frac{h_1}{h_2}\frac{\partial \phi}{\partial v}\right)=\frac{\partial^2 \phi}{\partial u^2}+\frac{\partial^2 \phi}{\partial v^2}=0 \tag{2.41}$$

The potential function ϕ satisfies the same equation in the u-v coordinate system as it does in the x-y coordinate system. The conductor contours S_1 and S_2 may be made to coincide with the constant coordinate curves in the u-v coordinate system as shown in Fig. 2.3b, so that the solution of Laplace's equation given by (2.41) will be much simpler as compared with the solution of Laplace's equation in the x-y coordinate system. It may be noted that in Fig. 2.3b, u = constant curves represent the equipotential lines, and the v = constant curves represent the electric flux lines. The energy stored in the electrostatic field in the x-y coordinate system is given by

$$W_e=\frac{1}{2}C(V_2-V_1)^2=\frac{\epsilon}{2}\iint\limits_{xy\text{-plane}}\left[\left(\frac{\partial \phi}{\partial x}\right)^2+\left(\frac{\partial \phi}{\partial y}\right)^2\right]dx\,dy$$

$$=\frac{\epsilon}{2}\iint\limits_{uv\text{-plane}}\left[\frac{1}{h_1^2}\left(\frac{\partial \phi}{\partial u}\right)^2+\frac{1}{h_2^2}\left(\frac{\partial \phi}{\partial v}\right)^2\right]h_1h_2\,du\,dv$$

$$=\frac{\epsilon}{2}\iint\limits_{uv\text{-plane}}\left[\left(\frac{\partial \phi}{\partial u}\right)^2+\left(\frac{\partial \phi}{\partial v}\right)^2\right]du\,dv \tag{2.42}$$

Thus the capacitance of the two-conductor transmission line in the u-v plane is the same as its capacitance in the x-y plane. In other words, the capacitance of the TEM line remains invariant under conformal transformation.

2.3.2 *The Schwarz-Christoffel Transformation* [1]

Planar transmission lines with which we are concerned in this book, have straight boundaries and can be treated as closed polygons. One of the most useful forms of conformal transformation that is suitable for such problems is the Schwarz-Christoffel transformation. It maps the interior of the polygon in one plane, say, the Z-plane, to the upper half of another plane, say, the T-plane. The sides of the polygon become the real axis in the T-plane. The upper half of the T-plane is not likely to be the final configuration possessing a known solution. A second Schwarz-Christoffel transformation may be used to convert the upper half of the T-plane to a polygon of simpler geometry in another plane, say, the W-plane. Sometimes, a combination of more than two transformations may be required to arrive at the desired configuration.

Consider the following differential equation defining the basic transformation:

$$\frac{dZ}{dT}=F'(T)=A(T-r_1)^{-k_1}(T-r_2)^{-k_2}\ldots(T-r_N)^{-k_N}=A\prod_{i=1}^{N}(T-r_i)^{-k_i} \tag{2.43}$$

where $Z=x+jy$ and $T=r+js$ are complex variables, A is a complex constant, r_i and k_i are real numbers, and $r_1 < r_2 < \ldots < r_N$. Since the argument of a product of complex numbers is equal to the sum of the arguments of the individual factors, we have

$$\angle \underline{F'(T)}=\angle \underline{A}-k_i\sum_{i=1}^{N}\angle \underline{(T-r_i)} \tag{2.44}$$

Let us now consider the problem of mapping a N-sided polygon in the Z-plane to the upper half of the T-plane as shown in Fig. 2.4. The real numbers $r_1, r_2, \ldots, r_N$ are plotted on the real axis of the T-plane. For any point T on the r-axis, the number $(T-r_i)$ is positive if T lies on the right side of r_i, and is negative if T lies on the left side. Hence,

$$\underline{\angle (T-r_i)} = \begin{cases} 0, & T > r_i \\ \pi, & T < r_i \end{cases} \tag{2.45}$$

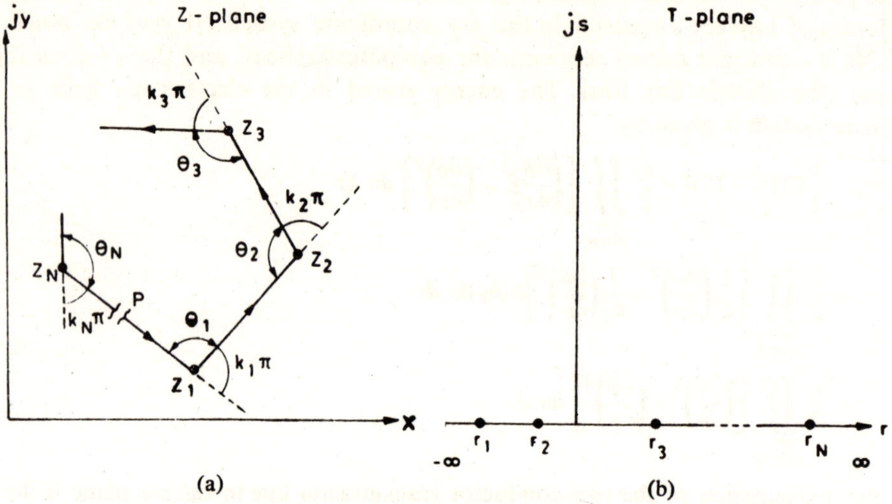

FIG. 2.4 Schwarz-Christoffel transformation: (a) Z-plane and (b) T-plane.

From (2.44) and (2.45), it can be seen that as T traverses on the real axis of the T-plane from left to right and as each point r_i is crossed, $\underline{\angle F'(T)}$ changes discontinuously by an amount $k_i\pi$ for the path traced in the Z-plane. Consequently, as T traverses the real axis of the T-plane, the corresponding point Z traverses a polygon in the Z-plane. The interior angle θ_i formed by two adjacent sides of the polygon is given by $\theta_i = \pi(1-k_i)$. For a N-sided closed polygon, the sum of all the exterior angles must be equal to 2π or equivalently, $\sum_{i=1}^{N} k_i = 2$.

Integrating the expression (2.43) with respect to T, we have

$$Z = A \int \prod_{i=1}^{N} (T-r_i)^{-k_i}\, dT + B \tag{2.46}$$

where B is an arbitrary constant. The modulus of the constant A determines the size of the polygon, the argument of A determines its orientation, and the constant B serves to locate its position in the W-plane.

In the Schwarz-Christoffel transformation, the polygon may be opened at some arbitrary point to suit the problem. In Fig. 2.4a, it is opened at a point P between Z_N and Z_1. One side of the opening becomes $-\infty$ and the other $+\infty$ in the T-plane as shown in Fig. 2.4b. If for the sake of convenience, the opening is made at a corner of the polygon, that corner does not take part in the transformation, and, there must be one factor less in the expression (2.43). It may be noted that dZ/dT must be either zero or infinite at each of the corners. This is satisfied in (2.43).

2.4 Variational Approach [2]

2.4.1 Basis of Variational Method

The variational method is an established mathematical technique, generally applied to those problems where the physical system under consideration so acts that some function of its behaviour attains the least or the greatest value. For example, in an electrostatic system, the well-known Thomson's theorem states that the charges which reside on conducting bodies and which give rise to the electric field **E** will distribute themselves in such a way that the energy function W_e is minimized.

Consider a system of perfect conductors S_1, S_2, . . . , S_N as shown in Fig. 2.5. Let Q_1, Q_2, . . . , Q_N be the charges on the conductors and let the conductors be held at potentials V_1, V_2, . . . , V_N. The potential function ϕ in the space surrounding the conductors is the solution of Laplace's equation $\nabla^2\phi=0$, subject to the boundary conditions $\phi=V_i$ on S_i with $i=1$, 2, . . . , N. Since the electric field **E** is obtained from the gradient of the potential function ϕ, the electrostatic energy stored is given by

$$W_e=\frac{\epsilon}{2}\int_{\text{vol}}\nabla\phi\cdot\nabla\phi\,dV \tag{2.47}$$

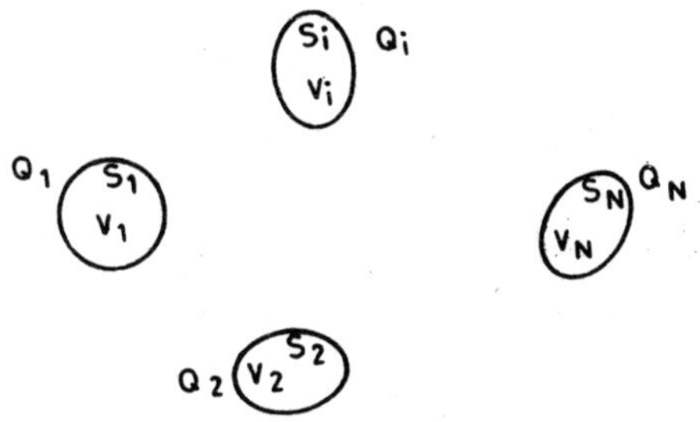

FIG. 2.5 System of conductors with surface charges Q_1, Q_2, . . ., Q_i, . . ., Q_N.

where the integration is carried over the entire volume containing the electric field. Supposing the charges on the conductors are slightly displaced from their equilibrium positions while keeping the potentials on the conductors constant, then the potential distribution in the surrounding space also changes. Let the potential function ϕ change to $\phi+\delta\phi$ where $\delta\phi$ is an incremental change in ϕ. Here, $\delta\phi$ is a function of the space coordinates but vanishes on the conductors S_i bacause V_i's are held constant. The change in the energy function is then given by

$$\delta W_e=\frac{\epsilon}{2}\left[\int_{\text{vol}}\nabla(\phi+\delta\phi)\cdot\nabla(\phi+\delta\phi)\,dV-\int_{\text{vol}}\nabla\phi\cdot\nabla\phi\,dV\right]$$

$$=\frac{\epsilon}{2}\left[\int_{\text{vol}}2\nabla\phi\cdot\nabla\delta\phi\,dV+\int_{\text{vol}}\nabla\delta\phi\cdot\nabla\delta\phi\,dV\right] \tag{2.48}$$

By considering a surface S enclosing the volume containing the electric field and applying the divergence theorem, it can be shown that the first integral vanishes. For a detailed derivation of this result, the reader is rererred to Collin [2]. The final expression for the change in the energy function is

$$\delta W_e = \frac{\epsilon}{2} \int_{\text{vol}} \nabla \delta\phi \cdot \nabla \delta\phi \, dV \tag{2.49}$$

If we insert a trial function for the potential distribution which differs by a small quantity $\delta\phi$ from the correct value, the resulting value of W_e will differ from its true value by an amount proportional to $(\delta\phi)^2$. Equivalently, for a first-order change in ϕ, the change in W_e is only of second order. The energy function W_e is therefore a stationary function for the equilibrium conditions. Further, from (2.49), it can be seen that W_e is a positive quantity. Hence, the true value of W_e is a minimum, since any change from the equilibrium increases the energy function W_e.

2.4.2 Variational Expression for Upper Bound on Capacitance

Consider a two-conductor transmission line as shown in Fig. 2.1. The energy stored in the electrostatic field per unit length along the line is given by

$$W_e = \frac{\epsilon}{2} \iint_{\substack{xy-\\ \text{plane}}} |\nabla_t \phi|^2 \, dx \, dy = \frac{1}{2} C V_0^2 \tag{2.50}$$

where C is the capacitance per unit length of the line and V_0 is the potential difference between the two conductors. The capacitance per unit length of the line is given by

$$C = \frac{\epsilon}{V_0^2} \iint_{\substack{xy-\\ \text{plane}}} |\nabla_t \phi|^2 \, dx \, dy = \frac{\epsilon \displaystyle\iint_{\substack{xy-\\ \text{plane}}} |\nabla_t \phi|^2 \, dx \, dy}{\left(\displaystyle\int_{S_1}^{S_2} \nabla_t \phi \cdot \mathbf{dl} \right)^2} \tag{2.51}$$

where V_0 is the line integral of $\nabla_t \phi$ from S_1 to S_2. Here, (2.51) is the variational expression for the capacitance C. If we substitute an approximate value for ϕ, the calculated value of C will always be larger than the true value. The variational method involves the determination of the capacitance by employing a trial function containing several variable parameters. The particular form of the trial function is based on some a priori guesses as to the nature of the exact solution. Let these parameters be α_i, with $i = 1, 2, \ldots, N$, so that $C = C(\alpha_1, \alpha_2 \ldots \alpha_N)$. The condition that the integral is stationary leads then to the simultaneous equations $\partial C/\partial \alpha_i = 0$ for $i = 1, 2, \ldots, N$. Solution of the N homogeneous equations yields the values of α_i and hence the minimum value of C.

2.4.3 Variational Expression for Lower Bound on Capacitance

The variational formula (2.51) expressed in terms of the potential distribution ϕ gives an upper bound on the capacitance. In this section, we present another variational expression which gives a lower bound on the capacitance. This is obtained by solving the electrostatic problem in terms of the unknown charge distribution by utilizing Green's function technique. The method has been described in detail in Collin [2], and hence only the salient steps leading to the variational formula for the capacitance are presented here.

Consider a two-conductor TEM transmission line, the cross-sectional view of which is shown in Fig. 2.6a. Conductor S_2 is at a positive potential V_0 and conductor S_1 is at zero potential. We formulate the problem in terms of the two-dimensional Poisson's equation:

$$\nabla_t^2 \phi = -\frac{\rho(x_0, y_0)}{\epsilon} \tag{2.52}$$

with the boundary condition that $\phi = 0$ on S_1. Here $\rho(x_0, y_0)$ is the unknown charge distribution on S_2, for which a suitable trial function would be substituted later.

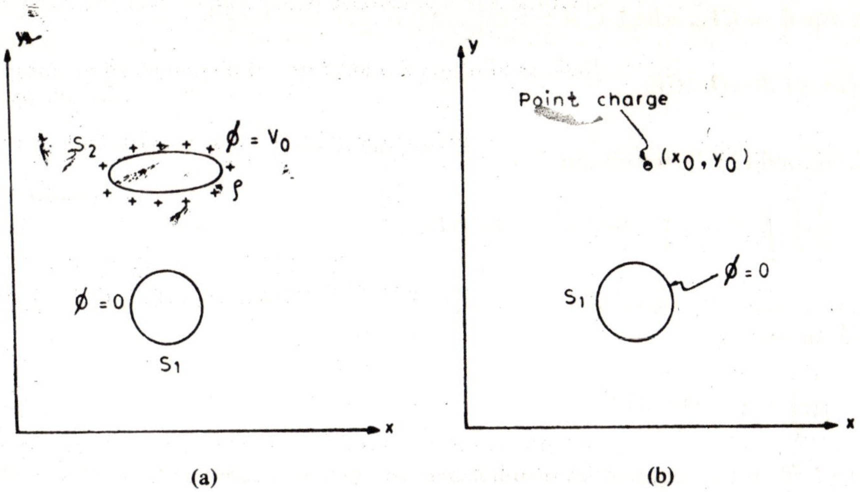

FIG. 2.6 (a) Two-dimensional electrostatic problem —— Formulation in terms of surface charge on conductor S_2 and (b) illustration for determining Green's function for (a).

Green's function G corresponding to this problem is first obtained by considering the potential due to a point charge at (x_0, y_0) in the presence of the conductor S_1 as illustrated in Fig. 2.6b. Now, G must satisfy the Poisson's equation

$$\nabla_t^2 G = -\frac{1}{\epsilon}\delta(x - x_0)\delta(y - y_0) \tag{2.53}$$

with $G = 0$ on S_1 and at infinity. Here, $\delta(x - x_0)$ and $\delta(y - y_0)$ are Dirac's delta functions. We represent the solution G by $G(x, y/x_0, y_0)$, because it is a function of both the point of observation and the source point. The solution of (2.52) is then obtained by applying the principle of superposition:

$$\phi(x, y) = \int_{S_2} G(x, y/x_0, y_0)\rho(x_0, y_0)\, dl_0 \tag{2.54}$$

where the integration is carried out over the contour of S_2 on which the charges are distributed. On the surface of S_2, the potential must reduce to V_0. We can therefore write

$$V_0 = \int_{S_2} G(x, y/x_0, y_0)\rho(x_0, y_0)\, dl_0, \qquad x, y \text{ on } S_2 \tag{2.55}$$

This is the basic equation which can, in principle, be used to determine the unknown charge distribution on the conductor S_2. Here, this equation is used to obtain a variational expression for the capacitance, in which the charge distribution becomes a trial function. Multiply both sides of (2.55) by $\rho(x, y)$ and integrate over S_2:

$$V_0 \int_{S_2} \rho(x, y)\, dl = \int_{S_2}\int_{S_2} G(x, y/x_0, y_0)\rho(x, y)\rho(x_0, y_0)\, dl\, dl_0 \tag{2.56}$$

Now, the integral on the left-hand side gives the total charge per unit length of conductor S_2 and is equal to CV_0, where C is the capacitance per unit length, i.e.,

$$\int_{S_2} \rho(x, y)\, dl = Q = CV_0 \tag{2.57}$$

Using (2.56) and (2.57), we obtain

$$\frac{1}{C} = \frac{1}{Q^2}\int_{S_2}\int_{S_2} G(x, y/x_0, y_0)\rho(x, y)\rho(x_0, y_0)\, dl\, dl_0 \tag{2.58}$$

Alternatively, using the definition of $\phi(x, y)$ as given in (2.54), the capacitance expression can be written in the form

$$\frac{1}{C} = \frac{1}{Q^2}\int_{S_2} \phi(x, y)\rho(x, y)\, dl \tag{2.59}$$

Equation (2.58) is the required variational expression for C. Collin [2] has shown that (2.58) is a stationary expression for C for arbitrary first-order changes in the functional form of ρ. Further, when ρ is changed by a small value $\delta\rho$, the first order change in $1/C$ vanishes and the second-order change is always positive. Thus, for any trial function $\rho(x_0, y_0)$, the calculated value of $1/C$ is always larger than the true value, and we have a lower bound on the capacitance.

2.5 Transverse Transmission Line Technique—Application to Planar Structures with Multilayer Dielectrics

In the preceding section, two separate variational expressions have been derived for determining the capacitance of a transmission line. In order to apply these expressions to a particular planar transmission structure, it is first necessary to determine either the potential distribution function or the Green's function. These functions can be derived by solving a set of algebraic equations obtained by applying the boundary conditions at the various boundaries and interfaces. The application of this method to a planar transmission line with multilayer dielectrics, however, becomes increasingly complicated with an increase in the number of dielectric layers. On the other hand, the 'transverse transmission line' technique, reported by Crampagne *et al.* [5], provides a simple method of determining the Green's function for MIC lines with multilayer dielectrics. This method makes use of the analogy between the basic equations governing multilayer MIC line and those of a two-wire transmission line excited by a current source and having stepped characteristic admittance along the line. The following sections describe the 'transverse transmission line' technique of determining the Green's function for a multilayer planar transmission line with side walls. The variational formula given by (2.58) can then be directly used to find the capacitance of such structures. If the side walls are separated by an infinite distance, it is much easier to determine the potential function in the Fourier transform domain using the 'transverse transmission line' technique and then apply the variational formula (2.59) in the Fourier domain.

2.5.1 Determination of Green's Function in Space Domain

We consider the problem of determining the Green's function for a general multilayer dielectric structure with an infinitesimally thin strip conductor located at $y = y_0$ and with side walls separated by a distance c as shown in Fig. 2.7a. For a point charge located at (x_0, y_0), Green's function should satisfy Poisson's differential equation in the xy-plane:

$$\nabla_t^2 G(x, y/x_0, y_0) = -\frac{1}{\epsilon}\delta(x - x_0)\cdot\delta(y - y_0) \tag{2.60}$$

The boundary condition on each of the side walls at $x = 0$ and c is either of Dirichlet type for an electric wall ($G = 0$) or of Neumann type for a magnetic wall (normal derivative of G, $\partial G/\partial n = 0$). For the structure considered, Green's function can be expressed as the sum of the product of elementary functions with separated variables:

$$G = \sum_n G_n(x) G_n(y) \tag{2.61}$$

In order to satisfy the boundary conditions on the vertical side walls, the following expressions are found for $G_n(x)$ for three separate cases.

Case a. Electric walls at $x = 0$ and c:

$$G_n(x) = \sin\frac{n\pi x}{c}, \qquad n = 1, 2, \ldots, \infty \tag{2.62}$$

Case b. Electric wall at $x = 0$ and magnetic wall at $x = c$:

$$G_n(x) = \sin\frac{(2n+1)\pi x}{2c}, \qquad n = 0, 1, 2, \ldots, \infty \tag{2.63}$$

Case c. Magnetic walls at $x = 0$ and c:

$$G_n(x) = \cos\frac{n\pi x}{c}, \qquad n = 1, 2, \ldots, \infty \tag{2.64}$$

Using these expressions in (2.60) and noting that the functions, $\sin(n\pi x/c)$, $\sin[(2n+1)\pi x/2c]$ and $\cos(n\pi x/c)$ are orthogonal in the interval $(0, c)$, the following differential equations are obtained:

Case a:

$$\left[\frac{d^2}{dy^2} - \left(\frac{n\pi}{c}\right)^2\right] G_n(y) = -\frac{2}{c\epsilon}\delta(y - y_0)\sin\frac{n\pi x_0}{c} \tag{2.65}$$

Case b:

$$\left[\frac{d^2}{dy^2} - \left\{\frac{(2n+1)\pi}{2c}\right\}^2\right] G_n(y) = -\frac{2}{c\epsilon}\delta(y - y_0)\sin\left[\frac{(2n+1)\pi x_0}{2c}\right] \tag{2.66}$$

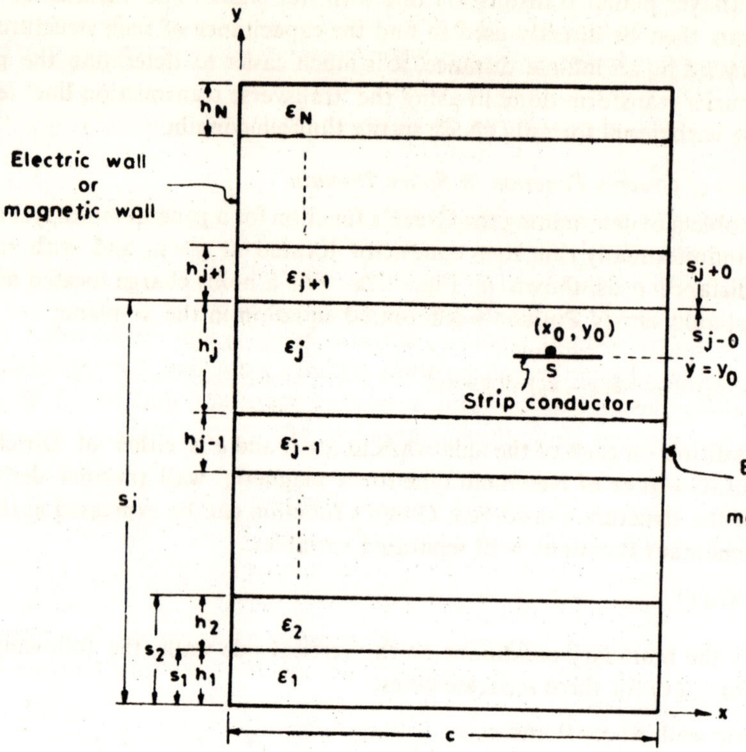

(a)

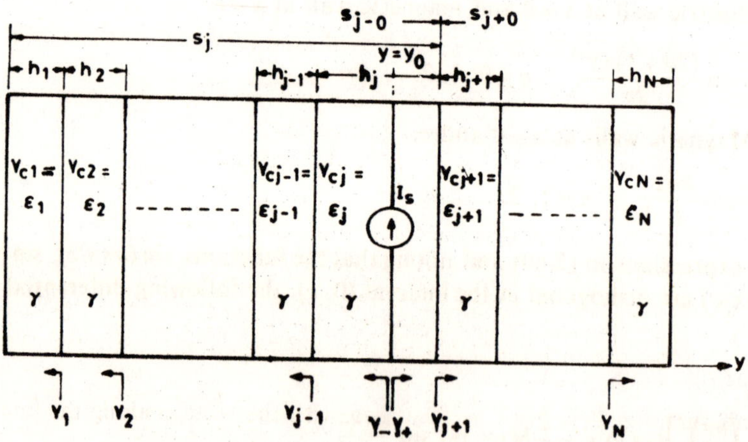

(b)

FIG. 2.7 (a) Geometry of N-layer dielectric with side walls and an infinitesimally thin strip conductor at $y=y_0$ and (b) two-wire transmission line equivalent: $\gamma=n\pi/c$ for cases (a) and (c); $\gamma=(2n+1)\pi/2c$ for case (b).

Case c:

$$\left[\frac{d^2}{dy^2}-\left(\frac{n\pi}{c}\right)^2\right]G_n(y)=-\frac{2}{c\epsilon}\delta(y-y_0)\cos\frac{n\pi x_0}{c} \tag{2.67}$$

Green's function G and hence $G_n(y)$ should satisfy the following boundary conditions at the various dielectric interfaces:

$$G_n(y)|_{y=s_j-0}=G_n(y)|_{y=s_j+0} \tag{2.68a}$$

$$\epsilon_j\left.\frac{\partial G_n(y)}{\partial y}\right|_{y=s_j-0}=\epsilon_{j+1}\left.\frac{\partial G_n(y)}{\partial y}\right|_{y=s_j+0},\qquad j=1,2,\ldots,N-1 \tag{2.68b}$$

We now identify the differential equation satisfied by $G_n(y)$ and the boundary conditions, with the equations satisfied by the voltage along a transmission line. Consider a transmission line extending along the y-direction with a current source of intensity I_s at $y=y_0$. The voltage and current relations along the line are [2, 5, 10]:

$$\frac{dV}{dy}=-\gamma Z_c I \tag{2.69}$$

$$\frac{dI}{dy}=-\frac{\gamma}{Z_c}V+I_s\delta(y-y_0) \tag{2.70}$$

where Z_c is the characteristic impedance of the line and γ is the propagation constant. Equations (2.69) and (2.70) can be solved simultaneously to lead to the following differential equation for the voltage:

$$\frac{d^2V}{dy^2}-\gamma^2V=-\gamma Z_c I_s\,\delta(y-y_0) \tag{2.71}$$

If the transmission line has stepped characteristic admittances and Y_{cj} is the characteristic admittance of the jth section of transmission line as shown in Fig. 2.7b, the continuity conditions at the interface between two different admittances are given by

$$V|_{y=s_j-0}=V|_{y=s_j+0} \tag{2.72a}$$

and $I_j=I_{j+1}$, which combined with (2.69) gives

$$Y_{cj}\left.\frac{\partial V}{\partial y}\right|_{y=s_j-0}=Y_{cj+1}\left.\frac{\partial V}{\partial y}\right|_{y=s_j+0} \tag{2.72b}$$

Comparing (2.71) satisfied by the voltage along the transmission line with equations (2.65), (2.66), and (2.67) satisfied by $G_n(y)$ for the three cases, and further comparing the boundary conditions (2.72a) and (2.72b) with (2.68a) and (2.68b), respectively, the following equivalences can be observed.

1. Green's function $G_n(y)$ can be identified with the voltage along the line

$$V\equiv G_n(y) \tag{2.73}$$

2. The dielectric constant of the layer can be identified with the characteristic admittance of the corresponding section of two wire line

$$Y_{cj}\equiv\epsilon_j \tag{2.74}$$

With these equivalences, the boundary conditions satisfied by the Green's function at the various interfaces are equivalent to the boundary conditions satisfied by the voltages at the interfaces between two dissimilar characteristic admittances. The voltage on the transmission line at $y = y_0$ is given by

$$V|_{y=y_0} = \frac{I_s}{Y} \tag{2.75}$$

where Y is the admittance at the plane $y = y_0$. In the subsequent expressions, we denote Y as Y_I. The subscript I is used to refer to the isotropic structure and to distinguish from the admittance Y_A of the anisotropic structure considered in Sec. 2.6. Referring to Fig. 2.7b, $Y_I = Y_+ + Y_-$, where Y_+ is the input admittance of the transmission line at $y = y_0$ looking in the positive y-direction and Y_- is the input admittance of the transmission line looking in the negative y-direction. Comparing the right-hand side of (2.65)-(2.67) with that of (2.71) and using (2.73)-(2.75), we obtain the necessary expressions for Green's function for the three cases.

Case a:

$$Z_c = \frac{1}{\epsilon}, \quad \gamma = \frac{n\pi}{c}, \quad \text{and} \quad I_s = \frac{2}{n\pi} \sin \frac{n\pi x_0}{c}$$

Thus

$$G_n(y)|_{y=y_0} = \frac{2}{n\pi Y_I} \sin \frac{n\pi x_0}{c} \tag{2.76}$$

Substituting (2.62) and (2.76) in (2.61), the expression for Green's function at the charge plane $y = y_0$ becomes

$$G(x, y/x_0, y_0)|_{y=y_0} = \sum_{n=1}^{\infty} \frac{2}{n\pi Y_I} \sin \frac{n\pi x}{c} \sin \frac{n\pi x_0}{c} \tag{2.77}$$

Case b:

$$Z_c = \frac{1}{\epsilon}, \quad \gamma = \frac{(2n+1)\pi}{2c}, \quad \text{and} \quad I_s = \frac{4}{(2n+1)\pi} \sin \left[\frac{(2n+1)\pi x_0}{2c}\right]$$

We thus have

$$G_n(y)|_{y=y_0} = \frac{4}{(2n+1)\pi Y_I} \sin \left[\frac{(2n+1)\pi x_0}{2c}\right] \tag{2.78}$$

Substituting (2.63) and (2.78) in (2.61), we obtain

$$G(x, y/x_0, y_0)|_{y=y_0} = \sum_{n=0}^{\infty} \frac{4}{(2n+1)\pi Y_I} \sin \left[\frac{(2n+1)\pi x}{2c}\right] \sin \left[\frac{(2n+1)\pi x_0}{2c}\right] \tag{2.79}$$

Case c:

$$Z_c = \frac{1}{\epsilon}, \quad \gamma = \frac{n\pi}{c}, \quad \text{and} \quad I_s = \frac{2}{n\pi} \cos \frac{n\pi x_0}{c}$$

Thus

$$G_n(y)|_{y=y_0} = \frac{2}{n\pi Y_I} \cos \frac{n\pi x_0}{c} \tag{2.80}$$

Using (2.64) and (2.80) in (2.61), we obtain

$$G(x, y/x_0, y_0)|_{y=y_0} = \sum_{n=1}^{\infty} \frac{2}{n\pi Y_I} \cos\frac{n\pi x}{c} \cos\frac{n\pi x_0}{c} \tag{2.81}$$

Thus, the problem of determining the Green's function reduces to that of determining the admittance at the charge plane $y = y_0$. The expression for Y_I is determined from the transmission line equivalent (Fig. 2.7b).

2.5.2 Determination of Potential Function in the Fourier Transform Domain

Consider a general N-layer dielectric structure as shown in Fig. 2.8a. The structure extends to infinity in the positive and negative x-directions. For an infinitesimally thin line charge situated at $y = y_0$, the potential function $\phi(x, y)$ satisfies Poisson's differential equation in the two-dimensional xy-plane:

$$\begin{aligned} \nabla_t^2 \phi &= -\frac{1}{\epsilon}\rho(x, y) \\ &= -\frac{1}{\epsilon} f(x)\delta(y - y_0) \end{aligned} \tag{2.82}$$

where $f(x)$ gives the distribution of charge in the x-direction. Applying the Fourier transform with respect to x, (2.82) can be rewritten as

$$\left[\frac{d^2}{dy^2} - \beta^2\right]\hat{\phi}(\beta, y) = -\frac{1}{\epsilon}\hat{f}(\beta)\delta(y - y_0) \tag{2.83}$$

where $\hat{\phi}(\beta, y)$ and $\hat{f}(\beta)$ are Fourier transforms of $\phi(x, y)$ and $f(x)$, respectively. They are defined as

$$\hat{\phi}(\beta, y) = \int_{-\infty}^{\infty} \phi(x, y)\, e^{j\beta x}\, dx \tag{2.84}$$

$$\hat{f}(\beta) = \int_{-\infty}^{\infty} f(x)\, e^{j\beta x}\, dx \tag{2.85}$$

The boundary conditions in the Fourier domain at the various dielectric interfaces are given by

$$\hat{\phi}(\beta, y)|_{y=s_j-0} = \hat{\phi}(\beta, y)|_{y=s_j+0} \tag{2.86a}$$

$$\epsilon_j \frac{\partial \hat{\phi}}{\partial y}(\beta, y)|_{y=s_j-0} = \epsilon_{j+1}\frac{\partial \hat{\phi}}{\partial y}(\beta, y)|_{y=s_j+0}, \qquad j = 1, 2, \ldots, N-1 \tag{2.86b}$$

The transmission line equivalent (Fig. 2.8b) is the same as that shown in Fig. 2.7b. Comparing (2.71), (2.72a), and (2.72b) satisfied by the voltage along the transmission line, with equations (2.83), (2.86a), and (2.86b) satisfied by the potential function, the following analogous relations can be identified:

1. The potential function can be identified with the voltage along the line

$$V \equiv \hat{\phi}(\beta, y) \tag{2.87}$$

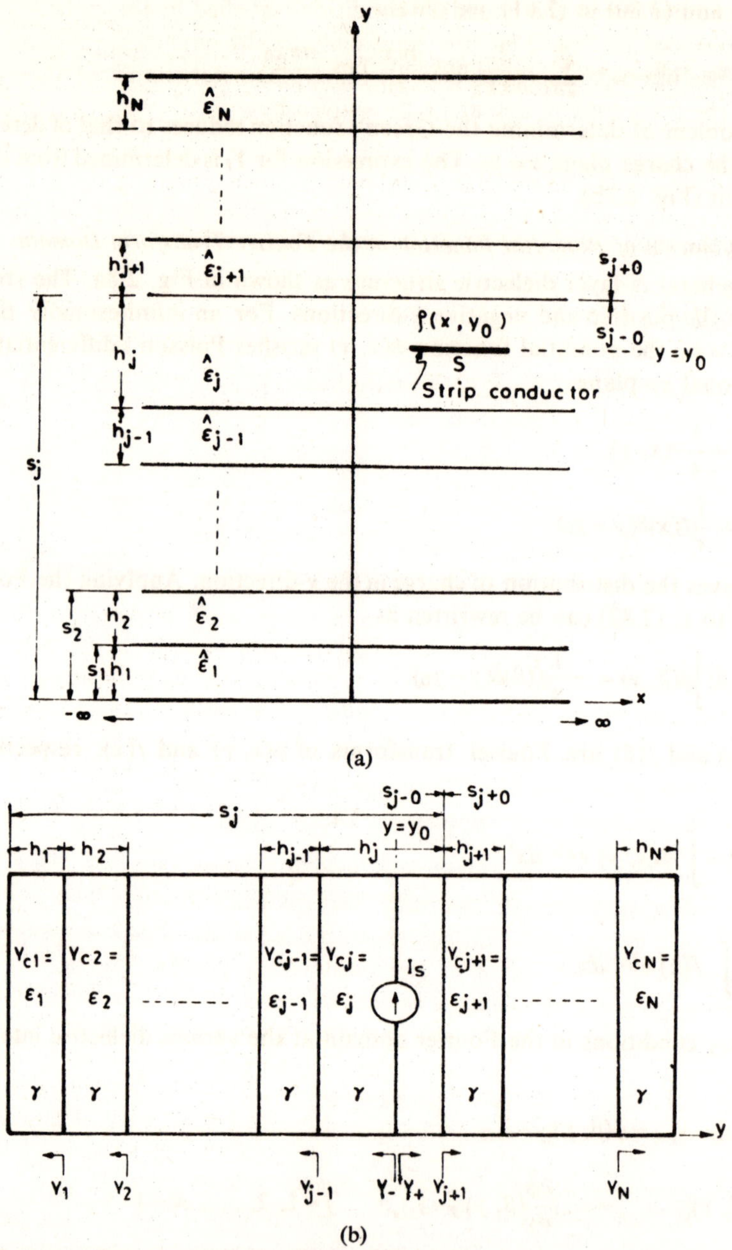

FIG. 2.8 (a) Geometry of N-layer dielectric with charge at $y = y_0$ and (b) two-wire transmission line equivalent ($\gamma = \beta$).

2. The dielectric constant of the jth layer can be identified with the characteristic admittance of the jth section of the two-wire transmission line

$$Y_{cj} \equiv \epsilon_j \tag{2.88}$$

With these equivalences, the boundary conditions satisfied by the potential function at the various interfaces in Fig. 2.8a are equivalent to the boundary conditions satisfied by the voltages at the interfaces between two dissimilar characteristic admittances in Fig. 2.8b. The voltage on the transmission line at $y=y_0$ is given by

$$V|_{y=y_0}=\frac{I_s}{Y_I}, \qquad Y_I=Y_+ + Y_- \tag{2.89}$$

Comparing the right-hand sides of (2.71) and (2.83), we have

$$\gamma Z_c I_s=\frac{\hat{f}(\beta)}{\epsilon} \tag{2.90}$$

Further, with $Z_c=1/\epsilon$ and $\gamma=\beta$, (2.90) reduces to

$$I_s=\frac{\hat{f}(\beta)}{\beta} \tag{2.91}$$

From (2.87), (2.89), and (2.91), we obtain

$$\hat{\phi}(\beta, y)|_{y=y_0}=\frac{\hat{f}(\beta)}{\beta Y_I} \tag{2.92}$$

The problem of determining $\hat{\phi}(\beta, y)$ at $y=y_0$ reduces to that of finding the Fourier transform of the charge distribution $f(x)$ and the admittance parameter Y_I for the particular structure.

2.6 Transverse Transmission Line Technique—Application to Planar Structures with Multilayer Anisotropic Dielectrics

Planar transmission lines with anisotropic dielectric substrates offer interesting possibilities in MIC applications. Of particular interest are uniaxially anisotropic substrates cut with their optical axis perpendicular to the plane of the substrate. Since the material permittivity is constant everywhere in the plane of the substrate, a propagating wave travelling along a stripline is not subjected to a change in permittivity at bends and other discontinuities in the line. In this section, we apply the 'transverse transmission line' technique to the anisotropic structure, and on the basis of the analogy with a two-wire transmission line consisting of sections of lengths with different propagation constants and characteristic admittances, obtain expressions for Green's function and the potential function [7].

2.6.1 Determination of Green's Function in Space Domain

Consider a multilayer structure with an infinitesimally thin strip conductor S located at $y=y_0$ as shown in Fig. 2.9a. The side walls are separated by a distance c. The substrates are assumed to be uniaxially anisotropic having a tensor permittivity given by

$$\hat{\epsilon}_j=\epsilon_0\begin{bmatrix}\epsilon_{xxj} & 0 & 0\\ 0 & \epsilon_{yyj} & 0\\ 0 & 0 & \epsilon_{xxj}\end{bmatrix}, \qquad j=1,2,\ldots,N \tag{2.93}$$

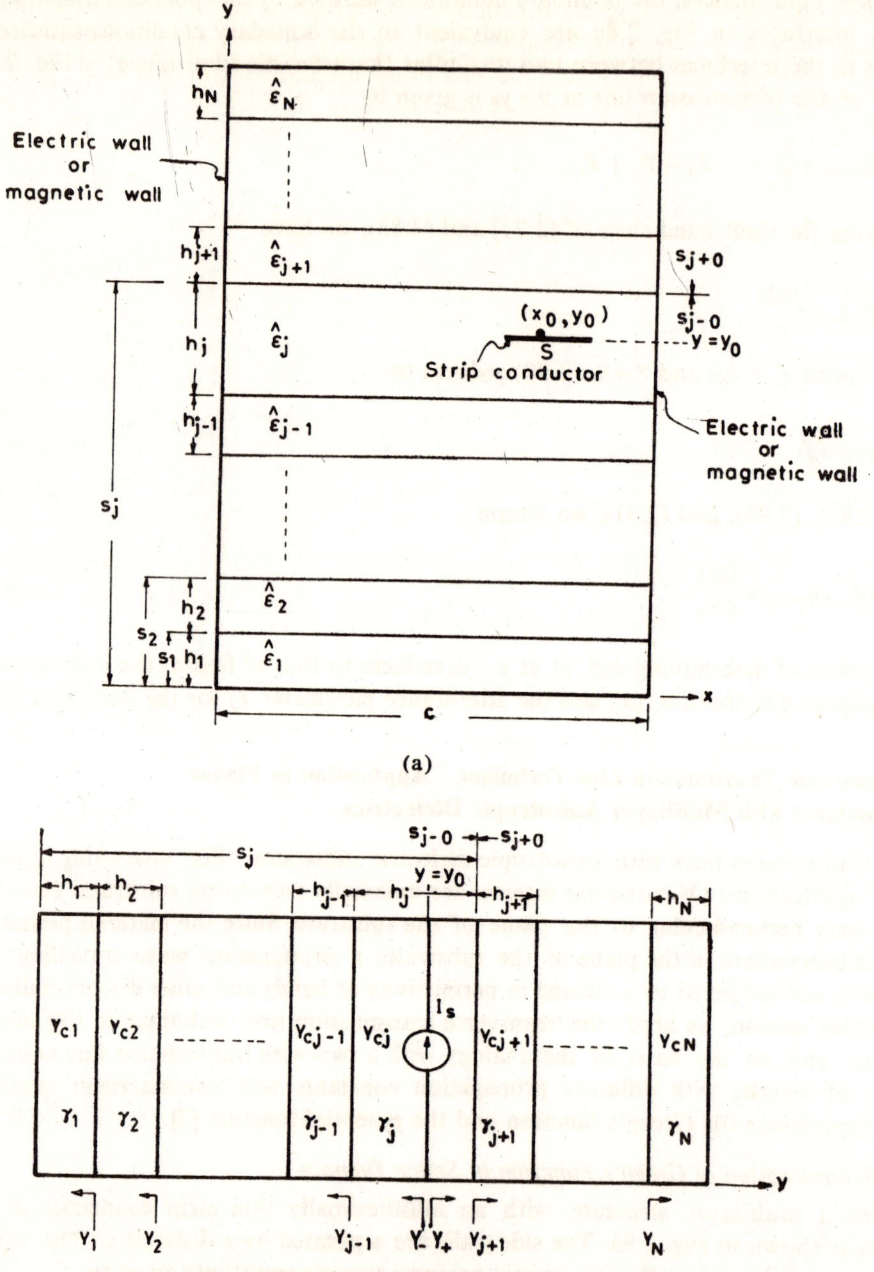

FIG. 2.9 (a) Geometry of N-layer anisotropic dielectric with side walls and an infinitesimally thin conductor at $y=y_0$ and (b) two-wire transmission line equivalent: $Y_{cj}=\epsilon_0\sqrt{\epsilon_{xxj}\,\epsilon_{yyj}}$; $\gamma_j=(n\pi/c)\sqrt{\epsilon_{xxj}/\epsilon_{yyj}}$ for cases (a) and (c); $\gamma_j=[(2n+1)\pi/2c]\sqrt{\epsilon_{xxj}/\epsilon_{yyj}}$ for case (b).

In order to determine the Green's function for the structure, we first consider a point charge at (x_0, y_0) instead of the strip conductor S. Under the quasi-static approximation, the Green's function should satisfy the following Poisson's differential equation in the xy-plane:

$$\nabla \cdot [\hat{\epsilon}_j \cdot \nabla G(x, y/x_0, y_0)] = -\delta(x-x_0)\delta(y-y_0) \tag{2.94}$$

Expanding and denoting $G(x, y/x_0, y_0)$ as G, we get

$$\epsilon_{xxj}\frac{\partial^2 G}{\partial x^2} + \epsilon_{yyj}\frac{\partial^2 G}{\partial y^2} = -\frac{1}{\epsilon_0}\delta(x-x_0)\delta(y-y_0) \tag{2.95}$$

For the structure considered, Green's function can be expressed as the sum of the product of elementary functions as for the isotropic case given by (2.61). The expressions for $G_n(x)$ for the three different combinations of side boundary walls are the same as those given in (2.62)-(2.64) for the isotropic case. Using (2.61)-(2.64) in (2.95) and noting that the functions $\sin(n\pi x/c)$, $\sin[(2n+1)\pi x/2c]$ and $\cos(n\pi x/c)$ are orthogonal in the interval $(0, c)$, the following differential equations are obtained for the three cases of boundary wall conditions:

Case a. Electric walls at $x=0$ and c:

$$\left[\frac{d^2}{dy^2} - \left\{\left(\frac{n\pi}{c}\right)\Big/\sqrt{\frac{\epsilon_{xxj}}{\epsilon_{yyj}}}\right\}^2\right]G_n(y) = -\frac{2}{c\epsilon_0\epsilon_{yyj}}\delta(y-y_0)\cdot\sin\frac{n\pi x_0}{c} \tag{2.96}$$

Case b. Electric wall at $x=0$ and magnetic wall at $x=c$:

$$\left[\frac{d^2}{dy^2} - \left\{\frac{(2n+1)\pi}{2c}\sqrt{\frac{\epsilon_{xxj}}{\epsilon_{yyj}}}\right\}^2\right]G_n(y) = -\frac{2}{c\epsilon_0\epsilon_{yyj}}\delta(y-y_0)\sin\frac{(2n+1)\pi x_0}{2c} \tag{2.97}$$

Case c. Magnetic walls at $x=0$ and c:

$$\left[\frac{d^2}{dy^2} - \left\{\left(\frac{n\pi}{c}\right)\Big/\sqrt{\frac{\epsilon_{xxj}}{\epsilon_{yyj}}}\right\}^2\right]G_n(y) = -\frac{2}{c\epsilon_0\epsilon_{yyj}}\delta(y-y_0)\cos\frac{n\pi x_0}{c} \tag{2.98}$$

Green's function G, and hence $G_n(y)$ should satisfy the following boundary conditions at the various interfaces:

$$G_n(y)\big|_{y=s_{j-0}} = G_n(y)\big|_{y=s_{j+0}} \tag{2.99}$$

$$\epsilon_0\epsilon_{yyj}\frac{\partial G_n(y)}{\partial y}\bigg|_{y=s_{j-0}} = \epsilon_0\epsilon_{yyj+1}\frac{\partial G_n(y)}{\partial y}\bigg|_{y=s_{j+0}}, \qquad j=1, 2, \ldots, N-1 \tag{2.100}$$

For a transmission line extending along the y-direction and excited by a current source of intensity I_s at $y=y_0$, the voltage along the line satisfies the following differential equation:

$$\left[\frac{d^2}{dy^2} - \gamma^2\right]V = -\frac{\gamma}{Y_c}I_s\delta(y-y_0) \tag{2.101}$$

Let the transmission line consist of N sections with Y_{cj} as the characteristic admittance and γ_j as the propagation constant of the jth section as shown in Fig 2.9b. At the junction between any two sections, the voltage must be continuous:

$$V\big|_{y=s_{j-0}} = V\big|_{y=s_{j+0}} \tag{2.102}$$

Using the continuity condition for the current, namely $I_j = I_{j+1}$, with (2.69) yields

$$\frac{Y_{cj}}{\gamma_j}\frac{\partial V}{\partial y}\bigg|_{y=s_{j-0}} = \frac{Y_{cj+1}}{\gamma_{j+1}}\frac{\partial V}{\partial y}\bigg|_{y=s_{j+0}} \tag{2.103}$$

We now consider the analogy between a N-layer anisotropic planar structure and a N-section two-wire transmission line.

Case a. Comparing (2.96) and (2.99) with (2.101) for the jth section and (2.102), respectively, we obtain the following equivalent relations:

$$V \equiv G_n(y) \tag{2.104}$$

$$\gamma_j = \frac{n\pi}{c}\sqrt{\frac{\epsilon_{xxj}}{\epsilon_{yyj}}} \tag{2.105}$$

$$I_s = \frac{Y_{cj}}{\gamma_j}\,\frac{2}{c\epsilon_0\epsilon_{yyj}}\sin\frac{n\pi x_0}{c} \tag{2.106}$$

Using (2.105) in (2.103) and normalizing with respect to $n\pi/c$ for mathematical convenience, we get

$$Y_{cj}\sqrt{\frac{\epsilon_{yyj}}{\epsilon_{xxj}}}\,\frac{\partial V}{\partial y}\bigg|_{y=s_{j-0}} = Y_{cj+1}\sqrt{\frac{\epsilon_{yyj+1}}{\epsilon_{xxj+1}}}\,\frac{\partial V}{\partial y}\bigg|_{y=s_{j+0}} \tag{2.107}$$

Comparing (2.100) with (2.107) gives

$$Y_{cj} = \epsilon_0\sqrt{\epsilon_{xxj}\epsilon_{yyj}} \tag{2.108}$$

Using (2.105) and (2.108) in (2.106), the expression for I_s simplifies to

$$I_s = \frac{2}{n\pi}\sin\frac{n\pi x_0}{c} \tag{2.109}$$

The voltage V on the transmission line at $y = y_0$ is given by

$$V|_{y=y_0} = \frac{I_s}{Y} \tag{2.110}$$

Using (2.104) and (2.109) in (2.110), and setting $Y = Y_A$ to denote the admittance at the charge plane for the anisotropic structure, $G_n(y)$ at $y = y_0$ is expressed as

$$G_n(y)|_{y=y_0} = \frac{2}{n\pi Y_A}\sin\frac{n\pi x_0}{c} \tag{2.111}$$

Substituting (2.62) and (2.111) in (2.61), the Green's function at the charge plane $y = y_0$ for case a becomes

$$G(x, y/x_0, y_0)|_{y=y_0} = \sum_{n=1}^{\infty}\frac{2}{n\pi Y_A}\sin\frac{n\pi x}{c}\sin\frac{n\pi x_0}{c} \tag{2.112}$$

The procedure for determining the Green's function for cases (b) and (c) is exactly the same as for case (a). The final expressions for the Green's function for these two cases are obtained as

Case b:

$$G(x, y/x_0, y_0)|_{y=y_0} = \sum_{n=0}^{\infty}\frac{4}{(2n+1)\pi Y_A}\sin\left\{\frac{(2n+1)\pi x}{2c}\right\}\sin\left\{\frac{(2n+1)\pi x_0}{2c}\right\} \tag{2.113}$$

Case c:

$$G(x, y/x_0, y_0)|_{y=y_0} = \sum_{n=1}^{\infty}\frac{2}{n\pi Y_A}\cos\frac{n\pi x}{c}\cos\frac{n\pi x_0}{c} \tag{2.114}$$

The Green's function expressions (2.112), (2.113), and (2.114) for the anisotropic structure are identical to (2.77), (2.79), and (2.81), respectively, for the isotropic structure except that the admittance expression at the charge plane now corresponds to that for the anisotropic case. While deriving the expressions for Y_A, the propagation constant and the characteristic admittance of each section are taken as per (2.105) and (2.108), respectively, for cases (a) and (c). For case (b), the propagation constant γ_j and the characteristic admittance Y_{cj} are given by $\gamma_j = \frac{(2n+1)\pi}{2c}\sqrt{\frac{\epsilon_{xxj}}{\epsilon_{yyj}}}$ and $Y_{cj} = \epsilon_0\sqrt{\epsilon_{xxj}\epsilon_{yyj}}$ respectively.

2.6.2 Determination of Potential Function in the Fourier Transform Domain

Consider a general N-layer anisotropic structure with an infinitesimally thin line charge $\rho(x, y)$ situated at $y = y_0$ as shown in Fig. 2.10a. The substrates are assumed to be uniaxially anisotropic with a dielectric permittivity tensor given as in (2.93). The structure extends to infinity in the positive and negative x-directions. Under the quasi-static approximation, the potential function $\phi(x, y)$ satisfies the two-dimensional anisotropic Poisson's differential equation:

$$\epsilon_{xxj}\frac{\partial^2\phi}{\partial x^2} + \epsilon_{yyj}\frac{\partial^2\phi}{\partial y^2} = -\frac{1}{\epsilon_0}\rho(x, y)$$

$$= -\frac{1}{\epsilon_0} f(x)\delta(y - y_0) \tag{2.115}$$

where $f(x)$ gives the distribution of charge in the x-direction. Applying the Fourier transform with respect to x, (2.115) can be written in the form

$$\left[\frac{d^2}{dy^2} - \left(\beta\sqrt{\frac{\epsilon_{xxj}}{\epsilon_{yyj}}}\right)^2\right]\hat{\phi}(\beta, y) = -\frac{1}{\epsilon_0\epsilon_{yyj}}\hat{f}(\beta)\delta(y - y_0) \tag{2.116}$$

where $\hat{\phi}(\beta, y)$ and $\hat{f}(\beta)$ are the Fourier transforms of $\phi(x, y)$ and $f(x)$, respectively, as defined in Sec. 2.5.2. The boundary conditions in the Fourier domain at the various interfaces are

$$\hat{\phi}(\beta, y)\big|_{y=s_{j-0}} = \hat{\phi}(\beta, y)\big|_{y=s_{j+0}} \tag{2.117a}$$

$$\epsilon_0\epsilon_{yyj}\frac{\partial\hat{\phi}}{\partial y}(\beta, y)\big|_{y=s_{j-0}} = \epsilon_0\epsilon_{yyj+1}\frac{\partial\hat{\phi}}{\partial y}(\beta, y)\big|_{y=s_{j+0}}, \quad j = 1, 2, \ldots, N-1 \tag{2.117b}$$

The transmission line equivalent (Fig. 2.10b) for this structure is the same as that shown in Fig. 2.9b. Thus the transmission line equation (2.101) and the continuity conditions (2.102) and (2.103) are applicable. Comparing (2.101) for the jth section with (2.116), the following equivalent relations are obtained:

$$V \equiv \hat{\phi}(\beta, y) \tag{2.118}$$

$$\gamma_j \equiv \beta\sqrt{\frac{\epsilon_{xxj}}{\epsilon_{yyj}}} \tag{2.119}$$

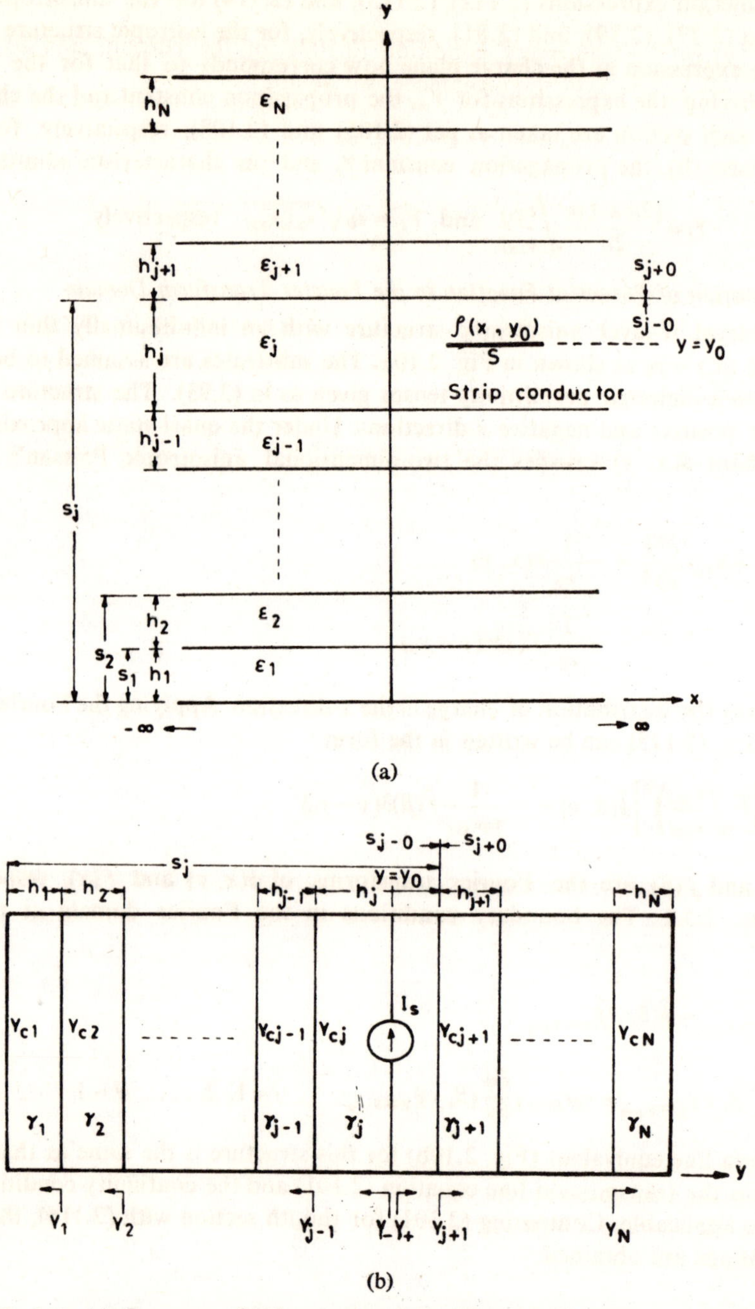

FIG. 2.10 (a) Geometry of N-layer anisotropic dielectric with charge at $y=y_0$ and (b) two-wire transmission line equivalent:

$$Y_{cj}=\epsilon_0\sqrt{\epsilon_{xxj}\,\epsilon_{yyj}},\quad \gamma_j=\beta\sqrt{\epsilon_{xxj}/\epsilon_{yyj}}.$$

$$I_s = \frac{Y_{cj}}{\gamma_j} \frac{\hat{f}(\beta)}{\epsilon_0 \epsilon_{yyj}} \tag{2.120}$$

Using (2.119) in the continuity condition (2.103) for the current, and normalizing with respect to β for mathematical convenience, we obtain

$$Y_{cj} \sqrt{\frac{\epsilon_{yyj}}{\epsilon_{xxj}}} \frac{\partial V}{\partial y}\bigg|_{y=s_{j-0}} = Y_{cj+1} \sqrt{\frac{\epsilon_{yyj+1}}{\epsilon_{xxj+1}}} \frac{\partial V}{\partial y}\bigg|_{y=s_{j+0}} \tag{2.121}$$

Comparing (2.121) with (2.117b) yields the expression for the characteristic admittance Y_{cj} in terms of the elements of the dielectric tensor:

$$Y_{cj} = \epsilon_0 \sqrt{\epsilon_{xxj} \epsilon_{yyj}} \tag{2.122}$$

When (2.119) and (2.122) are used in (2.120), the expression for I_s simplifies to

$$I_s = \frac{\hat{f}(\beta)}{\beta} \tag{2.123}$$

With (2.123) the voltage V on the transmission line at $y = y_0$ can be expressed in the form

$$V\big|_{y=y_0} = \frac{I_s}{Y_A} = \frac{\hat{f}(\beta)}{\beta Y_A} \tag{2.124}$$

where Y_A is the admittance at the charge plane $y = y_0$.

Since the voltage is equivalent to $\hat{\phi}(\beta, y)$ from (2.118), we can write

$$\hat{\phi}(\beta, y)\big|_{y=y_0} = \frac{\hat{f}(\beta)}{\beta Y_A} \tag{2.125}$$

Expression (2.125) is the same as that given in (2.92) for the isotropic structure, except that the admittance parameter Y_I in (2.92) is replaced by Y_A corresponding to the anisotropic structure.

2.7 Unified Approach to the Determination of Capacitance—Variational Method Combined with Transverse Transmission Line Technique

2.7.1 Unified Capacitance Formula for Planar Structures with Side Walls

The variational expression for the capacitance of any two-conductor transmission line having an arbitrary cross-section is given by (2.58). We now apply this formula to a multilayer planar transmission line with side walls as depicted in Figs. 2.7a and 2.9a. For an infinitesimally thin strip conductor S, the charge distribution can be assumed to be of the form

$$\rho(x, y) = f(x)\delta(y - y_0) \tag{2.126}$$

where $f(x)$ is the charge distribution in the x-direction. Substituting (2.126) in (2.58), the variational formula for the capacitance of a multilayer structure with side walls can be expressed as

$$\frac{1}{C} = \frac{\iint_S G(x, y_0/x_0, y_0) f(x) f(x_0)\, dx\, dx_0}{\left[\int_S f(x)\, dx\right]^2} \tag{2.127}$$

Expressions for the Green's function at the charge plane $G(x, y_0/x_0, y_0)$ for various boundary conditions at the side walls are derived in Secs. 2.5 and 2.6 using the 'transverse transmission line' technique. Substituting (2.77), (2.79), and (2.81) for the isotropic structure, and (2.112)-(2.114) for the anisotropic structure in (2.127), the expressions for the capacitance for the three cases of side wall conditions take the following form:

Case a. Electric walls at $x=0$ and c:

$$\frac{1}{C}=\frac{\sum_{n=1}^{\infty}\frac{2}{n\pi Y}\left[\int_S f(x)\sin\frac{n\pi x}{c}\,dx\right]^2}{\left[\int_S f(x)\,dx\right]^2} \tag{2.128}$$

Case b. Electric wall at $x=0$ and magnetic wall at $x=c$:

$$\frac{1}{C}=\frac{\sum_{n=0}^{\infty}\frac{4}{(2n+1)\pi Y}\left[\int_S f(x)\sin\left\{\frac{(2n+1)\pi x}{2c}\right\}dx\right]^2}{\left[\int_S f(x)\,dx\right]^2} \tag{2.129}$$

Case c. Magnetic walls at $x=0$ and c:

$$\frac{1}{C}=\frac{\sum_{n=1}^{\infty}\frac{2}{n\pi Y}\left[\int_S f(x)\cos\frac{n\pi x}{c}\,dx\right]^2}{\left[\int_S f(x)\,dx\right]^2} \tag{2.130}$$

where the admittance at the charge plane $Y=Y_+ + Y_-$ is represented as Y_I for the isotropic structure and Y_A for the anisotropic structure. The expression for the admittance is easily obtained by applying the standard transmission line formula for the input admittance Y_{in} of a section of transmission line. If l_j is the length of the jth section, its input admittance $Y_{in\,j}$ is given by

$$Y_{in\,j}=Y_{cj}\left[\frac{Y_{lj}+Y_{cj}\tanh(\gamma_j l_j)}{Y_{cj}+Y_{lj}\tanh(\gamma_j l_j)}\right] \tag{2.131}$$

where Y_{lj} is the load admittance of the jth section which is the same as the input admittance $Y_{in\,j+1}$ of the $(j+1)$th section. Y_{cj} and γ_j are the characteristic admittance and propagation constant of the jth section, respectively. If the dielectric layer is isotropic,

$$Y_{cj}=\epsilon_j \tag{2.132a}$$

and

$$\gamma_j=\gamma=\frac{n\pi}{c}, \quad \text{for cases (a) and (c)}$$

$$=\frac{(2n+1)\pi}{2c}, \quad \text{for case (b)} \tag{2.132b}$$

If the dielectric layer is uniaxially anisotropic with permittivity tensor as in (2.93), then

$$Y_{cj} = \epsilon_0 \sqrt{\epsilon_{xxj}\epsilon_{yyj}} \tag{2.133a}$$

and

$$\begin{aligned} \gamma_j &= \frac{n\pi}{c}\sqrt{\frac{\epsilon_{xxj}}{\epsilon_{yyj}}}, && \text{for cases (a) and (c)} \\ &= \frac{(2n+1)\pi}{2c}\sqrt{\frac{\epsilon_{xxj}}{\epsilon_{yyj}}}, && \text{for case (b)} \end{aligned} \tag{2.133b}$$

Expression (2.131) is iterated over each section of the transmission line to determine Y_+ and Y_- and summing them to obtain Y at the charge plane. With the substitution of Y, the capacitance C is evaluated by assuming a suitable charge distribution $f(x)$ on the strip conductor and performing the line integration over the width of the strip conductor as per (2.128)-(2.130) for the three cases. These expressions are valid for any value of c greater than the width w of the strip conductor. With increasing value of c, the effect of side walls on the capacitance of the structure reduces and for a sufficiently large value of c, the capacitance approaches that of the structure with no side walls.

2.7.2 *Unified Capacitance Formula for Planar Structures with Infinite Parallel Ground Planes*

Structures in which the side walls have practically no effect on the capacitance can be analysed in a simpler way in the Fourier transform domain by assuming the dielectric layers to extend to infinity in the x-direction as depicted in Figs. 2.8a and 2.10a. Consider the general variational expression for the capacitance given by (2.59). If the strip conductor S in Fig. 2.8a as well as in Fig. 2.10a is assumed to be infinitesimally thin, the charge distribution $\rho(x, y)$ on it can be expressed in the form $f(x)\delta(y-y_0)$. Substituting this in (2.59) reduces it to a line integral given by

$$\frac{1}{C} = \frac{1}{Q^2}\int_S \phi(x, y_0)\, f(x)\, dx \tag{2.134}$$

where

$$Q = \int_S f(x)\, dx$$

By applying Parseval's theorem in Fourier transform theory [11], (2.134) can be transformed into the Fourier-transformed coordinate, giving

$$\frac{1}{C} = \frac{1}{2\pi Q^2}\int_{-\infty}^{\infty} \hat{\phi}(\beta, y_0)\hat{f}(\beta)\, d\beta \tag{2.135}$$

For a symmetric charge distribution on the strip conductor about $x=0$, $f(x)=f(-x)$, and hence. $\hat{f}(\beta)=\hat{f}(-\beta)$ and $\hat{\phi}(\beta, y_0)=\hat{\phi}(-\beta, y_0)$. Expression (2.134) then reduces to

$$\frac{1}{C} = \frac{1}{\pi Q^2}\int_0^{\infty} \hat{\phi}(\beta, y_0)\hat{f}(\beta)\, d\beta \tag{2.136}$$

Using (2.92) and (2.125) in (2.136), the capacitance expression becomes

$$\frac{1}{C}=\frac{1}{\pi Q^2}\int_0^\infty \frac{[\hat{f}(\beta)]^2}{\beta Y}\,d\beta \tag{2.137}$$

where Y, the admittance at the charge plane $y=y_0$ is Y_I for the isotropic structure, and Y_A for the anisotropic structure. The admittance parameter Y is obtained by iterating (2.131) over each section of the transmission line to determine Y_+ and Y_- and summing them. In (2.131), Y_{cj} and γ_j are the characteristic admittance and propagation constant of the jth section, respectively, and Y_{lj} is the load admittance. If the dielectric layer is isotropic, then

$$Y_{cj}=\epsilon_j \tag{2.138a}$$

and

$$\gamma_j=\gamma=\beta \tag{2.138b}$$

If the dielectric layer is uniaxially anisotropic with permittivity tensor as in (2.93), then

$$Y_{cj}=\epsilon_0\sqrt{\epsilon_{xxj}\epsilon_{yyj}} \tag{2.139a}$$

and

$$\gamma_j=\beta\sqrt{\frac{\epsilon_{xxj}}{\epsilon_{yyj}}} \tag{2.139b}$$

By assuming a suitable charge distribution $f(x)$ on the strip conductor and from the knowledge of the admittance parameter Y at the charge plane, the capacitance of planar structures with dielectric substrates extending to infinity in the lateral direction can be determined.

Variational expressions (2.128)-(2.130) for the capacitance of planar structures with side walls for the three different side wall conditions are general and can be applied to any multilayer structure with iso/anisotropic substrates. The only parameter that needs to be changed is the admittance at the charge plane depending on the structure under consideration. The variational method in the space domain combined with the 'transverse transmission line' technique, therefore, offers a simple unified approach to solve a class of planar structures with side walls. The top and bottom sides may be bounded or unbounded. Similarly, expression (2.137) is a general variational expression for the capacitance of planar structures having multilayer iso/anisotropic substrates extending to infinity in the lateral ($\pm x$)-direction. For a given charge distribution, the only parameter that needs to be changed is the admittance at the charge plane corresponding to the structure considered. The variational method in the Fourier domain combined with the 'transverse transmission line' technique, therefore, provides an unified approach to solve single conductor planar structures with no effect of side walls on the propagation.

2.7.3 *Effect of Small but Finite Thickness of Strip Conductor*

In the unified analytical approach presented in the earlier sections, the strip conductor is assumed to be of zero thickness. In practice, the conductor thickness is so small (order of a few microns) that its effect on the characteristic impedance of the structure is negligible, and hence, the capacitance formulas presented in Secs. 2.7.1 and 2.7.2 give accurate results adequate for most applications. However, the effect of small but finite thickness t of the strip conductor can be taken into account by the following approximate procedure.

Consider a general multilayer dielectric structure with side walls, with a strip conductor of thickness t situated between $y=y_0$ and $y=y_0+t$ or between $y=y_0$ and $y=y_0-t$ (Fig. 2.11a). The dielectric layers may be isotropic or uniaxially anisotropic. The transmission line equivalent of this structure is shown in Fig. 2.11b. The effect of a small but finite thickness

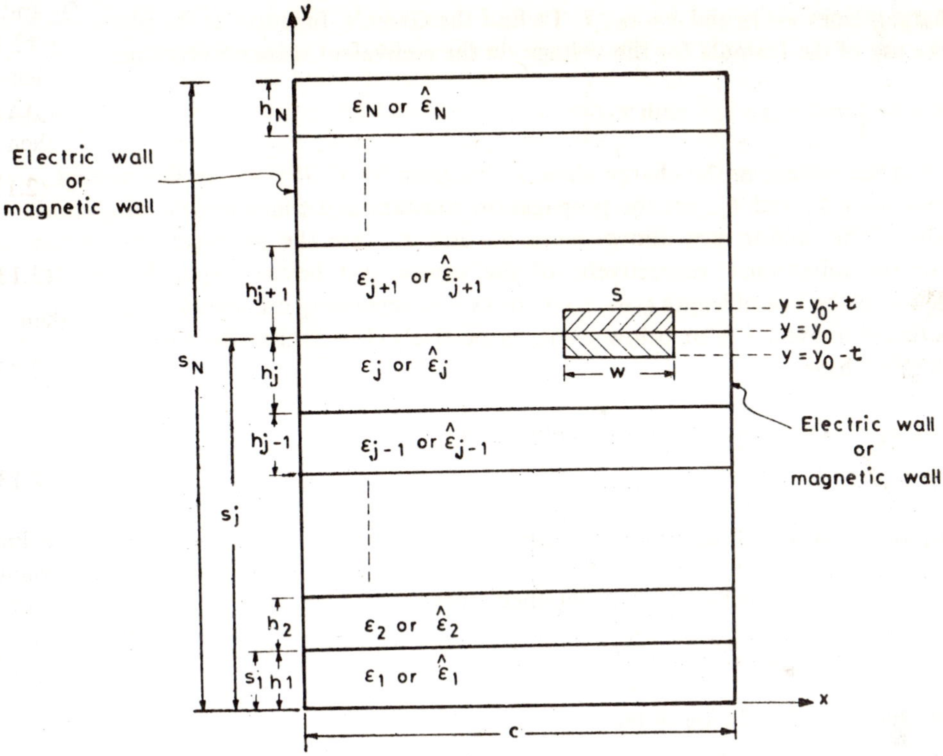

(a)

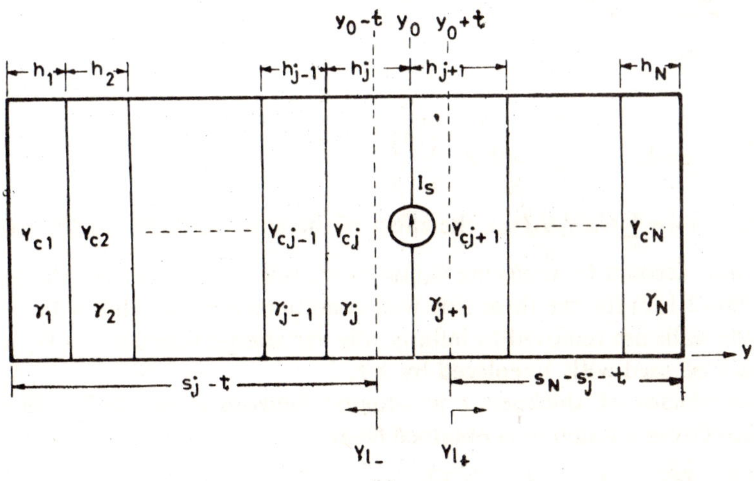

(b)

FIG. 2.11 (a) Multilayer isotropic/anisotropic structure with finite thickness strip conductor and (b) two-wire transmission line equivalent.

t of the strip conductor can be approximately taken into account by considering two layers of charge, one at the plane $y=y_0$ and the other at $y=y_0\pm t$, where the +ve sign applies if the strip conductor is situated above $y=y_0$ and the −ve sign applies if the strip conductor is situated below $y=y_0$. The Green's function at the charge plane $y=y_0$ in the variational expression for capacitance (2.127) is replaced by the average of Green's functions at the two charge planes $y=y_0$ and $y=y_0\pm t$. To find the Green's function at the plane $y=y_0\pm t$, we make use of the formula for the voltage on the equivalent transmission line:

$$V=V_{\pm}\left[\cosh \gamma_{\pm}t+\frac{Y_{l\pm}}{Y_{c\pm}}\sinh \gamma_{\pm}t\right] \tag{2.140}$$

where V is the voltage at the charge plane $y=y_0$ given by (2.75), $V_{\pm}$ is the voltage at the plane $y=y_0\pm t$, γ_+ and Y_{c+} are the propagation constant and characteristic admittance, respectively, of the section just above $y=y_0$, γ_- and Y_{c-} are the propagation constant and characteristic admittance, respectively, of the section just below $y=y_0$. $Y_{l\pm}$ is the input admittance of the line of length (s_N-s_j-t) or (s_j-t) depending on whether the strip conductor is situated above $y=y_0$ or below $y=y_0$, respectively (Fig. 2.11b). Because $G_n(y)$ is analogous to V, we have

$$G_n(y_0)=G_n(y_0\pm t)\left[\cosh \gamma_{\pm}t+\frac{Y_{l\pm}}{Y_{c\pm}}\sinh \gamma_{\pm}t\right] \tag{2.141}$$

or

$$G_n(y_0\pm t)=G_n(y_0)\left[\cosh \gamma_{\pm}t+\frac{Y_{l\pm}}{Y_{c\pm}}\sinh \gamma_{\pm}t\right]^{-1} \tag{2.142}$$

The average Green's function $G_{n\,\mathrm{av}}$ is obtained from

$$G_{n\,\mathrm{av}}=\frac{1}{2}\left[G_n(y_0)+G_n(y_0\pm t)\right] \tag{2.143}$$

Substituting from (2.142), we obtain

$$\begin{aligned} G_{n\,\mathrm{av}}&=G_n(y_0)\frac{1}{2}\left[1+\left\{\cosh \gamma_{\pm}t+\frac{Y_{l\pm}}{Y_{c\pm}}\sinh \gamma_{\pm}t\right\}^{-1}\right] \\ &=G_n(y_0)t_{n\pm} \end{aligned} \tag{2.144}$$

where

$$t_{n\pm}=\frac{1}{2}\left[1+\left\{\cosh \gamma_{\pm}t+\frac{Y_{l\pm}}{Y_{c\pm}}\sinh \gamma_{\pm}t\right\}^{-1}\right] \tag{2.145}$$

Since $G_n(x, y_0/x_0, y_0)=\sum_n G_n(x)G_n(y_0)$, the effect of the finite thickness of the strip conductor may be taken into account by replacing $G_n(y_0)$ by $G_n(y_0)t_{n\pm}$. Equivalently the space domain formulas (2.128)-(2.130) for the three side wall conditions remain valid with Y replaced by $Y/t_{n\pm}$. If the side walls are removed to infinity, the variational formula (2.137) in the Fourier domain can still be used with Y replaced by $Y/t_{n\pm}$.

If the strip conductor of thickness t is situated between $y=y_0+(t/2)$ and $y=y_0-(t/2)$, then the average Green's function is obtained from

$$G_{n\,\mathrm{av}}=\frac{1}{2}\left[G_n\left(y_0+\frac{t}{2}\right)+G_n\left(y_0-\frac{t}{2}\right)\right] \tag{2.146}$$

where

$$G_n\left(y_0+\frac{t}{2}\right)=G_n(y_0)\left[\cosh\left(\gamma_+\frac{t}{2}\right)+\frac{Y_{l+}}{Y_{c+}}\sinh\left(\gamma_+\frac{t}{2}\right)\right]^{-1} \quad (2.147a)$$

$$G_n\left(y_0-\frac{t}{2}\right)=G_n(y_0)\left[\cosh\left(\gamma_-\frac{t}{2}\right)+\frac{Y_{l-}}{Y_{c-}}\sinh\left(\gamma_-\frac{t}{2}\right)\right]^{-1} \quad (2.147b)$$

where Y_{l+} is the input admittance looking to the right of the line of length $[s_N-s_j-(t/2)]$ and Y_{l-} is the admittance looking to the left of the line of length $[s_j-(t/2)]$. Using (2.147) in (2.146), we obtain

$$G_{n\,\text{av}}=\frac{1}{2}G_n(y_0)t_n \quad (2.148)$$

where

$$t_n=\left[\left\{\cosh\left(\gamma_+\frac{t}{2}\right)+\frac{Y_{l+}}{Y_{c+}}\sinh\left(\gamma_+\frac{t}{2}\right)\right\}^{-1} +\left\{\cosh\left(\gamma_-\frac{t}{2}\right)+\frac{Y_{l-}}{Y_{c-}}\sinh\left(\gamma_-\frac{t}{2}\right)\right\}^{-1}\right] \quad (2.149)$$

Thus, in structures where the strip conductor of thickness t is symmetrically displaced about $y=y_0$, the admittance parameter Y in the variational formulas in the space domain given by (2.128)-(2.130), and in the variational formulas in the Fourier domain given by (2.137) gets replaced by Y/t_n.

2.8 Finite-Difference Method

The application of the finite-difference method to TEM transmission lines involves the solution of Laplace's equation in the cross-sectional plane subject to the boundary conditions on the inner and outer conductors. The domain between the conductors is divided into a finite set of mesh points and Laplace's equation is solved in the finite-difference form by digital computation. This method has been applied to TEM transmission lines and has also been extended to quasi-TEM transmission lines with limited amount of inhomogeneity [3], [4]. In the following, we present the procedure for the determination of propagation parameters of a homogeneous TEM structure such as the strip-transmission line. The modification required near an air-dielectric boundary, as is encountered in a suspended stripline, is also outlined.

2.8.1 Homogeneous TEM Structure

We consider the example of a homogeneous strip transmission line (Fig. 2.12) to illustrate the procedure involved in the application of the finite-difference method. Laplace's equation for the potential function $\phi(x, y)$, in the cross-sectional plane, and the boundary conditions are given by

$$\frac{\partial^2\phi}{\partial x^2}+\frac{\partial^2\phi}{\partial y^2}=0 \quad (2.150)$$

$$\phi(x,y)=\begin{cases}1, & \text{on the strip conductor (boundary } S_1)\\ 0, & \text{on the outer boundary (boundary } S_2)\end{cases} \quad (2.151)$$

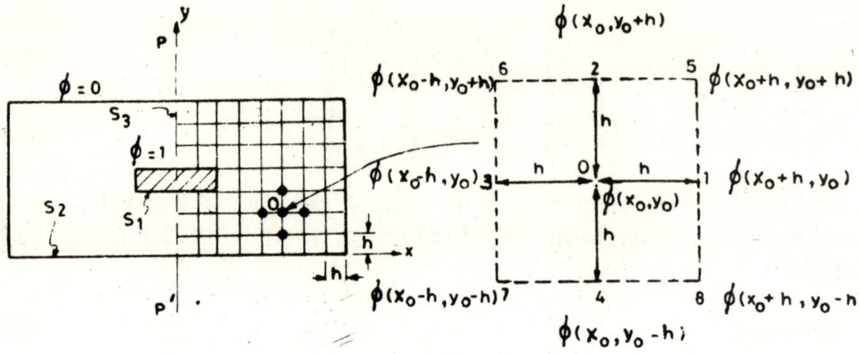

FIG. 2.12 Homogeneous stripline and superposition of square meshes on the right half subdomain.

Since the structure is symmetric about the line PP', the problem can be simplified by considering only one-half of the structure, say, to the right of PP'. The problem is thus transformed into a combination of the classical Dirichlet and Neumann problems, since the normal derivative $\partial\phi/\partial n$ is specified on the boundary S_3 defining the line of symmetry. The region is divided into square meshes of side h each. Consider a typical mesh point O located at (x_0, y_0). Laplace's equation is now reduced to a difference equation by considering it in terms of the potential at the point O and its four neighbouring points 1, 2, 3, and 4. Applying Taylor's theorem in the x-direction, we get

$$\phi(x_0+h, y_0)-\phi(x_0, y_0)=h\left.\frac{\partial\phi}{\partial x}\right|_0+\frac{h^2}{2!}\left.\frac{\partial^2\phi}{\partial x^2}\right|_0+\frac{h^3}{3!}\left.\frac{\partial^3\phi}{\partial x^3}\right|_0+\frac{h^4}{4!}\left.\frac{\partial^4\phi}{\partial x^4}\right|_0+\dots \quad (2.152)$$

$$\phi(x_0-h, y_0)-\phi(x_0, y_0)=-h\left.\frac{\partial\phi}{\partial x}\right|_0+\frac{h^2}{2!}\left.\frac{\partial^2\phi}{\partial x^2}\right|_0-\frac{h^3}{3!}\left.\frac{\partial^3\phi}{\partial x^3}\right|_0+\frac{h^4}{4!}\left.\frac{\partial^4\phi}{\partial x^4}\right|_0+\dots \quad (2.153)$$

Simple algebraic manipulation of (2.152) and (2.153) yields

$$\frac{\partial^2\phi}{\partial x^2}=\frac{\phi(x_0+h, y_0)+\phi(x_0-h, y_0)-2\phi(x_0, y_0)}{h^2}-\frac{2h^2}{4!}\frac{\partial^4\phi}{\partial x^4}-\dots \quad (2.154)$$

Similarly, applying Taylor's theorem in the y-direction, gives

$$\frac{\partial^2\phi}{\partial y^2}=\frac{\phi(x_0, y_0+h)+\phi(x_0, y_0-h)-2\phi(x_0, y_0)}{h^2}-\frac{2h^2}{4!}\frac{\partial^4\phi}{\partial y^4}-\dots \quad (2.155)$$

Using the expansions given by (2.154) and (2.155) in (2.150), we obtain

$$[\phi(x_0+h, y_0)+\phi(x_0-h, y_0)+\phi(x_0, y_0+h)+\phi(x_0, y_0-h)]-4\phi(x_0, y_0)=\delta(h^4)$$
$$+\text{higher order terms} \quad (2.156)$$

where $\delta(h^4)$ is an error term of the fourth order. The next error term is of the sixth order and so on. An alternate equation of the same type can also be written by expanding the potentials at the diagonal points 5, 6, 7, and 8, with respect to the point (x_0, y_0):

$$[\phi(x_0+h, y_0+h)+\phi(x_0-h, y_0+h)+\phi(x_0-h, y_0-h)+\phi(x_0+h, y_0-h)]$$
$$-4\phi(x_0, y_0)=-\delta(4h^4)+\text{higher order terms} \quad (2.157)$$

The error in (2.157) is four times the error in (2.156) and it is of the opposite sign. Combining (2.156) and (2.157) in the ratio four to one, we obtain a more accurate approximation to Laplace's equation:

$$[\phi(x_0+h, y_0+h)+\phi(x_0-h, y_0+h)+\phi(x_0-h, y_0-h)+\phi(x_0+h, y_0-h)] \\ +4[\phi(x_0+h, y_0)+\phi(x_0-h, y_0)+\phi(x_0, y_0+h)+\phi(x, y_0-h)] \\ -20\phi(x_0, y_0)=\delta(h^6) \tag{2.158}$$

The error term can be further reduced by including additional mesh points. If the error term can be neglected, the potential at the point (x_0, y_0) can be expressed as the algebraic mean of the potentials at the neighbouring points. For example, setting the error term equal to zero in (2.156), we obtain

$$\phi(x_0, y_0)=\frac{1}{4}\left[\phi(x_0+h, y_0)+\phi(x_0-h, y_0)+\phi(x_0, y_0+h)+\phi(x_0, y_0-h)\right] \tag{2.159}$$

In general, if ϕ_{ij} represents the potential at any interior mesh point in a homogeneous region, we can write

$$\phi_{ij}=\frac{1}{4}\left[\phi_{i+1,j}+\phi_{i-1,j}+\phi_{i,j+1}+\phi_{i,j-1}\right] \tag{2.160}$$

The next step is to solve these simultaneous difference equations for the potentials at all the mesh points. The numerical solution is obtained by applying the relaxation process. A constant potential $\phi=1$ is assigned to all the mesh points on the inner conductor boundary, $\phi=0$ to all the mesh points on the outer conductor boundary and an initial guess value ϕ_{ij}^0 to each of the interior mesh points. Starting from the inner boundary, the lattice is scanned through successive rows, and the old value for each mesh point is discarded and replaced by the new value. The relaxation formula for the new value $\phi_{i,j}^{n+1}$ in terms of its old value $\phi_{i,j}^n$ is

$$\phi_{i,j}^{n+1}=\phi_{i,j}^n-\Omega_0 R_{i,j}^n \tag{2.161}$$

$R_{i,j}^n$ is the residue at each point and is calculated from

$$R_{i,j}^n=\phi_{i,j}^n-\frac{1}{4}(\phi_{i+1,j}^n+\phi_{i-1,j}^n+\phi_{i,j+1}^n+\phi_{i,j-1}^n) \tag{2.162}$$

Ω_0 is the relaxation factor which determines the rate of convergence. The optimum value of Ω_0 is determined by trial and error and has been found to lie in general between 0 and 2 [4].

After scanning the whole net several times, it is further divided into finer meshes by setting $h_{new}=(1/2)\, h_{old}$ and the process is repeated. The relaxation process is continued until the difference

$$\Delta_{i,j}^n=\phi_{i,j}^{n+1}-\phi_{i,j}^n \tag{2.163}$$

for certain test points becomes negligibly small. With this, the process converges, thus providing potentials at each of the mesh points.

2.8.2 *Modification at an Air-Dielectric Boundary*

For a mesh point on an air-dielectric boundary (Fig. 2.13), the difference equation is derived by requiring that the normal component of the displacement density is continuous at the interface and that the finite-difference equation has to be identical with (2.160) when the dielectric is replaced by air. The resulting equation is given by

$$\phi_{i,j}=\frac{(\phi_{i,j+1}+\epsilon_r\phi_{i,j-1})}{2(1+\epsilon_r)}+\frac{(\phi_{i-1,j}+\phi_{i+1,j})}{4} \tag{2.164}$$

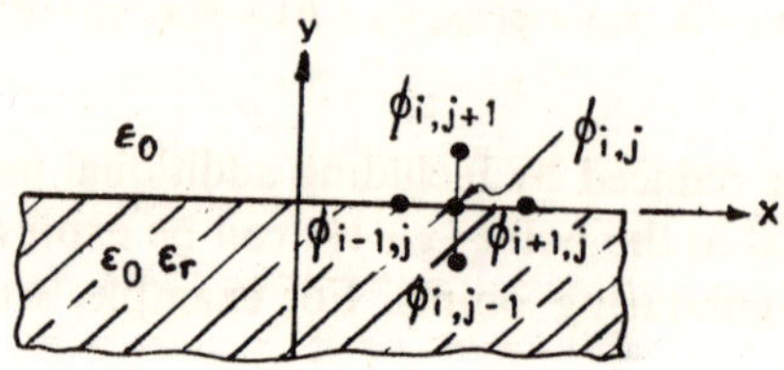

FIG. 2.13 Mesh point on air-dielectric boundary.

2.8.3 *Determination of Static Capacitance*

The capacitance of TEM or quasi-TEM lines can now be computed from the line integrals of the electric field normal to the boundary surfaces. The normal component of the electric field vector E_n is given by

$$E_n=-\frac{\partial\phi}{\partial n} \tag{2.165}$$

The derivative $\partial\phi/\partial n$ may be approximated by a difference formula in terms of the potentials at the adjacent mesh points. If we set the potential difference between the strip and the outer conductor equal to unity, the expression for the capacitance per unit length of the line becomes

$$C=\epsilon_0\oint_{S_2}\epsilon_r E_n\cdot ds \tag{2.166}$$

where S_2 is the inner surface of the shielding enclosure. The value of ϵ_r is taken as unity while carrying out the integration over that part of the conductor surface which is in contact with air. The strip conductor boundary S_1 could also be considered for this calculation. This is, however, not preferred, since the field varies rapidly near the strip. The effective dielectric constant ϵ_f, the phase velocity v, the guide wavelength λ, and the characteristic impedance Z can then be calculated using the formulas (2.23) to (2.26) and (2.20). In these formulas, C^a, which is the capacitance per unit length of the same structure with all dielectrics replaced by air is calculated using (2.166) with $\epsilon_r=1$.

In the finite-difference method outlined above, the accuracy of the solution depends on the fineness of mesh size. Particularly, in coupled strip transmission lines, extremely fine mesh sizes are necessary, since the charge and field singularities at the adjacent edges of the strip conductors dominate for small gaps. This results in a very large system of equations to be solved, leading to difficulties in convergence and accuracy.

2.9 Variational Series Based on Conformal Transformation

A combined approach which makes use of the variational series based on conformal transformation has been reported by Smith [8] for the characterization of capacitances of coupled lines. This method does not have the difficulties of convergence encountered in the finite difference method for closely spaced lines. Singularities are absorbed into the capacitance terms in such a way as to render them fast convergent.

The configuration analysed by Smith consists of an array of edge-coupled strip conductors in suspended stripline configuration. A typical cell of the array consists of a single strip conductor with electric or magnetic walls (Fig. 2.14). The capacitance is derived from the variational solution for the upper bound on the system energy. The trial function is the potential at the top air-dielectric interface which is assumed to be a magnetic wall. Applying conformal transformation method exclusively to region 3, a closed form expression is derived for the potential across the gap and also the capacitance C_3 of region 3. The potentials in the three regions are also obtained in terms of the Fourier series expansion and their Fourier coefficients are determined from the known interface potential at $y = h_2$ and also by matching the potentials at the interfaces $y = h_1$ and h_2. The components of the electric field are derived from the potentials from which a variational series for the energy $(1/2)CV^2$, of the system is derived. For $V = 1$, the series gives the capacitance. The capacitance of region 3 weighted by the factor $(\epsilon_r + 1)/2$ is subtracted term by term from the series to achieve rapid convergence. The known value of C_3 weighted by $(\epsilon_r + 1)/2$ is then added to the series to obtain the capacitance of the cell. For detailed derivation, the reader is referred to Smith [8].

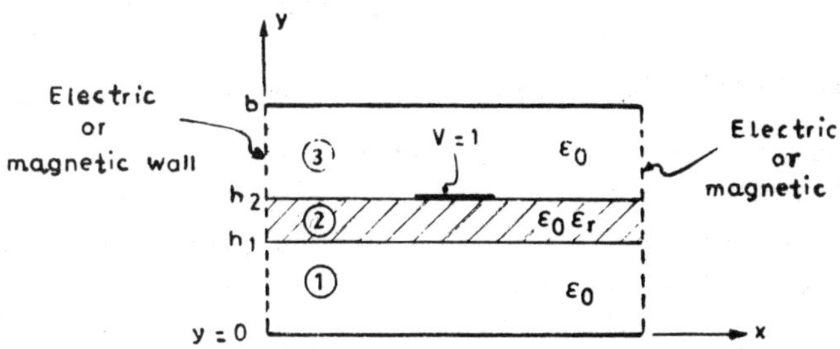

FIG. 2.14 Geometry of one cell of coupled lines in suspended stripline.

2.10 Concluding Remarks

Of the various quasi-static analytical techniques described in this chapter, the 'variational method in space domain' combined with the 'transverse transmission line' (TTL) technique offers the most general as well as the simplest approach. It can be used to solve a class of strip- and microstrip-like transmission structures with isotropic/uniaxially anisotropic dielectric substrates. In the case of coupled lines, any symmetric structure which can be resolved into even- and odd-mode configurations can be analysed. The capacitance expressions are general and can be used for a class of such structures irrespective of the number of dielectric layers. The only parameter that needs to be changed in the formula is the admittance at the charge plane. In the following chapters, this method is used to derive the propagation parameters of the stripline and a variety of stripline-like transmission lines. For the specific case of the homogeneous symmetric stripline, other methods are also equally simple to apply. In particular, since the conformal mapping method yields exact expression for its characteristic impedance, the procedure involved in the derivation of the impedance formula is also included in the next chapter.

REFERENCES

1. L.A. Pipes, *Applied Mathematics for Engineers and Physicists*, McGraw-Hill, New York, 1958.
2. R.E. Collin, *Field Theory of Guided Waves*, McGraw-Hill, New York, 1960.
3. H.E. Green, The numerical solution of some important transmission line problems, *IEEE Trans. Microwave Theory Tech.*, Vol. MTT-13, pp. 676-692, Sept. 1965.
4. M.V. Schneider, Computation of impedance and attenuation of TEM lines by finite difference methods, *IEEE Trans. Microwave Theory Tech.*, Vol. MTT-13, pp. 793-800, Nov. 1965.
5. R. Crampagne, M. Ahmadpanah, and J.L. Guiraud, A simple method for determining the Green's function for a large class of MIC lines having multilayered dielectric substrates, *IEEE Trans. Microwave Theory Tech.*, Vol. MTT-26, pp. 82-87, Feb. 1978.
6. B. Bhat and S.K. Koul, Unified approach to solve a class of strip and microstrip-like transmission lines, *IEEE Trans. Microwave Theory Tech.*, Vol. MTT-30, pp. 679-686, May 1982. Also correction *IEEE Trans. Microwave Theory Tech.*, Vol. MTT-30, p. 2067, Nov. 1982.
7. S.K. Koul and B. Bhat, Generalized analysis of microstrip-like transmission lines and coplanar strips with anisotropic substrates for MIC, electro-optic modulator and SAW applications, *IEEE Trans. Microwave Theory Tech.*, Vol. MTT-31, pp. 1051-1059, Dec. 1983.
8. J.I. Smith, The even- and odd-mode capacitance parameters for coupled lines in suspended substrate, *IEEE Trans. Microwave Theory Tech.*, Vol. MTT-19, pp. 424-431, May 1971.
9. R.V. Churchill, *Introduction to Complex Variables and Applications*, McGraw-Hill, New York, 1948.
10. Y. Chang and I.E. Chang, Simple method for the variational analysis of a generalized N-dielectric layer transmission line, *Electron. Lett.*, Vol. 6, No. 3, pp. 49-50, Feb. 1970.
11. A. Papoulis, *The Fourier Integral and its Applications*, McGraw-Hill, New York, 1962.

CHAPTER 3

Analysis of Striplines and Stripline-like Transmission Lines

3.1 Introduction

This chapter presents in detail the analyses of striplines and stripline-like transmission lines having a single strip conductor and single or multilayer iso/anisotropic dielectrics. First, we present the analysis of a homogeneous stripline using the conformal mapping method. This analysis was reported first by Cohn [1] who also derived an exact expression for the characteristic impedance of the stripline under the assumption that the strip conductor is of zero thickness. Next, we describe the variational method of analysing inhomogeneous lines, in which the potential function or the Green's function is determined by solving a set of algebraic equations obtained by matching the boundary conditions at the various dielectric interfaces [2-4]. We illustrate this method by considering a simple example of a two-layered stripline. The remaining part of the chapter deals with the application of the unified variational method which combines the 'transverse transmission line' (TTL) technique as described in Chapter 2. This method is used to analyse a whole class of stripline-like transmission lines including the simple homogeneous stripline.

Symmetric stripline, homogeneously filled with a dielectric material, is commonly used in the realization of passive microwave components such as filters, couplers, and power dividers. Homogeneous stripline with a rectangular shielding box finds applications in areas such as EM susceptibility and emission testing, biological effects of RF exposure, calibration of radiation survey meters, and electric probes [5, 6]. Such lines possess an air dielectric with a thin center conductor to maximize the space available between the conductors. In the case of striplines with off-set strip conductor, the off-set distance from the center serves as a variable, thus introducing some flexibility in the design [7]. Besides these homogeneous striplines, a variety of stripline-like transmission lines offer potential use in MIC design. Stripline-like transmission lines considered here are basically inhomogeneous striplines with two or more dielectric (including air) layers situated between the two ground planes. Some of the useful practical configurations are shown in Table 3.1 (Sec. 3.4). For example, the microstrip-like stripline is a convenient geometry for mounting active components [8]. The three-layer sandwiched stripline is specially useful in the design of passive components such as couplers. The thickness of the central layer offers considerable flexibility in the design. The suspended stripline is a special case of the sandwiched stripline. In this, the dielectric substrate carrying the strip conductor is suspended symmetrically between the top and the bottom ground planes. This is an excellent transmission medium for the supercomponent approach to circuit design, that is, large-scale integration of microwave devices on a single printed circuit substrate [9]. This transmission line also finds extensive applications at millimeter wave frequencies [10, 11]. The analysis of all these transmission lines including the homogeneous stripline in a rectangular shield is carried out in this chapter. The unified approach which uses the variational method in the space domain combined with the 'TTL' technique developed in Chapter 2 is applied to derive the capacitance formulas for structures with shielding side walls. For structures with no side wall effects, alternative formulas are also derived by applying the variational method in the Fourier domain combined with the 'TTL' technique. These results, however, would be the same as those computed from the space domain formulas when the side wall spacing is considered sufficiently large.

Stripline-like transmission lines using anisotropic substrates made of materials such as sapphire and pyrolytic boron nitride find useful applications in microwave integrated circuits. In particular, the single crystal sapphire is attractive because of its well-defined and repeatable electrical characteristics. The material is optically transparent, thus enabling mounting of chip devices accurately on the substrate. These crystalline materials have layered atomic planes with the optical axis perpendicular to the planes. They are usually cut into substrates with their plane surfaces perpendicular to the optical axis so that the material permittivity is constant everywhere in the plane of the substrate. A propagating wave travelling along the transmission line is then not subjected to a change in permittivity at bends and corners in the strip conductor. Anisotropic substrates can also be cut such that the optical axis is oriented at an arbitrary angle with respect to the vertical coordinate axis in the transverse plane (x-y plane) of the structure. Stripline-like transmission lines employing such substrates also find useful circuit applications, especially in improving the directivity of couplers. The variational capacitance formula for stripline-like transmission lines with isotropic substrates is shown to be applicable to structures with anisotropic substrates also. The principal axes of the anisotropic substrate may be either aligned, or oriented at an angle, with respect to the x-y coordinate axes taken in the transverse plane of the structure. The only parameter which needs to be changed is the admittance at the charge plane depending on the structure under consideration.

3.2 Symmetric Stripline with Zero Thickness Center Conductor—Conformal Transformation Method [1, 12]

A symmetric homogeneous strip transmission line is one structure which has been analysed rigorously [1] by the conformal transformation method to yield an exact expression for its capacitance. Figure 3.1a shows the geometry of the stripline. The center conductor is assumed to be of zero thickness and the two ground planes are considered to be of infinite width. The medium between the ground planes is filled with a homogeneous, lossless, isotropic dielectric having a relative dielectric constant ϵ_r. The electrostatic potential ϕ in the cross-sectional area of the stripline satisfies Laplace's equation subject to the boundary conditions that $\phi = V_0$ on the strip conductor and $\phi = 0$ on the two ground planes. In Fig. 3.1b is shown the symmetric one quarter of the stripline, where, along the dotted portion of the boundary, the normal gradient of the potential is zero. The capacitance of the stripline is four times

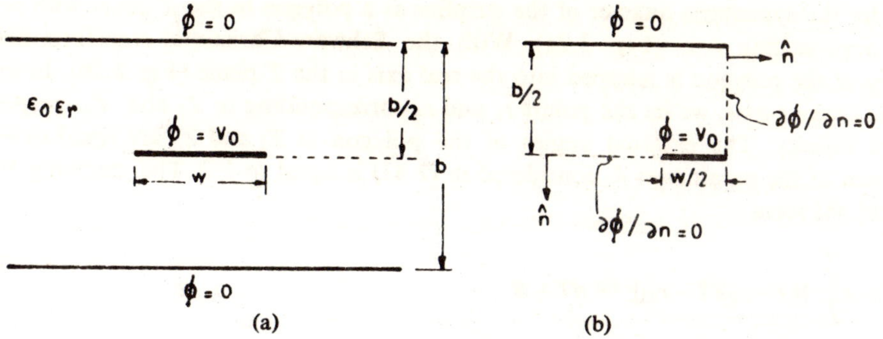

FIG. 3.1 (a) Geometry of the symmetric stripline and (b) one quarter of the stripline.

the capacitance of the quarter section. The Schwarz-Christoffel transformation discussed in Sec. 2.3 is applied twice to map this quarter section into a parallel plate geometry as shown in Fig. 3.2. Once the plate spacing is expressed in terms of the stripline parameters through the mapping function, the capacitance is obtained from the simple parallel plate capacitance formula.

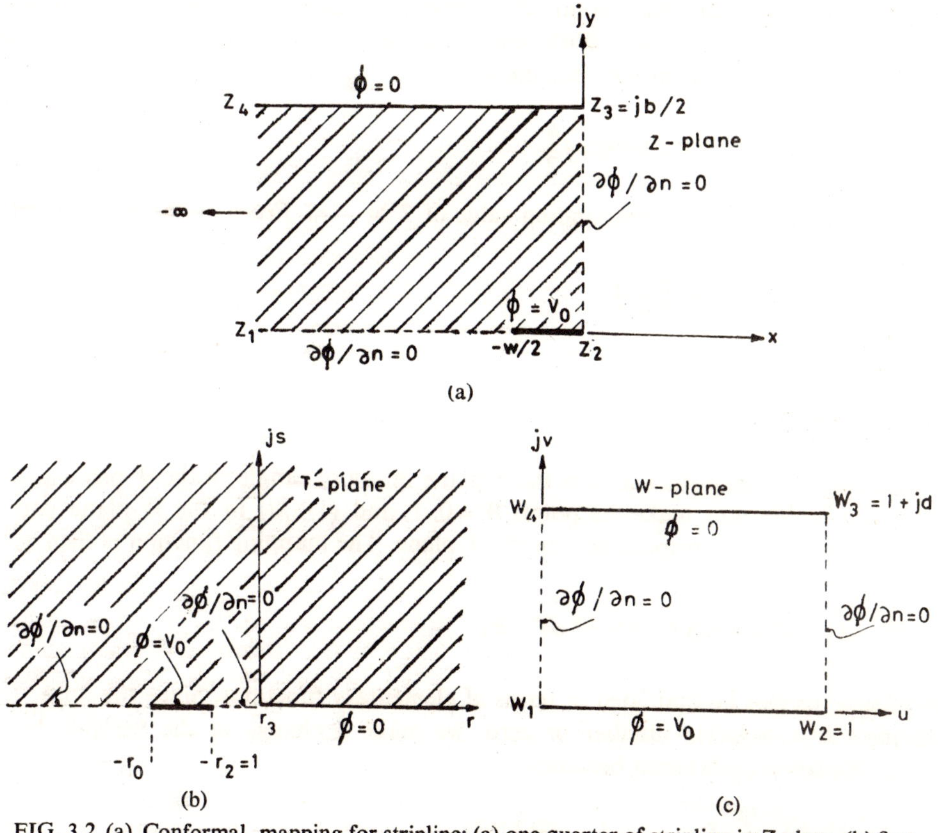

FIG. 3.2 (a) Conformal mapping for stripline: (a) one quarter of stripline in Z-plane (b) first mapping into upper half of *T*-plane, and (c) second mapping into parallel plate geometry.

Consider the symmetric quarter of the stripline as a polygon in the Z-plane with points Z_1 and Z_4 approaching $-\infty$ (Fig. 3.2a). With the Schwarz-Christoffel transformation, the boundary of the polygon is mapped into the real axis in the T-plane (Fig. 3.2b). In order to cover the entire r-axis, we let the points r_1 and r_4 corresponding to Z_1 and Z_4, respectively, approach infinity. The external angles of the polygon at Z_2 and Z_3 are equal to $\pi/2$ and, hence, each of the parameters k_i considered in (2.43) is equal to 1/2. The mapping function thus takes the form

$$Z = A\int^{T} [(T-r_2)(T-r_3)]^{-1/2}\, dT + B \tag{3.1}$$

where A and B are arbitrary constants. We now choose $r_2 = -1$ and $r_3 = 0$ so that the mapping function becomes

$$\begin{aligned} Z &= A\int^{T} [T(T+1)]^{-1/2}\, dT + B \\ &= 2A \ln [T^{1/2} + (T+1)^{1/2}] + B \end{aligned} \tag{3.2a}$$

Corresponding to the boundary points $Z=0$ and $(jb/2)$ in the Z-plane, we have $T=-1$ and 0, respectively, in the T-plane. With this substitution in (3.2a), we get $A=-b/2\pi$ and $B=(jb/2)$. The tranformation then becomes

$$Z = -\frac{b}{\pi} \ln [T^{1/2} + (T+1)^{1/2}] + j\frac{b}{2} \tag{3.2b}$$

When $Z=-(w/2)$, let the corresponding value of T be $-r_0$. From (3.2b), we then get

$$r_0^{1/2} + (r_0 - 1)^{1/2} = \exp\left(\frac{\pi w}{2b}\right)$$

or

$$r_0 = \cosh^2\left(\frac{\pi w}{2b}\right) \tag{3.3}$$

The next step is to map the r-axis in the T-plane into a rectangle in the W-plane as illustrated in Fig. 3.2c. We now make the points $W=0$, 1, and $(1+jd)$ in the W-plane correspond to $T=-r_0$, -1, and 0, respectively, in the T-plane. The mapping function is then given by

$$W = A_1\int^{T} [T(T+r_0)(T+1)]^{-1/2}\, dT + B_1 \tag{3.4a}$$

This integral cannot be evaluated in terms of elementary functions. Since the upper limit of integration is a negative number or zero, we make a change in the variable by setting $T=-\zeta^2$. Equation (3.4a) then becomes

$$W = \left(\frac{2jA_1}{r_0^{1/2}}\right) \int^{(-T)^{1/2}} \left[(1-\zeta^2)\left(1-\frac{\zeta^2}{r_0}\right)\right]^{-1/2} d\zeta + B_1 \tag{3.4b}$$

This integral represents an inverse elliptic function, and tables are available for its evaluation [13-15]. The inverse elliptic function is denoted as $\mathrm{sn}^{-1}(x, k)$ and is defined by the following integral:

$$\mathrm{sn}^{-1}(x, k) = \int_0^x [(1-\zeta^2)(1-k^2\zeta^2)]^{-1/2}\, d\zeta \tag{3.5}$$

where x may be a complex variable, and k is called the modulus of the elliptic function [14]. Since B_1 in (3.4b) is as yet arbitrary, we may put in a lower limit of integration, and, by introducing a new constant $A_2 = (2jA_1)/r_0^{1/2}$, we get

$$W = A_2\, \mathrm{sn}^{-1}[(-T)^{1/2}, r_0^{-1/2}] + B_1 \tag{3.6}$$

We now substitute the values of $T = -r_0$, -1, 0, and the corresponding values of $W = 0$, 1, $(1+jd)$, respectively, into (3.6). This results in three simultaneous equations involving three unknowns A_2, B_1 and d, which on solving gives

$$d = \frac{-j\, \mathrm{sn}^{-1}(1, r_0^{-1/2})}{[\mathrm{sn}^{-1}(r_0^{1/2}, r_0^{-1/2}) - \mathrm{sn}^{-1}(1, r_0^{-1/2})]} \tag{3.7}$$

The function $\mathrm{sn}(x+jy)$ is periodic in x with period $4K$, and periodic in y with period $2K'$ where

$$K = \int_0^1 \frac{d\zeta}{[(1-\zeta^2)(1-k^2\zeta^2)]^{1/2}} \tag{3.8a}$$

$$K' = \int_0^1 \frac{d\zeta}{[(1-\zeta^2)(1-\zeta^2+k^2\zeta^2)]^{1/2}} = \int_1^{1/k} \frac{d\zeta}{[(\zeta^2-1)(1-k^2\zeta^2)]^{1/2}} \tag{3.8b}$$

K as given by (3.8a) is called the complete elliptic integral of the first kind. From (3.8a) and (3.8b), we obtain

$$K - jK' = \int_0^{1/k} \frac{d\zeta}{[(1-\zeta^2)(1-k^2\zeta^2)]^{1/2}} \tag{3.8c}$$

Further, from (3.5), (3.8a), and (3.8c), we can write

$$\mathrm{sn}^{-1}(1, k) = K \tag{3.8d}$$

$$\mathrm{sn}^{-1}\left(\frac{1}{k}, k\right) = K - jK' \tag{3.8e}$$

The expression for d given in (3.7) can be written as

$$d = \frac{K}{K'} = \frac{K(k)}{K(k')} \tag{3.9a}$$

where

$$k = \mathrm{sech}\left(\frac{\pi w}{2b}\right) \tag{3.9b}$$

$$k' = \sqrt{1-k^2} = \tanh\left(\frac{\pi w}{2b}\right) \tag{3.9c}$$

The capacitance per unit length of the stripline is now simply four times the capacitance of the parallel plate capacitor shown in Fig. 3.2c and is given by, $C=(4\epsilon_0\epsilon_r/d)$. The characteristic impedance of the stripline then becomes

$$Z = \frac{\sqrt{\mu_0\epsilon_0\epsilon_r}}{C} = \frac{29.976\pi}{\sqrt{\epsilon_r}} \cdot \frac{K(k)}{K(k')} \tag{3.10}$$

(3.10) gives an exact formula for the characteristic impedance of a homogeneous symmetric stripline with zero thickness strip conductor.

Tables of $[K(k)/K(k')]$ versus k are available in the literature [15]. In addition, some simple, yet very accurate closed form formulas, are also reported for this ratio in terms of k [16]. They are now used to obtain 'ready-to-use formulas' for the characteristic impedance of the stripline.

For $w/b \leqslant 0.5$:

$$0.5 \leqslant k^2 \leqslant 1, \qquad 1 \leqslant \frac{K(k)}{K(k')} \leqslant \infty$$

$$\frac{K(k)}{K(k')} \simeq \frac{1}{\pi} \ln\left[\frac{2(1+\sqrt{k})}{(1-\sqrt{k})}\right] \tag{3.11}$$

With this approximation, the expression for the characteristic impedance given by (3.10) simplifies to

$$Z = \frac{29.976}{\sqrt{\epsilon_r}} \ln\left[\frac{2(1+\sqrt{k})}{(1-\sqrt{k})}\right] \tag{3.12a}$$

Let $p = Z\sqrt{\epsilon_r}/29.976$ so that $\sqrt{k} = (e^p-2)/(e^p+2)$. Using the expression (3.9b) for k, we can express w/b in terms of the characteristic impedance Z:

$$\frac{w}{b} = \frac{2}{\pi} \cosh^{-1}\left[\left(\frac{e^p+2}{e^p-2}\right)^2\right] \tag{3.12b}$$

For $w/b \geqslant 0.5$:

$$0 \leqslant k^2 \leqslant 0.5, \qquad 0 \leqslant \frac{K(k)}{K(k')} \leqslant 1$$

$$\frac{K(k)}{K(k')} \simeq \frac{\pi}{\ln\left[\dfrac{2(1+\sqrt{k'})}{(1-\sqrt{k'})}\right]} \tag{3.13}$$

Substituting this expression in (3.10), we obtain

$$Z = \frac{29.976\pi^2}{\sqrt{\epsilon_r}\ln\left[\dfrac{2(1+\sqrt{k'})}{(1-\sqrt{k'})}\right]} \tag{3.14a}$$

Let $q=(29.976\pi^2)/Z\sqrt{\epsilon_r}$, so that $\sqrt{k'}=(e^q-2)/(e^q+2)$. Using the expression (3.9c) for k', we can express w/b in the form

$$\frac{w}{b}=\frac{2}{\pi}\tanh^{-1}\left[\left(\frac{e^q-2}{e^q+2}\right)^2\right] \tag{3.14b}$$

The accuracy of $[K(k)/K(k')]$ as calculated using (3.11) and (3.13) is of the order of three parts per million. Hence, the simple closed form formulas for Z given by (3.12a) and (3.14a) yield virtually exact results. In (3.12b) and (3.14b), the parameter w/b is expressed in terms of Z and ϵ_r. These equations can be directly used in the determination of stripline dimensions for a specified value of the characteristic impedance.

3.3 Asymmetric Double Layer Stripline—Illustration of Variational Method

3.3.1 Analysis in the Fourier Domain

The application of the variational method to the simplest inhomogeneous planar transmission line, namely, the microstrip line was first reported by Yamashita and Mittra [2]. The method was later extended to other inhomogeneous planar structures, such as the suspended stripline and the double layer microstrip by Yamashita [3]. In this method, the line capacitance is formulated as a variational integral in the Fourier transformed coordinate. The potential function appearing in the integrand is obtained as the solution of Laplace's equation in the Fourier transformed domain, subject to the boundary conditions at the dielectric interfaces. The integration is then carried out by substituting a suitable trial function for the charge distribution on the strip conductor.

In this section, we demonstrate the above analytical approach by considering a double layer stripline as shown in Fig. 3.3. The strip conductor is assumed to be of negligible thickness and is placed asymmetrically between two large ground planes. The potential function $\phi(x, y)$ in the cross-sectional area of the stripline is governed by the Poisson's equation:

$$\frac{\partial^2\phi}{\partial x^2}+\frac{\partial^2\phi}{\partial y^2}=-\frac{\rho(x, y)}{\epsilon} \tag{3.15}$$

where ϵ is the permittivity of the medium, $\rho(x, y)$ is the charge distribution on the strip conductor which is expressed as

$$\rho(x, y)=f(x)\delta(y-h) \tag{3.16}$$

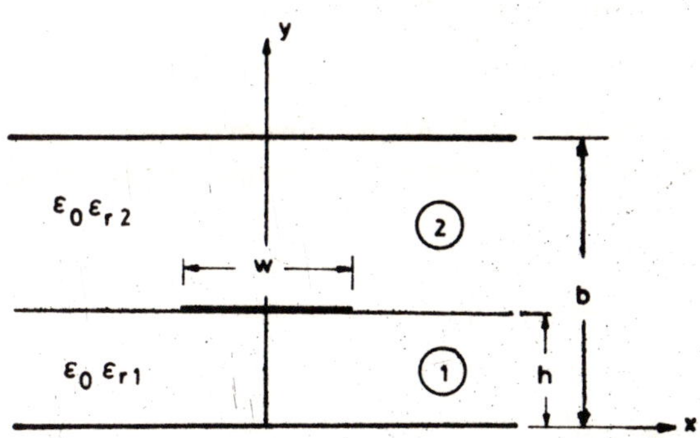

FIG. 3.3 Double layer stripline.

We now define the Fourier transform with respect to x in the form

$$\hat{f}(\beta)=\int_{-\infty}^{\infty} f(x)\, e^{j\beta x}\, dx \tag{3.17}$$

Taking the Fourier transform of (3.15), the equation governing the Fourier transformed potential function $\hat{\phi}(\beta, y)$ in the dielectric regions can be written as

$$\left(\frac{d^2}{dy^2}-\beta^2\right)\hat{\phi}(\beta, y)=0, \qquad y \neq 0, h, b \tag{3.18}$$

The solutions in the two dielectric regions marked 1 and 2 (Fig. 3.3) are

$$\hat{\phi}_1(\beta, y)=A \sinh \beta y, \qquad 0<y<h \tag{3.19a}$$

$$\hat{\phi}_2(\beta, y)=B \sinh \beta(b-y), \qquad h<y<b \tag{3.19b}$$

The conditions to be satisfied at the dielectric interface $y=h$ are

$$\hat{\phi}_2(\beta, y)|_{y=h}=\hat{\phi}_1(\beta, y)|_{y=h} \tag{3.20a}$$

$$\epsilon_{r2}\frac{d}{dy}\hat{\phi}_2(\beta, y)\bigg|_{y=h}=\epsilon_{r1}\frac{d}{dy}\hat{\phi}_1(\beta, y)\bigg|_{y=h}-\frac{\hat{f}(\beta)}{\epsilon_0} \tag{3.20b}$$

where ϵ_{r1} and ϵ_{r2} are the relative dielectric constants of regions 1 and 2, respectively. Using the conditions given in (3.20), the two unknown constants A and B can be solved and expressed in terms of $\hat{f}(\beta)$.

$$A=\frac{\hat{f}(\beta)}{\beta\epsilon_0[\epsilon_{r1} \coth \beta h+\epsilon_{r2} \coth \{\beta(b-h)\}]\cdot\sinh \beta h} \tag{3.21a}$$

$$B=\frac{A \sinh \beta h}{\sinh \{\beta(b-h)\}} \tag{3.21b}$$

The Fourier transformed potential function at the plane of the strip conductor is now obtained by substituting for A in (3.19a) and setting $y=h$. Thus,

$$\hat{\phi}(\beta, h)=\frac{\hat{f}(\beta)}{\beta\epsilon_0[\epsilon_{r1} \coth \beta h+\epsilon_{r2} \coth \{\beta(b-h)\}]} \tag{3.22}$$

where β is real and positive. It may be noted that $f(x)$ and $\phi(x, h)$ are symmetric in x, and hence, $\hat{f}(\beta)=\hat{f}(-\beta)$ and $\hat{\phi}(\beta, h)=\hat{\phi}(-\beta, h)$. We now make use of the variational expression (2.59) for the capacitance per unit length of a transmission line. Referring to Fig. 3.3, we can write

$$\frac{1}{C}=\frac{1}{Q^2}\int_{-\infty}^{\infty} \phi(x, h) f(x)\, dx$$

$$=\frac{1}{Q^2}\int_{-w/2}^{w/2} \phi(x, h) f(x)\, dx$$

where

$$Q = \int_{-w/2}^{w/2} f(x)\, dx \tag{3.24}$$

In order to circumvent the difficulty of performing the inverse Fourier transform of (3.22) for use in (3.23), we transform (3.23) into a variational form in the Fourier transformed coordinate. To this end, we make use of Parseval's theorem [17], which, in terms of our symmetric functions $f(x)$ and $\phi(x, h)$ can be stated as

$$\int_{-\infty}^{\infty} \phi(x, h) f(x)\, dx = \frac{1}{2\pi} \int_{-\infty}^{\infty} \hat{\phi}(\beta, h) \hat{f}(\beta)\, d\beta \tag{3.25}$$

With this relation, expression (3.23) for the capacitance becomes

$$\frac{1}{C} = \frac{1}{2\pi Q^2} \int_{-\infty}^{\infty} \hat{\phi}(\beta, h) \hat{f}(\beta)\, d\beta \tag{3.26}$$

Substituting for $\hat{\phi}(\beta, h)$ from (3.22), we obtain

$$\frac{1}{C} = \frac{1}{2\pi Q^2} \int_{-\infty}^{\infty} \frac{[\hat{f}(\beta)]^2}{\beta\epsilon_0[\epsilon_{r1} \coth \beta h + \epsilon_{r2} \coth \beta(b-h)]} \cdot d\beta$$

$$= \frac{1}{\pi Q^2} \int_{0}^{\infty} \frac{[\hat{f}(\beta)]^2}{\beta\epsilon_0[\epsilon_{r1} \coth \beta h + \epsilon_{r2} \coth \beta(b-h)]} \cdot d\beta \tag{3.27}$$

This is the desired integral form for the calculation of capacitance per unit length of the stripline. Since the expression is variational, the calculated value of C is relatively insensitive to the choice of the trial function for $f(x)$. Further, for a given set of trial functions, the one that maximizes the capacitance clearly gives the closest value to the true value of capacitance. The choice of the trial function for the charge distribution is discussed later in Sec. 3.4.

3.3.2 Analysis in Space Domain Using Green's Function

The variational expression for the capacitance given in (2.58) can be used directly provided Green's function for the structure can be derived explicitly. We illustrate this procedure by considering the example of a double layer stripline as in the preceding subsection but with the incorporation of shielding side walls as shown in Fig. 3.4. The potential function $\phi(x, y)$ in the cross-sectional area of the structure satisfies the same equation as in (3.15). The strip conductor is also assumed to be infinitesimally thin so that the charge distribution specified by (3.16) applies.

We determine the Green's function G due to a point charge located at $x = x_0$, $y = h$. G should satisfy Poisson's equation given by

$$\frac{\partial^2 G}{\partial x^2} + \frac{\partial^2 G}{\partial y^2} = -\frac{\delta(x-x_0)\delta(y-h)}{\epsilon} \tag{3.28}$$

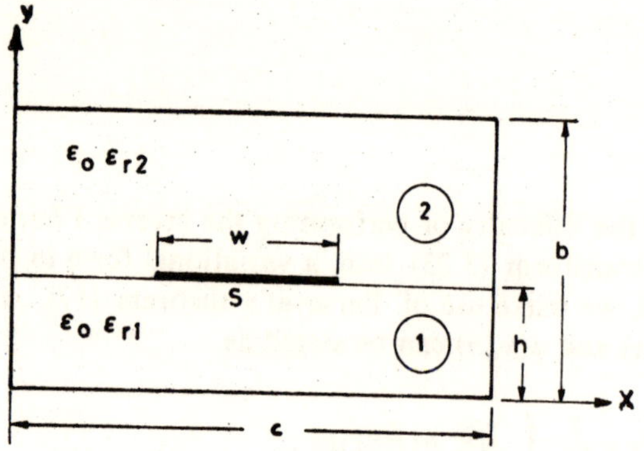

FIG. 3.4 Double layer stripline with side walls.

At the four conducting walls located at $x=0$, c, and $y=0$, b, Green's function should vanish. With this condition incorporated, the solution for G in the two dielectric regions marked 1 and 2 (Fig. 3.4) is

$$G_1(x, y/x_0, h) = \sum_{n=1}^{\infty} A_n \cdot \sinh\frac{n\pi y}{c} \sin\frac{n\pi x}{c} \sin\frac{n\pi x_0}{c}, \qquad 0 < y < h \tag{3.29a}$$

$$G_2(x, y/x_0, h) = \sum_{n=1}^{\infty} B_n \sinh\left\{\frac{n\pi}{c}(b-y)\right\} \cdot \sin\frac{n\pi x}{c} \cdot \sin\frac{n\pi x_0}{c}, \qquad h < y < b \tag{3.29b}$$

where A_n and B_n are unknown constants, which are to be determined by satisfying the following conditions at the interface $y=h$:

$$G_2(x, y)|_{y=h} = G_1(x, y)|_{y=h} \tag{3.30a}$$

$$\epsilon_{r2} \left.\frac{\partial G_2(x, y)}{\partial y}\right|_{y=h} = \epsilon_{r1} \left.\frac{\partial G_1(x, y)}{\partial y}\right|_{y=h} - \frac{\delta(x-x_0)}{\epsilon_0} \tag{3.30b}$$

Incorporating the spectral representation of $\delta(x-x_0)$ in the form [18]

$$\delta(x-x_0) = \frac{2}{c}\sum_{n=1}^{\infty} \sin\frac{n\pi x}{c} \cdot \sin\frac{n\pi x_0}{c} \tag{3.31}$$

we can solve for A_n and B_n using (3.29) in (3.30). Green's function at the plane $y=h$ is then obtained by substituting for A_n in (3.29a) and setting $y=h$. Thus,

$$G(x, h/x_0, h) = \sum_{n=1}^{\infty} \frac{2}{n\pi\epsilon_0} \cdot \frac{\sin\left(\frac{n\pi x}{c}\right) \cdot \sin\left(\frac{n\pi x_0}{c}\right)}{\left[\epsilon_{r1} \coth\left(\frac{n\pi h}{c}\right) + \epsilon_{r2} \coth\left\{\frac{n\pi}{c}(b-h)\right\}\right]} \tag{3.32}$$

Referring to Fig. 3.4, the variational expression for the capacitance given by (2.58) can be written in the form

$$\frac{1}{C} = \frac{1}{Q^2} \int_S \int_S G(x, h/x_0, h) f(x) f(x_0)\, dx\, dx_0 \tag{3.33a}$$

where

$$Q=\int_S f(x)\,dx \tag{3.33b}$$

The integration is to be carried out on the strip conductor from $x=(c-w)/2$ to $(c+w)/2$. Substituting for Green's function from (3.32), the final variational expression for the capacitance per unit length of the line becomes

$$\frac{1}{C}=\frac{1}{Q^2}\sum_{n=1}^{\infty}\frac{2}{n\pi\epsilon_0}\frac{\left[\int_{(c-w)/2}^{(c+w)/2} f(x)\sin\frac{n\pi x}{c}\,dx\right]^2}{\left[\epsilon_{r1}\coth\frac{n\pi h}{c}+\epsilon_{r2}\coth\left(\frac{n\pi}{c}(b-h)\right)\right]} \tag{3.34}$$

3.4 Striplines and Stripline-like Transmission Lines with Side Walls and Iso/Anisotropic Dielectric Layers—Unified Approach [19]

3.4.1 General Capacitance Formulas for a Specified Charge Distribution

Table 3.1 illustrates the cross-sectional views of striplines and a variety of stripline-like transmission lines with a rectangular outer conductor. It is assumed that the strip conductor is infinitesimally thin. Since these structures have electric side walls separated by a distance c, the capacitance expression (2.128) applies. It is given by

$$\frac{1}{C}=\frac{\left[\sum_{n=1}^{\infty}(2/n\pi Y)\left\{\int_S f(x)\sin(n\pi x/c)\,dx\right\}^2\right]}{\left[\int_S f(x)\,dx\right]^2} \tag{3.35}$$

We now have to specify a suitable charge distribution $f(x)$ before performing the integration. The charge density on an infinitely thin strip conductor in free space has been calculated based on the electrostatic theory [20], and the solution indicates a rapid increase of the charge density near the edge of the strip. Such a function is given by

$$\begin{aligned} f(x)&=[1-\{(2/w)\mid x-c/2\mid\}^2]^{-1}, \qquad (c-w)/2\leqslant x\leqslant(c+w)/2\\ &=0, \qquad \text{otherwise}\end{aligned} \tag{3.36}$$

where w is the width of the strip conductor. Experimental verification of a similar distribution on a homogeneous stripline has been reported by Dukes [21].

For the structures considered here, it would therefore be appropriate to assume a trial function of this type. Since (3.35) is variational, one may use an approximate trial function for $f(x)$ and incur only a second-order error in the value of capacitance. A trial function that maximizes the capacitance yields the closest value to the exact result. Approximating the charge distribution given in (3.36) in the form of a series, we can write

$$\begin{aligned} f(x)&=A_0+A_1\mid x-c/2\mid+A_2\mid x-c/2\mid^2+\ldots, \qquad (c-w)/2\leqslant x\leqslant(c+w)/2\\ &=0, \qquad \text{otherwise}\end{aligned} \tag{3.37}$$

Table 3.1 Admittance parameters of striplines and stripline-like transmission lines with side walls

Structure	*Admittance Y at $y=y_0$, $\beta_n=n\pi/c$*
A1. Symmetric stripline 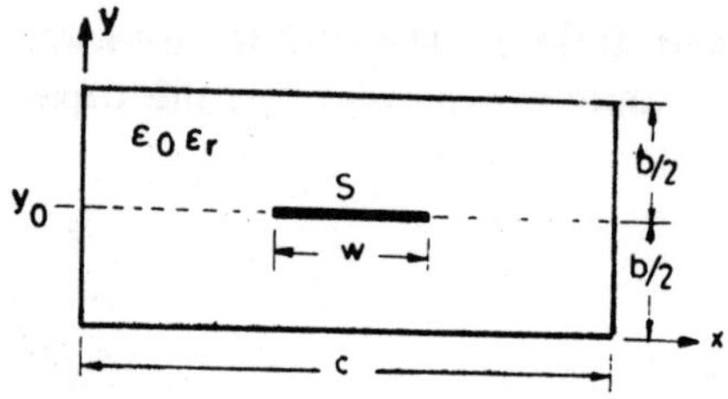	$2\epsilon_0\epsilon_r \coth(\beta_n b/2)$
A2. Stripline with off-set center conductor 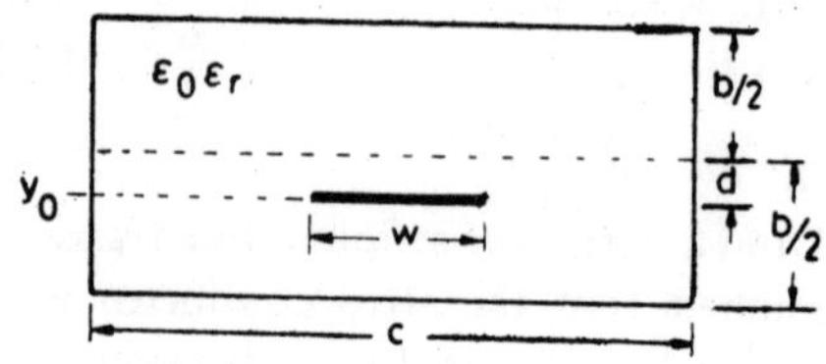	$\epsilon_0\epsilon_r[\coth\{\beta_n(b/2+d)\}+\coth\{\beta_n(b/2-d)\}]$
A3. Microstrip-like stripline 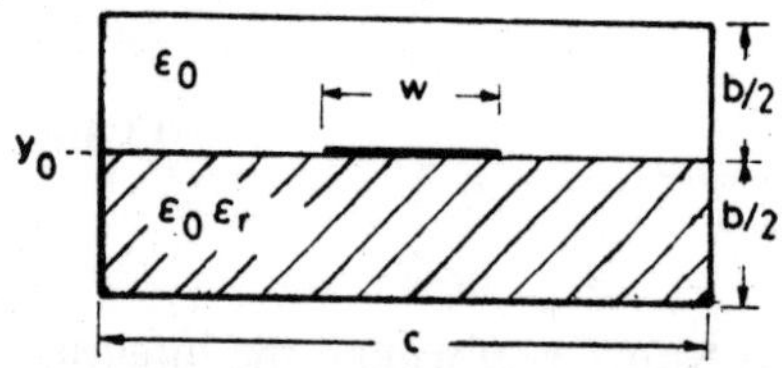	$\epsilon_0[\coth(\beta_n b/2)+\epsilon_r \coth(\beta_n b/2)]$
A4. Suspended stripline 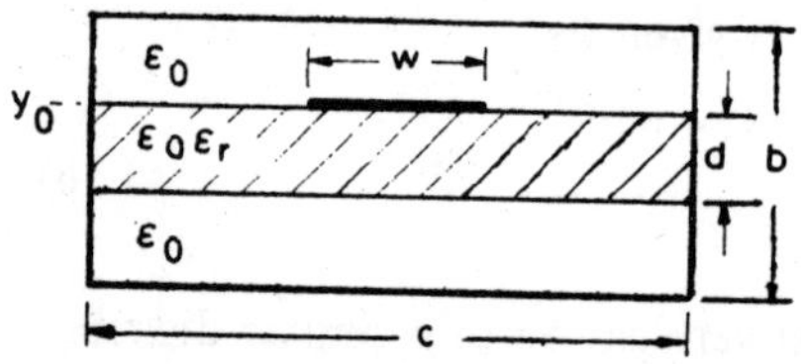	$\epsilon_0\left[\coth(\beta_n(b-d)/2)+\epsilon_r\left\{\frac{\coth\{\beta_n(b-d)/2\}\coth(\beta_n d)+\epsilon_r}{\epsilon_r\coth(\beta_n d)+\coth\{\beta_n(b-d)/2\}}\right\}\right]$
A5. Sandwiched stripline 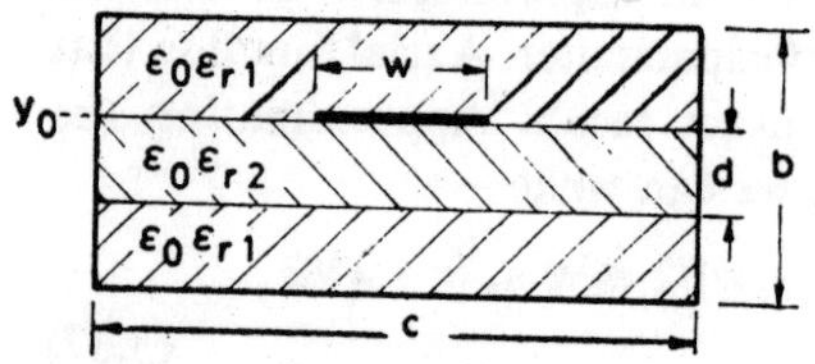	$\epsilon_0\left[\epsilon_{r1}\coth(\beta_n(b-d)/2)+\epsilon_{r2}\left\{\frac{\epsilon_{r1}\coth\{\beta_n(b-d)/2\}\coth(\beta_n d)+\epsilon_{r2}}{\epsilon_{r2}\coth(\beta_n d)+\epsilon_{r1}\coth\{\beta_n(b-d)/2\}}\right\}\right]$

When a finite series is chosen, the coefficients of the series should be adjusted to obtain the maximum value of the line capacitance. Taking more terms would mean more computing time. A compromise between the computing time and accuracy is therefore necessary in practice.

It is found that a trial function given by the following expression gives quite accurate results adequate for almost all practical purposes:

$$f(x) = (1/w)[1 + A \mid (2/w)(x - c/2) \mid^3], \qquad (c-w)/2 \leqslant x \leqslant (c+w)/2 \tag{3.38}$$

$$= 0, \qquad \text{otherwise}$$

where A is an arbitrary constant and is determined by maximizing the capacitance. Substituting (3.38) in (3.35), the expression for C becomes

$$C = \frac{\left[\int_{(c-w)/2}^{(c+w)/2} \frac{1}{w}\left\{1 + A \mid (2/w)(x - c/2) \mid^3\right\} dx \right]^2}{\sum_{n=1}^{\infty} [2/(n\pi Y w^2)] \left[\int_{(c-w)/2}^{(c+w)/2} \{1 + A \mid (2/w)(x-c/2) \mid^3\} \sin (n\pi x/c)\, dx \right]^2}$$

$$= \left[(1 + 0.25A)^2\right] \Big/ \left[\sum_{n=1}^{\infty} [2/(n\pi Y w^2)] I_n^2 \right] \tag{3.39}$$

where

$$I_n = \int_{(c-w)/2}^{(c+w)/2} \sin (n\pi x/c)\, dx + (8A/w^3) \int_{(c-w)/2}^{(c+w)/2} \mid x - c/2 \mid^3 \sin (n\pi x/c)\, dx$$

$$= (2c/n\pi) \sin (n\pi/2) [\sin (n\pi w/2c) + A(2c/n\pi w)^3 \{3[(n\pi w/2c)^2 - 2] \cos (n\pi w/2c)$$
$$+ (n\pi w/2c) [(n\pi w/2c)^2 - 6] \sin (n\pi w/2c) + 6\}] \tag{3.40}$$

Substituting (3.40) in (3.39) and noting that $\sin (n\pi/2)$ which appears in I_n vanishes for even values of n including $n = 0$, we obtain

$$C = [(1 + 0.25A)^2] / [\sum_{n \text{ odd}} (L_n + AM_n)^2 P_n / Y] \tag{3.41}$$

where

$$L_n = \sin (\beta_n w/2) \tag{3.42a}$$

$$M_n = (2/\beta_n w)^3 [3\{(\beta_n w/2)^2 - 2\} \cos (\beta_n w/2) + (\beta_n w/2)\{(\beta_n w/2)^2 - 6\} \cdot \sin (\beta_n w/2) + 6] \tag{3.42b}$$

$$P_n = (2/n\pi)(2/\beta_n w)^2 \tag{3.42c}$$

$$\beta_n = n\pi/c \tag{3.42d}$$

In order to determine the constant A, we maximize C by differentiating C with respect to A and setting it equal to zero. Solving $\partial C/\partial A = 0$, we obtain

$$A = -[\sum_{n \text{ odd}} (L_n - 4M_n) L_n P_n / Y] / \sum_{n \text{ odd}} (L_n - 4M_n) M_n P_n / Y] \tag{3.43}$$

For the charge distribution specified by (3.38), expression (3.41) combined with (3.42) and (3.43) can be used to evaluate the capacitance of any single conductor multilayer stripline-like transmission line with side walls separated by a distance c. The only parameter that needs to be changed is the admittance Y at the charge plane depending on the structure.

3.4.2 Admittance Parameter for Isotropic Structures

The admittance at the charge plane for isotropic structures is obtained by applying the transmission line formula (2.131), with $Y_{cj}=\epsilon_j=\epsilon_0\epsilon_{rj}$ and $\gamma_j=\beta_n=n\pi/c$. As an illustration, we derive below the admittance expression for a general three-layer structure. Figure 3.5 shows the geometry of a three-layer structure along with its transmission line equivalent circuit for determining the admittance parameter. The short circuits at either ends of the equivalent circuit shown in Fig. 3.5b correspond to the ground planes at $y=0$ and b in Fig. 3.5a. The various admittances marked in Fig. 3.5b are

$$Y_1=\epsilon_0\epsilon_{r1}\coth(\beta_n b_1) \tag{3.44a}$$

$$Y_-=\epsilon_0\epsilon_{r2}\left[\frac{Y_1+\epsilon_0\epsilon_{r2}\tanh(\beta_n b_2)}{\epsilon_0\epsilon_{r2}+Y_1\tanh(\beta_n b_2)}\right] \tag{3.44b}$$

$$Y_+=\epsilon_0\epsilon_{r3}\coth(\beta_n b_3) \tag{3.44c}$$

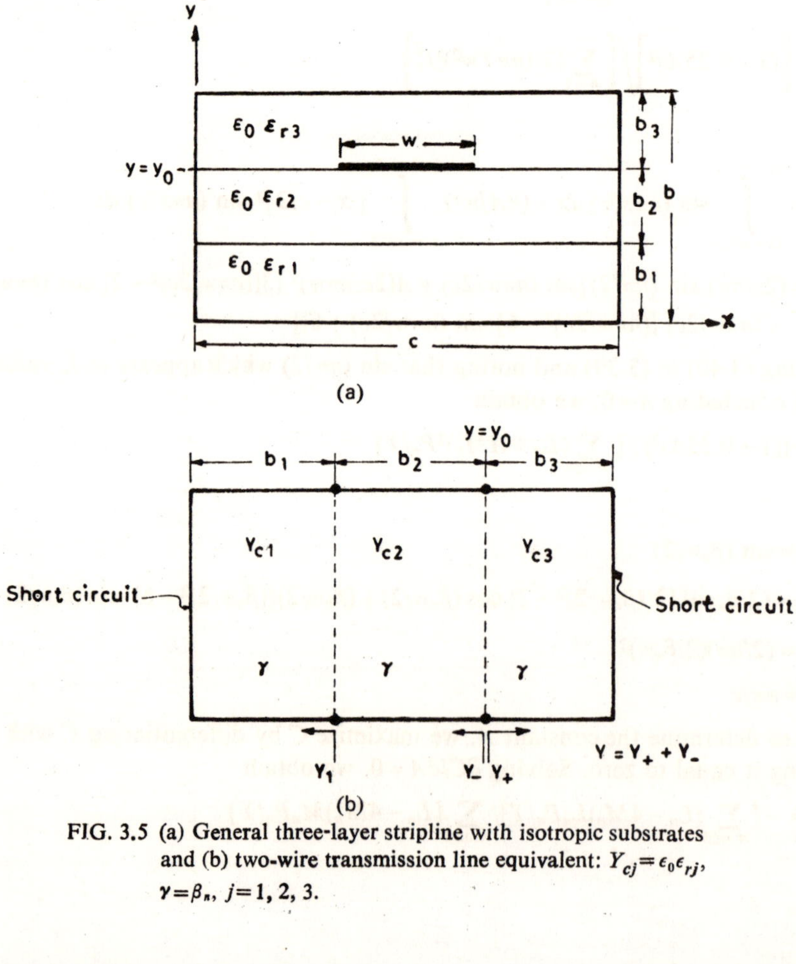

FIG. 3.5 (a) General three-layer stripline with isotropic substrates and (b) two-wire transmission line equivalent: $Y_{cj}=\epsilon_0\epsilon_{rj}$, $\gamma=\beta_n$, $j=1, 2, 3$.

The admittance parameter Y is obtained by summing the two admittances Y_+ and Y_-:

$$Y=\epsilon_0\left[\epsilon_{r3}\coth(\beta_n b_3)+\epsilon_{r2}\left\{\frac{\epsilon_{r1}\coth(\beta_n b_1)\coth(\beta_n b_2)+\epsilon_{r2}}{\epsilon_{r2}\coth(\beta_n b_2)+\epsilon_{r1}\coth(\beta_n b_1)}\right\}\right] \tag{3.45}$$

where $\beta_n=n\pi/c$. Expressions for the admittance parameters of some practical stripline and stripline-like transmission lines are listed in Table 3.1.

3.4.3 Admittance Parameter for Structures with Anisotropic Substrates

Table 3.2 shows transmission line structures corresponding to Table 3.1 with isotropic dielectric substrates replaced by anisotropic dielectric substrates. The substrates are cut such that the optical axis is oriented perpendicular to the plane of the substrate. The permittivity tensor then becomes a diagonal dyadic with respect to the coordinate system. Thus, while applying the formula (2.131), we use $Y_{cj}=\epsilon_0\sqrt{\epsilon_{xxj}\epsilon_{yyj}}$ and $\gamma_j=\beta_n\sqrt{\epsilon_{xxj}/\epsilon_{yyj}}$ with $\beta_n=n\pi/c$ for the anisotropic layer, and $Y_{cj}=\epsilon_0\epsilon_{rj}$ and $\gamma_j=\beta_n$ for the isotropic layer. Considering a general three-layer structure (Fig. 3.6a), the various admittances marked in its transmission line equivalent representation (Fig. 3.6b) are

$$Y_1=\epsilon_0\sqrt{\epsilon_{xx1}\epsilon_{yy1}}\coth\left(\beta_n b_1\sqrt{\epsilon_{xx1}/\epsilon_{yy1}}\right) \tag{3.46a}$$

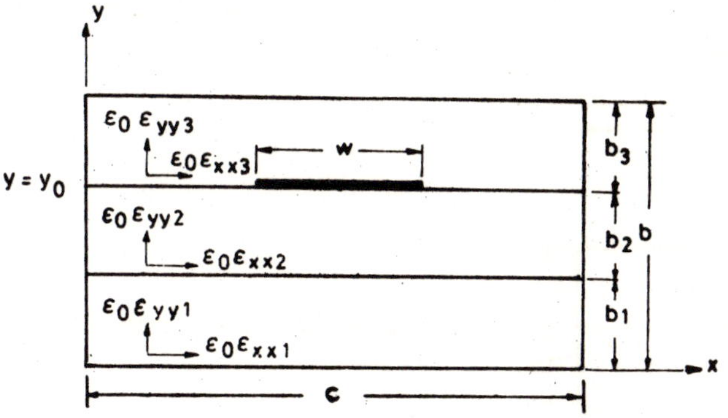

(a)

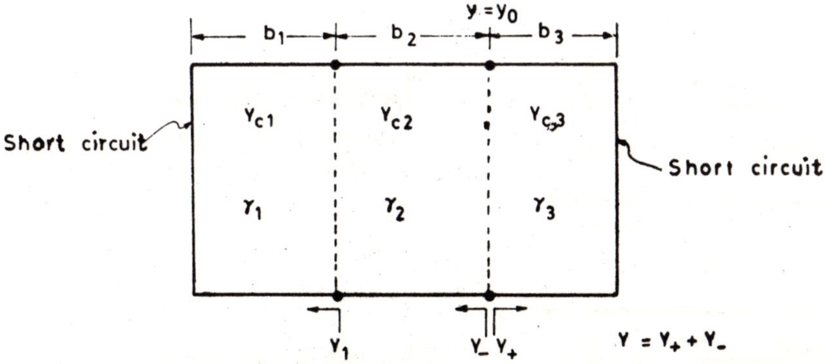

(b)

FIG. 3.6 (a) General three-layer stripline with anisotropic substrates and (b) two-wire transmission line equivalent: $Y_{cj}=\epsilon_0\sqrt{\epsilon_{xxj}\epsilon_{yyj}}$, $\gamma_j=\beta_n\sqrt{\epsilon_{xxj}/\epsilon_{yyj}}$, $j=1, 2, 3$.

Table 3.2 Admittance parameters of striplines and stripline-like transmission lines with side walls and anisotropic dielectric substrates—optical axis aligned along the *y*-axis

$$\epsilon_{rfj}=\sqrt{\epsilon_{xxj}\epsilon_{yyj}},\ \epsilon_{aj}=\sqrt{\epsilon_{xxj}/\epsilon_{yyj}},\ j=1,2$$

Structure	*Admittance Y at $y=y_0$, $\beta_n=n\pi/c$*
A1. Symmetric stripline 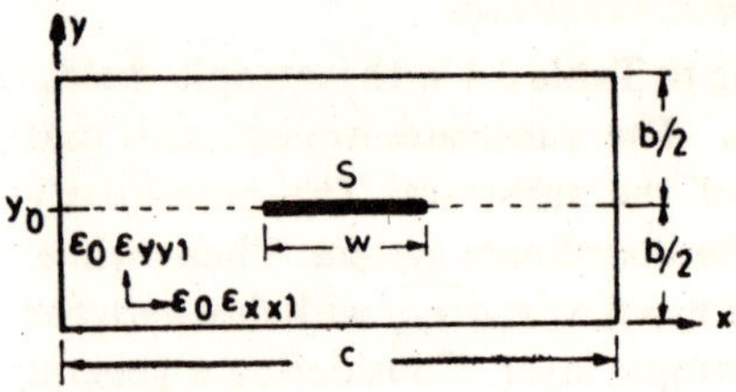 	$2\epsilon_0\epsilon_{rf1}\coth(\beta_n\epsilon_{a1}b/2)$
A2. Stripline with off-set center conductor 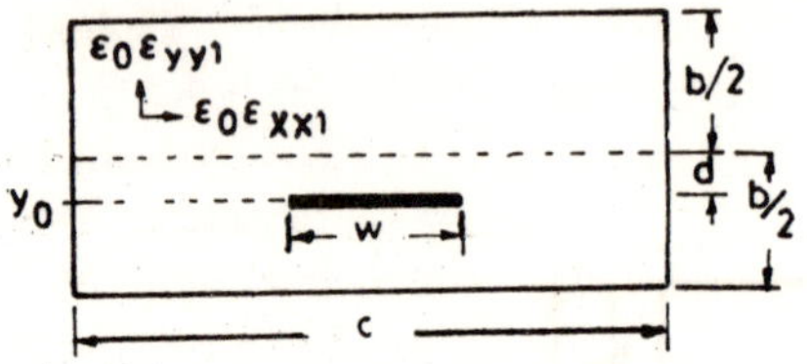 	$\epsilon_0\epsilon_{rf1}[\coth(\beta_n\epsilon_{a1}((b/2)+d))+\coth(\beta_n\epsilon_{a1}((b/2)-d))]$
A3. Microstrip-like stripline 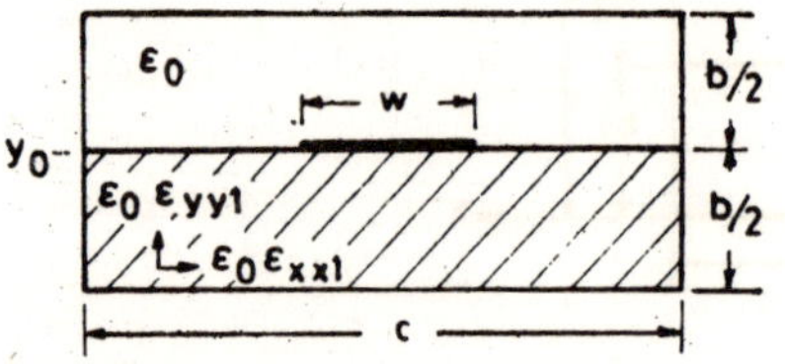 	$\epsilon_0[\coth(\beta_n b/2)+\epsilon_{rf1}\coth(\beta_n\epsilon_{a1}b/2)]$
A4. Suspended stripline 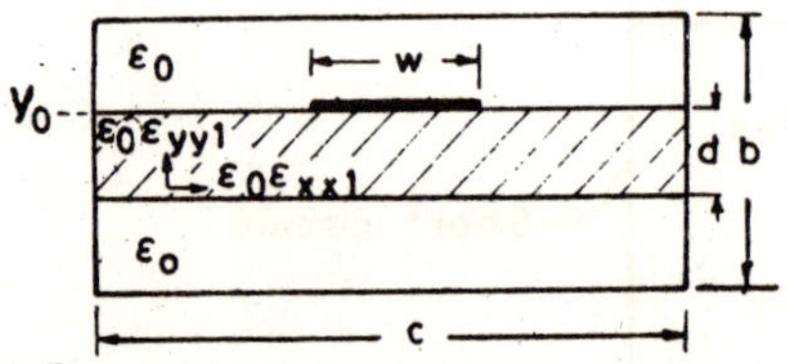 	$\epsilon_0\left[\coth(\beta_n((b-d)/2))+\epsilon_{rf1}\cdot\left\{\frac{\coth(\beta_n((b-d)/2))\coth(\beta_n\epsilon_{a1}d)+\epsilon_{rf1}}{\epsilon_{rf1}\coth(\beta_n\epsilon_{a1}d)+\coth(\beta_n((b-d)/2))}\right\}\right]$
A5. Sandwiched stripline 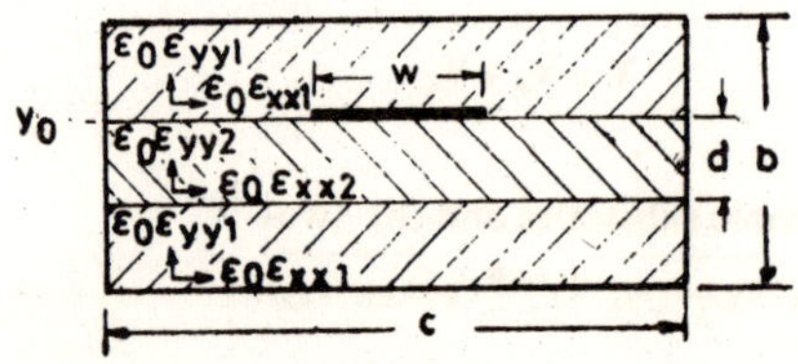 	$\epsilon_0\left[\epsilon_{rf1}\coth(\beta_n\epsilon_{a1}((b-d)/2))+\epsilon_{rf2}\cdot\left\{\frac{\epsilon_{rf1}\coth(\beta_n\epsilon_{a1}((b-d)/2))\coth(\beta_n\epsilon_{a2}d)+\epsilon_{rf2}}{\epsilon_{rf2}\coth(\beta_n\epsilon_{a2}d)+\epsilon_{rf1}\coth(\beta_n\epsilon_{a1}((b-d)/2))}\right\}\right]$

$$Y_- = \epsilon_0\sqrt{\epsilon_{xx2}\epsilon_{yy2}}\left[\frac{Y_1+\epsilon_0\sqrt{\epsilon_{xx2}\epsilon_{yy2}}\tanh\left(\beta_n b_2\sqrt{\epsilon_{xx2}/\epsilon_{yy2}}\right)}{\epsilon_0\sqrt{\epsilon_{xx2}\epsilon_{yy2}}+Y_1\tanh\left(\beta_n b_2\sqrt{\epsilon_{xx2}/\epsilon_{yy2}}\right)}\right] \quad (3.46b)$$

$$Y_+ = \epsilon_0\sqrt{\epsilon_{xx3}\epsilon_{yy3}}\coth\left(\beta_n b_3\sqrt{\epsilon_{xx3}/\epsilon_{yy3}}\right) \quad (3.46c)$$

The admittance parameter Y is obtained by summing Y_+ and Y_-:

$$Y = \epsilon_0\left[\sqrt{\epsilon_{xx3}\epsilon_{yy3}}\coth\left(\beta_n b_3\sqrt{\epsilon_{xx3}/\epsilon_{yy3}}\right)+\sqrt{\epsilon_{xx2}\epsilon_{yy2}}\cdot\left\{\frac{\sqrt{\epsilon_{xx1}\epsilon_{yy1}}\coth\left(\beta_n b_1\sqrt{\epsilon_{xx1}/\epsilon_{yy1}}\right)\coth\left(\beta_n b_2\sqrt{\epsilon_{xx2}/\epsilon_{yy2}}\right)+\sqrt{\epsilon_{xx2}\epsilon_{yy2}}}{\sqrt{\epsilon_{xx2}\epsilon_{yy2}}\coth\left(\beta_n b_2\sqrt{\epsilon_{xx2}/\epsilon_{yy2}}\right)+\sqrt{\epsilon_{xx1}\epsilon_{yy1}}\coth\left(\beta_n b_1\sqrt{\epsilon_{xx1}/\epsilon_{yy1}}\right)}\right\}\right] \quad (3.47)$$

The admittance expressions for some of the striplines and stripline-like transmission lines with uniaxially anisotropic substrates having permittivity tensor given in (2.93) are listed in Table 3.2.

3.5 Striplines and Stripline-like Transmission Lines with Infinite Ground Planes and Iso/Anisotropic Dielectric Substrates—Unified Approach

3.5.1 General Capacitance Formula for a Specified Charge Distribution

In the structures given in Table 3.1, if the side walls are removed sufficiently away from the strip conductor, they will have practically no effect on the propagation parameters. Since the unified capacitance expression (3.41) is valid for any value of c greater than w, it can be numerically solved by taking c sufficiently large so as to have a negligible effect of side walls on the capacitance of the structure. However, as c increases, the computational time also increases. An alternative way of analysing a structure with no effect of side walls is to consider the dielectric layers and the top and bottom ground planes to extend to infinity in the x-direction ($c=\infty$) as shown in Table 3.3. The unified capacitance expression (2.137) which is computationally faster can then be used. It is given by

$$\frac{1}{C} = (1/\pi)\int_0^\infty\left[\frac{\hat{f}(\beta)}{Q}\right]^2(1/\beta Y)\,d\beta \quad (3.48)$$

where

$$Q = \int_{-w/2}^{w/2} f(x)\,dx \quad (3.49)$$

We now choose a trial function for the charge distribution given by

$$f(x) = 1+|2x/w|^3, \qquad -w/2\leqslant x\leqslant w/2$$
$$= 0, \qquad \text{otherwise} \quad (3.50)$$

Taking the Fourier transform of (3.50), the ratio $\hat{f}(\beta)/Q$ in the integrand of (3.48) becomes

$$\hat{f}(\beta)/Q = \frac{8}{5}\{\sin(\beta w/2)/(\beta w/2)\}+\{12/(5(\beta w/2)^2)\}[\cos(\beta w/2)-2(\sin(\beta w/2)/(\beta w/2))+(\sin(\beta w/4)/(\beta w/4))^2] \quad (3.51)$$

Table 3.3 Admittance parameters of striplines and stripline-like transmission lines with infinite parallel ground planes

Structure	*Admittance Y at $y = y_0$*
• Symmetric stripline 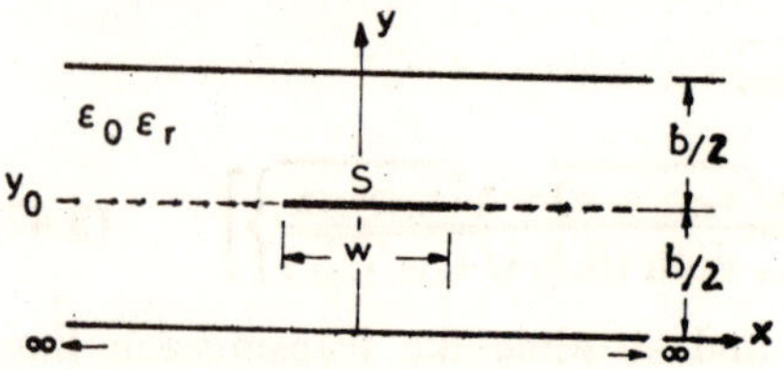	$2\epsilon_0\epsilon_r \coth(\beta b/2)$
• Stripline with off-set center conductor 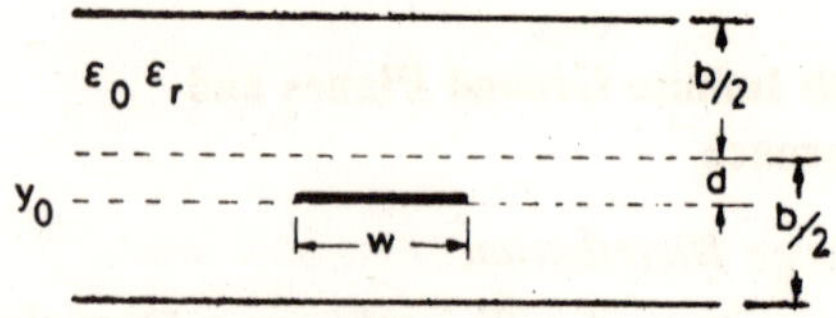	$\epsilon_0\epsilon_r[\coth(\beta(b/2+d)) + \coth(\beta(b/2-d))]$
• Microstrip-like stripline 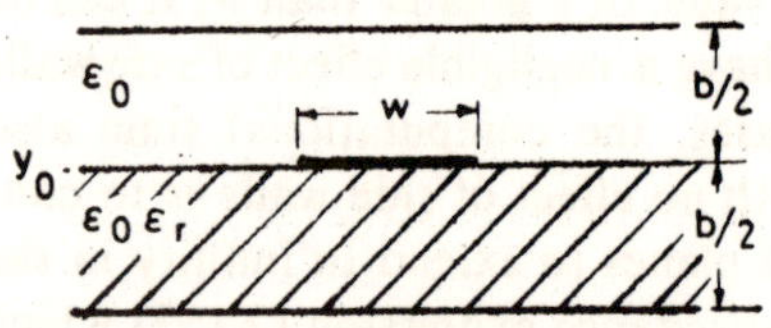	$\epsilon_0[\coth(\beta b/2) + \epsilon_r \coth(\beta b/2)]$
• Suspended stripline 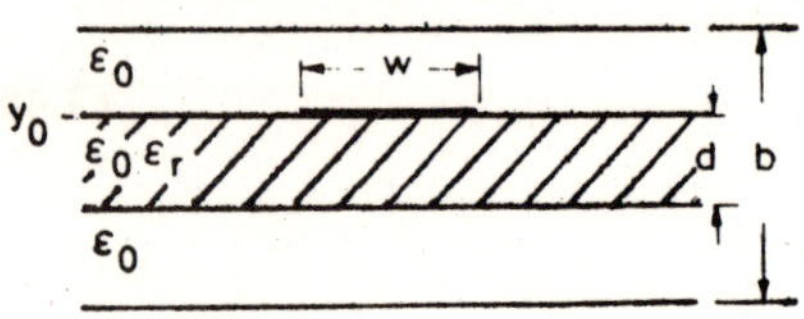	$\epsilon_0\left[\coth(\beta(b-d)/2) + \epsilon_r\left\{\dfrac{\coth(\beta(b-d)/2)\coth(\beta d) + \epsilon_r}{\epsilon_r \coth(\beta d) + \coth(\beta(b-d)/2)}\right\}\right]$
• Sandwiched stripline 	$\epsilon_0\left[\epsilon_{r1}\coth(\beta(b-d)/2) + \epsilon_{r2}\left\{\dfrac{\epsilon_{r1}\coth(\beta(b-d)/2)\coth(\beta d) + \epsilon_{r2}}{\epsilon_{r2}\coth(\beta d) + \epsilon_{r1}\coth(\beta(b-d)/2)}\right\}\right]$

For the charge distribution specified by (3.50), (3.48) combined with (3.51) provides an unified formula for evaluating the capacitance of single conductor striplines and stripline-like transmission lines with side walls separated by infinite distance. The only parameter that needs to be changed is the admittance parameter Y evaluated at the charge plane depending on the structure.

3.5.2 *Admittance Parameter for Isotropic Structures*

As discussed in Sec. 2.7, the admittance parameter Y for isotropic structures can be obtained by applying the transmission line formula (2.131), with $Y_{cj}=\epsilon_j=\epsilon_0\epsilon_{rj}$ and $\gamma_j=\beta$. It may be noted that the transmission line equivalent for a structure with side walls separated by infinite distance is the same as that for a corresponding structure with side walls separated by a distance c, when β_n is replaced by β in the expression for the propagation constant γ_j. Therefore, the admittance expressions for the structures shown in Table 3.3 are the same as those for the corresponding structures in Table 3.1 with β_n replaced by β.

3.5.3 *Admittance Parameter for Structures with Anisotropic Substrates*

Table 3.4 shows transmission line structures corresponding to Table 3.3 with isotropic dielectric substrates replaced by anisotropic dielectric substrates. In these, the optical axis is aligned along the y-axis of the coordinate system. Thus, for determining the admittance parameter Y, expression (2.131) can be used with $Y_{cj}=\epsilon_0\sqrt{\epsilon_{xxj}\,\epsilon_{yyj}}$ and $\gamma_j=\beta\sqrt{\epsilon_{xxj}/\epsilon_{yyj}}$ for the anisotropic layer, and $Y_{cj}=\epsilon_0\epsilon_{rj}$ and $\gamma_j=\beta$ for the isotropic layer. With these, the admittance parameters of the structures shown in Table 3.4 are the same as those of the corresponding structures with side walls shown in Table 3.2 with β_n replaced by β.

3.6 Striplines and Stripline-like Transmission Lines Using Anisotropic Substrates with Principal Axes Rotated with Respect to the Coordinate Axes—Variational Technique Combined with Spectral Domain Analysis [22]

3.6.1 *An Example of a Two-Layered Structure—Spectral Domain Analysis under Quasi-Static Approximation*

Consider a two-layered anisotropic stripline with a rectangular shielding box as shown in Fig. 3.7. The permittivity tensor of the two anisotropic regions is given by

$$\hat{\epsilon}_j'=\epsilon_0\begin{bmatrix}\epsilon_{\xi j} & 0 & 0\\ 0 & \epsilon_{\eta j} & 0\\ 0 & 0 & \epsilon_{zzj}\end{bmatrix} \tag{3.52}$$

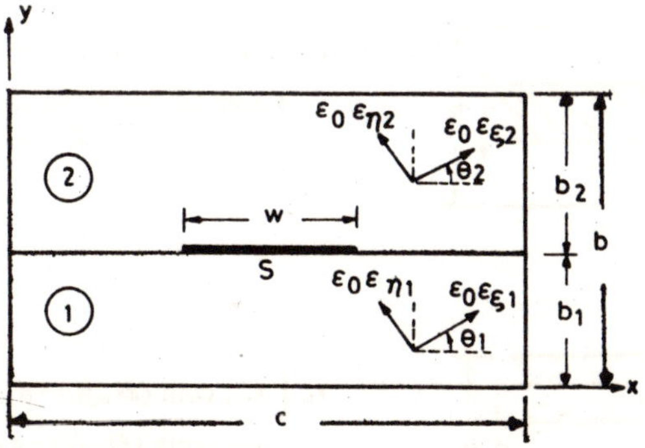

FIG. 3.7 Two-layered stripline with anisotropic substrates—principal axes arbitrarily rotated with respect to x-y coordinate axes.

Table 3.4 Admittance parameters of striplines and stripline-like transmission lines with anisotropic dielectric substrates with infinite parallel ground planes – optical axis aligned along the y-axis

$$\epsilon_{rfj}=\sqrt{\epsilon_{xxj}\epsilon_{yyj}},\quad \epsilon_{aj}=\sqrt{\epsilon_{xxj}/\epsilon_{yyj}},\quad j=1,\ 2$$

Structure	*Admittance Y at $y=y_0$*

• Symmetric stripline

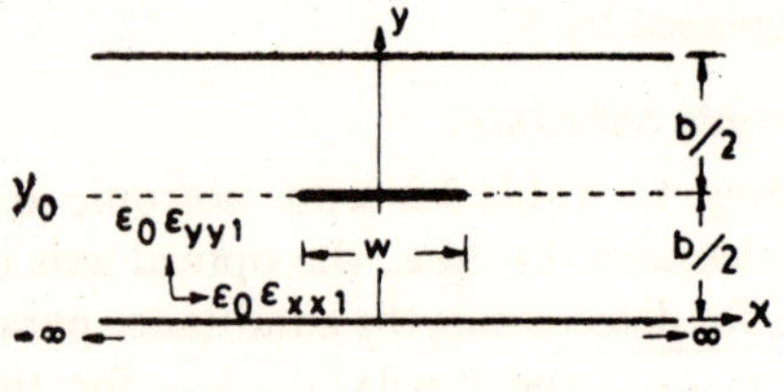

$$2\epsilon_0\epsilon_{rf1}\coth(\beta\epsilon_{a1}b/2)$$

• Stripline with off-set center conductor

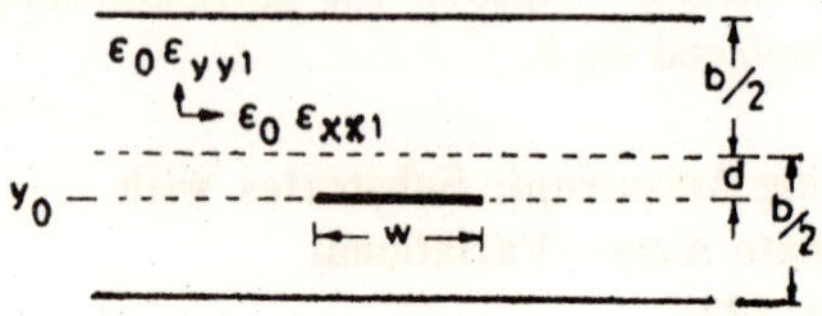

$$\epsilon_0\epsilon_{rf1}[\coth(\beta\epsilon_{a1}((b/2)+d))+\coth(\beta\epsilon_{a1}((b/2)-d))]$$

• Microstrip-like stripline

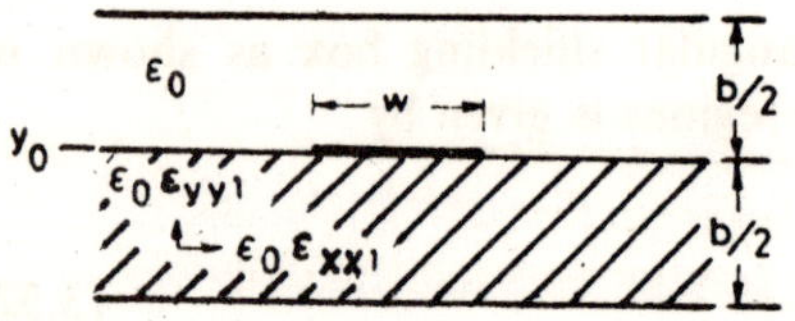

$$\epsilon_0[\coth(\beta b/2)+\epsilon_{rf1}\coth(\beta\epsilon_{a1}b/2)]$$

• Suspended stripline

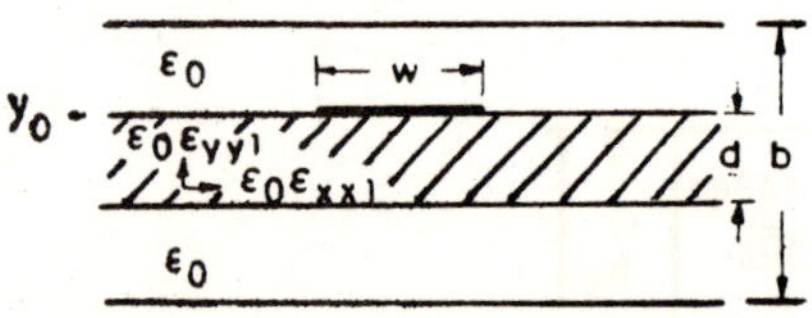

$$\epsilon_0\left[\coth(\beta((b-d)/2))+\epsilon_{rf1}\cdot\left\{\frac{\coth(\beta((b-d)/2))\coth(\beta\epsilon_{a1}d)+\epsilon_{rf1}}{\epsilon_{rf1}\coth(\beta\epsilon_{a1}d)+\coth(\beta((b-d)/2))}\right\}\right]$$

• Sandwiched stripline

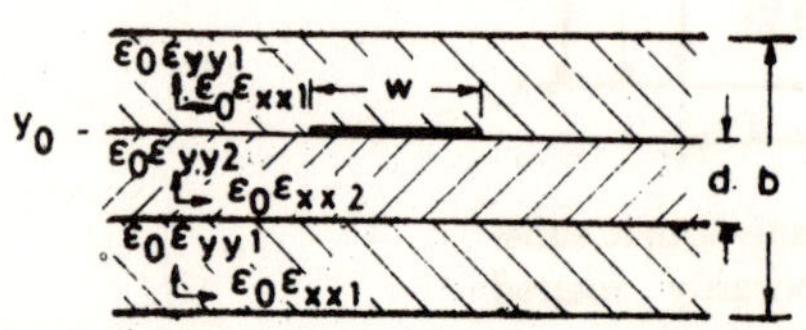

$$\epsilon_0\left[\epsilon_{rf1}\coth(\beta\epsilon_{a1}((b-d)/2))+\epsilon_{rf2}\cdot\left\{\frac{\epsilon_{rf1}\coth(\beta\epsilon_{a1}((b-d)/2))\coth(\beta\epsilon_{a2}d)+\epsilon_{rf2}}{\epsilon_{rf2}\coth(\beta\epsilon_{a2}d)+\epsilon_{rf1}\coth(\beta\epsilon_{a1}((b-d)/2))}\right\}\right]$$

where the subscript $j=1, 2$ represents regions 1 and 2 respectively, $\epsilon_{\xi j}$ and $\epsilon_{\eta j}$ are the relative dielectric constants along the principal axes ξj and ηj respectively, of the anisotropic medium. The axes ξj and ηj are oriented at angle θ_j with respect to the x and y axes, in the x-y plane. The permittivity tensor $\hat{\epsilon}_j$ in the x-y-z coordinate system is obtained by applying the following transformation [23]:

$$\hat{\epsilon}_j = (\hat{L}_j^t \cdot \hat{\epsilon}_j' \cdot \hat{L}_j)/\det \hat{L}_j \tag{3.53}$$

where

$$\hat{L}_j = \begin{bmatrix} \cos\theta_j & -\sin\theta_j & 0 \\ \sin\theta_j & \cos\theta_j & 0 \\ 0 & 0 & 1 \end{bmatrix} \tag{3.54}$$

Thus,

$$\hat{\epsilon}_j = \epsilon_0 \begin{bmatrix} \epsilon_{xxj} & \epsilon_{xyj} & 0 \\ \epsilon_{xyj} & \epsilon_{yyj} & 0 \\ 0 & 0 & \epsilon_{zzj} \end{bmatrix} \tag{3.55}$$

where

$$\epsilon_{xxj} = \epsilon_{\xi j} \cos^2\theta_j + \epsilon_{\eta j} \sin^2\theta_j \tag{3.56a}$$

$$\epsilon_{xyj} = (\epsilon_{\eta j} - \epsilon_{\xi j}) \sin\theta_j \cos\theta_j \tag{3.56b}$$

$$\epsilon_{yyj} = \epsilon_{\xi j} \sin^2\theta_j + \epsilon_{\eta j} \cos^2\theta_j \tag{3.56c}$$

We first determine the quasi-static potential distribution function in spectral domain, by solving Laplace's equation in two regions subject to the appropriate boundary conditions. The strip conductor is assumed to be of zero thickness and having a charge distribution

$$\rho(x, y) = f(x)\delta(y - b_1) \tag{3.57}$$

The Laplace's equation for the two regions in the x-y coordinate system is

$$\nabla_t \cdot [\hat{\epsilon}_j \cdot \nabla_t \phi_j(x, y)] = 0 \tag{3.58}$$

Expanding, we get

$$\epsilon_{xxj} \frac{\partial^2 \phi_j(x, y)}{\partial x^2} + 2\epsilon_{xyj} \frac{\partial^2 \phi_j(x, y)}{\partial x\, \partial y} + \epsilon_{yyj} \frac{\partial^2 \phi_j(x, y)}{\partial y^2} = 0 \tag{3.59}$$

Let $\hat{\phi}_j(\beta_n, y)$ be the finite Fourier transform of $\phi_j(x, y)$ in the interval 0 to c, where β_n is the discrete Fourier variable given by

$$\beta_n = n\pi/c \tag{3.60}$$

Taking the Fourier transform of (3.59), we obtain

$$\epsilon_{yyj} \frac{\partial^2 \hat{\phi}_j(\beta_n, y)}{\partial y^2} + 2j\beta_n \epsilon_{xyj} \frac{\partial \hat{\phi}_j(\beta_n, y)}{\partial y} - \beta_n^2 \epsilon_{xxj} \hat{\phi}_j(\beta_n, y) = 0 \tag{3.61}$$

Assuming a y variation of the form $\exp(-\zeta_j y)$ in (3.61), we obtain the following algebraic equation for ζ_j:

$$\epsilon_{yyj}\zeta_j^2 - 2j\beta_n\epsilon_{xyj}\zeta_j - \beta_n^2\epsilon_{xxj} = 0 \tag{3.62}$$

The solution for ζ_j can be expressed as

$$\zeta_j = \pm\gamma_j + j\alpha_j \tag{3.63}$$

where

$$\alpha_j = \beta_n\epsilon_{xyj}/\epsilon_{yyj} \tag{3.64a}$$

$$\gamma_j = \beta_n\epsilon_{rfj}/\epsilon_{yyj} \tag{3.64b}$$

$$\epsilon_{rfj} = \sqrt{\epsilon_{xxj}\epsilon_{yyj} - \epsilon_{xyj}^2} \tag{3.64c}$$

The general solution to the Fourier transformed potential distribution is

$$\hat{\phi}_j(\beta_n, y) = A_j \exp(-\zeta_j^+ y) + B_j \exp(-\zeta_j^- y) \tag{3.65}$$

where

$$\zeta_j^+ = \gamma_j + j\alpha_j \tag{3.66a}$$

$$\zeta_j^- = -\gamma_j + j\alpha_j \tag{3.66b}$$

We now apply the boundary conditions at the top and bottom ground planes and at the interface $y = b_1$. Since the potential is zero at $y = 0$, $B_1 = -A_1$. The potential distribution in region 1 takes the form

$$\hat{\phi}_1(\beta_n, y) = -2A_1 \exp(-j\alpha_1 y) \sinh(\gamma_1 y) \tag{3.67}$$

Setting the potential $\hat{\phi}_2(\beta_n, y) = 0$ at $y = b$ gives $B_2 = -A_2 \exp(-2\gamma_2 b)$. With this, the potential distribution in region 2 can be expressed as

$$\hat{\phi}_2(\beta_n, y) = 2A_2 \exp(-\gamma_2 b) \exp(-j\alpha_2 y) \sinh \gamma_2(b - y) \tag{3.68}$$

The electric displacement density in the transformed domain in two regions can be obtained from

$$\hat{D}_j(\beta_n, y) = \epsilon_0 \begin{pmatrix} \epsilon_{xxj} & \epsilon_{xyj} \\ \epsilon_{xyj} & \epsilon_{yyj} \end{pmatrix} \begin{pmatrix} -\partial\hat{\phi}_j(\beta_n, y)/\partial x \\ -\partial\hat{\phi}_j(\beta_n, y)/\partial y \end{pmatrix} \tag{3.69}$$

Using (3.67) and (3.68) in (3.69), the y-components of the displacement density in two regions are obtained as

$$\hat{D}_{y1}(\beta_n, y) = 2A_1\epsilon_0\gamma_1\epsilon_{yy1} \exp(-j\alpha_1 y) \cosh \gamma_1 y \tag{3.70}$$

$$\hat{D}_{y2}(\beta_n, y) = 2A_2\epsilon_0\gamma_2\epsilon_{yy2} \exp(-\gamma_2 b) \exp(-j\alpha_2 y) \cosh \gamma_2(b - y) \tag{3.71}$$

At the interface $y = b_1$, the normal component of $\hat{D}(\beta_n, y)$ should satisfy the condition

$$\hat{D}_{y2}(\beta_n, y)\big|_{y=b_1} - \hat{D}_{y1}(\beta_n, y)\big|_{y=b_1} = \hat{f}(\beta_n) \tag{3.72}$$

where $\hat{f}(\beta_n)$ is the Fourier transform of the charge density distribution. Substituting (3.70) and (3.71) in (3.72), we get

$$2\epsilon_0[A_2\gamma_2\epsilon_{yy2}\exp(-\gamma_2 b)\exp(-j\alpha_2 b_1)\cosh\gamma_2(b-b_1)-A_1\gamma_1\epsilon_{yy1}\exp(-j\alpha_1 b_1)\cosh\gamma_1 b_1]$$
$$=\hat{f}(\beta_n) \quad (3.73)$$

Since the potential must be continuous at the interface, we equate the potentials given by (3.67) and (3.68) at $y=b_1$ to give

$$-2A_1\exp(-j\alpha_1 b_1)\sinh\gamma_1 b_1=2A_2\exp(-\gamma_2 b)\exp(-j\alpha_2 b_1)\sinh\gamma_2(b-b_1) \quad (3.74)$$

Solving (3.73) and (3.74) for A_1, we have

$$A_1=\frac{-\hat{f}(\beta_n)}{2\epsilon_0\exp(-j\alpha_1 b_1)\sinh\gamma_1 b_1[\gamma_1\epsilon_{yy1}\coth\gamma_1 b_1+\gamma_2\epsilon_{yy2}\coth\gamma_2(b-b_1)]} \quad (3.75)$$

Substituting for A_1 in (3.67) gives the transformed potential distribution in region 1:

$$\hat{\phi}_1(\beta_n, y)=\frac{\hat{f}(\beta_n)\exp(-j\alpha_1 y)\sinh\gamma_1 y}{\epsilon_0\exp(-j\alpha_1 b_1)\sinh\gamma_1 b_1[\gamma_1\epsilon_{yy1}\coth\gamma_1 b_1+\gamma_2\epsilon_{yy2}\coth\gamma_2(b-b_1)]} \quad (3.76)$$

The potential in the Fourier domain at the charge plane $y=b_1$ is then

$$\hat{\phi}(\beta_n, b_1)=\frac{\hat{f}(\beta_n)}{\epsilon_0[\gamma_1\epsilon_{yy1}\coth(\gamma_1 b_1)+\gamma_2\epsilon_{yy2}\coth\gamma_2(b-b_1)]} \quad (3.77)$$

From (3.64b), $\gamma_j\epsilon_{yyj}=\beta_n\epsilon_{rfj}$. Using this, (3.77) can be written as

$$\hat{\phi}(\beta_n, b_1)=\frac{\hat{f}(\beta_n)}{\beta_n\epsilon_0[\epsilon_{rf1}\coth\gamma_1 b_1+\epsilon_{rf2}\coth\gamma_2(b-b_1)]} \quad (3.78)$$

It may be recognized that the factor multiplying β_n in the denominator of (3.78) is nothing but the admittance at the charge plane $y=b_1$ of an equivalent isotropic structure having a dielectric constant of $\epsilon_0\epsilon_{rf1}$ and a propagation constant γ_1 for region 1, and a dielectric constant $\epsilon_0\epsilon_{rf2}$ and a propagation constant γ_2 for region 2. Denoting this factor as Y (to conform to the notation used in the unified approach), we can write

$$\hat{\phi}(\beta_n, b_1)=\hat{f}(\beta_n)/\beta_n Y \quad (3.79)$$

where

$$Y=\epsilon_0[\epsilon_{rf1}\coth\gamma_1 b_1+\epsilon_{rf2}\coth\gamma_2(b-b_1)] \quad (3.80)$$

$$\epsilon_{rf1}=\sqrt{\epsilon_{xx1}\epsilon_{yy1}-\epsilon_{xy1}^2} \quad (3.81a)$$

$$\epsilon_{rf2}=\sqrt{\epsilon_{xx2}\epsilon_{yy2}-\epsilon_{xy2}^2} \quad (3.81b)$$

$$\gamma_1=(\beta_n\sqrt{\epsilon_{xx1}\epsilon_{yy1}-\epsilon_{xy1}^2})/\epsilon_{yy1} \quad (3.81c)$$

$$\gamma_2=(\beta_n\sqrt{\epsilon_{xx2}\epsilon_{yy2}-\epsilon_{xy2}^2})/\epsilon_{yy2} \quad (3.81d)$$

As a verification of the analogy stated above, we set $\theta_j = 0$ or $\pi/2$ in (3.56). The permittivity tensor for each of the two anisotropic regions then becomes diagonal with parameters

$$\epsilon_{rf1} = \sqrt{\epsilon_{xx1}\epsilon_{yy1}},\ \epsilon_{rf2} = \sqrt{\epsilon_{xx2}\epsilon_{yy2}},\ \gamma_1 = \beta_n\sqrt{\epsilon_{xx1}/\epsilon_{yy1}},\ \text{and}\ \gamma_2 = \beta_n\sqrt{\epsilon_{xx2}/\epsilon_{yy2}}.$$

With these parameters, it is seen that the admittance expression (3.80) is identical to the one that can be derived independently using the 'transverse transmission line' technique for the same two-layered uniaxially anisotropic structure, with the optical axis oriented perpendicular to the plane of the substrate.

In (3.79), the factor multiplying $\hat{f}(\beta_n)$ may be interpreted as the Fourier transform of Green's function $\hat{G}(\beta_n, b_1)$. Equation (3.79) is rewritten as

$$\hat{\phi}(\beta_n, b_1) = \hat{G}(\beta_n, b_1)\hat{f}(\beta_n) \tag{3.82}$$

where

$$\hat{G}(\beta_n, b_1) = 1/(\beta_n Y) \tag{3.83}$$

The potential $\hat{\phi}(\beta_n, b_1)$ is constituted by $\hat{\phi}_c(\beta_n, b_1)$, the transform of the potential distribution on the strip conductor and $\hat{\phi}_d(\beta_n, b_1)$ on the complementary region at the interface. It is therefore represented as

$$\hat{\phi}_c(\beta_n, b_1) + \hat{\phi}_d(\beta_n, b_1) = \hat{G}(\beta_n, b_1)\hat{f}(\beta_n) \tag{3.84}$$

From (3.84), $\hat{\phi}_d(\beta_n, b_1)$ can be eliminated through the use of Parseval's theorem [17, 24]. For this, we define an inner product as

$$\langle \psi_1(\beta_n), \psi_2(\beta_n) \rangle = \sum_{n=1}^{\infty} \psi_1(\beta_n)\psi_2(\beta_n) \tag{3.85}$$

Taking the inner product of (3.84) with $\hat{f}(\beta_n)$, we obtain

$$\langle \hat{\phi}_c(\beta_n, b_1), \hat{f}(\beta_n) \rangle = \langle \hat{G}(\beta_n, b_1)\hat{f}(\beta_n), \hat{f}(\beta_n) \rangle \tag{3.86}$$

The inner product of $\hat{\phi}_d(\beta_n, b_1)$ with $\hat{f}(\beta_n)$ vanishes, since they exist on complementary regions. By virtue of Parseval's theorem, we have

$$\sum_{n=1}^{\infty} \psi_1(\beta_n)\psi_2(\beta_n) = (c/2)\int_0^c \psi_1(x)\psi_2(x)\,dx \tag{3.87}$$

Using this relation in (3.86) and setting $\phi_c(x)$ equal to the potential V on the strip conductor, we obtain from (3.87)

$$\begin{aligned}(c/2)V\int_S f(x)\,dx &= \langle \hat{G}(\beta_n, b_1)\hat{f}(\beta_n), \hat{f}(\beta_n) \rangle \\ &= \sum_{n=1}^{\infty} \hat{G}(\beta_n, b_1)[\hat{f}(\beta_n)]^2 \end{aligned} \tag{3.88}$$

Noting that $V=Q/C$, where C is the capacitance per unit length of the line and Q is the total charge on the strip conductor per unit length given by

$$Q=\int_S f(x)\,dx \tag{3.89}$$

we obtain from (3.88), the following expression for capacitance:

$$\frac{1}{C}=\frac{\frac{2}{c}\sum_{n=1}^{\infty}\hat{G}(\beta_n, b_1)[\hat{f}(\beta_n)]^2}{\left[\int_S f(x)\,dx\right]^2} \tag{3.90}$$

Substituting for $\hat{G}(\beta_n, b_1)$ from (3.83) and expressing $\hat{f}(\beta_n)$ as the finite Fourier integral of $f(x)$, we obtain

$$\frac{1}{C}=\frac{\frac{2}{c}\sum_{n=1}^{\infty}(1/\beta_n Y)\left[\int_0^c f(x)\sin\beta_n x\,dx\right]^2}{\left[\int_S f(x)\,dx\right]^2} \tag{3.91}$$

Noting that $\beta_n = n\pi/c$ and $f(x)$ exists only on the strip conductor, we can write

$$\frac{1}{C}=\frac{\sum_{n=1}^{\infty}\frac{2}{n\pi Y}\left[\int_S f(x)\sin(n\pi x/c)\,dx\right]^2}{\left[\int_S f(x)\,dx\right]^2} \tag{3.92}$$

3.6.2 Generalized Capacitance Expression

The variational expression (3.92) derived above gives the capacitance of a two-layered stripline with anisotropic substrates in which the principal axes in the jth anisotropic substrate make an angle θ_j with respect to the x-y coordinate axes. It is interesting to note that (3.92) is the same as the capacitance expression (2.128) for single conductor multilayer dielectric structures. In deriving (2.128), the substrates were assumed to be either isotropic, or anisotropic, with a diagonal permittivity tensor as specified in (2.93). In both the equations, (2.128) and (3.92), Y is the only parameter which needs to be changed depending on the type of dielectric used. Equation (2.128) is therefore general because it is also applicable to anisotropic structures in which the principal axes of the anisotropic dielectric are oriented at an arbitrary angle with respect to the x-y coordinate axes. For the purpose of easy reference, we now consolidate the final results.

The unified capacitance expression is

$$\frac{1}{C}=\frac{\sum_{n=1}^{\infty}\frac{2}{n\pi Y}\left[\int_S f(x)\sin(n\pi x/c)\,dx\right]^2}{\left[\int_S f(x)\,dx\right]^2} \tag{3.93}$$

The admittance at the charge plane for any multilayer structure with anisotropic substrates having arbitrarily oriented principal axes in the x-y plane can also be determined by using a two-wire transmission line equivalent (Fig. 3.8). The characteristic admittance Y_{cj} and the

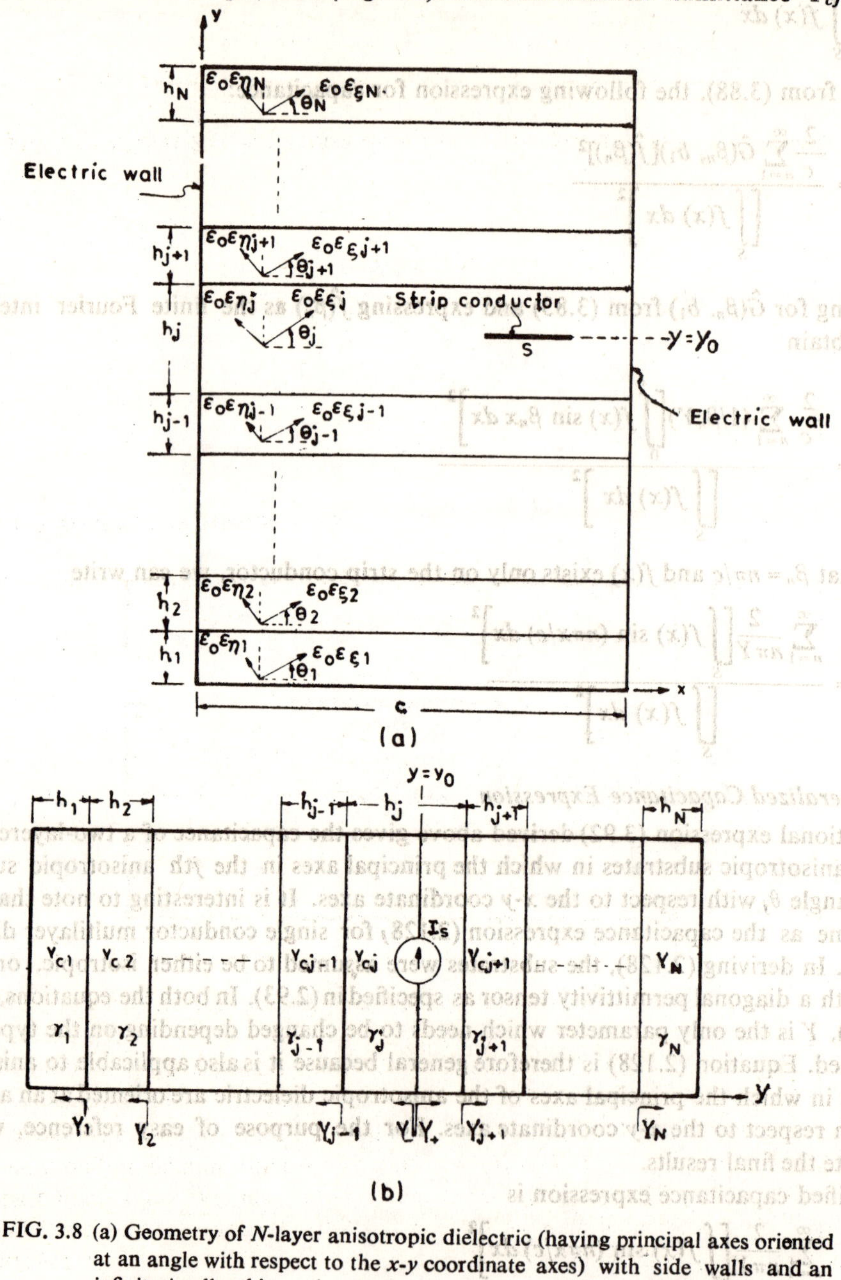

FIG. 3.8 (a) Geometry of N-layer anisotropic dielectric (having principal axes oriented at an angle with respect to the x-y coordinate axes) with side walls and an infinitesimally thin strip conductor at $y=y_0$ and (b) two-wire transmission line equivalent: $Y_{cj}=\epsilon_0\sqrt{\epsilon_{\eta j}\,\epsilon_{\xi j}}$, $\gamma_j=\beta_n\sqrt{\epsilon_{\eta j}/\epsilon_{\xi j}}\;[(\epsilon_{\eta j}/\epsilon_{\xi j}-1)\cos^2\theta_j+1]^{-1}$, $j=1$ to N, $\beta_n=n\pi/c$.

propagation constant γ_j of the jth section of the transmission line equivalent are given by

$$Y_{cj}=\epsilon_j=\epsilon_0\epsilon_{rfj}=\epsilon_0\sqrt{\epsilon_{xxj}\epsilon_{yyj}-\epsilon_{xyj}^2} \tag{3.94}$$

$$\gamma_j=(\beta_n\sqrt{\epsilon_{xxj}\epsilon_{yyj}-\epsilon_{xyj}^2})/\epsilon_{yyj}, \quad \beta_n=\frac{n\pi}{c} \tag{3.95}$$

where

$$\epsilon_{xxj}=\epsilon_{\xi j}\cos^2\theta_j+\epsilon_{\eta j}\sin^2\theta_j \tag{3.96a}$$

$$\epsilon_{xyj}=(\epsilon_{\eta j}-\epsilon_{\xi j})\sin\theta_j\cos\theta_j \tag{3.96b}$$

$$\epsilon_{yyj}=\epsilon_{\xi j}\sin^2\theta_j+\epsilon_{\eta j}\cos^2\theta_j \tag{3.96c}$$

Substituting these relations in (3.94), we get

$$Y_{cj}=\epsilon_0\sqrt{\epsilon_{\xi j}\epsilon_{\eta j}} \tag{3.97}$$

$$\gamma_j=\beta_n\sqrt{\epsilon_{\eta j}/\epsilon_{\xi j}}[(\epsilon_{\eta j}/\epsilon_{\xi j}-1)\cos^2\theta_j+1]^{-1}, \quad \beta_n=\frac{n\pi}{c} \tag{3.98}$$

While applying the transmission line formula (2.131) to any structure having mixed isotropic and anisotropic layers, the parameters to be used are

(i) $Y_{cj}=\epsilon_0\epsilon_{rj}$ and $\gamma_j=\beta_n$, $\beta_n=n\pi/c$ for isotropic layer;
(ii) $Y_{cj}=\epsilon_0\sqrt{\epsilon_{xxj}\epsilon_{yyj}}$ and $\gamma_j=\beta_n\sqrt{\epsilon_{xxj}/\epsilon_{yyj}}$ for uniaxially anisotropic substrates with optical axis aligned perpendicular to the substrate plane; and
(iii) Y_{cj} and γ_j given by (3.97) and (3.98), respectively, for anisotropic substrate with arbitrary orientation of the principal axes in the x-y plane.

Table 3.5 gives the admittance expressions for some of the typical stripline and stripline-like transmission lines with anisotropic substrates for arbitrary orientation of the principal axes in the x-y plane.

If the structure in Fig. 3.8a extends to infinity in the positive and negative x-directions, then the unified formula given by (2.137) applies. The admittance parameter is the same as that of the corresponding structure having side walls, with β_n replaced by β.

3.7 Conclusion

The unified capacitance formulas developed in Chapter 2 for multilayer planar structures are applied to stripline-like transmission structures by assuming an appropriate charge distribution on the strip conductor. Closed form analytical expressions are derived which can be directly used in the computation of capacitance of any single or multilayered transmission line. The dielectric layers may be isotropic or anisotropic. In the case of anisotropic substrates, the principal axes of the anisotropic medium may be either aligned or oriented at an angle with respect to the x-y coordinate axes. The only parameter that needs to be changed is the admittance Y evaluated at the charge plane of the structure.

Table 3.5 Admittance parameters of striplines and stripline-like transmission lines with side walls and anisotropic dielectric substrates—principal axes arbitrarily oriented in the *x-y* plane

$$\epsilon_{rfj}=\sqrt{\epsilon_{\xi j}\epsilon_{\eta j}}\,,\quad \epsilon_{ej}=\sqrt{\epsilon_{\eta j}/\epsilon_{\xi j}}\,[(\epsilon_{\eta j}/\epsilon_{\xi j}-1)\cos^2\theta_j+1]^{-1},\ j=1,2$$

Structure	*Admittance Y at $y=y_0$, $\beta_n=n\pi/c$*

- Symmetric stripline

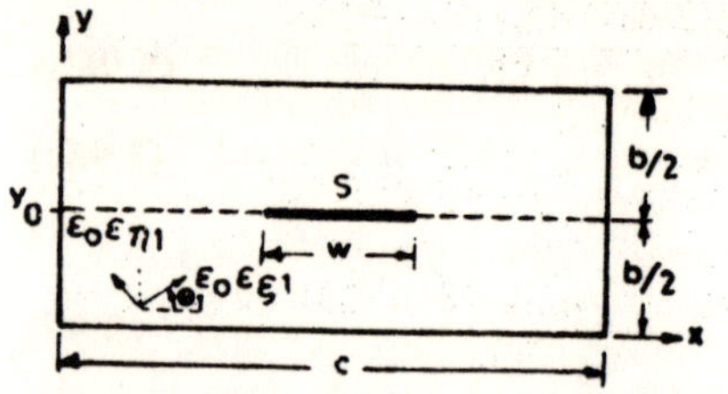

$$2\epsilon_0\epsilon_{rf1}\coth(\beta_n\epsilon_{e1}b/2)$$

- Stripline with off-set center conductor

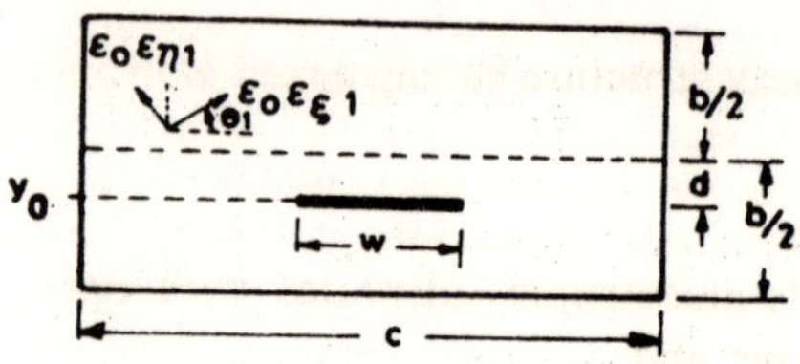

$$\epsilon_0\epsilon_{rf1}\,[\coth(\beta_n\epsilon_{e1}((b/2)+d))+\coth(\beta_n\epsilon_{e1}((b/2)-d))]$$

- Microstrip-like stripline

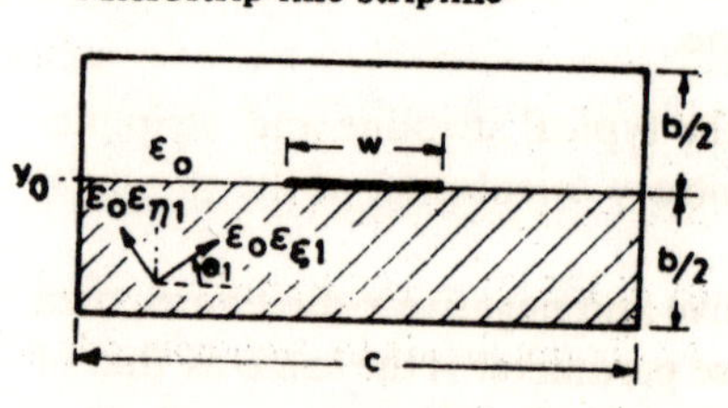

$$\epsilon_0[\coth(\beta_n b/2)+\epsilon_{rf1}\coth(\beta_n\epsilon_{e1}b/2)]$$

- Suspended stripline

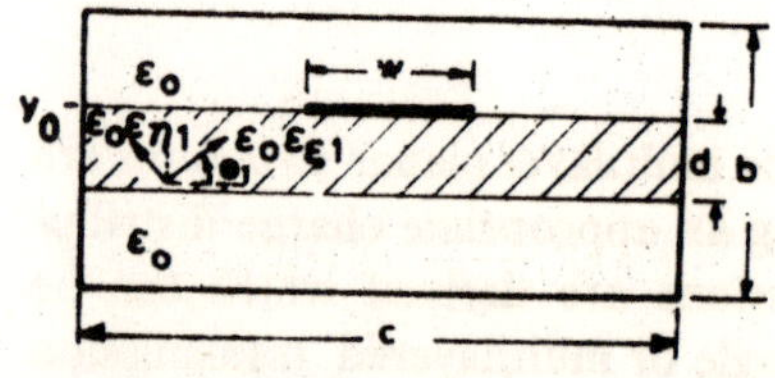

$$\epsilon_0\left[\coth(\beta_n((b-d)/2))+\epsilon_{rf1}\cdot\left\{\frac{\coth(\beta_n((b-d)/2))\coth(\beta_n\epsilon_{e1}d)+\epsilon_{rf1}}{\epsilon_{rf1}\coth(\beta_n\epsilon_{e1}d)+\coth(\beta_n((b-d)/2))}\right\}\right]$$

- Sandwiched stripline

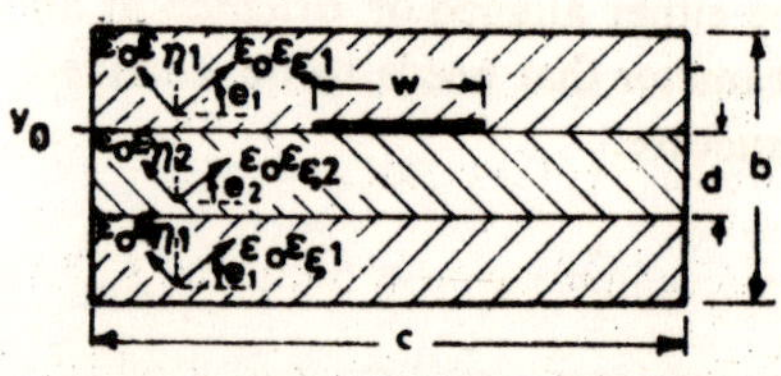

$$\epsilon_0\left[\epsilon_{rf1}\coth(\beta_n\epsilon_{e1}((b-d)/2))+\epsilon_{rf2}\cdot\left\{\frac{\epsilon_{rf1}\coth(\beta_n\epsilon_{e1}((b-d)/2))\coth(\beta_n\epsilon_{e2}d)+\epsilon_{rf2}}{\epsilon_{rf2}\coth(\beta_n\epsilon_{e2}d)+\epsilon_{rf1}\coth(\beta_n\epsilon_{e1}((b-d)/2))}\right\}\right]$$

REFERENCES

1. S.B. Cohn, Characteristic impedance of the shielded strip transmission line, *IRE Trans. Microwave Theory Tech.*, Vol. MTT-2, pp. 52-57, July 1954.
2. E. Yamashita and R. Mittra, Variational method for the analysis of microstrip lines, *IEEE Trans. Microwave Theory Tech.*, Vol. MTT-16, pp. 251-256, April 1968.
3. E. Yamashita, Variational method for the analysis of microstrip-like transmission lines, *IEEE Trans. Microwave Theory Tech.*, Vol. MTT-16, pp. 529-535, Aug. 1968.
4. E. Yamashita and K. Atsuki, Stripline with rectangular outer conductor and three dielectric layers, *IEEE Trans. Microwave Theory Tech.*, Vol. MTT-18, pp. 238-244, May 1970.
5. C.M. Weil, The characteristic impedance of rectangular transmission lines with thin center conductor and air dielectric, *IEEE Trans. Microwave Theory Tech.*, Vol. MTT-26, pp. 235-242, April 1978.
6. M.L. Crawford, Generation of standard EM fields using TEM transmission cells, *IEEE Trans. Electromagn. Compat.*, Vol. EMC-16, pp. 189-195, Nov. 1974.
7. S. Rimmon, Master the challenge of offset stripline design, *Microwaves*, Vol. 15, pp. 40-44, May 1976.
8. R. Garg, Stripline-like microstrip configuration, *Microwave J.*, Vol. 22, pp. 103-104, 116, April 1979.
9. B. Robinson, Supercomponents in ECM systems, *Microwave J.*, Vol. 19, pp. 47-50, March 1976.
10. K. Chang, K. Louie, A.J. Crote, R.S. Tahim, M.J. Mlinar, G.M. Hayashibara, and C. Sun, V-band low noise integrated circuit receiver, *IEEE Trans. Microwave Theory Tech.*, Vol. MTT-31, pp. 149-154, Feb. 1983.
11. R.S. Tahim, G.M. Hayashibara and K. Chang, Design and performance of W-band broadband integrated circuit mixers, *IEEE Trans. Microwave Theory Tech.*, Vol MTT-31, pp. 277-283, March 1983.
12. R.E. Collin, *Field Theory of Guided Waves*, McGraw-Hill, New York, 1960.
13. E. Jahnke and F. Emde, *Tables of Functions*, 4th Edn., Dover Publications, New York, 1945.
14. F. Bowmane, *Introduction to Elliptic Functions*, Dover Publications, New York, 1961.
15. M. Abramowitz and I.A. Stegun, *Handbook of Mathematical Functions*, Applied Mathematics Series, No. 55, NBS, Washington D.C., 1965.
16. M.A.R. Gunston, *Microwave Transmission Line Impedance Data*, Van Nostrand Reinhold, New York, 1972.
17. A. Papoulis, *The Fourier Integral and its Applications*, McGraw-Hill, New York, 1962.
18. B. Friedman, *Principles and Techniques of Applied Mathematics*, John Wiley & Sons, New York, 1956.
19. B. Bhat and S.K. Koul, Unified approach to solve a class of strip and microstrip-like transmission lines, *IEEE Trans. Microwave Theory Tech.*, Vol. MTT-30, pp. 679-686, May 1982. Also correction in *IEEE Trans. Microwave Theory Tech.*, Vol. MTT-30, p. 2067, Nov. 1982.
20. E. Hallen, *Electromagnetic Theory*, Capman and Hall, London, p. 64, 1962.
21. J.M.C. Dukes, An investigation into some fundamental properties of strip transmission lines with the aid of an electrostatic tank, *Proc. IEE* (London), Vol. 103, pt. B, pp. 319-333, May 1956.
22. B. Bhat and S.K. Koul, New approach to analyse strip- and microstrip-like transmission lines with anisotropic substrates having a tilted optical axis, *IEE Proc.*, MOA, Vol. 131, pt. H, No. 3, pp. 191-197, June 1984.
23. S. Kusase and R. Terakado, Mapping theory of two-dimensional anisotropic region, *Proc. IEEE*, Vol. 67, pp. 171-172, Jan. 1979.
24. J.W. Dettman, *Mathematical Methods in Physics and Engineering*, McGraw-Hill, New York, 1962.

CHAPTER 4

Analysis of Coupled Striplines and Stripline-like Transmission Lines

4.1 Definitions of Basic Parameters

The electromagnetic coupling effects between parallel coupled transmission lines form the basis of a variety of useful circuits such as directional couplers [1–5], filters [5–7], baluns [8] and phase shift networks [9].

A pair of strip conductors situated in a homogeneous, lossless dielectric medium can support two fundamental TEM modes, called the 'even-mode' and the 'odd-mode'. These terms refer to the two possible ways in which the electromagnetically coupled two-conductor strip transmission lines can be excited. In the even-mode, the two strip conductors are excited in equal amplitude and in-phase (both positive with respect to ground) and, in the odd-mode, they are excited in equal amplitude, but with opposite phase (one positive and the other negative with respect to ground). The electric field configurations for these two modes of excitation are illustrated schematically in Fig. 4.1 for the edge-coupled as well as the broadside-coupled strip transmission lines. In these illustrations, pp' represents the plane of symmetry. At the plane of symmetry, the normal component of the electric field is zero ($\hat{n}\cdot\mathbf{E}=0$) for the even-mode excitation, whereas the tangential component is zero ($\hat{n}\times\mathbf{E}=0$) for the odd-mode excitation, where $\hat{n}$ is the unit normal perpendicular to the plane of symmetry. This is equivalent to placing, at pp', a 'magnetic wall' and an 'electric wall' for the even- and odd-modes, respectively. In terms of the potential function $\phi(x,y)$, $\phi=0$ at an electric boundary, and $\partial\phi/\partial n=0$ at the magnetic boundary. The two modes have different characteristic impedances, but they propagate with the same phase velocity if the mode is pure TEM. If the medium is inhomogeneous, as would result when layered dielectrics with different dielectric constants are used, pure TEM mode cannot propagate. However, if the transverse dimensions are small in comparison with the half-wavelength in the structure, the dominant mode can be represented as quasi-TEM. In such structures, the two normal modes which are supported by a pair of coupled lines propagate with different phase velocities.

We designate the two characteristic impedances of the structure as Z_e and Z_o where the subscripts e and o represent the even- and odd-modes, respectively. Here, Z_e and Z_o are defined as the input impedances of an infinite length of one line, in the presence of the second one also of infinite length, when the two conductors are excited in even-mode and odd-mode,

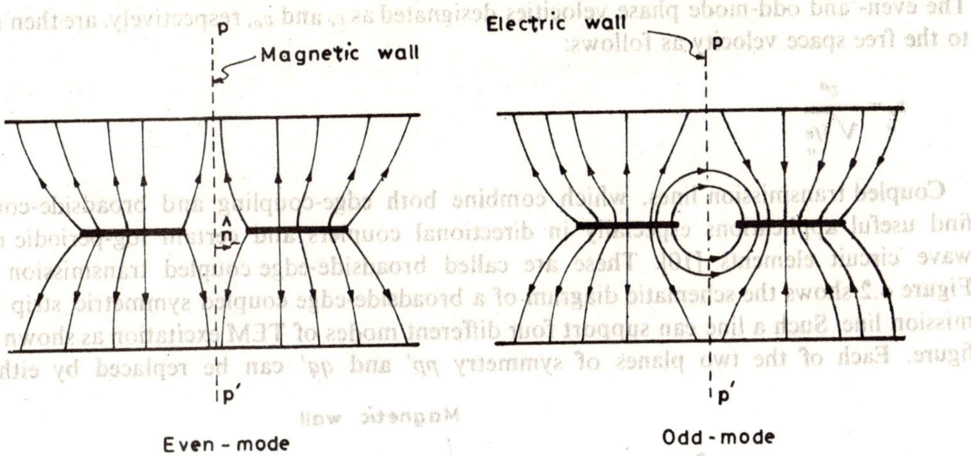

(a) Edge-coupled stripline

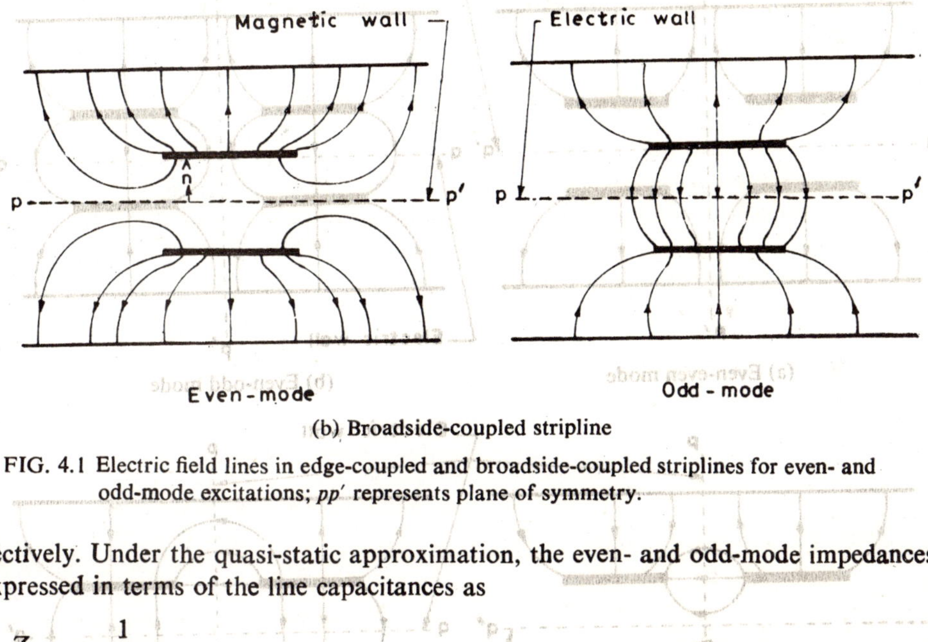

(b) Broadside-coupled stripline

FIG. 4.1 Electric field lines in edge-coupled and broadside-coupled striplines for even- and odd-mode excitations; *pp'* represents plane of symmetry.

respectively. Under the quasi-static approximation, the even- and odd-mode impedances can be expressed in terms of the line capacitances as

$$Z_{\substack{e\\o}} = \frac{1}{v^a \sqrt{C_{\substack{e\\o}} C^a_{\substack{e\\o}}}} \tag{4.1}$$

where v^a is the velocity of light in free space, C_e and C_o are the even- and odd-mode capacitances per unit length of the coupled line, and C_e^a and C_o^a are the even- and odd-mode capacitances per unit length of the coupled line when all dielectrics are replaced by free space. Similarly, the effective dielectric constants for the even- and odd-modes represented as ϵ_{fe} and ϵ_{fo}, respectively, can be expressed in terms of the capacitances as

$$\epsilon_{f\substack{e\\o}} = \frac{C_{\substack{e\\o}}}{C^a_{\substack{e\\o}}} \tag{4.2}$$

The even- and odd-mode phase velocities designated as v_e and v_o, respectively, are then related to the free space velocity as follows:

$$v_{\substack{e\\o}} = \frac{v^a}{\sqrt{\epsilon_{\substack{fe\\o}}}} \tag{4.3}$$

Coupled transmission lines, which combine both edge-coupling and broadside-coupling find useful applications especially in directional couplers and certain log-periodic microwave circuit elements [10]. These are called broadside-edge coupled transmission lines. Figure 4.2 shows the schematic diagram of a broadside-edge coupled symmetric strip transmission line. Such a line can support four different modes of TEM excitation as shown in the figure. Each of the two planes of symmetry pp' and qq' can be replaced by either an

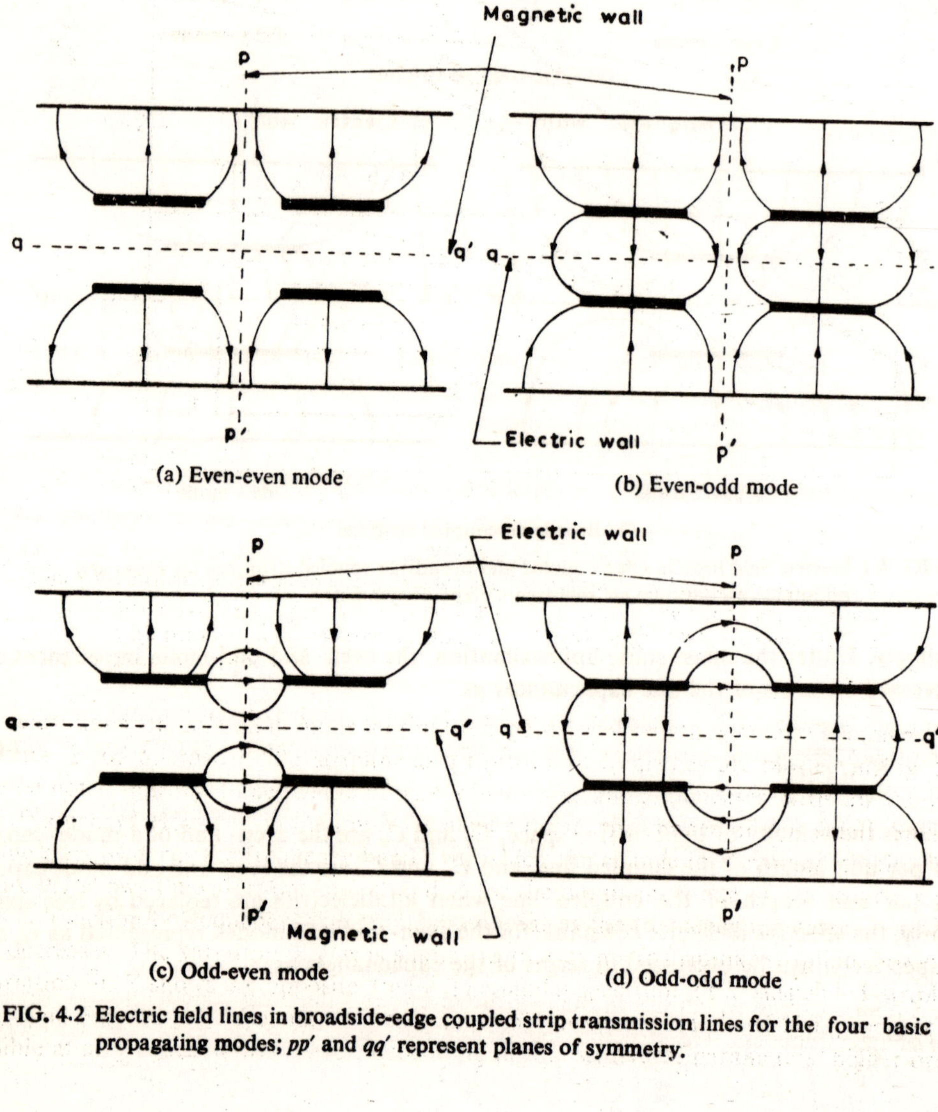

FIG. 4.2 Electric field lines in broadside-edge coupled strip transmission lines for the four basic propagating modes; pp' and qq' represent planes of symmetry.

electric wall or a magnetic wall depending on the type of excitation. The four modes are designated as:

(1) Even-even mode (*ee*): *pp'* magnetic wall, *qq'* magnetic wall,
(2) Even-odd mode (*eo*): *pp'* magnetic wall, *qq'* electric wall,
(3) Odd-even mode (*oe*): *pp'* electric wall, *qq'* magnetic wall,
(4) Odd-odd mode (*oo*): *pp'* electric wall, *qq'* electric wall,

The first index characterizes the excitation of the horizontal strip conductors with respect to *pp'* and the second one of the vertical strip conductors with respect to *qq'*. The characteristic impedances, effective dielectric constants, and phase velocities of the four modes in any broadside-edge coupled stripline-like transmission structure, under the quasi-static approximation, are given by

$$Z_{\substack{ee\\eo\\oe\\oo}} = \frac{1}{v^a\sqrt{C_{\substack{ee\\eo\\oe\\oo}}C^a_{\substack{ee\\eo\\oe\\oo}}}} \tag{4.4}$$

$$\epsilon_{f\substack{ee\\eo\\oe\\oo}} = C_{\substack{ee\\eo\\oe\\oo}}/C^a_{\substack{ee\\eo\\oe\\oo}} \tag{4.5}$$

$$v_{\substack{ee\\eo\\oe\\oo}} = \frac{v^a}{\sqrt{\epsilon_{f\substack{ee\\eo\\oe\\oo}}}} \tag{4.6}$$

where C_{ee}, C_{eo}, C_{oe}, and C_{oo} are the line capacitances per unit length of the structure for the four propagating modes, and C^a_{ee}, C^a_{eo}, C^a_{oe}, and C^a_{oo} are the corresponding capacitances of the structure when all dielectrics are replaced by air.

The design of circuits using coupled lines necessitates information on the propagation parameters corresponding to the various basic modes comprising the propagation in the transmission structure. The propagation parameters, namely, the characteristic impedance and the phase velocity can be determined from the knowledge of the capacitance of the structure corresponding to the various basic modes. This chapter presents the analyses of a variety of coupled stripline and stripline-like transmission lines under the quasi-static approximation. The edge-coupled homogeneous symmetric stripline is analysed using the conformal mapping method to obtain exact closed form expressions for the characteristic impedances of the two fundamental modes. This method, however, becomes prohibitively difficult to apply to inhomogeneous coupled structures. The unified approach which combines the variational technique with the 'transverse transmission line' method developed in Chapter 2 is applied to edge-coupled, broadside-coupled and broadside-edge coupled structures. These structures can have one or more dielectric layers, either isotropic or anisotropic.

4.2 Edge-Coupled Symmetric Stripline—Conformal Transformation Method [11]

Consider an edge-coupled homogeneous, symmetric stripline as shown in Fig. 4.3a. The two ground planes extend to infinity in the $\pm x$ directions and the strip conductors are assumed to be of zero thickness. The structure can be analysed in terms of the even- and odd-modes, by placing at the plane of symmetry, pp', a magnetic wall and an electric wall, respectively. In view of the symmetry of the structure, we may reduce the electrostatic boundary value problem to that of only one quarter of the structure as illustrated in Fig. 4.3b.

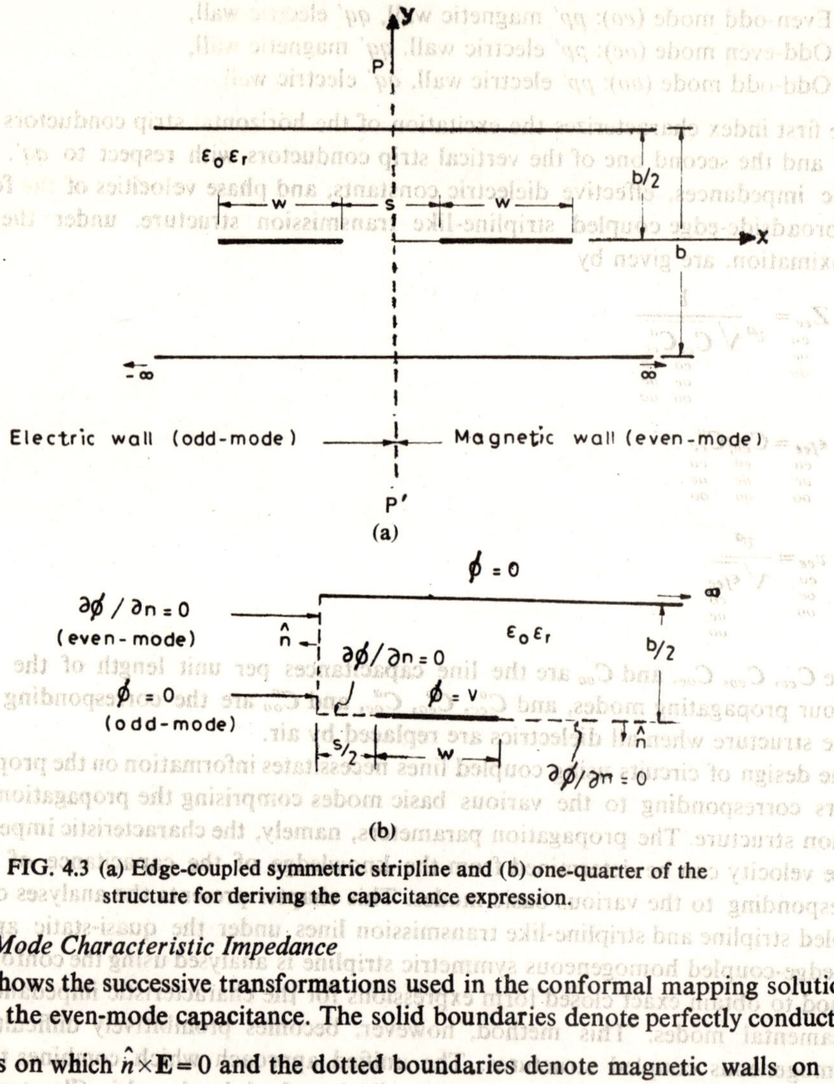

FIG. 4.3 (a) Edge-coupled symmetric stripline and (b) one-quarter of the structure for deriving the capacitance expression.

4.2.1 Even-Mode Characteristic Impedance

Figure 4.4 shows the successive transformations used in the conformal mapping solution for determining the even-mode capacitance. The solid boundaries denote perfectly conducting or electric walls on which $\hat{n} \times \mathbf{E} = 0$ and the dotted boundaries denote magnetic walls on which $\hat{n} \cdot \mathbf{E} = 0$.

As a first step towards the solution of this boundary value problem, we map the rectangular boundary formed by a parallel plate capacitor shown in Fig. 4.4a to the real axis in the ζ-plane as in Fig. 4.4b. Such a mapping can be achieved through the Schwarz-Christoffel transformation discussed in Chapter 2. The mapping function is of the form:

$$W = A_0 \int^{\zeta} \frac{d\zeta}{[(\zeta + 1/k)(\zeta + 1)(\zeta - 1)(\zeta - 1/k)]^{1/2}} + B$$

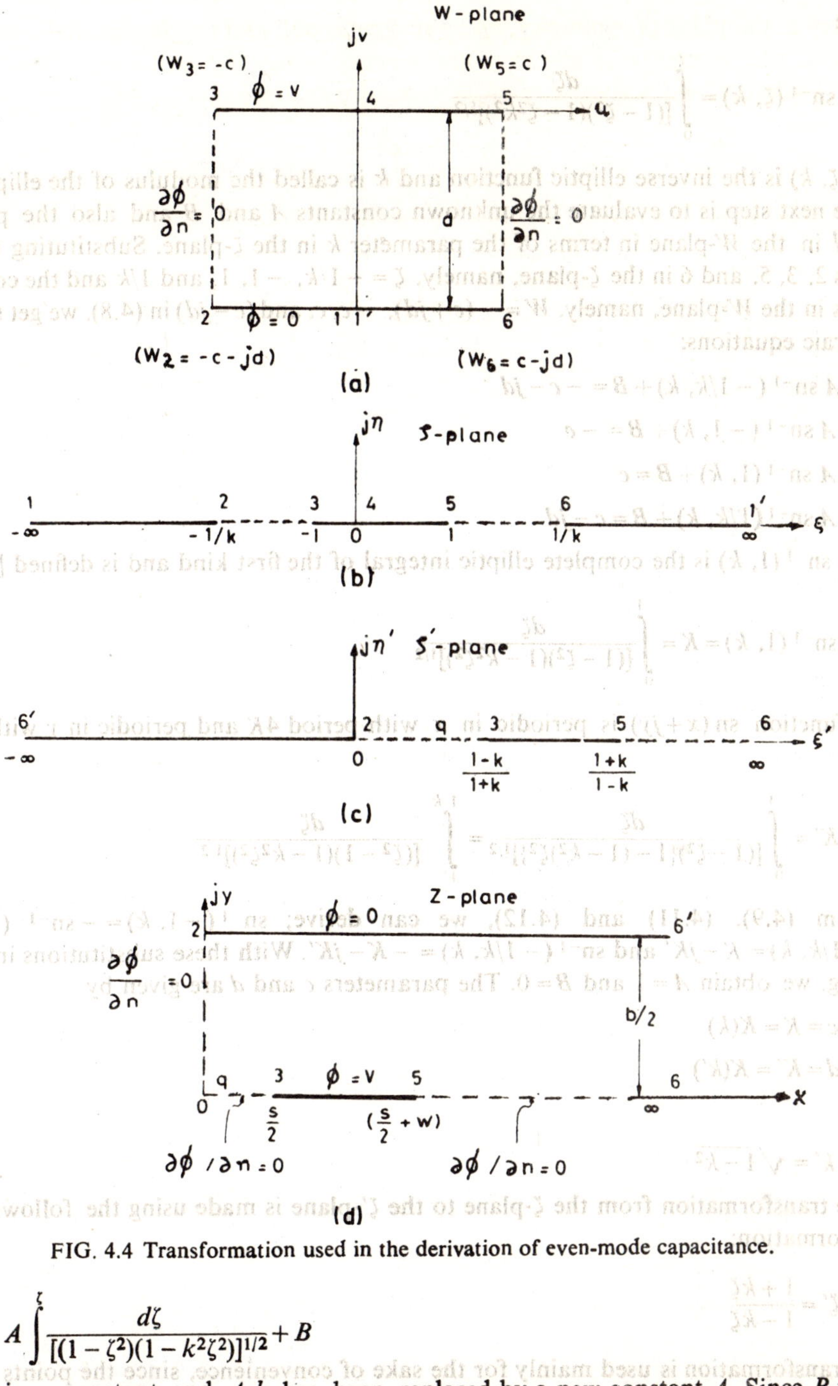

FIG. 4.4 Transformation used in the derivation of even-mode capacitance.

$$= A\int^{\zeta}\frac{d\zeta}{[(1-\zeta^2)(1-k^2\zeta^2)]^{1/2}}+B \tag{4.7}$$

where B is a constant and A_0k has been replaced by a new constant A. Since B is as yet arbitrary, we may introduce a lower limit of integration as zero. Equation (4.7) can then be expressed in the form

$$W = A\,\mathrm{sn}^{-1}(\zeta, k) + B \tag{4.8}$$

where

$$\mathrm{sn}^{-1}(\zeta, k) = \int_0^{\zeta} \frac{d\zeta}{[(1-\zeta^2)(1-\zeta^2 k^2)]^{1/2}} \tag{4.9}$$

$\mathrm{sn}^{-1}(\zeta, k)$ is the inverse elliptic function and k is called the modulus of the elliptic function.

The next step is to evaluate the unknown constants A and B and also the parameters c and d in the W-plane in terms of the parameter k in the ζ-plane. Substituting the values at points 2, 3, 5, and 6 in the ζ-plane, namely, $\zeta = -1/k$, -1, 1, and $1/k$ and the corresponding values in the W-plane, namely, $W = -(c+jd)$, $-c$, c, and $(c-jd)$ in (4.8), we get the following algebraic equations:

$$A\,\mathrm{sn}^{-1}(-1/k, k) + B = -c - jd \tag{4.10a}$$

$$A\,\mathrm{sn}^{-1}(-1, k) + B = -c \tag{4.10b}$$

$$A\,\mathrm{sn}^{-1}(1, k) + B = c \tag{4.10c}$$

$$A\,\mathrm{sn}^{-1}(1/k, k) + B = c - jd \tag{4.10d}$$

Here, $\mathrm{sn}^{-1}(1, k)$ is the complete elliptic integral of the first kind and is defined [12] as

$$\mathrm{sn}^{-1}(1, k) = K = \int_0^1 \frac{d\zeta}{[(1-\zeta^2)(1-k^2\zeta^2)]^{1/2}} \tag{4.11}$$

The function $\mathrm{sn}(x+jy)$ is periodic in x with period $4K$ and periodic in y with period $4K'$ where

$$K' = \int_0^1 \frac{d\zeta}{[(1-\zeta^2)\{1-(1-k^2)\zeta^2\}]^{1/2}} = \int_1^{1/k} \frac{d\zeta}{[(\zeta^2-1)(1-k^2\zeta^2)]^{1/2}} \tag{4.12}$$

From (4.9), (4.11) and (4.12), we can derive; $\mathrm{sn}^{-1}(-1, k) = -\mathrm{sn}^{-1}(1, k) = -K$, $\mathrm{sn}^{-1}(1/k, k) = K - jK'$ and $\mathrm{sn}^{-1}(-1/k, k) = -K - jK'$. With these substitutions in (4.10) and solving, we obtain $A = 1$ and $B = 0$. The parameters c and d are given by

$$c = K = K(k) \tag{4.13a}$$

$$d = K' = K(k') \tag{4.13b}$$

where

$$k' = \sqrt{1-k^2} \tag{4.13c}$$

The transformation from the ζ-plane to the ζ'-plane is made using the following bilinear transformation:

$$\zeta' = \frac{1+k\zeta}{1-k\zeta} \tag{4.14}$$

This transformation is used mainly for the sake of convenience, since the points mapped in the ζ'-plane are better suited for transforming to the desired configuration shown in Fig. 4.4d. The Schwarz-Christoffel transformation is again applied to map the real axis in the ζ'-plane to the boundary of the rectangular region shown in Fig. 4.4d. The transformation is given by

$$Z = A_1 \int_0^{\zeta'} \frac{d\zeta'}{[\zeta'(\zeta'-q)]^{1/2}} \tag{4.15}$$

Evaluating the integral in (4.15) gives

$$Z = 2A_1 \tanh^{-1} \sqrt{\frac{\zeta'-q}{\zeta'}} + B_1 \tag{4.16}$$

Because Z is required to be zero when $\zeta' = q$, we get $B_1 = 0$. Substituting $Z = j\ (b/2)$ corresponding to $\zeta' = 0$ gives $A_1 = b/2\pi$. Hence (4.16) reduces to

$$Z = (b/\pi)[\tanh^{-1} \sqrt{1-(q/\zeta')}] \tag{4.17}$$

We now substitute the values of Z at the edges 3 and 5 of the strip conductor and the corresponding values in the ζ'-plane, in (4.17), to give

$$s/2 = (b/\pi)[\tanh^{-1}\{1-((1+k)/(1-k))q\}^{1/2}] \tag{4.18a}$$

$$(s/2)+w = (b/\pi)[\tanh^{-1}\{1-((1-k)/(1+k))q\}^{1/2}] \tag{4.18b}$$

Using the hyperbolic function identities, q can be eliminated from (4.18) to yield the following expression:

$$k = k_e = \tanh(\pi w/2b) \cdot \tanh\{(\pi/2)\cdot(w+s)/b\} \tag{4.19}$$

where the subscript e is attached to k to denote the even-mode.

The capacitance C_1 of the structure in Fig. 4.4d is the same as that of the parallel plate structure shown in Fig. 4.4a. It is, therefore, given by

$$C_1 = \frac{2\epsilon_0\epsilon_r c}{d} \tag{4.20}$$

Substituting for c and d from (4.13) and noting that $k = k_e$ for the even-mode, (4.20) becomes

$$C_1 = 2\epsilon_0\epsilon_r \frac{K(k_e)}{K(k_e')} \tag{4.21}$$

where

$$k_e' = \sqrt{1-k_e^2} \tag{4.22}$$

The capacitance C_e of the coupled stripline for the even-mode excitation is then given by

$$C_e = 4\epsilon_0\epsilon_r \frac{K(k_e)}{K(k_e')} \tag{4.23}$$

With (4.23), the expression for the even-mode characteristic impedance Z_e becomes

$$Z_e = \frac{1}{v^a C_e} = \frac{30\pi}{\sqrt{\epsilon_r}} \frac{K(k_e')}{K(k_e)} \tag{4.24}$$

4.2.2 Odd-Mode Characteristic Impedance

The procedure and transformations for deriving the expression for odd-mode characteristic impedance Z_o are exactly the same as those used for deriving the even-mode characteristic impedance Z_e, except for the following:

(1) The point q is placed between the points 6′ and 2 instead of between 2 and 3 in Figs. 4.4c and d.

(2) Points 2 and q are made to coincide with $Z=0$ and $j(b/2)$, respectively in Fig. 4.4d.

These changes facilitate the vertical section of the boundary between $Z=0$ and $j(b/2)$ to be an electric wall as is required in the odd-mode case. When the above changes are incorporated, the expression for the odd-mode capacitance C_o of the stripline becomes

$$C_o = 4\epsilon_o\epsilon_r \frac{K(k_o)}{K(k_o')} \tag{4.25}$$

where

$$k_o = \tanh(\pi w/2b)\cdot\coth((\pi/2)\cdot(w+s)/b) \tag{4.26}$$

$$k_o' = \sqrt{1-k_o^2} \tag{4.27}$$

The subscript o is attached to k to denote the odd-mode case. The expression for the characteristic impedance is then given by

$$Z_o = \frac{30\pi}{\sqrt{\epsilon_r}}\frac{K(k_o')}{K(k_o)} \tag{4.28}$$

4.2.3 *Explicit Expressions for the Dimensional Parameters of the Edge-Coupled Symmetric Stripline* [13]

In practical designs, one is invariably required to determine the dimensional parameters of the circuit from the known characteristic impedances. Expressions (4.24) and (4.28) give impedances in terms of the geometrical dimensions. The determination of dimensions from these formulas involves the use of numerical procedures in which the geometrical parameters are iteratively changed until the desired impedance is obtained. In order to obtain inverse formulas which are versatile, we need the inverse function of the parameter K'/K. We first define two parameters ξ_e and ξ_o for the even- and odd-modes as

$$\xi_{\substack{e\\o}} = \frac{K(k'_{\substack{e\\o}})}{K(k_{\substack{e\\o}})} \tag{4.29}$$

The inverse function can be expressed in terms of Jacobi's theta function [14] as

$$k_{\substack{e\\o}} = \text{arm}\,\xi_{\substack{e\\o}} = \left[\frac{\nu_2(0/j\xi_{\substack{e\\o}})}{\nu_3(0/j\xi_{\substack{e\\o}})}\right]^2 \tag{4.30}$$

where the functions arm ξ_e and arm ξ_o are introduced for convenience. From (4.24), (4.28), and (4.29), we have

$$\xi_{\substack{e\\o}} = \frac{Z_{\substack{e\\o}}\sqrt{\epsilon_r}}{30\pi} \tag{4.31}$$

Further, from (4.19) and (4.26), we can write

$$w/b = (2/\pi)[\tanh^{-1}\sqrt{k_e k_o}\,] \tag{4.32}$$

$$(w+s)/b = (2/\pi)[\tanh^{-1}\sqrt{k_e/k_o}\,] \tag{4.33}$$

In a practical design, k_e and k_o are calculated from the knowledge of Z_e, Z_o and ϵ_r using (4.30) and (4.31). The dimensional parameters w/b and s/b are then obtained directly from (4.32) and (4.33). The numerical computation of Jacobi's theta functions in (4.30) can be simplified by using the following rapidly convergent series:

$$\nu_2(0/\tau) = 2q^{1/4} \sum_{n=0}^{\infty} q^{n^2-n} \tag{4.34a}$$

$$\nu_3(0/\tau) = 1 + 2 \sum_{n=1}^{\infty} q^{n^2} \tag{4.34b}$$

where

$$q = e^{j\pi\tau} \tag{4.34c}$$

It may be noted that the degenerate case $Z_e = Z_o = Z$ gives the formula for calculating the parameter w/b for a single conductor symmetric stripline having a strip width w and ground plane spacing b. It is given by

$$w/b = (2/\pi)\,[\tanh^{-1}\{\text{arm}\,(Z\sqrt{\epsilon_r}/30\pi)\}] \tag{4.35}$$

4.3 Edge-Coupled Stripline and Stripline-like Transmission Lines with Side Walls and Isotropic/Uniaxially Anisotropic Dielectric Layers—Unified Approach [15, 16]

4.3.1 General Capacitance Formula

Figure 4.5a shows the cross-sectional view of a multilayer edge-coupled transmission structure with a rectangular shielding enclosure. The dielectric layers may be isotropic or uniaxially anisotropic. The dielectric constant of the jth layer is ϵ_j or $\hat{\epsilon}_j$ depending on whether it is isotropic or uniaxially anisotropic. The expressions are

$$\epsilon_j = \epsilon_o \epsilon_{rj} \tag{4.36}$$

and

$$\hat{\epsilon}_j = \epsilon_o \begin{bmatrix} \epsilon_{xxj} & 0 & 0 \\ 0 & \epsilon_{yyj} & 0 \\ 0 & 0 & \epsilon_{xxj} \end{bmatrix} \tag{4.37}$$

For the purpose of analysis, the plane of symmetry pp' can be represented by a magnetic wall for the even-mode excitation and by an electric wall for the odd-mode excitation. It is then adequate to consider only half the structure between $x = 0$ and $c/2$ as shown in Fig. 4.5b for the determination of capacitance. In Chapter 2, general capacitance formulas have been derived for single conductor multilayer structures for the two side wall conditions; case (a) corresponding to the odd-mode when both the side walls are electric and case (b) corresponding to the even-mode when the left side wall is electric and the right side wall is magnetic. Since the side wall spacing is $c/2$ in the present structure (Fig. 4.5b), formulas (2.128) and (2.129) can be directly used with c replaced by $c/2$. Thus, the capacitances C_e and C_o for the

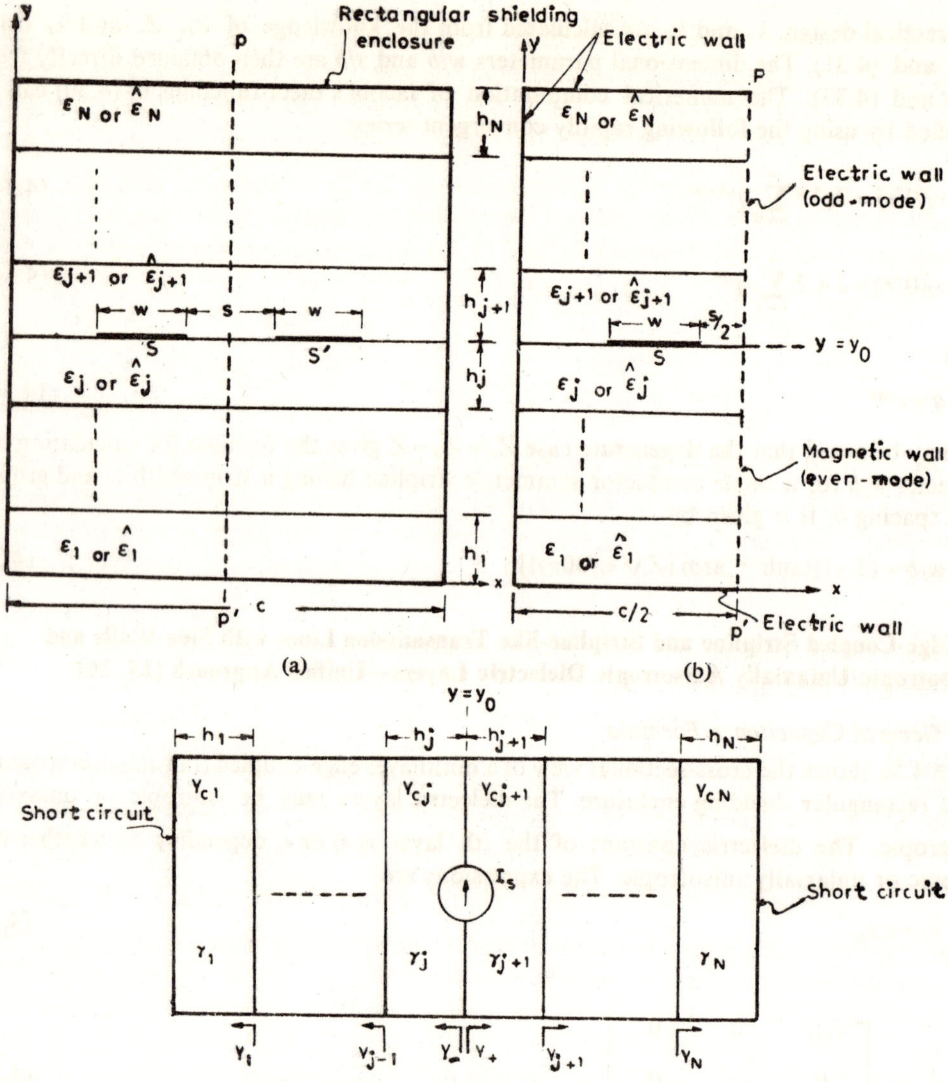

FIG. 4.5 (a) Multilayer edge-coupled transmission structure with isotropic/uniaxially anisotropic substrates, (b) half of the structure for even- and odd-mode analysis, and (c) two-wire transmission line equivalent for both even- and odd-modes. For isotropic layer: $Y_{ej}=\epsilon_j=\epsilon_0\epsilon_{rj}$, $\gamma_j=\gamma=\beta_n$, $\beta_n=n\pi/c$, $j=1, 2, \ldots, N$; For uniaxially anisotropic layer: $Y_{ej}=\epsilon_0\sqrt{\epsilon_{xxj}\epsilon_{yyj}}$, $\gamma_j=\beta_n\sqrt{\epsilon_{xxj}/\epsilon_{yyj}}$, $\beta_n=n\pi/c$; $j=1, 2, \ldots, N$.

even- and odd-modes, respectively, are given by

$$\frac{1}{C_e}=\frac{\displaystyle\sum_{n=0}^{\infty}\frac{4}{(2n+1)\pi Y}\left[\int_S f(x)_e \sin\left\{\frac{(2n+1)\pi x}{c}\right\}dx\right]^2}{\left[\displaystyle\int_S f(x)_e\,dx\right]^2} \tag{4.38}$$

$$\frac{1}{C_o}=\frac{\sum_{n=1}^{\infty}\frac{2}{n\pi Y}\left[\int_S f(x)_o \sin\frac{2n\pi x}{c}\,dx\right]^2}{\left[\int_S f(x)_o\,dx\right]^2} \tag{4.39}$$

If, in (4.38), $(2n+1)$ is replaced by n, the summation can be carried over odd values of n. Similarly, in (4.39), if $2n$ is replaced by n, the summation can be carried over even values of n excluding the value $n=0$. Thus, (4.38) and (4.39) can be together expressed as

$$\frac{1}{C_{\substack{e\\o}}}=\frac{\sum_{\substack{n\ \text{odd}\\ \text{even}}}\frac{4}{n\pi Y}\left[\int_S f(x)_{\substack{e\\o}} \sin\frac{n\pi x}{c}\,dx\right]^2}{\left[\int_S f(x)_{\substack{e\\o}}\,dx\right]^2} \tag{4.40}$$

Here, $f(x)_e$ and $f(x)_o$ are the charge distributions on the strip conductor S in Fig. 4.5b for the even- and odd-modes, respectively. The admittance parameter Y can be determined from the two-wire transmission line equivalent shown in Fig. 4.5c. It may be noted that the expression for Y is the same for both even- and odd-mode cases, since it does not depend on the boundary conditions at the side walls.

4.3.2 Closed Form Capacitance Expressions for Specified Even- and Odd-Mode Charge Distributions

Several techniques are available in the literature for the determination of charge distributions in coupled strip transmission lines [17, 18]. However, for the purpose of determining the capacitance values required for practical designs, there is no need to go through the difficulty of finding the actual charge distribution for each structure. It is adequate to choose trial functions for $f(x)_e$ and $f(x)_o$ which would best approximate the nature of charge density distribution so that the evaluation of the integral is mathematically tractable and computationally simple.

Figure 4.6 shows the nature of charge distribution on edge-coupled strip conductors for even- and odd-mode excitations. For loosely coupled lines, the charge distribution on each strip is nearly symmetric (Fig. 4.6a). As the spacing s between the strip conductors is reduced to increase the coupling, the distribution tends to be increasingly asymmetric (Fig. 4.6b). Referring to Fig. 4.5a, the trial function which approximates the charge distribution on the strip conductor S for the even- and odd-modes can be assumed to be of the form

$$\begin{aligned} f(x)_{\substack{e\\o}} &= \frac{1}{w}\left[1+A_{\substack{e\\o}}\left|\frac{2}{w}\left(x-\frac{c-s-w}{2}-g_{\substack{e\\o}}\right)\right|^3\right], \qquad \frac{c-s}{2}-w\leqslant x\leqslant\frac{c-s}{2} \\ &=0, \qquad \text{otherwise} \end{aligned} \tag{4.41}$$

where

$$0\leqslant g_e\leqslant\frac{w}{2}, \quad \text{for even-mode, and} \quad -\frac{w}{2}\leqslant g_o\leqslant 0, \quad \text{for odd-mode}$$

Here, A_e and A_o are arbitrary constants and are determined by maximizing C_e and C_o, respectively, and g_e and g_o are parameters which basically account for the degree of asymmetry

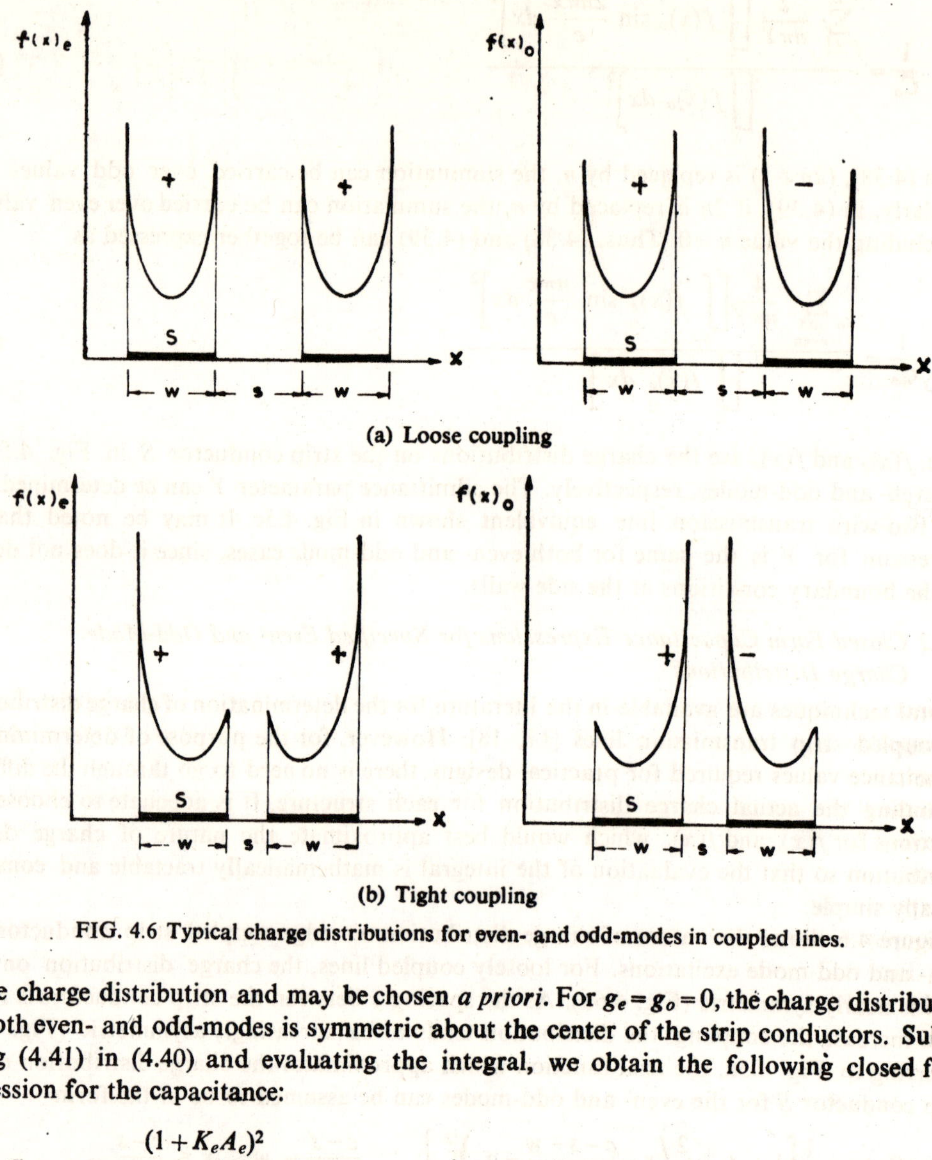

FIG. 4.6 Typical charge distributions for even- and odd-modes in coupled lines.

in the charge distribution and may be chosen *a priori*. For $g_e = g_o = 0$, the charge distribution for both even- and odd-modes is symmetric about the center of the strip conductors. Substituting (4.41) in (4.40) and evaluating the integral, we obtain the following closed form expression for the capacitance:

$$C_{\substack{e\\o}} = \frac{(1+K_{\substack{e\\o}}A_{\substack{e\\o}})^2}{\sum\limits_{\substack{n\ \mathrm{odd}\\ \mathrm{even}}} (L_n + M_{ne}A_{\substack{e\\o}})^2 P_n / Y} \tag{4.42}$$

where

$$K_{\substack{e\\o}} = \frac{2}{w^4}\left[\left(\frac{w}{2}+g_{\substack{e\\o}}\right)^4 + \left(\frac{w}{2}-g_{\substack{e\\o}}\right)^4\right] \tag{4.43a}$$

$$L_n = \sin\left(\frac{\beta_n w}{2}\right)\cdot \sin\left\{\beta_n\left(\frac{c-s-w}{2}\right)\right\}, \qquad \beta_n = \frac{n\pi}{c} \tag{4.43b}$$

$$M_{\substack{ne\\o}}=\left(\frac{2}{\beta_n w}\right)^3\cdot\frac{1}{2}\left[3\left\{\beta_n^2\left(\frac{w}{2}+g_{\substack{e\\o}}\right)^2-2\right\}\sin\left\{\beta_n\left(\frac{c-s}{2}-w\right)\right\}\right.$$

$$+\beta_n\left(\frac{w}{2}+g_{\substack{e\\o}}\right)\left\{\beta_n^2\left(\frac{w}{2}+g_{\substack{e\\o}}\right)^2-6\right\}\cos\left\{\beta_n\left(\frac{c-s}{2}-w\right)\right\}+3\left\{\beta_n^2\left(\frac{w}{2}-g_{\substack{e\\o}}\right)^2-2\right\}\cdot$$

$$\sin\left\{\beta_n\left(\frac{c-s}{2}\right)\right\}-\beta_n\left(\frac{w}{2}-g_{\substack{e\\o}}\right)\left\{\beta_n^2\left(\frac{w}{2}-g_{\substack{e\\o}}\right)^2-6\right\}\cos\left\{\beta_n\left(\frac{c-s}{2}\right)\right\}$$

$$\left.+12\sin\left\{\beta_n\left(\frac{c-s-w}{2}+g_{\substack{e\\o}}\right)\right\}\right] \tag{4.43c}$$

$$P_n=\frac{4}{n\pi}\left(\frac{2}{\beta_n w}\right)^2 \tag{4.43d}$$

By setting

$$\frac{\partial C_e}{\partial A_e}=0 \qquad \text{and} \qquad \frac{\partial C_o}{\partial A_o}=0$$

and solving the resulting equations, we obtain

$$A_{\substack{e\\o}}=-\frac{\sum\limits_{\substack{n\ \text{odd}\\ \text{even}}}(K_{\substack{e\\o}}L_n-M_{\substack{ne\\o}})L_nP_n/Y}{\sum\limits_{\substack{n\ \text{odd}\\ \text{even}}}(K_{\substack{e\\o}}L_n-M_{\substack{ne\\o}})M_{\substack{ne\\o}}P_n/Y} \tag{4.43e}$$

Since A_e and A_o given by (4.43e) are obtained by maximizing the capacitances C_e and C_o, respectively, the choice of g_e and g_o will not be critical. It is found that the symmetric distribution corresponding to $g_e=g_o=0$ yields quite accurate results of capacitance for loose to moderate couplings for which values of s/b lie generally above 0.2. For tight couplings involving highly asymmetric charge distribution (values of s/b generally less than about 0.2), one may choose $0<g_e\leqslant w/2$ for even-mode, and $-w/2\leqslant g_o<0$ for odd-mode. (This aspect is discussed in detail in Chapter 6 before presenting the numerical data on edge-coupled lines.)

For the symmetric case, setting $g_e=g_o=0$ in (4.41)-(4.43), the charge distribution and the resulting closed form expression for the capacitance are given by

$$f(x)_{\substack{e\\o}}=\frac{1}{w}\left[1+A_{\substack{e\\o}}\left|\frac{2}{w}\left(x-\frac{c-s-w}{2}\right)\right|^3\right], \qquad \frac{c-s}{2}-w\leqslant x\leqslant\frac{c-s}{2}$$

$$=0, \qquad \text{otherwise} \tag{4.44}$$

$$C_{\substack{e\\o}}=\frac{(1+0.25A_{\substack{e\\o}})^2}{\sum\limits_{\substack{n\ \text{odd}\\ \text{even}}}(L_n+M_nA_{\substack{e\\o}})^2P_n/Y} \tag{4.45}$$

where

$$L_n=\sin\left(\frac{\beta_n w}{2}\right)\sin\left\{\beta_n\left(\frac{c-s-w}{2}\right)\right\}, \qquad \beta_n=\frac{n\pi}{c} \tag{4.46a}$$

$$M_n = \left(\frac{2}{\beta_n w}\right)^3 \cdot \sin\left\{\beta_n\left(\frac{c-s-w}{2}\right)\right\} \cdot \left[3\left\{\left(\frac{\beta_n w}{2}\right)^2 - 2\right\}\cos\left(\frac{\beta_n w}{2}\right) + \left(\frac{\beta_n w}{2}\right)\left\{\left(\frac{\beta_n w}{2}\right)^2 - 6\right\}\cdot \sin\left(\frac{\beta_n w}{2}\right) + 6\right] \tag{4.46b}$$

$$P_n = \left(\frac{2}{\beta_n w}\right)^2 \cdot \left(\frac{4}{n\pi}\right) \tag{4.46c}$$

$$A_{\substack{e\\o}} = -\frac{\sum\limits_{\substack{n\text{ odd}\\\text{even}}} (L_n - 4M_n)L_n P_n / Y}{\sum\limits_{\substack{n\text{ odd}\\\text{even}}} (L_n - 4M_n)M_n P_n / Y} \tag{4.46d}$$

For charge distributions specified as in (4.41) and (4.44), the capacitance expressions given in (4.42) and (4.45), respectively, are general and can be used to determine the even- and odd-mode capacitances of any edge-coupled structure having isotropic/anisotropic dielectric layers. The only parameter that needs to be changed is the admittance Y at the charge plane depending on the structure.

4.3.3 Admittance Parameters

The transmission line equivalent of Fig. 4.5b for both even- and odd-modes is shown in Fig. 4.5c. The admittance Y for the edge-coupled structure is the same for both the modes. Thus, the procedure described in Sec. 3.4 is directly applicable. The expression for Y appearing in the capacitance formulas (4.42) and (4.45) are obtained by applying transmission line formula (2.131), with $Y_{cj} = \epsilon_j = \epsilon_0 \epsilon_{rj}$, $\gamma_j = \beta_n = n\pi/c$ for the isotropic layer, and $Y_{cj} = \epsilon_0 \epsilon_{rfj}$, $\gamma_j = \beta_n \epsilon_{aj}$ with $\epsilon_{rfj} = \sqrt{\epsilon_{xxj}\epsilon_{yyj}}$, $\epsilon_{aj} = \sqrt{\epsilon_{xxj}/\epsilon_{yyj}}$ and $\beta_n = n\pi/c$ for anisotropic layers in which the optical axis coincides with the y-axis. The admittance expressions for some practical edge-coupled stripline-like transmission lines using isotropic substrates are listed in Table 4.1 and those for structures using anisotropic substrates in which the optical axis is aligned along the y-axis are listed in Table 4.2. It may be noted that the expressions given in Tables 4.1 and 4.2 are the same as those given in Tables 3.1 and 3.2, respectively, for the single conductor configurations.

4.4 Edge-Coupled Stripline and Stripline-like Transmission Lines with Side Walls and Anisotropic Substrates Having Principal Axes Rotated with Respect to the *x-y* Coordinate Axes—Variational Technique Combined with Spectral Domain Analysis [19]

In Sec. 3.6, a detailed spectral domain analysis under quasi-static approximation has been presented for a two-layered single conductor stripline with anisotropic substrates in which the principal axes are rotated with respect to the x-y coordinate axes. Identification of the admittance parameter in the variational capacitance expression led to generalization of the formula to multilayer structures. In this section, we follow a similar procedure to evolve generalized variational capacitance expressions for edge-coupled multilayer structures having anisotropic substrates with principal axes of each substrate rotated with respect to the x-y coordinate axes.

Table 4.1 Admittance parameters of edge-coupled isotropic transmission structures with side walls

Structure	*Admittance Y at $y=y_0$, $\beta_n=n\pi/c$*
A1. Edge-coupled symmetric stripline 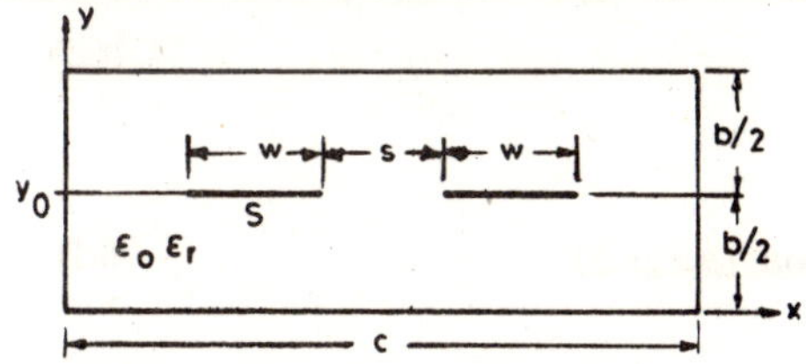	$2\epsilon_0\epsilon_r \coth(\beta_n b/2)$
A2. Edge-coupled stripline with off-set conductors 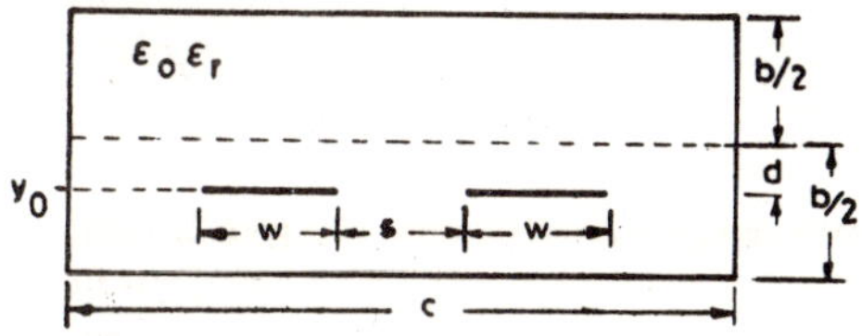	$\epsilon_0\epsilon_r[\coth(\beta_n((b/2)+d))+\coth(\beta_n((b/2)-d))]$
A3. Edge-coupled microstrip-like stripline 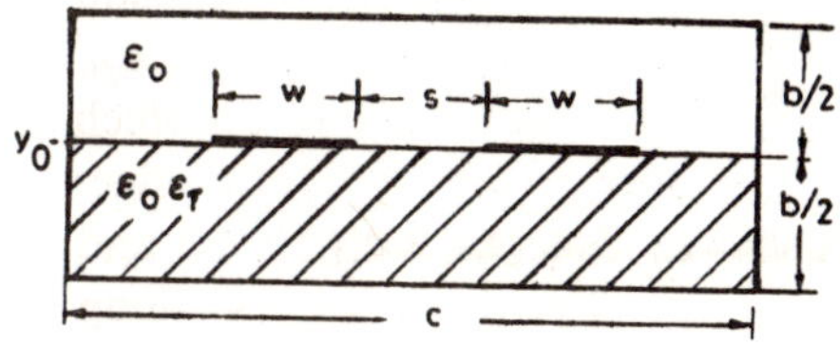	$\epsilon_0[\coth(\beta_n b/2)+\epsilon_r \coth(\beta_n b/2)]$
A4. Edge-coupled suspended stripline 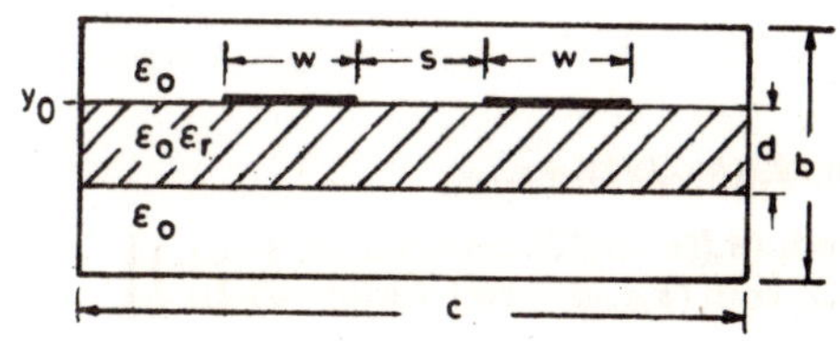	$\epsilon_0\left[\coth(\beta_n((b-d)/2)) + \epsilon_r\left(\frac{\coth(\beta_n((b-d)/2))\coth(\beta_n d)+\epsilon_r}{\epsilon_r\coth(\beta_n d)+\coth(\beta_n((b-d)/2))}\right)\right]$
A5. Edge-coupled sandwiched stripline 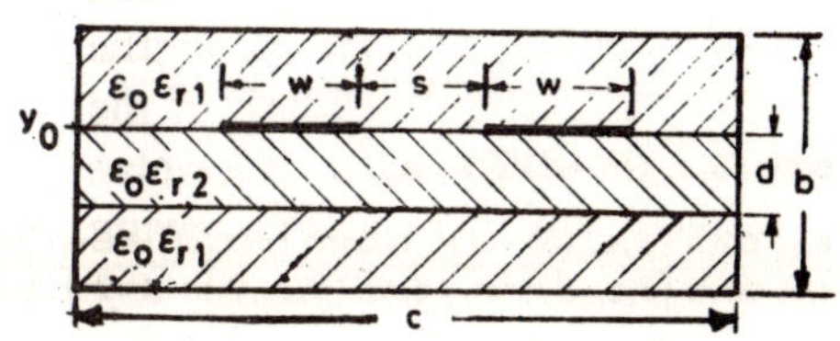	$\epsilon_0\left[\epsilon_{r1}\coth(\beta_n((b-d)/2)) + \epsilon_{r2}\left(\frac{\epsilon_{r1}\coth(\beta_n((b-d)/2))\coth(\beta_n d)+\epsilon_{r2}}{\epsilon_{r2}\coth(\beta_n d)+\epsilon_{r1}\coth(\beta_n((b-d)/2))}\right)\right]$

Table 4.2 Admittance parameters of edge-coupled transmission structures with side walls and having uniaxially anisotropic substrates—optical axis aligned along *y*-axis

$$\epsilon_{rfj}=\sqrt{\epsilon_{xxj}\epsilon_{yyj}},\ \epsilon_{aj}=\sqrt{\epsilon_{xxj}/\epsilon_{yyj}},\qquad j=1,2$$

Structure	*Admittance at* $y=y_0$, $\beta_n=n\pi/c$

A1. Edge-coupled symmetric stripline

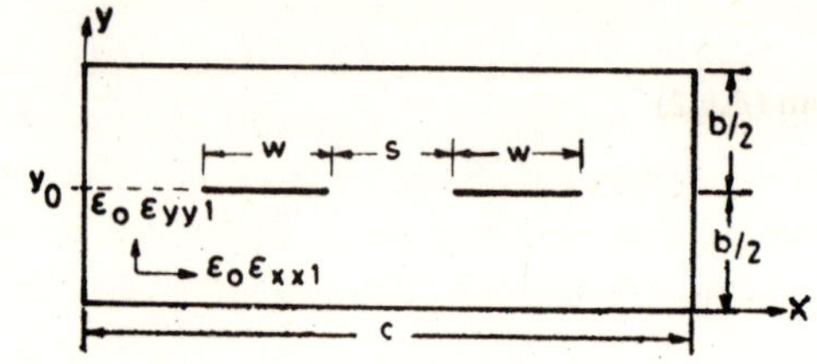

$$2\epsilon_0\epsilon_{rf1}\coth(\beta_n\epsilon_{a1}b/2)$$

A2. Edge-coupled stripline with off-set conductors

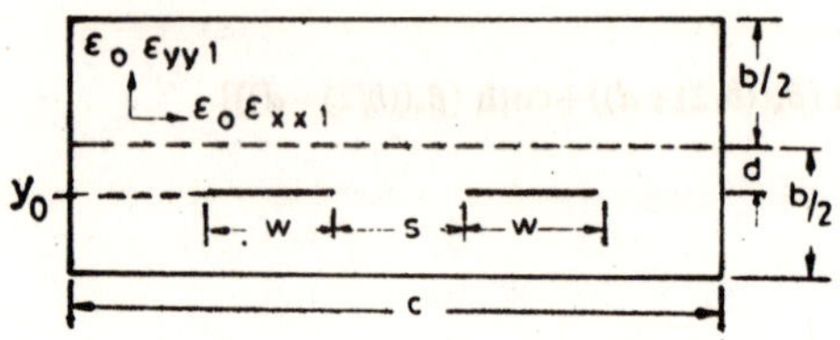

$$\epsilon_0\epsilon_{rf1}[\coth(\beta_n\epsilon_{a1}((b/2)+d))+\coth(\beta_n\epsilon_{a1}((b/2)-d))]$$

A3. Edge-coupled microstrip-like stripline

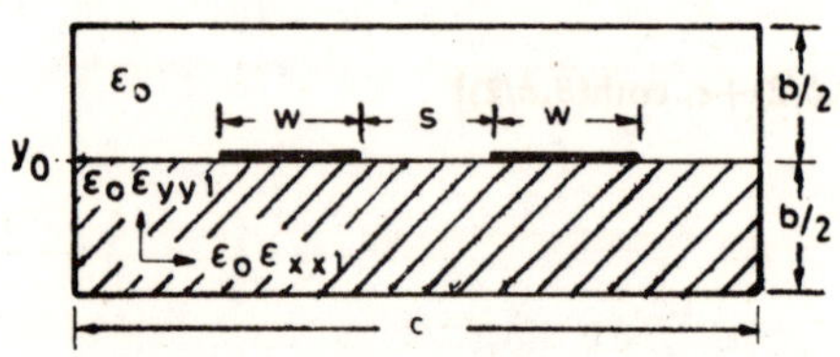

$$\epsilon_0[\coth(\beta_n b/2)+\epsilon_{rf1}\coth(\beta_n\epsilon_{a1}b/2)]$$

A4. Edge-coupled suspended stripline

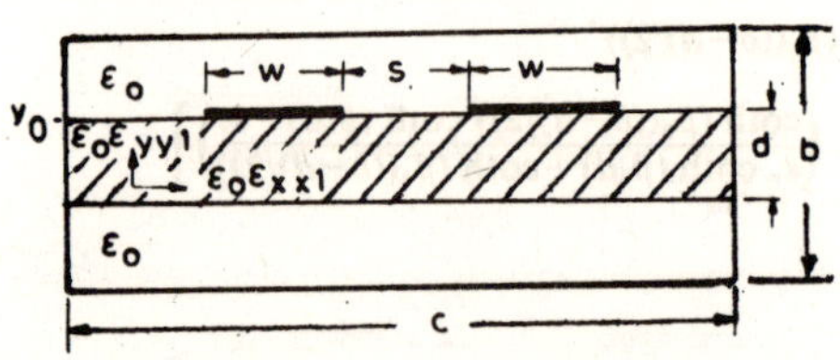

$$\epsilon_0\left[\coth(\beta_n((b-d)/2))+\epsilon_{rf1}\cdot\left(\frac{\coth(\beta_n((b-d)/2))\coth(\beta_n\epsilon_{a1}d)+\epsilon_{rf1}}{\epsilon_{rf1}\coth(\beta_n\epsilon_{a1}d)+\coth(\beta_n((b-d)/2))}\right)\right]$$

A5. Edge-coupled sandwiched stripline

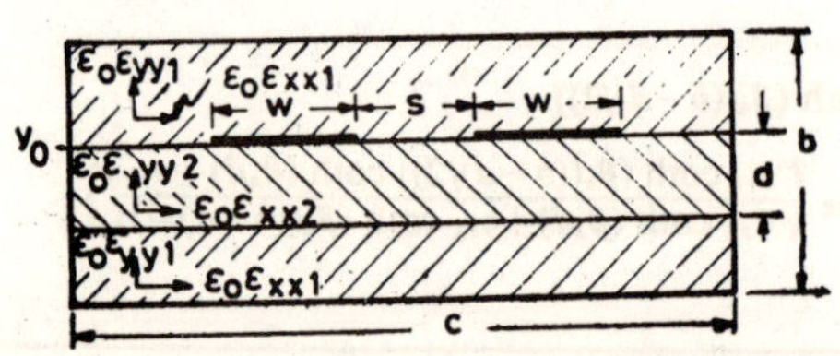

$$\epsilon_0\left[\epsilon_{rf1}\coth(\beta_n\epsilon_{a1}((b-d)/2))+\epsilon_{rf2}\cdot\left\{\frac{\epsilon_{rf1}\coth(\beta_n\epsilon_{a1}((b-d)/2))\coth(\beta_n\epsilon_{a2}d)+\epsilon_{rf2}}{\epsilon_{rf2}\coth(\beta_n\epsilon_{a2}d)+\epsilon_{rf1}\coth(\beta_n\epsilon_{a1}((b-d)/2))}\right\}\right]$$

4.4.1 Two-Layered Edge-Coupled Structure—Spectral Domain Analysis under Quasi-Static Approximation

Consider a two-layered edge-coupled anisotropic stripline with a rectangular shielding enclosure as shown in Fig. 4.7. As considered in Sec. 3.6.1, the permittivity tensor $\hat{\epsilon}_j$ for the two layers, $j = 1, 2$, in the x-y-z coordinate system is given by [20]:

$$\hat{\epsilon}_j = \epsilon_0 \begin{bmatrix} \epsilon_{xxj} & \epsilon_{xyj} & 0 \\ \epsilon_{xyj} & \epsilon_{yyj} & 0 \\ 0 & 0 & \epsilon_{xxj} \end{bmatrix}, \quad j = 1, 2 \tag{4.47}$$

where

$$\epsilon_{xxj} = \epsilon_{\xi j} \cos^2 \theta_j + \epsilon_{\eta j} \sin^2 \theta_j \tag{4.48a}$$

$$\epsilon_{xyj} = (\epsilon_{\eta j} - \epsilon_{\xi j}) \sin \theta_j \cos \theta_j \tag{4.48b}$$

$$\epsilon_{yyj} = \epsilon_{\xi j} \sin^2 \theta_j + \epsilon_{\eta j} \cos^2 \theta_j \tag{4.48c}$$

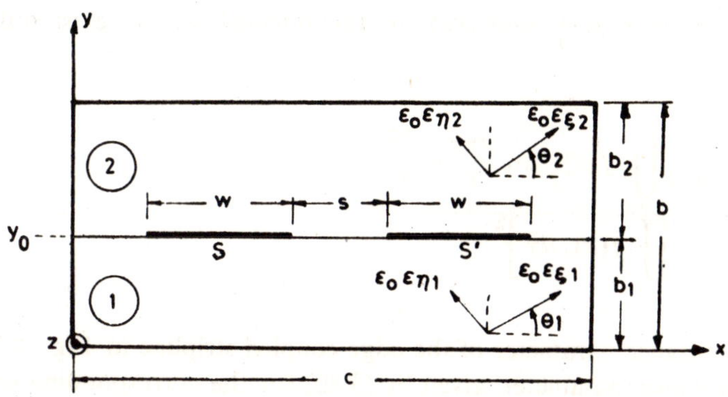

FIG. 4.7 Double layered edge-coupled stripline with anisotropic substrates.

The two strip conductors are assumed to be of zero thickness and spaced by a distance s. The structure is analysed in terms of the even- and odd-modes. Let $f(x)_e$ and $f(x)_o$ denote the even- and odd-mode charge distributions, respectively, on the strip conductor S and $\hat{f}(\beta_n)_e$ and $\hat{f}(\beta_n)_o$ denote their corresponding finite Fourier transforms with β_n as a discrete Fourier variable. The finite Fourier transforms $\hat{f}(\beta_n)_e$ and $\hat{f}(\beta_n)_o$ are defined as

$$\hat{f}(\beta_n)_{\substack{e\\o}} = \int_0^{c/2} f(x)_{\substack{e\\o}} \sin \beta_n x \, dx \tag{4.49}$$

where

$$\beta_n = \frac{n\pi}{c}, \quad \begin{matrix} n \text{ even for odd-mode} \\ n \text{ odd for even-mode} \end{matrix} \tag{4.50}$$

The analysis procedure for the present edge-coupled structure is the same as that described in Sec. 3.6.1 for the corresponding single conductor two-layer structure, except that, while applying Parseval's theorem, we consider the range of integration to be from 0 to $c/2$. Thus, equation (3.87) gets replaced by

$$\sum_{n=1}^{\infty} \psi_1(\beta_n)\psi_2(\beta_n) = (c/4) \int_0^{c/2} \psi_1(x)\psi_2(x)\, dx \tag{4.51}$$

With this, the capacitance expression in (3.91) gets replaced by

$$\frac{1}{C_{\substack{e\\o}}} = \frac{\dfrac{4}{c} \sum\limits_{\substack{n\ \text{odd}\\ \text{even}}} \dfrac{1}{\beta_n Y} \left[\displaystyle\int_0^{c/2} f(x)_{\substack{e\\o}} \sin \beta_n x\, dx \right]^2}{\left[\displaystyle\int_S f(x)_{\substack{e\\o}}\, dx \right]^2} \tag{4.52}$$

Noting that $\beta_n = n\pi/c$, and, $f(x)_e$ and $f(x)_o$ in the interval $(o, c/2)$ exist only on the strip conductor S, we get

$$\frac{1}{C_{\substack{e\\o}}} = \frac{\sum\limits_{\substack{n\ \text{odd}\\ \text{even}}} \dfrac{4}{n\pi Y} \left[\displaystyle\int_S f(x)_{\substack{e\\o}} \sin \frac{n\pi x}{c}\, dx \right]^2}{\left[\displaystyle\int_S f(x)_{\substack{e\\o}}\, dx \right]^2} \tag{4.53}$$

where Y is the admittance parameter of the edge-coupled stripline in Fig. 4.7 which is the same as the admittance parameter given by (3.80) for the corresponding single conductor structure.

4.4.2 Generalization to a Class of Edge-Coupled Transmission Structures

For specified even- and odd-mode charge distributions, the only parameter that depends on the medium properties in the capacitance formula (4.53) is the admittance Y. We also observe that (4.53) is precisely the same expression as (4.40) when the admittance parameter appropriate to the structure is substituted. Expression (4.40) is therefore general and can be applied to multilayer edge-coupled transmission lines with anisotropic dielectric substrates even when the optical axis is oriented at an arbitrary angle with respect to the vertical (y-) axis in the transverse plane. If the even- and odd-mode charge distributions are assumed to be of the form given in (4.41), the closed form formula given by (4.42) combined with (4.43) can be used to compute the even- and odd-mode capacitances.

Since the admittance parameter of the edge-coupled transmission line is the same as that of the corresponding single conductor line, the transmission line equivalent given in Fig. 3.8 and the procedure described in Sec. 3.6.2 are applicable. Table 4.3 lists the admittance parameters for some of the edge-coupled structures with anisotropic substrates having principal axes rotated with respect to the x-y coordinate axes.

Table 4.3 Admittance parameters of edge-coupled transmission structures with side walls and anisotropic substrates – Principal axes rotated with respect to the *x-y* coordinate axes

$$\epsilon_{rfj}=\sqrt{\epsilon_{\eta j}\epsilon_{\xi j}}\,;\quad \epsilon_{aj}=\sqrt{\epsilon_{\eta j}/\epsilon_{\xi j}}\,[(\epsilon_{\eta j}/\epsilon_{\xi j}-1)\cos^2\theta_j+1]^{-1};\quad j=1,2$$

Structure	*Admittance Y at* $y=y_0$, $\beta_n=n\pi/c$
• Edge-coupled symmetric stripline 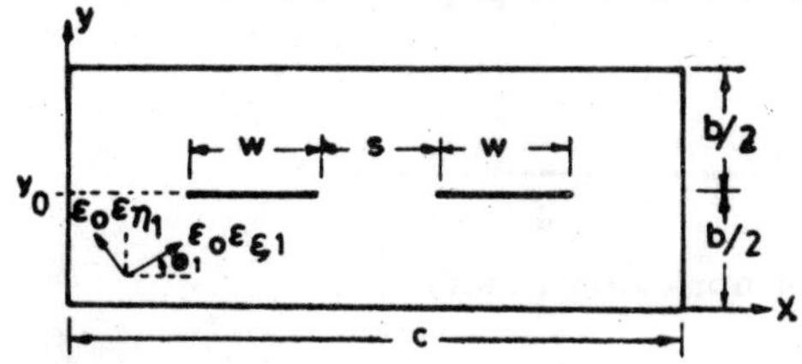	$2\epsilon_0\epsilon_{rf1}\coth(\beta_n\epsilon_{a1}b/2)$
• Edge-coupled stripline with off-set conductors 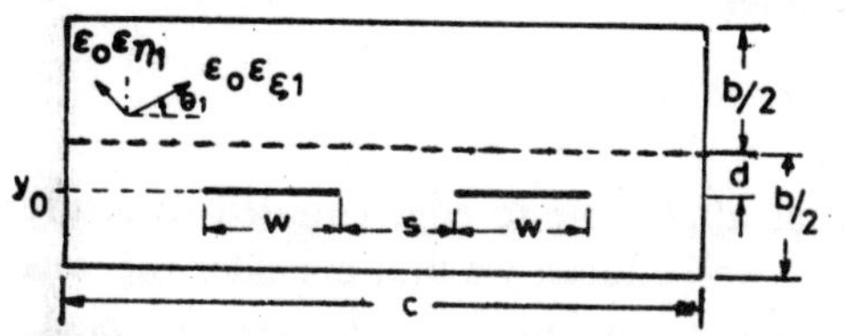	$\epsilon_0\epsilon_{rf1}[\coth(\beta_n\epsilon_{a1}((b/2)+d))+\coth(\beta_n\epsilon_{a1}((b/2)-d))]$
• Edge-coupled microstrip-like stripline 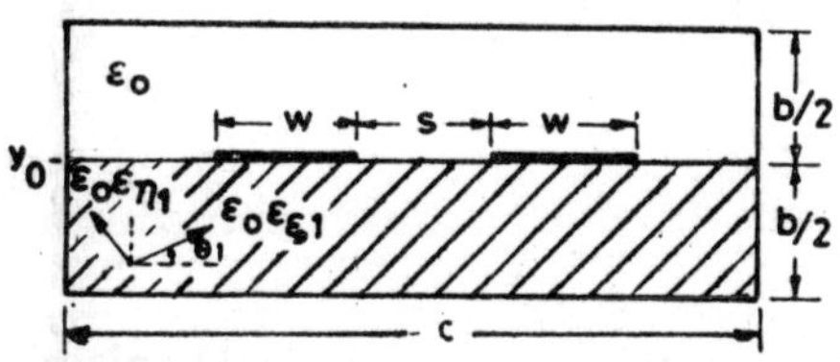	$\epsilon_0[\coth(\beta_n b/2)+\epsilon_{rf1}\coth(\beta_n\epsilon_{a1}b/2)]$
• Edge-coupled suspended stripline 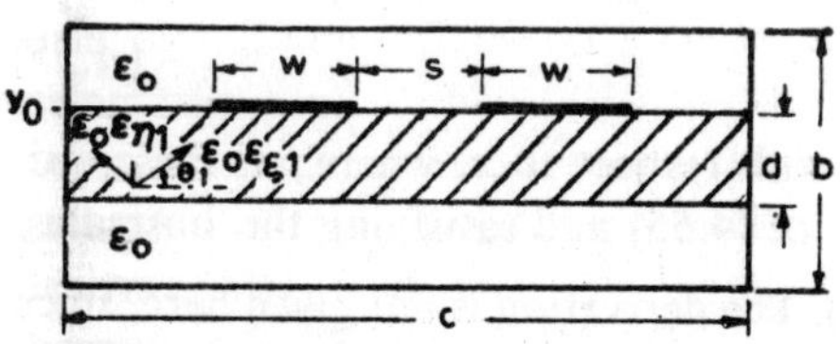	$\epsilon_0\left[\coth(\beta_n((b-d)/2))+\epsilon_{rf1}\cdot\left\{\dfrac{\coth(\beta_n((b-d)/2))\coth(\beta_n\epsilon_{a1}d)+\epsilon_{rf1}}{\epsilon_{rf1}\coth(\beta_n\epsilon_{a1}d)+\coth(\beta_n((b-d)/2))}\right\}\right]$
• Edge-coupled sandwiched stripline 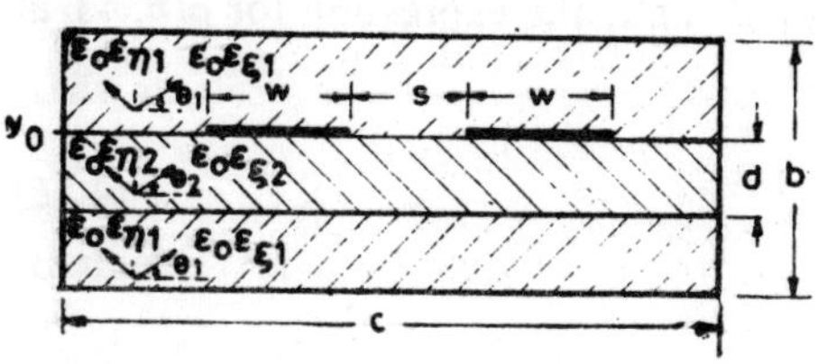	$\epsilon_0\left[\epsilon_{rf1}\coth(\beta_n\epsilon_{a1}((b-d)/2))+\epsilon_{rf2}\cdot\left\{\dfrac{\epsilon_{rf1}\coth(\beta_n\epsilon_{a1}((b-d)/2))\coth(\beta_n\epsilon_{a2}d)+\epsilon_{rf2}}{\epsilon_{rf2}\coth(\beta_n\epsilon_{a2}d)+\epsilon_{rf1}\coth(\beta_n\epsilon_{a1}((b-d)/2))}\right\}\right]$

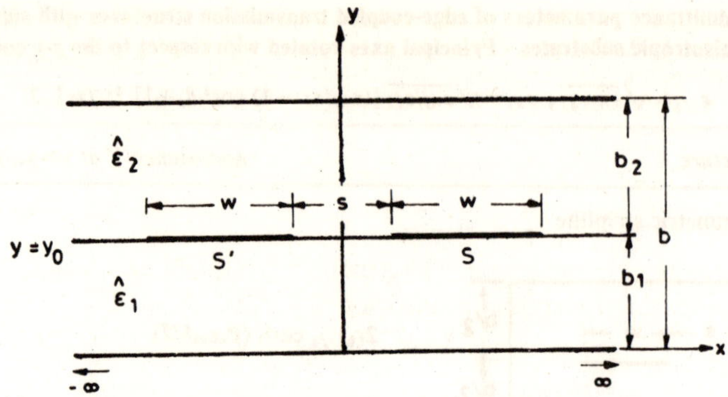

FIG. 4.8 Double layered edge-coupled anisotropic stripline with infinitely wide ground planes.

4.5 Edge-Coupled Stripline and Stripline-like Transmission Lines with Iso/Anisotropic Dielectric Layers and Infinitely Wide Ground Planes

4.5.1 Two-Layered Edge-Coupled Anisotropic Stripline with Infinitely Wide Ground Planes

In order to evolve generalized capacitance expressions applicable to edge-coupled structures with iso/anisotropic dielectric layers and with no side wall effects, we first consider the analysis of a simple double layer edge-coupled structure shown in Fig. 4.8. The ground planes are considered to be infinitely wide and the strip conductors are assumed to be of zero thickness. The two anisotropic regions, $j=1, 2$, have permittivities represented as

$$\hat{\epsilon}_j = \epsilon_0 \begin{bmatrix} \epsilon_{xxj} & \epsilon_{xyj} & 0 \\ \epsilon_{xyj} & \epsilon_{yyj} & 0 \\ 0 & 0 & \epsilon_{xxj} \end{bmatrix}, \quad j=1, 2 \tag{4.54}$$

The potential function $\phi_j(x, y)$ in the two regions satisfies the two-dimensional Laplace's equation given by

$$\nabla_t \cdot [\hat{\epsilon}_j \cdot \nabla_t \phi_j(x, y)] = 0 \tag{4.55}$$

We define $\hat{\phi}(\beta, y)$ as the Fourier transform of $\phi(x, y)$ with respect to x, where β is the continuous Fourier variable. Taking the Fourier transform of (4.55) and satisfying the boundary conditions at $y=0$, b_1, and b, we can solve for $\hat{\phi}_j(\beta, y)$. The derivation is not given here, since the steps involved are the same as those presented in Sec. 3.6.1 [from (3.61) to (3.81)] for a similar structure with side walls, when β_n is replaced by β. Thus the expression for $\hat{\phi}(\beta, y)$ at the charge plane $y=y_0$ can be obtained from (3.79) as

$$\hat{\phi}(\beta, y_0) = \frac{\hat{f}(\beta)}{\beta Y} \tag{4.56}$$

where

$$Y = \epsilon_0[\epsilon_{rf1} \coth(\gamma_1 b_1) + \epsilon_{rf2} \coth(\gamma_2(b-b_1))] \tag{4.57a}$$

$$\epsilon_{rf1} = \sqrt{\epsilon_{xx1}\epsilon_{yy1} - \epsilon_{xy1}^2} \tag{4.57b}$$

$$\epsilon_{rf2} = \sqrt{\epsilon_{xx2}\epsilon_{yy2} - \epsilon_{xy2}^2} \tag{4.57c}$$

$$\gamma_1 = \beta\sqrt{\epsilon_{xx1}\epsilon_{yy1} - \epsilon_{xy1}^2}/\epsilon_{yy1} \tag{4.57d}$$

$$\gamma_2 = \beta\sqrt{\epsilon_{xx2}\epsilon_{yy2} - \epsilon_{xy2}^2}/\epsilon_{yy2} \tag{4.57e}$$

Y is the admittance at the charge plane. The general variational expression for the capacitance of a single conductor transmission line (Sec. 2.7.2) is

$$\frac{1}{C} = \frac{1}{2\pi Q^2}\int_{-\infty}^{\infty} \hat{\phi}(\beta, y_0)\hat{f}(\beta)\, d\beta \tag{4.58}$$

For the edge-coupled structure, the even- and odd-mode capacitances are given by

$$\frac{1}{C_{e,o}} = \frac{1}{2\pi Q_{e,o}^2}\int_0^{\infty} \hat{\phi}_e(\beta, y_0)\hat{f}(\beta)_{e,o}\, d\beta \tag{4.59}$$

Using (4.56) in (4.59) and noting that the admittance parameter Y is the same for both even- and odd-modes, we can write

$$\frac{1}{C_{e,o}} = \frac{1}{2\pi}\int_0^{\infty}\left[\frac{\hat{f}(\beta)_{e,o}}{Q_{e,o}}\right]^2 \frac{1}{\beta Y}\, d\beta \tag{4.60}$$

where

$$\hat{f}(\beta)_e = 2\int_{s/2}^{(s/2)+w} f(x)_e \cos\beta x\, dx \tag{4.61a}$$

$$\hat{f}(\beta)_o = 2\int_{s/2}^{(s/2)+w} f(x)_o \sin\beta x\, dx \tag{4.61b}$$

$$Q_{e,o} = \int_{s/2}^{(s/2)+w} f(x)_{e,o}\, dx \tag{4.61c}$$

Here $f(x)_e$ and $f(x)_o$ are the charge distributions on the strip conductor S for the even- and odd-modes, respectively.

4.5.2 Generalization to Edge-Coupled Structures using Multilayer Iso/Anisotropic Substrates and Infinitely Wide Ground Planes

The variational expression given in (4.60) is general and can be applied to any multilayer edge-coupled transmission line with iso/anisotropic substrates by substituting the admittance parameter Y appropriate to the structure. A variety of edge-coupled stripline-like transmission lines with side walls and with substrates which are isotropic, uniaxially anisotropic with optical axis along the y-axis, and anisotropic with principal axes rotated with respect to the

x-y coordinate axes, have been listed in Tables 4.1, 4.2 and 4.3, respectively, along with their admittance parameters. The same admittance expressions, with β_n replaced by β, are valid for the corresponding structures with $c \to \infty$.

4.6 Broadside-Coupled Stripline and Stripline-like Transmission Lines with Side Walls and Anisotropic Dielectric Substrates—Unified Approach [15, 21, 22]

4.6.1 General Capacitance Formulas

Consider a multilayer, broadside-coupled symmetric stripline with side walls as shown in Fig. 4.9a. The dielectric layers may be isotropic or uniaxially anisotropic with permittivities defined as in (4.36) and (4.37), respectively. Using the symmetry about the plane qq' at $y = b/2$, the structure can be analysed using the even- and odd-mode method. For the even-mode, the plane qq' is replaced by a magnetic wall, and, for the odd-mode, it is replaced by an electric wall. The problem thus reduces to that of finding the capacitance of half the structure between $y=0$ and $b/2$ with appropriate boundary walls at $y=b/2$ as shown in Fig. 4.9b. Since Fig. 4.9b is a single conductor structure with electric walls at $x=0$ and c, the capacitance expression (2.128) applies. The expressions for the even- and odd-modes become

$$\frac{1}{C_{\substack{e\\o}}} = \frac{\displaystyle\sum_{n=1}^{\infty} \frac{2}{n\pi Y_{\substack{e\\o}}} \left[\int_S f(x)_{\substack{e\\o}} \sin(n\pi x/c)\, dx \right]^2}{\left[\displaystyle\int_S f(x)_{\substack{e\\o}}\, dx \right]^2} \tag{4.62}$$

where Y_e and Y_o are the even- and odd-mode admittances, respectively, at the charge plane $y = y_0$ of half the structure shown in Fig. 4.9b.

4.6.2 Closed Form Capacitance Expressions for Specified Even- and Odd-mode Charge Distributions

In symmetric broadside-coupled lines, the charge distribution on each strip conductor would be symmetrical with respect to the center of the strip. We therefore assume a charge distribution on the strip conductor S in Fig. 4.9b to be of the form

$$f(x)_{\substack{e\\o}} = 1/w \left[1 + A_{\substack{e\\o}} \,|\, (2/w)(x - c/2) \,|^3 \right], \quad \frac{c-w}{2} \leqslant x \leqslant \frac{c+w}{2}$$
$$= 0, \qquad\qquad \text{otherwise} \tag{4.63}$$

Evaluating (4.62) with (4.63) and maximizing the capacitances C_e and C_o with respect to the arbitrary constants A_e and A_o, respectively, we obtain

$$C_{\substack{e\\o}} = \frac{(1 + 0.25 A_{\substack{e\\o}})^2}{\displaystyle\sum_{n\ \text{odd}} (L_n + A_{\substack{e\\o}} M_n)^2 P_n / Y_{\substack{e\\o}}} \tag{4.64}$$

where

$$L_n = \sin(\beta_n w/2) \tag{4.65a}$$

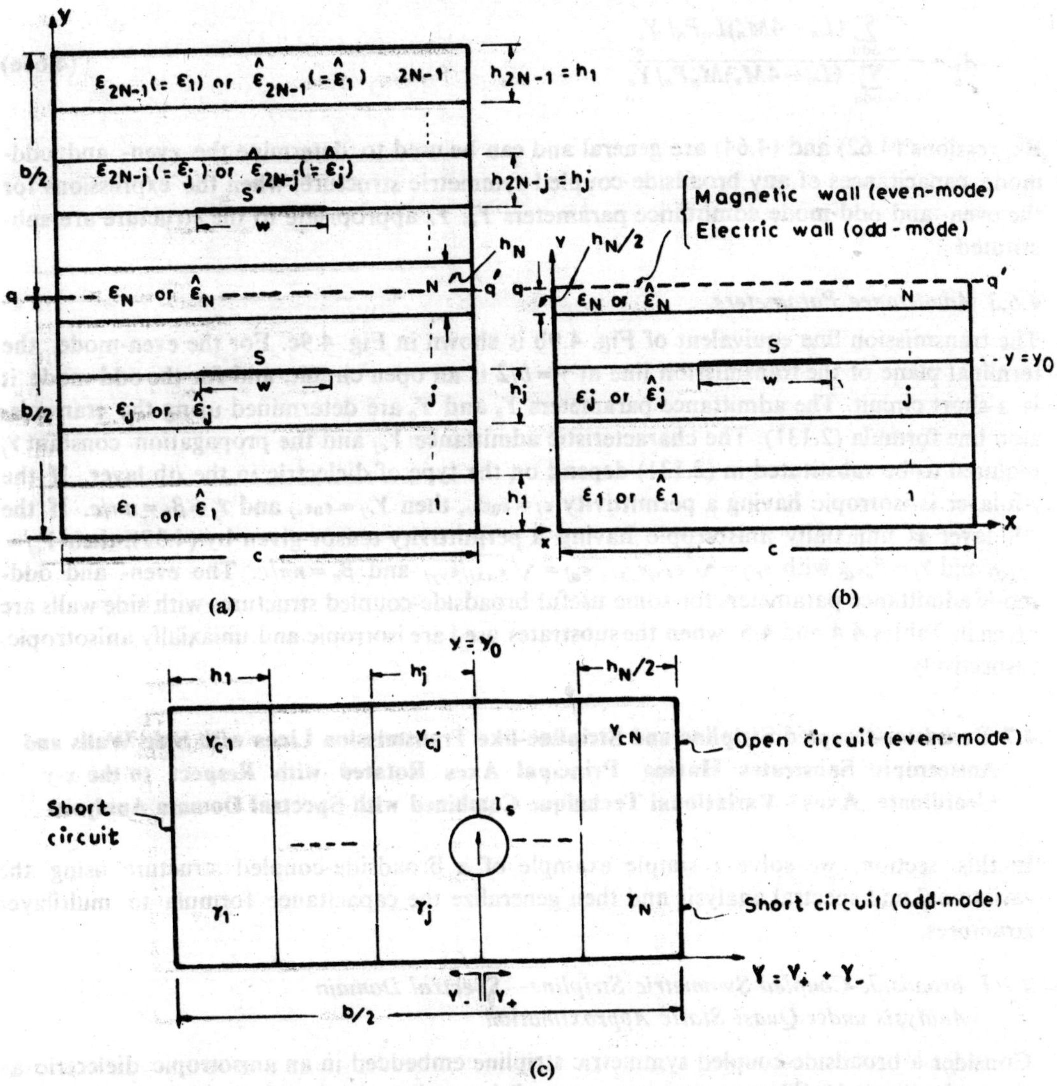

FIG. 4.9 (a) General broadside-coupled symmetric isotropic/uniaxially anisotropic structure, (b) half of the structure for even- and odd-mode analysis, (c) two-wire transmission line equivalent for determining the admittance parameter Y: For isotropic layer, $Y_{cj} = \epsilon_j = \epsilon_0\epsilon_{rj}$, $\gamma_j = \beta_n = n\pi/c$, $j = 1, \ldots, N$; For uniaxially anisotropic layer, $Y_{cj}=\epsilon_0\sqrt{\epsilon_{xxj}\epsilon_{yyj}}$, $\gamma_j=\beta_n\sqrt{\epsilon_{xxj}/\epsilon_{yyj}}$, $\beta_n=n\pi/c, j=1, \ldots, N$.

$$M_n=\left(\frac{2}{\beta_n w}\right)^3\left[\ 3\left\{\left(\frac{\beta_n w}{2}\right)^2-2\right\}\cos\left(\frac{\beta_n w}{2}\right)+\left(\frac{\beta_n w}{2}\right)\left\{\left(\frac{\beta_n w}{2}\right)^2-6\right\}\sin\left(\frac{\beta_n w}{2}\right)+6\right] \tag{4.65b}$$

$$P_n=(2/n\pi)(2/\beta_n w)^2 \tag{4.65c}$$

$$\beta_n=n\pi/c \tag{4.65d}$$

$$A_{\substack{e\\o}} = -\frac{\sum\limits_{n\text{ odd}} (L_n - 4M_n)L_nP_n/Y_{\substack{e\\o}}}{\sum\limits_{n\text{ odd}} (L_n - 4M_n)M_nP_n/Y_{\substack{e\\o}}} \tag{4.65e}$$

Expressions (4.62) and (4.64) are general and can be used to determine the even- and odd-mode capacitances of any broadside-coupled symmetric structure, when the expressions for the even- and odd-mode admittance parameters Y_e, Y_o appropriate to the structure are substituted.

4.6.3 Admittance Parameters

The transmission line equivalent of Fig. 4.9b is shown in Fig. 4.9c. For the even-mode, the terminal plane of the transmission line at $y = b/2$ is an open circuit, and for the odd-mode it is a short circuit. The admittance parameters Y_e and Y_o are determined using the transmission line formula (2.131). The characteristic admittance Y_{cj} and the propagation constant γ_j required to be substituted in (2.131) depend on the type of dielectric in the jth layer. If the jth layer is isotropic having a permittivity $\epsilon_j = \epsilon_0\epsilon_{rj}$, then $Y_{cj} = \epsilon_0\epsilon_{rj}$ and $\gamma_j = \beta_n = n\pi/c$. If the jth layer is uniaxially anisotropic having a permittivity tensor given by (4.37), then $Y_{cj} = \epsilon_0\epsilon_{rfj}$ and $\gamma_j = \beta_n\epsilon_{aj}$ with $\epsilon_{rfj} = \sqrt{\epsilon_{xxj}\epsilon_{yyj}}$, $\epsilon_{aj} = \sqrt{\epsilon_{xxj}/\epsilon_{yyj}}$ and $\beta_n = n\pi/c$. The even- and odd-mode admittance parameters for some useful broadside-coupled structures with side walls are given in Tables 4.4 and 4.5, when the substrates used are isotropic and uniaxially anisotropic, respectively.

4.7 Broadside-Coupled Stripline and Stripline-like Transmission Lines with Side Walls and Anisotropic Substrates Having Principal Axes Rotated with Respect to the *x-y* Coordinate Axes—Variational Technique Combined with Spectral Domain Analysis

In this section, we solve a simple example of a broadside-coupled structure using the variational and spectral analysis and then generalize the capacitance formula to multilayer structures.

4.7.1 Broadside-Coupled Symmetric Stripline—Spectral Domain Analysis under Quasi-Static Approximation

Consider a broadside-coupled symmetric stripline embedded in an anisotropic dielectric as shown in Fig. 4.10. The permittivity tensor of the medium is given by

$$\hat{\epsilon} = \epsilon_0 \begin{bmatrix} \epsilon_{xx} & \epsilon_{xy} & 0 \\ \epsilon_{xy} & \epsilon_{yy} & 0 \\ 0 & 0 & \epsilon_{xx} \end{bmatrix} \tag{4.66}$$

where

$$\epsilon_{xx} = \epsilon_\xi \cos^2\theta + \epsilon_\eta \sin^2\theta \tag{4.67a}$$

$$\epsilon_{xy} = (\epsilon_\eta - \epsilon_\xi)\sin\theta\cos\theta \tag{4.67b}$$

$$\epsilon_{yy} = \epsilon_\xi \sin^2\theta + \epsilon_\eta \cos^2\theta \tag{4.67c}$$

Table 4.4 Admittance parameters of broadside-coupled isotropic transmission structures with side walls

Structure	*Admittance Y at $y=y_0$*

A1. Broadside-coupled homogeneous stripline

$$\beta_n = n\pi/c, \qquad p = \begin{cases} 1 & \text{even-mode} \\ -1 & \text{odd-mode} \end{cases}$$

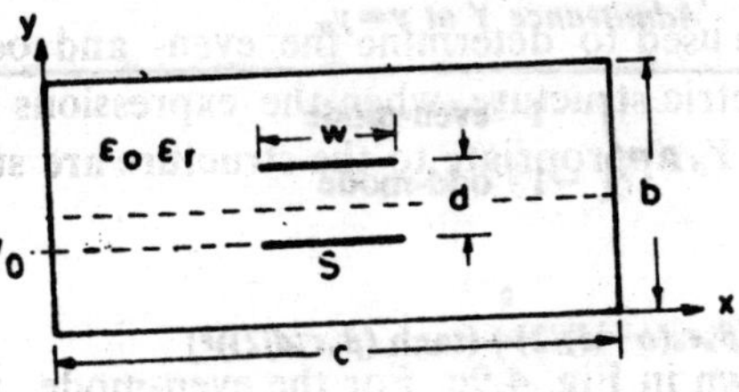

$$\epsilon_0\epsilon_r[\coth(\beta_n((b-d)/2)) + (\tanh(\beta_n d/2))^p]$$

A2. Broadside-coupled suspended stripline

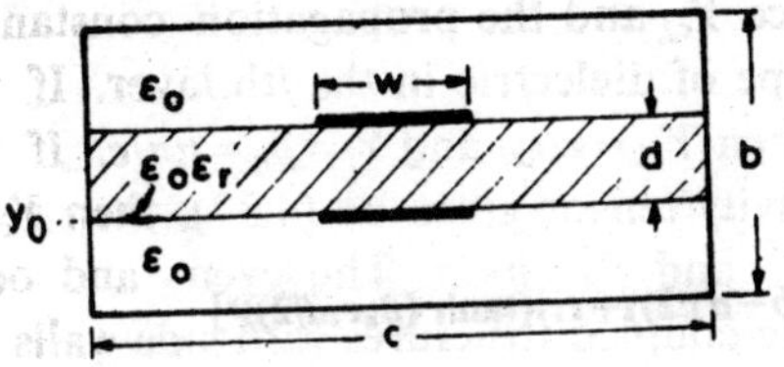

$$\epsilon_0[\coth(\beta_n((b-d)/2)) + \epsilon_r(\tanh(\beta_n d/2))^p]$$

A3. Broadside-coupled inverted stripline

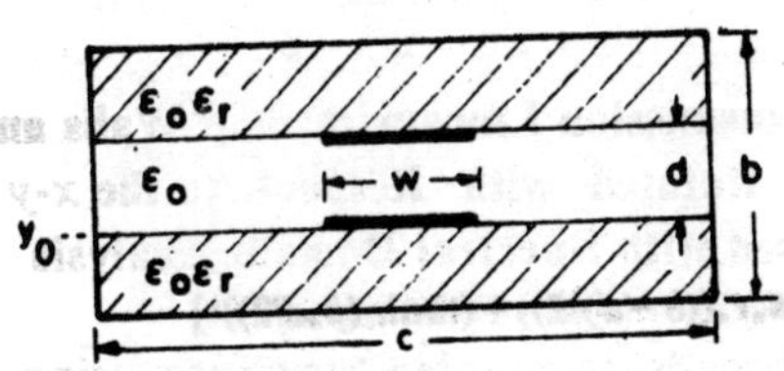

$$\epsilon_0[\epsilon_r \coth(\beta_n((b-d)/2)) + (\tanh(\beta_n d/2))^p]$$

A4. Broadside-coupled stripline with double suspended substrate

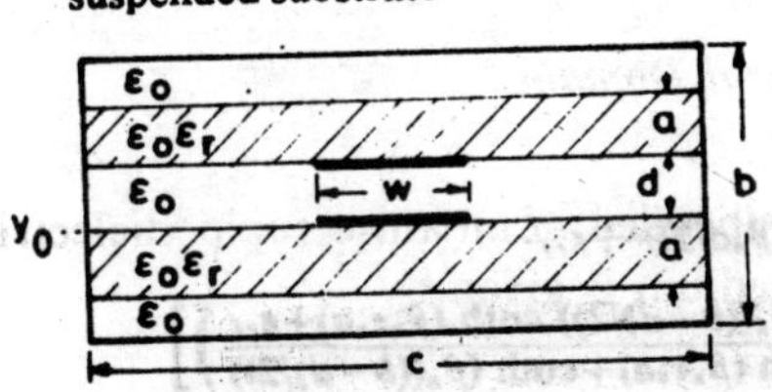

$$\epsilon_0\left[(\tanh(\beta_n d/2))^p + \epsilon_r\left(\frac{\coth(\beta_n((b-d)/2-a))\coth(\beta_n a)+\epsilon_r}{\epsilon_r\coth(\beta_n a)+\coth(\beta_n((b-d)/2-a))}\right)\right]$$

A5. Broadside-coupled stripline with double suspended inverted substrate

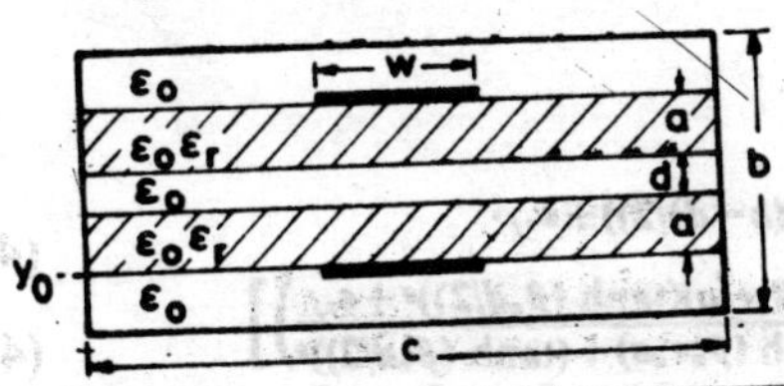

$$\epsilon_0\left[\coth(\beta_n((b-d)/2-a)) + \epsilon_r\left(\frac{\coth(\beta_n a)(\tanh(\beta_n d/2))^p+\epsilon_r}{\epsilon_r\coth(\beta_n a)+(\tanh(\beta_n d/2))^p}\right)\right]$$

Table 4.5 Admittance parameters of broadside-coupled transmission structures with side walls and uniaxially anisotropic substrates – optical axis aligned along the y-axis

$$\epsilon_{rf}=\sqrt{\epsilon_{xx}\epsilon_{yy}},\quad \epsilon_a=\sqrt{\epsilon_{xx}/\epsilon_{yy}}$$

Structure	*Admittance Y at $y=y_0$*
A1. Broadside-coupled homogeneous stripline 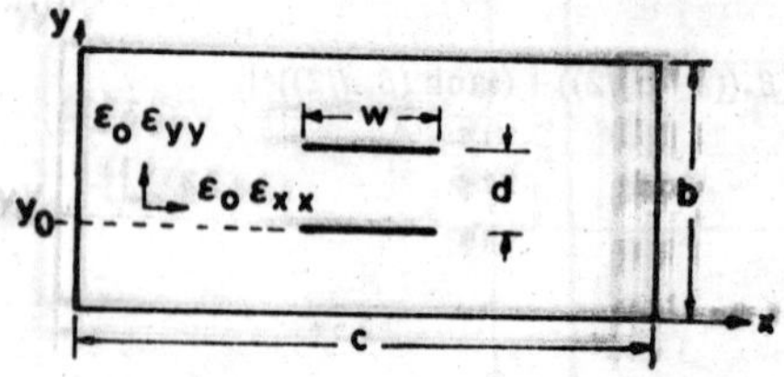	$\beta_n=n\pi/c,\quad p=\begin{cases}1 & \text{even-mode}\\ -1 & \text{odd-mode}\end{cases}$ $\epsilon_0\epsilon_{rf}[\coth(\beta_n\epsilon_a(b-d)/2)+(\tanh(\beta_n\epsilon_a d/2))^p]$
A2. Broadside-coupled suspended stripline 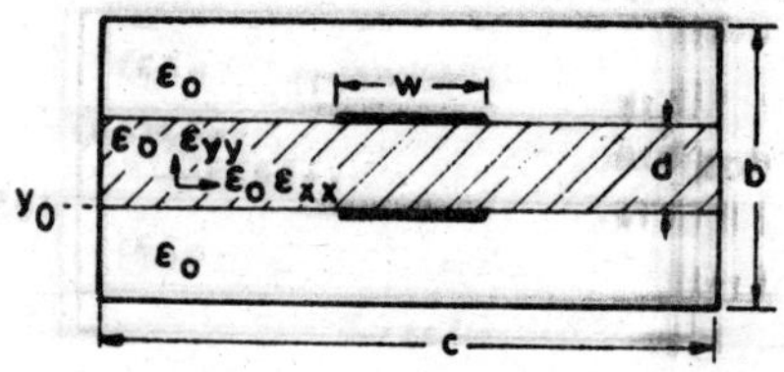	$\epsilon_0[\coth(\beta_n((b-d)/2))+\epsilon_{rf}(\tanh(\beta_n\epsilon_a d/2))^p]$
A3. Broadside-coupled inverted stripline 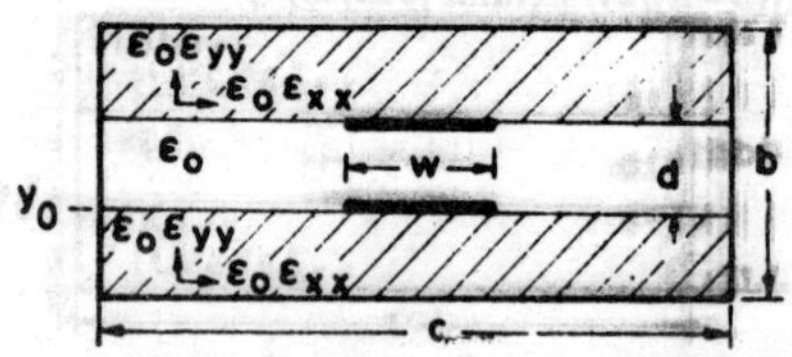	$\epsilon_0[\epsilon_{rf}\coth(\beta_n\epsilon_a((b-d)/2))+(\tanh(\beta_n d/2))^p]$
A4. Broadside-coupled stripline with double suspended substrate 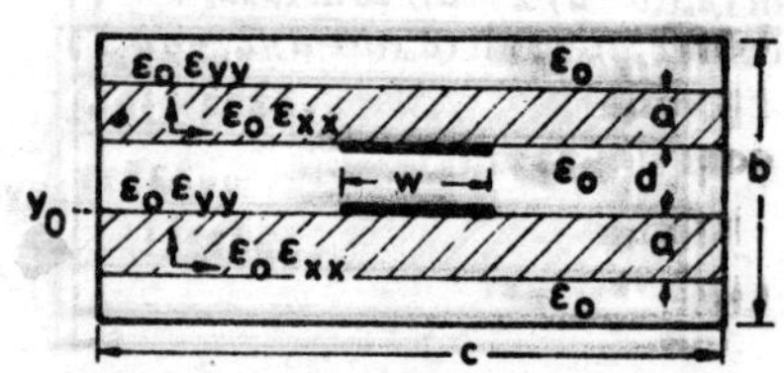	$\epsilon_0\left[(\tanh(\beta_n d/2))^p+\epsilon_{rf}\cdot\left(\dfrac{\coth(\beta_n((b-d)/2))\coth(\beta_n\epsilon_a a)+\epsilon_{rf}}{\epsilon_{rf}\coth(\beta_n\epsilon_a a)+\coth(\beta_n((b-d)/2))}\right)\right]$
A5. Broadside-coupled stripline with double suspended inverted substrate 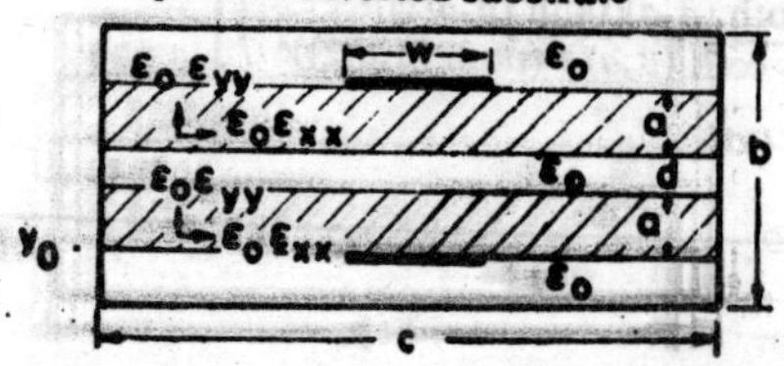	$\epsilon_0\left[\coth(\beta_n((b-d)/2))+\epsilon_{rf}\cdot\left(\dfrac{\coth(\beta_n\epsilon_a a)(\tanh(\beta_n d/2))^p+\epsilon_{rf}}{\epsilon_{rf}\coth(\beta_n\epsilon_a a)+(\tanh(\beta_n d/2))^p}\right)\right]$

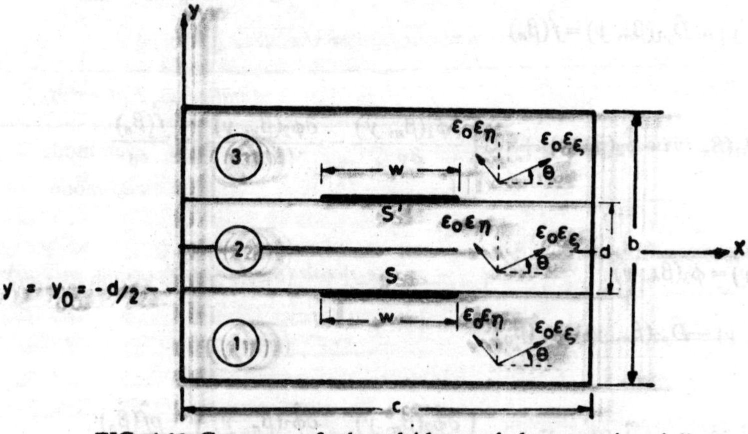

FIG. 4.10 Geometry of a broadside-coupled symmetric stripline with anisotropic substrates.

The two strip conductors are assumed to be of zero thickness and spaced (by a distance d) symmetrically with respect to the top and bottom walls of the structure. In order to derive the quasi-static capacitance of the structure, we first solve for the potential distribution in the spectral domain in terms of the charge distribution on the strip conductors for the even- and odd-mode excitations. Denoting $\hat{\phi}(\beta_n, y)$ as the Fourier transform of the potential function $\phi(x, y)$, Laplace's equation in the Fourier domain takes the form

$$\epsilon_{yy}\frac{\partial^2\hat{\phi}_j(\beta_n, y)}{\partial y^2}+2j\beta_n\epsilon_{xy}\frac{\partial\hat{\phi}_j(\beta_n, y)}{\partial y}-\beta_n^2\epsilon_{xx}\hat{\phi}_j(\beta_n, y)=0 \tag{4.68}$$

where the subscript j in $\hat{\phi}_j(\beta_n, y)$ refers to the three regions, $j=1$, 2, and 3, marked in Fig. 4.10. Assuming a y variation of the form $e^{-\zeta y}$ and solving (4.68), we obtain

$$\hat{\phi}_j(\beta_n, y)=A_j\exp(-\zeta^+y)+B_j\exp(-\zeta^-y) \tag{4.69}$$

where

$$\zeta^+=\gamma+j\alpha \tag{4.70a}$$

$$\zeta^-=-\gamma+j\alpha \tag{4.70b}$$

$$\alpha=\beta_n\epsilon_{xx}/\epsilon_{yy} \tag{4.70c}$$

$$\gamma=\beta_n\epsilon_{rf}/\epsilon_{yy} \tag{4.70d}$$

$$\epsilon_{rf}=\sqrt{\epsilon_{xx}\epsilon_{yy}-\epsilon_{xy}^2} \tag{4.70e}$$

A_j and B_j for $j=1$, 2, and 3 are arbitrary constants which are to be determined by applying the following boundary conditions:

At $y=-b/2$,

$$\hat{\phi}_1(\beta_n, y)=0 \tag{4.71a}$$

At $y=-d/2$,

$$\hat{\phi}_1(\beta_n, y)=\hat{\phi}_2(\beta_n, y) \tag{4.71b}$$

$$\hat{D}_{y2}(\beta_n, y) - \hat{D}_{y1}(\beta_n, y) = \hat{f}(\beta_n) \tag{4.71c}$$

or

$$j\beta_n\epsilon_{xy}[\hat{\phi}_1(\beta_n, y) - \hat{\phi}_2(\beta_n, y)] + \epsilon_{yy}\left[\frac{\partial\hat{\phi}_1(\beta_n, y)}{\partial y} - \frac{\partial\hat{\phi}_2(\beta_n, y)}{\partial y}\right] = \frac{\hat{f}(\beta_n)}{\epsilon_0} \tag{4.71d}$$

At $y = d/2$,

$$\hat{\phi}_2(\beta_n, y) = \hat{\phi}_3(\beta_n, y) \tag{4.71e}$$

$$\hat{D}_{y3}(\beta_n, y) - \hat{D}_{y2}(\beta_n, y) = p\hat{f}(\beta_n) \tag{4.71f}$$

or

$$j\beta_n\epsilon_{xy}[\hat{\phi}_2(\beta_n, y) - \hat{\phi}_3(\beta_n, y)] + \epsilon_{yy}\left[\frac{\partial\hat{\phi}_2(\beta_n, y)}{\partial y} - \frac{\partial\hat{\phi}_3(\beta_n, y)}{\partial y}\right] = \frac{p\hat{f}(\beta_n)}{\epsilon_0} \tag{4.71g}$$

At $y = b/2$,

$$\hat{\phi}_3(\beta_n, y) = 0 \tag{4.71h}$$

In (4.71c), (4.71d), (4.71f), and (4.71g), $\hat{f}(\beta_n) = \hat{f}(\beta_n)_e$ and $p = 1$ for the even- mode and $\hat{f}(\beta_n) = \hat{f}(\beta_n)_0$ and $p = -1$ for the odd-mode.

Here, $\hat{f}(\beta_n)_e$ and $\hat{f}(\beta_n)_o$ are the Fourier transforms of the even- and odd-mode charge distributions $f(x)_e$ and $f(x)_o$, respectively, and are defined as

$$\hat{f}(\beta_n)_{\substack{e\\o}} = \int_0^c f(x)_{\substack{e\\o}} \sin \beta_n x \, dx \tag{4.72a}$$

where

$$\beta_n = n\pi/c \tag{4.72b}$$

With the use of the boundary conditions (4.71), the potential distribution $\hat{\phi}_1(\beta_n, y)$ for the even- and odd-modes in region 1 is obtained as

$$\hat{\phi}_1(\beta_n, y)_{\substack{e\\o}} = -\left[\frac{\hat{f}(\beta_n)_{\substack{e\\o}}}{\epsilon_0\beta_n\epsilon_{rf}\sinh\gamma b}\right]\left[e^{j\alpha d/2}\sinh\left\{\gamma\left(\frac{b-d}{2}\right)\right\} + pe^{-j\alpha d/2}\sinh\{\gamma((b+d)/2)\}\right]$$
$$\times e^{-j\alpha y}\sinh\{\gamma((b/2)+y)\} \tag{4.73}$$

At $y = -d/2$, corresponding to the location of the lower strip conductor, the potential is

$$\hat{\phi}_1\left(\beta_n, -\frac{d}{2}\right)_{\substack{e\\o}} = -\left[\frac{\hat{f}(\beta_n)_{\substack{e\\o}}}{\epsilon_0\beta_n\epsilon_{rf}\sinh\gamma b}\right]\left[e^{j\alpha d}\sinh\left\{\gamma\left(\frac{b-d}{2}\right)\right\}\right.$$
$$\left. + p\sinh\left\{\gamma\left(\frac{b+d}{2}\right)\right\}\right]\sinh\left\{\gamma\left(\frac{b-d}{2}\right)\right\} \tag{4.74}$$

Denoting the magnitude of the potential $\hat{\phi}_1(\beta_n, -d/2)_{\substack{e\\o}}$ as $\hat{\phi}(\beta_n, -d/2)_{\substack{e\\o}}$, we have

$$\hat{\phi}(\beta_n, -d/2)_{\substack{e\\o}} = \frac{\hat{f}(\beta_n)_{\substack{e\\o}}}{\beta_n \epsilon_0 \epsilon_{rf}\left[\coth\left\{\gamma\left(\frac{b-d}{2}\right)\right\} + \left\{\tanh\left(\frac{\gamma d}{2}\right)\right\}^p\right]}, \quad p = \begin{cases} 1 & \text{even-mode} \\ -1 & \text{odd-mode} \end{cases} \tag{4.75}$$

We note that the factor multiplying β_n in the denominator of (4.75) is nothing but the admittance at the charge plane $p = -d/2$ of an equivalent isotropic structure having a relative dielectric constant ϵ_{rf} and propagation constant γ, with a magnetic wall at $y=0$ for the even-mode and an electric wall at $y=0$ for the odd-mode. Denoting these admittance parameters for even- and odd-modes as Y_e and Y_o respectively, the even- and odd-mode potentials at $y=-d/2$ are expressed as

$$\hat{\phi}(\beta_n, -d/2)_{\substack{e\\o}} = \hat{G}(\beta_n, -d/2)_{\substack{e\\o}} \hat{f}(\beta_n)_{\substack{e\\o}} \tag{4.76}$$

where

$$\hat{G}(\beta_n, -d/2)_{\substack{e\\o}} = \frac{1}{\beta_n Y_{\substack{e\\o}}} \tag{4.77}$$

$$Y_{\substack{e\\o}} = \epsilon_0 \epsilon_{rf}\left[\coth\left\{\gamma\left(\frac{b-d}{2}\right)\right\} + \left\{\tanh\left(\frac{\gamma d}{2}\right)\right\}^p\right], \quad p = \begin{cases} 1 & \text{even-mode} \\ -1 & \text{odd-mode} \end{cases} \tag{4.78}$$

The potential $\hat{\phi}(\beta_n, -d/2)$ can be expressed as the sum of two potentials, namely, the potential on the strip conductor $\hat{\phi}_c(\beta_n, -d/2)$, and the potential on the complementary dielectric interface $\hat{\phi}_d(\beta_n, -d/2)$. For the even- and odd-mode excitations, this is expressed as

$$\hat{\phi}(\beta_n, -d/2)_{\substack{e\\o}} = \hat{\phi}_{c\substack{e\\o}}(\beta_n, -d/2) + \hat{\phi}_{d\substack{e\\o}}(\beta_n, -d/2) \tag{4.79}$$

using (4.79) in (4.76), we can write

$$\hat{\phi}_{c\substack{e\\o}}(\beta_n, -d/2) + \hat{\phi}_{d\substack{e\\o}}(\beta_n, -d/2) = \hat{G}(\beta_n, -d/2)_{\substack{e\\o}} \hat{f}(\beta_n)_{\substack{e\\o}} \tag{4.80}$$

$\hat{\phi}_{d\substack{e\\o}}(\beta_n, -d/2)$ can now be eliminated through the use of Parseval's theorem. For this, we define an inner product as

$$\langle \psi_1(\beta_n), \psi_2(\beta_n) \rangle = \sum_{n=1}^{\infty} \psi_1(\beta_n)\psi_2(\beta_n) \tag{4.81}$$

Taking the inner product of (4.80) with $\hat{f}(\beta_n)_{\substack{e\\o}}$, we obtain

$$\langle \hat{\phi}_{c\substack{e\\o}}(\beta_n, -d/2), \hat{f}(\beta_n)_{\substack{e\\o}} \rangle = \langle \hat{G}(\beta_n, -d/2)_{\substack{e\\o}} \hat{f}(\beta_n)_{\substack{e\\o}}, \hat{f}(\beta_n)_{\substack{e\\o}} \rangle \tag{4.82}$$

The inner product of $\hat{\phi}_{d\substack{e\\o}}(\beta_n, -d/2)$ with $\hat{f}(\beta_n)_{\substack{e\\o}}$ vanishes, since they exist on complementary regions. By virtue of Parseval's theorem, we have

$$\sum_{n=1}^{\infty} \psi_1(\beta_n)\psi_2(\beta_n) = (c/2)\int_0^c \psi_1(x)\psi_2(x)\,dx \tag{4.83}$$

Using the relations (4.83) and (4.81) in (4.82) and setting $\phi_{ce}(x, -d/2) = V_e$ and $\phi_{co}(x, -d/2) = V_o$ for the even- and odd-modes, respectively, we obtain

$$(c/2)V_{\substack{e\\o}}\int_S f(x)_{\substack{e\\c}}\,dx = \langle \hat{G}(\beta_n, -d/2)_{\substack{e\\o}}\hat{f}(\beta_n)_{\substack{e\\o}}, \hat{f}(\beta_n)_{\substack{e\\o}} \rangle \tag{4.84}$$

Denoting Q_e as the total charge on the strip conductor S and C_e as its capacitance for even-mode excitation, and Q_o and C_o as the corresponding quantities for the odd-mode excitation, we can write

$$Q_{\substack{e\\o}} = C_{\substack{e\\o}}V_{\substack{e\\o}} = \int_S f(x)_{\substack{e\\o}}\,dx \tag{4.85}$$

Using (4.85) in (4.84) and with (4.81), we obtain the following expression for the even- and odd-mode capacitances:

$$\frac{1}{C_{\substack{e\\o}}} = \frac{(2/c)\sum_{n=1}^{\infty}\hat{G}(\beta_n, -d/2)_{\substack{e\\o}}\hat{f}(\beta_n)_{\substack{e\\o}}\hat{f}(\beta_n)_{\substack{e\\o}}}{\left[\int_S f(x)_{\substack{e\\o}}\,dx\right]^2} \tag{4.86}$$

Substituting for $\hat{f}(\beta_n)_{\substack{e\\o}}$ and $\hat{G}(\beta_n, -d/2)_{\substack{e\\o}}$ from (4.72) and (4.77), respectively, in (4.86), and noting that the integral of $f(x)_{\substack{e\\o}}$ from 0 to c is non-zero only on the strip conductor S, we obtain

$$\frac{1}{C_{\substack{e\\o}}} = \frac{\sum_{n=1}^{\infty}\frac{2}{n\pi Y_{\substack{e\\o}}}\left[\int_S f(x)_{\substack{e\\o}}\sin(n\pi x/c)\,dx\right]^2}{\left[\int_S f(x)_{\substack{e\\o}}\,dx\right]^2} \tag{4.87}$$

4.7.2 Generalization to a Class of Broadside-Coupled Symmetric Transmission Structures

The capacitance expression (4.87) obtained through the variational and spectral analysis of the structure shown in Fig. 4.10 is precisely the same as the earlier expression (4.62), when the admittance parameters corresponding to the structure under consideration are substituted for Y_e and Y_o. With the charge distribution specified as in (4.63), expression (4.64) combined with (4.65) offers a general closed form capacitance formula valid also for structures with anisotropic dielectric substrates in which the principal axes are rotated with respect to the x-y coordinate axes. For a general broadside-coupled symmetric structure having $(2N-1)$

layers as shown in Fig. 4.9a but with $\hat{\epsilon}_j$ for the jth substrate defined as in (4.47), the two-wire transmission line equivalent shown in Fig. 4.9c can be used to determine Y_e and Y_o by substituting for Y_{cj} and γ_j as given below.

$$Y_{cj} = \epsilon_0 \epsilon_{rfj} \tag{4.88a}$$

$$\gamma_j = \beta_n \epsilon_{aj} \tag{4.88b}$$

where

$$\epsilon_{rfj} = \sqrt{\epsilon_{xxj}\epsilon_{yyj} - \epsilon_{xyj}^2} = \sqrt{\epsilon_{\xi j}\epsilon_{\eta j}} \tag{4.88c}$$

$$\epsilon_{aj} = \epsilon_{rfj}/\epsilon_{yyj} = \sqrt{\epsilon_{\eta j}/\epsilon_{\xi j}}\,[(\epsilon_{\eta j}/\epsilon_{\xi j} - 1)\cos^2\theta_j + 1]^{-1} \tag{4.88d}$$

$$\beta_n = n\pi/c \tag{4.88e}$$

The expressions for the even- and odd-mode admittance parameters for a class of symmetrical broadside-coupled structures with anisotropic substrates in which the principal axes are rotated with respect to the x-y coordinate axes are listed in Table 4.6.

4.8 Broadside-Coupled Symmetric Stripline and Stripline-like Transmission Lines with Iso/Anisotropic Dielectric Layers and Infinitely Wide Ground Planes

In order to evolve general capacitance expressions for multilayer structures, we first consider a simple example of broadside-coupled symmetric structure shown in Fig. 4.11. The ground planes are of infinite extent and the strip conductors are assumed to be of zero thickness. With the permittivity tensor $\hat{\epsilon}_j$ of the medium defined as in (4.66) the potential function $\phi_j(x, y)$ in the three regions ($j = 1$, 2, and 3) satisfies Laplace's equation given by

$$\nabla_t \cdot [\hat{\epsilon}_j \cdot \nabla_t \phi_j(x, y)] = 0 \tag{4.89}$$

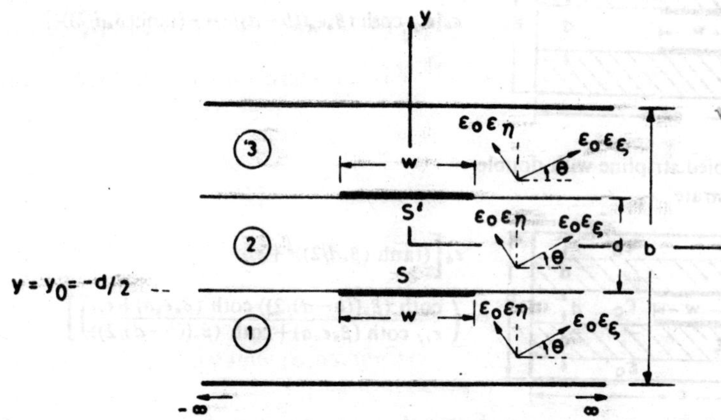

FIG. 4.11 Geometry of a broadside-coupled anisotropic symmetric stripline with infinitely wide ground planes.

We define $\hat{\phi}(\beta, y)$ as the Fourier transform of $\phi(x, y)$, where β is the continuous Fourier variable. The derivation of the potential function $\hat{\phi}(\beta, -d/2)$ at the charge plane $y = -d/2$ involves exactly the same steps as given in Sec. 4.7.1 from (4.68) to (4.75) except that, in (4.72), the integral limits 0 and c are replaced by $-w/2$ and $w/2$, respectively and β_n gets replaced

Table 4.6 Admittance parameters of broadside-coupled transmission structures with side walls and anisotropic substrates-Principal axes rotated with respect to the *x-y* coordinate axes

$$\epsilon_{rf}=\sqrt{\epsilon_\eta\epsilon_\xi},\quad \epsilon_a=\sqrt{\epsilon_\eta/\epsilon_\xi}[(\epsilon_\eta/\epsilon_\xi-1)\cos^2\theta+1]^{-1}$$

Structure	*Admittance Y at $y=y_0$*
• **Broadside-coupled homogeneous stripline** 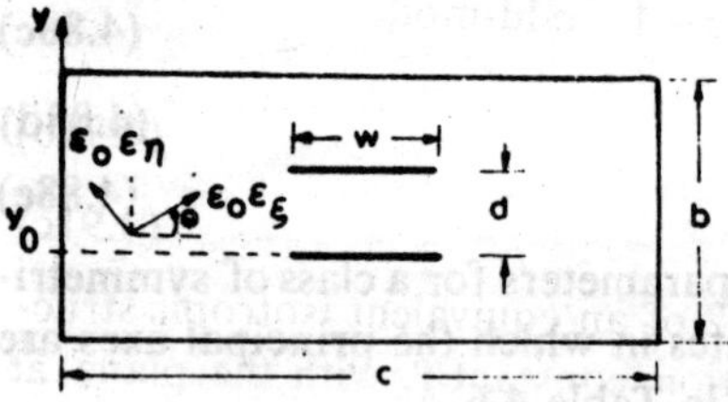	$\beta_n=n\pi/c$, $p=\begin{cases}1 & \text{even-mode}\\ -1 & \text{odd-mode}\end{cases}$ $\epsilon_0\epsilon_{rf}[\coth(\beta_n\epsilon_a((b-d)/2))+(\tanh(\beta_n\epsilon_a d/2))^p]$
• **Broadside-coupled suspended stripline** 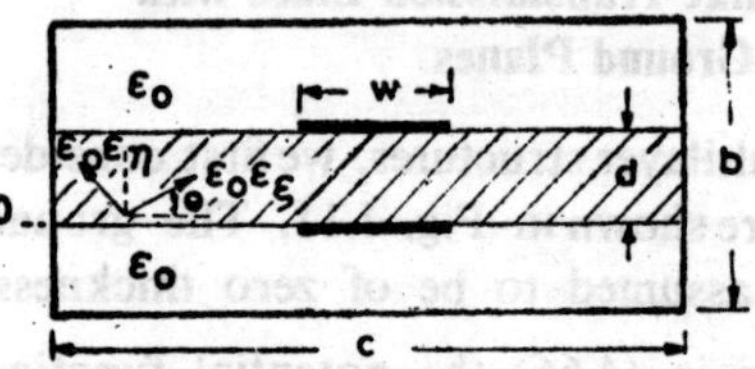	$\epsilon_0[\coth(\beta_n((b-d)/2))+\epsilon_{rf}(\tanh(\beta_n\epsilon_a d/2))^p]$
• **Broadside-coupled inverted stripline** 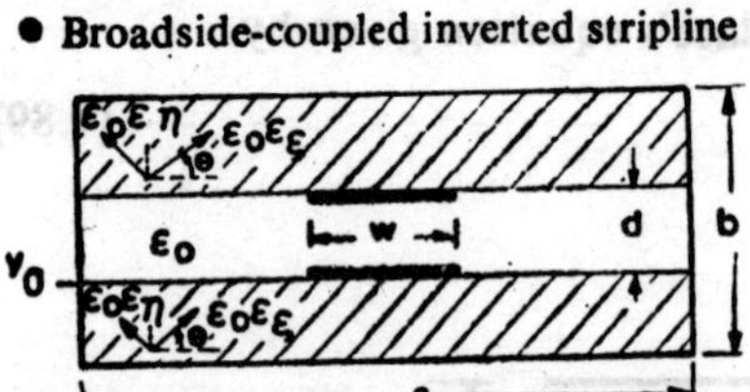	$\epsilon_0[\epsilon_{rf}\coth(\beta_n\epsilon_a((b-d)/2))+(\tanh(\beta_n d/2))^p]$
• **Broadside-coupled stripline with double suspended substrate** 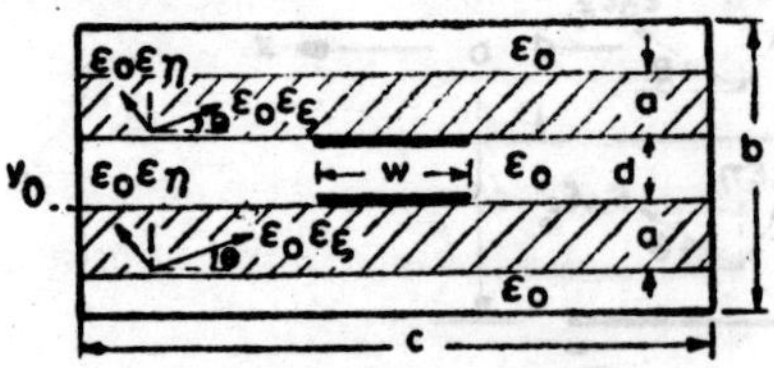	$\epsilon_0\left[(\tanh(\beta_n d/2))^p+\epsilon_{rf}\cdot\left(\frac{\coth(\beta_n((b-d)/2))\coth(\beta_n\epsilon_a a)+\epsilon_{rf}}{\epsilon_{rf}\coth(\beta_n\epsilon_a a)+\coth(\beta_n((b-d)/2))}\right)\right]$
• **Broadside-coupled stripline with double suspended inverted substrate** 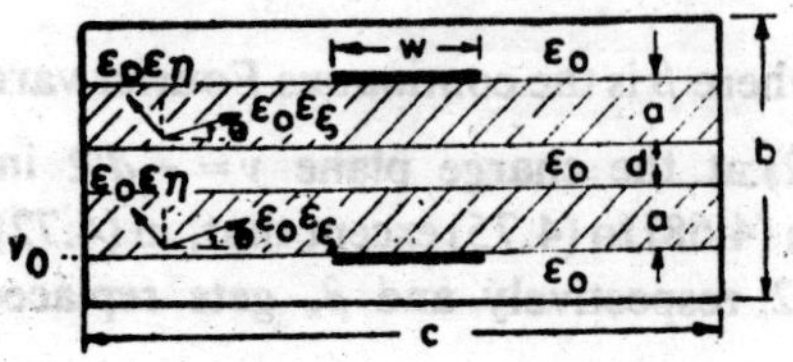	$\epsilon_0\left[\coth(\beta_n((b-d)/2))+\epsilon_{rf}\cdot\left(\frac{\coth(\beta_n\epsilon_a a)(\tanh(\beta_n d/2))^p+\epsilon_{rf}}{\epsilon_{rf}\coth(\beta_n\epsilon_a a)+(\tanh(\beta_n d/2))^p}\right)\right]$

by β throughout. The expression at the charge plane $y = y_0 = -d/2$ is obtained from (4.75) as

$$\hat{\phi}(\beta, -d/2)_{\substack{e\\o}} = \frac{\hat{f}(\beta)_{\substack{e\\o}}}{\beta Y_{\substack{e\\o}}} \tag{4.90}$$

where

$$Y_{\substack{e\\o}} = \epsilon_0 \epsilon_{rf}\left[\coth\left\{\gamma\left(\frac{b-d}{2}\right)\right\} + \left\{\tanh\left(\frac{\gamma d}{2}\right)\right\}^p\right], \quad p = \begin{cases} 1 & \text{even-mode} \\ -1 & \text{odd-mode} \end{cases} \tag{4.91a}$$

$$\epsilon_{rf} = \sqrt{\epsilon_{xx}\epsilon_{yy} - \epsilon_{xy}^2} \tag{4.91b}$$

$$\gamma = \beta\epsilon_{rf}/\epsilon_{yy} \tag{4.91c}$$

Y_e and Y_o are the admittances at the charge plane $y = -d/2$ of an equivalent isotropic structure having a relative dielectric constant ϵ_{rf} and propagation constant γ, with the plane at $y = 0$ replaced by a magnetic wall and an electric wall, respectively. The even- and odd-mode capacitances of the structure shown in Fig. 4.11 are then given by

$$\frac{1}{C_{\substack{e\\o}}} = \frac{1}{2\pi Q_{\substack{e\\o}}^2}\int_{-\infty}^{\infty} \hat{\phi}(\beta, -d/2)_{\substack{e\\o}} \hat{f}(\beta)_{\substack{e\\o}}\, d\beta \tag{4.92}$$

Using (4.90) in (4.92), we obtain

$$\frac{1}{C_{\substack{e\\o}}} = \frac{1}{\pi Q_{\substack{e\\o}}^2}\int_0^{\infty} \frac{\left[\hat{f}(\beta)_{\substack{e\\o}}\right]^2}{\beta Y_{\substack{e\\o}}}\, d\beta \tag{4.93}$$

where

$$Q_{\substack{e\\o}} = \int_S f(x)_{\substack{e\\o}}\, dx \tag{4.94}$$

is the total charge on the strip conductor S for the two excitations. The even- and odd-mode admittance parameters Y_e and Y_o in (4.93) can be obtained from the two-wire transmission line equivalent of the corresponding structure with side walls by replacing β_n by β. We note that (4.93) can be considered as a general capacitance expression applicable to any multilayer broadside-coupled symmetric structure with no side wall effects. Tables 4.4, 4.5, and 4.6 can be directly used for structures with $c \to \infty$, when β_n is replaced by β in the admittance expressions.

4.9 Broadside-Edge Coupled Symmetric Stripline and Stripline-like Transmission Lines with Side Walls and Iso/Anisotropic Dielectric Substrates

4.9.1 Capacitance Formulas for Structures with Isotropic/Uniaxially Anisotropic Substrates—Unified Approach [15, 23]

Consider the example of a broadside-edge coupled symmetric structure having side walls and three dielectric layers as shown in Fig. 4.12a. Using the symmetry with respect to the planes pp' and qq', the structure is analysed in terms of the following four basic modes:

(1) even-even mode (*ee*): pp' magnetic wall, qq' magnetic wall,
(2) even-odd mode (*eo*): pp' magnetic wall, qq' electric wall,
(3) odd-even mode (*oe*): pp' electric wall, qq' magnetic wall, and
(4) odd-odd mode (*oo*): pp' electric wall, qq' electric wall.

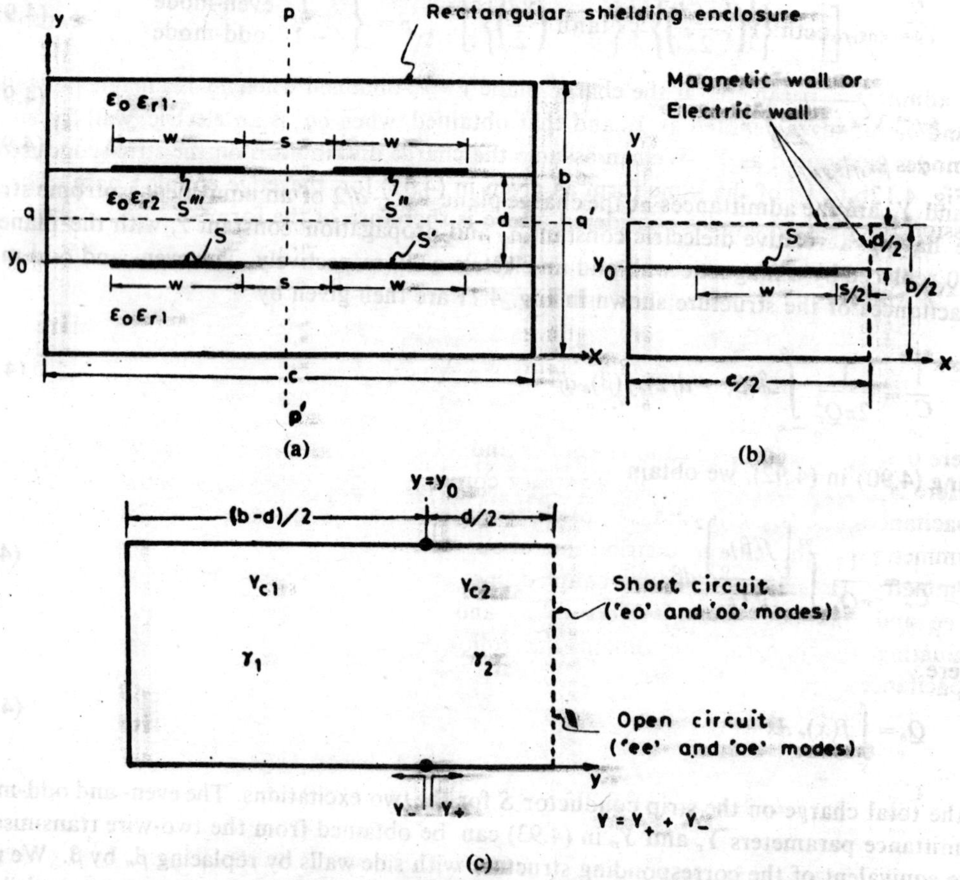

FIG. 4.12 (a) Three-layered broadside-edge coupled symmetric stripline, (b) one quarter of the structure for the purpose of analysis, (c) two-wire transmission line equivalent, $Y_{cj}=\epsilon_0\epsilon_{rj}$, $\gamma_j=\beta_n=n\pi/c$, $j=1, 2$; $Y=\epsilon_0[\epsilon_{r1}\coth(\beta_n((b-d)/2))+\epsilon_{r2}(\tanh(\beta_n d/2))^p]$, $p=1$ for *ee* and *oe* modes and -1 for *eo* and *oo* modes.

It suffices to analyse one quarter of the structure lying between $0\leqslant x\leqslant c/2$ and $0\leqslant y\leqslant b/2$ with appropriate boundary conditions corresponding to the four modes. For *ee* and *eo* modes (cases 1 and 2), the boundary at $x=0$ is an electric wall and that at $x=c/2$ is a magnetic wall; whereas for *oe* and *oo* modes (cases 3 and 4), both the boundaries at $x=0$ and $c/2$ are electric walls. Cases 1 and 2, therefore, correspond to the even-mode excitation and cases 3 and 4 correspond to the odd-mode excitation of an edge-coupled structure. Thus, the capacitance formula (4.40) derived in Sec. 4.3.1 for edge-coupled structures applies. Considering the admittance parameters corresponding to the boundaries at $y=0$ and $b/2$, the capacitance expression for the four modes can be expressed in the form

$$\frac{1}{C_{\substack{ee\\eo\\oe\\oo}}}=\frac{\displaystyle\sum_{\substack{n\text{ odd}\\ \text{odd}\\ \text{even}\\ \text{even}}}\frac{4}{n\pi Y_{\substack{e\\o\\e\\o}}}\left[\int_S f(x)_{\substack{ee\\eo\\oe\\oo}}\sin\left(\frac{n\pi x}{c}\right)dx\right]^2}{\left[\displaystyle\int_S f(x)_{\substack{ee\\eo\\oe\\oo}}\,dx\right]^2} \tag{4.95}$$

The admittance parameter at the charge plane $y=y_o$ obtained when qq' is a magnetic wall for *ee* and *oe* modes is denoted as Y_e and that obtained when qq' is an electric wall for *eo* and *oo* modes is denoted as Y_o. We can assume the charge distribution on the strip conductor S in Fig. 4.12b to be of the same form as given in (4.41) for the edge-coupled case. The expression for the four modes considered above is therefore of the form

$$f(x)_{\substack{ee\\eo\\oe\\oo}}=\left(\frac{1}{w}\right)\left[1+A_{\substack{ee\\eo\\oe\\oo}}\left|\left(\frac{2}{w}\right)\left(x-\frac{c-s-w}{2}-g_{\substack{e\\e\\o\\o}}\right)\right|^3\right],\qquad \frac{c-s}{2}-w\leqslant x\leqslant\frac{c-s}{2}$$

$$=0,\qquad \text{otherwise} \tag{4.96}$$

where $0\leqslant g_e\leqslant w/2$, for *ee* and *eo* modes and $-w/2\leqslant g_o\leqslant 0$, for *oe* and *oo* modes.

Here A_{ee}, A_{eo}, A_{oe}, and A_{oo} are arbitrary constants and are determined by maximizing the capacitances C_{ee}, C_{eo}, C_{oe}, and C_{oo}, respectively. The parameters g_e and g_o account for the asymmetry in the charge distribution about the center of the strip conductor. Since this asymmetry arises due to the edge-coupled lines and not due to broadside coupling, g_e refers to *ee* and *eo* modes and g_o refers to *oe* and *oo* modes. Substituting (4.96) in (4.95), and evaluating the integral, we obtain the following closed form expression for the four capacitances:

$$C_{\substack{ee\\eo\\oe\\oo}}=\frac{\left(1+K_{\substack{e\\e\\o\\o}}A_{\substack{ee\\eo\\oe\\oo}}\right)^2}{\displaystyle\sum_{\substack{n\text{ odd}\\ \text{odd}\\ \text{even}\\ \text{even}}}\left(L_n+M_{n\substack{e\\e\\o\\o}}A_{\substack{ee\\eo\\oe\\oo}}\right)^2 P_n/Y_{\substack{e\\o\\e\\o}}} \tag{4.97}$$

where

$$A_{\substack{ee\\eo\\oe\\oo}}=-\frac{\displaystyle\sum_{\substack{n\text{ odd}\\ \text{odd}\\ \text{even}\\ \text{even}}}\left(K_{\substack{e\\e\\o\\o}}L_n-M_{n\substack{e\\e\\o\\o}}\right)L_nP_n/Y_{\substack{e\\o\\e\\o}}}{\displaystyle\sum_{\substack{n\text{ odd}\\ \text{odd}\\ \text{even}\\ \text{even}}}\left(K_{\substack{e\\e\\o\\o}}L_n-M_{n\substack{e\\e\\o\\o}}\right)M_{n\substack{e\\e\\o\\o}}P_n/Y_{\substack{e\\o\\e\\o}}} \tag{4.98}$$

Here, the parameters K_e and M_{ne} refer to *ee* and *eo* modes and the parameters K_o and M_{no} refer to *oe* and *oo* modes. The expressions for $K_{\substack{e\\o}}$, L_n, $M_{n\substack{e\\o}}$ and P_n are the same as given in (4.43a)-(4.43d).

As discussed in Sec. 4.3.2, for the edge-coupled lines, the choice of g_e and g_o is not critical. Thus, except for a very small value of s/b, we may set $g_e = g_o = 0$ in the capacitance formula. For the symmetric charge distribution, we have

$$f(x)_{\substack{ee\\eo\\oe\\oo}} = \frac{1}{w}\left[1 + A_{\substack{ee\\eo\\oe\\oo}}\left|\frac{2}{w}\left(x - \frac{c-s-w}{2}\right)\right|^3\right], \quad \frac{c-s}{2} - w \leqslant x \leqslant \frac{c-s}{2} \tag{4.99}$$

$$= 0, \qquad \text{otherwise}$$

The capacitance expression (4.97) can then be written in the form

$$C_{\substack{ee\\eo\\oe\\oo}} = \frac{\left(\dfrac{1+0.25A_{\substack{ee\\eo\\oe\\oo}}}{}\right)^2}{\sum\limits_{\substack{n\,\text{odd}\\\text{odd}\\\text{even}\\\text{even}}} (L_n + M_n A_{\substack{ee\\eo\\oe\\oo}})^2 P_n / Y_{\substack{e\\o\\e\\o}}} \tag{4.100}$$

where

$$A_{\substack{ee\\eo\\oe\\oo}} = -\frac{\sum\limits_{\substack{n\,\text{odd}\\\text{odd}\\\text{even}\\\text{even}}} (L_n - 4M_n) L_n P_n / Y_{\substack{e\\o\\e\\o}}}{\sum\limits_{\substack{n\,\text{odd}\\\text{odd}\\\text{even}\\\text{even}}} (L_n - 4M_n) M_n P_n / Y_{\substack{e\\o\\e\\o}}} \tag{4.101}$$

The expressions for L_n, M_n and P_n are given in (4.46a)-(4.46c). For the charge distributions specified in (4.96) [or (4.99)] the corresponding closed form capacitance formula (4.97) [or (4.100)] applies to any multilayer symmetric broadside-edge coupled structure. It may be noted that the admittance parameter Y_e for the *ee* and *oe* modes is the same as that of the corresponding broadside-coupled structure for the even-mode. Similarly, the admittance parameter Y_o for the *eo* and *oo* modes is the same as that of the corresponding broadside-coupled structure for the odd-mode (Fig. 4.12c). Therefore, the admittance expressions presented in Tables 4.4 and 4.5 for the broadside-coupled structures are also applicable to the corresponding broadside-edge coupled structures. However, for the sake of completeness, a class of broadside-edge coupled structures and their admittance parameters are listed in Table 4.7 when the substrates used are isotropic, and, in Table 4.8, when the substrates used are uniaxially anisotropic with the optical axis aligned with the vertical (y-) axis.

4.9.2 General Capacitance Formula for Structures with Anisotropic Substrates having Principal Axes Rotated with Respect to the x-y Coordinate Axes—Spectral Domain Analysis under Quasi-Static Approximation

Consider first the example of a broadside-edge coupled symmetric structure with a rectangular shielding enclosure as shown in Fig. 4.13. The medium is anisotropic having the permittivity tensor defined as in (4.66). The strip couductors S, S', S'' and S''' are assumed to be of zero thickness and are placed symmetrically with respect to each other.

Table 4.7 Admittance parameters of broadside-edge coupled isotropic transmission structures with side walls

Structure	*Admittance Y at $y=y_0$*
A1. Broadside-edge coupled homogeneous stripline 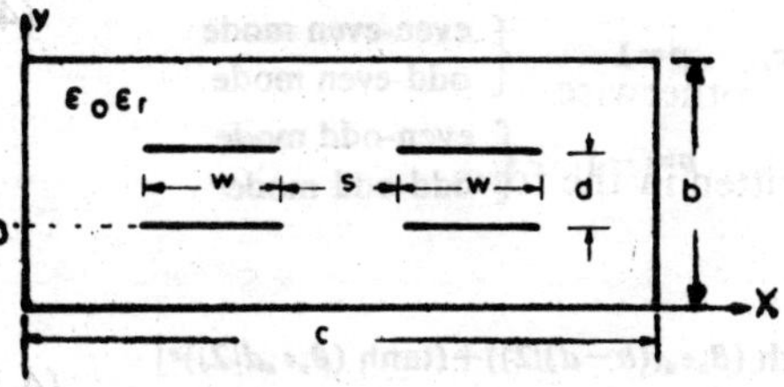	$\beta_n = n\pi/c$, $p=1$ {even-even mode, odd-even mode}; $p=-1$ {even-odd mode, odd-odd mode} $\epsilon_0\epsilon_r[\coth(\beta_n((b-d)/2))+(\tanh(\beta_n d/2))^p]$
A2. Broadside-edge coupled suspended stripline 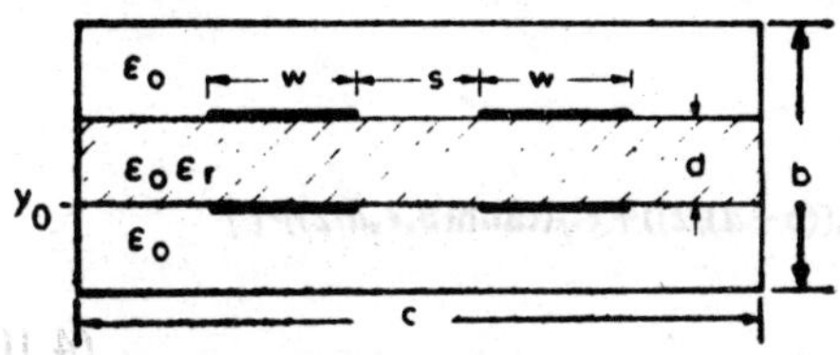	$\epsilon_0[\coth(\beta_n((b-d)/2))+\epsilon_r(\tanh(\beta_n d/2))^p]$
A3. Broadside edge-coupled inverted stripline 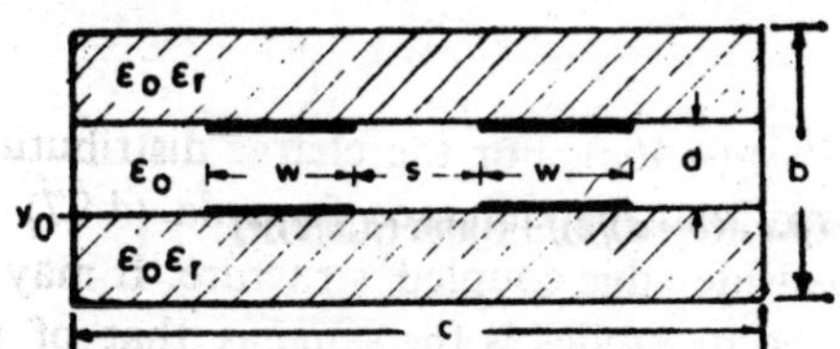	$\epsilon_0[\epsilon_r\coth(\beta_n((b-d)/2))+(\tanh(\beta_n d/2))^p]$
A4. Broadside-edge coupled stripline with double suspended substrate 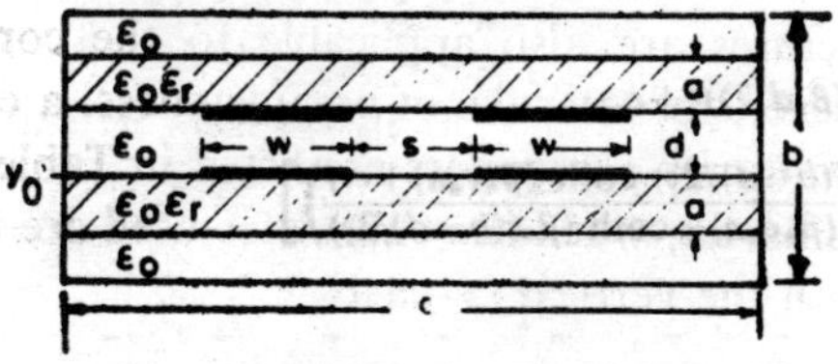	$\epsilon_0\left[(\tanh(\beta_n d/2))^p+\epsilon_r\left(\frac{\coth(\beta_n((b-d)/2-a))\coth(\beta_n a)+\epsilon_r}{\epsilon_r\coth(\beta_n a)+\coth(\beta_n((b-d)/2-a))}\right)\right]$
A5. Broadside-edge coupled stripline with double suspended inverted substrate 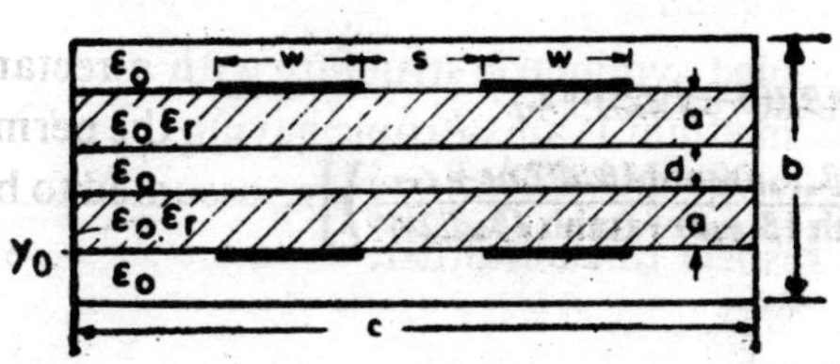	$\epsilon_0\left[\coth(\beta_n((b-d)/2-a))+\epsilon_r\left(\frac{\coth(\beta_n a)(\tanh(\beta_n d/2))^p+\epsilon_r}{\epsilon_r\coth(\beta_n a)+(\tanh(\beta_n d/2))^p}\right)\right]$

Table 4.8 Admittance parameters of broadside-edge coupled transmission structures with side walls and uniaxially anisotropic substrates—Optical axis aligned along the *y*-axis

$$\epsilon_{rf}=\sqrt{\epsilon_{xx}\epsilon_{yy}},\quad \epsilon_a=\sqrt{\epsilon_{xx}/\epsilon_{yy}}$$

Structure	*Admittance Y at* $y=y_0$
• **Broadside-edge coupled homogeneous stripline**	$\beta_n=n\pi/c,\quad p=1$ {even-even mode, odd-even mode}; $p=-1$ {even-odd mode, odd-odd mode}
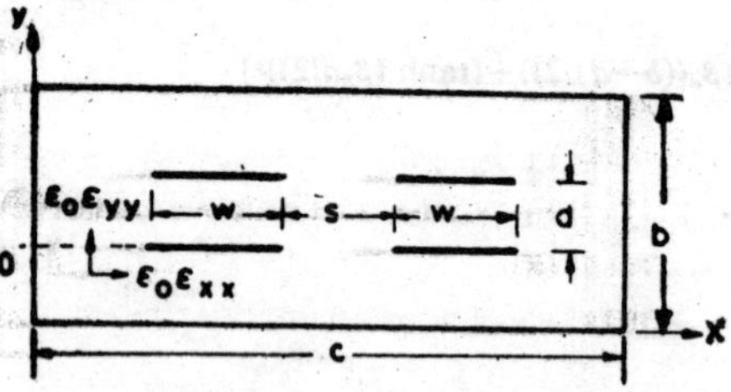	$\epsilon_0\epsilon_{rf}[\coth(\beta_n\epsilon_a((b-d)/2))+(\tanh(\beta_n\epsilon_a d/2))^p]$
• **Broadside-edge coupled suspended stripline**	
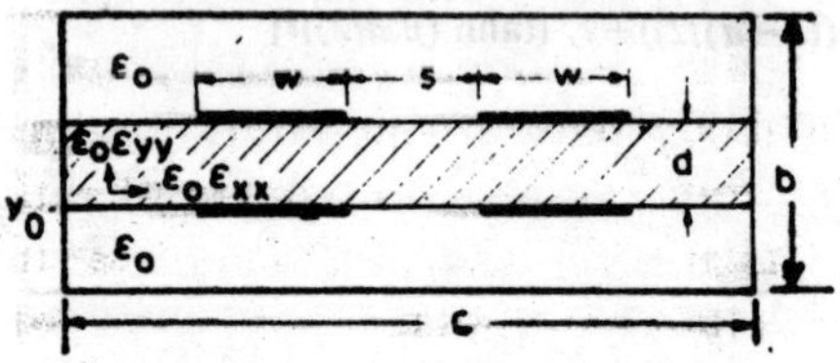	$\epsilon_0[\coth(\beta_n((b-d)/2))+\epsilon_{rf}(\tanh(\beta_n\epsilon_a d/2))^p]$
• **Broadside-edge coupled inverted stripline**	
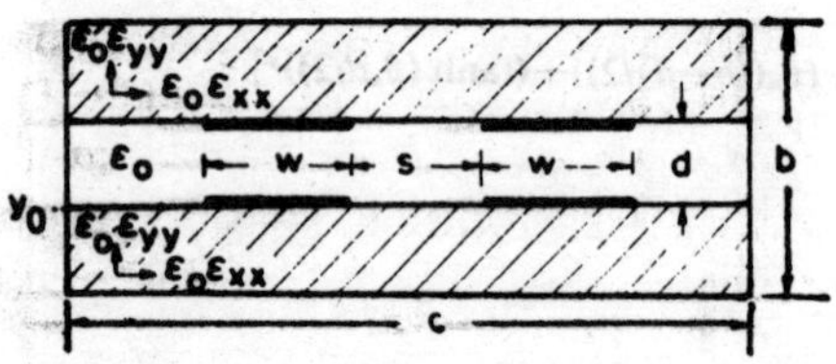	$\epsilon_0[\epsilon_{rf}\coth(\beta_n\epsilon_a((b-d)/2))+(\tanh(\beta_n d/2))^p]$
• **Broadside-edge coupled stripline with double suspended substrate**	
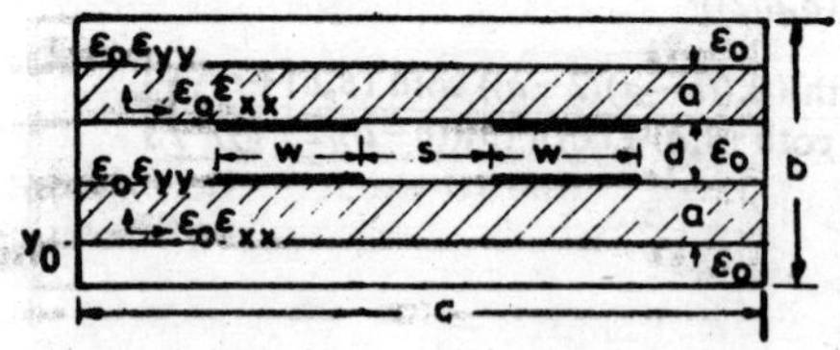	$\epsilon_0\left[(\tanh(\beta_n d/2))^p+\epsilon_{rf}\cdot\left(\frac{\coth(\beta_n((b-d)/2))\coth(\beta_n\epsilon_a a)+\epsilon_{rf}}{\epsilon_{rf}\coth(\beta_n\epsilon_a a)+\coth(\beta_n((b-d)/2))}\right)\right]$
• **Broadside-edge coupled stripline with double suspended inverted substrate**	
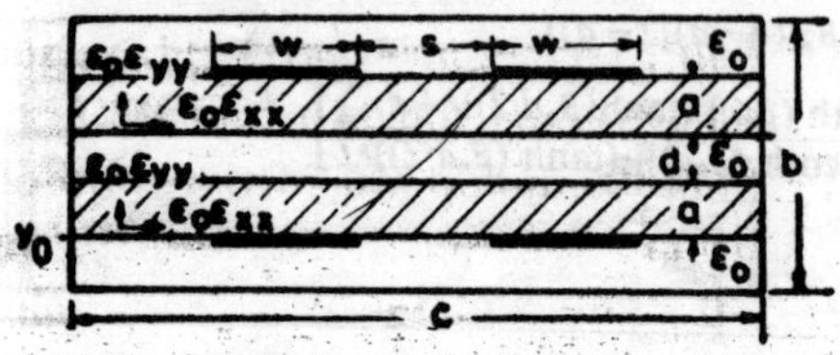	$\epsilon_0\left[\coth(\beta_n((b-d)/2))+\epsilon_{rf}\cdot\left(\frac{\coth(\beta_n\epsilon_a a)(\tanh(\beta_n d/2))^p+\epsilon_{rf}}{\epsilon_{rf}\coth(\beta_n\epsilon_a a)+(\tanh(\beta_n d/2))^p}\right)\right]$

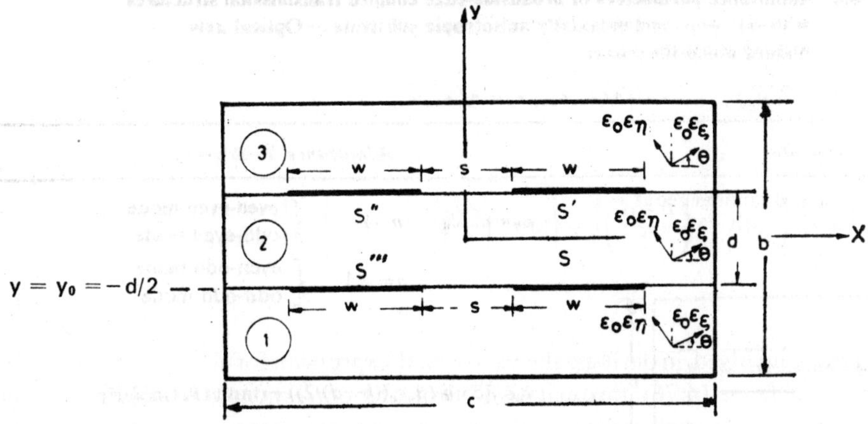

FIG. 4.13 Broadside-edge coupled symmetric stripline with anisotropic substrates.

In order to derive the quasi-static capacitance formula for the structure, we first solve for the potential distribution in the spectral domain in terms of the charge distributions on the strip conductors for the following four basic modes of excitation. The charge distributions on the four conductors are denoted as: (i) $f_1(x)|_S = f_1(x)|_{S'} = f_1(x)|_{S''} = f_1(x)|_{S'''} = f(x)_{ee}$ for the even-even (*ee*) mode excitation; (ii) $f_2(x)|_S = -f_2(x)|_{S'} = -f_2(x)|_{S''} = f_2(x)|_{S'''} = f(x)_{eo}$ for even-odd (*eo*) mode excitation; (iii) $f_3(x)|_S = f_3(x)|_{S'} = -f_3(x)|_{S''} = -f_3(x)|_{S'''} = f(x)_{oe}$ for odd-even (*oe*) mode excitation; and (iv) $f_4(x)|_S = -f_4(x)|_{S'} = f_4(x)|_{S''} = -f_4(x)|_{S'''} = f(x)_{oo}$ for odd-odd (*oo*) mode excitation. The Fourier transforms of these charge distributions are defined as

$$\hat{f}(\beta_n)_{\substack{ee\\eo\\oe\\oo}} = \int_0^{c/2} f(x)_{\substack{ee\\eo\\oe\\oo}} \sin \beta_n x \, dx \tag{4.102}$$

where

$$\beta_n = n\pi/c, \quad \begin{cases} n \text{ even for } oe \text{ and } oo \text{ modes} \\ n \text{ odd for } ee \text{ and } eo \text{ modes} \end{cases} \tag{4.103}$$

The analysis procedure for solving the present structure is the same as that described in Sec. 4.7.1 for the broadside-coupled structure. Laplace's equation (4.68) is solved subject to the boundary conditions specified in (4.71) with the following changes. In (4.71c), (4.71d), (4.71f), and (4.71g), $\hat{f}(\beta_n) = \hat{f}(\beta_n)_{ee}$ with $p = 1$ for the *ee* mode, $\hat{f}(\beta_n) = \hat{f}(\beta_n)_{eo}$ with $p = -1$ for the *eo* mode, $\hat{f}(\beta_n) = \hat{f}(\beta_n)_{oe}$ with $p = 1$ for the *oe* mode, and $\hat{f}(\beta_n) = \hat{f}(\beta_n)_{oo}$ with $p = -1$ for the *oo* mode. The magnitude of the transformed potential at the charge plane $y = -d/2$ for the four basic modes of excitation can be written from (4.76) as

$$\hat{\phi}(\beta_n, -d/2)_{\substack{ee\\eo\\oe\\oo}} = \hat{G}(\beta_n, -d/2)_{\substack{ee\\eo\\oe\\oo}} \hat{f}(\beta_n)_{\substack{ee\\eo\\oe\\oo}} \tag{4.104}$$

where

$$\hat{G}(\beta_n, -d/2)_{\substack{ee\\eo\\oe\\oo}} = \frac{1}{\beta_n Y_{\substack{e\\o\\e\\o}}} \tag{4.105}$$

$$Y_{\substack{e\\o\\e\\o}} = Y_{\substack{ee\\eo\\oe\\oo}} = \epsilon_0\epsilon_{rf}\left[\coth\left\{\gamma\left(\frac{b-d}{2}\right)\right\} + \left\{\tanh\left(\frac{\gamma d}{2}\right)\right\}^p\right], \quad p = \begin{cases} 1, & ee \text{ and } oe \text{ modes} \\ -1, & eo \text{ and } oo \text{ modes} \end{cases} \tag{4.106}$$

The analytical steps involved in deriving the variational expression for the capacitance are the same as those given from (4.76) through (4.87) for the corresponding broadside-coupled structure, except that, while applying Parseval's theorem, we consider the range of integration to be from 0 to $c/2$. Thus, in place of (4.83), we use

$$\sum_{n=1}^{\infty} \psi_1(\beta_n)\psi_2(\beta_n) = (c/4)\int_0^{c/2} \psi_1(x)\psi_2(x)\,dx \tag{4.107}$$

and, the subscript e is replaced by ee and oe, and o is replaced by eo and oo. With these changes, the expression for the capacitance given by (4.53) gets replaced by

$$\frac{1}{C_{\substack{ee\\eo\\oe\\oo}}} = \frac{\displaystyle\sum_{\substack{n\ \text{odd}\\ \text{odd}\\ \text{even}\\ \text{even}}} \frac{4}{n\pi Y_{\substack{e\\o\\e\\o}}}\left[\int_S f(x)_{\substack{ee\\eo\\oe\\oo}} \sin\frac{n\pi x}{c}\,dx\right]^2}{\left[\displaystyle\int_S f(x)_{\substack{ee\\eo\\oe\\oo}}\,dx\right]^2} \tag{4.108}$$

We note that (4.108) is the same as the capacitance expression (4.95) derived using the unified approach, when the admittance parameters appropriate for the structure under consideration are substituted. Thus, for the charge distributions specified as in (4.96), the closed form capacitance expression given by (4.97) applies to structures with anisotropic substrates in which the principal axes are rotated with respect to the x-y coordinate axes. The admittance parameters are the same as those for the corresponding broadside-coupled structures. They are listed in Table 4.9.

4.10 Broadside-Edge Coupled Symmetric Stripline and Stripline-like Transmission Lines with Iso/Anisotropic Dielectric Layers with Infinitely Wide Ground Planes

In order to obtain generalized capacitance expressions applicable to broadside-edge coupled structures with iso/anisotropic substrates and with no side wall effects, we first solve a simple three-layered structure shown in Fig. 4.14. The ground planes are considered to be infinitely wide and the strip conductors are assumed to be of zero thickness. Considering the four basic modes (ee, eo, oe, and oo modes) of excitation, the Fourier-transformed potential at the

Table 4.9 Admittance parameters of broadside-edge coupled transmission structures with side walls and anisotropic substrates—Principal axes rotated with respect to the *x-y* coordinate axes

$$\epsilon_{rf}=\sqrt{\epsilon_\xi\epsilon_\eta},\quad \epsilon_a=\sqrt{\epsilon_\eta/\epsilon_\xi}\,[(\epsilon_\eta/\epsilon_\xi-1)\cos^2\theta+1]^{-1}$$

Structure	*Admittance Y* at $y=y_0$
• Broadside-edge coupled homogeneous stripline 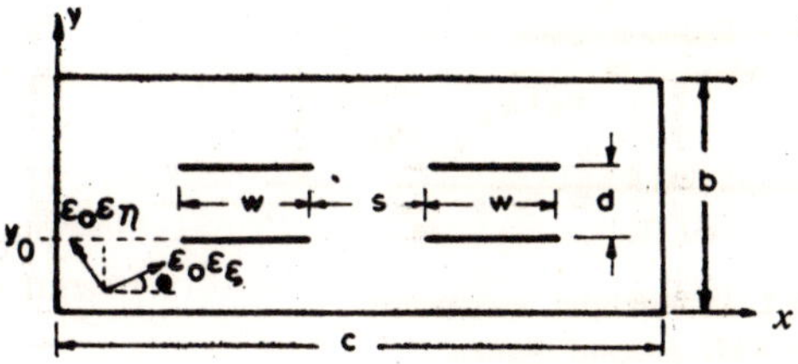	$\beta_n=n\pi/c$, $p=1$ {even-even mode, odd-even mode}; $p=-1$ {even-odd mode, odd-odd mode} $\epsilon_0\epsilon_{rf}\,[\coth(\beta_n\epsilon_a((b-d)/2))+(\tanh(\beta_n\epsilon_a d/2))^p]$
• Broadside-edge coupled suspended stripline 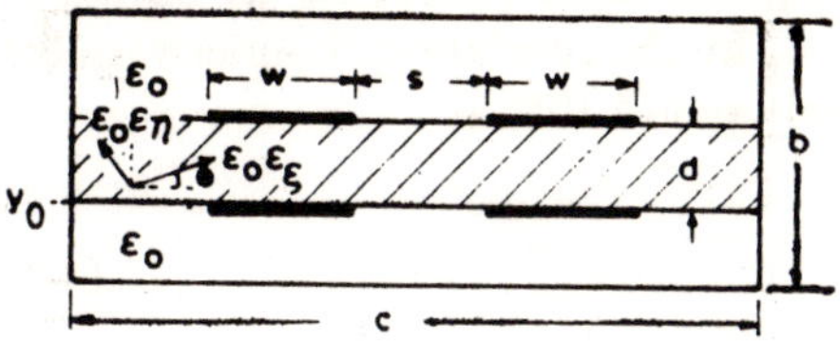	$\epsilon_0\,[\coth(\beta_n(b-d)/2)+\epsilon_{rf}\,(\tanh(\beta_n\epsilon_a d/2))^p]$
• Broadside-edge coupled inverted stripline 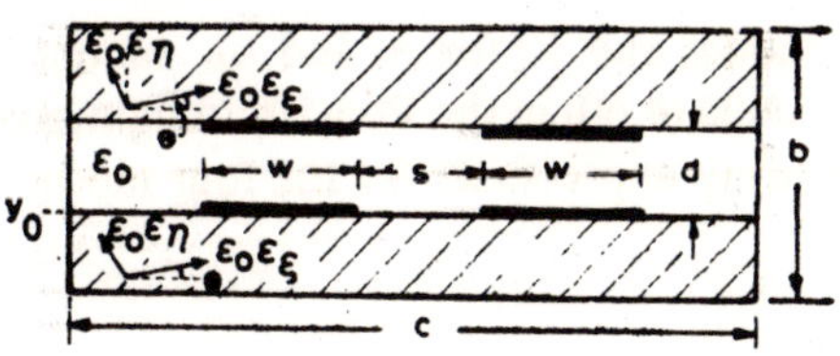	$\epsilon_0[\epsilon_{rf}\,[\coth(\beta_n\epsilon_a(b-d)/2)+(\tanh(\beta_n d/2))^p]$
• Broadside-edge coupled stripline with double suspended substrate 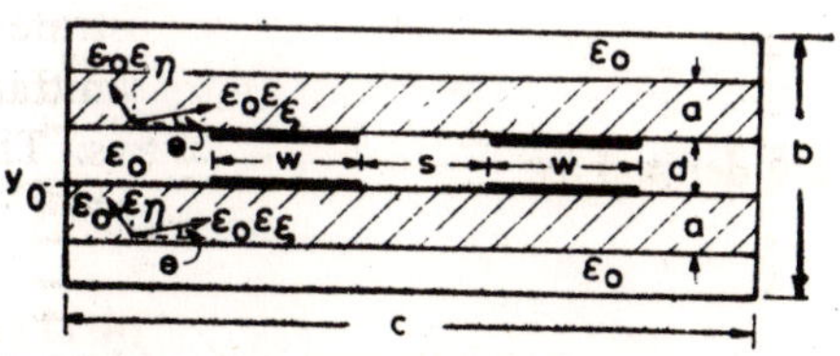	$\epsilon_0\left[(\tanh(\beta_n d/2))^p+\epsilon_{rf}\cdot\left(\frac{\coth(\beta_n(b-d)/2)\coth(\beta_n\epsilon_a a)+\epsilon_{rf}}{\epsilon_{rf}\coth(\beta_n\epsilon_a a)+\coth(\beta_n(b-d)/2)}\right)\right]$
• Broadside-edge coupled stripline with double suspended inverted substrate 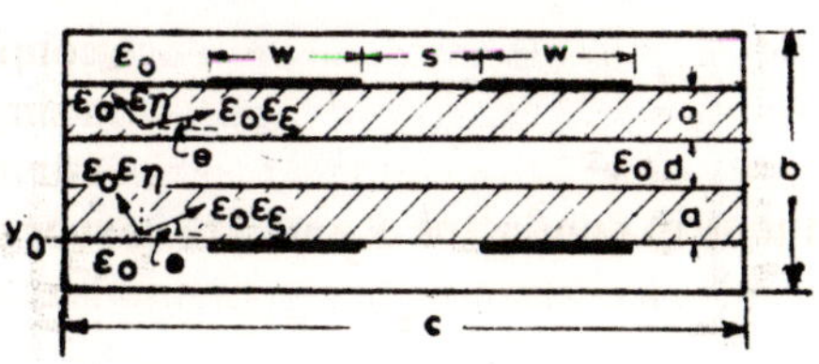	$\epsilon_0\left[\coth(\beta_n(b-d)/2)+\epsilon_{rf}\cdot\left(\frac{\coth(\beta_n\epsilon_a a)\,(\tanh(\beta_n d/2))^p+\epsilon_{rf}}{\epsilon_{rf}\coth(\beta_n\epsilon_a a)+(\tanh(\beta_n d/2))^p}\right)\right]$

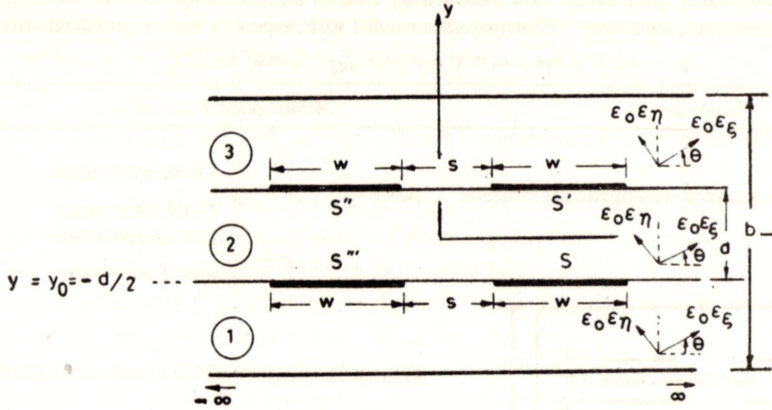

FIG. 4.14 Broadside-edge coupled anisotropic symmetric stripline with infinitely wide ground planes.

charge plane $y = -d/2$ can be determined following the same procedure as outlined in Sec. 4.9.2 by replacing the discrete Fourier variable β_n by the continuous Fourier variable β. The transformed potential can be written from (4.104) and (4.105) as

$$\hat{\phi}(\beta, -d/2)_{\substack{ee\\eo\\oe\\oo}} = \frac{\hat{f}(\beta)}{\beta Y_{\substack{e\\o\\e\\o}}} \tag{4.109}$$

where Y_e is the admittance parameter for ee and oe modes and Y_o for eo and oo modes as given by (4.106). We can now use the capacitance formula (4.60) in the Fourier domain given for the edge-coupled structures to write the corresponding expressions for the four modes of the present structure:

$$\frac{1}{C_{\substack{ee\\eo\\oe\\oo}}} = \frac{1}{2\pi} \int_0^{\infty} \left[\frac{\hat{f}(\beta)_{\substack{ee\\eo\\oe\\oo}}}{Q_{\substack{ee\\eo\\oe\\oo}}} \right]^2 \frac{1}{\beta Y_{\substack{e\\o\\e\\o}}} \, d\beta \tag{4.110}$$

The Fourier transforms of the charge distributions are defined as

$$\hat{f}(\beta)_{\substack{ee\\eo}} = 2 \int_{s/2}^{(s/2)+w} f(x)_{\substack{ee\\eo}} \cos \beta x \, dx \tag{4.111a}$$

$$\hat{f}(\beta)_{\substack{oe\\oo}} = 2 \int_{s/2}^{(s/2)+w} f(x)_{\substack{oe\\oo}} \sin \beta x \, dx \tag{4.111b}$$

Q_{ee}, Q_{eo}, Q_{oe}, and Q_{oo} represent the total charge on the strip conductor S and are given by the expression

$$Q_{\substack{ee\\eo\\oe\\oo}} = \int_{s/2}^{(s/2)+w} f(x)_{\substack{ee\\eo\\oe\\oo}}\, dx \tag{4.112}$$

We note that the capacitance expression (4.110), derived for the simple structure, offers a general formula and can be applied to any multilayer broadside-edge coupled symmetric structure by substituting the admittance parameters appropriate to the structure. The substrates used may be isotropic, or anisotropic with principal axes either aligned or rotated with respect to the x-y coordinate axes. The admittance expressions are the same as those for the corresponding structure with side walls when β_n is replaced by β. Hence, for the structures given in Tables 4.7, 4.8, and 4.9, if we let $c \to \infty$, the same admittance expressions listed in the tables can be used with β_n replaced by β.

4.11 Conclusion

The unified capacitance formulas developed in Chapter 2 for single conductor multilayer planar structures are utilized to derive general expressions for the capacitances of multilayer coupled transmission structures. By assuming appropriate charge distribution on the strip conductors corresponding to the various basic propagating modes, closed form expressions are derived for the line capacitances of edge-coupled, broadside-coupled, and broadside-edge coupled symmetric striplines and stripline-like transmission lines with side walls. The formulas are general and can be applied to any symmetric coupled structure, irrespective of the number of dielectric layers. Further, the dielectric layers may be isotropic or anisotropic. The only parameter that needs to be changed is the admittance parameter Y. This parameter can be easily determined using the isotropic equivalent of the structure and its transmission line equivalent for each of the basic modes of propagation.

For coupled structures in which the side walls have no effect on the propagation parameters, capacitance expressions are derived in the form of integrals in the Fourier domain. General formulas are presented for the edge-coupled, broadside-coupled, and broadside-edge coupled symmetric striplines and stripline-like transmission lines with no side wall effects. These formulas are applicable to a class of such symmetric structures having one or more dielectric layers, which may be isotropic, or anisotropic. The admittance parameters of these structures are the same as those of the corresponding structures with side walls when the discrete Fourier variable β_n is replaced by the continuous Fourier variable β.

REFERENCES

1. J.K. Shimizu and E.M.T. Jones, Coupled transmission line directional couplers, *IRE Trans. Microwave Theory Tech.*, Vol. MTT-6, pp. 403-410, Oct. 1958.
2. P.P. Toulios and A.C. Todd, Synthesis of symmetrical TEM-mode directional couplers, *IEEE Trans. Microwave Theory Tech.*, Vol. MTT-13, pp. 536-544, Sept. 1965.
3. J.E. Dalley, A stripline directional coupler utilizing a non-homogeneous dielectric medium, *IEEE Trans. Microwave Theory Tech.*, Vol. MTT-17, pp. 706-711, Sept. 1969.
4. Y. Tajima and S. Kamihashi, Multiconductor couplers, *IEEE Trans. Microwave Theory Tech.*, Vol. MTT-26, pp. 795-801, Oct. 1978.
5. E.M.T Jones and J.T. Bolljahn, Coupled strip-transmission line filters and couplers, *IRE Trans. Microwave Theory Tech.*, Vol. MTT-4, pp. 75-81, April 1956.

6. S.B. Cohn, Parallel-coupled transmission line resonator filters, *IRE Trans. Microwave Theory Tech*, Vol. MTT-6, pp. 223-231, April 1958.
7. J.L. Allen, Inhomogeneous coupled line filter with large mode velocity ratios, *IEEE Trans. Microwave Theory Tech.*, Vol. MTT-22, pp. 1182-1186, Dec. 1974.
8. E.M.T. Jones and J.K. Shimizu, A wide band stripline balun, *IRE Trans. Microwave Theory Tech.*, Vol. MTT-7, pp. 128-134, Jan. 1959.
9. B.M. Schiffman, A new class of broad-band microwave 90-degree phase shifters, *IRE Trans. Microwave Theory Tech.*, Vol. MTT-6, pp. 232-237, April 1958.
10. J.W. Duncan, Characteristic impedances of multiconductor strip transmission lines, *IEEE Trans. Microwave Theory Tech.*, Vol. MTT-13, pp. 107-118, Jan. 1965.
11. S.B. Cohn, Shielded strip transmission lines, *IRE Trans. Microwave Theory Tech.*, Vol. MTT-3, pp. 29-38, Oct. 1955.
12. E. Jahnke and F. Emde, *Tables of Functions*, 4th Edn., Dover Publications, New York, 1945.
13. L.J.P. Linner, Explicit expressions for geometrical dimensions of symmetrical coupled striplines, *Electron. Lett.*, Vol. 10, No. 4, pp. 45-46, Feb. 1974.
14. F. Tricomi, *Elliptische Funktionen*, Liepzig, 1948.
15. B. Bhat and S.K. Koul, Unified approach to solve a class of strip and microstrip-like transmission lines, *IEEE Trans. Microwave Theory Tech.*, Vol. MTT-30, pp. 679-686, May 1982. Also correction in *IEEE Trans. Microwave Theory Tech.*, Vol. MTT-30, p. 2067, Nov. 1982.
16. S.K. Koul and B. Bhat, Generalized analysis of microstrip-like transmission lines and coplanar strips with anisotropic substrates for MIC, electro-optic modulator and SAW application, *IEEE Trans. Microwave Theory Tech.*, Vol. MTT-31, pp. 1051-1059, Dec. 1983.
17. E. Yamashita and S Yamazaki, Parallel-stripline embedded in or printed on a dielectric sheet, *IEEE Trans. Microwave Theory Tech.*, Vol. MTT-16, pp. 972-973, Nov. 1968.
18. T. Itoh and R. Mittra, An accurate method for calculating charge and potential distributions in coupled microstrip lines, *Proc. IEEE*, Vol. 59, pp. 332-334, Feb. 1971.
19. B. Bhat and S.K. Koul, A new approach to analyse microstrip-like transmission lines with anisotropic substrates having tilted optical axis, *IEE Proc.*, MOA, Vol. 131, pt H, No. 3, pp. 191-197, June 1984. Also correction in *IEE Proc.*, MOA, Vol. 131, p. 280, Oct. 1984.
20. S. Kusase and R. Terakado, Mapping theory of two-dimensional anisotropic regions, *Proc. IEEE*, Vol. 67, pp. 171-172, Jan. 1979.
21. S.K. Koul and B. Bhat, Transverse transmission line method for the analysis of broadside-coupled microstrip lines with anisotropic substrates, *Arch. Elek. Ubertragung*, Vol. 37, pp. 59-64, Jan./Feb. 1983.
22. S.K. Koul and B. Bhat, A generalized TEM analysis of broadside-coupled planar transmission lines with isotropic and anisotropic substrates, *Arch. Elek. Ubertragung*, Vol. 38, pp. 37-45, Jan./Feb. 1984.
23. S.K. Koul and B. Bhat, Broadside, edge-coupled symmetric strip transmission lines, *IEEE Trans. Microwave Theory Tech.*, Vol. MTT-30, pp. 1874-1880, Nov. 1982.

CHAPTER 5

Design Data on Single Strip Conductor Stripline-Like Transmission Lines—Characteristic Impedance and Effective Dielectric Constant

5.1 Introduction

The two most important parameters of a transmission line required in the design of any microwave integrated circuit are the characteristic impedance Z and the effective dielectrie constant ϵ_f. With the knowledge of ϵ_f, the guide wavelength λ and the phase velocity v of the transmission structure can be readily determined.

This chapter presents extensive design data on Z and ϵ_f of a variety of single strip conductor stripline-like structures. Unified formulae derived in Chapter 3 for a class of single conductor striplines with multilayer iso/anisotropic dielectric substrates are listed first. For some simple structures, closed form expressions are available in the literature. These are listed along with their range of validity. Formulas are quoted without proof for some special structures. Interested readers may obtain details of the analytical derivations from the references listed in this chapter.

Few graphs are plotted basically to explain the propagation characteristics as a function of various parameters of the structure and to point out the special features. Considerable emphasis has been placed on providing accurate data in the form of tables for structures of practical importance. Tabulation is made in sufficiently small intervals of the parameters, to enable the reader to obtain any intermediate value accurately by direct linear interpolation between any two successive values. All tables of data are included at the end of this chapter. In figures as well as in tables, the values of t/b (thickness of the strip conductor normalized with respect to the ground plane spacing) is specified wherever necessary; where the value is not specified, it is presumed that $t/b=0$.

5.2 Design Equations

5.2.1 *Characteristic Impedance, Effective Dielectric Constant and Guide Wavelength*

Under the quasi-TEM approximation, the expressions for the characteristic impedance Z, effective dielectric constant ϵ_f, and guide wavelength λ of a stripline-like transmission line are given by

$$Z=\frac{1}{v^a\sqrt{CC^a}} \tag{5.1}$$

$$\epsilon_f=C/C^a \tag{5.2}$$

$$\lambda=\lambda^a/\sqrt{\epsilon_f} \tag{5.3}$$

In these formulas, C is the capacitance per unit length of the structure, C^a is the capacitance per unit length of the structure with all dielectrics replaced by air, v^a is the velocity of propagation in air, and λ^a is the free space wavelength. The evaluation of Z, ϵ_f and λ requires first the determination of the capacitance C of the transmission structure.

5.2.2 Unified Variational Capacitance Formula for Structures with Iso/Anisotropic Substrates

Table 3.1 (Chapter 3) gives some of the practical stripline-like transmission lines. Referring to the coordinate system marked in structure Al of Table 3.1, the charge distribution on the strip conductor (assumed to be of zero thickness) is specified as

$$f(x)=(1/w)\left[1+A\,|\,\frac{2}{w}(x-c/2)\,|^3\right],\qquad \frac{c-w}{2}\leqslant x\leqslant\frac{c+w}{2} \tag{5.4}$$

The unified variational expression for the capacitance is given by

$$C=\frac{(1+0.25A)^2}{\sum\limits_{n\ \mathrm{odd}}(L_n+AM_n)^2P_n/Y} \tag{5.5a}$$

where

$$L_n=\sin\beta_n w/2,\qquad \beta_n=n\pi/c \tag{5.5b}$$

$$M_n=(2/\beta_n w)^3\left[3\left\{\left(\frac{\beta_n w}{2}\right)^2-2\right\}\cos\left(\frac{\beta_n w}{2}\right)+\left(\frac{\beta_n w}{2}\right)\left\{\left(\frac{\beta_n w}{2}\right)^2-6\right\}\sin\left(\frac{\beta_n w}{2}\right)+6\right] \tag{5.5c}$$

$$P_n=(2/n\pi)(2/\beta_n w)^2 \tag{5.5d}$$

$$A=-\frac{\sum\limits_{n\ \mathrm{odd}}(L_n-4M_n)L_nP_n/Y}{\sum\limits_{n\ \mathrm{odd}}(L_n-4M_n)M_nP_n/Y} \tag{5.5e}$$

The capacitance formula (5.5) is applicable to any single conductor stripline-like transmission line with one or more dielectric layers. The dielectric layers may be isotropic or anisotropic. The only factor that needs to be changed is the admittance parameter Y depending on the structure. The admittance parameters of some of the useful structures with isotropic and anisotropic structures are listed in Tables 3.1, 3.2 and 3.5 of Chapter 3.

If the strip conductor has a small but finite thickness t, (5.5) can still be used by replacing Y by $Y/h(\beta_n, t)$. The expressions $h(\beta_n, t)$ for the structures considered in Tables 3.1, 3.2 and 3.5 are now given.

Symmetric stripline:

$$h(\beta_n, t)=\frac{\sinh\left\{\beta_n\left(\frac{b-t}{2}\right)\right\}}{\sinh\left(\beta_n\frac{b}{2}\right)} \tag{5.6a}$$

Stripline with off-set centre conductor:

$$h(\beta_n, t) = \frac{1}{2}\left[\frac{\sinh\left\{\beta_n\left(\frac{b-t}{2}+d\right)\right\}}{\sinh\left\{\beta_n\left(\frac{b}{2}+d\right)\right\}} + \frac{\sinh\left\{\beta_n\left(\frac{b-t}{2}-d\right)\right\}}{\sinh\left\{\beta_n\left(\frac{b}{2}-d\right)\right\}}\right] \tag{5.6b}$$

Microstrip-like stripline:

$$h(\beta_n, t) = \frac{1}{2}\left[1 + \frac{\sinh\left\{\beta_n\left(\frac{b}{2}-t\right)\right\}}{\sinh\left(\beta_n \frac{b}{2}\right)}\right] \tag{5.6c}$$

Suspended and sandwiched striplines:

$$h(\beta_n, t) = \frac{1}{2}\left[1 + \frac{\sinh\left\{\beta_n\left(\frac{b-d}{2}-t\right)\right\}}{\sinh\left\{\beta_n\left(\frac{b-d}{2}\right)\right\}}\right] \tag{5.6d}$$

In deriving (5.6a) and (5.6b), it is assumed that the strip conductor is assumed to lie between $y = y_0 + (t/2)$ and $y_0 - (t/2)$, and in deriving (5.6c) and (5.6d), it is assumed to be between $y = y_0$ and $y_0 + t$.

5.2.3 Other Closed Form Expressions

(i) Symmetric stripline with zero thickness strip conductor: The exact expression for the characteristic impedance Z of a symmetric stripline with zero thickness strip conductor has been reported by Cohn [1]. The formula due to Cohn is now given with reference to the dimensional parameters marked in Fig. 5.1 and with $t/b = 0$.

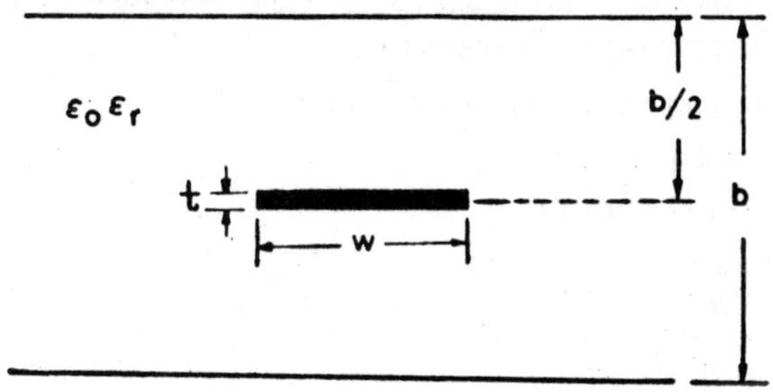

FIG. 5.1 Symmetric stripline with rectangular strip conductor.

Exact formula:

$$Z\sqrt{\epsilon_r} = 29.976\pi \frac{K(k)}{K(k')} \tag{5.7a}$$

where

$$k = \text{sech}\,(\pi w/2b) \tag{5.7b}$$

$$k' = \tanh\,(\pi w/2b) \tag{5.7c}$$

Tables of data on elliptic integrals are available in the literature [2, 3]. Making use of the elliptic integral approximation [4], (5.7a) can be expressed in simple closed form as follows:

Empirical formulas:

For $w/b \leqslant 0.5$:

$$Z\sqrt{\epsilon_r} = 29.976 \ln\left[2\left(\frac{1+\sqrt{k}}{1-\sqrt{k}}\right)\right] \tag{5.8a}$$

For $w/b \geqslant 0.5$:

$$Z\sqrt{\epsilon_r} = \frac{29.976\pi^2}{\ln\left[2\left(\frac{1+\sqrt{k'}}{1-\sqrt{k'}}\right)\right]} \tag{5.8b}$$

These empirical formulas yield results which are virtually exact.

(*ii*) *Symmetric stripline with rectangular strip conductor:* Practical circuits are frequently etched from copper-clad printed circuit boards or gold metallized dielectric substrates. The thickness t of the strip conductor is generally two to three times the skin depth, and hence, is usually very small in comparison with the ground plane spacing. The approximation $t/b = 0$ is therefore reasonable from the practical point of view. However, for special applications, such as when high power transmission is involved, a more rugged structure which utilizes a centre conductor of appreciable thickness is required. Striplines with rectangular strip conductor and air dielectric are frequently used at UHF and L bands in the realization of high directivity couplers, interdigital and combline type of filters.

The analysis of a symmetric stripline with a rectangular centre conductor (shown in Fig. 5.1) has been carried out by Bates [5], Waldron [6, 7] and Ikeda and Sato [8] using conformal transformation technique, and by Collin [9] using the variational method. Some design data generated using Waldron's formula are reported in Gunston [4]. Cohn [1] has developed simple approximate formulas which yield impedance data within one per cent accuracy for the following range of parameters.

For $w/b \geqslant 0.35(1 - t/b)$ and $t/b \leqslant 0.25$ (Ref. Fig. 5.1)

$$Z\sqrt{\epsilon_r} = \frac{94.172}{X(w/b) + (1/\pi)F(X)} \tag{5.9a}$$

$$F(X) = \frac{(X+1)^{X+1}}{(X-1)^{X-1}} \tag{5.9b}$$

$$X = (1 - t/b) \tag{5.9c}$$

(*iii*) *Homogeneous stripline with off-set rectangular conductor:* Figure 5.2 shows the cross-section of a homogeneous stripline with rectangular strip conductor placed at arbitrary distances from the two ground planes. Structures of this type have been analyzed by approximate methods by Rosenzweig [10] and Guckel [11]. The formulas developed by Rosenzweig relate the characteristic impedance to the various shunt capacities of the line as shown in Fig. 5.2. These formulas have been published in Howe [12]:

$$Z\sqrt{\epsilon_r} = \frac{376.7}{(C_t/\epsilon_r)} \tag{5.10a}$$

where

$$C_t = (C_{p1} + C_{p2} + 2C_{f1} + 2C_{f2}) \tag{5.10b}$$

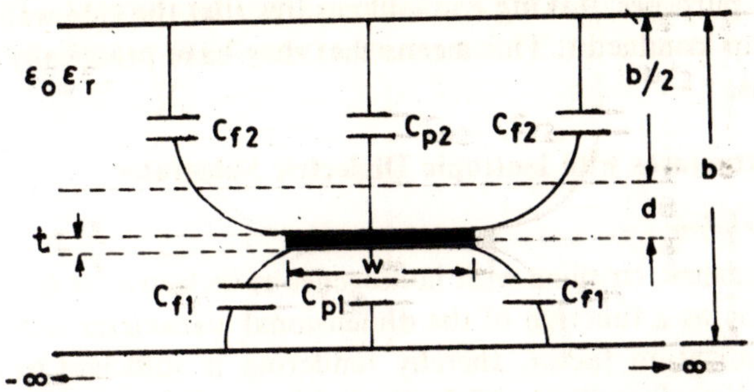

FIG 5.2 Stripline with off-set rectangular strip conductor.

For $w/(b-t) \geqslant 0.35$:

$$C_{p1} = \epsilon_r \left[\frac{2w/(b-2d)}{1-t/(b-2d)} \right] \tag{5.10c}$$

$$C_{p2} = \epsilon_r \left[\frac{2w/(b+2d)}{1-t/(b+2d)} \right] \tag{5.10d}$$

$$C_{f1} = (\epsilon_r/\pi) \left[\frac{1}{1-t/(b-2d)} \ln \left\{ \frac{1}{1-t/(b-2d)} + 1 \right\} - \left\{ \frac{1}{1-t/(b-2d)} - 1 \right\} \ln \left\{ \frac{1}{(1-t(b-2d))^2} - 1 \right\} \right] \tag{5.10e}$$

$$C_{f2} = (\epsilon_r/\pi) \left[\frac{2}{1-t/(b+2d)} \ln \left\{ \frac{1}{1-t/(b+2d)} + 1 \right\} - \left\{ \frac{1}{1-t/(b+2d)} - 1 \right\} \ln \left\{ \frac{1}{(1-t/(b+2d))^2} - 1 \right\} \right] \tag{5.10f}$$

A correction for the case where $w/(b-t) \leqslant 0.35$ may be made by the method of Getsinger [13] by defining a new value of w/b as

$$\frac{w_N}{b} = \left[\frac{0.07(1-t/b) + w/b}{1.2} \right] \tag{5.10g}$$

which provides a reasonable correction to the fringing capacity for the condition

$$0.1 < \frac{w_N/b}{(1-t/b)} < 0.35 \tag{5.10h}$$

In the following sections, detailed design data are presented for the five structures listed in Tables 3.1, 3.2 and 3.5. Data for symmetric stripline and stripline with off-set centre conductor for zero thickness strip conductor as well as for rectangular strip conductor are computed using the closed form expressions given in this section. For microstrip-like stripline, suspended stripline, and sandwiched stripline, the unified formula for C given by (5.5) is used for computing the characteristic impedance Z and the effective dielectric constant ϵ_f. Detailed data are presented in the form of tables for $t/b = 0$ and $c/w = 40$. Since the conducting patterns in these structures are realized using vacuum evaporation and photolithographic techniques, the conductor thickness is so small that t/b lies between 0.0005 and 0.005 in the majority of cases. Hence, the zero thickness approximation for the strip conductor is sufficiently accurate for practical design purposes. Taking $c/w=40$ ensures that the side walls are sufficiently far away from the strip conductor. This means that they have practically no effect on the propagation parameters.

5.3 Transmission Structures with Isotropic Dielectric Substrates

5.3.1 Symmetric Stripline

In the case of a symmetric stripline with homogeneous dielectric, it is convenient to present design data for $Z\sqrt{\epsilon_r}$ as a function of the dimensional parameter w/b. Here $\sqrt{\epsilon_r}$ serves as a convenient normalization factor, thereby rendering it sufficient to use a single graph to present the characteristic impedance of the line with any dielectric filling.

The variational formula (5.5) derived using the unified approach permits calculation of the characteristic impedance for any spacing c between the side walls. Over the lower range of values of c/w, as c/w is increased, the impedance of the stripline increases sharply and then saturates (Fig. 5.3). For a typical set of parameters, $w/b=0.1$ and $t/b=0$ considered in Fig. 5.3, it is seen that the effect of side walls is insignificant for $c/w > 18$. Thus, in all the computations, c/w is taken as 40 for all structures to ensure that the side walls have no effect on the propagation parameters. Figure 5.4 shows the variation in $Z\sqrt{\epsilon_r}$ computed using (5.5) as a function of the parameter A in the assumed charge distribution. Impedance close to the true value is obtained with the optimum value of A, at which the capacitance given by (5.5) is maximum, and hence, $Z\sqrt{\epsilon_r}$ is minimum. The expression for optimum A is given by (5.5e). It can be seen that for $w/b=0.1$, the error in $Z\sqrt{\epsilon_r}$ is within 2.5 Ω when A is varied between 1 and 10 about its optimum value. This corresponds to a maximum error of about 1.2 per cent in $Z\sqrt{\epsilon_r}$. Computations for higher values of w/b show that the variation in $Z\sqrt{\epsilon_r}$ as a function of A becomes smaller with an increase in the width of the strip conductor. Thus, except for high impedance values of the order of 100 Ω or more, a simpler charge distribution with $A=1$ gives fairly accurate results. However, all the design data presented in this book using the unified approach are generated using the optimum value of A for the corresponding structure.

Figure 5.5 shows the variation of $Z\sqrt{\epsilon_r}$ of a symmetric stripline as a function of w/b for $t/b=0$. The solid line shows results computed from the exact formula (5.7) derived using the conformal mapping method by assuming $c=\infty$. The dots show the values computed using the unified approach formula (5.5) with the optimum value of A and $c/w=40$. There is good agreement between the two; the unified approach yields results which are accurate to within 1 per cent.

Table 5.1 lists the normalized characteristic impedance $Z\sqrt{\epsilon_r}$ of a symmetric stripline as a function of w/b for various ratios of t/b. For $t/b=0$, the values are computed from the exact formula (5.7), and for t/b finite, the values are computed using (5.9) which is accurate to within one per cent.

5.3.2 Stripline with Off-set Strip Conductor

As in the case of a homogeneous symmetric stripline, the characteristic impedance of a homogeneous stripline with off-set centre conductor can also be normalized with respect to $\sqrt{\epsilon_r}$ to render the expression for $Z\sqrt{\epsilon_r}$ independent of ϵ_r. This can be easily verified by substituting the capacitance expression (5.5) in (5.1). A typical plot of the variation in $Z\sqrt{\epsilon_r}$ as a function of w/b for various values of the normalized off-set distance d/b is shown in Fig. 5.6. The results are computed using (5.5) by assuming $t/b=0$ and $c/w=40$. It can be seen that

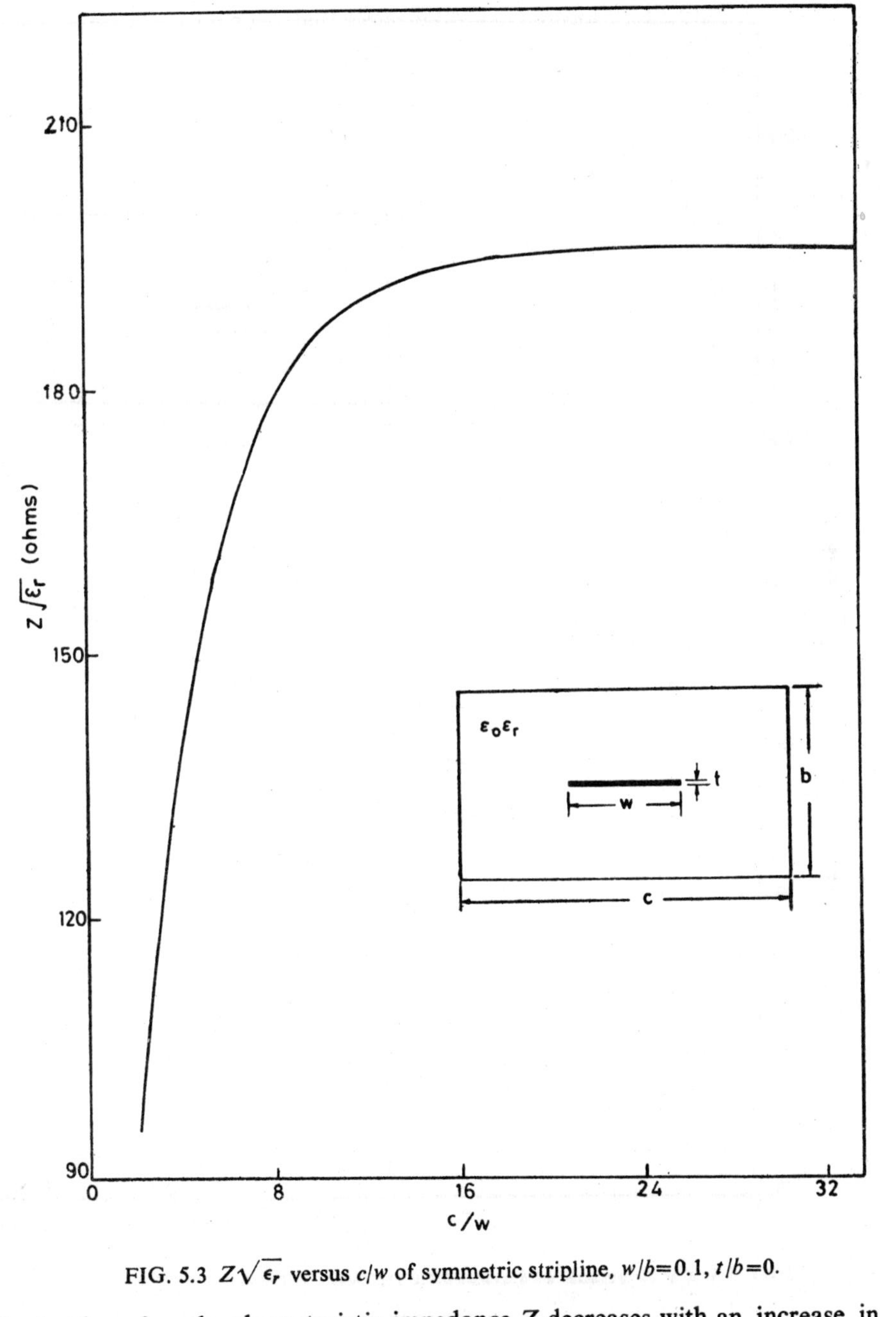

FIG. 5.3 $Z\sqrt{\epsilon_r}$ versus c/w of symmetric stripline, $w/b=0.1$, $t/b=0$.

for a fixed value of ϵ_r, the characteristic impedance Z decreases with an increase in w/b as well as in d/b. Table 5.2 gives detailed data for various values of d/b.

Closed form empirical formula (5.10) due to Rosenzweig [10] enables computation of the characteristic impedance of the stripline as a function of the off-set distance d as well as the finite thickness t of the strip conductor. These formulas are used to generate data for several combinations of d/b and t/b, and the data are listed in Tables 5.3 through 5.10.

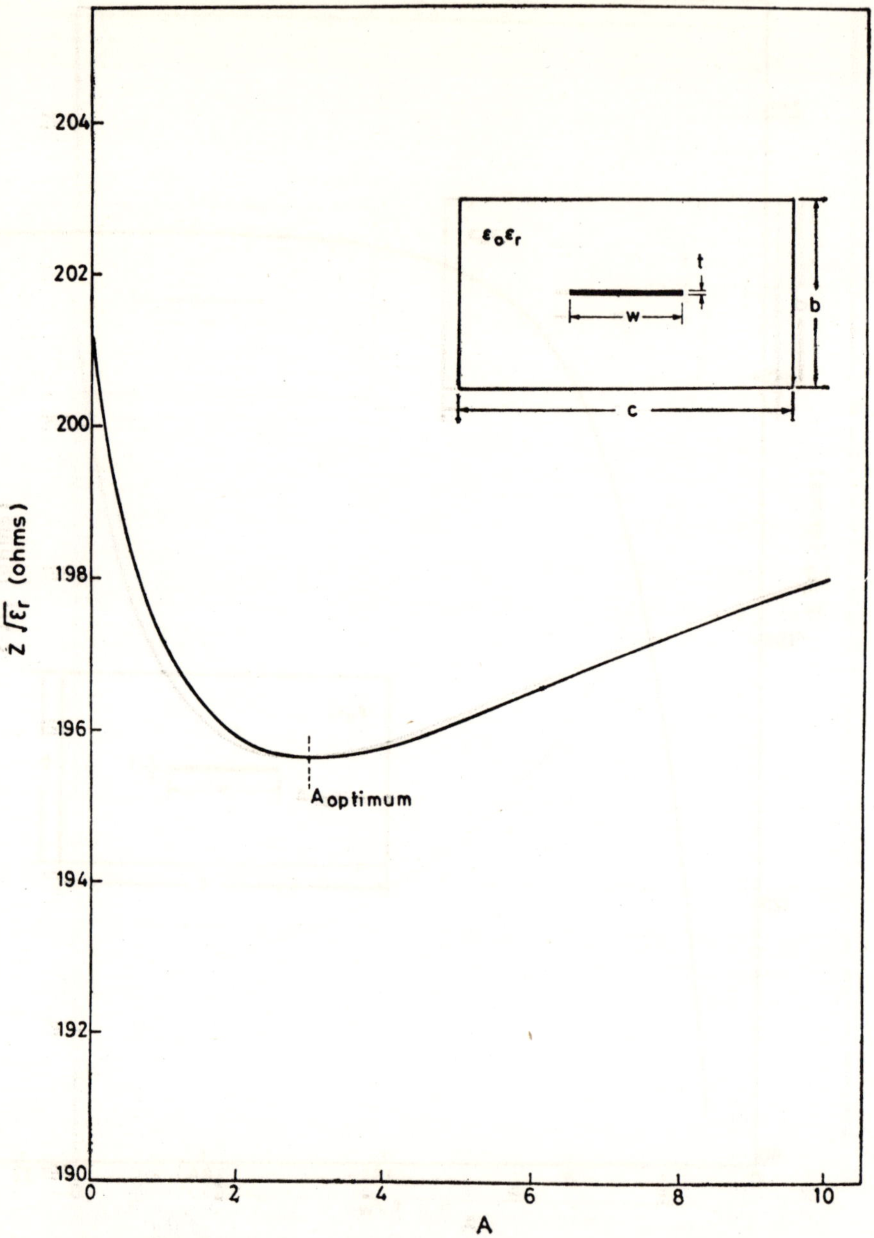

FIG. 5.4 $Z\sqrt{\epsilon_r}$ versus A of symmetric stripline, $w/b=0.1$, $c/w=40$, $t/b=0$.

5.3.3 Microstrip-like Stripline

An interesting property of the microstrip-like stripline (structure A_3 of Table 3.1) is that the effective dielectric constant ϵ_f, and hence, the ratio of guide wavelength to free space wavelength λ/λ^a is independent of the dimensional parameters of the structure. The value of ϵ_f is $(\epsilon_r+1)/2$ where ϵ_r is the relative dielectric constant of the substrate. It behaves like a homo-

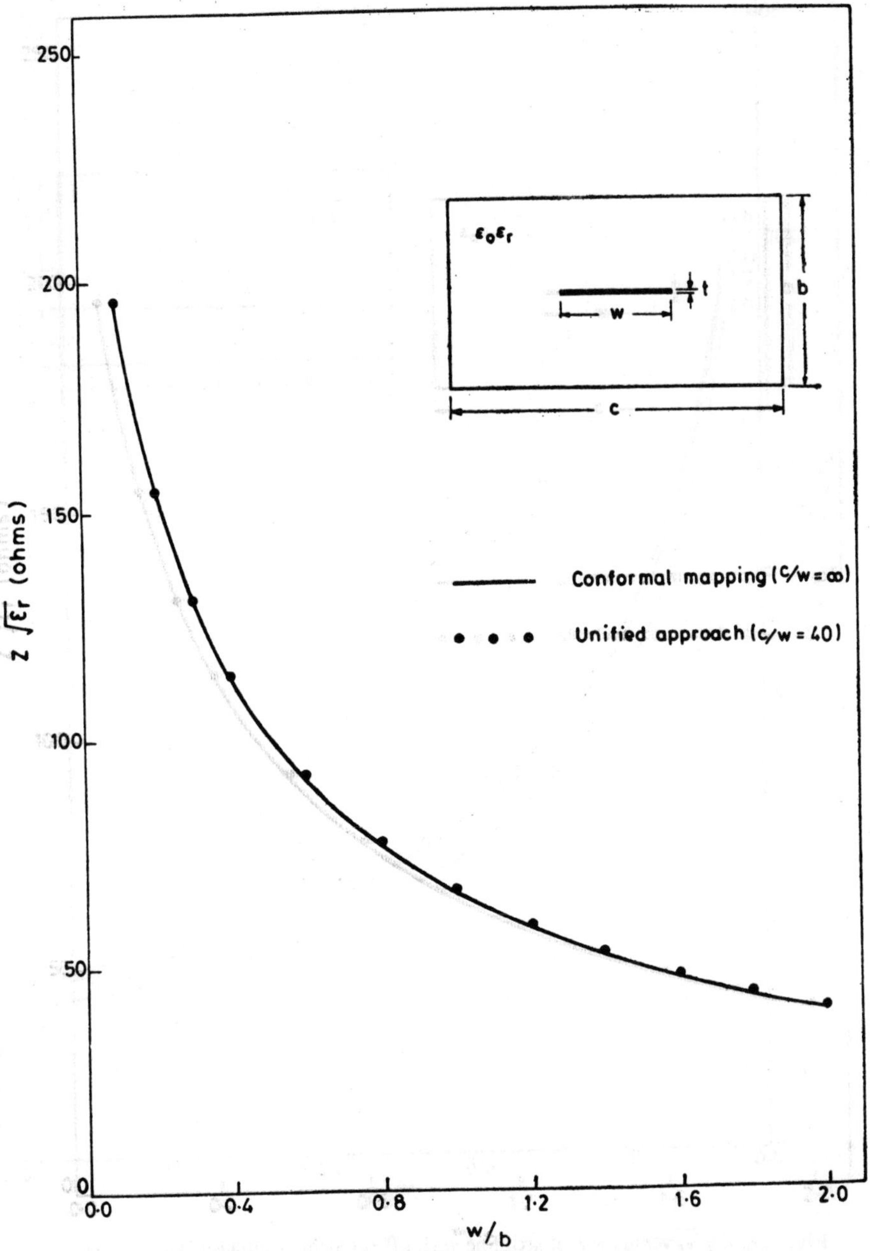

FIG. 5.5 $Z\sqrt{\epsilon_r}$ versus w/b of symmetric stripline—comparison of computed results using unified approach formula [(5.5) with (5.1), $c/w=40$, $t/b=0$] with conformal mapping method [(5.8), $c/w=\infty$, $t/b=0$].

geneous symmetric stripline completely filled with a dielectric having a relative dielectric constant $(\epsilon_r+1)/2$. Hence, the impedance data of the symmetric stripline presented in Table 5.1 in terms of $Z\sqrt{\epsilon_r}$ for $t/b=0$ can be directly used by replacing the normalization factor ϵ_r

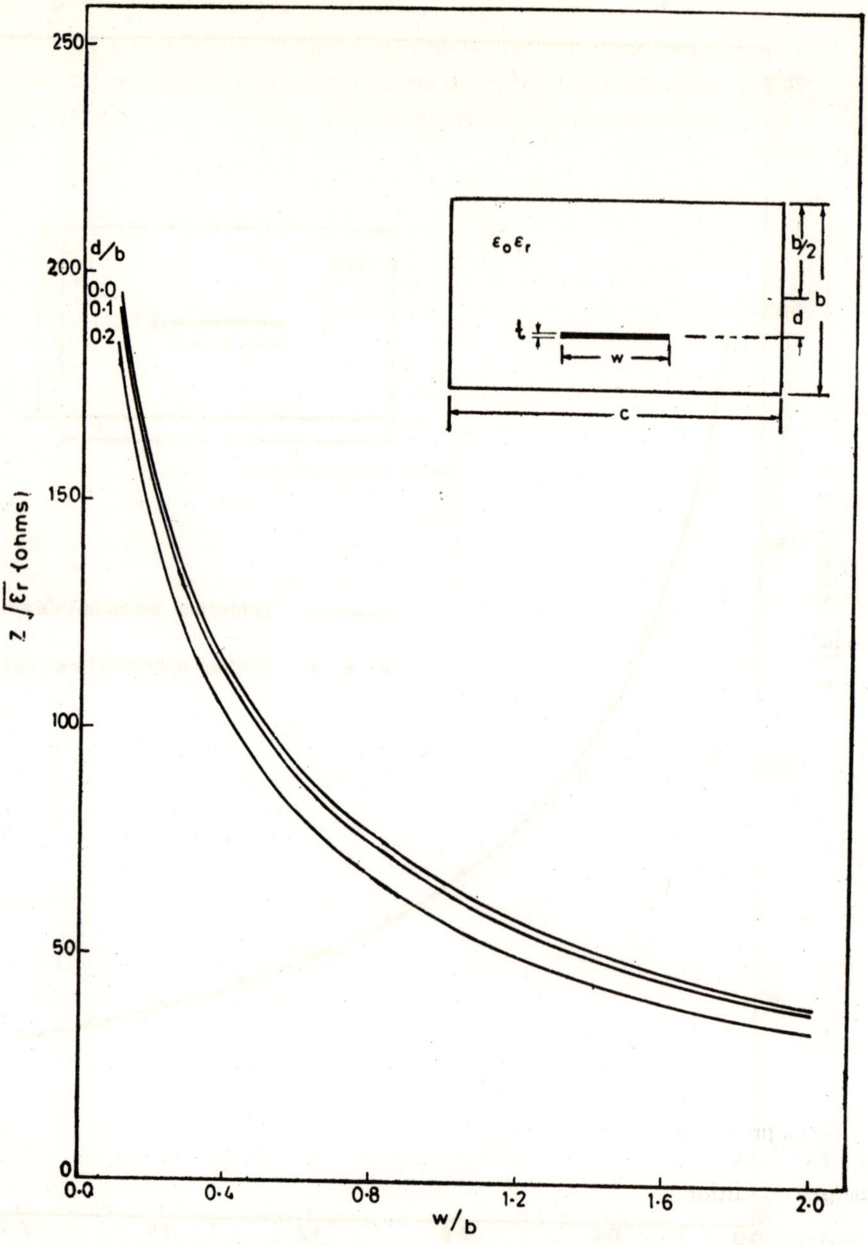

FIG. 5.6 $Z\sqrt{\epsilon_r}$ versus w/b of stripline with off-set strip conductor with d/b as parameter, $t/b=0$.

by $(\epsilon_r+1)/2$. In other words, the characteristic impedance Z of the microstrip-like stripline for a zero thickness strip conductor can be obtained by simply dividing the tabulated values of $Z\sqrt{\epsilon_r}$ by $\sqrt{(\epsilon_r+1)/2}$.

In Fig. 5.7, the plots of Z and λ/λ^a for the case $b_1/b_2=1$ correspond to the microstrip-like stripline. When the height of the air region is greater than that of the dielectric substrate

($b_1/b_2 > 1$), ϵ_f increases or equivalently λ/λ^a decreases with an increase in w/b_2. On the other hand, for $b_1/b_2 < 1$, ϵ_f decreases and consequently λ/λ^a increases with an increase in w/b_2. The characteristic impedance Z, however, decreases with increase in w/b_2 for any fixed value of the ratio b_1/b_2.

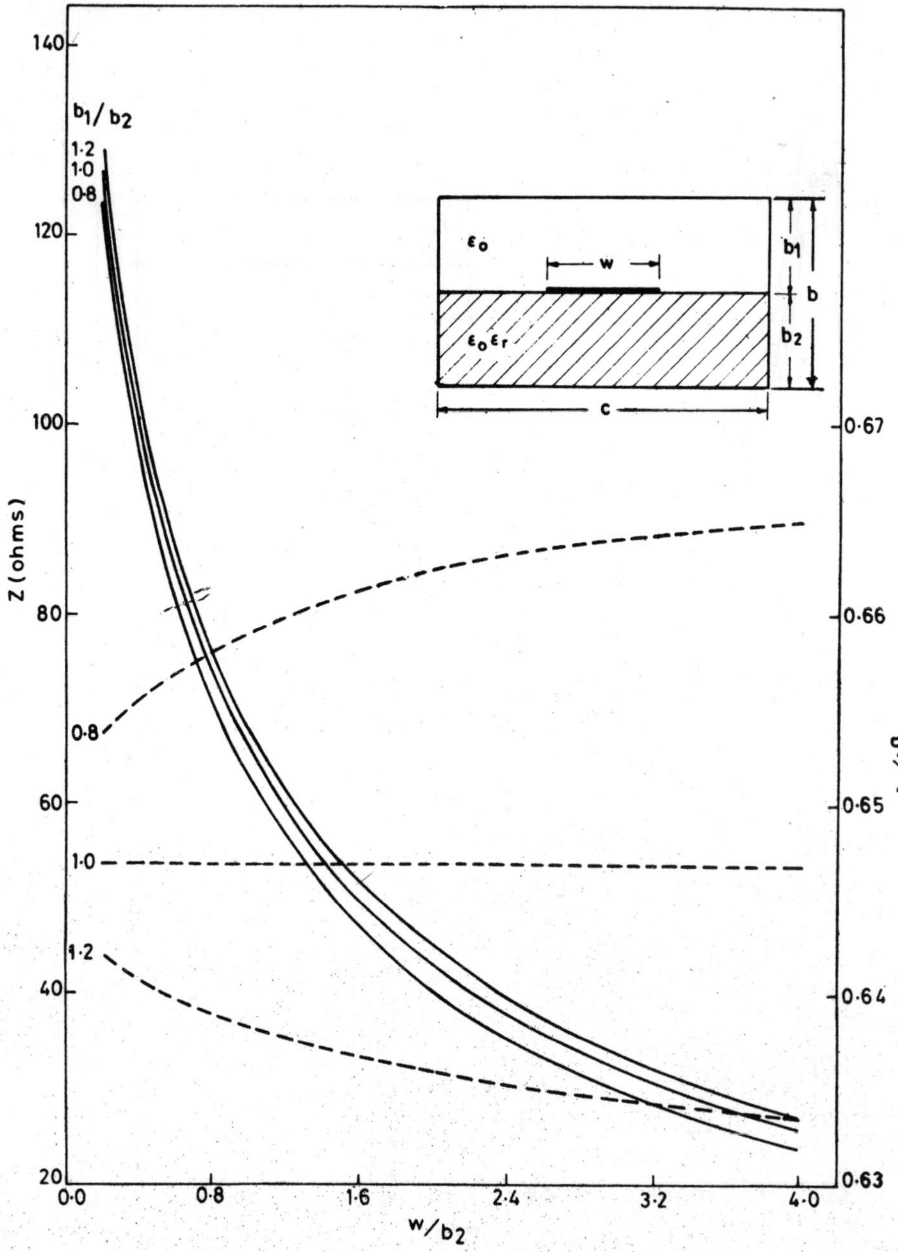

FIG. 5.7 Z and λ/λ^a versus w/b_2 of microstrip-like stripline for various values of b_1/b_2, ϵ_r=3.78, c/w=40: ——— Z, — — — λ/λ^a.

5.3.4 *Suspended Stripline*

In the suspended stripline, the dielectric substrate is placed symmetrically between the two ground planes with an equal amount of air gap on either side of the substrate (structure A4 of Table 3.1). Figure 5.8 depicts the variation in Z and λ/λ^a as a function of w/b for

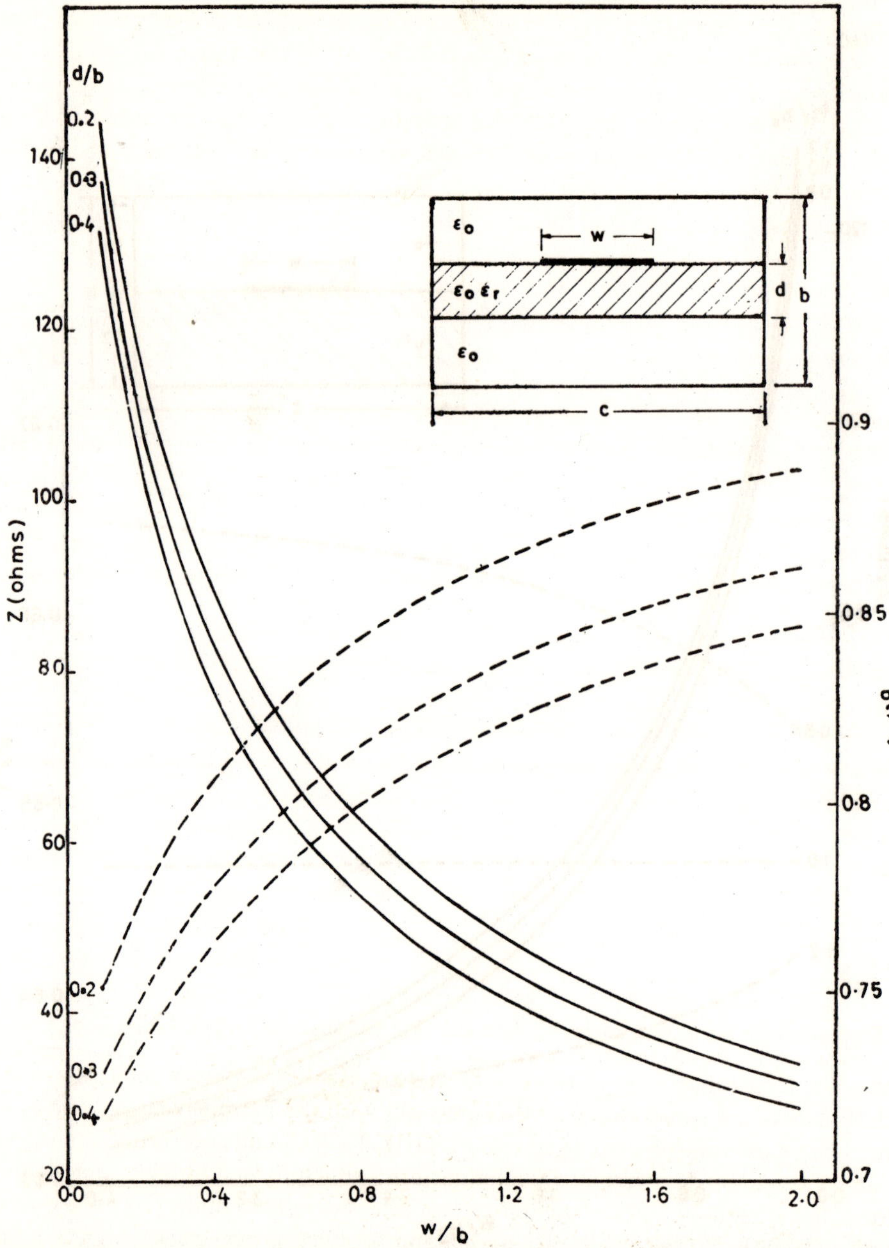

FIG. 5.8 Z and λ/λ^a versus w/b of suspended stripline with d/b as parameter, $\epsilon_r = 3.78$, $c/w = 40$; ——— Z, — — — λ/λ^a.

three different thicknesses of the substrate. For a fixed value of w/b, both Z and λ/λ^a decrease with an increase in d/b, whereas, for a fixed value of d/b, Z decreases but λ/λ^a inareases with an increase in w/b.

Tables 5.11-5.13 give detailed design data on Z and ϵ_f of the suspended stripline for three commonly used substrates, namely, duroid ($\epsilon_r = 2.22$), quartz ($\epsilon_r = 3.78$), and alumina ($\epsilon_r = 9.6$). Unified capacitance formula (5.5) is used for computing the results and the data are presented for values of d/b in the range 0.1 to 0.5 and w/b in the range 0.1 to 2.5. In all the computations, c/w is set equal to 40, and $t/b = 0$ is implicit in the formula itself.

An important property of the suspended stripline is its low effective dielectric constant. For example, in a suspended stripline using duroid substrate, ϵ_f lies between 1.085 and 1.456 for structural parameters in the range $0.1 \leqslant w/b \leqslant 2.5$ and $0.1 \leqslant d/b \leqslant 0.5$ (Table 5.11); specifically, the lowest value of ϵ_f equal to 1.085 is for $w/b = 2.5$ and $d/b = 0.1$, and the highest value 1.456 is for $w/b = 0.1$ and $d/b = 0.5$. Lower the value of ϵ_f, closer would be the mode of propagation to the TEM mode and weaker would be the dispersion. The dispersion characteristics are discussed in detail in Chapter 9. Another important advantage of the suspended stripline is its low attenuation constant. These distinguishing features of the suspended stripline make it eminently suitable for use at millimeter wave frequencies.

Practical suspended striplines employ grooves in the side walls of the metallic enclosure to support the dielectric substrate (inset to Figs. 5.9a and b). Yamashita *et al.* [14] have analyzed the effect of grooves on the characteristic impedance and guide wavelength of the line. Their results are reproduced in Figs. 5.9a and b. Clearly, the introduction of grooves increases Z as well as λ/λ^a. The deeper the grooves, the higher are the values of Z and λ/λ^a for normalized groove depth g/b up to about 0.1, and for values of g/b beyond 0.1, the effect is almost negligible.

5.3.5 Sandwiched Stripline

The sandwiched stripline considered here consists of a three-layered structure, in which, a substrate of high dielectric constant is sandwiched symmetrically between the ground planes with another dielectric layer of lower dielectric constant on either side (structure A5 of Table 3.1). Figure 5.10 shows the variation of Z and λ/λ^a as a function of w/b for a stripline with Epsilam-10 substrate ($\epsilon_{r2} = 10.2$) sandwiched between two duroid substrates ($\epsilon_{r1} = 2.22$). For $\epsilon_{r2} > \epsilon_{r1}$ as in the present case, the nature of variation of Z and λ/λ^a as a function of w/b and d/b is similar to that observed in the suspended stripline case. The detailed design data for Z and ϵ_f are presented in Table 5.14 for a practical structure with parameters $\epsilon_{r1} = 2.22$ and $\epsilon_{r2} = 2.56$, and in Table 5.15 for a structure with parameters $\epsilon_{r1} = 2.22$ and $\epsilon_{r2} = 10.2$.

5.4 Transmission Structures with Uniaxially Anisotropic Dielectric Substrates

In practice, the permittivity of alumina ceramic substrates is known to vary from batch to batch. Further, the substrates exhibit dispersion, with the permittivity varying as a function of frequency (about 2 per cent over 1 to 10 GHz) [15]. Circuits fabricated in suspended stripline and microstrip-like stripline configurations, which demand highly repeatable electrical performance, therefore, utilize sapphire as an alternative to alumina [16]. The dielectric constant of sapphire is close to that of alumina and it offers lower losses, higher homogeneity, and highly repeatable electrical characteristics from specimen to specimen. Sapphire, how-

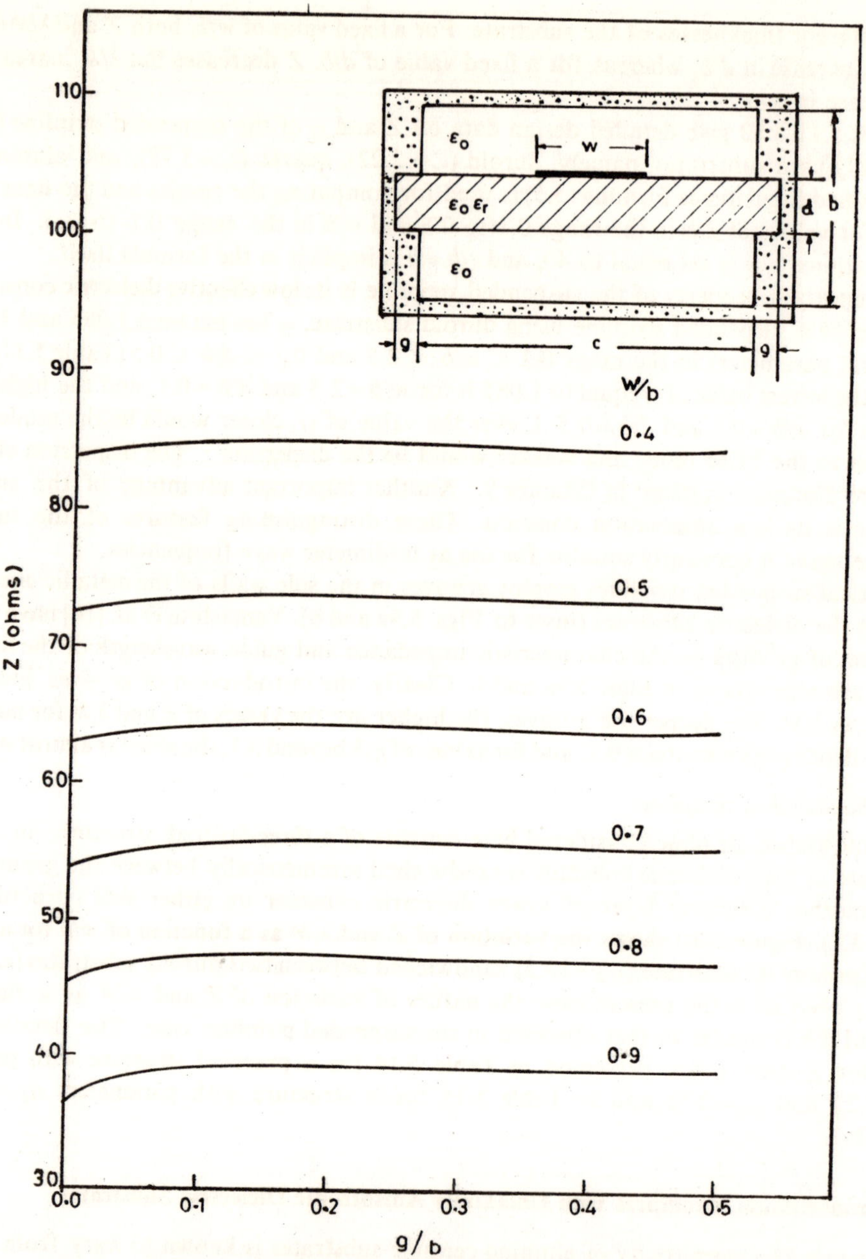

FIG. 5.9 (a) Effect of side wall grooves on characteristic impedance of suspended stripline, ϵ_r=2.22, d/b=0.2, c/b=1.0 (From Yamashita et al [14], Copyright © 1985 IEEE, reprinted with permission).

ever, is anisotropic. It is normally cut with the optical axis perpendicular to the plane of the substrate (*x-z* plane) and has a permittivity given by $\epsilon_{xx}=\epsilon_{zz}=9.4$ and $\epsilon_{yy}=11.6$. Pyrolytic boron nitride is another useful crystalline substrate possessing anisotropic tensor permittivity

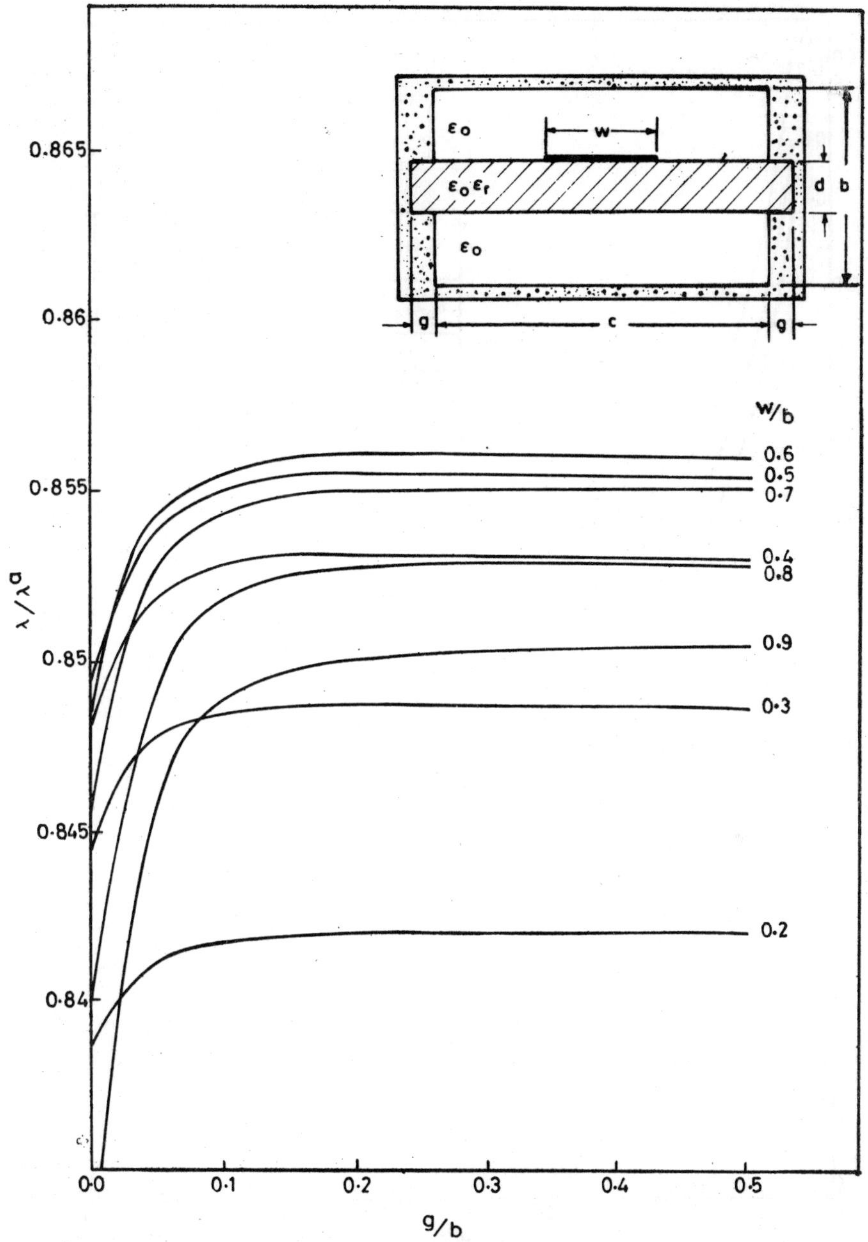

FIG. 5.9 (b) Effect of side wall grooves on normalized guide wavelength of suspended stripline, $\epsilon_r = 2.22$, $d/b = 0.2$, $c/b = 1.0$ (From Yamashita et al [14], Copyright © 1985 IEEE, reprinted with permission).

with the elements of the tensor given by $\epsilon_{xx} = \epsilon_{zz} = 5.12$ and $\epsilon_{yy} = 3.4$ [17]. This can be used as an alternative to the isotropic quartz substrate having a relative dielectric constant $\epsilon_r = 3.78$. Epsilam-10 being a flexible substrate is useful for homogeneous striplines and this material is known to exhibit a certain amount of anisotropy given by $\epsilon_{xx} = \epsilon_{zz} = 13$ and $\epsilon_{yy} = 10.2$.

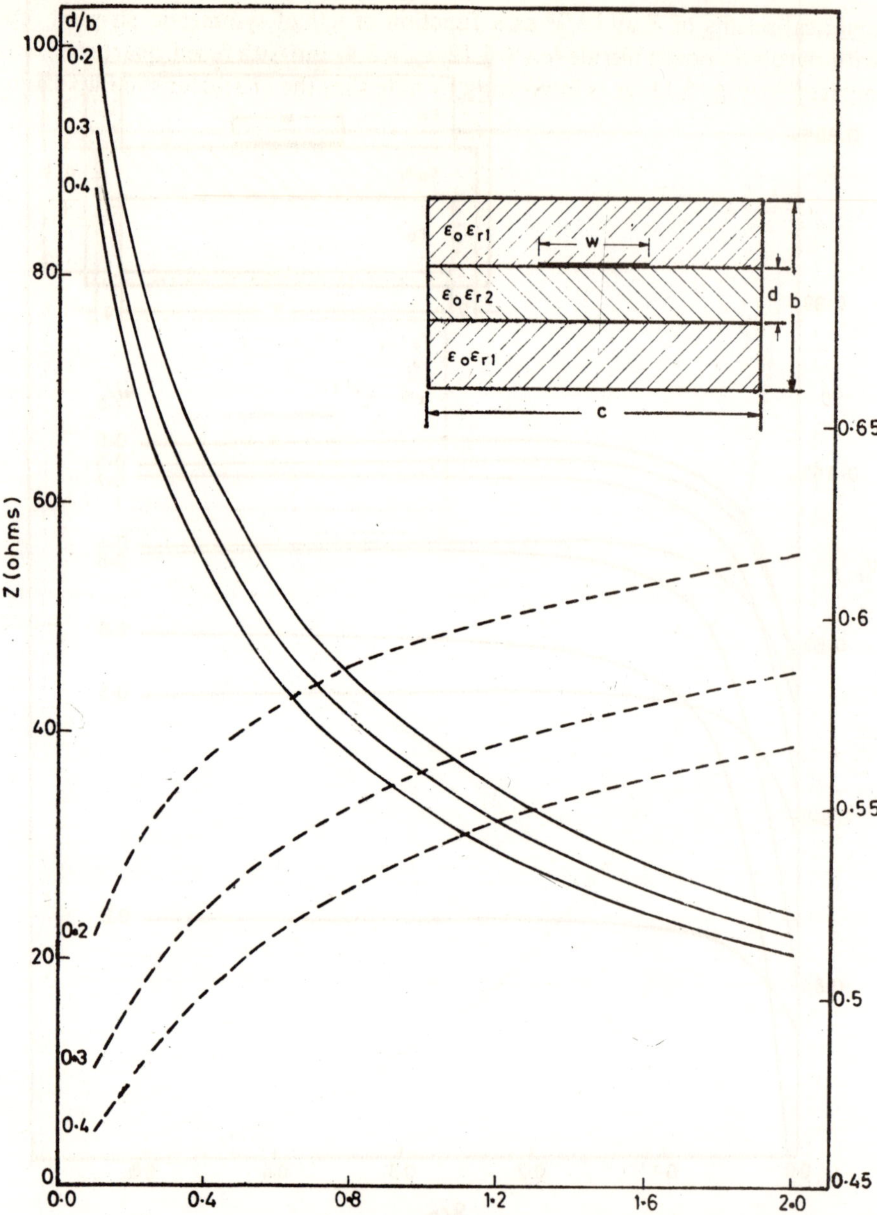

FIG. 5.10 Z and λ/λ^a versus w/b of sandwiched stripline with d/b as parameter, $\epsilon_{r1}=2.22$, $\epsilon_{r2}=10.2$, $c/w=40$; ———— Z , — — — λ/λ^a.

In this section, we present design data on homogeneous striplines, microstrip-like striplines, and suspended striplines using sapphire and pyrolytic boron nitride substrates. Design data are also presented for homogeneous striplines with Epsilam-10 dielectric taking into account its anisotropy. Unified capacitance formula (5.5) is used to generate the data.

5.4.1 Symmetric Stripline

The numerical results of Z and λ/λ^a as a function of w/b of symmetric stripline completely filled with pyrolytic boron nitride ($\epsilon_{xx} = 5.12$, $\epsilon_{yy} = 3.4$) and with fused quartz ($\epsilon_{xx} = \epsilon_{yy} = 3.78$) are compared in Fig. 5.11. It is interesting to note that the characteristic impedances in both

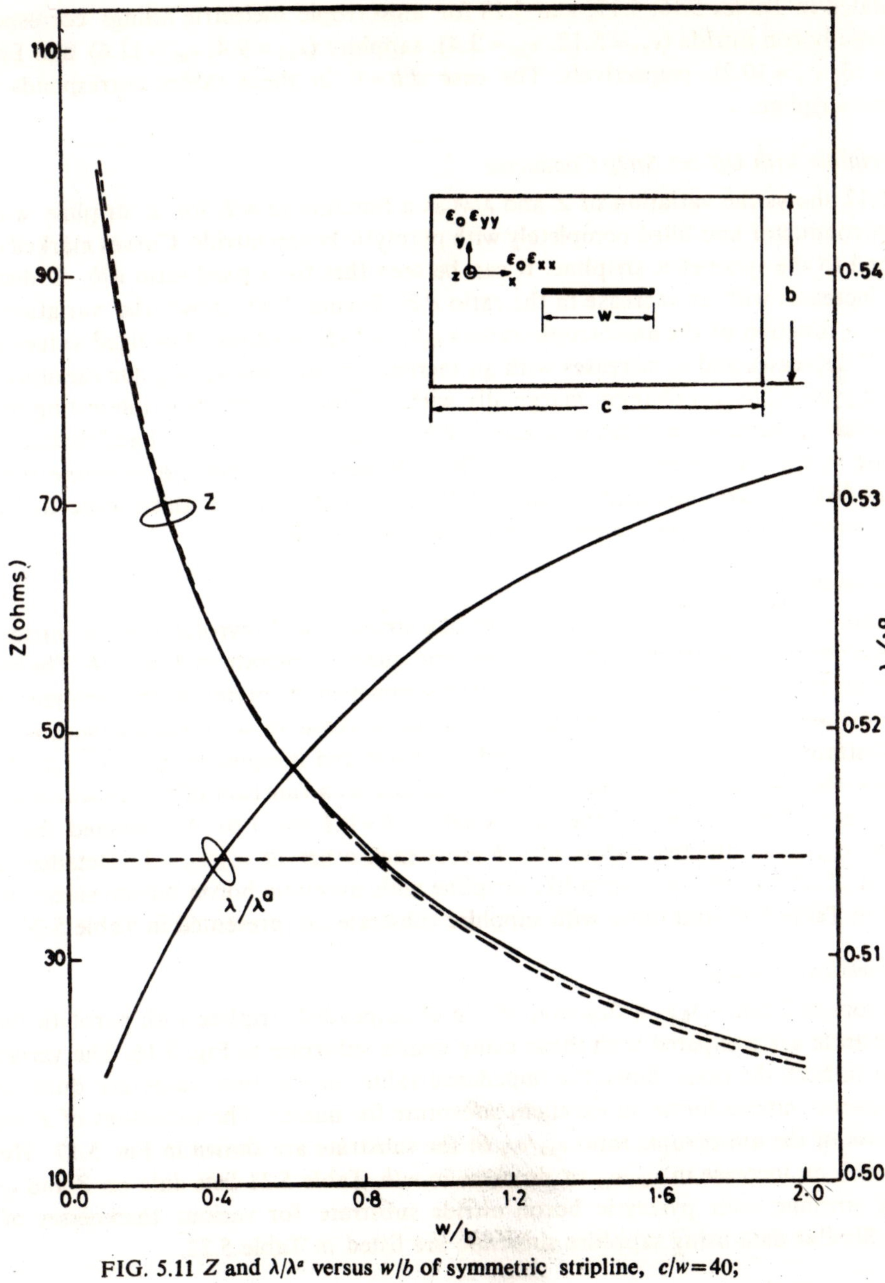

FIG. 5.11 Z and λ/λ^a versus w/b of symmetric stripline, $c/w=40$;
——— $\epsilon_{xx} = \epsilon_{zz} = 5.12$, $\epsilon_{yy} = 3.4$, — — — $\epsilon_{xx} = \epsilon_{yy} = \epsilon_{zz} = 3.78$.

cases are nearly the same, while the normalized guide wavelength varies quite differently. This is because, with isotropic dielectric, the effective dielectric constant ϵ_f which is the same as the relative dielectric constant ϵ_r of the dielectric is independent of w/b, whereas with the anisotropic dielectric, it changes with a change in w/b. In the case of pyrolytic boron nitride, ϵ_f decreases and, hence, λ/λ^a increases with an increase in w/b. Detailed data on Z and ϵ_f are tabulated in Tables 5.16, 5.17, and 5.18 for anisotropic dielectric fillings corresponding to pyrolytic boron nitride ($\epsilon_{xx}=5.12$, $\epsilon_{yy}=3.4$), sapphire ($\epsilon_{xx}=9.4$, $\epsilon_{yy}=11.6$) and Epsilam-10 ($\epsilon_{xx}=13$, $\epsilon_{yy}=10.2$), respectively. The case $d/b=0$ in these tables corresponds to the symmetric stripline.

5.4.2 Stripline with Off-set Strip Conductor

Figure 5.12 shows the variation of Z and λ/λ^a as a function of w/b for a stripline with off-set centre conductor and filled completely with pyrolytic boron nitride. Curves marked $d/b=0$ correspond to the symmetric stripline. It can be seen that for a fixed ratio w/b, Z decreases and λ/λ^a increases with an increase in the ratio d/b. Figure 5.13 shows the variation of Z and ϵ_f as a function of the anisotropic ratio $\epsilon_{xx}/\epsilon_{yy}$ of the medium. For fixed values of w/b and d/b, Z decreases and ϵ_f increases with an increase in the ratio $\epsilon_{xx}/\epsilon_{yy}$. On the other hand, for a fixed ratio $\epsilon_{xx}/\epsilon_{yy}$, ϵ_f changes marginally with variation in d/b. It is interesting to note, however, that ϵ_f increases with an increase in d/b for values of $\epsilon_{xx}/\epsilon_{yy} < 1$ and decreases with an increase in d/b for values of $\epsilon_{xx}/\epsilon_{yy} > 1$. Detailed data on Z and ϵ_f of stripline for various off-set distances are tabulated in Tables 5.16, 5.17 and 5.18 for anisotropic dielectrics; pyrolytic boron nitride, sapphire and Epsilam-10, respectively.

5.4.3 Microstrip-like Stripline

The variations of Z and λ/λ^a of microstrip-like stripline with pyrolytic boron nitride substrate are compared with those using the isotropic quartz substrate in Fig. 5.14. The behaviour is similar to that shown in Fig. 5.11 for the symmetric stripline. With pyrolytic boron nitride substrate, the ratio λ/λ^a increases with an increase in w/b, whereas with isotropic quartz substrate ($\epsilon_r=\epsilon_{xx}=\epsilon_{yy}$), it is independent of w/b and is equal to $\sqrt{2/(1+\epsilon_r)}$. Figure 5.15 shows the variation of Z and ϵ_f of the structure as a function of the anisotropic ratio $\epsilon_{xx}/\epsilon_{yy}$ for two different values of the strip width ($w/b=0.1$ and 1.0). As expected, the curves of ϵ_f versus $\epsilon_{xx}/\epsilon_{yy}$ for the two values of w/b cross each other at $\epsilon_{xx}/\epsilon_{yy}=1$. Detailed tables of data on Z and ϵ_f of microstrip-like stripline with pyrolytic boron nitride substrate are presented in Table 5.19 and those with sapphire substrate are presented in Table 5.20.

5.4.4 Suspended Stripline

The variations of Z and λ/λ_0 as a function of w/b of suspended stripline with pyrolytic boron nitride substrate are compared with those using quartz substrate in Fig. 5.16. The variations are similar in both the cases. Since the impedance values in the two cases are quite close, pyrolytic boron nitride forms an excellent substitute for quartz. The variations of Z and ϵ_f as a function of the anisotropic ratio $\epsilon_{xx}/\epsilon_{yy}$ of the substrate are shown in Fig. 5.17. Here ϵ_f increases with an increase in $\epsilon_{xx}/\epsilon_{yy}$ or decrease in w/b. Table 5.21 lists data on Z and ϵ_f for suspended stripline with pyrolytic boron nitride substrate for various thicknesses of the substrate. Similar data using sapphire substrate are listed in Table 5.22.

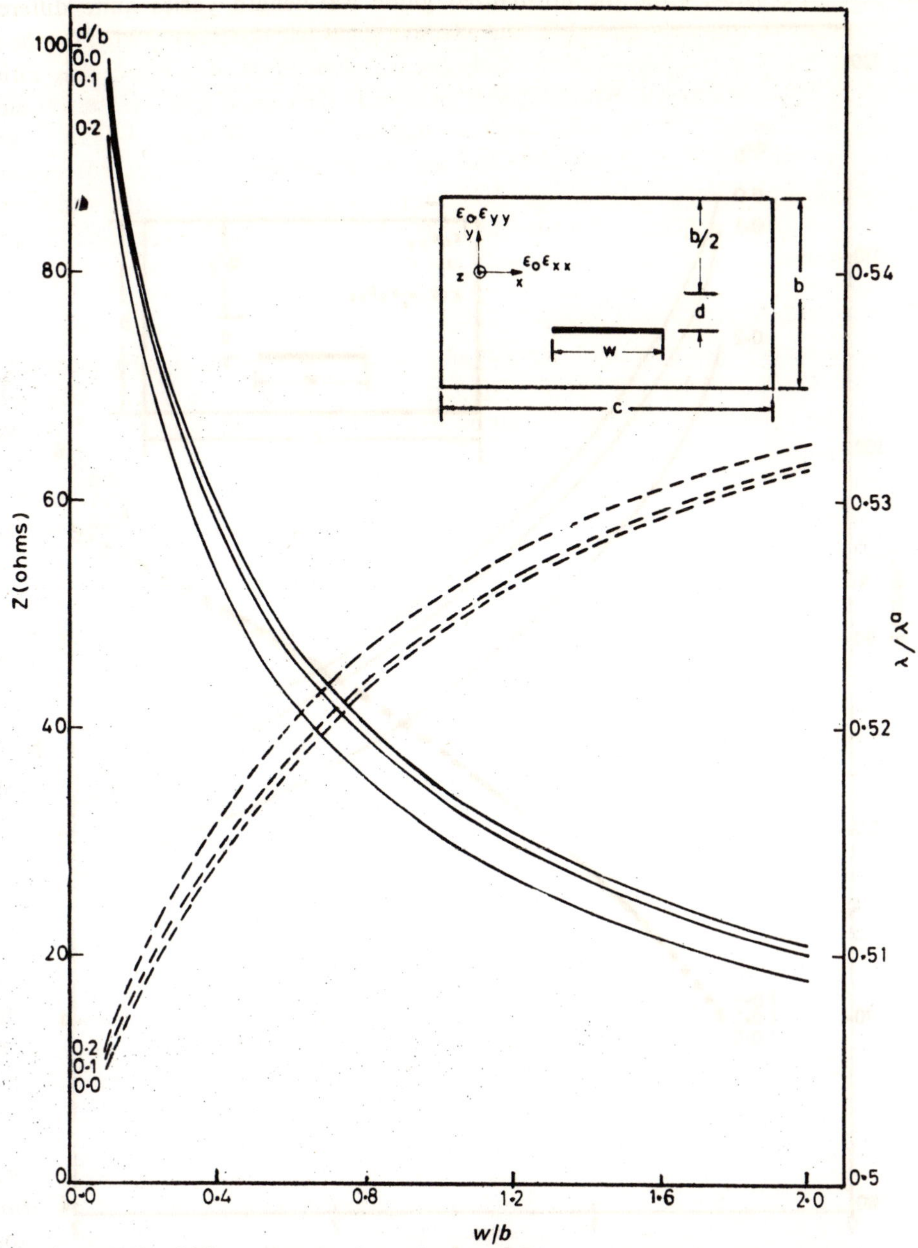

FIG. 5.12 Z and λ/λ^e versus w/b of stripline with off-set strip conductor with uniaxially anistoropic dielectric substrate (pyrolytic boron nitride), $\epsilon_{xx}=\epsilon_{zz}=5.12$, $\epsilon_{yy}=3.4$, $c/w=40$; ——— Z, — — — λ/λ^e.

5.5 Transmission Structures using Anisotropic Dielectric Substrate with Principal Axes Rotated with Respect to the Coordinate Axes

Anisotropic pyrolytic boron nitride and sapphire substrates can be cut with the optical axis lying in the transverse plane (x-y plane) but oriented at an arbitrary angle θ with respect to

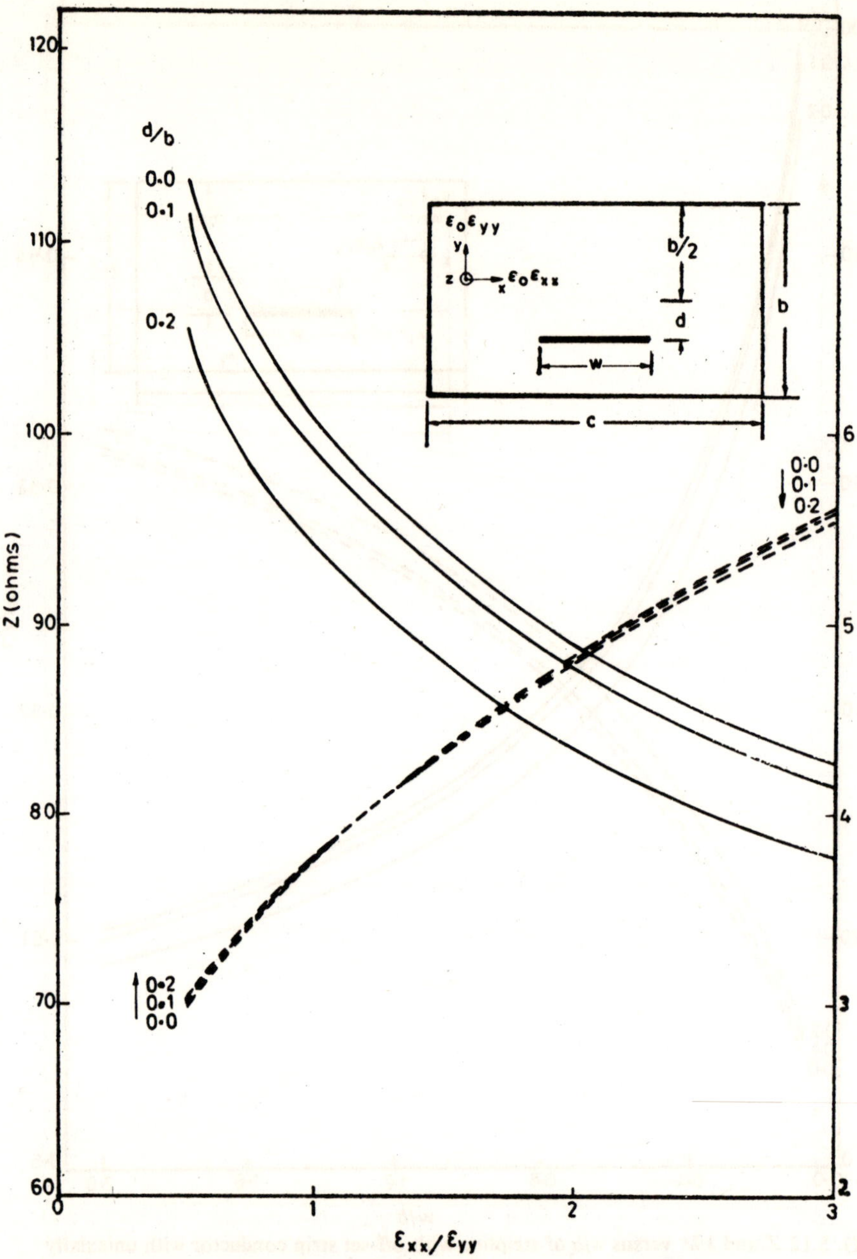

FIG. 5.13 Z and ϵ_f versus anisotropic ratio $\epsilon_{xx}/\epsilon_{yy}$ of stripline with off-set strip conductor, $\epsilon_{yy}=3.78$, $w/b=0.1$, $c/w=40$; ———, Z — — — ϵ_f.

the plane of the substrate (*x-z* plane). The tilt angle θ, then, serves as an additional parameter which can be used for controlling the characteristic impedance Z and effective dielectric constant ϵ_f of the transmission line. Typical stripline-like transmission structures using

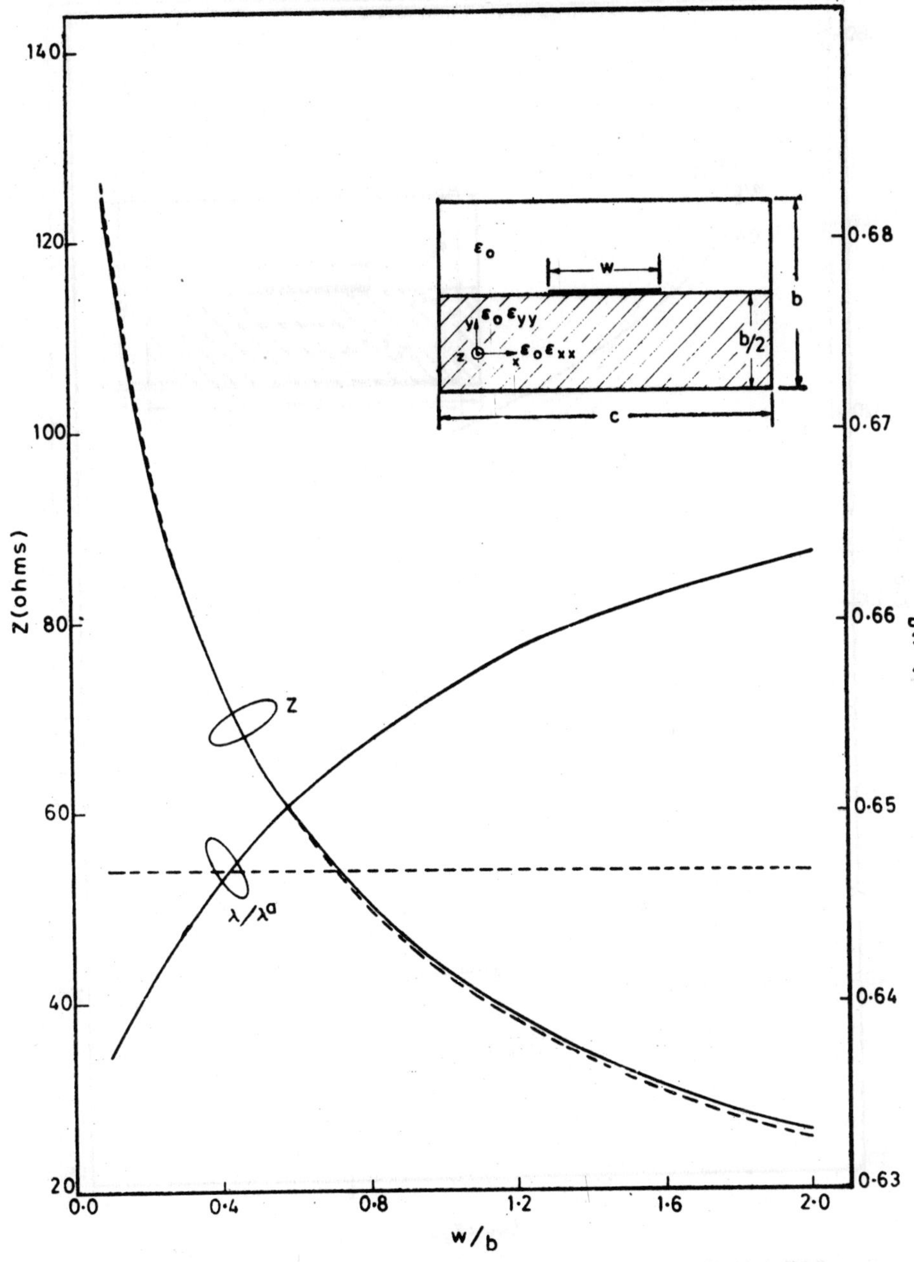

FIG. 5.14 Z and λ/λ^a versus w/b of microstrip-like stripline, $c/w=40$; ——— $\epsilon_{xx}=\epsilon_{zz}=5.12$, $\epsilon_{yy}=3.4$; — — —$\epsilon_{xx}=\epsilon_{yy}=\epsilon_{zz}=3.78$.

such anisotropic substrates are shown in Table 3.5 along with their admittance parameters. The unified capacitance formula (5.5) is used to generate characteristics of Z and ϵ_f as a

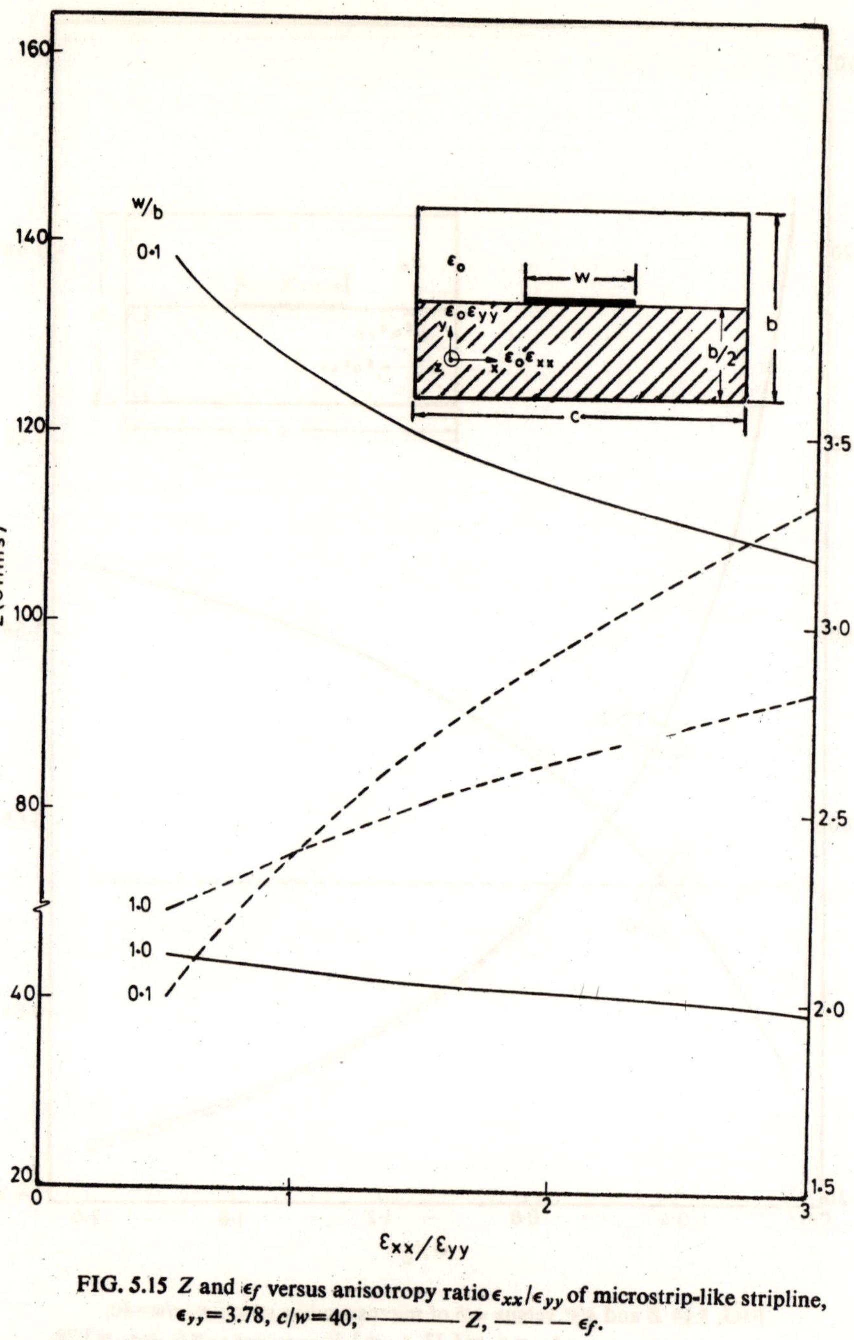

FIG. 5.15 Z and ϵ_f versus anisotropy ratio $\epsilon_{xx}/\epsilon_{yy}$ of microstrip-like stripline, ϵ_{yy}=3.78, c/w=40; ———— Z, — — — ϵ_f.

function of the tilt angle θ. The characteristics are presented in graphical form in order to illustrate the effect of the tilt angle θ on the propagation parameters of various structures.

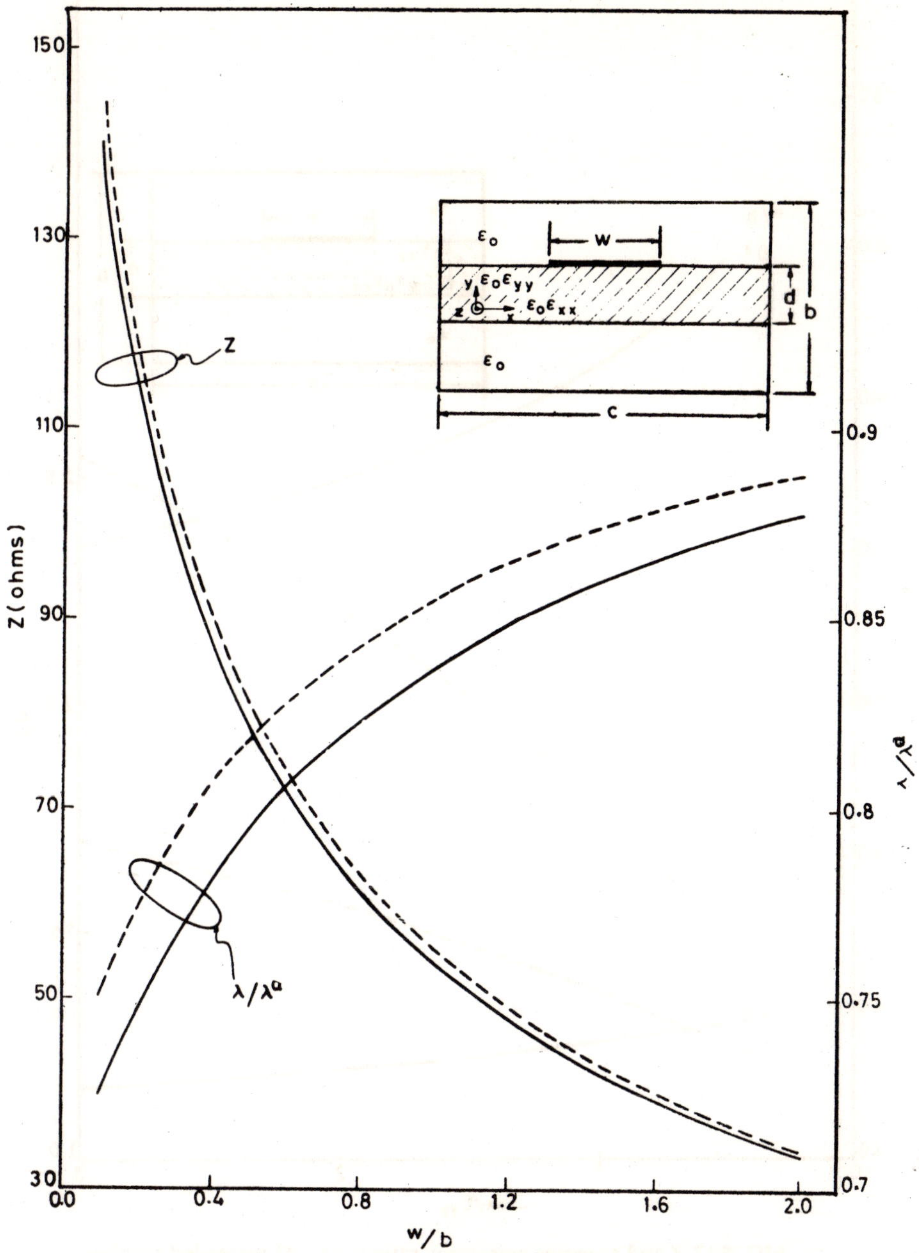

FIG. 5.16 Z and λ/λ^a versus w/b of suspended stripline, $d/b=0.2$;
———— $\epsilon_{xx}=\epsilon_{zz}=5.12$, $\epsilon_{yy}=3.4$; — — — $\epsilon_{xx}=\epsilon_{yy}=\epsilon_{zz}=3.78$.

Figures 5.18 and 5.19 illustrate the variations in Z and ϵ_f as a function of the tilt angle θ for symmetric stripline filled with pyrolytic boron nitride and sapphire dielectric, respectively. It is seen from Fig. 5.18, that with pyrolytic boron nitride, Z decreases and ϵ_f increases

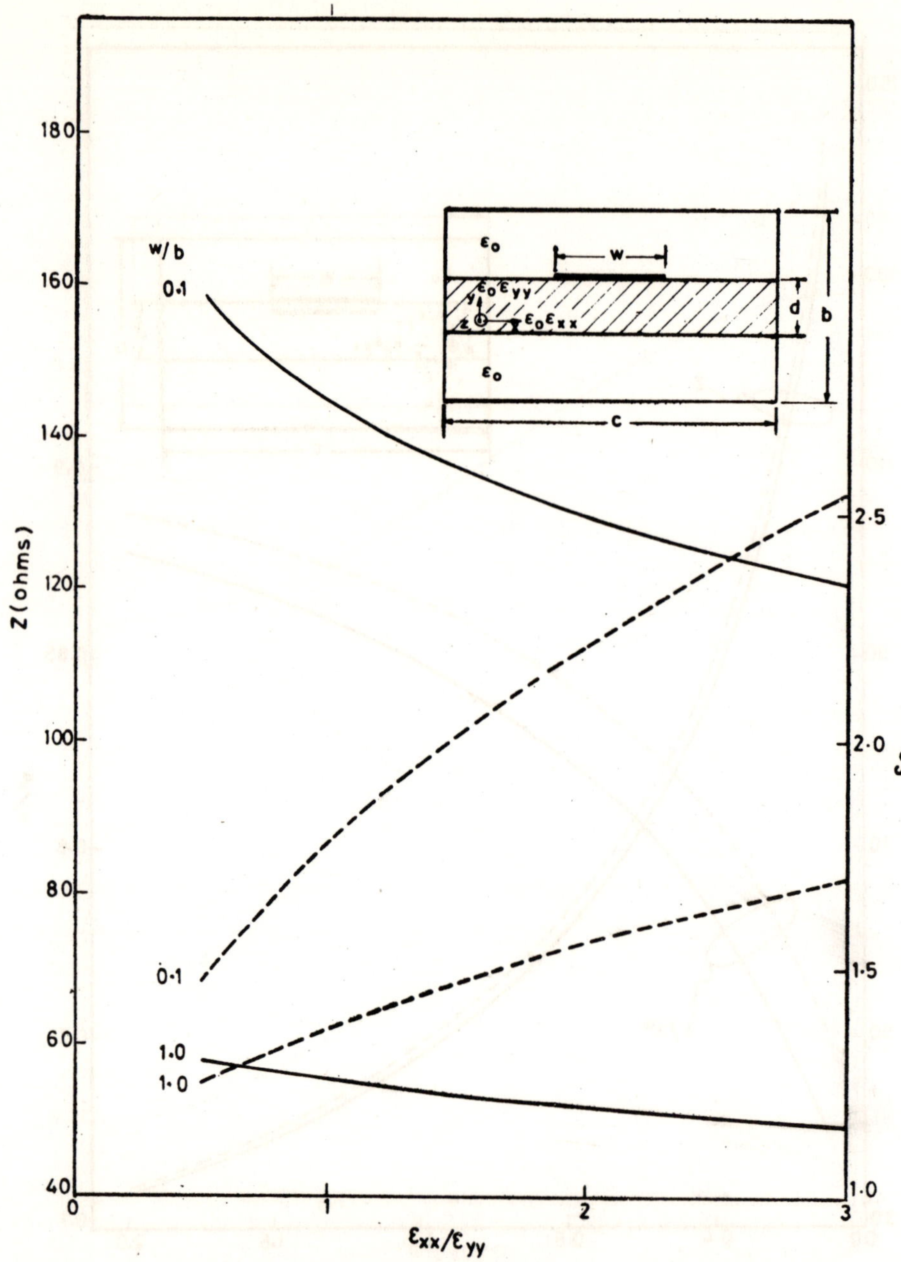

FIG. 5.17 Z and ϵ_f versus anisotropy ratio $\epsilon_{xx}/\epsilon_{yy}$ of suspended stripline, $\epsilon_{yy}=3.78$, $d/b=0.2$, $c/w=40$; ———— Z, — — — ϵ_f.

when the tilt angle θ is increased from 0° to 90°. This behaviour can be explained as follows. For $\theta=0$, the horizontal component of the permittivity ($\epsilon_{xx}=\epsilon_\xi=5.12$) is higher than the vertical component ($\epsilon_{yy}=\epsilon_\eta=3.4$). As θ is increased, the value of the vertical component increases and reaches a maximum value 5.12 at $\theta=90°$, whereas, the value of the

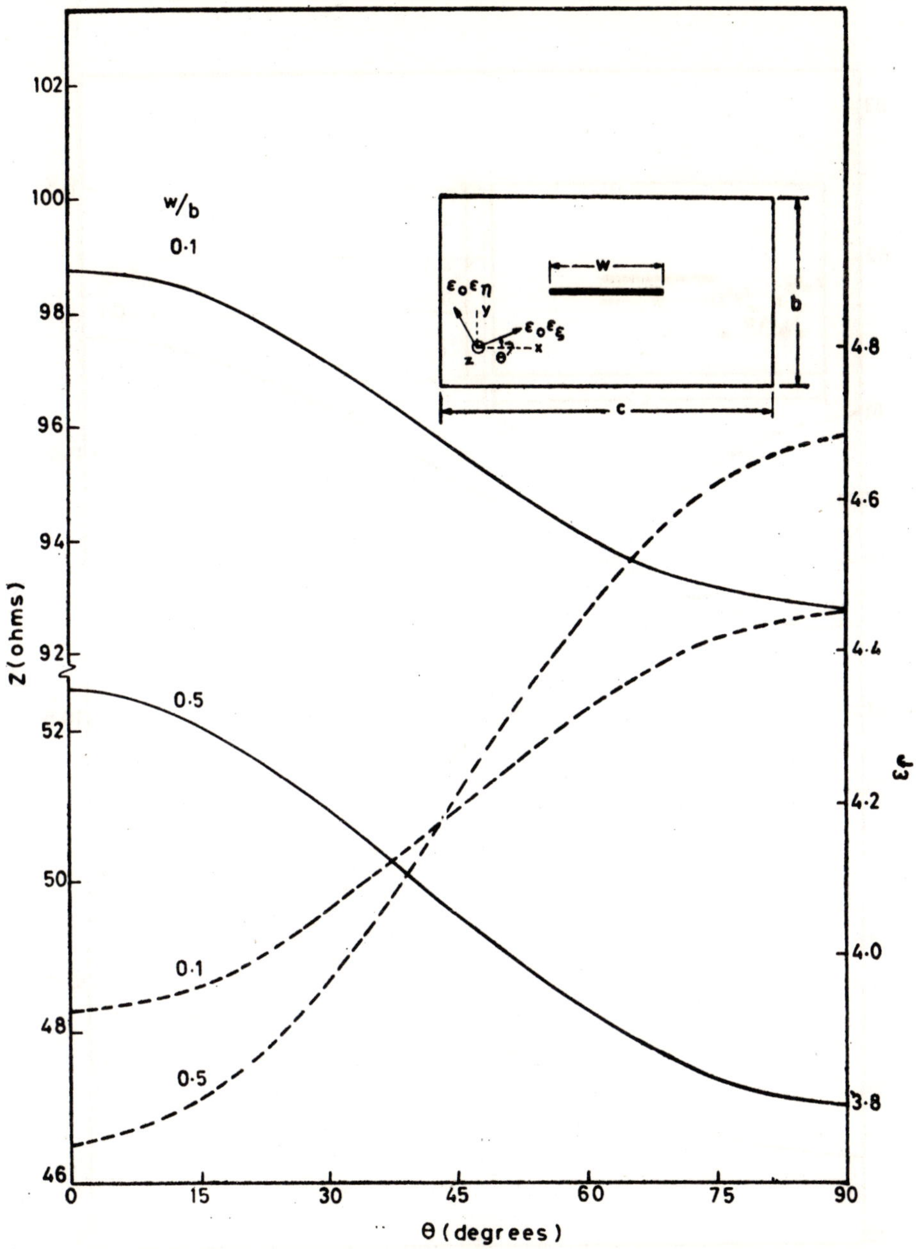

FIG. 5.18 Z and ϵ_f versus θ of symmetric stripline with anisotropic dielectric substrate (pyrolytic boron nitride), $\epsilon_\xi=5.12$, $\epsilon_\eta=3.4$ at $\theta=0$, $c/w=40$; ——— Z, – – – ϵ_f.

horizontal component decreases and reaches a minimum value 3.4 at $\theta=90°$. Since the vertical component of the electric field is stronger than the horizontal component in the dielectric, ϵ_f increases with increasing θ. In the case of stripline with sapphire substrate (Fig. 5.19), the trend in Z and ϵ_f is opposite to that observed in Fig. 5.18. This is expected,

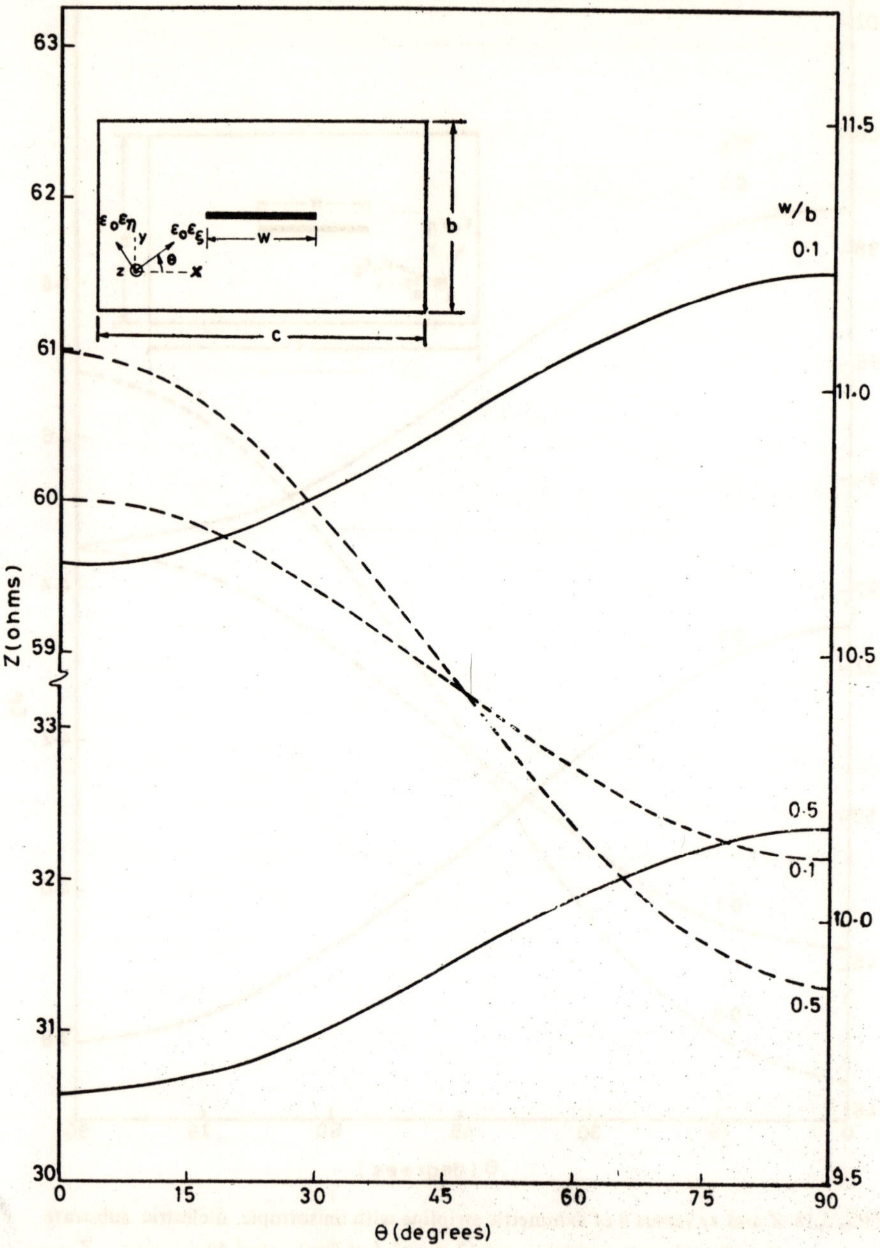

FIG. 5.19 Z and ϵ_f versus θ of symmetric stripline with anisotropic dielectric substrate (sapphire), $\epsilon_\xi=9.4$, $\epsilon_\eta=11.6$ at $\theta=0$, $c/w=40$; ——— Z, — — — ϵ_f.

since, for sapphire at $\theta=0$, the vertical component of the permittivity ($\epsilon_{yy}=\epsilon_{\eta}=11.6$) is higher than the horizontal component ($\epsilon_{xx}=\epsilon_{\xi}=9.4$), and, with increasing θ, the vertical component decreases. Similar behaviour can be observed in the variation of Z and ϵ_f as a

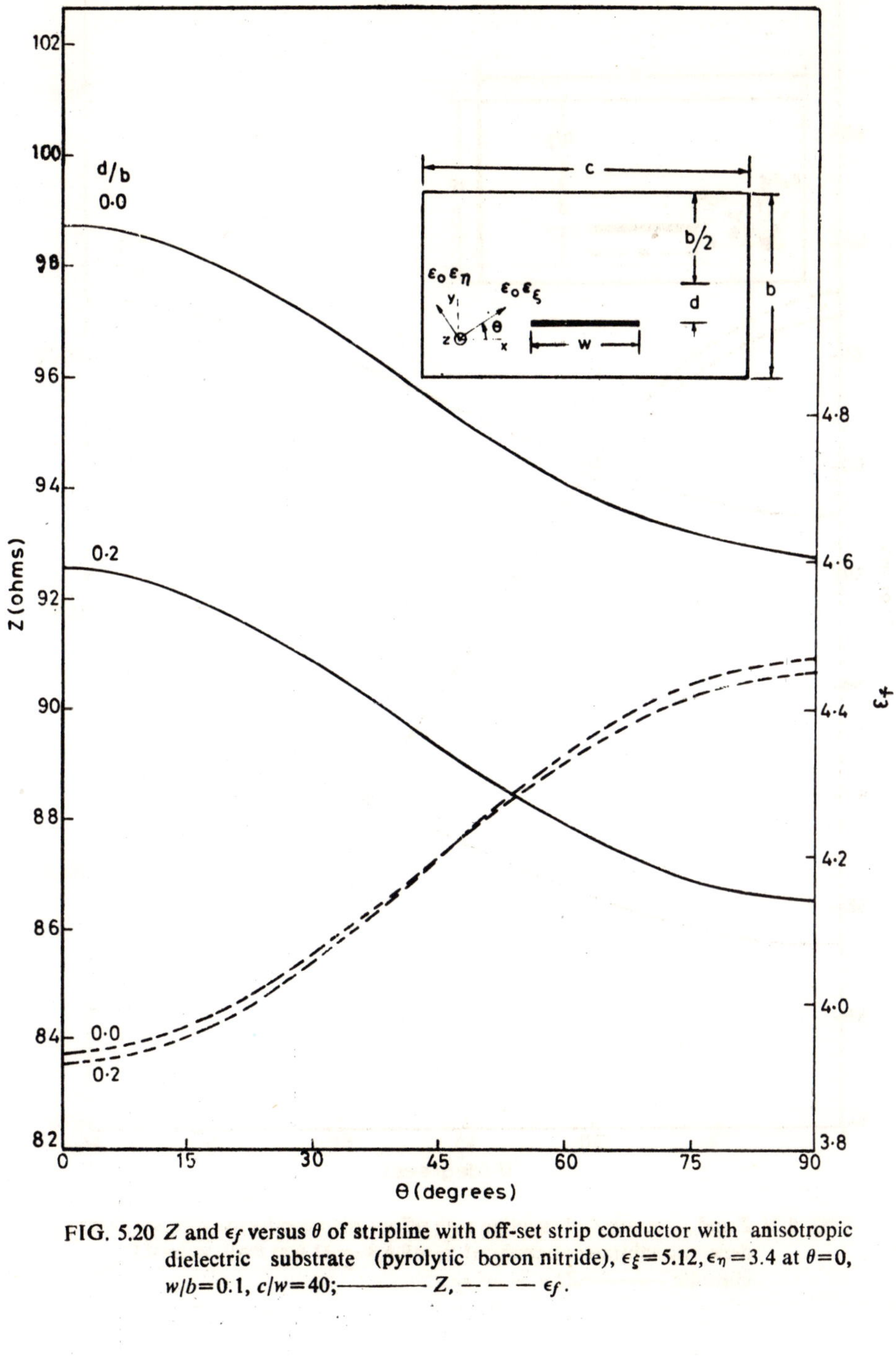

FIG. 5.20 Z and ϵ_f versus θ of stripline with off-set strip conductor with anisotropic dielectric substrate (pyrolytic boron nitride), $\epsilon_{\xi}=5.12, \epsilon_{\eta}=3.4$ at $\theta=0$, $w/b=0.1$, $c/w=40$;——— Z, — — — ϵ_f.

function of tilt angle θ in the case of stripline with off-set strip conductor (Figs. 5.20 and 5.21) and microstrip-like stripline (Figs. 5.22 and 5.23).

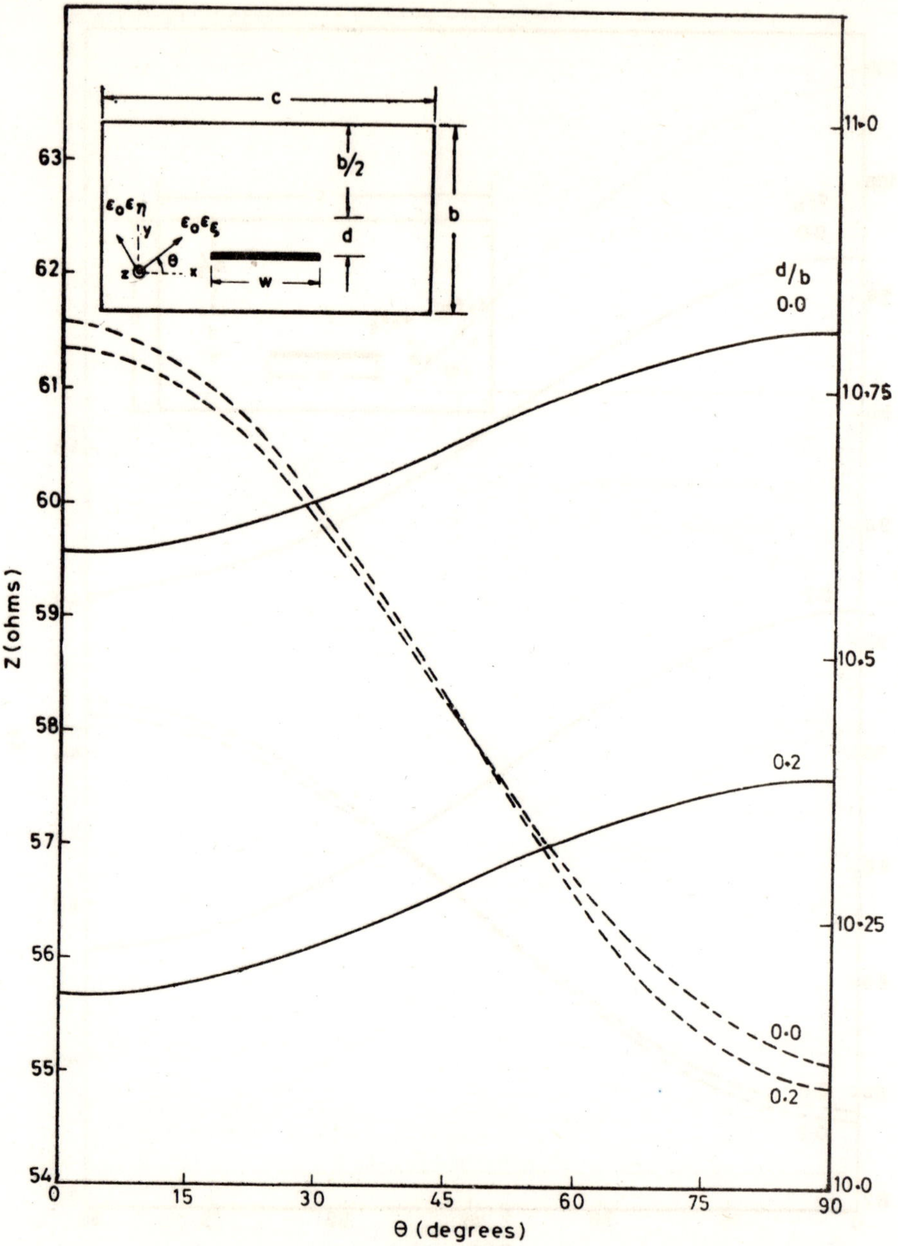

FIG. 5.21 Z and ϵ_f versus θ of stripline with off-set strip conductor with anisotropic dielectric substrates (sapphire), $\epsilon_\xi = 9.4$, $\epsilon_\eta = 11.6$ at $\theta = 0$, $w/b = 0.1$, $c/w = 40$;———— Z; — — — ϵ_f.

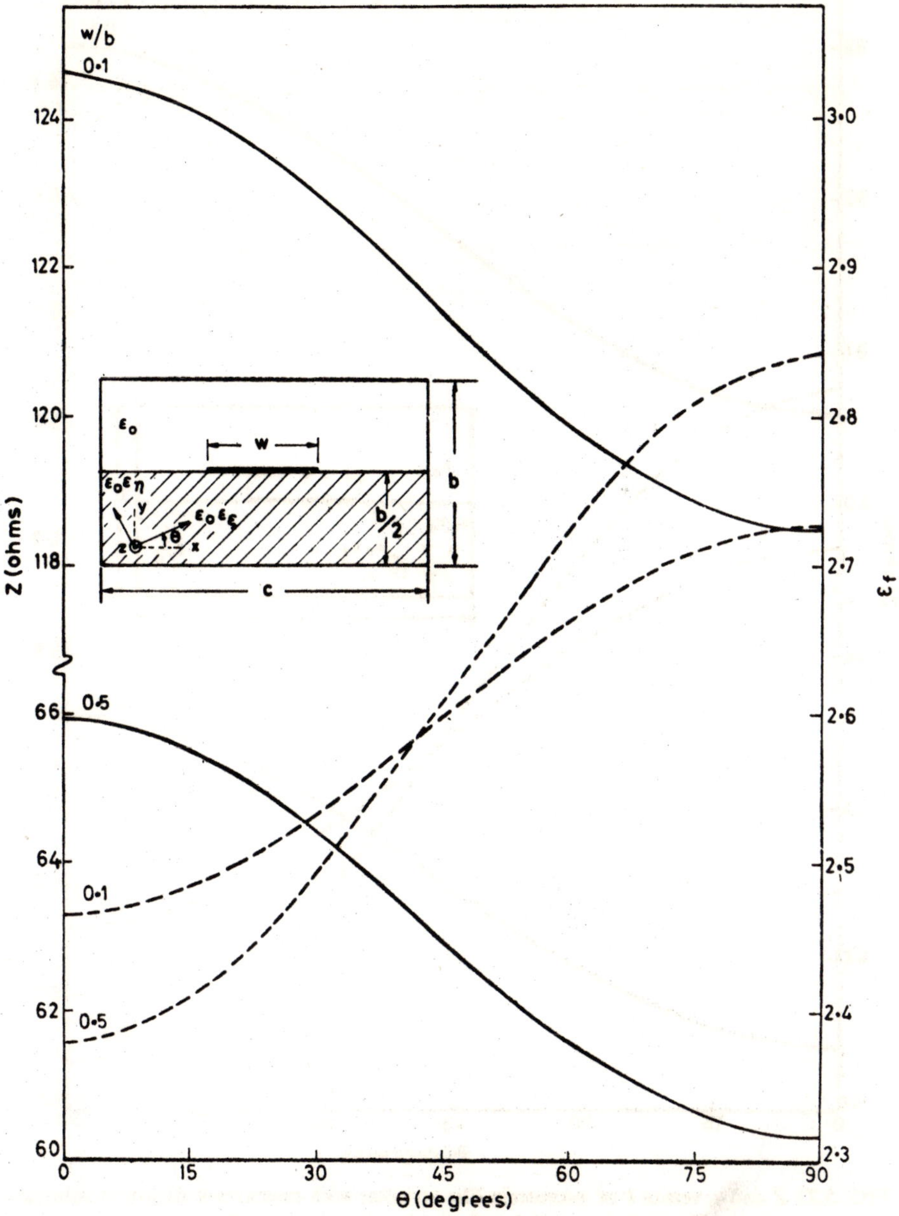

FIG. 5.22 Z and ϵ_f versus θ of microstrip-like stripline with anisotropic dielectric substrate (pyrolytic boron nitride), $\epsilon_\xi=5.12$, $\epsilon_\eta=3.4$ at $\theta=0$, $c/w=40$; ——— Z, — — — ϵ_f.

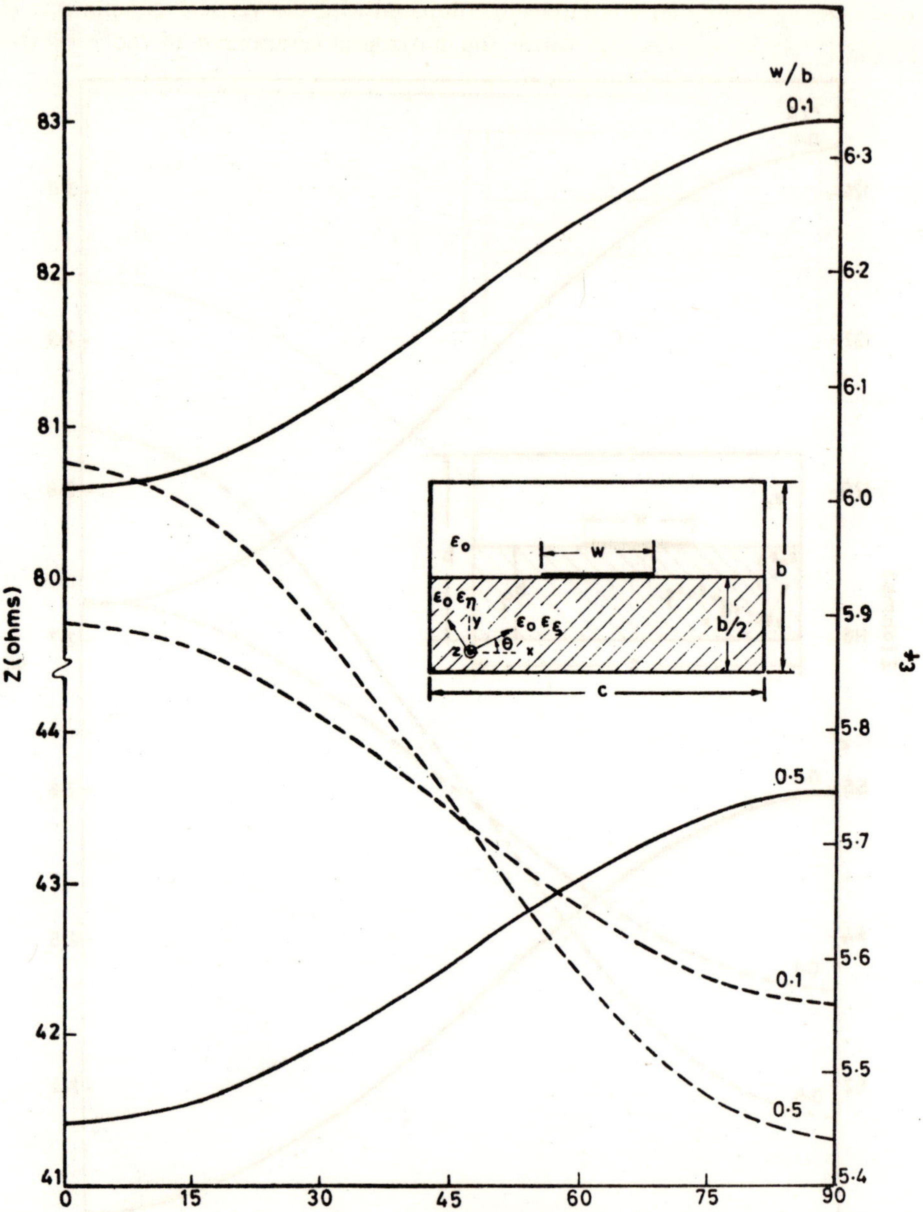

FIG. 5.23 Z and ϵ_f versus θ of microstrip-like stripline with anisotropic dielectric substrate (sapphire), $\epsilon_\xi = 9.4$, $\epsilon_\eta = 11.6$ at $\theta = 0$, $c/w = 40$; ———— Z, — — — ϵ_f.

Figures 5.24 and 5.25 show the variations in Z and ϵ_f as a function of θ for suspended stripline with pyrolytic boron nitride and sapphire substrates, respectively. The plots are drawn for a fixed value of the dielectric thickness ($d/b = 0.2$) and for two strip widths ($w/b = 0.1$ and 0.5).

When the substrate thickness is small compared to the air gap, the electric field in the substrate has a stronger horizontal component compared with the vertical component. Hence, with pyrolytic boron nitride, for which the horizontal component of the permittivity is

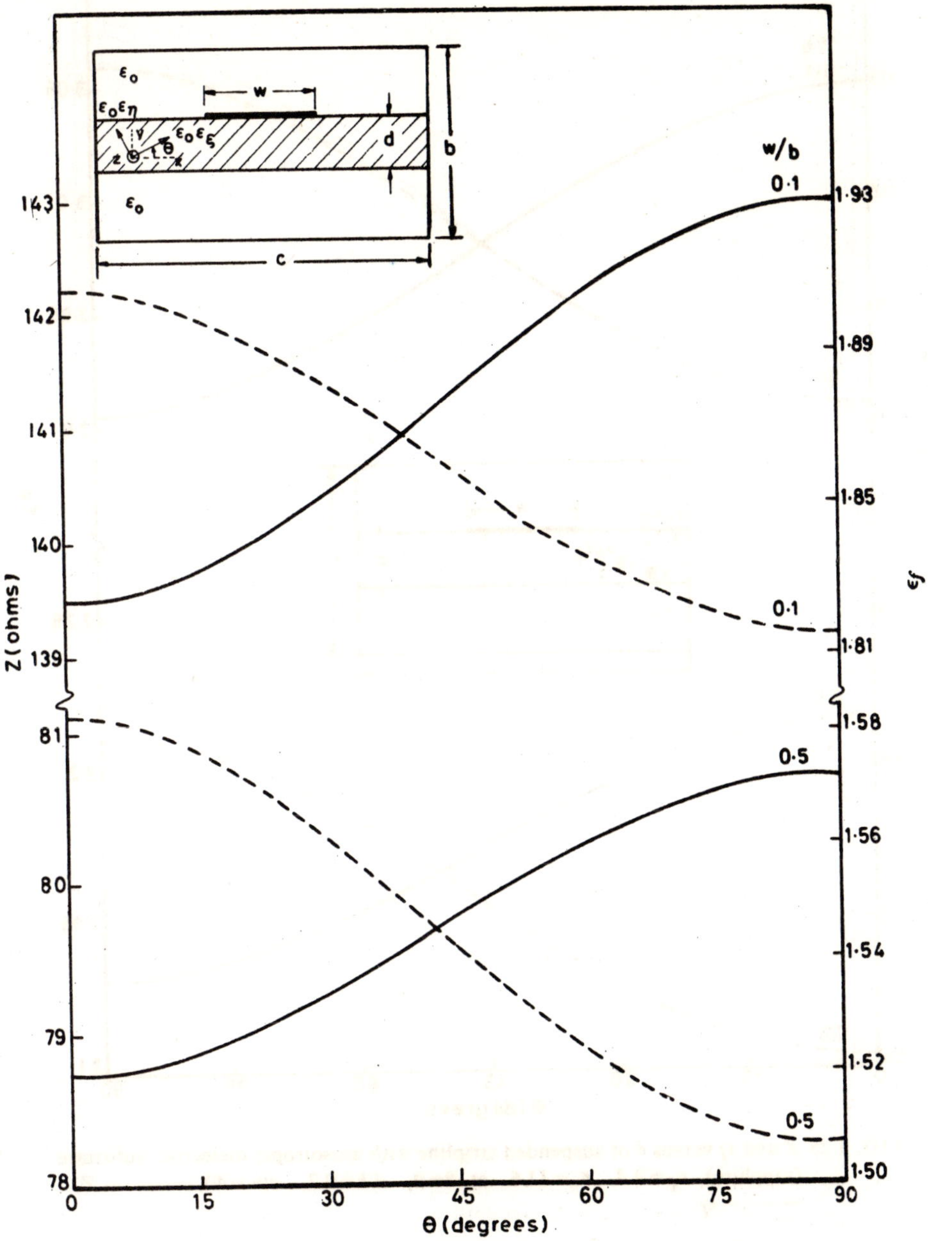

FIG. 5.24 Z and ϵ_f versus θ of suspended stripline with anisotropic dielectric substrate (pyrolytic boron nitride), $\epsilon_\xi=5.12$, $\epsilon_\eta=3.4$ at $\theta=0$, $d/b=0.2$. $c/w=40$; ——— Z, – – – ϵ_f.

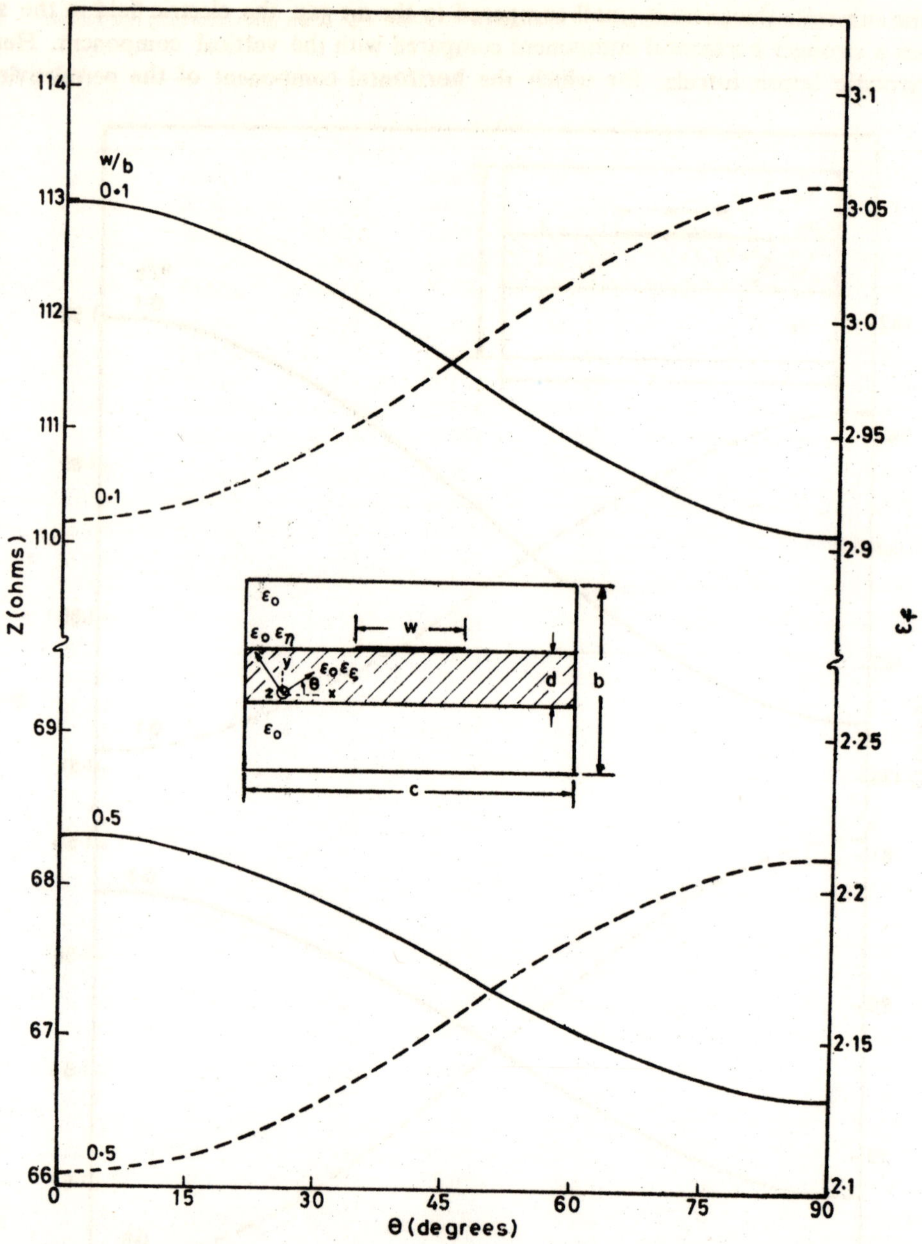

FIG. 5.25 Z and ϵ_f versus θ of suspended stripline with anisotropic dielectric substrate (sapphire), $\epsilon_\xi=9.4$, $\epsilon_\eta=11.6$ at $\theta=0$, $d/b=0.2$, $c/w=40$;———— Z, – – – ϵ_f.

maximum at $\theta=0$ and minimum at $\theta=90°$, Z increases and ϵ_f decreases with an increase in θ (Fig. 5.24). In the suspended stripline with sapphire substrate, the trend in Z and ϵ_f with an increase in θ is opposite (Fig. 5.25) to that observed in Fig. 5.24.

Table 5.1 Characteristic impedance of symmetric stripline with finite thickness strip conductor (See Fig. 5.1)

t/b → / w/b ↓	0.0	0.01	0.02	0.04	0.06	0.10	0.15	0.20	0.25
					$Z\sqrt{\epsilon_r}$ (ohms)				
0.10	194.20	183.49	176.23	165.10	156.48	142.81	129.53	119.07	110.12
0.11	188.51	178.56	171.73	161.16	153.04	139.83	127.05	116.98	108.24
0.12	183.32	174.01	167.55	157.45	149.72	137.07	124.76	114.90	106.51
0.13	178.56	169.78	163.64	153.96	146.52	134.37	122.60	112.88	104.83
0.14	174.15	165.82	160.00	150.73	143.56	131.91	120.45	111.06	103.17
0.15	170.04	162.11	156.58	147.68	140.77	129.53	118.49	109.17	101.51
0.16	166.21	158.62	153.34	144.78	138.11	127.27	116.45	107.50	99.89
0.17	162.62	155.32	150.27	142.08	135.57	125.13	114.58	105.88	98.39
0.18	159.23	152.19	147.35	139.50	133.12	123.06	112.74	104.26	96.90
0.19	156.04	149.22	144.57	137.02	130.78	121.04	110.93	102.67	95.44
0.20	153.01	146.38	141.91	134.64	128.52	119.07	109.17	101.23	94.11
0.21	150.13	143.74	139.39	132.35	126.40	117.13	107.58	99.67	92.82
0.22	147.39	141.21	136.98	130.14	124.35	115.25	105.99	98.25	91.51
0.23	144.78	138.78	134.66	128.02	122.37	113.48	104.42	96.86	90.23
0.24	142.29	136.44	132.42	125.96	120.45	111.78	102.94	95.48	88.99
0.25	139.90	134.19	130.27	123.98	118.64	110.12	101.51	94.11	87.84
0.26	137.61	132.03	128.19	122.05	116.88	108.51	100.11	92.78	86.59
0.27	135.41	129.94	126.18	120.20	115.17	106.95	98.74	91.50	85.15
0.28	133.29	127.92	124.23	118.41	113.50	105.42	97.38	90.45	84.14
0.29	131.25	125.96	122.35	116.68	111.89	103.93	96.04	89.38	83.15
0.30	129.29	124.07	120.52	114.99	110.31	102.47	95.20	88.33	82.18
0.31	127.39	122.23	118.75	113.35	108.77	101.05	94.08	87.30	81.23
0.32	125.56	120.45	117.03	111.76	107.28	100.67	92.99	86.30	80.31
0.33	123.79	118.73	115.36	110.20	106.65	99.49	91.92	85.33	79.41
0.34	122.08	117.04	113.73	109.47	105.38	98.34	90.87	84.37	78.52
0.35	120.42	115.54	112.81	108.16	104.14	97.21	89.85	83.44	77.66
0.36	118.81	114.12	111.45	106.89	102.93	96.11	88.86	82.52	76.82
0.37	117.25	112.74	110.12	105.64	101.75	95.03	87.88	81.63	75.99
0.38	115.73	111.40	108.82	104.42	100.59	93.97	86.93	80.75	75.18
0.39	114.26	110.08	107.55	103.22	99.46	92.94	85.99	79.90	74.39
0.40	112.83	108.80	106.31	102.06	98.35	91.94	85.08	79.06	73.61
0.41	111.44	107.54	105.10	100.92	97.27	90.95	84.18	78.24	72.85
0.42	110.08	106.32	103.92	99.81	96.22	89.98	83.31	77.43	72.11
0.43	108.77	105.12	102.76	98.72	95.18	89.04	82.45	76.64	71.38
0.44	107.48	103.94	101.63	97.65	94.17	88.11	81.61	75.87	70.67
0.45	106.23	102.80	100.52	96.61	93.18	87.21	80.79	75.12	69.97
0.46	105.01	101.68	99.44	95.58	92.21	86.32	79.98	74.37	69.28
0.47	103.82	100.58	98.38	94.58	91.26	85.45	79.19	73.65	68.61
0.48	102.66	99.51	97.34	93.60	90.32	84.59	78.41	72.93	67.95
0.49	101.52	98.46	96.33	92.64	89.41	83.76	77.65	72.23	67.30
0.50	100.42	97.43	95.33	91.71	88.52	82.94	76.60	71.55	66.66
0.51	99.33	96.42	94.36	90.78	87.64	82.13	76.17	70.88	66.04
0.52	98.28	95.43	93.40	89.88	86.78	81.35	75.45	70.21	65.43
0.53	97.24	94.46	92.47	89.00	85.94	80.57	74.75	69.57	64.83
0.54	96.23	93.52	91.55	88.13	85.11	79.81	74.06	68.93	64.24

(Contd.)

w/b \ t/b	0.0	0.01	0.02	0.04	0.06	0.10	0.15	0.20	0.25
					$Z\sqrt{\epsilon_r}$ (ohms)				
0.55	95.24	92.59	90.65	87.28	84.30	79.07	73.38	68.30	63.66
0.56	94.27	91.68	89.77	86.44	83.51	78.34	72.71	67.69	63.09
0.57	93.32	90.78	88.90	85.62	82.73	77.62	72.06	67.09	62.53
0.58	92.40	89.91	88.05	84.82	81.96	76.92	71.41	66.50	61.98
0.59	91.49	89.05	87.22	84.03	81.21	76.22	70.78	65.91	61.44
0.60	90.60	88.21	86.41	83.26	80.47	75.54	70.16	65.34	60.91
0.61	89.72	87.38	85.60	82.50	79.75	74.88	69.55	64.78	60.39
0.62	88.87	86.57	84.82	81.75	79.03	74.22	68.95	64.23	59.88
0.63	88.03	85.77	84.04	81.02	78.33	73.58	68.36	63.68	59.38
0.64	87.21	84.99	83.29	80.30	77.65	72.94	67.79	63.15	58.88
0.65	86.40	84.22	82.54	79.59	76.97	72.32	67.22	62.63	58.39
0.66	85.61	83.47	81.81	78.90	76.31	71.71	66.66	62.11	57.92
0.67	84.83	82.73	81.09	78.22	75.66	71.11	66.11	61.60	57.44
0.68	84.07	82.00	80.38	77.54	75.01	70.52	65.56	61.10	56.98
0.69	83.32	81.29	79.69	76.89	74.38	69.93	65.03	60.61	56.53
0.70	82.58	80.58	79.01	76.24	73.76	69.36	64.51	60.13	56.08
0.71	81.86	79.89	78.34	75.60	73.15	68.80	63.99	59.65	55.63
0.72	81.15	79.21	77.68	74.97	72.55	68.24	63.48	59.18	55.20
0.73	80.45	78.55	77.03	74.36	71.96	67.70	62.98	58.72	54.77
0.74	79.77	77.89	76.39	73.75	71.38	67.16	62.49	58.27	54.35
0.75	79.09	77.24	75.76	73.15	70.81	66.63	62.01	57.82	53.93
0.76	78.43	76.61	75.15	72.56	70.25	66.11	61.53	57.38	53.53
0.77	77.78	75.98	74.54	71.99	69.70	65.60	61.06	56.94	53.12
0.78	77.14	75.37	73.94	71.42	69.15	65.10	60.60	56.52	52.73
0.79	76.51	74.77	73.35	70.86	68.62	64.60	60.14	56.10	52.34
0.80	75.89	74.17	72.78	70.31	68.09	64.11	59.70	55.68	51.95
0.81	75.29	73.59	72.21	69.76	67.57	63.63	59.25	55.27	51.57
0.82	74.69	73.01	71.65	69.23	67.06	63.16	58.82	54.87	51.20
0.83	74.10	72.44	71.09	68.70	66.55	62.69	58.39	54.47	50.83
0.84	73.52	71.88	70.55	68.19	66.06	62.23	57.97	54.08	50.47
0.85	72.95	71.33	70.02	67.68	65.57	61.78	57.55	53.70	50.11
0.86	72.38	70.79	69.49	67.17	65.09	61.33	57.14	53.32	49.75
0.87	71.83	70.26	68.97	66.68	64.61	60.89	56.73	52.94	49.41
0.88	71.29	69.73	68.46	66.19	64.14	60.45	56.33	52.57	49.06
0.89	70.75	69.21	67.95	65.71	63.68	60.03	55.94	52.21	48.72
0.90	70.22	68.70	67.46	65.23	63.23	59.60	55.55	51.85	48.39
0.91	69.70	68.20	66.97	64.77	62.78	59.19	55.17	51.49	48.06
0.92	69.19	67.71	66.48	64.30	62.34	58.78	54.79	51.14	47.74
0.93	68.68	67.22	66.01	63.85	61.90	58.37	54.42	50.80	47.42
0.94	68.18	66.74	65.54	63.40	61.47	57.97	54.05	50.46	47.10
0.95	67.69	66.26	65.08	62.96	61.05	57.58	53.69	50.12	46.79
0.96	67.21	65.79	64.62	62.53	60.63	57.19	53.33	49.79	46.48
0.97	66.73	65.33	64.17	62.10	60.21	56.81	52.98	49.47	46.18
0.98	66.26	64.88	63.73	61.67	59.81	56.43	52.63	49.14	45.88
0.99	65.80	64.43	63.29	61.25	59.41	56.06	52.29	48.82	45.58

(Contd.)

w/b \ t/b	0.0	0.01	0.02	0.04	0.06	0.10	0.15	0.20	0.25
					$Z\sqrt{\epsilon_r}$ (ohms)				
1.02	64.45	63.12	62.02	60.03	58.23	54.96	51.28	47.89	44.71
1.04	63.58	62.28	61.19	59.25	57.48	54.26	50.63	47.29	44.15
1.06	62.73	61.46	60.39	58.48	56.74	53.57	50.00	46.70	43.61
1.08	61.90	60.66	59.61	57.73	56.02	52.91	49.38	46.13	43.08
1.10	61.10	59.88	58.85	57.01	55.32	52.25	48.78	45.57	42.56
1.12	60.32	59.12	58.11	56.30	54.64	51.62	48.19	45.03	42.05
1.14	59.55	58.38	57.39	55.60	53.97	51.00	47.62	44.50	41.56
1.16	58.81	57.66	56.68	54.93	53.32	50.39	47.06	43.98	41.07
1.18	58.08	56.95	55.99	54.27	52.69	49.80	46.51	43.47	40.60
1.20	57.37	56.26	55.32	53.62	52.07	49.22	45.98	42.97	40.14
1.22	56.68	55.59	54.67	52.99	51.46	48.65	45.46	42.49	39.69
1.24	56.01	54.94	54.03	52.38	50.87	48.10	44.94	42.02	39.25
1.26	55.35	54.30	53.40	51.78	50.29	47.56	44.45	41.55	38.81
1.28	54.71	53.67	52.79	51.19	49.73	47.03	43.96	41.10	38.39
1.30	54.08	53.06	52.19	50.62	49.17	46.52	43.48	40.65	37.98
1.32	53.46	52.46	51.61	50.06	48.63	46.01	43.01	40.22	37.58
1.34	52.86	51.88	51.04	49.51	48.11	45.52	42.55	39.80	37.18
1.36	52.28	51.31	50.48	48.97	47.59	45.03	42.11	39.38	36.79
1.38	51.70	50.75	49.93	48.45	47.08	44.56	41.67	38.97	36.41
1.40	51.14	50.20	49.40	47.94	46.59	44.10	41.24	38.57	36.04
1.42	50.59	49.67	48.88	47.43	46.10	43.64	40.82	38.18	35.68
1.44	50.05	49.14	48.36	46.94	45.63	43.20	40.41	37.80	35.32
1.46	49.53	48.63	47.86	46.46	45.16	42.76	40.00	37.42	34.97
1.48	49.01	48.13	47.37	45.99	44.70	42.33	39.61	37.05	34.63
1.50	48.51	47.64	46.89	45.52	44.26	41.91	39.22	36.69	34.29
1.52	48.01	47.16	46.42	45.07	43.82	41.50	38.84	36.34	33.96
1.54	47.53	46.68	45.95	44.62	43.39	41.10	38.46	35.99	33.64
1.56	47.05	46.22	45.50	44.19	42.97	40.71	38.10	35.65	33.32
1.58	46.59	45.77	45.06	43.76	42.55	40.32	37.74	35.32	33.01
1.60	46.13	45.32	44.62	43.34	42.15	39.94	37.39	34.99	32.70
1.62	45.68	44.88	44.19	42.93	41.75	39.57	37.04	34.67	32.40
1.64	45.24	44.46	43.77	42.52	41.36	39.20	36.70	34.35	32.11
1.66	44.81	44.04	43.36	42.13	40.98	38.84	36.37	34.04	31.82
1.68	44.39	43.62	42.96	41.74	40.60	38.49	36.04	33.74	31.53
1.70	43.97	43.22	42.56	41.36	40.23	38.14	35.72	33.44	31.26
1.72	43.57	42.82	42.17	40.98	39.87	37.80	35.40	33.14	30.98
1.74	43.17	42.43	41.79	40.61	39.51	37.47	35.09	32.85	30.71
1.76	42.78	42.05	41.42	40.25	39.16	37.14	34.79	32.57	30.45
1.78	42.39	41.67	41.05	39.90	38.82	36.82	34.49	32.29	30.19
1.80	42.01	41.30	40.69	39.55	38.48	36.50	34.19	32.01	29.93
1.82	41.64	40.94	40.33	39.20	38.15	36.19	33.90	31.74	29.68
1.84	41.28	40.58	39.98	38.87	37.83	35.88	33.62	31.48	29.43
1.86	40.92	40.23	39.64	38.53	37.50	35.58	33.34	31.22	29.19
1.88	40.56	39.89	39.30	38.21	37.19	35.28	33.06	30.96	28.95
1.90	40.22	39.55	38.97	37.89	36.88	34.99	32.79	30.71	28.71

(Contd.)

w/b \ t/b	0.0	0.01	0.02	0.04	0.06	0.10	0.15	0.20	0.25
					$Z\sqrt{\epsilon_r}$ (ohms)				
1.92	39.88	39.22	38.64	37.57	36.57	34.70	32.52	30.46	28.48
1.94	39.54	38.89	38.32	37.26	36.27	34.42	32.26	30.22	28.25
1.96	39.21	38.57	38.00	36.96	35.98	34.14	32.00	29.98	28.03
1.98	38.89	38.25	37.69	36.66	35.69	33.87	31.75	29.74	27.81
2.00	38.57	37.94	37.39	36.36	35.40	33.60	31.50	29.51	27.59
2.05	37.80	37.18	36.65	35.65	34.71	32.95	30.89	28.94	27.06
2.10	37.05	36.46	35.93	34.96	34.04	32.32	30.31	28.39	26.55
2.15	36.34	35.76	35.25	34.29	33.40	31.72	29.74	27.87	26.06
2.20	35.65	35.09	34.59	33.66	32.78	31.13	29.20	27.36	25.59
2.25	34.99	34.44	33.95	33.04	32.18	30.57	28.68	26.87	25.14
2.30	34.35	33.81	33.34	32.45	31.61	30.03	28.17	26.40	24.70
2.35	33.73	33.21	32.74	31.88	31.06	29.51	27.69	25.95	24.27
2.40	33.14	32.63	32.17	31.32	30.52	29.00	27.21	25.51	23.86
2.45	32.57	32.07	31.62	30.79	30.00	28.51	26.76	25.08	23.46
2.50	32.01	31.52	31.09	30.27	29.50	28.04	26.32	24.67	23.08

Table 5.2 Characteristic impedance of stripline with off-set centre conductor $c/w=40$ (See structure A2 of Table 3.1)

d/b / w/b	0.0	0.01	0.02	0.04	0.06	0.08	0.10	0.15	0.20
					$Z\sqrt{\epsilon_r}$ (ohms)				
0.10	195.64	195.61	195.52	195.17	194.57	193.73	192.65	188.76	183.01
0.11	189.94	189.92	189.83	189.47	188.88	188.04	186.96	183.07	177.34
0.12	184.75	184.72	184.63	184.28	183.69	182.85	181.77	177.89	172.17
0.13	179.98	179.95	179.86	179.51	178.92	178.08	177.00	173.13	167.41
0.14	175.56	175.53	175.44	175.09	174.50	173.67	172.59	168.72	163.02
0.15	171.46	171.43	171.34	170.99	170.40	169.57	168.49	164.63	158.94
0.16	167.62	167.59	167.50	167.15	166.56	165.73	164.65	160.81	155.14
0.17	164.02	163.99	163.90	163.55	162.96	162.14	161.06	157.22	151.56
0.18	160.63	160.60	160.51	160.16	159.58	158.75	157.68	153.85	148.20
0.19	157.43	157.40	157.31	156.96	156.38	155.56	154.48	150.67	145.03
0.20	154.40	154.37	154.28	153.93	153.35	152.53	151.46	147.65	142.03
0.21	151.52	151.49	151.40	151.05	150.47	149.65	148.58	144.78	139.19
0.22	148.78	148.75	148.66	148.31	147.73	146.91	145.85	142.06	136.48
0.23	146.16	146.13	146.04	145.70	145.12	144.30	143.24	139.46	133.90
0.24	143.66	143.63	143.54	143.20	142.62	141.81	140.75	136.98	131.44
0.25	141.26	141.24	141.15	140.81	140.23	139.42	138.36	134.61	129.09
0.26	138.97	138.94	138.85	138.51	137.94	137.13	136.08	132.34	126.86
0.27	136.76	136.74	136.65	136.31	135.74	134.93	133.88	130.15	124.69
0.28	134.64	134.61	134.53	134.19	133.62	132.81	131.77	128.06	122.61
0.29	132.60	132.57	132.49	132.15	131.58	130.78	129.74	126.05	120.62
0.30	130.63	130.61	130.52	130.18	129.62	128.82	127.78	124.11	118.69
0.31	128.74	128.71	128.63	128.29	127.72	126.94	125.91	122.23	116.84
0.32	126.92	126.89	126.81	126.47	125.91	125.12	124.08	120.42	115.05
0.33	125.15	125.12	125.04	124.70	124.14	123.35	122.32	118.67	113.32
0.34	123.43	123.40	123.32	122.98	122.42	121.63	120.61	116.97	111.65
0.35	121.77	121.74	121.65	121.32	120.76	119.98	118.96	115.33	110.03
0.36	120.15	120.12	120.04	119.71	119.15	118.37	117.35	113.75	108.47
0.37	118.59	118.56	118.48	118.15	117.59	116.18	115.80	112.21	106.95
0.38	117.07	117.04	116.96	116.63	116.08	115.30	114.29	110.71	105.48
0.39	115.59	115.56	115.48	115.15	114.60	113.83	112.82	109.26	104.06
0.40	114.16	114.13	114.05	113.72	113.17	112.40	111.40	107.85	102.67
0.42	111.40	111.37	111.29	110.97	110.43	109.66	108.67	105.15	100.02
0.44	108.79	108.76	108.68	108.36	107.82	107.06	106.08	102.59	97.51
0.46	106.31	106.28	106.20	105.88	105.35	104.60	103.62	100.17	95.14
0.48	103.95	103.92	103.84	103.52	102.99	102.25	101.28	97.86	92.88
0.50	101.70	101.67	101.59	101.28	100.75	100.01	99.06	95.67	90.74
0.52	99.55	99.52	99.44	99.13	98.61	97.88	96.93	93.57	88.70
0.54	97.49	97.46	97.39	97.08	96.56	95.84	94.90	91.57	86.75
0.56	95.52	95.50	95.42	95.12	94.60	93.89	92.95	89.66	84.89
0.58	93.64	93.61	93.54	93.23	92.73	92.02	91.09	87.83	83.12
0.60	91.83	91.80	91.73	91.43	90.93	90.22	89.31	86.08	81.41
0.62	90.09	90.06	89.99	89.69	89.20	88.50	87.59	84.40	79.78
0.64	88.42	88.39	88.32	88.02	87.53	86.84	85.95	82.78	78.22

(Contd.)

w/b \ d/b	0.0	0.01	0.02	0.04	0.06	0.08	0.10	0.15	0.20
					$Z\sqrt{\epsilon_r}$ (ohms)				
0.66	86.81	86.78	86.71	86.42	85.93	85.25	84.36	81.23	76.71
0.68	85.26	85.23	85.16	84.87	84.39	83.71	82.84	79.74	75.27
0.70	83.76	83.74	83.67	83.38	82.91	82.23	81.37	78.30	73.88
0.72	82.32	82.30	82.23	81.95	81.47	80.81	79.95	76.91	72.54
0.74	80.93	80.91	80.84	80.56	80.09	79.43	78.58	75.58	71.25
0.76	79.59	79.57	79.50	79.22	78.75	78.10	77.26	74.29	70.01
0.78	78.29	78.27	78.20	77.92	77.46	76.82	75.98	73.04	68.81
0.80	77.03	77.01	76.94	76.67	76.21	75.57	74.75	71.84	67.65
0.82	75.81	75.97	75.73	75.46	75.01	74.37	73.55	70.67	66.53
0.84	74.64	74.61	74.55	74.28	73.83	73.21	72.40	69.55	65.45
0.86	73.49	73.47	73.41	73.14	72.70	72.08	71.28	68.46	64.40
0.88	72.39	72.37	72.30	72.04	71.60	70.99	70.19	67.40	63.39
0.90	71.31	71.29	71.23	70.97	70.53	69.93	69.14	66.37	62.40
0.92	70.27	70.25	70.19	69.63	69.50	68.90	68.12	65.38	61.45
0.94	69.26	69.24	69.17	68.92	68.49	67.90	67.13	64.42	60.53
0.96	68.28	68.25	68.19	67.94	67.52	66.93	66.16	63.48	59.63
0.98	67.32	67.30	67.24	66.99	66.57	65.99	65.23	62.57	58.76
1.00	66.39	66.37	66.31	66.06	65.65	65.07	64.32	61.69	57.92
1.05	64.18	64.16	64.10	63.86	63.45	62.89	62.16	59.59	55.91
1.10	62.11	62.09	62.03	61.79	61.40	60.85	60.13	57.63	54.04
1.15	60.17	60.15	60.09	59.86	59.47	58.93	58.24	55.79	52.29
1.20	58.34	58.32	58.27	58.04	57.67	57.14	56.46	54.07	50.65
1.25	56.63	56.61	56.55	56.33	55.97	55.45	54.78	52.45	49.11
1.30	55.01	54.99	54.94	54.72	54.36	53.86	53.21	50.92	47.66
1.35	53.48	53.47	53.41	53.20	52.85	52.36	51.72	49.48	46.29
1.40	52.04	52.02	51.97	51.76	51.42	50.93	50.31	48.12	45.00
1.45	50.67	50.65	50.60	50.40	50.06	49.59	48.98	46.84	43.78
1.50	49.37	49.35	49.30	49.11	48.78	48.31	47.71	45.62	42.63
1.55	48.14	48.12	48.07	47.89	47.55	47.10	46.51	44.46	41.53
1.60	46.96	46.95	46.90	46.71	46.39	45.94	45.37	43.36	40.49
1.65	45.84	45.83	45.78	45.59	45.28	44.85	44.28	42.31	39.50
1.70	44.78	44.76	44.72	44.53	44.23	43.80	43.25	41.31	38.55
1.75	43.76	43.74	43.70	43.52	43.22	42.80	42.26	40.36	37.65
1.80	42.79	42.77	42.73	42.55	42.26	41.84	41.31	39.45	36.79
1.85	41.86	41.84	41.80	41.62	41.34	40.93	40.41	38.58	35.97
1.90	40.96	40.95	40.91	40.74	40.45	40.05	39.54	37.74	35.19
1.95	40.11	40.10	40.05	39.89	39.61	39.22	38.71	36.94	34.44
2.00	39.29	39.28	39.24	39.07	38.80	38.41	37.92	36.18	33.72
2.05	38.50	38.49	38.45	38.29	38.02	37.64	37.15	35.45	33.02
2.10	37.75	37.73	37.69	37.54	37.27	36.90	36.42	34.74	32.36
2.15	37.02	37.01	36.97	36.81	36.55	36.19	35.71	34.06	31.72
2.20	36.32	36.31	36.27	36.12	35.86	35.50	35.03	33.41	31.11
2.25	35.65	35.63	35.60	35.45	35.19	34.84	34.38	32.78	30.52
2.30	35.00	34.99	34.95	34.80	34.55	34.20	33.75	32.18	29.95
2.35	34.37	34.36	34.32	34.18	33.93	33.59	33.14	31.60	29.40
2.40	33.77	33.75	33.72	33.57	33.33	32.99	32.56	31.03	28.88
2.45	33.18	33.17	33.14	32.99	32.76	32.42	31.99	30.49	28.37
2.50	32.62	32.61	32.57	32.43	32.20	31.87	31.45	29.97	27.87

Table 5.3 Characteristic impedance of stripline with off-set centre conductor and t/b finite; $d/b = 0.05$ (See Fig. 5.2)

w/b \ t/b	0.001	0.01	0.05	0.10	0.15	0.20	0.25
			$Z\sqrt{\epsilon_r}$ (ohms)				
0.10	190.38	183.44	163.29	146.00	132.64	121.58	112.06
0.12	181.48	175.08	156.37	140.13	127.50	116.98	107.88
0.14	173.37	167.46	150.00	134.72	122.74	112.71	104.00
0.16	165.96	160.47	144.14	129.71	118.32	108.75	100.39
0.18	159.15	154.04	138.72	125.06	114.22	105.05	97.02
0.20	152.88	148.11	133.68	120.73	110.38	101.60	93.87
0.22	147.09	142.61	129.01	116.69	106.80	98.36	90.91
0.24	141.72	137.51	124.64	112.91	103.44	95.33	88.14
0.26	136.72	132.77	120.57	109.37	100.29	92.47	85.53
0.28	132.07	128.33	116.75	106.05	97.32	89.79	83.42
0.30	127.72	124.19	113.17	102.92	94.58	87.67	81.47
0.32	123.65	120.30	109.80	100.09	92.37	85.64	79.60
0.34	119.84	116.65	106.81	97.75	90.25	83.71	77.81
0.36	116.55	113.60	104.29	95.52	88.24	81.86	76.10
0.38	113.70	110.87	101.89	93.39	86.31	80.09	74.47
0.40	110.99	108.26	99.60	91.35	84.46	78.40	72.90
0.42	108.41	105.78	97.41	89.40	82.69	76.77	71.40
0.44	105.94	103.41	95.31	87.53	80.99	75.22	69.96
0.46	103.59	101.14	93.30	85.74	79.36	73.72	68.58
0.48	101.33	98.98	91.37	84.02	77.80	72.28	67.25
0.50	99.17	96.90	89.52	82.36	76.30	70.90	65.97
0.52	97.11	94.90	87.75	80.77	74.85	69.57	64.74
0.54	95.12	92.99	86.04	79.24	73.46	68.29	63.55
0.56	93.22	91.15	84.40	77.77	72.11	67.06	62.41
0.58	91.39	89.39	82.82	76.35	70.82	65.87	61.31
0.60	89.63	87.69	81.30	74.99	69.57	64.72	60.24
0.62	87.94	86.05	79.83	73.67	68.37	63.61	59.21
0.64	86.31	84.48	78.41	72.39	67.20	62.53	58.22
0.66	84.74	82.96	77.05	71.16	66.08	61.50	57.26
0.68	83.22	81.49	75.73	69.97	64.99	60.49	56.33
0.70	81.76	80.08	74.45	68.82	63.94	59.52	55.43
0.72	80.35	78.71	73.22	67.71	62.92	58.58	54.56
0.74	78.99	77.39	72.03	66.63	61.93	57.67	53.71
0.76	77.67	76.12	70.88	65.59	60.97	56.79	52.89
0.78	76.40	74.88	69.76	64.57	60.05	55.93	52.10
0.80	75.16	73.69	68.67	63.59	59.15	55.10	51.33
0.82	73.97	72.53	67.62	62.64	58.27	54.29	50.58
0.84	72.81	71.41	66.61	61.72	57.43	53.51	49.85
0.86	71.69	70.32	65.62	60.82	56.60	52.75	49.15
0.88	70.60	69.26	64.66	59.95	55.80	52.01	48.46
0.90	69.55	68.24	63.73	59.10	55.02	51.29	47.79
0.92	68.53	57.24	62.82	58.28	54.27	50.59	47.14
0.94	67.53	66.28	61.94	57.48	53.53	49.91	46.51

(Contd.)

w/b \ t/b	0.001	0.01	0.05	0.10	0.15	0.20	0.25
				$Z\sqrt{\epsilon_r}$ (ohms)			
0.96	66.57	65.34	61.09	56.70	52.82	49.25	45.90
0.98	65.63	64.43	60.26	55.94	52.12	48.60	45.30
1.00	64.72	63.54	59.45	55.20	51.44	47.97	44.71
1.05	62.54	61.42	57.51	53.44	49.82	46.47	43.32
1.10	60.51	59.45	55.70	51.79	48.29	45.06	42.01
1.15	58.61	57.59	54.00	50.24	46.86	43.73	40.77
1.20	56.82	55.85	52.40	48.77	45.51	42.48	39.61
1.25	55.14	54.21	50.90	47.39	44.23	41.29	38.51
1.30	53.55	52.66	49.47	46.09	43.03	40.18	37.47
1.35	52.06	51.20	48.13	44.85	41.89	39.12	36.48
1.40	50.64	49.82	46.85	43.68	40.80	38.11	35.55
1.45	49.30	48.51	45.65	42.57	39.78	37.16	34.66
1.50	48.03	47.27	44.50	41.52	38.80	36.25	33.82
1.55	46.82	46.09	43.41	40.51	37.87	35.38	33.01
1.60	45.68	44.96	42.37	39.55	36.98	34.56	32.24
1.65	44.58	43.89	41.38	38.64	36.13	33.77	31.51
1.70	43.54	42.87	40.43	37.77	35.33	33.02	30.81
1.75	42.55	41.90	39.53	36.94	34.55	32.30	30.14
1.80	41.60	40.97	38.66	36.14	33.81	31.61	29.50
1.85	40.69	40.08	37.84	35.38	33.10	30.95	28.89
1.90	39.82	39.23	37.04	34.64	32.42	30.32	28.30
1.95	38.98	38.41	36.29	33.94	31.77	29.71	27.73
2.00	38.18	37.63	35.56	33.27	31.14	29.13	27.19
2.05	37.42	36.88	34.86	32.62	30.54	28.57	26.67
2.10	36.68	36.16	34.18	32.00	29.96	28.03	26.16
2.15	35.97	35.46	33.53	31.40	29.40	27.61	25.68
2.20	35.29	34.79	32.91	30.82	28.87	27.01	25.21
2.25	34.63	34.15	32.31	30.26	28.35	26.52	24.76
2.30	34.00	33.53	31.73	29.72	27.85	26.06	24.33
2.35	33.39	32.93	31.17	29.20	27.37	25.61	23.91
2.40	32.80	32.35	30.63	28.70	26.90	25.17	23.50
2.45	32.24	31.80	30.11	28.22	26.45	24.75	23.11
2.50	31.69	31.26	29.61	27.75	26.01	24.35	22.73

Table 5.4 Characteristic impedance of stripline with off-set centre conductor and t/b finite; $d/b=0.1$ (See Fig. 5.2)

w/b \ t/b	0.001	0.01	0.05	0.10	0.15	0.20	0.25
				$Z\sqrt{\epsilon_r}$ (ohms)			
0.10	189.75	182.67	162.21	144.68	131.12	119.88	110.17
0.12	180.64	174.14	155.15	138.70	125.88	115.19	105.90
0.14	172.36	166.36	148.68	133.20	121.05	110.85	101.95
0.16	164.82	159.25	142.72	128.11	116.57	106.83	98.29
0.18	157.90	152.72	137.22	123.40	112.41	103.08	94.88
0.20	151.54	146.71	132.13	119.03	108.54	99.60	91.69
0.22	145.67	141.15	127.41	114.95	104.92	96.34	88.72
0.24	140.24	136.00	123.01	111.15	101.54	93.28	85.93
0.26	135.21	131.21	118.90	107.58	98.37	90.42	83.31
0.28	130.52	126.75	115.06	104.24	95.39	87.72	81.20
0.30	126.14	122.58	111.46	101.10	92.65	85.60	79.24
0.32	122.05	118.68	108.08	98.27	90.43	83.57	77.37
0.34	118.21	115.01	105.09	95.93	88.32	81.64	75.58
0.36	114.92	111.95	102.57	93.70	86.30	79.79	73.88
0.38	112.06	109.21	100.17	91.57	84.38	78.03	72.25
0.40	109.35	106.61	97.87	89.53	82.53	76.34	70.70
0.42	106.76	104.12	95.68	87.59	80.77	74.73	69.21
0.44	104.30	101.76	93.59	85.72	79.08	73.18	67.78
0.46	101.94	99.49	91.58	83.94	77.46	71.69	66.40
0.48	99.69	97.33	89.66	82.22	75.91	70.27	65.09
0.50	97.54	95.25	87.82	80.58	74.41	68.90	63.82
0.52	95.48	93.27	86.05	79.00	72.98	67.58	62.60
0.54	93.50	91.36	84.35	77.48	71.59	66.31	61.43
0.56	91.60	89.53	82.72	76.02	70.26	65.09	60.30
0.58	89.78	87.77	81.15	74.61	68.98	63.91	59.21
0.60	88.03	86.08	79.64	73.25	67.75	62.77	58.16
0.62	86.35	84.46	78.18	71.94	66.55	61.68	57.15
0.64	84.73	82.89	76.78	70.68	65.40	60.62	56.17
0.66	83.17	81.38	75.43	69.46	64.29	59.60	55.23
0.68	81.66	79.93	74.12	68.29	63.21	58.61	54.31
0.70	80.21	78.52	72.85	67.15	62.17	57.65	53.43
0.72	78.81	77.17	71.63	66.05	61.17	56.72	52.57
0.74	77.46	75.86	70.45	64.98	60.20	55.83	51.74
0.76	76.15	74.59	69.31	63.95	59.25	54.96	50.94
0.78	74.89	73.37	68.21	62.95	58.34	54.12	50.16
0.80	73.67	72.19	67.13	61.98	57.45	53.30	49.40
0.82	72.49	71.04	66.10	61.04	56.59	52.51	48.67
0.84	71.34	69.93	65.09	60.13	55.76	51.74	47.96
0.86	70.23	68.85	64.11	59.25	54.95	50.99	47.27
0.88	69.16	67.81	63.17	58.39	54.16	50.27	46.60
0.90	68.11	66.79	62.25	57.55	53.39	49.56	45.94
0.92	67.10	65.81	61.35	56.74	52.65	48.87	45.31
0.94	66.12	64.86	60.49	55.96	51.93	48.21	44.69
0.96	65.16	63.93	59.64	55.19	51.22	47.56	44.09
0.98	64.24	63.03	58.82	54.44	50.54	46.93	43.51

(Contd.)

t/b \ w/b	0.001	0.01	0.05	0.10	0.15	0.20	0.25
				$Z\sqrt{\epsilon_r}$ (ohms)			
1.00	63.34	62.15	58.02	53.72	49.87	46.31	42.94
1.05	61.19	60.07	56.12	51.98	48.28	44.84	41.58
1.10	59.19	58.11	54.34	50.36	46.79	43.46	40.30
1.15	57.31	56.28	52.66	48.83	45.38	42.16	39.10
1.20	55.55	54.57	51.09	47.40	44.06	40.94	37.97
1.25	53.89	52.95	49.61	46.04	42.81	39.79	36.90
1.30	52.33	51.43	48.21	44.76	41.63	38.70	35.89
1.35	50.85	49.99	46.89	43.55	40.52	37.67	34.93
1.40	49.46	48.63	45.63	42.41	39.46	36.69	34.03
1.45	48.14	47.35	44.45	41.32	38.46	35.76	33.17
1.50	46.89	46.12	43.32	40.29	37.50	34.88	32.35
1.55	45.71	44.96	42.26	39.30	36.60	34.04	31.57
1.60	44.58	43.86	41.23	38.37	35.73	33.24	30.83
1.65	43.60	42.81	40.26	37.48	34.91	32 47	30.12
1.70	42.48	41.81	39.34	36.63	34.12	31.74	29.45
1.75	41.51	40.86	38.45	35.81	33.37	31.04	28.80
1.80	40.57	39.94	37.61	35.03	32.65	30.38	28.18
1.85	39.68	39.07	36.80	34.29	31.96	29.74	27.59
1.90	38.83	38.24	36.02	33.57	31.30	29.12	27.02
1.95	38.01	37.44	35.28	32.89	30.66	28.54	26.48
2.00	37.23	36.67	34.57	32.23	30.05	27.97	25.95
2.05	36.48	35.93	33.88	31.60	29.47	27.43	25.45
2.10	35.75	35.22	33.22	30.99	28.90	26.91	24.97
2.15	35.06	34.54	32.59	30.41	28.36	26.40	24.50
2.20	34.39	33.89	31.98	29.84	27.84	25.92	24.05
2.25	33.75	33.26	31.39	29.30	27.34	25.45	23.62
2.30	33.13	32.65	30.83	28.78	26.85	25.00	23.20
2.35	32.53	32.07	30.28	28.27	26.38	24.57	22.80
2.40	31.96	31.50	29.75	27.78	25.93	24.15	22.41
2.45	31.40	30.96	29.25	27.31	25.49	23.74	22.03
2.50	30.87	30.43	28.75	26.86	25.07	23.35	21.67

Table 5.5 Characteristic impedance of stripline with off-set centre conductor and t/b finite; d/b=0.15 (See Fig. 5.2)

w/b \ t/b	0.001	0.01	0.05	0.10	0.15	0.20	0.25
			$Z\sqrt{\epsilon_r}$ (ohms)				
0.10	188.61	181.31	160.29	142.32	128.41	116.83	106.76
0.12	179.13	172.44	152.99	136.15	123.00	111.98	102.34
0.14	170.57	164.41	146.32	130.49	118.03	107.52	98.28
0.16	162.78	157.09	140.20	125.28	113.44	103.40	94.52
0.18	155.68	150.39	134.58	120.47	109.20	99.58	91.04
0.20	149.17	144.24	129.39	116.02	105.26	96.04	87.80
0.22	143.18	138.58	124.59	111.88	101.60	92.74	84.79
0.24	137.65	133.34	120.13	108.03	98.18	89.65	81.98
0.26	132.54	128.48	115.98	104.44	94.99	86.77	79.35
0.28	127.79	123.97	112.10	101.08	92.00	84.07	77.23
0.30	123.37	119.76	108.48	97.92	89.24	81.94	75.27
0.32	119.24	115.83	105.08	95.08	87.03	79.92	73.41
0.34	115.38	112.15	102.09	92.74	84.92	77.99	71.64
0.36	112.07	109.07	99.56	90.51	82.91	76.16	69.95
0.38	109.21	106.33	97.16	88.39	80.99	74.41	68.34
0.40	106.49	103.72	94.87	86.36	79.16	72.73	66.80
0.42	103.91	101.24	92.69	84.43	77.41	71.14	65.33
0.44	101.45	98.88	90.60	82.58	75.74	69.61	63.92
0.46	99.10	96.62	88.61	80.80	74.14	68.14	62.57
0.48	96.85	94.47	86.70	79.11	72.60	66.74	61.28
0.50	94.71	92.41	84.87	77.48	71.13	65.39	60.04
0.52	92.66	90.43	83.12	75.92	69.71	64.09	58.85
0.54	90.70	88.54	81.44	74.42	68.35	62.85	57.70
0.56	88.81	86.73	79.83	72.98	67.04	61.65	56.60
0.58	87.01	84.99	78.28	71.59	65.79	60.50	55.54
0.60	85.28	83.31	76.78	70.26	64.57	59.39	54.52
0.62	83.61	81.71	75.35	68.97	63.41	58.32	53.54
0.64	82.01	80.16	73.96	67.73	62.28	57.29	52.59
0.66	80.47	78.67	72.63	66.53	61.19	56.29	51.67
0.68	78.98	77.20	71.34	65.38	60.14	55.33	50.79
0.70	77.55	75.65	70.10	64.26	59.13	54.40	49.93
0.72	76.17	74.51	68.90	63.18	58.14	53.50	49.11
0.74	74.84	73.23	67.74	62.14	57.20	52.63	48.31
0.76	73.55	71.98	66.62	61.13	56.28	51.79	47.53
0.78	72.31	70.78	65.54	60.16	55.39	50.97	46.78
0.80	71.11	69.61	64.49	59.21	54.53	50.18	46.06
0.82	69.95	68.49	63.47	58.30	53.69	49.42	45.35
0.84	68.82	67.40	62.49	57.41	52.88	48.67	44.67
0.86	67.73	66.34	61.53	56.55	52.09	47.95	44.01
0.88	66.68	65.32	60.61	55.71	51.33	47.25	43.36

(Contd.)

w/b \ t/b	0.001	0.01	0.05	0.10	0,15	0.20	0.25
			$Z\sqrt{\epsilon_r}$ (ohms)				
0.90	65.65	64.33	59.71	54.90	50.59	46.57	42.74
0.92	64.66	-63.36	58.84	54.11	49.87	45.91	42.13
0.94	63.70	62.43	57.99	53.34	49.17	45.27	41.54
0.96	62.76	61.52	57.16	52.60	48.49	44.65	40.97
0.98	61.86	60.64	56.36	51.87	47.83	44.C4	40.41
1.00	60.98	59.78	55.59	51.17	47.18	43.45	39.87
1.05	58.88	57.74	53.73	49.49	45.65	42.04	38.57
1.10	56.92	55.84	52.00	47.91	44.21	40.72	37.36
1.15	55.09	54.06	50.37	46.44	42.85	39.47	36.22
1.20	53.37	52.39	48.84	45.05	41.58	38.31	35.14
1.25	51.76	50.81	47.41	43.74	40.38	37 21	34.13
1.30	50.24	49.33	46.05	42.51	39.25	36.17	33.18
1.35	48.81	47.94	44.77	41.34	38.18	35.18	32.28
1.40	47.45	46.62	43.56	40.24	37.17	34.25	31.42
1.45	46.17	45.37	42.41	39.19	36.21	33.37	30.61
1.50	44.96	44.18	41.33	38.20	35.30	32.53	29.84
1.55	43.81	43.06	40.29	37.25	34.43	31.74	29.11
1.60	42.72	41.99	39.31	36.35	33.61	30.98	28.41
1.65	41.68	40.98	38.37	35.50	32.82	30.25	27.75
1.70	40.69	40.01	37.48	34.68	32.07	29.56	27.11
1.75	39.74	39.09	36.63	33.90	31.35	28.90	26.51
1.80	38.84	38.20	35.81	33.15	30.67	28.27	25.93
1.85	37.98	37.36	35.03	32.44	30.01	27.67	25.37
1.90	37.16	36.55	34.29	31.76	29.38	27.09	24.84
1.95	36.37	35.78	33.57	31.10	28.78	26.53	24.33
2.00	35.61	35.04	32.89	30.47	28.20	26.00	23.84
2.05	34.88	34.33	32.23	29.87	27.64	25.49	23.37
2.10	34.19	33.65	31.60	29.29	27.11	25.00	22.92
2.15	33.52	32.99	30.99	28.73	26.59	24.52	22.49
2.20	32.87	32.36	30.41	28.19	26.10	24.07	22.07
2.25	32.25	31.76	29.84	27.67	25.62	23.63	21.67
2.30	31.66	31.17	29.30	27.18	25.16	23.20	21.28
2.35	31.08	30.61	28.78	26.69	24.72	22.80	20.90
2.40	30.53	30.06	28.27	26.23	24.29	22.40	20.54
2.45	29.99	29.54	27.78	25.78	23.87	22.02	20.19
2.50	29.48	29.03	27.31	25.35	23.47	21.65	19.85

Table 5.6 Characteristic impedance of stripline with off-set centre conductor and t/b finite; $d/b=0.2$ (See Fig. 5.2)

w/b \ t/b	0.001	0.01	0.05	0.10	0.15	0.20	0.25
			$Z\sqrt{\epsilon_r}$ (ohms)				
0.10	186.81	179.16	157.30	138.65	124.18	112.04	101.35
0.12	176.78	169.80	149.63	132.18	118.51	106.95	96.70
0.14	167.77	161.37	142.66	126.29	113.34	102.30	92.45
0.16	159.63	153.74	136.32	120.91	108.60	98.04	88.56
0.18	152.25	146.80	130.52	115.96	104.24	94.12	84.99
0.20	145.52	140.45	125.19	111.40	100.22	90.50	81.69
0.22	139.36	134.64	120.28	107.19	96.49	87.15	78.64
0.24	133.70	129.28	115.74	103.28	93.04	84.04	75.81
0.26	128.48	124.34	111.53	99.65	89.82	81.15	73.18
0.28	123.65	119.76	107.62	96.26	86.81	78.44	71.07
0.30	119.17	115.50	103.97	93.10	84.06	76.32	69.12
0.32	115.01	111.54	100.56	90.26	81.85	74.32	67.28
0.34	111.13	107.84	97.56	87.93	79.76	72.41	65.54
0.36	107.80	104.75	95.04	85.71	77.77	70.60	63.88
0.38	104.94	102.01	92.65	83.60	75.87	68.88	62.31
0.40	102.22	99.41	90.38	81.60	74.07	67.24	60.81
0.42	99.64	96.94	88.21	79.69	72.35	65.68	59.38
0.44	97.19	94.59	86.15	77.86	70.71	64.19	58.01
0.46	94.86	92.35	84.17	76.12	69.14	62.76	56.71
0.48	92.63	90.22	82.29	74.45	67.64	61.40	55.47
0.50	90.51	88.18	80.49	72.86	66.20	60.09	54.28
0.52	88.48	86.23	78.77	71.33	64.83	58.84	53.13
0.54	86.54	84.36	77.12	69.86	63.50	57.64	52.04
0.56	84.69	82.58	75.54	68.46	62.24	56.49	50.99
0.58	82.91	80.86	74.02	67.11	61.02	55.38	49.98
0.60	81.21	79.22	72.56	65.81	59.85	54.31	49.01
0.62	79.57	77.65	71.16	64.56	58.72	53.29	48.08
0.64	78.00	76.13	69.81	63.36	57.63	52.30	47.18
0.66	76.49	74.67	68.51	62.20	56.59	51.35	46.32
0.68	75.04	73.27	67.25	61.08	55.58	50.44	45.48
0.70	73.64	71.92	66.05	60.00	54.60	49.55	44.68
0.72	72.29	70.62	64.88	58.96	53.66	48.70	43.90
0.74	70.99	69.36	63.76	57.96	52.75	47.87	43.15
0.76	69.74	68.15	62.67	56.99	51.88	47.07	42.43
0.78	68.53	66.98	61.62	56.05	51.03	46.30	41.73
0.80	67.36	65.85	60.61	55.14	50.20	45.56	41.05
0.82	66.23	64.76	59.63	54.26	49.41	44.83	40.39
0.84	65.14	63.70	58.68	53.41	48.64	44.13	39.76
0.86	64.08	62.67	57.76	52.58	47.89	43.45	39.14
0.88	63.06	61.68	56.86	51.78	47.16	42.80	38.55

(Contd.)

w/b \ t/b	0.001	0.01	0.05	0.10	0.15	0.20	0.25
			$Z\sqrt{\epsilon_r}$ (ohms)				
0.90	62.07	60.72	56.00	51.00	46.46	42.16	37.97
0.92	61.11	59.79	55.16	50.25	45.78	41.54	37.40
0.94	60.17	58.89	54.34	49.52	45.12	40.94	36.86
0.96	59.27	58.01	53.55	48.81	44.47	40.35	36.33
0.98	58.40	57.16	52.79	48.12	43.85	39.78	35.81
1.00	57.54	56.34	52.04	47.45	43.24	39.23	35.31
1.05	55.52	54.37	50.26	45.85	41.79	37.91	34.12
1.10	53.64	52.54	48.60	44.35	40.43	36.68	33.01
1.15	51.88	50.83	47.05	42.95	39.16	35.53	31.96
1.20	50.23	49.23	45.59	41.64	37.97	34.45	30.98
1.25	48.68	47.72	44.22	40.40	36.85	33.43	30.06
1.30	47.22	46.31	42.93	39.23	35.79	32.47	29.19
1.35	45.85	44.97	41.72	38.13	34.79	31.56	28.37
1.40	44.56	43.71	40.57	37.09	33.85	30.70	27.60
1.45	43.34	42.52	39.48	36.11	32.95	29.89	26.87
1.50	42.18	41.39	38.45	35.18	32.10	29.12	26.17
1.55	41.08	40.32	37.47	34.29	31.30	28.39	25.51
1.60	40.04	39.31	36.54	33.45	30.53	27.69	24.88
1.65	39.05	38.34	35.65	32.64	29.80	27.03	24.28
1.70	38.11	37.42	34.81	31.88	29.10	26.40	23.71
1.75	37.21	36.54	34.01	31.15	28.44	25.80	23.17
1.80	36.36	35.71	33.24	30.45	27.81	25.22	22.65
1.85	35.54	34.91	32.50	29.78	27.20	24.67	22.16
1.90	34.76	34.15	31.80	29.15	26.62	24.14	21.68
1.95	34.01	33.41	31.13	28.54	26.06	23.64	21.23
2.00	33.29	32.71	30.49	27.95	25.53	23.15	20.79
2.05	32.61	32.04	29.87	27.39	25.02	22.69	20.37
2.10	31.95	31.40	29.27	26.85	24.52	22.24	19.97
2.15	31.31	30.78	28.70	26.33	24.05	21.81	19.58
2.20	30.70	30.18	28.15	25.83	23.59	21.40	19.21
2.25	30.12	29.61	27.63	25.35	23.16	21.00	18.85
2.30	29.56	29.06	27.12	24.88	22.73	20.62	18.51
2.35	29.01	28.53	26.63	24.43	22.33	20.25	18.17
2.40	28.49	28.01	26.15	24.00	21.93	19.89	17.85
2.45	27.99	27.52	25.70	23.59	21.55	19.55	17.54
2.50	27.50	27.04	25.26	23.19	21.19	19.22	17.24

Table 5.7 Characteristic impedance of stripline with off-set centre conductor and t/b finite; $d/b = 0.25$ (See Fig. 5.2)

w/b \ t/b	0.001	0.01	0.05	0.10	0.15	0.20	0.25
			$Z\sqrt{\epsilon_r}$ (ohms)				
0.10	184.07	175.92	152.82	133.16	117.79	104.71	92.89
0.12	173.22	165.84	144.62	126.27	111.76	99.27	87.90
0.14	163.58	156.85	137.25	120.06	106.31	94.37	83.41
0.16	154.95	148.78	130.60	114.43	101.37	89.94	79.36
0.18	147.19	141.51	124.56	109.31	96.87	85.90	75.69
0.20	140.17	134.91	119.06	104.63	92.75	82.21	72.34
0.22	133.78	128.90	114.02	100.33	88.97	78.82	69.27
0.24	127.96	123.41	109.40	96.37	85.48	75.70	66.46
0.26	122.62	118.36	105.13	92.71	82.26	72.82	63.86
0.28	117.70	113.71	101.18	89.32	79.27	70.15	61.80
0.30	113.17	109.41	97.52	86.17	76.54	68.07	59.91
0.32	108.97	105.42	94.12	83.35	74.37	66.11	58.14
0.34	105.08	101.72	91.13	81.04	72.31	64.26	56.46
0.36	101.75	98.64	88.63	78.86	70.37	62.51	54.89
0.38	98.90	95.92	86.27	76.79	68.53	60.85	53.39
0.40	96.20	93.34	84.03	74.83	66.78	59.28	51.98
0.42	93.64	90.89	81.90	72.97	65.12	57.79	50.64
0.44	91.22	88.58	79.88	71.19	63.54	56.37	49.36
0.46	88.92	86.37	77.95	69.51	62.03	55.02	48.15
0.48	86.73	84.28	76.12	67.90	60.60	53.73	47.00
0.50	84.65	82.28	74.37	66.36	59.22	52.50	45.90
0.52	82.66	80.38	72.70	64.89	57.91	51.33	44.85
0.54	80.77	78.56	71.10	63.48	56.66	50.20	43.85
0.56	78.96	76.82	69.57	62.14	55.46	49.13	42.89
0.58	77.23	75.16	68.10	60.85	54.31	48.10	41.97
0.60	75.58	73.56	66.70	59.61	53.21	47.11	41.09
0.62	73.99	72.04	65.35	58.42	52.15	46.17	40.25
0.64	72.47	70.57	64.06	57.28	51.13	45.26	39.44
0.66	71.01	69.17	62.81	56.18	50.15	44.38	38.66
0.68	69.61	67.82	61.62	55.12	49.20	43.54	37.91
0.70	68.26	66.52	60.46	54.10	48.30	42.73	37.20
0.72	66.96	65.27	59.35	53.12	47.42	41.95	36.50
0.74	65.72	64.06	58.29	52.18	46.58	41.19	35.84
0.76	64.51	62.90	57.25	51.26	45.76	40.47	35.19
0.78	63.35	61.78	56.26	50.38	44.98	39.77	34.57
0.80	62.24	60.70	55.30	49.53	44.22	39.09	33.98
0.82	61.16	59.66	54.37	48.71	43.48	38.44	33.40
0.84	60.11	58.65	53.47	47.91	42.77	37.80	32.84
0.86	59.11	57.68	52.60	47.14	42.08	37.19	32.30
0.88	58.13	56.74	51.76	46.39	41.42	36.60	31.77

(*Contd.*)

w/b \ t/b	0.001	0.01	0.05	0.10	0.15	0.20	0.25
			$Z\sqrt{\epsilon_r}$ (ohms)				
0.90	57.19	55.83	50.94	45.67	44.77	36.02	31.27
0.92	56.28	54.94	50.15	44.97	40.15	35.47	30.78
0.94	55.39	54.09	49.39	44.29	39.54	34.93	30.30
0.96	54.53	53.26	48.64	43.63	38.95	34.40	29.84
0.98	53.70	52.45	47.92	42.99	38.38	33.90	29.39
1.00	52.90	51.67	47.22	42.37	37.83	33.40	28.96
1.05	50.99	49.82	45.56	40.89	36.51	32.23	27.98
1.10	49.21	48.09	44.01	39.51	35.28	31.14	26.97
1.15	47.55	46.48	42.56	38.22	34.13	30.12	26.07
1.20	46.00	44.98	41.21	37.02	33.05	29.16	25.24
1.25	44.54	43.57	39.93	35.88	32.04	28.26	24.45
1.30	43.18	42.24	38.74	34.82	31.09	27.42	23.71
1.35	41.90	41.00	37.61	33.81	30.19	26.62	23.02
1.40	40.69	39.82	36.55	32.86	29.35	25.87	22.36
1.45	39.55	38.71	35.54	31.97	28.55	25.16	21.74
1.50	38.47	37.66	34.59	31.12	27.79	24.49	21.15
1.55	37.45	36.67	33.69	30.31	27.07	23.86	20.60
1.60	36.48	35.72	32.84	29.55	26.39	23.25	20.07
1.65	35.56	34.83	32.03	28.82	25.74	22.68	19.57
1.70	34.68	33.98	31.25	28.13	25.12	22.13	19.10
1.75	33.85	33.17	30.51	27.47	24.53	21.61	18.64
1.80	33.06	32.39	29.81	26.84	23.97	21.11	18.21
1.85	32.30	31.65	29.14	26.24	23.43	20.64	17.80
1.90	31.58	30.95	28.50	25.67	22.92	20.19	17.40
1.95	30.89	30.27	27.88	25.12	22.43	19.75	17.02
2.00	30.22	29.63	27.30	24.59	21.96	19.33	16.66
2.05	29.59	29.01	26.73	24.08	21.51	18.94	16.32
2.10	28.98	28.42	26.19	23.60	21.08	18.55	15.98
2.15	28.40	27.85	25.67	23.13	20.66	18.19	15.66
2.20	27.84	27.30	25.17	22.69	20.26	17.83	15.36
2.25	27.30	26.77	24.69	22.25	19.88	17.49	15.06
2.30	26.78	26.27	24.23	21.84	19.51	17.17	14.78
2.35	26.28	25.78	23.78	21.44	19.15	16.85	14.51
2.40	25.80	25.31	23.35	21.06	18.80	16.55	14.24
2.45	25.34	24.86	22.94	20.68	18.47	16.25	13.99
2.50	24.49	24.42	22.54	20.32	18.15	15.97	13.74

Table 5.8 Characteristic impedance of stripline with off-set centre conductor and t/b finite; $d/b = 0.3$ (See Fig. 5.2)

w/b \ t/b	0.001	0.01	0.05	0.10	0.15	0.20	0.25
				$Z\sqrt{\epsilon_r}$ (ohms)			
0.10	179.85	170.96	146.04	124.83	107.96	93.10	78.85
0.12	167.80	159.84	137.11	117.36	101.42	87.19	73.39
0.14	157.27	150.08	129.20	110.74	95.62	81.98	68.64
0.16	147.99	141.44	122.16	104.82	90.45	77.36	64.46
0.18	139.73	133.75	115.84	99.50	85.81	73.24	60.76
0.20	132.35	126.84	110.15	94.70	81.63	69.53	57.47
0.22	125.71	120.62	104.99	90.34	77.83	66.18	54.51
0.24	119.71	114.98	100.29	86.36	74.37	63.14	51.84
0.26	114.25	109.84	95.99	82.72	71.21	60.36	49.43
0.28	109.27	105.14	92.05	79.38	68.30	57.82	47.53
0.30	104.70	100.83	88.42	76.29	65.68	55.86	45.82
0.32	100.51	96.85	85.06	73.55	63.60	54.03	44.22
0 34	96.63	93.18	82.14	71.32	61.65	52.32	42.74
0.36	93.34	90.15	79.70	69.22	59.81	50.71	41.35
0.38	90.53	87.47	77.41	67.25	58.08	49.19	40.05
0.40	87.89	84.95	75.24	65.38	56.45	47.77	38.82
0.42	85.39	82.58	73.19	63.61	54.90	46.42	37.67
0.44	83.03	80.33	71.25	61.94	53.44	45.15	36.59
0.46	80.80	78.20	69.41	60.35	52.06	43.95	35.56
0.48	78.69	76.18	67.67	58.84	50.74	42.80	34.60
0.50	76.68	74.26	66.01	57.41	49.49	41.72	33.68
0.52	74.77	72.44	64.43	56.04	48.30	40.69	32.81
0.54	72.96	70.70	62.92	54.74	47.17	39.71	31.99
0.56	71.23	69.04	61.48	53.50	46.08	38.78	31.20
0.58	69.58	67.47	60.11	52.31	45.05	37.88	30.45
0.60	68.01	65.96	58.79	51.17	44.06	37.03	29.74
0.62	66.51	64.51	57.53	50.08	43.12	36.22	29.06
0.64	65.07	63.13	56.33	49.04	42.21	35.44	28.41
0.66	63.69	61.81	55.17	48.04	41.34	34.69	27.79
0.68	62.37	60.54	54.06	47.08	40.51	33.98	27.20
0.70	61.10	59.32	53.00	46.15	39.71	33.29	26.63
0.72	59.88	58.15	51.97	45.27	38.94	32.64	26.08
0.74	58.71	57.03	50.99	44.41	38.20	32.00	25.56
0.76	57.59	55.95	50.04	43.59	37.48	31.39	25.05
0.78	56.51	54.91	49.12	42.80	36.80	30.80	24.57
0.80	55.47	53.90	48.24	42.03	36.13	30.24	24.10
0.82	54.46	52.94	47.39	41.30	35.49	29.69	23.65
0.84	53.49	52.00	46.57	40.58	34.88	29.17	23.22
0.86	52.56	51.10	45.78	39.90	34.28	28.66	22.81
0.88	51.65	50.23	45.01	39.23	33.71	28.17	22.40

(Contd.)

w/b \ t/b	0.001	0.01	0.05	0.10	0.15	0.20	0.25
			$Z\sqrt{\epsilon_r}$ (ohms)				
0.90	50.78	49.39	44.27	38.59	33.15	27.70	22.02
0.92	49.94	48.58	43.55	37.97	32.61	27.24	21.64
0.94	49.12	47.79	42.86	37.36	32.09	26.80	21.28
0.96	48.33	47.03	42.19	36.78	31.59	26.37	20.93
0.98	47.57	46.29	41.54	36.21	31.10	25.95	20.59
1.00	46.83	45.57	40.90	35.67	30.62	25.55	20.26
1.05	45.07	43.88	39.41	34.36	29.50	24.60	19.49
1.10	43.45	42.31	38.01	33.16	28.45	23.71	18.77
1.15	41.93	40.84	36.72	32.03	27.48	22.89	18.10
1.20	40.52	39.47	35.50	30.97	26.57	22.12	17.48
1.25	39.20	38.20	34.37	29.99	25.72	21.40	16.90
1.30	37.96	37.00	33.30	29.06	24.92	20.73	16.35
1.35	36.80	35.87	32.30	28.19	24.17	20.10	15.84
1.40	35.71	34.81	31.36	27.37	23.47	19.50	15.36
1.45	34.68	33.82	30.47	26.60	22.80	18.94	14.91
1.50	33.71	32.87	29.63	25.87	22.17	18.41	14.49
1.55	32.79	31.98	28.84	25.18	21.58	17.91	14.09
1.60	31.92	31.14	28.09	24.52	21.01	17.44	13.71
1.65	31.09	30.34	27.37	23.90	20.48	16.99	13.35
1.70	30.31	29.58	26.69	23.31	19.97	16.56	13.01
1.75	29.56	28.85	26.05	22.75	19.48	16.16	12.68
1.80	28.86	28.17	25.43	22.21	19.02	15.77	12.37
1.85	28.18	27.51	24.84	21.70	18.58	15.40	12.08
1.90	27.53	26.88	24.28	21.21	18.16	15.05	11.80
1.95	26.92	26.28	23.75	20.74	17.76	14.71	11.53
2.00	26.33	25.71	23.23	20.29	17.38	14.39	11.28
2.05	25.76	25.16	22.74	19.87	17.01	14.09	11.03
2.10	25.22	24.64	22.27	19.46	16.65	13.79	10.80
2.15	24.71	24.13	21.82	19.06	16.32	13.51	10.57
2.20	24.21	23.65	21.39	18.68	15.99	13.24	10.36
2.25	23.73	23.18	20.97	18.32	15.68	12.98	10.15
2.30	23.27	22.74	20.57	17.97	15.38	12.73	9.95
2.35	22.83	22.31	20.18	17.63	15.09	12.49	9.76
2.40	22.41	21.89	19.81	17.31	14.81	12.25	9.58
2.45	22.00	21.49	19.45	17.00	14.54	12.03	9.40
2.50	21.60	21.11	19.11	16.70	14.29	11.81	9.23

Table 5.9 Characteristic impedance of stripline with off-set centre conductor and t/b finite; $d/b = 0.35$ (See Fig. 5.2)

w/b \ t/b	0.001	0.01	0.05	0.10	0.15	0.20
			$Z\sqrt{\epsilon_r}$ (ohms)			
0.10	172.95	162.95	135.27	111.42	91.54	72.26
0.12	159.15	150.34	125.32	103.18	84.35	65.80
0.14	147.40	139.54	116.73	96.07	78.20	60.40
0.16	137.26	130.19	109.25	89.88	72.89	55.82
0.18	128.42	122.01	102.67	84.44	68.26	51.88
0.20	120.66	114.80	96.83	79.62	64.17	48.46
0.22	113.78	108.40	91.62	75.32	66.55	45.47
0.24	107.64	102.67	86.95	71.47	57.32	42.82
0.26	102.13	97.51	82.73	67.98	54.41	40.47
0.28	97.16	92.85	78.90	64.82	51.79	38.36
0.30	92.65	88.62	75.40	61.95	49.45	36.76
0.32	88.54	84.75	72.21	59.42	47.62	35.29
0.34	84.77	81.21	69.45	57.38	45.93	33.93
0.36	81.61	78.30	67.17	55.48	44.34	32.68
0.38	78.92	75.76	65.03	53.70	42.87	31.51
0.40	76.40	73.37	63.03	52.03	41.49	30.43
0.42	74.04	71.13	61.14	50.46	40.19	29.41
0.44	71.82	69.03	59.37	48.99	38.98	28.46
0.46	69.72	67.04	57.69	47.59	37.83	27.57
0.48	67.75	65.17	56.11	46.28	36.75	26.74
0.50	65.89	63.40	54.61	45.03	35.73	25.95
0.52	64.12	61.72	53.19	43.85	34.77	25.21
0.54	62.45	60.13	51.84	42.73	33.85	24.51
0.56	60.86	58.61	50.56	41.67	32.99	23.85
0.58	59.35	57.18	49.34	40.66	32.16	23.22
0.60	57.92	55.81	48.18	39.70	31.38	22.63
0.62	56.55	54.50	47.07	38.78	30.63	22.06
0.64	55.25	53.26	46.01	37.90	29.92	21.52
0.66	54.00	52.07	45.00	37.06	29.24	21.01
0.68	52.81	50.93	44.03	36.26	28.59	20.52
0.70	51.67	49.84	43.10	35.49	27.97	20.06
0.72	50.58	48.80	42.21	34.75	27.38	19.61
0.74	49.53	47.80	41.35	34.04	26.81	19.19
0.76	48.53	46.84	40.53	33.37	26.26	18.78
0.78	47.56	45.91	39.75	32.71	25.74	18.39
0.80	46.64	45.03	38.99	32.09	25.23	18.01
0.82	45.75	44.17	38.26	31.48	24.75	17.65
0.84	44.89	43.35	37.56	30.90	24.28	17.31
0.86	44.06	42.56	36.88	30.34	23.83	16.97
0.88	43.27	41.80	36.23	29.80	23.40	16.65

(Contd.)

w/b \ t/b	0.001	0.01	0.05	0.10	0.15	0.20
			$Z\sqrt{\epsilon_r}$ (ohms)			
0.90	42.50	41.06	35.60	29.28	22.98	16.34
0.92	41.76	40.35	34.99	28.78	22.58	16.05
0.94	41.04	39.66	34.40	28.29	22.19	15.76
0.96	40.35	39.00	33.83	27.82	21.81	15.48
0.98	39.68	38.36	33.28	27.37	21.45	15.22
1.00	39.03	37.74	32.74	26.93	21.10	14.96
1.05	37.51	36.27	31.48	25.89	20.27	14.35
1.10	36.09	34.91	30.32	24.92	19.50	13.79
1.15	34.78	33.65	29.23	24.03	18.79	13.27
1.20	33.56	32.48	28.22	23.20	18.13	12.79
1.25	32.43	31.39	27.28	22.42	17.51	12.35
1.30	31.37	30.36	26.40	21.69	16.94	11.93
1.35	30.37	29.41	25.58	21.01	16.40	11.54
1.40	29.44	28.51	24.80	20.37	15.89	11.17
1.45	28.56	27.66	24.07	19.77	15.42	10.83
1.50	27.73	26.86	23.38	19.20	14.97	10.51
1.55	26.95	26.11	22.73	18.67	14.55	10.20
1.60	26.21	25.40	22.12	18.16	14.15	9.92
1.65	25.52	24.73	21.54	17.68	13.77	9.65
1.70	24.85	24.09	20.98	17.23	13.41	9.39
1.75	24.23	23.48	20.46	16.80	13.07	9.15
1.80	23.63	22.91	19.96	16.39	12.75	8.92
1.85	23.06	22.36	19.48	15.99	12.44	8.70
1.90	22.52	21.83	19.03	15.62	12.15	8.49
1.95	22.00	21.33	18.60	15.27	11.87	8.29
2.00	21.51	20.86	18.18	14.92	11.60	8.10
2.05	21.03	20.40	17.79	14.60	11.34	7.92
2.10	20.58	19.96	17.41	14.29	11.10	7.74
2.15	20.15	19.54	17.05	13.99	10.86	7.58
2.20	19.73	19.14	16.70	13.70	10.64	7.42
2.25	19.33	18.76	16.36	13.43	10.43	7.27
2.30	18.95	18.39	16.04	13.16	10.22	7.12
2.35	18.58	18.03	15.73	12.91	10.02	6.98
2.40	18.23	17.69	15.44	12.67	9.83	6.84
2.45	17.89	17.36	15.15	12.43	9.64	6.71
2.50	17.56	17.04	14.88	12.20	9.47	6.59

Table 5.10 Characteristic impedance of stripline with off-set centre conductor and t/b finite; d/b = 0.4 (See Fig. 5.2)

w/b \ t/b	0.001	0.01	0.05	0.10	0.15
			$Z\sqrt{\epsilon_r}$ (ohms)		
0.10	160.33	148.56	116.11	86.44	56.74
0.12	143.92	133.81	104.88	77.41	49.39
0.14	130.56	121.73	95.63	70.09	43.72
0.16	119.47	111.65	87.88	64.04	39.22
0.18	110.11	103.11	81.29	58.94	35.55
0.20	102.11	95.78	75.62	54.60	32.52
0.22	92.20	89.43	70.69	50.86	29.96
0.24	89.16	83.86	66.36	47.59	27.78
0.26	83.85	78.95	62.54	44.72	25.89
0.28	79.13	74.58	59.13	42.17	24.24
0.30	74.91	70.67	56.07	39.90	22.82
0.32	71.12	67.15	53.31	37.95	21.73
0.34	67.70	63.97	50.96	36.39	20.75
0.36	64.85	61.38	49.04	34.96	19.85
0.38	62.45	59.14	47.26	33.64	19.02
0.40	60.22	57.05	45.60	32.41	18.26
0.42	58.14	55.11	44.06	31.27	17.56
0.44	56.21	53.29	42.62	30.21	16.91
0.46	54.40	51.59	41.26	29.22	16.31
0.48	52.70	50.00	40.00	28.29	15.74
0.50	51.10	48.50	38.80	27.42	15.22
0.52	49.60	47.09	37.68	26.60	14.73
0.54	48.18	45.76	36.62	25.83	14.27
0.56	46.85	44.50	35.62	25.10	13.84
0.58	45.58	43.31	34.67	24.41	13.43
0.60	44.38	42.18	33.77	23.76	13.05
0.62	43.24	41.10	32.91	23.14	12.68
0.64	42.16	40.09	32.10	22.55	12.34
0.66	41.14	39.12	31.33	21.99	12.02
0.68	40.16	38.19	30.59	21.46	11.71
0.70	39.22	37.31	29.89	20.96	11.42
0.72	38.33	36.47	29.22	20.48	11.14
0.74	37.48	35.67	28.58	20.01	10.87
0.76	36.67	34.90	27.96	19.57	10.62
0.78	35.89	34.16	27.37	19.15	10.38
0.80	35.14	33.45	26.81	18.75	10.15
0.82	34.42	32.78	26.27	18.36	9.93
0.84	33.73	32.13	25.75	17.99	9.72
0.86	33.07	31.50	25.25	17.63	9.52
0.88	32.44	30.90	24.77	17.29	9.32

(Contd.)

t/b \ w/b	0.001	0.01	0.05	0.10	0.15
			$Z\sqrt{\epsilon_r}$ (ohms)		
0.90	31.83	30.32	24.31	16.96	9.14
0.92	31.24	29.76	23.86	16.64	8.96
0.94	30.67	29.22	23.43	16.34	8.78
0.96	30.12	28.70	23.02	16.04	8.62
0.98	29.59	28.20	22.62	15.76	8.46
1.00	29.08	27.72	22.23	15.48	8.31
1.05	27.88	26.58	21.32	14.84	7.94
1.10	26.78	25.53	20.48	14.24	7.61
1.15	25.75	24.56	19.70	13.70	7.31
1.20	24.81	23.67	18.99	13.19	7.03
1.25	23.93	22.83	18.32	12.72	6.77
1.30	23.11	22.05	17.69	12.28	6.53
1.35	22.34	21.33	17.11	11.87	6.30
1.40	21.63	20.65	16.57	11.49	6.09
1.45	20.96	20.01	16.06	11.13	5.89
1.50	20.33	19.41	15.58	10.79	5.71
1.55	19.73	18.84	15.12	10.47	5.54
1.60	19.17	18.31	14.70	10.17	5.37
1.65	18.64	17.80	14.29	9.89	5.22
1.70	18.14	17.33	13.91	9.62	5.08
1.75	17.67	16.88	13.55	9.37	4.94
1.80	17.21	16.45	13.21	9.13	4.81
1.85	16.79	16.04	12.88	8.90	4.69
1.90	16.38	15.65	12.57	8.68	4.57
1.95	15.99	15.28	12.27	8.48	4.46
2.00	15.62	14.93	11.99	8.28	4.35
2.05	15.27	14.59	11.72	8.09	4.25
2.10	14.93	14.27	11.46	7.91	4.15
2.15	14.61	13.96	11.21	7.74	4.06
2.20	14.30	13.67	10.98	7.57	3.97
2.25	14.00	13.38	10.75	7.42	3.89
2.30	13.72	13.11	10.53	7.26	3.81
2.35	13.44	12.85	10.32	7.12	3.73
2.40	13.18	12.60	10.12	6.98	3.66
2.45	12.93	12.36	9.93	6.84	3.58
2.50	12.69	12.13	9.74	6.72	3.51

Table 5.11 Characteristic impedance and effective dielectric constant of suspended stripline; $\epsilon_r = 2.22$, $c/w = 40$ (See structure A4 of Table 3.1)

d/b → w/b ↓	0.1		0.2		0.3		0.4		0.5	
	Z (ohms)	ϵ_f	Z (ohms)	ϵ_f	Z (ohms)	ϵ_f	Z (ohms)	ϵ_f	Z (ohms)	ϵ_f
0.10	171.38	1.293	163.71	1.385	157.92	1.429	152.01	1.450	145.07	1.456
0.12	162.62	1.280	155.08	1.374	149.33	1.419	143.45	1.440	136.52	1.447
0.14	155.19	1.269	147.78	1.364	142.08	1.410	136.22	1.432	129.32	1.439
0.16	148.74	1.259	141.46	1.355	135.81	1.402	129.97	1.425	123.11	1.431
0.18	143.03	1.250	135.90	1.346	130.28	1.395	124.48	1.418	117.65	1.424
0.20	137.90	1.242	130.92	1.338	125.35	1.387	119.58	1.411	112.80	1.417
0.22	133.26	1.234	126.43	1.331	120.90	1.381	115.17	1.404	108.44	1.410
0.24	129.02	1.228	122.33	1.324	116.86	1.374	111.16	1.398	104.48	1.404
0.26	125.10	1.221	118.56	1.317	113.14	1.368	107.50	1.393	100.86	1.398
0.28	121.48	1.215	115.08	1.311	109.72	1.362	104.12	1.387	97.53	1.393
0.30	118.11	1.210	111.84	1.305	106.55	1.357	100.99	1.381	94.45	1.387
0.32	114.97	1.205	108.83	1.300	103.59	1.351	98.07	1.376	91.60	1.382
0.34	112.01	1.201	106.00	1.295	100.82	1.346	95.35	1.371	88.93	1.377
0.36	109.22	1.196	103.34	1.290	98.22	1.341	92.80	1.366	86.44	1.372
0.38	106.58	1.192	100.83	1.285	95.77	1.336	90.40	1.361	84.10	1.367
0.40	104.09	1.188	98.46	1.280	93.46	1.332	88.14	1.357	81.90	1.363
0.45	98.38	1.180	93.04	1.270	88.20	1.321	83.01	1.346	76.92	1.352
0.50	93.32	1.172	88.25	1.260	83.56	1.311	78.50	1.336	72.57	1.342
0.55	88.78	1.166	83.95	1.251	79.42	1.302	74.49	1.327	68.71	1.333
0.60	84.69	1.160	80.08	1.244	75.69	1.293	70.89	1.319	65.27	1.325
0.65	80.97	1.154	76.57	1.236	72.32	1.286	67.65	1.311	62.18	1.317
0.70	77.57	1.150	73.36	1.230	69.25	1.279	64.70	1.304	59.38	1.310
0.75	74.45	1.145	70.42	1.224	66.43	1.272	62.01	1.297	56.83	1.303
0.80	71.58	1.141	67.71	1.219	63.85	1.266	59.54	1.291	54.49	1.297
0.85	68.93	1.137	65.21	1.214	61.46	1.260	57.27	1.285	52.35	1.291
0.90	66.46	1.134	62.88	1.209	59.25	1.255	55.17	1.280	50.37	1.286
0.95	64.17	1.131	60.72	1.205	57.19	1.250	53.22	1.275	48.54	1.281
1.00	62.03	1.128	58.71	1.200	55.27	1.246	51.40	1.270	46.84	1.276
1.05	60.04	1.125	56.82	1.197	53.48	1.241	49.71	1.265	45.26	1.271
1.10	58.16	1.122	55.05	1.193	51.81	1.237	48.12	1.261	43.78	1.267
1.15	56.40	1.120	53.39	1.190	50.23	1.234	46.64	1.257	42.40	1.263
1.20	54.75	1.117	51.83	1.187	48.75	1.230	45.24	1.254	41.10	1.260
1.25	53.19	1.115	50.35	1.184	47.35	1.227	43.92	1.250	39.88	1.256
1.30	51.71	1.113	48.96	1.181	46.04	1.224	42.68	1.247	38.73	1.253
1.35	50.32	1.111	47.65	1.178	44.79	1.221	41.51	1.244	37.65	1.250
1.40	49.00	1.109	46.40	1.176	43.61	1.218	40.40	1.241	36.62	1.247
1.45	47.74	1.108	45.21	1.173	42.49	1.215	39.35	1.238	35.65	1.244
1.50	46.55	1.106	44.09	1.171	41.43	1.213	38.36	1.235	34.73	1.241
1.55	45.42	1.104	43.02	1.169	40.41	1.210	37.41	1.233	33.85	1.239
1.60	44.34	1.103	42.00	1.167	39.45	1.208	36.50	1.230	33.02	1.236

(Contd.)

w/b \ d/b	0.1 Z (ohms)	ϵ_f	0.2 Z (ohms)	ϵ_f	0.3 Z (ohms)	ϵ_f	0.4 Z (ohms)	ϵ_f	0.5 Z (ohms)	ϵ_f
1.65	43.31	1.102	41.03	1.165	38.53	1.206	35.64	1.228	32.23	1.234
1.70	42.33	1.100	40.10	1.163	37.65	1.203	34.82	1.226	31.48	1.232
1.75	41.39	1.099	39.21	1.161	36.82	1.201	34.04	1.224	30.75	1.230
1.80	40.49	1.098	38.36	1.160	36.02	1.200	33.29	1.222	30.07	1.228
1.85	39.63	1.096	37.55	1.158	35.25	1.198	32.57	1.220	29.41	1.226
1.90	38.80	1.095	36.77	1.156	34.51	1.196	31.89	1.218	28.78	1.224
1.95	38.01	1.094	36.02	1.155	33.81	1.194	31.23	1.216	28.17	1.222
2.00	37.25	1.093	35.30	1.153	33.13	1.193	30.60	1.214	27.60	1.220
2.05	36.52	1.092	34.61	1.152	32.48	1.191	29.99	1.213	27.04	1.219
2.10	35.82	1.091	33.95	1.151	31.85	1.190	29.40	1.211	26.51	1.217
2.15	35.14	1.090	33.31	1.149	31.25	1.188	28.84	1.210	25.99	1.216
2.20	34.49	1.089	32.70	1.148	30.67	1.187	28.30	1.208	25.50	1.214
2.25	33.86	1.089	32.10	1.147	30.11	1.185	27.78	1.207	25.02	1.213
2.30	33.26	1.088	31.53	1.146	29.57	1.184	27.28	1.205	24.57	1.212
2.35	32.68	1.087	30.98	1.145	29.05	1.183	26.80	1.204	24.12	1.210
2.40	32.11	1.086	30.44	1.144	28.55	1.182	26.33	1.203	23.70	1.209
2.45	31.57	1.085	29.93	1.143	28.07	1.180	25.88	1.202	23.29	1.208
2.50	31.04	1.085	29.43	1.142	27.60	1.179	25.44	1.200	22.89	1.207

Table 5.12 Characteristic impedance and effective dielectric constant of suspended stripline; $\epsilon_r = 3.78$, $c/w = 40$ (See structure A4 of Table 3.1)

w/b \ d/b	0.1		0.2		0.3		0.4		0.5	
	Z (ohms)	ϵ_f	Z (ohms)	ϵ_f	Z (ohms)	ϵ_f	Z (ohms)	ϵ_f	Z (ohms)	ϵ_f
0.10	155.41	1.573	144.57	1.776	137.52	1.884	131.29	1.943	124.72	1.969
0.12	148.04	1.545	137.41	1.750	130.43	1.860	124.24	1.920	117.70	1.946
0.14	141.76	1.521	131.34	1.727	124.43	1.839	118.27	1.900	111.76	1.926
0.16	136.27	1.500	126.08	1.706	119.23	1.819	113.11	1.881	106.63	1.907
0.18	131.40	1.481	121.43	1.686	114.65	1.801	108.56	1.864	102.12	1.890
0.20	127.01	1.464	117.26	1.668	110.56	1.784	104.51	1.847	98.10	1.873
0.22	123.01	1.449	113.49	1.652	106.86	1.767	100.85	1.831	94.48	1.858
0.24	119.35	1.434	110.04	1.636	103.48	1.752	97.52	1.817	91.21	1.843
0.26	115.96	1.422	106.87	1.621	100.39	1.738	94.48	1.803	88.20	1.829
0.28	112.80	1.410	103.93	1.607	97.53	1.724	91.67	1.789	85.43	1.815
0.30	109.86	1.399	101.20	1.594	94.88	1.711	89.06	1.776	82.87	1.802
0.32	107.10	1.389	98.65	1.582	92.40	1.698	86.63	1.764	80.49	1.790
0.34	104.50	1.379	96.24	1.571	90.08	1.686	84.36	1.752	78.27	1.778
0.36	102.04	1.371	93.98	1.559	87.89	1.675	82.22	1.740	76.19	1.766
0.38	99.71	1.362	91.83	1.549	85.83	1.664	80.22	1.729	74.23	1.755
0.40	97.50	1.354	89.80	1.539	83.88	1.653	78.32	1.718	72.39	1.744
0.45	92.43	1.337	85.15	1.516	79.43	1.628	74.01	1.693	68.22	1.719
0.50	87.90	1.321	81.02	1.495	75.49	1.606	70.21	1.670	64.55	1.696
0.55	83.82	1.308	77.29	1.476	71.96	1.585	66.82	1.649	61.30	1.675
0.60	80.12	1.296	73.92	1.459	68.78	1.567	63.77	1.630	58.39	1.656
0.65	76.75	1.285	70.85	1.444	65.88	1.549	61.01	1.612	55.76	1.638
0.70	73.66	1.275	68.03	1.430	63.23	1.533	58.49	1.596	53.37	1.621
0.75	70.82	1.266	65.44	1.418	60.80	1.519	56.18	1.580	51.19	1.606
0.80	68.19	1.257	63.04	1.406	58.55	1.505	54.06	1.566	49.19	1.591
0.85	65.75	1.250	60.82	1.395	56.47	1.493	52.10	1.553	47.35	1.578
0.90	63.49	1.243	58.75	1.385	54.54	1.481	50.28	1.540	45.65	1.566
0.95	61.38	1.236	56.82	1.376	52.74	1.470	48.59	1.529	44.07	1.554
1.00	59.40	1.230	55.01	1.367	51.05	1.460	47.01	1.518	42.60	1.543
1.05	57.55	1.224	53.32	1.359	49.48	1.451	45.53	1.508	41.22	1.533
1.10	55.81	1.219	51.73	1.351	47.90	1.442	44.15	1.498	39.94	1.523
1.15	54.18	1.214	50.23	1.344	46.60	1.433	42.85	1.489	38.73	1.514
1.20	52.64	1.209	48.82	1.337	45.28	1.426	41.62	1.481	37.59	1.506
1.25	51.18	1.205	47.48	1.331	44.04	1.418	40.46	1.473	36.52	1.498
1.30	49.80	1.200	46.22	1.325	42.87	1.411	39.37	1.466	35.51	1.490
1.35	48.50	1.196	45.02	1.320	41.75	1.405	38.33	1.459	34.56	1.483
1.40	47.26	1.193	43.89	1.314	40.70	1.398	37.35	1.452	33.66	1.476
1.45	46.08	1.189	42.81	1.309	39.69	1.392	36.42	1.446	32.80	1.470
1.50	44.96	1.186	41.78	1.304	38.74	1.387	35.53	1.439	31.98	1.464
1.55	43.90	1.182	40.80	1.300	37.82	1.381	34.68	1.434	31.21	1.458
1.60	42.88	1.179	39.86	1.295	36.96	1.376	33.88	1.428	30.47	1.452

(Contd.)

w/b \ d/b	0.1 Z (ohms)	ϵ_f	0.2 Z (ohms)	ϵ_f	0.3 Z (ohms)	ϵ_f	0.4 Z (ohms)	ϵ_f	0.5 Z (ohms)	ϵ_f
1.65	41.91	1.176	38.97	1.291	36.13	1.372	33.11	1.423	29.77	1.447
1.70	40.98	1.173	38.12	1.287	35.33	1.367	32.38	1.418	29.09	1.442
1.75	40.10	1.171	37.30	1.283	34.57	1.362	31.67	1.413	28.45	1.437
1.80	39.25	1.168	36.52	1.280	33.85	1.358	31.00	1.409	27.84	1.432
1.85	38.43	1.166	35.77	1.276	33.15	1.354	30.36	1.404	27.25	1.428
1.90	37.65	1.163	35.05	1.273	32.48	1.350	29.74	1.400	26.68	1.424
1.95	36.90	1.161	34.35	1.270	31.84	1.347	29.14	1.396	26.14	1.420
2.00	36.18	1.159	33.69	1.267	31.22	1.343	28.57	1.392	25.62	1.416
2.05	35.48	1.157	33.05	1.264	30.63	1 339	28.02	1.389	25.12	1.412
2.10	34.82	1.155	32.43	1.261	30.05	1.336	27.50	1.385	24.64	1.408
2.15	34.17	1.153	31.84	1.258	29.50	1.333	26.99	1.382	24.18	1.405
2.20	33.55	1.151	31.27	1.255	28.97	1.330	26.50	1.379	23.73	1.402
2.25	32.95	1.150	30.72	1.253	28.46	1.327	26.02	1.375	23.30	1.399
2.30	32.38	1.148	30.18	1.250	27.97	1.324	25.57	1.372	22.89	1.395
2.35	31.82	1.146	29.67	1.248	27.49	1.321	25.13	1.369	22.49	1.392
2.40	31.28	1.145	29.17	1.246	27.03	1.319	24.70	1.366	22.10	1.390
2.45	30.76	1.143	28.69	1.244	26.58	1.316	24.29	1.364	21.73	1.387
2.50	30.26	1.142	28.22	1.241	26.15	1.314	23.89	1.361	21.37	1.384

Table 5.13 Characteristic impedance and effective dielectric constant of suspended stripline; $\epsilon_r = 9.6$, $c/w = 40$ (See structure A4 of Table 3.1)

d/b	0.1		0.2		0.3		0.4		0.5	
w/b	Z (ohms)	ϵ_f	Z (ohms)	ϵ_f	Z (ohms)	ϵ_f	Z (ohms)	ϵ_f	Z (ohms)	ϵ_f
0.10	128.30	2.308	114.04	2.854	105.61	3.195	99.03	3.415	92.95	3.545
0.12	123.02	2.237	109.18	2.772	100.93	3.106	94.43	3.324	88.36	3.454
0.14	118.45	2.178	104.96	2.704	96.85	3.035	90.41	3.251	84.37	3.380
0.16	114.43	2.127	101.26	2.644	93.26	2.973	86.88	3.188	80.87	3.315
0.18	110.84	2.081	97.97	2.590	90.08	2.917	83.75	3.131	77.78	3.257
0.20	107.58	2.040	95.01	2.541	87.22	2.866	80.95	3.079	75.01	3.204
0.22	104.60	2.004	92.31	2.496	84.63	2.818	78.41	3.030	72.52	3.154
0.24	101.84	1.970	89.84	2.454	82.26	2.773	76.10	2.984	70.25	3.107
0.26	99.27	1.940	87.56	2.415	80.08	2.731	73.97	2.941	68.16	3.062
0.28	96.87	1.911	85.44	2.379	78.05	2.692	72.01	2.900	66.23	3.020
0.30	94.62	1.885	83.46	2.344	76.18	2.654	70.18	2.861	64.45	2.980
0.32	92.51	1.862	81.60	2.312	74.41	2.619	68.47	2.823	62.78	2.941
0.34	90.50	1.839	79.84	2.282	72.75	2.585	66.87	2.788	61.23	2.905
0.36	88.60	1.818	78.18	2.253	71.19	2.553	65.36	2.754	59.77	2.870
0.38	86.79	1.798	76.60	2.226	69.71	2.522	63.94	2.722	58.39	2.836
0.40	85.06	1.780	75.11	2.200	68.31	2.493	62.59	2.691	57.09	2.804
0.45	81.07	1.738	71.65	2.141	65.10	2.425	59.52	2.168	54.14	2.730
0.50	77.48	1.701	68.55	2.088	62.23	2.364	56.79	2.553	51.53	2.662
0.55	74.22	1.668	65.74	2.041	59.64	2.308	54.35	2.493	49.20	2.600
0.60	71.24	1.639	63.18	1.998	57.29	2.258	52.14	2.438	47.10	2.544
0.65	68.50	1.613	60.82	1.960	55.14	2.212	50.12	2.388	45.20	2.492
0.70	65.98	1.589	58.65	1.925	53.16	2.170	48.27	2.342	43.47	2.444
0.75	63.64	1.567	56.64	1.893	51.33	2.131	46.57	2.300	41.87	2.400
0.80	61.48	1.547	54.76	1.863	49.63	2.095	44.99	2.261	40.40	2.359
0.85	59.46	1.528	53.01	1.836	48.04	2.062	43.53	2.224	39.04	2.321
0.90	57.57	1.511	51.38	1.811	46.56	2.032	42.17	2.190	37.78	2.286
0.95	55.80	1.495	49.85	1.787	45.18	2.003	40.89	2.159	36.60	2.253
1.00	54.14	1.481	48.40	1.766	43.88	1.977	39.69	2.129	35.50	2.222
1.05	52.58	1.467	47.05	1.745	42.65	1.952	38.57	2.102	34.46	2.193
1.10	51.10	1.454	45.76	1.726	41.49	1.929	37.51	2.076	33.49	2.166
1.15	49.71	1.442	44.55	1.709	40.40	1.907	36 51	2.052	32.57	2.141
1.20	48.39	1.430	43.40	1.692	39.36	1.887	35.56	2.029	31.71	2.117
1.25	47.15	1.419	42.32	1.676	38.38	1.868	34.66	2.007	30.89	2.094
1.30	45.96	1.409	41.28	1.661	37.44	1.849	33.81	1.987	30.11	2.073
1.35	44.84	1.400	40.30	1.647	36.56	1.832	33.00	1.968	29.38	2.052
1.40	43.77	1.391	39.36	1.634	35.71	1.816	32.23	1.950	28.68	2.033
1.45	42.74	1.382	38.47	1.621	34.90	1.801	31.50	1.932	28.01	2.015
1.50	41.77	1.374	37.61	1.609	34.13	1.786	30.80	1.916	27.37	1.998
1.55	40.84	1.366	36.80	1.598	33.39	1.772	30.13	1.901	26.77	1.982
1.60	39.95	1.359	36.01	1.587	32.69	1.759	29.48	1.886	26.18	1.966

(Contd.)

w/b \ d/b	0.1 Z (ohms)	ϵ_f	0.2 Z (ohms)	ϵ_f	0.3 Z (ohms)	ϵ_f	0.4 Z (ohms)	ϵ_f	0.5 Z (ohms)	ϵ_f
1.65	39.10	1.351	35.27	1.577	32.01	1.746	28.87	1.872	25.63	1.951
1.70	38.29	1.345	34.55	1.567	31.37	1.734	28.28	1.858	25.10	1.937
1.75	37.50	1.338	33.86	1.557	30.74	1.723	27.72	1.845	24.59	1.924
1.80	36.75	1.332	33.20	1.548	30.15	1.712	27.18	1.833	24.10	1.911
1.85	36.03	1.326	32.57	1.540	29.57	1.701	26.65	1.822	23.63	1.899
1.90	35.34	1.321	31.95	1.531	29.02	1.691	26.15	1.810	23.18	1.887
1.95	34.67	1.315	31.37	1.523	28.49	1.682	25.67	1.800	22.74	1.875
2.00	34.03	1.310	30.80	1.516	27.98	1.672	25.21	1.789	22.33	1.865
2.05	33.41	1.305	30.25	1.508	27.48	1.664	24.76	1.779	21.92	1.854
2.10	32.81	1.300	29.72	1.501	27.01	1.655	24.33	1.770	21.53	1.844
2.15	32.24	1.296	29.22	1.494	26.54	1.647	23.91	1.761	21.16	1.835
2.20	31.68	1.291	28.72	1.488	26.10	1.639	23.51	1.752	20.80	1.825
2.25	31.15	1.287	28.25	1.481	25.67	1.631	23.12	1.743	20.45	1.816
2.30	30.63	1.283	27.79	1.475	25.25	1.624	22.74	1.735	20.11	1.808
2.35	30.12	1.279	27.34	1.469	24.85	1.617	22.37	1.727	19.78	1.799
2.40	29.64	1.275	26.91	1.464	24.46	1.610	22.02	1.719	19.47	1.792
2.45	29.17	1.271	26.49	1.458	24.08	1.603	21.68	1.712	19.16	1.784
2.50	28.71	1.268	26.09	1.453	23.72	1.597	21.35	1.705	18.86	1.776

Table 5.14 Characteristic impedance and effective dielectric constant of sandwiched stripline; $\epsilon_{r1} = 2.22$, $\epsilon_{r2} = 2.56$, $c/w = 40$ (See structure A5 of Table 3.1)

d/b → / w/b ↓	0.1		0.2		0.3		0.4		0.5	
	Z (ohms)	ϵ_f	Z (ohms)	ϵ_f	Z (ohms)	ϵ_f	Z (ohms)	ϵ_f	Z (ohms)	ϵ_f
0.10	129.17	2.290	128.05	2.317	125.91	2.341	123.11	2.351	119.27	2.355
0.12	122.08	2.286	120.99	2.313	118.87	2.338	116.08	2.349	112.25	2.352
0.14	116.08	2.282	115.03	2.310	112.93	2.336	110.15	2.347	106.33	2.350
0.16	110.90	2.280	109.88	2.307	107.79	2.333	105.02	2.345	101.23	2.349
0.18	106.33	2.277	105.34	2.304	103.27	2.331	100.52	2.343	96.74	2.347
0.20	102.24	2.275	101.29	2.302	99.24	2.329	96.50	2.341	92.75	2.345
0.22	98.56	2.273	97.63	2.300	95.61	2.327	92.88	2.339	89.15	2.344
0.24	95.20	2.271	94.30	2.298	92.30	2.325	89.59	2.338	85.89	2.342
0.26	92.12	2.270	91.24	2.296	89.27	2.324	86.58	2.336	82.91	2.341
0.28	89.28	2.268	88.42	2.294	86.47	2.322	83.81	2.335	80.16	2.340
0.30	86.64	2.267	85.81	2.293	83.88	2.321	81.24	2.334	77.62	2.338
0.32	84.19	2.266	83.38	2.292	81.47	2.319	78.85	2.332	75.26	2.337
0.34	81.89	2.265	81.10	2.290	79.22	2.318	76.62	2.331	73.06	2.336
0.36	79.73	2.264	78.96	2.289	77.10	2.317	74.52	2.330	70.99	2.334
0.38	77.70	2.263	76.94	2.288	75.11	2.315	72.55	2.328	69.06	2.333
0.45	71.41	2.261	70.71	2.284	68.96	2.311	66.49	2.324	63.11	2.329
0.50	67.55	2.259	66.89	2.282	65.19	2.309	62.78	2.322	59.49	2.327
0.55	64.12	2.257	63.49	2.280	61.84	2.306	59.49	2.320	56.29	2.324
0.60	61.03	2.256	60.43	2.278	58.83	2.304	56.55	2.317	53.42	2.322
0.65	58.24	2.255	57.66	2.276	56.12	2.302	53.89	2.315	50.85	2.320
0.70	55.70	2.254	55.14	2.275	53.65	2.300	51.48	2.313	48.52	2.318
0.75	53.37	2.253	52.84	2.274	51.39	2.299	49.28	2.312	46.40	2.317
0.80	51.24	2.252	50.73	2.272	49.32	2.297	47.26	2.310	44.46	2.315
0.85	49.27	2.251	48.78	2.271	47.41	2.296	45.41	2.309	42.68	2.313
0.90	47.45	2.250	46.97	2.270	45.64	2.295	43.70	2.307	41.04	2.312
0.95	45.76	2.250	45.30	2.269	44.01	2.293	42.11	2.306	39.52	2.311
1.00	44.18	2.249	43.74	2.268	42.48	2.292	40.64	2.305	38.11	2.310
1.05	42.72	2.249	42.29	2.267	41.06	2.291	39.26	2.304	36.80	2.308
1.10	41.34	2.248	40.93	2.267	39.73	2.290	37.98	2.303	35.58	2.307
1.15	40.05	2.247	39.65	2.266	38.49	2.289	36.77	2.302	34.43	2.306
1.20	38.84	2.247	38.45	2.265	37.32	2.288	35.65	2.301	33.36	2.305
1.25	37.70	2.247	37.33	2.265	36.22	2.288	34.59	2.300	32.35	2.304
1.30	36.63	2.246	36.26	2.264	35.18	2.287	33.59	2.299	31.40	2.303
1.35	35.62	2.246	35.26	2.263	34.21	2.286	32.64	2.298	30.51	2.303
1.40	34.66	2.245	34.31	2.263	33.28	2.285	31.75	2.297	29.66	2.302
1.45	33.75	2.245	33.41	2.262	32.40	2.285	30.91	2.297	28.86	2.301
1.50	32.88	2.245	32.55	2.262	31.57	2 284	30.10	2.296	28.10	2.301
1.55	32.06	2.244	31.74	2.261	30.78	2.284	29.34	2.295	27.39	2.300
1.60	31.28	2.244	30.97	2.261	30.03	2.283	28.62	2.295	26.70	2.299
1.65	30.54	2.244	30.23	2.260	29.31	2.282	27.93	2.294	26.05	2.299

(Contd.)

d/b → w/b ↓	0.1 Z (ohms)	ϵ_f	0.2 Z (ohms)	ϵ_f	0.3 Z (ohms)	ϵ_f	0.4 Z (ohms)	ϵ_f	0.5 Z (ohms)	ϵ_f
1.70	29.83	2.243	29.53	2.260	28.63	2.282	27.28	2.294	25.43	2.298
1.75	29.15	2.243	28.86	2.260	27.98	2.281	26.65	2.293	24.84	2.297
1.80	28.51	2.243	28.22	2.259	27.35	2.281	26.05	2.292	24.28	2.297
1.85	27.89	2.243	27.61	2.259	26.76	2.280	25.48	2.292	23.74	2.296
1.90	27.30	2.242	27.02	2.259	26.19	2.280	24.93	2.292	23.22	2.296
1.95	26.73	2.242	26.46	2.258	25.64	2.280	24.41	2.291	22.73	2.295
2.00	26.18	2.242	25.92	2.258	25.11	2.279	23.90	2.291	22.26	2.295
2.05	25.66	2.242	25.40	2.258	24.61	2.279	23.42	2.290	21.80	2.294
2.10	25.16	2.242	24.90	2.257	24.13	2.278	22.96	2.290	21.37	2.294
2.15	24.67	2.241	24.42	2.257	23.66	2.278	22.51	2.289	20.95	2.293
2.20	24.21	2.241	23.96	2.257	23.21	2.278	22.08	2.289	20.54	2.293
2.25	23.76	2.241	23.52	2.257	22.78	2.277	21.67	2.288	20.16	2.293
2.30	23.33	2.241	23.09	2.256	22.37	2.277	21.27	2.288	19.78	2.292
2.35	22.91	2.241	22.68	2.256	21.97	2.276	20.89	2.287	19.42	2.292
2.40	22.51	2.241	22.28	2.256	21.58	2.276	20.52	2.287	19.08	2.292
2.45	22.12	2.240	21.90	2.255	21.21	2.276	20.16	2.287	18.74	2.291
2.50	21.75	2.240	21.53	2.255	20.85	2.275	19.82	2.287	18.42	2.291

Table 5.15 Characteristic impedance and effective dielectric constant of sandwiched stripline; $\epsilon_{r1} = 2.22$, $\epsilon_{r2} = 10.2$, $c/w = 40$ (See structure A5 of Table 3.1)

d/b → w/b ↓	0.1		0.2		0.3		0.4		0.5	
	Z (ohms)	ϵ_f	Z (ohms)	ϵ_f	Z (ohms)	ϵ_f	Z (ohms)	ϵ_f	Z (ohms)	ϵ_f
0.10	108.20	3.263	100.42	3.767	92.48	4.339	87.50	4.654	83.24	4.834
0.12	103.23	3.196	95.79	3.690	88.03	4.264	83.09	4.583	78.87	4.765
0.14	98.96	3.141	91.83	3.624	84.24	4.198	79.36	4.521	75.16	4.705
0.16	95.19	3.094	88.38	3.566	80.95	4.138	76.12	4.463	71.94	4.650
0.18	91.82	3.053	85.30	3.514	78.04	4.083	73.26	4.411	69.11	4.599
0.20	88.77	3.018	82.53	3.467	75.43	4.032	70.70	4.361	66.58	4.550
0.22	85.99	2.986	80.00	3.425	73.06	3.985	68.39	4.315	64.30	4.505
0.24	83.42	2.958	77.67	3.386	70.90	3.941	66.28	4.271	62.23	4.462
0.26	81.04	2.933	75.52	3.351	68.91	3.899	64.35	4.229	60.33	4.421
0.28	78.82	2.910	73.52	3.319	67.06	3.861	62.56	4.190	58.57	4.382
0.30	76.75	2.889	71.64	3.289	65.34	3.824	60.90	4.153	56.94	4.345
0.32	74.81	2.870	69.89	3.262	63.74	3.790	59.35	4.117	55.42	4.309
0.34	72.98	2.852	68.23	3.236	62.22	3.757	57.89	4.083	54.00	4.274
0.36	71.25	2.836	66.66	3.212	60.80	3.726	56.52	4.050	52.67	4.241
0.38	69.91	2.820	65.17	3.189	59.44	3.697	55.23	4.018	51.41	4.209
0.40	68.05	2.806	63.76	3.168	58.16	3.669	54.01	3.988	50.23	4.179
0.45	64.47	2.773	60.50	3.120	55.22	3.604	51.21	3.918	47.52	4.107
0.50	61.28	2.745	57.60	3.078	52.60	3.546	48.73	3,854	45.14	4.042
0.55	58.42	2.720	54.97	3.040	50.24	3.494	46.50	3.796	43.01	3.982
0.60	55.82	2.697	52.59	3.007	48.10	3.448	44.49	3.743	41.09	3.926
0.65	53.45	2.677	50.42	2.977	46.14	3.405	42.66	3.695	39.35	3.875
0.70	51.28	2.659	48.42	2.950	44.34	3.367	40.98	3.650	37.76	3.828
0.75	49.28	2.643	46.58	2.925	42.69	3.331	39.44	3.609	36.30	3.785
0.80	47.44	2.627	44.88	2.902	41.16	3.299	38.02	3.571	34.96	3.744
0.85	45.73	2.613	43.30	2.881	39.73	3.269	36.69	3.536	33.72	3.707
0.90	44.14	2.601	41.84	2.862	38.41	3.241	35.46	3.503	32.57	3.672
0.95	42.66	2.589	40.46	2.844	37.17	3.215	34.31	3.473	31.49	3.639
1.00	41.27	2.577	39.18	2.827	36.01	3.191	33.24	3.444	30.49	3.608
1.05	39.98	2.567	37.98	2.811	34.92	3.169	32.23	3.418	29.55	3.580
1.10	38.76	2.557	36.85	2.796	33.89	3.148	31.29	3.393	28.67	3.553
1.15	37.61	2.548	35.78	2.783	32.93	3.128	30.39	3.369	27.82	3.528
1.20	36.54	2.540	34.78	2.770	32.02	3.109	29.55	3.348	27.06	3.504
1.25	35.52	2.532	33.83	2.757	31.16	3.092	28.75	3.327	26.32	3.482
1.30	34.55	2.524	32.93	2.746	30.34	3.075	28.00	3.307	25.62	3.460
1.35	33.64	2.517	32.07	2.735	29.57	3.060	27.29	3.289	24.96	3.440
1.40	32.77	2.511	31.27	2.725	28.83	3.045	26.61	3.271	24.33	3.422
1.45	31.95	2.504	30.50	2.715	28.13	3.031	25.96	3.255	23.73	3.404
1.50	31.17	2.498	29.76	2.705	27.47	3.017	25.35	3.239	23.16	3.387
1.55	30.43	2.493	29.07	2.697	26.83	3.005	24.76	3.224	22.62	3.370
1.60	29.72	2.487	28.40	2.688	26.23	2.993	24.20	3.210	22.11	3.355

(Contd.)

w/b \ d/b	0.1 Z (ohms)	ϵ_f	0.2 Z (ohms)	ϵ_f	0.3 Z (ohms)	ϵ_f	0.4 Z (ohms)	ϵ_f	0.5 Z (ohms)	ϵ_f
1.65	29.04	2.482	27.77	2.680	25.65	2.981	23.66	3.196	21.61	3.340
1.70	28.39	2.477	27.16	2.672	25.09	2.970	23.15	3.183	21.14	3.326
1.75	27.77	2.472	26.58	2.665	24.56	2.960	22.66	3.171	20.69	3.313
1.80	27.18	2.468	26.02	2.658	24.05	2.950	22.19	3.159	20.26	3.300
1.85	26.61	2.464	25.48	2.651	23.57	2.940	21.74	3.148	19.84	3.288
1.90	26.06	2.460	24.97	2.645	23.10	2.931	21.31	3.137	19.44	3.276
1.95	25.54	2.456	24.48	2.639	22.65	2.922	20.89	3.126	19.06	3.264
2.00	25.04	2.452	24.00	2.633	22.21	2.913	20.49	3.116	18.69	3.253
2.05	24.55	2.449	23.55	2.627	21.80	2.905	20.11	3.107	18.34	3.243
2.10	24.09	2.445	23.11	2.622	21.40	2.897	19.74	3.098	18.00	3.233
2.15	23.64	2.442	22.69	2.617	21.01	2.890	19.38	3.089	17.67	3.223
2.20	23.21	2.439	22.28	2.612	20.64	2.882	19.04	3.080	17.35	3.214
2.25	22.79	2.436	21.88	2.607	20.28	2.875	18.71	3.072	17.05	3.205
2.30	22.39	2.433	21.50	2.602	19.93	2.868	18.39	3.064	16.75	3.196
2.35	22.00	2.430	21.14	2.597	19.59	2.862	18.08	3.055	16.47	3.188
2.40	21.62	2.427	20.78	2.593	19.27	2.855	17.78	3.048	16.19	3.180
2.45	21.26	2.425	20.44	2.589	18.95	2.849	17.49	3.041	15.93	3.172
2.50	20.91	2.422	20.11	2.585	18.65	2.843	17.21	3.034	15.67	3.164

Table 5.16 Characteristic impedance and effective dielectric constant of stripline with anisotropic dielectric (pyrolytic boron nitride); $\epsilon_{xx1} = 5.12$, $\epsilon_{yy1} = 3.40$, $c/w = 40$ (See structure A2 of Table 3.2)

d/b	0.0		0.05		0.10		0.15		0.20	
w/b	Z (ohms)	ϵ_f	Z (ohms)	ϵ_f	Z (ohms)	ϵ_f	Z (ohms)	ϵ_f	Z (ohms)	ϵ_f
0.10	98.73	3.926	98.37	3.926	97.27	3.923	95.36	3.918	92.54	3.911
0.12	93.40	3.913	93.04	3.912	91.93	3.909	90.03	3.904	87.22	3.896
0.14	88.89	3.901	88.53	3.900	87.43	3.897	85.54	3.891	82.74	3.883
0.16	85.00	3.889	84.64	3.888	83.54	3.885	81.65	3.879	78.86	3.870
0.18	81.57	3.878	81.21	3.877	80.11	3.874	78.23	3.867	75.46	3.858
0.20	78.50	3.868	78.15	3.867	77.06	3.863	75.19	3.857	72.42	3.846
0.22	75.74	3.858	75.39	3.857	74.30	3.853	72.44	3.846	69.69	3.836
0.24	73.23	3.849	72.87	3.847	71.79	3.843	69.94	3.836	67.20	3.825
0.26	70.92	3.840	70.57	3.838	69.49	3.834	67.65	3.827	64.94	3.816
0.28	68.79	3.831	68.44	3.829	67.37	3.825	65.54	3.818	62.84	3.807
0.30	66.82	3.822	66.47	3.821	65.41	3.817	63.59	3.809	60.91	3.798
0.32	64.98	3.815	64.64	3.813	63.58	3.809	61.77	3.801	59.10	3.789
0.34	63.26	3.807	62.92	3.805	61.87	3.801	60.06	3.793	57.42	3.781
0.36	61.64	3.799	61.30	3.798	60.26	3.793	58.47	3.785	55.84	3.773
0.38	60.12	3.792	59.78	3.790	58.74	3.786	56.96	3.777	54.36	3.765
0.40	58.68	3.785	58.34	3.783	57.31	3.778	55.55	3.770	52.97	3.758
0.45	55.40	3.767	55.07	3.766	54.06	3.761	52.33	3.753	49.80	3.740
0.50	52.51	3.751	52.18	3.750	51.19	3.745	49.49	3.737	47.02	3.724
0.55	49.92	3.736	49.60	3.735	48.63	3.730	46.97	3.722	44.56	3.709
0.60	47.59	3.723	47.28	3.721	46.33	3.716	44.70	3.708	42.35	3.696
0.65	45.49	3.710	45.18	3.708	44.25	3.703	42.66	3.695	40.36	3.683
0.70	43.56	3.697	43.26	3.696	42.35	3.691	40.80	3.683	38.56	3.672
0.75	41.80	3.686	41.51	3.685	40.62	3.680	39.10	3.672	36.21	3.661
0.80	40.18	3.675	39.89	3.674	39.02	3.670	37.54	3.662	35.40	3.651
0.85	38.68	3.666	38.40	3.664	37.55	3.660	36.10	3.653	34.02	3.642
0.90	37.30	3.656	37.02	3.655	36.19	3.651	34.77	3.644	32.74	3.633
0.95	36.00	3.647	35.74	3.646	34.92	3.642	33.54	3.635	31.55	3.625
1.00	34.80	3.639	34.54	3.638	33.74	3.634	32.39	3.627	30.45	3.617
1.05	33.68	3.632	33.42	3.630	32.64	3.626	31.32	3.620	29.43	3.610
1.10	32.62	3.624	32.37	3.623	31.61	3.619	30.32	3.613	28.47	3.603
1.15	31.63	3.617	31.39	3.616	30.64	3.613	29.38	3.606	27.57	3.597
1.20	30.70	3.611	30.46	3.610	29.73	3.606	28.50	3.600	26.73	3.591
1.25	29.83	3.605	29.59	3.604	28.87	3.600	27.66	3.594	25.94	3.585
1.30	29.00	3.599	28.76	3.598	28.06	3.595	26.88	3.589	25.19	3.580
1.35	28.21	3.594	27.99	3.593	27.30	3.589	26.14	3.593	24.48	3.575
1.40	27.47	3.588	27.25	3.587	26.58	3.584	25.44	3.578	23.82	3.570
1.45	26.77	3.584	26.55	3.582	25.89	3.579	24.78	3.574	23.19	3.565
1.50	26.10	3.579	25.88	3.578	25.24	3.575	24.15	3.569	22.59	3.561
1.55	25.46	3.574	25.25	3.573	24.62	3.570	23.55	3.565	22.02	3.556
1.60	24.85	3.570	24.65	3.569	24.03	3.566	22.98	3.561	21.48	3.552

(Contd.)

w/b \ d/b	0.0		0.05		0.10		0.15		0.20	
	Z (ohms)	ϵ_f	Z (ohms)	ϵ_f	Z (ohms)	ϵ_f	Z (ohms)	ϵ_f	Z (ohms)	ϵ_f
1.65	24.28	3.566	24.07	3.565	23.46	3.562	22.43	3.556	20.97	3.549
1.70	23.72	3.562	23.53	3.561	22.93	3.558	21.92	3.552	20.48	3.545
1.75	23.20	3.558	23.00	3.557	22.41	3.554	21.42	3.549	20.01	3.541
1.80	22.69	3.554	22.50	3.553	21.92	3.550	20.95	3.545	19.56	3.538
1.85	22.21	3.551	22.02	3.550	21.45	3.547	20.50	3.542	19.13	3.535
1.90	21.75	3.548	21.56	3.547	21.00	3.544	20.06	3.539	18.72	3.532
1.95	21.31	3.544	21.12	3.543	20.57	3.541	19.65	3.536	18.33	3.529
2.00	20.88	3.541	20.70	3.540	20.16	3.538	19.25	3.533	17.96	3.526
2.05	20.47	3.538	20.29	3.538	19.76	3.535	18.87	3.530	17.59	3.523
2.10	20.08	3.536	19.90	3.535	19.38	3.532	18.50	3.527	17.25	3.521
2.15	19.70	3.533	19.53	3.532	19.01	3.529	18.14	3.525	16.91	3.518
2.20	19.33	3.530	19.16	3.529	18.66	3.527	17.80	3.522	16.59	3.516
2.25	18.98	3.528	18.81	3.527	18.31	3.524	17.47	3.520	16.28	3.514
2.30	18.64	3.525	18.48	3.525	17.98	3.522	17.16	3.518	15.98	3.511
2.35	18.31	3.523	18.15	3.522	17.67	3.520	16.85	3.515	15.70	3.509
2.40	18.00	3.521	17.84	3.520	17.36	3.517	16.56	3.513	15.42	3.507
2.45	17.69	3.518	17.53	3.518	17.06	3.515	16.27	3.511	15.15	3.505
2.50	17.40	3.516	17.24	3.516	16.78	3.513	16.00	3.509	14.89	3.503

Table 5.17 Characteristic impedance and effective dielectric constant of stripline with anisotropic dielectric (sapphire); $\epsilon_{xx1} = 9.4$, $\epsilon_{yy1} = 11.6$, $c/w = 40$ (See structure A2 of Table 3.2)

d/b →	0.0		0.05		0.10		0.15		0.20	
w/b ↓	Z (ohms)	ϵ_f	Z (ohms)	ϵ_f	Z (ohms)	ϵ_f	Z (ohms)	ϵ_f	Z (ohms)	ϵ_f
0.10	59.56	10.788	59.34	10.789	58.64	10.793	57.44	10.801	55.66	10.812
0.12	56.20	10.808	55.97	10.810	55.27	10.814	54.08	10.822	52.31	10.834
0.14	53.35	10.827	53.13	10.829	52.43	10.834	51.24	10.842	49.48	10.855
0.16	50.90	10.845	50.67	10.847	49.98	10.852	48.79	10.861	47.04	10.875
0.18	48.74	10.862	48.51	10.864	47.83	10.870	46.64	10.879	44.90	10.894
0.20	46.81	10.879	46.59	10.880	45.90	10.886	44.73	10.896	43.00	10.912
0.22	45.07	10.894	44.85	10.896	44.17	10.902	43.00	10.913	41.28	10.929
0.24	43.49	10.909	43.27	10.911	42.60	10.917	41.44	10.928	39.73	10.946
0.26	42.05	10.924	41.83	10.926	41.16	10.932	40.00	10.944	38.31	10.963
0.28	40.71	10.937	40.49	10.940	39.83	10.946	38.68	10.959	37.01	10.978
0.30	39.48	10.951	39.26	10.954	38.60	10.961	37.46	10.974	35.80	10.992
0.32	38.33	10.966	38.11	10.968	37.45	10.975	36.33	10.988	34.68	11.006
0.34	37.25	10.979	37.04	10.981	36.38	10.988	35.27	11.001	33.63	11.019
0.36	36.24	10.991	36.03	10.994	35.38	11.001	34.27	11.013	32.66	11.032
0.38	35.29	11.003	35.08	11.006	34.44	11.013	33.34	11.025	31.74	11.045
0.40	34.40	11.015	34.19	11.017	33.55	11.024	32.46	11.037	30.88	11.056
0.45	32.36	11.042	32.16	11.045	31.53	11.052	30.47	11.065	28.93	11.082
0.50	30.57	11.068	30.37	11.070	29.76	11.077	28.73	11.088	27.23	11.107
0.55	28.98	11.089	28.78	11.092	28.19	11.099	27.18	11.112	25.72	11.130
0.60	27.55	11.111	27.36	11.114	26.78	11.121	25.80	11.133	24.38	11.151
0.65	26.26	11.132	26.07	11.134	25.51	11.141	24.55	11.153	23.18	11.171
0.70	25.08	11.151	24.90	11.153	24.36	11.160	23.43	11.172	22.09	11.189
0.75	24.01	11.168	23.84	11.171	23.30	11.177	22.40	11.189	21.10	11.205
0.80	23.03	11.185	22.36	11.187	22.34	11.194	21.46	11.205	20.20	11.221
0.85	22.13	11.200	21.96	11.202	21.46	11.209	20.60	11.219	19.37	11.235
0.90	21.29	11.215	21.13	11.217	20.64	11.223	19.80	11.233	18.61	11.249
0.95	20.52	11.228	20.36	11.230	19.88	11.236	19.07	11.246	17.90	11.261
1.00	19.80	11.241	19.65	11.243	19.18	11.248	18.39	11.258	17.25	11.273
1.05	19.13	11.252	18.98	11.254	18.52	11.260	17.75	11.270	16.64	11.284
1.10	18.51	11.264	18.36	11.265	17.91	11.271	17.16	11.280	16.08	11.294
1.15	17.92	11.274	17.77	11.276	17.34	11.281	16.60	11.291	15.55	11.304
1.20	17.37	11.284	17.23	11.286	16.80	11.291	16.08	11.300	15.06	11.314
1.25	16.85	11.293	16.71	11.295	16.30	11.300	15.60	11.309	14.59	11.322
1.30	16.36	11.302	16.23	11.303	15.82	11.309	15.14	11.317	14.16	11.330
1.35	15.90	11.310	15.77	11.312	15.37	11.317	14.70	11.326	13.75	11.338
1.40	15.47	11.318	15.34	11.320	14.95	11.325	14.29	11.333	13.36	11.346
1.45	15.06	11.325	14.93	11.327	14.55	11.332	13.91	11.340	12.99	11.353
1.50	14.67	11.333	14.54	11.334	14.17	11.339	13.54	11.347	12.65	11.359
1.55	14.29	11.339	14.17	11.341	13.81	11.346	13.19	11.354	12.32	11.366
1.60	13.94	11.346	13.82	11.347	13.47	11.352	12.86	11.360	12.01	11.372

(*Contd.*)

w/b \ d/b	0.0 Z (ohms)	ϵ_f	0.05 Z (ohms)	ϵ_f	0.10 Z (ohms)	ϵ_f	0.15 Z (ohms)	ϵ_f	0.20 Z (ohms)	ϵ_f
1.65	13.61	11.352	13.49	11.353	13.14	11.358	12.55	11.366	11.71	11.378
1.70	13.29	11.358	13.17	11.359	12.83	11.364	12.25	11.372	11.43	11.383
1.75	12.98	11.364	12.87	11.365	12.53	11.369	11.96	11.377	11.16	11.388
1.80	12.69	11.369	12.58	11.370	12.25	11.375	11.69	11.382	10.90	11.393
1.85	12.41	11.374	12.30	11.376	11.98	11.380	11.43	11.387	10.66	11.398
1.90	12.14	11.379	12.04	11.381	11.72	11.385	11.18	11.392	10.42	11.403
1.95	11.89	11.384	11.78	11.385	11.47	11.390	10.94	11.397	10.20	11.407
2.00	11.64	11.389	11.54	11.390	11.23	11.394	10.72	11.401	9.98	11.412
2.05	11.41	11.393	11.31	11.394	11.00	11.398	10.50	11.405	9.77	11.416
2.10	11.18	11.397	11.08	11.399	10.79	11.403	10.29	11.410	9.58	11.419
2.15	10.96	11.401	10.87	11.403	10.57	11.407	10.08	11.413	9.39	11.423
2.20	10.75	11.405	10.66	11.407	10.37	11.411	9.89	11.417	9.20	11.427
2.25	10.55	11.409	10.46	11.410	10.18	11.414	9.70	11.421	9.03	11.430
2.30	10.36	11.413	10.27	11.414	9.99	11.418	9.52	11.424	8.86	11.434
2.35	10.17	11.416	10.08	11.418	9.81	11.421	9.35	11.428	8.69	11.437
2.40	9.99	11.420	9.90	11.421	9.63	11.425	9.18	11.431	8.54	11.440
2.45	9.82	11.423	9.73	11.424	9.46	11.428	9.02	11.434	8.39	11.443
2.50	9.65	11.426	9.56	11.428	9.30	11.431	8.86	11.437	8.24	11.446

Table 5.18 Characteristic impedance and effective dielectric constant of stripline with anisotropic dielectric (Epsilam-10); $\epsilon_{xx1} = 13.0$, $\epsilon_{yy1} = 10.2$, $c/w = 40$ (See structure A2 of Table 3.2)

d/b → w/b ↓	0.0		0.05		0.10		0.15		0.20	
	Z(ohms)	ϵ_f	Z(ohms)	ϵ_f	Z(ohms)	ϵ_f	Z(ohms)	ϵ_f	Z(ohms)	ϵ_f
0.10	58.71	11.103	58.49	11.102	57.83	11.097	56.68	11.089	54.99	11.077
0.12	55.50	11.081	55.28	11.079	54.62	11.074	53.48	11.065	51.79	11.052
0.14	52.79	11.059	52.57	11.058	51.91	11.052	50.77	11.043	49.09	11.028
0.16	50.45	11.040	50.23	11.038	49.57	11.032	48.44	11.022	46.76	11.006
0.18	48.39	11.021	48.17	11.019	47.51	11.013	46.38	11.002	44.71	10.986
0.20	46.55	11.003	46.33	11.001	45.68	10.995	44.55	10.984	42.89	10.966
0.22	44.89	10.986	44.67	10.984	44.02	10.977	42.90	10.966	41.25	10.948
0.24	43.37	10.970	43.16	10.968	42.51	10.961	41.40	10.948	39.76	10.930
0.26	41.99	10.954	41.78	10.952	41.13	10.944	40.02	10.932	38.40	10.915
0.28	40.71	10.939	40.50	10.936	39.86	10.929	38.76	10.917	37.14	10.899
0.30	39.52	10.924	39.31	10.922	38.68	10.915	37.59	10.903	35.98	10.883
0.32	38.42	10.912	38.21	10.910	37.58	10.902	36.49	10.888	34.90	10.868
0.34	37.39	10.898	37.18	10.896	36.55	10.888	35.47	10.874	33.89	10.854
0.36	36.42	10.885	36.21	10.883	35.59	10.875	34.51	10.861	32.94	10.840
0.38	35.50	10.872	35.30	10.870	34.68	10.862	33.61	10.848	32.06	10.827
0.40	34.64	10.860	34.44	10.857	33.82	10.849	32.77	10.835	31.22	10.814
0.45	32.68	10.830	32.48	10.827	31.87	10.819	30.84	10.805	29.33	10.784
0.50	30.94	10.803	30.75	10.800	30.15	10.792	29.14	10.777	27.67	10.754
0.55	29.39	10.777	29.20	10.774	28.62	10.766	27.64	10.750	26.20	10.729
0.60	28.00	10.752	27.82	10.748	27.25	10.740	26.28	10.726	24.88	10.705
0.65	26.75	10.728	26.56	10.726	26.01	10.718	25.06	10.704	23.70	10.684
0.70	25.60	10.707	25.42	10.705	24.88	10.697	23.96	10.683	22.62	10.664
0.75	24.55	10.688	24.37	10.685	23.84	10.678	22.94	10.664	21.65	10.645
0.80	23.58	10.669	23.41	10.667	22.89	10.660	22.02	10.647	20.75	10.628
0.85	22.69	10.652	22.52	10.650	22.02	10.643	21.16	10.630	19.93	10.612
0.90	21.87	10.636	21.70	10.634	21.21	10.627	20.37	10.615	19.17	10.597
0.95	21.10	10.621	20.94	10.619	20.46	10.612	19.64	10.600	18.47	10.583
1.00	20.38	10.607	20.23	10.605	19.76	10.598	18.96	10.587	17.82	10.570
1.05	19.72	10.594	19.57	10.592	19.10	10.585	18.32	10.574	17.21	10.558
1.10	19.09	10.582	18.94	10.579	18.49	10.573	17.73	10.562	16.64	10.546
1.15	18.51	10.570	18.36	10.568	17.92	10.562	17.18	10.551	16.11	10.535
1.20	17.96	10.559	17.81	10.557	17.38	10.551	16.65	10.540	15.61	10.525
1.25	17.44	10.548	17.30	10.546	16.87	10.540	16.16	10.530	15.14	10.515
1.30	16.95	10.538	16.81	10.536	16.40	10.531	15.70	10.521	14.70	10.506
1.35	16.48	10.529	16.35	10.527	15.94	10.521	15.26	10.512	14.29	10.497
1.40	16.04	10.520	15.91	10.518	15.52	10.513	14.85	10.503	13.90	10.489
1.45	15.63	10.512	15.50	10.510	15.11	10.504	14.46	10.495	13.52	10.481
1.50	15.23	10.504	15.11	10.502	14.73	10.496	14.09	10.487	13.17	10.473
1.55	14.86	10.496	14.73	10.494	14.36	10.489	13.73	10.480	12.84	10.466
1.60	14.50	10.489	14.38	10.487	14.01	10.482	13.40	10.473	12.52	10.459

(Contd.)

w/b \ d/b	0.0 Z(ohms)	ϵ_f	0.05 Z(ohms)	ϵ_f	0.10 Z(ohms)	ϵ_f	0.15 Z(ohms)	ϵ_f	0.20 Z(ohms)	ϵ_f
1.65	14.16	10.482	14.04	10.480	13.68	10.475	13.08	10.466	12.22	10.453
1.70	13.83	10.475	13.72	10.473	13.37	10.468	12.77	10.460	11.93	10.447
1.75	13.52	10.469	13.41	10.467	13.06	10.462	12.48	10.453	11.65	10.441
1.80	13.23	10.463	13.12	10.461	12.78	10.456	12.20	10.447	11.39	10.435
1.85	12.94	10.457	12.83	10.455	12.50	10.450	11.94	10.442	11.14	10.430
1.90	12.67	10.451	12.56	10.450	12.23	10.445	11.68	10.436	10.90	10.424
1.95	12.41	10.446	12.30	10.444	11.98	10.439	11.44	10.431	10.67	10.419
2.00	12.16	10.441	12.05	10.439	11.74	10.434	11.20	10.426	10.45	10.414
2.05	11.92	10.436	11.82	10.434	11.50	10.429	10.98	10.422	10.24	10.410
2.10	11.69	10.431	11.59	10.429	11.28	10.425	10.76	10.417	10.03	10.405
2.15	11.47	10.426	11.37	10.425	11.06	10.420	10.56	10.412	9.84	10.401
2.20	11.25	10.422	11.15	10.420	10.86	10.416	10.36	10.408	9.65	10.397
2.25	11.04	10.417	10.95	10.416	10.66	10.412	10.16	10.404	9.47	10.393
2.30	10.85	10.416	10.75	10.412	10.46	10.407	9.98	10.400	9.29	10.389
2.35	10.65	10.409	10.56	10.408	10.28	10.403	9.80	10.396	9.12	10.386
2.40	10.47	10.405	10.38	10.404	10.10	10.400	9.63	10.392	8.96	10.382
2.45	10.29	10.401	10.20	10.400	9.92	10.396	9.46	10.389	8.81	10.378
2.50	10.12	10.398	10.03	10.397	9.75	10.392	9.30	10.385	8.65	10.375

Table 5.19 Characteristic impedance and effective dielectric constant of microstrip-like stripline with anisotropic dielectric (pyrolytic boron nitride); $\epsilon_{xx1} = 5.12$, $\epsilon_{yy1} = 3.4$, $c/w = 40$ (See structure A3 of Table 3.2)

w/b	Z(ohms)	ϵ_f	w/b	Z(ohms)	ϵ_f	w/b	Z(ohms)	ϵ_f
0.10	124.60	2.465	0.60	59.73	2.363	1.55	31.82	2.288
0.12	117.83	2.459	0.65	57.07	2.357	1.60	31.06	2.286
0.14	112.10	2.453	0.70	54.64	2.350	1.65	30.33	2.284
0.16	107.16	2.447	0.75	52.41	2.345	1.70	29.64	2.282
0.18	102.80	2.441	0.80	50.36	2.339	1.75	28.98	2.280
0.20	98.92	2.436	0.85	48.47	2.334	1.80	28.35	2.278
0.22	95.41	2.431	0.90	46.72	2.330	1.85	27.74	2.277
0.24	92.22	2.426	0.95	45.09	2.325	1.90	27.16	2.275
0.26	89.30	2.422	1.00	43.58	2.321	1.95	26.60	2.273
0.28	86.60	2.418	1.05	42.16	2.317	2.00	26.07	2.272
0.30	84.09	2.413	1.10	40.83	2.313	2.05	25.55	2.270
0.32	81.76	2.410	1.15	39.59	2.310	2.10	25.06	2.269
0.34	79.58	2.406	1.20	38.41	2.307	2.15	24.58	2.268
0.36	77.53	2.402	1.25	37.31	2.304	2.20	24.13	2.666
0.38	75.60	2.398	1.30	36.27	2.301	2.25	23.69	2.265
0.40	73.77	2.394	1.35	35.28	2.298	2.30	23.26	2.264
0.45	69.62	2.386	1.40	34.35	2.295	2.35	22.85	2.263
0.50	65.95	2.378	1.45	33.46	2.293	2.40	22.45	2.261
0.55	62.68	2.370	1.50	32.62	2.290	2.45	22.07	2.260
						2.50	21.70	2.259

Table 5.20 Characteristic impedance and effective dielectric constant of microstrip-like stripline with anisotropic dielectric (sapphire); $\epsilon_{xx1} = 9.40$, $\epsilon_{yy1} = 11.60$, $c/w = 40$ (See structure A3 of Table 3.2)

w/b	Z(ohms)	ϵ_f	w/b	Z(ohms)	ϵ_f	w/b	Z(ohms)	ϵ_f
0.10	80.58	5.894	0.60	37.31	6.057	1.55	19.38	6.170
0.12	76.03	5.905	0.65	37.57	6.067	1.60	18.90	6.173
0.14	72.19	5.914	0.70	33.98	6.077	1.65	18.45	6.176
0.16	68.87	5.923	0.75	32.53	6.086	1.70	18.01	6.179
0.18	65.95	5.931	0.80	31.20	6.094	1.75	17.60	6.182
0.20	63.35	5.940	0.85	29.98	6.102	1.80	17.20	6.185
0.22	61.01	5.947	0.90	28.85	6.109	1.85	16.83	6.187
0.24	58.87	5.955	0.95	27.81	6.115	1.90	16.47	6.190
0.26	56.91	5.962	1.00	26.83	6.122	1.95	16.12	6.192
0.28	55.11	5.969	1.05	25.93	6.127	2.00	15.79	6.194
0.30	53.44	5.976	1.10	25.08	6.132	2.05	15.47	6.196
0.32	51.89	5.983	1.15	24.29	6.137	2.10	15.16	6.199
0.34	50.43	5.990	1.20	23.54	6.142	2.15	14.87	6.201
0.36	49.07	5.996	1.25	22.84	6.147	2.20	14.58	6.203
0.38	47.78	6.002	1.30	22.18	6.151	2.25	14.31	6.205
0.40	46.57	6.008	1.35	21.56	6.155	2.30	14.05	6.206
0.45	43.82	6.021	1.40	20.97	6.159	2.35	13.79	6.208
0.50	41.40	6.034	1.45	20.41	6.163	2.40	13.55	6.210
0.55	39.24	6.046	1.50	19.88	6.166	2.45	13.31	6.212
						2.50	13.09	6.213

Table 5.21 Characteristic impedance and effective dielectric constant of suspended stripline with anisotropic dielectric (pyrolytic boron nitride); $\epsilon_{xx1} = 5.12$, $\epsilon_{yy1} = 3.4$, $c/w = 40$ (See structure A4 of Table 3.2)

d/b → w/b ↓	0.1 Z(ohms)	0.1 ϵ_f	0.2 Z(ohms)	0.2 ϵ_f	0.3 Z(ohms)	0.3 ϵ_f	0.4 Z(ohms)	0.4 ϵ_f	0.5 Z(ohms)	0.5 ϵ_f
0.10	149.996	1.688	139.499	1.907	132.909	2.017	127.106	2.073	120.940	2.094
0.12	142.951	1.657	132.648	1.878	126.128	1.989	120.357	2.046	114.209	2.067
0.14	136.950	1.630	126.828	1.852	120.366	1.965	114.626	2.023	108.503	2.043
0.16	131.719	1.605	121.779	1.828	115.371	1.943	109.663	2.001	103.567	2.022
0.18	127.077	1.583	117.321	1.806	110.969	1.922	105.292	1.981	99.227	2.001
0.20	122.901	1.563	113.330	1.786	107.036	1.903	101.392	1.962	95.361	1.982
0.22	119.102	1.545	109.719	1.767	103.483	1.885	97.875	1.945	91.880	1.964
0.24	115.617	1.528	106.421	1.749	100.245	1.867	94.675	1.928	88.724	1.948
0.26	112.396	1.513	103.388	1.732	97.275	1.851	91.750	1.912	85.830	1.931
0.28	109.401	1.499	100.580	1.716	94.531	1.835	89.044	1.896	83.165	1.915
0.30	106.604	1.485	97.968	1.701	91.988	1.820	86.535	1.881	80.699	1.900
0.32	103.987	1.473	95.531	1.687	89.610	1.806	84.199	1.867	78.406	1.886
0.34	101.513	1.462	93.237	1.673	87.382	1.792	82.014	1.853	76.266	1.872
0.36	99.176	1.451	91.076	1.660	85.288	1.779	79.964	1.840	74.262	1.859
0.38	96.962	1.441	89.033	1.648	83.313	1.766	78.034	1.827	72.379	1.846
0.40	94.859	1.431	87.099	1.636	81.446	1.754	76.212	1.815	70.605	1.834
0.45	90.027	1.409	82.668	1.608	77.186	1.725	72.068	1.786	66.582	1.805
0.50	85.708	1.390	78.724	1.583	73.412	1.698	68.413	1.759	63.049	1.778
0.55	81.815	1.373	75.179	1.561	70.034	1.674	65.154	1.735	59.913	1.753
0.60	78.280	1.357	71.966	1.540	66.985	1.651	62.224	1.712	57.105	1.731
0.65	75.052	1.344	69.036	1.521	64.213	1.631	59.570	1.691	54.571	1.710
0.70	72.089	1.331	66.348	1.504	61.679	1.612	57.151	1.671	52.269	1.690
0.75	69.358	1.320	63.871	1.488	59.349	1.594	54.934	1.653	50.167	1.672
0.80	66.832	1.309	61.579	1.473	57.198	1.577	52.894	1.636	48.239	1.655
0.85	64.487	1.299	59.452	1.460	55.205	1.562	51.008	1.620	46.461	1.639
0.90	62.304	1.290	57.470	1.447	53.352	1.548	49.258	1.605	44.816	1.624
0.95	60.266	1.282	55.619	1.436	51.623	1.534	47.630	1.591	43.289	1.610
1.00	58.359	1.274	53.886	1.425	50.006	1.522	46.109	1.578	41.866	1.597
1.05	56.571	1.267	52.259	1.415	48.490	1.510	44.686	1.566	40.537	1.585
1.10	54.889	1.260	50.729	1.405	47.065	1.499	43.351	1.554	39.293	1.573
1.15	53.306	1.254	49.288	1.396	45.723	1.489	42.095	1.543	38.125	1.563
1.20	51.813	1.248	47.927	1.388	44.457	1.479	40.911	1.533	37.026	1.552
1.25	50.401	1.242	46.640	1.380	43.260	1.470	39.794	1.523	35.990	1.542
1.30	49.064	1.237	45.420	1.372	42.126	1.461	38.736	1.514	35.012	1.533
1.35	47.798	1.232	44.264	1.365	41.051	1.453	37.735	1.505	34.086	1.524
1.40	46.595	1.227	43.165	1.358	40.031	1.445	36.785	1.497	33.208	1.516
1.45	45.451	1.222	42.120	1.352	39.060	1.438	35.881	1.489	32.376	1.508
1.50	44.363	1.218	41.125	1.346	38.135	1.431	35.022	1.481	31.584	1.501
1.55	43.325	1.214	40.175	1.340	37.254	1.424	34.203	1.474	30.831	1.494
1.60	42.335	1.210	39.269	1.335	36.412	1.418	33.422	1.468	30.113	1.487

(Contd.)

w/b \ d/b	0.1 Z(ohms)	0.1 ϵ_f	0.2 Z(ohms)	0.2 ϵ_f	0.3 Z(ohms)	0.3 ϵ_f	0.4 Z(ohms)	0.4 ϵ_f	0.5 Z(ohms)	0.5 ϵ_f
1.65	41.390	1.206	38.403	1.330	35.608	1.412	32.676	1.461	29.428	1.480
1.70	40.485	1.203	37.575	1.325	34.839	1.406	31.963	1.455	28.774	1.474
1.75	39.620	1.199	36.781	1.320	34.102	1.400	31.280	1.449	28.148	1.468
1.80	38.790	1.196	36.021	1.315	33.396	1.395	30.626	1.443	27.549	1.462
1.85	37.995	1.193	35.292	1.311	32.719	1.390	29.999	1.438	26.975	1.457
1.90	37.232	1.190	34.591	1.307	32.069	1.385	29.397	1.433	26.425	1.452
1.95	36.498	1.187	33.918	1.303	31.441	1.380	28.819	1.428	25.897	1.447
2.00	35.793	1.184	33.270	1.299	30.843	1.376	28.263	1.423	25.389	1.442
2.05	35.115	1.181	32.647	1.295	30.065	1.372	27.729	1.418	24.901	1.437
2.10	34.461	1.179	32.047	1.291	29.708	1.368	27.214	1.414	24.432	1.433
2.15	33.832	1.176	31.469	1.288	29.171	1.364	26.718	1.410	23.980	1.428
2.20	33.225	1.174	30.910	1.285	28.654	1.360	26.239	1.406	23.544	1.424
2.25	32.639	1.172	30.372	1.281	28.154	1.356	25.778	1.402	23.124	1.420
2.30	32.074	1.170	29.852	1.278	27.671	1.352	25.332	1.398	22.718	1.417
2.35	31.528	1.167	29.349	1.275	27.205	1.349	24.902	1.394	22.327	1.413
2.40	31.000	1.165	28.863	1.272	26.754	1.346	24.486	1.391	21.949	1.409
2 45	30.489	1.163	28.393	1.270	26.318	1.342	24.084	1.387	21.583	1.406
2.50	29.995	1.162	27.938	1.267	25.896	1.339	23.694	1.384	21.229	1.403

Table 5.22 Characteristic impedance and effective dielectric constant of suspended stripline with anisotropic dielectric (sapphire) $\epsilon_{xx1} = 9.4$, $\epsilon_{yy1} = 11.60$, $c/w = 40$ (See structure A4 of Table 3.2)

w/b \ d/b	0.1		0.2		0.3		0.4		0.5	
	Z (ohms)	ϵ_f	Z (ohms)	ϵ_f	Z (ohms)	ϵ_f	Z (ohms)	ϵ_f	Z (ohms)	ϵ_f
0.10	127.848	2.324	112.973	2.908	104.119	3.287	97.238	3.542	90.954	3.703
0.12	122.654	2.251	108.243	2.820	99.587	3.191	92.790	3.443	86.520	3.602
0.14	118.155	2.189	104.128	2.747	95.624	3.113	88.900	3.363	82.662	3.521
0.16	114.188	2.136	100.513	2.684	92.139	3.046	85.482	3.294	79.282	3.450
0.18	110.631	2.089	97.295	2.626	89.042	2.985	82.448	3.231	76.289	3.386
0.20	107.395	2.047	94.396	2.574	86.259	2.930	79.727	3.174	73.611	3.327
0.22	104.432	2.010	91.756	2.527	83.734	2.878	77.263	3.120	71.192	3.272
0.24	101.695	1.976	89.331	2.482	81.423	2.830	75.014	3.070	68.992	3.221
0.26	99.148	1.944	87.088	2.441	79.295	2.785	72.952	3.024	66.969	3.172
0.28	96.764	1.916	85.000	2.403	77.320	2.743	71.039	2.979	65.102	3.126
0.30	94.525	1.889	83.047	2.368	75.484	2.703	69.259	2.937	63.369	3.082
0.32	92.420	1.865	81.217	2.334	73.758	2.665	67.596	2.897	61.754	3.040
0.34	90.419	1.842	79.484	2.303	72.136	2.630	66.036	2.859	60.242	3.001
0.36	88.520	1.821	77.843	2.273	70.605	2.595	64.568	2.822	58.823	2.963
0.38	86.714	1.801	76.286	2.245	69.156	2.563	63.182	2.787	57.486	2.927
0.40	84.991	1.793	74.804	2.218	67.780	2.532	61.869	2.754	56.223	2.892
0.45	81.008	1.740	71.384	2.157	64.621	2.461	58.867	2.677	53.347	2.811
0.50	77.421	1.703	68.310	2.103	61.797	2.397	56.199	2.607	50.804	2.738
0.55	74.165	1.670	65.521	2.055	59.248	2.339	53.803	2.544	48.532	2.672
0.60	71.191	1.641	62.973	2.011	56.928	2.287	51.632	2.486	46.485	2.612
0.65	68.460	1.615	60.632	1.972	54.802	2.239	49.653	2.434	44.627	2.557
0.70	65.942	1.591	58.471	1.936	52.845	2.195	47.836	2.385	42.928	2.506
0.75	63.610	1.569	56.467	1.904	51.034	2.156	46.161	2.341	41.369	2.459
0.80	61.443	1.549	54.603	1.874	49.351	2.119	44.610	2.300	39.929	2.416
0.85	59.425	1.530	52.864	1.846	47.783	2.085	43.167	2.262	38.595	2.375
0.90	57.543	1.513	51.236	1.821	46.316	2.054	41.821	2.227	37.354	2.338
0.95	55.776	1.497	49.709	1.797	44.941	2.025	40.562	2.194	36.195	2.303
1.00	54.116	1.482	48.273	1.775	43.649	1.997	39.380	2.163	35.111	2.271
1.05	52.555	1.468	46.921	1.755	42.432	1.972	38.269	2.135	34.094	2.241
1.10	51.082	1.455	45.644	1.736	41.283	1.948	37.221	2.108	33.137	2.212
1.15	49.692	1.443	44.437	1.718	40.197	1.926	36.232	2.083	32.234	2.186
1.20	48.376	1.431	43.293	1.701	39.168	1.905	35.295	2.059	31.382	2.161
1.25	47.129	1.420	42.209	1.685	38.192	1.886	34.407	2.037	30.575	2.137
1.30	45.945	1.410	41.178	1.670	37.264	1.867	33.564	2.016	29.809	2.115
1.35	44.821	1.401	40.197	1.655	36.382	1.850	32.763	1.997	29.083	2.094
1.40	43.750	1.391	39.263	1.642	35.541	1.833	32.000	1.978	28.392	2.074
1.45	42.730	1.383	38.373	1.629	34.739	1.818	31.272	1.960	27.734	2.055
1.50	41.757	1.375	37.522	1.617	33.974	1.803	30.578	1.943	27.106	2.037
1.55	40.827	1.367	36.709	1.605	33.241	1.789	29.914	1.927	26.507	2.021
1.60	39.939	1.359	35.931	1.594	32.541	1.775	29.279	1.912	25.934	2.004

(Contd.)

w/b \ d/b	0.1 Z (ohms)	ϵ_f	0.2 Z (ohms)	ϵ_f	0.3 Z (ohms)	ϵ_f	0.4 Z (ohms)	ϵ_f	0.5 Z (ohms)	ϵ_f
1.65	39.088	1.352	35.186	1.584	31.869	1.762	28.671	1.898	25.386	1.989
1.70	38.273	1.346	34.471	1.574	31.226	1.750	28.088	1.884	24.861	1.974
1.75	37.492	1.339	33.786	1.564	30.608	1.738	27.529	1.871	24.358	1.960
1.80	36.741	1.333	33.127	1.555	30.014	1.727	26.991	1.858	23.875	1.947
1.85	36.021	1.327	32.494	1.546	29.444	1.716	26.475	1.846	23.411	1.934
1.90	35.328	1.321	31.885	1.538	28.895	1.706	25.979	1.835	22.965	1.922
1.95	34.661	1.316	31.298	1.530	28.366	1.696	25.500	1.824	22.536	1.910
2.00	34.091	1.311	31.733	1.522	27.857	1.687	25.040	1.813	22.123	1.899
2.05	33.401	1.306	30.188	1.515	27.365	1.678	24.596	1.803	21.725	1.888
2.10	32.804	1.301	29.662	1.507	26.891	1.669	24.167	1.793	21.341	1.878
2.15	32.229	1.296	29.155	1.501	26.433	1.661	23.753	1.784	20.970	1.868
2.20	31.673	1.292	28.664	1.494	25.991	1.653	23.354	1.775	20.612	1.858
2.25	31.136	1.288	28.190	1.487	25.564	1.645	22.968	1.766	20.267	1.849
2.30	30.617	1.284	27.731	1.481	25.150	1.637	22.594	1.757	19.933	1.840
2.35	30.115	1.280	27.288	1.475	24.750	1.630	22.233	1.749	19.609	1.832
2.40	29.629	1.276	26.858	1.470	24.362	1.623	21.883	1.741	19.297	1.823
2.45	29.158	1.272	26.441	1.464	23.986	1.616	21.543	1.734	18.994	1.815
2.50	28.703	1.268	26.038	1.459	23.622	1.609	21.215	1.726	18.700	1.808

REFERENCES

1. S.B. Cohn, Characteristic impedance of shielded strip transmission line, *IRE Trans. Microwave Theory Tech.*, Vol. MTT-2, pp. 52-55, July 1954.
2. E. Jahnke and F. Emde, *Tables of Functions*, 4th Edn., Dover Publications, New York, 1945.
3. M. Abramowitz and I.A. Stegun, *Handbook of Mathematical Functions*, Applied Mathematics Series, No. 55, NBS, Washington D.C., 1965.
4. M.A.R. Gunston, *Microwave Transmission-Line Impedance Data*, Van Nostrand Reinhold Company, London, 1972.
5. R.H.T. Bates, Characteristic impedance of the shielded slab line, *IRE Trans. Microwave Theory Tech.*, Vol. MTT-4, pp. 28-33, Jan. 1956.
6. R.A. Waldron, *Theory of Guided Electromagnetic Waves*, Van Nostrand Reinhold Company, London, 1970 (pp. 187-191).
7. R.A. Waldron, Theory of a stripline cavity for measurement of dielectric constants and gyromagnetic resonance line widths, *IEEE Trans. Microwave Theory Tech.*, Vol. MTT-12, pp. 123-131, Jan. 1964.
8. T. Ikeda and R. Sato, Analysis of the characteristic impedance and the effective dielectric resistance of strip transmission lines with a rectangular inner conductor by the use of conformal mappings, *Electronics and Comm.*, Japan, Vol. 50, pp. 111-117, March 1967.
9. R.E. Collin, *Field Theory of Guided Waves*, McGraw-Hill, New York, 1960.
10. A. Rosenzweig, Determining the characteristic impedance of a nonsymmetrical strip transmission line, *LEL Technical Memo*, No. TM 0014, Varian Associates Copiague, New York.
11. H. Guckel, Characteristic impedance of generalized rectangular transmission line, *IEEE Trans. Microwave Theory Tech.*, Vol. MTT-13, pp. 270-274, May 1965.
12. H. Howe, *Stripline Circuit Design*. Artech House, Dedham, Mass., 1974.
13. W. Getsinger, Coupled rectangular bars between parallel plates, *IEEE Trans. Microwave Theory Tech.*, Vol. MTT-10, pp. 65-72, Jan. 1972.
14. E. Yamashita, B.Y. Wang and K. Atsuki, Effects of side-wall grooves on transmission characteristics of suspended striplines, *IEEE MTT-S Int. Microwave Symp. Digest*, pp. 145-148, June 1985.
15. S. Arnold, Dispersive effects in microstrip on alumina substrates, *Electron. Lett.*, Vol. 5, No. 26, pp. 673-674, Dec. 1969.
16. R.P. Owens, J.E. Aitken and T.C. Edwards, Quasi-static characteristics of microstrip on an anisotropic sapphire substrate, *IEEE Trans. Microwave Theory Tech.*, Vol. MTT-24, pp. 499-505, Aug. 1976.
17. C.M. Krowne, Microstrip transmission lines on pyrolytic boron nitride, *Electron. Lett.*, Vol. 12, No. 24, pp. 642-643, Nov. 1976.

CHAPTER 6

Design Data on Edge-coupled Stripline-Like Transmission Lines—Characteristic Impedances and Effective Dielectric Constants

6.1 Introduction

Of the various types of coupled lines reported in the literature, the edge-coupled parallel conductor configuration is most extensively used in practical circuits, both passive and active. While the commonly encountered applications have been in directional couplers [1–4] and filters [2–5], they are also used in a variety of other circuits such as phase shifters [6, 7], d.c. blocks [8], wide-band magic *T*[9] and interdigital capacitors [10]. A majority of these circuits adopt either the edge-coupled homogeneous stripline or edge-coupled microstrip line and a few adopt edge-coupled suspended stripline especially at higher microwave frequencies. The application potentials of edge-coupled lines can be enhanced by exploiting the characteristics of a variety of other structural configurations. Tables 4.1–4.3 (of Chapter 4) show a class of edge-coupled stripline-like transmission lines employing iso/anisotropic substrates. In this chapter, we present detailed design data on the even- and odd-mode characteristic impedances and effective dielectric constants of some of the useful structures.

For the edge-coupled homogeneous symmetric stripline (structure Al of Table 4.1), Cohn [11] has derived formulas for the even- and odd-mode characteristic impedances using the conformal mapping method. The expressions are exact for zero-thickness strip conductors. Extensive design data computed using Cohn's exact impedance formulas are presented in the form of tables. Normalized impedances $Z_e\sqrt{\epsilon_r}$ and $Z_o\sqrt{\epsilon_r}$ are tabulated as functions of w/b and s/b. Since this structure is extensively used in couplers, tables of data listing dimensional parameters as a function of coupling in *dB* are also provided. These data are generated using the formulas reported by Shelton [12]. Homogeneous strip transmission lines with edge-coupled thick rectangular strips are useful when tight coupling between the lines is desired. Such structures with air dielectric are useful in the realization of very low loss filters and couplers at lower microwave frequencies. Getsinger [13] has derived approximate impedance expressions for coupled rectangular bars centered between parallel plates by directly applying the Schwarz-Christoffel transformation. Based on Getsinger's analysis, Matthaei *et al.* [3] have provided complete design curves characterizing the various fringe capacitances and also formulas for determining the structural parameters from the even- and odd-mode impedances. Design data on edge-coupled homogeneous striplines with finite thickness strip conductors are therefore not given here.

The edge-coupled microstrip-like stripline (structure A3 of Table 4.1) behaves exactly like an edge-coupled symmetric stripline homogeneously filled with a dielectric having a relative dielectric constant $(\epsilon_r + 1)/2$. Thus, the tables of data provided for the edge-coupled symmetric stripline are also applicable to edge-coupled microstrip-like stripline when ϵ_r in the normalized impedances $Z_e\sqrt{\epsilon_r}$ and $Z_o\sqrt{\epsilon_r}$ is replaced by $(\epsilon_r + 1)/2$.

The even- and odd-mode capacitance parameters for edge-coupled lines in suspended substrate stripline have been reported by Smith [14]. The formulas are derived from variational series using conformal transformations. A simpler approach which uses the variational method combined with the 'transverse transmission line' technique has been presented by Bhat and Koul [15], which can be applied to a class of edge-coupled multilayer stripline-like transmission lines. The same technique has been extended by the authors to structures using anisotropic substrates [16, 17]. A complete description of this analysis is presented in Chapters 2 and 4 and unified capacitance formulas applicable to a class of edge-coupled structures such as those listed in Tables 4.1–4.3 are derived. These formulas enable computation of data on the even- and odd-mode impedances and effective dielectric constants of the various structures, using a single computer program. In order to verify the accuracy of the results, the computed impedance characteristics for the edge-coupled homogeneous symmetric stripline are compared with those computed using conformal mapping formulas. Unified capacitance formulas are used to present design data on the even- and odd-mode characteristic impedances and effective dielectric constants of other useful edge-coupled structures for which exact formulas are not available. These include the homogeneous stripline with off-set conductors and suspended stripline employing isotropic dielectrics; and the homogeneous symmetric stripline, homogeneous stripline with off-set strip conductors, microstrip-like stripline and suspended stripline, all employing anisotropic dielectrics. The design parameters can be extracted from the graphical plots and also the tables to a reasonable accuracy adequate for most practical purposes. All tables of data are included at the end of this chapter.

6.2 Design Equations

6.2.1 Even- and Odd-Mode Propagation Parameters

Under the quasi-TEM approximation, the even- and odd-mode propagation parameters of edge-coupled stripline-like transmission line are given by

$$Z_{\substack{e\\o}} = \frac{1}{v^a\sqrt{C_{\substack{e\\o}}C^a_{\substack{e\\o}}}} \tag{6.1}$$

$$\epsilon_{\substack{fe\\o}} = \frac{C_{\substack{e\\o}}}{C^a_{\substack{e\\o}}} \tag{6.2}$$

$$v_{\substack{e\\o}} = \frac{v^a}{\sqrt{\epsilon_{\substack{fe\\o}}}} \tag{6.3}$$

$$\lambda_{\substack{e\\o}} = \frac{\lambda^a}{\sqrt{\epsilon_{\substack{fe\\o}}}} \tag{6.4}$$

In these formulas, C_e, Z_e, ϵ_{fe}, v_e and λ_e are the capacitance per unit length, characteristic impedance, effective dielectric constant, phase velocity and guide wavelength, respectively, of the structure in the even-mode excitation, and C_o, Z_o, ϵ_{fo}, v_o and λ_o are the corresponding parameters in the odd-mode excitation. C_e^a and C_o^a are the capacitances per unit length for even- and odd-modes respectively, when all the dielectrics in the structure are replaced by air. The parameters v^a and λ^a represent the free space velocity and free space wavelength, respectively. The determination of the even- and odd-mode characteristic impedances, phase velocities and guide wavelengths, therefore, involves, first the evaluation of the capacitance of the structure. We summarize below the unified variational capacitance formulas derived in Chapter 4 as well as other closed form exact/empirical formulas available in the literature for some specific structures.

6.2.2 Unified Variational Capacitance Formula for Structures with Iso/Anisotropic Substrates

Tables 4.1–4.3 list a class of edge-coupled stripline-like transmission lines. Referring to the coordinate system marked in structure A1 of Table 4.1, the charge distribution on the strip conductor S (assumed to be of zero thickness) is specified as

$$f(x)_{\substack{e\\o}} = \frac{1}{w}\left[1 + A_{\substack{e\\o}}\left|\frac{2}{w}\left(x - \frac{c-s-w}{2} - g_{\substack{e\\o}}\right)\right|^3\right]; \quad \frac{c-s}{2} - w \leqslant x \leqslant \frac{c-s}{2} \tag{6.5}$$

$$= 0; \qquad \text{otherwise}$$

where

$$0 \leqslant g_e \leqslant \frac{w}{2}, \quad \text{for even-mode, and} \quad -\frac{w}{2} \leqslant g_o \leqslant 0, \quad \text{for odd-mode}$$

The subscripts e and o in (6.5) refer to the even- and odd-mode excitations, respectively. The unified capacitance formula for the two excitations is given by

$$C_{\substack{e\\o}} = \frac{(1 + K_{\substack{e\\o}}A_{\substack{e\\o}})^2}{\sum\limits_{\substack{n\ \text{odd}\\ \text{even}}} (L_n + M_{\substack{ne\\o}}A_{\substack{e\\o}})^2 P_n/Y} \tag{6.6}$$

where

$$K_{\substack{e\\o}} = \frac{2}{w^4}\left[\left(\frac{w}{2} + g_{\substack{e\\o}}\right)^4 + \left(\frac{w}{2} - g_{\substack{e\\o}}\right)^4\right] \tag{6.7a}$$

$$L_n = \sin\left(\frac{\beta_n w}{2}\right)\cdot\sin\left\{\beta_n\left(\frac{c-s-w}{2}\right)\right\}, \qquad \beta_n = \frac{n\pi}{c} \tag{6.7b}$$

$$\begin{aligned} M_{\substack{ne\\o}} = \left(\frac{2}{\beta_n w}\right)^3\cdot\frac{1}{2}\Bigg[&3\left\{\beta_n^2\left(\frac{w}{2} + g_{\substack{e\\o}}\right)^2 - 2\right\}\sin\left\{\beta_n\left(\frac{c-s}{2} - w\right)\right\} \\ &+ \beta_n\left(\frac{w}{2} + g_{\substack{e\\o}}\right)\left\{\beta_n^2\left(\frac{w}{2} + g_{\substack{e\\o}}\right)^2 - 6\right\}\cos\left\{\beta_n\left(\frac{c-s}{2} - w\right)\right\} \\ &+ 3\left\{\beta_n^2\left(\frac{w}{2} - g_{\substack{e\\o}}\right)^2 - 2\right\}\cdot\sin\left\{\beta_n\left(\frac{c-s}{2}\right)\right\} - \beta_n\left(\frac{w}{2} - g_{\substack{e\\o}}\right)\left\{\beta_n^2\left(\frac{w}{2} - g_{\substack{e\\o}}\right)^2 - 6\right\} \\ &\cos\left\{\beta_n\left(\frac{c-s}{2}\right)\right\} + 12\sin\left\{\beta_n\left(\frac{c-s-w}{2} + g_{\substack{e\\o}}\right)\right\}\Bigg] \end{aligned} \tag{6.7c}$$

$$P_n = \left(\frac{4}{n\pi}\right)\left(\frac{2}{\beta_n w}\right)^2 \tag{6.7d}$$

$$A_{\substack{e\\o}} = -\frac{\sum\limits_{\substack{n\text{ odd}\\ \text{even}}} (K_{\substack{e\\o}} L_n - M_{n\substack{e\\o}}) L_n P_n / Y}{\sum\limits_{\substack{n\text{ odd}\\ \text{even}}} (K_{\substack{e\\o}} L_n - M_{n\substack{e\\o}}) M_{n\substack{e\\o}} P_n / Y} \tag{6.7e}$$

The choice of g_e and g_o in the above formulas is not critical, since the capacitance formula (6.6) is derived by maximizing C_e and C_o with respect to A_e and A_o, respectively. This aspect is demonstrated numerically in Sec. 6.3. The design data presented in this chapter are generated by setting $g_e = g_o = 0$. For this case, the capacitance formula reduces to

$$C_{\substack{e\\o}} = \frac{(1 + 0.25 A_{\substack{e\\o}})^2}{\sum\limits_{\substack{n\text{ odd}\\ \text{even}}} (L_n + M_n A_{\substack{e\\o}})^2 P_n / Y} \tag{6.8}$$

where

$$L_n = \sin\left(\frac{\beta_n w}{2}\right) \sin\left\{\beta_n\left(\frac{c - s - w}{2}\right)\right\}, \quad \beta_n = \frac{n\pi}{c} \tag{6.9a}$$

$$M_n = \left(\frac{2}{\beta_n w}\right)^3 \sin\left\{\beta_n\left(\frac{c - s - w}{2}\right)\right\}\left[3\left\{\left(\frac{\beta_n w}{2}\right)^2 - 2\right\}\cos\left(\frac{\beta_n w}{2}\right)\right.$$
$$\left. + \left(\frac{\beta_n w}{2}\right)\left\{\left(\frac{\beta_n w}{2}\right)^2 - 6\right\}\sin\left(\frac{\beta_n w}{2}\right) + 6\right] \tag{6.9b}$$

$$P_n = \frac{4}{n\pi}\left(\frac{2}{\beta_n w}\right)^2 \tag{6.9c}$$

$$A_{\substack{e\\o}} = -\frac{\sum\limits_{\substack{n\text{ odd}\\ \text{even}}} (L_n - 4M_n) L_n P_n / Y}{\sum\limits_{\substack{n\text{ odd}\\ \text{even}}} (L_n - 4M_n) M_n P_n / Y} \tag{6.9d}$$

The capacitance formula (6.6) or (6.8) is applicable to any edge-coupled stripline-like transmission line possessing a plane of symmetry between the strip conductors as considered in Fig. 4.1. The structure may consist of one or more dielectric layers which may be isotropic or anisotropic. The only parameter that needs to be changed in the computation is the admittance Y depending on the structure under consideration. The expressions for Y for some of the useful edge-coupled structures with iso/anisotropic substrates are listed in Tables 4.1–4.3.

6.2.3 Other Closed Form Expressions

(i) *Characteristic impedances of edge-coupled symmetric striplines with zero thickness strip conductors*: Exact expressions for the even- and odd-mode characteristic impedances of edge-coupled homogeneous symmetric stripline with zero thickness strip conductors (inset in Fig. 6.1) have been reported by Cohn [11]. The detailed derivation which uses the conformal mapping method is presented in Chapter 4. The closed form expression for Z_e and Z_o are

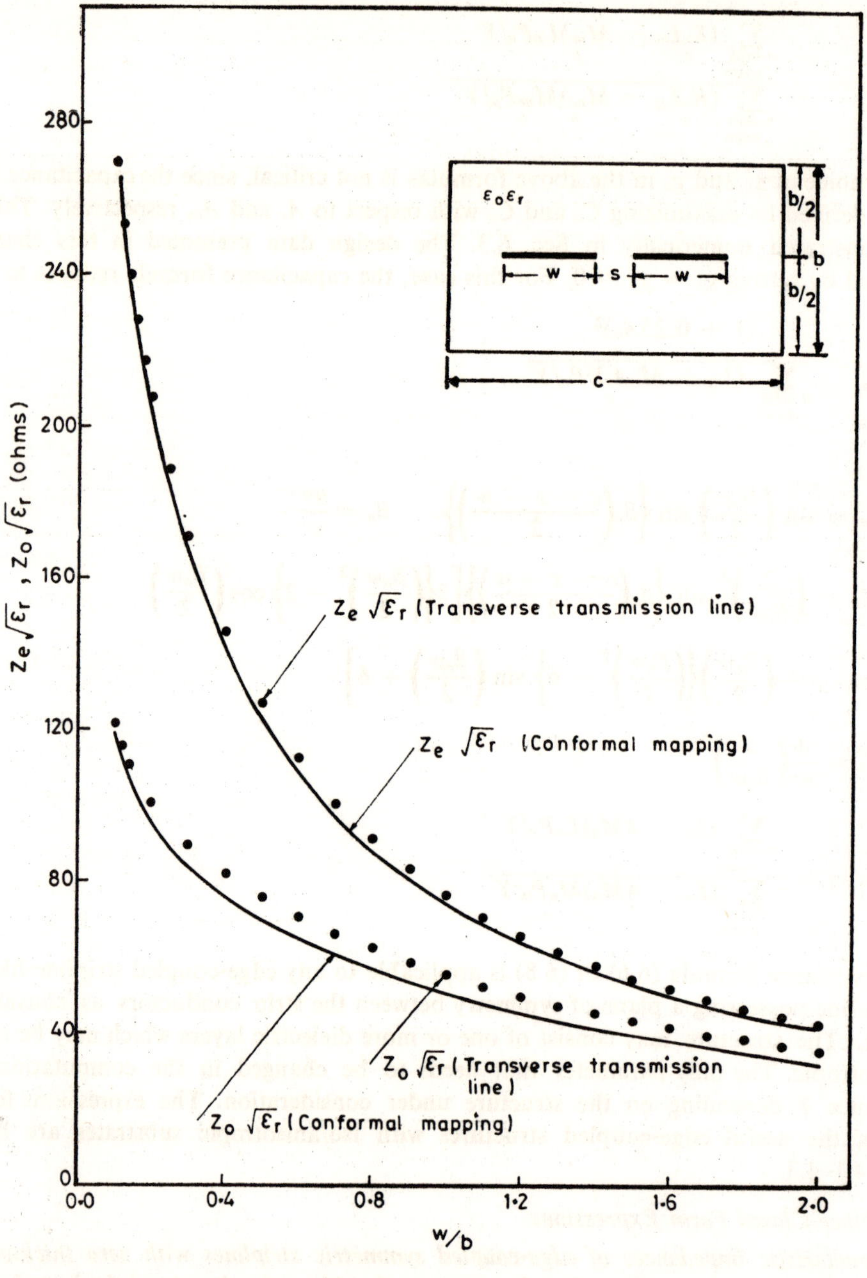

FIG. 6.1 $Z_e\sqrt{\epsilon_r}$ and $Z_o\sqrt{\epsilon_r}$ versus w/b of edge-coupled symmetric stripline—comparison of computed results using unified approach formula [(6.8) with (6.1), $c/(2w+s) = 20$] with conformal mapping method [(6.10), $c/(2w+s) = \infty$]; $s/b = 0.1$.

given as follows:

$$Z_{\substack{e\\o}}\sqrt{\epsilon_r} = \frac{30\pi K(k'_{\substack{e\\o}})}{K(k_{\substack{e\\o}})} \tag{6.10}$$

where

$$k_e = \tanh\left(\frac{\pi w}{2b}\right)\tanh\left(\frac{\pi}{2}\cdot\frac{w+s}{b}\right) \tag{6.11a}$$

$$k_o = \tanh\left(\frac{\pi w}{2b}\right)\coth\left(\frac{\pi}{2}\cdot\frac{w+s}{b}\right) \tag{6.11b}$$

$$k'_{\substack{e\\o}} = \sqrt{1 - k^2_{\substack{e\\o}}} \tag{6.11c}$$

In (6.10), $K(k)$ and $K(k')$ are complete elliptic integrals of the first kind. Closed form formulas for the quantity $[K(k')/K(k)]$ are available in [18].

(ii) *Explicit expressions for the dimensional parameters of edge-coupled symmetric stripline in terms of coupling:* In Sec. 4.2.3, explicit expressions derived by Linner [19] for determining the dimensional parameters of the edge-coupled symmetric stripline from its even- and odd-mode impedances are summarized. The computation, however, involves the evaluation of Jacobi's theta function. Shelton [12] has analysed a homogeneous parallel off-set strip transmission line using the conformal mapping technique. He has presented comprehensive design equations for determining the structural parameters directly from the knowledge of the even- and odd-mode impedances. In the special case for loose coupling, if the vertical off-set distance between the strip conductors is set equal to zero, the resulting formulas apply to edge-coupled symmetric stripline. The formulas are simpler than those presented by Linner [19] and can be directly used to determine the dimensional parameters of a coupler from the coupling value. If C_v denotes the voltage coupling coefficient and Z_e and Z_o represent the even- and odd-mode impedances, respectively, then,

$$C_v = \frac{\zeta - 1}{\zeta + 1}, \qquad \zeta = \frac{Z_e}{Z_o} \tag{6.12}$$

In order that the coupler is perfectly matched to its terminating transmission line having a characteristic impedance Z_t, it is necessary that

$$Z_t = \sqrt{Z_e Z_o} \tag{6.13}$$

With (6.13) in (6.12), we can express Z_e and Z_o in terms of the coupling coefficient C_v:

$$Z_e = Z_t\sqrt{\frac{1 + C_v}{1 - C_v}} \tag{6.14a}$$

$$Z_o = Z_t\sqrt{\frac{1 - C_v}{1 + C_v}}. \tag{6.14b}$$

Referring to the edge-coupled symmetric stripline shown as an inset in Fig. 6.1, the normalized width w/b and spacing s/b of the structure can be calculated using the following set of equations [12]:

$$C_o = \frac{120\pi\sqrt{\zeta}}{\sqrt{\epsilon_r}Z_e} \tag{6.15a}$$

$$\Delta C = \frac{120\pi(\zeta - 1)}{Z_t\sqrt{\epsilon_r\zeta}} \tag{6.15b}$$

$$k = \frac{1}{(e^{0.5\pi\Delta c} - 1)} \tag{6.15c}$$

$$a = \sqrt{k^2 + k} + k \tag{6.15d}$$

$$q = k/a \tag{6.15e}$$

$$C_{fo} = (2/\pi)\left[\ln\left\{\frac{(1+a)}{a(1-q)}\right\} - \ln(q)\right] \tag{6.15f}$$

$$w/b = 0.25(C_o - C_{fo} - 0.8825424) \tag{6.15g}$$

$$s/b = (1/\pi)\ln\left(\frac{1-q}{1+a}\right) \tag{6.15h}$$

In the following sections, we present computed design data on some of the useful edge-coupled structures listed in Tables 4.1–4.3. All the computations are carried out for zero-thickness strip conductor ($t/b = 0$). As discussed in Chapter 5 (Sec. 5.2), this approximation gives fairly accurate values of impedances valid for most practical purposes. Further, the side walls are chosen sufficiently away from the strip conductors [$c/(2w + s) = 20$] so that, they have practically no effect on the propagation parameters.

6.3 Edge-coupled Transmission Structures with Isotropic Dielectric Substrates

6.3.1 Edge-coupled Symmetric Stripline

For the structure completely filled with a homogeneous isotropic dielectric, the even- and odd-mode impedances can be normalized with respect to $\sqrt{\epsilon_r}$ so as to render the expressions for $Z_e\sqrt{\epsilon_r}$ and $Z_o\sqrt{\epsilon_r}$ independent of ϵ_r. Thus, a single graph or table giving variations of $Z_e\sqrt{\epsilon_r}$ and $Z_o\sqrt{\epsilon_r}$ as a function of the dimensional parameters of the edge-coupled stripline can be used with any dielectric filling.

In Figs. 6.1 and 6.2, the variations of $Z_e\sqrt{\epsilon_r}$ and $Z_o\sqrt{\epsilon_r}$ computed using the conformal mapping formula (6.10) are compared with those computed using the unified variational formula (6.8) derived by assuming a symmetric charge distribution ($g_e = g_o = 0$). The conformal mapping formula yields exact results, whereas the results computed from the unified formula are slightly higher. For a small value of the normalized spacing, $s/b = 0.1$, it can be seen from Fig. 6.1 that the maximum error incurred in using the unified formula (6.8) is about 3 ohms in $Z_e\sqrt{\epsilon_r}$ and 5 ohms in $Z_o\sqrt{\epsilon_r}$. The same error is reflected in Z_e and Z_o for an air-filled line. For a fixed value of w/b, the error reduces with increasing s/b and gives fairly accurate results for $s/b > 0.2$, adequate for most practical applications (Fig. 6.2). The accuracy of the results in the range $s/b \leqslant 0.2$ can be improved by using the formula (6.6) with appropriate values of g_e and g_o. As depicted in Figs. 6.3a and 6.3b, for $s/b = 0.1$, an improvement of the order of 1 to 2 ohms in $Z_e\sqrt{\epsilon_r}$ and 2 to 3 ohms in $Z_o\sqrt{\epsilon_r}$ is achieved. Further, for $s/b = 0.5$, the improvement in $Z_e\sqrt{\epsilon_r}$ and $Z_o\sqrt{\epsilon_r}$ is marginal. Use of (6.6) is

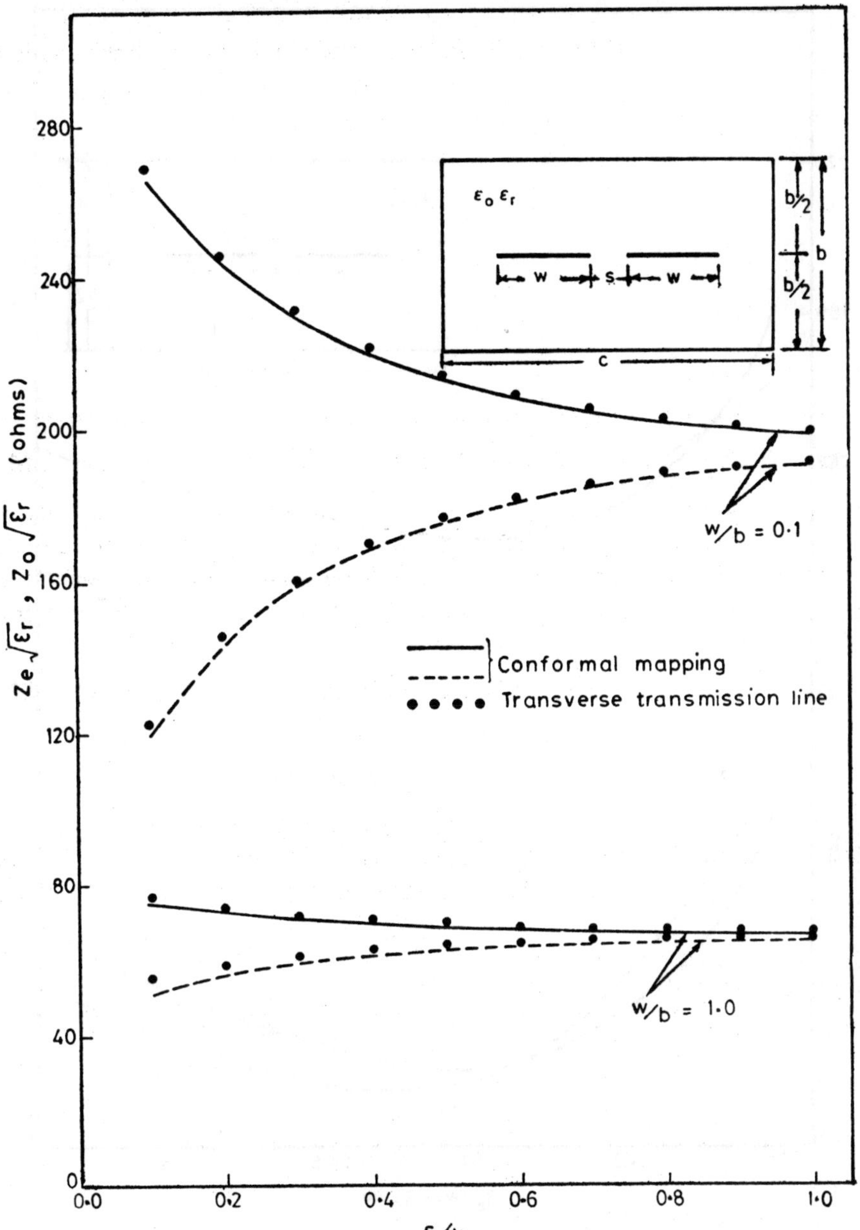

FIG. 6.2 $Z_e \sqrt{\epsilon_r}$ and $Z_o \sqrt{\epsilon_r}$ versus s/b of edge-coupled symmetric stripline—comparison of computed results using unified approach formula [(6.8) with (6.1), $c/(2w + s) = 20$] with conformal mapping method [(6.10), $c/(2w + s) = \infty$]; ——— $Z_e \sqrt{\epsilon_r}$, – – – $Z_o \sqrt{\epsilon_r}$

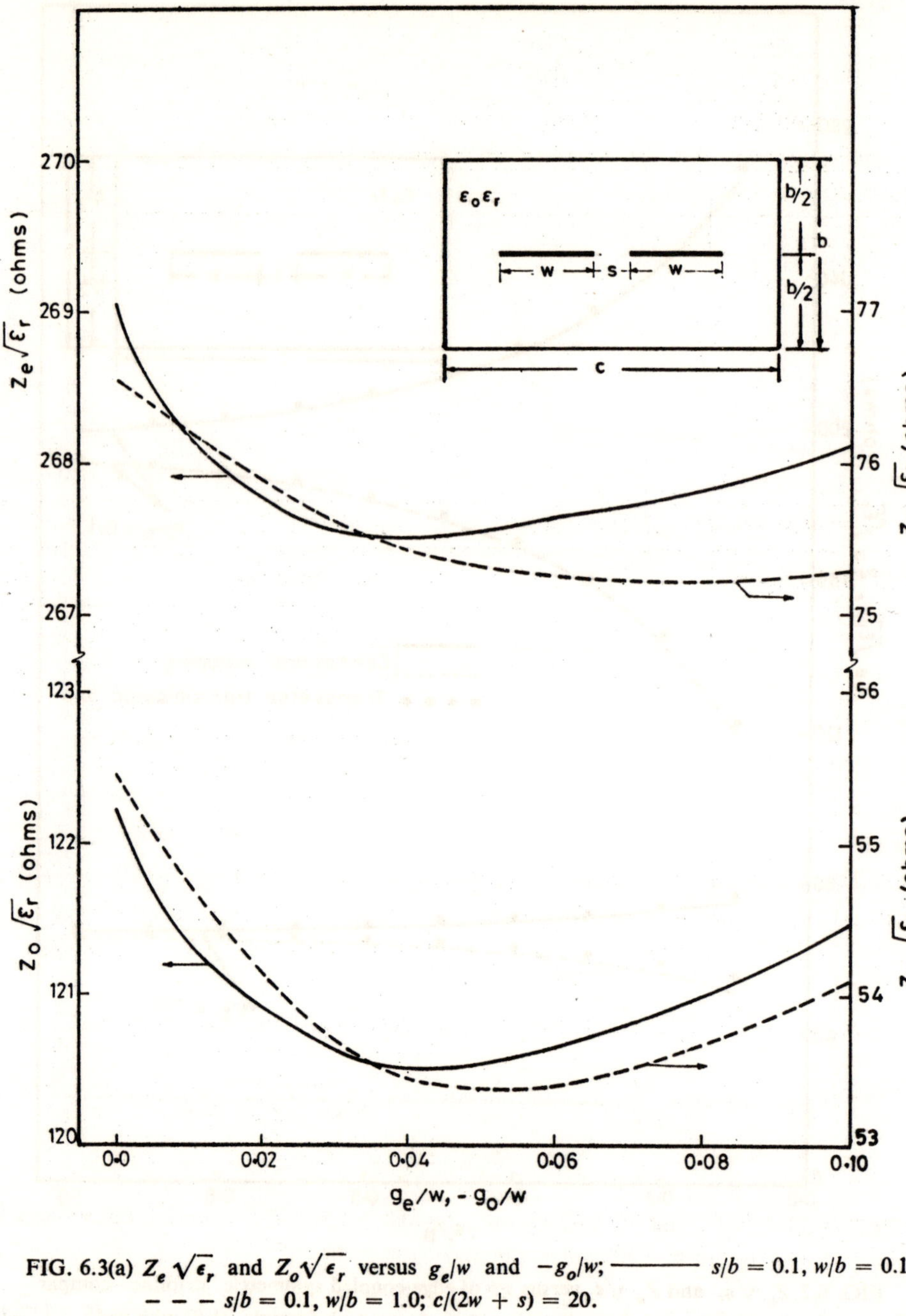

FIG. 6.3(a) $Z_e\sqrt{\epsilon_r}$ and $Z_o\sqrt{\epsilon_r}$ versus g_e/w and $-g_o/w$; ——— $s/b = 0.1$, $w/b = 0.1$, – – – $s/b = 0.1$, $w/b = 1.0$; $c/(2w + s) = 20$.

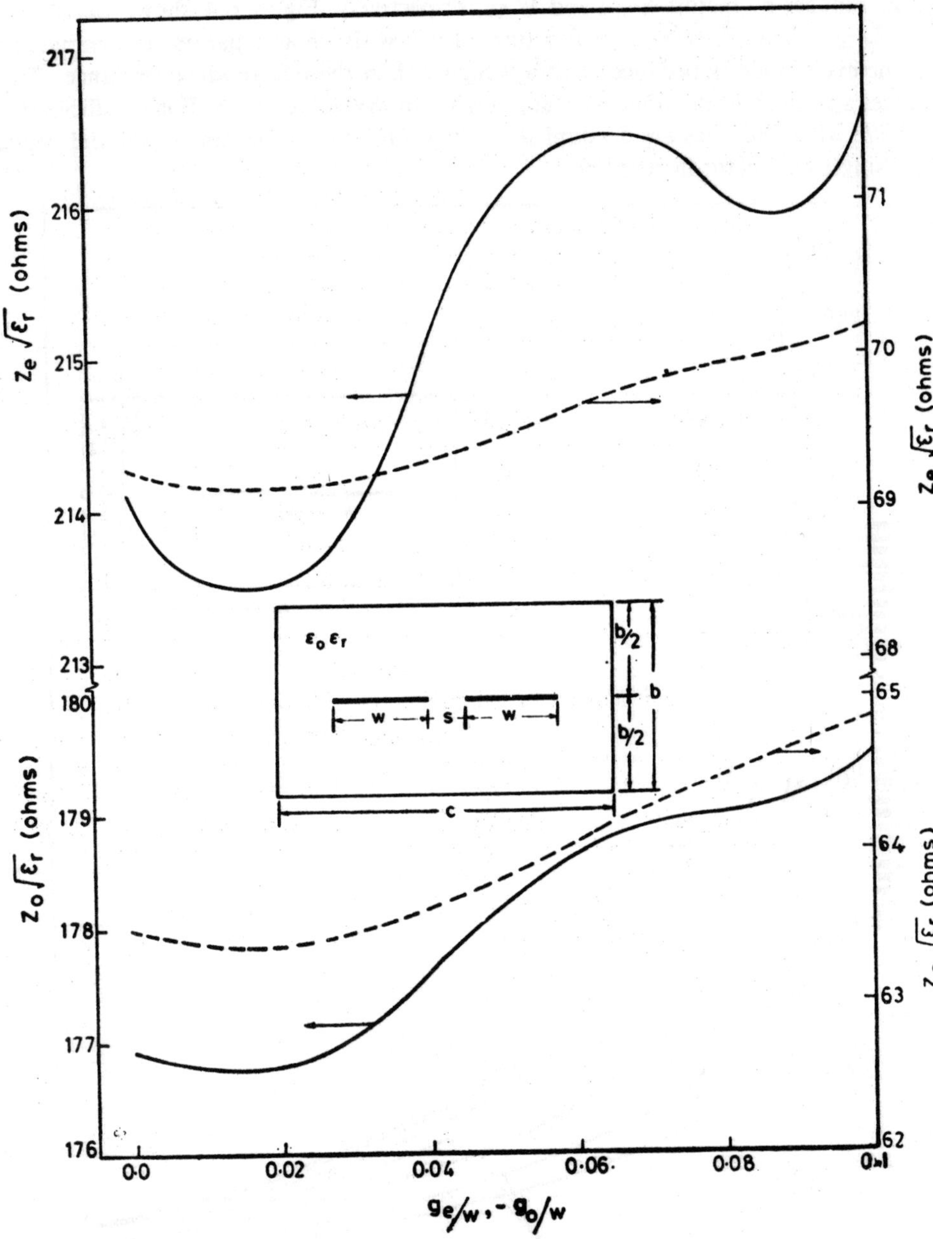

FIG. 6.3(b) $Z_e\sqrt{\epsilon_r}$ and $Z_o\sqrt{\epsilon_r}$ versus g_e/w and $-g_o/w$; ——— $s/b = 0.5$, $w/b = 0.1$, – – – $s/b = 0.5$, $w/b = 1.0$; $c/(2w + s) = 20$.

therefore recommended for calculating the propagation parameters of air stripline for $s/b \leqslant 0.2$. For $s/b > 0.2$ (6.8) gives equally accurate results.

For a dielectric-filled stripline with $\epsilon_r > 1$, the error reduces by a factor $\sqrt{\epsilon_r}$ as compared with the air stripline. Similarly, for striplines with partial dielectric fillings, the error in

Z_e and Z_o reduces by factors $\sqrt{\epsilon_{fe}}$ and $\sqrt{\epsilon_{fo}}$, respectively. Figure 6.4 shows detailed variations of $Z_e\sqrt{\epsilon_r}$ and $Z_o\sqrt{\epsilon_r}$ as a function of w/b with s/b as a parameter computed using (6.10). The even-mode impedance is always higher than the odd-mode impedance. The two impedances approach each other with an increase in the value of s/b. For a sufficiently large value of s/b, when the lines are decoupled, the two impedances become equal and reduce to that of a single conductor in stripline.

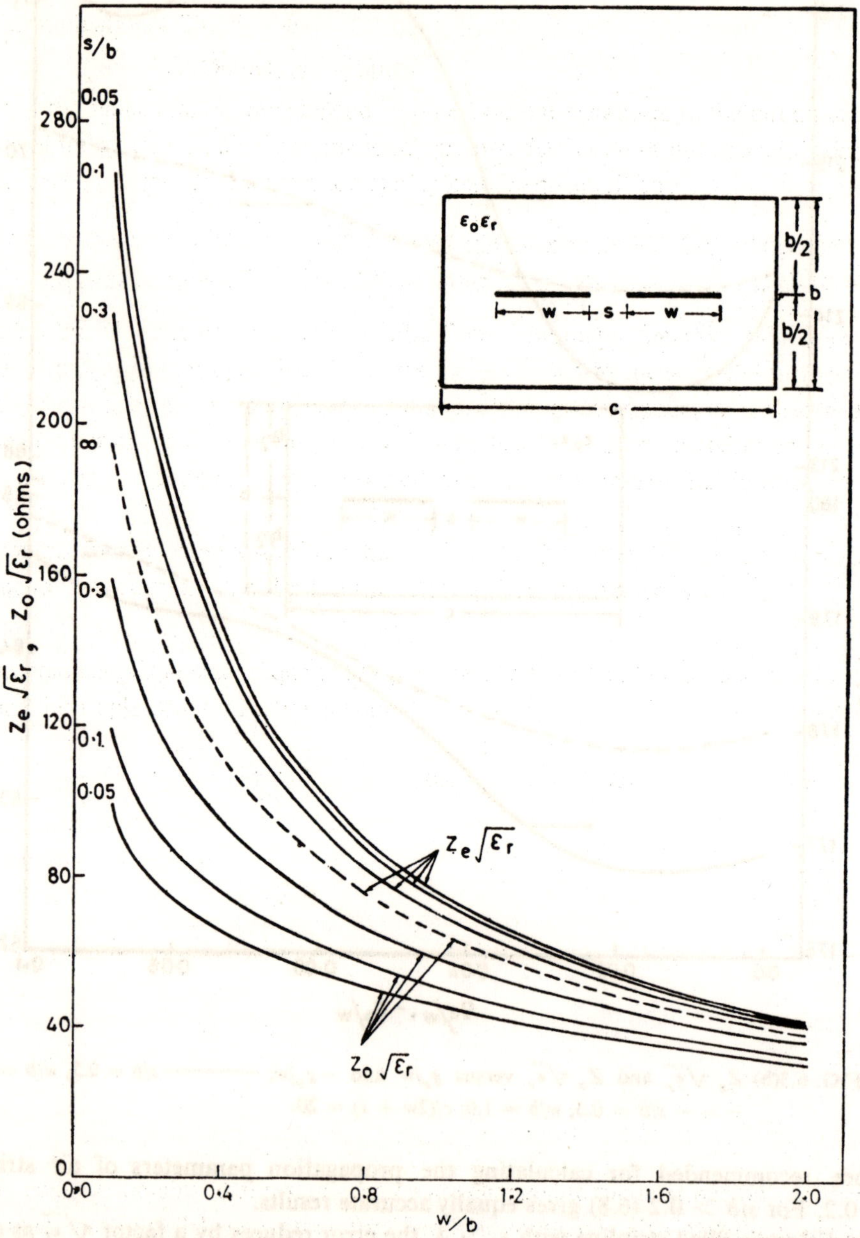

FIG. 6.4 $Z_e\sqrt{\epsilon_r}$ and $Z_o\sqrt{\epsilon_r}$ versus w/b of edge-coupled symmetric stripline with s/b as parameter; $c/(2w + s) = \infty$.

Since edge-coupled stripline filled with homogeneous dielectric supports pure TEM modes, the even- and odd-mode phase velocities are equal. This is an advantageous feature which is utilized in the design of a number of passive components especially for achieving high directivity in couplers. Extensive impedance data are generated using the conformal mapping formula (6.10) for values of w/b ranging from 0.1 to 2.5 and s/b ranging from 0.001 to 1. The results are tabulated in Table 6.1 over small intervals of w/b and s/b so that the impedances for any intermediate value of w/b or s/b can be obtained accurately by linear interpolation. Tables 6.2–6.4 provide data on the dimensional parameters of the edge-coupled stripline as a function of coupling, computed using the formulas (6.12)–(6.15). Three different ground plane spacings that are normally used in practice at microwave frequencies are considered. Since edge-coupled lines are used in couplers requiring coupling of the order of 8 dB or more, data are presented for coupling values in the range 8 dB to 40 dB in steps of 0.5 dB.

6.3.2 *Edge-coupled Stripline with Off-set Conductors*

Structure A2 in Table 4.1 shows the cross-sectional view of edge-coupled stripline with off-set strip conductors. The medium is completely filled with a homogeneous isotropic dielectric and the strip conductors are off-set by a distance d from the symmetry plane with respect to the top and bottom ground planes.

Figure 6.5 depicts the variations in $Z_e \sqrt{\epsilon_r}$ and $Z_o \sqrt{\epsilon_r}$ as a function of d/b for a fixed value of $s/b = 0.3$ and with w/b as a parameter. Both the impedances decrease with an increase in the off-set distance d. Detailed data on the impedances as a function of w/b for various values of d/b and s/b are tabulated in Tables 6.5–6.12. All the data presented in Fig. 6.5 and Tables 6.5–6.12 are computed using the unified formula (6.8). The data for $d/b = 0$ corresponds to the edge-coupled symmetric stripline and can be compared with the values of Table 6.1 computed using the exact conformal mapping formulas (6.10). It is observed that the tabulated impedances $Z_e \sqrt{\epsilon_r}$ and $Z_o \sqrt{\epsilon_r}$ in Tables 6.5–6.12 are in agreement (within 2.5 ohms) with the corresponding values in Table 6.1. For values of $s/b \leqslant 0.2$, it is recommended to use (6.6) with appropriate values of g_e and g_o.

6.3.3 *Edge-coupled Microstrip-like Stripline*

The edge-coupled microstrip-like stripline (structure A3 in Table 4.1) behaves like a symmetric stripline homogeneously filled with a dielectric having a relative dielectric constant $[(\epsilon_r + 1)/2]$. The even- and odd-mode phase velocities are equal and hence the structure possesses all the advantages of a homogeneous symmetric stripline. In addition, since the dielectric above the strip conductors is air, it is amenable for mounting chip devices. Figure 6.6 shows the variations in Z_e and Z_o as a function of w/b of an edge-coupled microstrip-like stripline with s/b as a parameter. The characteristics are similar to those shown in Fig. 6.4 for an edge-coupled symmetric stripline. The complete design data presented in Table 6.1 are applicable to edge-coupled microstrip-like stripline when ϵ_r in $Z_e \sqrt{\epsilon_r}$ and $Z_o \sqrt{\epsilon_r}$ is replaced by $(\epsilon_r + 1)/2$.

6.3.4 *Edge-coupled Suspended Stripline*

The edge-coupled suspended stripline (structure A4 in Table 4.1) makes use of a dielectric substrate suspended symmetrically between the top and bottom ground planes. The commonly used substrate materials are quartz ($\epsilon_r = 3.78$) and alumina ($\epsilon_r = 9.6$). Detailed data on the even- and odd-mode impedances and effective dielectric constants of the structure using quartz

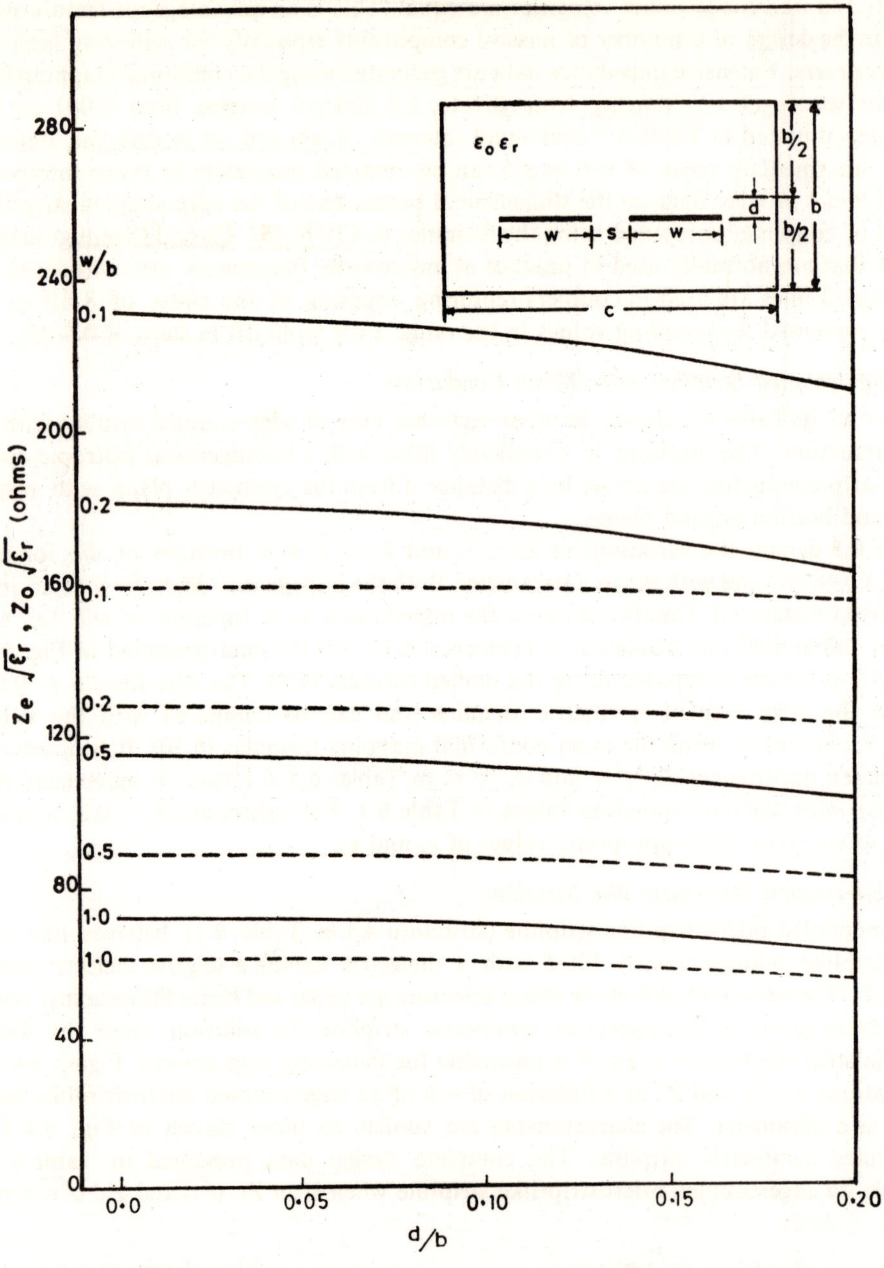

FIG. 6.5 $Z_e\sqrt{\epsilon_r}$ and $Z_o\sqrt{\epsilon_r}$ versus d/b of edge-coupled stripline with off-set strip conductor and with w/b as parameter; $c/(2w + s) = 20$, $s/b = 0.3$; ——— $Z_e\sqrt{\epsilon_r}$, – – – $Z_o\sqrt{\epsilon_r}$.

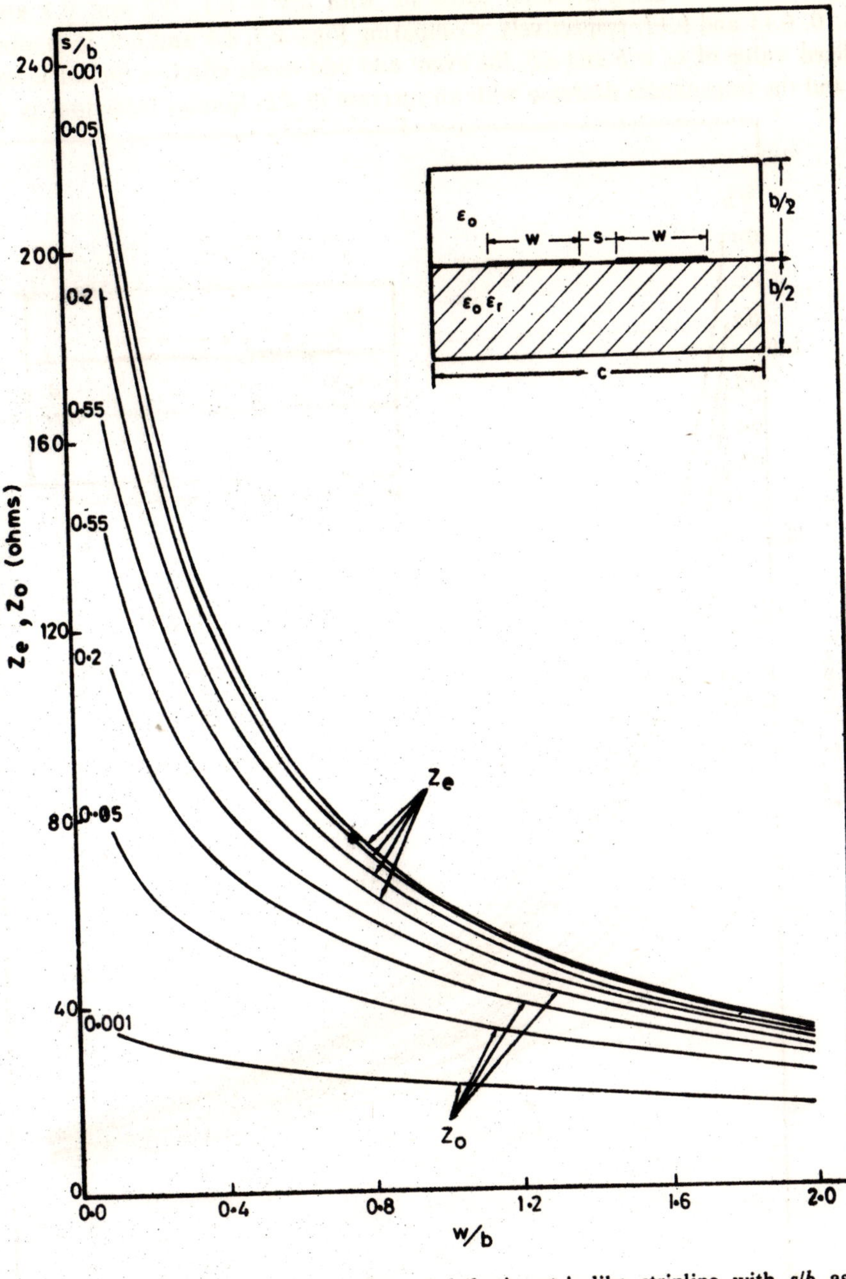

FIG. 6.6 Z_e and Z_o versus w/b of edge-coupled microstrip-like stripline with s/b as parameter; $c/(2w + s) = \infty$, $\epsilon_r = 2.22$.

substrate are plotted in Figs. 6.7, 6.8 and 6.9 for $d/b = 0.1$, 0.2 and 0.3, respectively. Corresponding design data using alumina substrate with $d/b = 0.1$, 0.2 and 0.3 are plotted in Figs. 6.10, 6.11 and 6.12, respectively. Comparing Figs. 6.7, 6.8 and 6.9, it is observed that for a fixed value of ϵ_r, w/b and s/b, the even- and odd-mode effective dielectric constants increase and the impedances decrease with an increase in d/b. Similar behaviour is depicted in

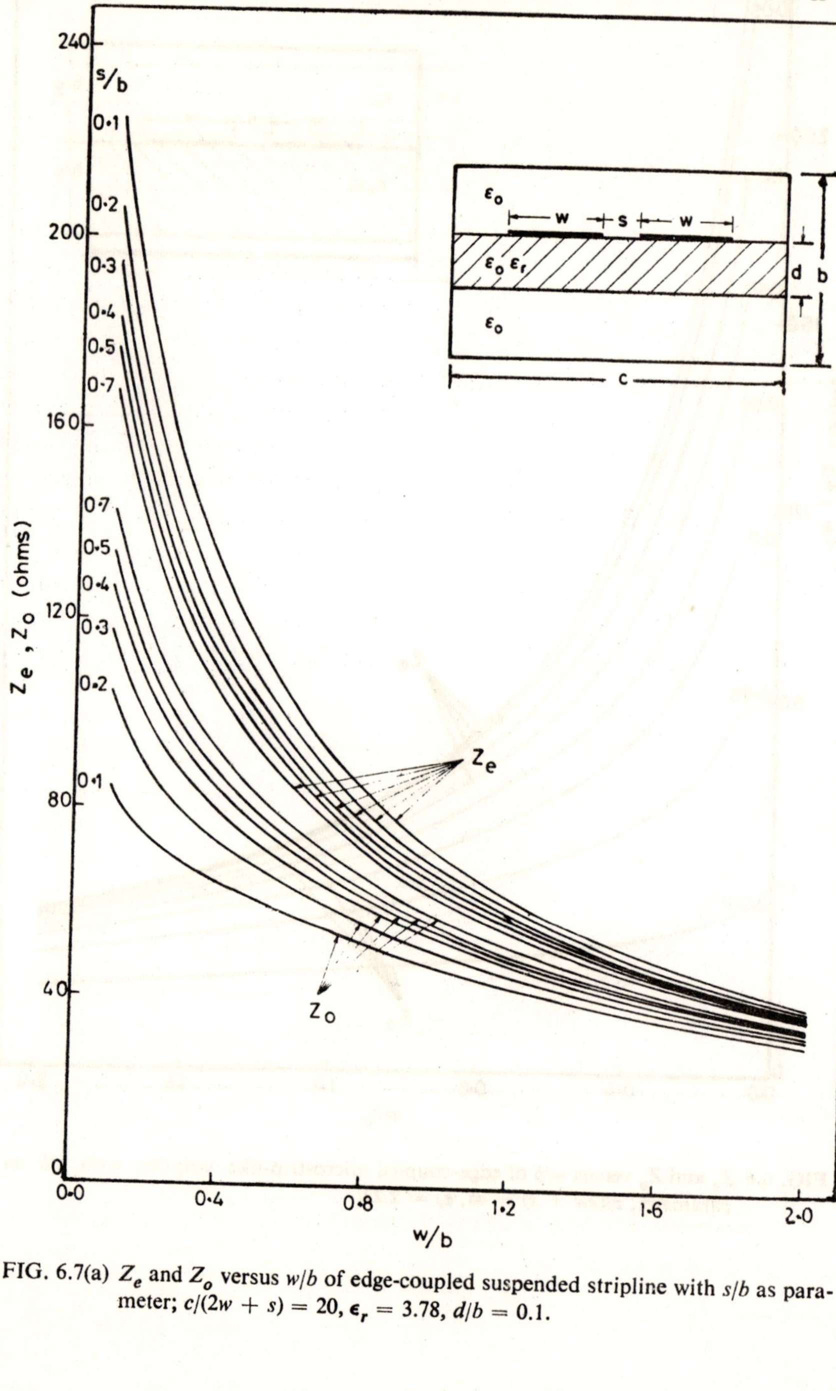

FIG. 6.7(a) Z_e and Z_o versus w/b of edge-coupled suspended stripline with s/b as parameter; $c/(2w + s) = 20$, $\epsilon_r = 3.78$, $d/b = 0.1$.

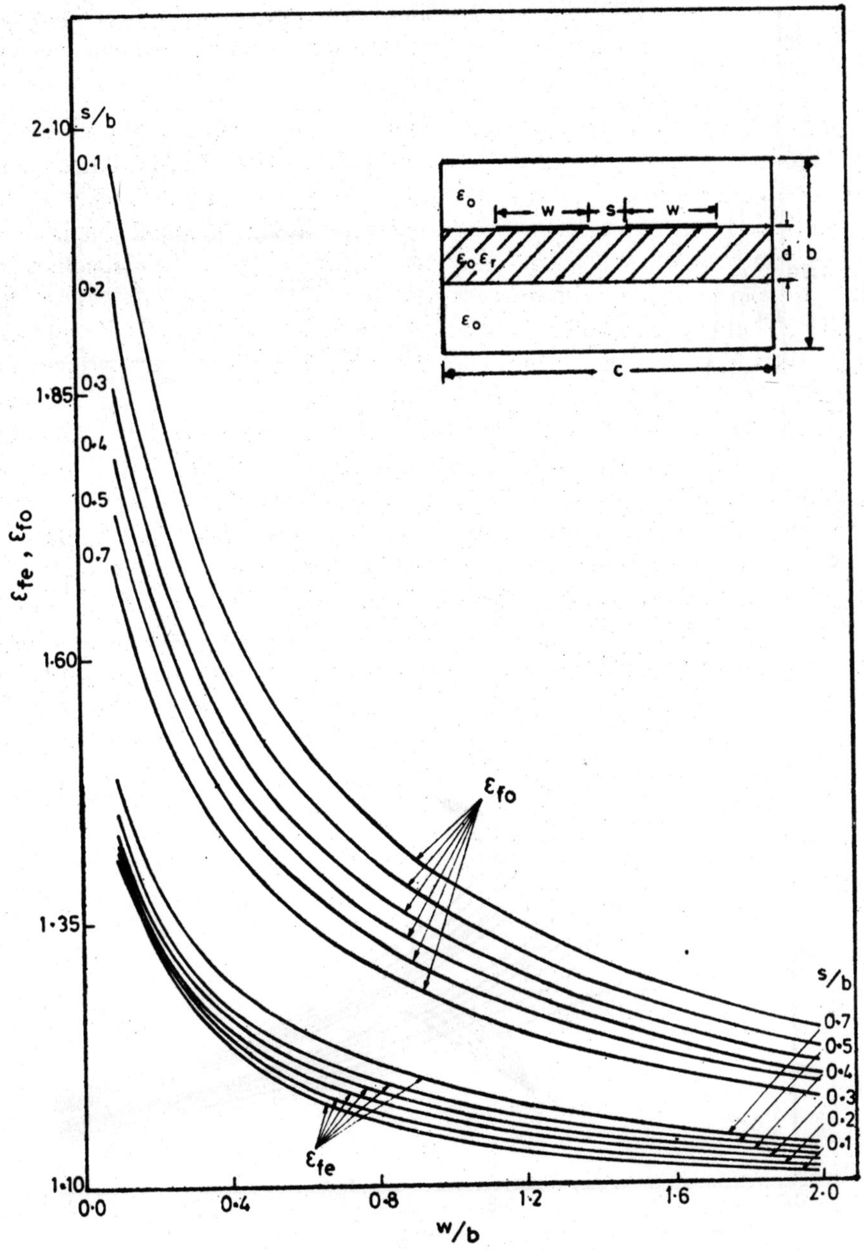

FIG. 6.7(b) ϵ_{fe} and ϵ_{fo} versus w/b of edge-coupled suspended stripline with s/b as parameter; $c/(2w + s) = 20$, $\epsilon_r = 3.78$, $d/b = 0.1$.

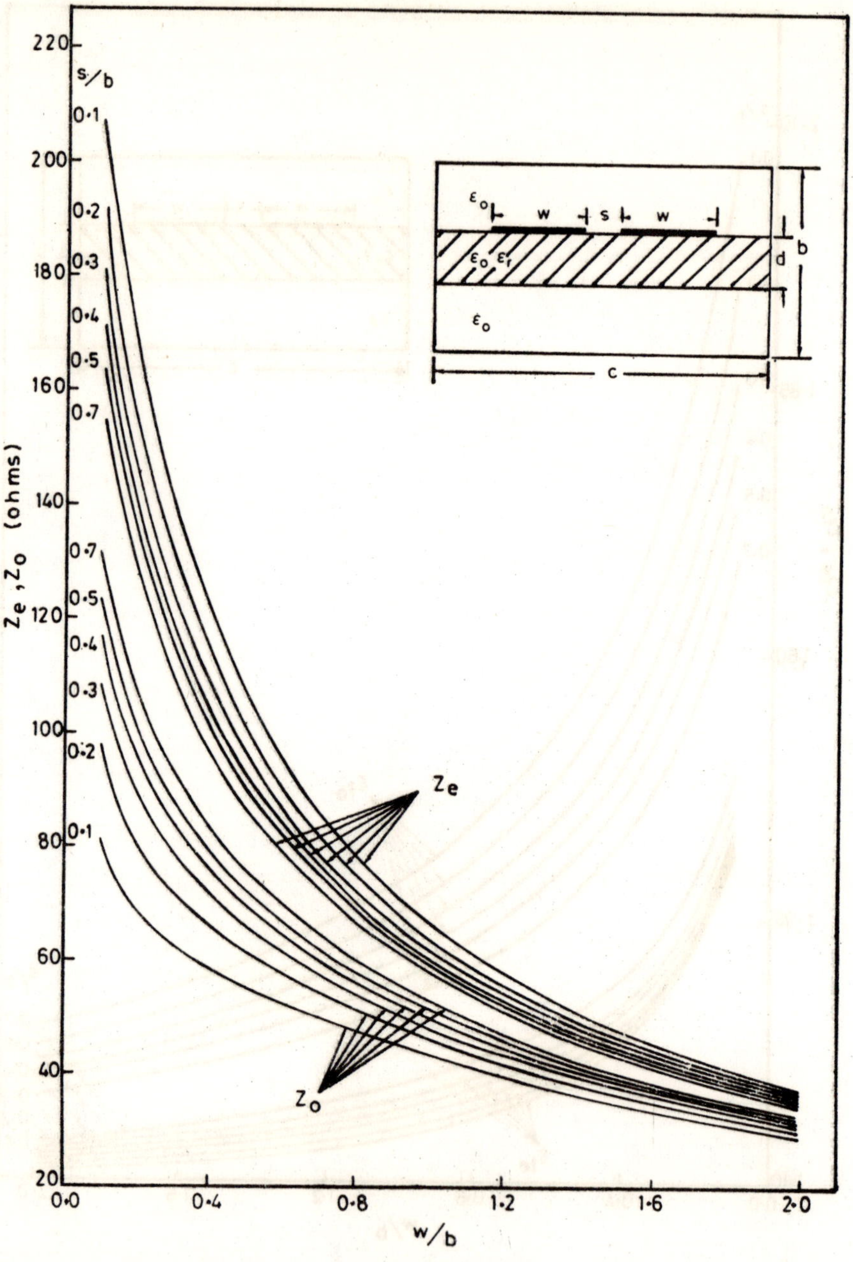

FIG. 6.8(a) Z_e and Z_o versus w/b of edge-coupled suspended stripline with s/b as parameter; $c/(2w+s) = 20$, $\epsilon_r = 3.78$, $d/b = 0.2$.

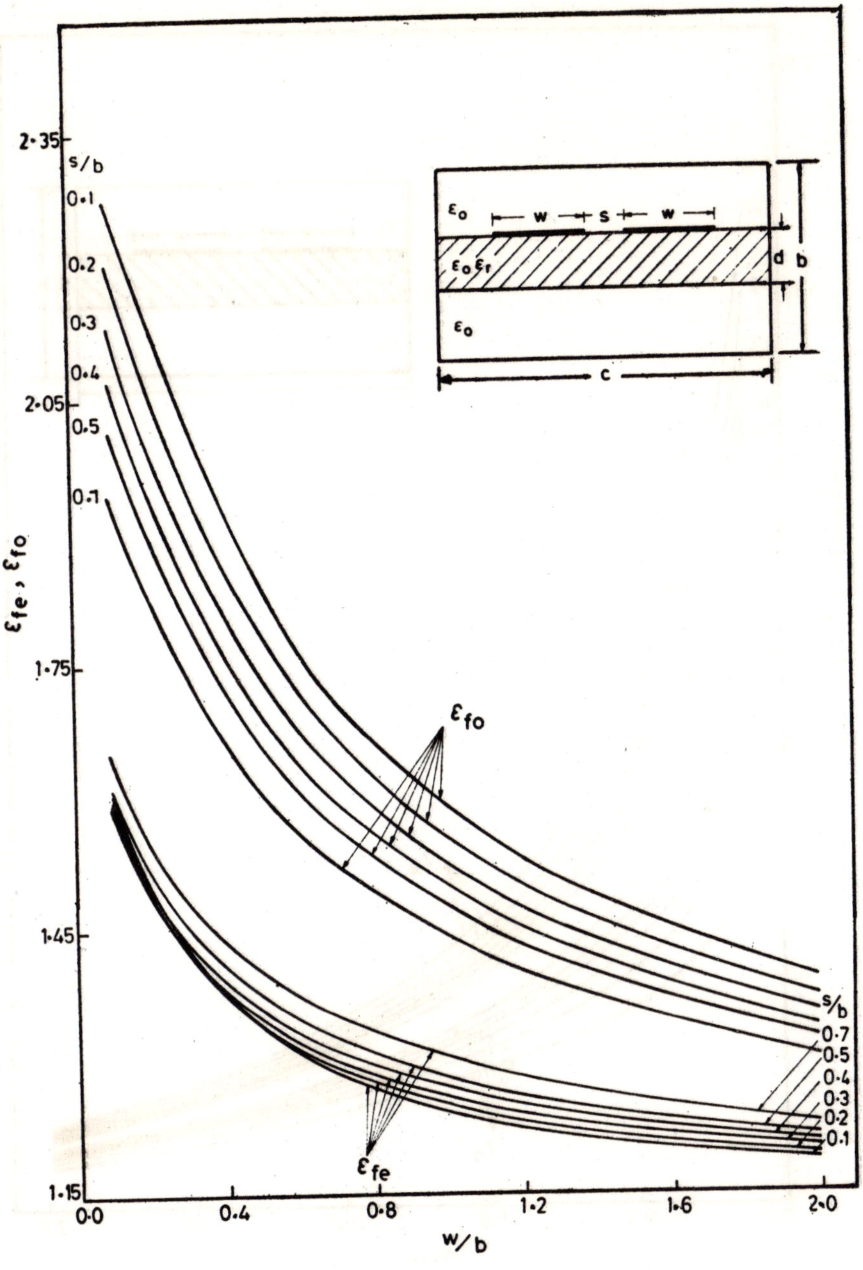

FIG. 6.8(b) ϵ_{fe} and ϵ_{fo} versus w/b of edge-coupled suspended stripline with s/b as parameter; $c/(2w + s) = 20$, $\epsilon_r = 3.78$, $d/b = 0.2$.

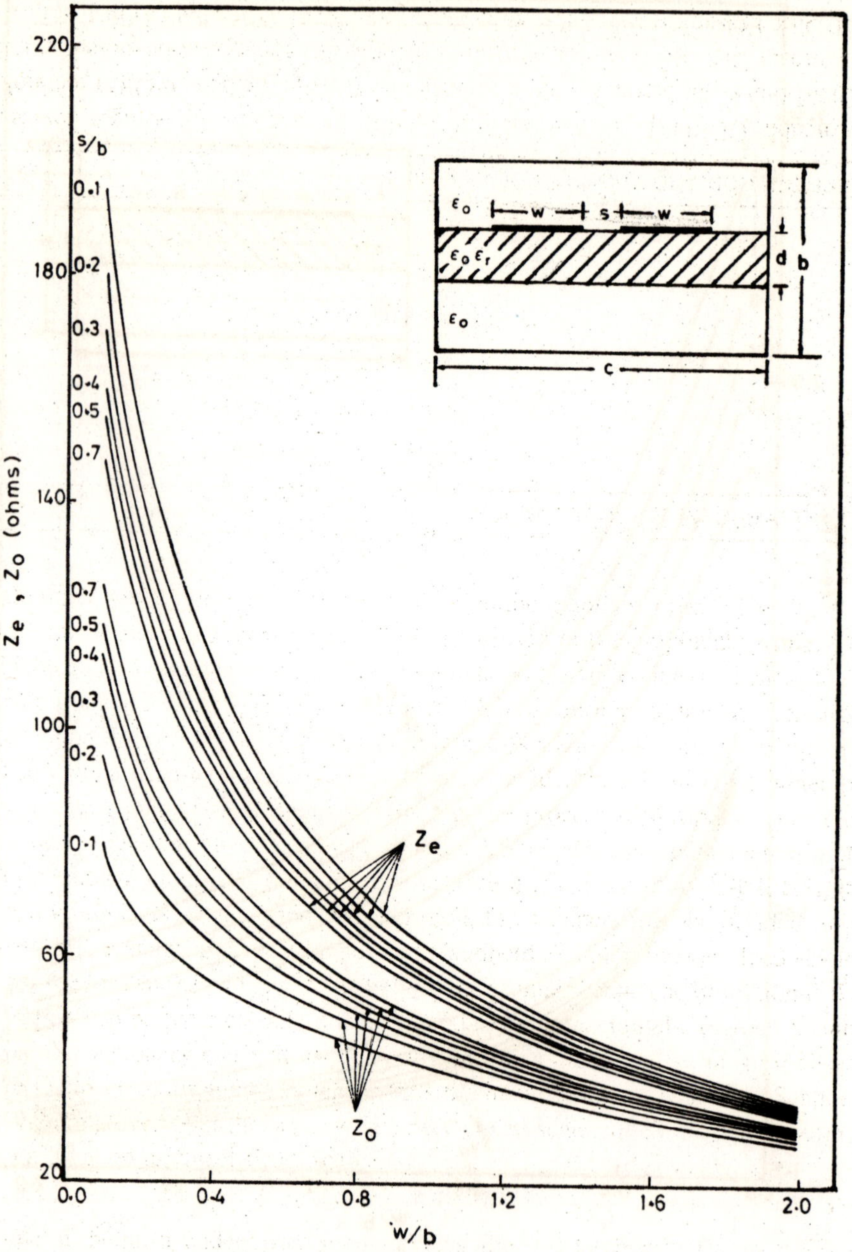

FIG. 6.9(a) Z_e and Z_o versus w/b of edge-coupled suspended stripline with s/b as parameter; $c/(2w + s) = 20$, $\epsilon_r = 3.78$, $d/b = 0.3$.

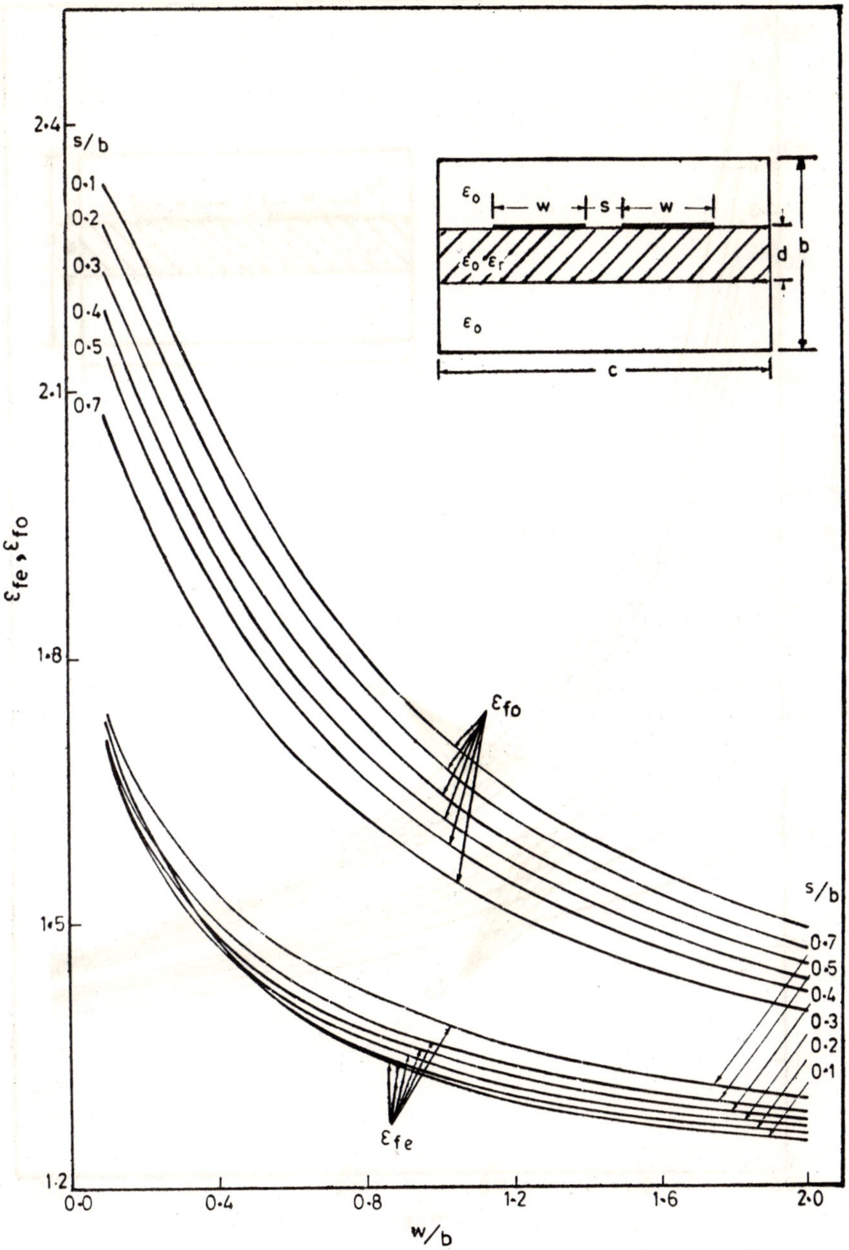

FIG. 6.9(b) ϵ_{fe} and ϵ_{fo} versus w/b of edge-coupled suspended stripline with s/b as parameter; $c/(2w+s) = 20$, $\epsilon_r = 3.78$, $d/b = 0.3$.

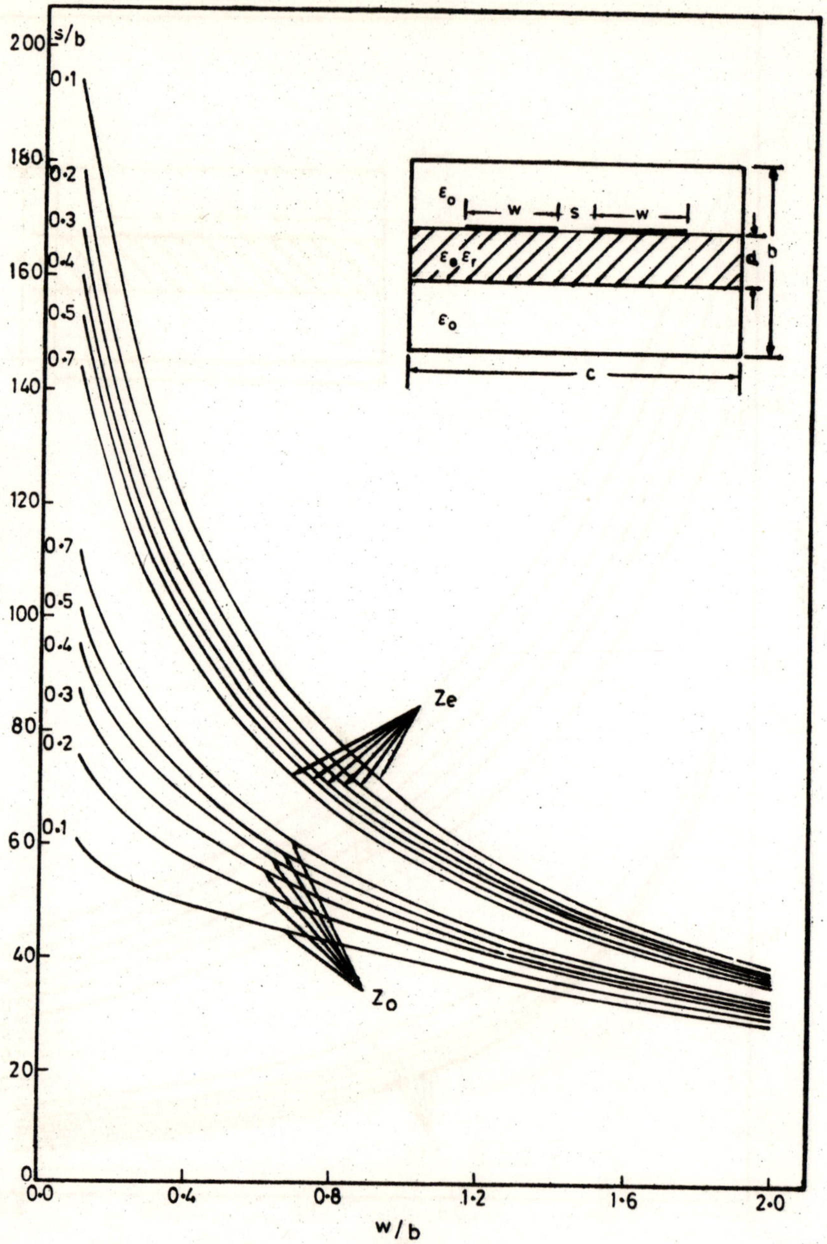

FIG. 6.10(a) Z_e and Z_o versus w/b of edge-coupled suspended stripline with s/b as parameter; $c/(2w + s) = 20$, $\epsilon_r = 9.6$, $d/b = 0.1$.

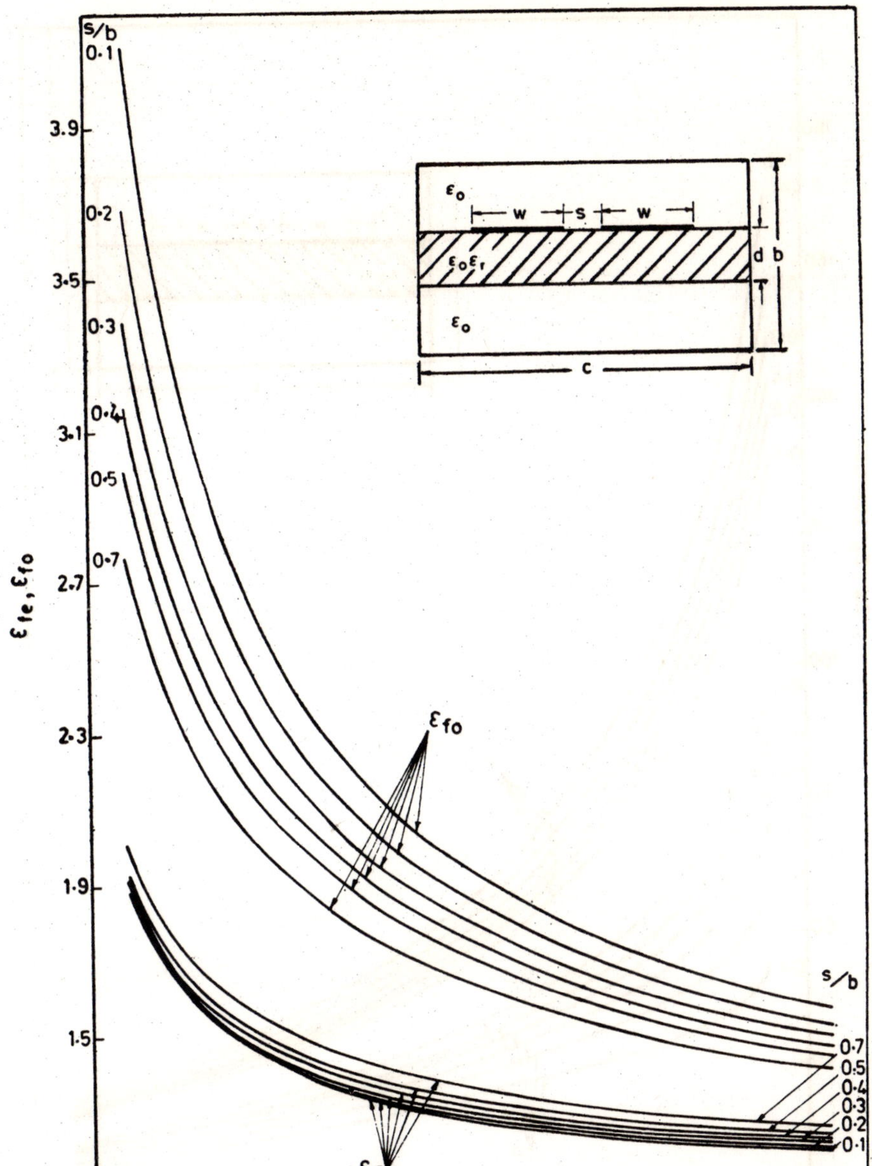

FIG. 6.10(b) ϵ_{fe} and ϵ_{fo} versus w/b of edge-coupled suspended stripline with s/b as parameter; $c/(2w + s) = 20$, $\epsilon_r = 9.6$, $d/b = 0.1$.

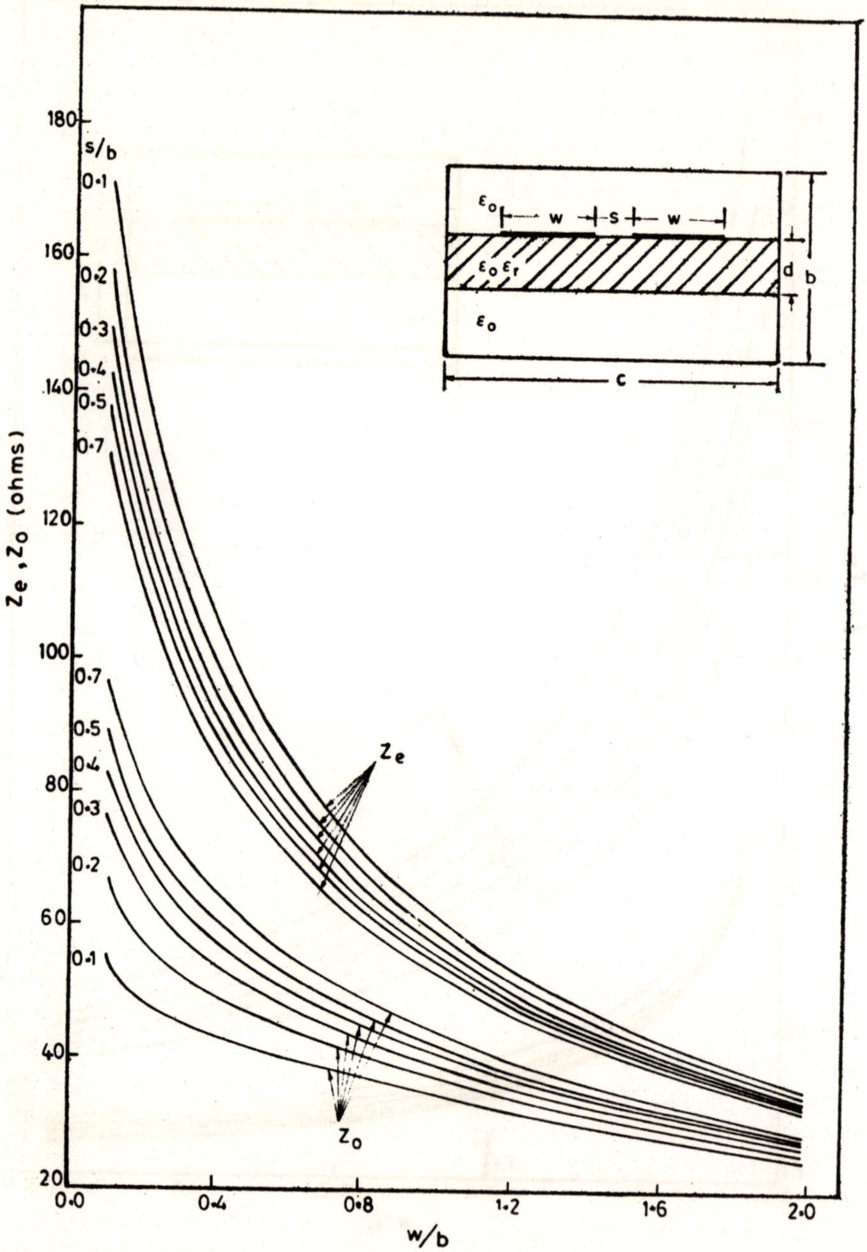

FIG. 6.11(a) Z_e and Z_o versus w/b of edge-coupled suspended stripline with s/b as parameter; $c/(2w + s) = 20$, $\epsilon_r = 9.6$, $d/b = 0.2$.

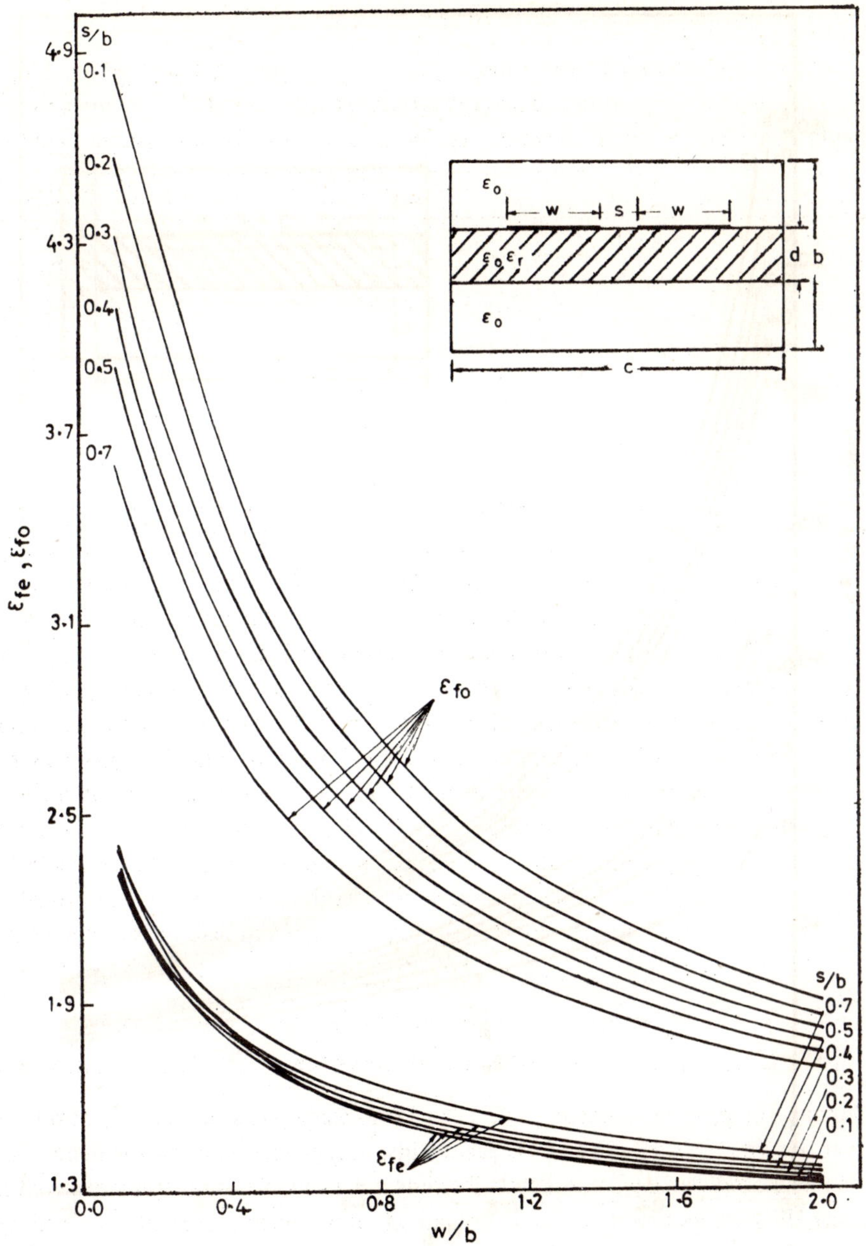

FIG. 6.11(b) ϵ_{fe} and ϵ_{fo} versus w/b of edge-coupled suspended stripline with s/b as parameter; $c/(2w+s) = 20$, $\epsilon_r = 9.6$, $d/b = 0.2$.

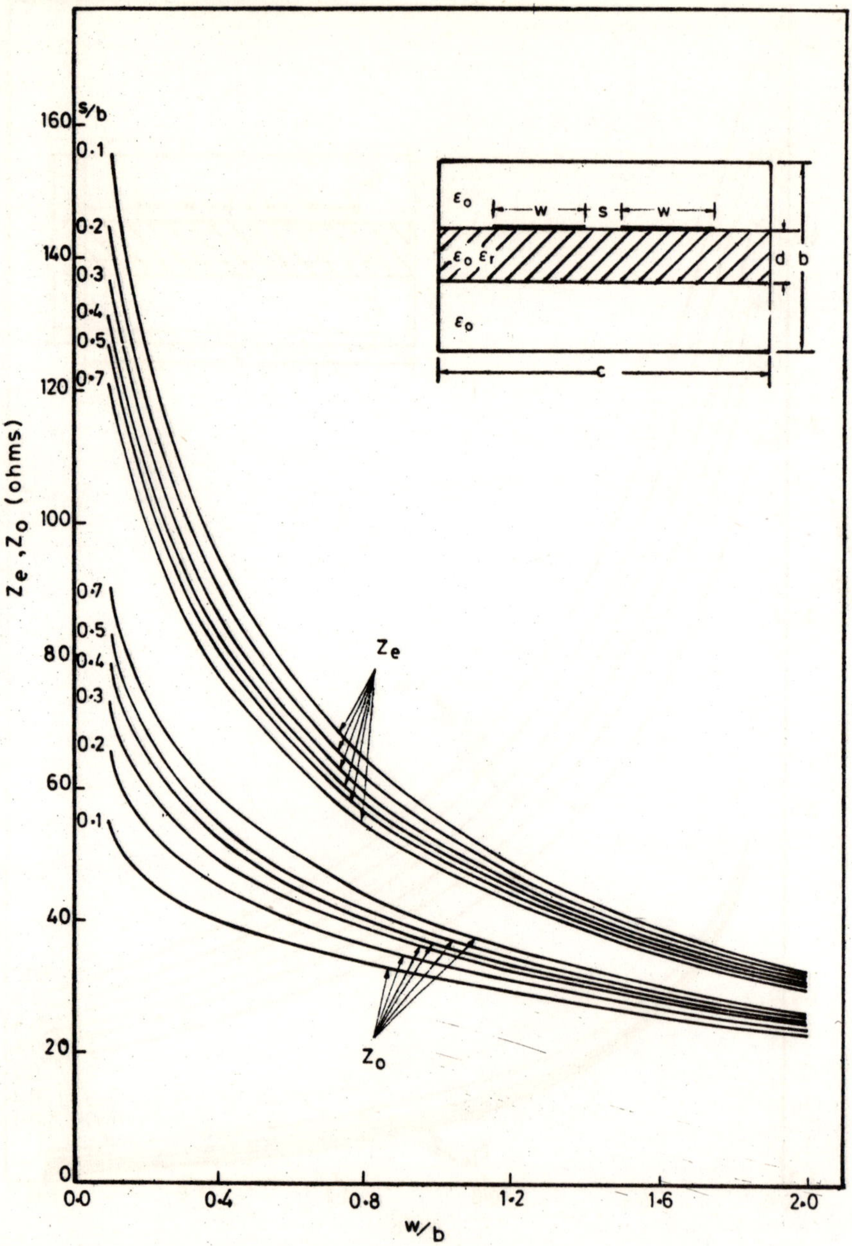

FIG. 6.12(a) Z_e and Z_o versus w/b of edge-coupled suspended stripline with s/b as parameter; $c/(2w + s) = 20$, $\epsilon_r = 9.6$, $d/b = 0.3$.

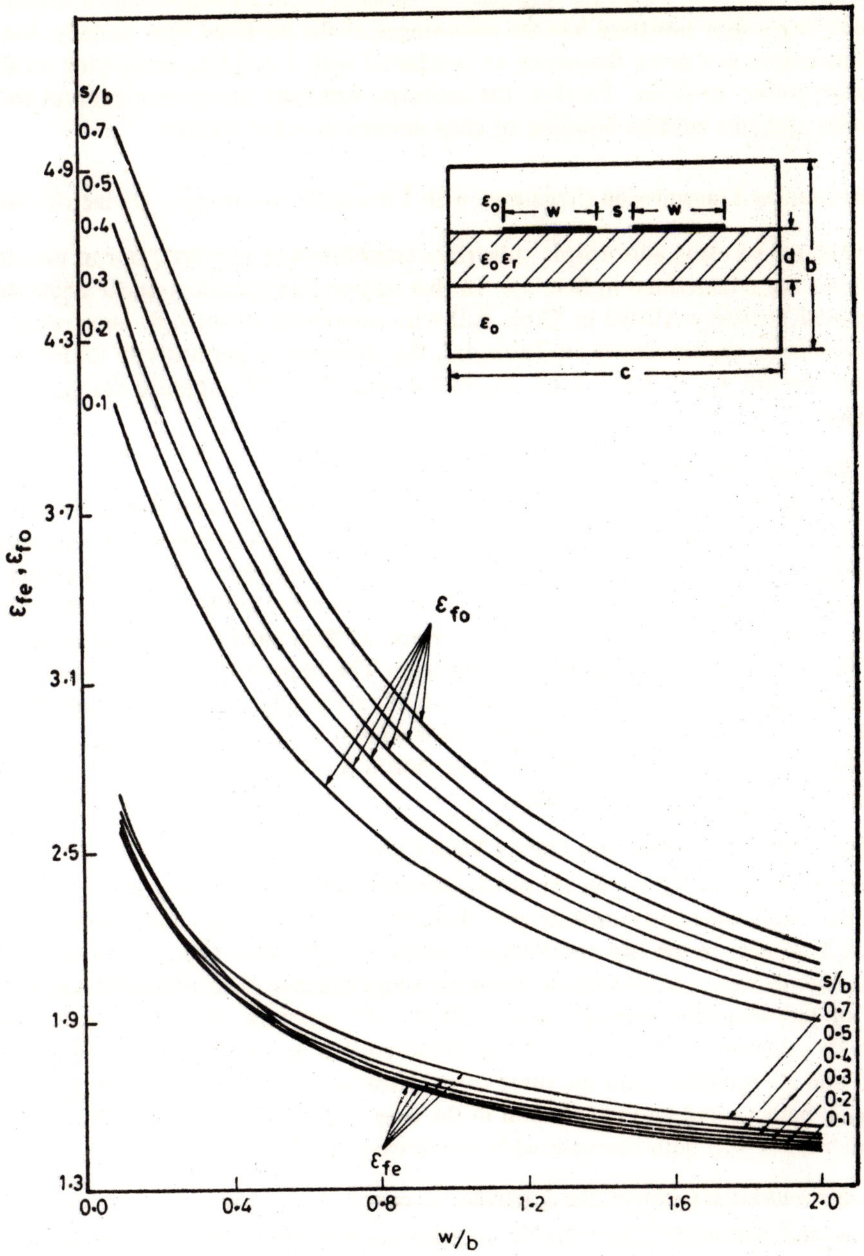

FIG. 6.12(b) ϵ_{fe} and ϵ_{fo} versus w/b of edge-coupled suspended stripline with s/b as parameter; $c/(2w + s) = 20$, $\epsilon_r = 9.6$, $d/b = 0.3$.

Figs. 6.10, 6.11 and 6.12. The characteristics of the suspended stripline approach that of the air-filled stripline as the substrate thickness is reduced. Thus, an edge-coupled suspended stripline employing a thin substrate has the advantages of the air-filled line, namely, low loss and larger dimensions at a given frequency as compared with a stripline employing solid dielectric or microstrip-like stripline. Further, the dielectric substrate serves as a support for the strip conductors and also enables bonding of chip devices in active circuits.

6.4 Edge-coupled Transmission Structures with Uniaxially Anisotropic Dielectric Substrates

The advantages of using anisotropic substrates (sapphire and pyrolytic boron nitride) in MIC applications are enumerated in Sec. 5.4. In this section, we present typical characteristics of edge-coupled structures (listed in Table 4.2) with uniaxially anisotropic substrates. Referring to the coordinate system shown in Table 4.2, the elements of permittivity tensor of pyrolytic boron nitride are $\epsilon_{xx} = \epsilon_{zz} = 5.12$, $\epsilon_{yy} = 3.4$ and those of sapphire are $\epsilon_{xx} = \epsilon_{zz} = 9.4$, $\epsilon_{yy} = 11.6$.

6.4.1 Edge-coupled Symmetric Stripline

Figure 6.13 shows the variations in characteristic impedances Z_e and Z_o and the phase velocity ratio v_e/v_o as a function of w/b for an edge-coupled symmetric stripline homogeneously filled with pyrolytic boron nitride. As in the case of isotropic structure, Z_e is greater than Z_o and both Z_e and Z_o decrease with an increase in w/b. However, unlike the isotropic counterpart, v_e is not equal to v_o. The ratio v_e/v_o is slightly greater than unity and increases with an increase in w/b for small values of w/b and then decreases. The impedance characteristics of the structure homogeneously filled with sapphire (Fig 6.14) are similar to those displayed in Fig. 6.13. However, the variation in v_o/v_e in Fig. 6.14 is similar to that in v_e/v_o in Fig. 6.13. This is an expected trend, since, in the case of pyrolytic boron nitride ϵ_{xx} is greater than ϵ_{yy} whereas the converse is true in the case of sapphire.

6.4.2 Edge-coupled Stripline with Off-set Strip Conductor

Structure A2 of Table 4.2 shows the cross-sectional view of edge-coupled stripline homogeneously filled with uniaxially anisotropic dielectric and with strip conductors asymmetrically placed with respect to the top and bottom conductors. Figures 6.15a, 6.15b and 6.15c show the variations in even- and odd-mode impedances and effective dielectric constants as a function of d/b of the structure homogeneously filled with sapphire, pyrolytic boron nitride and epsilam-10 substrates, respectively. It is observed that as the off-set distance d is increased, both Z_e and Z_o decrease in all the three cases. However, in the case of sapphire (for which $\epsilon_{xx} < \epsilon_{yy}$), both ϵ_{fe} and ϵ_{fo} increase and in the case of pyrolytic boron nitride and epsilam-10 (for which $\epsilon_{xx} > \epsilon_{yy}$) both decrease with an increase in d/b.

6.4.3 Edge-coupled Microstrip-like Stripline

The even- and odd-mode characteristic impedances and effective dielectric constants as a function of w/b of edge-coupled microstrip-like stripline employing uniaxially anisotropic dielectric substrate are plotted in Figs. 6.16a and 6.16b, respectively, with pyrolytic boron nitride as the dielectric and in Figs. 6.17a and 6.17b, respectively, with sapphire as the dielectric. In both the cases, for a fixed value of s/b, both Z_e and Z_o decrease with an increase in w/b and for a fixed value of w/b, Z_e decreases and Z_o increases with an increase in s/b. In the case of pyrolytic

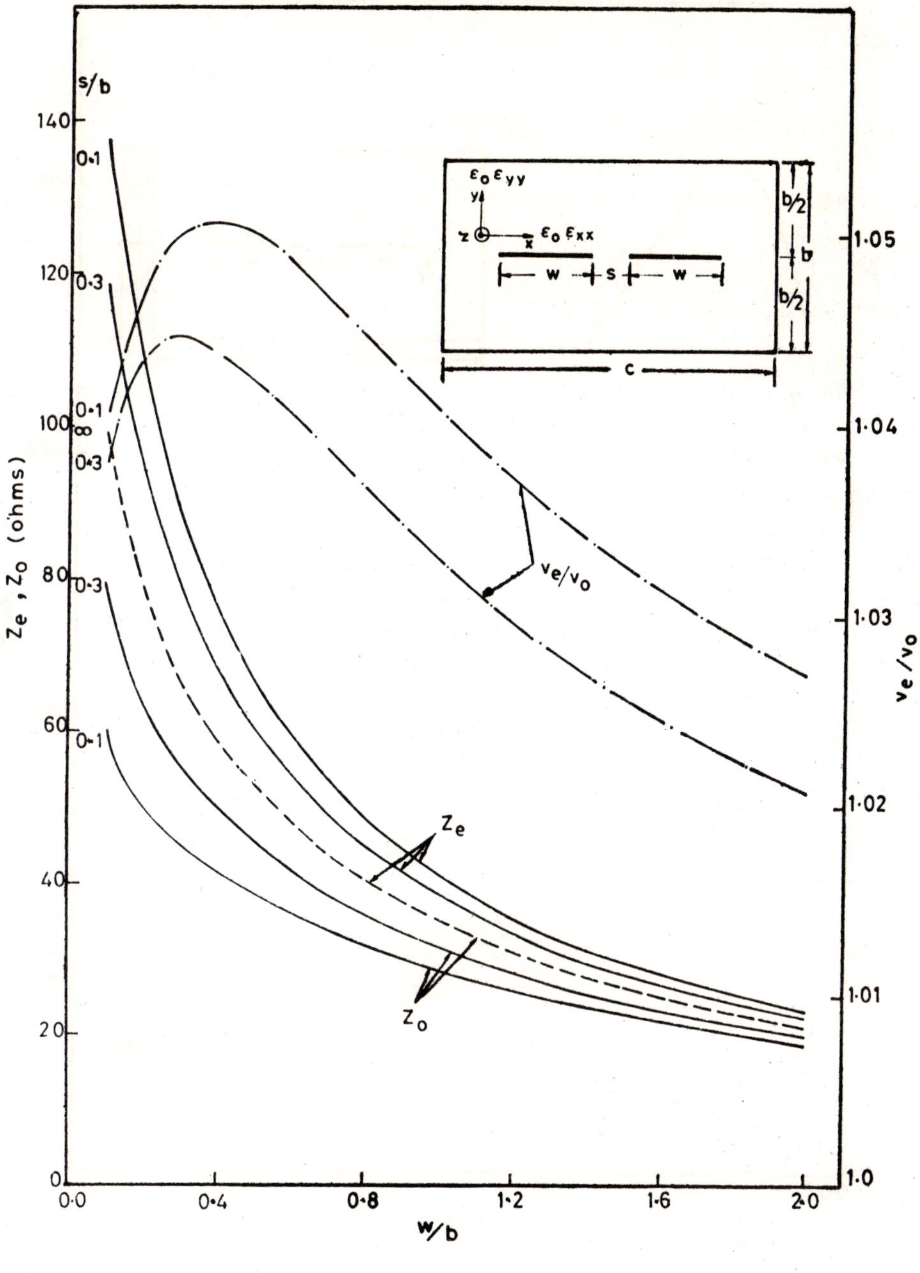

FIG. 6.13 Characteristic impedances Z_e and Z_o and phase velocity ratio v_e/v_o versus w/b of edge-coupled symmetric stripline with uniaxially anisotropic dielectric (pyrolytic boron nitride); $c/(2w + s) = 20$, $\epsilon_{xx} = \epsilon_{zz} = 5.12$, $\epsilon_{yy} = 3.4$.

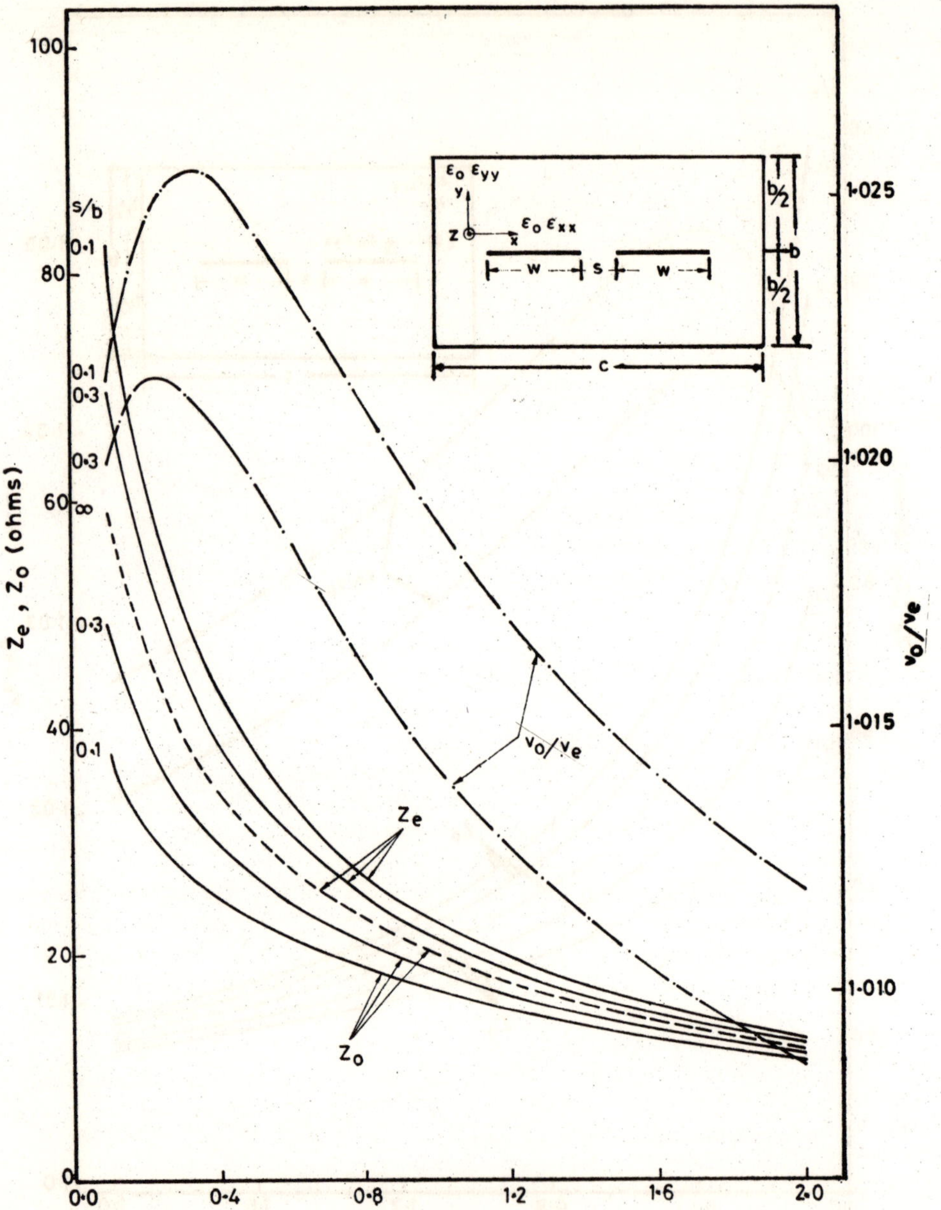

FIG. 6.14 Characteristic impedances Z_e and Z_o and phase velocity ratio v_o/v_e versus w/b of edge-coupled symmetric stripline with uniaxially anisotropic dielectric (sapphire); $c/(2w + s) = 20$, $\epsilon_{xx} = \epsilon_{zz} = 9.4$, $\epsilon_{yy} = 11.6$.

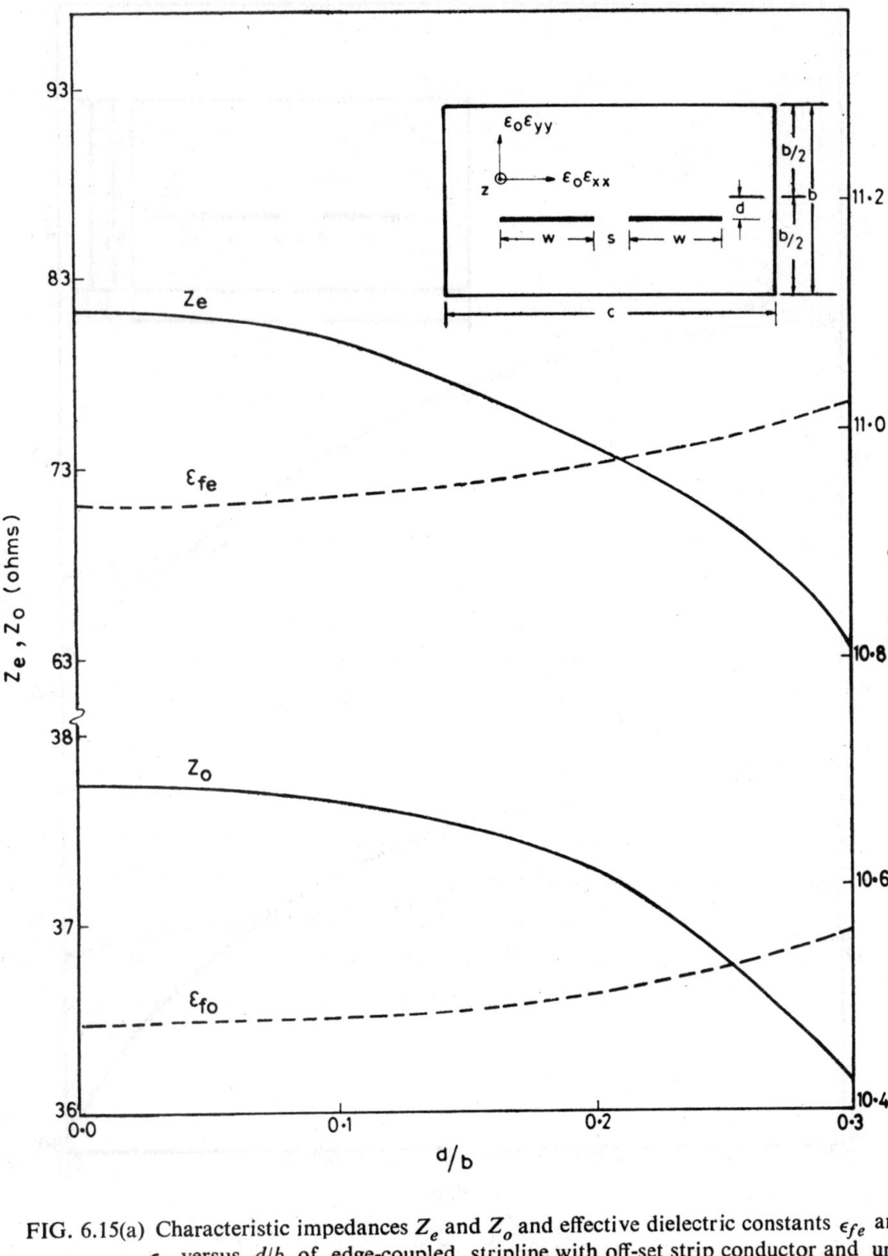

FIG. 6.15(a) Characteristic impedances Z_e and Z_o and effective dielectric constants ϵ_{fe} and ϵ_{fo} versus d/b of edge-coupled stripline with off-set strip conductor and uniaxially anisotropic dielectric substrate (sapphire); $c/(2w+s) = 20$, $s/b = 0.1$, $w/b = 0.1$, $\epsilon_{xx} = \epsilon_{zz} = 9.4$, $\epsilon_{yy} = 11.6$.

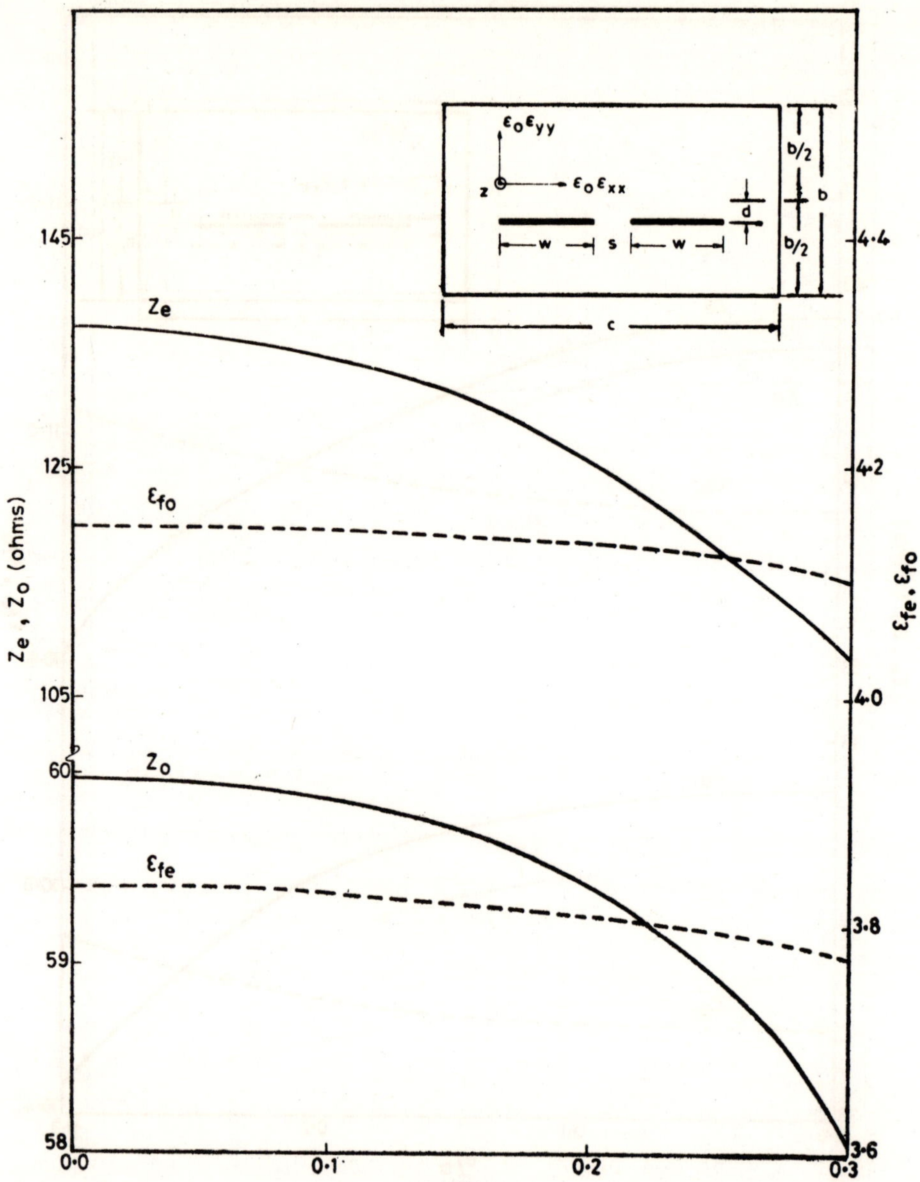

FIG. 6.15 (b) Characteristic impedances Z_e and Z_o and effective dielectric constants ϵ_{fe} and ϵ_{fo} versus d/b of edge-coupled stripline with off-set strip conductor and uniaxially anisotropic dielectric substrate (pyrolytic boron nitride); $c/(2w + s) = 20$, $s/b = 0.1$, $w/b = 0.1$, $\epsilon_{xx} = \epsilon_{zz} = 5.12$, $\epsilon_{yy} = 3.4$.

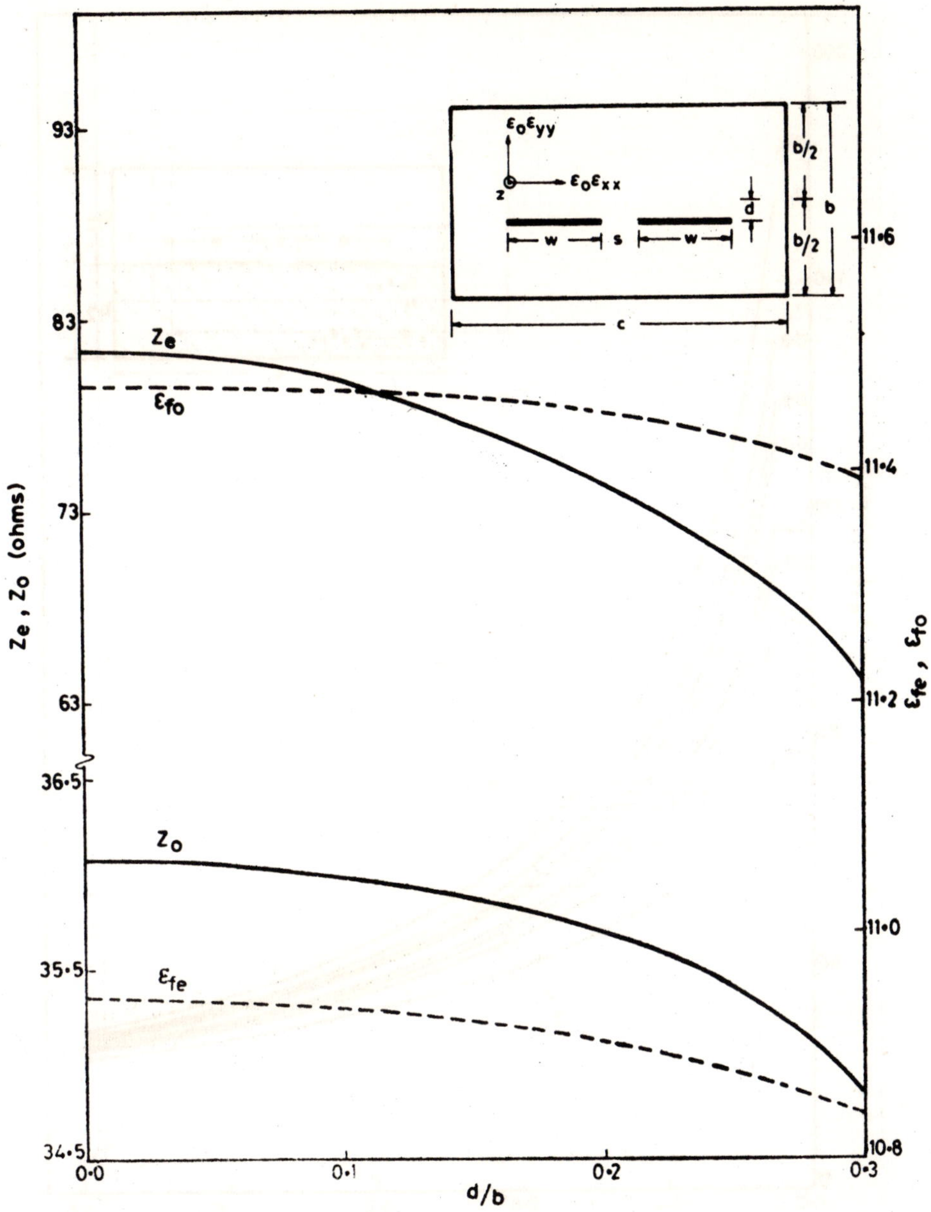

FIG. 6.15(c) Characteristic impedances Z_e and Z_o and effective dielectric constants ϵ_{fe} and ϵ_{fo} versus d/b of edge-coupled stripline with off-set strip conductor and uniaxially anisotropic dielectric substrate (epsilam-10); $c/(2w + s) = 20$, $s/b = 0.1$, $w/b = 0.1$, $\epsilon_{xx} = \epsilon_{zz} = 13$, $\epsilon_{yy} = 10.2$.

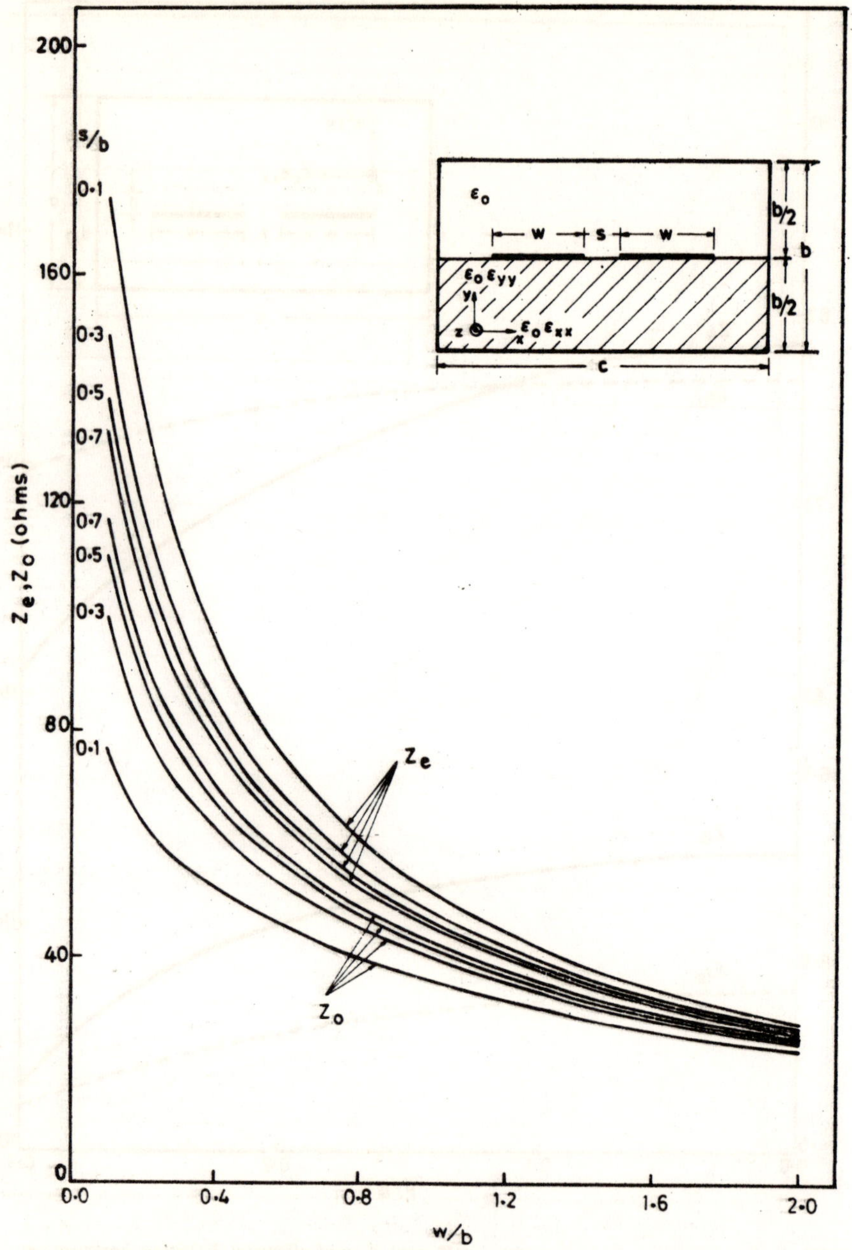

FIG. 6.16(a) Z_e and Z_o versus w/b of edge-coupled microstrip-like stripline with uniaxially anisotropic dielectric substrate (pyrolytic boron nitride); $c/(2w + s) = 20$, $\epsilon_{xx} = \epsilon_{zz} = 5.12$, $\epsilon_{yy} = 3.4$.

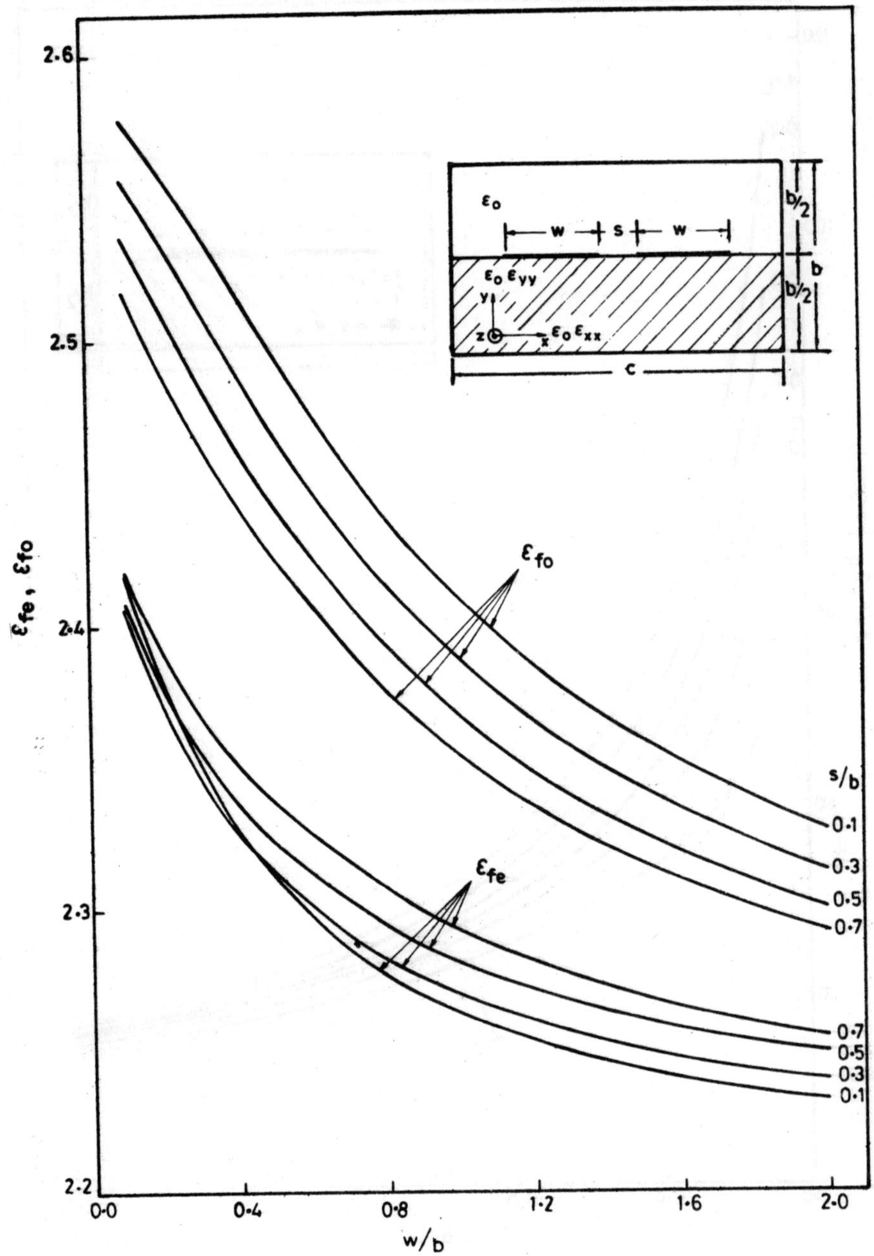

FIG. 6.16(b) ϵ_{fe} and ϵ_{fo} versus w/b of edge-coupled microstrip-like stripline with uniaxially anisotropic dielectric substrate (pyrolytic boron nitride); $c/(2w + s) = 20$, $\epsilon_{xx} = \epsilon_{zz} = 5.12$, $\epsilon_{yy} = 3.4$.

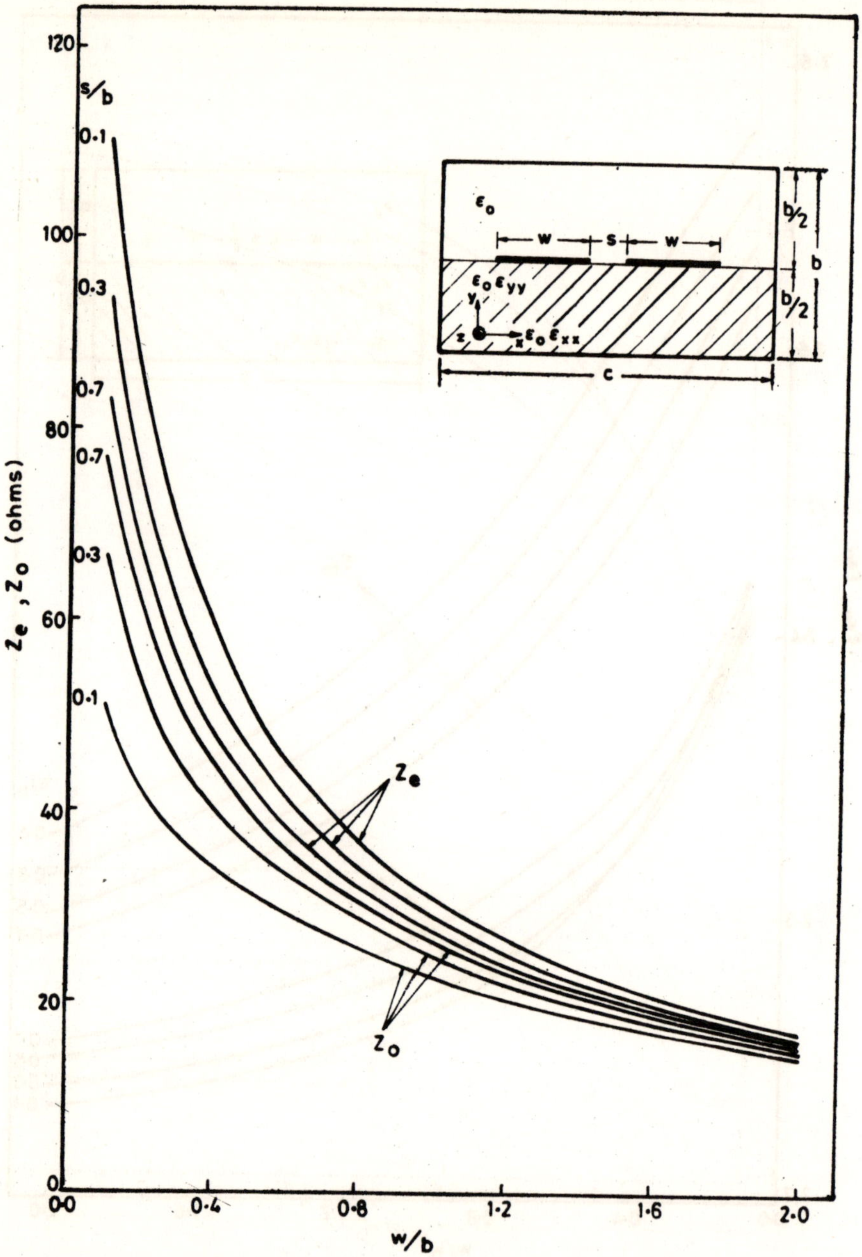

FIG. 6.17(a) Z_e and Z_o versus w/b of edge-coupled microstrip-like stripline with uniaxially anisotropic dielectric substrate (sapphire); $c/(2w + s) = 20$, $\epsilon_{xx} = \epsilon_{zz} = 9.4$, $\epsilon_{yy} = 11.6$.

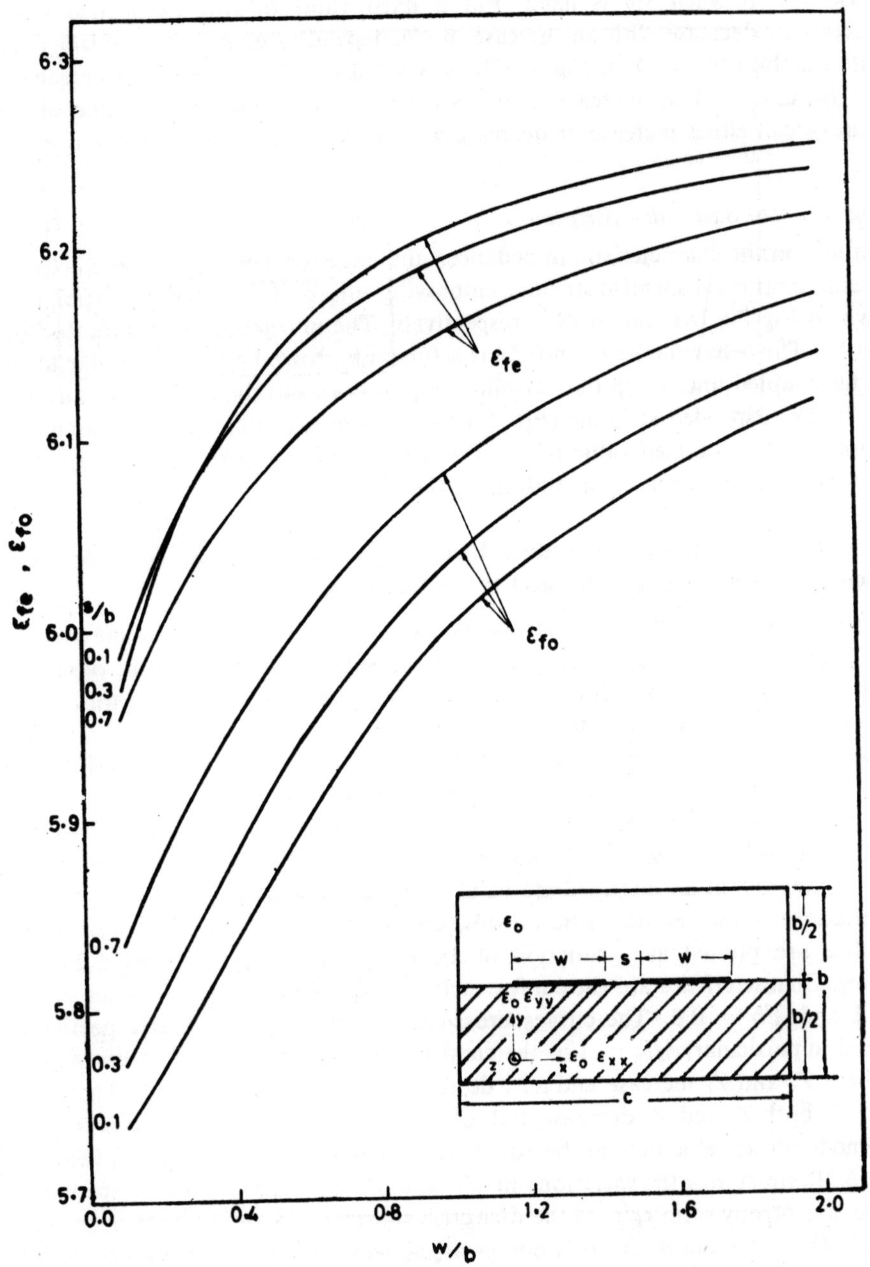

FIG. 6.17(b) ϵ_{fe} and ϵ_{fo} versus w/b of edge-coupled microstrip-like stripline with uniaxially anisotropic dielectric substrate (sapphire); $c/(2w + s) = 20$, $\epsilon_{xx} = \epsilon_{zz} = 9.4$, $\epsilon_{yy} = 11.6$.

boron nitride ($\epsilon_{xx} > \epsilon_{yy}$), Fig. 6.16b shows that ϵ_{fo} is greater than ϵ_{fe} and both decrease with an increase in w/b when s/b is fixed. For a fixed value of w/b, ϵ_{fo} decreases, but ϵ_{fe} can either increase or decrease with an increase in s/b, depending on the value of w/b. On the other hand, with sapphire ($\epsilon_{xx} < \epsilon_{yy}$), Fig. 6.17b shows that ϵ_{fe} of the structure is greater than ϵ_{fo} and both increase with an increase in w/b when s/b is fixed. For a fixed value of w/b, ϵ_{fo} increases but ϵ_{fe} can either increase or decrease with an increase in s/b depending on the value of w/b.

6.4.4 *Edge-coupled Suspended Stripline*

The variations in the characteristic impedances and effective dielectric constants as a function of w/b of edge-coupled suspended stripline employing uniaxially anisotropic sapphire substrate are shown in Figs. 6.18a and 6.18b, respectively. The normalized substrate thickness d/b is chosen as 0.1. The behaviour of Z_e and Z_o as a function of w/b and s/b is similar to that in the case of edge-coupled microstrip-like stripline (Fig. 6.17a), but the trend in ϵ_{fe} and ϵ_{fo} is quite different. In the suspended stripline (Fig. 6.18b), ϵ_{fo} is greater than ϵ_{fe} and both decrease with an increase in w/b for a fixed value of s/b. For a fixed value of w/b, ϵ_{fo} decreases, whereas ϵ_{fe} can either increase or decrease, depending on the value of w/b.

6.5 Edge-coupled Transmission Structures Using Anisotropic Dielectric Substrates with Principal Axes Rotated with Respect to the Coordinate Axes

Anisotropic dielectric substrates with principal axes rotated with respect to the coordinate axes offer an important advantage in edge-coupled transmission lines, in terms of equalizing the even- and odd-mode phase velocities. Characteristics demonstrating this feature are discussed here for some of the structures listed in Table 4.3.

Figure 6.19 illustrates the variations in Z_e, Z_o and v_e/v_o of edge-coupled stripline filled with pyrolytic boron nitride as a function of the tilt angle θ. Plots are shown for stripline with symmetrically placed strip conductors ($d/b = 0$) as well as with off-set strip conductors ($d/b = 0.2$). In both the cases, Z_e, Z_o and v_e/v_o decrease as θ is varied from 0° to 90° and the curves for v_e/v_o cross at an intermediate value of θ. At the crossing $v_e/v_o = 1$.

The variations in the characteristic impedances and phase velocity ratio v_o/v_e of microstrip-like stripline are plotted as a function of anisotropy ratio $\epsilon_\xi/\epsilon_\eta$ in Figs. 6.20a and 6.20b, respectively. Characteristics are shown for various values of θ, for fixed values of $\epsilon_\eta = 3.78$, $w/b = 0.1$ and $s/b = 0.3$. The curves cross each other at $\epsilon_\xi/\epsilon_\eta = 1$, irrespective of the tilt angle θ and in particular, $v_o/v_e = 1$ at the crossing. Further, for a fixed value of $\epsilon_\xi/\epsilon_\eta$ as θ increases, both Z_e and Z_o increase and v_o/v_e decreases in the range $\epsilon_\xi/\epsilon_\eta < 1$, whereas for values of $\epsilon_\xi/\epsilon_\eta > 1$, both Z_e and Z_o decrease and v_o/v_e increases. For values of $\epsilon_\xi/\epsilon_\eta \neq 1$, the even- and odd-mode phase velocities can be equalized by choosing an appropriate value of θ.

Figure 6.21a illustrates the variations in Z_e and Z_o of edge-coupled suspended stripline versus the anisotropy ratio $\epsilon_\xi/\epsilon_\eta$ of the dielectric substrate. With the parameters, $\epsilon_\eta = 3.78$, $w/b = 0.5$, $s/b = 0.3$ and $d/b = 0.2$, curves are shown for different values of θ. The impedance curves are similar to those for the edge-coupled microstrip-like stripline (Fig. 6.20a). The variation in the phase velocity ratio v_e/v_o of edge-coupled suspended stripline as a function of $\epsilon_\xi/\epsilon_\eta$ (Fig. 6.21b) is different compared to that for the edge-coupled microstrip-like stripline (Fig. 6.20b). It can be seen from Fig. 6.21b that the ratio v_e/v_o is not unity at $\epsilon_\xi/\epsilon_\eta = 1$. The even- and odd-mode phase velocities can, however, be equalized by choosing values of $\epsilon_\xi/\epsilon_\eta \leqslant 1$ and an appropriate value of θ.

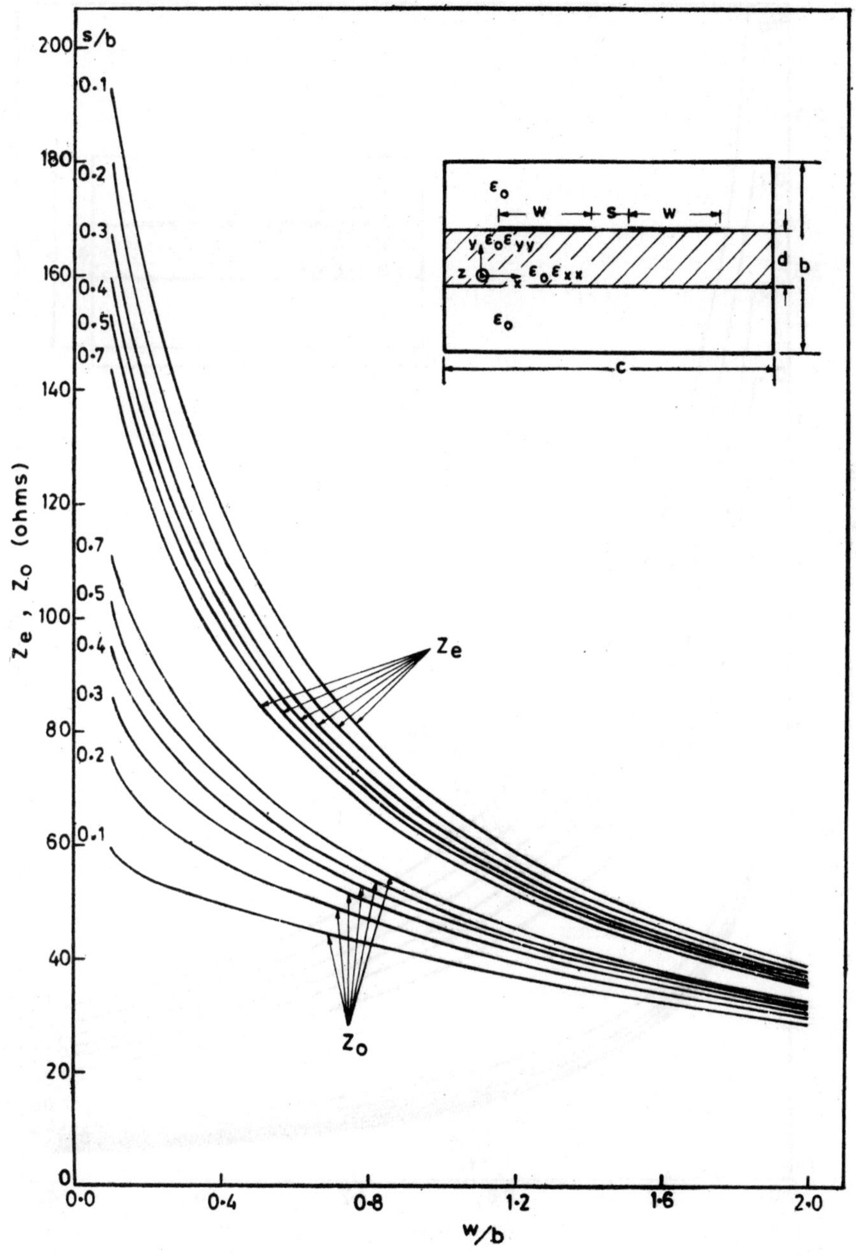

FIG. 6.18(a) Z_e and Z_o versus w/b of edge-coupled suspended stripline with uniaxially anisotropic dielectric substrate (sapphire); $c/(2w + s) = 20$, $\epsilon_{xx} = \epsilon_{zz} = 9.4$, $\epsilon_{yy} = 11.6$, $d/b = 0.1$.

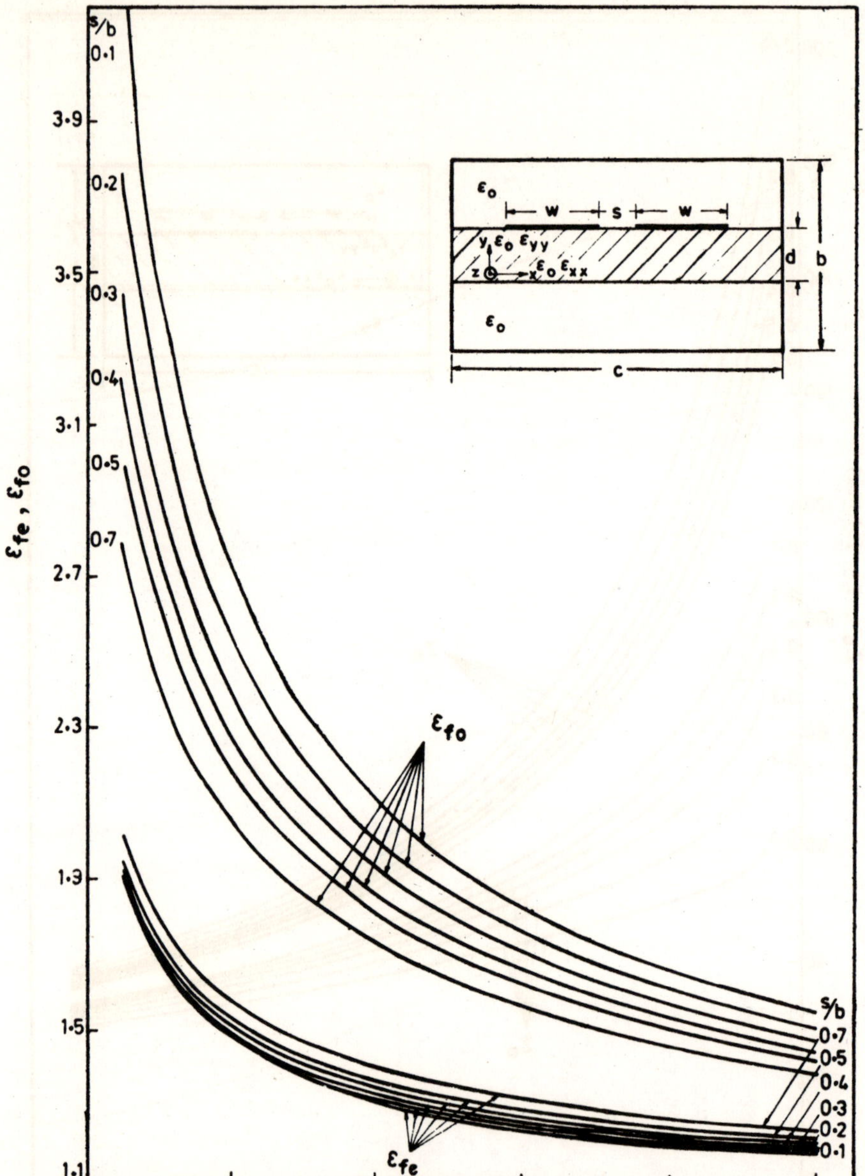

FIG. 6.18(b) ϵ_{fe} and ϵ_{fo} versus w/b of edge-coupled suspended stripline with uniaxially anisotropic dielectric substrate (sapphire); $c/(2w + s) = 20$, $\epsilon_{xx} = \epsilon_{zz} = 9.4$, $\epsilon_{yy} = 11.6$, $d/b = 0.1$.

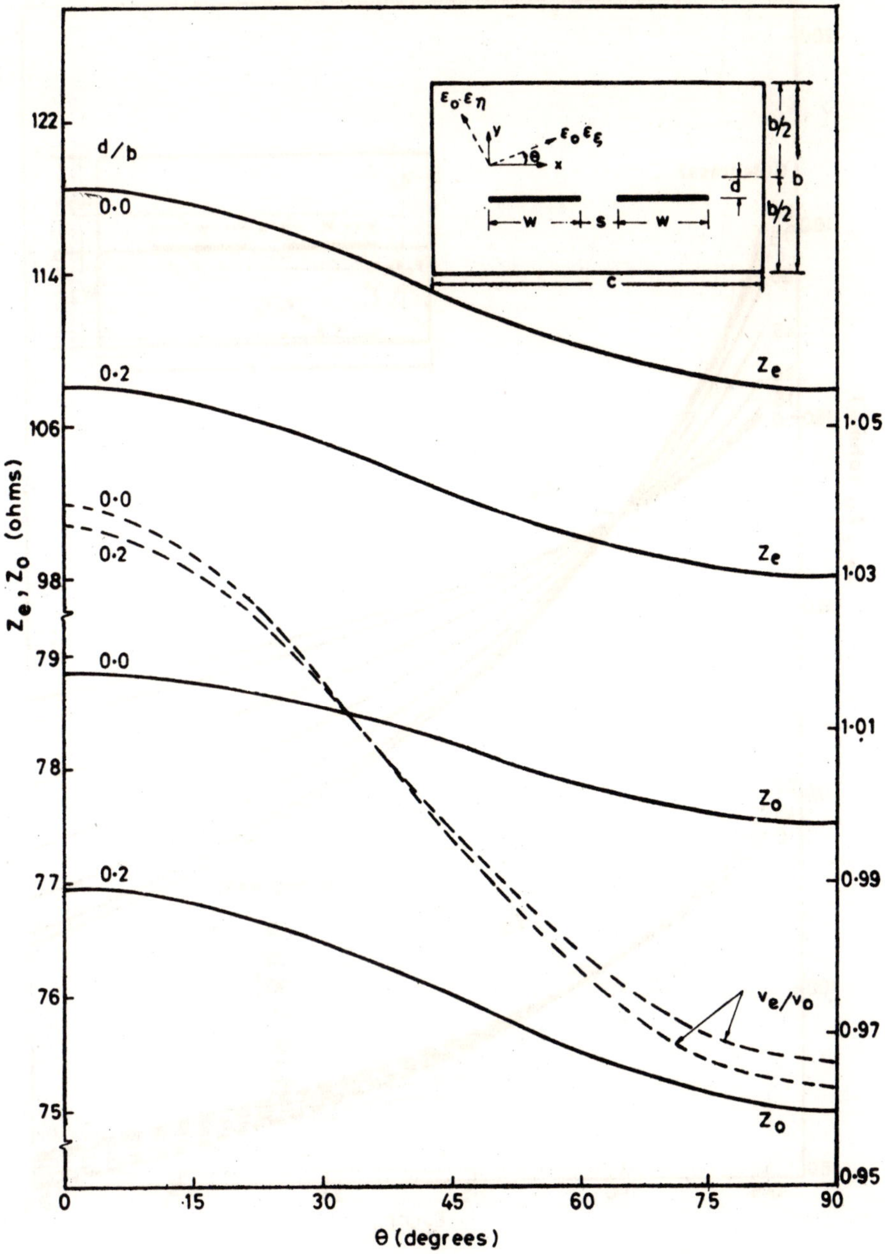

FIG. 6.19 Characteristic impedances Z_e and Z_o and phase velocity ratio v_e/v_o versus angle θ of edge-coupled stripline filled with pyrolytic boron nitride; $c/(2w + s) = 20$, $\epsilon_\xi = 5.12$, $\epsilon_\eta = 3.4$ at $\theta = 0$, $w/b = 0.1$, $s/b = 0.3$.

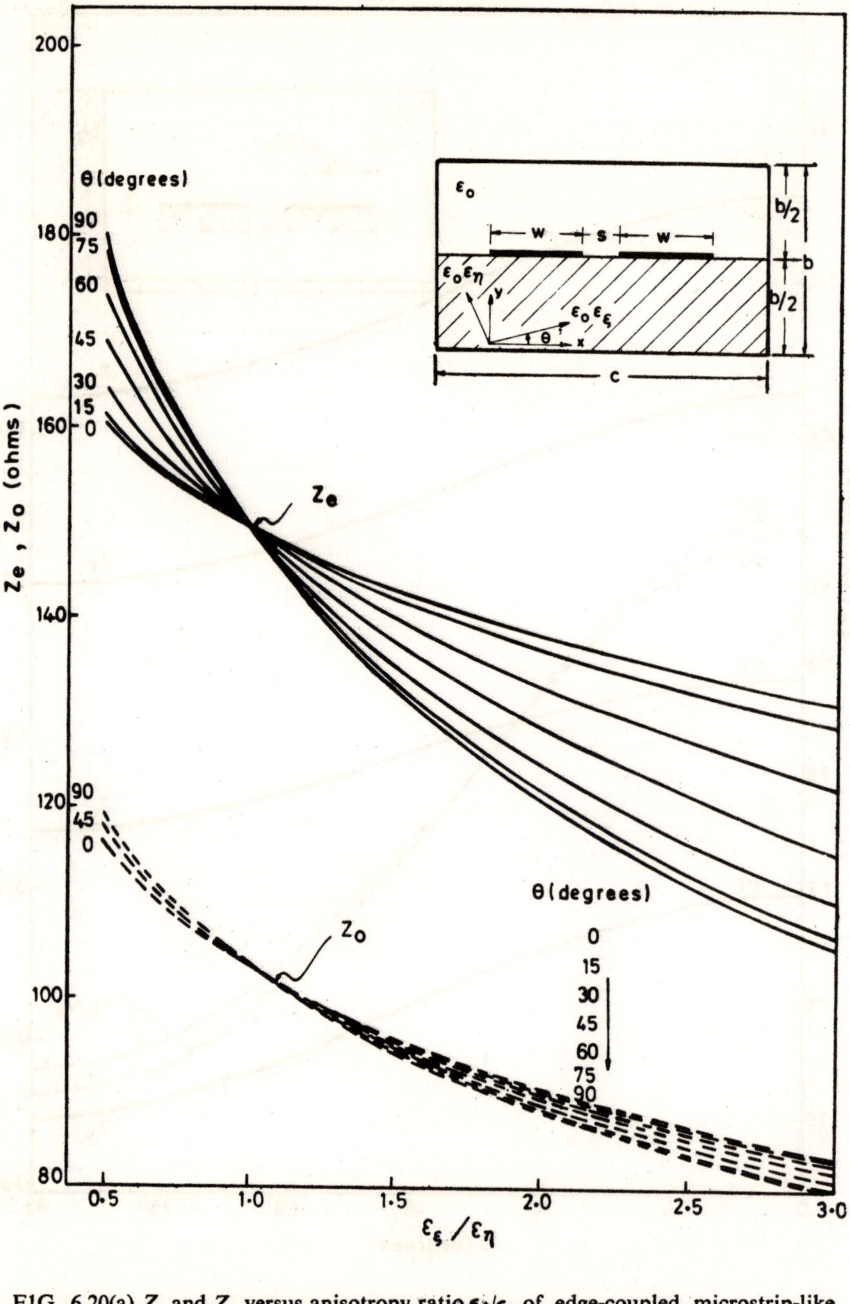

FIG. 6.20(a) Z_e and Z_o versus anisotropy ratio $\epsilon_\xi/\epsilon_\eta$ of edge-coupled microstrip-like stripline with anisotropic dielectric substrate; $c/(2w + s) = 20$, $\epsilon_\eta = 3.78$, $w/b = 0.1$, $s/b = 0.3$.

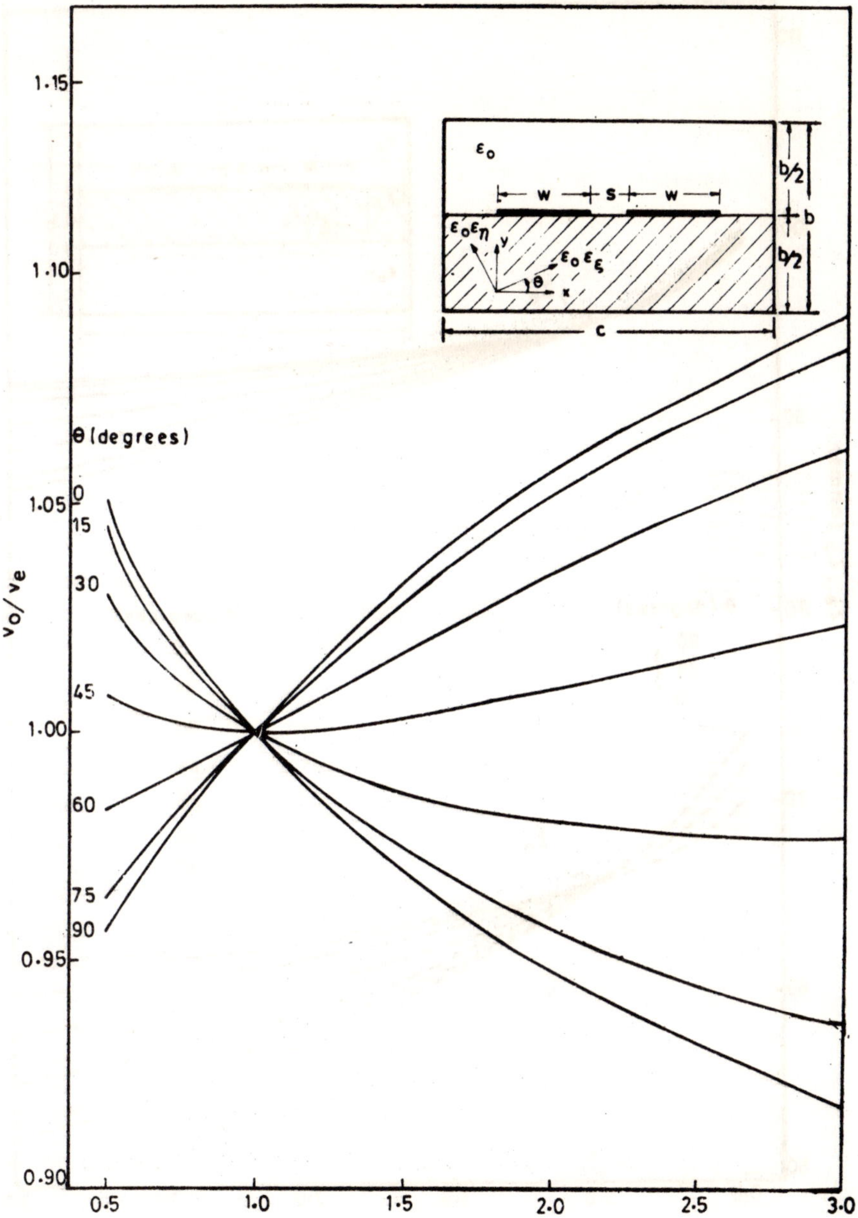

FIG. 6.20(b) Phase velocity ratio v_o/v_e versus anisotropy ratio $\epsilon_\xi/\epsilon_\eta$ of edge-coupled microstrip-like stripline with anisotropic dielectric substrate; $c/(2w + s) = 20$, $\epsilon_\eta = 3.78$, $w/b = 0.1$, $s/b = 0.3$.

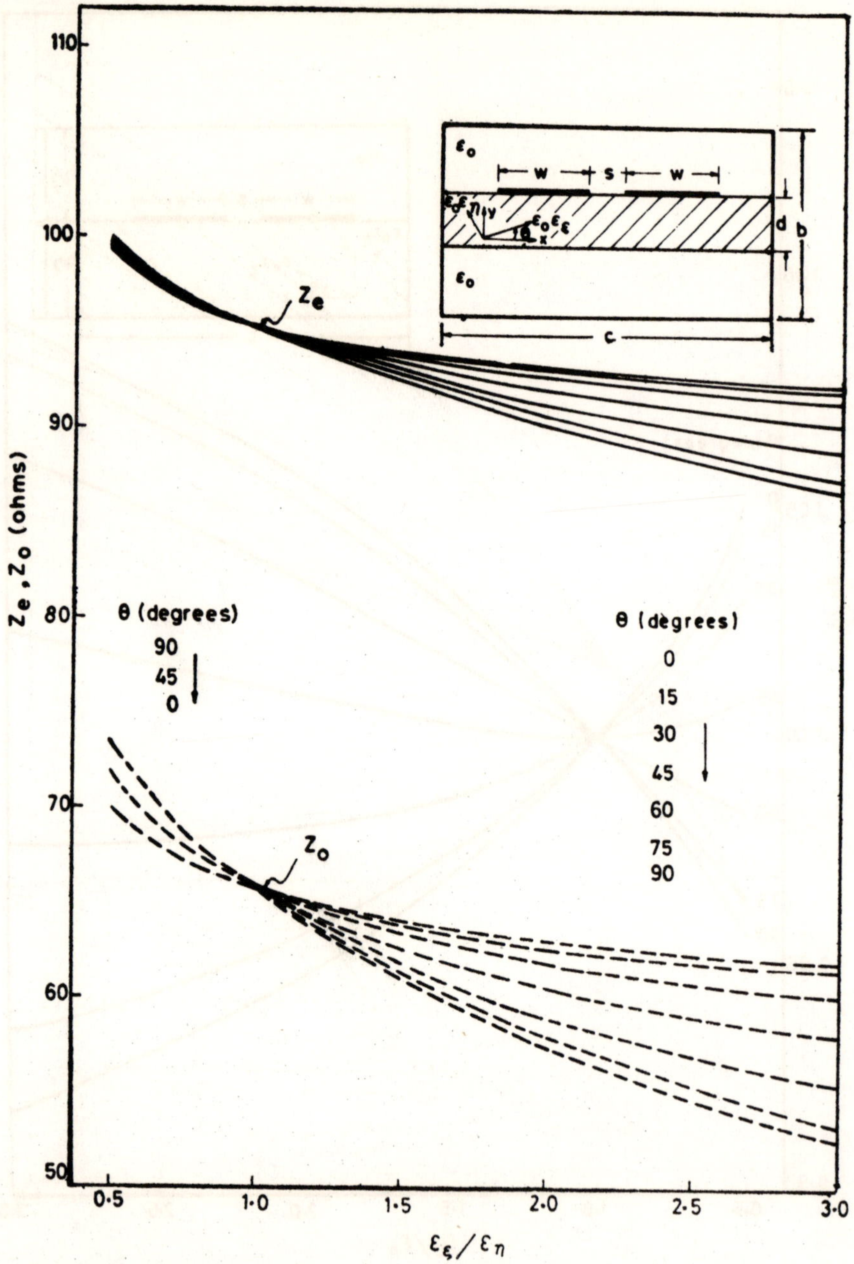

FIG. 6.21(a) Z_e and Z_o versus anisotropy ratio $\epsilon_\xi/\epsilon_\eta$ of edge-coupled suspended stripline with anisotropic dielectric substrate; $c/(2w + s) = 20$, $\epsilon_\eta = 3.78$, $w/b = 0.5$, $s/b = 0.3$, $d/b = 0.2$.

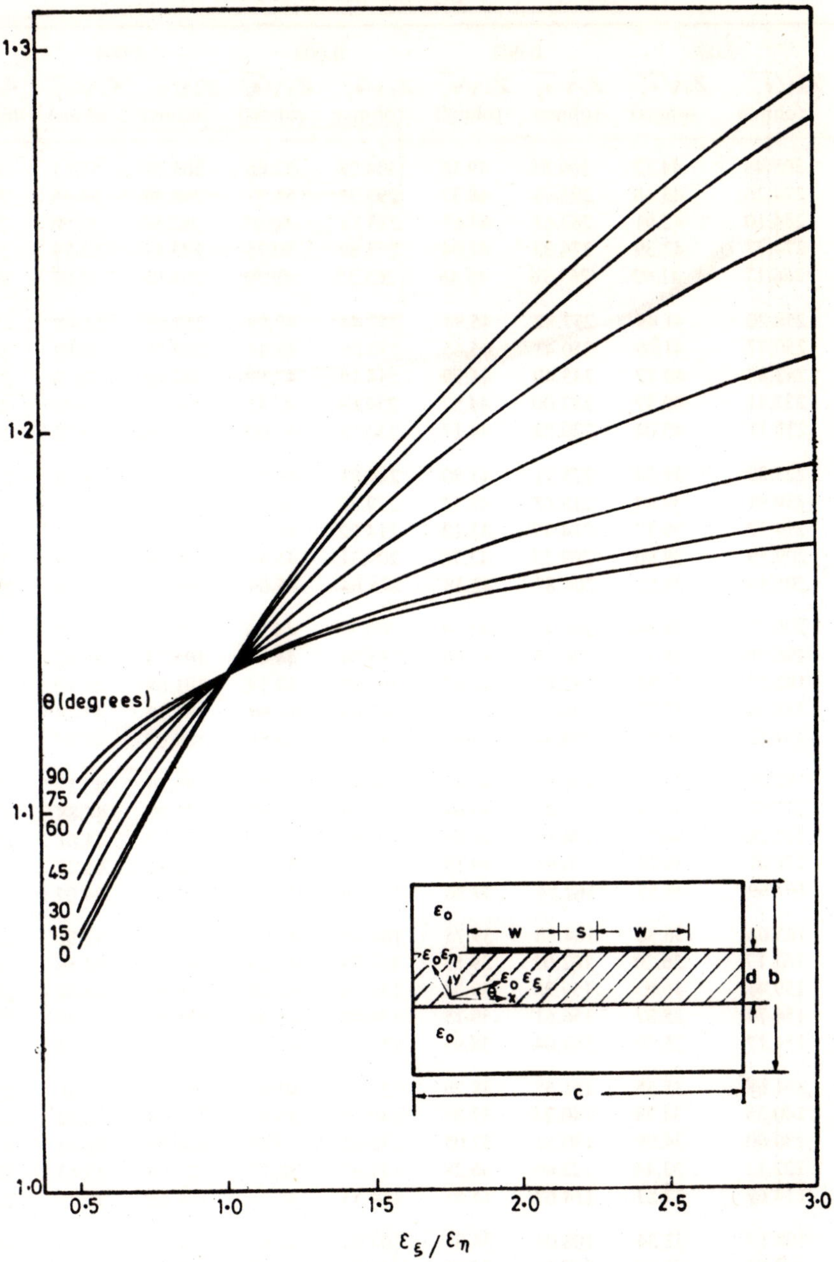

FIG. 6.21(b) v_e/v_o versus anisotropy ratio $\epsilon_\xi/\epsilon_\eta$ of edge-coupled suspended stripline with anisotropic dielectric substrate; $c/(2w + s) = 20$, $\epsilon_\eta = 3.78$, $w/b = 0.5$, $s/b = 0.3$, $d/b = 0.2$.

Table 6.1 Characteristic impedances of edge-coupled symmetric stripline; $c/(2w + s) = \infty$ (See structure A1 of Table 4.1)

s/b / w/b	0.001		0.002		0.003		0.004		0.005	
	$Z_e\sqrt{\epsilon_r}$ (ohms)	$Z_o\sqrt{\epsilon_r}$ (ohms)	$Z_e\sqrt{\epsilon_r}$ (ohms)	$Z_o\sqrt{\epsilon_r}$ (ohms)	$Z_e\sqrt{\epsilon_r}$ (ohms)	$Z_o\sqrt{\epsilon_r}$ (ohms)	$Z_e\sqrt{\epsilon_r}$ (ohms)	$Z_o\sqrt{\epsilon_r}$ (ohms)	$Z_e\sqrt{\epsilon_r}$ (ohms)	$Z_o\sqrt{\epsilon_r}$ (ohms)
0.10	305.44	44.12	304.85	49.16	304.28	52.66	303.70	55.45	303.15	57.82
0.11	294.26	43.48	293.73	48.37	293.21	51.77	292.70	54.46	292.19	56.75
0.12	284.10	42.91	283.61	47.67	283.14	50.97	282.66	53.58	282.19	55.79
0.13	274.77	42.39	274.33	47.04	273.89	50.25	273.45	52.79	273.02	54.94
0.14	266.17	41.92	265.76	46.46	265.35	49.59	264.95	52.07	264.55	54.16
0.15	258.20	41.49	257.82	45.94	257.44	48.99	257.06	51.41	256.69	53.45
0.16	250.77	41.09	250.41	45.45	250.06	48.44	249.71	50.80	249.36	52.80
0.17	243.83	40.72	243.49	44.99	243.16	47.92	242.83	50.24	242.50	52.19
0.18	237.31	40.37	237.00	44.57	236.68	47.45	236.37	49.71	236.07	51.62
0.19	231.11	40.04	230.88	44.17	230.59	47.00	230.29	49.22	230.00	51.09
0.20	225.39	39.74	225.11	43.80	224.83	46.57	224.55	48.76	224.28	50.60
0.21	219.91	39.44	219.65	43.44	219.38	46.17	219.12	48.32	218.86	50.13
0.22	214.72	39.17	214.47	43.10	214.22	45.79	213.97	47.91	213.72	49.68
0.23	209.79	38.90	209.55	42.78	209.31	45.43	209.08	47.51	208.84	49.26
0.24	205.10	38.65	204.87	42.48	204.64	45.09	204.42	47.14	204.19	48.85
0.25	200.62	38.40	200.41	42.18	200.19	44.76	199.98	46.78	199.76	48.47
0.26	196.35	38.17	196.15	41.90	195.94	44.44	195.73	46.43	195.53	48.10
0.27	192.27	37.94	192.07	41.63	191.88	44.14	191.68	46.10	191.48	47.74
0.28	188.36	37.72	188.17	41.37	187.98	43.84	187.80	45.78	187.61	47.40
0.29	184.62	37.51	184.44	41.12	184.25	43.56	184.07	45.47	183.89	47.07
0.30	181.03	37.31	180.85	40.87	180.68	43.29	180.50	45.17	180.33	46.75
0.31	177.58	37.11	177.41	40.64	177.24	43.02	177.07	44.88	176.91	46.44
0.32	174.26	36.92	174.10	40.41	173.99	42.76	173.77	44.61	173.61	46.14
0.33	171.07	36.73	170.91	40.18	170.75	42.51	170.60	44.33	170.44	45.85
0.34	167.99	36.55	167.84	39.96	167.69	42.27	167.54	44.07	167.39	45.57
0.35	165.03	36.38	164.88	39.75	164.74	42.03	164.59	43.81	164.45	45.30
0.36	162.17	36.20	162.03	39.55	161.89	41.80	161.75	43.56	161.61	45.03
0.37	159.41	36.03	159.28	39.34	159.14	41.58	159.01	43.32	158.87	44.77
0.38	156.74	35.87	156.61	39.15	156.48	41.36	156.35	43.08	156.23	44.51
0.39	154.17	35.70	154.04	38.95	153.92	41.14	153.79	42.84	153.67	44.26
0.40	151.68	35.55	151.55	38.76	151.43	40.93	151.31	42.61	151.19	44.02
0.45	140.35	34.79	140.24	37.87	140.14	39.93	140.03	41.54	139.93	42.87
0.50	130.60	34.09	130.51	37.05	130.42	39.02	130.33	40.55	130.24	41.82
0.55	122.13	33.44	122.05	36.28	121.97	38.17	121.89	39.63	121.81	40.84
0.60	114.69	32.83	114.62	35.55	114.55	37.37	114.48	38.77	114.41	39.93
0.65	108.10	32.24	108.04	34.87	107.98	36.61	107.92	37.96	107.86	39.07
0.70	102.24	31.69	102.18	34.22	102.13	35.90	102.07	37.19	102.02	38.25
0.75	96.97	31.15	96.92	33.60	96.87	35.21	96.82	36.45	96.77	37.47
0.80	92.22	30.64	92.18	33.00	92.13	34.56	92.09	35.75	92.04	36.73
0.85	87.92	30.14	87.88	32.42	87.84	33.93	87.80	35.08	87.76	36.02

(Contd.)

w/b \ s/b	0.001		0.002		0.003		0.004		0.005	
	$Z_e\sqrt{\epsilon_r}$ (ohms)	$Z_o\sqrt{\epsilon_r}$ (ohms)	$Z_e\sqrt{\epsilon_r}$ (ohms)	$Z_o\sqrt{\epsilon_r}$ (ohms)	$Z_e\sqrt{\epsilon_r}$ (ohms)	$Z_o\sqrt{\epsilon_r}$ (ohms)	$Z_e\sqrt{\epsilon_r}$ (ohms)	$Z_o\sqrt{\epsilon_r}$ (ohms)	$Z_e\sqrt{\epsilon_r}$ (ohms)	$Z_o\sqrt{\epsilon_r}$ (ohms)
0.90	84.00	29.66	83.96	31.87	83.92	33.32	83.89	34.43	83.85	35.34
0.95	80.41	29.20	80.38	31.34	80.34	32.74	80.31	33.81	80.27	34.69
1.00	77.12	28.75	77.09	30.82	77.06	32.18	77.02	33.21	76.99	34.06
1.05	74.09	28.32	74.06	30.32	74.03	31.63	74.00	32.63	73.97	33.45
1.10	71.28	27.91	71.25	29.84	71.23	31.11	71.20	32.08	71.17	32.87
1.15	68.68	27.50	68.66	29.38	68.63	30.61	68.61	31.54	68.58	32.30
1.20	66.27	27.10	66.24	28.93	66.22	30.11	66.20	31.02	66.17	31.76
1.25	64 01	26.72	63.99	28.49	63.97	29.64	63.95	30.52	63.93	31.23
1.30	61.91	26.35	61.89	28.07	61.87	29.19	61.85	30.03	61.83	30.72
1.35	59.94	25.98	59.92	27.65	59.90	28.74	59.88	29.56	59.86	30.23
1.40	58.09	25.63	58.07	27.26	58.05	28.31	58.04	29.10	58.02	29.75
1.45	56.35	25.28	56.34	26.88	56.32	27.89	56.30	28.66	56.29	29.28
1.50	54.72	24.97	54.70	26.50	54.68	27.48	54.67	28.23	54.65	28.84
1.55	53.17	24.62	53.16	26.14	53.14	27.08	53.13	27.81	53.11	28.41
1.60	51.71	24.32	51.70	25.78	51.68	26.71	51.67	27.41	51.65	27.99
1.65	50.33	24.00	50.32	25.44	50.30	26.34	50.29	27.03	50.28	27.57
1.70	49.02	23.70	49.01	25.09	48.99	25.98	48.98	26.64	48.97	27.19
1.75	47.78	23.44	47.76	24.77	47.75	25.63	47.74	26.28	47.73	26 80
1.80	46.59	23.15	46.58	24.45	46.57	25.28	46.56	25.91	46.55	26.42
1.85	45.47	22.90	45.46	24.12	45.45	24.95	45.44	25.57	45.43	26.06
1.90	44.40	22.62	44.39	23.86	44.38	24.61	44.37	25.22	44.36	25.71
1.95	43.37	22.28	43.36	23.56	43.35	24.30	43.34	24.91	43.34	25.37
2.00	42.40	22.07	42.35	23.22	42.38	24.02	42.37	24.58	42.36	25.02
2.05	41.46	21.90	41.46	22.96	41.45	23.73	41.44	24.26	41.43	24.69
2.10	40.57	21.59	40.56	22.70	40.55	23.41	40.55	23.96	40.54	24.40
2.15	39.72	21.43	39.71	22.41	39.70	23.14	39.69	23.69	39.68	24.07
2.20	38.90	21.22	38.89	22.16	38.88	22.87	38.87	23.34	38.86	23.79
2.25	38.11	20.94	38.10	22.00	38 09	22.58	38.09	23.11	38.08	23.47
2.30	37.35	20.68	37.35	21.71	37.34	22.37	37.33	22.77	37.32	23.18
2.35	36.63	20.64	36.62	21.52	36.61	22.12	36.61	22.57	36.60	22.94
2.40	35.93	20.37	35.92	21.16	35.91	21.86	35.91	22.25	35.90	22.68
2.45	35.26	20.32	35.25	20.91	35.24	21.52	35.24	21.98	35.23	22.35
2.50	34.61	20.09	34.60	20.75	34.60	21.41	34.59	21.75	34.58	22.16

(Contd.)

w/b \ s/b	0.01		0.02		0.03		0.04		0.05	
	$Z_e\sqrt{\epsilon_r}$ (ohms)	$Z_o\sqrt{\epsilon_r}$ (ohms)	$Z_e\sqrt{\epsilon_r}$ (ohms)	$Z_o\sqrt{\epsilon_r}$ (ohms)	$Z_e\sqrt{\epsilon_r}$ (ohms)	$Z_o\sqrt{\epsilon_r}$ (ohms)	$Z_e\sqrt{\epsilon_r}$ (ohms)	$Z_o\sqrt{\epsilon_r}$ (ohms)	$Z_e\sqrt{\epsilon_r}$ (ohms)	$Z_o\sqrt{\epsilon_r}$ (ohms)
0.10	300.40	66.51	295.30	77.82	290.62	85.99	286.31	92.62	282.31	98.27
0.11	289.69	65.12	285.01	76.00	280.70	83.85	276.69	90.21	272.97	95.64
0.12	279.90	63.89	275.58	74.38	271.59	81.95	267.86	88.08	264.38	93.32
0.13	270.90	62.79	266.90	72.94	263.18	80.25	259.69	86.18	256.43	91.23
0.14	262.58	61.79	258.86	71.64	255.38	78.72	252.11	84.45	249.04	89.34
0.15	254.86	60.88	251.38	70.45	248.11	77.32	245.04	82.88	242.14	87.62
0.16	247.65	60.05	244.38	69.36	241.31	76.04	238.41	81.44	235.67	86.05
0.17	240.89	59.27	237.82	68.35	234.92	74.85	232.18	80.11	229.58	84.59
0.18	234.55	58.55	231.65	67.41	228.91	73.75	226.31	78.87	223.84	83.24
0.19	228.57	57.88	225.83	66.54	223.23	72.73	220.76	77.72	218.42	81.98
0.20	222.93	57.25	220.32	65.72	217.86	71.77	215.51	76.64	213.27	80.79
0.21	217.58	56.65	215.11	64.95	212.76	70.86	210.52	75.63	208.39	79.68
0.22	212.50	56.09	210.15	64.22	207.92	70.01	205.78	74.67	203.74	78.63
0.23	207.68	55.55	205.44	63.53	203.30	69.20	201.26	73.76	199.31	77.64
0.24	203.09	55.04	200.95	62.87	198.91	68.43	196.95	72.90	195.09	76.70
0.25	198.70	54.56	196.66	62.25	194.71	67.70	192.84	72.08	191.04	75.80
0.26	194.52	54.09	192.56	61.65	190.69	67.00	188.90	71.30	187.18	74.94
0.27	190.51	53.65	188.64	61.08	186.85	66.33	185.12	70.55	183.47	74.12
0.28	186.68	53.22	184.88	60.53	183.16	65.69	181.50	69.83	179.91	73.34
0.29	183.00	52.80	181.28	60.00	179.62	65.07	178.03	69.14	176.50	72.58
0.30	179.47	52.40	177.81	59.49	176.22	64.48	174.69	68.47	173.21	71.86
0.31	176.08	52.02	174.49	59.00	172.95	63.91	171.48	67.83	170.05	71.16
0.32	172.82	51.64	171.29	58.52	169.81	63.35	168.38	67.21	167.01	70.48
0.33	169.68	51.28	168.20	58.06	166.78	62.82	165.40	66.61	164.08	69.83
0.34	166.66	50.93	165.23	57.61	163.85	62.30	162.53	66.03	161.25	69.19
0.35	163.74	50.59	162.36	57.18	161.04	61.79	159.75	65.47	158.52	68.58
0.36	160.93	50.26	159.60	56.76	158.31	61.30	157.07	64.93	155.88	67.98
0.37	158.21	49.93	156.93	56.34	155.68	60.83	154.49	64.39	153.33	67.40
0.38	155.59	49.62	154.34	55.94	153.14	60.36	151.98	63.88	150.86	66.84
0.39	153.05	49.31	151.84	55.55	150.68	59.91	149.56	63.37	148.47	66.29
0.40	150.59	49.00	149.43	55.17	148.30	59.47	147.21	62.88	146.16	65.76
0.45	139.42	47.59	138.42	53.39	137.45	57.41	136.52	60.59	135.61	63.27
0.50	129.80	46.30	128.93	51.78	128.09	55.55	127.28	58.53	126.49	61.03
0.55	121.43	45.11	120.67	50.29	119.93	53.85	119.21	56.65	118.52	58.99
0.60	114.07	43.95	113.40	48.92	112.75	52.28	112.11	54.92	111.50	57.12
0.65	107.55	42.95	106.96	47.63	106.38	50.82	105.81	53.31	105.27	55.38
0.70	101.74	41.97	101.21	46.43	100.69	49.45	100.18	51.81	99.69	53.76
0.75	96.53	41.03	96.05	45.29	95.58	48.16	95.12	50.39	94.68	52.24
0.80	91.82	40.15	91.39	44.21	90.96	46.94	90.55	49.06	90.15	50.81
0.85	87.55	39.30	87.16	43.19	86.77	45.79	86.40	47.81	86.03	49.47

(Contd.)

s/b → w/b ↓	0.01		0.02		0.03		0.04		0.05	
	$Z_e\sqrt{\epsilon_r}$ (ohms)	$Z_o\sqrt{\epsilon_r}$ (ohms)	$Z_e\sqrt{\epsilon_r}$ (ohms)	$Z_o\sqrt{\epsilon_r}$ (ohms)	$Z_e\sqrt{\epsilon_r}$ (ohms)	$Z_o\sqrt{\epsilon_r}$ (ohms)	$Z_e\sqrt{\epsilon_r}$ (ohms)	$Z_o\sqrt{\epsilon_r}$ (ohms)	$Z_e\sqrt{\epsilon_r}$ (ohms)	$Z_o\sqrt{\epsilon_r}$ (ohms)
0.90	83.66	38.49	83.30	42.21	82.95	44.70	82.61	46.61	82.27	48.19
0.95	80.11	37.72	79.77	41.28	79.45	43.66	79.14	45.48	78.83	46.98
1.00	76.84	36.97	76.53	40.39	76.24	42.66	75.95	44.41	75.66	45.83
1.05	73.83	36.26	73.54	39.54	73.27	41.72	73.00	43.38	72.74	44.74
1.10	71.04	35.57	70.78	38.73	70.53	40.81	70.28	42.40	70.03	43.70
1.15	68.46	34.91	68.22	37.94	67.98	39.94	67.75	41.47	67.52	42.71
1.20	66.06	34.28	65.83	37.20	65.61	39.11	65.40	40.57	65.19	41.76
1.25	63.82	33.66	63.61	36.47	63.40	38.32	63.20	39.72	63.01	40.86
1.30	61.73	33.07	61.53	35.78	61.34	37.55	61.15	38.90	60.97	39.99
1.35	59.77	32.50	59.58	35.11	59.40	36.82	59.23	38.11	59.06	39.16
1.40	57.93	31.95	57.76	34.47	57.59	36.11	57.42	37.35	57.26	38.36
1.45	56.20	31.42	56.04	33.85	55.88	35.43	55.72	36.63	55.57	37.59
1.50	54.57	30.90	54.42	33.25	54.27	34.78	54.12	35.93	53.98	36.86
1.55	53.04	30.41	52.89	32.68	52.75	34.15	52.61	35.25	52.47	36.15
1.60	51.58	29.92	51.45	32.12	51.31	33.54	51.18	34.61	51.05	35.47
1.65	50.21	29.46	50.08	31.58	49.95	32.95	49.83	33.98	49.70	34.82
1.70	48.91	29.00	48.78	31.06	48.66	32.39	48.54	33.38	48.43	34.18
1.75	47.67	28.56	47.55	30.56	47.44	31.84	47.32	32.80	47.21	33.57
1.80	46.49	28.13	46.38	30.07	46.27	31.31	46.16	32.24	46.06	32.99
1.85	45.37	27.72	45.26	29.60	45.16	30.80	45.06	31.70	44.96	32.42
1.90	44.30	27.33	44.20	29.14	44.10	30.31	44.01	31.17	43.91	31.87
1.95	43.29	26.93	43.19	28.70	43.09	29.83	43.00	30.66	42.91	31.34
2.00	42.31	26.55	42.22	28.27	42.13	29.36	42.04	30.17	41.95	30.83
2.05	41.38	26.19	41.30	27.86	41.21	28.91	41.12	29.70	41.04	30.33
2.10	40.49	25.84	40.41	27.45	40.33	28.47	40.24	29.24	40.16	29.85
2.15	39.64	25.49	39.56	27.05	39.48	28.05	39.40	28.79	39.33	29.39
2.20	38.82	25.15	38.75	26.68	38.67	27.64	38.60	28.36	38.52	28.93
2.25	38.04	24.83	37.97	26.30	37.89	27.24	37.82	27.95	37.75	28.49
2.30	37.29	24.51	37.22	25.93	37.14	26.86	37.08	27.54	37.01	28.08
2.35	36.56	24.20	36.49	25.59	36.43	26.48	36.36	27.14	36.29	27.66
2.40	35.87	23.90	35.80	25.25	35.74	26.12	35.67	26.75	35.61	27.27
2.45	35.20	23.60	35.13	24.94	35.07	25.75	35.01	26.38	34.95	26.88
2.50	34.55	23.28	34.49	24.61	34.43	25.41	34.37	26.01	34.31	26.51

(Contd.)

w/b \ s/b	0.10		0.15		0.20		0.25		0.30	
	$Z_e\sqrt{\epsilon_r}$ (ohms)	$Z_o\sqrt{\epsilon_r}$ (ohms)	$Z_e\sqrt{\epsilon_r}$ (ohms)	$Z_o\sqrt{\epsilon_r}$ (ohms)	$Z_e\sqrt{\epsilon_r}$ (ohms)	$Z_o\sqrt{\epsilon_r}$ (ohms)	$Z_e\sqrt{\epsilon_r}$ (ohms)	$Z_o\sqrt{\epsilon_r}$ (ohms)	$Z_e\sqrt{\epsilon_r}$ (ohms)	$Z_o\sqrt{\epsilon_r}$ (ohms)
0.10	265.88	118.80	253.53	132.75	243.83	143.25	235.99	151.55	229.55	158.27
0.11	257.53	115.41	245.79	128.87	236.50	139.04	228.97	147.08	222.77	153.60
0.12	249.82	112.38	238.65	125.41	229.75	135.26	222.51	143.05	216.53	149.39
0.13	242.67	109.66	232.02	122.28	223.49	131.83	216.52	139.40	210.75	145.55
0.14	236.01	107.19	225.83	119.43	217.65	128.71	210.94	136.06	205.37	142.04
0.15	229.77	104.94	220.04	116.82	212.18	125.83	205.71	132.99	200.33	138.80
0.16	223.91	102.86	214.60	114.41	207.04	123.18	200.80	130.14	195.60	135.80
0.17	218.38	100.94	209.46	112.18	202.19	120.71	196.17	127.49	191.15	133.00
0.18	213.16	99.16	204.60	110.10	197.60	118.41	191.79	125.01	186.93	130.39
0.19	208.21	97.49	199.99	108.15	193.25	116.25	187.64	122.69	182.94	127.93
0.20	203.51	95.93	195.62	106.32	189.11	114.22	183.70	120.50	179.14	125.61
0.21	199.04	94.46	191.45	104.60	185.18	112.31	179.94	118.44	175.53	123.42
0.22	194.78	93.07	187.47	102.97	181.42	110.50	176.35	116.48	172.08	121.34
0.23	190.72	91.75	183.67	101.43	177.83	108.78	172.92	114.62	168.78	119.37
0.24	186.83	90.50	180.04	99.96	174.39	107.15	169.64	112.85	165.63	117.49
0.25	183.11	89.31	176.56	98.56	171.09	105.59	166.50	111.16	162.61	115.70
0.26	179.54	88.18	173.22	97.23	167.93	104.10	163.48	109.55	159.70	113.98
0.27	176.11	87.09	170.01	95.95	164.89	102.67	160.58	108.00	156.91	112.33
0.28	172.82	86.05	166.93	94.72	161.97	101.30	157.78	106.52	154.23	110.75
0.29	169.66	85.05	163.96	93.55	159.16	99.99	155.10	105.09	151.64	109.24
0.30	166.62	84.08	161.10	92.41	156.45	98.72	152.50	103.72	149.15	107.77
0.31	163.68	83.16	158.34	91.32	153.83	97.50	150.00	102.39	146.74	106.36
0.32	160.85	82.26	155.68	90.27	151.31	96.32	147.59	101.12	144.42	105.00
0.33	158.12	81.40	153.11	89.25	148.87	95.19	145.26	99.88	142.17	103.69
0.34	155.49	80.56	150.63	88.27	146.51	94.09	143.00	98.69	140.00	102.42
0.35	152.94	79.75	148.23	87.31	144.23	93.02	140.82	97.53	137.90	101.18
0.36	150.48	78.96	145.91	86.39	142.02	91.99	138.70	96.41	135.86	99.99
0.37	148.09	78.20	143.66	85.49	139.88	90.99	136.65	95.32	133.89	98.83
0.38	145.79	77.46	141.48	84.62	137.80	90.01	134.66	94.26	131.97	97.70
0.39	143.55	76.74	139.37	83.77	135.79	89.07	132.74	93.24	130.11	96.61
0.40	141.38	76.03	137.32	82.95	133.84	88.15	130.86	92.24	128.31	95.55
0.45	131.47	72.77	127.93	79.13	124.89	83.88	122.27	87.62	120.01	90.62
0.50	122.87	69.86	119.76	75.73	117.07	80.09	114.76	83.51	112.75	86.26
0.55	115.33	67.22	112.58	72.66	110.19	76.69	108.13	79.83	106.34	82.35
0.60	108.67	64.81	106.22	69.86	104.08	73.59	102.23	76.49	100 63	78.81
0.65	102.74	62.59	100.54	67.30	98.62	70.76	96.96	73.45	95.51	75.59
0.70	97.42	60.54	95.44	64.94	93.71	68.16	92.20	70.65	90.89	72.63
0.75	92.63	58.62	90.83	62.74	89.26	65.75	87.89	68.07	86.69	69.91
0.80	88.29	56.83	86.65	60.70	85.22	63.51	83.97	65.68	82.87	67.39
0.85	84.33	55.16	82.84	58.79	81.53	61.43	80.38	65.45	79.38	65.05

(Contd.)

w/b \ s/b	0.10		0.15		0.20		0.25		0.30	
	$Z_e\sqrt{\epsilon_r}$ (ohms)	$Z_o\sqrt{\epsilon_r}$ (ohms)	$Z_e\sqrt{\epsilon_r}$ (ohms)	$Z_o\sqrt{\epsilon_r}$ (ohms)	$Z_e\sqrt{\epsilon_r}$ (ohms)	$Z_o\sqrt{\epsilon_r}$ (ohms)	$Z_e\sqrt{\epsilon_r}$ (ohms)	$Z_o\sqrt{\epsilon_r}$ (ohms)	$Z_e\sqrt{\epsilon_r}$ (ohms)	$Z_o\sqrt{\epsilon_r}$ (ohms)
0.90	80.72	53.58	79.35	57.00	78.14	59.48	77.09	61.37	76.17	62.87
0.95	77.40	52.09	76.14	55.32	75.03	57.65	74.06	59.43	73.20	60.83
1.00	74.34	50.68	73.18	53.74	72.15	55.93	71.25	57.61	70.46	58.93
1.05	71.52	49.35	70.44	52.24	69.49	54.32	68.66	55.90	67.92	57.14
1.10	68.90	48.09	67.90	50.83	67.02	52.79	66.24	54.28	65.56	55.45
1.15	66.47	46.89	65.54	49.50	64.72	51.35	63.99	52.76	63.35	53.86
1.20	64.21	45.75	63.34	48.23	62.57	49.99	61.89	51.32	61.25	52.37
1.25	62.09	44.67	61.28	47.02	60.55	48.70	59.92	49.96	59.36	50.95
1.30	60.11	43.63	59.34	45.88	58.67	47.47	58.07	48.67	57.54	49.61
1.35	58.25	42.64	57.53	44.78	59.90	46.30	56.33	47.44	55.84	48.33
1.40	56.50	41.70	55.83	43.74	55.23	45.19	54.70	46.28	54.23	47.12
1.45	54.86	40.79	54.22	42.75	53.65	44.13	53.15	45.17	52.71	45.97
1.50	53.30	39.93	52.70	41.80	52.17	43.12	51.69	44.11	51.28	44.88
1.55	51.84	39.10	51.27	40.90	50.76	42.16	50.31	43.10	49.92	43.83
1.60	50.45	38.31	49.91	40.03	49.43	41.23	49.00	42.14	48.63	42.84
1.65	49.13	37.54	48.62	39.19	48.17	40.35	47.76	41.21	47.41	41.88
1.70	47.88	36.81	47.40	38.39	46.96	39.50	46.58	40.33	46.24	40.97
1.75	46.70	36.10	46.23	37.63	45.82	38.69	45.46	39.49	45.13	40.10
1.80	45.57	35.42	45.13	36.89	44.73	37.91	44.39	38.67	44.08	39.26
1.85	44.49	34.77	44.07	36.18	43.70	37.16	43.36	37.90	43.07	38.46
1.90	43.46	34.14	43.06	35.50	42.71	36.45	42.39	37.15	42.11	37.69
1.95	42.48	33.53	42.10	34.84	41.76	35.75	41.45	36.43	41.19	36.95
2.00	41.55	32.95	41.18	34.21	40.85	35.09	40.56	35.74	40.30	36.24
2.05	40.65	32.38	40.30	33.60	39.99	34.45	39.71	35.07	39.46	35.56
2.10	39.79	31.83	39.45	33.01	39.15	33.83	38.89	34.43	38.65	34.90
2.15	38.97	31.30	38.64	32.44	38.36	33.23	38.10	33.81	37.87	34.27
2.20	38.18	30.79	37.87	31.90	37.59	32.65	37.34	33.22	37.13	33.65
2.25	37.42	30.30	37.12	31.36	36.86	32.10	36.62	32.64	36.41	33.06
2.30	36.69	29.82	36.40	30.85	36.15	31.56	35.92	32.09	35.72	32.49
2.35	35.99	29.36	35.71	30.35	35.47	31.04	35.25	31.55	35.05	31.94
2.40	35.31	28.91	35.05	29.88	34.81	30.54	34.60	31.03	34.41	31.41
2.45	34.66	28.48	34.41	29.41	34.18	30.05	33.98	30.53	33.80	30.89
2.50	34.04	28.05	33.79	28.95	33.57	29.58	33.37	30.04	33.20	30.40

(Contd.)

w/b \ s/b	0.35		0.40		0.45		0.50		0.55	
	$Z_e\sqrt{\epsilon_r}$ (ohms)	$Z_o\sqrt{\epsilon_r}$ (ohms)	$Z_e\sqrt{\epsilon_r}$ (ohms)	$Z_o\sqrt{\epsilon_r}$ (ohms)	$Z_e\sqrt{\epsilon_r}$ (ohms)	$Z_o\sqrt{\epsilon_r}$ (ohms)	$Z_e\sqrt{\epsilon_r}$ (ohms)	$Z_o\sqrt{\epsilon_r}$ (ohms)	$Z_e\sqrt{\epsilon_r}$ (ohms)	$Z_o\sqrt{\epsilon_r}$ (ohms)
0.10	224.20	163.80	219.71	168.41	215.93	172.28	212.72	175.54	210.00	178.30
0.11	217.60	158.97	213.26	163.45	209.59	167.21	206.49	170.38	203.85	173.06
0.12	211.53	154.61	207.23	158.96	203.78	162.61	200.78	165.69	198.22	168.30
0.13	205.92	150.63	201.85	154.86	198.42	158.41	195.50	161.41	193.02	163.95
0.14	200.70	146.98	196.76	151.10	193.43	154.55	190.60	157.47	188.20	159.94
0.15	195.82	143.61	192.00	147.61	188.78	150.98	186.03	153.82	183.70	156.22
0.16	191.23	140.48	187.54	144.38	184.41	147.66	181.75	150.42	179.49	152.76
0.17	186.92	137.56	183.34	141.36	180.30	144.55	177.72	147.24	175.52	149.52
0.18	182.84	134.83	179.37	138.53	176.42	141.64	173.92	144.26	171.78	146.48
0.19	178.97	132.26	175.61	135.87	172.75	138.90	170.32	141.45	168.24	143.61
0.20	175.30	129.83	172.03	133.35	169.26	136.31	166.90	138.80	164.88	140.91
0.21	171.80	127.54	168.63	130.97	165.94	133.85	163.64	136.28	161.68	138.34
0.22	168.46	125.36	165.39	128.71	162.78	131.52	160.54	133.89	158.64	135.90
0.23	165.28	123.29	162.29	126.56	159.75	129.31	157.58	131.62	155.73	133.58
0.24	162.23	121.32	159.33	124.51	156.86	127.19	154.75	129.45	152.95	131.36
0.25	159.30	119.44	156.49	122.50	154.08	125.17	152.03	127.37	150.28	129.24
0.26	156.49	117.64	153.76	120.68	151.42	123.24	149.43	125.39	147.72	127.21
0.27	153.79	115.91	151.14	118.88	148.87	121.38	146.92	123.48	145.26	125.26
0.28	151.20	114.25	148.61	117.16	146.40	119.60	144.51	121.65	142.90	123.38
0.29	148.70	112.65	146.18	115.50	144.03	117.88	142.19	119.80	140.62	121.58
0.30	146.29	111.12	143.84	113.90	141.75	116.23	139.95	118.19	138.42	119.84
0.31	143.96	109.63	141.58	112.35	139.54	114.63	137.79	116.55	136.30	118.17
0.32	141.71	108.20	139.39	110.86	137.41	113.09	135.71	114.97	134.25	116.58
0.33	139.54	106.82	137.28	109.43	135.35	111.61	133.69	113.44	132.27	114.99
0.34	137.44	105.48	135.24	108.03	133.35	110.16	131.74	111.96	130.35	113.47
0.35	135.40	104.19	133.26	106.68	131.42	108.77	129.85	110.53	128.50	112.01
0.36	133.43	102.93	131.34	105.37	129.55	107.42	128.01	109.14	126.70	110.58
0.37	131.52	101.71	129.48	104.11	127.74	106.11	126.24	107.79	124.95	109.21
0.38	129.66	100.53	127.68	102.87	125.98	104.83	124.51	106.48	123.26	107.87
0.39	127.86	99.38	125.93	101.68	124.27	103.59	122.84	105.21	121.61	106.56
0.40	126.11	98.26	124.23	100.51	122.61	102.39	121.21	103.97	120.02	105.30
0.45	118.07	93.09	116.40	95.13	114.96	96.83	113.72	98.26	112.66	99.46
0.50	111.03	88.51	109.54	90.37	108.25	91.91	107.14	93.21	106.19	94.30
0.55	104.80	84.41	103.46	86.11	102.31	87.52	101.31	88.70	100.45	89.69
0.60	99.24	80.70	98.04	82.26	96.99	83.55	96.10	84.63	95.32	85.54
0.65	94.25	77.33	93.16	78.76	92.22	79.95	91.40	80.94	90.70	81.77
0.70	89.74	74.24	88.75	75.56	87.90	76.66	87.15	77.57	86.51	78.34
0.75	85.65	71.40	84.75	72.62	83.97	73.64	83.29	74.48	82.70	75.19
0.80	81.92	68.78	81.09	69.91	80.37	70.85	79.75	71.60	79.21	72.28
0.85	78.50	66.34	77.74	67.40	77.08	68.27	76.51	69.00	76.01	69.60

(*Contd.*)

s/b → w/b ↓	0.35		0.40		0.45		0.50		0.55	
	$Z_e\sqrt{\epsilon_r}$ (ohms)	$Z_o\sqrt{\epsilon_r}$ (ohms)	$Z_e\sqrt{\epsilon_r}$ (ohms)	$Z_o\sqrt{\epsilon_r}$ (ohms)	$Z_e\sqrt{\epsilon_r}$ (ohms)	$Z_o\sqrt{\epsilon_r}$ (ohms)	$Z_e\sqrt{\epsilon_r}$ (ohms)	$Z_o\sqrt{\epsilon_r}$ (ohms)	$Z_e\sqrt{\epsilon_r}$ (ohms)	$Z_o\sqrt{\epsilon_r}$ (ohms)
0.90	75.36	64.08	74.66	65.06	74.05	65.88	73.52	66.55	73.06	67.12
0.95	72.46	61.96	71.81	62.88	71.24	63.64	70.75	64.27	70.33	64.80
1.00	69.77	59.99	69.17	60.85	68.64	61.56	68.19	62.15	67.79	62.64
1.05	67.28	58.13	66.72	58.94	66.23	59.61	65.80	60.16	65.44	60.62
1.10	64.96	56.39	64.44	57.15	63.98	57.78	63.58	58.30	63.24	58.73
1.15	62.79	54.75	62.30	55.47	61.88	56.06	61.51	56.55	61.18	56.96
1.20	60.77	53.20	60.31	53.88	59.91	54.44	59.56	54.90	59.26	55.28
1.25	58.87	51.74	58.44	52.38	58.06	52.91	57.73	53.34	57.45	53.71
1.30	57.08	50.36	56.68	50.96	56.32	51.46	56.02	51.87	55.75	52.22
1.35	55.40	49.04	55.02	49.62	54.69	50.09	54.40	50.48	54.15	50.81
1.40	53.82	47.80	53.46	48.35	53.15	48.80	52.87	49.17	52.63	49.47
1.45	52.32	46.62	51.98	47.14	51.69	47.56	51.43	47.91	51.20	48.21
1.50	50.91	45.49	50.59	45.95	50.31	46.39	50.06	46.73	49.85	47.00
1.55	49.57	44.42	49.27	44.89	49.00	45.28	48.77	45.59	48.56	45.86
1.60	48.30	43.39	48.01	43.84	47.76	44.21	47.53	44.52	47.34	44.77
1.65	47.09	42.42	46.82	42.85	46.57	43.20	46.36	43.49	46.18	43.73
1.70	45.94	41.48	45.68	41.89	45.45	42.23	45.25	42.51	45.08	42.74
1.75	44.85	40.59	44.60	40.98	44.38	41.30	44.19	41.57	44.02	41.79
1.80	43.81	39.73	43.57	40.11	43.36	40.42	43.18	40.67	43 02	40.88
1.85	42.81	38.91	42.58	39.27	42.38	39.57	42.21	39.81	42.06	40.01
1.90	41.86	38.12	41.64	38.47	41.45	38.76	41.28	38.99	41.14	39.18
1.95	40.95	37.37	40.74	37.70	40.56	37.97	40.40	38.20	40.26	38.38
2.00	40.08	36.64	39.88	36.96	39.70	37.22	39.55	37.44	39.42	37.62
2.05	39.24	35.94	39.05	36.25	38.88	36.50	38.74	36.71	38.61	36.88
2.10	38.44	35.27	38.26	35.57	38.10	35.81	37.96	36.01	37.83	36.17
2.15	37.67	34.62	37.50	34.91	37.34	35.14	37.21	35.33	37.09	35.49
2.20	36.93	34.00	36.76	34.27	36.61	34.50	36.48	34.68	36.37	34.83
2.25	36.22	33.40	36.06	33.66	35.92	33.88	35.79	34.05	35.68	34.20
2.30	35.54	32.81	35.38	33.07	35.24	33.28	35.12	33.45	35.03	33.59
2.35	34.88	32.25	34.73	32.50	34.60	32.70	34.48	32.87	34.38	33.00
2.40	34.25	31.71	34.10	31.95	33.97	32.14	33.86	32.30	33.76	32.43
2.45	33.64	31.18	33.49	31.41	33.37	31.61	33.26	31.76	33.17	31.89
2.50	33.05	30.67	32.91	30.90	32.79	31.08	32.69	31.23	32.59	31.36

(Contd.)

s/b → w/b ↓	0.60		0.65		0.70		0.75		0.80	
	$Z_e\sqrt{\epsilon_r}$ (ohms)	$Z_o\sqrt{\epsilon_r}$ (ohms)	$Z_e\sqrt{\epsilon_r}$ (ohms)	$Z_o\sqrt{\epsilon_r}$ (ohms)	$Z_e\sqrt{\epsilon_r}$ (ohms)	$Z_o\sqrt{\epsilon_r}$ (ohms)	$Z_e\sqrt{\epsilon_r}$ (ohms)	$Z_o\sqrt{\epsilon_r}$ (ohms)	$Z_e\sqrt{\epsilon_r}$ (ohms)	$Z_o\sqrt{\epsilon_r}$ (ohms)
0.10	207.69	180.64	205.72	182.63	204.04	184.33	202.61	185.77	201.39	187.00
0.11	201.61	175.34	199.70	177.27	198.07	178.92	196.68	180.32	195.49	181.52
0.12	196.04	170.52	194.19	172.40	192.61	174.00	191.26	175.36	190.10	176.53
0.13	190.91	166.10	189.11	167.93	187.58	169.49	186.26	170.81	185.14	171.95
0.14	186.15	162.03	184.40	163.81	182.91	165.33	181.64	166.62	180.55	167.72
0.15	181.71	158.26	180.01	159.99	178.57	161.47	177.33	162.72	176.27	163.79
0.16	177.55	154.74	175.90	156.43	174.50	157.86	173.29	159.09	172.27	160.13
0.17	173.64	151.45	172.04	153.09	170.67	154.49	169.51	155.68	168.51	156.70
0.18	169.96	148.36	168.40	149.96	167.07	151.32	165.93	152.48	164.96	153.47
0.19	166.47	145.45	164.95	147.01	163.66	148.33	162.56	149.46	161.61	150.42
0.20	163.16	142.69	161.69	144.21	160.43	145.50	159.35	146.60	158.43	147.54
0.21	160.01	140.08	158.58	141.56	157.36	142.82	156.31	143.89	155.41	144.81
0.22	157.01	137.60	155.62	139.04	154.43	140.27	153.41	141.32	152.54	142.21
0.23	154.15	135.23	152.79	136.64	151.63	137.84	150.64	138.86	149.79	139.73
0.24	151.41	132.98	150.09	134.35	148.96	135.52	147.99	136.51	147.17	137.36
0.25	148.78	130.82	147.50	132.16	146.40	133.30	145.46	134.27	144.65	135.09
0.26	146.26	128.75	145.01	130.05	143.94	131.17	143.03	132.11	142.24	132.92
0.27	143.84	126.76	142.62	128.04	141.58	129.12	140.69	130.05	139.93	130.83
0.28	141.51	124.85	140.33	126.10	139.31	127.16	138.44	128.06	137.70	128.83
0.29	139.27	123.02	138.11	124.23	137.12	125.27	136.28	126.15	135.55	126.90
0.30	137.11	121.25	135.98	122.44	135.01	123.45	134.19	124.31	133.48	125.04
0.31	135.02	119.54	133.92	120.70	132.98	121.69	132.17	122.53	131.48	123.24
0.32	133.00	117.89	131.93	119.03	131.01	119.99	130.23	120.81	129.55	121.51
0.33	131.05	116.30	130.01	117.41	129.11	118.35	128.35	119.15	127.69	119.83
0.34	129.16	114.75	128.15	115.84	127.27	116.76	126.52	117.54	125.88	118.21
0.35	127.34	113.26	126.34	114.32	125.49	115.22	124.76	115.99	124.14	116.64
0.36	125.57	111.81	124.60	112.85	123.76	113.73	123.05	114.48	122.44	115.12
0.37	123.85	110.40	122.90	111.42	122.09	112.28	121.39	113.01	120.80	113.64
0.38	122.18	109.04	121.26	110.03	120.46	110.88	119.78	111.59	119.20	112.20
0.39	120.56	107.71	119.66	108.69	118.88	109.51	118.22	110.21	117.65	110.81
0.40	118.99	106.42	118.11	107.38	117.35	108.18	116.70	108.87	116.14	109.45
0.45	111.74	100.47	110.95	101.33	110.28	102.06	109.70	102.67	109.20	103.20
0.50	105.37	95.21	104.66	95.99	104.06	96.65	103.53	97.21	103.09	97.68
0.55	99.71	90.52	99.08	91.23	98.53	91.83	98.06	92.33	97.66	92.76
0.60	94.65	86.30	94.08	86.94	93.58	87.49	93.15	87.95	92.79	88.34
0.65	90.09	82.47	89.57	83.06	89.12	83.56	88.73	83.98	88.40	84.34
0.70	85.96	78.98	85.48	79.52	85.07	79.98	84.71	80.37	84.41	80.70
0.75	82.19	75.78	81.75	76.28	81.38	76.70	81.05	77.06	80 77	77.36
0.80	78.75	72.83	78.34	73.29	78.00	73.69	77.70	74.02	77.44	74.30
0.85	75.58	70.11	75.21	70.54	74.89	70.90	74.61	71.21	74.38	71.47

(Contd.)

s/b →	0.60		0.65		0.70		0.75		0.80	
w/b ↓	$Z_e\sqrt{\epsilon_r}$ (ohms)	$Z_o\sqrt{\epsilon_r}$ (ohms)	$Z_e\sqrt{\epsilon_r}$ (ohms)	$Z_o\sqrt{\epsilon_r}$ (ohms)	$Z_e\sqrt{\epsilon_r}$ (ohms)	$Z_o\sqrt{\epsilon_r}$ (ohms)	$Z_e\sqrt{\epsilon_r}$ (ohms)	$Z_o\sqrt{\epsilon_r}$ (ohms)	$Z_e\sqrt{\epsilon_r}$ (ohms)	$Z_o\sqrt{\epsilon_r}$ (ohms)
0.90	72.66	67.59	72.32	67.99	72.02	68.32	71.77	68.61	71.55	68.85
0.95	69.96	65.24	69.64	65.61	69.37	65.93	69.13	66.19	68.92	66.42
1.00	67.45	63.06	67.15	63.40	66.90	63.69	66.68	63.94	66.49	64.15
1.05	65.12	61.01	64.84	61.34	64.60	61.61	64.40	61.84	64.22	62.04
1.10	62.94	59.09	62.68	59.40	62.46	59.66	62.27	59.87	62.10	60.06
1.15	60.90	57.30	60.66	57.58	60.45	57.82	60.27	58.03	60.12	58.20
1.20	59.00	55.60	58.77	55.87	58.57	56.10	58.40	56.29	58.26	56.45
1.25	57.20	54.01	56.99	54.26	56.81	54.48	56.65	54.66	56.51	54.81
1.30	55.52	52.50	55.32	52.74	55.14	52.95	54.99	53.12	54.86	53.26
1.35	53.93	51.08	53.74	51.31	53.57	51.50	53.43	51.66	53.31	51.80
1.40	52.43	49.73	52.25	49.95	52.09	50.13	51.96	56.28	51.84	50.41
1.45	51.01	48.45	50.84	48.66	50.69	48.83	50.56	48.97	50.45	49.10
1.50	49.66	47.24	49.50	47.43	49.36	47.59	49.24	47.73	49.14	47.85
1.55	48.39	46.08	48.23	46.27	48.10	46.42	47.99	46.55	47.89	46.66
1.60	47.17	44.98	47.03	45.16	46.90	45.30	46.80	45.43	46.70	45.53
1.65	46.02	43.93	45.88	44.10	45.76	44.24	45.66	44.36	45.57	44.46
1.70	44.92	42.93	44.79	43.09	44.68	43.22	44.58	43.34	44.49	43.43
1.75	43.88	41.97	43.75	42.13	43.64	42.26	43.55	42.36	43.47	42.46
1.80	42.88	41.06	42.76	41.21	42.65	41.33	42.56	41.43	42.49	41.52
1.85	41.92	40.18	41.81	40.32	41.71	40.44	41.62	40.54	41.55	40.62
1.90	41.01	39.34	40.90	39.48	40.81	39.59	40.72	39.69	40.65	39.77
1.95	40.14	38.54	40.03	38.67	39.94	38.78	39.86	38.87	39.79	38.94
2.00	39.30	37.77	39.20	37.89	39.11	37.99	39.04	38.08	38.97	38.16
2.05	38.50	37.02	38.40	37.14	38.32	37.24	38.24	37.33	38.18	37.40
2.10	37.73	36.31	37.63	36.42	37.55	36.52	37.48	36.60	37.42	36.67
2.15	36.98	35.62	36.90	35.73	36.82	35.83	36.75	35.90	36.69	35.97
2.20	36.27	34.96	36.19	35.07	36.11	35.16	36.05	35.23	35.99	35.30
2.25	35.59	34.32	35.50	34.43	35.43	34.51	35.37	34.59	35.32	34.65
2.30	34.93	33.71	34.85	33.81	34.78	33.89	34.72	33.96	34.67	34.02
2.35	34.29	33.12	34.21	33.21	34.15	33.29	34.09	33.36	34.04	33.42
2.40	33.68	32.55	33.60	32.64	33.54	32.72	33.48	32.78	33.44	32.84
2.45	33.09	31.99	33.01	32.08	32.95	32.16	32.90	32.22	32.85	32.27
2.50	32.51	31.46	32.45	31.55	32.39	31.62	32.33	31.68	32.29	31.73

(Contd.)

s/b → / w/b ↓	0.85		0.90		0.95		1.00	
	$Z_e\sqrt{\epsilon_r}$ (ohms)	$Z_o\sqrt{\epsilon_r}$ (ohms)	$Z_e\sqrt{\epsilon_r}$ (ohms)	$Z_o\sqrt{\epsilon_r}$ (ohms)	$Z_e\sqrt{\epsilon_r}$ (ohms)	$Z_o\sqrt{\epsilon_r}$ (ohms)	$Z_e\sqrt{\epsilon_r}$ (ohms)	$Z_o\sqrt{\epsilon_r}$ (ohms)
0.10	200.34	188.05	199.45	188.95	198.69	189.72	198.04	190.37
0.11	194.48	182.54	193.61	183.41	192.87	184.15	192.24	184.79
0.12	189.12	177.52	188.28	178.37	187.56	179.09	186.94	179.71
0.13	184.19	172.91	183.37	173.73	182.67	174.44	182.07	175.04
0.14	179.62	168.66	178.82	169.46	178.15	170.14	177.57	170.73
0.15	175.37	164.71	174.59	165.48	173.93	166.15	173.37	166.72
0.16	171.39	161.02	170.64	161.78	170.00	162.42	169.45	162.98
0.17	167.65	157.56	166.92	158.30	166.30	158.93	165.76	159.47
0.18	164.13	154.31	163.42	155.03	162.81	155.64	162.30	156.17
0.19	160.80	151.24	160.11	151.95	159.52	152.54	159.02	153.05
0.20	157.65	148.34	156.98	149.02	156.40	149.61	155.91	150.10
0.21	154.65	145.59	154.00	146.25	153.44	146.82	152.96	147.30
0.22	151.79	142.97	151.16	143.61	150.61	144.17	150.15	144.64
0.23	149.07	140.47	148.45	141.10	147.92	141.64	147.46	142.10
0.24	146.46	138.08	145.86	138.70	145.34	139.22	144.90	139.67
0.25	143.97	135.79	143.38	136.40	142.87	136.91	142.44	137.34
0.26	141.57	133.61	141.00	134.19	140.51	134.69	140.09	135.12
0.27	139.27	131.50	138.71	132.07	138.23	132.56	137.83	132.98
0.28	137.06	129.48	136.51	130.04	136.05	130.51	135.65	130.92
0.29	134.93	127.54	134.40	128.08	133.94	128.54	133.55	128.94
0.30	132.88	125.66	132.36	126.19	131.91	126.65	131.53	127.03
0.31	130.89	123.85	130.39	124.37	129.95	124.81	129.58	125.19
0.32	128.98	122.10	128.49	122.61	128.06	123.04	127.70	123.41
0.33	127.13	120.41	126.65	120.91	126.23	121.33	125.88	121.69
0.34	125.33	118.78	124.86	119.26	124.46	119.67	124.12	120.03
0.35	123.60	117.19	123.14	117.67	122.75	118.07	122.41	118.41
0.36	121.92	115.66	121.47	116.12	121.08	116.51	120.76	116.85
0.37	120.28	114.17	119.85	114.62	119.47	115.01	119.15	115.53
0.38	118.70	112.72	118.27	113.16	117.91	113.54	117.59	113.86
0.39	117.16	111.32	116.74	111.75	116.39	112.12	116.08	112.43
0.40	115.67	109.95	115.26	110.37	114.91	110.73	114.61	111.04
0.45	108.77	103.65	108.41	104.03	108.09	104.35	107.83	104.63
0.50	102.70	98.09	102.38	98.43	102.09	98.72	101.85	98.97
0.55	97.31	93.13	97.01	93.44	96.76	93.71	96.54	93.93
0.60	92.47	88.68	92.21	88.96	91.97	89.20	91.78	89.41
0.65	88.11	84.64	87.87	84.90	87.66	85.12	87.47	85.31
0.70	84.15	80.97	83.92	81.21	83.73	81.41	83.57	81.59
0.75	80.53	77.62	80.33	77.84	80.15	78.02	80.00	78.18
0.80	77.22	74.53	77.03	74.73	76.87	74.91	76.73	75.05
0.85	74.17	71.69	74.00	71.87	73.85	72.03	73.72	72.17

(*Contd.*)

s/b → w/b ↓	0.85		0.90		0.95		1.00	
	$Z_e\sqrt{\epsilon_r}$ (ohms)	$Z_o\sqrt{\epsilon_r}$ (ohms)	$Z_e\sqrt{\epsilon_r}$ (ohms)	$Z_o\sqrt{\epsilon_r}$ (ohms)	$Z_e\sqrt{\epsilon_r}$ (ohms)	$Z_o\sqrt{\epsilon_r}$ (ohms)	$Z_e\sqrt{\epsilon_r}$ (ohms)	$Z_o\sqrt{\epsilon_r}$ (ohms)
0.90	71.36	69.05	71 20	69.23	71.06	69.37	70.94	69.50
0.95	68.75	66.61	68.60	66.77	68.47	66.90	68.36	67.02
1.00	66.33	64.33	66.19	64.48	66.07	64.61	65.96	64.72
1.05	64.07	62.20	63.94	62.34	63.83	62.46	63.73	62.56
1.10	61.96	60.21	61.84	60.34	61.73	60.46	61.64	60.55
1.15	59.99	58.35	59.87	58.47	59.77	58.58	59.69	58.66
1.20	58.13	56.59	58.03	56.71	57.93	56.81	57.85	56.89
1.25	56.39	54.94	56.29	55.05	56.20	55.14	56.13	55.22
1.30	54.75	53.38	54.66	53.49	54.58	53.58	54.50	53.65
1.35	53.21	51.91	53.12	52.01	53.04	52.09	52.97	52.16
1.40	51.74	50.52	51.66	50.61	51.59	50.69	51.52	50.76
1.45	50.36	49.20	50.28	49.29	50.21	49.36	50.15	49.43
1.50	49.05	47.95	48.97	48.03	48.91	48.10	48.85	48.16
1.55	47.80	46.76	47.73	46.84	47.67	46.90	47.61	46.96
1.60	46.62	45.62	46.55	45.70	46.49	45.76	46.44	45.82
1.65	45.49	44.55	45.43	44.62	45.37	44.68	45.32	44.73
1.70	44.42	43.52	44.36	43.59	44.30	43.64	44.26	43.69
1.75	43.40	42.53	43.34	42.60	43.29	42.66	43.24	42.70
1.80	42.42	41.59	42.36	41.66	42.31	41.71	42.27	41.76
1.85	41.49	40.70	41.23	40.76	41.38	40.81	41.34	40.85
1.90	40.59	39.83	40.54	39.89	40.49	39.34	40.46	39.98
1.95	39.74	39.01	39.69	39.06	39.64	39.11	39.60	39.15
2.00	38.91	38.22	38.87	38.27	38.82	38.32	38.79	38.35
2.05	38.13	37.48	38.08	37.51	38.04	37.55	38.01	37.59
2.10	37.37	36.73	37.33	36.78	37.29	36.82	37.25	36.85
2.15	36.64	36.03	36.60	36.07	36.56	36.11	36.53	36.15
2.20	35.94	35.35	35.90	35.40	35.87	35.43	35.84	35.47
2.25	35.27	34.70	35.23	34.74	35.20	34.78	35.17	34.81
2.30	34.62	34.07	34.58	34.11	34.55	34.15	34.52	34.18
2.35	34.00	33.47	33.96	33.51	33.93	33.54	33.90	33.57
2.40	33.40	32.88	33.36	32.92	33.33	32.95	33.30	32.98
2.45	32.81	32.32	32.78	32.36	32.75	32.39	32.72	32.41
2.50	32.25	31.77	32.22	31.81	32.19	31.84	32.16	31.87

Table 6.2 Design data on edge-coupled stripline: Dielectric constant $\epsilon_r = 2.22$, ground plane spacing $b = 0.062''$, $c/(2w + s) = \infty$, $Z_t = 50$ ohms (See structure A1 of Table 4.1)

C_v(dB)	w(mm)	s(mm)	C_v(dB)	w(mm)	s(mm)	C_v(dB)	w(mm)	s(mm)
8.0	0.9441	0.0318	13.5	1.1774	0.1817	27.0	1.2915	0.8712
8.1	0.9504	0.0334	14.0	1.1894	0.2014	27.5	1.2921	0.8996
8.2	0.9567	0.0349	14.5	1.2003	0.2218	28.0	1.2927	0.9280
8.3	0.9628	0.0366	15.0	1.2101	0.2430	28.5	1.2932	0.9564
8.4	0.9689	0.0382	15.5	1.2191	0.2649	29.0	1.2936	0.9849
8.5	0.9749	0.0399	16.0	1.2271	0.2875	29.5	1.2940	1.0135
8.6	0.9807	0.0417	16.5	1.2344	0.3107	30.0	1.2944	1.0420
8.7	0.9865	0.0435	17.0	1.2409	0.3344	30.5	1.2947	1.0706
8.8	0.9923	0.0453	17.5	1.2468	0.3586	31.0	1.2950	1.0992
8.9	0.9979	0.0472	18.0	1.2521	0.3833	31.5	1.2953	1.1279
9.0	1.0034	0.0491	18.5	1.2569	0.4083	32.0	1.2955	1.1566
9.1	1.0089	0.0511	19.0	1.2612	0.4338	32.5	1.2957	1.1853
9.2	1.0142	0.0531	19.5	1.2650	0.4596	33.0	1.2959	1.2140
9.3	1.0195	0.0552	20.0	1.2684	0.4857	33.5	1.2960	1.2427
9.4	1.0247	0.0573	20.5	1.2715	0.5121	34.0	1.2962	1.2715
9.5	1.0299	0.0595	21.0	1.2743	0.5388	34.5	1.2963	1.3002
9.6	1.0349	0.0617	21.5	1.2767	0.5657	35.0	1.2964	1.3290
9.7	1.0399	0.0639	22.0	1.2789	0.5927	35.5	1.2965	1.3577
9.8	1.0447	0.0662	22.5	1.2809	0.6200	36.0	1.2966	1.3865
9.9	1.0495	0.0686	23.0	1.2827	0.6475	36.5	1.2967	1.4153
10.0	1.0543	0.0710	23.5	1.2843	0.6750	37.0	1.2968	1.4441
10.5	1.0768	0.0836	24.0	1.2857	0.7028	37.5	1.2968	1.4729
11.0	1.0974	0.0974	24.5	1.2869	0.7306	38.0	1.2969	1.5017
11.5	1.1164	0.1122	25.0	1.2881	0.7586	38.5	1.2969	1.5305
12.0	1.1338	0.1281	25.5	1.2891	0.7866	39.0	1.2970	1.5594
12.5	1.1497	0.1450	26.0	1.2900	0.8148	39.5	1.2970	1.5882
13.0	1.1642	0.1629	26.5	1.2908	0.8430	40.0	1.2971	1.6170

Table 6.3 Design data on edge-coupled stripline: Dielectric constant $\epsilon_r = 2.22$, ground plane spacing $b = 0.124''$, $c/(2w + s) = \infty$, $Z_t = 50$ ohms (See structure A1 of Table 4.1)

C_v(dB)	w(mm)	s(mm)	C_v(dB)	w(mm)	s(mm)	C_v(dB)	w(mm)	s(mm)
8.0	1.8881	0.0637	13.5	2.3548	0.3634	27.0	2.5830	1.7425
8.1	1.9008	0.0667	14.0	2.3788	0.4027	27.5	2.5842	1.7992
8.2	1.9133	0.0699	14.5	2.4006	0.4436	28.0	2.5854	1.8560
8.3	1.9256	0.0731	15.0	2.4203	0.4861	28.5	2.5864	1.9129
8.4	1.9378	0.0764	15.5	2.4381	0.5299	29.0	2.5873	1.9699
8.5	1.9497	0.0798	16.0	2.4542	0.5750	29.5	2.5881	2.0269
8.6	1.9615	0.0833	16.5	2.4687	0.6213	30.0	2.5888	2.0841
8.7	1.9731	0.0869	17.0	2.4818	0.6688	30.5	2.5895	2.1412
8.8	1.9845	0.0906	17.5	2.4936	0.7172	31.0	2.5900	2.1985
8.9	1.9958	0.0944	18.0	2.5042	0.7665	31.5	2.5905	2.2558
9.0	2.0068	0.0983	18.5	2.5138	0.8167	32.0	2.5910	2.3131
9.1	2.0177	0.1022	19.0	2.5223	0.8676	32.5	2.5914	2.3705
9.2	2.0285	0.1063	19.5	2.5300	0.9192	33.0	2.5918	2.4280
9.3	2.0391	0.1104	20.0	2.5369	0.9714	33.5	2.5921	2.4854
9.4	2.0495	0.1146	20.5	2.5430	1.0242	34.0	2.5924	2.5429
9.5	2.0597	0.1190	21.0	2.5485	1.0775	34.5	2.5926	2.6004
9.6	2.0698	0.1234	21.5	2.5535	1.1313	35.0	2.5928	2.6579
9.7	2.0797	0.1279	22.0	2.5579	1.1855	35.5	2.5931	2.7155
9.8	2.0895	0.1325	22.5	2.5618	1.2400	36.0	2.5932	2.7730
9.9	2.0991	0.1372	23.0	2.5654	1.2949	36.5	2.5934	2.8306
10.0	2.1085	0.1420	23.5	2.5685	1.3501	37.0	2.5935	2.8882
10.5	2.1535	0.1672	24.0	2.5714	1.4055	37.5	2.5937	2.9458
11.0	2.1949	0.1947	24.5	2.5739	1.4612	38.0	2.5938	3.0035
11.5	2.2329	0.2244	25.0	2.5761	1.5171	38.5	2.5939	3.0611
12.0	2.2676	0.2562	25.5	2.5781	1.5732	39.0	2.5940	3.1187
12.5	2.2994	0.2901	26.0	2.5799	1.6295	39.5	2.5941	3.1764
13.0	2.3284	0.3258	26.5	2.5815	1.6859	40.0	2.5941	3.2340

Table 6.4 Design data on edge-coupled stripline: Dielectric constant $\epsilon_r = 2.22$, ground plane spacing $b = 0.25''$, $c/(2w+s) = \infty$, $Z_t = 50$ ohms (See structure A1 of Table 4.1)

C_v(dB)	w(mm)	s(mm)	C_v(dB)	w(mm)	s(mm)	C_v(dB)	w(mm)	s(mm)
8.0	3.8067	0.1284	13.5	4.7476	0.7327	27.0	5.2076	3.5131
8.1	3.8323	0.1346	14.0	4.7960	0 8119	27.5	5.2101	3.6274
8.2	3.8575	0.1409	14.5	4.8398	0.8944	28.0	5.2124	3.7419
8.3	3.8823	0.1474	15.0	4.8796	0.9800	28.5	5.2145	3.8566
8.4	3.9068	0.1541	15.5	4.9155	1.0683	29.0	5.2163	3.9715
8.5	3.9309	0.1610	16.0	4.9480	1.1593	29.5	5.2179	4.0865
8.6	3.9546	0.1680	16.5	4.9773	1.2527	30.0	5.2194	4.2017
8.7	3.9780	0.1753	17.0	5.0037	1.3483	30.5	5.2207	4.3170
8.8	4.0010	0.1827	17.5	5.0275	1.4459	31.0	5.2218	4.4324
8.9	4.0237	0.1903	18.0	5.0489	1.5454	31.5	5.2229	4.5480
9.0	4.0461	0.1981	18.5	5.0681	1.6465	32.0	5.2238	4.6636
9.1	4.0680	0.2061	19.0	5.0853	1.7492	32.5	5.2246	4.7793
9.2	4.0897	0.2142	19.5	5.1008	1.8532	33.0	5.2253	4.8951
9.3	4.1110	0.2226	20.0	5.1146	1.9586	33.5	5.2260	5.0109
9.4	4.1320	0.2311	20.5	5.1271	2.0650	34.0	5.2266	5.1268
9.5	4.1526	0.2398	21.0	5.1382	2.1725	34.5	5.2271	5.2428
9.6	4.1730	0.2487	21.5	5.1481	2.2809	35.0	5.2275	5.3588
9.7	4.1929	0.2578	22.0	5.1570	2.3901	35.5	5.2279	5.4748
9.8	4.2126	0.2671	22.5	5.1650	2.5001	36.0	5.2283	5.5908
9.9	4.2320	0.2766	23.0	5.1721	2.6107	36.5	5.2286	5.7069
10.0	4.2511	0.2862	23.5	5.1785	2.7220	37.0	5.2289	5.8231
10.5	4.3418	0.3371	24.0	5.1842	2.8338	37.5	5.2292	5.9392
11.0	4.4252	0.3926	24.5	5.1893	2.9460	38.0	5.2294	6.0554
11.5	4.5017	0.4525	25.0	5.1938	3.0588	38.5	5.2296	6.1715
12.0	4.5719	0.5166	25.5	5.1979	3.1718	39.0	5.2298	6.2877
12.5	4.6359	0.5848	26.0	5.2015	3.2853	39.5	5.2300	6.4040
13.0	4.6944	0.6569	26.5	5.2047	3.3991	40.0	5.2301	6.5203

Table 6.5 Characteristic impedance of edge-coupled off-set stripline, $c/(2w + s) = 20$, $s/b = 0.3$ (See structure A2 of Table 4.1)

d/b → / w/b ↓	0.0		0.05		0.10		0.15		0.20	
	$Z_e\sqrt{\epsilon_r}$ (ohms)	$Z_o\sqrt{\epsilon_r}$ (ohms)	$Z_e\sqrt{\epsilon_r}$ (ohms)	$Z_o\sqrt{\epsilon_r}$ (ohms)	$Z_e\sqrt{\epsilon_r}$ (ohms)	$Z_o\sqrt{\epsilon_r}$ (ohms)	$Z_e\sqrt{\epsilon_r}$ (ohms)	$Z_o\sqrt{\epsilon_r}$ (ohms)	$Z_e\sqrt{\epsilon_r}$ (ohms)	$Z_o\sqrt{\epsilon_r}$ (ohms)
0.10	231.23	159.94	229.98	159.71	226.19	158.99	219.75	157.66	210.43	155.49
0.12	218.27	151.15	217.04	150.91	213.32	150.14	206.98	148.73	197.83	146.43
0.14	207.17	143.90	205.96	143.64	202.29	142.82	196.07	141.32	187.09	138.90
0.16	197.46	137.74	196.27	137.46	192.67	136.60	186.55	135.03	177.03	132.49
0.18	188.83	132.39	187.66	132.11	184.12	131.20	178.11	129.55	169.46	126.91
0.20	181.08	127.68	179.92	127.38	176.44	126.44	170.54	124.72	162.05	121.99
0.22	174.04	123.47	172.91	123.16	169.49	122.18	163.70	120.40	155.36	117.61
0.24	167.61	119.67	166.50	119.35	163.14	118.33	157.45	116.52	149.27	113.62
0.26	161.70	116.21	160.61	115.88	157.31	114.85	151.72	112.96	143.70	109.98
0.28	156.24	113.05	155.17	112.70	151.92	111.62	146.43	109.68	138.60	106.63
0.30	151.16	110.10	150.11	109.74	146.92	108.63	141.53	106.64	133.84	103.54
0.32	146.44	107.36	145.40	106.99	142.27	105.85	137.01	103.82	129.43	100.67
0.34	142.06	104.79	141.04	104.41	137.96	103.25	132.75	101.18	125.31	97.99
0.36	137.91	102.38	136.91	101.99	133.88	100.81	128.77	98.71	121.46	95.47
0.38	134.01	100.10	133.02	99.71	130.05	98.51	125.02	96.38	117.85	93.11
0.40	130.33	97.95	129.36	97.56	126.43	96.34	121.50	94.18	114.45	90.88
0.42	126.85	95.91	125.90	95.51	123.03	94.28	118.18	92.10	111.25	88.77
0.44	123.57	93.97	122.64	93.57	119.81	92.32	115.04	90.12	108.24	86.78
0.46	120.46	92.12	119.53	91.71	116.76	90.45	112.07	88.24	105.39	84.88
0.48	117.49	90.35	116.59	89.94	113.85	88.67	109.25	86.44	102.68	83.07
0.50	114.68	88.66	113.78	88.24	111.10	86.97	106.57	84.73	100.11	81.35
0.60	102.42	81.15	101.59	80.72	99.11	79.42	94.94	77.16	89.00	73.78
0.70	92.54	74.88	91.77	74.45	89.48	73.15	85.61	70.90	80.13	67.57
0.80	84.39	69.54	83.69	69.12	81.55	67.83	77.69	65.62	72.87	62.36
0.90	77.57	64.92	76.90	64.51	74.91	63.25	71.56	61.09	66.81	57.92
1.00	71.76	60.88	71.14	60.47	69.27	59.25	66.13	57.15	61.69	54.07
1.10	66.75	57.31	66.17	56.92	64.41	55.73	61.46	53.68	57.29	50.71
1.20	62.40	54.14	61.85	53.76	60.16	52.60	57.41	50.62	53.48	47.75
1.30	58.58	51.30	58.06	50.93	56.46	49.80	53.86	47.89	50.14	45.11
1.40	55.20	48.74	54.70	48.38	53.21	47.29	50.72	45.43	47.19	42.75
1.50	52.18	46.42	51.71	46.08	50.30	45.02	47.92	43.22	44.57	40.62
1.60	49.48	44.32	49.03	43.98	47.68	42.95	45.42	41.21	42.23	38.69
1.70	47.04	42.39	46.61	42.07	45.32	41.07	43.16	39.38	40.11	36.94
1.80	44.84	40.63	44.42	40.31	43.19	39,34	41.12	37.70	38.20	35.34
1.90	42.82	39.00	42.43	38.69	41.24	37.75	39.26	36.16	36.46	33.87
2.00	40.98	37.50	40.61	37.20	39.47	36.29	37.56	34.74	34.88	32.51
2.10	39.30	36.11	38.93	35.82	37.84	34.93	36.00	33.42	33.42	31.26
2.20	37.74	34.82	37.39	34.54	36.33	33.67	34.57	32.20	32.08	30.11
2.30	36.31	33.62	35.97	33.34	34.95	32.50	33.25	31.07	30.85	29.03
2.40	34.98	32.50	34.65	32.23	33.67	31.41	32.02	30.02	29.71	28.03
2.50	33.74	31.45	33.42	31.18	32.47	30.38	30.88	29.03	28.64	27.10

Table 6.6 Characteristic impedance of edge-coupled off-set stripline, $c/(2w + s) = 20$, $s/b = 0.4$ (See structure A2 of Table 4.1)

d/b → / w/b ↓	0.0		0.05		0.10		0.15		0.20	
	$Z_e\sqrt{\epsilon_r}$ (ohms)	$Z_o\sqrt{\epsilon_r}$ (ohms)	$Z_e\sqrt{\epsilon_r}$ (ohms)	$Z_o\sqrt{\epsilon_r}$ (ohms)	$Z_e\sqrt{\epsilon_r}$ (ohms)	$Z_o\sqrt{\epsilon_r}$ (ohms)	$Z_e\sqrt{\epsilon_r}$ (ohms)	$Z_o\sqrt{\epsilon_r}$ (ohms)	$Z_e\sqrt{\epsilon_r}$ (ohms)	$Z_o\sqrt{\epsilon_r}$ (ohms)
0.10	221.20	169.90	220.04	169.58	216.52	168.59	210.56	166.78	201.97	163.87
0.12	208.86	160.51	207.72	160.17	204.27	159.14	198.41	157.25	189.97	154.24
0.14	198.33	152.69	197.21	152.35	193.81	151.27	188.05	149.31	179.76	146.19
0.16	189.14	146.02	188.04	145.66	184.69	144.54	179.03	142.51	170.89	139.31
0.18	180.99	140.21	179.91	139.84	176.62	138.68	171.05	136.59	163.05	133.30
0.20	173.68	135.07	172.61	134.69	169.37	133.49	163.90	131.35	156.04	127.98
0.22	167.05	130.46	165.99	130.07	162.82	128.84	157.44	126.65	149.71	123.22
0.24	160.99	126.29	159.96	125.89	156.83	124.63	151.54	122.39	143.95	118.94
0.26	155.42	122.48	154.41	122.07	151.33	120.79	146.13	118.54	138.68	114.99
0.28	150.28	118.98	149.28	118.59	146.26	117.29	141.15	114.96	133.87	111.37
0.30	145.51	115.77	144.52	115.34	141.55	114.01	136.55	111.66	129.37	108.03
0.32	141.06	112.75	140.06	112.31	137.19	110.97	132.27	108.58	125.19	104.92
0.34	136.94	109.92	135.99	109.48	133.11	108.12	128.25	105.71	121.29	102.02
0.36	133.03	107.27	132.10	106.82	129.27	105.45	124.49	103.01	117.65	99.30
0.38	129.36	104.77	128.44	104.32	125.65	102.93	120.95	100.48	114.22	96.75
0.40	125.90	102.41	124.99	101.66	122.25	100.55	117.62	98.09	111.01	94.35
0.42	122.62	100.18	121.73	99.72	119.03	98.30	114.48	95.83	107.98	92.08
0.44	119.53	98.05	118.65	97.59	116.00	96.17	111.52	93.68	105.12	89.93
0.46	116.59	96.03	115.72	95.56	113.11	94.14	108.70	91.64	102.41	87.89
0.48	113.80	94.10	112.94	93.63	110.37	92.20	106.03	89.70	99.84	85.95
0.50	111.14	92.25	110.30	91.78	107.76	90.35	103.49	87.85	97.40	84.10
0.60	99.54	84.09	98.76	83.62	96.41	82.19	92.45	79.70	86.81	76.01
0.70	90.16	77.34	89.43	76.87	87.25	75.46	83.57	73.02	78.34	69.42
0.80	82.41	71.61	81.73	71.16	79.69	69.78	76.25	67.40	71.38	63.91
0.90	75.88	66.69	75.25	66.25	73.33	64.91	70.12	62.60	65.56	59.23
1.00	70.32	62.41	69.72	61.98	67.92	60.68	64.90	58.45	60.62	55.20
1.10	65.51	58.65	64.95	58.23	63.25	56.97	60.40	54.82	56.37	51.69
1.20	61.32	55.31	60.78	54.91	59.18	53.69	56.49	51.61	52.68	48.60
1.30	57.63	52.34	57.12	51.95	55.60	50.77	53.05	48.76	49.45	45.86
1.40	54.36	49.67	53.88	49.29	52.43	48.15	50.01	46.21	46.58	43.41
1.50	51.44	47.25	50.98	46.89	49.60	45.78	47.29	43.91	44.03	41.21
1.60	48.81	45.06	48.38	44.71	47.06	43.64	44.86	41.83	41.74	39.22
1.70	46.44	43.07	46.03	42.72	44.77	41.69	42.66	39.94	39.68	37.42
1.80	44.29	41.24	43.89	40.91	42.69	39.90	40.67	38.21	37.81	35.77
1.90	42.33	39.56	41.95	39.24	40.79	38.26	38.85	36.62	36.11	34.26
2.00	40.54	38.01	40.17	37.70	39.05	36.75	37.19	35.16	34.56	32.87
2.10	38.89	36.58	38.53	36.27	37.46	35.36	35.66	33.81	33.13	31.59
2.20	37.37	35.25	37.02	34.95	35.99	34.06	34.26	32.55	31.81	30.40
2.30	35.97	34.01	35.64	33.73	34.64	32.86	33.96	31.40	30.61	29.31
2.40	34.66	32.86	34.34	32.58	33.38	31.74	31.76	30.31	29.48	28.28
2.50	33.45	31.78	33.14	31.51	32.20	30.69	30.64	29.30	28.43	27.33

Table 6.7 Characteristic impedance of edge-coupled off-set stripline, $c/(2w+s) = 20$, $s/b = 0.5$ (See structure A2 of Table 4.1)

d/b → / w/b ↓	0.0		0.05		0.10		0.15		0.20	
	$Z_e\sqrt{\epsilon_r}$ (ohms)	$Z_o\sqrt{\epsilon_r}$ (ohms)	$Z_e\sqrt{\epsilon_r}$ (ohms)	$Z_o\sqrt{\epsilon_r}$ (ohms)	$Z_e\sqrt{\epsilon_r}$ (ohms)	$Z_o\sqrt{\epsilon_r}$ (ohms)	$Z_e\sqrt{\epsilon_r}$ (ohms)	$Z_o\sqrt{\epsilon_r}$ (ohms)	$Z_e\sqrt{\epsilon_r}$ (ohms)	$Z_o\sqrt{\epsilon_r}$ (ohms)
0.10	214.10	176.91	213.02	176.51	209.76	175.26	204.23	173.02	196.28	169.48
0.12	202.19	167.12	201.13	166.70	197.92	165.42	192.48	163.11	184.46	159.48
0.14	192.15	158.98	191.11	158.55	187.95	157.23	182.59	154.86	174.89	151.16
0.16	183.34	151.96	182.31	151.52	179.20	150.17	173.93	147.75	166.35	143.97
0.18	175.48	145.82	174.47	145.38	171.41	143.99	166.22	141.52	158.77	137.68
0.20	168.47	140.38	167.47	139.92	164.46	138.51	159.35	136.00	152.01	132.11
0.22	162.11	135.50	161.13	135.03	158.16	133.60	153.14	131.04	145.92	127.10
0.24	156.32	131.06	155.35	130.59	152.43	129.13	147.48	126.55	140.38	122.57
0.26	150.99	127.01	150.04	146.54	147.16	125.06	142.29	122.44	135.30	118.43
0.28	146.07	123.28	145.13	122.80	142.30	121.30	137.50	118.66	130.62	114.62
0.30	141.51	119.83	140.59	119.34	137.80	117.83	133.08	115.17	126.30	111.11
0.32	137.25	116.61	136.34	116.12	133.59	114.60	128.94	111.91	122.27	107.84
0.34	133.27	113.61	132.38	113.11	129.67	111.57	125.09	108.88	118.53	104.79
0.36	129.54	110.79	128.66	110.28	125.99	108.74	121.48	106.03	115.02	101.94
0.38	126.03	108.13	125.16	107.62	122.53	106.07	118.09	103.36	111.73	99.26
0.40	122.72	105.62	121.86	105.11	119.27	103.55	114.90	100.83	108.63	96.74
0.42	119.59	103.24	118.74	102.73	116.19	101.17	111.88	98.45	105.71	94.36
0.44	116.62	100.98	115.79	100.47	113.27	98.91	109.33	96.19	102.95	92.11
0.46	113.80	98.83	112.98	98.32	110.51	96.76	106.32	94.04	100.34	89.97
0.48	111.13	96.79	110.32	96.28	107.88	94.71	103.75	91.99	97.86	87.94
0.50	101.58	94.83	107.78	94.32	105.38	92.76	101.31	90.04	95.51	86.01
0.60	97.45	86.21	96.71	85.70	94.46	84.16	90.68	81.49	85.29	77.56
0.70	88.43	79.10	87.74	78.60	85.64	77.09	82.11	74.50	77.08	70.69
0.80	80.96	73.10	80.31	72.62	78.34	71.15	75.03	68.65	70.33	64.98
0.90	74.66	67.96	74.04	67.50	72.20	66.08	69.09	63.66	64.68	60.14
1.00	69.26	63.51	68.69	63.06	66.95	61.69	64.02	59.36	59.87	55.98
1.10	64.60	59.60	64.05	59.17	62.41	57.85	59.64	55.61	55.72	52.37
1.20	60.53	56.15	60.01	55.73	58.45	54.46	55.83	52.31	52.12	49.19
1.30	56.93	53.08	56.44	52.67	54.96	51.45	52.47	49.37	48.95	46.38
1.40	53.74	50.32	53.27	49.93	51.86	48.75	49.50	46.75	46.15	43.87
1.50	50.89	47.84	50.44	47.46	49.10	46.33	46.84	44.40	43.64	41.62
1.60	48.33	45.59	47.90	45.23	46.61	44.13	44.45	42.26	41.40	39.59
1.70	46.01	43.54	45.60	43.19	44.36	42.13	42.29	40.33	39.37	37.75
1.80	43.90	41.67	43.50	41.33	42.32	40.30	40.34	38.56	37.53	36.07
1.90	41.97	39.95	41.60	39.62	40.46	38.62	38.55	36.94	35.86	34.53
2.00	40.21	38.37	39.85	38.05	38.75	37.08	36.92	35.45	34.33	33.12
2.10	38.59	36.90	38.24	36.59	37.18	35.66	35.42	34.07	32.92	31.81
2.20	37.09	35.55	36.76	35.25	35.74	34.34	34.03	32.80	31.62	30.61
2.30	35.71	34.29	35.38	33.99	34.40	33.11	32.75	31.62	30.42	29.49
2.40	34.42	33.11	34.11	32.83	33.16	31.97	31.56	30.52	29.31	28.45
2.50	33.23	32.01	32.92	31.74	32.00	30.90	30.46	29.49	28.28	27.84

Table 6.8 Characteristic impedance of edge-coupled offset stripline, $c/(2w + s) = 20$, $s/b = 0.5$ (See structure A2 of Table 4.1)

d/b	0.0		0.05		0.10		0.15		0.20	
w/b	$Z_e\sqrt{\epsilon_r}$ (ohms)	$Z_o\sqrt{\epsilon_r}$ (ohms)	$Z_e\sqrt{\epsilon_r}$ (ohms)	$Z_o\sqrt{\epsilon_r}$ (ohms)	$Z_e\sqrt{\epsilon_r}$ (ohms)	$Z_o\sqrt{\epsilon_r}$ (ohms)	$Z_e\sqrt{\epsilon_r}$ (ohms)	$Z_o\sqrt{\epsilon_r}$ (ohms)	$Z_e\sqrt{\epsilon_r}$ (ohms)	$Z_o\sqrt{\epsilon_r}$ (ohms)
0.10	208.98	181.93	207.97	181.45	204.93	179.99	199.77	177.73	192.35	137.31
0.12	197.38	171.85	196.39	171.37	193.39	169.87	188.31	167.21	180.99	163.08
0.14	187.52	163.41	186.54	162.92	183.59	161.39	178.58	158.68	171.36	154.49
0.16	178.94	156.15	177.98	155.65	175.07	154.09	170.13	151.34	163.01	147.11
0.18	171.35	149.78	170.40	149.27	167.53	147.70	162.66	144.91	155.65	140.63
0.20	164.56	144.13	163.62	143.61	160.80	142.02	156.00	139.20	149.08	134.89
0.22	158.42	139.05	157.50	138.53	154.70	136.91	149.97	134.07	143.15	129.73
0.24	152.81	134.43	151.90	133.90	149.15	132.27	144.48	129.41	137.77	125.06
0.26	147.67	130.21	146.77	129.68	144.06	128.04	139.46	125.16	132.84	120.82
0.28	142.91	126.32	142.03	125.79	139.35	124.14	134.82	121.27	128.33	116.88
0.30	138.51	122.73	137.63	122.22	135.00	120.56	130.56	117.65	124.12	113.25
0.32	134.39	119.40	133.54	118.86	130.97	117.19	126.56	114.27	120.22	109.88
0.34	130.59	116.27	129.74	115.72	127.17	114.05	122.82	111.12	116.57	106.73
0.36	126.99	113.32	126.15	112.77	123.62	111.10	119.33	108.17	113.16	103.79
0.38	123.59	110.55	122.77	110.00	120.27	108.32	116.03	105.40	109.96	101.02
0.40	120.39	107.93	119.57	107.38	117.11	105.70	112.93	102.78	106.94	98.42
0.42	117.36	105.45	116.56	104.90	114.13	103.22	110.01	100.31	104.10	95.96
0.44	114.50	103.10	113.71	102.55	111.31	100.88	107.25	97.97	101.42	93.64
0.46	111.78	100.86	111.00	100.31	108.63	98.64	104.62	95.74	98.88	91.43
0.48	109.19	98.73	108.42	98.18	106.08	96.51	102.13	93.62	96.47	89.34
0.50	106.73	96.69	105.96	96.15	103.66	94.49	99.76	91.61	94.17	87.34
0.60	95.94	87.74	95.22	87.20	93.06	85.57	89.42	82.77	84.20	78.65
0.70	87.18	80.37	86.51	79.85	84.48	78.27	81.06	75.56	76.19	71.59
0.80	79.91	74.17	79.27	73.67	77.37	72.14	74.16	69.53	69.59	65.74
0.90	73.76	68.88	73.17	68.40	71.37	66.93	68.35	64.42	64.05	60.78
1.00	68.50	64.30	67.93	63.83	66.24	62.42	63.39	60.01	59.33	56.53
1.10	63.94	60.29	63.40	59.84	61.80	58.48	59.10	56.17	55.26	52.84
1.20	59.94	56.75	59.44	56.32	57.91	55.02	55.35	52.80	51.72	49.61
1.30	56.42	53.61	55.94	53.20	54.49	51.94	52.05	49.81	48.60	46.74
1.40	53.29	50.80	52.83	50.40	51.45	49.19	49.13	47.14	45.83	44.19
1.50	50.49	48.26	50.05	47.88	48.73	46.71	46.51	44.74	43.37	41.91
1.60	47.97	45.97	47.54	45.60	46.28	44.47	44.16	42.57	41.15	39.85
1.70	45.68	43.89	45.28	43.53	44.07	42.44	42.03	40.60	39.15	37.98
1.80	43.60	41.98	43.22	41.63	42.05	40.58	40.10	38.81	37.33	36.28
1.90	41.71	40.23	41.34	39.90	40.22	38.88	38.34	37.17	35.68	34.72
2.00	39.97	38.63	39.61	38.30	38.53	37.32	36.72	35.66	34.16	33.29
2.10	38.37	37.14	38.02	36.82	36.98	35.87	35.24	34.26	32.77	31.97
2.20	36.89	35.76	36.56	35.46	35.55	34.53	33.87	32.97	31.49	30.75
2.30	35.53	34.49	35.21	34.19	34.23	33.30	32.61	31.78	30.30	29.63
2.40	34.26	33.30	33.95	33.01	33.01	32.14	31.43	30.67	29.20	28.58
2.50	33.08	32.19	32.77	31.91	31.86	31.06	30.33	29.63	28.18	27.60

Table 6.9 Characteristic impedance of edge-coupled off-set stripline, $c/(2w + s) = 20$, $s/b = 0.7$ (See structure A2 of Table 4.1)

d/b →	0.0		0.05		0.10		0.15		0.20	
w/b ↓	$Z_e\sqrt{\epsilon_r}$ (ohms)	$Z_o\sqrt{\epsilon_r}$ (ohms)	$Z_e\sqrt{\epsilon_r}$ (ohms)	$Z_o\sqrt{\epsilon_r}$ (ohms)	$Z_e\sqrt{\epsilon_r}$ (ohms)	$Z_o\sqrt{\epsilon_r}$ (ohms)	$Z_e\sqrt{\epsilon_r}$ (ohms)	$Z_o\sqrt{\epsilon_r}$ (ohms)	$Z_e\sqrt{\epsilon_r}$ (ohms)	$Z_o\sqrt{\epsilon_r}$ (ohms)
0.10	205.27	185.55	204.33	185.02	201.47	183.37	196.61	180.45	189.60	175.98
0.12	193.88	175.27	192.95	174.73	190.12	173.06	185.33	170.11	178.41	165.58
0.14	184.22	166.64	183.30	166.09	180.51	164.40	175.78	161.41	168.94	156.84
0.16	175.83	159.21	174.92	158.66	172.17	156.94	167.50	153.93	160.74	149.33
0.18	168.42	152.68	167.52	152.12	164.80	150.40	160.18	147.35	153.51	142.73
0.20	161.78	146.88	160.89	146.31	158.21	144.57	153.65	141.51	147.06	136.87
0.22	155.79	141.26	154.91	141.08	152.26	139.33	147.76	136.26	141.26	131.60
0.24	150.31	136.90	149.45	136.33	146.83	134.56	142.39	131.48	135.97	126.82
0.26	145.30	132.56	144.44	131.98	141.86	130.21	137.47	127.12	131.14	122.50
0.28	140.66	128.56	139.82	127.98	137.26	126.20	132.93	123.14	126.72	118.48
0.30	136.36	124.85	135.53	124.27	133.01	122.53	128.77	119.42	122.60	114.77
0.32	132.36	121.43	131.54	120.85	129.09	119.07	124.86	115.96	118.76	111.32
0.34	128.65	118.20	127.84	117.61	125.38	115.83	121.21	112.73	115.19	108.10
0.36	125.14	115.16	124.33	114.58	121.91	112.80	117.78	109.70	111.85	105.09
0.38	121.82	112.31	121.03	111.73	118.63	109.95	114.57	106.86	108.70	102.26
0.40	118.70	109.61	117.92	109.03	115.55	107.25	111.53	104.17	105.75	99.60
0.42	115.75	107.06	114.98	106.48	112.64	104.71	108.67	101.64	102.96	97.09
0.44	112.96	104.64	112.20	104.06	109.89	102.29	105.97	99.24	100.34	94.72
0.46	110.30	102.33	109.55	101.76	107.27	100.00	103.40	96.96	97.84	92.46
0.48	107.78	100.14	107.03	99.57	104.78	97.82	100.96	94.79	95.47	90.33
0.50	105.37	98.05	104.63	97.48	102.41	95.73	98.64	92.73	93.22	88.29
0.60	94.83	88.85	94.14	88.29	92.04	86.59	88.50	83.68	83.43	79.42
0.70	86.26	81.29	85.61	80.75	83.64	79.12	80.31	76.32	75.55	72.23
0.80	79.13	74.95	78.52	74.43	76.66	72.86	73.53	70.17	69.05	66.27
0.90	73.11	69.54	72.52	69.05	70.77	67.53	67.81	64.96	63.60	61.24
1.00	67.94	64.87	67.38	64.39	65.72	62.94	62.93	60.48	58.95	56.92
1.10	63.45	60.79	62.93	60.33	61.35	58.94	58.70	56.58	54.93	53.18
1.20	59.52	57.19	59.02	56.75	57.52	55.42	55.01	53.15	51.43	49.90
1.30	56.05	54.00	55.58	53.57	54.15	52.29	51.75	50.12	48.34	47.00
1.40	52.96	51.14	52.51	50.73	51.15	49.50	48.86	47.41	45.61	44.42
1.50	50.19	48.57	49.76	48.18	48.46	46.99	46.27	44.98	43.17	42.11
1.60	47.70	46.25	47.29	45.87	46.04	44.72	43.94	42.79	40.97	40.03
1.70	45.44	44.13	45.05	43.77	43.85	42.67	41.84	40.80	38.99	38.14
1.80	43.39	42.20	43.01	41.85	41.86	40.79	39.93	38.99	37.19	36.42
1.90	41.51	40.44	41.15	40.10	40.04	39.07	38.18	37.33	35.55	34.85
2.00	39.79	38.81	39.44	38.48	38.37	37.48	36.58	35.81	34.04	33.41
2.10	38.21	37.31	37.86	36.99	36.83	36.03	35.11	34.40	32.66	32.08
2.20	36.74	35.92	36.41	35.61	35.42	34.67	33.75	33.10	31.39	30.86
2.30	35.39	34.64	35.07	34.33	34.11	33.43	32.50	31.90	30.21	29.73
2.40	34.13	33.43	33.82	33.14	32.89	32.26	31.33	30.78	29.12	28.67
2.50	32.96	32.31	32.66	32.03	31.75	31.17	30.24	29.73	28.10	27.68

Table 6.10 Characteristic impedance of edge-coupled offset stripline, $c/(2w+s) = 20$, $s/b = 0.8$ (See structure A2 of Table 4.1)

d/b →	0.0		0.05		0.10		0.15		0.20	
w/b ↓	$Z_e\sqrt{\epsilon_r}$ (ohms)	$Z_o\sqrt{\epsilon_r}$ (ohms)	$Z_e\sqrt{\epsilon_r}$ (ohms)	$Z_o\sqrt{\epsilon_r}$ (ohms)	$Z_e\sqrt{\epsilon_r}$ (ohms)	$Z_o\sqrt{\epsilon_r}$ (ohms)	$Z_e\sqrt{\epsilon_r}$ (ohms)	$Z_o\sqrt{\epsilon_r}$ (ohms)	$Z_e\sqrt{\epsilon_r}$ (ohms)	$Z_o\sqrt{\epsilon_r}$ (ohms)
0.10	202.53	188.15	201.64	187.57	198.93	185.77	194.31	182.62	187.62	177.82
0.12	191.34	177.76	190.45	177.17	187.77	175.36	183.20	172.18	176.58	167.35
0.14	181.81	168.99	180.94	168.39	178.28	166.57	173.77	163.36	167.22	158.51
0.16	173.55	161.42	172.68	160.82	170.06	158.98	165.59	155.76	159.12	150.89
0.18	166.27	154.79	165.42	154.19	162.82	152.34	158.40	149.10	151.99	144.22
0.20	159.74	148.87	158.90	148.27	156.33	146.41	151.96	143.16	145.62	138.27
0.22	153.86	143.55	153.02	142.94	150.48	141.08	146.16	137.82	139.90	132.93
0.24	148.49	138.70	147.66	138.09	145.15	136.22	140.87	132.96	134.68	128.08
0.26	143.56	134.27	142.74	133.66	140.26	131.79	136.04	128.53	129.92	123.69
0.28	139.01	130.18	138.20	129.57	135.75	127.70	131.57	124.48	125.56	119.61
0.30	134.80	126.39	134.00	125.79	131.57	123.95	127.48	120.70	121.50	115.84
0.32	130.87	122.90	130.09	122.29	127.72	120.42	123.63	117.17	117.72	112.34
0.34	127.23	119.60	126.45	118.99	124.08	117.12	120.04	113.89	114.20	109.07
0.36	123.78	116.50	123.01	115.89	120.66	114.03	116.67	110.81	110.90	106.02
0.38	120.53	113.59	119.77	112.98	117.45	111.12	113.50	107.91	107.81	103.15
0.40	117.47	110.83	116.71	110.23	114.42	108.38	110.52	105.18	104.89	100.45
0.42	114.57	108.23	113.82	107.62	111.56	105.78	107.70	102.60	102.15	97.90
0.44	111.84	105.76	111.10	105.16	108.85	103.32	105.05	100.16	99.56	95.49
0.46	109.23	103.41	108.50	102.81	106.28	100.98	102.52	97.84	97.10	93.20
0.48	106.75	101.17	106.03	100.57	103.83	98.76	100.12	95.63	94.76	91.03
0.50	104.38	99.03	103.66	98.44	101.50	96.64	97.82	93.53	92.54	88.97
0.60	94.82	89.65	93.34	89.08	91.38	87.33	87.84	83.34	82.88	79.97
0.70	85.59	81.96	84.95	81.41	83.02	79.73	79.76	76.86	75.09	72.69
0.80	78.57	75.52	77.97	74.99	76.15	73.38	73.07	70.63	68.67	66.66
0.90	72.63	70.03	72.06	69.52	70.33	67.97	67.42	65.35	63.27	61.56
1.00	67.52	65.29	66.98	64.80	65.35	63.32	62.59	60.81	58.67	57.20
1.10	63.09	61.15	62.58	60.68	61.03	59.27	58.41	56.87	54.69	53.62
1.20	59.21	57.51	58.72	57.06	57.24	55.70	54.76	53.41	51.22	50.11
1.30	55.77	54.28	55.31	53.85	53.90	52.54	51.53	50.34	48.16	47.19
1.40	52.72	51.39	52.27	50.97	50.92	49.72	48.66	47.61	45.45	44.59
1.50	49.98	48.79	49.55	48.39	48.26	47.19	46.10	45.16	43.02	42.26
1.60	47.51	46.44	47.10	46.06	45.86	44.90	43.79	42.95	40.84	40.16
1.70	45.27	44.32	44.88	43.95	43.70	42.83	41.70	40.95	38.88	38.26
1.80	43.24	42.37	42.86	42.02	41.72	40.94	39.81	39.12	37.09	36.54
1.90	41.38	40.59	41.01	40.24	39.91	39.21	38.07	37.45	35.46	34.95
2.00	39.67	38.95	39.32	38.62	38.26	37.61	36.48	35.92	33.96	33.50
2.10	38.09	37.44	37.75	37.12	36.73	36.14	35.02	34.50	32.59	32.17
2.20	36.64	36.04	36.31	35.73	35.32	34.78	33.67	33.19	31.32	30.94
2.30	35.29	34.74	34.97	34.44	34.02	33.52	32.41	31.98	30.15	29.79
2.40	34.04	33.53	33.73	33.24	32.81	32.35	31.25	30.85	29.06	28.73
2.50	32.87	32.40	32.58	32.11	31.68	31.25	30.17	29.80	28.04	27.74

Table 6.11 Characteristic impedance of edge-coupled off-set stripline, $c/(2w + s) = 20$ $s/b = 0.9$ (See structure A2 of Table 4.1)

d/b → / w/b ↓	0.0		0.05		0.10		0.15		0.20	
	$Z_e\sqrt{\epsilon_r}$ (ohms)	$Z_o\sqrt{\epsilon_r}$ (ohms)	$Z_e\sqrt{\epsilon_r}$ (ohms)	$Z_o\sqrt{\epsilon_r}$ (ohms)	$Z_e\sqrt{\epsilon_r}$ (ohms)	$Z_o\sqrt{\epsilon_r}$ (ohms)	$Z_e\sqrt{\epsilon_r}$ (ohms)	$Z_o\sqrt{\epsilon_r}$ (ohms)	$Z_e\sqrt{\epsilon_r}$ (ohms)	$Z_o\sqrt{\epsilon_r}$ (ohms)
0.10	200.54	190.04	199.68	189.42	197.09	187.51	192.65	184.17	186.21	179.13
0.12	189.42	179.52	188.58	178.89	186.00	176.98	181.61	173.62	175.22	168.56
0.14	180.05	170.68	179.21	170.06	176.66	168.12	172.31	164.75	165.98	159.68
0.16	171.89	163.04	171.06	162.41	168.53	160.47	164.23	157.09	157.96	152.01
0.18	164.69	156.31	163.86	155.67	161.36	153.73	157.10	150.34	150.88	145.26
0.20	158.27	150.33	157.45	149.69	154.97	147.74	150.74	144.35	144.60	139.27
0.22	152.44	144.92	151.63	144.28	149.18	142.33	145.00	138.94	138.91	133.87
0.24	147.16	140.01	146.36	139.37	143.93	137.42	139.78	134.04	133.77	128.68
0.26	142.29	135.51	141.50	134.87	139.10	132.92	134.99	129.44	129.05	124.54
0.28	137.81	131.36	137.03	130.73	134.65	128.78	130.59	125.44	124.74	120.42
0.30	133.66	127.52	132.88	126.88	130.52	124.93	126.55	121.61	120.71	116.61
0.32	129.78	123.97	129.02	123.34	126.72	121.40	122.74	118.05	116.97	113.07
0.34	126.20	120.62	125.44	119.99	123.13	118.06	119.19	114.72	113.49	109.77
0.36	122.79	117.48	122.04	116.85	119.75	114.92	115.86	111.60	110.22	106.68
0.38	119.59	114.52	118.85	113.89	116.58	111.97	112.74	108.66	107.16	103.78
0.40	116.57	111.72	115.83	111.10	113.60	109.19	109.79	105.90	104.28	101.05
0.42	113.71	109.08	112.98	108.45	110.77	106.56	107.00	103.29	101.56	98.47
0.44	111.00	106.57	110.28	105.95	108.09	104.06	104.37	100.81	98.98	96.04
0.46	108.43	104.18	107.71	103.57	105.55	101.69	101.87	98.47	96.55	93.73
0.48	105.98	101.91	105.27	101.30	103.13	99.44	99.49	96.23	94.24	91.54
0.50	103.65	99.75	102.95	99.14	100.83	97.29	97.23	94.11	92.04	89.45
0.60	93.43	90.24	92.76	89.65	90.76	87.87	87.36	84.81	82.47	80.37
0.70	85.10	82.45	84.47	81.89	82.58	80.17	79.37	77.25	74.76	73.02
0.80	78.16	75.93	77.57	75.39	75.77	73.75	72.74	70.96	68.39	66.93
0.90	72.28	70.38	71.71	69.86	70.01	68.29	67.14	65.63	63.04	61.80
1.00	67.22	65.59	66.69	65.09	65.07	63.60	62.35	61.06	58.47	57.40
1.10	62.83	61.41	62.32	60.94	60.79	59.51	58.20	57.08	54.52	53.59
1.20	58.98	57.74	58.49	57.28	57.03	55.91	54.57	53.59	51.07	50.26
1.30	55.57	54.48	55.11	54.04	53.72	52.73	51.37	50.51	48.03	47.32
1.40	52.54	51.57	52.10	51.15	50.76	49.89	48.52	47.76	45.33	44.71
1.50	49.82	48.95	49.39	48.55	48.12	47.34	45.97	45.29	42.92	42.36
1.60	47.36	46.59	46.96	46.20	45.73	45.03	43.67	43.06	40.75	40.25
1.70	45.14	44.44	44.75	44.07	43.57	42.94	41.60	41.05	38.79	38.34
1.80	43.12	42.48	42.74	42.12	41.61	41.04	39.71	39.21	37.01	36.61
1.90	41.26	40.69	40.90	40.34	39.81	39.30	37.98	37.53	35.38	35.02
2.00	39.57	39.04	39.22	38.71	38.16	37.69	36.40	35.99	33.90	33.56
2.10	38.00	37.52	37.66	37.20	36.65	36.22	34.94	34.57	32.53	32.22
2.20	36.55	36.11	36.23	35.80	35.24	34.85	33.60	33.25	31.26	30.98
2.30	35.21	34.81	34.90	34.50	33.95	33.58	32.35	32.03	30.09	29.84
2.40	33.97	33.59	33.66	33.30	32.74	32.40	31.19	30.90	29.01	28.77
2.50	32.80	32.46	32.51	32.17	31.61	31.30	30.11	29.85	28.00	27.78

Table 6.12 Characteristic impedance of edge-coupled off-set stripline; $c/(2w + s) = 20$, $s/b = 1.0$ (See structure A2 of Table 4.1)

d/b → / w/b ↓	0.0		0.05		0.10		0.15		0.20	
	$Z_e\sqrt{\epsilon_r}$ (ohms)	$Z_o\sqrt{\epsilon_r}$ (ohms)	$Z_e\sqrt{\epsilon_r}$ (ohms)	$Z_o\sqrt{\epsilon_r}$ (ohms)	$Z_e\sqrt{\epsilon_r}$ (ohms)	$Z_o\sqrt{\epsilon_r}$ (ohms)	$Z_e\sqrt{\epsilon_r}$ (ohms)	$Z_o\sqrt{\epsilon_r}$ (ohms)	$Z_e\sqrt{\epsilon_r}$ (ohms)	$Z_o\sqrt{\epsilon_r}$ (ohms)
0.10	199.00	191.34	198.17	190.69	195.66	188.69	191.37	185.21	185.10	179.98
0.12	188.07	180.83	187.25	180.18	184.76	178.17	180.50	174.68	174.29	169.45
0.14	178.74	171.91	177.93	171.25	175.46	169.24	171.23	165.74	165.07	160.50
0.16	170.66	164.19	169.85	163.53	167.40	161.52	163.21	158.02	157.10	152.78
0.18	163.55	157.42	162.74	156.76	160.31	154.75	156.16	151.25	150.10	146.02
0.20	157.16	151.37	156.37	150.71	153.96	148.69	149.84	145.19	143.83	139.97
0.22	151.42	145.92	150.63	145.26	148.25	143.25	144.16	139.75	138.21	134.55
0.24	146.17	140.95	145.39	140.30	143.02	138.28	138.98	134.80	133.09	129.61
0.26	141.37	136.41	140.60	136.76	138.25	133.74	134.24	130.27	128.42	125.15
0.28	136.93	132.22	136.17	131.56	133.84	129.56	129.81	126.11	124.13	121.00
0.30	132.81	128.33	132.05	127.68	129.75	125.70	125.86	122.27	120.14	117.16
0.32	128.99	124.76	128.25	124.10	125.99	122.11	122.10	118.68	116.44	113.60
0.34	125.43	121.36	124.69	120.71	122.43	118.73	118.57	115.32	112.97	110.27
0.36	122.07	118.19	121.33	117.54	119.09	115.57	115.28	112.17	109.73	107.16
0.38	118.90	115.20	118.17	114.55	115.96	112.59	112.18	109.21	106.69	104.24
0.40	115.91	112.37	115.19	111.73	112.99	109.78	109.25	106.42	103.83	101.48
0.42	113.08	109.70	112.36	109.06	110.19	107.12	106.49	103.79	101.13	98.89
0.44	110.40	107.16	109.69	106.63	107.54	104.60	103.88	101.29	98.58	96.44
0.46	107.85	104.75	107.15	104.12	105.02	102.21	101.40	98.92	96.16	94.11
0.48	105.43	102.46	104.73	101.83	102.62	99.93	99.04	96.67	93.86	91.90
0.50	103.12	100.27	102.43	99.65	100.34	97.76	96.80	94.53	91.68	89.80
0.60	93.00	90.67	92.34	90.07	90.37	88.26	87.02	85.16	82.19	80.65
0.70	84.74	82.81	84.12	82.23	82.25	80.50	79.08	77.54	74.52	73.25
0.80	77.86	76.23	77.27	75.68	75.50	74.02	72.49	71.20	68.19	67.13
0.90	72.02	70.63	71.46	70.11	69.78	68.53	66.94	65.84	62.87	61.96
1.00	67.00	65.81	66.47	65.31	64.87	63.80	62.18	61.23	58.32	57.55
1.10	62.64	61.60	62.13	61.12	60.61	59.68	58.05	57.23	54.39	53.72
1.20	58.81	57.90	58.33	57.45	56.88	56.06	54.44	53.72	50.96	50.37
1.30	55.42	54.63	54.97	54.19	53.58	52.86	51.25	50.62	47.93	47.42
1.40	52.41	51.70	51.97	51.28	50.64	50.01	48.42	47.86	45.24	44.79
1.50	49.70	49.07	49.28	48.66	48.01	47.44	45.88	45.38	42.84	42.44
1.60	47.26	46.69	46.86	46.30	45.64	45.13	43.59	43.15	40.68	40.32
1.70	45.05	44.54	44.66	44.16	43.49	43.03	41.52	41.12	38.73	38.40
1.80	43.03	42.57	42.66	42.21	41.53	41.12	39.64	39.28	36.95	36.66
1.90	41.19	40.77	40.83	40.42	39.74	39.37	37.92	37.59	35.33	35.07
2.00	39.50	39.11	39.15	38.78	38.10	37.76	36.34	36.04	33.85	33.61
2.10	37.94	37.59	37.60	37.26	36.59	36.28	34.89	34.62	32.48	32.26
2.20	36.49	36.17	36.17	35.86	35.19	34.90	33.55	33.30	31.22	31.02
2.30	35.16	34.86	34.84	34.56	33.90	33.63	32.31	32.08	30.06	29.87
2.40	33.92	33.64	33.61	33.35	32.69	32.45	31.15	30.94	28.97	28.80
2.50	32.76	32.51	32.46	32.32	31.57	31.35	30.08	29.88	27.97	27.81

REFERENCES

1. L. Young (ed.), *Parallel-Coupled Lines and Directional Couplers*, Artech House, Dedham, Mass., 1972 (a reprint volume).
2. H. Howe Jr., *Stripline Circuit Design*, Artech House, Dedham, Mass., 1974.
3. G.L. Matthaei, L. Young and E.M.T. Jones, *Microwave Filters, Impedance Matching Networks and Coupling Structures*, McGraw-Hill, New York, 1964.
4. L. Young (ed.), *Microwave Filters Using Parallel-Coupled Lines*, Artech House, Dedham, Mass., 1972 (a reprint volume).
5. A. Matsumoto, *Microwave Filters and Circuits*, Academic Press, New York, 1970.
6. B.M. Schiffman, A new class of broadband microwave 90° phase shifters, *IRE Trans. Microwave Theory Tech.*, Vol. MTT-6, pp. 232-237, April 1958.
7. J.P. Shelton and J.A. Mosko, Synthesis and design of equi-ripple TEM directional couplers and fixed phase shifters, *IEEE Trans. Microwave Theory Tech.*, Vol. MTT-14, pp. 462-473, Oct. 1966.
8. D. Lacombe and J. Cohen, Octave-band microstrip d.c. blocks, *IEEE Trans. Microwave Theory Tech.*, Vol. MTT-20, pp. 555-556, Aug. 1972.
9. E.M.T. Jones, Wide-band stripline magic T, *IRE Trans. Microwave Theory Tech.*, Vol. MTT-8, pp. 160-168, March. 1960,
10. G.D. Alley, Interdigital capacitors and their application to lumped element microwave integrated circuits, *IEEE Trans. Microwave Theory Tech.*, Vol. MTT-18, pp. 1028-1033, Dec. 1970.
11. S.B. Cohn, Shielded coupled strip transmission lines, *IRE Trans. Microwave Theory Tech.*, Vol. MTT-3, pp. 29-38, Oct. 1955.
12. J.P. Shelton, Impedances of offset parallel-coupled strip transmission lines, *IEEE Trans. Microwave Theory Tech.*, Vol. MTT-14, pp. 7-14, Jan. 1966.
13. W.J. Getsinger, Coupled rectangular bars between parallel plates, *IRE Trans. Microwave Theory Tech.*, Vol. MTT-10, pp. 65-72, Jan. 1962.
14. J.I. Smith, The even- and odd-mode capacitance parameters for coupled lines in suspended substrate, *IEEE Trans. Microwave Theory Tech.*, Vol. MTT-19, pp. 424-431, May 1971.
15. B. Bhat and S.K. Koul, Unified approach to solve a class of strip and microstrip-like transmission lines, *IEEE Trans. Microwave Theory Tech.*, Vol. MTT-30, pp. 679-686, May 1982. Also correction, *IEEE Trans., Microwave Theory Tech.*, Vol. MTT-30, p. 2067, Nov. 1982.
16. S.K. Koul and B. Bhat, Generalized analysis of Microstrip-like transmission lines and coplanar strips with anisotropic substrates for MIC, electrooptic modulator and SAW application, *IEEE Trans. Microwave Theory Tech.*, Vol. MTT-31, pp. 1051-1059, Dec. 1983.
17. B. Bhat and S.K. Koul, A new approach to analyse strip and microstrip-like transmission lines with anisotropic substrates having tilted optical axis, *IEE Proc.*, Vol. 131, pt. H, No. 3, pp. 191-197, June 1984.
18. M.A.R. Gunston, *Microwave Transmission Line Impedance Data*, Van Nostrand Reinhold Company, London, 1972.
19. L.J.P. Linner, Explicit expressions for geometrical dimensions of symmetric coupled striplines, *Electron. Lett.*, Vol. 10, No. 4, pp. 44-45, Feb. 1974.

CHAPTER 7

Design Data on Broadside-coupled Stripline-Like Transmission Lines—Characteristic Impedances and Effective Dielectric Constants

7.1 Introduction

Broadside-coupled strip transmission lines are particularly useful in circuits where tight coupling is desired. Some of the specific circuits include wide-band filters and 3 dB directional couplers [1–4]. One of the simplest structures, namely, the broadside-coupled homogeneous stripline (Fig. 7.1) has been analysed by Cohn [1] using the Schwarz-Christoffel transformation method. Besides this, the analysis of a few other simple structures using the variational technique has been reported. For example, variational expressions for the even- and odd-mode capacitances have been reported for (i) broadside-coupled suspended stripline using isotropic dielectric (structure A2 of Table 4.4) by Allen and Estes [5] and Gish and Graham [6], (ii) broadside-coupled suspended stripline using uniaxially anisotropic substrate (structure A2 of Table 4.5) by Bahl and Bhartia [7] and D'Assuncao *et al.* [8], (iii) broadside-coupled inverted stripline using isotropic dielectric (structure A3 of Table 4.4) by Allen and Estes [5] and Bahl and Bhartia [9] and (iv) broadside-coupled inverted stripline using uniaxially anisotropic dielectric (structure A3 of Table 4.5) by Bahl and Bhartia [7] and D'Assuncao *et al.* [10]. The variational analysis combined with the 'transmission line technique' presented in Chapters 2 and 4 leading to the unified capacitance formula (4.64) is the most comprehensive and is based on the publications by the authors [11–13].

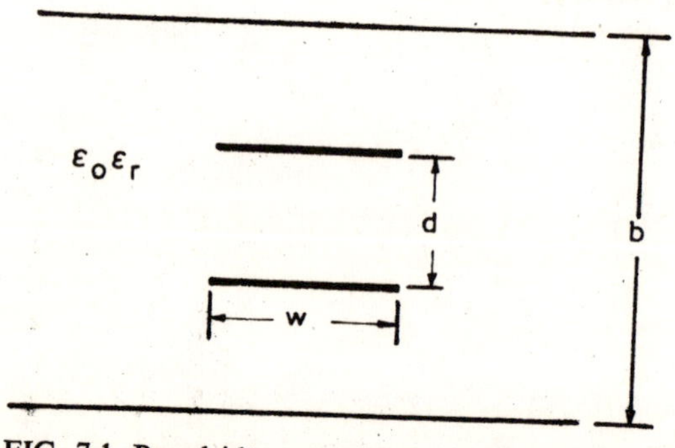

FIG. 7.1 Broadside-coupled homogeneous stripline with infinitely large ground planes.

In the following sections, we first summarize the characteristic impedance formulas reported by Cohn [1] for the broadside-coupled homogeneous stripline and also the unified capacitance expressions derived in Sec. 4.6 for a class of broadside-coupled symmetric structures. Tables 4.4–4.6 list a variety of broadside-coupled structures with iso/anisotropic substrates. Design data are generated for all the structures using the unified capacitance formula (4.64). In all the computations, the side walls are chosen sufficiently away from the strip conductor ($c/w = 40$), so that they have practically no effect on the propagation parameters. The propagation characteristics of the structures are described, supported by graphical illustrations. The graphs provided are mostly for commercially available dielectric substrates and can be read to an accuracy which is adequate for most engineering requirements. For the commonly used structures, data are also tabulated at sufficiently small intervals of the parameters. All the impedance data presented in this chapter are believed to be accurate within 1-2%. All tables of data are included towards the end of this chapter.

7.2 Design Equations

7.2.1 Unified Capacitance Formulas for Structures with Iso/Anisotropic Substrates

The unified variational capacitance expressions derived in Sec. 4.6 for the even- and odd-mode excitations of broadside-coupled structures are summarized below. The even- and odd-mode characteristic impedances, effective dielectric constants, phase velocities and guide wavelengths can then be determined using (6.1)–(6.4), respectively.

Referring to the coordinate system marked in structure A1 of Table 4.4, the charge distribution on the strip conductor S is assumed to be of the form

$$f(x)_{\substack{e\\o}} = (1/w)[1 + A_{\substack{e\\o}} \,|\, (2/w)(x - c/2) \,|^3], \qquad \frac{c-w}{2} \leqslant x \leqslant \frac{c+w}{2}$$
$$= 0, \qquad \text{otherwise} \tag{7.1}$$

For this distribution, the even- and odd-mode capacitances are expressed as:

$$C_{\substack{e\\o}} = \frac{(1 + 0.25A_{\substack{e\\o}})^2}{\sum\limits_{n\ \text{odd}} (L_n + A_{\substack{e\\o}} M_n)^2 P_n / Y_{\substack{e\\o}}} \tag{7.2}$$

where

$$L_n = \sin\left(\frac{\beta_n w}{2}\right); \qquad \beta_n = \frac{n\pi}{c} \tag{7.3a}$$

$$M_n = \left(\frac{2}{\beta_n w}\right)^3 \left[3\left\{\left(\frac{\beta_n w}{2}\right)^2 - 2\right\} \cos\left(\frac{\beta_n w}{2}\right) + \left(\frac{\beta_n w}{2}\right)\left\{\left(\frac{\beta_n w}{2}\right)^2 - 6\right\} \sin\left(\frac{\beta_n w}{2}\right) + 6\right] \tag{7.3b}$$

$$P_n = \left(\frac{2}{n\pi}\right)\left(\frac{2}{\beta_n w}\right)^2 \tag{7.3c}$$

$$A_{\substack{e\\o}} = -\frac{\sum\limits_{n\ \text{odd}} (L_n - 4M_n) L_n P_n / Y_{\substack{e\\o}}}{\sum\limits_{n\ \text{odd}} (L_n - 4M_n) M_n P_n / Y_{\substack{e\\o}}} \tag{7.3d}$$

The expressions for the even- and odd-mode admittances Y_e and Y_o, for some useful structures are listed in Tables 4.4 to 4.6.

7.2.2 *Characteristic Impedance of Broadside-coupled Homogeneous Stripline—Conformal Mapping Formulas*

Cohn [1] has derived formulas for the even- and odd-mode characteristic impedances of broadside-coupled homogeneous stripline (Fig. 7.1) by applying the Schwarz-Christoffel transformation method. These formulas are valid for $w/d \geqslant 0.35$ and are given as follows:

$$Z_e\sqrt{\epsilon_r} = 59.952\pi \frac{K(k')}{K(k)} \tag{7.4a}$$

$$Z_o\sqrt{\epsilon_r} = \frac{94.172\pi(d/b)}{\tanh^{-1} k} \tag{7.4b}$$

where

$$k' = \sqrt{1 - k^2} \tag{7.4c}$$

$$w/b = (2/\pi)\left[\tanh^{-1}\sqrt{\frac{k\left(k - \frac{d}{b}\right)}{\left(1 - k\frac{d}{b}\right)}} - d/b \tanh^{-1}\sqrt{\frac{\left(k - \frac{d}{b}\right)}{k\left(1 - k\frac{d}{b}\right)}}\right] \tag{7.4d}$$

In (7.4a), $K(k)$ and $K(k')$ are complete elliptic integrals of the first kind. The quantity $[K(k')/K(k)]$ as a function of k has been tabulated in Oberhettinger and Magnus [14]. With the aid of these tables, formula (7.4) can be used to determine Z_e and Z_o for specified w/b and d/b. The quantity $[K(k')/K(k)]$ can also be expressed in simpler closed form formulas as follows [15].

For $0.5 \leqslant k^2 \leqslant 1;\ 1 \leqslant \dfrac{K(k)}{K(k')} \leqslant \infty$

$$\frac{K(k)}{K(k')} \simeq \frac{1}{\pi} \ln\left(\frac{2(1 + \sqrt{k})}{(1 - \sqrt{k})}\right) \tag{7.5a}$$

For $0 \leqslant k^2 \leqslant 0.5;\ 0 \leqslant \dfrac{K(k)}{K(k')} \leqslant 1$

$$\frac{K(k)}{K(k')} \simeq \frac{\pi}{\ln\left(\dfrac{2(1 + \sqrt{k'})}{(1 - \sqrt{k'})}\right)} \tag{7.5b}$$

The accuracy of $[K(k)/K(k')]$ as calculated using (7.5) is of the order of three parts per million.

Cohn [1] has also reported another set of simple formulas which are valid under the condition $w/d \geqslant 0.35$ and $(w/b)/(1 - d/b) \geqslant 0.35$. The expressions are in terms of the fringe capacitances and are given by

$$Z_e\sqrt{\epsilon_r} = \frac{59.952\pi}{\left[\dfrac{w/b}{(1 - d/b)} + \dfrac{C'_{fe}}{\epsilon}\right]} \tag{7.6a}$$

$$Z_o\sqrt{\epsilon_r} = \frac{59.952\pi}{\left[\frac{w/b}{(1 - d/b)} + \frac{w}{d} + \frac{C'_{fo}}{\epsilon}\right]} \tag{7.6b}$$

where

$$\frac{C'_{fe}}{\epsilon} = 0.4413 + 1/\pi\left[\ln\left(\frac{1}{1 - d/b}\right) + \frac{d/b}{(1 - d/b)}\ln(b/d)\right] \tag{7.7a}$$

$$\frac{C'_{fo}}{\epsilon} = (1/\pi)(b/d)\left[\ln\left(\frac{1}{1 - d/b}\right) + \frac{d/b}{(1 - d/b)}\ln(b/d)\right] \tag{7.7b}$$

In (7.6) and (7.7), C'_{fe} and C'_{fo} are the fringe capacitances per unit length of the line for the even- and odd-modes, respectively, and ϵ is the absolute dielectric constant of the medium. The error in the computed values of characteristic impedances using the above formulas is reported to be less than 0.1% for close couplings and between 0.1% and 1% for loose couplings of practical interest [1]. When broadside-coupled lines are fabricated using high dielectric constant substrates, it is recommended to use (7.4) instead of (7.6).

7.2.3 Explicit Expressions for the Dimensional Parameters of Broadside-coupled Symmetric Homogeneous Stripline in Terms of Coupling

Consider the broadside-coupled symmetric stripline shown in Fig. 7.1. If C_v denotes the voltage coupling coefficient and Z_e and Z_o represent the even- and odd-mode characteristic impedances, then

$$C_v = \frac{\zeta - 1}{\zeta + 1}, \qquad \zeta = \frac{Z_e}{Z_o} \tag{7.8a}$$

In order that the coupler is perfectly matched to its terminating transmission line having a characteristic impedance Z_t, it is necessary that

$$Z_t = \sqrt{Z_e Z_o} \tag{7.8b}$$

With (7.8b) in (7.8a), we can express Z_e and Z_o in terms of coupling coefficient C_v as:

$$Z_e = Z_t\sqrt{\frac{1 + C_v}{1 - C_v}} \tag{7.9a}$$

$$Z_o = Z_t\sqrt{\frac{1 - C_v}{1 + C_v}} \tag{7.9b}$$

From (7.4a) and (7.9a), we have

$$\frac{K(k')}{K(k)} = \frac{59.952\pi}{Z_t\sqrt{\epsilon_r}}\left(\frac{1 - C_v}{1 + C_v}\right)^{1/2} \tag{7.10}$$

From (7.5a) and (7.5b), k can be expressed as follows:

For $0.5 \leqslant k^2 \leqslant 1$; $1 \leqslant \frac{K(k)}{K(k')} \leqslant \infty$

$$k = \left(\frac{0.5\exp[\pi K(k)/K(k')] - 1}{0.5\exp[\pi K(k)/K(k')] + 1}\right)^2 \tag{7.11a}$$

For $0 \leqslant k^2 \leqslant 0.5; 0 \leqslant \dfrac{K(k)}{K(k')} \leqslant 1$

$$k = \left[1 - \left(\frac{0.5 \exp\,[\pi K(k')/K(k)] - 1}{0.5 \exp\,[\pi K(k')/K(k)] + 1}\right)^4\right]^{1/2} \tag{7.11b}$$

From (7.4b) and (7.9b), we have

$$d/b = \frac{Z_t}{94.172\pi}\sqrt{\frac{1 - C_v}{1 + C_v}}\sqrt{\epsilon_r}\,\tanh^{-1}(k) \tag{7.12}$$

For a given coupling coefficient C_v, ϵ_r, and Z_t, the values of d/b and w/b can be obtained from (7.12) and (7.4d), respectively.

7.3 Broadside-coupled Transmission Structures with Isotropic Dielectric Substrates

7.3.1 Broadside-coupled Homogeneous Stripline

For the broadside-coupled homogeneous stripline filled with isotropic dielectric, Fig. 7.2a illustrates variations in $Z_e\sqrt{\epsilon_r}$ and $Z_o\sqrt{\epsilon_r}$ as a function of the normalized spacing d/b. The same graph applies for any value of ϵ_r. For a fixed value of w/b, as the spacing d between the conductors is increased, the even-mode impedance Z_e reduces, whereas the odd-mode impedance Z_o increases. Table 7.1 provides detailed data on $Z_e\sqrt{\epsilon_r}$ and $Z_o\sqrt{\epsilon_r}$ as a function of w/b for d/b from 0.05 to 0.5 in steps of 0.05. Since the even- and odd-mode phase velocities are equal, this configuration is ideal for realizing high directivity backward-wave couplers.

The variations of d/b and w/b versus voltage coupling coefficient C_v in decibels are shown in Fig. 7.2b. The terminating line impedance Z_t is chosen as 50 ohms. For a fixed value of $\epsilon_r = 2.56$, d/b increases as C_v (dB) increases from 1 to 30 dB. On the other hand, w/b first increases and then decreases as C_v (dB) increases. Table 7.2 provides design data on the dimensional parameters of a broadside-coupled stripline as a function of coupling, computed using the formulas (7.4d) and (7.12). The substrate chosen is duroid ($\epsilon_r = 2.22$). Data are presented for coupling values in the range 1 to 10 dB in steps of 0.1 dB and 10.5 to 30 dB in steps of 0.5 dB.

7.3.2 Broadside-coupled Suspended Stripline

Figure 7.3a illustrates variations in the even- and odd-mode impedances of the structure as a function of w/b for various values of d/b with duroid ($\epsilon_r = 2.22$) as the dielectric substrate. For the same set of parameters, the variations in the even- and odd-mode effective dielectric constants are depicted in Fig. 7.3b. Increasing the width of the strip conductors with all other parameters held fixed decreases Z_e, Z_o and ϵ_{fe} and increases ϵ_{fo}. ϵ_{fo} is greater than ϵ_{fe} and hence the even-mode phase velocity v_e is greater than the odd-mode phase velocity v_o. Further, by adjusting the parameters, a wide range of mode phase velocities can be achieved. In particular, the structure with a large phase velocity ratio can be advantageously utilized in the design of high directivity forward-wave tight couplers [16]. Detailed data on Z_e, Z_o, ϵ_{fe} and ϵ_{fo} of broadside-coupled suspended stripline are tabulated as functions of w/b and d/b in Tables 7.3–7.5 with duroid ($\epsilon_r = 2.22$), quartz ($\epsilon_r = 3.78$) and alumina ($\epsilon_r = 9.6$), respectively, as substrate materials. Comparing the data presented for the three values of ϵ_r, it is observed that as the dielectric constant of the centre slab is increased, both Z_e and Z_o decrease and the phase velocity ratio v_e/v_o increases.

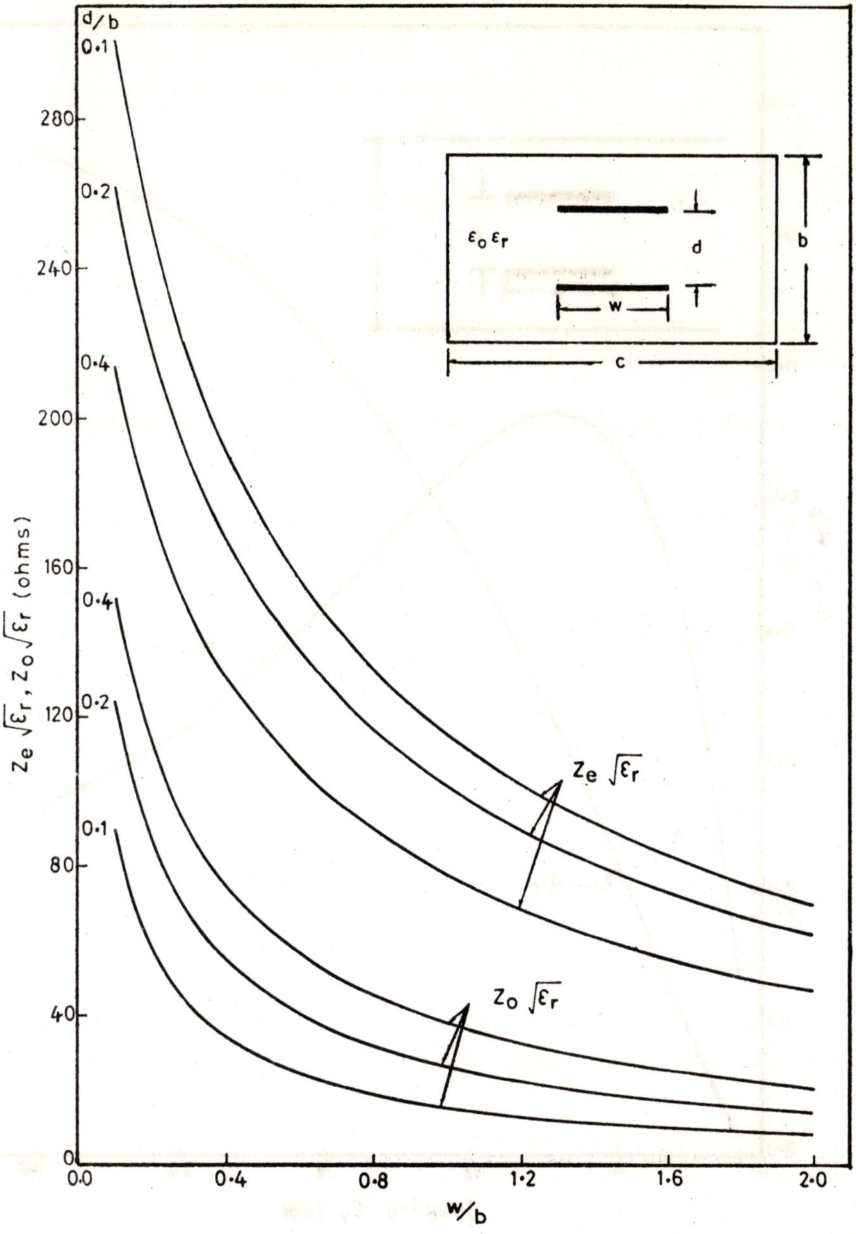

FIG. 7.2a $Z_e\sqrt{\epsilon_r}$ and $Z_o\sqrt{\epsilon_r}$ versus w/b and d/b as parameter for broadside-coupled homogeneous stripline.

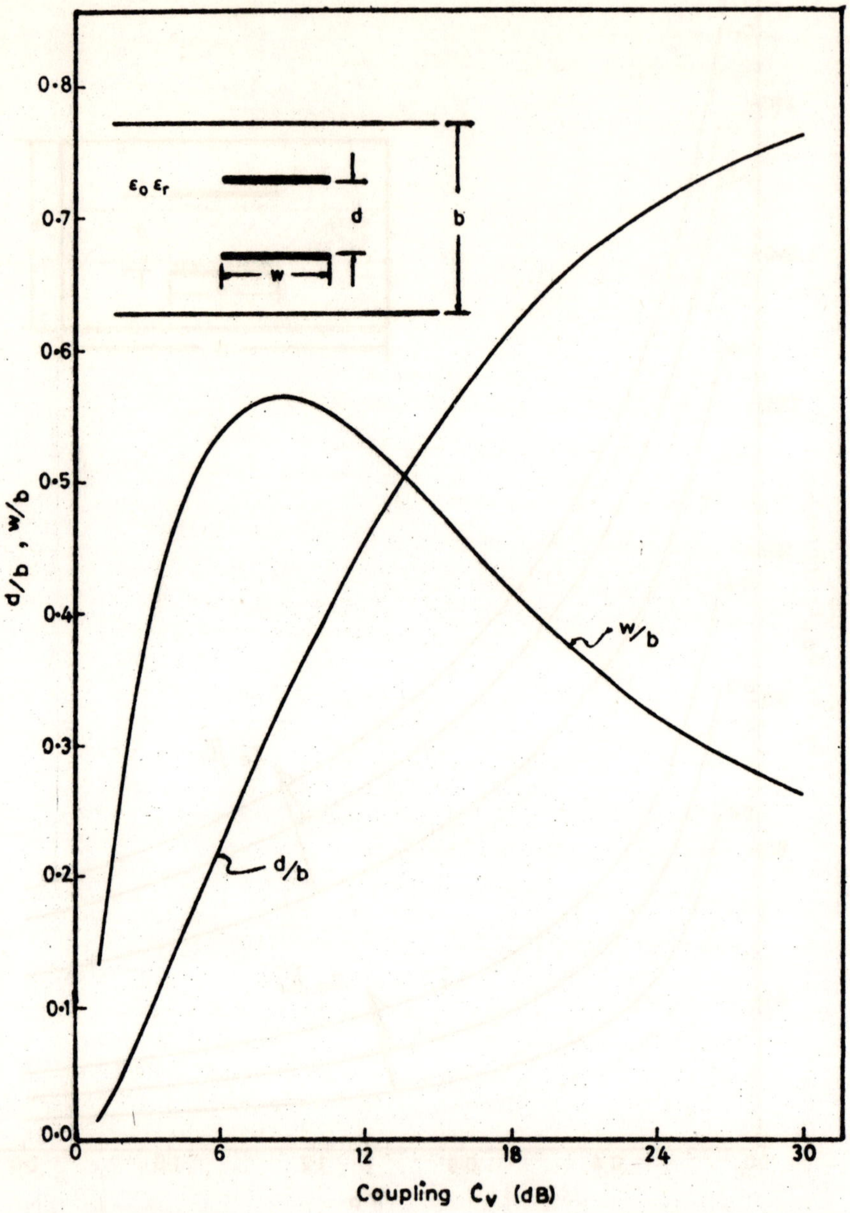

FIG. 7.2b Variations of d/b and w/b with voltage coupling coefficient C_v; $\epsilon_r = 2.56$, $Z_t = 50$ ohms.

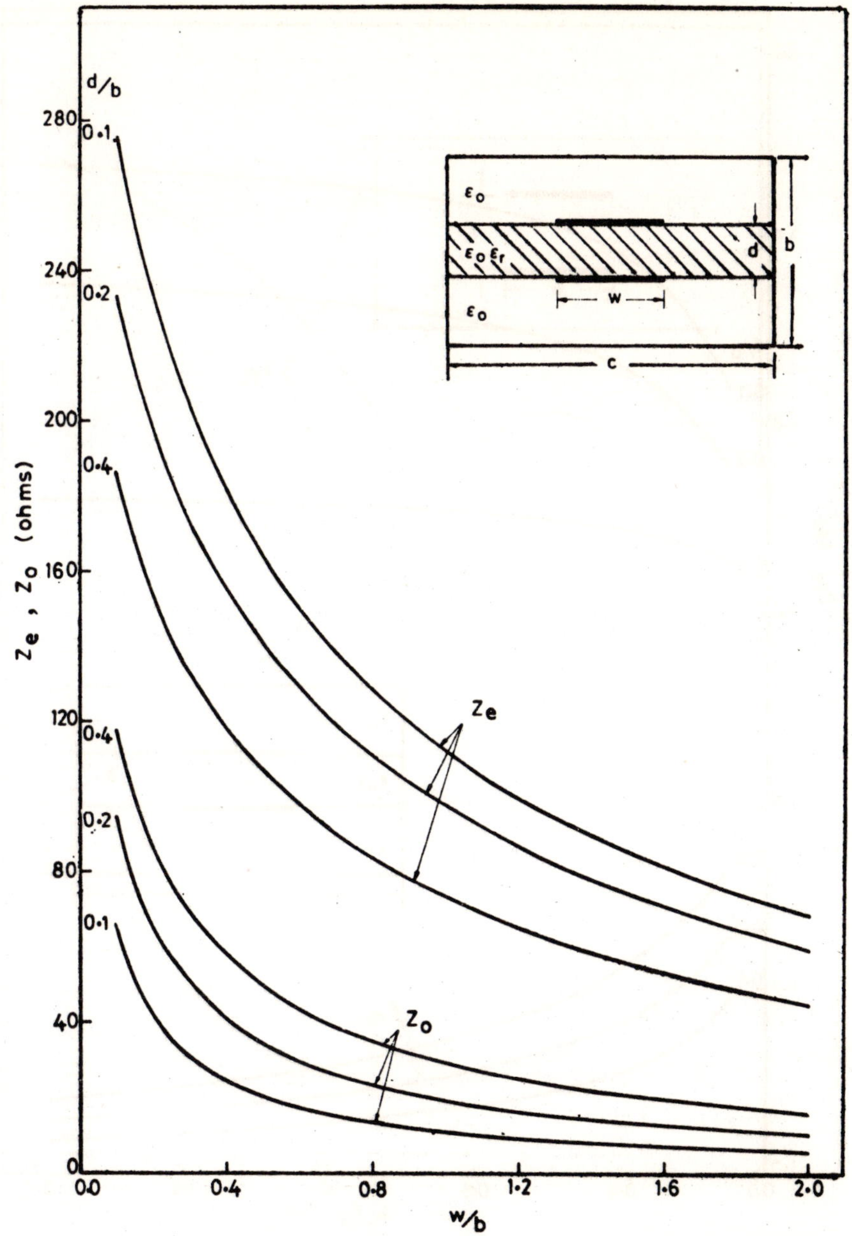

FIG. 7.3a Z_e and Z_o versus w/b with d/b as parameter for broadside-coupled suspended stripline; $\epsilon_r = 2.22$, $c/w = 40$.

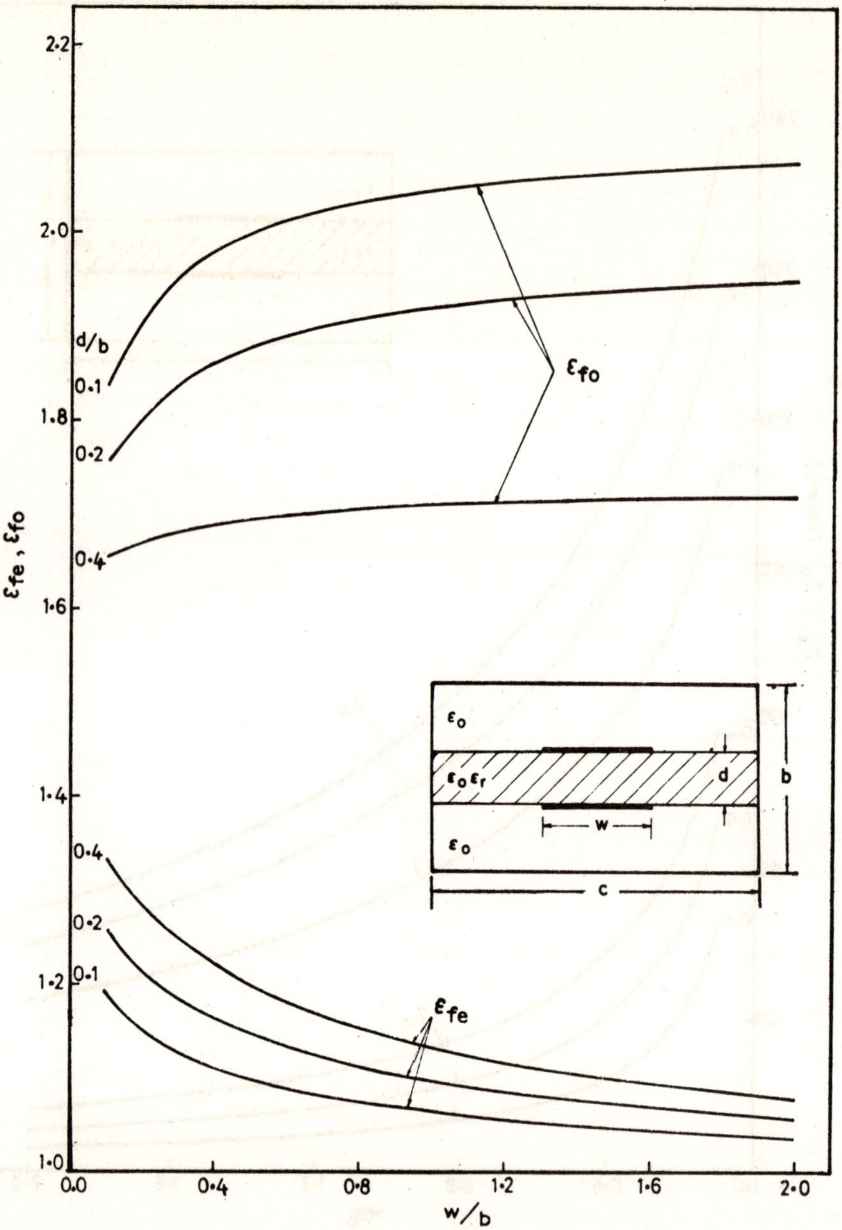

FIG. 7.3b ϵ_{fe} and ϵ_{fo} versus w/b with d/b as parameter for broadside-coupled suspended stripline; $\epsilon_r = 2.22$, $c/w = 40$.

7.3.3 Broadside-coupled Inverted Stripline

The impedance characteristics of the broadside-coupled inverted stripline (Fig. 7.4a) are similar to those of the broadside-coupled suspended stripline (Fig. 7.3a). However, since the dielectric regions are interchanged in the two structures, the variations in the effective dielectric constants are complementary (Figs. 7.4b and 7.3b). As is apparent from Fig. 7.4b, in the broadside-coupled inverted stripline, the odd-mode phase velocity v_o is greater than the even-mode phase velocity and the ratio v_o/v_e can be increased by decreasing d/b and increasing w/b. Detailed data on Z_e, Z_o, ϵ_{fe} and ϵ_{fo} of this structure are tabulated as functions of w/b and d/b in Tables 7.6–7.8 for $\epsilon_r = 2.22$, 3.78 and 9.6, respectively. From these data, it can be seen that as ϵ_r increases, both Z_e and Z_o decrease and the phase velocity ratio v_o/v_e increases.

7.3.4 Broadside-coupled Stripline with Double Suspended Substrates

This structure employs two dielectric substrates suspended symmetrically with the strip conductors facing each other as shown in structure A4 of Table 4.4. For this, the variations in the characteristic impedances and effective dielectric constants as a function of w/b are illustrated in Figs. 7.5a and 7.5b, respectively. Characteristics are plotted for $\epsilon_r = 3.78$, $a/b = 0.1$ with d/b as a parameter. For a larger substrate thickness, i.e., $a/b = 0.2$ and with all other parameters same as in Figs. 7.5a and 7.5b, the impedance characteristics of the structure are plotted in Fig. 7.6a and the variations in the effective dielectric constants are shown in Fig. 7.6b. As expected, with an increase in the air gap d, Z_e decreases. For the odd-mode, the structure is similar to the conventional inverted microstrip for small values of d and is similar to the suspended microstrip [17] for large values of d. Consequently, with increasing d, the odd-mode impedance increases initially and then decreases. For the structural parameters considered in Figs. 7.5a and 7.6a, Z_o increases with an increase in d/b. Referring to Figs. 7.5b and 7.6b, it is seen that as d/b is increased from 0.05 to 0.3, ϵ_{fo} increases, whereas ϵ_{fe} can either increase or decrease depending on the value of w/b. However, ϵ_{fe} is always greater than ϵ_{fo}. Further, phase velocity ratio close to unity is possible by choosing a larger air gap d and thinner substrates.

7.3.5 Broadside-coupled Stripline with Double Suspended Inverted Substrates

The cross-sectional view of broadside-coupled stripline with double suspended inverted substrates is shown as structure A5 in Table 4.4. It is similar to structure A4 except that the strip conductors on the two substrates face the top and bottom conducting planes. The impedance characteristics of the structure as a function of w/b with d/b as a parameter are plotted in Fig. 7.7a for $a/b = 0.05$ and in Fig. 7.8a for $a/b = 0.1$. The variations in the effective dielectric constants for the same set of parameters are shown in Fig. 7.7b for $a/b = 0.05$ and in Fig. 7.8b for $a/b = 0.1$. In all the cases, fused quartz having $\epsilon_r = 3.78$ is chosen as the substrate material. The variations in Z_e and Z_o (Figs. 7.7a and 7.8a) are similar to those of the broadside-coupled stripline with double suspended substrates (Figs. 7.5a and 7.6a). In the present structure, however, ϵ_{fo} is greater than ϵ_{fe}. Further, for a fixed value of w/b, as d/b is increased from a small value, the change in ϵ_{fe} is rather small, whereas ϵ_{fo} decreases considerably. With the appropriate choice of parameters, the phase velocity ratio can be nearly equalized.

7.3.6 Broadside-coupled Sandwiched Stripline

In the broadside-coupled sandwiched stripline, the strip conductors are sandwiched between

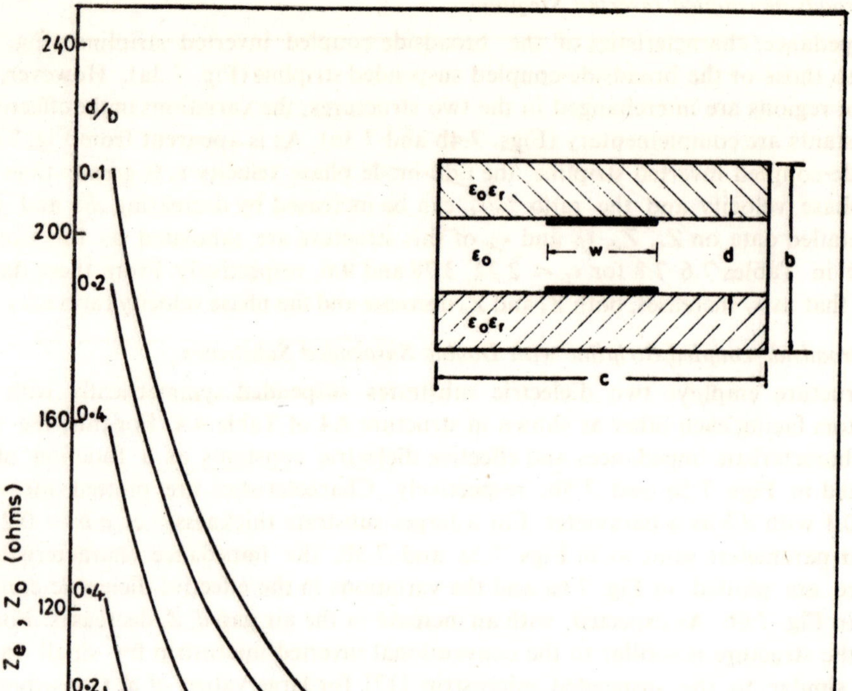

FIG. 7.4(a) Z_e and Z_o versus w/b with d/b as parameter for broadside-coupled inverted stripline; $\epsilon_r = 2.22$, $c/w = 40$.

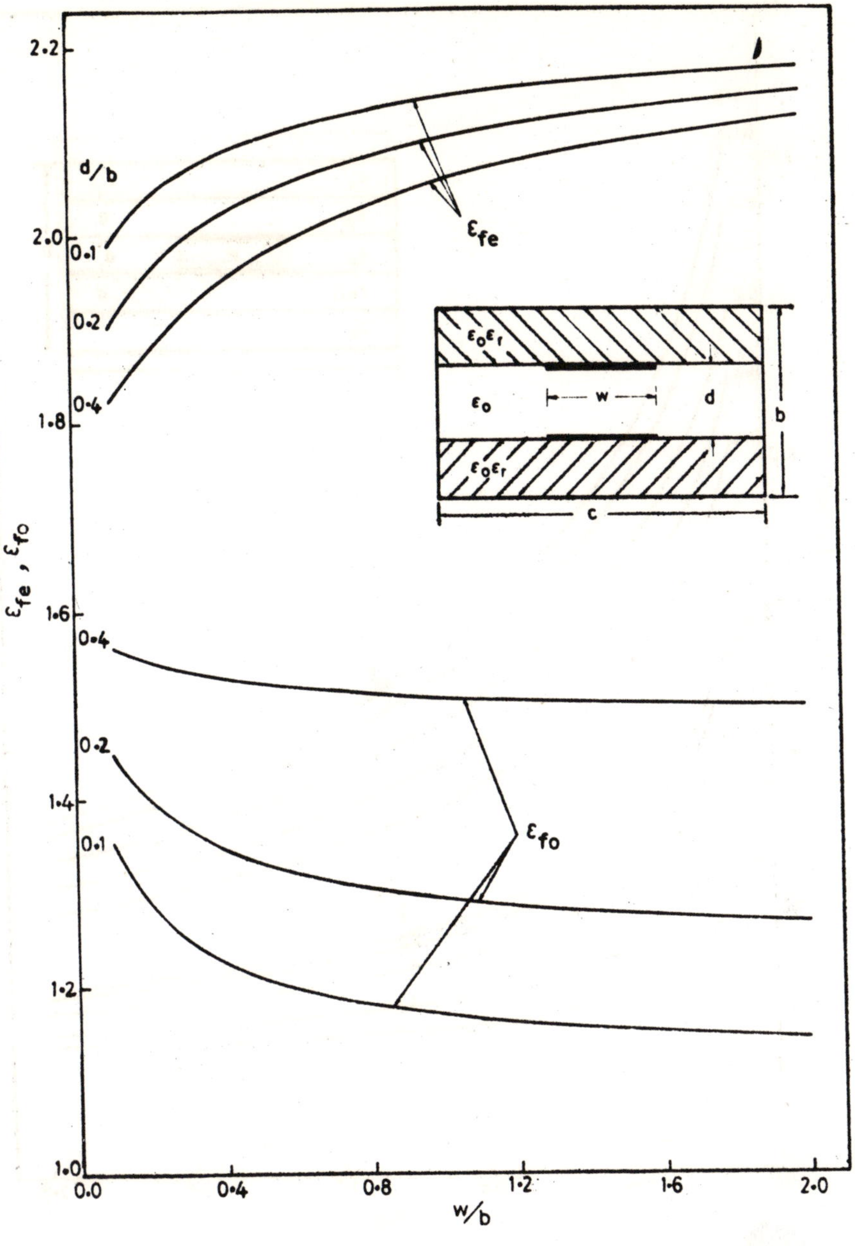

FIG. 7.4(b) ϵ_{fe} and ϵ_{fo} versus w/b with d/b as parameter for broadside-coupled inverted stripline; $\epsilon_r = 2.22$, $c/w = 40$.

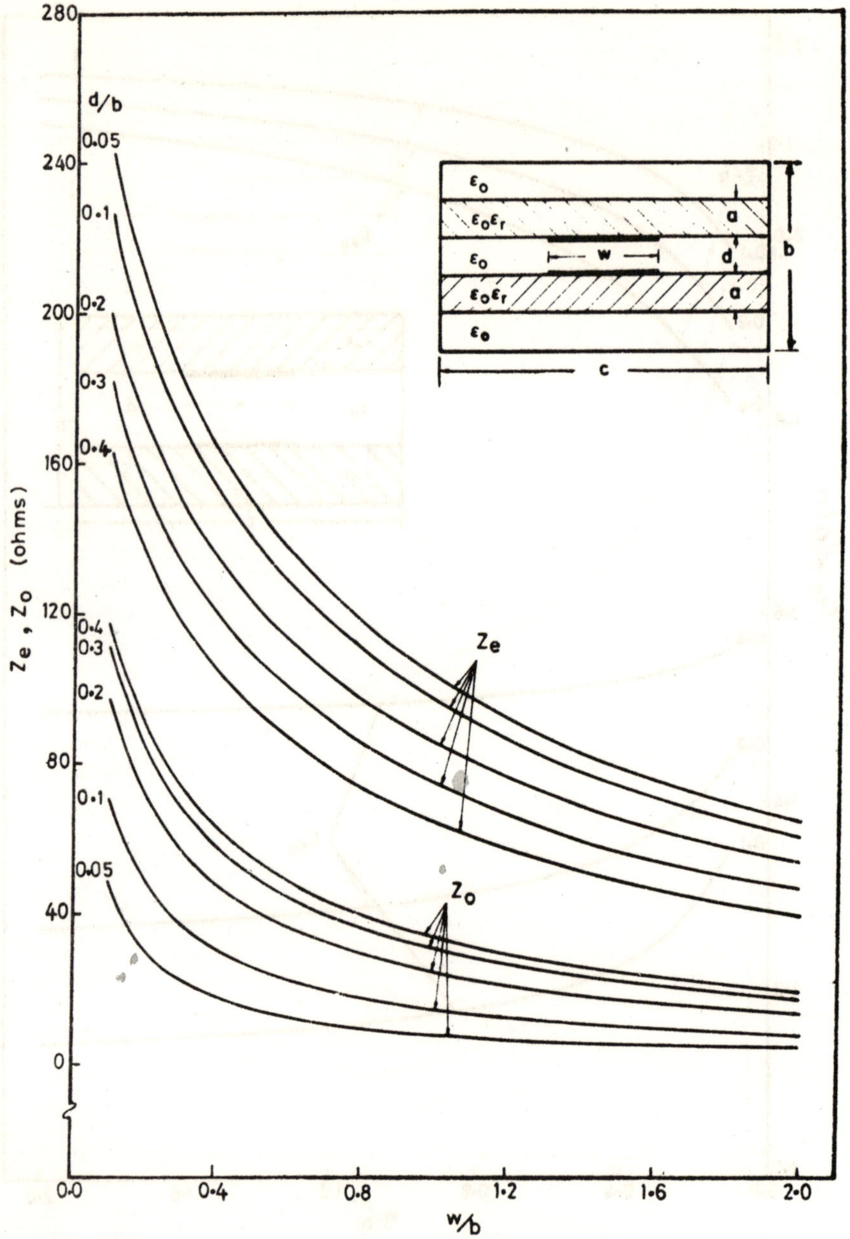

FIG. 7.5(a) Z_e and Z_o versus w/b with d/b as parameter for broadside-coupled stripline with double suspended dielectric substrates; $\epsilon_r = 3.78$, $a/b = 0.1$, $c/w = 40$.

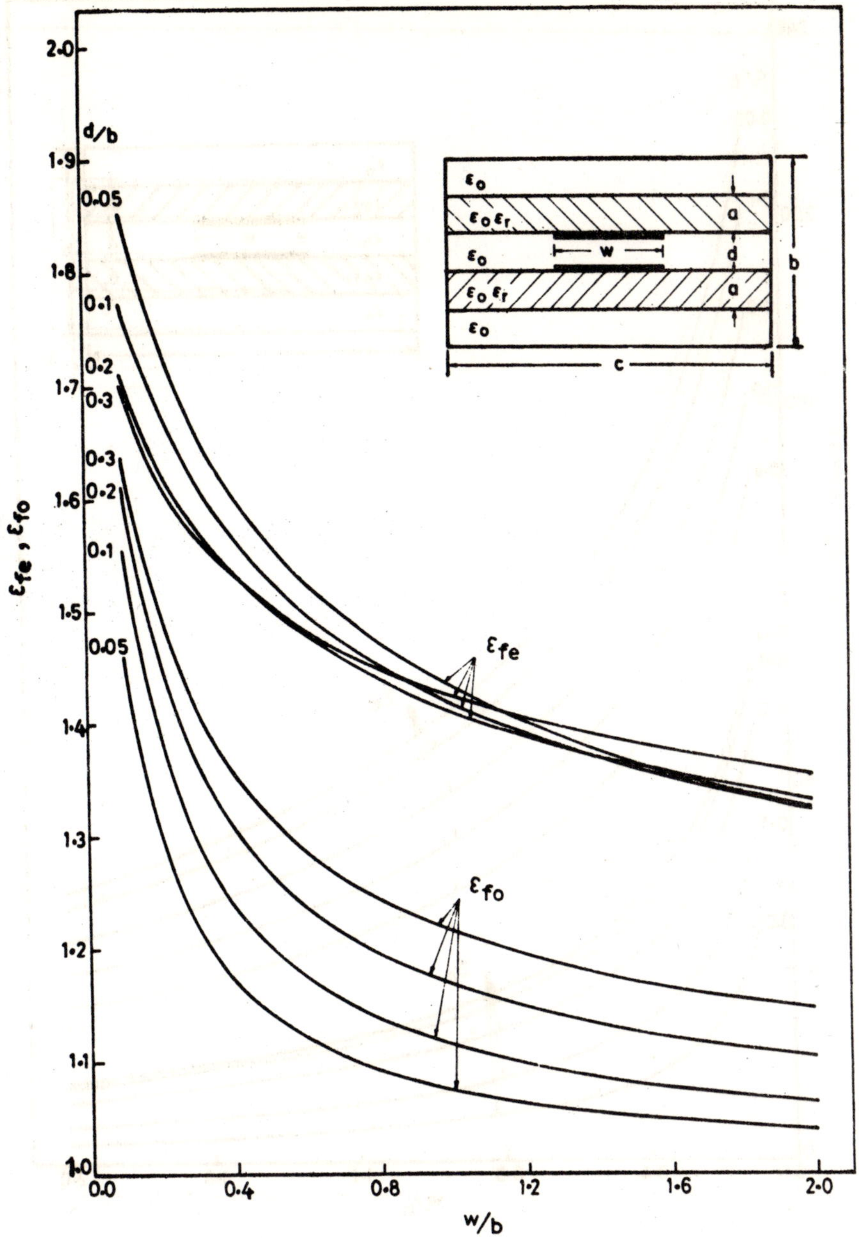

FIG. 7.5(b) ϵ_{fe} and ϵ_{fo} versus w/b with d/b as parameter for broadside-coupled stripline with double suspended dielectric substrates; $\epsilon_r = 3.78$, $a/b = 0.1$, $c/w = 40$.

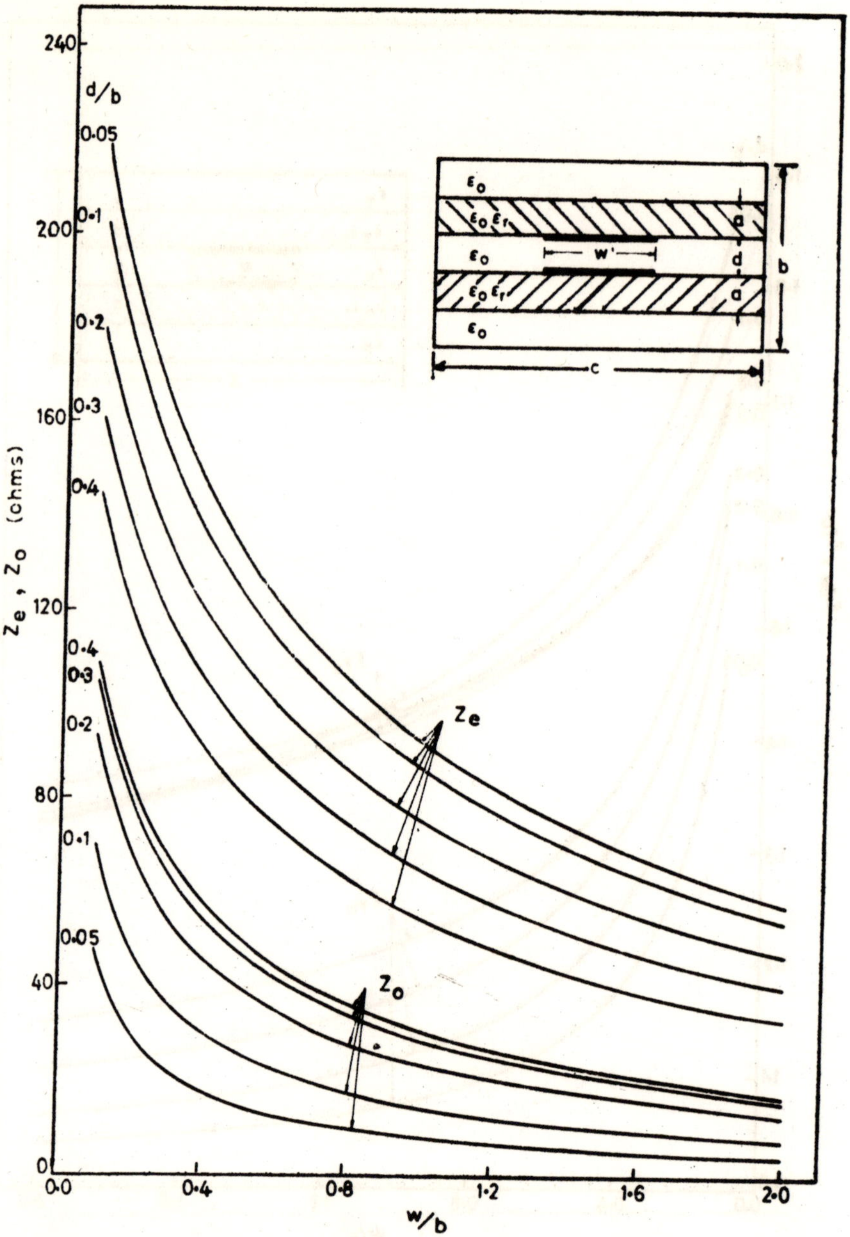

FIG. 7.6(a) Z_e and Z_o versus w/b with d/b as parameter for broadside-coupled stripline with double suspended dielectric substrates; $\epsilon_r = 3.78$, $a/b = 0.2$, $c/w = 40$.

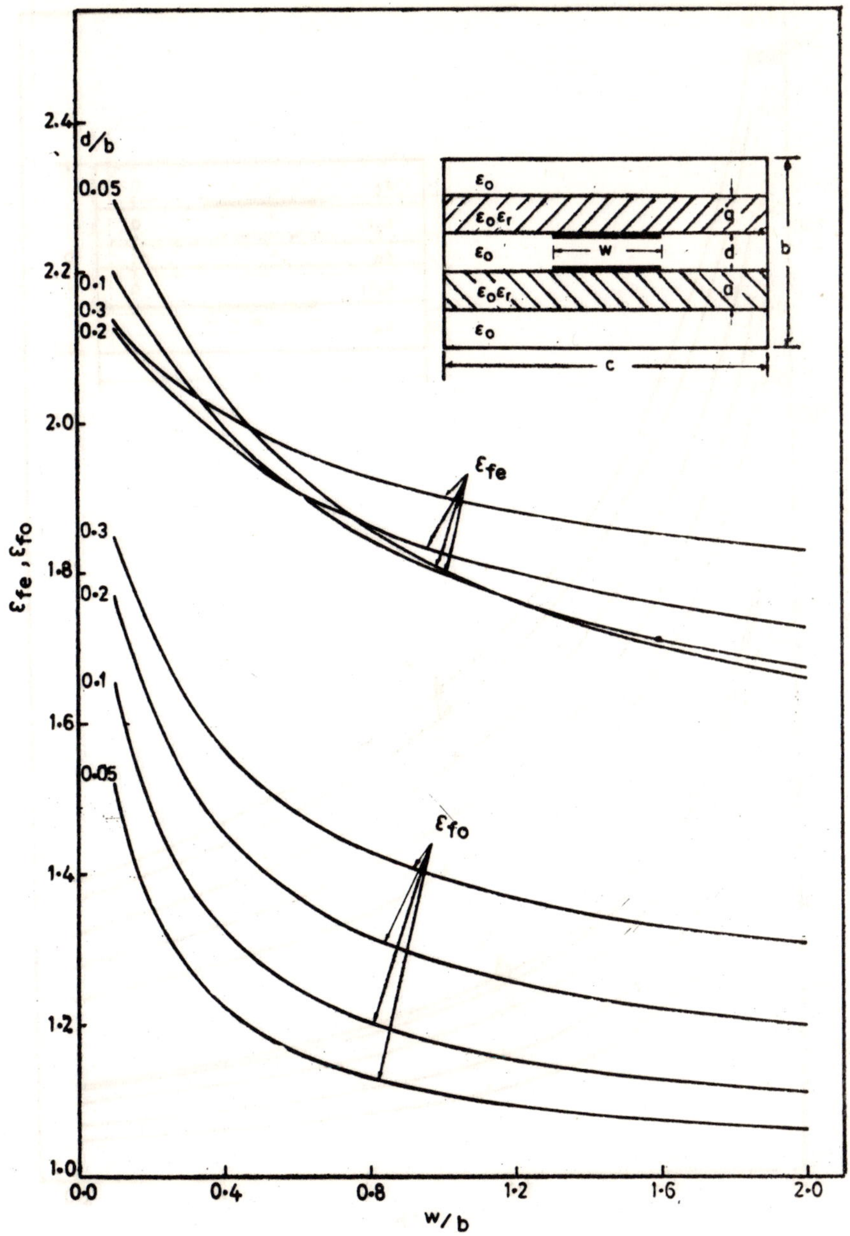

FIG. 7.6(b) ϵ_{fe} and ϵ_{fo} versus w/b with d/b as parameter for broadside-coupled stripline with double suspended dielectric substrates; $\epsilon_r = 3.78$, $a/b = 0.2$, $c/w = 40$.

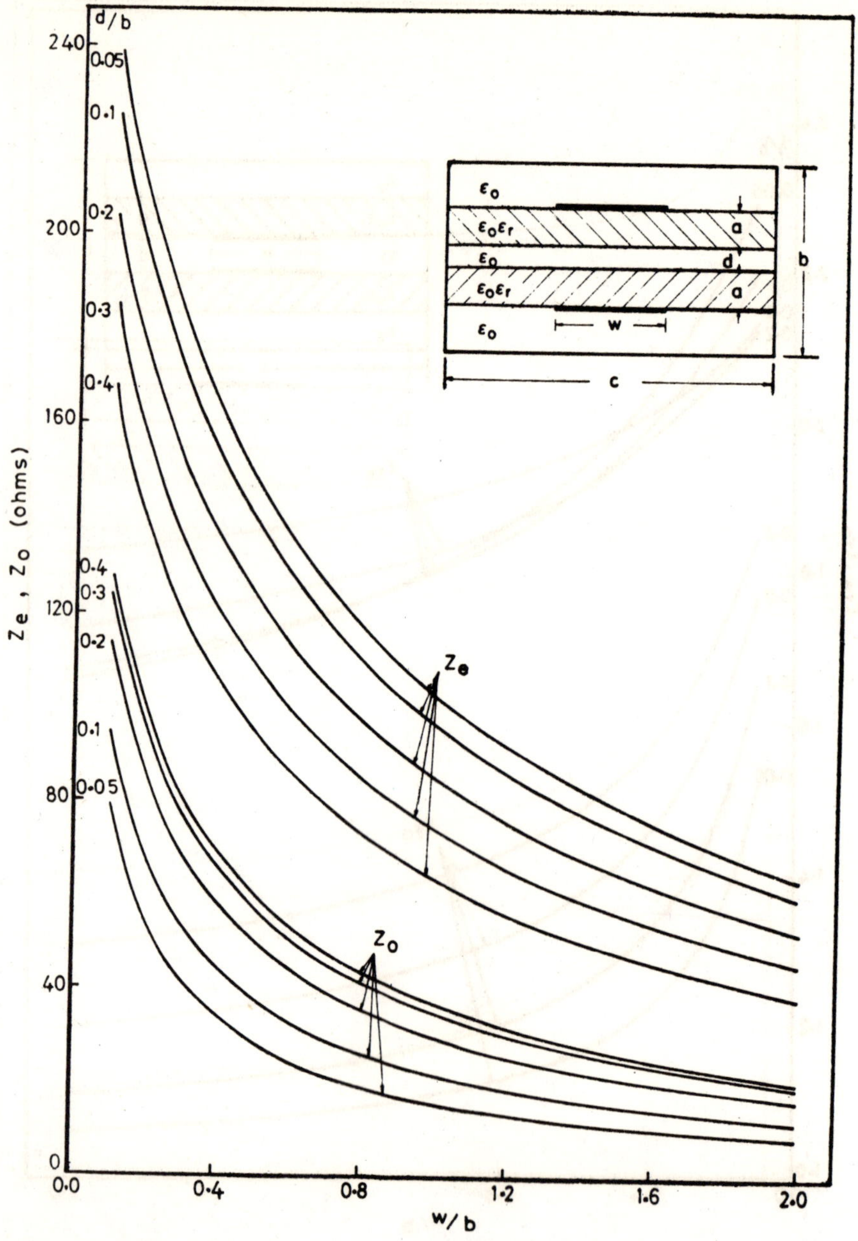

FIG. 7.7(a) Z_e and Z_o versus w/b with d/b as parameter for broadside-coupled stripline with double suspended inverted dielectric substrates; $\epsilon_r = 3.78$, $a/b = 0.05$, $c/w = 40$.

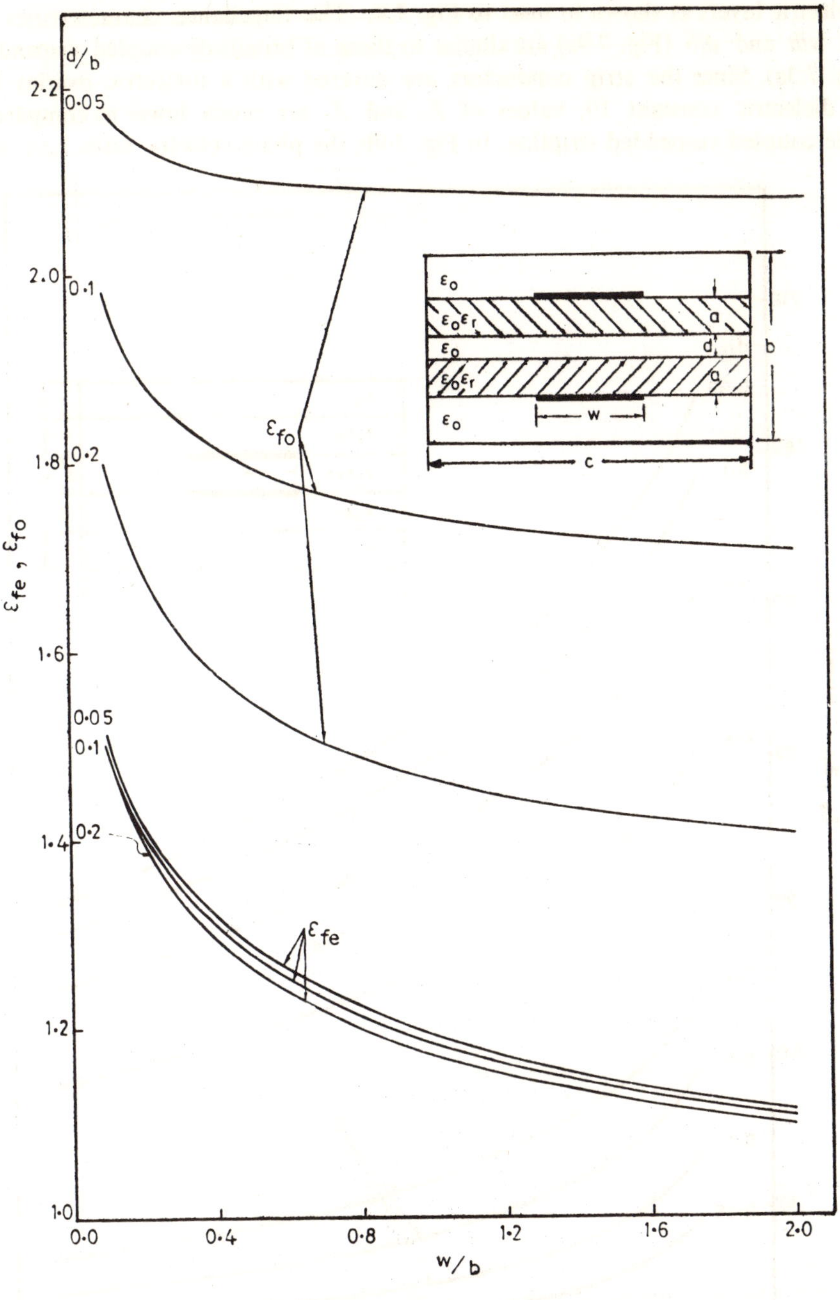

FIG. 7.8(b) ϵ_{fe} and ϵ_{fo} versus w/b with d/b as parameter for broadside-coupled stripline with double-suspended inverted dielectric substrates; $\epsilon_r = 3.78$, $a/b = 0.1$, $c/w = 40$.

solid dielectric layers as shown in inset to Fig. 7.9. The impedance characteristics as fun tions of w/b and d/b (Fig. 7.9a) are similar to those of broadside-coupled suspended stri line (Fig. 7.3a). Since the strip conductors are covered with a dielectric overlay having relative dielectric constant 10, values of Z_e and Z_o are much lower as compared to tl broadside-coupled suspended stripline. In Fig. 7.9b, the phase velocity ratio v_o/v_e is plott

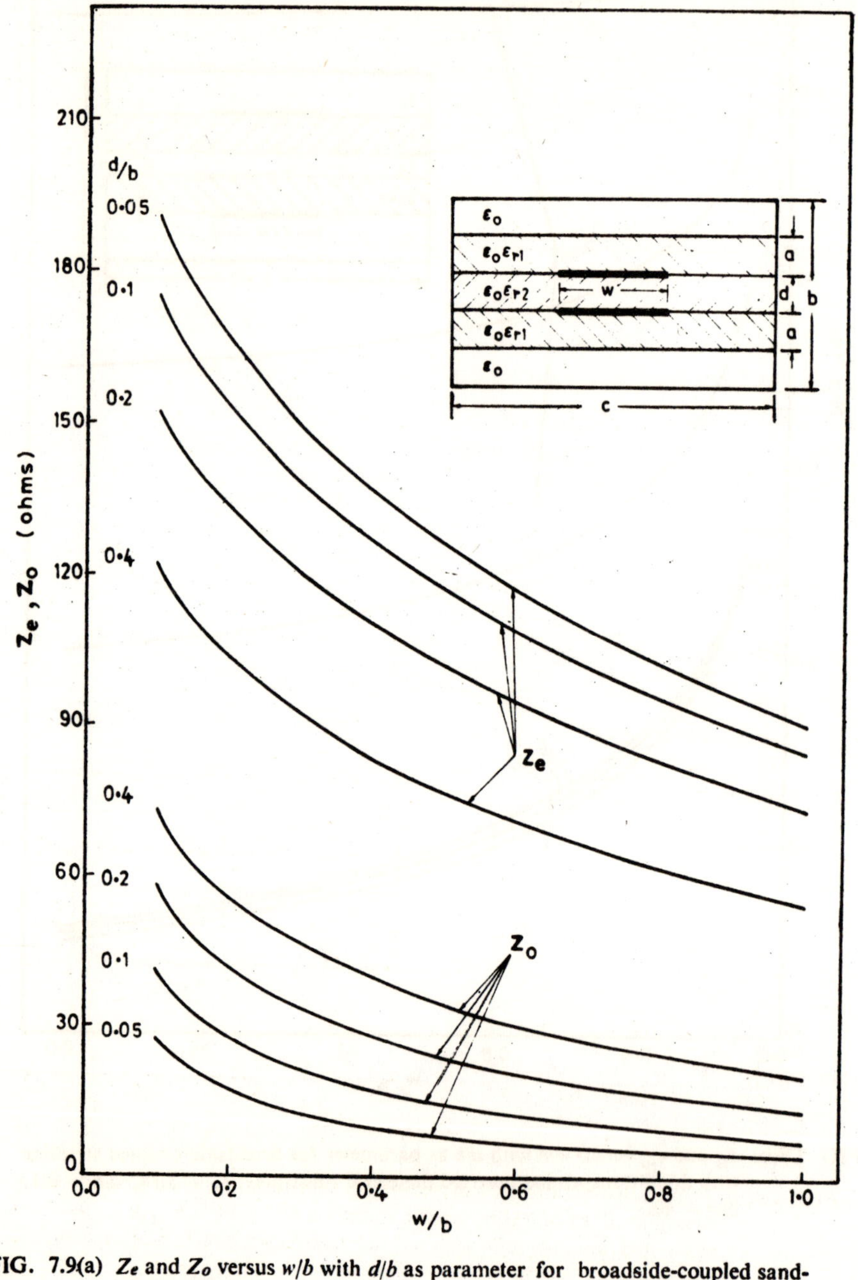

FIG. 7.9(a) Z_e and Z_o versus w/b with d/b as parameter for broadside-coupled sandwiched stripline; $\epsilon_{r1} = 10.0$, $\epsilon_{r2} = 3.78$, $a/b = 0.1$, $c/w = 40$.

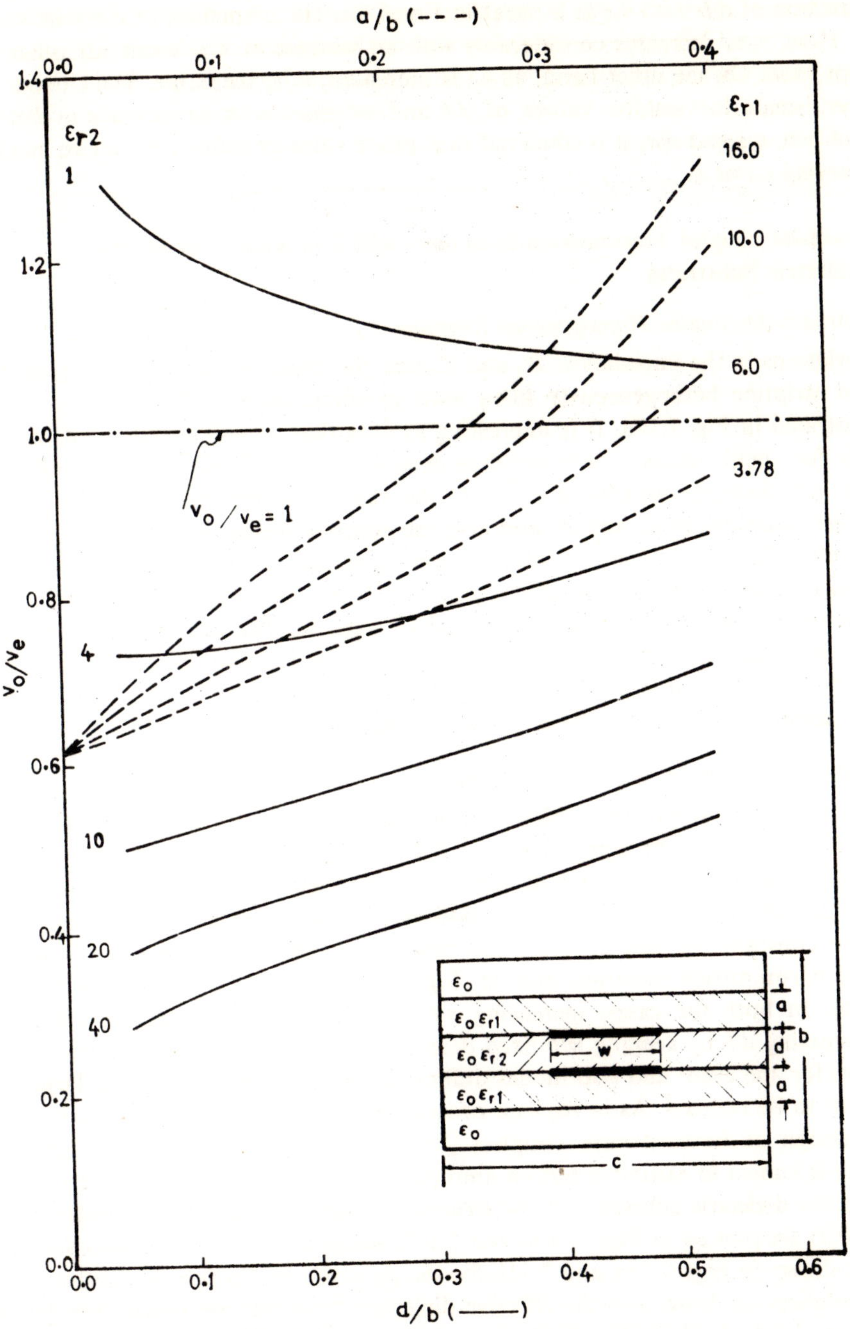

FIG. 7.9(b) Phase velocity ratio v_o/v_e of broadside-coupled sandwiched stripline as a function of (i) d/b with ϵ_{r2} as parameter; $\epsilon_{r1} = 10$, $a/b = 0.1$, $c/w = 40$, $w/b = 0.5$ and (ii) a/b with ϵ_{r1} as parameter; $\epsilon_{r2} = 3.78$, $d/b = 0.1$, $c/w = 40$, $w/b = 0.5$.

as a function of d/b with ϵ_{r2} as a parameter and also as a function of a/b with ϵ_{r1} as a parameter. Here, v_o/v_e decreases considerably with an increase in ϵ_{r2} when all other parameters are kept fixed. On the other hand, as ϵ_{r1} is increased, v_o/v_e increases. The change in v_o/v_e is, however, small for smaller values of d/b and increases with an increase in d/b. For a fixed value of w/b, ϵ_{r2} and d/b, it is observed that phase velocity ratio v_o/v_e can be made equal to 1 by varying ϵ_{r1} or a.

7.4 Broadside-coupled Transmission Structures with Uniaxially Anisotropic Dielectric Substrates

7.4.1 Broadside-coupled Homogeneous Stripline

The variations in the impedances Z_e and Z_o and the phase velocity ratio v_o/v_e of a broadside-coupled stripline homogeneously filled with pyrolytic boron nitride ($\epsilon_{xx} = 5.12$, $\epsilon_{yy}=3.4$) are illustrated in Fig. 7.10a. It is interesting to note that as w/b is increased, v_o/v_e increases initially for small values of w/b and then decreases. For a fixed value of w/b, as d/b is increased, v_o/v_e decreases and for $d/b = 0.4$, the ratio is nearly equal to unity. Figure 7.10b shows the variations in Z_e and Z_o and also the effective dielectric constants ϵ_{fe} and ϵ_{fo} of the structure as a function of the anisotropy ratio $\epsilon_{xx}/\epsilon_{yy}$ normalized with respect to $\epsilon_{yy} = 3.78$. The dimensional parameters are $d/b = w/b = 0.1$. As the ratio $\epsilon_{xx}/\epsilon_{yy}$ is increased from 0.5 to 3, both Z_e and Z_o decrease and ϵ_{fe} and ϵ_{fo} increase. The decrease in Z_o is negligible as compared to that in Z_e. As expected, at $\epsilon_{xx}/\epsilon_{yy} = 1$, ϵ_{fe} is equal to ϵ_{fo}. Detailed design data on Z_e, Z_o, ϵ_{fe} and ϵ_{fo} of the structure are tabulated in Tables 7.9–7.11, for porolytic boron nitride, sapphire and Epsilam-10 substrates, respectively.

7.4.2 Broadside-coupled Suspended Stripline

Broadside-coupled suspended stripline using uniaxially anisotropic pyrolytic boron nitride ($\epsilon_{xx} = 5.12$, $\epsilon_{yy} = 3.4$) or sapphire ($\epsilon_{xx} = 9.4$, $\epsilon_{yy} = 11.6$) as the suspended dielectric layer is attractive in the design of filters and high directivity tight couplers demanding highly reproducible characteristics. The even- and odd-mode characteristic impedances and effective dielectric constants of the structure are plotted in Figs. 7.11a and 7.11b, respectively, for pyrolytic boron nitride substrate and in Figs. 7.12a and 7.12b, respectively, for sapphire substrate. In both the cases, increasing w/b by keeping d/b fixed decreases both Z_e and Z_o, and, increasing d/b by keeping w/b fixed decreases Z_e and increases Z_o. The effective dielectric constants for the even- and odd-modes differ considerably, thus permitting large phase velocity ratios to be realized. As in the case of isotropic counterpart, this property can be used in the design of forward-wave couplers [16]. Since the commercially available Epsilam-10 substrate is known to exhibit a certain amount of anisotropy, the characteristic impedance and effective dielectric constants of the structure using its anisotropic permittivity ($\epsilon_{xx} = 13$, $\epsilon_{yy} = 10.2$) are plotted in Figs. 7.13a and 7.13b, respectively. The characteristics are similar to those shown in Figs. 7.11a and 7.11b for the pyrolytic boron nitride substrate except that the impedances are lower and the effective dielectric constants are higher for the same set of parameters w/b and d/b. In Fig. 7.14, the impedances and effective dielectric constants of the broadside-coupled suspended stripline are plotted as a function of the anisotropy ratio $\epsilon_{xx}/\epsilon_{yy}$ with $\epsilon_{yy} = 3.78$ and $w/b = d/b = 0.1$. Increasing the ratio $\epsilon_{xx}/\epsilon_{yy}$ decreases Z_e and Z_o and increases ϵ_{fe} and ϵ_{fo}.

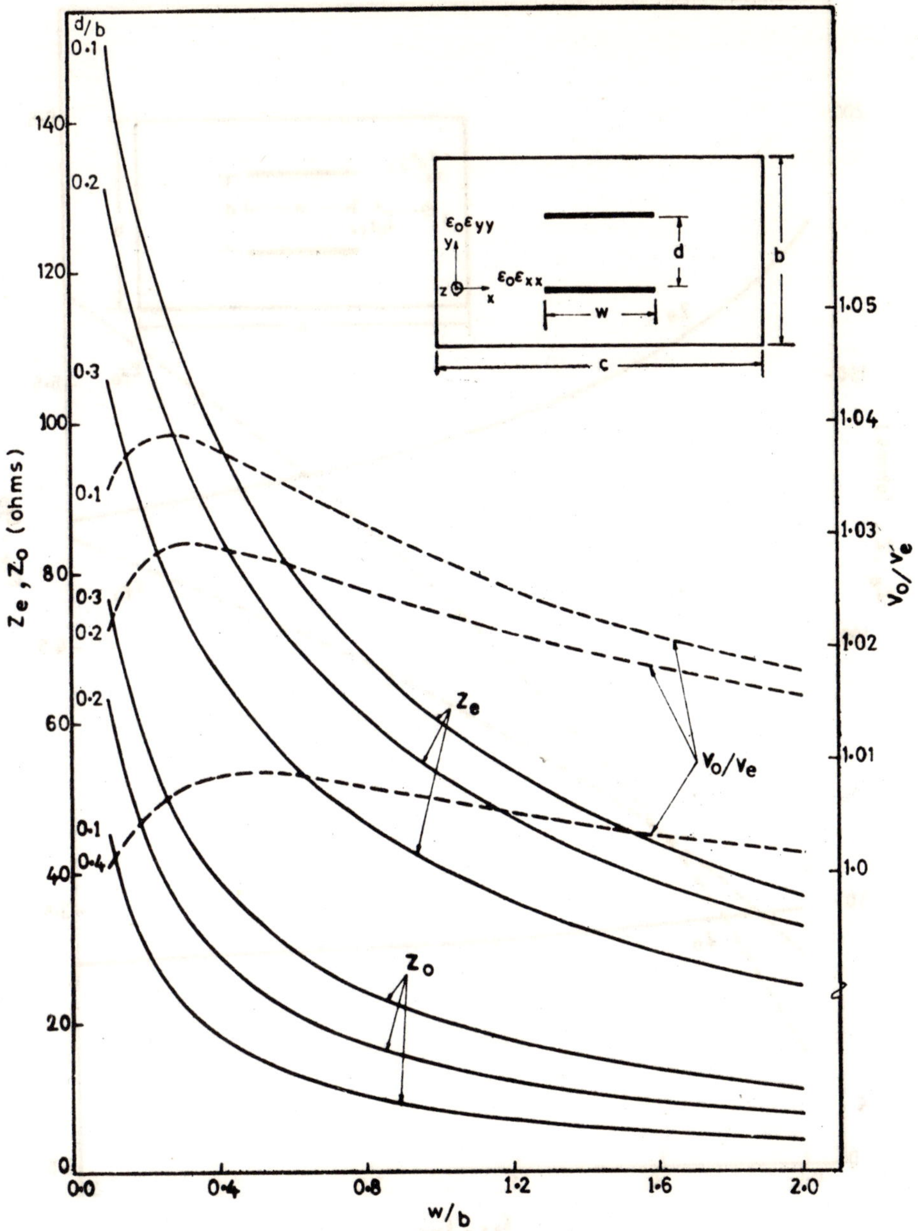

FIG. 7.10(a) Characteristic impedances Z_e and Z_o and phase velocity ratio v_o/v_e versus w/b with d/b as parameter of broadside-coupled homogeneous stripline filled with uniaxially anisotropic dielectric substrate (pyrolytic boron nitride); $\epsilon_{xx} = \epsilon_{zz} = 5.12$, $\epsilon_{yy} = 3.4$, $c/w = 40$.

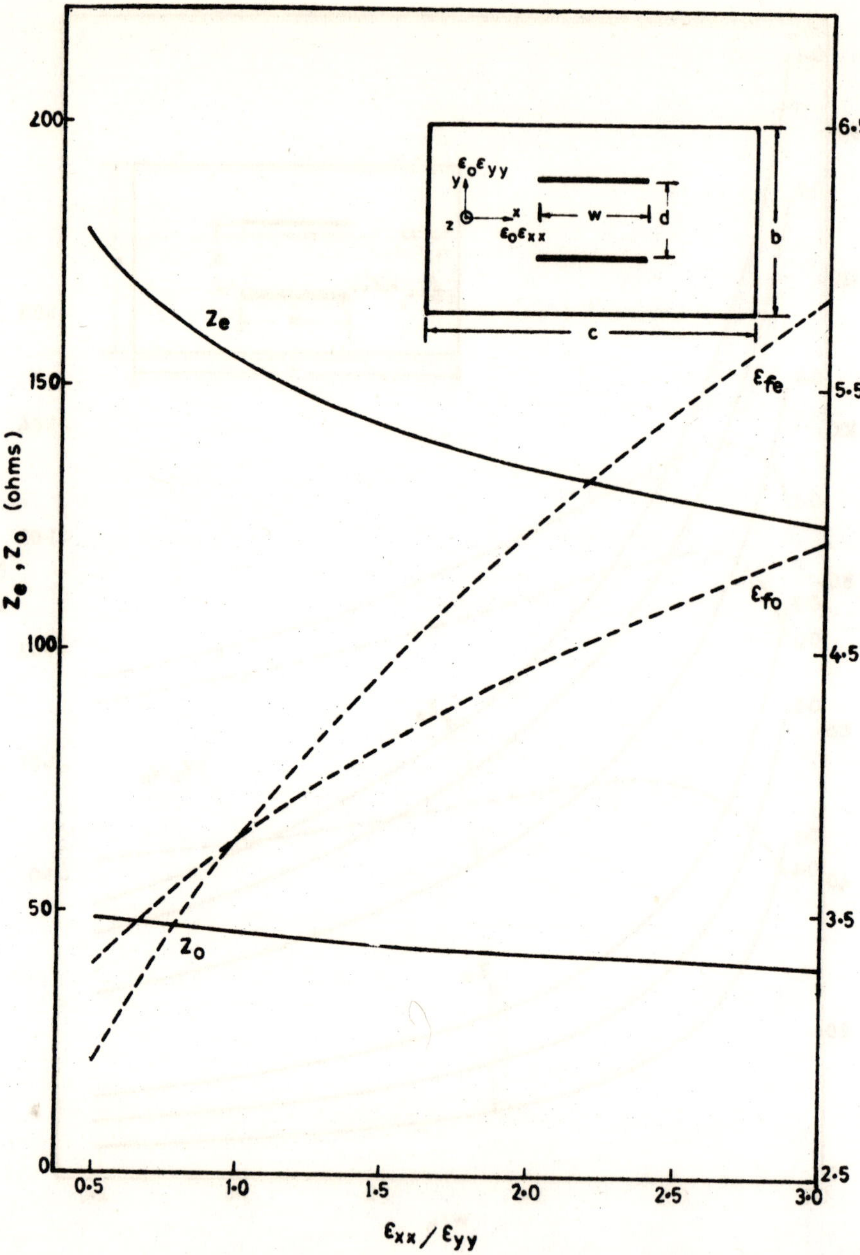

FIG. 7.10(b) Characteristic impedances Z_e and Z_o and effective dielectric constants ϵ_{fe} and ϵ_{fo} versus anisotropy ratio $\epsilon_{xx}/\epsilon_{yy}$ of broadside-coupled homogeneous strip-line filled with uniaxially anisotropic dielectric; $\epsilon_{yy} = 3.78$, $w/b = d/b = 0.1$, $c/w = 40$.

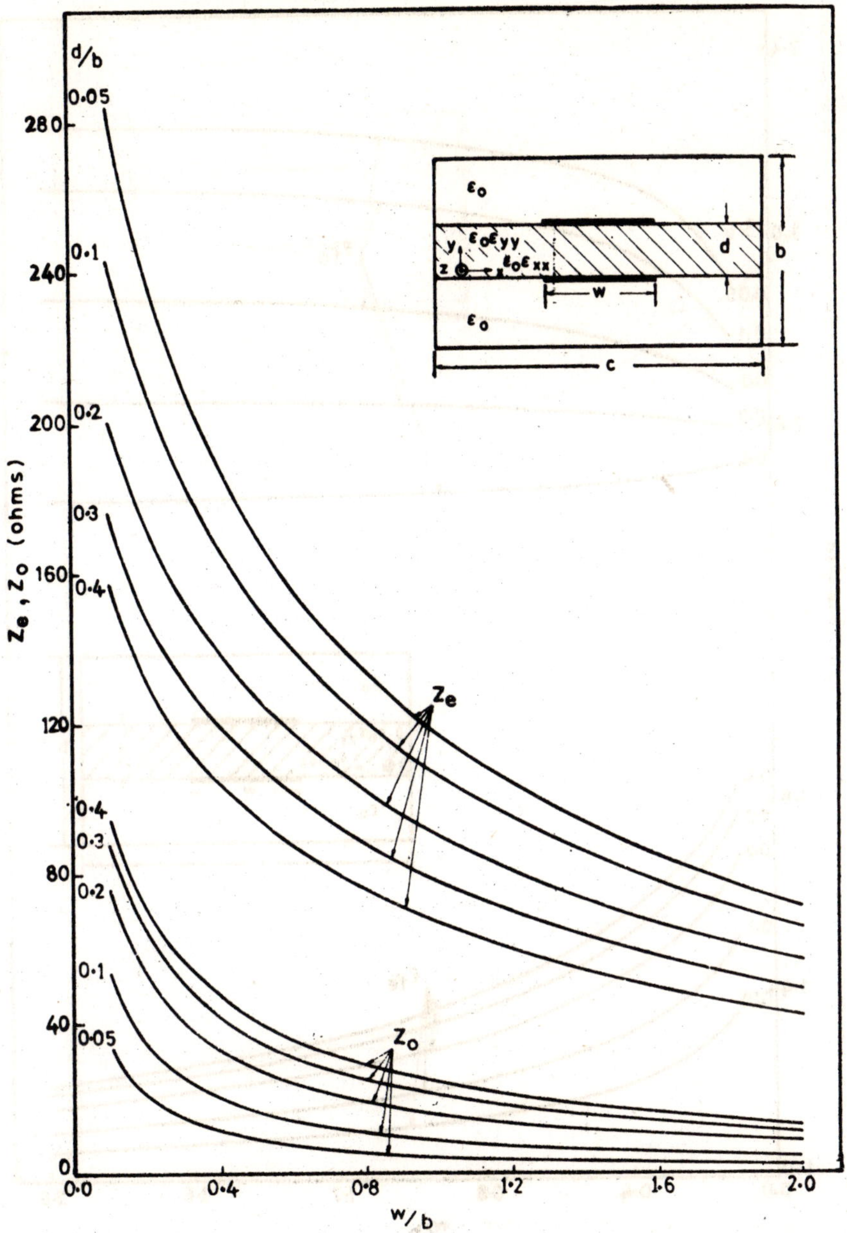

FIG. 7.11(a) Z_e and Z_o versus w/b with d/b as parameter of broadside-coupled suspended stripline with uniaxially anisotropic dielectric substrate (pyrolytic boron nitride); $\epsilon_{xx} = \epsilon_{zz} = 5.12$, $\epsilon_{yy} = 3.4$, $c/w = 40$.

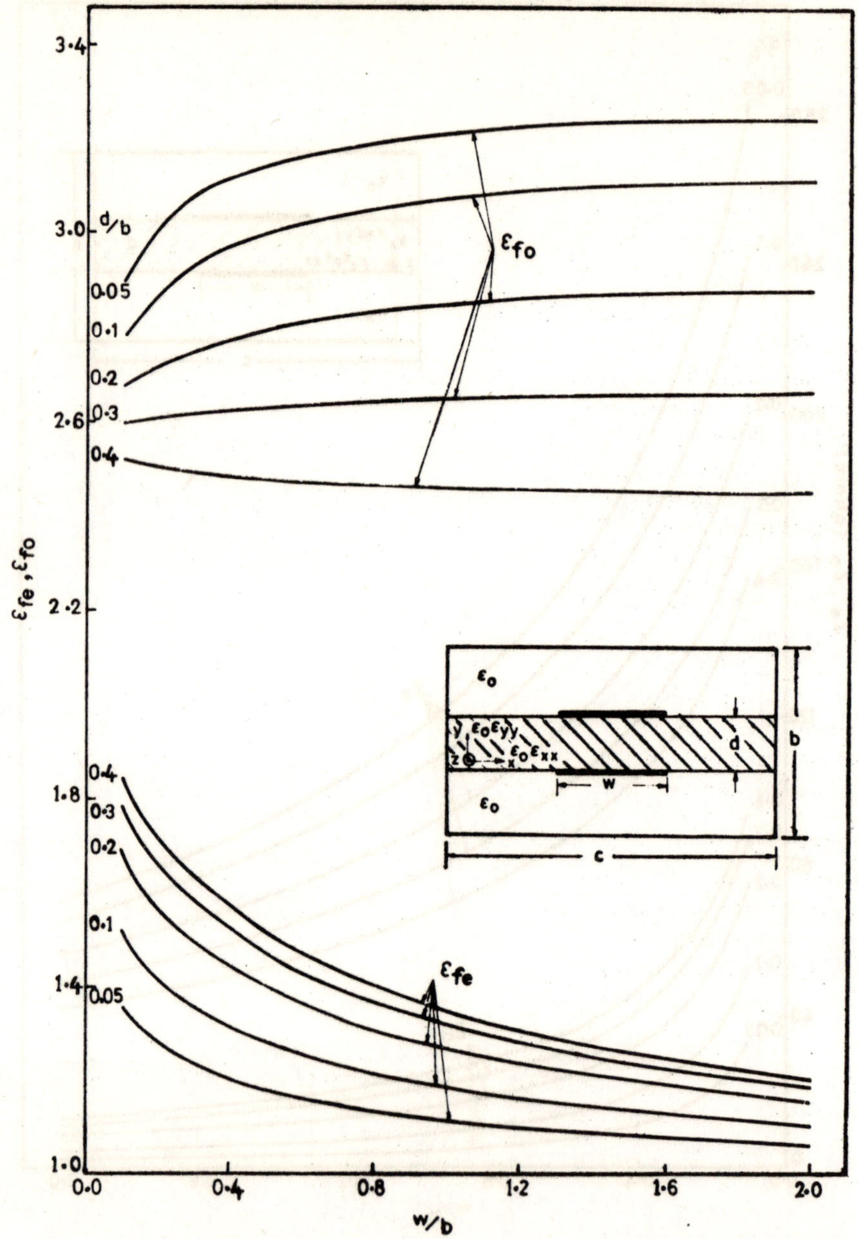

FIG. 7.11(b) ϵ_{fe} and ϵ_{fo} versus w/b with d/b as parameter of broadside-coupled suspended stripline with uniaxially anisotropic dielectric substrate (pyrolytic boron nitride); $\epsilon_{xx} = \epsilon_{zz} = 5.12$, $\epsilon_{yy} = 3.4$, $c/w = 40$.

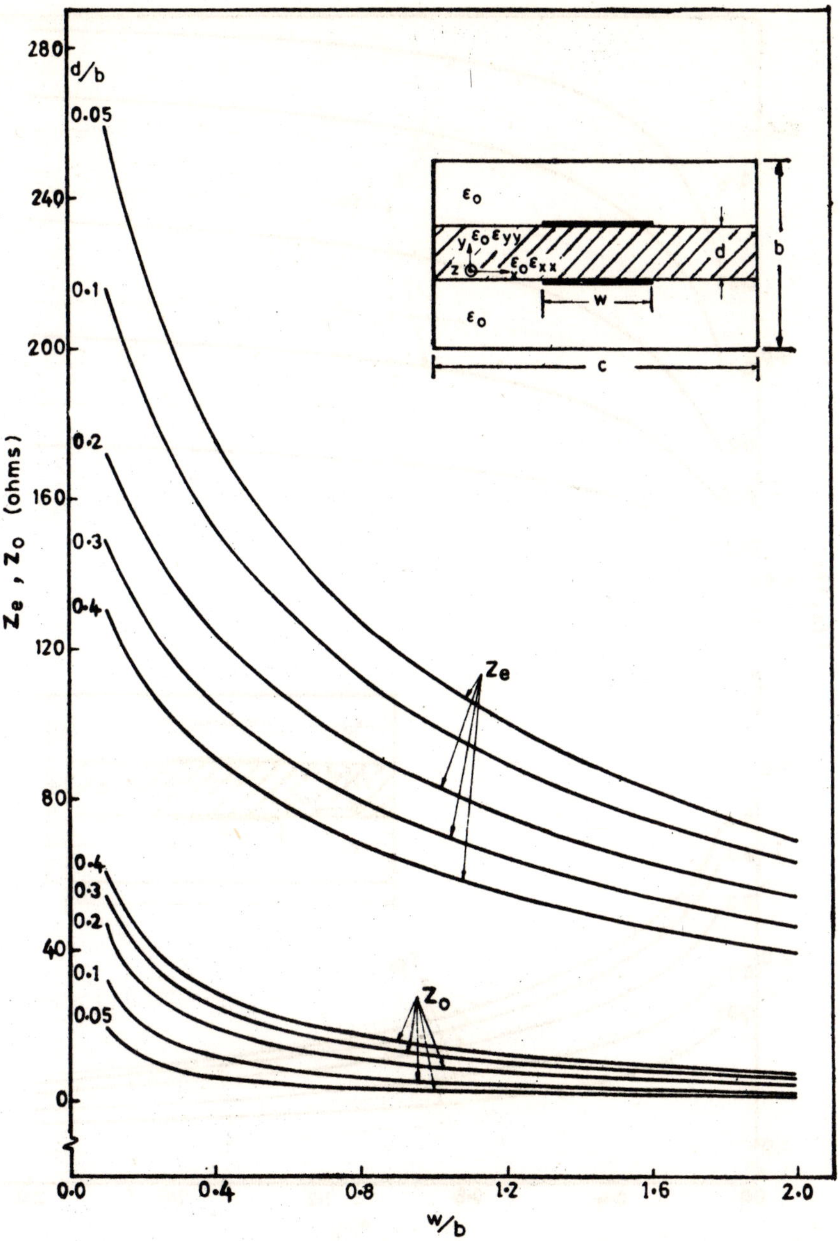

FIG. 7.12(a) Z_e and Z_o versus w/b with d/b as parameter of broadside-coupled suspended stripline with uniaxially anisotropic dielectric substrate (sapphire); $\epsilon_{xx} = \epsilon_{zz} = 9.4$, $\epsilon_{yy} = 11.6$, $c/w = 40$.

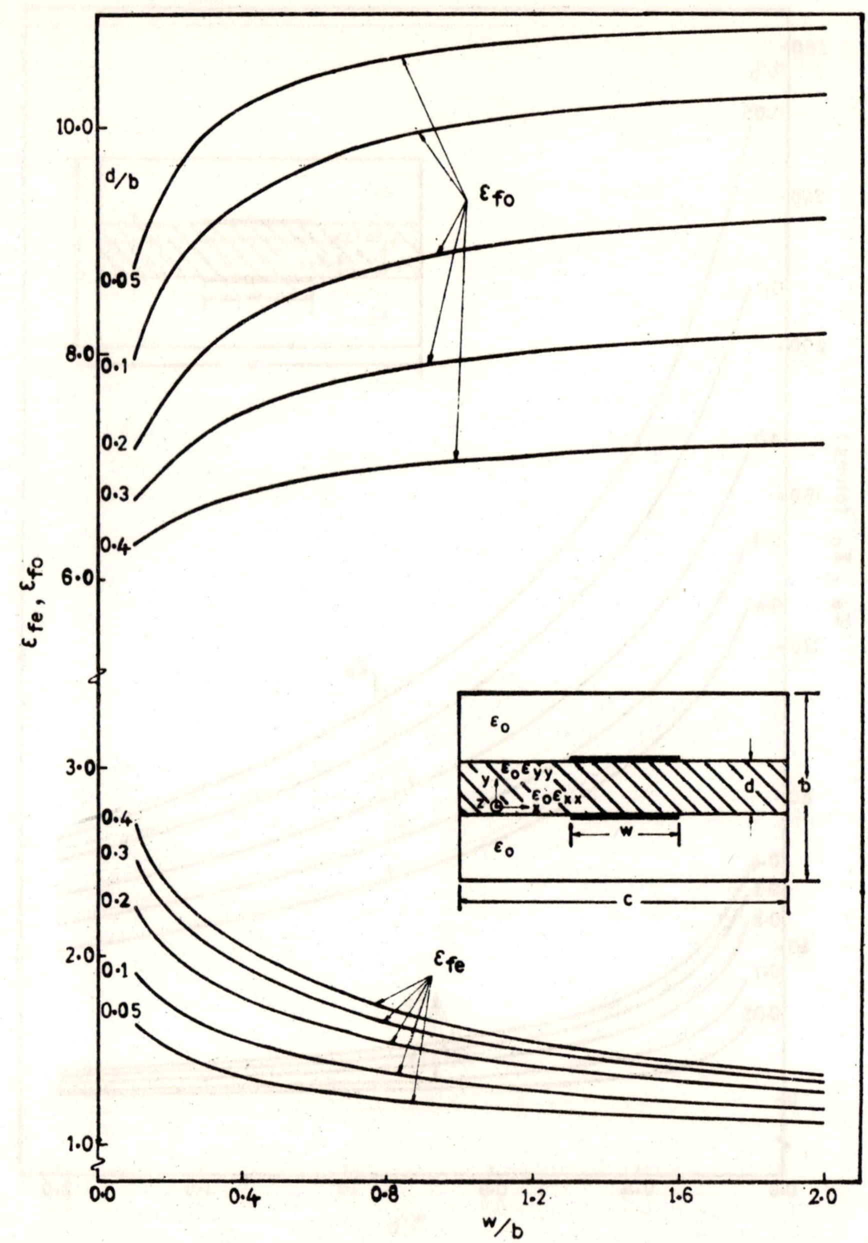

FIG. 7.12(b) ϵ_{fe} and ϵ_{fo} versus w/b with d/b as parameter of broadside-coupled suspended stripline with uniaxially anisotropic dielectric substrate (sapphire); $\epsilon_{xx} = \epsilon_{zz} = 9.4$, $\epsilon_{yy} = 11.6$, $c/w = 40$.

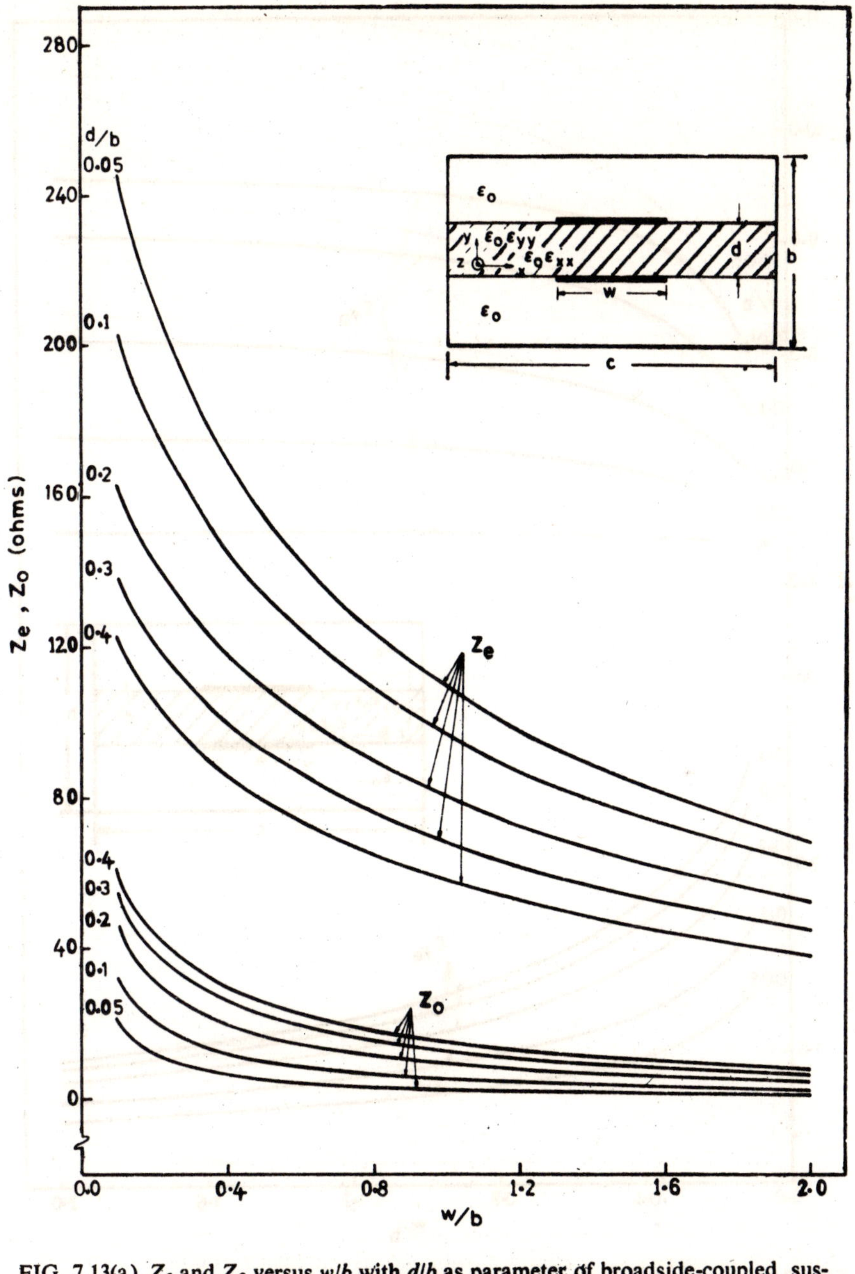

FIG. 7.13(a) Z_e and Z_o versus w/b with d/b as parameter of broadside-coupled suspended stripline with uniaxially anisotropic dielectric substrate (epsilam-10); $\epsilon_{xx} = \epsilon_{zz} = 13.0$, $\epsilon_{yy} = 10.2$, $c/w = 40$.

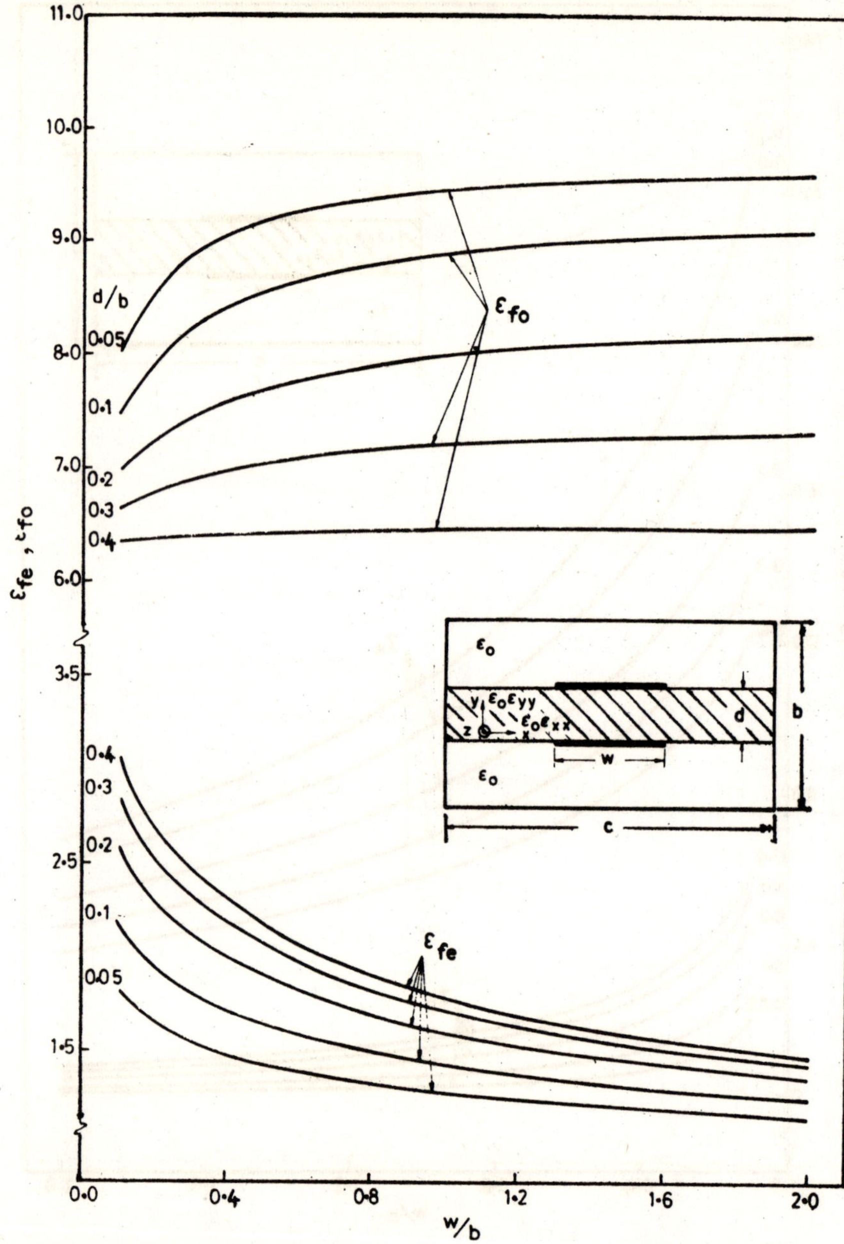

FIG. 7.13(b) ϵ_{fe} and ϵ_{fo} versus w/b with d/b as parameter of broadside-coupled suspended stripline with uniaxially anisotropic dielectric substrate (epsilam-10); $\epsilon_{xx} = \epsilon_{zz} = 13.0$, $\epsilon_{yy} = 10.2$, $c/w = 40$.

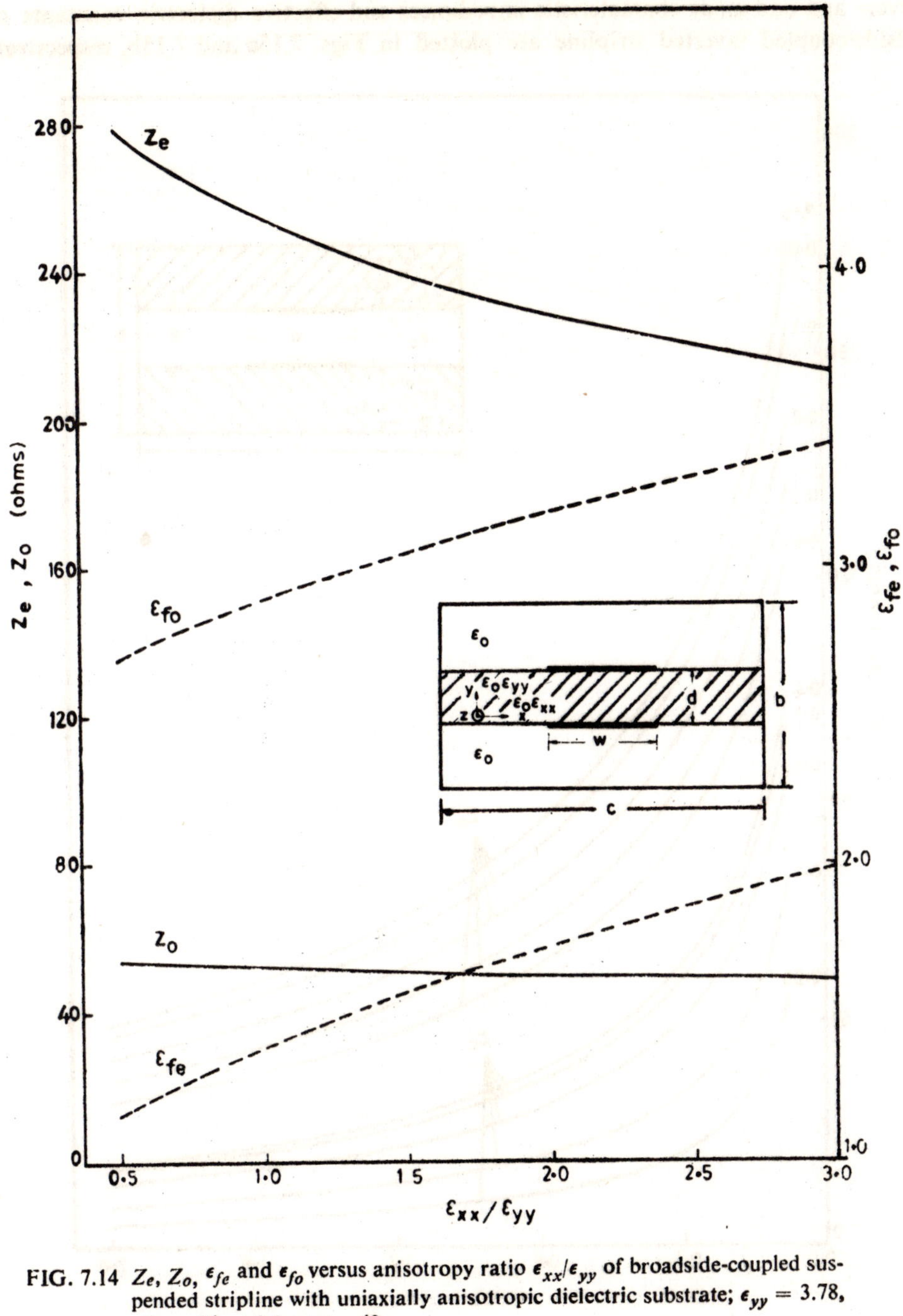

FIG. 7.14 Z_e, Z_o, ϵ_{fe} and ϵ_{fo} versus anisotropy ratio $\epsilon_{xx}/\epsilon_{yy}$ of broadside-coupled suspended stripline with uniaxially anisotropic dielectric substrate; $\epsilon_{yy} = 3.78$, $w/b = d/b = 0.1$, $c/w = 40$.

7.4.3 *Broadside-coupled Inverted Stripline*

The even- and odd-mode characteristic impedances and effective dielectric constants of the broadside-coupled inverted stripline are plotted in Figs. 7.15a and 7.15b, respectively, for

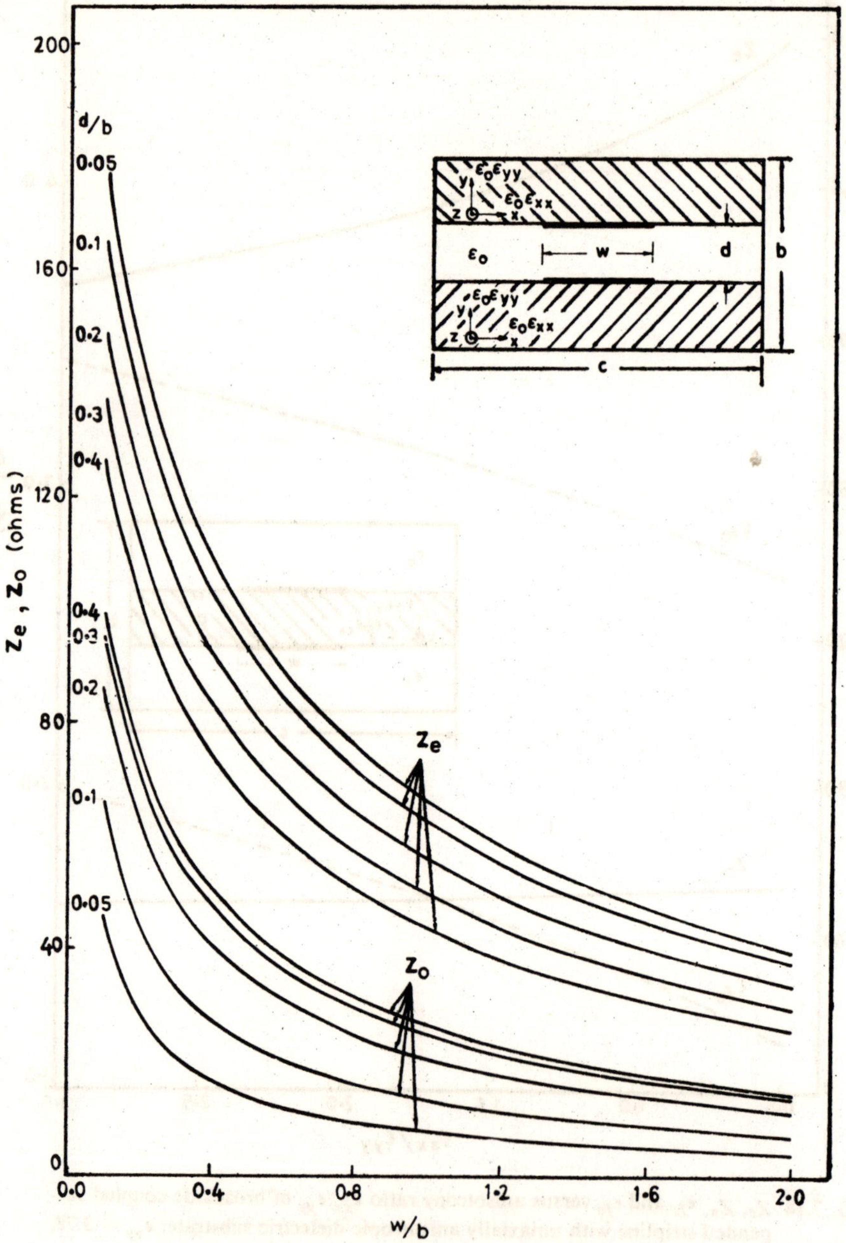

FIG. 7.15(a) Z_e and Z_o versus w/b with d/b as parameter of broadside-coupled inverted stripline with uniaxially anisotropic dielectric substrates (pyrolytic boron nitride); $\epsilon_{xx} = \epsilon_{zz} = 5.12$, $\epsilon_{yy} = 3.4$, $c/w = 40$.

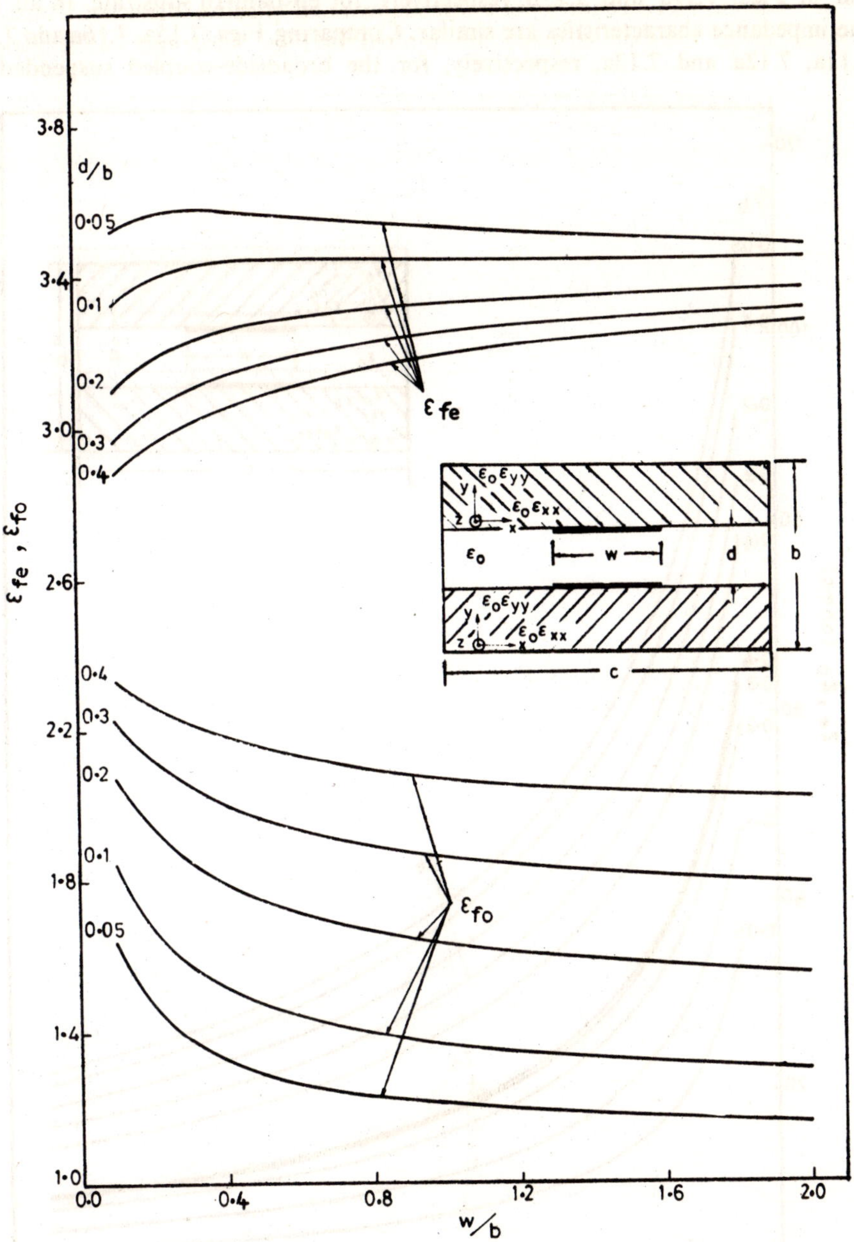

FIG. 7.15(b) ϵ_{fe} and ϵ_{fo} versus w/b with d/b as parameter of broadside-coupled inverted stripline with uniaxially anisotropic dielectric substrates (pyrolytic boron nitride); $\epsilon_{xx} = \epsilon_{zz} = 5.12$, $\epsilon_{yy} = 3.4$, $c/w = 40$.

pyrolytic boron nitride substrate, in Figs. 7.16a and 7.16b, respectively, for sapphire substrate and in Figs. 7.17a and 7.17b, respectively, for epsilam-10 substrate. In all the three cases, the impedance characteristics are similar. Comparing Figs. 7.15a, 7.16a and 7.17a with Figs. 7.11a, 7.12a and 7.13a, respectively, for the broadside-coupled suspended stripline

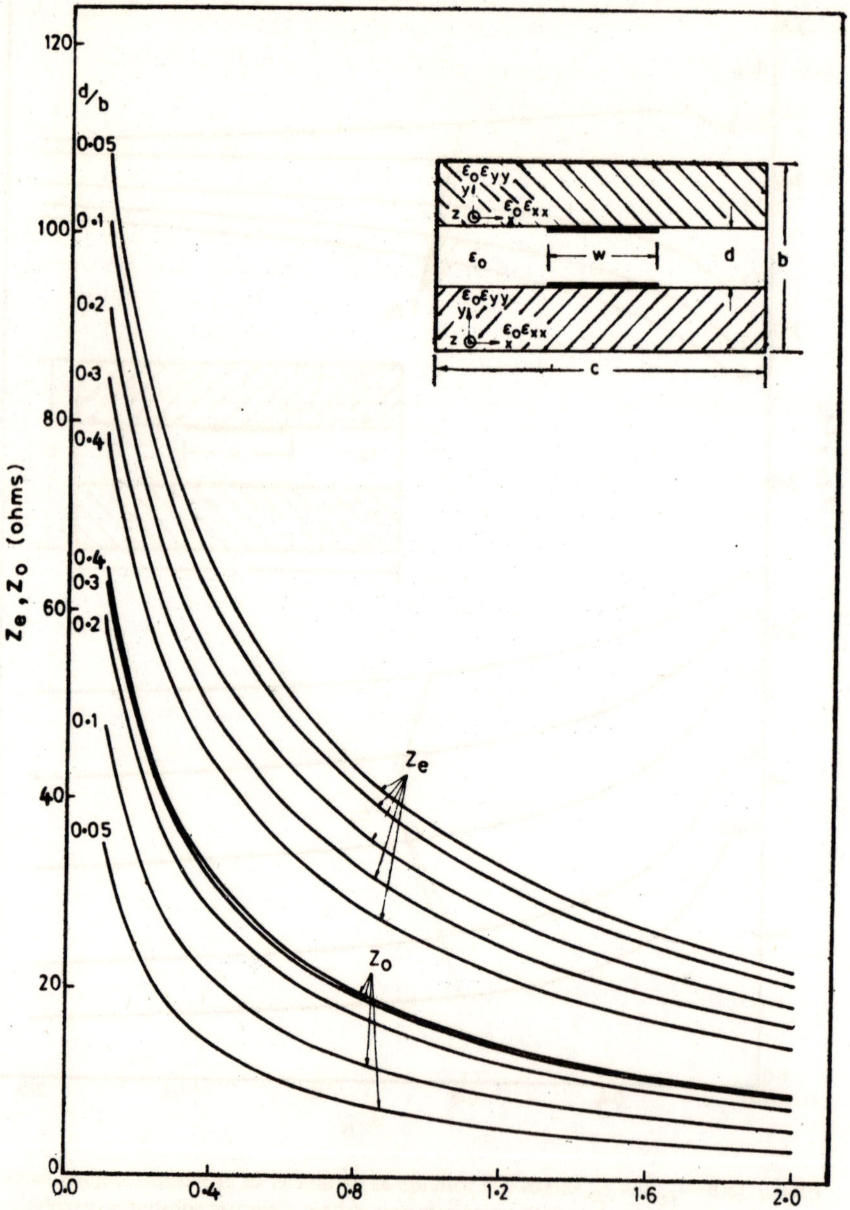

FIG. 7.16(a) Z_e and Z_o versus w/b with d/b as parameter of broadside-coupled inverted stripline with uniaxially anisotropic dielectric substrates; (sapphire); $\epsilon_{xx} = \epsilon_{zz} = 9.4$, $\epsilon_{yy} = 11\ 6$, $c/w = 40$.

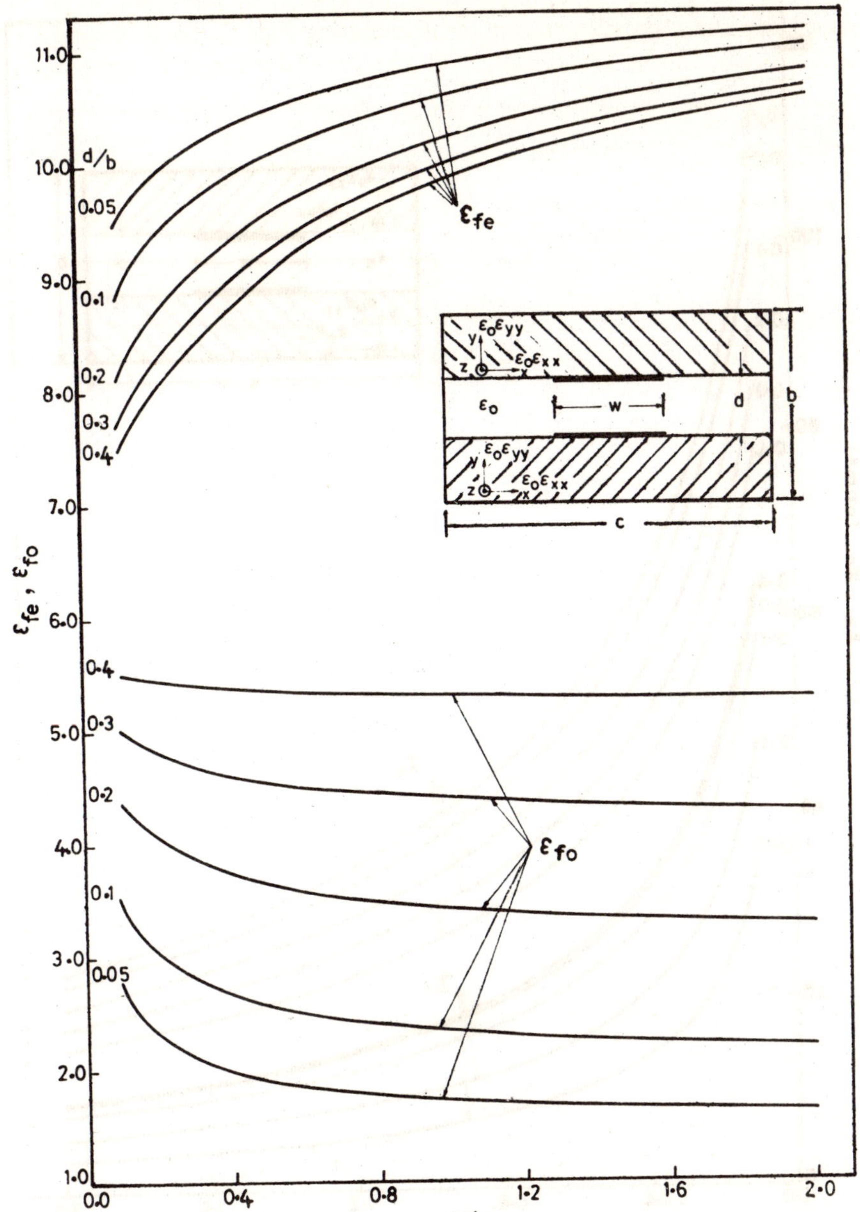

FIG. 7.16(b) ϵ_{fe} and ϵ_{fo} versus w/b with d/b as parameter of broadside-coupled inverted stripline with uniaxially anisotropic dielectric substrates (sapphire); $\epsilon_{xx} = \epsilon_{zz} = 9.4$, $\epsilon_{yy} = 11.6$, $c/w = 40$.

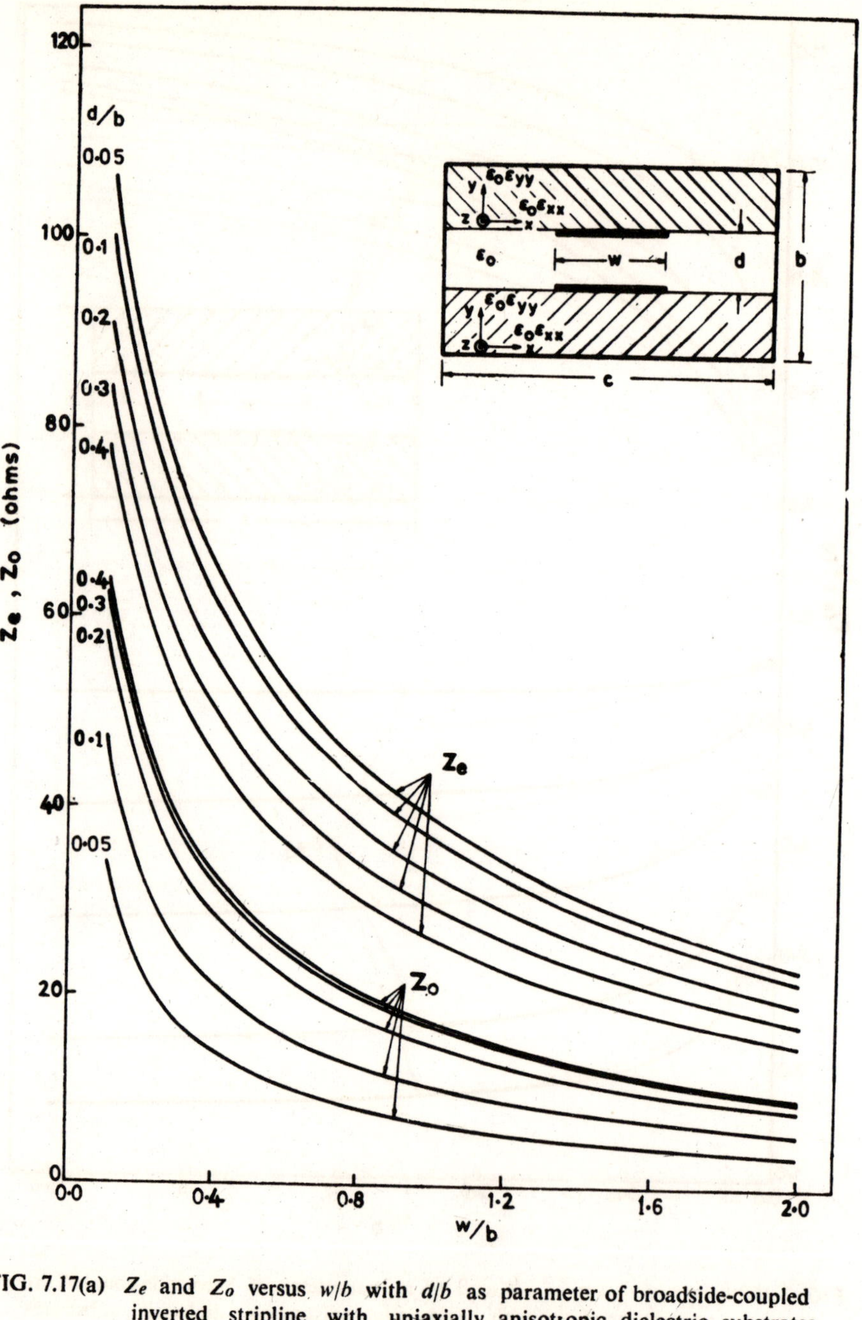

FIG. 7.17(a) Z_e and Z_o versus w/b with d/b as parameter of broadside-coupled inverted stripline with uniaxially anisotropic dielectric substrates (epsilam-10); $\epsilon_{xx} = \epsilon_{zz} = 13.0$, $\epsilon_{yy} = 10.2$, $c/w = 40$.

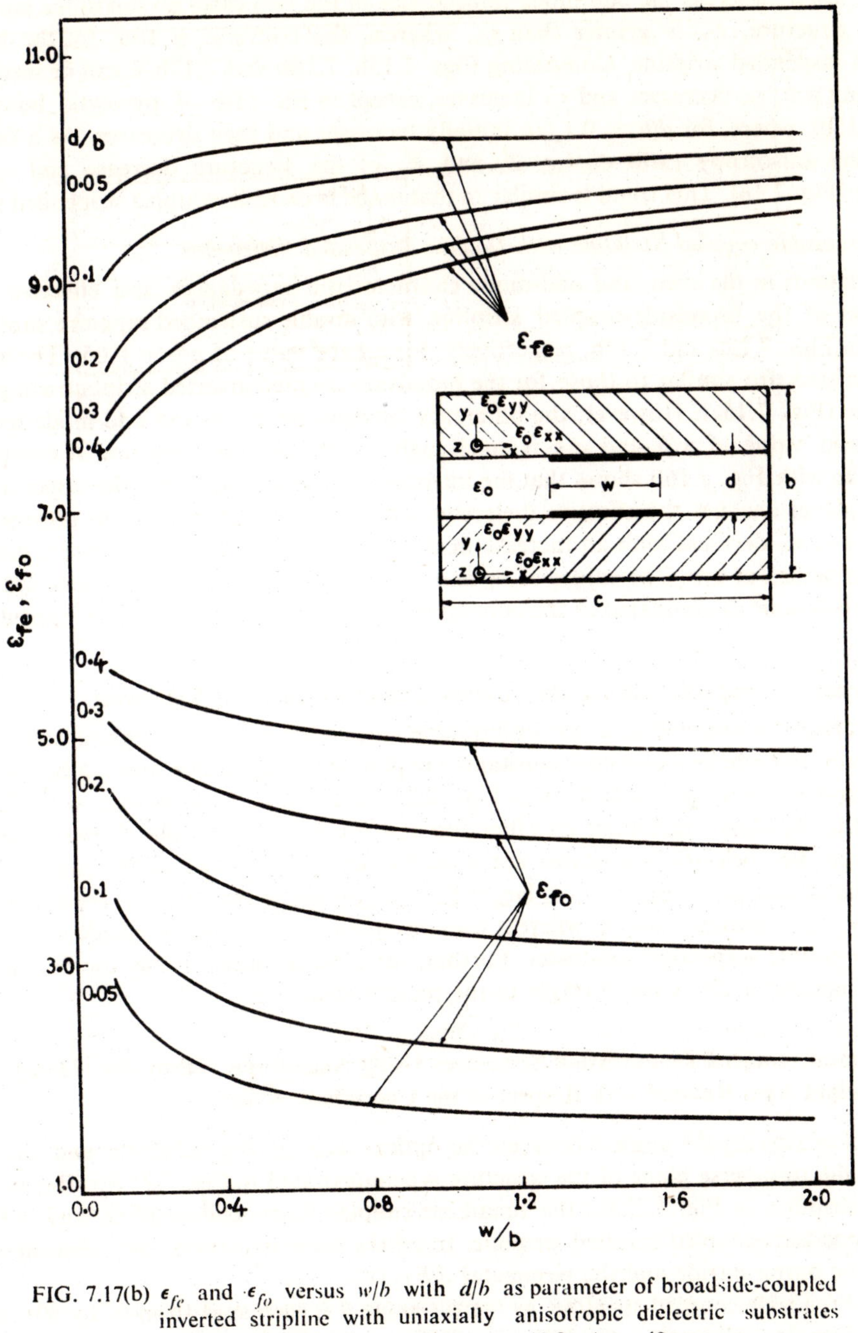

FIG. 7.17(b) ϵ_{fe} and ϵ_{fo} versus w/b with d/b as parameter of broadside-coupled inverted stripline with uniaxially anisotropic dielectric substrates (epsilam-10); $\epsilon_{xx} = \epsilon_{zz} = 13.0$, $\epsilon_{yy} = 10.2$, $c/w = 40$.

shows that the even-mode impedances in the broadside-coupled inverted stripline are considerably lower, whereas the odd-mode impedances in the two cases do not differ much. In the present structure, ϵ_{fe} is greater than ϵ_{fo}, whereas the converse is true for the broadside-coupled suspended stripline. Comparing Figs. 7.15b, 7.16b and 7.17b, it can be seen that with increasing w/b, ϵ_{fo} decreases and ϵ_{fe} increases, except in the case of pyrolytic boron nitride (Fig. 7.15b), where, for $d/b \leqslant 0.1$, ϵ_{fe} initially increases and then decreases. As a function of increasing anisotropy ratio $\epsilon_{xx}/\epsilon_{yy}$, Z_e and Z_o of the structure decrease and ϵ_{fe} and ϵ_{fo} increase (Fig. 7.18). This trend is similar to that in the broadside-coupled suspended stripline.

7.4.4 *Broadside-coupled Stripline with Double Suspended Substrates*

The variations in the even- and odd-mode characteristic impedances and effective dielectric constants of the broadside-coupled stripline with double suspended sapphire substrates are shown in Figs. 7.19a and 7.19b, respectively, for a fixed value of $a/b = 0.05$. The impedance characteristics are similar to those for the broadside-coupled inverted stripline using the same dielectric (Fig. 7.16a). However, the difference between the even- and odd-mode impedances for a fixed value of w/b and d/b is much higher in the present structure. A comparison of Fig. 7.19b with Fig. 7.16b shows that the introduction of air gaps near the upper and lower ground planes changes the effective dielectric constants considerably. In the present structure both ϵ_{fe} and ϵ_{fo} decrease with an increase in w/b (Fig. 7.19b), and the decrease is rather sharp for lower values of w/b. The values of ϵ_{fe} and ϵ_{fo} for a fixed set of parameters w/b and d/b are much lower as compared to the corresponding values in the broadside-coupled inverted stripline.

7.4.5 *Broadside-coupled Stripline with Double Suspended Inverted Substrates*

For the broadside-coupled stripline with double-suspended inverted sapphire substrates, the impedances and effective dielectric constants are plotted in Figs. 7.20a and 7.20b, respectively. The parameter a/b is fixed at 0.05 as in Figs. 7.19a and 7.19b. Comparison with Fig. 7.19a shows that impedance variations as a function of w/b and d/b are similar. In the present structure, since the dielectric substrates are situated between the two strip conductors, field concentration due to the dielectric is larger for the odd mode than for the even mode. Consequently, ϵ_{fo} is greater than ϵ_{fe}, whereas the converse is true for the broadside-coupled stripline with double suspended substrates. Further, for a fixed value of w/b, the decrease in ϵ_{fo} with an increase in d/b is much larger in the present structure.

7.5 Broadside-coupled Transmission Structures Using Anisotropic Substrates Having Principal Axes Rotated with Respect to the Coordinate Axes

The effect of varying the angle θ between the optical axis of the substrate and the vertical y-axis in the transverse plane of the structure is demonstrated in Fig. 7.21 for the broadside-coupled stripline; in Fig. 7.22 for the broadside-coupled suspended stripline; and in Fig. 7.23 for the broadside-coupled inverted stripline. In all the three structures, the substrate material is pyrolytic boron nitride and the parameter d/b is 0.1.

In the broadside-coupled stripline, as the tilt angle θ is increased from 0° to 90°, both Z_e and Z_o decrease, whereas ϵ_{fe} and ϵ_{fo} increase. ϵ_{fo} is smaller than ϵ_{fe} for smaller values of θ, but increases at a much faster rate than ϵ_{fe} as θ increases. The two curves cross each other at some intermediate value of θ. Equalization of phase velocities can therefore be achieved by

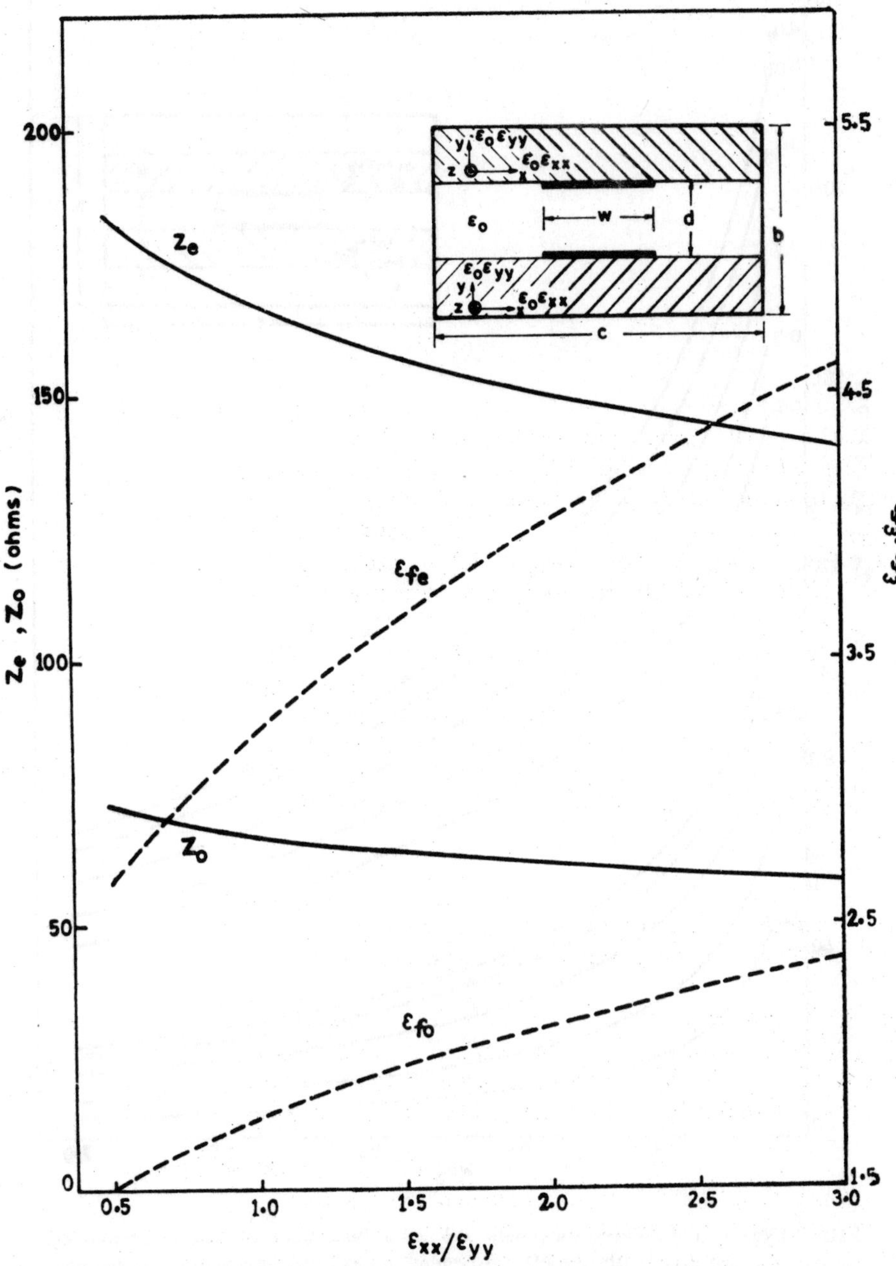

FIG. 7.18 Z_e, Z_o, ϵ_{fe} and ϵ_{fo} versus anisotropy ratio $\epsilon_{xx}/\epsilon_{yy}$ of broadside-coupled inverted stripline with uniaxially anisotropic dielectric substrates; $\epsilon_{yy} = 3.78$, $w/b = d/b = 0.1$, $c/w = 40$

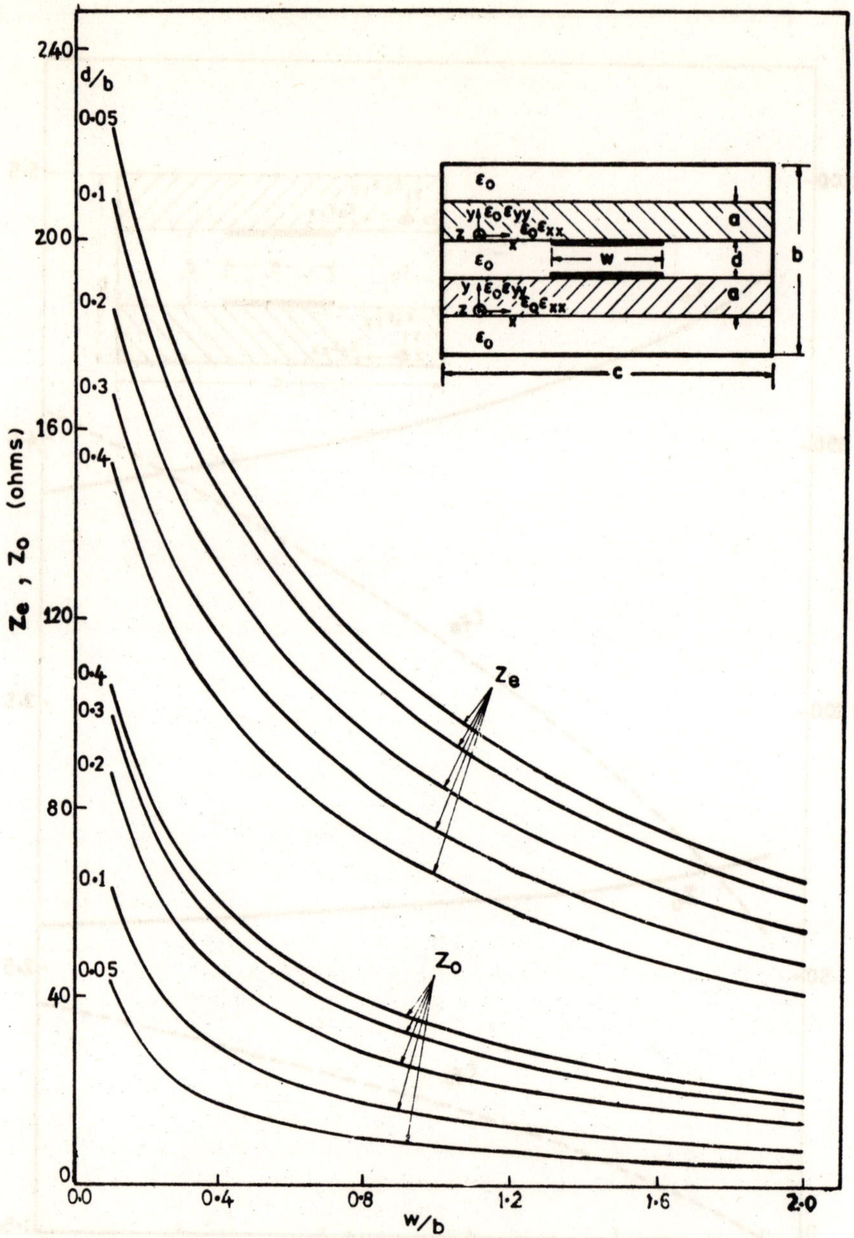

FIG. 7.19(a) Z_e and Z_o versus w/b with d/b as a parameter of broadside-coupled stripline with double suspended uniaxially anisotropic substrates (sapphire); $\epsilon_{xx} = \epsilon_{zz} = 9.4$, $\epsilon_{yy} = 11.6$, $a/b = 0.05$, $c/w = 40$.

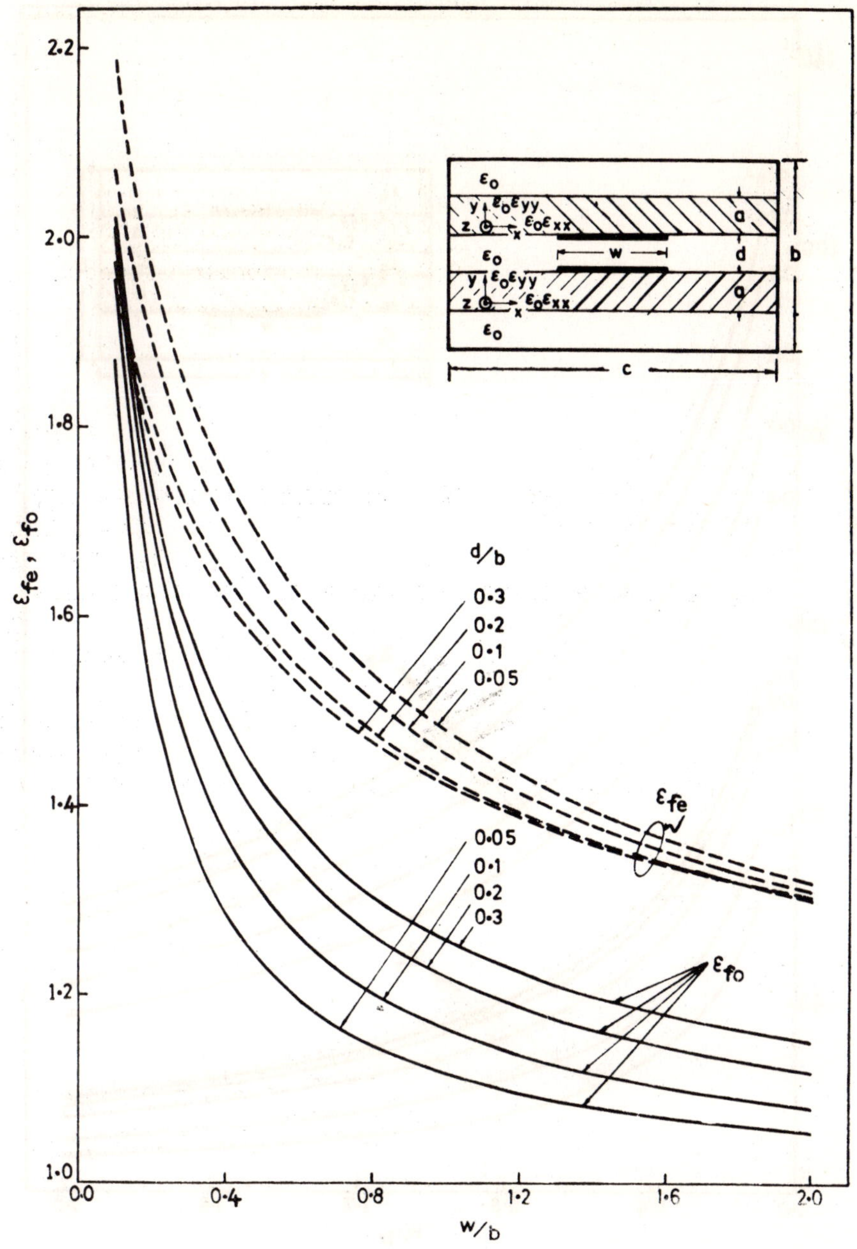

FIG. 7.19(b) ϵ_{fe} and ϵ_{fo} versus w/b with d/b as parameter of broadside-coupled stripline with double suspended uniaxially anisotropic substrates (sapphire); $\epsilon_{xx} = \epsilon_{zz} = 9.4$, $\epsilon_{yy} = 11.6$, $a/b = 0.05$, $c/w = 40$.

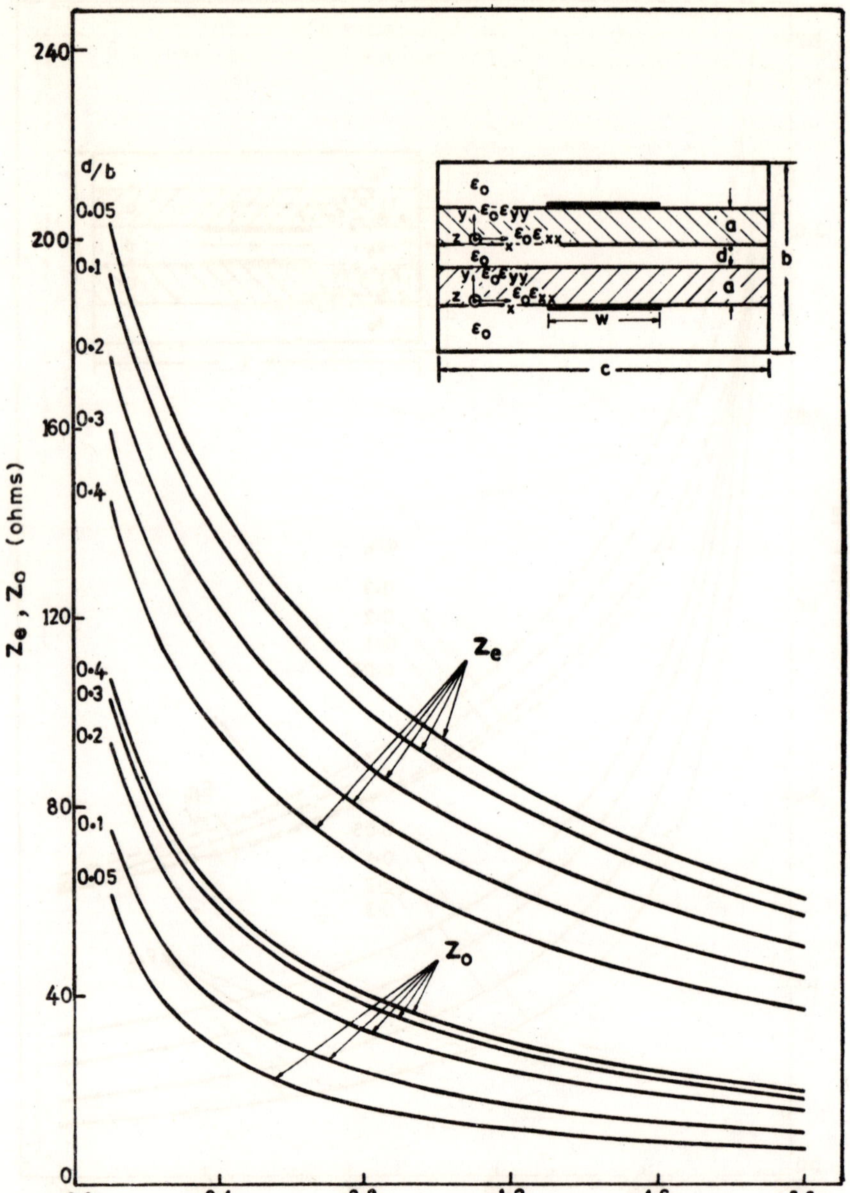

FIG. 7.20(a) Z_e and Z_o versus w/b with d/b as parameter of broadside-coupled stripline with double suspended inverted uniaxially anisotropic substrates (sapphire); $\epsilon_{xx} = \epsilon_{zz} = 9.4$, $\epsilon_{yy} = 11.6$, $a/b = 0.05$, $c/w = 40$.

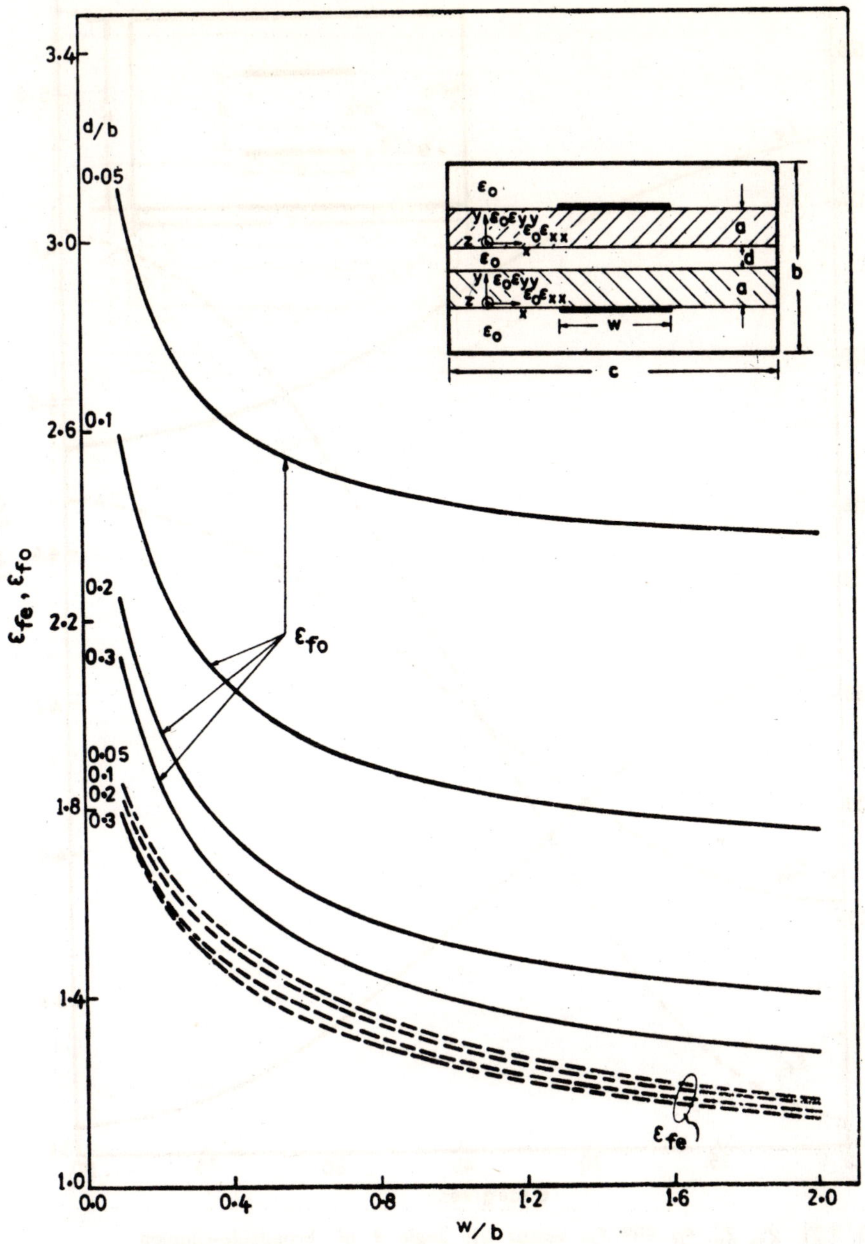

FIG. 7.20(b) ϵ_{fe} and ϵ_{fo} versus w/b with d/b as parameter of broadside-coupled stripline with double suspended inverted uniaxially anisotropic substrates (sapphire); $\epsilon_{xx} = \epsilon_{zz} = 9.4$, $\epsilon_{yy} = 11.6$, $a/b = 0.05$, $c/w = 40$.

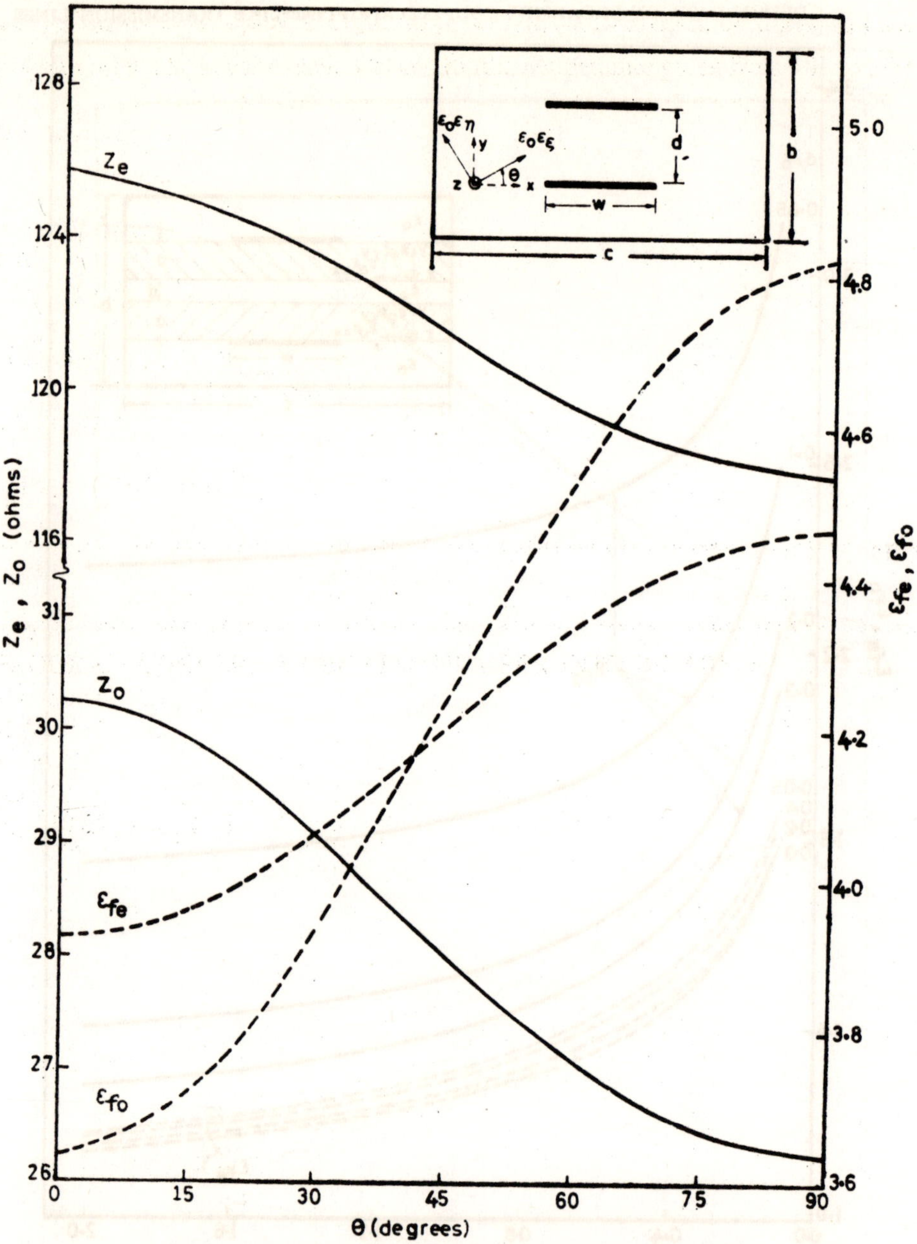

FIG 7.21 Z_e, Z_o, ϵ_{fe} and ϵ_{fo} versus tilt angle θ of broadside-coupled homogeneous stripline filled with pyrolytic boron nitride substrate; $\epsilon_\xi = 5.12$, $\epsilon_\eta = 3.4$ at $\theta = 0$, $w/b = 0.2$, $d/b = 0.1$, $c/w = 40$.

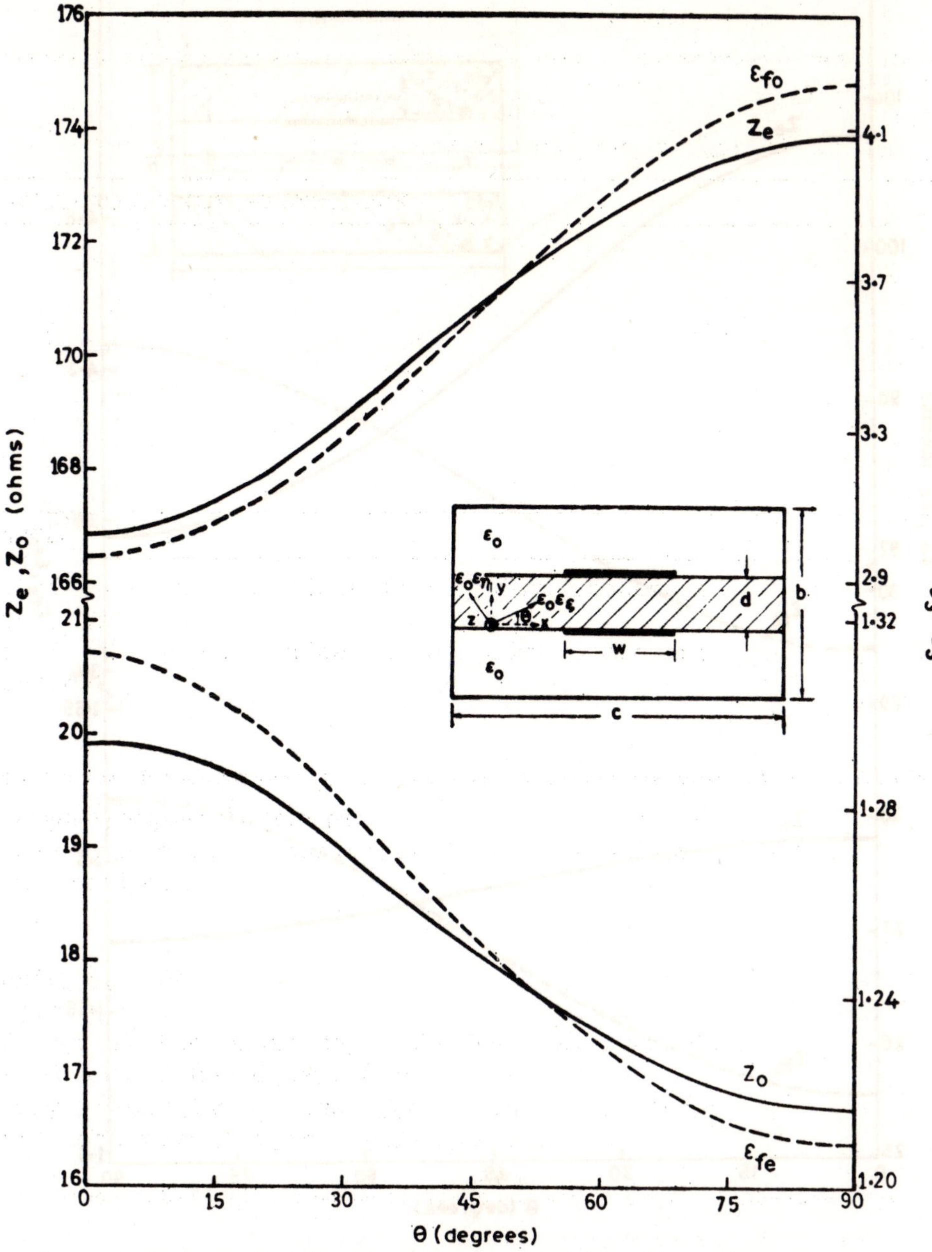

FIG. 7.22 Z_e, Z_o, ϵ_{fe} and ϵ_{fo} versus tilt angle θ of broadside-coupled suspended stripline with pyrolytic boron nitride substrate; $\epsilon_\xi = 5.12$, $\epsilon_\eta = 3.4$ at $\theta = 0$, $w/b = 0.4$, $d/b = 0.1$, $c/w = 40$.

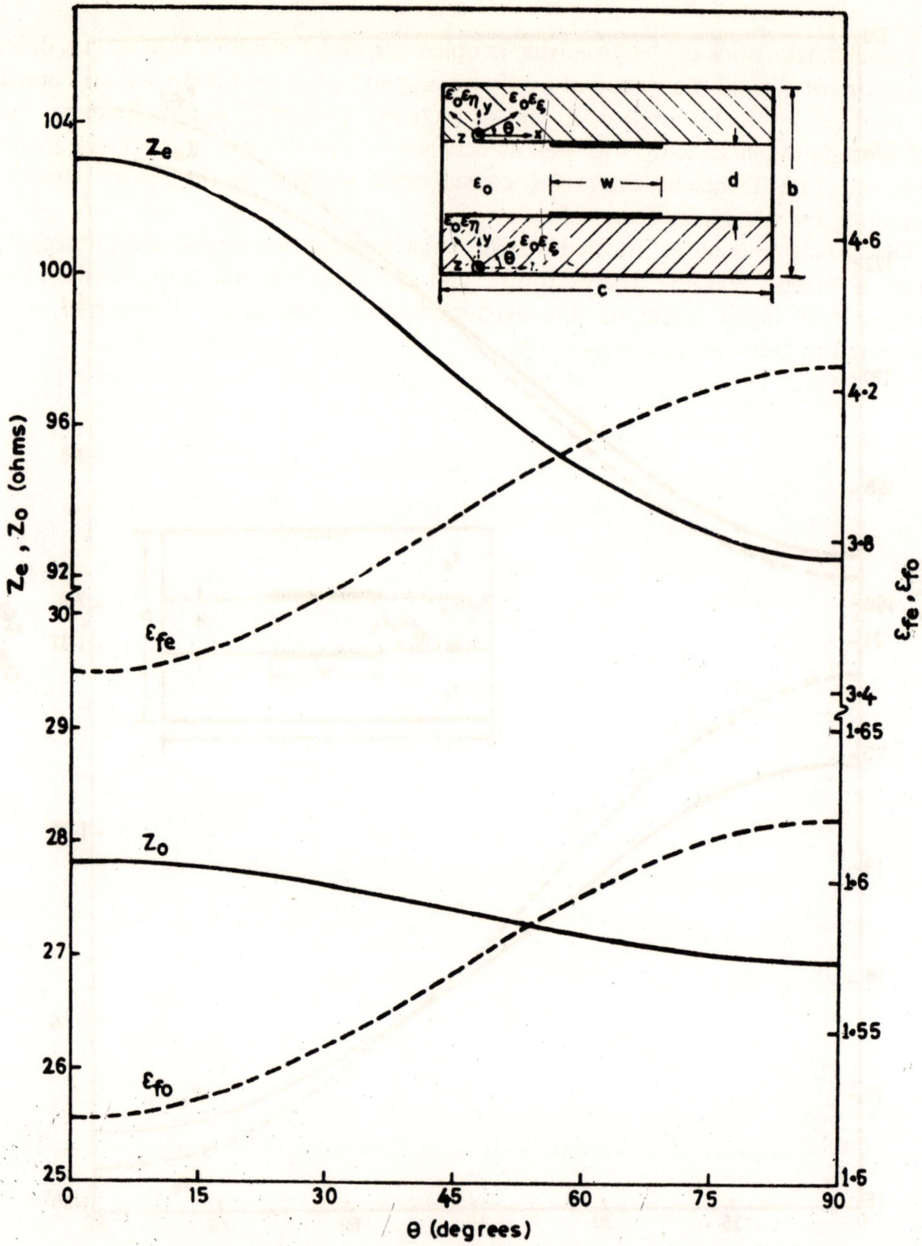

FIG. 7.23 Z_e, Z_o, ϵ_{fe} and ϵ_{fo} versus tilt angle θ of broadside coupled inverted stripline with pyrolytic boron nitride substrates; $\epsilon_\xi = 5.12$, $\epsilon_\eta = 3.4$ at $\theta = 0$, $w/b = 0.4$, $d/b = 0.1$, $c/w = 40$.

appropriately cutting the anisotropic substrate so as to orient the optical axis at the desired angle.

The characteristics of the broadside-coupled suspended stripline behave in a different way as the orientation of the principal axes with respect to the coordinate axes is changed. With an increase in the tilt angle θ, Z_e and ϵ_{fo} increase, whereas Z_o and ϵ_{fe} decrease. Further, the difference between Z_e and Z_o as well as between ϵ_{fe} and ϵ_{fo} increases with an increase in θ. This structure therefore caters for considerable increase in the phase velocity ratio by increasing the tilt angle θ.

The broadside-coupled inverted stripline behaves somewhat similar to the broadside-coupled stripline except that the curves for ϵ_{fe} and ϵ_{fo} versus θ do not cross. Further, the effect of varying θ is much larger on the even-mode indicating larger changes in Z_e and ϵ_{fe} as compared to those in Z_o and ϵ_{fo}.

Table 7.1 Characteristic impedances of broadside-coupled homogeneous stripline; $c/w = 40$ (See structure A1 of Table 4.4)

d/b	0.05		0.10		0.15		0.20		0.25	
w/b	$Z_e\sqrt{\epsilon_r}$ (ohms)	$Z_o\sqrt{\epsilon_r}$ (ohms)	$Z_e\sqrt{\epsilon_r}$ (ohms)	$Z_o\sqrt{\epsilon_r}$ (ohms)	$Z_e\sqrt{\epsilon_r}$ (ohms)	$Z_o\sqrt{\epsilon_r}$ (ohms)	$Z_e\sqrt{\epsilon_r}$ (ohms)	$Z_o\sqrt{\epsilon_r}$ (ohms)	$Z_e\sqrt{\epsilon_r}$ (ohms)	$Z_o\sqrt{\epsilon_r}$ (ohms)
0.10	331.70	58.39	300.45	89.23	278.50	109.45	261.55	123.79	247.58	134.28
0.12	316.65	51.41	287.58	80.21	266.55	99.59	250.04	113.53	236.30	123.85
0.14	303.51	45.97	276.33	72.96	256.21	91.52	240.15	105.04	226.66	115.13
0.16	291.84	41.61	266.29	66.98	247.03	84.76	231.44	97.86	218.22	107.70
0.18	281.34	38.02	257.21	61.95	238.76	78.99	223.63	91.68	210.69	101.27
0.20	271.80	35.01	248.92	57.66	231.21	74.00	216.54	86.28	203.89	95.63
0.22	263.06	32.45	241.28	53.94	224.27	69.64	210.03	81.53	197.66	90.63
0.24	255.01	30.25	234.20	50.69	217.83	65.78	204.01	77.30	191.92	86.16
0.26	247.54	28.33	227.61	47.82	211.82	62.34	198.41	73.50	186.59	82.13
0.28	240.58	26.64	221.45	45.26	206.20	59.26	193.16	70.08	181.62	78.48
0.30	234.06	25.14	215.66	42.97	200.91	56.48	188.23	66.97	176.95	75.15
0.32	227.95	23.80	210.21	40.90	195.92	53.95	183.58	64.13	172.56	72.10
0.34	222.19	22.60	205.07	39.03	191.20	51.64	179.18	61.53	168.40	69.29
0.36	216.75	21.52	200.19	37.32	186.72	49.53	175.01	59.14	164.47	66.71
0.38	211.59	20.54	195.57	35.76	182.47	47.59	171.05	56.93	160.73	64.31
0.40	206.71	19.64	191.18	34.32	178.43	45.79	167.27	54.88	157.17	62.08
0.50	185.51	16.12	172.04	28.59	160.76	38.54	150.77	46.54	141.62	52.93
0.55	176.54	14.80	163.90	26.39	153.23	35.71	143.72	43.26	134.98	49.31
0.60	168.42	13.67	156.53	24.50	146.39	33.28	137.32	40.41	128.95	46.15
0.65	161.04	12.71	149.80	22.86	140.15	31.15	131.48	37.91	123.45	43.37
0.70	154.28	11.87	143.64	21.43	134.43	29.28	126.12	35.70	118.40	40.90
0.75	148.08	11.13	137.98	20.17	129.17	27.62	121.19	33.74	113.76	38.71
0.80	142.36	10.48	132.74	19.05	124.31	26.14	116.63	31.98	109.50	36.73
0.85	137.07	9.91	127.90	18.04	119.80	24.80	112.44	30.39	105.52	34.95
0.90	132.20	9.39	123.40	17.14	115.64	23.60	108.51	28.95	101.82	33.33
0.95	127.63	8.92	119.23	16.33	111.74	22.51	104.85	27.65	98.38	31.85
1.00	123.36	8.50	115.31	15.58	108.09	21.51	101.43	26.45	95.16	30.50
1.05	119.37	8.11	111.64	14.90	104.67	20.60	98.23	25.35	92.15	29.26
1.10	115.64	7.76	108.19	14.28	101.47	19.76	95.23	24.34	89.33	28.11
1.15	112.12	7.44	104.96	13.71	98.45	18.99	92.40	23.41	86.67	27.05
1.20	108.82	7.15	101.90	13.18	95.61	18.27	89.74	22.55	84.17	26.07
1.25	105.70	6.87	99.03	12.69	92.93	17.61	87.22	21.74	81.80	25.15
1.30	102.76	6.62	96.31	12.23	90.39	17.00	84.85	20.99	79.57	24.30
1.35	99.97	6.38	93.73	11.81	87.99	16.42	82.60	20.30	77.46	23.50
1.40	97.33	6.16	91.29	11.41	85.71	15.88	80.46	19.64	75.45	22.76
1.45	94.83	5.96	88.97	11.04	83.55	15.38	78.44	19.03	73.55	22.05
1.50	92.45	5.77	86.76	10.70	81.49	14.91	76.51	18.45	71.74	21.39
1.55	90.19	5.59	84.67	10.37	79.53	14.46	74.67	17.91	70.02	20.77
1.60	88.03	5.42	82.67	10.07	77.66	14.04	72.92	17.40	68.37	20.19
1.65	85.98	5.26	80.76	9.78	75.88	13.65	71.25	16.92	66.81	19.63

(Contd.)

d/b → w/b ↓	0.05		0.10		0.15		0.20		0.25	
	$Z_e\sqrt{\epsilon_r}$ (ohms)	$Z_o\sqrt{\epsilon_r}$ (ohms)	$Z_e\sqrt{\epsilon_r}$ (ohms)	$Z_o\sqrt{\epsilon_r}$ (ohms)	$Z_e\sqrt{\epsilon_r}$ (ohms)	$Z_o\sqrt{\epsilon_r}$ (ohms)	$Z_e\sqrt{\epsilon_r}$ (ohms)	$Z_o\sqrt{\epsilon_r}$ (ohms)	$Z_e\sqrt{\epsilon_r}$ (ohms)	$Z_o\sqrt{\epsilon_r}$ (ohms)
1.70	84.01	5.11	78.93	9.50	74.18	13.27	69.66	16.46	65.31	19.11
1.75	82.14	4.97	77.19	9.25	72.55	12.92	68.13	16.03	63.88	18.61
1.80	80.35	4.83	75.52	9.00	70.99	12.58	66.67	15.62	62.51	18.14
1.85	78.63	4.71	73.93	8.77	69.50	12.26	65.27	15.23	61.19	17.69
1.90	76.98	4.59	72.39	8.55	68.07	11.96	63.93	14.85	59.93	17.26
1.95	75.40	4.47	70.92	8.34	66.69	11.67	62.64	14.50	58.72	16.86
2.00	73.89	4.36	69.51	8.14	65.37	11.39	61.40	14.16	57.56	16.47

d/b → w/b ↓	0.30		0.35		0.40		0.45		0.50	
	$Z_e\sqrt{\epsilon_r}$ (ohms)	$Z_o\sqrt{\epsilon_r}$ (ohms)	$Z_e\sqrt{\epsilon_r}$ (ohms)	$Z_o\sqrt{\epsilon_r}$ (ohms)	$Z_e\sqrt{\epsilon_r}$ (ohms)	$Z_o\sqrt{\epsilon_r}$ (ohms)	$Z_e\sqrt{\epsilon_r}$ (ohms)	$Z_o\sqrt{\epsilon_r}$ (ohms)	$Z_e\sqrt{\epsilon_r}$ (ohms)	$Z_o\sqrt{\epsilon_r}$ (ohms)
0.10	235.50	142.05	224.64	147.67	214.59	151.48	205.01	153.69	195.65	154.41
0.12	224.35	131.46	213.59	137.00	203.60	140.76	194.07	142.96	184.77	143.67
0.14	214.87	122.64	204.21	128.08	194.30	131.79	184.83	133.95	175.58	134.66
0.16	206.60	115.07	196.05	120.45	186.22	124.11	176.83	126.25	167.64	126.95
0.18	199.25	108.49	188.83	113.77	179.02	117.38	169.77	119.49	160.65	120.18
0.20	192.63	102.70	182.34	107.88	172.70	111.43	163.46	113.50	154.41	114.18
0.22	186.60	97.54	176.45	102.62	166.91	106.11	157.76	108.14	148.79	108.82
0.24	181.06	92.91	171.04	97.89	161.62	101.31	152.56	103.31	143.67	103.97
0.26	175.92	88.72	166.06	93.60	156.75	96.95	147.78	98.92	138.98	99.57
0.28	171.14	84.92	161.43	89.69	152.23	92.98	143.37	94.91	134.66	95.55
0.30	166.67	81.44	157.10	86.11	148.03	89.33	139.26	91.23	130.65	91.85
0.32	162.46	78.24	153.05	82.81	144.09	85.97	135.43	87.83	126.25	88.44
0.34	158.50	75.29	149.23	79.76	140.40	82.86	131.84	84.68	123.46	85.28
0.36	154.75	72.56	145.62	76.94	136.91	79.97	128.47	81.76	120.18	82.35
0.38	151.19	70.03	142.21	74.31	133.62	77.28	125.29	70.03	117.09	79.61
0.40	147.80	67.67	138.97	71.86	130.50	74.77	122.30	76.49	114.18	77.05
0.50	133.05	57.94	124.88	61.71	117.04	64.34	109.34	65.90	101.72	66.41
0.55	126.76	54.06	118.91	57.65	111.34	60.15	103.90	61.64	96.52	62.13
0.60	121.06	50.67	113.53	54.09	106.19	56.48	98.99	57.89	91.85	58.36
0.65	115.86	47.68	108.60	50.94	101.51	53.23	94.55	54.58	87.63	55.03
0.70	111.12	45.02	104.08	48.14	97.23	50.33	90.49	51.63	83.79	52.06
0.75	106.73	42.64	99.93	45.63	93.30	47.73	86.77	48.97	80.28	49.39
0.80	102.68	40.50	96.10	43.37	89.68	45.39	83.35	46.58	77.05	46.98
0.85	98.93	38.57	92.56	41.32	86.34	43.26	80.20	44.41	74.08	44.79
0.90	95.44	36.81	89.27	39.46	83.23	41.33	77.27	42.43	71.33	42.80
0.95	92.20	35.20	86.21	37.76	80.35	39.56	74.55	40.62	68.78	40.98
1.00	89.17	33.73	83.35	36.19	77.65	37.93	72.02	38.96	66.41	39.31
1.05	86.33	32.37	80.68	34.75	75.14	36.43	69.66	37.43	64.20	37.76
1.10	83.67	31.12	78.17	33.42	72.78	35.05	67.44	36.01	62.13	36.33
1.15	81.16	29.96	75.81	32.19	70.56	33.76	65.37	34.70	60.19	35.01

(Contd.)

d/b → w/b ↓	0.30		0.35		0.40		0.45		0.50	
	$Z_e\sqrt{\epsilon_r}$ (ohms)	$Z_o\sqrt{\epsilon_r}$ (ohms)	$Z_e\sqrt{\epsilon_r}$ (ohms)	$Z_o\sqrt{\epsilon_r}$ (ohms)	$Z_e\sqrt{\epsilon_r}$ (ohms)	$Z_o\sqrt{\epsilon_r}$ (ohms)	$Z_e\sqrt{\epsilon_r}$ (ohms)	$Z_o\sqrt{\epsilon_r}$ (ohms)	$Z_e\sqrt{\epsilon_r}$ (ohms)	$Z_o\sqrt{\epsilon_r}$ (ohms)
1.20	78.81	28.89	73.60	31.05	68.48	32.57	63.41	33.48	58.36	33.78
1.25	76.59	27.89	71.51	29.98	66.51	31.46	61.57	32.34	56.65	32.63
1.30	74.49	26.95	69.53	28.98	64.66	30.42	59.84	31.27	55.03	31.56
1.35	72.50	26.08	67.66	28.05	62.91	29.45	58.20	30.28	53.50	30.55
1.40	70.61	25.26	65.89	27.18	61.25	28.53	56.65	29.34	52.06	29.61
1.45	68.82	24.49	64.21	26.35	59.67	27.67	55.17	28.46	50.69	28.72
1.50	67.12	23.76	62.61	25.58	58.18	26.87	53.78	27.63	49.39	27.89
1.55	65.50	23.08	61.09	24.85	56.75	26.10	52.45	26.85	48.15	27.10
1.60	63.96	22.43	59.65	24.16	55.40	25.38	51.18	26.11	46.98	26.35
1.65	62.49	21.82	58.27	23.50	54.11	24.70	49.98	25.41	45.86	25.65
1.70	61.08	21.24	56.95	22.89	52.87	24.05	48.83	24.74	44.79	24.98
1.75	59.74	20.69	55.69	22.30	51.69	23.44	47.73	24.11	43.77	24.34
1.80	58.45	20.17	54.48	21.74	50.57	22.85	46.68	23.51	42.80	23.74
1.85	57.22	19.68	53.33	21.21	49.49	22.30	45.67	22.94	41.87	23.16
1.90	56.04	19.20	52.22	20.70	48.45	21.77	44.71	22.40	40.98	22.61
1.95	54.91	18.75	51.16	20.22	47.46	21.26	43.79	21.88	40.12	22.09
2.00	53.82	18.32	50.14	19.76	46.51	20.78	42.90	21.39	39.31	21.59

Table 7.2 Design data on broadside-coupled stripline; Dielectric constant $\epsilon_r = 2.22$, $Z_t = 50$ ohms, $c/w = \infty$ (See structure A1 of Table 4.4)

C_v (dB)	w/b	d/b	C_v (dB)	w/b	d/b	C_v (dB)	w/b	d/b
1.0	0.1621	0.0184	5.4	0.5894	0.2070	9.8	0.6230	0.3890
1.1	0.1817	0.0218	5.5	0.5928	0.2113	9.9	0.6218	0.3927
1.2	0.2007	0.0253	5.6	0.5960	0.2158	10.0	0.6207	0.3964
1.3	0.2190	0.0290	5.7	0.5990	0.2201	10.5	0.6141	0.4148
1.4	0.2366	0.0327	5.8	0.6019	0.2246	11.0	0.6065	0.4326
1.5	0.2536	0.0366	5.9	0.6046	0.2289	11.5	0.5980	0.4499
1.6	0.2699	0.0405	6.0	0.6072	0.2333	12.0	0.5887	0.4667
1.7	0.2856	0.0445	6.1	0.6096	0.2377	12.5	0.5788	0.4829
1.8	0.3008	0.0485	6.2	0.6119	0.2420	13.0	0.5684	0.4986
1.9	0.3153	0.0526	6.3	0.6140	0.2464	13.5	0.5576	0.5138
2.0	0.3293	0.0567	6.4	0.6161	0.2508	14.0	0.5466	0.5284
2.1	0.3428	0.0609	6.5	0.6179	0.2551	14.5	0.5354	0.5425
2.2	0.3558	0.0651	6.6	0.6197	0.2594	15.0	0.5240	0.5560
2.3	0.3683	0.0694	6.7	0.6213	0.2637	15.5	0.5126	0.5691
2.4	0.3803	0.0737	6.8	0.6228	0.2679	16.0	0.5012	0.5816
2.5	0.3919	0.0779	6.9	0.6241	0.2722	16.5	0.4899	0.5937
2.6	0.4030	0.0823	7.0	0.6254	0.2765	17.0	0.4788	0.6052
2.7	0.4137	0.0866	7.1	0.6265	0.2807	17.5	0.4678	0.6163
2.8	0.4241	0.0910	7.2	0.6275	0.2850	18.0	0.4569	0.6269
2.9	0.4341	0.0954	7.3	0.6285	0.2892	18.5	0.4463	0.6370
3.0	0.4436	0.0998	7.4	0.6293	0.2933	19.0	0.4360	0.6467
3.1	0.4529	0.1043	7.5	0.6299	0.2975	19.5	0.4259	0.6560
3.2	0.4618	0.1087	7.6	0.6306	0.3017	20.0	0.4161	0.6649
3.3	0.4703	0.1131	7.7	0.6311	0.3059	20.5	0.4066	0.6733
3.4	0.4786	0.1176	7.8	0.6315	0.3100	21.0	0.3975	0.6814
3.5	0.4865	0.1220	7.9	0.6318	0.3141	21.5	0.3886	0.6891
3.6	0.4941	0.1265	8.0	0.6320	0.3182	22.0	0.3800	0.6964
3.7	0.5014	0.1309	8.1	0.6321	0.3223	22.5	0.3718	0.7034
3.8	0.5085	0.1354	8.2	0.6322	0.3264	23.0	0.3638	0.7100
3.9	0.5153	0.1399	8.3	0.6322	0.3304	23.5	0.3562	0.7164
4.0	0.5218	0.1444	8.4	0.6320	0.3344	24.0	0.3490	0.7224
4.1	0.5281	0.1489	8.5	0.6318	0.3885	24.5	0.3419	0.7281
4.2	0.5341	0.1534	8.6	0.6316	0.3425	25.0	0.3352	0.7336
4.3	0.5399	1.1578	8.7	0.6312	0.3464	25.5	0.3288	0.7387
4.4	0.5454	0.1623	8.8	0.6308	0.3504	26.0	0.3226	0.7436
4.5	0.5508	0.1668	8.9	0.6303	0.3543	26.5	0.3168	0.7483
4.6	0.5559	0.1713	9.0	0.6297	0.3583	27.0	0.3112	0.7527
4.7	0.5608	0.1758	9.1	0.6291	0.3622	27.5	0.3058	0.7569
4.8	0.5654	0.1802	9.2	0.6284	0.3661	28.0	0.3008	0.7609
4.9	0.5699	0.1847	9.3	0.6276	0.3699	28.5	0.2959	0.7649
5.0	0.5742	0.1892	9.4	0.6268	0.3738	29.0	0.2913	0.7683
5.1	0.5783	0.1936	9.5	0.6259	0.3776	29.5	0.2869	0.7717
5.2	0.5821	0.1980	9.6	0.6250	0.3814	30.0	0.2828	0.7749
5.3	0.5859	0.2025	9.7	0.6240	0.3852			

Table 7.3 Characteristic impedance and effective dielectric constants of broadside-coupled suspended stripline; $c/w = 40$, $\epsilon_r = 2.22$ (Duroid) (See structure A2 of Table 4.4)

d/b = **0.05**

w/b	Z_e(ohms)	Z_o(ohms)	ϵ_{fe}	ϵ_{fo}	w/b	Z_e(ohms)	Z_o(ohms)	ϵ_{fe}	ϵ_{fo}
0.10	312.005	42.125	1.130	1.921	0.90	129.464	6.454	1.043	2.116
0.12	298.946	36.890	1.122	1.942	0.95	125.091	6.130	1.041	2.118
0.14	287.422	32.840	1.115	1.960	1.00	121.020	5.837	1.039	2.120
0.16	277.100	29.608	1.109	1.975	1.05	117.191	5.570	1.038	2.122
0.18	267.750	26.965	1.104	1.988	1.10	113.597	5.327	1.036	2.124
0.20	259.203	24.761	1.100	2.000	1.15	110.215	5.104	1.035	2.126
0.22	251.334	22.894	1.096	2.010	1.20	107.029	4.899	1.034	2.127
0.24	244.044	21.291	1.092	2.019	1.25	104.021	4.710	1.033	2.129
0.26	237.257	19.900	1.089	2.027	1.30	101.176	4.535	1.031	2.130
0.28	230.909	18.681	1.086	2.034	1.35	98.482	4.372	1.030	2.131
0.30	224.950	17.604	1.083	2.040	1.40	95.928	4.221	1.030	2.132
0.32	219.338	16.644	1.080	2.046	1.45	93.501	4.080	1.029	2.133
0.34	214.037	15.784	1.078	2.052	1.50	91.194	3.947	1.028	2.134
0.36	209.017	15.009	1.076	2.057	1.55	88.997	3.823	1.027	2.135
0.38	204.252	14.307	1.073	2.061	1.60	86.903	3.707	1.026	2.136
0.40	199.721	13.667	1.071	2.066	1.65	84.905	3.598	1.025	2.137
0.50	179.931	11.171	1.063	2.082	1.70	82.995	3.494	1.025	2.137
0.55	171.515	10.237	1.059	2.089	1.75	81.170	3.397	1.024	2.138
0.60	163.876	9.446	1.056	2.094	1.80	79.422	3.305	1.023	2.139
0.65	156.903	8.769	1.053	2.099	1.85	77.747	3.217	1.023	2.139
0.70	150.509	8.182	1.051	2.103	1.90	76.141	3.135	1.022	2.140
0.75	144.621	7.669	1.048	2.107	1.95	74.599	3.056	1.022	2.140
0.80	139.180	7.216	1.046	2.110	2.00	73.119	2.981	1.021	2.141
0.85	134.136	6.814	1.044	2.113					

(*Contd.*)

$d/b = \mathbf{0.10}$

w/b	Z_e (ohms)	Z_o (ohms)	ϵ_{fe}	ϵ_{fo}	w/b	Z_e (ohms)	Z_o (ohms)	ϵ_{fe}	ϵ_{fo}
0.10	275.493	65.854	1.189	1.836	0.90	119.315	12.006	1.070	2.039
0.12	264.811	58.909	1.179	1.854	0.95	115.412	11.424	1.067	2.042
0.14	255.378	53.355	1.171	1.870	1.00	111.744	10.896	1.065	2.045
0.16	246.892	48.799	1.163	1.884	1.05	108.319	10.415	1.062	2.048
0.18	239.157	44.986	1.157	1.897	1.10	105.084	9.974	1.060	2.050
0.20	232.040	41.742	1.151	1.908	1.15	102.037	9.569	1.058	2.052
0.22	225.446	38.947	1.145	1.918	1.20	99.160	9.195	1.056	2.054
0.24	219.300	36.511	1.141	1.927	1.25	96.441	8.849	1.054	2.056
0.26	213.549	34.367	1.136	1.936	1.30	93.867	8.528	1.053	2.058
0.28	208.144	32.466	1.132	1.943	1.35	91.426	8.230	1.051	2.059
0.30	203.050	30.766	1.128	1.950	1.40	89.108	7.952	1.050	2.061
0.32	198.235	29.238	1.124	1.957	1.45	86.905	7.692	1.048	2.062
0.34	193.671	27.855	1.121	1.963	1.50	84.807	7.448	1.047	2.063
0.36	189.337	26.599	1.118	1.968	1.55	82.807	7.219	1.045	2.064
0.38	185.212	25.452	1.115	1.974	1.60	80.899	7.004	1.044	2.066
0.40	181.279	24.400	1.112	1.978	1.65	79.076	6.801	1.043	2.067
0.50	164.029	20.227	1.100	1.997	1.70	77.333	6.610	1.042	2.068
0.55	156.638	18.634	1.095	2.005	1.75	75.665	6.429	1.041	2.069
0.60	149.906	17.272	1.090	2.012	1.80	74.067	6.258	1.040	2.069
0.65	143.744	16.096	1.086	2.018	1.85	72.534	6.095	1.039	2.070
0.70	138.077	15.070	1.082	2.023	1.90	71.062	5.941	1.038	2.071
0.75	132.849	14.166	1.079	2.028	1.95	69.649	5.794	1.037	2.072
0.80	128.004	13.364	1.075	2.032	2.00	68.291	5.654	1.036	2.073
0.85	123.505	12.648	1.072	2.035					

(Contd.)

$d/b = 0.15$

w/b	Z_e (ohms)	Z_o (ohms)	ϵ_{fe}	ϵ_{fo}	w/b	Z_e (ohms)	Z_o (ohms)	ϵ_{fe}	ϵ_{fo}
0.10	251.183	81.854	1.229	1.788	0.90	110.772	16.800	1.090	1.973
0.12	241.465	74.179	1.219	1.803	0.95	107.192	16.010	1.087	1.976
0.14	232.987	67.919	1.209	1.816	1.00	103.837	15.290	1.084	1.979
0.16	225.417	62.695	1.201	1.828	1.05	100.700	14.632	1.080	1.982
0.18	218.549	58.256	1.194	1.839	1.10	97.736	14.028	1.078	1.984
0.20	212.243	54.430	1.187	1.848	1.15	94.942	13.472	1.075	1.986
0.22	206.403	51.095	1.181	1.857	1.20	92.304	12.958	1.073	1.989
0.24	200.957	48.157	1.175	1.866	1.25	89.808	12.483	1.071	1.991
0.26	195.852	45.549	1.170	1.873	1.30	87.444	12.040	1.069	1.993
0.28	191.048	43.215	1.165	1.880	1.35	85.200	11.628	1.067	1.994
0.30	186.511	41.114	1.160	1.887	1.40	83.069	11.242	1.065	1.996
0.32	182.213	39.212	1.156	1.893	1.45	81.042	10.881	1.063	1.997
0.34	178.132	37.480	1.152	1.899	1.50	79.110	10.543	1.061	1.999
0.36	174.248	35.897	1.148	1.904	1.55	77.269	10.225	1.059	2.000
0.38	170.546	34.444	1.145	1.909	1.60	75.511	9.925	1.058	2.001
0.40	167.011	33.105	1.141	1.913	1.65	73.830	9.643	1.056	2.002
0.50	151.444	27.724	1.127	1.932	1.70	72.223	9.376	1.055	2.004
0.55	144.748	25.641	1.121	1.940	1.75	70.683	9.123	1.054	2.005
0.60	138.638	23.845	1.115	1.947	1.80	69.207	8.884	1.052	2.006
0.65	133.035	22.295	1.110	1.952	1.85	67.792	8.656	1.051	2.006
0.70	127.879	20.927	1.105	1.957	1.90	66.432	8.440	1.050	2.007
0.75	123.114	19.717	1.101	1.962	1.95	65.126	8.235	1.049	2.008
0.80	118.696	18.693	1.097	1.966	2.00	63.869	8.039	1.048	2.009
0.85	114.588	17.672	1.093	1.970					

(Contd.)

$d/b = \mathbf{0.20}$

w/b	Z_e(ohms)	Z_o(ohms)	ϵ_{fe}	ϵ_{fo}	w/b	Z_e (ohms)	Z_o(ohms)	ϵ_{fe}	ϵ_{fo}
0.10	233.067	93.469	1.259	1.754	0.90	103.208	20.926	1.105	1.914
0.12	223.800	85.442	1.248	1.766	0.95	99.896	19.966	1.102	1.917
0.14	215.793	78.818	1.238	1.776	1.00	96.804	19.089	1.098	1.920
0.16	208.706	73.229	1.230	1.786	1.05	93.888	18.286	1.095	1.922
0.18	202.320	68.430	1.222	1.795	1.10	91.144	17.547	1.092	1.925
0.20	196.490	64.257	1.214	1.803	1.15	88.556	16.866	1.089	1.927
0.22	191.111	60.587	1.208	1.811	1.20	86.112	16.235	1.086	1.929
0.24	186.110	57.330	1.202	1.818	1.25	83.800	15.649	1.083	1.931
0.26	181.433	54.417	1.196	1.824	1.30	81.609	15.104	1.081	1.932
0.28	177.034	51.795	1.190	1.831	1.35	79.530	14.595	1.079	1.934
0.30	172.884	49.421	1.185	1.836	1.40	77.554	14.119	1.076	1.935
0.32	168.952	47.260	1.181	1.842	1.45	75.674	13.673	1.074	1.937
0.34	165.218	45.283	1.176	1.846	1.50	73.883	13.255	1.072	1.938
0.36	161.664	43.468	1.172	1.851	1.55	72.175	12.863	1.070	1.940
0.38	158.273	41.796	1.168	1.855	1.60	70.543	12.491	1.069	1.941
0.40	155.034	40.248	1.164	1.860	1.65	68.984	12.141	1.067	1.942
0.50	140.741	33.972	1.148	1.877	1.70	67.492	11.809	1.065	1.943
0.55	134.578	31.517	1.140	1.884	1.75	66.062	11.495	1.064	1.944
0.60	128.947	29.393	1.134	1.890	1.80	64.692	11.197	1.062	1.945
0.65	123.780	27.537	1.128	1.895	1.85	63.377	10.915	1.061	1.946
0.70	119.020	25.901	1.123	1.900	1.90	62.114	10.646	1.059	1.947
0.75	114.618	24.448	1.118	1.904	1.95	60.901	10.389	1.058	1.947
0.80	110.535	23.149	1.113	1.908	2.00	59.733	10.145	1.057	1.948
0.85	106.752	21.983	1.109	1.911					

(Contd.)

$d/b = \mathbf{0.25}$

w/b	Z_e(ohms)	Z_o(ohms)	ϵ_{fe}	ϵ_{fo}	w/b	Z_e(ohms)	Z_o(ohms)	ϵ_{fe}	ϵ_{fo}
0.10	218.567	102.210	1.283	1.726	0.85	99.611	25.644	1.122	1.857
0.12	209.538	94.009	1.272	1.735	0.90	96.311	24.437	1.118	1.860
0.14	201.790	87.183	1.262	1.744	0.95	93.238	23.338	1.113	1.863
0.16	194.976	81.382	1.253	1.751	1.00	90.345	22.334	1.110	1.865
0.18	188.876	76.368	1.244	1.759	1.05	87.628	21.414	1.106	1.867
0.20	183.336	71.978	1.237	1.765	1.10	85.071	20.564	1.103	1.869
0.22	178.249	68.094	1.230	1.772	1.15	82.660	19.779	1.099	1.870
0.24	173.540	64.627	1.223	1.777	1.20	80.383	19.052	1.096	1.872
0.26	169.146	61.511	1.217	1.783	1.25	78.299	18.375	1.094	1.874
0.28	165.029	58.692	1.211	1.788	1.30	76.187	17.745	1.091	1.875
0.30	161.148	56.127	1.206	1.793	1.35	74.250	17.156	1.088	1.877
0.32	157.482	53.783	1.201	1.797	1.40	72.409	16.605	1.086	1.878
0.34	154.002	51.632	1.196	1.801	1.45	70.658	16.088	1.084	1.879
0.36	150.694	49.649	1.191	1.805	1.50	68.989	15.602	1.081	1.880
0.38	147.540	47.815	1.187	1.809	1.55	67.398	15.144	1.079	1.881
0.40	144.527	46.114	1.183	1.812	1.60	65.878	14.712	1.077	1.882
0.50	131.242	39.161	1.164	1.827	1.65	64.425	14.304	1.075	1.883
0.55	125.512	36.420	1.157	1.833	1.70	63.035	13.918	1.073	1.884
0.60	120.275	34.037	1.149	1.838	1.75	61.703	13.552	1.072	1.885
0.65	115.466	31.948	1.143	1.843	1.80	60.426	13.206	1.070	1.886
0.70	111.035	30.099	1.137	1.847	1.85	59.201	12.877	1.068	1.887
0.75	106.937	28.452	1.132	1.851	1.90	58.025	12.563	1.067	1.888
0.80	103.145	26.975	1.127	1.854	1.95	56.893	12.264	1.065	1.889
					2.00	55.805	11.978	1.064	1.890

(Contd.)

$d/b = 0.30$

w/b	Z_e(ohms)	Z_o(ohms)	ϵ_{fe}	ϵ_{fo}	w/b	Z_e(ohms)	Z_o(ohms)	ϵ_{fe}	ϵ_{fo}
0.10	206.344	108.904	1.303	1.701	0.85	92.976	28.704	1.132	1.805
0.12	197.456	100.582	1.291	1.708	0.90	89.889	27.377	1.127	1.808
0.14	189.865	93.649	1.281	1.715	0.95	87.002	26.166	1.123	1.810
0.16	183.222	87.719	1.271	1.721	1.00	84.296	25.058	1.119	1.812
0.18	177.301	82.571	1.263	1.726	1.05	81.756	24.039	1.115	1.814
0.20	171.949	78.043	1.255	1.732	1.10	79.365	23.100	1.111	1.815
0.22	167.056	74.020	1.248	1.736	1.15	77.111	22.231	1.108	1.817
0.24	162 541	70.413	1.241	1.741	1.20	74.982	21.428	1.105	1.818
0.26	158.346	67.159	1.234	1.745	1.25	72.969	20.676	1.102	1.819
0.28	154.427	64.203	1.228	1.749	1.30	71.062	19.976	1.099	1.820
0.30	150.743	61.506	1.222	1.753	1.35	69.252	19.321	1.096	1.821
0.32	147.268	59.032	1.217	1.757	1.40	67.532	18.708	1.093	1.823
0.34	143.979	56.754	1.212	1.760	1.45	65.897	18.132	1.091	1.824
0.36	140.856	54.649	1.207	1.763	1.50	64.338	17.591	1.088	1.825
0.38	137.885	52.697	1.202	1.766	1.55	62.852	17.081	1.086	1.826
0.40	135.050	50.881	1.198	1.769	1.60	61.433	16.599	1.084	1.826
0.50	122.575	43.419	1.178	1.781	1.65	60.077	16.143	1.082	1.827
0.55	117.205	40.456	1.170	1.786	1.70	58.779	15.712	1.080	1.828
0.60	112.300	37.873	1.162	1.790	1.75	57.536	15 303	1.078	1.829
0.65	107.799	35.600	1.155	1.794	1.80	56.345	14.914	1.076	1.829
0.70	103.666	33.584	1.149	1.797	1.85	55.202	14.545	1.075	1.830
0.75	99.830	31.784	1.143	1.800	1.90	54.104	14.193	1.073	1.831
0.80	96.274	30.166	1.138	1.803	1.95	53.049	13.858	1.071	1.831
					2.00	52.034	13.539	1.070	1.832

(Contd.)

$d/b = 0.35$

w/b	Z_e(ohms)	Z_o(ohms)	ϵ_{fe}	ϵ_{fo}	w/b	Z_e(ohms)	Z_o(ohms)	ϵ_{fe}	ϵ_{fo}
0.10	195.621	113.989	1.319	1.678	0.85	86.674	31.190	1.140	1.755
0.12	186.826	105.601	1.307	1.683	0.90	83.779	29.768	1.135	1.757
0.14	179.340	98.587	1.297	1.688	0.95	81.073	28.470	1.131	1.759
0.16	172.811	92.585	1.287	1.692	1.00	78.537	27.280	1.126	1.760
0.18	167.011	87.351	1.278	1.696	1.05	76.157	26.185	1.122	1.762
0.20	161.788	82.734	1.270	1.700	1.10	73.919	25.173	1.118	1.763
0.22	157.029	78.618	1.263	1.704	1.15	71.808	24.237	1.115	1.764
0.24	152.654	74.919	1.255	1.707	1.20	69.816	23.368	1.111	1.765
0.26	148.601	71.570	1.249	1.710	1.25	67.932	22.558	1.108	1.766
0.28	144.825	68.521	1.242	1.713	1.30	66.148	21.803	1.105	1.767
0.30	141.287	65.731	1.236	1.716	1.35	64.456	21.098	1.102	1.768
0.32	137.959	63.165	1.231	1.719	1.40	62.848	20.435	1.099	1.769
0.34	134.815	60.797	1.225	1.721	1.45	61.319	19.812	1.097	1.769
0.36	131.837	58.604	1.220	1.724	1.50	59.863	19.226	1.094	1.770
0.38	129.009	56.567	1.215	1.726	1.55	58.474	18.673	1.092	1.771
0.40	126.315	54.668	1.210	1.728	1.60	57.149	18.151	1.089	1.771
0.50	114.507	46.826	1.189	1.737	1.65	55.882	17.657	1.087	1.772
0.55	109.442	43.697	1.181	1.740	1.70	54.671	17.189	1.085	1.773
0.60	104.839	40.960	1.173	1.744	1.75	53.510	16.745	1.083	1.773
0.65	100.619	38.546	1.165	1.746	1.80	52.398	16.324	1.081	1.774
0.70	96.721	36.400	1.158	1.749	1.85	51.332	15.922	1.079	1.774
0.75	93.119	34.481	1.152	1.751	1.90	50.308	15.540	1.078	1.765
0.80	89.779	32.753	1.146	1.754	1.95	49.323	15.176	1.076	1.775
					2.00	48.377	14.828	1.074	1.776

(Contd.)

$d/b = 0.40$

w/b	Z_e(ohms)	Z_o(ohms)	ϵ_{fe}	ϵ_{fo}	w/b	Z_e(ohms)	Z_o(ohms)	ϵ_{fe}	ϵ_{fo}
0.10	185.902	117.730	1.332	1.655	0.85	80.615	33.116	1.147	1.707
0.12	177.173	109.301	1.321	1.659	0.90	77.897	31.623	1.142	1.708
0.14	169.761	102.237	1.310	1.662	0.95	75.359	30.259	1.137	1.709
0.16	163.313	96.189	1.300	1.665	1.00	72.982	29.007	1.132	1.710
0.18	157.604	90.901	1.291	1.667	1.05	70.751	27.854	1.128	1.711
0.20	152.475	86.227	1.283	1.670	1.10	68.654	26.789	1.124	1.712
0.22	147.816	82.051	1.275	1.672	1.15	66.679	25.802	1.120	1.712
0.24	143.546	78.291	1.268	1.674	1.20	64.814	24.884	1.116	1.713
0.26	139.602	74.880	1.261	1.677	1.25	63.052	24.030	1.113	1.714
0.28	135.937	71.768	1.254	1.678	1.30	61.384	23.231	1.110	1.715
0.30	132.513	68.914	1.248	1.680	1.35	59.802	22.484	1.107	1.715
0.32	129.299	66.286	1.242	1.682	1.40	58.299	21.783	1.104	1.716
0.34	126.271	63.857	1.236	1.684	1.45	56.871	21.127	1.101	1.716
0.36	123.410	61.603	1.231	1.685	1.50	55.511	20.507	1.098	1.716
0.38	120.696	59.505	1.226	1.687	1.55	54.215	19.921	1.096	1.717
0.40	118.120	57.548	1.221	1.688	1.60	52.979	19.368	1.093	1.717
0.50	106.902	49.434	1.199	1.694	1.65	51.797	18.845	1.091	1.718
0.55	102.100	46.183	1.189	1.696	1.70	50.667	18.348	1.089	1.718
0.60	97.732	43.333	1.181	1.699	1.75	49.585	17.878	1.087	1.718
0.65	93.736	40.815	1.173	1.701	1.80	48.549	17.430	1.085	1.719
0.70	90.063	38.574	1.166	1.702	1.85	47.555	17.004	1.083	1.719
0.75	86.672	36.565	1.159	1.704	1.90	46.601	16.599	1.081	1.719
0.80	83.532	34.756	1.153	1.705	1.95	45.684	16.212	1.079	1.720
					2.00	44.803	15.843	1.078	1.720

(Contd.)

d/b = **0.45**

w/b	Z_e(ohms)	Z_o(ohms)	ϵ_{fe}	ϵ_{fo}	w/b	Z_e(ohms)	Z_o(ohms)	ϵ_{fe}	ϵ_{fo}
0.10	176.842	120.278	1.344	1.633	0.85	74.720	34.488	1.152	1.658
0.12	168.162	111.826	1.332	1.634	0.90	72.169	32.946	1.146	1.659
0.14	160.806	104.734	1.321	1.636	0.95	69.788	31.536	1.141	1.659
0.16	154.424	98.657	1.311	1.638	1.00	67.561	30.242	1.136	1.660
0.18	148.784	93.337	1.302	1.639	1.05	65.473	29.049	1.132	1.660
0.20	143.730	88.627	1.293	1.640	1.10	63.511	27.945	1.128	1.661
0.22	139.152	84.416	1.285	1.641	1.15	61.663	26.923	1.124	1.661
0.24	134.966	80.617	1.278	1.642	1.20	59.921	25.972	1.120	1.662
0.26	131.117	77.168	1.270	1.643	1.25	58.275	25.085	1.116	1.662
0.28	127.553	74.017	1.263	1.644	1.30	56.718	24.257	1.113	1.662
0.30	124.221	71.124	1.257	1.645	1.35	55.242	23.482	1.110	1.663
0.32	121.103	68.456	1.251	1.646	1.40	53.841	22.753	1.107	1.663
0.34	118.172	65.987	1.245	1.647	1.45	52.510	22.069	1.104	1.663
0.36	115.408	63.694	1.239	1.648	1.50	51.243	21.425	1.101	1.663
0.38	112.795	61.558	1.234	1.648	1.55	50.036	20.819	1.099	1.663
0.40	110.328	59.562	1.229	1.649	1.60	48.885	20.244	1.096	1.663
0.50	99.566	51.269	1.206	1.652	1.65	47.785	19.700	1.094	1.664
0.55	95.000	47.937	1.196	1.653	1.70	46.734	19.184	1.092	1.664
0.60	90.858	45.011	1.187	1.654	1.75	45.728	18.694	1.089	1.664
0.65	87.078	42.423	1.179	1.655	1.80	44.764	18.228	1.087	1.664
0.70	83.610	40.116	1.171	1.656	1.85	43.841	17.785	1.085	1.664
0.75	80.416	38.047	1.164	1.657	1.90	42.954	17.362	1.084	1.664
0.80	77.461	36.180	1.158	1.658	1.95	42.103	16.960	1.082	1.665
					2.00	41.285	16.575	1.080	1.665

(Contd.)

$d/b = 0.50$

w/b	Z_e(ohms)	Z_o(ohms)	ϵ_{fe}	ϵ_{fo}	w/b	Z_e(ohms)	Z_o(ohms)	ϵ_{fe}	ϵ_{fo}
0.10	168.173	121.709	1.354	1.610	0.85	68.923	35.301	1.155	1.610
0.12	159.535	113.245	1.341	1.610	0.90	66.531	33.731	1.150	1.610
0.14	152.228	106.140	1.330	1.610	0.95	64.303	32.295	1.144	1.610
0.16	145.899	100.049	1.320	1.610	1.00	62.220	30.976	1.139	1.610
0.18	140.320	94.713	1.311	1.610	1.05	60.269	29.760	1.135	1.610
0.20	135.350	89.987	1.302	1.610	1.10	58.438	28.635	1.130	1.610
0.22	130.840	85.758	1.293	1.610	1.15	56.715	27.591	1.126	1.610
0.24	126.726	81.941	1.285	1.610	1.20	55.092	26.621	1.122	1.610
0.26	122.949	78.472	1.278	1.610	1.25	53.559	25.716	1.119	1.610
0.28	119.451	75.302	1.271	1.610	1.30	52.110	24.871	1.115	1.610
0.30	116.202	72.388	1.264	1.610	1.35	50.737	24.078	1.112	1.610
0.32	113.182	69.701	1.258	1.610	1.40	49.435	23.335	1.109	1.610
0.34	110.336	67.211	1.252	1.610	1.45	48.199	22.635	1.106	1.610
0.36	107.659	64.898	1.246	1.610	1.50	47.022	21.977	1.103	1.610
0.38	105.131	62.742	1.241	1.610	1.55	45.902	21.355	1.100	1.610
0.40	102.739	60.727	1.235	1.610	1.60	44.834	20.770	1.098	1.610
0.50	92.420	52.339	1.211	1.610	1.65	43.815	20.213	1.096	1.610
0.55	88.069	48.962	1.201	1.610	1.70	42.841	19.685	1.093	1.610
0.60	84.136	45.996	1.192	1.610	1.75	41.910	19.184	1.091	1.610
0.65	80.558	43.368	1.183	1.610	1.80	41.018	18.708	1.089	1.610
0.70	77.284	41.025	1.175	1.610	1.85	40.163	18.254	1.087	1.610
0.75	74.274	38.921	1.168	1.610	1.90	39.343	17.822	1.085	1.610
0.80	71.496	37.023	1.162	1.610	1.95	38.556	17.410	1.083	1.610
					2.00	37.800	17.016	1.081	1.610

Table 7.4 Characteristic impedances and effective dielectric constants of broadside-coupled suspended stripline; $c/w = 40$, $\epsilon_r = 3.78$ (fused quartz) (See structure A2 of Table 4.4)

$d/b = 0.05$

w/b	Z_e(ohms)	Z_o(ohms)	ϵ_{fe}	ϵ_{fo}	w/b	Z_e(ohms)	Z_o(ohms)	ϵ_{fe}	ϵ_{fo}
0.10	294.592	33.231	1.268	3.087	0.85	131.075	5.268	1.094	3.535
0.12	283.047	29.030	1.252	3.136	0.90	126.619	4.989	1.090	3.541
0.14	272.786	25.795	1.238	3.177	0.95	122.442	4.737	1.087	3.546
0.16	263.541	23.217	1.226	3.212	1.00	118.531	4.510	1.083	3.551
0.18	255.124	21.114	1.216	3.242	1.05	114.862	4.303	1.080	3.556
0.20	247.397	19.364	1.207	3.269	1.10	111.429	4.115	1.077	3.560
0.22	240.254	17.885	1.199	3.292	1.15	108.180	3.942	1.074	3.564
0.24	233.613	16.617	1.192	3.313	1.20	105.116	3.783	1.072	3.567
0.26	227.411	15.518	1.185	3.332	1.25	102.219	3.637	1.069	3.570
0.28	221.593	14.557	1.179	3.349	1.30	99.477	3.501	1.067	3.573
0.30	216.117	13.708	1.173	3.364	1.35	96.877	3.375	1.065	3.576
0.32	210.947	12.952	1.168	3.378	1.40	94.409	3.258	1.063	3.579
0.34	206.053	12.276	1.163	3.390	1.45	92.063	3.149	1.061	3.581
0.36	201.407	11.668	1.158	3.402	1.50	89.830	3.046	1.059	3.583
0.38	196.990	11.117	1.154	3.412	1.55	87.702	2.950	1.057	3.585
0.40	192.781	10.615	1.150	3.422	1.60	85.672	2.860	1.056	3.587
0.50	174.339	8.664	1.132	3.462	1.65	83.733	2.776	1.054	3.589
0.55	166.443	7.934	1.125	3.478	1.70	81.880	2.696	1.053	3.591
0.60	159.254	7.317	1.118	3.491	1.75	80.106	2.621	1.051	3.591
0.65	152.675	6.789	1.113	3.502	1.80	78.407	2.550	1.050	3.593
0.70	146.627	6.333	1.107	3.512	1.85	76.777	2.482	1.049	3.594
0.75	141.045	5.933	1.102	3.520	1.90	75.214	2.418	1.048	3.596
0.80	135.875	5.581	1.098	3.528	1.95	73.712	2.357	1.046	3.597
					2.00	72.269	2.299	1.045	3.598

(Contd.)

$d/b = \mathbf{0.10}$

w/b	Z_e(ohms)	Z_o(ohms)	ϵ_{fe}	ϵ_{fo}	w/b	Z_e(ohms)	Z_o(ohms)	ϵ_{fe}	ϵ_{fo}
0.10	255.229	52.476	1.386	2.891	0.85	119.255	9.849	1.150	3.356
0.12	246.062	46.835	1.366	2.933	0.90	115.347	9.346	1.144	3.365
0.14	237.914	42.336	1.349	2.970	0.95	111.700	8.891	1.139	3.372
0.16	230.544	38.654	1.334	3.003	1.00	108.264	8.477	1.134	3.379
0.18	223.791	35.580	1.321	3.032	1.05	105.036	8.101	1.130	3.384
0.20	217.551	32.970	1.309	3.058	1.10	101.993	7.756	1.125	3.390
0.22	211.745	30.726	1.298	3.082	1.15	99.124	7.439	1.121	3.395
0.24	206.313	28.773	1.289	3.103	1.20	96.423	7.147	1.117	3.399
0.26	201.213	27.057	1.280	3.123	1.25	93.854	6.877	1.113	3.404
0.28	196.404	25.541	1.271	3.140	1.30	91.419	6.627	1.110	3.408
0.30	191.858	24.185	1.264	3.156	1.35	89.106	6.394	1.106	3.411
0.32	187.549	22.967	1.256	3.171	1.40	86.907	6.177	1.103	3.415
0.34	183.455	21.867	1.249	3.185	1.45	84.813	5.974	1.100	3.418
0.36	179.557	20.868	1.243	3.198	1.50	82.817	5.784	1.098	3.421
0.38	175.839	19.956	1.237	3.210	1.55	80.913	5.606	1.095	3.424
0.40	172.286	19.121	1.231	3.221	1.60	79.094	5.438	1.092	3.426
0.50	156.618	15.816	1.207	3.267	1.65	77.354	5.280	1.090	3.429
0.55	149.862	14.557	1.196	3.285	1.70	75.689	5.131	1.088	3.431
0.60	143.685	13.484	1.187	3.301	1.75	74.093	4.990	1.085	3.433
0.65	138.013	12.558	1.178	3.115	1.80	72.563	4.857	1.083	3.435
0.70	132.782	11.751	1.170	3.327	1.85	71.095	4.730	1.081	3.437
0.75	127.941	11.041	1.163	3.338	1.90	69.684	4.610	1.079	3.439
0.80	123.443	10.411	1.156	3.348	1.95	68.327	4.496	1.077	3.441
					2.00	67.022	4.387	1.076	3.443

(Contd.)

$d/b = 0.15$

w/b	Z_e(ohms)	Z_o(ohms)	ϵ_{fe}	ϵ_{fo}	w/b	Z_e(ohms)	Z_o(ohms)	ϵ_{fe}	ϵ_{fo}
0.10	230.004	65.601	1.466	2.784	0.85	109.772	13.851	1.191	3.207
0.12	221.782	59.336	1.445	2.817	0.90	106.264	13.163	1.184	3.215
0.14	214.567	54.236	1.426	2.848	0.95	102.962	12.539	1.178	3.222
0.16	208.096	49.987	1.409	2.875	1.00	99.862	11.972	1.172	3.229
0.18	202.201	46.383	1.394	2.900	1.05	96.944	11.453	1.166	3.235
0.20	196.766	43.283	1.381	2.923	1.10	94.193	10.978	1.160	3.240
0.22	191.715	40.584	1.368	2.944	1.15	91.595	10.540	1.155	3.246
0.24	186.986	38.212	1.357	2.963	1.20	89.150	10.135	1.150	3.250
0.26	182.540	36.108	1.347	2.981	1.25	86.821	9.762	1.146	3.256
0.28	178.341	34.228	1.337	2.998	1.30	84.611	9.414	1.141	3.260
0.30	174.365	32.538	1.328	3.013	1.35	82.511	9.090	1.137	3.264
0.32	170.586	31.010	1.319	3.027	1.40	80.512	8.787	1.133	3.267
0.34	166.989	29.620	1.311	3.040	1.45	78.608	8.504	1.130	3.271
0.36	163.558	28.351	1.303	3.052	1.50	76.792	8.238	1.126	3.274
0.38	160.279	27.188	1.296	3.064	1.55	75.058	7.988	1.123	3.277
0.40	157.140	26.117	1.289	3.074	1.60	73.400	7.753	1.120	3.280
0.50	143.236	21.826	1.260	3.118	1.65	71.814	7.531	1.117	3.283
0.55	137.213	20.169	1.247	3.136	1.70	70.294	7.322	1.114	3.285
0.60	131.697	18.746	1.236	3.151	1.75	68.837	7.124	1.111	3.287
0.65	126.619	17.509	1.225	3.165	1.80	67.439	6.936	1.108	3.290
0.70	121.930	16.426	1.216	3.177	1.85	66.096	6.758	1.106	3.292
0.75	117.583	15.468	1.207	3.188	1.90	64.806	6.589	1.103	3.294
0.80	113.542	14.615	1.199	3.198	1.95	63.564	6.428	1.101	3.296
					2.00	62.369	6.275	1.099	3.298

(Contd.)

$d/b = \mathbf{0.20}$

w/b	Z_e(ohms)	Z_o(ohms)	ϵ_{fe}	ϵ_{fo}	w/b	Z_e(ohms)	Z_o(ohms)	ϵ_{fe}	ϵ_{fo}
0.10	211.634	75.215	1.527	2.709	0.85	101.677	17.336	1.223	3.073
0.12	203.858	68.648	1.504	2.735	0.90	98.451	16.497	1.215	3.081
0.14	197.102	63.236	1.484	2.759	0.95	95.426	15.734	1.207	3.087
0.16	191.098	58.674	1.467	2.782	1.00	92.585	15.038	1.200	3.094
0.18	185.671	54.764	1.451	2.802	1.05	89.909	14.401	1.194	3.099
0.20	180.700	51.367	1.436	2.822	1.10	87.386	13.816	1.187	3.105
0.22	176.098	48.384	1.423	2.839	1.15	85.003	13.276	1.182	3.110
0.24	171.807	45.741	1.410	2.856	1.20	82.757	12.776	1.176	3.114
0.26	167.779	43.380	1.398	2.871	1.25	80.618	12.313	1.171	3.118
0.28	163.982	41.257	1.388	2.885	1.30	78.588	11.881	1.166	3.122
0.30	160.388	39.337	1.377	2.898	1.35	76.658	11.479	1.161	3.126
0.32	156.973	37.591	1.368	2.911	1.40	74.821	11.103	1.157	3.129
0.34	153.722	35.996	1.359	2.922	1.45	73.070	10.751	1.152	3.133
0.36	150.620	34.533	1.350	2.933	1.50	71.399	10.420	1.148	3.136
0.38	147.654	33.186	1.342	2.943	1.55	69.804	10.110	1.144	3.140
0.40	144.813	31.941	1.334	2.953	1.60	68.277	9.817	1.141	3.142
0.50	132.200	26.903	1.301	2.993	1.65	66.817	9.541	1.137	3.145
0.55	126.721	24.941	1.286	3.008	1.70	65.417	9.279	1.134	3.147
0.60	121.694	23.243	1.273	3.022	1.75	64.075	9.031	1.131	3.150
0.65	117.063	21.762	1.261	3.035	1.80	62.786	8.796	1.128	3.152
0.70	112.782	20.457	1.250	3.046	1.85	61.549	8.573	1.125	3.154
0.75	108.809	19.300	1.240	3.056	1.90	60.358	8.361	1.122	3.156
0.80	105.113	18.266	1.231	3.065	1.95	59.213	8.159	1.119	3.158
					2.00	58.110	7.967	1.116	3.159

(Contd.)

$d/b = 0.25$

w/b	Z_e(ohms)	Z_o(ohms)	ϵ_{fe}	ϵ_{fo}	w/b	Z_e(ohms)	Z_o(ohms)	ϵ_{fe}	ϵ_{fo}
0.10	197.179	82.527	1.577	2.648	0.85	94.459	20.348	1.248	2.950
0.12	189.644	75.809	1.553	2.669	0.90	91.476	19.384	1.239	2.956
0.14	183.142	70.223	1.532	2.688	0.95	88.679	18.506	1.231	2.962
0.16	177.405	65.480	1.513	2.706	1.00	86.050	17.704	1.223	2.968
0.18	172.253	61.384	1.496	2.722	1.05	83.576	16.969	1.216	2.973
0.20	167.560	57.802	1.481	2.737	1.10	81.252	16.291	1.209	2.977
0.22	163.240	54.636	1.466	2.752	1.15	79.046	15.666	1.202	2.982
0.24	159.228	51.814	1.453	2.765	1.20	76.958	15.086	1.196	2.986
0.26	155.476	49.279	1.440	2.778	1.25	74.978	14.548	1.190	2.989
0.28	151.950	46.988	1.429	2.789	1.30	73.099	14.046	1.185	2.993
0.30	148.619	44.906	1.418	2.800	1.35	71.313	13.578	1.180	2.996
0.32	145.463	43.005	1.407	2.811	1.40	69.612	13.139	1.175	2.999
0.34	142.461	41.261	1.397	2.820	1.45	67.991	12.728	1.170	3.002
0.36	139.598	39.656	1.388	2.830	1.50	66.444	12.342	1.166	3.005
0.38	136.864	38.172	1.379	2.838	1.55	64.966	11.978	1.162	3.007
0.40	134.245	36.797	1.371	2.846	1.60	63.553	11.635	1.157	3.010
0.50	122.627	31.190	1.334	2.880	1.65	62.200	11.311	1.154	3.012
0.55	117.578	28.983	1.318	2.894	1.70	60.903	11.005	1.150	3.014
0.60	112.943	27.069	1.304	2.906	1.75	59.660	10.714	1.146	3.016
0.65	108.670	25.392	1.290	2.917	1.80	58.466	10.439	1.143	3.019
0.70	104.717	23.914	1.278	2.926	1.85	57.319	10.178	1.140	3.021
0.75	101.049	22.595	1.267	2.935	1.90	56.216	9.929	1.137	3.023
0.80	97.647	21.413	1.257	2.942	1.95	55.155	9.692	1.134	3.024
					2.00	54.133	9.466	1.131	3.026

(*Contd.*)

$d/b=0.30$

w/b	Z_e(ohms)	Z_o(ohms)	ϵ_{fe}	ϵ_{fo}	w/b	Z_e(ohms)	Z_o(ohms)	ϵ_{fe}	ϵ_{fo}
0.10	185.169	88.204	1.617	2.594	0.85	87.849	22.916	1.268	2.833
0.12	177.780	81.383	1.593	2.609	0.90	85.075	21.850	1.259	2.838
0.14	171.433	75.704	1.571	2.624	0.95	82.475	20.878	1.250	2.843
0.16	165.861	70.850	1.552	2.638	1.00	80.042	19.989	1.241	2.847
0.18	160.882	66.638	1.534	2.651	1.05	77.740	19.172	1.233	2.851
0.20	156.369	62.937	1.518	2.663	1.10	75.569	18.418	1.226	2.855
0.22	152.233	59.651	1.502	2.674	1.15	73.518	17.722	1.219	2.859
0.24	148.409	56.708	1.488	2.684	1.20	71.576	17.075	1.212	2.862
0.26	144.847	54.054	1.475	2.694	1.25	69.735	16.474	1.206	2.865
0.28	141.510	51.646	1.463	2.703	1.30	67.988	15.914	1.200	2.868
0.30	138.366	49.450	1.451	2.712	1.35	66.327	15.390	1.195	2.871
0.32	135.394	47.437	1.440	2.720	1.40	64.746	14.900	1.189	2.873
0.34	132.574	45.585	1.429	2.728	1.45	63.239	14.439	1.184	2.876
0.36	129.891	43.875	1.419	2.735	1.50	61.801	14.006	1.180	2.878
0.38	127.330	42.290	1.410	2.742	1.55	60.428	13.598	1.175	2.880
0.40	124.883	40.818	1.401	2.749	1.60	59.114	13.213	1.171	2.882
0.50	114.049	34.774	1.361	2.776	1.65	57.857	12.849	1.167	2.884
0.55	109.349	32.380	1.344	2.787	1.70	56.652	12.504	1.163	2.886
0.60	105.038	30.295	1.328	2.797	1.75	55.496	12.178	1.159	2.888
0.65	101.065	28.462	1.314	2.806	1.80	54.387	11.867	1.155	2.889
0.70	97.402	26.838	1.301	2.814	1.85	53.321	11.573	1.152	2.891
0.75	93.990	25.389	1.289	2.821	1.90	52.297	11.292	1.148	2.892
0.80	90.814	24.091	1.278	2.827	1.95	51.310	11.024	1.145	2.894
					2.00	50.361	10.769	1.142	2.895

(Contd.)

$d/b = 0.35$

w/b	Z_e(ohms)	Z_o(ohms)	ϵ_{fe}	ϵ_{fo}	w/b	Z_e(ohms)	Z_o(ohms)	ϵ_{fe}	ϵ_{fo}
0.10	174.770	92.605	1.652	2.543	0.85	81.682	25.055	1.284	2.720
0.12	167.478	85.726	1.626	2.554	0.90	79.093	23.910	1.274	2.724
0.14	161.235	79.977	1.604	2.565	0.95	76.666	22.863	1.264	2.727
0.16	155.773	75.060	1.584	2.575	1.00	74.386	21.903	1.256	2.731
0.18	150.910	70.775	1.566	2.584	1.05	72.239	21.019	1.247	2.734
0.20	146.521	66.996	1.549	2.593	1.10	70.214	20.204	1.239	2.737
0.22	142.513	69.629	1.553	2.601	1.15	68.302	19.450	1.232	2.739
0.24	138.821	60.605	1.518	2.609	1.20	66.492	18.749	1.225	2.742
0.26	135.393	57.869	1.504	2.616	1.25	64.776	18 097	1.219	2.744
0.28	132.191	55.380	1.491	2.623	1.30	63.148	17.488	1.212	2.747
0.30	129.186	53.102	1.479	2.629	1.35	61.601	16.919	1.206	2.749
0.32	126.352	51.010	1.467	2.635	1.40	60.128	16.386	1.201	2.751
0.34	123.670	49.080	1.456	2.641	1.45	58.724	15.884	1.196	2.753
0.36	121.123	47.294	1.445	2.646	1.50	57.385	15.413	1.191	2.754
0.38	118.698	45.634	1.435	2.652	1.55	56.107	14.968	1.186	2.756
0.40	116.384	44.089	1.426	2.656	1.60	54.884	14.548	1.181	2.757
0.50	106.181	37.716	1.383	2.677	1.65	53.714	14.151	1.177	2.759
0.55	101.772	35.177	1.365	2.685	1.70	52.593	13.775	1.173	2.760
0.60	97.748	32.958	1.349	2.693	1.75	51.517	13.418	1.169	2.762
0.65	94.030	31.003	1.334	2.700	1.80	50.485	13.079	1.165	2.763
0.70	90.593	29.267	1.320	2.706	1.85	49 494	12.757	1.161	2.764
0.75	87.405	27.714	1.307	2.711	1.90	48.541	12.450	1.157	2.765
0.80	84.450	26.318	1.295	2.716	1.95	47.624	12.157	1.154	2.766
					2.00	46.741	11.878	1.151	2.767

(Contd.)

$d/b = 0.40$

w/b	Z_e(ohms)	Z_o(ohms)	ϵ_{fe}	ϵ_{fo}	w/b	Z_e(ohms)	Z_o(ohms)	ϵ_{fe}	ϵ_{fo}
0.10	165.461	95.947	1.682	2.492	0.85	75.811	26.780	1.297	2.610
0.12	158.237	89.033	1.656	2.500	0.90	73.389	25.569	1.286	2.612
0.14	152.068	83.240	1.632	2.507	0.95	71.120	24.462	1.276	2.615
0.16	146.688	78.282	1.612	2.514	1.00	68.989	23.449	1.267	2.616
0.18	141.926	73.949	1.592	2.520	1.05	66.984	22.514	1.258	2.618
0.20	137.626	70.119	1.575	2.525	1.10	65.094	21.651	1.250	2.620
0.22	133.713	66.700	1.558	2.531	1.15	63.309	20.851	1.242	2.622
0.24	130.118	63.621	1.543	2.536	1.20	61.620	20.107	1.235	2.624
0.26	126.791	60.830	1.528	2.540	1.25	60.020	19.415	1.228	2.626
0.28	123.693	58.284	1.515	2.545	1.30	58.502	18.768	1.222	2.627
0.30	120.793	55.951	1.502	2.549	1.35	57.059	18.163	1.215	2.628
0.32	118.065	53.803	1.489	2.553	1.40	55.687	17.595	1.210	2.630
0.34	115.490	51.818	1.478	2.557	1.45	54.380	17.061	1.204	2.631
0.36	113.051	49.977	1.467	2.561	1.50	53.133	16.559	1.199	2.632
0.38	110.734	48.265	1.456	2.564	1.55	51.942	16.085	1.194	2.633
0.40	108.527	46.667	1.446	2.567	1.60	50.804	15.637	1.189	2.634
0.50	98.859	40.051	1.402	2.581	1.65	49.715	15.214	1.184	2.635
0.55	94.696	37.403	1.383	2.586	1.70	48.672	14.813	1.180	2.636
0.60	90.892	35.084	1.365	2.591	1.75	47.672	14.432	1.176	2.637
0.65	87.396	33.036	1.349	2.596	1.80	46.713	14.070	1.172	2.638
0.70	84.169	31.214	1.334	2.600	1.85	45.791	13.725	1.168	2.639
0.75	81.180	29.582	1.321	2.603	1.90	44.906	13.397	1.164	2.639
0.80	78.401	28.111	1.309	2.607	1.95	44.053	13.084	1.161	2.640
					2.00	43.233	12.786	1.157	2.641

(Contd.)

$d/b = 0.45$

w/b	Z_e(ohms)	Z_o(ohms)	ϵ_{fe}	ϵ_{fo}	w/b	Z_e(ohms)	Z_o(ohms)	ϵ_{fe}	ϵ_{fo}
0.10	156.893	98.355	1.707	2.442	0.85	70.154	28.089	1.307	2.500
0.12	149.732	91.419	1.680	2.445	0.90	67.886	26.830	1.296	2.501
0.14	143.617	85.602	1.656	2.449	0.95	65.764	25.680	1.285	2.503
0.16	138.295	80.617	1.635	2.452	1.00	63.772	24.624	1.275	2.504
0.18	133.583	76.254	1.615	2.455	1.05	61.900	23.651	1.266	2.505
0.20	129.354	72.392	1.597	2.458	1.10	60.136	22.754	1.258	2.505
0.22	125.514	68.939	1.580	2.461	1.15	58.470	21.920	1.250	2.506
0.24	121.997	65.826	1.564	2.463	1.20	56.896	21.145	1.242	2.507
0.26	118.751	62.999	1.549	2.466	1.25	55.405	20 422	1.235	2.508
0.28	115.736	60.417	1.534	2.468	1.30	53.991	19.747	1.228	2.508
0.30	112.921	58.047	1.521	2.470	1.35	52.647	19.115	1.222	2.509
0.32	110.280	55.862	1.508	2.472	1.40	51.370	18.521	1.216	2.510
0.34	107.793	53.840	1.496	2.474	1.45	50.154	17.964	1.210	2.510
0.36	105.444	51.963	1.484	2.476	1.50	48.994	17.438	1.205	2.511
0.38	103.218	50.214	1.473	2.477	1.55	47.888	16.943	1.199	2.511
0.40	101.113	48.581	1.463	2.479	1.60	46.830	16.474	1.195	2.512
0.50	91.877	41.798	1.416	2.486	1.65	45.818	16 030	1.190	2.512
0.55	87.928	39.073	1.396	2.488	1.70	44.850	15.610	1.185	2.513
0.60	84.331	36.682	1.378	2.491	1.75	43.921	15.211	1.181	2.513
0.65	81.034	34.568	1.361	2.493	1.80	43.031	14.831	1.177	2.514
0.70	77.997	32.683	1.346	2.495	1.85	42.176	14.470	1.173	2.514
0.75	75.189	30.994	1.332	2.497	1.90	41.354	14.126	1.169	2.514
0.80	72.581	29.470	1.319	2.498	1.95	40.564	13.798	1.165	2.515
					2.00	39.804	13.485	1.162	2.515

(Contd.)

$d/b = 0.50$

w/b	Z_e(ohms)	Z_o(ohms)	ϵ_{fe}	ϵ_{fo}	w/b	Z_e(ohms)	Z_o(ohms)	ϵ_{fe}	ϵ_{fo}
0.10	148.800	99.892	1.729	2.390	0.85	64.638	28.973	1.314	2.390
0.12	141.664	92.946	1.701	2.390	0.90	62.516	27.684	1.302	2.390
0.14	135.594	87.115	1.677	2.390	0.95	60.533	26.505	1.291	2.390
0.16	130.323	82.115	1.655	2.390	1.00	58.673	25.422	1.281	2.390
0.18	125.666	77.736	1.634	2.390	1.05	56.927	24.424	1.272	2.390
0.20	121.496	73.857	1.615	2.390	1.10	55.282	23.501	1.263	2.390
0.22	117.718	70.386	1.598	2.390	1.15	53.732	22.645	1.255	2.390
0.24	114.267	67.253	1.581	2.390	1.20	52.267	21.852	1.247	2.390
0.26	111.089	64.406	1.565	2.390	1.25	50.880	21.109	1.240	2.390
0.28	108.146	61.803	1.550	2.390	1.30	49.566	20.415	1.233	2.390
0.30	105.404	59.413	1.536	2.390	1.35	48.319	19.765	1.226	2.390
0.32	102.852	57.207	1.523	2.390	1.40	47.133	19.155	1.220	2.390
0.34	100.442	55.163	1.511	2.390	1.45	46.005	18.581	1.214	2.390
0.36	98.169	53.265	1.499	2.390	1.50	44.929	18.040	1.208	2.390
0.38	96.020	51.495	1.487	2.390	1.55	43.903	17.529	1.203	2.390
0.40	93.982	49.841	1.476	2.390	1.60	42.923	17.047	1.198	2.390
0.50	85.141	42.957	1.427	2.390	1.65	41.987	16.590	1.193	2.390
0.55	81.387	40.185	1.407	2.390	1.70	41.090	16.157	1.188	2.390
0.60	77.978	37.750	1.387	2.390	1.75	40.231	15.746	1.184	2.390
0.65	74.863	35.594	1.370	2.390	1.80	39.408	15.355	1.180	2.390
0.70	72.001	33.670	1.354	2.390	1.85	38.617	14.982	1.176	2.390
0.75	69.360	31.944	1.340	2.390	1.90	37.858	14.627	1.172	2.390
0.80	66.913	30.386	1.326	2.390	1.95	37.128	14.289	1.168	2.390
					2.00	36.426	13.966	1.164	2.390

Table 7.5 Characteristic impedances and effective dielectric constants of broadside-coupled suspended stripline; $c/w = 40$, $\epsilon_r = 9.6$ (Alumina) (See structure A2 of Table 4.4)

$d/b = 0.05$

w/b	Z_e(ohms)	Z_o(ohms)	ϵ_{fe}	ϵ_{fo}	w/b	Z_e(ohms)	Z_o(ohms)	ϵ_{fe}	ϵ_{fo}
0.10	257.818	21.441	1.665	7.415	0.85	123.037	3.333	1.241	8.832
0.12	248.908	18.685	1.618	7.570	0.90	119.104	3.155	1.232	8.851
0.14	240.875	16.568	1.588	7.700	0.95	115.403	2.996	1.223	8.869
0.16	233.603	14.887	1.561	7.811	1.00	111.926	2.851	1.215	8.884
0.18	226.931	13.520	1.537	7.908	1.05	108.652	2.720	1.207	8.898
0.20	220.763	12.384	1.516	7.992	1.10	105.565	2.601	1.200	8.911
0.22	215.025	11.426	1.497	8.066	1.15	102.648	2.491	1.193	8.923
0.24	209.659	10.606	1.479	8.132	1.20	99.889	2.391	1.187	8.934
0.26	204.620	9.897	1.463	8.192	1.25	97.273	2.298	1.181	8.944
0.28	199.870	9.277	1.449	8 245	1.30	94.792	2.212	1.175	8.953
0.30	195.379	8.730	1.435	8.293	1.35	92.445	2.132	1.169	8.962
0.32	191.119	8.244	1.423	8.337	1.40	90.200	2.058	1.164	8.970
0.34	187.071	7.810	1.411	8.377	1.45	88.062	1.989	1.160	8.978
0.36	183.213	7.419	1.400	8.414	1.50	86.022	1.924	1.155	8.985
0.38	179.554	7.065	1.389	8.448	1.55	84.074	1.863	1.151	8.991
0.40	176.035	6.744	1.379	8.479	1.60	82.212	1.806	1.147	8.997
0.50	160.474	5.495	1.336	8.606	1.65	80.430	1.753	1.143	9.003
0.55	153.740	5.029	1.319	8.654	1.70	78.723	1.702	1.139	9.008
0.60	147.573	4.636	1.303	8.695	1.75	77.087	1.654	1.135	9.014
0.65	141.895	4.300	1.288	8.731	1.80	75.517	1.609	1.132	9.018
0.70	136.649	4.009	1.275	8.762	1.85	74.009	1.567	1.129	9.023
0.75	131.784	3.755	1.263	8.787	1.90	72.560	1.526	1.126	9.027
0.80	127.258	3.532	1.251	8.811	1.95	71.166	1.488	1.123	9.031
					2.00	69.825	1.451	1.120	9.035

(Contd.)

d/b = **0.10**

w/b	Z_e(ohms)	Z_o(ohms)	ϵ_{fe}	ϵ_{fo}	w/b	Z_e(ohms)	Z_o(ohms)	ϵ_{fe}	ϵ_{fo}
0.10	216.259	34.194	1.930	6.809	0.85	109.178	6.271	1.372	8.280
0.12	209.538	30.446	1.884	6.941	0.90	105.873	5.949	1.358	8.307
0.14	203.464	27.469	1.844	7.055	0.95	102.776	5.657	1.346	8.329
0.16	197.919	25.037	1.810	7.158	1.00	99.847	5.393	1.334	8.350
0.18	192.802	23.010	1.780	7.250	1.05	97.082	5.152	1.322	8.369
0.20	188.045	21.294	1.752	7.332	1.10	94.468	4.931	1.312	8.386
0.22	183.594	19.820	1.727	7.407	1.15	91.992	4.729	1.302	8.402
0.24	179.408	18.540	1.704	7.474	1.20	89.644	4.542	1.292	8.416
0.26	175.457	17.418	1.683	7.536	1.25	87.413	4.370	1.283	8.430
0.28	171.715	16.425	1.663	7.592	1.30	85.291	4.210	1.275	8.442
0.30	168.161	15.541	1.645	7.644	1.35	83.271	4.062	1.267	8.454
0.32	164.778	14.747	1.627	7.692	1.40	81.344	3.923	1.259	8.464
0.34	161.551	14.031	1.611	7.736	1.45	79.505	3.794	1.252	8.472
0.36	158.466	13.382	1.596	7.777	1.50	77.757	3.673	1.245	8.482
0.38	155.513	12.790	1.582	7.815	1.55	76.076	3.560	1.239	8.491
0.40	152.681	12.249	1.568	7.851	1.60	74.465	3.453	1.232	8.499
0.50	140.073	10.109	1.508	7.997	1.65	72.922	3.352	1.226	8.507
0.55	134.576	9.297	1.483	8.054	1.70	71.441	3.257	1.221	8.514
0.60	129.516	8.606	1.461	8.105	1.75	70.018	3.168	1.215	8.521
0.65	124.842	8.010	1.440	8.149	1.80	68.651	3.083	1.210	8.527
0.70	120.505	7.491	1.421	8.187	1.85	67.337	3.002	1.205	8.534
0.75	116.470	7.035	1.403	8.221	1.90	66.071	2.926	1.201	8.540
0.80	112.702	6.631	1.387	8.252	1.95	64.852	2.853	1.196	8.545
					2.00	63.677	2.784	1.192	8.550

(Contd.)

d/b = **0.15**

w/b	Z_e(ohms)	Z_o(ohms)	ϵ_{fe}	ϵ_{fo}	w/b	Z_e(ohms)	Z_o(ohms)	ϵ_{fe}	ϵ_{fo}
0.10	191.237	42.992	2.121	6.481	0.85	98.986	8.872	1.465	7.815
0.12	185.426	38.808	2.066	6.586	0.90	96.093	8.428	1.448	7.841
0.14	180.187	35.409	2.022	6.681	0.95	93.357	8.026	1.433	7.864
0.16	175.434	32.582	1.983	6.767	1.00	90.777	7.660	1.418	7.885
0.18	171.070	30.189	1.948	6.846	1.05	88.338	7.327	1.404	7.905
0.20	167.025	28.139	1.916	6.916	1.10	86.030	7.020	1.391	7.923
0.22	163.246	26.354	1.887	6.982	1.15	83.841	6.739	1.379	7.939
0.24	159.692	24.787	1.861	7.043	1.20	81.763	6.479	1.367	7.955
0.26	156.334	23.399	1.836	7.099	1.25	79.786	6.239	1.357	7.971
0.28	153.150	22.162	1.813	7.150	1.30	77.905	6.015	1.346	7.984
0.30	150.120	21.051	1.791	7.198	1.35	76.111	5.807	1.336	7.997
0.32	147.231	20.047	1.771	7.242	1.40	74.398	5.613	1.327	8.008
0.34	144.469	19.136	1.752	7.284	1.45	72.762	5.431	1.318	8.019
0.36	141.824	18.304	1.733	7.322	1.50	71.197	5.260	1.310	8.029
0.38	139.288	17.543	1.716	7.358	1.55	69.708	5.100	1.302	8.039
0.40	136.851	16.842	1.700	7.392	1.60	68.271	4.950	1.294	8.048
0.50	125.955	14.042	1.629	7.533	1.65	66.893	4.808	1.287	8.056
0.55	121.181	12.964	1.599	7.589	1.70	65.570	4.673	1.280	8.064
0.60	116.779	12.040	1.571	7.639	1.75	64.298	4.546	1.273	8.072
0.65	112.701	11.238	1.546	7.683	1.80	63.075	4.426	1.267	8.079
0.70	108.912	10.536	1.524	7.721	1.85	61.897	4.312	1.261	8.086
0.75	105.381	9.917	1.502	7.756	1.90	60.763	4.204	1.255	8.092
0.80	102.080	9.366	1.483	7.787	1.95	59.669	4.101	1.249	8.098
					2.00	58.614	4.003	1.244	8.104

(Contd.)

d/b = **0.20**

w/b	Z_e(ohms)	Z_o(ohms)	ϵ_{fe}	ϵ_{fo}	w/b	Z_e(ohms)	Z_o(ohms)	ϵ_{fe}	ϵ_{fo}
0.10	173.580	49.492	2.270	6.256	0.85	90.729	11.170	1.536	7.403
0.12	168.248	45.096	2.209	6.338	0.90	88.112	10.625	1.517	7.427
0.14	163.440	41.478	2.159	6.414	0.95	85.646	10.130	1.499	7.448
0.16	159.100	38.432	2.116	6.484	1.00	83.320	9.679	1.482	7.468
0.18	155.142	35.828	2.078	6.549	1.05	81.120	9.266	1.466	7.486
0.20	151.496	33.564	2.043	6.609	1.10	79.036	8.887	1.452	7.503
0.22	148.106	31.582	2.011	6.664	1.15	77.059	8.538	1.438	7.519
0.24	144.929	29.827	1.982	6.716	1.20	75.181	8.214	1.425	7.534
0.26	141.937	28.266	1.954	6.762	1.25	73.395	7.915	1.412	7.547
0.28	139.103	26.861	1.928	6.806	1.30	71.693	7.636	1.401	7.560
0.30	136.410	25.591	1.904	6.848	1.35	70.069	7.376	1.390	7.571
0.32	133.844	24.438	1.881	6.887	1.40	68.519	7.133	1.379	7.582
0.34	131.390	23.386	1.860	6.923	1.45	67.036	6.906	1.369	7.593
0.36	129.014	22.422	1.839	6.957	1.50	65.618	6.692	1.359	7.602
0.38	126.785	21.535	1.820	6.989	1.55	64.268	6.492	1.350	7.614
0.40	124.618	20.717	1.802	7.019	1.60	62.965	6.303	1.341	7.622
0.50	114.906	17.411	1.722	7.145	1.65	61.715	6.125	1.333	7.630
0.55	110.639	16.125	1.687	7.196	1.70	60.513	5.956	1.325	7.638
0.60	106.697	15.016	1.656	7.241	1.75	59.358	5.797	1.317	7.645
0.65	103.042	14.049	1.628	7.281	1.80	58.246	5.645	1.310	7.652
0.70	99.642	13.199	1.602	7.316	1.85	57.176	5.501	1.303	7.659
0.75	96.469	12.446	1.578	7.348	1.90	56.144	5.365	1.297	7.665
0.80	93.501	11.774	1.556	7.377	1.95	55.149	5.235	1.290	7.671
					2.00	54.189	5.111	1.284	7.677

(Contd.)

$d/b = \mathbf{0.25}$

w/b	Z_e(ohms)	Z_o(ohms)	ϵ_{fe}	ϵ_{fo}	w/b	Z_e(ohms)	Z_o(ohms)	ϵ_{fe}	ϵ_{fo}
0.10	160.004	54.483	2.394	6.075	0.85	83.634	13.187	1.592	7.023
0.12	154.960	49.980	2.325	6.140	0.90	81.244	12.558	1.571	7.043
0.14	150.405	46.240	2.271	6.199	0.95	78.992	11.985	1.551	7.062
0.16	146.288	43.067	2.225	6.254	1.00	76.865	11.462	1.533	7.079
0.18	142.572	40.330	2.184	6.306	1.05	74.853	10.983	1.516	7.096
0.20	139.166	37.939	2.146	6.354	1.10	72.948	10.542	1.499	7.110
0.22	136.016	35.828	2.112	6.399	1.15	71.140	10.135	1.484	7.124
0.24	133.079	33.948	2.080	6.441	1.20	69.422	9.758	1.470	7.137
0.26	130.321	32.262	2.050	6.481	1.25	67.787	9.407	1.456	7.149
0.28	127.720	30.740	2.022	6.518	1.30	66.229	9.081	1.443	7.160
0.30	125.254	29.358	1.996	6.552	1.35	64.743	8.777	1.431	7.170
0.32	122.908	28.101	1.971	6.583	1.40	63.323	8.492	1.420	7.180
0.34	120.669	26.946	1.948	6.613	1.45	61.966	8.225	1.409	7.189
0.36	118.527	25.883	1.925	6.642	1.50	60.675	7.974	1.398	7.198
0.38	116.473	24.902	1.904	6.669	1.55	59.430	7.738	1.388	7.206
0.40	114.501	23.994	1.884	6.694	1.60	58.237	7.516	1.378	7.214
0.50	105.667	20.297	1.796	6.801	1.65	57.090	7.305	1.369	7.221
0.55	101.785	18.847	1.759	6.844	1.70	55.989	7.107	1.361	7.228
0.60	98.196	17.590	1.724	6.883	1.75	54.930	6.918	1.352	7.234
0.65	94.866	16.490	1.693	6.917	1.80	53.911	6.740	1.344	7.241
0.70	91.767	15.519	1.665	6.948	1.85	52.929	6.571	1.337	7.248
0.75	88.873	14.656	1.668	6.975	1.90	51.982	6.409	1.329	7.254
0.80	86.176	13.883	1.614	7.000	1.95	51.070	6.256	1.322	7.259
					2.00	50.188	6.109	1.315	7.265

(Contd.)

$d/b = \mathbf{0.30}$

w/b	Z_e(ohms)	Z_o(ohms)	ϵ_{fe}	ϵ_{fo}	w/b	Z_e(ohms)	Z_o(ohms)	ϵ_{fe}	ϵ_{fo}
0.10	148.914	58.408	2.501	5.915	0.85	77.319	14.942	1.637	6.663
0.12	144.056	53.833	2.425	5.963	0.90	75.117	14.242	1.614	6.680
0.14	139.665	50.027	2.367	6.010	0.95	73.041	13.605	1.593	6.695
0.16	135.721	46.776	2.317	6.052	1.00	71.080	13.021	1.574	6.709
0.18	132.158	43.958	2.273	6.092	1.05	69.227	12.486	1.555	6.722
0.20	128.908	41.483	2.233	6.129	1.10	67.470	11.993	1.538	6.735
0.22	125.915	39.288	2.196	6.164	1.15	65.804	11.537	1.521	6.746
0.24	123.136	37.323	2.162	6.197	1.20	64.220	11.114	1.506	6.757
0.26	120.538	35.553	2.130	6.228	1.25	62.713	10.720	1.491	6.766
0.28	118.081	33.948	2.101	6.257	1.30	61.276	10.354	1.478	6.776
0.30	115.773	32.486	2.073	6.284	1.35	59.907	10.012	1.465	6.784
0.32	113.583	31.147	2.046	6.310	1.40	58.599	9.691	1.452	6.792
0.34	111.498	29.916	2.021	6.334	1.45	57.355	9.390	1.440	6.800
0.36	109.508	28.780	1.997	6.357	1.50	56.157	9.107	1.429	6.807
0.38	107.603	27.732	1.974	6.377	1.55	55.009	8.841	1.418	6.814
0.40	105.777	26.756	1.952	6.397	1.60	53.908	8.589	1.408	6.820
0.50	97.617	22.755	1.858	6.483	1.65	52.852	8.352	1.398	6.826
0.55	94.037	21.174	1.817	6.518	1.70	51.836	8.127	1.389	6.832
0.60	90.732	19.799	1.780	6.549	1.75	50.859	7.914	1.380	6.837
0.65	87.665	18.591	1.747	6.577	1.80	49.920	7.711	1.371	6.843
0.70	84.822	17.522	1.716	6.602	1.85	49.014	7.519	1.363	6.848
0.75	82.156	16.568	1.688	6.624	1.90	48.142	7.336	1.355	6.852
0.80	79.661	15.713	1.661	6.644	1.95	47.300	7.162	1.347	6.857
					2.00	46.487	6.996	1.340	6.861

(Contd.)

$d/b = 0.35$

w/b	Z_e(ohms)	Z_o(ohms)	ϵ_{fe}	ϵ_{fo}	w/b	Z_e(ohms)	Z_o(ohms)	ϵ_{fe}	ϵ_{fo}
0.10	139.479	61.509	2.594	5.763	0.85	71.545	16.443	1.674	6.316
0.12	134.740	56.894	2.513	5.798	0.90	69.504	15.686	1.650	6.328
0.14	130.459	53.039	2.450	5.831	0.95	67.581	14.995	1.627	6.340
0.16	126.623	49.743	2.397	5.863	1.00	65.765	14.362	1.606	6.351
0.18	123.165	46.872	2.351	5.892	1.05	64.047	13.780	1.587	6.361
0.20	120.023	44.343	2.308	5.919	1.10	62.420	13.243	1.568	6.370
0.22	117.140	42.090	2.269	5.944	1.15	60.877	12.746	1.551	6.379
0.24	114.473	40.068	2.233	5.968	1.20	59.411	12.285	1.535	6.387
0.26	111.989	38.240	2.199	5.991	1.25	58.023	11.856	1.519	6.394
0.28	109.660	36.577	2.167	6.012	1.30	56.694	11.456	1.504	6.401
0.30	107.467	35.057	2.137	6.033	1.35	55.426	11.081	1.490	6.408
0.32	105.392	33.661	2.109	6.052	1.40	54.214	10.731	1.477	6.414
0.34	103.422	32.375	2.082	6.070	1.45	53.056	10.401	1.465	6.420
0.36	101.546	31.185	2.057	6.087	1.50	51.948	10.091	1.453	6.425
0.38	99.753	30.080	2.032	6.103	1.55	50.886	9.799	1.442	6.430
0.40	98.038	29.052	2.009	6.118	1.60	49.867	9.523	1.431	6.435
0.50	90.395	24.822	1.909	6.181	1.65	48.889	9.262	1.420	6.440
0.55	87.062	23.138	1.865	6.207	1.70	47.950	9.015	1.411	6.444
0.60	84.000	21.668	1.827	6.230	1.75	47.047	8.781	1.401	6.448
0.65	81.152	20.374	1.791	6.251	1.80	46.177	8.559	1.392	6.452
0.70	78.502	19.226	1.758	6.270	1.85	45.340	8.347	1.383	6.456
0.75	76.030	18.199	1.728	6.287	1.90	44.533	8.146	1.375	6.460
0.80	73.716	17.277	1.700	6.302	1.95	43.754	7.954	1.367	6.463
					2.00	43.003	7.770	1.359	6.466

(Contd.)

$d/b = 0.40$

w/b	Z_e(ohms)	Z_o(ohms)	ϵ_{fe}	ϵ_{fo}	w/b	Z_e(ohms)	Z_o(ohms)	ϵ_{fe}	ϵ_{fo}
0.10	131.180	63.936	2.676	5.613	0.85	66.168	17.697	1.703	5.976
0.12	126.514	59.296	2.590	5.636	0.90	64.269	16.894	1.677	5.984
0.14	122.308	55.409	2.524	5.657	0.95	62.481	16.159	1.654	5.992
0.16	118.547	52.085	2.468	5.678	1.00	60.793	15.486	1.632	5.999
0.18	115.167	49.179	2.418	5.697	1.05	59.198	14.867	1.611	6.006
0.20	112.103	46.613	2.373	5.714	1.10	57.687	14.294	1.592	6.012
0.22	109.302	44.322	2.332	5.731	1.15	56.254	13.764	1.573	6.017
0.24	106.718	42.260	2.294	5.747	1.20	54.892	13.272	1.556	6.023
0.26	104.318	40.392	2.258	5.762	1.25	53.597	12.813	1.540	6.028
0.28	102.076	38.689	2.224	5.776	1.30	52.364	12.385	1.522	6.032
0.30	99.971	37.138	2.192	5.789	1.35	51.187	11.985	1.510	6.037
0.32	97.984	35.693	2.163	5.801	1.40	50.064	11.609	1.497	6.041
0.34	96.102	34.366	2.134	5.813	1.45	48.990	11.256	1.484	6.045
0.36	94.315	33.136	2.107	5.824	1.50	47.962	10.924	1.471	6.048
0.38	92.611	31.993	2.082	5.835	1.55	46.978	10.610	1.459	6.052
0.40	90.985	30.927	2.057	5.845	1.60	46.034	10.314	1.448	6.055
0.50	83.796	26.520	1.951	5.886	1.65	45.128	10.034	1.437	6.058
0.55	80.666	24.757	1.905	5.904	1.70	44.258	9.769	1.427	6.061
0.60	77.787	23.214	1.864	5.919	1.75	43.421	9.517	1.417	6.064
0.65	75.124	21.852	1.826	5.933	1.80	42.616	9.278	1.408	6.067
0.70	72.650	20.641	1.791	5.945	1.85	41.841	9.050	1.399	6.069
0.75	70.345	19.557	1.759	5.956	1.90	41.094	8.833	1.390	6.072
0.80	68.189	18.581	1.730	5.966	1.95	40.374	8.626	1.382	6.074
					2.00	39.679	8.429	1.374	6.076

(Contd.)

$d/b = 0.45$

w/b	Z_e(ohms)	Z_o(ohms)	ϵ_{fe}	ϵ_{fo}	w/b	Z_e(ohms)	Z_o(ohms)	ϵ_{fe}	ϵ_{fo}
0.10	123.671	65.776	2.748	5.459	0.85	61.061	18.704	1.725	5.638
0.12	119.042	61.121	2.658	5.470	0.90	59.291	17.864	1.698	5.642
0.14	114.887	57.216	2.588	5.481	0.95	57.625	17.097	1.674	5.646
0.16	111.181	53.871	2.529	5.492	1.00	56.054	16.392	1.651	5.650
0.18	107.860	50.943	2.477	5.501	1.05	54.569	15.743	1.629	5.653
0.20	104.857	48.353	2.430	5.510	1.10	53.164	15.143	1.609	5.656
0.22	102.118	46.037	2.387	5.518	1.15	51.832	14.587	1.590	5.659
0.24	99.600	43.949	2.346	5.526	1.20	50.567	14.070	1.573	5.661
0.26	97.267	42.054	2.308	5.533	1.25	49.365	13.589	1.556	5.664
0.28	95.094	40.323	2.273	5.540	1.30	48.220	13.139	1.540	5.666
0.30	93.058	38.735	2.240	5.547	1.35	47.128	12.717	1.525	5.668
0.32	91.142	37.272	2.208	5.553	1.40	46.086	12.322	1.511	5.670
0.34	89.333	35.917	2.178	5.559	1.45	45.090	11.950	1.497	5.672
0.36	87.618	34.660	2.150	5.564	1.50	44.138	11.600	1.484	5.674
0.38	85.989	33.489	2.123	5.569	1.55	43.226	11.270	1.472	5.676
0.40	84.444	32.396	2.098	5.574	1.60	42.351	10.958	1.461	5.677
0.50	77.604	27.858	1.985	5.596	1.65	41.512	10.662	1.449	5.679
0.55	74.648	26.040	1.937	5.603	1.70	40.706	10.382	1.439	5.680
0.60	71.937	24.442	1.894	5.610	1.75	39.932	10.117	1.429	5.682
0.65	69.436	23.029	1.854	5.617	1.80	39.187	9.864	1.419	5.683
0.70	67.118	21.771	1.818	5.623	1.85	38.470	9.624	1.410	5.684
0.75	64.961	20.643	1.784	5.619	1.90	37.779	9.394	1.401	5.685
0.80	62.947	19.626	1.753	5.634	1.95	37.113	9.176	1.392	5.687
					2.00	36.470	8.967	1.384	5.688

(Contd.)

$d/b = \mathbf{0.50}$

w/b	Z_e(ohms)	Z_o(ohms)	ϵ_{fe}	ϵ_{fo}	w/b	Z_e(ohms)	Z_o(ohms)	ϵ_{fe}	ϵ_{fo}
0.10	116.698	67.079	2.811	5.300	0.85	56.149	19.458	1.741	5.300
0.12	112.081	62.414	2.718	5.300	0.90	54.496	18.592	1.713	5.300
0.14	107.959	58.498	2.645	5.300	0.95	52.944	17.801	1.688	5.300
0.16	104.296	55.141	2.583	5.300	1.00	51.481	17.074	1.664	5.300
0.18	101.022	52.200	2.529	5.300	1.05	50.100	16.403	1.642	5.300
0.20	98.069	49.595	2.479	5.300	1.10	48.794	15.780	1.621	5.300
0.22	95.383	47.264	2.433	5.300	1.15	47.556	15.208	1.602	5.300
0.24	92.918	45.160	2.391	5.300	1.20	46.383	14.674	1.583	5.300
0.26	90.642	43 248	2.351	5.300	1.25	45.267	14.175	1.566	5.300
0.28	88.526	41.501	2.314	5.300	1.30	44.205	13.709	1.550	5.300
0.30	86.549	39.895	2.279	5.300	1.35	43.193	13.272	1.534	5.300
0.32	84.704	38.414	2.246	5.300	1.40	42.228	12.863	1.520	5.300
0.34	82.955	37.042	2.215	5.300	1.45	41.306	12.477	1.506	5.300
0.36	81.302	35.767	2.185	5.300	1.50	40.425	12.114	1.493	5.300
0.38	79.733	34.578	2.157	5.300	1.55	39.581	11.771	1.480	5.300
0.40	78.241	33.467	2.130	5.300	1.60	38.772	11.447	1.468	5.300
0.50	71.714	28.844	2.012	5.300	1.65	37.997	11.140	1.457	5.300
0.55	68.912	26.983	1.962	5.300	1.70	37.252	10.849	1.446	5.300
0.60	66.351	25.352	1.916	5.300	1.75	36.537	10.573	1.435	5.300
0.65	63.995	23.904	1.875	5.300	1.80	35.849	10.311	1.425	5.300
0.70	61.818	22.612	1.837	5.300	1.85	35.187	10.061	1.416	5.300
0.75	59.795	21.453	1.802	5.300	1.90	34.550	9.822	1.407	5.300
0.80	57.911	20.407	1.770	5.300	1.95	33.935	9.595	1.398	5.300
					2.00	33.343	9.378	1.390	5.300

Table 7.6 Characteristic impedances and effective dielectric constants of broadside-coupled inverted stripline; $c/w = 40$, $\epsilon_r = 2.22$ (Duroid) (See structure A3 of Table 4.4)

$d/b = 0.05$

w/b	Z_e(ohms)	Z_o(ohms)	ϵ_{fe}	ϵ_{fo}	w/b	Z_e(ohms)	Z_o(ohms)	ϵ_{fe}	ϵ_{fo}
0.10	230.93	51.88	2.063	1.267	0.85	93.06	9.44	2.170	1.101
0.12	219.87	46.02	2.074	1.248	0.90	89.69	8.96	2.172	1.098
0.14	210.29	41.41	2.083	1.232	0.95	86.55	8.52	2.174	1.096
0.16	201.83	37.68	2.090	1.219	1.00	83.63	8.12	2.176	1.095
0.18	194.27	34.59	2.097	1.208	1.05	80.89	7.76	2.178	1.093
0.20	187.43	31.99	2.103	1.198	1.10	78.33	7.43	2.189	1.091
0.22	181.18	29.76	2.108	1.189	1.15	75.92	7.13	2.181	1.090
0.24	175.44	27.83	2.112	1.182	1.20	73.66	6.85	2.182	1.089
0.26	170.14	26.13	2.117	1.175	1.25	71.53	6.59	2.184	1.088
0.28	165.20	24.64	2.120	1.168	1.30	69.52	6.35	2.185	1.086
0.30	160.60	23.31	2.124	1.163	1.35	67.61	6.13	2.186	1.085
0.32	156.29	22.12	2.127	1.158	1.40	65.81	5.92	2.187	1.084
0.34	152.24	21.05	2.130	1.153	1.45	64.10	5.72	2.188	1.084
0.36	148.43	20.08	2.133	1.149	1.50	62.48	5.54	2.189	1.083
0.38	144.81	19.19	2.135	1.145	1.55	60.94	5.37	2.190	1.082
0.40	141.39	18.38	2.138	1.141	1.60	59.47	5.21	2.191	1.081
0.45	133.55	16.63	2.143	1.134	1.65	58.07	5.06	2.192	1.081
0.50	126.58	15.18	2.148	1.127	1.70	56.73	4.92	2.193	1.080
0.55	120.34	13.97	2.152	1.122	1.75	55.46	4.78	2.193	1 079
0.60	114.71	12.94	2.156	1.117	1.80	54.24	4.65	2.194	1.079
0.65	109.60	12.04	2.159	1.113	1.85	53.07	4.53	2.195	1.078
0.70	104.93	11.27	2.162	1.109	1.90	51.95	4.42	2.196	1.078
0.75	100.65	10.59	2.165	1.106	1.95	50.88	4.31	2.196	1.077
0.80	96.70	9.98	2.168	1.103	2.00	49.85	4.20	2.197	1.077

(Contd.)

d/b = **0.10**

w/b	Z_e(ohms)	Z_o(ohms)	ϵ_{fe}	ϵ_{fo}	w/b	Z_e(ohms)	Z_o(ohms)	ϵ_{fe}	ϵ_{fo}
0.10	213.08	76.74	1.988	1.352	0.85	87.53	16.64	2.135	1.176
0.12	203.27	69.49	2.001	1.335	0.90	84.39	15.83	2.139	1.173
0.14	194.77	63.49	2.013	1.321	0.95	81.47	15.09	2.142	1.170
0.16	187.24	58.57	2.022	1.308	1.00	78.73	14.42	2.145	1.168
0.18	180.47	54.41	2.031	1.297	1.05	76.18	13.80	2.148	1.166
0.20	174.33	50.83	2.038	1.287	1.10	73.78	13.24	2.150	1.164
0.22	168.70	47.72	2.045	1.278	1.15	71.53	12.72	2.153	1.162
0.24	163.50	44.99	2.051	1.269	1.20	69.42	12.23	2.155	1.160
0.26	158.68	42.56	2.057	1.262	1.25	67.42	11.79	2.157	1.159
0.28	154.19	40.39	2.062	1.255	1.30	65.54	11.37	2.159	1.157
0.30	149.99	38.44	2.067	1.249	1.35	63.76	10.98	2.161	1.156
0.32	146.04	36.67	2.072	1.244	1.40	62.07	10.62	2.163	1.155
0.34	142.34	35.07	2.076	1.239	1.45	60.47	10.28	2.165	1.153
0.36	138.83	33.60	2.080	1.234	1.50	58.95	9.97	2.166	1.152
0.38	135.51	32.25	2.083	1.229	1.55	57.50	9.67	2.168	1.151
0.40	132.35	31.01	2.087	1.225	1.60	56.13	9.39	2.169	1.150
0.45	125.12	28.28	2.095	1.216	1.65	54.81	9.12	2.171	1.149
0.50	118.68	26.01	2.102	1.208	1.70	53.56	8.87	2.172	1.149
0.55	112.90	24.07	2.108	1.202	1.75	52.36	8.63	2.173	1.148
0.60	107.68	22.40	2.113	1.196	1.80	51.21	8.41	2.174	1.147
0.65	102.93	20.95	2.119	1.191	1.85	50.12	8.19	2.176	1.146
0.70	98.59	19.68	2.123	1.186	1.90	49.07	7.99	2.177	1.146
0.75	94.60	18.55	2.128	1.182	1.95	48.06	7.79	2.178	1.145
0.80	90.93	17.54	2.132	1.179	2.00	47.09	7.61	2.179	1.144

(Contd.)

d/b = **0.20**

w/b	Z_e(ohms)	Z_o(ohms)	ϵ_{fe}	ϵ_{fo}	w/b	Z_e(ohms)	Z_o(ohms)	ϵ_{fe}	ϵ_{fo}
0.10	189.63	102.99	1.902	1.445	0.85	77.82	26.64	2.088	1.301
0.12	180.64	94.82	1.916	1.434	0.90	75.01	25.41	2.093	1.298
0.14	172.96	88.04	1.928	1.424	0.95	72.40	24.29	2.097	1.296
0.16	166.22	82.29	1.938	1.414	1.00	69.97	23.26	2.102	1.293
0.18	160.20	77.32	1.948	1.406	1.05	67.70	22.31	2.106	1.291
0.20	154.79	72.97	1.957	1.398	1.10	65.57	21.44	2.109	1.289
0.22	149.82	69.12	1.966	1.391	1.15	63.57	20.63	2.113	1.287
0.24	145.24	65.68	1.973	1.385	1.20	61.68	19.88	2.116	1.286
0.26	140.99	62.59	1.981	1.379	1.25	59.91	19.19	2.119	1.284
0.28	137.03	59.80	1.987	1.373	1.30	58.24	18.54	2.122	1.283
0.30	133.33	57.25	1.993	1.368	1.35	56.65	17.93	2.125	1.281
0.32	129.84	54.92	1.999	1.363	1.40	55.16	17.36	2.128	1.280
0.34	126.56	52.78	2.005	1.359	1.45	53.73	16.83	2.131	1.279
0.36	123.45	50.81	2.010	1.355	1.50	52.38	16.33	2.133	1.277
0.38	120.50	48.98	2.015	1.351	1.55	51.10	15.86	2.135	1.276
0.40	117.71	47.28	2.020	1.347	1.60	49.87	15.41	2.138	1.275
0.45	111.29	43.52	2.031	1.339	1.65	48.71	14.99	2.140	1.274
0.50	105.56	40.32	2.040	1.332	1.70	47.59	14.59	2.142	1.273
0.55	100.41	37.56	2.049	1.326	1.75	46.53	14.21	2.144	1.273
0.60	95.76	35.16	2.057	1.321	1.80	45.51	13.85	2.146	1.272
0.65	91.52	33.05	2.064	1.316	1.85	44.54	13.51	2.148	1.271
0.70	87.65	31.17	2.071	1.312	1.90	43.60	13.18	2.149	1.270
0.75	84.10	29.50	2.077	1.308	1.95	42.71	12.87	2.151	1.269
0.80	80.83	28.00	2.082	1.304	2.00	41.85	12.57	2.152	1.269

(Contd.)

$d/b = 0.30$

w/b	Z_e(ohms)	Z_o(ohms)	ϵ_{fe}	ϵ_{fo}	w/b	Z_e(ohms)	Z_o(ohms)	ϵ_{fe}	ϵ_{fo}
0.10	172.99	115.68	1.853	1.508	0.85	68.97	32.47	2.057	1.411
0.12	164.27	107.29	1.866	1.502	0.90	66.45	31.01	2.063	1.409
0.14	156.84	100.29	1.877	1.495	0.95	64.10	29.68	2.069	1.407
0.16	150.37	94.28	1.888	1.490	1.00	61.92	28.45	2.074	1.405
0.18	144.65	89.05	1.898	1.484	1.05	59.88	27.33	2.079	1.403
0.20	139.50	84.44	1.907	1.479	1.10	57.97	26.28	2.083	1.402
0.22	134.83	80.32	1.916	1.475	1.15	56.18	25.32	2.087	1.400
0.24	130.54	76.62	1.924	1.470	1.20	54.49	24.42	2.091	1.399
0.26	126.59	73.27	1.932	1.466	1.25	52.91	23.59	2.095	1.398
0.28	122.91	70.21	1.939	1.463	1.30	51.41	22.80	2.099	1.397
0.30	119.49	67.42	1.946	1.459	1.35	50.00	22.07	2.102	1.396
0.32	116.27	64.84	1.953	1.456	1.40	48.66	21.39	2.106	1.395
0.34	113.25	62.47	1.959	1.453	1 45	47.39	20.74	2.109	1.394
0.36	110.40	60.26	1.965	1.450	1.50	46.19	20.13	2.112	1.393
0.38	107.71	58.22	1.971	1.447	1.55	45.05	19.56	2.114	1.392
0.40	105.15	56.30	1.976	1.444	1.60	43.96	19.02	2.117	1.391
0.45	99.29	52.04	1.989	1.439	1.65	42.92	18.51	2.120	1.390
0.50	94.08	48.39	2.000	1.434	1.70	41.93	18.02	2.122	1.390
0.55	89.40	45.22	2.011	1.429	1.75	40.98	17.56	2.125	1.389
0.60	85.18	42.44	2.020	1.425	1.80	40.08	17.12	2.127	1.388
0.65	81.35	39.99	2.029	1.422	1.85	39.21	16.71	2.129	1.388
0.70	77.86	37.80	2.037	1.419	1.90	38.39	16.31	2.131	1.387
0.75	74.65	35.84	2.044	1.416	1.95	37.59	15.93	2.133	1.387
0.80	71.70	34.07	2.051	1.413	2.00	36.83	15.57	2.135	1.386

(*Contd.*)

d/b = **0.40**

w/b	Z_e(ohms)	Z_o(ohms)	ϵ_{fe}	ϵ_{fo}	w/b	Z_e(ohms)	Z_o(ohms)	ϵ_{fe}	ϵ_{fo}
0.10	159.02	121.22	1.821	1.562	0.85	60.48	35.18	2.038	1.513
0.12	150.37	112.77	1.833	1.558	0.90	58.21	33.61	2.044	1.511
0.14	143.07	105.68	1.845	1.556	0.95	56.11	32.19	2.050	1.510
0.16	136.73	99.60	1.855	1.553	1.00	54.15	30.87	2.056	1.509
0.18	131.15	94.28	1.865	1.550	1.05	52.33	29.66	2.062	1.509
0.20	126.16	89.56	1.874	1.548	1.10	50.62	28.54	2.067	1.508
0.22	121.64	85.35	1.883	1.545	1.15	49.03	27.50	2.071	1.507
0.24	117.52	81.55	1.892	1.543	1.20	47.53	26.54	2.076	1.506
0.26	113.74	78.09	1.900	1.541	1.25	46.12	25.64	2.080	1.506
0.28	110.24	74.94	1.907	1.539	1.30	44.79	24.80	2.084	1.505
0.30	106.99	72.04	1.915	1.538	1.35	43.53	24.01	2.088	1.504
0.32	103.95	69.37	1.922	1.536	1.40	42.35	23.27	2.092	1.504
0.34	101.11	66.89	1.928	1.534	1.45	41.22	22.57	2.095	1.503
0.36	98.44	64.59	1.935	1.533	1.49	40.37	22.04	2.098	1.503
0.38	95.92	62.45	1.941	1.531	1.55	39.15	21.29	2.102	1.502
0.40	93.53	60.45	1.947	1.530	1.59	38.37	20.82	2.104	1.502
0.45	88.10	55.97	1.961	1.527	1.65	37.27	20.15	2.107	1.501
0.50	83.31	52.11	1.974	1.524	1.70	36.40	19.63	2.110	1.501
0.55	79.02	48.76	1.985	1.522	1.75	35.56	19.13	2.113	1.501
0.60	75.17	45.81	1.996	1.520	1.80	34.77	18.66	2.115	1.500
0.65	71.68	43.20	2.006	1.518	1.85	34.00	18.21	2.118	1.500
0.70	68.50	40.87	2.015	1.517	1.90	33.28	17.78	2.120	1.500
0.75	65.60	38.78	2.023	1.515	1.95	32.58	17.36	2.122	1.499
0.80	62.94	36.89	2.231	1.514	2.00	31.91	16.97	2.124	1.499

(*Contd.*)

$d/b = 0.50$

w/b	Z_e(ohms)	Z_o(ohms)	ϵ_{fe}	ϵ_{fo}	w/b	Z_e(ohms)	Z_o(ohms)	ϵ_{fe}	ϵ_{fo}
0.10	145.81	121.71	1.801	1.610	0.85	52.04	35.30	2.027	1.610
0.12	137.24	113.25	1.813	1.610	0.90	50.02	33.73	2.034	1.610
0.14	130.02	106.14	1.824	1.610	0.95	48.15	32.29	2.041	1.610
0.16	123.78	100.05	1.835	1.610	1.00	46.42	30.98	2.047	1.610
0.18	118.30	94.71	1.844	1.610	1.05	44.81	29.76	2.053	1.610
0.20	113.43	89.99	1.854	1.610	1.10	43.30	28.63	2.058	1.610
0.22	109.02	85.76	1.863	1.610	1.15	41.90	27.59	2.063	1.610
0.24	105.03	81.94	1.873	1.610	1.20	40.58	26.62	2.068	1.610
0.26	101.38	78.47	1.880	1.610	1.25	39.35	25.72	2.073	1.610
0.28	98.02	75.30	1.888	1.610	1.30	38.18	24.87	2.077	1.610
0.30	94.91	72.39	1.896	1.610	1.35	37.09	24.08	2.081	1.610
0.32	92.03	69.70	1.903	1.610	1.40	36.05	23.33	2.085	1.610
0.34	89.33	67.21	1.910	1.610	1.45	35.07	22.64	2.089	1.610
0.36	86.80	64.90	1.917	1.610	1.50	34.33	22.11	2.091	1.610
0.38	84.43	62.74	1.923	1.610	1.55	33.26	21.36	2.095	1.610
0.40	82.19	60.73	1.930	1.610	1.60	32.59	20.88	2.098	1.610
0.45	77.13	56.22	1.945	1.610	1.65	31.63	20.21	2.102	1.610
0.50	72.69	52.34	1.958	1.610	1.70	30.88	19.69	2.104	1.610
0.55	68.76	48.96	1.971	1.610	1.75	30.15	19.18	2.107	1.610
0.60	65.24	46.00	1.982	1.610	1.80	29.46	18.71	2.110	1.610
0.65	62.08	43.37	1.992	1.610	1.85	28.81	18.25	2.112	1.610
0.70	59.22	41.02	2.002	1.610	1.90	28.18	17.82	2.115	1.610
0.75	56.61	38.92	2.011	1.610	1.95	27.57	17.41	2.117	1.610
0.80	54.22	37.02	2.019	1.610	2.00	27.00	17.02	2.120	1.610

Table 7.7 Characteristic impedances and effective dielectric constants of broadside-coupled inverted stripline; $c/w = 40$, $\epsilon_r = 3.78$ (Fused quartz) (See structure A3 of Table 4.4)

$d/b = 0.05$

w/b	Z_e(ohms)	Z_o(ohms)	ϵ_{fe}	ϵ_{fo}	w/b	Z_e(ohms)	Z_o(ohms)	ϵ_{fe}	ϵ_{fo}
0.10	179.59	46.53	3.412	1.574	0.85	71.61	8.96	3.664	1.223
0.12	170.79	41.51	3.438	1.534	0.90	69.01	8.51	3.669	1.218
0.14	163.19	37.53	3.459	1.501	0.95	66.58	8.10	3.674	1.214
0.16	156.51	34.29	3.477	1.473	1.00	64.32	7.73	3.678	1.210
0.18	150.54	31.59	3.493	1.449	1.05	62.21	7.39	3.682	1.206
0.20	145.16	29.30	3.506	1.429	1.10	60.23	7.08	3.686	1.203
0.22	140.25	27.34	3.518	1.409	1.15	58.37	6.79	3.690	1.200
0.24	135.75	25.63	3.529	1.393	1.20	56.62	6.53	3.693	1.198
0.26	131.59	24.13	3.539	1.378	1.25	54.98	6.29	3.696	1.195
0.28	127.73	22.80	3.548	1.365	1.30	53.43	6.06	3.699	1.193
0.30	124.13	21.61	3.556	1.354	1.35	51.96	5.85	3.702	1.191
0.32	120.76	20.54	3.563	1.343	1.40	50.57	5.65	3.704	1.189
0.34	117.59	19.58	3.570	1.333	1.45	49.25	5.47	3.707	1.187
0.36	114.61	18.70	3.577	1.324	1.50	48.00	5.30	3.709	1.185
0.38	111.79	17.90	3.583	1.316	1.55	46.81	5.14	3.711	1.184
0.40	109.12	17.17	3.589	1.309	1.60	45.68	4.98	3.713	1.182
0.45	103.02	15.58	3.601	1.292	1.65	44.60	4.84	3.715	1.181
0.50	97.61	14.26	3.613	1.278	1.70	43.57	4.70	3.717	1.179
0.55	92.76	13.15	3.622	1.267	1.75	42.59	4.58	3.719	1.178
0.60	88.39	12.19	3.631	1.257	1.80	41.65	4.46	3.720	1.177
0.65	84.42	11.37	3.639	1.248	1.85	40.75	4.34	3.722	1.176
0.70	80.80	10.65	3.646	1.241	1.90	39.89	4.23	3.723	1.175
0.75	77.49	10.02	3.653	1.234	1.95	39.07	4.13	3.725	1.174
0.80	74.43	9.46	3.659	1.228	2.00	38.27	4.03	3.726	1.173

(Contd.)

$d/b = 0.10$

w/b	Z_e(ohms)	Z_o(ohms)	ϵ_{fe}	ϵ_{fo}	w/b	Z_e(ohms)	Z_o(ohms)	ϵ_{fe}	ϵ_{fo}
0.10	167.07	67.07	3.234	1.770	0.85	67.58	15.29	3.582	1.392
0.12	159.14	60.92	3.266	1.733	0.90	65.13	14.56	3.590	1.386
0.14	152.28	55.93	3.293	1.702	0.95	62.86	13.89	3.598	1.381
0.16	146.24	51.77	3.316	1.674	1.00	60.73	13.29	3.605	1.376
0.18	140.82	48.23	3.336	1.650	1.05	58.74	12.73	3.611	1.371
0.20	135.91	45.18	3.354	1.628	1.10	56.88	12.21	3.617	1.367
0.22	131.43	42.52	3.371	1.609	1.15	55.14	11.74	3.623	1.363
0.24	127.30	40.18	3.385	1.592	1.20	53.50	11.30	3.628	1.359
0.26	123.47	38.09	3.399	1.576	1.25	51.95	10.90	3.633	1.356
0.28	119.91	36.22	3.411	1.562	1.30	50.49	10.52	3.638	1.353
0.30	116.59	34.53	3.422	1.549	1.35	49.11	10.16	3.643	1.350
0.32	113.47	32.99	3.433	1.537	1.40	47.80	9.83	3.647	1.347
0.34	110.53	31.60	3.442	1.526	1.45	46.56	9.52	3.651	1.345
0.36	107.76	30.32	3.452	1.515	1.50	45.38	9.23	3.655	1.343
0.38	105.14	29.14	3.460	1.506	1.55	44.26	8.96	3.658	1.340
0.40	102.66	28.05	3.468	1.497	1.60	43.20	8.70	3.662	1.338
0.45	96.97	25.66	3.487	1.478	1.65	42.18	8.46	3.665	1.336
0.50	91.92	23.65	3.503	1.461	1.70	41.21	8.23	3.668	1.335
0.55	87.39	21.93	3.518	1.447	1.75	40.28	8.01	3.671	1.333
0.60	83.30	20.45	3.531	1.435	1.80	39.40	7.80	3.674	1.331
0.65	79.59	19.16	3.543	1.424	1.85	38.55	7.61	3.677	1.330
0.70	76.20	18.02	3.554	1.415	1.90	37.74	7.42	3.679	1.328
0.75	73.09	17.01	3.564	1.406	1.95	36.96	7.24	3.682	1.327
0.80	70.22	16.11	3.574	1.399	2.00	36.21	7.07	3.684	1.325

(*Contd.*)

d/b = **0.20**

w/b	Z_e(ohms)	Z_o(ohms)	ϵ_{fe}	ϵ_{fo}	w/b	Z_e(ohms)	Z_o(ohms)	ϵ_{fe}	ϵ_{fo}
0.10	150.19	87.64	3.033	1.995	0.85	60.36	23.44	3.470	1.680
0.12	142.84	80.88	3.064	1.970	0.90	58.16	22.38	3.481	1.674
0.14	136.57	75.26	3.092	1.948	0.95	56.11	21.40	3.492	1.669
0.16	131.08	70.49	3.118	1.928	1.00	54.20	20.51	3.502	1.664
0.18	126.19	66.35	3.141	1.909	1.05	52.42	19.68	3.512	1.659
0.20	121.77	62.73	3.162	1.892	1.10	50.75	18.93	3.521	1.654
0.22	117.75	59.51	3.182	1.877	1.15	49.18	18.22	3.529	1.650
0.24	114.05	56.64	3.200	1.862	1.20	47.71	17.57	3.537	1.647
0.26	110.62	54.05	3.217	1.849	1.25	46.33	16.96	3.544	1.643
0.28	107.43	51.70	3.233	1.837	1.30	45.02	16.40	3.552	1.640
0.30	104.45	49.56	3.248	1.826	1.35	43.78	15.87	3.558	1.637
0.32	101.66	47.60	3.262	1.816	1.40	42.61	15.37	3.565	1.634
0.34	99.02	45.79	3.275	1.806	1.45	41.51	14.90	3.571	1.631
0.36	96.54	44.12	3.287	1.797	1.50	40.45	14.46	3.577	1.629
0.38	94.18	42.57	3.299	1.788	1.55	39.45	14.05	3.582	1.626
0.40	91.95	41.13	3.310	1.780	1.60	38.50	13.65	3.587	1.624
0.45	86.83	37.93	3.335	1.763	1.65	37.59	13.28	3.592	1.622
0.50	82.28	35.21	3.358	1.747	1.70	36.72	12.93	3.597	1.620
0.55	78.20	32.85	3.379	1.734	1.75	35.90	12.60	3.602	1.618
0.60	74.51	30.79	3.397	1.722	1.80	35.11	12.28	3.606	1.616
0.65	71.16	28.97	3.414	1.712	1.85	34.35	11.98	3.610	1.615
0.70	68.11	27.36	3.430	1.703	1.90	33.62	11.69	3.614	1.613
0.75	65.31	25.92	3.444	1.695	1.95	32.93	11.42	3.618	1.611
0.80	62.73	24.62	3.457	1.687	2.00	32.26	11.16	3.622	1.610

(*Contd.*)

$d/b = 0.30$

w/b	Z_e(ohms)	Z_o(ohms)	ϵ_{fe}	ϵ_{fo}	w/b	Z_e(ohms)	Z_o(ohms)	ϵ_{fe}	ϵ_{fo}
0.10	137.85	96.87	2.919	2.151	0.85	53.67	27.74	3.397	1.933
0.12	130.68	89.97	2.948	2.136	0.90	51.68	26.50	3.411	1.929
0.14	124.59	84.20	2.975	2.122	0.95	49.82	25.37	3.424	1.925
0.16	119.30	79.24	2.999	2.109	1.00	48.10	24.34	3.436	1.921
0.18	114.61	74.93	3.023	2.097	1.05	46.49	23.38	3.447	1.917
0.20	110.41	71.11	3.044	2.086	1.10	44.99	22.50	3.458	1.914
0.22	106.60	67.71	3.065	2.075	1.15	43.58	21.68	3.468	1.910
0.24	103.11	64.64	3.084	2.066	1.20	42.26	20.92	3.478	1.908
0.26	99.89	61.87	3.102	2.057	1.25	41.01	20.20	3.487	1.905
0.28	96.91	59.33	3.119	2.048	1.30	39.84	19.54	3.495	1.902
0.30	94.13	57.01	3.135	2.040	1.35	38.73	18.92	3.503	1.900
0.32	91.53	54.87	3.151	2.033	1.40	37.68	18.34	3.511	1.898
0.34	89.09	52.89	3.166	2.026	1.45	36.69	17.79	3.518	1.895
0.36	86.79	51.06	3.180	2.020	1.50	35.75	17.27	3.525	1.893
0.38	84.61	49.35	3.193	2.014	1.55	34.85	16.78	3.532	1.892
0.40	82.55	47.76	3.206	2.008	1.60	34.00	16.32	3.538	1.890
0.45	77.84	44.20	3.236	1.995	1.65	33.19	15.88	3.544	1.888
0.50	73.66	41.14	3.263	1.984	1.70	32.42	15.47	3.550	1.886
0.55	69.92	38.48	3.288	1.974	1.75	31.68	15.07	3.556	1.885
0.60	66.55	36.14	3.310	1.965	1.80	30.97	14.70	3.561	1.884
0.65	63.49	34.08	3.330	1.957	1.85	30.30	14.34	3.566	1.882
0.70	60.72	32.23	3.349	1.950	1.90	29.65	14.00	3.571	1.881
0.75	58.17	30.58	3.366	1.944	1.95	29.03	13.68	3.576	1.880
0.80	55.83	29.09	3.382	1.939	2.00	28.44	13.37	3.586	1.878

(*Contd.*)

$d/b = \mathbf{0.40}$

w/b	Z_e(ohms)	Z_o(ohms)	ϵ_{fe}	ϵ_{fo}	w/b	Z_e(ohms)	Z_o(ohms)	ϵ_{fe}	ϵ_{fo}
0.10	127.19	100.37	2.847	2.278	0.85	47.17	29.39	3.350	2.167
0.12	120.10	93.42	2.874	2.271	0.90	45.37	28.09	3.366	2.165
0.14	114.11	87.59	2.900	2.264	0.95	43.70	26.90	3.380	2.162
0.16	108.91	82.59	2.924	2.258	1.00	42.15	25.81	3.394	2.160
0.18	104.34	78.22	2.947	2.252	1.05	40.71	24.80	3.407	2.158
0.20	100.25	74.34	2.968	2.247	1.10	39.36	23.87	3.419	2.156
0.22	96.56	70.87	2.989	2.241	1.15	38.10	23.00	3.430	2.155
0.24	93.20	67.74	3.008	2.237	1.20	36.92	22.20	3.441	2.153
0.26	90.11	64.89	3.027	2.232	1.25	35.80	21.45	3.451	2.152
0.28	87.26	62.29	3.045	2.228	1.30	34.76	20.75	3.460	2.150
0.30	84.61	59.90	3.062	2.224	1.35	33.77	20.09	3.469	2.149
0.32	82.14	57.70	3.078	2.220	1.40	32.84	19.47	3.478	2.148
0.34	79.83	55.66	3.094	2.216	1.45	31.96	18.89	3.486	2.146
0.36	77.66	53.76	3.109	2.213	1.50	31.12	18.34	3.494	2.145
0.38	75.61	51.99	3.124	2.210	1.55	30.33	17.83	3.501	2.144
0.40	73.68	50.33	3.138	2.207	1.60	29.57	17.34	3.508	2.143
0.45	69.29	46.63	3.170	2.200	1.65	28.86	16.88	3.515	2.142
0.50	65.43	43.44	3.200	2.194	1.70	28.17	16.44	3.521	2.141
0.55	61.98	40.66	3.227	2.189	1.75	27.52	16.02	3.528	2.141
0.60	58.88	38.21	3.252	2.184	1.80	26.90	15.62	3.534	2.140
0.65	56.09	36.05	3.275	2.180	1.85	26.31	15.25	3.539	2.139
0.70	53.55	34.11	3.296	2.176	1.90	25.74	14.89	3.545	2.138
0.75	51.24	32.38	3.316	2.173	1.95	25.19	14.54	3.550	2.138
0.80	49.12	30.81	3.334	2.170	2.00	24.67	14.22	3.555	2.137

(Contd.)

$d/b = 0.50$

w/b	Z_e (ohms)	Z_o (ohms)	ϵ_{fe}	ϵ_{fo}	w/b	Z_e(ohms)	Z_o(ohms)	ϵ_{fe}	ϵ_{fo}
0.10	116.92	99.89	2.801	2.390	0.85	40.63	28.97	3.325	2.390
0.12	109.88	92.95	2.828	2.390	0.90	39.02	27.68	3.342	2.390
0.14	103.96	87.11	2.853	2.390	0.95	37.54	26.51	3.357	2.390
0.16	98.84	82.12	2.877	2.390	1.00	36.16	25.42	3.372	2.390
0.18	94.35	77.74	2.900	2.390	1.05	34.89	24.42	3.385	2.390
0.20	90.35	73.86	2.921	2.390	1.10	33.70	23.50	3.398	2.390
0.22	86.75	70.39	2.942	2.390	1.15	32.59	22.65	3.410	2.390
0.24	83.49	67.25	2.962	2.390	1.20	31.55	21.85	3.422	2.390
0.26	80.50	64.41	2.981	2.390	1.25	30.57	21.11	3.433	2.390
0.28	77.76	61.80	3.000	2.390	1.30	29.66	20.42	3.443	2.390
0.30	75.22	59.41	3.017	2.390	1.35	28.79	19.76	3.452	2 390
0.32	72.88	57.21	3.034	2.390	1.40	27.98	19.15	3.461	2.390
0.34	70.68	55.16	3.051	2.390	1.45	27.21	18.58	3.470	2.390
0.36	68.62	53.26	3.067	2.390	1.50	26.48	18.04	3.478	2.390
0.38	66.70	51.50	3.082	2.390	1.55	25.79	17.53	3.486	2.390
0.40	64.88	49.84	3.097	2.390	1.60	25.13	17.05	3.494	2.390
0.45	60.78	46.14	3.131	2.390	1.65	24.15	16.59	3.501	2.390
0.50	57.19	42.96	3.163	2.390	1.70	23.92	16.16	3.508	2.390
0.55	54.02	40.18	3.192	2.390	1.75	23.35	15.75	3.514	2.390
0.60	51.19	37.75	3.219	2.390	1.80	22.18	15.35	3.521	2.390
0.65	48.65	35.59	3.244	2.390	1.85	22.30	14.98	3.527	2.390
0.70	46.36	33.67	3.266	2.390	1.90	21.80	14.63	3.533	2.390
0.75	44.27	31.94	3.287	2.390	1.95	21.33	14.29	3.538	2.390
0.80	42.37	30.39	3.307	2.390	2.00	20.88	13.97	3.543	2.390

Table 7.8 Characteristic impedances and effective dielectric constants of broadside-coupled inverted stripline; $c/w = 40$, $\epsilon_r = 9.6$ (Alumina) (See structure A3 of Table 4.4)

$d/b = 0.05$

w/b	Z_e(ohms)	Z_o(ohms)	ϵ_{fe}	ϵ_{fo}	w/b	Z_e(ohms)	Z_o(ohms)	ϵ_{fe}	ϵ_{fo}
0.10	114.26	36.23	8.428	2.597	0.85	45.10	7.70	9.236	1.654
0.12	108.53	32.60	8.513	2.486	0.90	43.46	7.33	9.252	1.641
0.14	103.61	29.70	8.582	2.395	0.95	41.92	6.99	9.267	1.630
0.16	99.29	27.32	8.640	2.319	1.00	40.49	6.68	9.281	1.620
0.18	95.94	25.32	8.690	2.254	1.05	39.15	6.39	9.293	1.611
0.20	91.97	23.62	8.733	2.197	1.10	37.90	6.13	9.305	1.602
0.22	88.82	22.14	8.772	2.148	1.15	36.73	5.89	9,316	1.594
0.24	85.93	20.85	8.806	2.104	1.20	35.63	5.67	9.317	1.587
0.26	83.27	19.71	8.837	2.066	1.25	34.59	5.47	9.336	1.581
0.28	80.80	18.69	8.866	2.031	1.30	33.61	5.27	9.345	1.574
0.30	78.49	17.78	8.892	1.999	1.35	32.69	5.10	9.354	1.569
0.32	76.34	16.95	8.916	1.971	1.40	31.81	4.93	9.362	1.563
0.34	74.32	16.21	8.938	1.945	1.45	30.98	4.77	9.370	1.559
0.36	72.42	15.53	8.959	1.922	1.50	30.19	4.63	9.377	1.554
0.38	70.62	14.90	8.978	1.900	1.55	29.44	4 49	9.384	1.550
0.40	68.92	14.32	8.996	1.880	1.60	28.73	4.36	9.390	1.546
0.45	65.03	13.07	9.037	1.837	1.65	28.05	4.24	9.396	1.542
0.50	61.59	12.01	9.073	1.800	1.70	27.40	4.12	9.402	1.538
0.55	58.51	11.12	9.104	1.770	1.75	26.78	4.01	9.408	1.535
0.60	55.74	10.35	9.132	1.743	1.80	26.19	3.91	9.413	1.531
0.65	53.22	9.69	9.157	1.720	1.85	25.62	3.81	9.418	1.528
0.70	50.92	9.10	9.180	1.701	1.90	25.08	3.71	9.423	1.526
0.75	48.82	8.58	9.200	1.683	1.95	24.56	3.62	9.427	1.523
0.80	46.89	8.12	9.219	1.668	2.00	24.06	3.54	9.432	1.520

(Contd.)

d/b = **0.10**

w/b	Z_e(ohms)	Z_o(ohms)	ϵ_{fe}	ϵ_{fo}	w/b	Z_e(ohms)	Z_o(ohms)	ϵ_{fe}	ϵ_{fo}
0.10	107.17	49.67	7.860	3.227	0.85	42.69	12.24	8.975	2.177
0.12	101.92	45.40	7.962	3.122	0.90	41.13	11.67	9.000	2.158
0.14	97.40	41.90	8.049	3.032	0.95	39.69	11.15	9.024	2.143
0.16	93.43	38.97	8.124	2.953	1.00	38.33	10.68	9.046	2.129
0.18	89.88	36.47	8.189	2.885	1.05	37.07	10.24	9.067	2.116
0.20	86.68	34.31	8.247	2.824	1.10	35.89	9.84	9.086	2.105
0.22	83.75	32.41	8.299	2.771	1.15	34.78	9.47	9.104	2.094
0.24	81.07	30.72	8.346	2.722	1.20	33.74	9.13	9.121	2.084
0.26	78.59	29.22	8.389	2.678	1.25	32.76	8.81	9.137	2.075
0.28	76.28	27.86	8.428	2.639	1.30	31.83	8.51	9.152	2.067
0.30	74.13	26.63	8.464	2.603	1.35	30.96	8.23	9.166	2.059
0.32	72.11	25.11	8.498	2.570	1.40	30.13	7.97	9.180	2.052
0.34	70.22	24.49	8.529	2.539	1.45	29.34	7.72	9.193	2.045
0.36	68.43	23.55	8.558	2.511	1.50	28.60	7.49	9.205	2.039
0.38	66.75	22.68	8.586	2.485	1.55	27.89	7.28	9.216	2.033
0.40	65.15	21.88	8.612	2.461	1.60	27.21	7.07	9.227	2.027
0.45	61.49	20.10	8.671	2.408	1.65	26.57	6.88	9.238	2.021
0.50	58.25	18.60	8.724	2.363	1.70	25.96	6.69	9.248	2.016
0.55	55.35	17.31	8.770	2.324	1.75	25.37	6.52	9.257	2.012
0.60	52.73	16.19	8.812	2.291	1.80	24.81	6.35	9.266	2.007
0.65	50.35	15.21	8.851	2.261	1.85	24.27	6.20	9.275	2.003
0.70	48.19	14.34	8.886	2.236	1.90	23.76	6.05	9.283	1.999
0.75	46.20	13.56	8.918	2.213	1.95	23.27	5.90	9.291	1.995
0.80	44.38	12.87	8.948	2.192	2.00	22.80	5.77	9.298	1.991

(Contd.)

$d/b = 0.20$

w/b	Z_e(ohms)	Z_o(ohms)	ϵ_{fe}	ϵ_{fo}	w/b	Z_e(ohms)	Z_o(ohms)	ϵ_{fe}	ϵ_{fo}
0.10	97.35	61.89	7.219	4.001	0.85	38.31	17.32	8.613	3.079
0.12	92.43	57.30	7.319	3.926	0.90	36.89	16.55	8.650	3.061
0.14	88.23	53.17	7.408	3.859	0.95	35.58	15.84	8.684	3.045
0.16	84.57	50.21	7.490	3.798	1.00	34.35	15.20	8.717	3.031
0.18	81.32	47.38	7.564	3.743	1.05	33.21	14.60	8.747	3.017
0.20	78.39	44.89	7.632	3.694	1.10	32.15	14.05	8.775	3.005
0.22	75.72	42.68	7.695	3.648	1.15	31.14	13.53	8.802	2.993
0.24	73.27	40.70	7.753	3.606	1.20	30.20	13.06	8.828	2.982
0.26	71.01	38.91	7.807	3.568	1.25	29.32	12.61	8.851	2.972
0.28	68.91	37.28	7.857	3.533	1.30	28.48	12.20	8.874	2.963
0.30	66.95	35.79	7.905	3.500	1.35	27.69	11.81	8.896	2.954
0.32	65.12	34.43	7.949	3.470	1.40	26.95	11.45	8.916	2.946
0.34	63.39	33.17	7.991	3.441	1.45	26.24	11.10	8.936	2.938
0.36	61.76	32.00	8.030	3.415	1.50	25.57	10.78	8.954	2.931
0.38	60.22	30.92	8.068	3.391	1.55	24.93	10.48	8.972	2.924
0.40	58.76	29.91	8.103	3.368	1.60	24.32	10.19	8.989	2.918
0.45	55.43	27.66	8.184	3.316	1.65	23.75	9.92	9.005	2.911
0.50	52.47	25.73	8.257	3.272	1.70	23.19	9.66	9.020	2.906
0.55	49.82	24.05	8.322	3.234	1.75	22.67	9.41	9.034	2.900
0.60	47.43	22.59	8.381	3.200	1.80	22.16	9.18	9.049	2.895
0.65	45.27	21.29	8.436	3.170	1.85	21.68	8.96	9.062	2.890
0.70	43.30	20.14	8.485	3.144	1.90	21.22	8.74	9.075	2.885
0.75	41.49	19.10	8.531	3.120	1.95	20.78	8.54	9.087	2.881
0.80	39.83	18.17	8.574	3.098	2.00	20.36	8.35	9.099	2.877

(Contd.)

$d/b = \mathbf{0.30}$

w/b	Z_e(ohms)	Z_o(ohms)	ϵ_{fe}	ϵ_{fo}	w/b	Z_e(ohms)	Z_o(ohms)	ϵ_{fe}	ϵ_{fo}
0.10	89.91	66.74	6.861	4.531	0.85	34.17	19.59	8.378	3.877
0.12	85.09	62.09	6.952	4.484	0.90	32.88	18.73	8.423	3.863
0.14	81.01	58.19	7.036	4.441	0.95	31.69	17.94	8.464	3.850
0.16	77.46	54.85	7.114	4.402	1.00	30.58	17.22	8.503	3.838
0.18	74.33	51.93	7.186	4.365	1.05	29.54	16.55	8.539	3.827
0.20	71.52	49.34	7.255	4.332	1.10	28.57	15.93	8.573	3.817
0.22	68.98	47.03	7.319	4.301	1.15	27.67	15.36	8.605	3.808
0.24	66.65	44.95	7.380	4.272	1.20	26.82	14.82	8.636	3.799
0.26	64.51	43.06	7.438	4.245	1.25	26.02	14.32	8.665	3.791
0.28	62.53	41.34	7.492	4.219	1.30	25.26	13.86	8.692	3.784
0.30	60.68	39.75	7.544	4.196	1.35	24.55	13.42	8.718	3.776
0.32	58.96	38.29	7.594	4.174	1.40	23.88	13.01	8.743	3.770
0.34	57.34	36.94	7.641	4.153	1.45	23.25	12.62	8.766	3.763
0.36	55.82	35.69	7.686	4.134	1.50	22.64	12.26	8.789	3.757
0.38	54.38	34.52	7.729	4.116	1.55	22.07	11.92	8.810	3.752
0.40	53.03	33.42	7.770	4.098	1.60	21.52	11.59	8.830	3.747
0.45	49.93	30.98	7.865	4.060	1.65	21.01	11.28	8.850	3.742
0.50	47.19	28.47	7.951	4.026	1.70	20.51	10.99	8.868	3.737
0.55	44.74	27.04	8.029	3.997	1.75	20.04	10.71	8.886	3.732
0.60	42.54	25.42	8.100	3.971	1.80	19.59	10.45	8.903	3.728
0.65	40.55	23.99	8.165	3.948	1.85	19.16	10.20	8.920	3.724
0.70	38.74	22.72	8.225	3.927	1.90	18.75	9.96	8.936	3.720
0.75	37.09	21.57	8.280	3.906	1.95	18.35	9.73	8.951	3.716
0.80	35.57	20.53	8.331	3.892	2.00	17.97	9.51	8.965	3.713

(Contd.)

$d/b = \mathbf{0.40}$

w/b	Z_e(ohms)	Z_o(ohms)	ϵ_{fe}	ϵ_{fo}	w/b	Z_e(ohms)	Z_o(ohms)	ϵ_{fe}	ϵ_{fo}
0.10	83.29	68.10	6.638	4.948	0.85	30.10	20.15	8.227	4.608
0.12	78.53	63.43	6.723	4.925	0.90	28.93	19.27	8.277	4.600
0.14	74.50	59.51	6.802	4.905	0.95	27.85	18.46	8.323	4.593
0.16	71.02	56.15	6.877	4.885	1.00	26.85	17.71	8.367	4.587
0.18	67.95	53.20	6.947	4.867	1.05	25.91	17.02	8.407	4.581
0.20	65.21	50.59	7.051	4.851	1.10	25.04	16.39	8.446	4.575
0.22	62.74	48.25	7.079	4.835	1.15	24.23	15.80	8.482	4.570
0.24	60.49	46.14	7.140	4.820	1.20	23.47	15.25	8.516	4.565
0.26	58.43	44.22	7.199	4.806	1.25	22.75	14.73	8.548	4.560
0.28	56.52	42.47	7.255	4.793	1.30	22.08	14.25	8.579	4.556
0.30	54.76	40.85	7.309	4.781	1.35	21.44	13.80	8.607	4.552
0.32	53.11	39.36	7.361	4.769	1.40	20.84	13.38	8.635	4.548
0.34	51.58	37.98	7.411	4.758	1.45	20.28	12.98	8.661	4.545
0.36	50.13	36.70	7.459	4.748	1.50	19.74	12.61	8.686	4.542
0.38	48.78	35.50	7.505	4.738	1.55	19.23	12.25	8.710	4.538
0.40	47.50	34.38	7.549	4.729	1.60	18.75	11.92	8,732	4.535
0.45	44.59	31.87	7.653	4.708	1.65	18.29	11.60	8.754	4.533
0.50	42.04	29.71	7.748	4.690	1.70	17.85	11.30	8.775	4.530
0.55	39.71	27.82	7.835	4.674	1.75	17.43	11.02	8.795	4.527
0.60	37.74	26.16	7.914	4.660	1.80	17.03	10.74	8.814	4.525
0.65	35.91	24.69	7.987	4.647	1.85	16.65	10.48	8.832	4.522
0.70	34.26	23.38	8.054	4.636	1.90	16.29	10.24	8.850	4.520
0.75	32.75	22.19	8.116	4.626	1.95	15.94	10.00	8.866	4.518
0.80	31.37	21.12	8.174	4.616	2.00	15.61	9.78	8.883	4.516

(*Contd.*)

$d/b = 0.50$

w/b	Z_e(ohms)	Z_o(ohms)	ϵ_{fe}	ϵ_{fo}	w/b	Z_e(ohms)	Z_o(ohms)	ϵ_{fe}	ϵ_{fo}
0.10	76.75	67.08	6.499	5.300	0.85	25.96	19.46	8.145	5.300
0.12	72.02	62.41	6.582	5.300	0.90	24.91	18.59	8.199	5.300
0.14	68.04	58.50	6.659	5.300	0.95	23.95	17.80	8.249	5.300
0.16	64.61	55.14	6.733	5.300	1.00	23.06	17.07	8.295	5.300
0.18	61.60	52.20	6.803	5.300	1.05	22.23	16.40	8.339	5.300
0.20	58.92	49.59	6.870	5.300	1.10	21.46	15.78	8.380	5.300
0.22	56.51	47.26	6.934	5.300	1.15	20.74	15.21	8.419	5.300
0.24	54.32	45.16	6.996	5.300	1.20	20.07	14.67	8.455	5.300
0.26	52.33	43.25	7.056	5.300	1.25	19.44	14.18	8.490	5.300
0.28	50.49	41.50	7.114	5.300	1.30	18.85	13.71	8.522	5.300
0.30	48.80	39.90	7.169	5.300	1.35	18.29	13.27	8.553	5.300
0.32	47.23	38.41	7.223	5.300	1.40	17.77	12.86	8.582	5.300
0.34	45.77	37.04	7.275	5.300	1.45	17.27	12.41	8.610	5.300
0.36	44.40	35.77	7.325	5.300	1.50	16.81	12.11	8.637	5.300
0.38	43.12	34.58	7.373	5.300	1.55	16.36	11.77	8.662	5.300
0.40	41.91	33.47	7.420	5.300	1.60	15.94	11.45	8.686	5.300
0.45	39.19	30.98	7.530	5.300	1.65	15.54	11.14	8.709	5.300
0.50	36.82	28.84	7.631	5.300	1.70	15.16	10.85	8.731	5.300
0.55	34.73	26.98	7.723	5.300	1.75	14.80	10.75	8.752	5.300
0.60	32.87	25.35	7.808	5.300	1.80	14.45	10.31	8.772	5.300
0.65	31.20	23.90	7.887	5.300	1.85	14.12	10.06	8.791	5.300
0.70	29.70	22.61	7.959	5.300	1.90	13.81	9.82	8.810	5.300
0.75	28.34	21.45	8.026	5.300	1.95	13.51	9.60	8.828	5.300
0.80	27.09	20.41	8.088	5.300	2.00	13.22	9.38	8.845	5.300

Table 7.9 Characteristic impedances and effective dielectric constants of broadside-coupled stripline with uniaxially anisotropic (Pyrolytic boron nitride) substrate; $\epsilon_{xx} = 5.12$, $\epsilon_{yy} = 3.4$, $c/w = 40$ (See structure A1 of Table 4.5)

$d/b = 0.05$

w/b	Z_e(ohms)	Z_o(ohms)	ϵ_{fe}	ϵ_{fo}	w/b	Z_e(ohms)	Z_o(ohms)	ϵ_{fe}	ϵ_{fo}
0.10	166.300	30.597	3.978	3.641	1.05	62.499	4.375	3.648	3.440
0.15	149.880	23.047	3.940	3.592	1.10	60.606	4.187	3.640	3.438
0.20	137.500	18.555	3.907	3.560	1.15	58.824	4.014	3.633	3.436
0.25	127.531	15.554	3.880	3.537	1.20	57.144	3.855	3.626	3.435
0.30	119.207	13.400	3.855	3.520	1.25	55.557	3.709	3.620	3.433
0.35	112.080	11.774	3.833	3.507	1.30	54.056	3.572	3.614	3.432
0.40	105.865	10.503	3.812	3.496	1.35	52.633	3.446	3.608	3.431
0.45	100.371	9.482	3.794	3.488	1.40	51.284	3.328	3.602	3.430
0.50	95.464	8.614	3.776	3.480	1.45	50.001	3.218	3.597	3.429
0.55	91.043	7.938	3.760	3.474	1.50	48.781	3.115	3.592	3.428
0.60	87.032	7.341	3.745	3.468	1.55	47.619	3.018	3.587	3.427
0.65	83.372	6.827	3.731	3.464	1.60	46.510	2.927	3.582	3.426
0.70	80.016	6.380	3.718	3.459	1.65	45.452	2.841	3.578	3.425
0.75	76.926	5.989	3.705	3.456	1.70	44.441	2.761	3.574	3.424
0.80	74.069	5.642	3.694	3.452	1.75	43.473	2.684	3.570	3.424
0.85	71.420	5.333	3.683	3.449	1.80	42.547	2.612	3.566	3.423
0.90	68.965	5.057	3.674	3.447	1.85	41.659	2.544	3.562	3.422
0.95	66.665	4.807	3.665	3.444	1.90	40.807	2.479	3.559	3.422
1.00	64.515	4.581	3.656	3.442	1.95	39.989	2.418	3.556	3.420
					2.00	39.203	2.359	3.552	3.420

(Contd.)

$d/b = 0.10$

w/b	Z_e(ohms)	Z_o(ohms)	ϵ_{fe}	ϵ_{fo}	w/b	Z_e(ohms)	Z_o(ohms)	ϵ_{fe}	ϵ_{fo}
0.10	150.470	46.207	3.987	3.729	1.05	58.384	8.005	3.656	3.466
0.15	136.405	36.436	3.952	3.674	1.10	56.644	7.674	3.648	3.463
0.20	125.713	30.239	3.921	3.636	1.15	55.004	7.369	3.641	3.460
0.25	117.013	25.908	3.892	3.607	1.20	53.457	7.087	3.634	3.458
0.30	109.674	22.696	3.867	3.584	1.25	51.994	6.826	3.627	3.455
0.35	103.339	20.204	3.844	3.566	1.30	50.610	6.583	3.621	3.453
0.40	97.782	18.213	3.823	3.551	1.35	49.297	6.357	3.615	3.451
0.45	92.847	16.583	3.803	3.538	1.40	48.050	6.146	3.609	3.449
0.50	88.422	15.222	3.786	3.527	1.45	46.864	5.948	3.604	3.448
0.55	84.425	14.069	3.769	3.518	1.50	45.735	5.763	3.599	3.446
0.60	80.789	13.079	3.754	3.509	1.55	44.660	5.589	3.594	3.444
0.65	77.465	12.219	3.740	3.502	1.60	43.633	5.425	3.589	3.443
0.70	74.410	11.465	3.726	3.495	1.65	42.653	5.270	3.585	3.442
0.75	71.593	10.798	3.714	3.489	1.70	41.715	5.124	3.581	3.440
0.80	68.984	10.205	3.703	3.484	1.75	40.817	4.986	3.576	3.439
0.85	66.563	9.673	3.692	3.480	1.80	39.957	4.855	3.572	3.438
0.90	64.308	9.195	3.682	3.476	1.85	39.133	4.730	3.569	3.437
0.95	62.208	8.761	3.673	3.472	1.90	38.341	4.612	3.565	3.436
1.00	60.235	8.366	3.665	3.469	1.95	37.580	4.500	3.562	3.435
					2.00	36.849	4.393	3.558	3.434

(Contd.)

$d/b = 0.20$

w/b	Z_e(ohms)	Z_o(ohms)	ϵ_{fe}	ϵ_{fo}	w/b	Z_e(ohms)	Z_o(ohms)	ϵ_{fe}	ϵ_{fo}
0.10	131.132	63.411	3.978	3.811	1.05	51.344	13.552	3.660	3.500
0.15	118.566	52.265	3.951	3.757	1.10	49.827	13.021	3.653	3.496
0.20	109.314	44.759	3.924	3.716	1.15	48.396	12.529	3.645	3.492
0.25	101.886	39.259	3.898	3.684	1.20	47.046	12.072	3.638	3.488
0.30	95.639	35.019	3.874	3.657	1.25	45.770	11.648	3.632	3.484
0.35	90.239	31.634	3.851	3.635	1.30	44.561	11.252	3.625	3.481
0.40	85.484	28.862	3.829	3.616	1.35	43.415	10.882	3.620	3.478
0.45	81.247	26.544	3.809	3.600	1.40	42.326	10.536	3.614	3.475
0.50	77.435	24.576	3.791	3.586	1.45	41.290	10.211	3.609	3.473
0.55	73.981	22.885	3.774	3.572	1.50	40.304	9.905	3.603	3.470
0.60	70.833	21.411	3.758	3.561	1.55	39.364	9.618	3.599	3.469
0.65	67.950	20.116	3.744	3.552	1.60	38.467	9.346	3.594	3.467
0.70	65.298	18.968	3.730	3.543	1.65	37.609	9.089	3.589	3.465
0.75	62.848	17.945	3.718	3.535	1.70	36.789	8.846	3.585	3.463
0.80	60.579	17.026	3.707	3.527	1.75	36.004	8.615	3.581	3.461
0.85	58.479	16.196	3.697	3.521	1.80	35.251	8.396	3.577	3.459
0.90	56.514	15.444	3.687	3.515	1.85	34.529	8.188	3.573	3.458
0.95	54.678	14.757	3.677	3.509	1.90	33.837	7.990	3.570	3.456
1.00	52.958	14.129	3.669	3.504	1.95	33.171	7.801	3.566	3.455
					2.00	32.530	7.629	3.563	3.453

(Contd.)

$d/b = 0.30$

w/b	Z_e(ohms)	Z_o(ohms)	ϵ_{fe}	ϵ_{fo}	w/b	Z_e(ohms)	Z_o(ohms)	ϵ_{fe}	ϵ_{fo}
0.10	118.302	72.431	3.963	3.846	1.05	45.148	17.253	3.656	3.521
0.15	106.151	60.917	3.936	3.798	1.10	43.802	16.598	3.648	3.516
0.20	97.399	52.979	3.911	3.758	1.15	42.534	15.991	3.641	3.511
0.25	90.498	47.033	3.888	3.724	1.20	41.339	15.426	3.634	3.507
0.30	84.773	42.359	3.865	3.696	1.25	40.209	14.900	3.628	3.503
0.35	79.872	38.566	3.844	3.672	1.30	39.139	14.408	3.622	3.499
0.40	75.588	35.414	3.824	3.651	1.35	38.125	13.947	3.616	3.496
0.45	71.785	32.749	3.804	3.633	1.40	37.162	13.515	3.610	3.492
0.50	68.374	30.462	3.786	3.618	1.45	36.247	13.108	3.605	3.489
0.55	65.289	28.477	3.770	3.604	1.50	35.376	12.726	3.600	3.486
0.60	62.481	26.736	3.754	3.591	1.55	34.546	12.364	3.595	3.484
0.65	59.911	25.197	3.740	3.580	1.60	33.754	12.023	3.591	3.481
0.70	57.557	23.826	3.727	3.570	1.65	32.997	11.700	3.586	3.479
0.75	55.377	22.599	3.715	3.560	1.70	32.274	11.393	3.582	3.476
0.80	53.358	21.490	3.703	3.552	1.75	31.581	11.102	3.578	3.474
0.85	51.483	20.485	3.692	3.545	1.80	30.918	10.826	3.574	3.472
0.90	49.737	19.569	3.682	3.538	1.85	30.282	10.563	3.571	3.470
0.95	48.106	18.731	3.673	3.532	1.90	29.671	10.312	3.567	3.468
1.00	46.580	17.962	3.664	3.526	1.95	29.085	10.073	3.564	3.466
					2.00	28.521	9.845	3.560	3.465

(Contd.)

$d/b = 0.40$

w/b	Z_e(ohms)	Z_o(ohms)	ϵ_{fe}	ϵ_{fo}	w/b	Z_e(ohms)	Z_o(ohms)	ϵ_{fe}	ϵ_{fo}
0.10	108.032	77.070	3.945	3.863	1.05	39.346	19.385	3.647	3.532
0.15	96.064	65.420	3.917	3.817	1.10	38.150	18.663	3.639	3.527
0.20	87.539	57.324	3.892	3.778	1.15	37.026	17.992	3.632	3.522
0.25	80.904	51.199	3.869	3.745	1.20	35.965	17.367	3.625	3.517
0.30	75.467	46.339	3.847	3.716	1.25	34.965	16.784	3.619	3.513
0.35	70.864	42.361	3.827	3.691	1.30	34.018	16.239	3.613	3.509
0.40	66.877	39.031	3.808	3.670	1.35	33.122	15.728	3.607	3.505
0.45	63.368	36.196	3.790	3.651	1.40	32.272	15.247	3.602	3.502
0.50	60.250	33.751	3.774	3.634	1.45	31.464	14.796	3.597	3.498
0.55	57.438	31.617	3.758	3.620	1.50	30.696	14.369	3.592	3.495
0.60	54.891	29.739	3.743	3.606	1.55	29.964	13.967	3.587	3.492
0.65	52.568	28.073	3.729	3.595	1.60	29.267	13.587	3.583	3.490
0.70	50.439	26.583	3.716	3.584	1.65	28.601	13.226	3.579	3.487
0.75	48.480	25.244	3.704	3.575	1.70	27.965	12.884	3.574	3.484
0.80	46.670	24.033	3.693	3.566	1.75	27.357	12.559	3.571	3.482
0.85	44.993	22.933	3.682	3.558	1.80	26.774	12.250	3.567	3.480
0.90	43.433	21.932	3.672	3.550	1.85	26.216	11.956	3.563	3.477
0.95	41.979	21.012	3.663	3.544	1.90	25.680	11.675	3.560	3.475
1.00	40.619	20.166	3.655	3.538	1.95	25.166	11.408	3.557	3.473
					2.00	24.672	11.152	3.553	3.471

(Contd.)

$d/b = 0.50$

w/b	Z_e(ohms)	Z_o(ohms)	ϵ_{fe}	ϵ_{fo}	w/b	Z_e(ohms)	Z_o(ohms)	ϵ_{fe}	ϵ_{fo}
0.10	98.742	78.515	3.926	3.868	1.05	33.687	20.083	3.632	3.536
0.15	86.888	66.828	3.895	3.822	1.10	32.633	19.338	3.624	3.530
0.20	78.515	58.693	3.868	3.785	1.15	31.643	18.647	3.618	3.525
0.25	72.061	52.518	3.844	3.752	1.20	30.712	18.003	3.611	3.521
0.30	66.828	47.605	3.822	3.723	1.25	29.834	17.402	3.605	3.516
0.35	62.452	43.574	3.803	3.698	1.30	29.006	16.839	3.599	3.512
0.40	58.693	40.191	3.785	3.676	1.35	28.222	16.312	3.594	3.508
0.45	55.414	37.306	3.768	3.656	1.40	27.479	15.816	3.589	3.505
0.50	52.518	34.812	3.752	3.639	1.45	26.774	15.350	3.584	3.501
0.55	49.932	32.633	3.737	3.624	1.50	26.105	14.910	3.579	3.498
0.60	47.605	30.712	3.723	3.611	1.55	25.468	14.494	3.575	3.495
0.65	45.496	29.006	3.710	3.599	1.60	24.862	14.101	3.570	3.492
0.70	43.574	27.479	3.698	3.589	1.65	24.284	13.729	3.566	3.489
0.75	41.812	26.105	3.686	3.579	1.70	23.732	13.375	3.562	3.487
0.80	40.191	24.862	3.676	3.570	1.75	23.205	13.039	3.559	3.484
0.85	38.694	23.732	3.666	3.562	1.80	22.703	12.720	3.554	3.482
0.90	37.306	22.703	3.656	3.554	1.85	22.220	12.415	3.551	3.480
0.95	36.015	21.757	3.648	3.548	1.90	21.757	12.125	3.548	3.478
1.00	34.812	20.887	3.639	3.541	1.95	21.313	11.848	3.544	3.476
					2.00	20.887	11.583	3.541	3.474

Table 7.10 Characteristic impedances and effective dielectric constants of broadside-coupled stripline with uniaxially anisotropic substrate (sapphire); $\epsilon_{xx} = 9.4$, $\epsilon_{yy} = 11.6$, $c/w = 40$ (See structure A1 of Table 4.5)

$d/b = 0.05$

w/b	Z_e(ohms)	Z_o(ohms)	ϵ_{fe}	ϵ_{fo}	w/b	Z_e(ohms)	Z_o(ohms)	ϵ_{fe}	ϵ_{fo}
0.10	101.308	17.423	10.720	11.230	1.05	35.620	2.389	11.231	11.541
0.15	90.623	12.990	10.778	11.305	1.10	34.487	2.285	11.243	11.544
0.20	82.607	10.390	10.826	11.354	1.15	33.424	2.190	11.251	11.547
0.25	76.198	8.668	10.868	11.389	1.20	32.426	2.103	11.262	11.549
0.30	70.875	7.441	10.906	11.416	1.25	31.484	2.022	11.272	11.551
0.35	66.338	6.520	10.941	11.436	1.30	30.594	1.947	11.281	11.553
0.40	62.401	5.802	10.973	11.453	1.35	29.753	1.878	11.290	11.555
0.45	58.937	5.228	11.003	11.471	1.40	28.957	1.813	11.298	11.557
0.50	55.857	4.757	11.030	11.482	1.45	28.202	1.753	11.306	11.558
0.55	53.095	4.364	11.055	11.492	1.50	27.485	1.696	11.314	11.560
0.60	50.601	4.032	11.078	11.498	1.55	26.804	1.643	11.321	11.561
0.65	48.335	3.746	11.100	11.505	1.60	26.155	1.593	11.328	11.563
0.70	46.266	3.498	11.120	11.512	1.65	25.537	1.546	11.335	11.564
0.75	44.369	3.280	11.139	11.518	1.70	24.948	1.502	11.341	11.565
0.80	42.622	3.088	11.156	11.523	1.75	24.384	1.461	11.347	11.566
0.85	41.008	2.917	11.172	11.527	1.80	23.846	1.421	11.353	11.567
0.90	39.518	2.764	11.191	11.531	1.85	23.330	1.384	11.358	11.568
0.95	38.127	2.627	11.205	11.535	1.90	22.836	1.348	11.364	11.569
1.00	36.831	2.502	11.219	11.538	1.95	22.363	1.314	11.369	11.570
					2.00	21.908	1.282	11.374	11.571

(Contd.)

d/b=**0.10**

w/b	Z_e(ohms)	Z_o(ohms)	ϵ_{fe}	ϵ_{fo}	w/b	Z_e(ohms)	Z_o(ohms)	ϵ_{fe}	ϵ_{fo}
0.10	91.833	26.787	10.704	11.096	1.05	33.330	4.394	11.219	11.506
0.15	82.673	20.885	10.759	11.181	1.10	32.284	4.209	11.231	11.510
0.20	75.715	17.198	10.808	11.240	1.15	31.306	4.040	11.239	11.511
0.25	70.078	14.649	10.852	11.284	1.20	30.382	3.883	11.250	11.515
0.30	65.349	12.771	10.891	11.319	1.25	29.511	3.738	11.260	11.519
0.35	61.291	11.326	10.926	11.347	1.30	28.688	3.604	11.270	11.522
0.40	57.750	10.178	10.959	11.371	1.35	27.909	3.479	11.279	11.525
0.45	54.623	9.242	10.989	11.391	1.40	27.172	3.362	11.287	11.528
0.50	51.834	8.464	11.016	11.407	1.45	26.472	3.253	11.295	11.530
0.55	49.326	7.807	11.041	11.422	1.50	25.807	3.150	11.303	11.533
0.60	47.056	7.245	11.065	11.435	1.55	25.174	3.054	11.311	11.535
0.65	44.991	6.758	11.087	11.446	1.60	24.572	2.964	11.318	11.537
0.70	43.101	6.333	11.107	11.456	1.65	23.998	2.878	11.325	11.539
0.75	41.365	5.957	11.126	11.465	1.70	23.449	2.798	11.331	11.541
0.80	39.765	5.624	11.143	11.473	1.75	22.925	2.721	11.337	11.543
0.85	38.285	5.326	11.160	11.480	1.80	22.424	2.649	11.343	11.545
0.90	36.912	5.058	11.176	11.490	1.85	21.944	2.581	11.349	11.546
0.95	35.638	4.815	11.193	11.496	1.90	21.484	2.516	11.355	11.548
1.00	34.445	4.595	11.206	11.501	1.95	21.042	2.454	11.360	11.549
					2.00	20.619	2.395	11.365	11.551

(Contd.)

$d/b = 0.20$

w/b	Z_e(ohms)	Z_o(ohms)	ϵ_{fe}	ϵ_{fo}	w/b	Z_e(ohms)	Z_o(ohms)	ϵ_{fe}	ϵ_{fo}
0.10	79.911	37.374	10.713	10.971	1.05	29.340	7.493	11.210	11.540
0.15	71.853	30.469	10.757	11.056	1.10	28.428	7.192	11.221	11.457
0.20	65.890	25.878	10.800	11.117	1.15	27.570	6.915	11.232	11.463
0.25	61.094	22.547	10.841	11.168	1.20	26.763	6.658	11.243	11.468
0.30	57.065	20.003	10.880	11.209	1.25	26.002	6.419	11.253	11.473
0.35	53.593	17.987	10.917	11.244	1.30	25.283	6.197	11.262	11.478
0.40	50.550	16.347	10.950	11.273	1.35	24.602	5.989	11.271	11.482
0.45	47.853	14.983	10.981	11.299	1.40	23.958	5.795	11.280	11.486
0.50	45.439	13.832	11.009	11.321	1.45	23.346	5.613	11.288	11.491
0.55	43.265	12.845	11.035	11.340	1.50	22.764	5.443	11.296	11.494
0.60	41.294	11.990	11.059	11.357	1.55	22.211	5.282	11.303	11.501
0.65	39.497	11.241	11.080	11.372	1.60	21.683	5.131	11.311	11.504
0.70	37.853	10.581	11.101	11.386	1.65	21.180	4.987	11.317	11.507
0.75	36.342	9.993	11.120	11.398	1.70	20.700	4.852	11.324	11.510
0.80	34.948	9.467	11.137	11.409	1.75	20.241	4.724	11.330	11.513
0.85	33.662	8.994	11.157	11.419	1.80	19.802	4.602	11.336	11.516
0.90	32.465	8.565	11.172	11.428	1.85	19.381	4.486	11.342	11.518
0.95	31.350	8.175	11.186	11.436	1.90	18.977	4.376	11.348	11.521
1.00	30.310	7.819	11.200	11.443	1.95	18.590	4.271	11.353	11.523
					2.00	18.219	4.171	11.358	11.525

(*Contd.*)

$d/b = 0.30$

w/b	Z_e(ohms)	Z_o(ohms)	ϵ_{fe}	ϵ_{fo}	w/b	Z_e(ohms)	Z_o(ohms)	ϵ_{fe}	ϵ_{fo}
0.10	71.879	43.003	10.734	10.912	1.05	25.778	9.580	11.215	11.419
0.15	64.156	35.806	10.776	10.993	1.10	24.970	9.207	11.227	11.427
0.20	58.574	30.885	10.815	11.057	1.15	24.211	8.862	11.238	11.433
0.25	54.165	27.235	10.853	11.107	1.20	23.498	8.541	11.248	11.440
0.30	50.507	24.387	10.890	11.151	1.25	22.825	8.243	11.258	11.446
0.35	47.379	22.094	10.924	11.188	1.30	22.190	7.964	11.267	11.452
0.40	44.652	20.202	10.957	11.221	1.35	21.589	7.704	11.276	11.457
0.45	42.241	18.612	10.987	11.249	1.40	21.020	7.460	11.285	11.462
0.50	40.087	17.257	11.015	11.273	1.45	20.480	7.231	11.293	11.466
0.55	38.149	16.086	11.041	11.294	1.50	19.968	7.016	11.300	11.471
0.60	36.393	15.064	11.065	11.313	1.55	19.480	6.813	11.308	11.475
0.65	34.794	14.164	11.087	11.330	1.60	19.015	6.621	11.315	11.479
0.70	33.336	13.366	11.111	11.345	1.65	18.572	6.440	11.321	11.482
0.75	31.993	12.652	11.130	11.359	1.70	18.149	6.268	11.328	11.486
0.80	30.754	12.011	11.147	11.372	1.75	17.745	6.105	11.334	11.489
0.85	29.609	11.432	11.164	11.383	1.80	17.358	5.950	11.340	11.492
0.90	28.546	10.905	11.179	11.393	1.85	16.988	5.803	11.346	11.495
0.95	27.562	10.425	11.190	11.403	1.90	16.633	5.663	11.351	11.498
1.00	26.640	9.985	11.203	11.411	1.95	16.293	5.530	11.356	11.501
					2.00	15.966	5.403	11.361	11.504

(Contd.)

d/b = **0.40**

w/b	Z_e(ohms)	Z_o(ohms)	ϵ_{fe}	ϵ_{fo}	w/b	Z_e(ohms)	Z_o(ohms)	ϵ_{fe}	ϵ_{fo}
0.10	65.420	45.910	10.759	10.886	1.05	22.421	10.789	11.230	11.402
0.15	57.846	38.604	10.803	10.961	1.10	21.706	10.376	11.241	11.410
0.20	52.449	33.558	10.842	11.025	1.15	21.036	9.992	11.252	11.418
0.25	48.247	29.768	10.879	11.078	1.20	20.405	9.636	11.262	11.424
0.30	44.807	26.788	10.914	11.121	1.25	19.812	9.305	11 272	11.431
0.35	41.899	24.363	10.947	11.160	1.30	19.252	8.995	11.281	11.437
0.40	39.387	22.349	10.978	11.193	1.35	18.723	8.705	11.290	11.443
0.45	37.183	20.645	11.007	11.223	1.40	18.222	8.433	11.298	11.448
0.50	35.230	19.184	11.037	11.248	1.45	17.747	8.177	11.306	11.453
0.55	33.477	17.918	11.062	11.271	1.50	17.296	7.937	11.313	11.458
0.60	31.895	16.808	11.086	11.291	1.55	16.868	7.710	11.320	11.462
0.65	30.459	15.828	11.107	11.308	1.60	16.460	7.495	11.327	11.466
0.70	29.148	14.956	11.127	11.324	1.65	16.072	7.292	11.333	11.470
0.75	27.951	14.174	11.143	11.339	1.70	15.701	7.100	11.339	11.474
0.80	26.846	13.471	11.160	11.352	1.75	15.347	6.917	11.345	11.478
0.85	25.826	12.833	11.176	11.364	1.80	15.009	6.744	11.351	11.481
0.90	24.881	12.253	11.191	11.374	1.85	14.685	6.579	11.357	11.484
0.95	24.033	11.723	11.205	11.385	1.90	14.375	6.422	11.362	11.487
1.00	23.185	11.237	11.218	11.394	1.95	14.077	6.272	11.367	11.490
					2.00	13.792	6.129	11.372	11.493

(Contd.)

$d/b = 0.50$

w/b	Z_e(ohms)	Z_o(ohms)	ϵ_{fe}	ϵ_{fo}	w/b	Z_e(ohms)	Z_o(ohms)	ϵ_{fe}	ϵ_{fo}
0.10	59.569	46.817	10.788	10.878	1.05	19.138	11.186	11.252	11.397
0.15	52.090	39.481	10.836	10.951	1.10	18.512	10.759	11.263	11.405
0.20	46.817	34.403	10.878	11.015	1.15	17.925	10.364	11.274	11.413
0.25	42.761	30.575	10.916	11.068	1.20	17.375	9.996	11.284	11.420
0.30	39.481	27.555	10.951	11.111	1.25	16.857	9.654	11.293	11.426
0.35	36.745	25.092	10.986	11.151	1.30	16.369	9.334	11.302	11.432
0.40	34.403	23.040	11.015	11.185	1.35	15.909	9.034	11.310	11.438
0.45	32.367	21.302	11.043	11.214	1.40	15.474	8.753	11.318	11.443
0.50	30.575	19.809	11.068	11.240	1.45	15.061	8.489	11.325	11.448
0.55	28.985	18.512	11.089	11.263	1.50	14.671	8.240	11.332	11.453
0.60	27.555	17.375	11.111	11.284	1.55	14.300	8.005	11.339	11.458
0.65	26.265	16.369	11.132	11.302	1.60	13.947	7.783	11.346	11.462
0.70	25.092	15.474	11.151	11.318	1.65	13.611	7.573	11.352	11.466
0.75	24.022	14.671	11.168	11.332	1.70	13.291	7.374	11.358	11.470
0.80	23.040	13.947	11.185	11.346	1.75	12.986	7.186	11.363	11.474
0.85	22.136	13.291	11.200	11.358	1.80	12.694	7.006	11.369	11.477
0.90	21.302	12.694	11.214	11.369	1.85	12.415	6.835	11.374	11.481
0.95	20.528	12.148	11.228	11.379	1.90	12.148	6.672	11.379	11.484
1.00	19.809	11.647	11.240	11.388	1.95	11.892	6.517	11.384	11.487
					2.00	11.647	6.369	11.388	11.490

Table 7.11 Characteristic impedances and effective dielectric constants of broadside-coupled stripline with uniaxially anisotropic substrate (epsilam-10); ϵ_{xx} = 13.0, ϵ_{yy} = 10.2, c/w = 40 (See structure A1 of Table 4.5)

d/b = **0.05**

w/b	Z_e(ohms)	Z_o(ohms)	ϵ_{fe}	ϵ_{fo}	w/b	Z_e(ohms)	Z_o(ohms)	ϵ_{fe}	ϵ_{fo}
0.10	99.168	17.922	11.188	10.613	1.05	36.623	2.533	10.625	10.267
0.15	89.208	13.461	11.122	10.529	1.10	35.498	2.423	10.611	10.264
0.20	81.698	10.818	11.068	10.474	1.15	34.440	2.323	10.599	10.261
0.25	75.668	9.056	11.021	10.435	1.20	33.443	2.231	10.587	10.258
0.30	70.641	7.794	10.979	10.406	1.25	32.502	2.146	10.576	10.256
0.35	66.340	6.843	10.940	10.383	1.30	31.643	2.067	10.565	10.253
0.40	62.595	6.100	10.905	10.364	1.35	30.774	1.993	10.553	10.251
0.45	59.288	5.504	10.873	10.351	1.40	29.976	1.925	10.543	10.249
0.50	56.337	5.014	10.843	10.339	1.45	29.217	1.861	10.534	10.247
0.55	53.682	4.604	10.815	10.328	1.50	28.496	1.802	10.526	10.246
0.60	51.276	4.256	10.789	10.319	1.55	27.809	1.745	10.517	10.244
0.65	49.082	3.957	10.765	10.310	1.60	27.155	1.693	10.510	10.243
0.70	47.073	3.698	10.742	10.300	1.65	26.530	1.643	10.502	10.241
0.75	45.225	3.470	10.721	10.294	1.70	25.934	1.596	10.495	10.240
0.80	43.518	3.268	10.702	10.288	1.75	25.363	1.552	10.488	10.239
0.85	41.936	3.089	10.683	10.283	1.80	24.817	1.510	10.482	10.237
0.90	40.472	2.928	10.669	10.278	1.85	24.294	1.471	10.475	10.236
0.95	39.102	2.783	10.653	10.274	1.90	23.792	1.433	10.469	10.235
1.00	37.822	2.652	10.639	10.270	1.95	23.310	1.398	10.463	10.234
					2.00	22.848	1.364	10.458	10.233

(Contd.)

$d/b = 0.10$

w/b	Z_e(ohms)	Z_o(ohms)	ϵ_{fe}	ϵ_{fo}	w/b	Z_e(ohms)	Z_o(ohms)	ϵ_{fe}	ϵ_{fo}
0.10	89.759	27.197	11.205	10.763	1.05	34.227	4.641	10.638	10.314
0.15	81.230	21.381	11.145	10.668	1.10	33.192	4.448	10.625	10.309
0.20	74.746	17.707	11.090	10.603	1.15	33.218	4.270	10.612	10.304
0.25	69.473	15.147	11.041	10.554	1.20	31.299	4.106	10.601	10.300
0.30	65.031	13.250	10.998	10.515	1.25	30.431	3.954	10.589	10.296
0.35	61.202	11.784	10.958	10.484	1.30	29.610	3.813	10.578	10.292
0.40	57.847	10.613	10.922	10.458	1.35	28.836	3.682	10.565	10.286
0.45	54.873	9.655	10.889	10.436	1.40	28.097	3.560	10.556	10.283
0.50	52.209	8.858	10.858	10.417	1.45	27.395	3.445	10.547	10.280
0.55	49.805	8.182	10.830	10.401	1.50	26.728	3.337	10.538	10.277
0.60	47.621	7.602	10.804	10.386	1.55	26.092	3.236	10.529	10.275
0.65	45.627	7.099	10.780	10.374	1.60	25.485	3.141	10.521	10.272
0.70	43.796	6.658	10.757	10.362	1.65	24.906	3.051	10.514	10.270
0.75	42.110	6.269	10.736	10.352	1.70	24.352	2.966	10.506	10.268
0.80	40.550	5.923	10.716	10.344	1.75	23.823	2.886	10.499	10.266
0.85	39.103	5.613	10.698	10.336	1.80	23.315	2.810	10.492	10.264
0.90	37.757	5.334	10.681	10.331	1.85	22.829	2.738	10.486	10.262
0.95	36.505	5.081	10.668	10.325	1.90	22.363	2.669	10.480	10.260
1.00	35.229	4.851	10.652	10.319	1.95	21.915	2.604	10.474	10.258
					2.00	21.484	2.542	10.468	10.257

(*Contd.*)

$d/b =$ **0.20**

w/b	Z_e(ohms)	Z_o(ohms)	ϵ_{fe}	ϵ_{fo}	w/b	Z_e(ohms)	Z_o(ohms)	ϵ_{fe}	ϵ_{fo}
0.10	78.184	37.484	11.191	10.907	1.05	30.107	7.873	10.646	10.370
0.15	70.597	30.807	11.143	10.815	1.10	29.204	7.562	10.633	10.362
0.20	65.003	26.328	11.097	10.741	1.15	28.354	7.275	10.620	10.356
0.25	60.509	23.052	11.052	10.685	1.20	27.552	7.008	10.608	10.349
0.30	56.729	20.532	11.010	10.639	1.25	26.798	6.761	10.594	10.344
0.35	53.463	18.525	10.970	10.601	1.30	26.081	6.530	10.583	10.338
0.40	50.590	16.883	10.933	10.568	1.35	25.402	6.314	10.573	10.333
0.45	48.033	15.513	10.899	10.540	1.40	24.756	6.112	10.564	10.328
0.50	45.735	14.351	10.867	10.516	1.45	24.143	5.922	10.554	10.324
0.55	43.656	13.353	10.838	10.494	1.50	23.560	5.744	10.546	10.320
0.60	41.763	12.484	10.811	10.475	1.55	23.004	5.577	10.537	10.319
0.65	40.032	11.722	10.787	10.458	1.60	22.473	5.418	10.529	10.315
0.70	38.441	11.048	10.764	10.443	1.65	21.967	5.269	10.521	10.312
0.75	36.974	10.447	10.743	10.429	1.70	21.482	5.127	10.514	10.308
0.80	35.617	9.908	10.723	10.417	1.75	21.019	4.993	10.507	10.305
0.85	34.361	9.421	10.708	10.406	1.80	20.575	4.865	10.500	10.302
0.90	33.187	8.980	10.691	10.396	1.85	20.149	4.744	10.494	10.300
0.95	32.092	8.578	10.675	10.386	1.90	19.741	4.629	10.487	10.297
1.00	31.068	8.211	10.660	10.378	1.95	19.348	4.519	10.481	10.295
					2.00	18.971	4.414	10.475	10.292

(Contd.)

$d/b = 0.30$

w/b	Z_e(ohms)	Z_o(ohms)	ϵ_{fe}	ϵ_{fo}	w/b	Z_e(ohms)	Z_o(ohms)	ϵ_{fe}	ϵ_{fo}
0.10	70.479	42.896	11.165	10.966	1.05	26.467	10.036	10.639	10.405
0.15	63.158	35.985	11.119	10.884	1.10	25.670	9.652	10.623	10.397
0.20	57.879	31.229	11.070	10.815	1.15	24.917	9.296	10.610	10.389
0.25	53.715	27.678	11.036	10.754	1.20	24.207	8.966	10.599	10.382
0.30	50.260	24.890	10.997	10.705	1.25	23.537	8.658	10.588	10.375
0.35	47.303	22.631	10.959	10.664	1.30	22.903	8.370	10.577	10.369
0.40	44.719	20.758	10.924	10.628	1.35	22.302	8.101	10.567	10.363
0.45	42.427	19.176	10.891	10.597	1.40	21.732	7.848	10.558	10.357
0.50	40.373	17.821	10.860	10.570	1.45	21.190	7.611	10.549	10.352
0.55	38.517	16.647	10.831	10.546	1.50	20.675	7.387	10.540	10.347
0.60	36.830	15.618	10.804	10.525	1.55	20.184	7.176	10.532	10.342
0.65	35.287	14.709	10.780	10.506	1.60	19.716	6.977	10.524	10.338
0.70	33.876	13.901	10.760	10.489	1.65	19.270	6.788	10.517	10.334
0.75	32.570	13.177	10.738	10.473	1.70	18.843	6.610	10.509	10.330
0.80	31.363	12.524	10.719	10.459	1.75	18.434	6.440	10.502	10.326
0.85	30.243	11.633	10.700	10.447	1.80	18.043	6.279	10.496	10.322
0.90	29.200	11.395	10.683	10.435	1.85	17.668	6.125	10.489	10.316
0.95	28.228	10.903	10.668	10.424	1.90	17.308	5.979	10.483	10.316
1.00	27.319	10.452	10.653	10.415	1.95	16.963	5.840	10.477	10.312
					2.00	16.631	5.707	10.472	10.310

(Contd.)

$d/b = \mathbf{0.40}$

w/b	Z_e(ohms)	Z_o(ohms)	ϵ_{fe}	ϵ_{fo}	w/b	Z_e(ohms)	Z_o(ohms)	ϵ_{fe}	ϵ_{fo}
0.10	64.304	45.683	11.136	10.995	1.05	23.056	11.284	10.620	10.425
0.15	57.099	38.685	11.087	10.915	1.10	22.346	10.860	10.607	10.416
0.20	51.967	33.828	11.044	10.850	1.15	21.679	10.466	10.594	10.407
0.25	47.972	30.160	11.004	10.793	1.20	21.050	10.100	10.583	10.400
0.30	44.699	27.259	10.967	10.740	1.25	20.457	9.758	10.572	10.392
0.35	41.929	24.885	10.931	10.697	1.30	19.896	9.439	10.562	10.386
0.40	39.531	22.901	10.898	10.660	1.35	19.366	9.140	10.552	10.379
0.45	37.422	21.216	10.867	10.627	1.40	18.863	8.859	10.543	10.373
0.50	35.549	19.764	10.840	10.598	1.45	18.385	8.595	10.534	10.367
0.55	33.861	18.499	10.813	10.573	1.50	17.931	8.346	10.526	10.362
0.60	32.333	17.387	10.787	10.551	1.55	17.499	8.111	10.518	10.357
0.65	30.941	16.402	10.763	10.531	1.60	17.088	7.888	10.510	10.352
0.70	29.668	15.522	10.741	10.513	1.65	16.695	7.678	10.503	10.348
0.75	28.496	14.732	10.721	10.496	1.70	16.320	7.478	10.496	10.343
0.80	27.416	14.019	10.701	10.482	1.75	15.961	7.288	10.489	10.339
0.85	26.418	13.371	10.680	10.468	1.80	15.618	7.108	10.483	10.335
0.90	25.488	12.780	10.664	10.456	1.85	15.289	6.936	10.477	10.331
0.95	24.622	12.239	10.648	10.445	1.90	14.974	6.773	10.471	10.328
1.00	23.813	11.742	10.633	10.434	1.95	14.671	6.617	10.465	10.324
					2.00	14.381	6.467	10.460	10.321

(*Contd.*)

$d/b = 0.50$

w/b	Z_e(ohms)	Z_o(ohms)	ϵ_{fe}	ϵ_{fo}	w/b	Z_e(ohms)	Z_o(ohms)	ϵ_{fe}	ϵ_{fo}
0.10	58.717	46.551	11.103	11.003	1.05	19.724	11.692	10.594	10.431
0.15	51.586	39.529	11.049	10.924	1.10	19.099	11.255	10.582	10.422
0.20	46.551	34.647	11.003	10.861	1.15	18.512	10.850	10.570	10.413
0.25	42.672	30.948	10.962	10.803	1.20	17.961	10.472	10.559	10.405
0.30	39.529	28.013	10.924	10.751	1.25	17.442	10.120	10.548	10.398
0.35	36.902	25.606	10.893	10.707	1.30	16.951	9.790	10.538	10.391
0.40	34.647	23.590	10.861	10.670	1.35	16.418	9.481	10.529	10.384
0.45	32.682	21.873	10.831	10.636	1.40	16.049	9.191	10.520	10.378
0.50	30.948	20.391	10.803	10.607	1.45	15.633	8.918	10.512	10.372
0.55	29.401	19.099	10.778	10.582	1.50	15.239	8.661	10.504	10.367
0.60	28.013	17.961	10.751	10.559	1.55	14.863	8.418	10.496	10.362
0.65	26.754	16.951	10.728	10.538	1.60	14.506	8.188	10.489	10.357
0.70	25.606	16.049	10.707	10.520	1.65	14.165	7.971	10.482	10.352
0.75	24.556	15.239	10.688	10.504	1.70	13.840	7.766	10.475	10.348
0.80	23.590	14.506	10.670	10.489	1.75	13.529	7.568	10.469	10.344
0.85	22.698	13.840	10.652	10.475	1.80	13.232	7.381	10.463	10.339
0.90	21.873	13.232	10.636	10.463	1.85	12.948	7.204	10.457	10.336
0.95	21.106	12.676	10.621	10.451	1.90	12.676	7.035	10.451	10.332
1.00	20.391	12.164	10.607	10.441	1.95	12.415	6.873	10.446	10.328
					2.00	12.164	6.718	10.441	10.325

REFERENCES

1. S.B. Cohn, Characteristic impedances of broadside-coupled strip transmission lines, *IRE Trans. Microwave Theory Tech.*, Vol. MTT-8, pp. 633-637, Nov. 1960.
2. J.K. Shimizu, Stripline 3-dB directional couplers, *IRE WESCON Convention Record*, Pt. 1, pp. 4-15, 1957.
3. L.C. Gunderson and A. Guida, Stripline coupler design, *Microwave J.*, Vol. 8, pp. 97-101, June 1965.
4. C.P. Tresselt, The design and construction of broadband, high directivity, 90-degree couplers using nonuniform line techniques, *IEEE Trans. Microwave Theory Tech.*, Vol. MTT-14, pp. 647-656, Dec. 1966.
5. J.L. Allen and M.F. Estes, Broadside-coupled strips in a layered dielectric medium, *IEEE Trans. Microwave Theory Tech.*, Vol. MTT-20, pp. 662-669, Oct. 1972. Also correction *IEEE Trans. Microwave Theory Tech.*, Vol. MTT-23, p. 779, Sept. 1975.
6. D.L. Gish and O. Graham, Characteristic impedance and phase velocity of a dielectric-supported air-strip transmission line with side walls, *IEEE Trans. Microwave Theory Tech.*, Vol. MTT-18, pp. 131-146, March 1970.
7. I.J. Bahl and P. Bhartia, Analysis of inhomogeneous broadside-coupled striplines employing anisotropic substrates, *Arch. Elek. Ubertragung*, Vol. 35, pp. 492-495, Dec. 1981.
8. A.G. d'Assuncao, A.J. Giarola and D.A. Rogers, Inhomogeneous broadside-coupled striplines, *IEEE MTT-S, Int. Microwave Symp. Digest*, pp. 218–220, June 1981.
9. I.J. Bahl and P. Bhartia, Characteristics of inhomogeneous coupled striplines, *IEEE Trans. Microwave Theory Tech.*, Vol. MTT-28, pp. 529-535, June 1980.
10. A.G.d'Assuncao, A.J. Giarola and D.A. Rogers, Characteristics of broadside-coupled microstrip lines with iso/anisotropic substrates, *Electron Lett.*, Vol. 17, No. 7, pp. 264-265, April 1981.
11. B. Bhat and S.K. Koul, Unified approach to solve a class of strip and microstrip-like transmission lines, *IEEE Trans. Microwave Theory Tech.*, Vol. MTT-30, pp. 679-686, May 1982. Also correction *IEEE Trans. Microwave Theory Tech.* Vol. MTT-30, p. 2067, Nov. 1982.
12. S.K. Koul and B. Bhat, Transverse transmission line method for the analysis of broadside-coupled microstrip lines with anisotropic substrates, *Arch. Elek. Ubertragung*, Vol. 37, pp. 59-64, Jan./Feb. 1983.
13. S.K. Koul and B. Bhat, A generalized TEM-analysis of broadside-coupled planar transmission lines with isotropic and anisotropic substrates, *Arch. Elek. Ubertragung*, Vol. 38, pp. 37-45, Jan./Feb. 1984.
14. F. Oberhettinger and W. Magnus, *Anwendung der Elliptischen Funktionen in Physik and Technik*, Springer-Verlag, 1949.
15. M.A.R. Gunston, *Microwave Transmission Line Impedance Data,* Van Nostrand Reinhold Company, London, 1972.
16. J.E. Dalley, A stripline directional coupler utilizing a nonhomogeneous dielectric medium, *IEEE Trans. Microwave Theory Tech.*, Vol. MTT-17, pp 706-712, Sept. 1969.
17. S.K. Koul and B. Bhat, Characteristic impedance of microstrip-like transmission lines for millimeter wave applications, *Arch. Elek. Ubertragung*, Vol. 35, pp. 253-258, June 1981.

CHAPTER 8

Design Data on Broadside-Edge Coupled Stripline-Like Transmission Lines and some Special Structures

8.1 Introduction

Multiconductor coupled transmission lines are generally used in interdigital couplers [1, 2], meander-folded couplers [3], and certain log-periodic microwave circuit elements [4]. Depending on the terminal conditions, these coupled transmission lines can support several propagating modes. In order to study the electric response produced by these lines for various excitations, a knowledge of the characteristic impedances and the effective dielectric constants for all the possible modes of excitation is essential. Duncan [4] has analysed a system of four broadside-edge coupled strip conductors situated between two parallel ground planes by using the variational method. Using the self consistent field method, Ivanov [5] has presented formulas for the characteristic impedances of four rectangular bars symmetrically situated in a rectangular metal shield. A generalised analysis which combines the variational method with the transverse transmission line technique has been applied by Bhat and Koul [6] to solve a class of broadside-edge coupled symmetric structures. The impedance characteristics of some of the structures using isotropic as well as anisotropic dielectric substrates have been reported by Koul and Bhat [7, 8]. A detailed analysis leading to unified capacitance formulas valid for multilayer structures using iso/anisotropic substrates is presented in Chapter 4. Tables 4.7–4.9 list some broadside-edge coupled structures using iso/anisotropic substrates. In this chapter, we present design curves on the characteristic impedances and effective dielectric constants for three structures, namely, (a) broadside-edge coupled homogeneous striplines, (b) broadside-edge coupled suspended striplines, and (c) broadside-edge coupled inverted striplines. Both isotropic and uniaxially anisotropic cases are considered.

A few special structures of practical importance, which cannot be classified under symmetric edge-coupled, broadside-coupled or broadside-edge coupled structures are reported in the last section. Design equations reported in the literature are presented in consolidated form. For the design data already published, readers are referred to the appropriate references cited. Design data which are likely to be of practical utility but are not available in the literature are presented towards the end of this chapter.

8.2 Design Equations

8.2.1 Quasi-Static Propagation Parameters

Broadside-edge coupled strip transmission structures can support four orthogonal modes of excitation. They are designated as even-even (*ee*), even-odd (*eo*), odd-even (*oe*) and odd-odd (*oo*) modes. A detailed description of these modes is available in Chapter 4. The propagation parameters of the structure, namely, the characteristic impedance, effective dielectric constants, phase velocities and guide wavelengths for the four modes of excitation are given by,

$$Z_{\substack{ee\\eo\\oe\\oo}} = \frac{1}{v^a \sqrt{C_{\substack{ee\\eo\\oe\\oo}} C^a_{\substack{ee\\eo\\oe\\oo}}}} \tag{8.1}$$

$$\epsilon_{f\substack{ee\\eo\\oe\\oo}} = C_{\substack{ee\\eo\\oe\\oo}} / C^a_{\substack{ee\\eo\\oe\\oo}} \tag{8.2}$$

$$v_{\substack{ee\\eo\\oe\\oo}} = \frac{v^a}{\sqrt{\epsilon_{f\substack{ee\\eo\\oe\\oo}}}} \tag{8.3}$$

$$\lambda_{\substack{ee\\eo\\oe\\oo}} = \frac{\lambda^a}{\sqrt{\epsilon_{f\substack{ee\\eo\\oe\\oo}}}} \tag{8.4}$$

where C_{ee}, C_{eo}, C_{oe} and C_{oo} are the line capacitances per unit length of the structure for the *ee*, *eo*, *oe* and *oo* modes respectively. C^a_{ee}, C^a_{eo}, C^a_{oe} and C^a_{oo} are the corresponding capacitances of the structure wherr all dielectrics are replaced by air. v^a and λ^a represent the free space velocity and free space wavelength respectively.

8.2.2. Unified Variational Capacitance Formula for Structures with Iso/ Anisotropic Substrates

The determination of the quasi-static propagation parameters defined in (8.1)–(8.4) reduces to the evaluation of the capacitances for the four modes of excitation. The unified capacitance formulas derived in Chapter 4, are summarized below.

Tables 4.7–4.9 list a class of broadside-edge coupled symmetric stripline-like transmission structures. Referring to the coordinate system marked in structure A1 of Table 4.7, the charge distribution on the strip conductors (assumed to be of zero thickness) is chosen as,

$$f(x)_{\substack{ee\\eo\\oe\\oo}} = \left(\frac{1}{w}\right)\left[1 + A_{\substack{ee\\eo\\oe\\oo}} \left|\left(\frac{2}{w}\right)\left(x - \frac{c-s-w}{2} - g_{\substack{e\\e\\o\\o}}\right)\right|^3\right], \quad \frac{c-s}{2} - w \leqslant x \leqslant \frac{c-s}{2}$$

$$= 0 \qquad \text{, otherwise} \tag{8.5}$$

where $0 \leqslant g_e \leqslant w/2$, for *ee* and *eo* modes and $-w/2 \leqslant g_o \leqslant 0$, for *oe* and *oo* modes.

The variational capacitance formula for the four excitations is given by

$$C_{\substack{ee\\eo\\oe\\oo}} = \frac{\left(1 + K_{\substack{e\\e\\o\\o}} A_{\substack{ee\\eo\\oe\\oo}}\right)^2}{\sum\limits_{\substack{n\ \text{odd}\\ \text{odd}\\ \text{even}\\ \text{even}}} (L_n + M_{\substack{ne\\e\\o\\o}} A_{\substack{ee\\eo\\oe\\oo}})^2 P_n / Y_{\substack{e\\o\\e\\o}}} \tag{8.6}$$

where

$$K_{\substack{e\\o}} = \frac{2}{w^4}\left[\left(\frac{w}{2} + g_{\substack{e\\o}}\right)^4 + \left(\frac{w}{2} - g_{\substack{e\\o}}\right)^4\right] \tag{8.7a}$$

$$L_n = \sin\left(\frac{\beta_n w}{2}\right) \sin\left\{\beta_n \left(\frac{c - s - w}{2}\right)\right\}, \quad \beta_n = \frac{n\pi}{c} \tag{8.7b}$$

$$\begin{aligned} M_{\substack{ne\\o}} = \left(\frac{2}{\beta_n w}\right)^3 \cdot \frac{1}{2}\Bigg[& 3\left\{\beta_n^2\left(\frac{w}{2} + g_{\substack{e\\o}}\right)^2 - 2\right\}\cdot\sin\left\{\beta_n\left(\frac{c-s}{2} - w\right)\right\} \\ & + \beta_n\left(\frac{w}{2} + g_{\substack{e\\o}}\right)\cdot\left\{\beta_n^2\left(\frac{w}{2} + g_{\substack{e\\o}}\right)^2 - 6\right\}\cdot\cos\left\{\beta_n\left(\frac{c-s}{2} - w\right)\right\} \\ & + 3\left\{\beta_n^2\left(\frac{w}{2} - g_{\substack{e\\o}}\right)^2 - 2\right\}\cdot\sin\left\{\beta_n\left(\frac{c-s}{2}\right)\right\} - \beta_n\left(\frac{w}{2} - g_{\substack{e\\o}}\right) \\ & \left\{\beta_n^2\left(\frac{w}{2} - g_{\substack{e\\o}}\right)^2 - 6\right\}\cdot\cos\left\{\beta_n\left(\frac{c-s}{2}\right)\right\} + 12\sin\left\{\beta_n\left(\frac{c-s-w}{2} + g_{\substack{e\\o}}\right)\right\}\Bigg] \end{aligned} \tag{8.7c}$$

$$P_n = \left(\frac{4}{n\pi}\right)\left(\frac{2}{\beta_n w}\right)^2 \tag{8.7d}$$

$$A_{\substack{ee\\eo\\oe\\oo}} = -\frac{\sum\limits_{\substack{n\ \text{odd}\\ \text{odd}\\ \text{even}\\ \text{even}}} (K_{\substack{e\\o\\o}} L_n - M_{\substack{ne\\o\\o}}) L_n P_n / Y_{\substack{o\\e\\o}}}{\sum\limits_{\substack{n\ \text{odd}\\ \text{odd}\\ \text{even}\\ \text{even}}} (K_{\substack{e\\o\\o}} L_n - M_{\substack{ne\\o\\o}}) M_{\substack{ne\\o\\o}} P_n / Y_{\substack{o\\e\\o}}} \tag{8.7e}$$

The same arguments forwarded in Chapter 6 for the choice of g_e and g_o apply here also. Setting $g_e = g_o = 0$ corresponds to symmetric charge distribution.

$$\begin{aligned} f(x)_{\substack{ee\\eo\\oe\\oo}} &= (1/w)\left[1 + A_{\substack{ee\\eo\\oe\\oo}}\left|(2/w)\left(x - \frac{c-s-w}{2}\right)\right|^3\right], \quad \frac{c-s}{2} - w \leqslant x \leqslant \frac{c-s}{2} \\ &= 0 \qquad\qquad\qquad\qquad\qquad\qquad\qquad\qquad, \text{otherwise} \end{aligned} \tag{8.8}$$

The variational capacitance formula then simplifies to the following

$$C_{\substack{ee\\eo\\oe\\oo}} = \frac{\left(1 + 0.25 A_{\substack{ee\\eo\\oe\\oo}}\right)^2}{\sum\limits_{\substack{n\ \text{odd}\\ \text{odd}\\ \text{even}\\ \text{even}}} (L_n + M_n A_{\substack{ee\\eo\\oe\\oo}})^2 P_n / Y_{\substack{e\\o\\e\\o}}} \tag{8.9}$$

where

$$L_n = \sin\left(\frac{\beta_n w}{2}\right) \sin\left\{\beta_n\left(\frac{c-s-w}{2}\right)\right\}, \; \beta_n = \frac{n\pi}{c} \tag{8.10a}$$

$$M_n = \left(\frac{2}{\beta_n w}\right)^3 \sin\left\{\beta_n\left(\frac{c-s-w}{2}\right)\right\}\left[3\left\{\left(\frac{\beta_n w}{2}\right)^2 - 2\right\}\cos\left(\frac{\beta_n w}{2}\right) + \left(\frac{\beta_n w}{2}\right)\left\{\left(\frac{\beta_n w}{2}\right)^2 - 6\right\}\sin\left(\frac{\beta_n w}{2}\right) + 6\right] \tag{8.10b}$$

$$P_n = \frac{4}{n\pi}\left(\frac{2}{\beta_n w}\right)^2 \tag{8.10c}$$

$$A_{\substack{ee\\eo\\oe\\oo}} = -\frac{\displaystyle\sum_{\substack{n\ \text{odd}\\ \text{odd}\\ \text{even}\\ \text{even}}} (L_n - 4M_n)L_n P_n / Y_{\substack{e\\o\\e\\o}}}{\displaystyle\sum_{\substack{n\ \text{odd}\\ \text{odd}\\ \text{even}\\ \text{even}}} (L_n - 4M_n)M_n P_n / Y_{\substack{e\\o\\e\\o}}} \tag{8.10d}$$

The admittance parameter Y_e which is applicable to *ee* and *oe* modes and Y_o applicable to *eo* and *oo* modes are listed in Tables 4.7-4.9 along with the structural configurations.

In the following sections, the numerical results evaluated using (8.6) are presented. First, the effect of shielding side walls on the computed characteristic impedances were investigated by increasing the ratio $c/(2w + s)$. It was found that for $c/(2w + s) \geqslant 20$, the change in the various impedances is negligible. Hence, in all the computations, $c/(2w + s)$ was set equal to 20 so that the side walls have practically no effect on the propagation parameters.

8.3 Broadside-edge Coupled Transmission Structures Using Isotropic Dielectric Substrates

8.3.1 Broadside-edge Coupled Homogeneous Stripline

For the broadside-edge coupled homogeneous stripline, the normalized characteristic impedances $Z_{ee}\sqrt{\epsilon_r}$ and $Z_{eo}\sqrt{\epsilon_r}$ are plotted in Fig. 8.1a and $Z_{oe}\sqrt{\epsilon_r}$ and $Z_{oo}\sqrt{\epsilon_r}$ are plotted in Fig. 8.1b. The characteristics are shown as a function of w/b, with s/b as a parameter for a fixed value of $d/b = 0.1$. Similar characteristics of the structure computed for larger spacings between the broadside-coupled strip conductors are shown in Fig. 8.2a and b for $d/b = 0.2$ and in Fig. 8.3a and b for $d/b = 0.3$. These normalized curves are valid for any value of ϵ_r.

For a fixed value of s/b and d/b, increasing w/b, decreases all the four impedances. Increasing s/b, for a fixed value of w/b and d/b decreases Z_{ee} and Z_{eo} but increases Z_{oe} and Z_{oo}. In all the cases, the variations in Z_{eo} and Z_{oo}, with increasing s/b, are much less as compared with those in Z_{ee} and Z_{oe}. For a fixed set of parameters w/b and s/b, as d/b is increased, Z_{ee} and Z_{oe} decrease whereas Z_{eo} and Z_{oo} increase. Figures 8.1 to 8.3 show that a wide range of impedance values from 10 ohms to 400 ohms is possible for the structural parameters in the range $0.1 \leqslant w/b \leqslant 1.0$, $0.2 \leqslant s/b \leqslant 1.0$ and $0.1 \leqslant d/b \leqslant 0.3$.

8.3.2 Broadside-edge Coupled Suspended Stripline

The variations in the characteristic impedances (Z_{ee}, Z_{eo}, Z_{oe}, Z_{oo}) and effective dielectric

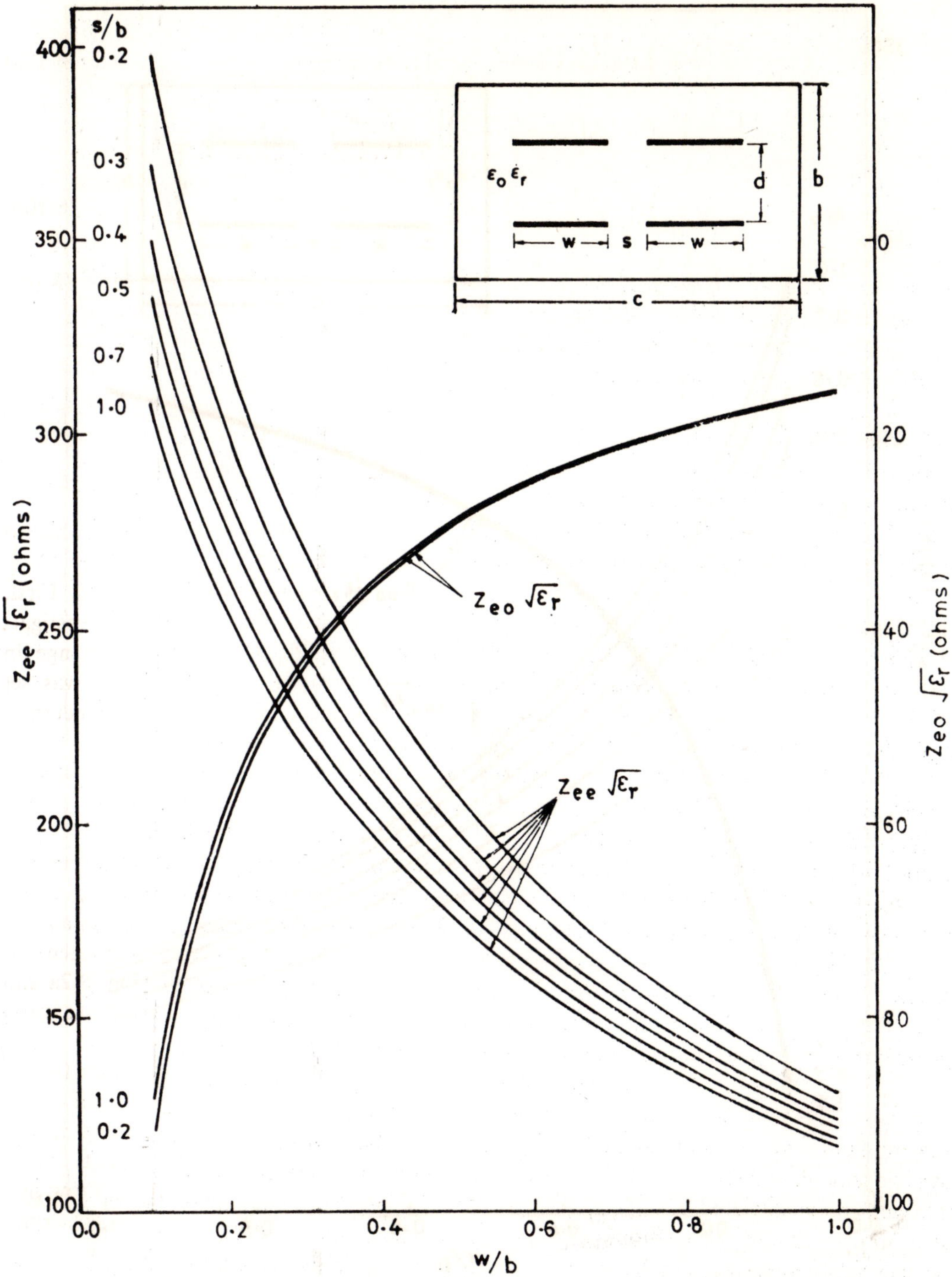

FIG. 8.1(a) The normalized characteristic impedances $Z_{ee}\sqrt{\epsilon_r}$ and $Z_{eo}\sqrt{\epsilon_r}$ versus w/b with s/b as parameter of broadside-edge coupled homogeneous stripline; $d/b = 0.1$, $c/(2w + s) = 20$.

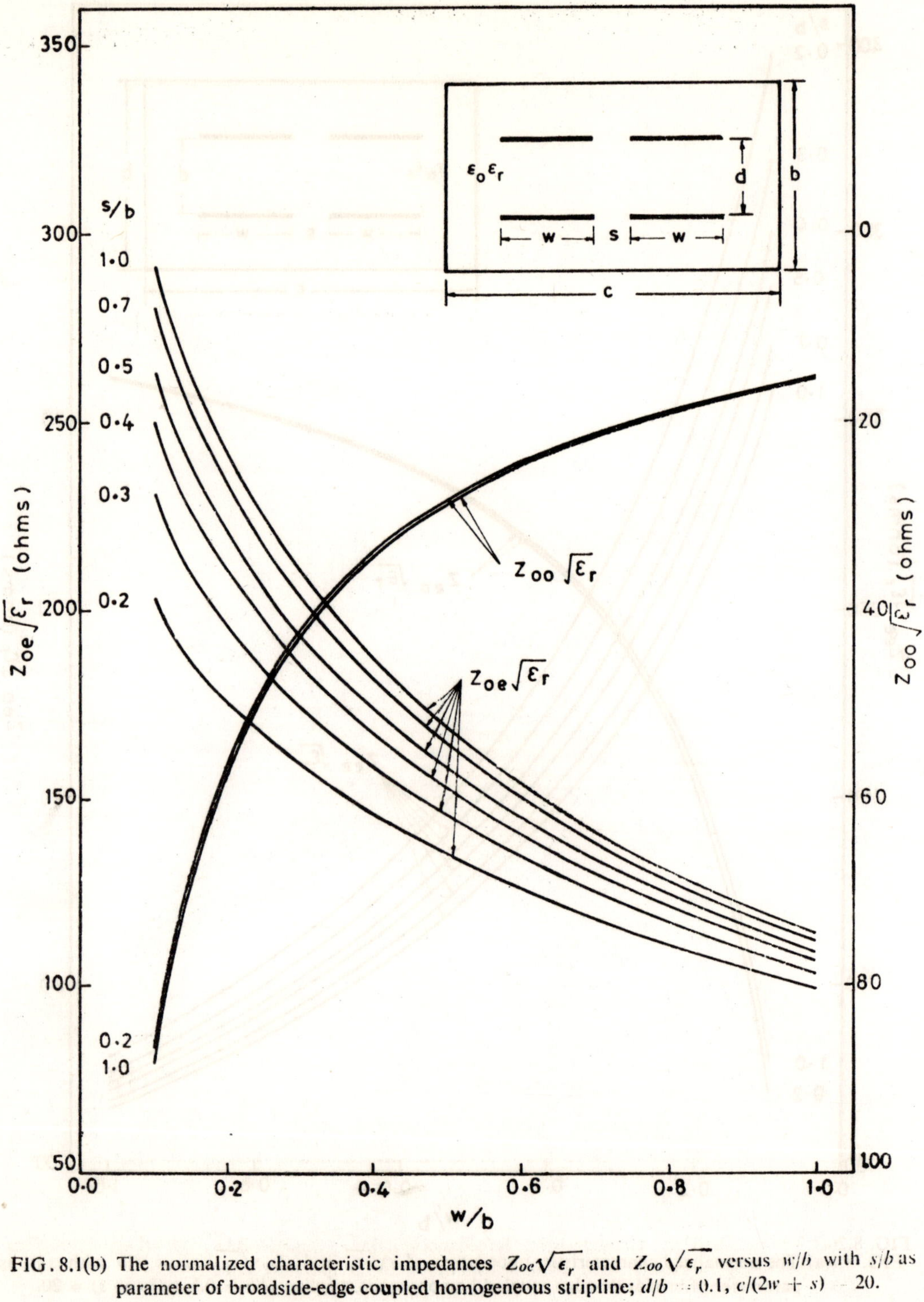

FIG. 8.1(b) The normalized characteristic impedances $Z_{oe}\sqrt{\epsilon_r}$ and $Z_{oo}\sqrt{\epsilon_r}$ versus w/b with s/b as parameter of broadside-edge coupled homogeneous stripline; $d/b = 0.1$, $c/(2w + s) = 20$.

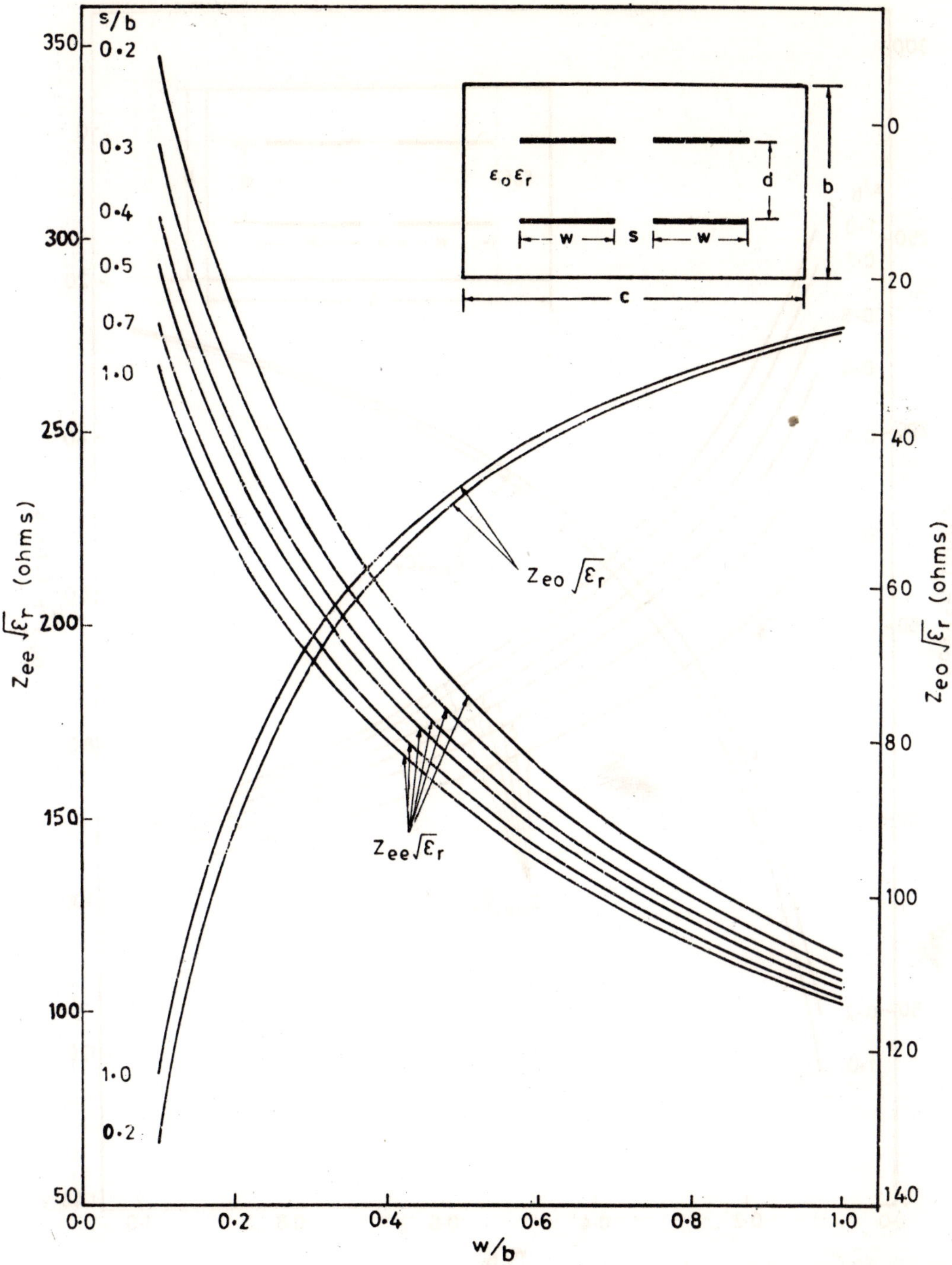

FIG. 8.2(a) The normalized characteristic impedances $Z_{ee}\sqrt{\epsilon_r}$ and $Z_{eo}\sqrt{\epsilon_r}$ versus w/b with s/b as parameter of broadside-edge coupled homogeneous stripline; $d/b = 0.2$, $c/(2w + s) = 20$.

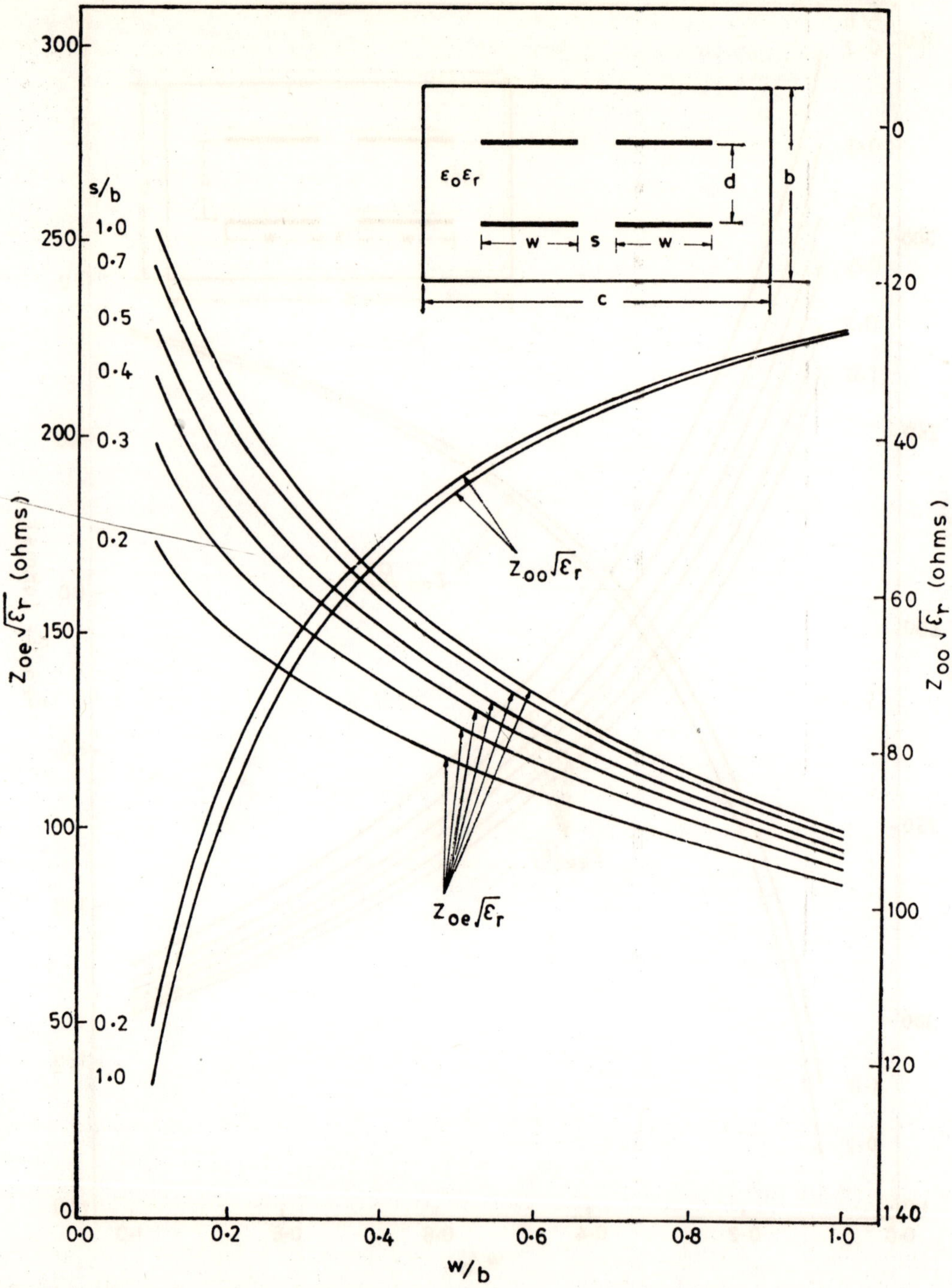

FIG. 8.2(b) The normalized characteristic impedances $Z_{oe}\sqrt{\epsilon_r}$ and $Z_{oo}\sqrt{\epsilon_r}$ versus w/b with s/b as parameter of broadside-edge coupled homogeneous stripline; $d/b = 0.2$, $c/(2w + s) = 20$.

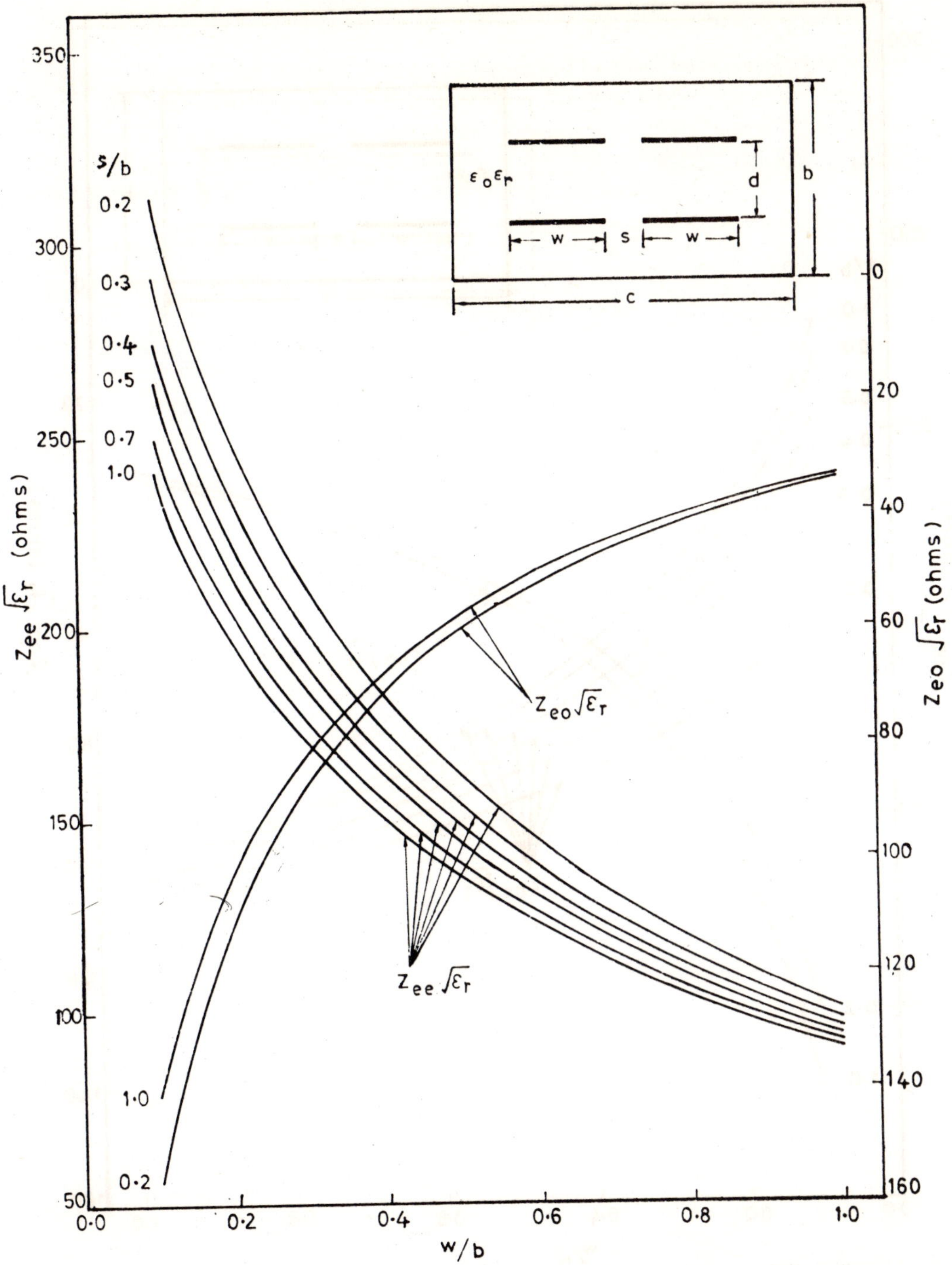

FIG. 8.3(a) The normalized characteristic impedances $Z_{ee}\sqrt{\epsilon_r}$ and $Z_{eo}\sqrt{\epsilon_r}$ versus w/b with s/b as parameter of broadside-edge coupled homogeneous stripline; $d/b = 0.3$, $c/(2w + s) = 20$.

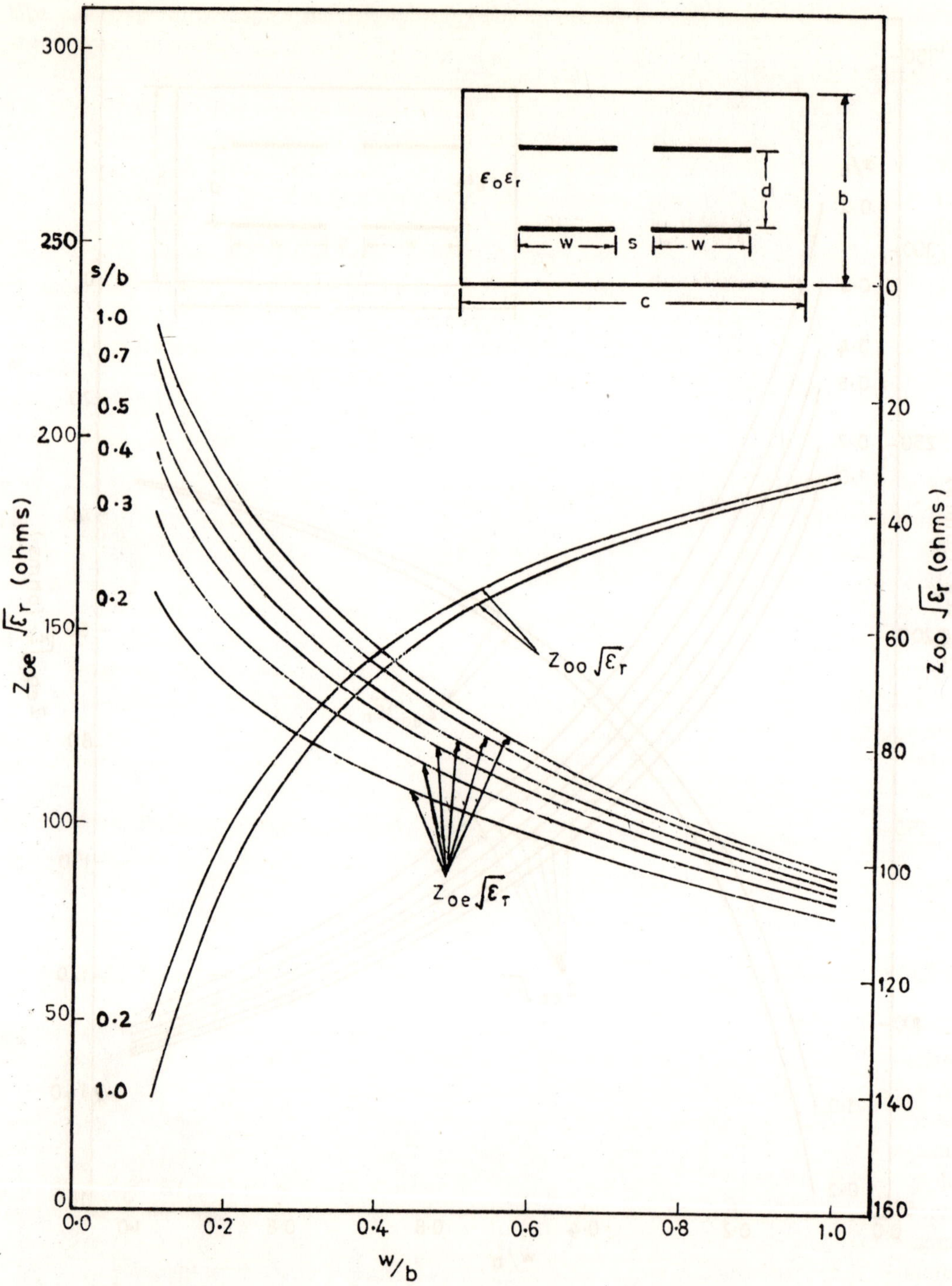

FIG. 8.3(b) The normalized characteristic impedances $Z_{oe}\sqrt{\epsilon_r}$ and $Z_{oo}\sqrt{\epsilon_r}$ versus w/b with s/b as parameter of broadside-edge coupled homogeneous stripline; $d/b = 0.3$, $c/(2w + s) = 20$.

constants (ϵ_{fee}, ϵ_{feo}, ϵ_{foe}, ϵ_{foo}) of the broadside-edge coupled suspended stripline versus w/b with s/b as a parameter are illustrated in Figs. 8.4 and 8.5 respectively. All the characteristics are plotted for a fixed value of $d/b = 0.1$ and with alumina ($\epsilon_r = 9.6$) as the substrate material. It is seen that the variations of Z_{ee}, Z_{eo}, Z_{oe} and Z_{oo} (Fig. 8.4a and b) are similar to those of the broadside-edge coupled homogeneous stripline (Fig. 8.1a and b). From Fig. 8.5a and b, it is observed that, with an increase in w/b, ϵ_{feo} and ϵ_{foo} increase whereas ϵ_{fee} and ϵ_{foe} decrease. The values of ϵ_{feo} and ϵ_{foo} are considerably higher than those of ϵ_{fee} and ϵ_{foe}.

Figures 8.6 and 8.7 show the variations in the characteristic impedances and effective dielectric constants respectively of the structure for the same range of parameters as considered in Fig. 8.4 and 8.5 but for a higher value of d/b ($= 0.2$). As in the case of broadside-edge coupled homogeneous stripline, as d/b is increased, Z_{ee} and Z_{oe} decrease whereas Z_{eo} and Z_{oo} increase. Comparing Fig. 8.7a and b with Fig. 8.5a and b, it is seen that as d/b is increased, with all other parameters fixed, ϵ_{fee} and ϵ_{foe} increase whereas ϵ_{feo} and ϵ_{foo} decrease. For the structural parameters $0.1 \leqslant w/b \leqslant 1.0$, $0.2 \leqslant s/b \leqslant 1.0$, $0.1 \leqslant d/b \leqslant 0.2$ and $\epsilon_r = 9.6$, the impedance values obtained are in the range 10 to 310 ohms.

8.3.3 Broadside-edge Coupled Inverted Stripline

For the broadside-edge coupled inverted stripline the characteristic impedances are plotted in Fig. 8.8a and b and the effective dielectric constants are plotted in Fig. 8.9a and b. The characteristics are shown as a function of w/b, with s/b as a parameter for a fixed value of $d/b = 0.1$ and $\epsilon_r = 9.6$. For a higher value of $d/b = 0.2$ and for the same set of other parameters, the characteristic impedances are plotted in Fig. 8.10a and b and the effective dielectric constants are plotted in Fig. 8.11a and b. It is observed that the variations of Z_{ee}, Z_{eo}, Z_{oe} and Z_{oo} are similar to those of the broadside-edge coupled suspended stripline. Impedance values in the range 10 to 150 ohms are obtained for the structural parameters in the range $0.1 \leqslant w/b \leqslant 1.0$, $0.1 \leqslant s/b \leqslant 1.0$, $0.1 \leqslant d/b \leqslant 0.2$ and for $\epsilon_r = 9.6$. From Fig. 8.9a and b, it is observed that as w/b is increased, ϵ_{fee} and ϵ_{foe} increase whereas ϵ_{feo} and ϵ_{foo} decrease. Further, the values of ϵ_{fee} and ϵ_{foe} are considerably higher than those of ϵ_{feo} and ϵ_{foo}. Comparing Fig. 8.9a and b with Fig. 8.11a and b it is seen that, as d/b increases with all other parameters fixed, ϵ_{fee} and ϵ_{foe} decrease whereas ϵ_{feo} and ϵ_{foo} increase. As expected, this trend is opposite to that observed in the broadside-edge coupled suspended stripline.

8.4 Effect of Dielectric Anisotropy on Broadside-edge Coupled Strip Transmission Structures

The effect of dielectric anisotropy on the characteristic impedances and effective dielectric constants is shown in Fig. 8.12 for the broadside-edge coupled homogeneous stripline, in Fig. 8.13 for the broadside-edge coupled suspended stripline and in Fig. 8.14 for the broadside-edge coupled inverted stripline. In all the three figures, the characteristics are plotted as a function of the anisotropy ratio $\epsilon_{xx}/\epsilon_{yy}$ for fixed parameters $w/b = d/b = 0.1$, $s/b = 0.2$, and $\epsilon_{yy} = 3.78$. With an increase in the anisotropy ratio $\epsilon_{xx}/\epsilon_{yy}$, the variations in the impedances Z_{ee}, Z_{eo}, Z_{oe} and Z_{oo}, in the three structures, are similar. All the four impedances decrease as $\epsilon_{xx}/\epsilon_{yy}$ is increased from 0.5 to 3. However, the change in Z_{eo} and Z_{oo} is small as compared to the change in Z_{ee} and Z_{oe}. The curves of effective dielectric constants ϵ_{fee}, ϵ_{feo}, ϵ_{foe} and ϵ_{foo} differ considerably for the three structures, although all the four quantities increase with an increase in $\epsilon_{xx}/\epsilon_{yy}$. In the broadside-coupled homogeneous stripline, as $\epsilon_{xx}/\epsilon_{yy}$ is increased, ϵ_{foe} increases by the largest amount followed by ϵ_{fee}, ϵ_{foo} and ϵ_{feo} in

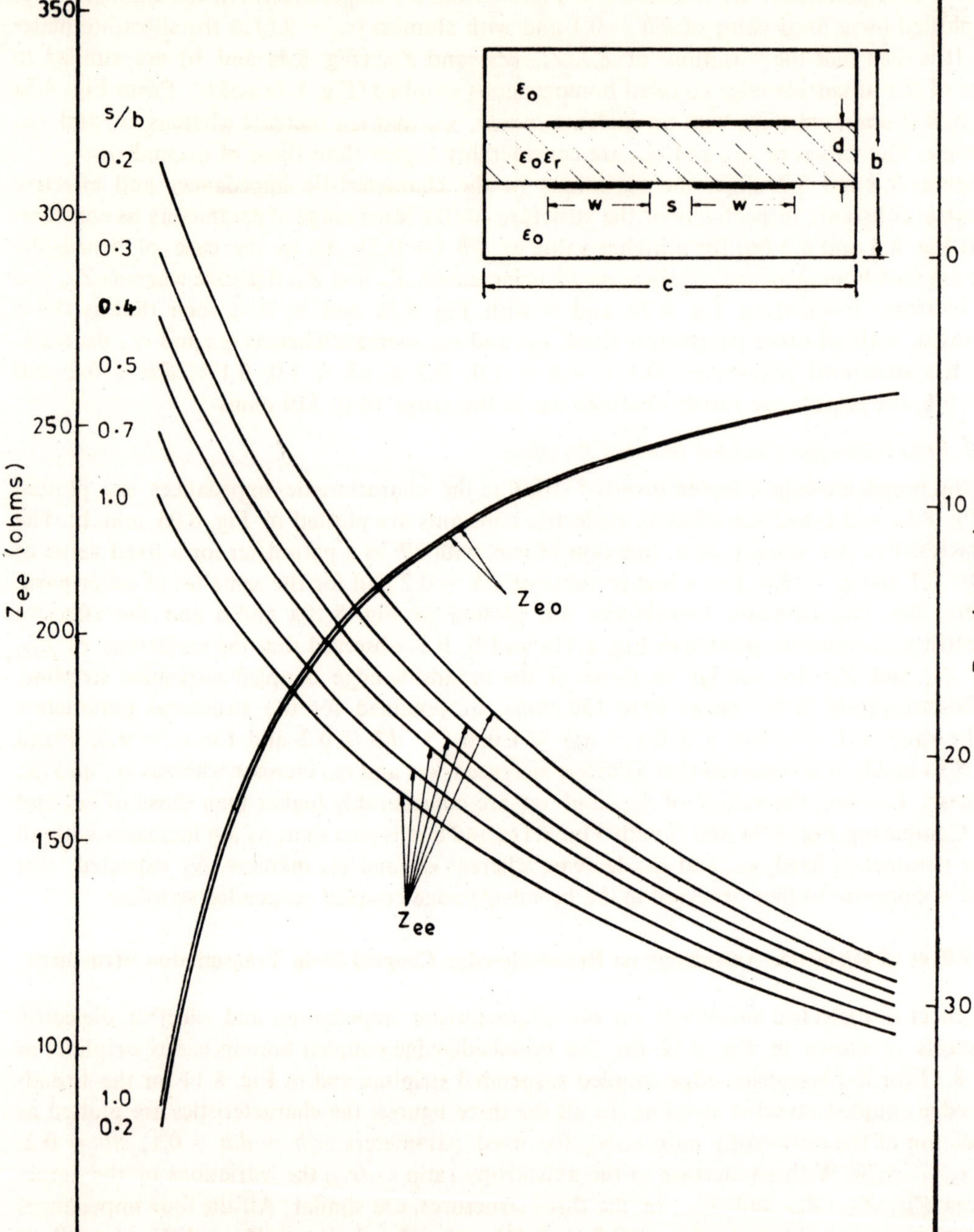

FIG. 8.4(a) Z_{ee} and Z_{eo} versus w/b with s/b as parameter of broadside-edge coupled suspended stripline; $\epsilon_r = 9.6$, $d/b = 0.1$, $c/(2w + s) = 20$.

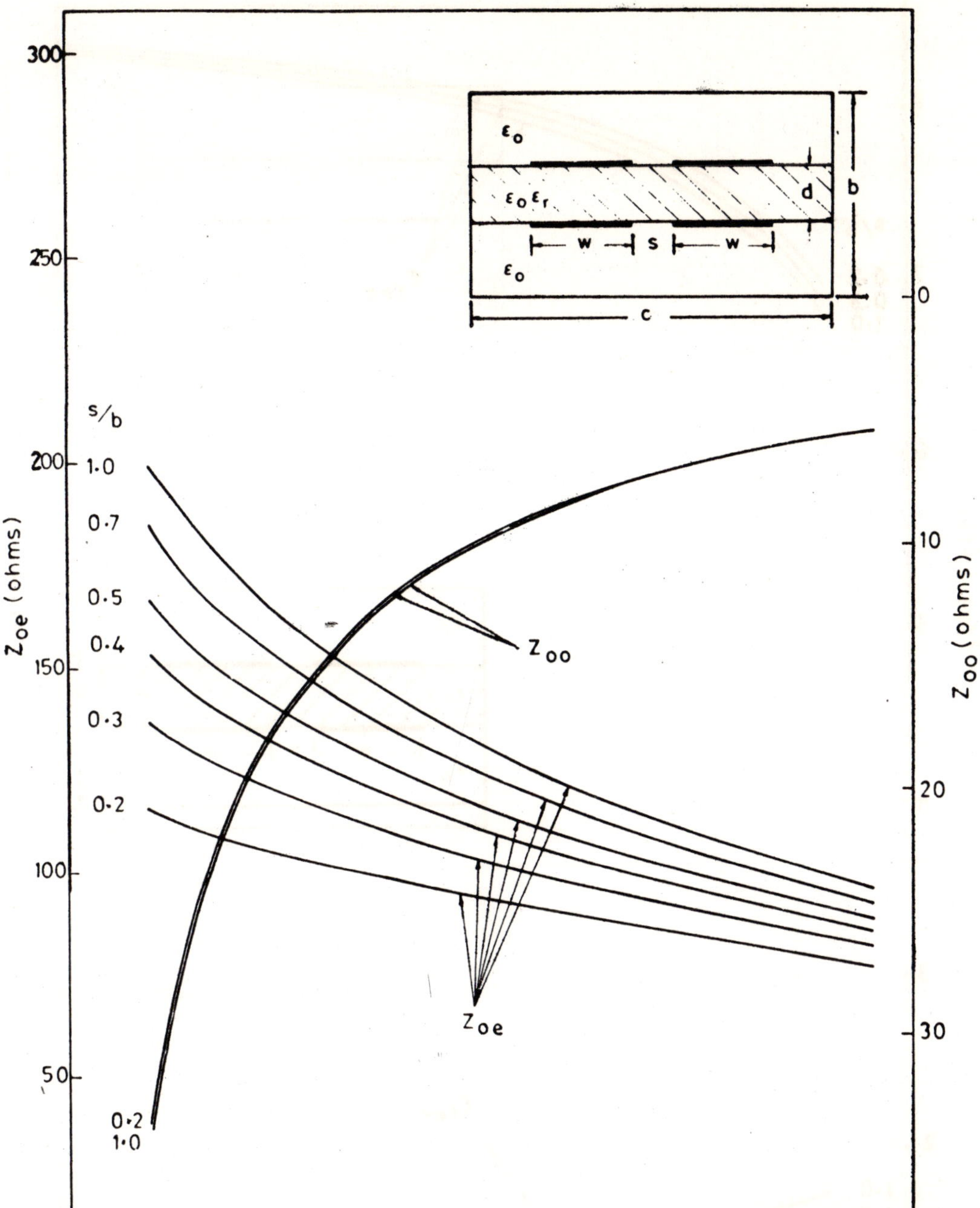

FIG. 8.4(b) Z_{oe} and Z_{oo} versus w/b with s/b as parameter of broadside-edge coupled suspended stripline; $\epsilon_r = 9.6$, $d/b = 0.1$, $c/(2w + s) = 20$.

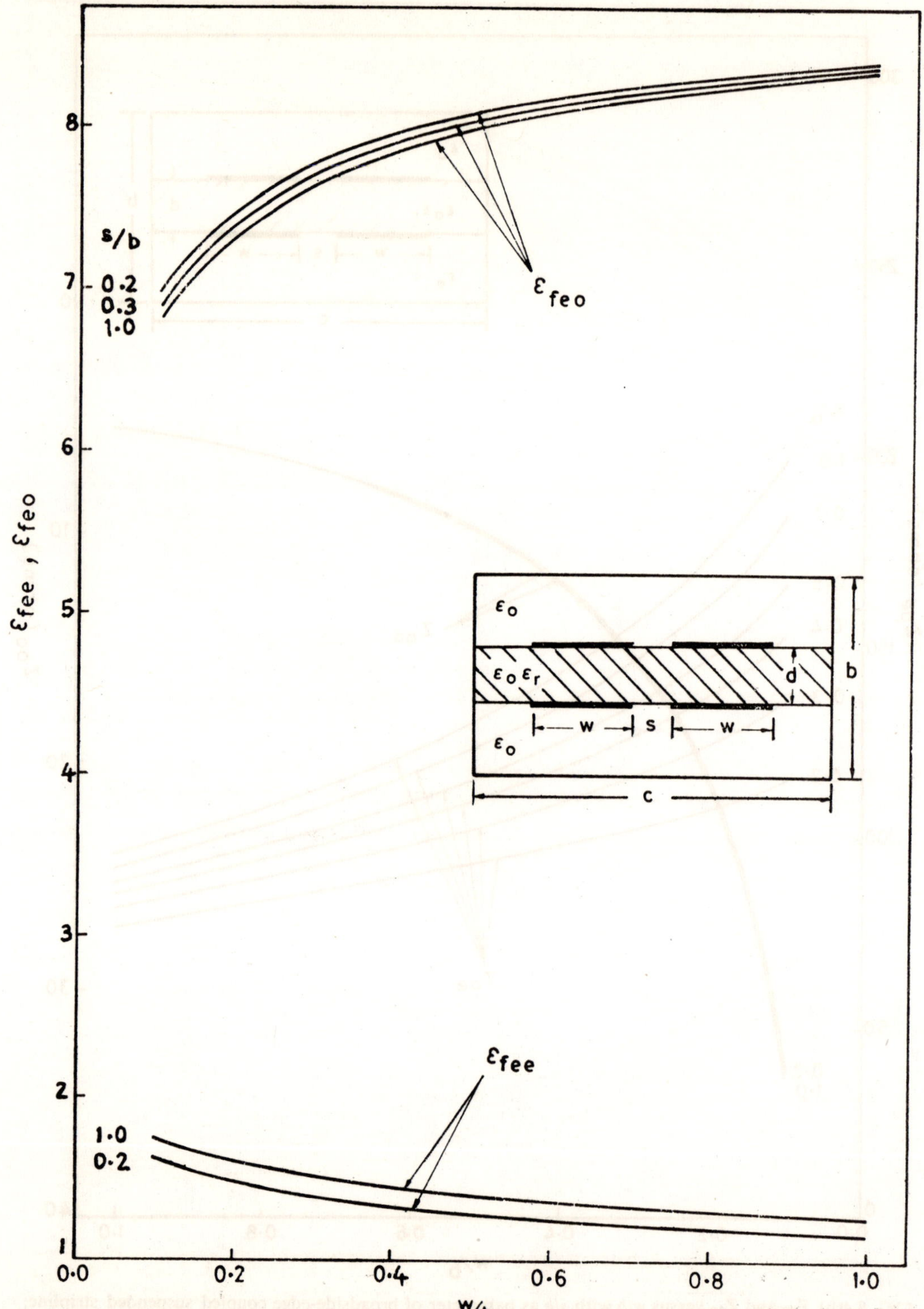

FIG. 8.5(a) ϵ_{fee} and ϵ_{feo} versus w/b with s/b as parameter of broadside-edge coupled suspended stripline; $\epsilon_r = 9.6$, $d/b = 0.1$, $c/(2w + s) = 20$.

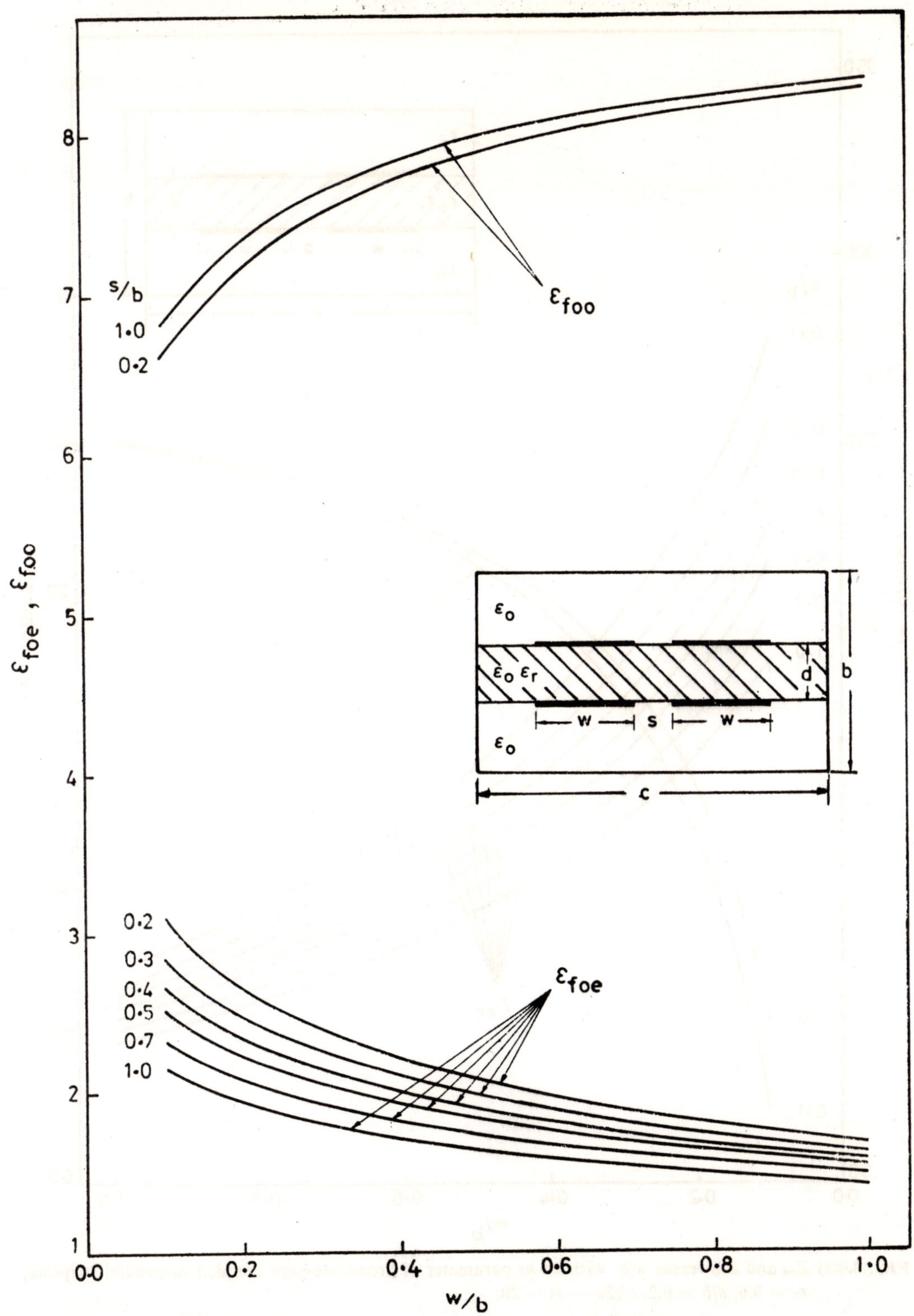

FIG. 8.5(b) ϵ_{foe} and ϵ_{foo} versus w/b with s/b as parameter of broadside-edge coupled suspended stripline; $\epsilon_r = 9.6$, $d/b = 0.1$, $c/(2w + s) = 20$.

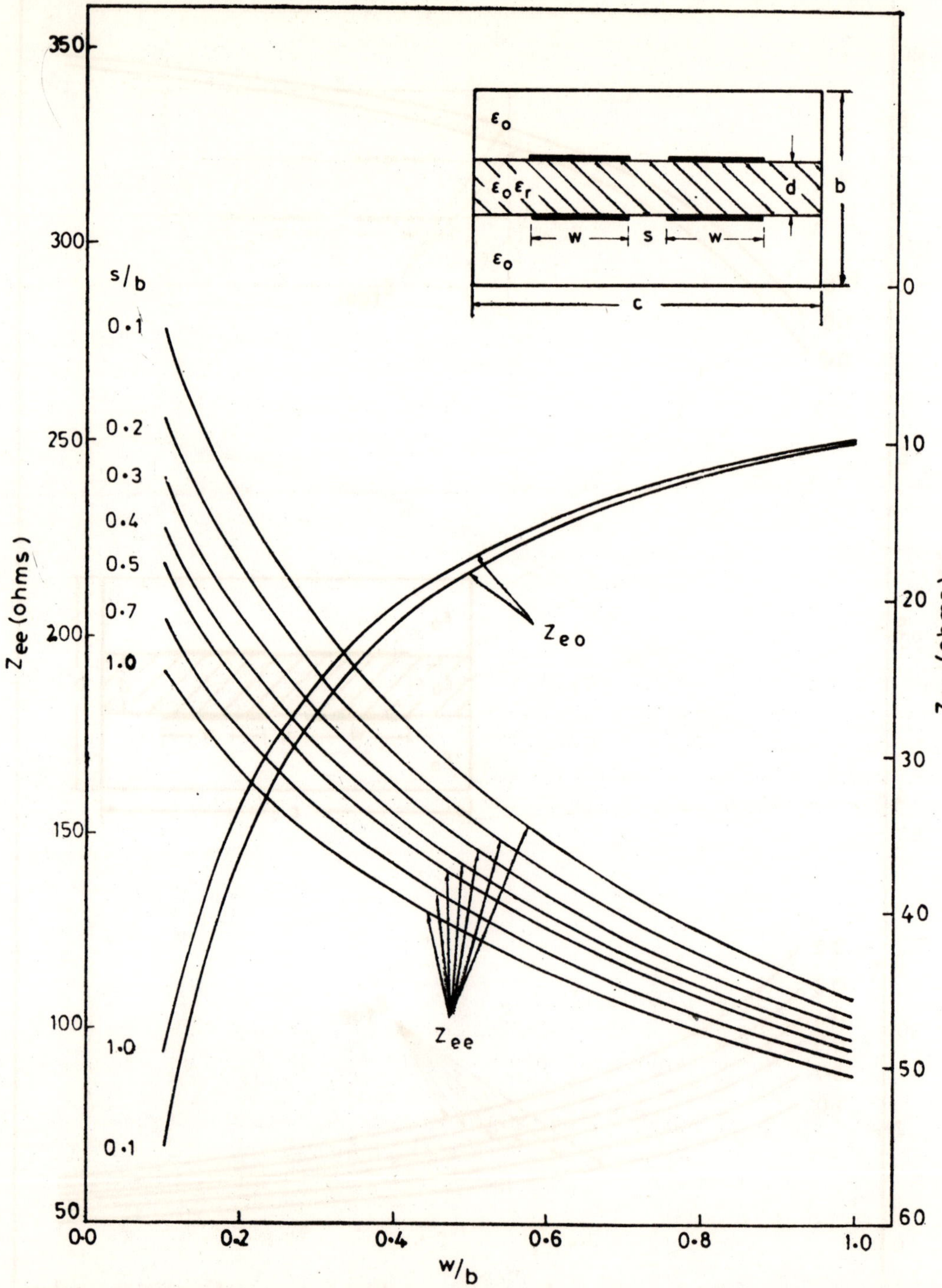

FIG. 8.6(a) Z_{ee} and Z_{eo} versus w/b with s/b as parameter of broadside-edge coupled suspended stripline; $\epsilon_r = 9.6$, $d/b = 0.2$, $c/(2w + s) = 20$.

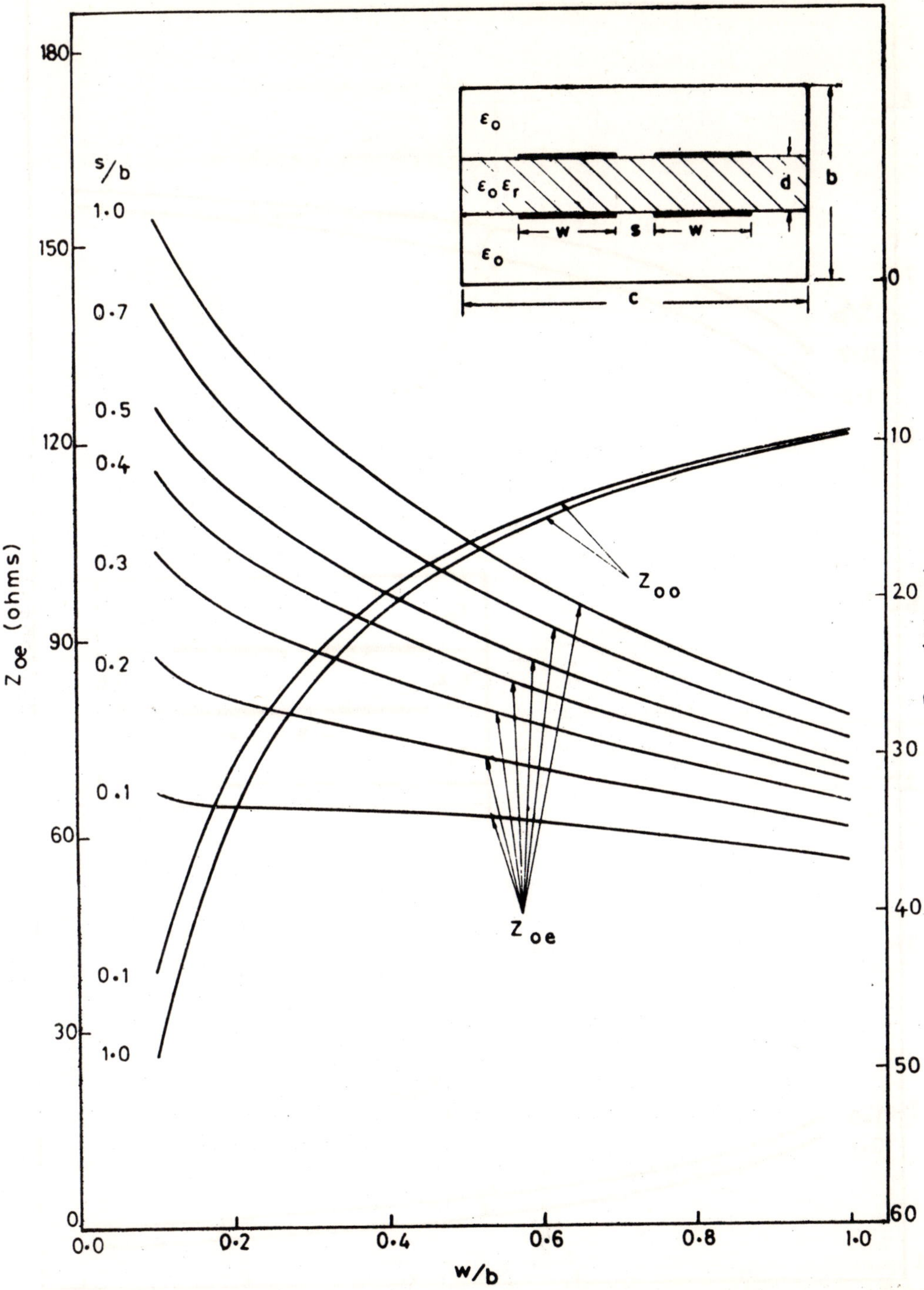

FIG. 8.6(b) Z_{oe} and Z_{oo} versus w/b with s/b as parameter of broadside-edge coupled suspended stripline; $\epsilon_r = 9.6$, $d/b = 0.2$, $c/(2w + s) = 20$.

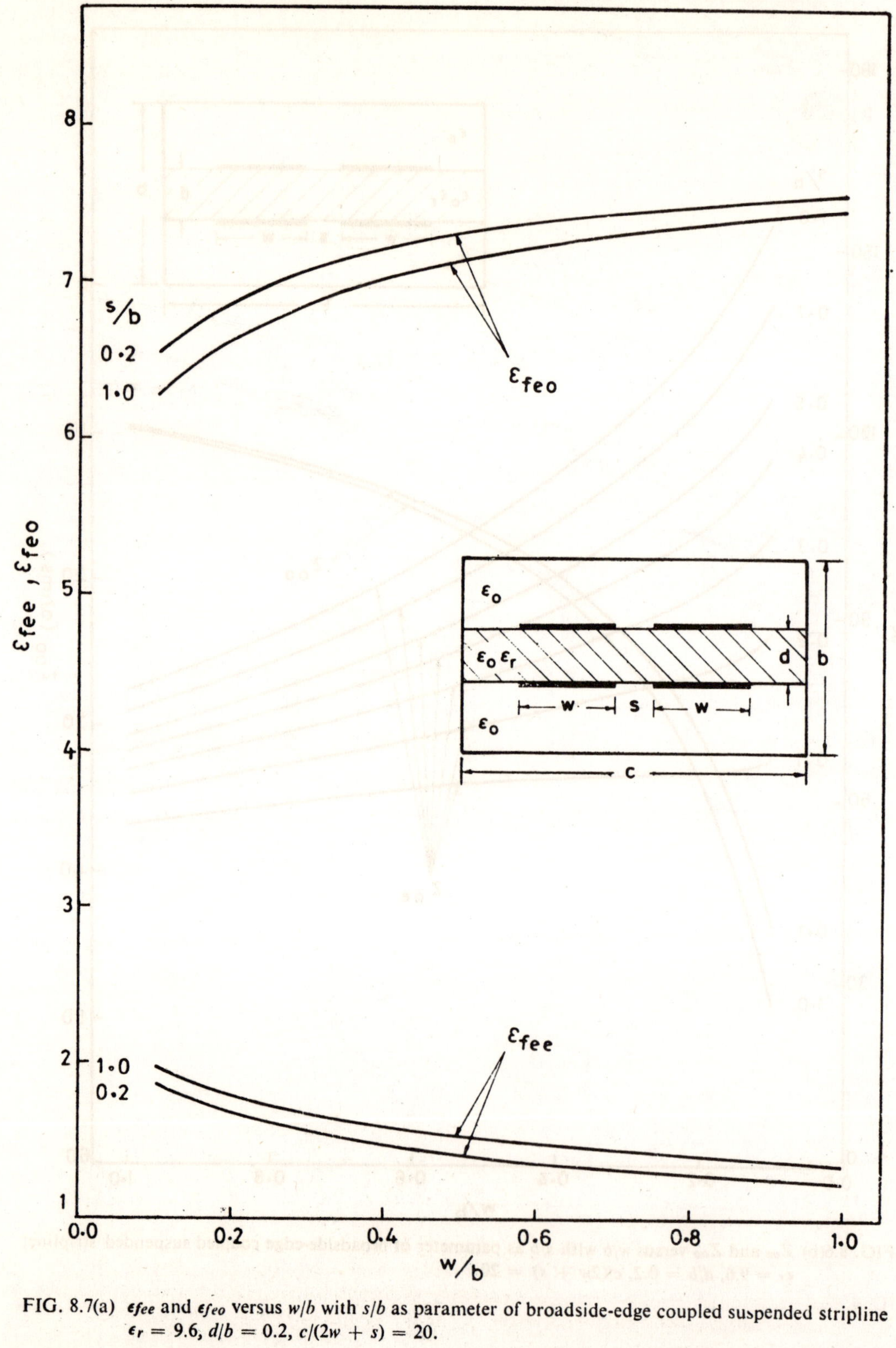

FIG. 8.7(a) ϵ_{fee} and ϵ_{feo} versus w/b with s/b as parameter of broadside-edge coupled suspended stripline $\epsilon_r = 9.6$, $d/b = 0.2$, $c/(2w + s) = 20$.

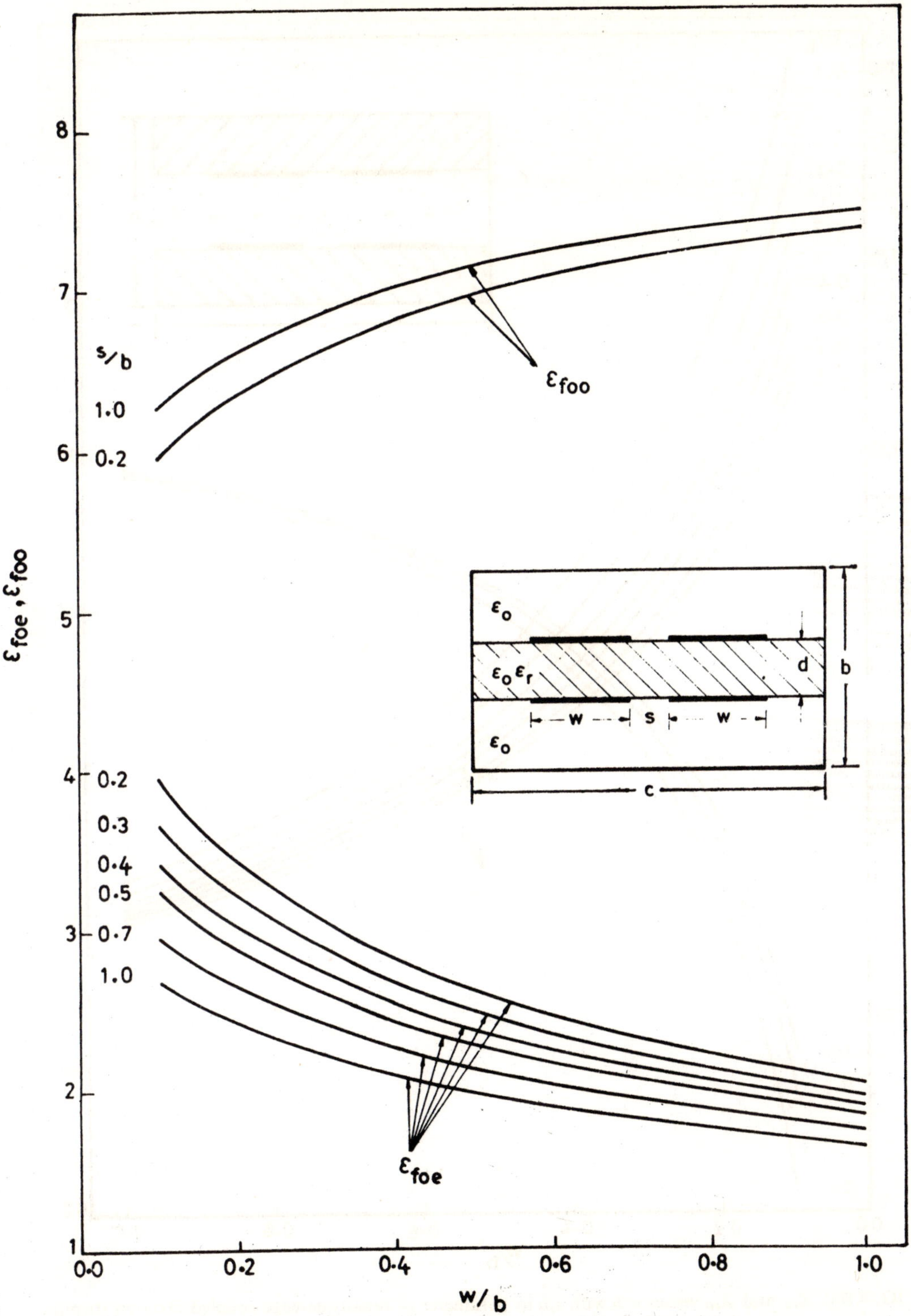

FIG. 8.7(b) ϵ_{foe} and ϵ_{foo} versus w/b with s/b as parameter of broadside-edge coupled suspended stripline; $\epsilon_r = 9.6$, $d/b = 0.2$, $c/(2w + s) = 20$.

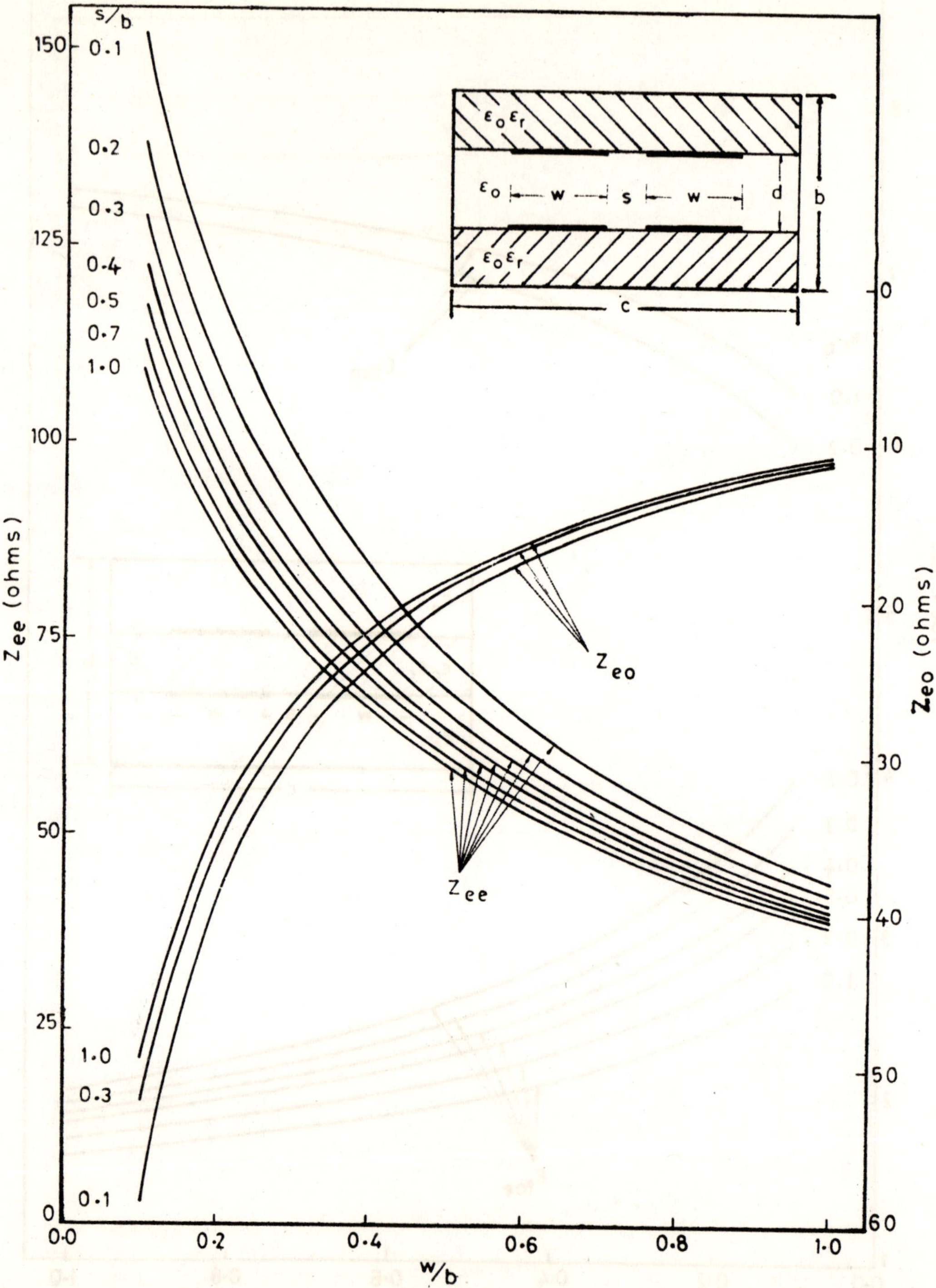

FIG. 8.8(a) Z_{ee} and Z_{eo} versus w/b with s/b as parameter of broadside-edge coupled inverted stripline; $\epsilon_r = 9.6$, $d/b = 0.1$, $c/(2w + s) = 20$.

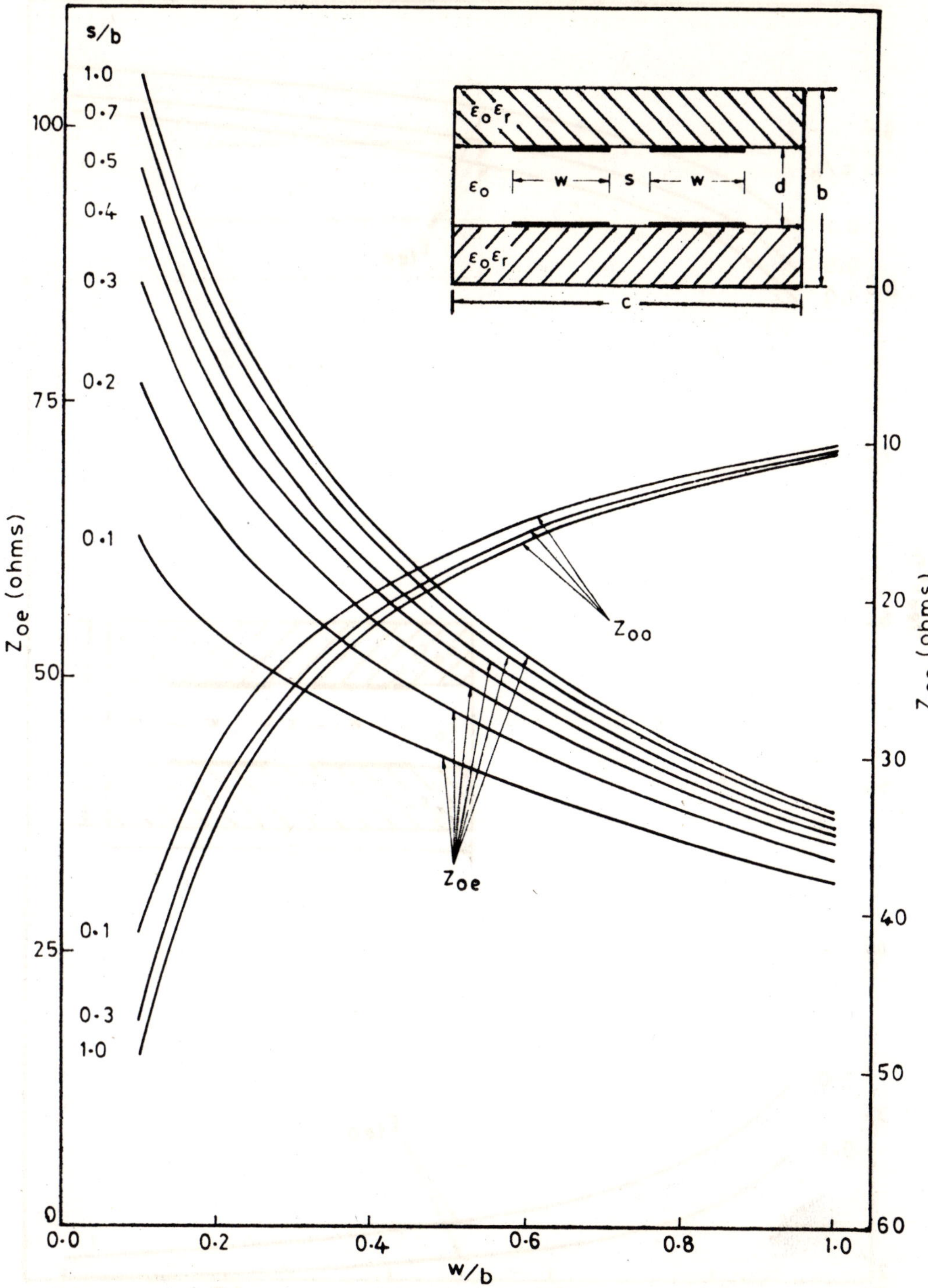

FIG. 8.8(b) Z_{oe} and Z_{oo} versus w/b with s/b as parameter of broadside-edge coupled inverted stripline; $\epsilon_r = 9.6$, $d/b = 0.1$, $c/(2w + s) = 20$.

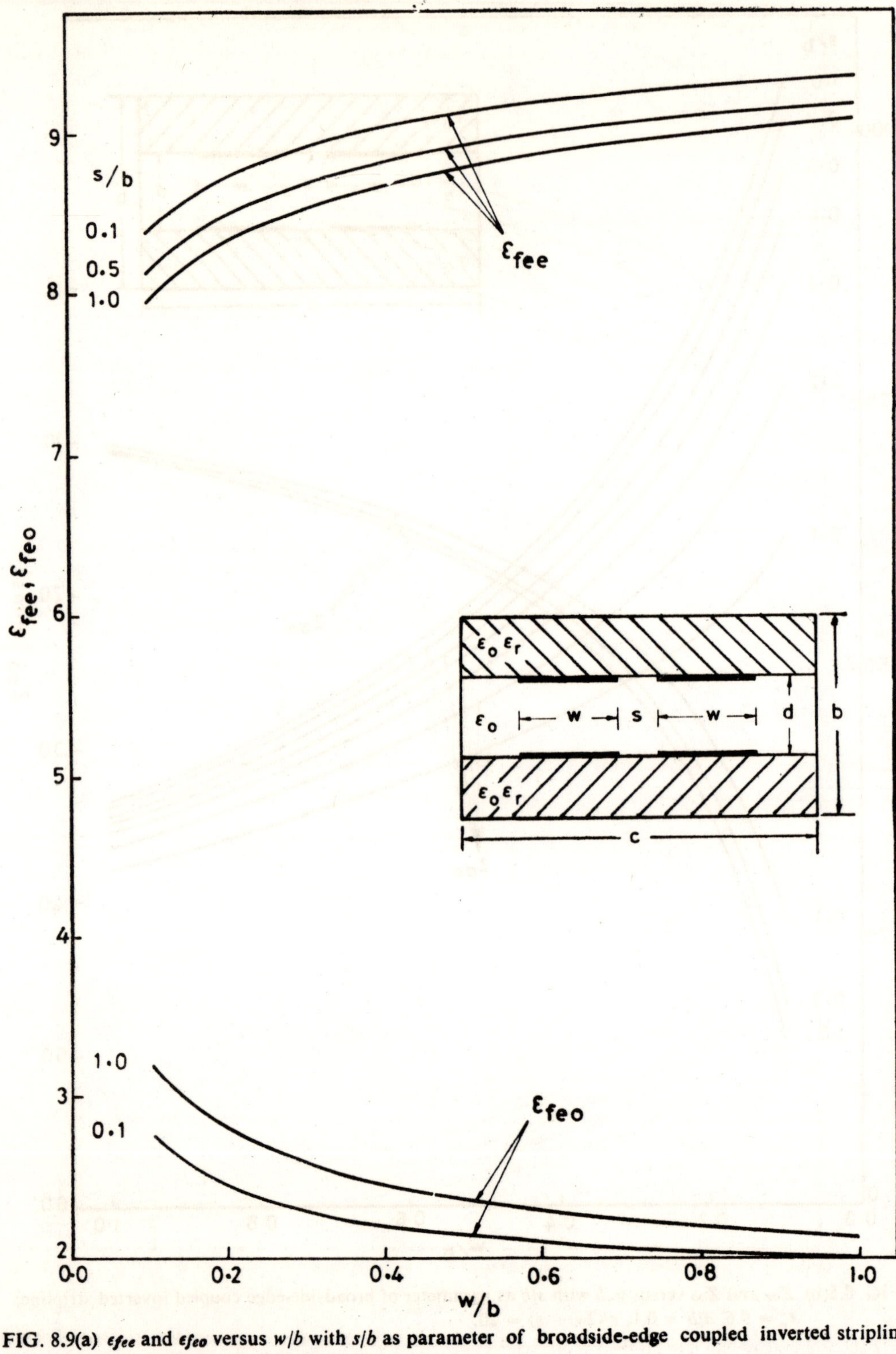

FIG. 8.9(a) ϵ_{fee} and ϵ_{feo} versus w/b with s/b as parameter of broadside-edge coupled inverted stripline; $\epsilon_r = 9.6$, $d/b = 0.1$, $c/(2w + s) = 20$.

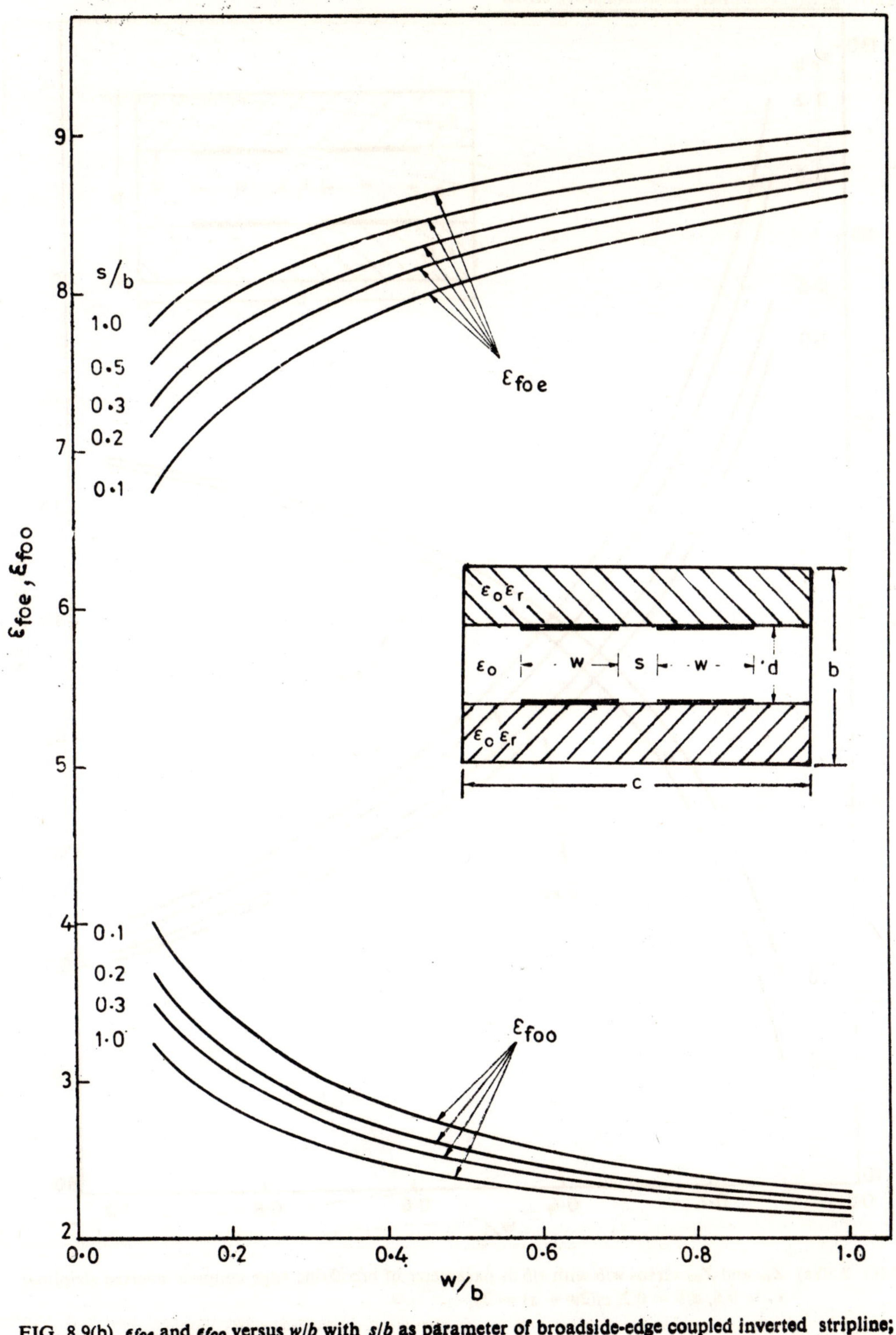

FIG. 8.9(b) ϵ_{foe} and ϵ_{foo} versus w/b with s/b as parameter of broadside-edge coupled inverted stripline; $\epsilon_r = 9.6$, $d/b = 0.1$, $c/(2w + s) = 20$.

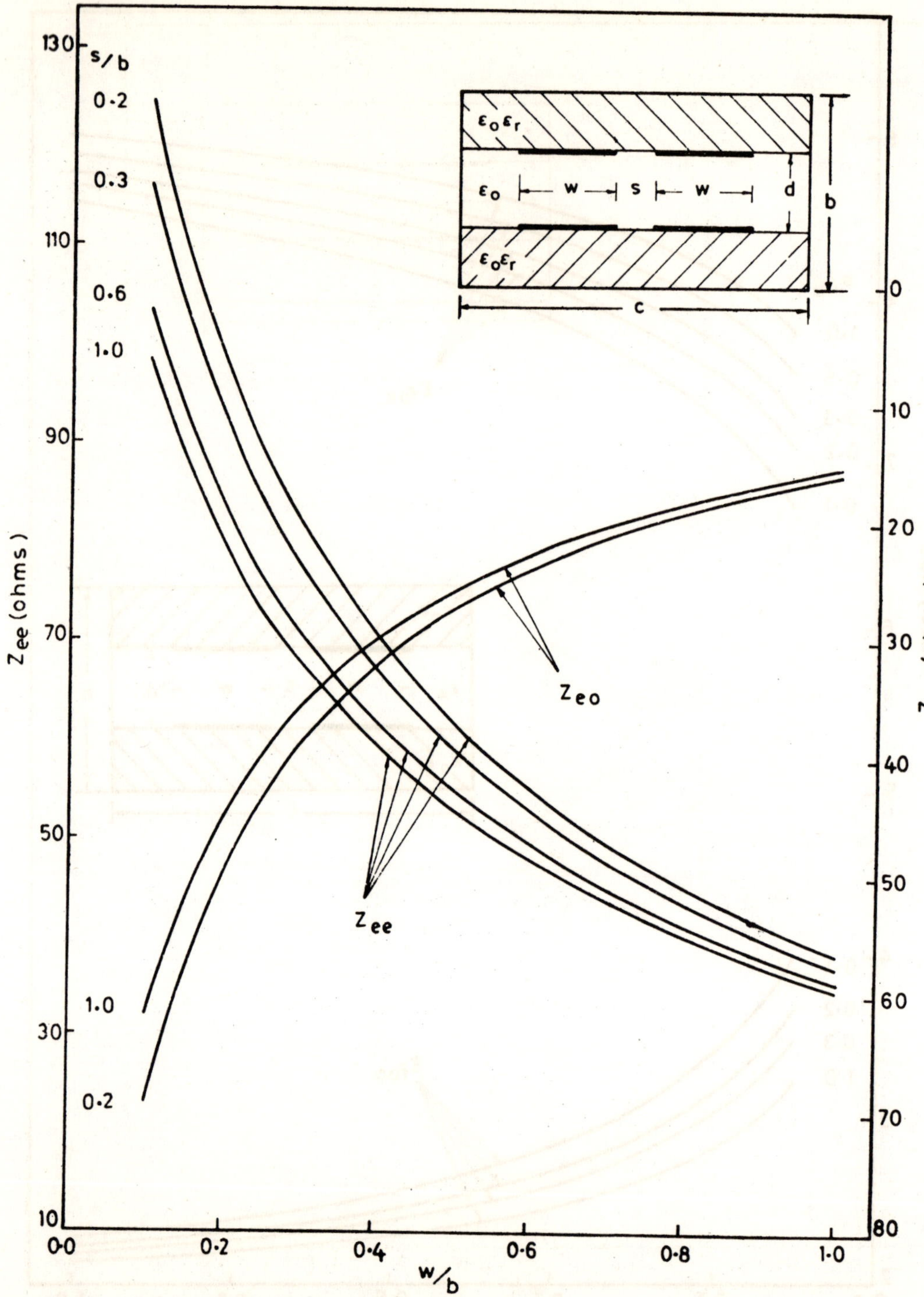

FIG. 8.10(a) Z_{ee} and Z_{eo} versus w/b with s/b as parameter of broadside-edge coupled inverted stripline; $\epsilon_r = 9.6$, $d/b = 0.2$, $c/(2w + s) = 20$.

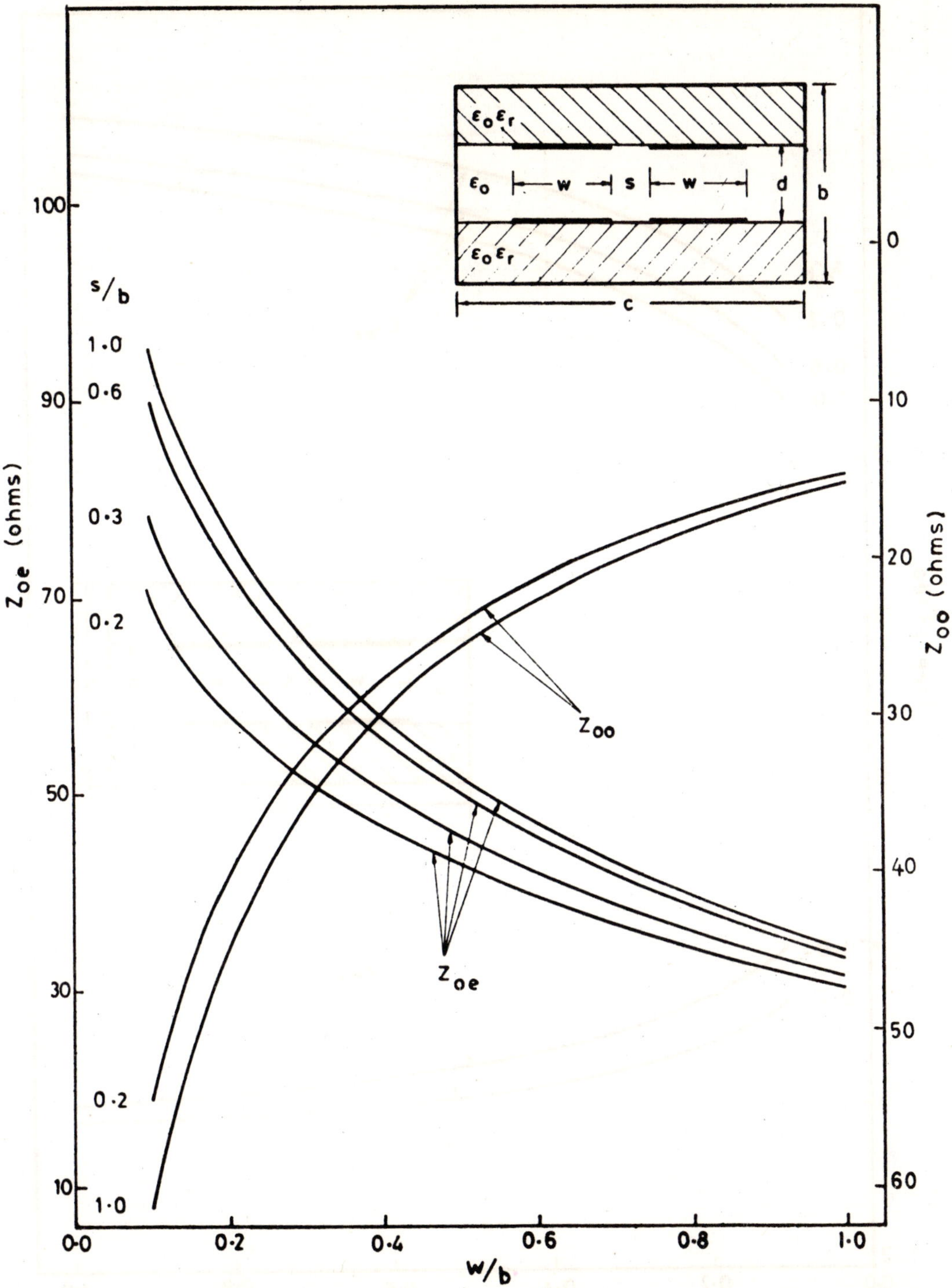

FIG. 8.10(b) Z_{oe} and Z_{oo} versus w/b with s/b as parameter of broadside-edge coupled inverted stripline; $\epsilon_r = 9.6$, $d/b = 0.2$, $c/(2w + s) = 20$.

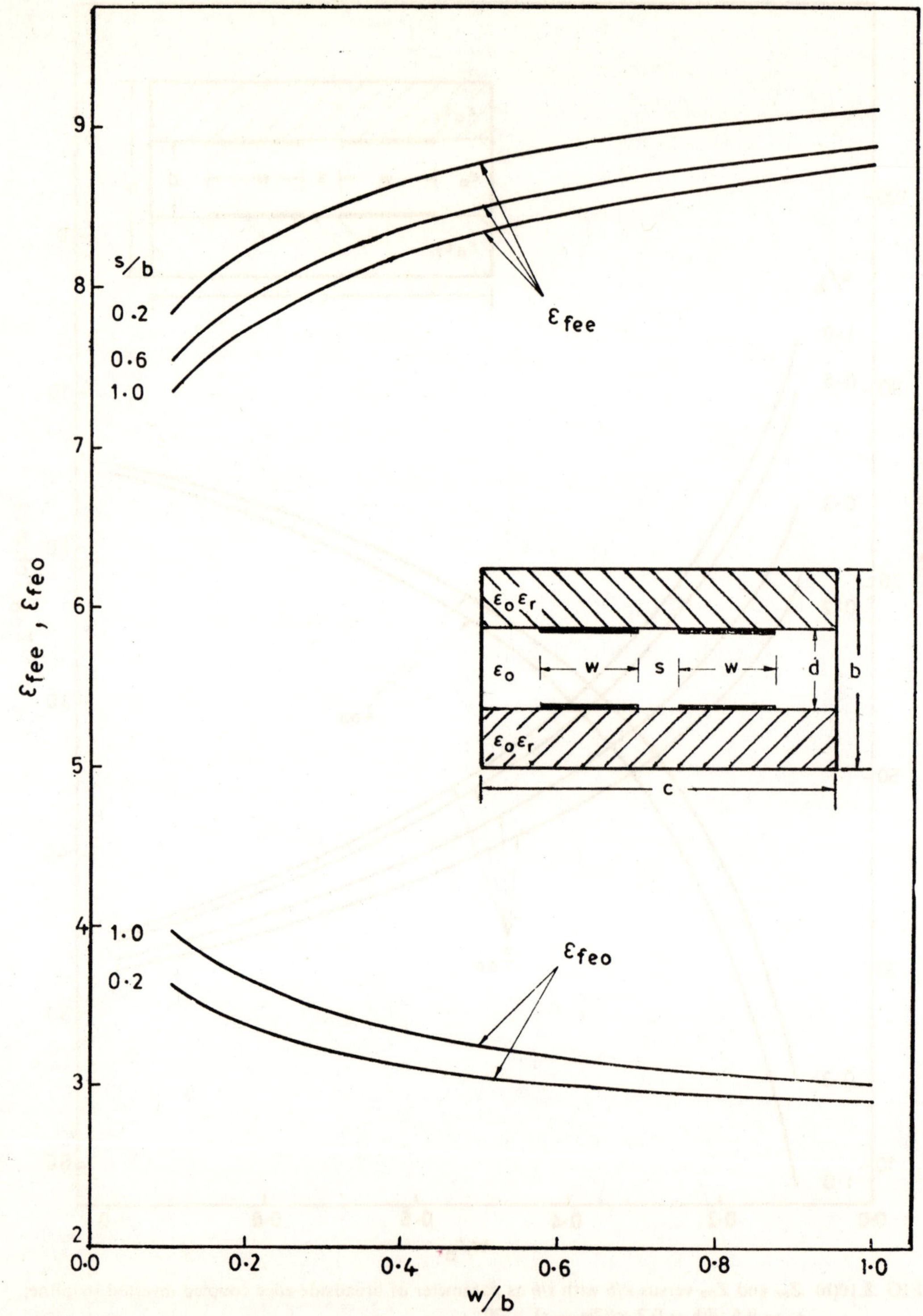

FIG. 8.11(a) ϵ_{fee} and ϵ_{feo} versus w/b with s/b as parameter of broadside-edge coupled inverted stripline; $\epsilon_r = 9.6$, $d/b = 0.2$, $c/(2w + s) = 20$.

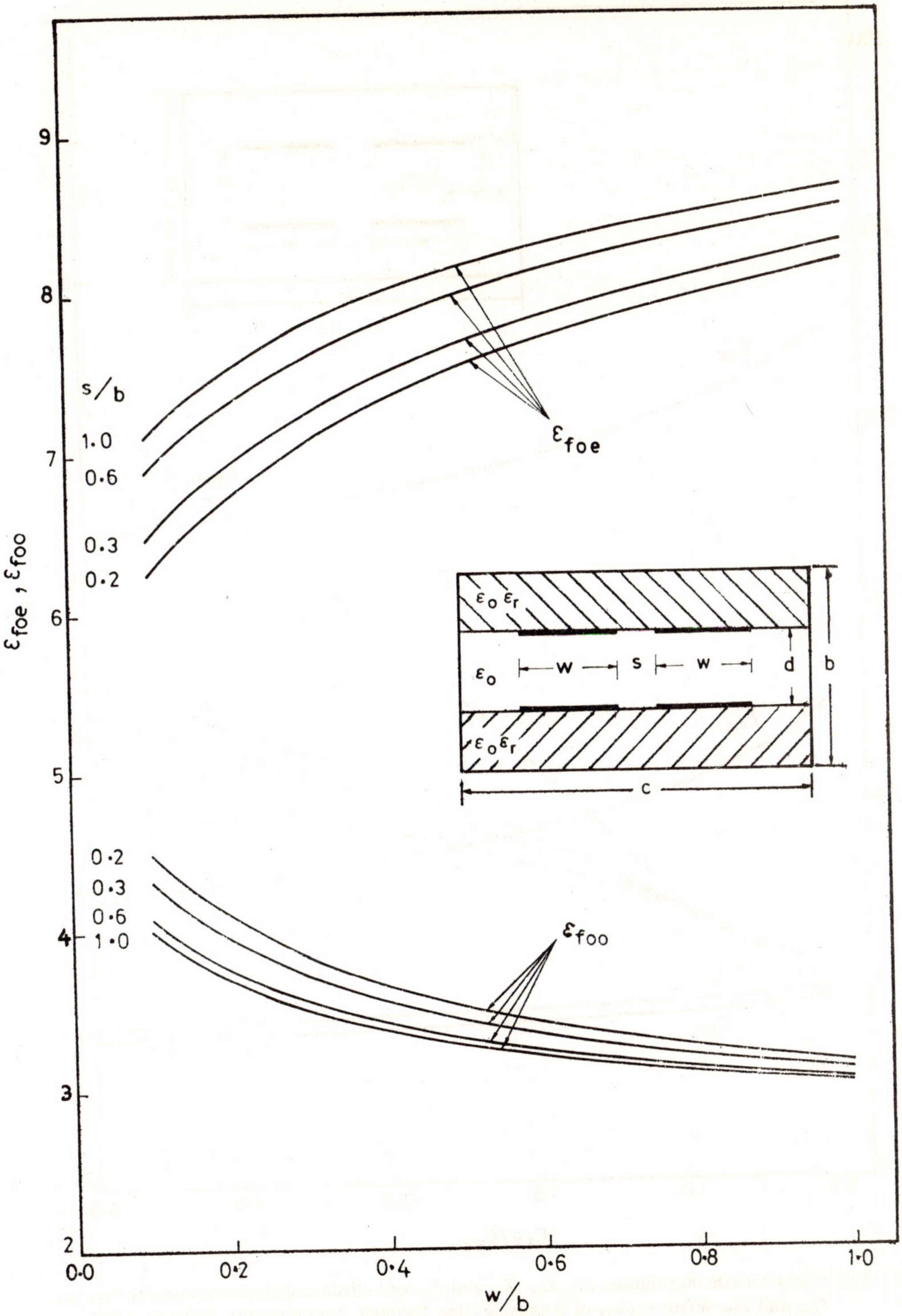

FIG. 8.11(b) ϵ_{foe} and ϵ_{foo} versus w/b with s/b as parameter of broadside-edge coupled inverted stripline; $\epsilon_r = 9.6$, $d/b = 0.2$, $c/(2w+s) = 20$.

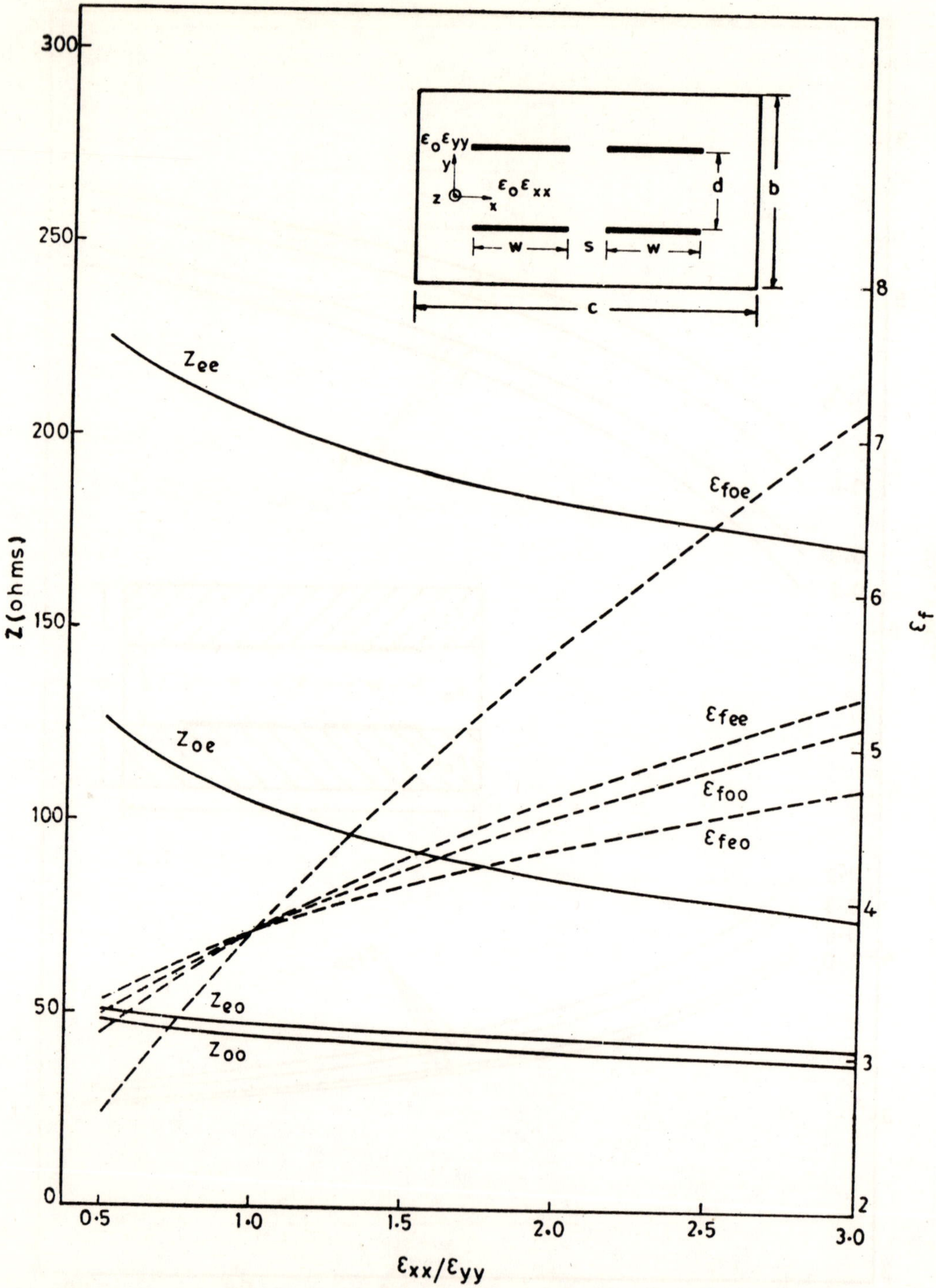

FIG. 8.12 Characteristic impedances Z_{ee}, Z_{eo}, Z_{oe} and Z_{oo} and effective dielectric constants ϵ_{fee}, ϵ_{feo}, ϵ_{foe} and ϵ_{foo} versus $\epsilon_{xx}/\epsilon_{yy}$ of broadside-edge coupled homogeneous stripline filled with uniaxially anisotropic dielectric; $\epsilon_{yy} = 3.78$, $w/b = d/b = 0.1$, $s/b = 0.2$, $c/(2w + s) = 20$.

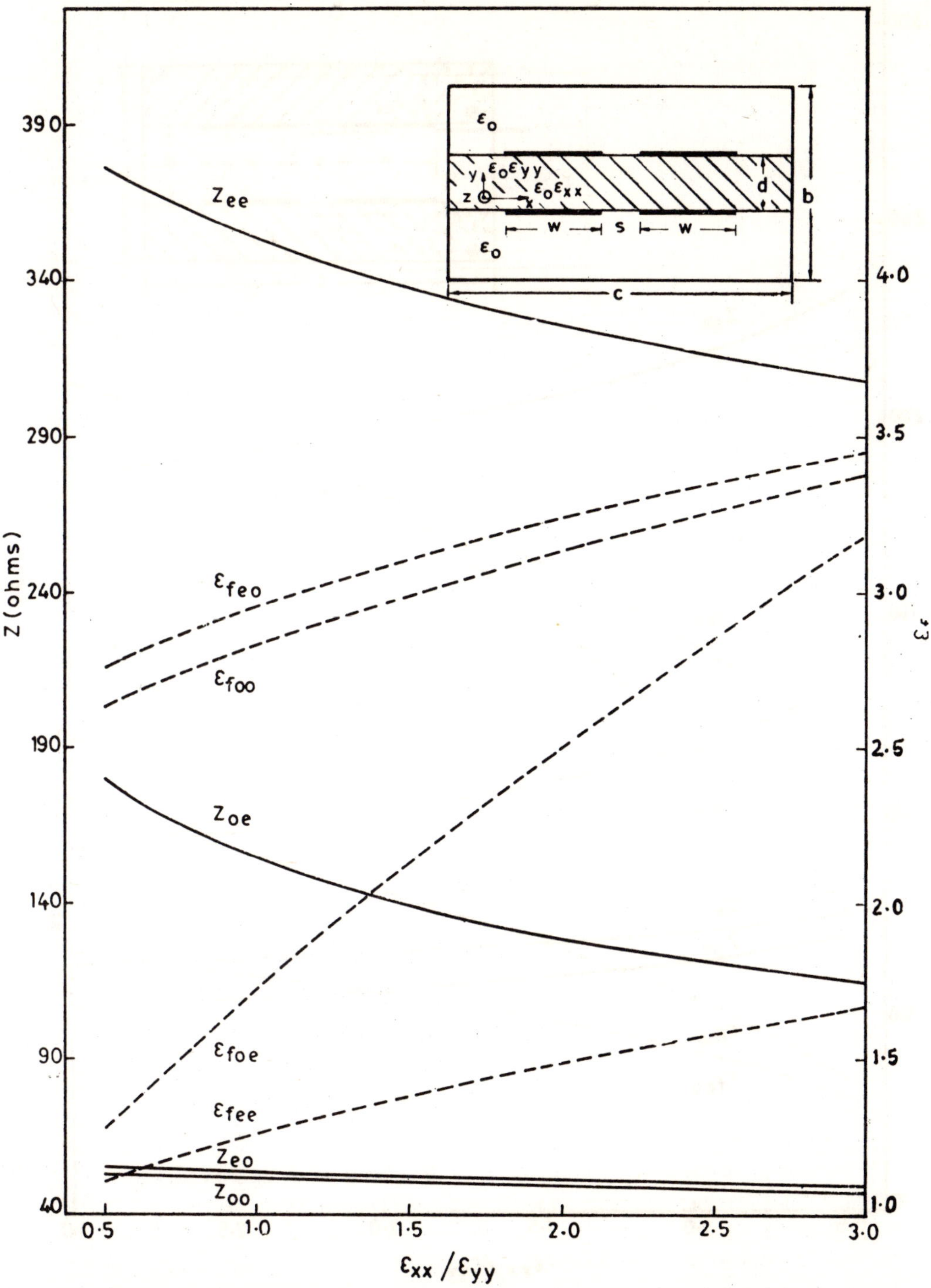

FIG. 8.13 Characteristic impedances Z_{ee}, Z_{eo}, Z_{oe} and Z_{oo} and effective dielectric constants ϵ_{fee}, ϵ_{feo}, ϵ_{foe} and ϵ_{foo} versus $\epsilon_{xx}/\epsilon_{yy}$ of broadside-edge coupled suspended stripline using uniaxially anisotropic dielectric substrate; $\epsilon_{yy} = 3.78$, $w/b = d/b = 0.1$, $s/b = 0.2$, $c/(2w + s) = 20$.

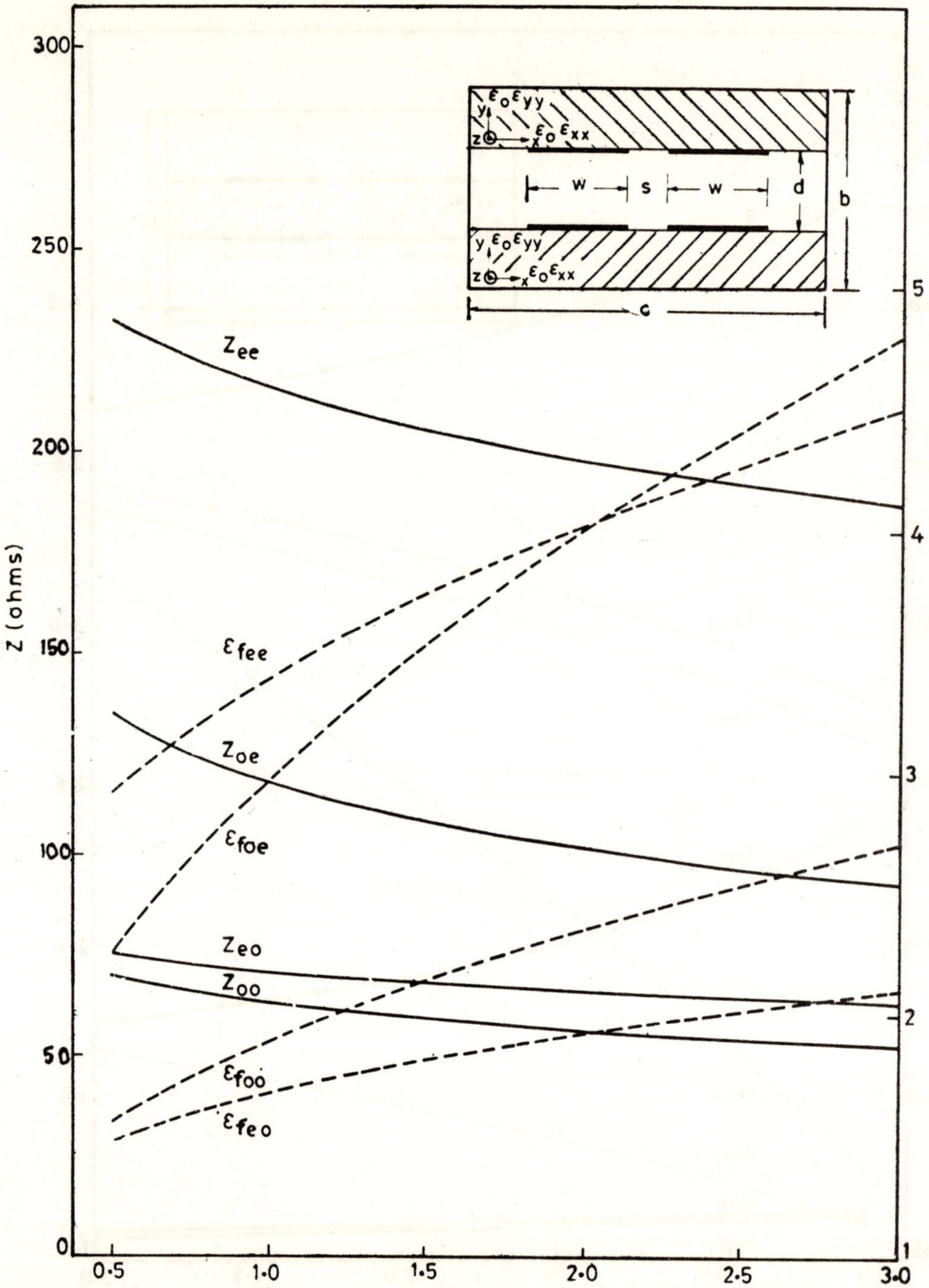

FIG. 8.14 Characteristic impedances Z_{ee}, Z_{eo}, Z_{oe} and Z_{oo} and effective dielectric constants ϵ_{fee}, ϵ_{feo}, ϵ_{foe} and ϵ_{foo} versus $\epsilon_{xx}/\epsilon_{yy}$ of broadside-edge coupled inverted stripline using uniaxially anisotropic dielectric substrates; $\epsilon_{yy} = 3.78$, $w/b = d/b = 0.1$, $s/b = 0.2$, $c/(2w + s) = 20$.

decreasing order (Fig. 8.12). As expected for a homogeneous structure, the four curves cross each other at $\epsilon_{xx}/\epsilon_{yy} = 1$. In the broadside-edge coupled suspended stripline, the values of ϵ_{feo} and ϵ_{foo} are higher than those of ϵ_{foe} and ϵ_{fee} for $\epsilon_{xx}/\epsilon_{yy}$ in the range 0.5 and 3.0 (Fig. 8.13), whereas the converse is true for the broadside-edge coupled inverted stripline (Fig. 8.14).

8.5 Special Structures

8.5.1 Broadside-Offset Coupled Homogeneous Stripline

The edge-coupled transmission structures described in Chapter 6 are generally used for loose couplings of the order of 8 db or more. For tighter couplings, the gap between the adjacent edges of the two conductors becomes too small to be fabricated accurately. For very tight coupling of the order of 3 db, the broadside-coupled structures described in Chapter 7 are ideal. Intermediate values can be achieved by increasing the spacing between the broadside-coupled strip conductors. This, however, may not always be convenient in practice, especially when it is to be integrated with other circuit components requiring smaller spacing between the conductors. An easy alternative proposed by Shelton [9] is to offset the broadside-coupled strip conductors while maintaining the vertical spacing between them constant as shown in Fig. 8.15. The structure is referred here as the broadside-offset coupled stripline. In the tight coupling condition, the coupled strips overlap over a width w_c as shown in Fig. 8.15a. The tightest coupling occurs when the two strips overlap com-

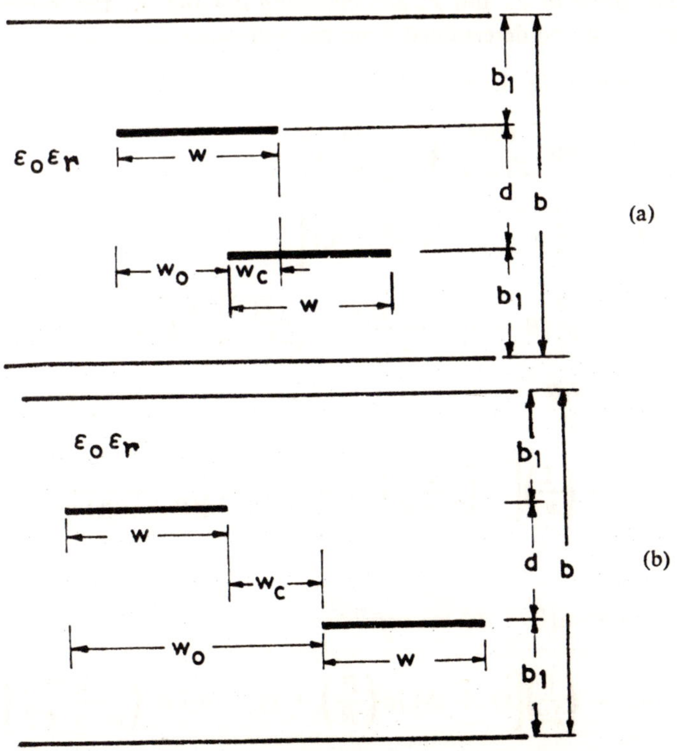

FIG. 8.15 Broadside-offset coupled homogeneous stripline. (a) Tight coupling, (b) Loose coupling

pletely ($w_c = w$). For this special case, the structure reduces to the broadside-coupled homogeneous stripline. In the loose coupling conditions, the two strip conductors are displaced outward so that a gap exists between the inner edges of the two strips (Fig. 8.15b). Since the structure offers considerable flexibility in varying the degree of coupling, it is ideal in the design of ultra-broadband couplers using either multi-section coupled lines with varied degrees of coupling [10] or tapered coupled lines [11].

Design Equations: For a broadside-offset coupled stripline homogeneously filled with an isotropic dielectric (Fig. 8.15), Shelton [9] has derived approximate expressions for the even- and odd-mode impedances in terms of the dimensional parameters. In the following, we summarize the design equations reported by Shelton for both tight and loose coupling cases. We first define,

$$\rho = Z_e/Z_o \tag{8.11a}$$

$$C' = 120\pi\sqrt{\rho}/(\sqrt{\epsilon_r}Z_0) \tag{8.11b}$$

$$Z_t = \sqrt{Z_e Z_o} \tag{8.11c}$$

where Z_e and Z_o are the even- and odd-mode impedances of the structure and ϵ_r is the relative dielectric constant of the dielectric medium.

Let

$$d_1 = d/b,\ w_1 = w/b,\ w_{c1} = w_c/b \tag{8.12}$$

For specified values of Z_e and Z_o and choosing d/b and ϵ_r, the normalized strip width w_1 and overlap w_{c1} can be determined from the following set of equations.

(i) Tight Coupling (Fig. 8.15a)

$$A = \exp\left[\frac{60\pi^2}{Z_t\sqrt{\epsilon_r}}\left(\frac{1 - \rho d_1}{\sqrt{\rho}}\right)\right] \tag{8.13a}$$

$$B = \frac{1}{2}\left[A - 2 + \sqrt{A^2 - 4A}\right] \tag{8.13b}$$

$$P = \frac{1}{2}\left[(B - 1)\frac{(1 + d_1)}{2} + \sqrt{\left(\frac{1 + d_1}{2}\right)^2 (B - 1)^2 + 4d_1 B}\right] \tag{8.13c}$$

$$R = \frac{d_1 B}{P} \tag{8.13d}$$

$$C_{fo} = \left(\frac{1}{\pi}\right)\left[-\left(\frac{2}{1 - d_1}\right)\ln d_1 + \frac{1}{d_1}\ln\left\{\frac{PR}{(P + d_1)(1 + P)(R - d_1)(1 - R)}\right\}\right] \tag{8.13e}$$

$$w_1 = d_1(1 - d_1)\frac{(C' - C_{fo})}{2} \tag{8.13f}$$

$$w_{o1} = \left(\frac{1}{2\pi}\right)\left[(1 + d_1)\ln\left(\frac{P}{R}\right) + (1 - d_1)\ln\left(\frac{1 + P}{d_1 + P}\right)\left(\frac{R - d_1}{1 - R}\right)\right] \tag{8.13g}$$

$$w_{c1} = w_1 - w_{o1} \tag{8.13h}$$

These formulas are valid under the approximations given by,

$$\frac{w_1}{1-d_1} \geqslant 0.35, \qquad \frac{w_{c1}}{d_1} \geqslant 0.7 \tag{8.13i}$$

(ii) Loose Coupling (Fig. 8.15b)

$$\Delta C = \frac{120\pi(\rho - 1)}{Z_t\sqrt{\epsilon_r \rho}} \tag{8.14a}$$

$$k = \frac{1}{\left[\exp\left(\frac{\pi \Delta C}{2}\right) - 1\right]} \tag{8.14b}$$

$$a = \sqrt{\left(\frac{d_1 - k}{d_1 + 1}\right)^2 + k} - \left(\frac{d_1 - k}{d_1 + 1}\right) \tag{8.14c}$$

$$q = k/a \tag{8.14d}$$

$$C_{fo} = \left(\frac{2}{\pi}\right)\left[\left(\frac{1}{1+d_1}\right)\ln\left\{\frac{(1+a)}{a(1-q)}\right\} - \left(\frac{1}{1-d_1}\right)\ln q\right] \tag{8.14e}$$

$$C_f(a=\infty) = -\left(\frac{2}{\pi}\right)\left[\left(\frac{1}{1+d_1}\right)\ln\left(\frac{1-d_1}{2}\right) + \left(\frac{1}{1-d_1}\right)\ln\left(\frac{1+d_1}{2}\right)\right] \tag{8.14f}$$

$$w_{c1} = \left(\frac{1}{\pi}\right)\left[d_1 \ln\left(\frac{q}{a}\right) + (1-d_1)\ln\left(\frac{1-q}{1+a}\right)\right] \tag{8.14g}$$

$$w_1 = \frac{(1-d_1^2)}{4}[C' - C_{fo} - C_f(a=\infty)] \tag{8.14h}$$

These equations are valid under the approximations given by

$$w_1/(1-d_1) \geqslant 0.35, \qquad 2(w_1 - w_{c1})/(1+d_1) \geqslant 0.85 \tag{8.14i}$$

In the case of a coupler, the voltage coupling coefficient C_v between the coupled lines is related to the even- and odd-mode impedances via the equations

$$Z_e = Z_t\sqrt{\frac{1+C_v}{1-C_v}} \tag{8.15a}$$

$$Z_o = Z_t\sqrt{\frac{1-C_v}{1+C_v}} \tag{8.15b}$$

where $Z_t = \sqrt{Z_e Z_o}$ is the characteristic impedance of the input and output lines which are matched to the coupler. If the coupling is specified in db, the coupling coefficient C_v is obtained from,

$$C_v = 10^{-\left(\frac{C_v \text{ in db}}{20}\right)} \tag{8.16}$$

Design data: Using Shelton's formulas, Mosko [10] has presented design curves showing the voltage coupling coefficient C_v versus w/b and w_c/b for various values of d/b and Z_t. Plots are shown for two dielectric materials; namely, polyolefin having $\epsilon_r \doteq 2.32$ and rexolite having $\epsilon_r = 1.65$. Howe [12] has also presented design data relating the even-mode impedances to the strip conductor width and the overlap for three different values of d and b. Data are tabulated for various dielectric materials, namely, Polyolefin ($\epsilon_r = 2.32$), microfibre teflon fibre-glass ($\epsilon_r = 2.4$), cross-linked polystyrene ($\epsilon_r = 2.53$), polyphenelene oxide or *ppo*($\epsilon_r = 2.55$) and glass filled polystyrene ($\epsilon_r = 2.62$).

Almost all of the above-mentioned materials are presently being replaced by the flexible glass reinforced teflon trade named as RT-duroid. Homogeneously filled strip transmission lines invariably make use of RT-duroid having $\epsilon_r = 2.22$. Design data generated using formulas (8.11) through (8.16) with $\epsilon_r = 2.22$ are tabulated in Tables 8.1-8.12 for 12 different values of d/b. The dimensional parameters w and w_c are specified for coupling values ranging from about 3 dB upto 40 dB. In the tables, positive values of w_c refer to tight coupling condition for which the coupled strips overlap (Fig. 8.15a) and negative values of w_c refer to loose coupling condition as shown in Fig. 8.15b.

8.5.2 Broadside-Offset Coupled Inhomogeneous Stripline

The configuration of the broadside-offset coupled inhomogeneous stripline is shown in Fig. 8.16. This structure offers additional flexibility in the sense that coupling can be varied by varying the spacing s between the strip conductors, keeping all other structural parameters fixed. In the limiting case ($s \to 0$), this structure reduces to the broadside-coupled suspended stripline (structure A2 of Table 4.4). Extensive design data for this structure has already been presented in Chapter 7. The broadside-offset coupled inhomogeneous stripline shown in Fig. 8.16 (with $c \to \infty$) has been analysed by Kitazawa et al. [13] using the variational method in space domain. Figure 8.17a and b show the variations of the even-and odd-mode effective dielectric constants and characteristic impedances, respectively as a function of s/w. We may observe that the even-mode impedance decreases and odd-mode impedance increases as s/w is increased. Further, the even-mode dielectric constant increases and odd-mode effective dielectric constant decreases as s/w is increased. However, the change in even-mode effective dielectric constant as s/w is increased is negligible.

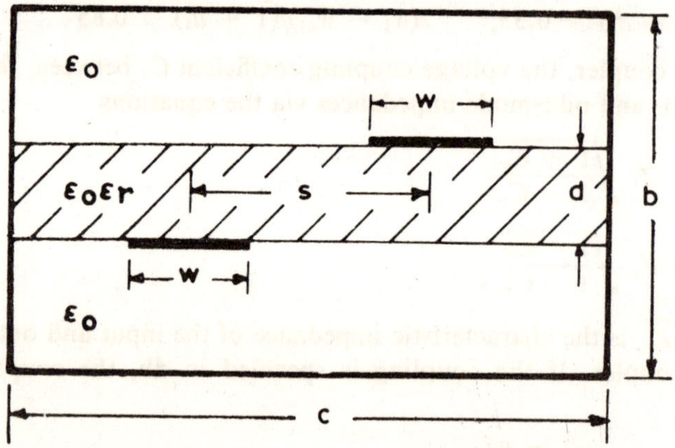

FIG. 8.16 Broadside-offset coupled inhomogeneous stripline

8.5.3 Unsymmetric Broadside-Coupled Inhomogeneous Stripline

The cross-section of an unsymmetric broadside-coupled inhomogeneous stripline is shown in Fig. 8.18. This structure is believed to be useful in the design of unsymmetrical broadside-

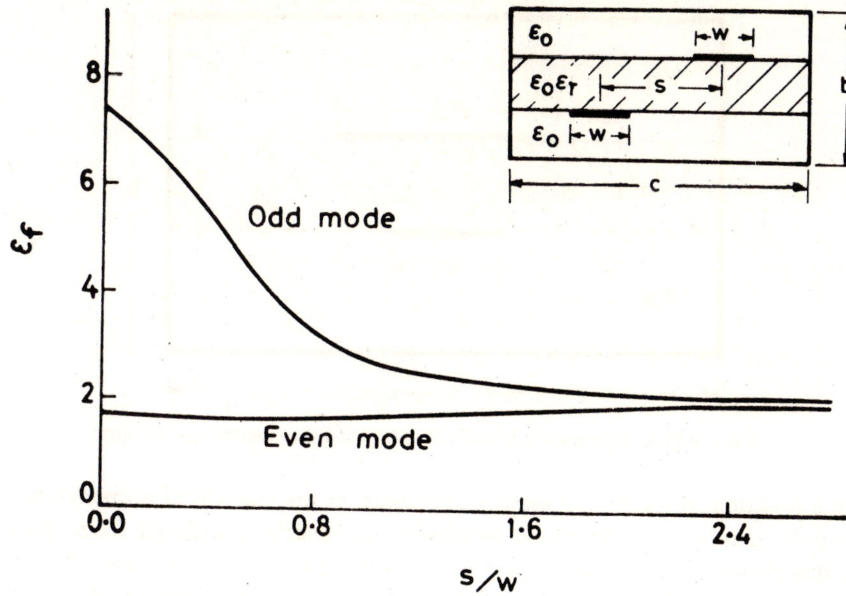

(a)

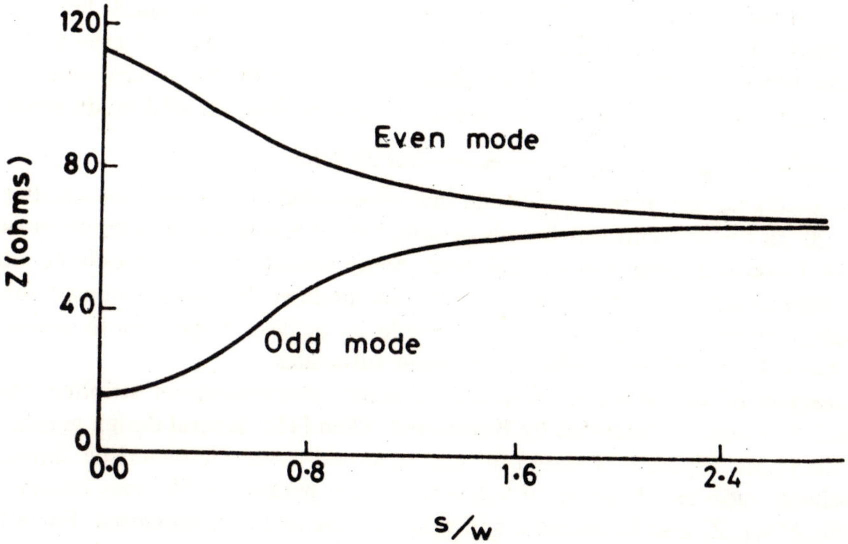

(b)

FIG. 8.17(a) Even-and odd-mode effective dielectric constants versus s/w
(b) Even-and odd-mode characteristic impedances versus s/w
$b/w = 2.0$, $d/w = 0.4$, $c/w = \infty$.
(From Kitazawa et al. [13], Electronics Letters, 1982, Vol. 10, pp. 425-426, reprinted with permission of IEE)

coupled line directional couplers in suspended substrate geometry. A method for determining the propagation parameters of this structure is reported by Chao [14]. After determining

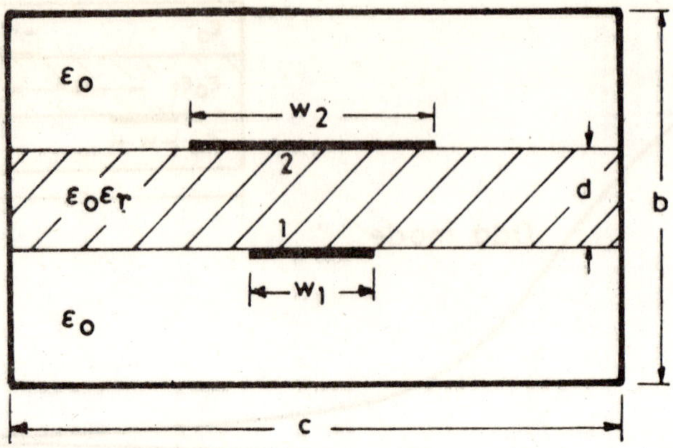

FIG. 8.18 Unsymmetric broadside-coupled inhomogeneous stripline.

the Green's function, the self and mutual capacitances are calculated numerically. The even- and odd-mode admittances and phase velocities are finally calculated from these self and mutual capacitances. For a fixed value of d, ϵ_r and w_2/b, Fig. 8.19 a shows the variations of the even-and odd-mode admittances as a function of w_1/b. The superscripts 1 and 2 refer to the two strip conductors. We may observe that the admittances $Y_e^1(=1/Z_e^1)$, $Y_o^1(=1/Z_o^1)$ and $Y_o^2(=1/Z_o^2)$ increase, whereas $Y_e^2(=1/Z_e^2)$ decreases as w_1/b is increased. For $w_1/b = w_2/b$, the admittance $Y_e^1 = Y_e^2$ and $Y_o^1 = Y_o^2$. The variations of normalized even-and odd-mode phase velocities as a function of w_1/b are plotted in Fig. 8.19b. For a fixed value of d, ϵ_r and w_2/b, the normalized even-mode phase velocity is greater than the odd-mode phase velocity.

8.5.4 Modified Broadside-Coupled Inhomogeneous Stripline

The basic configuration of the modified broadside-coupled inhomogeneous stripline is shown in Fig. 8.20. In the conventional broadside-coupled inhomogeneous stripline (structures A2 and A3 of Table 4.4), large even-to odd-mode phase velocity ratios can only be obtained by utilizing high dielectric constant substrates. In the modified broadside-coupled inhomogeneous stripline shown in Fig. 8.20, it is possible to achieve large even-to odd-mode phase velocity ratios by using lower dielectric constant substrates.

The analysis of the modified broadside-coupled inhomogeneous stripline based on the variational technique is presented by Barnes and Allen [15]. Several design graphs are available in this paper. Figure 8.21 shows the even-and odd-mode characteristic impedances and phase velocity ratio as a function of w/b with ϵ_{r1} as a parameter. We may observe that for a fixed value of ϵ_{r1}, Z_e and Z_o decrease and v_e/v_o increase as w/b is increased. For a fixed value of w/b, Z_o decreases and v_e/v_o increases as ϵ_{r1} increases from 2.35 to 40. For the structural parameters considered in Fig. 8.21, it may be observed that Z_e is invariant to a change in ϵ_{r1} for $w/b > 0.05$. For $w/b < 0.05$, Z_e decreases as ϵ_{r1} is increased.

8.5.5 Coupled Stripline with Interleaved Strip Conductors

The configuration of the coupled stripline with interleaved strip conductors (Fig. 8.22) was first proposed by Getsinger [16] for achieving close but variable coupling. In this, the two strip conductors of width w are maintained at the same potential. These form a single transmission line coupled to the transmission line formed by the interleaving strip conductor S_1.

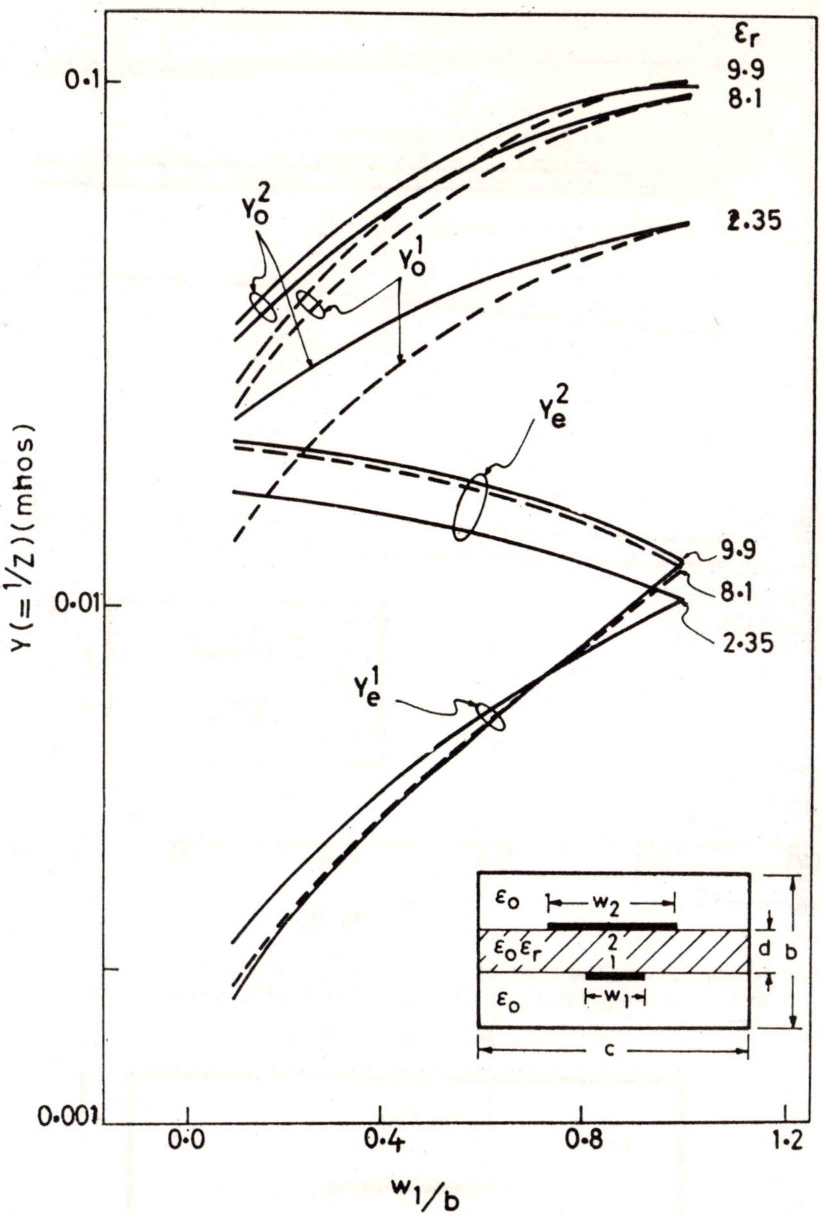

FIG. 8.19(a) Even-and odd-mode admittances as a function of w_1/b; $d/b = 0.2$, $c/b = 5.0$, $w_2/b = 1.0$.
(From Chao [14], Copyright © 1975 IEEE, reprinted with permission).

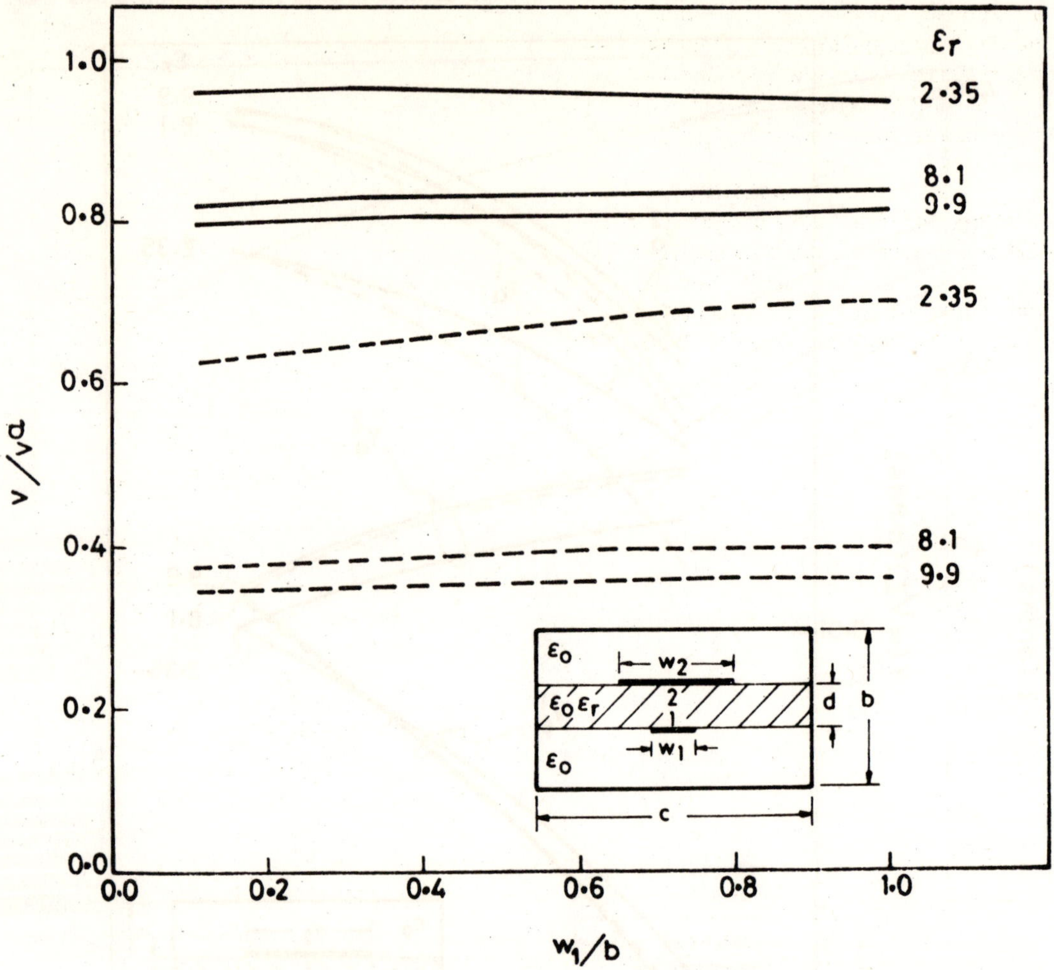

FIG. 8.19(b) Normalized even-and odd-mode phase velocity as a function of w_1/b; $d/b = 0.2$, $c/b = 5.0$, $w_2/b = 1.0$. (From Chao [14], Copyright © 1975 IEEE, reprinted with permission)

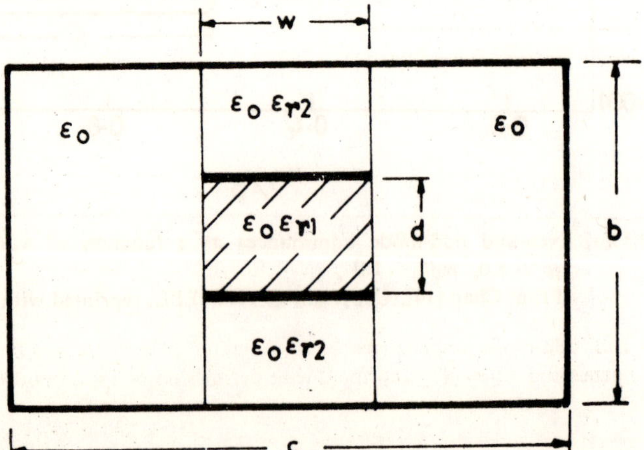

FIG. 8.20 Modified broadside-coupled inhomogeneous stripline

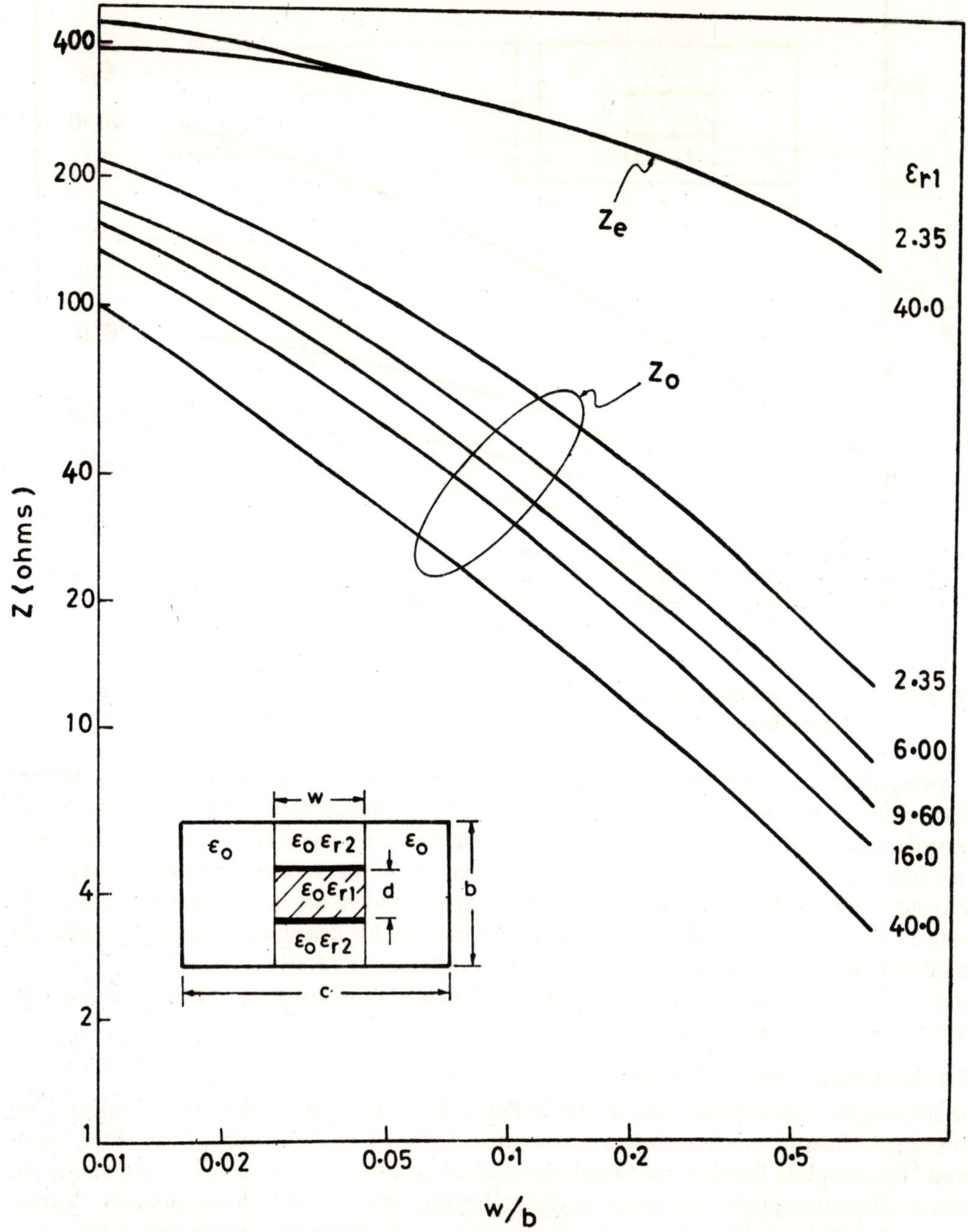

FIG. 8.21(a) Even- and odd-mode impedances as a function of w/b; $c/b = 4.0$, $d/b = 0.1$, $\epsilon_{r2} = 1.0$ (From Barnes and Allen [15], reprinted with permission of Taylor and Francis Ltd).

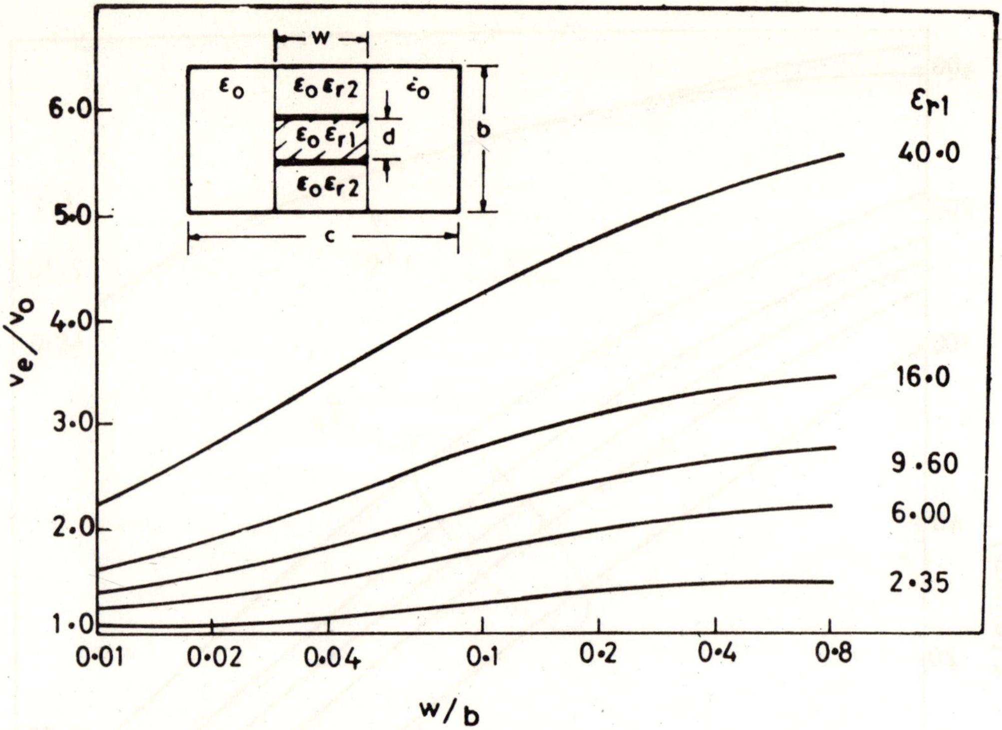

FIG. 8.21(b) Even-to odd-mode phase velocity ratio as a function of w/b; $c/b = 4.0$, $d/b = 0.1$, $\epsilon_{r2} = 1.0$. (From Barnes and Allen [15], reprinted with permission of Taylor and Francis Ltd)

A symmetrical construction with respect to the ground plane is used, in order to prevent radiation into a parallel-plate mode between the ground planes. Using the conformal mapping technique, Getsinger [16] has derived simple design formulas relating the even- and odd-mode impedances/capacitances to the dimensional parameters. Detailed design graphs of even- and odd-mode capacitances have been given which are to be used along with the formulas. Getsinger's formulas in consolidated form and his design data on the capacitances reproduced accurately in graphical form are available in Matthaei et al. [17]. Since the design curves cover almost the entire range of practical interest, no attempt is made to provide additional data.

8.5.6 *Slit-Coupled Strip Transmission Line*

The slit-coupled transmission line shown in Fig. 8.23 is basically a broadside-coupled strip transmission line, in which an intermediate ground plane having a longitudinal slit is introduced. The coupling between the broadside-coupled strips can be varied by varying the slit width *d*. This structure is very useful in the realization of multi-section components. Yamamoto et al. [18] have derived formulas for the even- and odd-mode impedances of the structure using the conformal mapping technique. The formulas are exact and are expressed in terms of the elliptic integral and Jacobian functions. A much simpler set of formulas for the special case when the spacing between the broadside-coupled strips is equal to half-the ground plane spacing as shown in Fig. 8.23 has been reported by Matthaei et al. [17]. They are reproduced below.

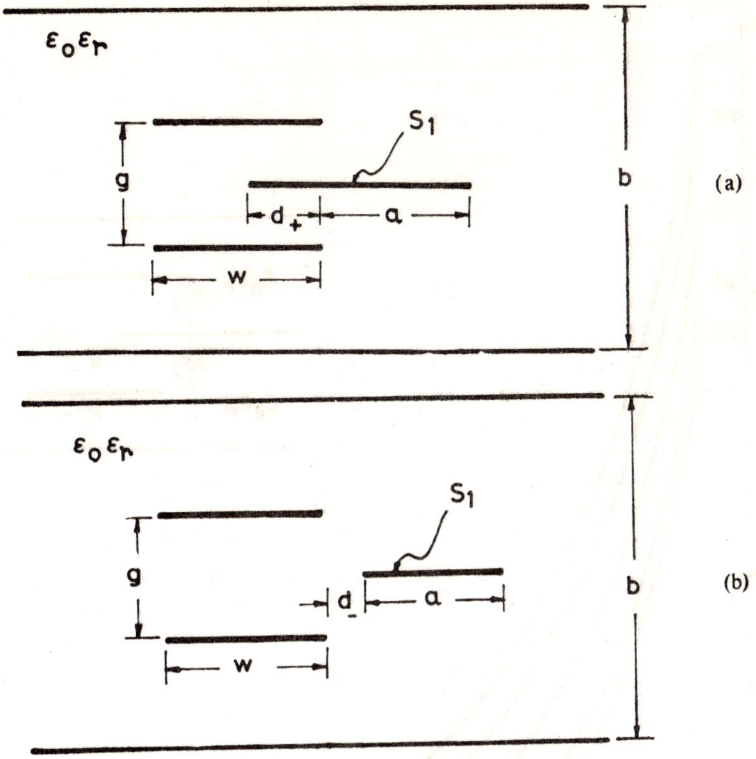

FIG. 8.22 Coupled stripline with interleaved strip conductors.

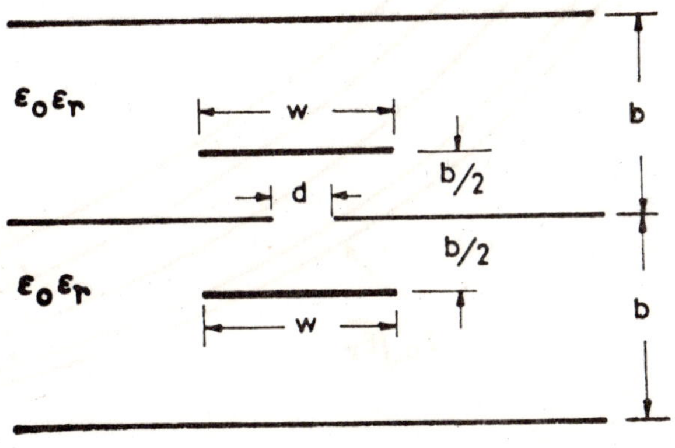

FIG. 8.23 Slit-coupled strip transmission line.

The odd-mode impedance Z_o of the structure shown in Fig. 8.24 is nearly the same as the characteristic impedance of a single conductor symmetric stripline with ground plane spacing b. This is given by

$$Z_o = \frac{30\pi}{\sqrt{\epsilon_r}} \frac{K(k_o)}{K(k_o')} \tag{8.17}$$

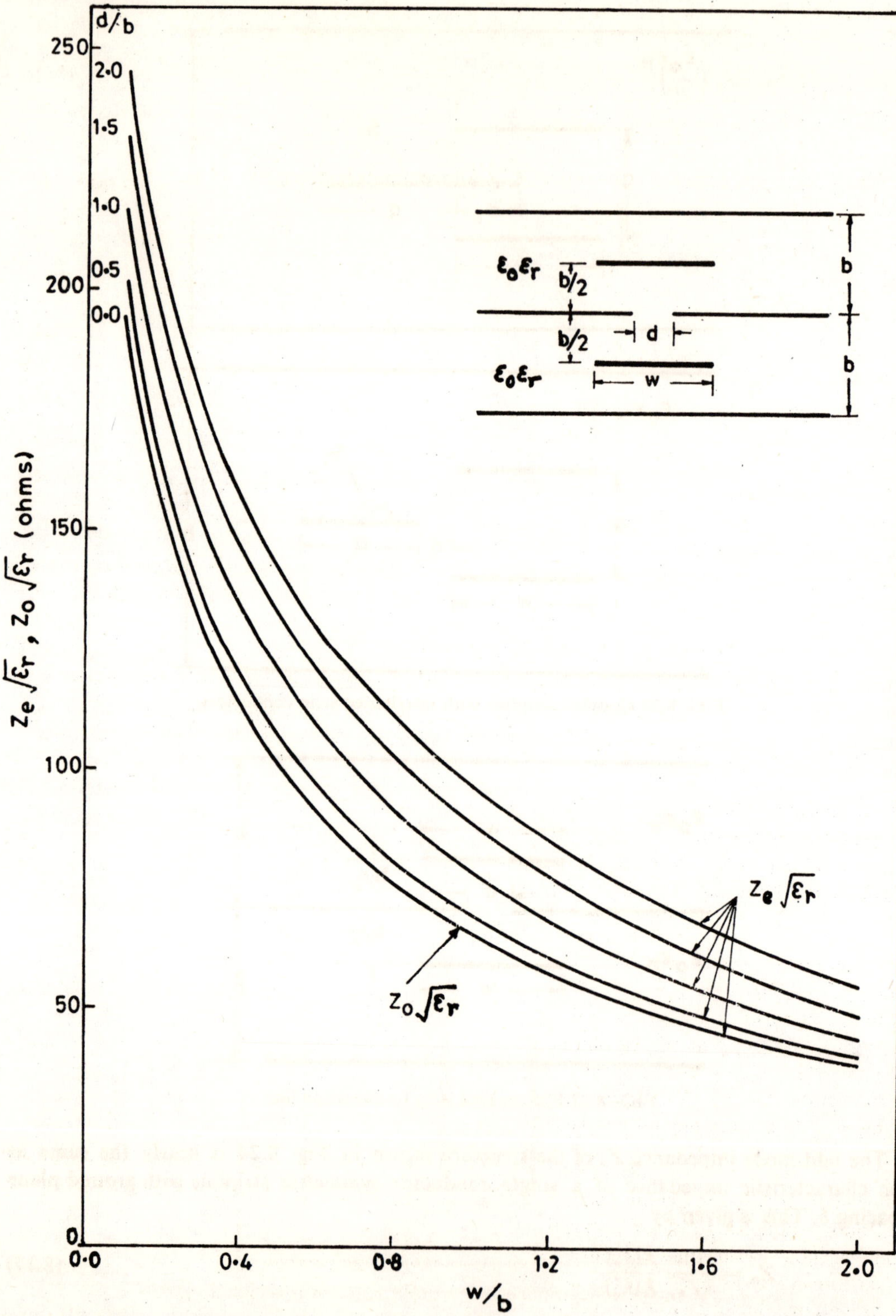

FIG. 8.24 The normalized characteristic impedances $Z_e\sqrt{\epsilon_r}$ and $Z_0\sqrt{\epsilon_r}$ versus w/b of slit-coupled strip transmission line.

where,

$$k_o = \text{sech}\left(\frac{\pi w}{2b}\right), \tag{8.18a}$$

$$k'_o = \sqrt{1 - k_o^2} \tag{8.18b}$$

The even-mode characteristic impedance Z_e is approximately given by,

$$\frac{1}{Z_e} = \frac{1}{2}\left[\frac{1}{Z_o} + \frac{1}{Z}\right] \tag{8.19}$$

where

$$Z = \frac{30\pi}{\sqrt{\epsilon_r}} \frac{K(k')}{K(k)} \tag{8.20a}$$

$$k = \sqrt{\frac{\cosh\left(\frac{\pi w}{b}\right) - 1}{\cosh \frac{\pi w}{b} + \cosh \frac{\pi d}{b}}} \tag{8.20b}$$

$$k' = \sqrt{1 - k^2} \tag{8.20c}$$

Figure 8.24 shows $Z_e\sqrt{\epsilon_r}$ as a function of w/b for various values of d/b. Since $Z_o\sqrt{\epsilon_r}$ is essentially independent of d/b, a single curve is superposed to show its variation with w/b.

8.5.7 Homogeneous Stripline with Coupled Vertical Strip Conductors

The cross-section of the homogeneous stripline with coupled vertical strip conductors is shown in Fig. 8.25. Cohn [19] has analysed the structure using the conformal mapping technique. The approximate formulas derived by Cohn for the even- and odd-mode characteristic impedances are given by,

$$Z_e\sqrt{\epsilon_r} = 188.344 \frac{K(k)}{K(k')} \tag{8.21a}$$

$$Z_o\sqrt{\epsilon_r} = \frac{295.849}{\left(\frac{b}{s}\cos^{-1} k + \ln \frac{1}{k}\right)} \tag{8.21b}$$

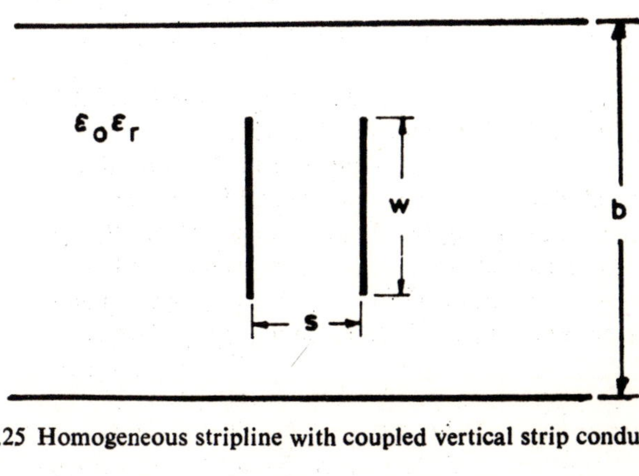

FIG. 8.25 Homogeneous stripline with coupled vertical strip conductors.

where $$k' = \sqrt{1 - k^2} \tag{8.21c}$$

and $$w/b = (2/\pi)\left[\tan^{-1}\left\{\frac{k'}{k}\sqrt{\frac{1-\left(\frac{k}{k'}\frac{s}{b}\right)}{1+\left(\frac{k'}{k}\frac{s}{b}\right)}}\right\} - \frac{s}{b}\tanh^{-1}\sqrt{\frac{1-\left(\frac{k}{k'}\frac{s}{b}\right)}{1+\left(\frac{k'}{k}\frac{s}{b}\right)}}\right] \tag{8.21d}$$

The inverse cosine and tangent functions are evaluated in radians between 0 and $\pi/2$. The above formulas are accurate for all values of s/b and w/b for values of w/s greater than 1.

Figure 8.26 shows the variations in $Z_e\sqrt{\epsilon_r}$ and $Z_o\sqrt{\epsilon_r}$ as a function of w/b for various values of s/b.

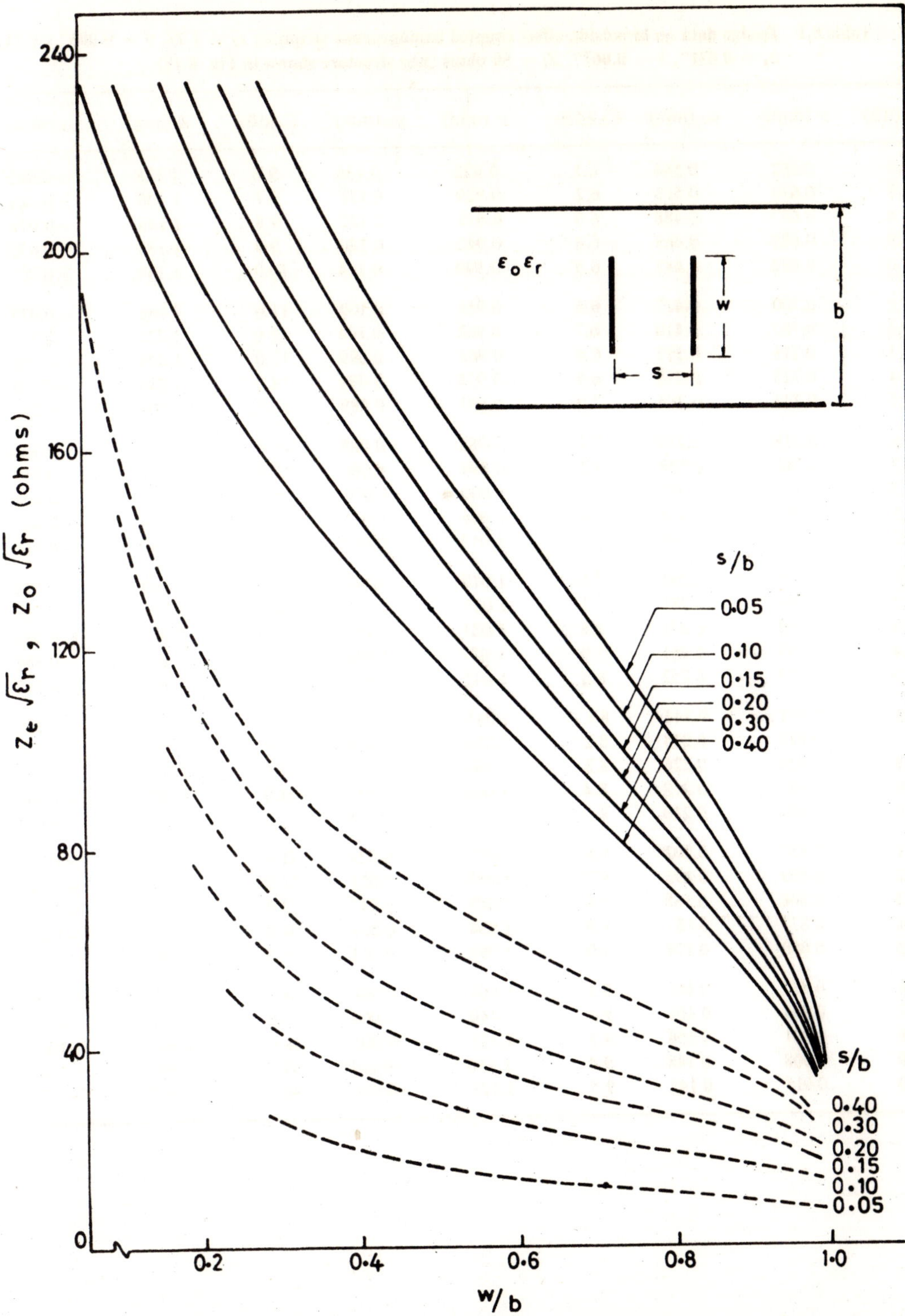

FIG. 8.26 The normalized characteristic impedances $Z_e\sqrt{\epsilon_r}$ and $Z_o\sqrt{\epsilon_r}$ versus w/b with s/b as parameter of homogeneous stripline with coupled vertical strip conductors; ———— $Z_e\sqrt{\epsilon_r}$, – – – $Z_o\sqrt{\epsilon_r}$

Table 8.1 Design data on broadside-offset coupled homogeneous stripline; $\epsilon_r = 2.22$, $d = 0.005''$, $b_1 = 0.031''$, $b = 0.067''$, $Z_t = 50$ ohms (See structure shown in Fig. 8.15)

C_v (dB)	w (mm)	w_c (mm)	C_v (dB)	w (mm)	w_c (mm)	C_v(dB)	w (mm)	w_c (mm)
2.6	0.662	0.544	6.1	0.922	0.136	9.6	1.125	−0.022
2.7	0.670	0.513	6.2	0.929	0.131	9.7	1.130	−0.026
2.8	0.677	0.488	6.3	0.935	0.125	9.8	1.135	−0.029
2.9	0.685	0.465	6.4	0.942	0.120	9.9	1.139	−0.033
3.0	0.692	0.445	6.5	0.949	0.114	10.0	1.144	−0.037
3.1	0.700	0.427	6.6	0.955	0.109	11.0	1.186	−0.075
3.2	0.707	0.410	6.7	0.962	0.104	12.0	1.223	−0.115
3.3	0.715	0.393	6.8	0.968	0.099	13.0	1.254	−0.158
3.4	0.723	0.378	6.9	0.975	0.094	14.0	1.280	−0.203
3.5	0.730	0.364	7.0	0.981	0.089	15.0	1.301	−0.250
3.6	0.738	0.350	7.1	0.988	0.085	16.0	1.319	−0.300
3.7	0.746	0.338	7.2	0.994	0.080	17.0	1.333	−0.352
3.8	0.754	0.325	7.3	1.000	0.076	18.0	1.345	−0.406
3.9	0.761	0.314	7.4	1.006	0.071	19.0	1.355	−0.461
4.0	0.769	0.302	7.5	1.013	0.067	20.0	1.362	−0.518
4.1	0.777	0.291	7.6	1.019	0.063	21.0	1.369	−0.576
4.2	0.784	0.281	7.7	1.025	0.058	22.0	1.373	−0.635
4.3	0.792	0.271	7.8	1.031	0.054	23.0	1.377	−0.694
4.4	0.799	0.261	7.9	1.037	0.050	24.0	1.381	−0.754
4.5	0.807	0.252	8.0	1.043	0.046	25.0	1.383	−0.815
4.6	0.814	0.243	8.1	1.048	0.043	26.0	1.385	−0.876
4.7	0.822	0.234	8.2	1.054	0.039	27.0	1.387	−0.937
4.8	0.829	0.226	8.3	1.060	0.035	28.0	1.388	−0.999
4.9	0.837	0.218	8.4	1.066	0.031	29.0	1.389	−1.060
5.0	0.844	0.210	8.5	1.071	0.028	30.0	1.390	−1.122
5.1	0.851	0.202	8.6	1.077	0.024	31.0	1.391	−1.184
5.2	0.859	0.195	8.7	1.083	0.021	32.0	1.391	−1.246
5.3	0.866	0.188	8.8	1.088	0.017	33.0	1.391	−1.308
5.4	0.873	0.181	8.9	1.094	0.014	34.0	1.392	−1.370
5.5	0.880	0.174	9.0	1.099	0.011	35.0	1.392	−1.433
5.6	0.877	0.167	9.1	1.104	0.008	36.0	1.392	−1.495
5.7	0.894	0.161	9.2	1.110	0.004	37.0	1.392	−1.557
5.8	0.901	0.154	9.3	1.115	0.001	38.0	1.393	−1.619
5.9	0.908	0.148	9.4	1.116	−0.014	39.0	1.393	−1.682
6.0	0.915	0.142	9.5	1.121	−0.018	40.0	1.393	−1.744

Table 8.2 Design data on broadside-offset coupled homogeneous stripline; $\epsilon_r = 2.22$, $d = 0.005''$, $b_1 = 0.062''$, $b = 0.129''$, $Z_t = 50$ ohms (See structure shown in Fig. 8.15)

C_v (dB)	w (mm)	w_c (mm)	C_v (dB)	w (mm)	w_c (mm)	C_v (dB)	w (mm)	w_c (mm)
1.6	0.851	0.819	5.4	1.602	0.125	9.2	2.116	−0.082
1.7	0.873	0.722	5.5	1.619	0.118	9.3	2.127	−0.087
1.8	0.895	0.671	5.6	1.635	0.112	9.4	2.137	−0.093
1.9	0.917	0.630	5.7	1.651	0.106	9.5	2.148	−0.098
2.0	0.939	0.595	5.8	1.667	0.099	9.6	2.158	−0.104
2.1	0.961	0.564	5.9	1.683	0.093	9.7	2.168	−0.109
2.2	0.983	0.536	6.0	1.699	0.087	9.8	2.177	−0.115
2.3	1.005	0.510	6.1	1.714	0.082	9.9	2.187	−0.120
2.4	1.027	0.486	6.2	1.730	0.076	10.0	2.197	−0.126
2.5	1.049	0.464	6.3	1.745	0.071	11.0	2.284	−0.187
2.6	1.071	0.443	6.4	1.760	0.065	12.0	2.358	−0.254
2.7	1.092	0.423	6.5	1.775	0.060	13.0	2.420	−0.329
2.8	1.113	0.405	6.6	1.790	0.055	14.0	2.472	−0.411
2.9	1.134	0.388	6.7	1.805	0.050	15.0	2.514	−0.499
3.0	1.155	0.371	6.8	1.819	0.045	16.0	2.549	−0.593
3.1	1.176	0.356	6.9	1.834	0.040	17.0	2.578	−0.691
3.2	1.196	0.341	7.0	1.848	0.035	18.0	2.601	−0.793
3.3	1.217	0.327	7.1	1.862	0.031	19.0	2.620	−0.899
3.4	1.237	0.313	7.2	1.876	0.026	20.0	2.635	−1.007
3.5	1.257	0.300	7.3	1.890	0.022	21.0	2.647	−1.118
3.6	1.277	0.288	7.4	1.904	0.017	22.0	2.657	−1.230
3.7	1.296	0.276	7.5	1.917	0.013	23.0	2.664	−1.344
3.8	1.316	0.264	7.6	1.931	0.009	24.0	2.670	−1.460
3.9	1.335	0.253	7.7	1.944	0.005	25.0	2.675	−1.576
4.0	1.354	0.243	7.8	1.958	0.001	26.0	2.679	−1.693
4.1	1.373	0.232	7.9	1.965	−0.016	27.0	2.682	−1.811
4.2	1.392	0.222	8.0	1.978	−0.021	28.0	2.685	−1.929
4.3	1.410	0.213	8.1	1.990	−0.026	29.0	2.687	−2.047
4.4	1.428	0.204	8.2	2.003	−0.031	30.0	2.689	−2.166
4.5	1.447	0.195	8.3	2.015	−0.036	31.0	2.690	−2.285
4.6	1.465	0.186	8.4	2.027	−0.041	32.0	2.691	−2.404
4.7	1.482	0.178	8.5	2.038	−0.046	33.0	2.692	−2.524
4.8	1.500	0.169	8.6	2.050	−0.051	34.0	2.692	−2.644
4.9	1.517	0.162	8.7	2.061	−0.056	35.0	2.693	−2.763
5.0	1.535	0.154	8.8	2.073	−0.061	36.0	2.693	−2.883
5.1	1.552	0.146	8.9	2.084	−0.066	37.0	2.693	−3.003
5.2	1.569	0.139	9.0	2.095	−0.072	38.0	2.694	−3.123
5.3	1.586	0.132	9.1	2.106	−0.077	39.0	2.694	−3.243
						40.0	2.694	−3.363

Table 8.3 Design data on broadside-offset coupled homogeneous stripline; $\epsilon_r = 2.22$, $d = 0.005''$, $b_1 = 0.125''$, $b = 0.255''$, $Z_t = 50$ ohms (See structure shown in Fig. 8.15)

C_v (dB)	w (mm)	w_c (mm)	C_v (dB)	w (mm)	w_c (mm)	C_v(dB)	w (mm)	w_c (mm)
1.2	1.109	0.883	5.2	3.024	0.083	9.2	4.175	−0.204
1.3	1.172	0.808	5.3	3.060	0.076	9.3	4.196	−0.213
1.4	1.234	0.750	5.4	3.096	0.069	9.4	4.217	−0.222
1.5	1.294	0.700	5.5	3.132	0.063	9.5	4.238	−0.231
1.6	1.354	0.656	5.6	3.167	0.056	9.6	4.259	−0.241
1.7	1.413	0.617	5.7	3.202	0.050	9.7	4.279	−0.251
1.8	1.471	0.582	5.8	3.236	0.044	9.8	4.299	−0.261
1.9	1.529	0.549	5.9	3.270	0.038	9.9	4.318	−0.271
2.0	1.585	0.520	6.0	3.304	0.032	10.0	4.338	−0.281
2.1	1.640	0.492	6.1	3.337	0.026	11.0	4.514	−0.392
2.2	1.695	0.467	6.2	3.370	0.020	12.0	4.663	−0.521
2.3	1.749	0.443	6.3	3.403	0.015	13.0	4.787	−0.665
2.4	1.802	0.421	6.4	3.436	0.010	14.0	4.890	−0.824
2.5	1.854	0.400	6.5	3.468	0.004	15.0	4.975	−0.996
2.6	1.905	0.380	6.6	3.494	−0.014	16.0	5.045	−1.180
2.7	1.956	0.362	6.7	3.525	−0.020	17.0	5.102	−1.373
2.8	2.005	0.344	6.8	3.555	−0.026	18.0	5.148	−1.574
2.9	2.054	0.327	6.9	3.585	−0.033	19.0	5.185	−1.782
3.0	2.103	0.311	7.0	3.615	−0.039	20.0	5.215	−1.996
3.1	2.151	0.296	7.1	3.644	−0.045	21.0	5.239	−2.214
3.2	2.198	0.282	7.2	3.673	−0.052	22.0	5.258	−2.436
3.3	2.244	0.268	7.3	3.701	−0.058	23.0	5.273	−2.662
3.4	2.290	0.255	7.4	3.729	−0.065	24.0	5.286	−2.889
3.5	2.335	0.242	7.5	3.757	−0.072	25.0	5.295	−3.119
3.6	2.380	0.230	7.6	3.785	−0.079	26.0	5.303	−3.350
3.7	2.424	0.218	7.7	3.812	−0.085	27.0	5.309	−3.582
3.8	2.467	0.207	7.8	3.838	−0.092	28.0	5.314	−3.816
3.9	2.510	0.196	7.9	3.865	−0.100	29.0	5.318	−4.050
4.0	2.552	0.185	8.0	3.891	−0.107	30.0	5.321	−4.285
4.1	2.594	0.175	8.1	3.916	−0.114	31.0	5.324	−4.520
4.2	2.636	0.165	8.2	3.941	−0.122	32.0	5.326	−4.756
4.3	2.677	0.156	8.3	3.966	−0.129	33.0	5.327	−4.992
4.4	2.717	0.147	8.4	3.991	−0.137	34.0	5.329	−5.228
4.5	2.757	0.138	8.5	4.015	−0.145	35.0	5.330	−5.465
4.6	2.796	0.129	8.6	4.039	−0.153	36.0	5.331	−5.702
4.7	2.835	0.121	8.7	4.062	−0.161	37.0	5.331	−5.939
4.8	2.874	0.113	8.8	4.085	−0.169	38.0	5.332	−6.176
4.9	2.912	0.105	8.9	4.108	−0.177	39.0	5.332	−6.413
5.0	2.949	0.098	9.0	4.131	−0.186	40.0	5.332	−6.650
5.1	2.987	0.090	9.1	4.153	−0.195			

Table 8.4 Design data on broadside-offset coupled homogeneous stripline; $\epsilon_r = 2.22$, $d = 0.01''$, $b_1 = 0.031''$, $b = 0.072''$, $Z_t = 50$ ohms (See structure shown in Fig. 8.15)

C_v(dB)	w (mm)	w_c (mm)	C_v (dB)	w (mm)	w_c (mm)	C_v (dB)	w (mm)	w_c (mm)
4.0	0.947	0.827	7.0	1.098	0.287	10.0	1.238	0.065
4.1	0.952	0.778	7.1	1.103	0.277	11.0	1.279	0.013
4.2	0.956	0.741	7.2	1.108	0.268	12.0	1.306	−0.067
4.3	0.961	0.709	7.3	1.113	0.258	13.0	1.335	−0.123
4.4	0.966	0.681	7.4	1.118	0.249	14.0	1.360	−0.179
4.5	0.971	0.655	7.5	1.122	0.240	15.0	1.381	−0.236
4.6	0.975	0.632	7.6	1.127	0.231	16.0	1.398	−0.294
4.7	0.980	0.610	7.7	1.132	0.223	17.0	1.413	−0.354
4.8	0.985	0.589	7.8	1.137	0.215	18.0	1.424	−0.414
4.9	0.990	0.569	7.9	1.142	0.206	19.0	1.434	−0.476
5.0	0.995	0.550	8.0	1.147	0.198	20.0	1.442	−0.539
5.1	1.000	0.532	8.1	1.152	0.190	21.0	1.448	−0.603
5.2	1.005	0.515	8.2	1.156	0.183	22.0	1.453	−0.667
5.3	1.010	0.499	8.3	1.161	0.175	23.0	1.457	−0.732
5.4	1.016	0.483	8.4	1.166	0.168	24.0	1.460	−0.798
5.5	1.021	0.468	8.5	1.171	0.160	25.0	1.463	−0.863
5.6	1.026	0.453	8.6	1.175	0.153	26.0	1.465	−0.929
5.7	1.031	0.439	8.7	1.180	0.146	27.0	1.467	−0.996
5.8	1.036	0.425	8.8	1.185	0.139	28.0	1.468	−1.062
5.9	1.041	0.412	8.9	1.189	0.132	29.0	1.469	−1.129
6.0	1.046	0.399	9.0	1.194	0.126	30.0	1.470	−1.196
6.1	1.052	0.386	9.1	1.198	0.119	31.0	1.471	−1.262
6.2	1.057	0.374	9.2	1.203	0.113	32.0	1.471	−1.329
6.3	1.062	0.362	9.3	1.207	0.106	33.0	1.472	−1.396
6.4	1.067	0.351	9.4	1.212	0.100	34.0	1.472	−1.463
6.5	1.072	0.339	9.5	1.216	0.094	35.0	1.472	−1.530
6.6	1.077	0.328	9.6	1.221	0.088	36.0	1.472	−1.597
6.7	1.082	0.318	9.7	1.225	0.082	37.0	1.473	−1.664
6.8	1.087	0.307	9.8	1.229	0.076	38.0	1.473	−1.731
6.9	1.092	0.297	9.9	1.234	0.070	39.0	1.473	−1.799
						40.0	1.473	−1.866

Table 8.5 Design data on broadside-offset coupled homogeneous stripline; $\epsilon_r = 2.22$, $d = 0.01''$, $b_1 = 0.062''$, $b = 0.134''$, $Z_t = 50$ ohms (See structure shown in Fig. 8.15)

C_v (dB)	w (mm)	w_c (mm)	C_v (dB)	w (mm)	w_c (mm)	C_v (dB)	w (mm)	w_c (mm)
2.6	1.325	1.088	6.1	1.844	0.273	9.6	2.251	−0.044
2.7	1.339	1.027	6.2	1.857	0.261	9.7	2.260	−0.051
2.8	1.354	0.976	6.3	1.871	0.250	9.8	2.270	−0.059
2.9	1.369	0.931	6.4	1.884	0.239	9.9	2.272	−0.066
3.0	1.384	0.891	6.5	1.898	0.229	10.0	2.288	−0.074
3.1	1.399	0.853	6.6	1.911	0.218	11.0	2.373	−0.151
3.2	1.415	0.819	6.7	1.924	0.208	12.0	2.446	−0.231
3.3	1.430	0.787	6.8	1.937	0.198	13.0	2.508	−0.315
3.4	1.446	0.757	6.9	1.950	0.188	14.0	2.559	−0.405
3.5	1.461	0.728	7.0	1.963	0.179	15.0	2.602	−0.500
3.6	1.476	0.701	7.1	1.975	0.169	16.0	2.638	−0.600
3.7	1.492	0.675	7.2	1.988	0.160	17.0	2.667	−0.704
3.8	1.507	0.651	7.3	2.000	0.151	18.0	2.690	−0.812
3.9	1.523	0.627	7.4	2.013	0.142	19.0	2.709	−0.923
4.0	1.538	0.605	7.5	2.025	0.134	20.0	2.725	−1.036
4.1	1.553	0.583	7.6	2.037	0.125	21.0	2.737	−1.152
4.2	1.569	0.562	7.7	2.049	0.117	22.0	2.747	−1.270
4.3	1.584	0.542	7.8	2.061	0.109	23.0	2.755	−1.389
4.4	1.599	0.523	7.9	2.073	0.101	24.0	2.761	−1.509
4.5	1.614	0.504	8.0	2.085	0.093	25.0	2.766	−1.630
4.6	1.629	0.486	8.1	2.097	0.085	26.0	2.770	−1.752
4.7	1.644	0.469	8.2	2.108	0.078	27.0	2.774	−1.874
4.8	1.659	0.452	8.3	2.120	0.070	28.0	2.776	−1.997
4.9	1.674	0.436	8.4	2.131	0.063	29.0	2.778	−2.121
5.0	1.688	0.420	8.5	2.143	0.056	30.0	2.780	−2.244
5.1	1.703	0.405	8.6	2.154	0.049	31.0	2.781	−2.368
5.2	1.717	0.390	8.7	2.165	0.042	32.0	2.782	−2.492
5.3	1.732	0.375	8.8	2.176	0.035	33.0	2.783	−2.617
5.4	1.746	0.361	8.9	2.187	0.028	34.0	2.784	−2.741
5.5	1.760	0.348	9.0	2.198	0.022	35.0	2.784	−2.865
5.6	1.775	0.334	9.1	2.209	0.015	36.0	2.784	−2.990
5.7	1.789	0.321	9.2	2.220	0.009	37.0	2.785	−3.114
5.8	1.803	0.309	9.3	2.230	0.003	38.0	2.785	−3.239
5.9	1.816	0.296	9.4	2.231	−0.029	39.0	2.785	−3.364
6.0	1.830	0.284	9.5	2.241	−0.036	40.0	2.785	−3.488

Table 8.6 Design data on broadside-offset coupled homogeneous stripline; $\epsilon_r = 2.22$, $d = 0.01''$, $b_1 = 0.125''$, $b = 0.26''$, $Z_t = 50$ ohms (See structure shown in Fig. 8.15)

C_v (dB)	w (mm)	w_c (mm)	C_v (dB)	w (mm)	w_c (mm)	C_v (dB)	w (mm)	w_c (mm)
1.6	1.710	1.612	5.4	3.228	0.249	9.2	4.265	−0.166
1.7	1.755	1.439	5.5	3.261	0.236	9.3	4.287	−0.177
1.8	1.799	1.339	5.6	3.294	0.223	9.4	4.308	−0.188
1.9	1.844	1.258	5.7	3.327	0.210	9.5	4.328	−0.199
2.0	1.889	1.188	5.8	3.359	0.198	9.6	4.349	−0.210
2.1	1.933	1.126	5.9	3.391	0.185	9.7	4.369	−0.221
2.2	1.978	1.070	6.0	3.423	0.174	9.8	4.388	−0.232
2.3	2.022	1.018	6.1	3.454	0.162	9.9	4.408	−0.243
2.4	2.067	0.970	6.2	3.486	0.151	10.0	4.427	−0.255
2.5	2.111	0.926	6.3	3.516	0.140	11.0	4.603	−0.377
2.6	2.154	0.884	6.4	3.547	0.129	12.0	4.752	−0.513
2.7	2.198	0.845	6.5	3.577	0.119	13.0	4.878	−0.664
2.8	2.241	0.809	6.6	3.607	0.108	14.0	4.982	−0.829
2.9	2.283	0.774	6.7	3.637	0.098	15.0	5.068	−1.006
3.0	2.325	0.741	6.8	3.666	0.088	16.0	5.139	−1.195
3.1	2.367	0.710	6.9	3.695	0.079	17.0	5.196	−1.393
3.2	2.409	0.680	7.0	3.724	0.069	18.0	5.243	−1.599
3.3	2.450	0.652	7.1	3.753	0.060	19.0	5.280	−1.812
3.4	2.491	0.625	7.2	3.781	0.051	20.0	5.311	−2.030
3.5	2.531	0.599	7.3	3.809	0.042	21.0	5.335	−2.253
3.6	2.571	0.574	7.4	3.837	0.033	22.0	5.354	−2.480
3.7	2.611	0.550	7.5	3.864	0.025	23.0	5.370	−2.710
3.8	2.650	0.527	7.6	3.892	0.016	24.0	5.383	−2.942
3.9	2.689	0.505	7.7	3.919	0.008	25.0	5.392	−3.176
4.0	2.727	0.484	7.8	3.935	−0.024	26.0	5.400	−3.412
4.1	2.765	0.463	7.9	3.961	−0.034	27.0	5.407	−3.649
4.2	2.803	0.443	8.0	3.986	−0.044	28.0	5.412	−3.887
4.3	2.840	0.424	8.1	4.011	−0.054	29.0	5.416	−4.126
4.4	2.877	0.406	8.2	4.036	−0.064	30.0	5.419	−4.366
4.5	2.914	0.388	8.3	4.060	−0.074	31.0	5.421	−4.606
4.6	2.950	0.371	8.4	4.048	−0.084	32.0	5.424	−4.846
4.7	2.986	0.354	8.5	4.108	−0.094	33.0	5.425	−5.087
4.8	3.022	0.338	8.6	4.131	−0.104	34.0	5.426	−5.328
4.9	3.057	0.322	8.7	4.154	−0.114	35.0	5.427	−5.570
5.0	3.092	0.306	8.8	4.177	−0.125	36.0	5.428	−5.811
5.1	3.126	0.291	8.9	4.200	−0.135	37.0	5.429	−6.053
5.2	3.161	0.277	9.0	4.222	−0.145	38.0	5.429	−6.294
5.3	3.195	0.263	9.1	4.244	−0.156	39.0	5.430	−6.536
						40.0	5.430	−6.778

Table 8.7 Design data on broadside-offset coupled homogeneous stripline; $\epsilon_r = 2.22$, $d = 0.015''$, $b_1 = 0.031''$, $b = 0.077''$, $Z_t = 50$ ohms (See structure shown in Fig. 8.15)

C_v (dB)	w (mm)	w_c (mm)	C_v (dB)	w (mm)	w_c (mm)	C_v (dB)	w (mm)	w_c (mm)
5.2	1.140	1.048	7.8	1.236	0.419	14.0	1.430	−0.138
5.3	1.143	0.973	7.9	1.240	0.406	15.0	1.450	−0.206
5.4	1.146	0.923	8.0	1.244	0.393	16.0	1.467	−0.275
5.5	1.150	0.882	8.1	1.248	0.381	17.0	1.481	−0.343
5.6	1.153	0.846	8.2	1.252	0.368	18.0	1.492	−0.412
5.7	1.156	0.815	8.3	1.256	0.356	19.0	1.502	−0.481
5.8	1.160	0.786	8.4	1.260	0.345	20.0	1.509	−0.551
5.9	1.164	0.759	8.5	1.264	0.333	21.0	1.516	−0.621
6.0	1.167	0.734	8.6	1.268	0.322	22.0	1.521	−0.692
6.1	1.171	0.710	8.7	1.271	0.311	23.0	1.525	−0.763
6.2	1.175	0.687	8.8	1.275	0.300	24.0	1.528	−0.834
6.3	1.178	0.666	8.9	1.279	0.290	25.0	1.531	−0.905
6.4	1.182	0.645	9.0	1.283	0.279	26.0	1.533	−0.977
6.5	1.186	0.625	9.1	1.287	0.269	27.0	1.535	−1.048
6.6	1.190	0.606	9.2	1.290	0.259	28.0	1.536	−1.120
6.7	1.194	0.588	9.3	1.294	0.249	29.0	1.537	−1.192
6.8	1.197	0.570	9.4	1.298	0.239	30.0	1.538	−1.264
6.9	1.201	0.553	9.5	1.302	0.230	31.0	1.538	−1.335
7.0	1.205	0.537	9.6	1.305	0.221	32.0	1.539	−1.407
7.1	1.209	0.520	9.7	1.309	0.211	33.0	1.539	−1.479
7.2	1.213	0.505	9.8	1.313	0.202	34.0	1.540	−1.551
7.3	1.217	0.490	9.9	1.316	0.194	35.0	1.540	−1.623
7.4	1.221	0.475	10.0	1.320	0.185	36.0	1.540	−1.695
7.5	1.225	0.460	11.0	1.355	0.105	37.0	1.540	−1.767
7.6	1.229	0.446	12.0	1.388	0.037	38.0	1.541	−1.839
7.7	1.233	0.432	13.0	1.407	−0.068	39.0	1.541	−1.911
						40.0	1.541	−1.983

Table 8.8 Design data on broadside-offset coupled homogeneous stripline; $\epsilon_r = 2.22$, $d = 0.015''$, $b_1 = 0.062''$, $b = 0.139''$, $Z_t = 50$ ohms (See structure shown in Fig. 8.15)

C_v (dB)	w (mm)	w_c (mm)	C_v (dB)	w (mm)	w_c (mm)	C_v (dB)	w (mm)	w_c (mm)
3.3	1.639	1.427	6.6	2.034	0.423	9.9	2.380	0.042
3.4	1.650	1.341	6.7	2.046	0.407	10.0	2.389	0.033
3.5	1.662	1.274	6.8	2.057	0.392	11.0	2.460	−0.093
3.6	1.673	1.217	6.9	2.069	0.377	12.0	2.530	−0.189
3.7	1.684	1.167	7.0	2.080	0.362	13.0	2.591	−0.286
3.8	1.696	1.121	7.1	2.091	0.348	14.0	2.642	−0.387
3.9	1.708	1.078	7.2	2.103	0.334	15.0	2.684	−0.491
4.0	1.720	1.039	7.3	2.114	0.320	16.0	2.720	−0.598
4.1	1.732	1.001	7.4	2.125	0.307	17.0	2.749	−0.709
4.2	1.744	0.966	7.5	2.136	0.294	18.0	2.773	−0.823
4.3	1.756	0.933	7.6	2.147	0.281	19.0	2.792	−0.941
4.4	1.768	0.901	7.7	2.158	0.268	20.0	2.807	−1.060
4.5	1.781	0.871	7.8	2.169	0.256	21.0	2.820	−1.181
4.6	1.793	0.842	7.9	2.180	0.244	22.0	2.830	−1.304
4.7	1.805	0.814	8.0	2.190	0.232	23.0	2.838	−1.429
4.8	1.817	0.787	8.1	2.201	0.220	24.0	2.844	−1.554
4.9	1.830	0.761	8.2	2.212	0.209	25.0	2.850	−1.680
5.0	1.842	0.736	8.3	2.222	0.197	26.0	2.854	−1.807
5.1	1.854	0.712	8.4	2.232	0.186	27.0	2.857	−1.935
5.2	1.866	0.688	8.5	2.243	0.176	28.0	2.860	−2.063
5.3	1.879	0.666	8.6	2.253	0.165	29.0	2.862	−2.191
5.4	1.891	0.644	8.7	2.263	0.154	30.0	2.863	−2.319
5.5	1.903	0.622	8.8	2.273	0.144	31.0	2.865	−2.448
5.6	1.915	0.602	8.9	2.283	0.134	32.0	2.866	−2.577
5.7	1.927	0.581	9.0	2.293	0.124	33.0	2.866	−2.706
5.8	1.939	0.562	9.1	2.303	0.114	34.0	2.867	−2.835
5.9	1.951	0.543	9.2	2.313	0.105	35.0	2.868	−2.964
6.0	1.963	0.524	9.3	2.323	0.095	36.0	2.868	−3.094
6.1	1.975	0.506	9.4	2.333	0.086	37.0	2.868	−3.223
6.2	1.987	0.489	9.5	2.342	0.077	38.0	2.869	−3.352
6.3	1.999	0.472	9.6	2.352	0.068	39.0	2.869	−3.482
6.4	2.011	0.455	9.7	2.361	0.059	40.0	2.869	−3.611
6.5	2.022	0.438	9.8	2.371	0.050			

Table 8.9 Design data on broadside-offset coupled homogeneous stripline; $\epsilon_r = 2.22$, $d = 0.015''$, $b_1 = 0.125''$, $b = 0.265''$, $Z_t = 50$ ohms (See Structure shown in Fig. 8.15)

C_v (dB)	w (mm)	w_c (mm)	C_v (dB)	w (mm)	w_c (mm)	C_v (dB)	w (mm)	w_c (mm)
2.1	2.230	1.940	5.8	3.493	0.393	9.5	4.421	−0.145
2.2	2.265	1.801	5.9	3.523	0.374	9.6	4.441	−0.158
2.3	2.301	1.697	6.0	3.552	0.357	9.7	4.461	−0.171
2.4	2.337	1.609	6.1	3.582	0.339	9.8	4.480	−0.184
2.5	2.373	1.531	6.2	3.611	0.322	9.9	4.499	−0.197
2.6	2.409	1.461	6.3	3.640	0.305	10.0	4.518	−0.211
2.7	2.445	1.396	6.4	3.669	0.289	11.0	4.692	−0.347
2.8	2.482	1.336	6.5	3.697	0.273	12.0	4.841	−0.494
2.9	2.518	1.280	6.6	3.726	0.258	13.0	4.966	−0.654
3.0	2.555	1.228	6.7	3.754	0.243	14.0	5.071	−0.826
3.1	2.591	1.178	6.8	3.782	0.228	15.0	5.158	−1.010
3.2	2.627	1.132	6.9	3.809	0.213	16.0	5.229	−1.204
3.3	2.663	1.087	7.0	3.836	0.199	17.0	5.287	−1.408
3.4	2.699	1.045	7.1	3.863	0.185	18.0	5.334	−1.619
3.5	2.735	1.005	7.2	3.890	0.171	19.0	5.372	−1.837
3.6	2.770	0.966	7.3	3.917	0.158	20.0	5.402	−2.061
3.7	2.805	0.929	7.4	3.943	0.145	21.0	5.427	−2.289
3.8	2.841	0.894	7.5	3.969	0.132	22.0	5.447	−2.521
3.9	2.875	0.860	7.6	3.995	0.119	23.0	5.463	−2.755
4.0	2.910	0.828	7.7	4.021	0.107	24.0	5.475	−2.992
4.1	2.945	0.796	7.8	4.047	0.095	25.0	5.485	−3.232
4.2	2.979	0.766	7.9	4.072	0.083	26.0	5.493	−3.472
4.3	3.013	0.737	8.0	4.097	0.071	27.0	5.500	−3.714
4.4	3.047	0.709	8.1	4.122	0.059	28.0	5.505	−3.957
4.5	3.080	0.681	8.2	4.146	0.048	29.0	5.509	−4.201
4.6	3.113	0.655	8.3	4.171	0.037	30.0	5.512	−4.445
4.7	3.146	0.630	8.4	4.195	0.026	31.0	5.515	−4.690
4.8	3.179	0.605	8.5	4.219	0.015	32.0	5.517	−4.935
4.9	3.212	0.581	8.6	4.243	0.005	33.0	5.518	−5.181
5.0	3.244	0.557	8.7	4.251	−0.043	34.0	5.520	−5.426
5.1	3.276	0.535	8.8	4.274	−0.056	35.0	5.521	−5.672
5.2	3.308	0.513	8.9	4.295	−0.069	36.0	5.522	−5.918
5.3	3.339	0.491	9.0	4.317	−0.081	37.0	5.522	−6.165
5.4	3.370	0.471	9.1	4.338	−0.094	38.0	5.523	−6.411
5.5	3.401	0.450	9.2	4.359	−0.107	39.0	5.523	−6.657
5.6	3.432	0.431	9.3	4.380	−0.120	40.0	5.524	−6.904
5.7	3.463	0.411	9.4	4.401	−0.132			

Table 8.10 Design data on broadside-offset coupled homogeneous stripline; $\epsilon_r = 2.22$, $d = 0.02''$, $b_1 = 0.031''$, $b = 0.082''$, $Z_t = 50$ ohms (See structure shown in Fig. 8.15)

C_v (dB)	w (mm)	w_c (mm)	C_v (dB)	w (mm)	w_c (mm)	C_v(dB)	w (mm)	w_c (mm)
6.4	1.286	1.112	8.6	1.351	0.526	18.0	1.551	−0.401
6.5	1.288	1.051	8.7	1.354	0.510	19.0	1.560	−0.478
6.6	1.291	1.003	8.8	1.357	0.494	20.0	1.568	−0.555
6.7	1.294	0.962	8.9	1.360	0.479	21.0	1.574	−0.633
6.8	1.296	0.926	9.0	1.363	0.464	22.0	1.579	−0.710
6.9	1.299	0.893	9.1	1.366	0.450	23.0	1.583	−0.787
7.0	1.302	0.862	9.2	1.369	0.435	24.0	1.586	−0.864
7.1	1.305	0.834	9.3	1.373	0.421	25.0	1.589	−0.941
7.2	1.308	0.807	9.4	1.376	0.408	26.0	1.591	−1.019
7.3	1.311	0.781	9.5	1.379	0.395	27.0	1.592	−1.096
7.4	1.314	0.757	9.6	1.382	0.382	28.0	1.594	−1.173
7.5	1.317	0.734	9.7	1.385	0.369	29.0	1.595	−1.250
7.6	1.320	0.712	9.8	1.388	0.356	30.0	1.596	−1.327
7.7	1.323	0.690	9.9	1.391	0.344	31.0	1.596	−1.404
7.8	1.326	0.670	10.0	1.394	0.332	32.0	1.597	−1.481
7.9	1.329	0.650	11.0	1.425	0.222	33.0	1.597	−1.558
8.0	1.332	0.630	12.0	1.454	0.129	34.0	1.598	−1.635
8.1	1.335	0.612	13.0	1.481	0.049	35.0	1.598	−1.712
8.2	1.338	0.593	14.0	1.493	−0.083	36.0	1.598	−1.789
8.3	1.341	0.576	15.0	1.511	−0.165	37.0	1.598	−1.865
8.4	1.344	0.559	16.0	1.527	−0.245	38.0	1.598	−1.942
8.5	1.348	0.542	17.0	1.540	−0.323	39.0	1.599	−2.019
						40.0	1.599	−2.096

Table 8.11 Design data on broadside-offset coupled homogeneous stripline; $\epsilon_r = 2.22$, $d = 0.02''$, $b_1 = 0.062''$, $b = 0.144''$, $Z_t = 50$ ohms (See structure shown in Fig 8.15)

C_v (dB)	w (mm)	w_c (mm)	C_v (dB)	w (mm)	w_c (mm)	C_v (dB)	w (mm)	w_c (mm)
4.0	1.895	1.654	7.0	2.195	0.574	10.0	2.476	0.129
4.1	1.904	1.557	7.1	2.205	0.554	11.0	2.558	0.025
4.2	1.913	1.482	7.2	2.215	0.535	12.0	2.611	−0.134
4.3	1.922	1.419	7.3	2.225	0.517	13.0	2.670	−0.246
4.4	1.932	1.362	7.4	2.235	0.498	14.0	2.719	−0.358
4.5	1.941	1.311	7.5	2.245	0.480	15.0	2.761	−0.472
4.6	1.951	1.263	7.6	2.255	0.463	16.0	2.796	−0.589
4.7	1.961	1.219	7.7	2.265	0.446	17.0	2.825	−0.708
4.8	1.970	1.178	7.8	2.274	0.429	18.0	2.849	−0.829
4.9	1.980	1.138	7.9	2.284	0.413	19.0	2.868	−0.953
5.0	1.990	1.101	8.0	2.294	0.397	20.0	2.883	−1.078
5.1	2.001	1.065	8.1	2.303	0.381	21.0	2.896	−1.206
5.2	2.011	1.031	8.2	2.313	0.365	22.0	2.906	−1.334
5.3	2.021	0.998	8.3	2.322	0.350	23.0	2.914	−1.464
5.4	2.031	0.966	8.4	2.332	0.335	24.0	2.921	−1.595
5.5	2.041	0.936	8.5	2.341	0.320	25.0	2.926	−1.727
5.6	2.052	0.906	8.6	2.351	0.306	26.0	2.930	−1.859
5.7	2.062	0.878	8.7	2.360	0.292	27.0	2.934	−1.991
5.8	2.072	0.850	8.8	2.369	0.278	28.0	2.936	−2.124
5.9	2.083	0.824	8.9	2.378	0.264	29.0	2.938	−2.258
6.0	2.093	0.798	9.0	2.388	0.251	30.0	2.940	−2.391
6.1	2.103	0.772	9.1	2.397	0.238	31.0	2.941	−2.525
6.2	2.113	0.748	9.2	2.406	0.225	32.0	2.942	−2.659
6.3	2.124	0.724	9.3	2.415	0.212	33.0	2.943	−2.793
6.4	2.134	0.701	9.4	2.424	0.200	34.0	2.944	−2.927
6.5	2.144	0.679	9.5	2.432	0.188	35.0	2.944	−3.061
6.6	2.154	0.657	9.6	2.441	0.175	36.0	2.945	−3.195
6.7	2.165	0.635	9.7	2.450	0.164	37.0	2.945	−3.329
6.8	2.175	0.614	9.8	2.459	0.152	38.0	2.945	−3.463
6.9	2.185	0.594	9.9	2.467	0.140	39.0	2.946	−3.597
						40.0	2.946	−3.731

Table 8.12 Design data on broadside-offset coupled homogeneous stripline; $\epsilon_r = 2.22$, $d = 0.02''$, $b_1 = 0.125''$, $b = 0.270''$, $Z_t = 50$ ohms (See structure shown in Fig. 8.15)

C_v (dB)	w (mm)	w_c (mm)	C_v (dB)	w (mm)	w_c (mm)	C_v (dB)	w (mm)	w_c (mm)
2.6	2.663	2.166	6.1	3.713	0.543	9.6	4.535	−0.091
2.7	2.692	2.046	6.2	3.741	0.520	9.7	4.554	−0.106
2.8	2.722	1.945	6.3	3.768	0.498	9.8	4.573	−0.121
2.9	2.753	1.857	6.4	3.795	0.476	9.9	4.591	−0.136
3.0	2.784	1.776	6.5	3.822	0.455	10.0	4.610	−0.151
3.1	2.814	1.703	6.6	3.848	0.434	11.0	4.781	−0.305
3.2	2.845	1.634	6.7	3.875	0.413	12.0	4.928	−0.466
3.3	2.877	1.570	6.8	3.901	0.393	13.0	5.053	−0.636
3.4	2.908	1.501	6.9	3.927	0.374	14.0	5.157	−0.817
3.5	2.939	1.453	7.0	3.953	0.355	15.0	5.244	−1.008
3.6	2.970	1.399	7.1	3.979	0.336	16.0	5.315	−1.209
3.7	3.001	1.347	7.2	4.004	0.318	17.0	5.374	−1.419
3.8	3.033	1.298	7.3	4.029	0.300	18.0	5.421	−1.636
3.9	3.064	1.251	7.4	4.054	0.282	19.0	5.460	−1.860
4.0	3.095	1.206	7.5	4.079	0.265	20.0	5.490	−2.089
4.1	3.126	1.163	7.6	4.104	0.248	21.0	5.515	−2.322
4.2	3.157	1.121	7.7	4.128	0.231	22.0	5.535	−2.559
4.3	3.187	1.081	7.8	4.152	0.215	23.0	5.551	−2.799
4.4	3.218	1.043	7.9	4.177	0.199	24.0	5.564	−3.041
4.5	3.249	1.006	8.0	4.200	0.183	25.0	5.574	−3.285
4.6	3.279	0.970	8.1	4.224	0.168	26.0	5.582	−3.530
4.7	3.309	0.935	8.2	4.248	0.153	27.0	5.589	−3.777
4.8	3.339	0.902	8.3	4.271	0.138	28.0	5.594	−4.025
4.9	3.369	0.869	8.4	4.294	0.123	29.0	5.598	−4.273
5.0	3.399	0.837	8.5	4.317	0.109	30.0	5.601	−4.522
5.1	3.428	0.807	8.6	4.340	0.095	31.0	5.604	−4.772
5.2	3.458	0.777	8.7	4.362	0.081	32.0	5.606	−5.022
5.3	3.487	0.748	8.8	4.384	0.067	33.0	5.608	−5.272
5.4	3.516	0.720	8.9	4.407	0.054	34.0	5.609	−5.523
5.5	3.545	0.693	9.0	4.429	0.041	35.0	5.610	−5.774
5.6	3.573	0.666	9.1	4.450	0.028	36.0	5.611	−6.024
5.7	3.602	0.640	9.2	4.472	0.015	37.0	5.612	−6.275
5.8	3.630	0.615	9.3	4.493	0.003	38.0	5.612	−6.527
5.9	3.658	0.590	9.4	4.495	−0.060	39.0	5.613	−6.778
6.0	3.686	0.566	9.5	4.515	−0.075	40.0	5.613	−7.029

REFERENCES

1. A. Presser, Interdigitated microstrip coupler design, *IEEE Trans. Microwave Theory Tech.*, Vol. MTT-26, pp. 801-805, Oct. 1978.
2. D. Kajfez, Z. Paunovic and S. Pavlin, Simplified design of Lange coupler, *IEEE Trans. Microwave Theory Tech.*, Vol. MTT-26, pp. 806-808, Oct. 1978.
3. S. Rehnmark, Meander-folded coupled lines, *IEEE Trans. Microwave Theory Tech.*, Vol. MTT-26, pp. 225-231, April 1978.
4. J.W. Duncan, Characteristic impedances of multiconductor strip transmission lines, *IEEE Trans. Microwave Theory Tech.*, Vol. MTT-13, pp. 107-118, Jan. 1965.
5. S.A. Ivanov, Characteristic impedances of four-conductor transmission lines, *IEEE Trans. Microwave Theory Tech.*, Vol. MTT-29, pp. 611-615, June, 1981.
6. B. Bhat and S.K. Koul, Unified approach to solve a class of strip and microstrip-like transmission lines, *IEEE Trans. Microwave Theory Tech.*, Vol. MTT-30, pp. 679-686, May 1982. Also correction *IEEE Trans. Microwave Theory Tech.*, Vol. MTT-30, p. 2067, Nov. 1982.
7. S.K. Koul and B. Bhat, Broadside-edge Coupled symmetric strip transmission lines, *IEEE Trans. Microwave Theory Tech.*, Vol. MTT-30, pp. 1874-1880, Nov. 1982.
8. S.K. Koul and B. Bhat, Simplified analysis of stripline, microstripline and coplanar strips with anisotropic substrates for MIC and SAW applications, *IEEE MTT-S Int. Microwave Symp.* Digest, pp. 236-238, June 1983.
9. J.P. Shelton, Impedances of offset parallel-coupled strip transmission lines, *IEEE Trans. Microwave Theory Tech.*, Vol. MTT-14, pp. 7-15, Jan. 1966.
10. J.A. Mosko, Coupling curves for offset parallel-coupled strip transmission lines, *Microwave J.*, Vol. 10, pp. 35-37, April 1967.
11. C.P. Tresselt, The design and construction of broadband high directivity, 90-degree couplers using nonuniform line techniques, *IEEE Trans. Microwave Theory Tech.*, Vol. MTT-14, pp. 647-656, Dec. 1966.
12. H. Howe Jr., *Stripline Circuit Design*, Artech House, Dedham, Mass., 1974.
13. T. Kitazawa, H. Kumagams and Y. Hayashi, Unsymmetric broadside-coupled striplines, *Electron. Lett.*, Vol. 18, No. 10, pp. 425-426, May 1982.
14. C.L. Chao, Characteristics of unsymmetrical broadside-coupled strips in an inhomogeneous dielectric medium, *IEEE MTT-S Int. Microwave Symp.* Digest, pp. 119-121, June 1975.
15. W.J. Barnes and J.L. Allen, Modified broadside-coupled strips in a layered dielectric medium, *Int. J. Electronics*, Vol. 40, No. 4. pp. 377-391, 1976.
16. W.J. Getsinger, A coupled stripline configuration using printed-circuit construction that allows very close coupling, *IRE Trans. Microwave Theory Tech.*, Vol. MTT-9, pp. 535-544, Nov. 1961.
17. G.L. Matthaei, L. Young and E.M.T Jones, *Microwave Filters, Impedance Matching Networks and Coupling Structures*, McGraw-Hill, New York, 1964, (Sec. 5.05)
18. S. Yamamoto, T. Azakami and T. Itakura, Slit-coupled strip transmission lines, *IEEE Trans. Microwave Theory Tech.*, Vol. MTT-14, pp. 542-553, Nov. 1966.
19. S.B. Cohn. Characteristic impedances of broadside-coupled strip transmission lines, *IRE Trans. Microwave Theory Tech.*, Vol. MTT-8, pp. 633-637, Nov. 1960.

CHAPTER 9

Hybrid-Mode Analysis and Dispersion Data on Stripline-like Transmission Lines

9.1 Introduction

The TEM and quasi-TEM mode analyses and design data on single and coupled stripline-like transmission lines are presented in Chapters 2–8. Each of these analyses has a certain range of validity depending on the structural parameters. For example, in a homogeneous stripline, the fundamental mode of propagation is a pure TEM mode. However, at higher frequencies when the wavelength in the medium becomes comparable to the transverse dimensions of the line, higher order modes get generated. The determination of the frequency above which higher order modes start propagating in rectangular striplines is essential for several applications such as in large scale TEM cells [1].

In a stripline-like transmission line employing layered dielectrics, the field distribution at lower frequencies essentially corresponds to quasi-TEM mode. Design information on quasi-static propagation parameters should therefore suffice if the circuits are designed at lower frequencies. As the frequency increases, the deviation of fields from the quasi-TEM mode results in dispersion. Since stripline-like transmission lines are finding increasing use at higher microwave and millimeter wave frequencies, it is essential to study the frequency dependent behaviour of the basic propagation parameters and also have a knowledge of the higher order mode propagation. This information is useful for determining the single-mode operating bandwidth of the transmission structure.

An important frequency dependent analytical procedure is the spectral domain method [2, 3]. This is a rigorous and numerically efficient method for the analysis of stripline and stripline-like transmission lines. In this method, the information on propagation constant at a given frequency is extracted from a set of algebraic equations that relate the Fourier transforms of the currents on the stripconductors to those of the electric fields at the complementary dielectric interface in the plane of the strip conductor. One advantage of this method is that its accuracy can be checked by comparing the results obtained by using different basis functions. For multilayer structures and some special geometries, spectral domain immittance approach [4] and spectral domain method using the concept of chain matrices [5] are preferred. The other useful frequency dependent analytical techniques are the integral equation method [6], finite difference techniques [7, 8], modal analysis [9, 10], method of lines [11, 12] and the planar waveguide model [13].

In this chapter, we first derive general expressions for the electromagnetic fields in terms of scalar electric and magnetic potentials. Next, we present the application of the spectral domain technique using the concept of chain matrix to analyze the frequency dependent propagation characteristics of a strip conductor embedded in a general N-layered dielectric structure. Formulas to compute the dispersion characteristics of various single and coupled stripline-like structures shown in Tables 3.1, 4.1, 4.4 and 4.7 can then be derived from the formulas presented for the N-layered dielectric structure. As an application of the above technique, the elements of the dyadic Green's function are derived for a microstrip-like stripline (structure A3 in Table 3.1) and broadside-coupled suspended stripline (structure A2 in Table 4.4). Lastly, dispersion characteristics of most of the structures shown in Tables 3.1, 4.1, 4.4 and 4.7 are presented.

9.2 Frequency Dependent Propagation Parameters

The frequency dependent propagation parameters which characterize a stripline-like transmission line are the propagation constant β and the characteristic impedance Z. The propagation constant β can be directly obtained by seeking the root of the characteristic equation derived in section 9.4. Once the propagation constant is obtained, the effective dielectric constant can be determined using the formula

$$\epsilon_f = \beta^2/k_0^2 \tag{9.1}$$

where $k_0 = (2\pi/\lambda^a)$ is the free space propagation constant and λ^a is the free space wavelength.

There is no unique definition for the characteristic impedance of stripline-like transmission lines. Three commonly used definitions of the characteristic impedance (Z) are based on [14, 15]

1. Power-current, $Z = 2P/II^*$; (9.2a)
2. Voltage-current, $Z = V/I$; (9.2b)
3. Power-voltage, $Z = VV^*/2P$; (9.2c)

where P denotes the average power transported by the fundamental mode in stripline-like transmission line, I is the longitudinal strip current and V is the centre strip voltage. However, the most commonly used definition for stripline-like transmission line is given by

$$Z = 2P_{avg}/II^* \tag{9.3}$$

where

$$I = \int_{strip} J_z\, dx \tag{9.4a}$$

and

$$P_{avg} = \frac{1}{2} R_e \iint_S (E_x H_y^* - E_y H_x^*)\, dx\, dy \tag{9.4b}$$

Here E_x, E_y, H_x and H_y are the transverse electric and magnetic fields and J_z is the current density in the longitudinal direction.

For the coupled stripline-like transmission lines, the effective dielectric constants and characteristic impedances are defined separately for the even- and odd-modes as

$$\epsilon_{fi} = \beta_i^2/k_0^2 \tag{9.5a}$$

$$Z_i = 2P_{avgi}/I_iI_i^* \tag{9.5b}$$

In the case of edge-coupled and broadside-coupled stripline-like transmission lines, $i = e$ denotes the even-mode, whereas, $i = o$ denotes the odd-mode case. For the broadside-edge coupled stripline-like transmission lines, $i = ee, eo, oe, oo$ denotes the even- even, even-odd, odd-even and odd-odd modes, respectively.

9.3 Derivation of General Electromagnetic F'elds

In a source free region, the electromagnetic fields satisfy the following Maxwell's equations:

$$\nabla \times \vec{E} = -\frac{\partial \vec{B}}{\partial t} \tag{9.6a}$$

$$\nabla \times \vec{H} = \frac{\partial \vec{D}}{\partial t} \tag{9.6b}$$

$$\nabla \cdot \vec{D} = 0 \tag{9.6c}$$

$$\nabla \cdot \vec{B} = 0 \tag{9.6d}$$

where

$$\vec{B} = \mu \vec{H} \quad \text{and} \quad \vec{D} = \epsilon \vec{E}$$

Electric and Magnetic Potentials

From (9.6d), we can write

$$\vec{B} = \nabla \times \vec{A^e} \tag{9.7a}$$

Substituting (9.7a) in (9.6a) we get

$$\nabla \times \left(\vec{E} + \frac{\partial \vec{A^e}}{\partial t}\right) = 0 \tag{9.7b}$$

or

$$\vec{E} = -\nabla \Phi^e - \frac{\partial \vec{A^e}}{\partial t} \tag{9.7c}$$

$\vec{A^e}$ and Φ^e are known as the electric vector potential and electric scalar potential, respectively. Substituting (9.7a) and (9.7c) in (9.6b), and simplifying, we get

$$-\nabla^2 \vec{A^e} + \mu\epsilon \frac{\partial^2 \vec{A^e}}{\partial t^2} + \nabla\left[\nabla \cdot \vec{A^e} + \mu\epsilon \frac{\partial \Phi^e}{\partial t}\right] = 0 \tag{9.8}$$

Imposing the Lorentz condition, namely,

$$\nabla \cdot \vec{A^e} + \mu\epsilon \frac{\partial \Phi^e}{\partial t} = 0 \tag{9.9}$$

(9.8) simplifies to

$$\nabla^2 \vec{A^e} - \mu\epsilon \frac{\partial^2 \vec{A^e}}{\partial t^2} = 0 \tag{9.10}$$

In a similar way it can be shown that

$$\vec{H} = -\nabla\Phi^h - \frac{\partial \vec{A}^h}{\partial t} \tag{9.11}$$

and

$$\nabla^2\vec{A}^h - \mu\epsilon\frac{\partial^2\vec{A}^h}{\partial t^2} = 0 \tag{9.12}$$

where $\vec{A}^h$ and Φ^h are known as the magnetic vector potential and the magnetic scalar potential, respectively.

Hertz Potential

Define

$$\vec{A}^e = \mu\epsilon\frac{\partial\vec{\pi}^e}{\partial t} \tag{9.13a}$$

and

$$\vec{A}^h = \mu\epsilon\frac{\partial\vec{\pi}^h}{\partial t} \tag{9.13b}$$

where $\vec{\pi}^e$ and $\vec{\pi}^h$ are known as the electric Hertz potential and magnetic Hertz potential, respectively.

With (9.13), (9.9) can be rewritten as

$$\mu\epsilon\frac{\partial}{\partial t}\nabla\cdot\vec{\pi}^e = -\mu\epsilon\frac{\partial\Phi^e}{\partial t} \tag{9.14a}$$

or

$$\Phi^e = -\nabla\cdot\vec{\pi}^e \tag{9.14b}$$

With (9.14b) and (9.13a), (9.7a) and (9.7c) can be written as

$$\vec{B} = \mu\epsilon\frac{\partial}{\partial t}\nabla\times\vec{\pi}^e \tag{9.15a}$$

$$\vec{E} = \nabla(\nabla\cdot\vec{\pi}^e) - \mu\epsilon\frac{\partial^2\vec{\pi}^e}{\partial t^2} \tag{9.15b}$$

Similarly, it can be shown that

$$\vec{D} = -\mu\epsilon\frac{\partial}{\partial t}\nabla\times\vec{\pi}^h \tag{9.16a}$$

$$\vec{H} = \nabla(\nabla\cdot\vec{\pi}^h) - \mu\epsilon\frac{\partial^2\vec{\pi}^h}{\partial t^2} \tag{9.16b}$$

The Hertz potentials $\vec{\pi}^e$ and $\vec{\pi}^h$ satisfy the wave equations:

$$\nabla^2\vec{\pi}^e - \mu\epsilon\frac{\partial^2\vec{\pi}^e}{\partial t^2} = 0 \tag{9.17a}$$

$$\nabla^2 \vec{\pi}^h - \mu\epsilon \frac{\partial^2 \vec{\pi}^h}{\partial t^2} = 0 \tag{9.17b}$$

Assuming a time dependence of $e^{j\omega t}$ and setting $k^2 = \omega^2\mu\epsilon$, the field and wave equations can be written as

$$\vec{E}^e = \nabla(\nabla \cdot \vec{\pi}^e) + k^2\vec{\pi}^e \tag{9.18a}$$

$$\vec{H}^e = j\omega\epsilon\nabla \times \vec{\pi}^e \tag{9.18b}$$

$$\vec{E}^h = -j\omega\mu\nabla \times \vec{\pi}^h \tag{9.18c}$$

$$\vec{H}^h = \nabla(\nabla \cdot \vec{\pi}^h) + k^2\vec{\pi}^h \tag{9.18d}$$

and

$$\nabla^2\vec{\pi}^e + k^2\vec{\pi}^e = 0 \tag{9.19a}$$

$$\nabla^2\vec{\pi}^h + k^2\vec{\pi}^h = 0 \tag{9.19b}$$

TE and TM Waves

The electromagnetic fields in general can be expressed as the superposition of TE and TM fields. For ease of mathematical analysis, the fields can be expressed in terms of the scalar potential functions, $\phi^e(x, y, z)$ and $\phi^h(x, y, z)$. Consider TE and TM modes with respect to z-direction. We can then define

$$\vec{\pi}^e = \hat{z}\phi^e(x, y, z), \quad \text{for TM or } E\text{-Waves} \tag{9.20a}$$

$$\vec{\pi}^h = \hat{z}\phi^h(x, y, z), \quad \text{for TE or } H\text{-Waves} \tag{9.20b}$$

Then, from (9.19), we get the following wave equations in terms of ϕ^e and ϕ^h:

$$\nabla^2\phi^e + k^2\phi^e = 0 \tag{9.21a}$$

$$\nabla^2\phi^h + k^2\phi^h = 0 \tag{9.21b}$$

The electric and magnetic fields are expressed as

$$\vec{E}^e = \left(\nabla \frac{\partial}{\partial z} + k^2\hat{z}\right)\phi^e \tag{9.22a}$$

$$\vec{H}^e = j\omega\epsilon\nabla \times \phi^e\hat{z} \tag{9.22b}$$

$$\vec{E}^h = -j\omega\mu\nabla \times \phi^h\hat{z} \tag{9.22c}$$

$$\vec{H}^h = \left(\nabla \frac{\partial}{\partial z} + k^2\hat{z}\right)\phi^h \tag{9.22d}$$

Expanding (9.22), we get

$$\vec{E}^e = \hat{x}\frac{\partial^2\phi^e}{\partial x\,\partial z} + \hat{y}\frac{\partial^2\phi^e}{\partial y\,\partial z} + \hat{z}\left(\frac{\partial^2\phi^e}{\partial z^2} + k^2\phi^e\right) \tag{9.23a}$$

$$\vec{H}^e = j\omega\epsilon\left(\hat{x}\frac{\partial\phi^e}{\partial y} - \hat{y}\frac{\partial\phi^e}{\partial x}\right) \tag{9.23b}$$

$$\vec{E^h} = -j\omega\mu\left(\hat{x}\frac{\partial\phi^h}{\partial y} - \hat{y}\frac{\partial\phi^h}{\partial x}\right) \tag{9.23c}$$

$$\vec{H^h} = \hat{x}\frac{\partial^2\phi^h}{\partial x\,\partial z} + \hat{y}\frac{\partial^2\phi^h}{\partial y\,\partial z} + \hat{z}\left(\frac{\partial^2\phi^h}{\partial z^2} + k^2\phi^h\right) \tag{9.23d}$$

The total $\vec{E}$ and $\vec{H}$ fields are then given by

$$\vec{E} = \vec{E^e} + \vec{E^h} \tag{9.24a}$$

$$\vec{H} = \vec{H^e} + \vec{H^h} \tag{9.24b}$$

9.4 Spectral Domain Technique Using the Concept of Chain Matrix

The spectral domain technique using the concept of chain matrix is the most general method of analyzing single and coupled stripline-like transmission lines [5]. In this method, the potential function coefficients of one layer are easily related to those of subsequent layers through a chain matrix. Because of this, the overall computing time is linearly dependent on the number of dielectric layers.

In the following, the analysis of a general N-layer stripline-like transmission structure is carried out using the above technique.

9.4.1 Analysis of General N-layer Stripline-Like Transmission Structure

Consider a general N-layer stripline-like transmission structure as shown in Fig. 9.1. We assume the strip conductor to be infinitesimally thin and perfectly conducting and dielectrics to be homogeneous, isotropic and lossless. The electric and magnetic fields in the ith region can be expressed in terms of $\phi^e(x, y, z)$ and $\phi^h(x, y, z)$ (derived from (9.23)) as follows:

$$E_{zi}(x, y, z) = \left(\frac{\partial^2}{\partial z^2} + k_i^2\right)\phi_i^e(x, y, z) \tag{9.25a}$$

$$H_{zi}(x, y, z) = \left(\frac{\partial^2}{\partial z^2} + k_i^2\right)\phi_i^h(x, y, z) \tag{9.25b}$$

$$E_{xi}(x, y, z) = \frac{\partial^2}{\partial x\,\partial z}\phi_i^e(x, y, z) - j\omega\mu_i\frac{\partial}{\partial y}\phi_i^h(x, y, z) \tag{9.25c}$$

$$H_{xi}(x, y, z) = j\omega\epsilon_i\frac{\partial}{\partial y}\phi_i^e(x, y, z) + \frac{\partial^2}{\partial x\,\partial z}\phi_i^h(x, y, z) \tag{9.25d}$$

$$E_{yi}(x, y, z) = \frac{\partial^2}{\partial y\,\partial z}\phi_i^e(x, y, z) + j\omega\mu_i\frac{\partial}{\partial x}\phi_i^h(x, y, z) \tag{9.25e}$$

$$H_{yi}(x, y, z) = -j\omega\epsilon_i\frac{\partial}{\partial x}\phi_i^e(x, y, z) + \frac{\partial^2}{\partial y\,\partial z}\phi_i^h(x, y, z) \tag{9.25f}$$

$\phi_i^e(x, y, z)$ and $\phi_i^h(x, y, z)$ satisfy the Helmholtz equation,

$$(\nabla^2 + k_i^2)\phi_i^e(x, y, z) = 0 \tag{9.26a}$$

$$(\nabla^2 + k_i^2)\phi_i^h(x, y, z) = 0 \tag{9.26b}$$

where $k_i^2 = \omega^2\mu_i\epsilon_i$

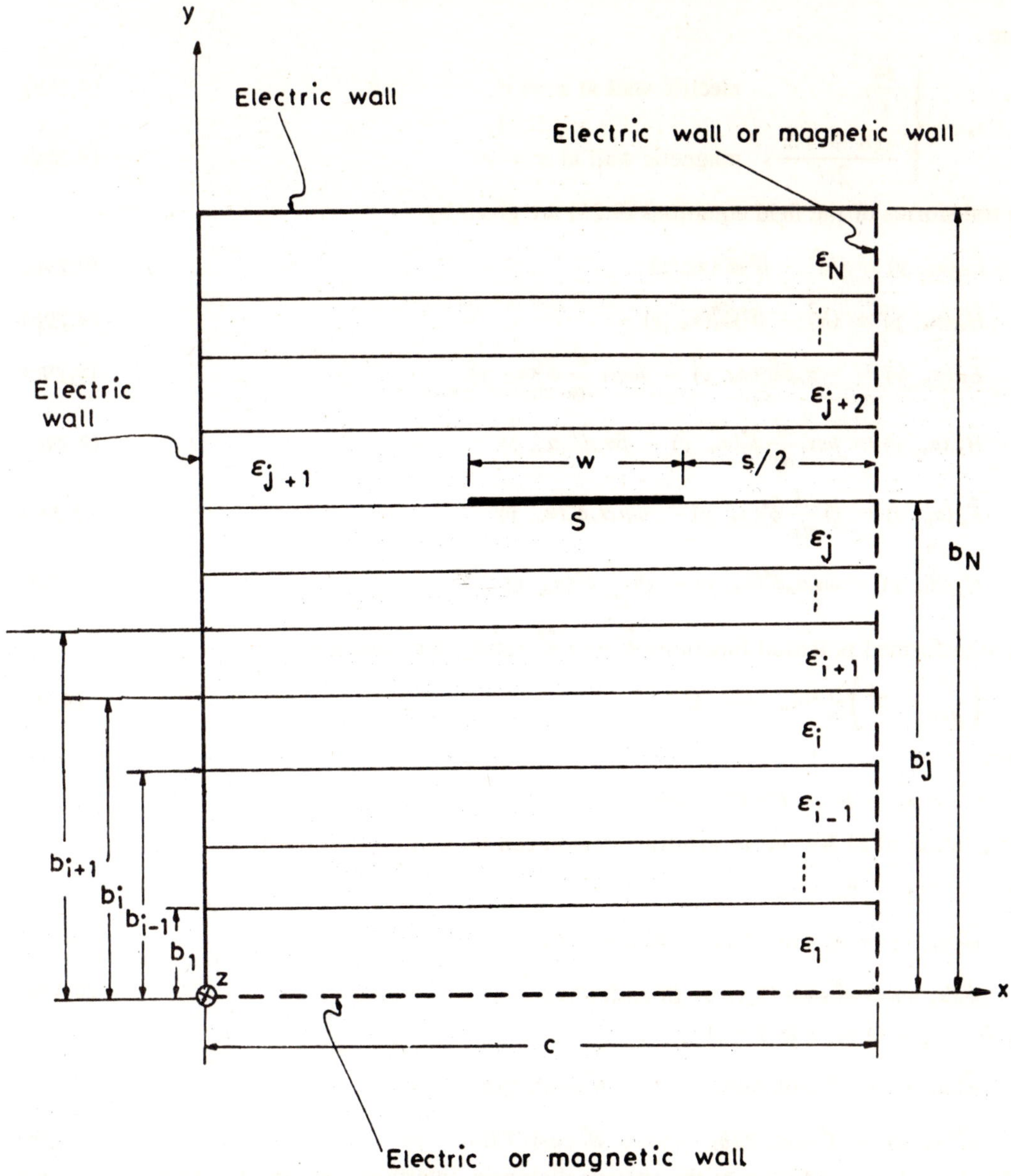

Fig. 9.1 A general N-layer stripline-like transmission structure for analysis.

For a time dependence of $e^{j\omega t}$ and propagation along the z-direction, we can write

$$\phi_i^{e,h}(x, y, z) = \phi_i^{e,h}(x, y)e^{-j\beta z} \tag{9.27}$$

where β is the propagation constant. The factor $e^{j\omega t}$ has been omitted for convenience. In order to convert (9.25) and (9.26) into the spectral domain, we define the Fourier transform of $\phi_i^{e,h}(x, y)$ as

$$\hat{\phi}_i^{e,h}(\alpha_n, y) = \int_0^c \phi_i^{e,h}(x, y)e^{j\alpha_n x}\, dx \tag{9.28a}$$

where

$$\alpha_n = \begin{cases} \dfrac{n\pi}{c}, & \text{electric wall at } x = c \quad (9.28b) \\[2ex] \dfrac{(2n+1)\pi}{2c}, & \text{magnetic wall at } x = c \quad (9.28c) \end{cases}$$

The transforms of the field equations (9.25) are given by

$$\hat{E}_{zi}(\alpha_n, y) = (k_i^2 - \beta^2)\hat{\phi}_i^e(\alpha_n, y) \tag{9.29a}$$

$$\hat{H}_{zi}(\alpha_n, y) = (k_i^2 - \beta^2)\hat{\phi}_i^h(\alpha_n, y) \tag{9.29b}$$

$$\hat{E}_{xi}(\alpha_n, y) = -\alpha_n\beta\hat{\phi}_i^e(\alpha_n, y) - j\omega\mu_i \frac{d}{dy}\hat{\phi}_i^h(\alpha_n, y) \tag{9.29c}$$

$$\hat{H}_{xi}(\alpha_n, y) = j\omega\epsilon_i \frac{d}{dy}\hat{\phi}_i^e(\alpha_n, y) - \beta\alpha_n\hat{\phi}_i^h(\alpha_n, y) \tag{9.29d}$$

$$\hat{E}_{yi}(\alpha_n, y) = j\beta\frac{d}{dy}\hat{\phi}_i^e(\alpha_n, y) - \omega\mu_i\alpha_n\hat{\phi}_i^h(\alpha_n, y) \tag{9.29e}$$

$$\hat{H}_{yi}(\alpha_n, y) = \omega\epsilon_i\alpha_n\hat{\phi}_i^e(\alpha_n, y) + j\beta\frac{d}{dy}\hat{\phi}_i^h(\alpha_n, y) \tag{9.29f}$$

The transformed potential functions $\hat{\phi}_i^e$ and $\hat{\phi}_i^h$ satisfy the equation

$$\left(\frac{d^2}{dy^2} - \gamma_i^2\right)\hat{\phi}_i^{e,h}(\alpha_n, y) = 0 \tag{9.30}$$

where

$$\gamma_i^2 = \alpha_n^2 + \beta^2 - k_i^2,\ k_i^2 = \omega^2\mu_i\epsilon_i \tag{9.31}$$

The solution of (9.30) for an arbitrary region can be written as follows:

For $b_{i-1} < y < b_i;\ i \leqslant j$

$$\hat{\phi}_i^e(\alpha_n, y) = A_i^e \sinh(\gamma_i y) + B_i^e \cosh(\gamma_i y) \tag{9.32a}$$

$$\hat{\phi}_i^h(\alpha_n, y) = A_i^h \sinh(\gamma_i y) + B_i^h \cosh(\gamma_i y) \tag{9.32b}$$

For $b_{i-1} < y < b_i;\ i \geqslant j + 1$

$$\hat{\phi}_i^e(\alpha_n, y) = A_i^e \sinh \gamma_i(b_i - y) + B_i^e \cosh \gamma_i(b_i - y) \tag{9.33a}$$

$$\hat{\phi}_i^h(\alpha_n, y) = A_i^h \sinh \gamma_i(b_i - y) + B_i^h \cosh \gamma_i(b_i - y) \tag{9.33b}$$

The boundary conditions at the various dielectric interfaces, except for those with strip sandwiched between them, are given by the following:

At $y = b_i$

$$\hat{E}_{zi}(\alpha_n) = \hat{E}_{zi+1}(\alpha_n) \tag{9.34a}$$

$$\hat{H}_{zi}(\alpha_n) = \hat{H}_{zi+1}(\alpha_n) \tag{9.34b}$$

$$\hat{E}_{xi}(\alpha_n) = \hat{E}_{xi+1}(\alpha_n) \tag{9.34c}$$

$$\hat{H}_{xi}(\alpha_n) = \hat{H}_{xi+1}(\alpha_n) \tag{9.34d}$$

At $y = b_j$, the interface where the strip is situated, we have

$$\hat{E}_{zj}(\alpha_n) = \hat{E}_{zj+1}(\alpha_n) = \hat{E}_z(\alpha_n) \tag{9.35a}$$

$$\hat{E}_{xj}(\alpha_n) = \hat{E}_{xj+1}(\alpha_n) = \hat{E}_x(\alpha_n) \tag{9.35b}$$

$$\hat{H}_{zj}(\alpha_n) - \hat{H}_{zj+1}(\alpha_n) = \hat{J}_x(\alpha_n) \tag{9.35c}$$

$$\hat{H}_{xj+1}(\alpha_n) - \hat{H}_{xj}(\alpha_n) = \hat{J}_z(\alpha_n) \tag{9.35d}$$

By substituting (9.32) and (9.33) in (9.29) and using (9.34), the coefficients of the potential function for any layer can be related to those of the subsequent one, except for the layers sandwiching the strip conductor. The relation is as follows:

$$\begin{bmatrix} A^e_{i+1} \\ B^e_{i+1} \\ A^h_{i+1} \\ B^h_{i+1} \end{bmatrix} = [P_{i+1,i}] \begin{bmatrix} A^e_i \\ B^e_i \\ A^h_i \\ B^h_i \end{bmatrix} \tag{9.36}$$

where

$$[P_{i+1,i}] = [\gamma_{i+1}u_i]^{-1}[\gamma_i v_i] \tag{9.37}$$

$$[\gamma_i v_i] = \begin{bmatrix} (k_i^2-\beta^2)\cdot\sinh(\gamma_i v_i) & (k_i^2-\beta^2)\cdot\cosh(\gamma_i v_i) & 0 & 0 \\ 0 & 0 & (k_i^2-\beta^2)\cdot\sinh(\gamma_i v_i) & (k_i^2-\beta^2)\cdot\cosh(\gamma_i v_i) \\ \alpha_n\beta\sinh(\gamma_i v_i) & \alpha_n\beta\cosh(\gamma_i v_i) & \mp j\omega\mu_i\gamma_i\cdot\cosh(\gamma_i v_i) & \mp j\omega\mu_i\gamma_i\cdot\sinh(\gamma_i v_i) \\ \pm j\omega\epsilon_i\gamma_i\cdot\cosh(\gamma_i v_i) & \pm j\omega\epsilon_i\gamma_i\cdot\sinh(\gamma_i v_i) & \alpha_n\beta\sinh(\gamma_i v_i) & \alpha_n\beta\cosh(\gamma_i v_i) \end{bmatrix} \tag{9.38}$$

$$[\gamma_{i+1}u_i] = \begin{bmatrix} (k_{i+1}^2-\beta^2)\cdot\sinh(\gamma_{i+1}u_i) & (k_{i+1}^2-\beta^2)\cdot\cosh(\gamma_{i+1}u_i) & 0 & 0 \\ 0 & 0 & (k_{i+1}^2-\beta^2)\cdot\sinh(\gamma_{i+1}u_i) & (k_{i+1}^2-\beta^2)\cdot\cosh(\gamma_{i+1}u_i) \\ \alpha_n\beta\sinh(\gamma_{i+1}u_i) & \alpha_n\beta\cosh(\gamma_{i+1}u_i) & \mp j\omega\mu_{i+1}\gamma_{i+1}\cdot\cosh(\gamma_{i+1}u_i) & \mp j\omega\mu_{i+1}\gamma_{i+1}\cdot\sinh(\gamma_{i+1}u_i) \\ \pm j\omega\epsilon_{i+1}\gamma_{i+i}\cdot\cosh(\gamma_{i+1}u_i) & \pm j\omega\epsilon_{i+1}\gamma_{i+i}\cdot\sinh(\gamma_{i+1}u_i) & \alpha_n\beta\sinh(\gamma_{i+1}u_i) & \alpha_n\beta\cosh(\gamma_{i+1}u_i) \end{bmatrix} \tag{9.39}$$

In the above matrices, $v_i = u_i = b_i$ for $i \leqslant j$, and $v_i = 0$, $u_i = b_{i+1} - b_i$ for $i \geqslant j + 1$. The upper signs (plus or minus) are for $i \leqslant j$, whereas the lower signs holds for $i \geqslant j + 1$. The matrix $[P_{i+1,i}]$ is called the chain matrix. By using (9.36), the coefficients of potential function for any layer (except the layers sandwiching the strip conductor) can be obtained; e.g. the relation between the $(i + 3)$th layer and ith layer is writen as

$$\begin{bmatrix} A^e_{i+3} \\ B^e_{i+3} \\ A^h_{i+3} \\ B^h_{i+3} \end{bmatrix} = [P_{i+3,i+2}]\,[P_{i+2,i+1}]\,[P_{i+1,i}] \begin{bmatrix} A^e_i \\ B^e_i \\ A^h_i \\ B^h_i \end{bmatrix} \tag{9.40}$$

The fields below and above the strip conductors can be related to the coefficients of the potential function by using the relation.

$$\begin{bmatrix} \hat{E}_{zj}(\alpha_n) \\ \hat{H}_{zj}(\alpha_n) \\ \hat{E}_{xj}(\alpha_n) \\ \hat{H}_{xj}(\alpha_n) \end{bmatrix} = [\gamma_j b_j] \begin{bmatrix} A^e_j \\ B^e_j \\ A^h_j \\ B^h_j \end{bmatrix} \tag{9.41}$$

and

$$\begin{bmatrix} \hat{E}_{zj+1}(\alpha_n) \\ \hat{H}_{zj+1}(\alpha_n) \\ \hat{E}_{xj+1}(\alpha_n) \\ \hat{H}_{xj+1}(\alpha_n) \end{bmatrix} = [\gamma_{j+1}(b_{j+1} - b_j)] \begin{bmatrix} A^e_{j+1} \\ B^e_{j+1} \\ A^h_{j+1} \\ B^h_{j+1} \end{bmatrix} \tag{9.42}$$

By using the chain matrix (9.36), we can rewrite (9.41) and (9.42) as

$$\begin{bmatrix} \hat{E}_{zj}(\alpha_n) \\ \hat{H}_{zj}(\alpha_n) \\ \hat{E}_{xj}(\alpha_n) \\ \hat{H}_{xj}(\alpha_n) \end{bmatrix} = [\gamma_j b_j][P_{j,j-1}] \ldots [P_{2,1}] \begin{bmatrix} A^e_1 \\ B^e_1 \\ A^h_1 \\ B^h_1 \end{bmatrix} \tag{9.43}$$

and

$$\begin{bmatrix} \hat{E}_{zj+1}(\alpha_n) \\ \hat{H}_{zj+1}(\alpha_n) \\ \hat{E}_{xj+1}(\alpha_n) \\ \hat{H}_{xj+1}(\alpha_n) \end{bmatrix} = [\gamma_{j+1}(b_{j+1} - b_j)]\,[P_{j+2,j+1}]^{-1} \ldots [P_{N,N-1}]^{-1} \begin{bmatrix} A^e_N \\ B^e_N \\ A^h_N \\ B^h_N \end{bmatrix} \tag{9.44}$$

At this stage, we shall consider two different cases separately

Case 1: Electric walls at $y = 0$ and $y = b_N$

$$B_1^e = B_N^e = 0 \tag{9.45}$$

$$A_1^h = A_N^h = 0 \tag{9.46}$$

By subtracting (9.44) from (9.43) and using (9.35), (9.45) and (9.46), we get

$$\begin{bmatrix} 0 \\ \hat{J}_x(\alpha_n) \\ 0 \\ -\hat{J}_z(\alpha_n) \end{bmatrix} = [Q] \begin{bmatrix} A_1^e \\ 0 \\ 0 \\ B_1^h \end{bmatrix} - [R] \begin{bmatrix} A_N^e \\ 0 \\ 0 \\ B_N^h \end{bmatrix} \tag{9.47}$$

where

$$[Q] = [\gamma_j b_j][P_{j,j-1}][P_{j-1,j-2}] \ldots [P_{2,1}] \tag{9.48a}$$

and

$$[R] = [\gamma_{j+1}(b_{j+1} - b_j)][P_{j+2,j+1}]^{-1} \ldots [P_{N,N-1}]^{-1} \tag{9.48b}$$

Equation (9.47) can be rewritten as

$$[U] \begin{bmatrix} 0 \\ \hat{J}_x(\alpha_n) \\ 0 \\ -\hat{J}_z(\alpha_n) \end{bmatrix} = \begin{bmatrix} A_1^e \\ B_1^h \\ A_N^c \\ B_N^h \end{bmatrix} \tag{9.49}$$

The matrix $[U] = [T]^{-1}$, where the elements of $[T]$ are $T_{mn} = M_{pq}$, $q \neq 2, 3, 6, 7$. The matrix $[M]$ is obtained from matrices $[Q]$ and $[R]$ by using the relation:

$$[M] = \{[Q], -[R]\} \tag{9.50}$$

From (9.49), we have

$$A_1^e = U_{12}\hat{J}_x(\alpha_n) - U_{14}\hat{J}_z(\alpha_n) \tag{9.51a}$$

$$B_1^h = U_{22}\hat{J}_x(\alpha_n) - U_{24}\hat{J}_z(\alpha_n) \tag{9.51b}$$

Substituting (9.45), (9.46) and (9.51) in (9.43) and simplifying, we get

$$\begin{bmatrix} \hat{E}_z(\alpha_n) \\ \hat{E}_x(\alpha_n) \end{bmatrix} = \begin{bmatrix} \hat{G}_{11}(\alpha_n, \beta) & \hat{G}_{12}(\alpha_n, \beta) \\ \hat{G}_{21}(\alpha_n, \beta) & \hat{G}_{22}(\alpha_n, \beta) \end{bmatrix} \begin{bmatrix} \hat{J}_x(\alpha_n) \\ \hat{J}_z(\alpha_n) \end{bmatrix} \tag{9.52}$$

where $\hat{G}_{11}(\alpha_n, \beta)$, $\hat{G}_{12}(\alpha_n, \beta)$, $\hat{G}_{21}(\alpha_n, \beta)$ and $\hat{G}_{22}(\alpha_n, \beta)$ are the elements of the dyadic Green's function in Fourier transformed domain given by

$$\hat{G}_{11}(\alpha_n, \beta) = (Q_{11}U_{12} + Q_{14}U_{22}) \tag{9.53a}$$

$$\hat{G}_{12}(\alpha_n, \beta) = -(Q_{11}U_{14} + Q_{14}U_{24}) \tag{9.53b}$$

$$\hat{G}_{21}(\alpha_n, \beta) = (Q_{31}U_{12} + Q_{34}U_{22}) \tag{9.53c}$$

$$\hat{G}_{22}(\alpha_n, \beta) = -(Q_{31}U_{14} + Q_{34}U_{24}) \tag{9.54d}$$

Case 2: Magnetic wall at $y = 0$ and Electric wall at $y = b_N$

$$A_1^e = B_1^h = 0 \tag{9.54a}$$

$$B_N^e = A_N^h = 0 \tag{9.54b}$$

Subtracting (9.44) from (9.43) and using (9.35) and (9.54), we get

$$[U]\begin{bmatrix} 0 \\ \hat{J}_x(\alpha_n) \\ 0 \\ -\hat{J}_z(\alpha_n) \end{bmatrix} = \begin{bmatrix} B_1^e \\ A_1^h \\ A_N^e \\ B_N^h \end{bmatrix} \tag{9.55}$$

The matrix $[U] = [T]^{-1}$, where elements of $[T]$ are $T_{mn} = M_{pq}$, $q \neq 1, 4, 6, 7$. The matrix $[M]$ is obtained from (9.50). Following the procedure given above, we obtain

$$\hat{G}_{11}(\alpha_n, \beta) = (Q_{12}U_{12} + Q_{13}U_{22}) \tag{9.56a}$$

$$\hat{G}_{12}(\alpha_n, \beta) = -(Q_{12}U_{14} + Q_{13}U_{24}) \tag{9.56b}$$

$$\hat{G}_{21}(\alpha_n, \beta) = (Q_{32}U_{12} + Q_{33}U_{22}) \tag{9.56c}$$

$$\hat{G}_{22}(\alpha_n, \beta) = -(Q_{32}U_{14} + Q_{33}U_{24}) \tag{9.56d}$$

9.4.2 Characteristic Equation

The set of algebraic equations given by (9.52) involve four unknown quantities; $\hat{J}_x(\alpha_n)$, $\hat{J}_z(\alpha_n)$, $\hat{E}_z(\alpha_n)$ and $\hat{E}_x(\alpha_n)$. Because the electric fields and currents are nonzero in complementary regions at the interface $y = b_j$, it is possible to eliminate $\hat{E}_z(\alpha_n)$ and $\hat{E}_x(\alpha_n)$ by using Galerkin's method and Parseval's theorem [16, 17] and solve for $\hat{J}_x(\alpha_n)$ and $\hat{J}_z(\alpha_n)$. In order to solve (9.52), we first expand $\hat{J}_x(\alpha_n)$ and $\hat{J}_z(\alpha_n)$ in terms of known basis functions $\hat{J}_{xm}(\alpha_n)$ and $\hat{J}_{zm}(\alpha_n)$, respectively

$$\hat{J}_x(\alpha_n) = \sum_{m=1}^{M} c_m \hat{J}_{xm}(\alpha_n) \tag{9.57a}$$

$$\hat{J}_z(\alpha_n) = \sum_{m=1}^{N} d_m \hat{J}_{zm}(\alpha_n) \tag{9.57b}$$

where c_m and d_m are unknown constants. Substituting (9.57) in (9.52), we get

$$\hat{G}_{11}(\alpha_n, \beta) \sum_{m=1}^{M} c_m \hat{J}_{xm}(\alpha_n) + \hat{G}_{12}(\alpha_n, \beta) \sum_{m=1}^{N} d_m \hat{J}_{zm}(\alpha_n) = \hat{E}_z(\alpha_n) \tag{9.58a}$$

$$\hat{G}_{21}(\alpha_n, \beta) \sum_{m=1}^{M} c_m \hat{J}_{xm}(\alpha_n) + \hat{G}_{22}(\alpha_n, \beta) \sum_{m=1}^{N} d_m \hat{J}_{zm}(\alpha_n) = \hat{E}_x(\alpha_n) \tag{9.58b}$$

We now make use of the definition of the inner product of two functions $\hat{f}(\alpha_n)$ and $\hat{g}(\alpha_n)$ in the form

$$\langle \hat{f}(\alpha_n), \hat{g}^*(\alpha_n) \rangle = \sum_{n=0}^{\infty} \hat{f}(\alpha_n)\hat{g}^*(\alpha_n) \tag{9.59}$$

According to Parseval's theorem, we have

$$\sum_{n=0}^{\infty} \hat{f}(\alpha_n)\hat{g}^*(\alpha_n) = \frac{c}{2}\int_0^c f(x)g^*(x)\,dx \tag{9.60}$$

Taking the inner product of (9.58a) with $\hat{J}_{zi}(\alpha_n)$ and (9.58b) with $\hat{J}_{xj}(\alpha_n)$, $(i = 1, \ldots N$ and $j = 1, \ldots M)$ and applying Parseval's theorem, we obtain the following set of equations:

$$\sum_{m=1}^{M} c_m P_{im} + \sum_{m=1}^{N} d_m Q_{im} = 0, \qquad i = 1, \ldots N \tag{9.61a}$$

$$\sum_{m=1}^{M} c_m R_{jm} + \sum_{m=1}^{N} d_m S_{jm} = 0, \qquad j = 1, \ldots M \tag{9.61b}$$

where

$$P_{im} = \sum_{n=0}^{\infty} \hat{J}^*_{zi}(\alpha_n)\hat{G}_{11}(\alpha_n, \beta)\hat{J}_{xm}(\alpha_n), \qquad i = 1, \ldots N \tag{9.62a}$$

$$Q_{im} = \sum_{n=0}^{\infty} \hat{J}^*_{zi}(\alpha_n)\hat{G}_{12}(\alpha_n, \beta)\hat{J}_{zm}(\alpha_n), \qquad i = 1, \ldots N \tag{9.62b}$$

$$R_{jm} = \sum_{n=0}^{\infty} \hat{J}^*_{xj}(\alpha_n)\hat{G}_{21}(\alpha_n, \beta)\hat{J}_{xm}(\alpha_n), \qquad j = 1, \ldots M \tag{9.62c}$$

$$S_{jm} = \sum_{n=0}^{\infty} \hat{J}^*_{xj}(\alpha_n)\hat{G}_{22}(\alpha_n, \beta)\hat{J}_{zm}(\alpha_n), \qquad j = 1, \ldots M \tag{9.62d}$$

The inner products on the right hand side of (9.61) vanish with the application of Parseval's theorem because the inverse transforms of $\hat{E}_x(\alpha_n)$, $\hat{E}_z(\alpha_n)$ and $\hat{J}_x(\alpha_n)$, $\hat{J}_z(\alpha_n)$ are nonzero in the complementary regions at the interface $y = b_j$. The propagation constants of the dominant and higher order modes are obtained by setting the determinant of the coefficient matrix in (9.61) equal to zero and seeking the roots of the resulting equations.

9.4.3 Choice of Basis Functions

The numerical solution of (9.61) is obtained by introducing a known set of basis functions. The accuracy of the final result depends on the accuracy with which the basis functions represent the true current distribution on the stripconductor S. In the following, simple as well as more accurate basis functions are considered.

Simple Basis Functions: An important feature of the spectral domain method as applied here is that quite accurate solutions are possible, even if a single term approximation to the current distribution $(N = 1, M = 0)$ is assumed. Furthermore, this results in simplicity as well as numerical efficiency. For a single term approximation to the current distribution $(N = 1, M = 0)$, the characteristic equation (9.61) simplifies to

$$\sum_{n=0}^{\infty} \hat{G}_{12}(\alpha_n, \beta) \mid \hat{J}_z(\alpha_n) \mid^2 = 0 \tag{9.63}$$

where $\hat{J}_{z1}(\alpha_n)$ is replaced by $\hat{J}_z(\alpha_n)$. The expression for $J_z(x)$ is chosen to be of the form

$$J_z(x) = \begin{cases} \dfrac{1}{2w}\left(1 + \left|\dfrac{2}{w}\left(x - \dfrac{2c - s - w}{2}\right)\right|^3\right); & c - \dfrac{s}{2} - w \leqslant x \leqslant c - \dfrac{s}{2} \\ 0 & ; \text{Otherwise} \end{cases} \tag{9.64}$$

If has been reported in the literature [18] that the zeroth order solution of propagation constant β obtained from (9.63) is accurate enough for most practical applications. However, it is desirable to check the convergence behaviour of the solution by comparing the zeroth order solution with the first order solution ($M = N = 1$). For the first order solution, we assume $J_z(x)$ as given in (9.64), and $J_x(x)$, to be of the form

$$J_x(x) = \begin{cases} \dfrac{1}{w}\sin\left[\dfrac{2\pi}{w}\left(x - \dfrac{2c - s - w}{2}\right)\right]; & c - \dfrac{s}{2} - w \leqslant x \leqslant c - \dfrac{s}{2} \\ 0 & ; \quad \text{Otherwise} \end{cases} \tag{9.65}$$

Using Fourier transforms of (9.64) and (9.65) in (9.62), and setting the determinant of coefficient matrix in (9.61) equal to zero yields the first order soution of the propagation constant β. The difference between the zeroth order and the first order results is usually very small.

More Accurate Basis Functions: For very accurate numerical results it is essential to increase the number of terms in the basis function or choose a more complicated but realistic basis function. The following choice of sinusoidal functions modified by an edge-condition term is reported to yield very accurate results [19].

$$J_{zi}(x) = \frac{\cos\left[\dfrac{i\pi}{w}\left(x - \dfrac{2c - s - 2w}{c}\right)\right]}{\sqrt{1 - \left[\dfrac{2}{w}\left(x - \dfrac{2c - s - 2w}{2}\right) - 1\right]^2}}; \quad c - \frac{s}{2} - w \leqslant x \leqslant c - \frac{s}{2} \tag{9.66}$$

$$= 0 \qquad ; \quad \text{Otherwise}$$

$$J_{xj}(x) = \frac{\sin\left[\dfrac{j\pi}{w}\left(x - \dfrac{2c - s - 2w}{2}\right)\right]}{\sqrt{1 - \left[\dfrac{2}{w}\left(x - \dfrac{2c - s - 2w}{2}\right) - 1\right]^2}}; \quad c - \frac{s}{2} - w \leqslant x \leqslant c - \frac{s}{2} \tag{9.67}$$

$$= 0 \qquad ; \quad \text{Otherwise}$$

These equations can be easily Fourier transformed and substituted in the characteristic equation. The accuracy of numerical results can be improved by increasing the number of terms in (9.57). This however increases the computational time. For most practical applications a choice of $N = M = 1$ is found to be adequate.

The formulas given by (9.61) and (9.62) are general and can be readily used to obtain propagation constants of a variety of stripline-like transmission lines provided the dyadic Green's functions for the corresponding structure are known. The generalized method of obtaining these functions is described in section 9.4. In the following, the application of this method for obtaining the dyadic Green's functions in the case of microstrip-like stripline and broadside-coupled suspended stripline is presented.

9.4.4 Application to Microstrip-Like Stripline

Consider the microstrip-like stripline structure shown in Fig. 9.2. The boundaries at $y = 0$ and $y = b$ represent electric walls. Hence the formula given in (9.53) is applicable for the determination of elements of the dyadic Green's function. Furthermore, since the boundaries at $x = 0$ and $x = c$ represent electric walls, α_n is given by (9.28b). In order to determine the various values of Q_{ij} and U_{ij} appearing in (9.53), we proceed as follows:

$$[Q] = [\gamma_1 h], \qquad h = b/2 \tag{9.68}$$

$$[R] = [\gamma_2 h], \qquad h = b/2 \tag{9.69}$$

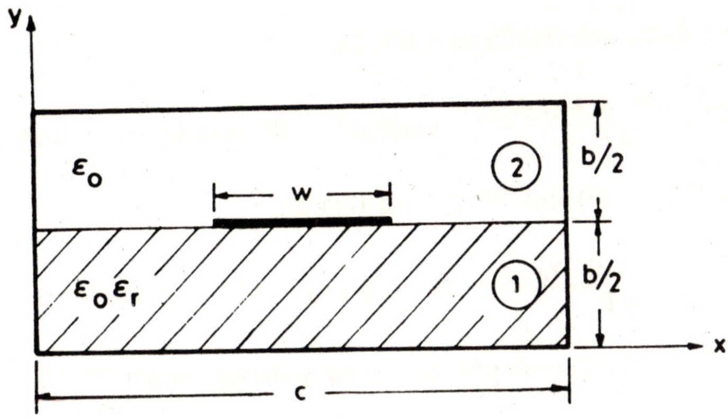

Fig. 9.2 Microstrip-like stripline structure for analysis.

The matrices $[\gamma_1 h]$ and $[\gamma_2 h]$ can be obtained from (9.38) and (9.39), respectively. Further, we have $[U] = [T]^{-1}$, where the elements of $[T]$ are $T_{mn} = M_{pq}$, $q \neq 2, 3, 6, 7$. Matrix $[M]$ is obtained from (9.50). Matrix $[T]$ for microstrip-like stripline is given by

$$[T] = \begin{bmatrix} (k^2-\beta^2)\cdot\sinh(\gamma b/2) & 0 & -(k_0^2-\beta^2)\cdot\sinh(\gamma_0 b/2) & 0 \\ 0 & (k^2-\beta^2)\cdot\cosh(\gamma b/2) & 0 & -(k_0^2-\beta^2)\cdot\cosh(\gamma_0 b/2) \\ \alpha_n\beta\sinh(\gamma b/2) & -j\omega\mu_0\gamma\cdot\sinh(\gamma b/2) & -\alpha_n\beta\sinh(\gamma_0 b/2) & -j\omega\mu_0\gamma_0\cdot\sinh(\gamma_0 b/2) \\ j\omega\epsilon_0\epsilon_r\gamma\cdot\cosh(\gamma b/2) & \alpha_n\beta\cosh(\gamma b/2) & j\omega\epsilon_0\gamma_0\cdot\cosh(\gamma_0 b/2) & -\alpha_n\beta\cosh(\gamma_0 b/2) \end{bmatrix} \tag{9.70}$$

where

$$\gamma_1 = \gamma = \sqrt{\alpha_n^2 + \beta^2 - k^2}; \qquad k^2 = \omega^2\mu_0\epsilon_0\epsilon_r \tag{9.71}$$

$$\gamma_2 = \gamma_0 = \sqrt{\alpha_n^2 + \beta^2 - k_0^2}; \qquad k_0^2 = \omega^2\mu_0\epsilon_0 \tag{9.72}$$

From (9.38) and (9.68), we have

$$Q_{11} = (k^2 - \beta^2) \sinh(\gamma b/2) \tag{9.73a}$$

$$Q_{14} = 0 \tag{9.73b}$$

$$Q_{31} = \alpha_n \beta \sinh(\gamma b/2) \tag{9.73c}$$

$$Q_{34} = -j\omega\mu_0\gamma \sinh(\gamma b/2) \tag{9.73d}$$

The elements of matrix $[U]$ can be obtained from the matrix $[T]^{-1}$. These are given by

$$U_{12} = -\frac{(k_0^2 - \beta^2)\sinh(\gamma_0 b/2)}{\Delta T}[j\omega\mu_0\gamma\alpha_n\beta \sinh(\gamma b/2)\cdot\cosh(\gamma_0 b/2) + j\omega\alpha_n\beta\mu_0\gamma_0 \sinh(\gamma_0 b/2)\cosh(\gamma b/2)] \tag{9.74a}$$

$$U_{14} = -\frac{(k_0^2 - \beta^2)\sinh(\gamma_0 b/2)}{\Delta T}[-j\omega\mu_0\gamma_0(k^2 - \beta^2)\cosh(\gamma b/2)\cdot\sinh(\gamma_0 b/2) - j\omega\mu_0\gamma(k_0^2 - \beta^2)\sinh(\gamma b/2)\cosh(\gamma_0 b/2)] \tag{9.74b}$$

$$U_{22} = -\frac{(k_0^2 - \beta^2)\sinh(\gamma_0 b/2)}{\Delta T}[-(k^2 - \beta^2)(k_0^2 - \alpha_n^2)\cdot\sinh(\gamma b/2)\cosh(\gamma_0 b/2) + \alpha_n^2\beta^2 \sinh(\gamma b/2)\cosh(\gamma_0 b/2) + k^2\gamma_0\gamma \sinh(\gamma_0 b/2)\cosh(\gamma b/2)] \tag{9.74c}$$

$$U_{24} = -\frac{(k_0^2 - \beta^2)\sinh(\gamma_0 b/2)}{\Delta T}[(k^2 - k_0^2)\alpha_n\beta\cdot\sinh(\gamma b/2)\cosh(\gamma_0 b/2)] \tag{9.74d}$$

where

$$\Delta T = -(k^2 - \beta^2)(k_0^2 - \beta^2)k_0^2 \sinh(\gamma b/2)\cosh(\gamma b/2)\sinh(\gamma_0 b/2)\cosh(\gamma_0 b/2) \cdot[\{\gamma \tanh(\gamma b/2) + \gamma_0\epsilon_r \tanh(\gamma_0 b/2)\}\{\gamma \coth(\gamma b/2) + \gamma_0 \coth(\gamma_0 b/2)\}] \tag{9.75}$$

Substituting (9.73) – (9.75) in (9.53) and simplifying, we obtain the following expressions for the elements of the dyadic Green's function

$$\hat{G}_{11}(\alpha_n, \beta) = \frac{j\alpha_n\beta}{\omega\epsilon_0\Delta}[\gamma_0 \tanh(\gamma_0 b/2) + \gamma \tanh(\gamma b/2)] \tag{9.76a}$$

$$\hat{G}_{12}(\alpha_n, \beta) = +\frac{j}{\omega\epsilon_0\Delta}[\gamma(k_0^2 - \beta^2)\tanh(\gamma b/2) + \gamma_0(k^2 - \beta^2)\cdot\tanh(\gamma_0 b/2)] \tag{9.76b}$$

$$\hat{G}_{21}(\alpha_n, \beta) = \frac{j}{\omega\epsilon_0\Delta}[\gamma_0(k_0^2 - \alpha_n^2)\tanh(\gamma b/2) + \gamma_0(k^2 - \alpha_n^2)\tanh(\gamma_0 b/2)] \tag{9.76c}$$

$$\hat{G}_{22}(\alpha_n, \beta) = +\hat{G}_{11}(\alpha_n, \beta) \tag{9.76d}$$

and

$$\Delta = [\gamma \tanh(\gamma b/2) + \gamma_0\epsilon_r \tanh(\gamma_0 b/2)] \cdot[\gamma \coth(\gamma b/2) + \gamma_0 \coth(\gamma_0 b/2)] \tag{9.77}$$

9.4.5 *Application to Broadside–Coupled Suspended Stripline*

Consider a broadside-coupled suspended stripline as shown in Fig. 9.3. This structure can support two orthogonal modes (even and odd) of excitation. In the odd-mode case, the symmetry plane $y = 0$ is replaced by an electric wall (Fig. 9.3b), whereas, in the even-mode case the symmetry plane $y = 0$ is replaced by a magnetic wall (Fig. 9.3c). In the following, elements of the dyadic Green's function are derived for the odd-mode and even-mode cases separately.

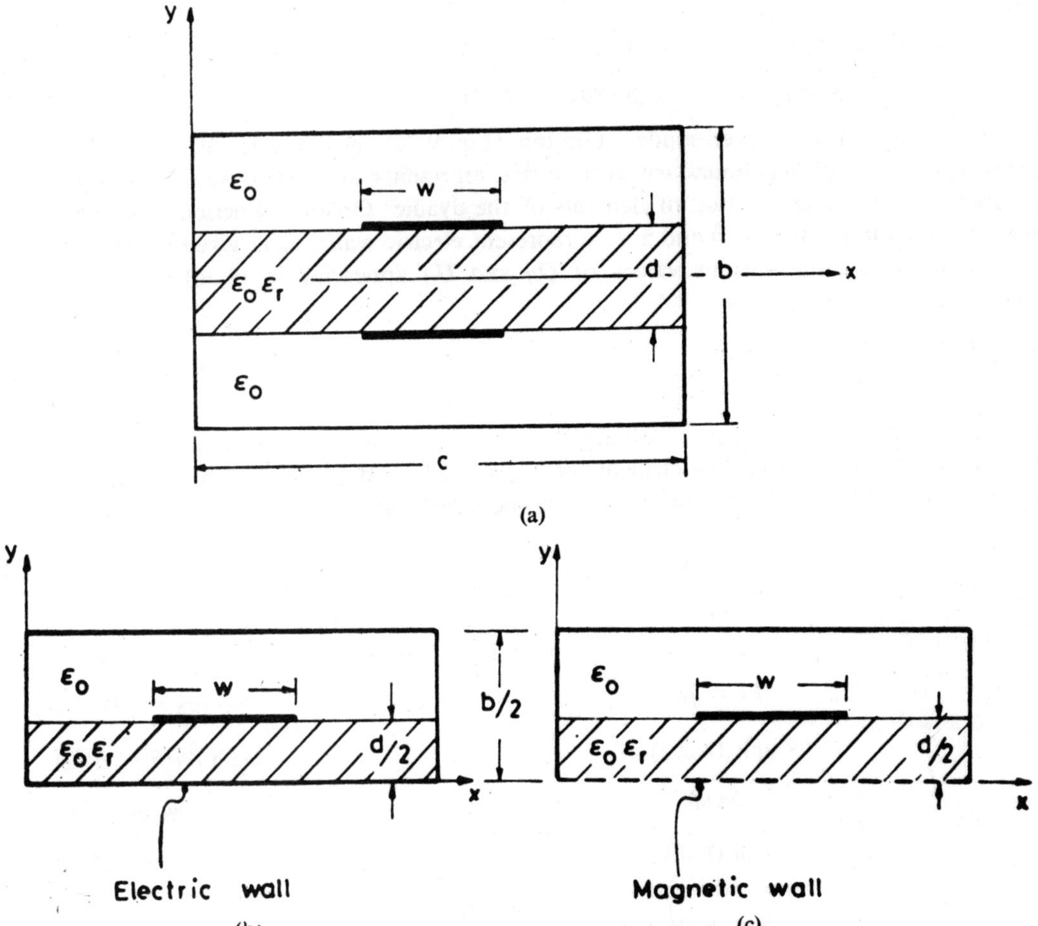

Fig. 9.3 (a) Broadside-coupled suspended stripline for analysis, (b) Cross-section of the structure for odd-mode excitation, (c) Cross-section of the structure for even-mode excitation.

Odd-Mode Case: For the odd-mode excitation the structure (Fig. 9.3b) resembles the microstrip-like stripline shown in Fig. 9.2 with substrate height equal to $d/2$ and air gap height equal to $(b - d)/2$. Incorporation of these changes in (9.76) and (9.77) gives

$$\hat{G}_{11}(\alpha_n, \beta) = \frac{j\alpha_n\beta}{\omega\epsilon_0\Delta}[\gamma_0 \tanh(\gamma_0(b - d)/2) + \gamma \tanh(\gamma d/2)] \tag{9.78a}$$

$$\hat{G}_{12}(\alpha_n, \beta) = + \frac{j}{\omega\epsilon_0\Delta}[\gamma(k_0^2 - \beta^2)\tanh(\gamma d/2) + \gamma_0(k^2 - \beta^2)\tanh(\gamma_0(b - d)/2)] \tag{9.78b}$$

$$\hat{G}_{21}(\alpha_n, \beta) = \frac{j}{\omega\epsilon_0\Delta}[\gamma_0(k_0^2 - \alpha_n^2)\tanh(\gamma\, d/2) + \gamma_0(k^2 - \alpha_n^2)\tanh(\gamma_0(b - d)/2)] \tag{9.78c}$$

$$\hat{G}_{22}(\alpha_n, \beta) = +\hat{G}_{11}(\alpha_n, \beta) \tag{9.78d}$$

and

$$\Delta = [\gamma\tanh(\gamma\, d/2) + \gamma_0\epsilon_r\tanh(\gamma_0(b - d)/2)] \cdot[\gamma\coth(\gamma\, d/2) + \gamma_0\coth(\gamma_0(b - d)/2] \tag{9.79}$$

Even-mode case: For the even-mode excitation, (Fig. 9.3c), the boundary at $y = 0$ represents a magnetic wall and the boundary at $y = b/2$ represents an electric wall. Hence (9.56) is applicable for the determination of elements of the dyadic Green's function. Furthermore, since the boundaries at $x = 0$ and $x = c$ represent electric walls, α_n is given by (9.28b). In order to determine the various values of Q_{ij} and U_{ij} appearing in (9.56), we proceed as follows:

$$[Q] = [\gamma_1 h], \qquad h = d/2 \tag{9.80}$$

$$[R] = [\gamma_2 h], \qquad h = (b - d)/2 \tag{9.81}$$

The matrics $[\gamma_1 h]$ and $[\gamma_2 h]$ can be obtained from (9.38) and (9.39), respectively. Further, we have $[U] = [T]^{-1}$, where the elements of $[T]$ are $T_{mn} = M_{pq}$, $q \neq 1, 4, 6, 7$. After obtaining the matrix $[M]$ from (9.50), matrix $[T]$ can be written as

$$[T] = \begin{bmatrix} (k^2 - \beta^2)\cdot\cosh(\gamma\, d/2) & 0 & -(k_0^2 - \beta^2)\cdot\sinh(\gamma_0(b - d)/2 & 0 \\ 0 & (k^2 - \beta^2)\cdot\sinh(\gamma\, d/2) & 0 & -(k_0^2 - \beta^2)\cdot\cosh(\gamma_0(b - d)/2) \\ \alpha_n\beta\cosh(\gamma\, d/2) & -j\omega\mu_0\gamma\cdot\cosh(\gamma\, d/2) & -\alpha_n\beta\sinh(\gamma_0(b - d)/2) & -j\omega\mu_0\gamma_0\cdot\sinh(\gamma_0(b - d)/2) \\ j\omega\epsilon_0\epsilon_r\gamma\cdot\sinh(\gamma\, d/2) & \alpha_n\beta\sinh(\gamma\, d/2) & j\omega\epsilon_0\gamma_0\cdot\cosh(\gamma_0(b - d)/2) & -\alpha_n\beta\cosh(\gamma_0(b-d)/2) \end{bmatrix} \tag{9.82}$$

where

$$\gamma_1 = \gamma = \sqrt{\alpha_n^2 + \beta^2 - k^2}; \qquad k^2 = \omega^2\mu_0\epsilon_0\epsilon_r \tag{9.83}$$

$$\gamma_2 = \gamma_0 = \sqrt{\alpha_n^2 + \beta^2 - k_0^2}; \qquad k_0^2 = \omega^2\mu_0\epsilon_0 \tag{9.84}$$

From (9.38) and (9.80), we have

$$Q_{12} = (k^2 - \beta^2)\cosh(\gamma\, d/2) \tag{9.85a}$$

$$Q_{13} = 0 \tag{9.85b}$$

$$Q_{32} = \alpha_n\beta \cosh(\gamma\, d/2) \tag{9.85c}$$

$$Q_{33} = -j\omega\mu_0\gamma \cosh(\gamma\, d/2) \tag{9.85d}$$

After obtaining the elements of matrix $[U]$ from the matrix $[T]^{-1}$ and then substituting various Q_{ij} and U_{ij} values in (9.56), we get the following expressions for the elements of the dyadic Green's function.

$$\hat{G}_{11}(\alpha_n, \beta) = \frac{j\alpha_n\beta}{\omega\epsilon_0\Delta}[\gamma_0 \tanh(\gamma_0(b-d)/2) + \gamma \coth(\gamma\, d/2)] \tag{9.86a}$$

$$\hat{G}_{12}(\alpha_n, \beta) = +\frac{j}{\omega\epsilon_0\Delta}[\gamma(k_0^2 - \beta^2)\coth(\gamma\, d/2) + \gamma_0(k^2 - \beta^2)\tanh(\gamma_0(b-d)/2)] \tag{9.86b}$$

$$\hat{G}_{21}(\alpha_n, \beta) = \frac{j}{\omega\epsilon_0\Delta}[\gamma_0(k_0^2 - \alpha_n^2)\coth(\gamma\, d/2) + \gamma_0(k^2 - \alpha_n^2)\tanh(\gamma_0(b-d)/2)] \tag{9.86c}$$

$$\hat{G}_{22}(\alpha_n, \beta) = +\hat{G}_{11}(\alpha_n, \beta) \tag{9.86d}$$

and

$$\Delta = [\gamma \coth(\gamma\, d/2) + \gamma_0\epsilon_r \tanh(\gamma_0(b-d)/2)] \cdot [\gamma \tanh(\gamma\, d/2) + \gamma_0 \coth(\gamma_0(b-d)/2)] \tag{9.87}$$

9.5 Numerical Results on Frequency Dependent Propagation Characteristics

In this section the numerical results on the frequency dependent propagation characteristics are presented in graphical form for a variety of single and coupled stripline-like transmission lines. Most of the results reported pertain to the dominant propagating mode. A few graphs showing the higher order mode behaviour are also included.

9.5.1 Single Conductor Stripline-Like Transmission Lines

The dominant mode in a rectangular stripline is the TEM mode. As the frequency increases, higher order modes can also propagate. For practical applications, it is essential to know the cut-off frequencies of these higher order modes. Figure 9.4 shows the normalized cut-off frequency versus w/c for different TE-modes in rectangular stripline [1]. It is observed that the TE_{10} mode cut-off remains unaltered by the presence of the stripconductor and continues to be governed by the criterion $\lambda_c = 2c$. Furthermore, the characteristics for TE_{01} mode cut-off cross the TE_{10} mode cut-off characteristics, for all values of $c/b > 1.0$. For $c/b = 1.0$, the TE_{01} mode cut-off characteristics are below those of the TE_{10} mode. It may be inferred that for $c/b = 1.0$, TE_{01} mode is always the first higher order mode. Data on TE_{11} and TE_{20} mode cut-off characteristics indicate that these modes can never propagate at frequencies below the cut-off frequency of the TE_{10} mode.

Figure 9.5 depicts the variations of normalized cut-off frequency versus c/b for the TE_{01} mode in rectangular striplines of different characteristic impedances [1]. For a rectangular stripline of characteristic impedance 50 ohms, we may observe that for $c/b < 1.93$, the first higher order mode is always TE_{01} whereas, for $c/b > 1.93$, the first higher order mode is the TE_{10} mode. The study of higher order mode behaviour in rectangular striplines is extremely important for proper use of the TEM transmission cells [1].

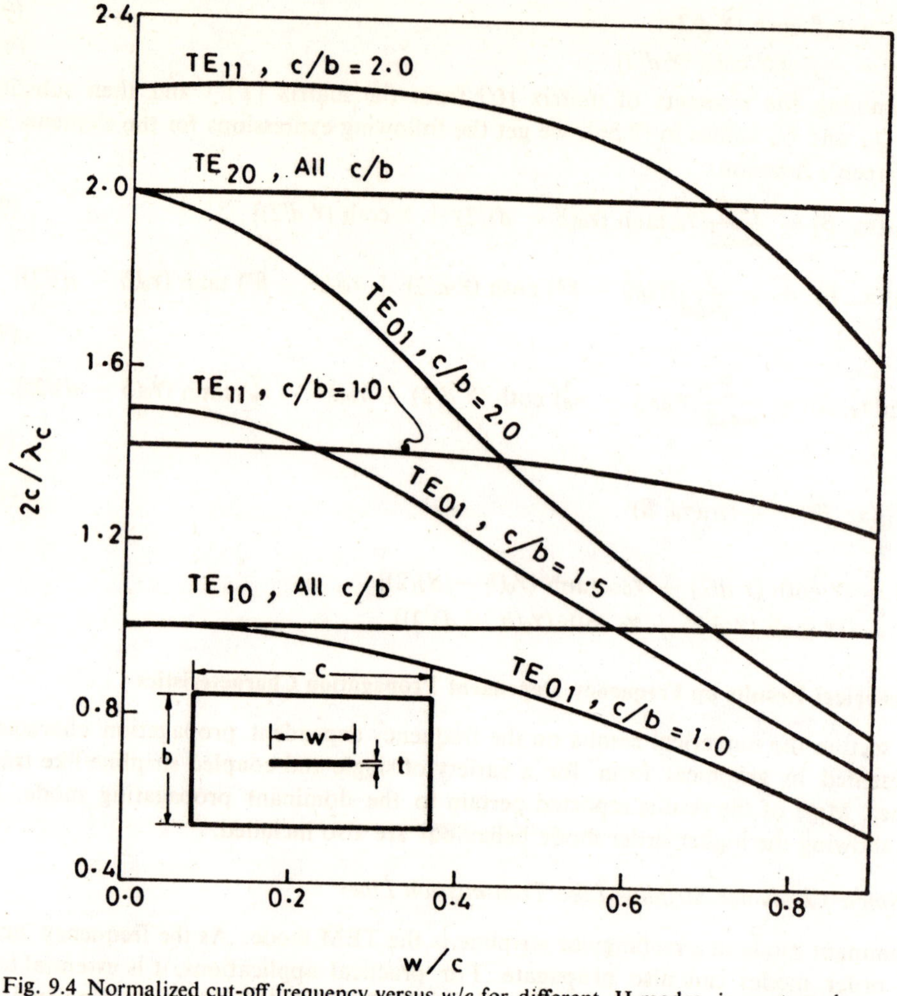

Fig. 9.4 Normalized cut-off frequency versus w/c for different H-modes in rectangular striplines with $c/b = 1.0$, 1.5 and 2.0; $t/b = 0$. (From Weil et al [1], reprinted with permission of Microwave Journal, Copyright © 1981 Horizon House-Microwave, Inc.)

The variations of dominant mode normalized guide-wavelength λ/λ^a and characteristic impedance Z of microstrip-like stripline as a function of frequency are plotted in Fig. 9.6. It may be noted that for a fixed value of w, λ/λ^a decreases and Z increases with an increase in the frequency. However, in the lower frequency range, the change in Z is small. Smaller the value of w, wider is the frequency band over which the change in Z remains small.

Figure 9.7 shows the variations of λ/λ^a and Z for the dominant mode of suspended stripline as a function of frequency. The graph shows that as the frequency increases, both λ/λ^a and Z decreases. The variation of λ/λ^a is similar to that of microstrip-like stripline, however, the variation in Z shows an opposite trend. The modal spectrum of a suspended stripline is shown in Fig. 9.8 [10]. We may observe that the dominant mode (EH_0) propagation is possible at all frequencies. However, depending on the structural parameters, higher order

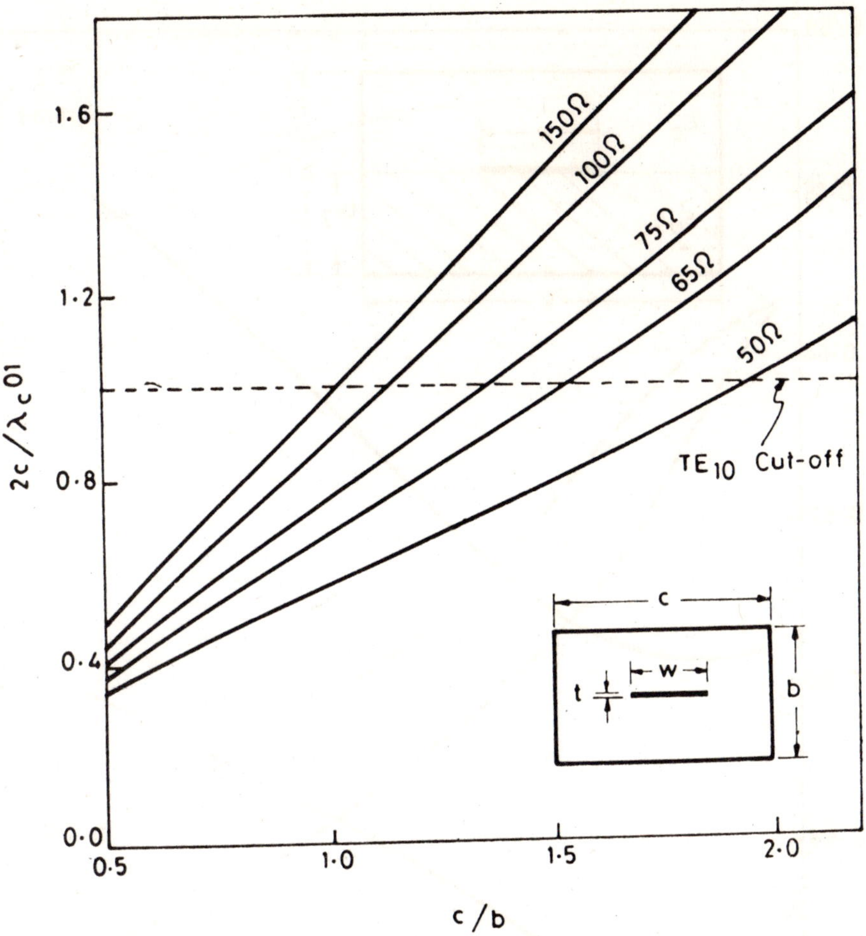

Fig. 9.5 Normalized cut-off frequency versus c/b for the TE_{01} mode in rectangular strip-lines of different characteristic impedance; $t/b = 0$. (From Weil et al [1], reprinted with permission of Microwave Journal, Copyright © 1981 Horizon House-Microwave, Inc:)

modes can also propagate, thereby limiting the dominant mode operating bandwidth. For the structural parameters chosen in Fig. 9.8, the first higher order mode can propagate above a frequency of 19 GHz. Figure 9.8 depicts the propagation characteristics of first seven higher order modes.

9.5.2 *Edge-Coupled Stripline-Like Transmission Lines*

The dominant even- and odd-mode guide wavelengths and characteristic impedances of an edge-coupled microstrip-like stripline as a function of frequency are plotted in Figs. 9.9 and 9.10 respectively. It may be observed from Fig. 9.9 that λ/λ^a for both even- and odd-modes decrease as the frequency increases. Further, the change in λ/λ^a for odd-mode is negligibly small especially for smaller values of w. From Fig. 9.10, it can be observed that for small values of w, the odd-mode characteristic impedance is insensitive to change in the frequency. For large values of w, the odd-mode characteristic impedance slightly

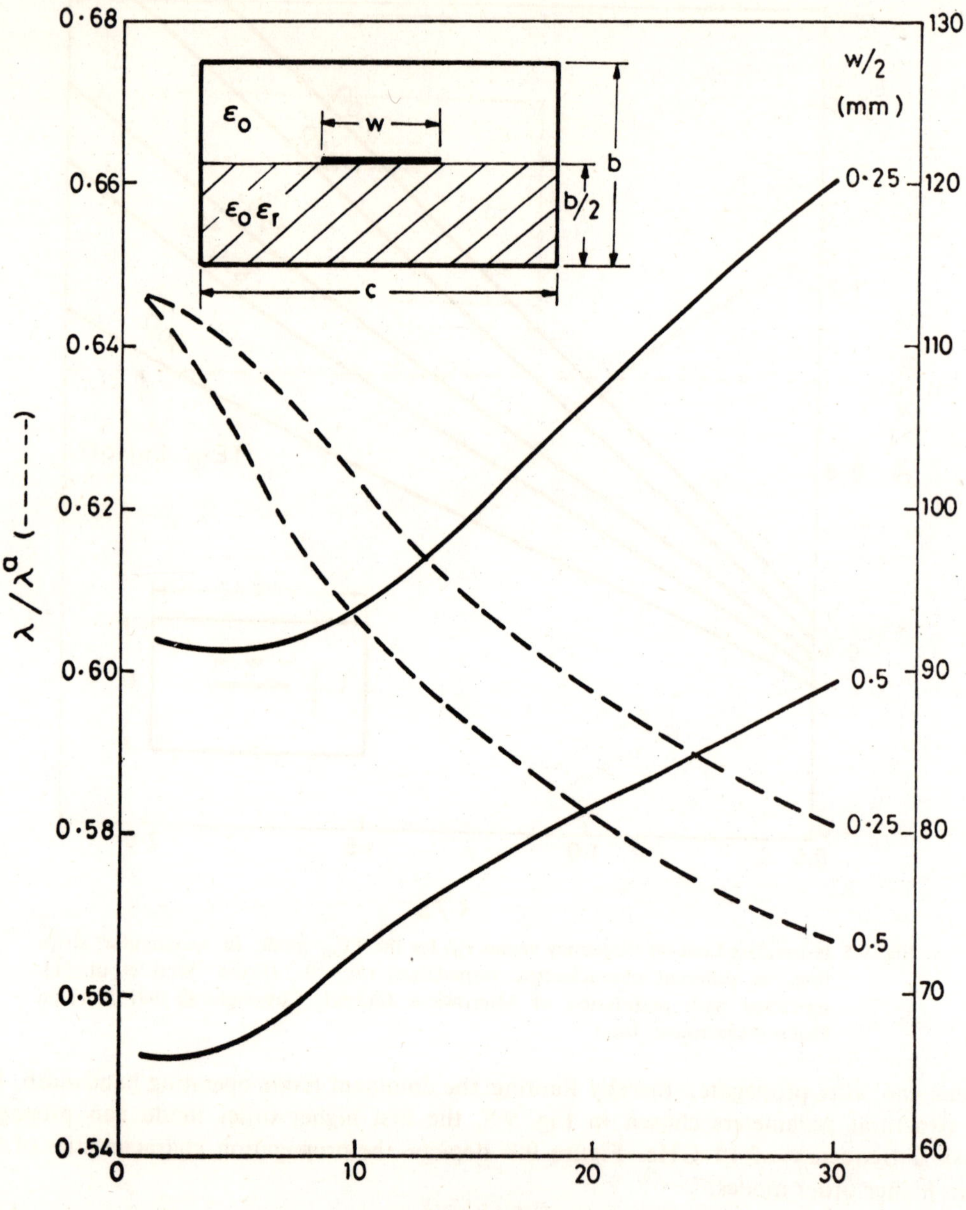

Fig. 9.6 Variations of dominant mode normalized guide wavelength λ/λ^a and characteristic impedance Z of microstrip-like stripline as a function of frequency ; $\epsilon_r = 3.78$, $b = 2$ mm, $c/w = 20$.

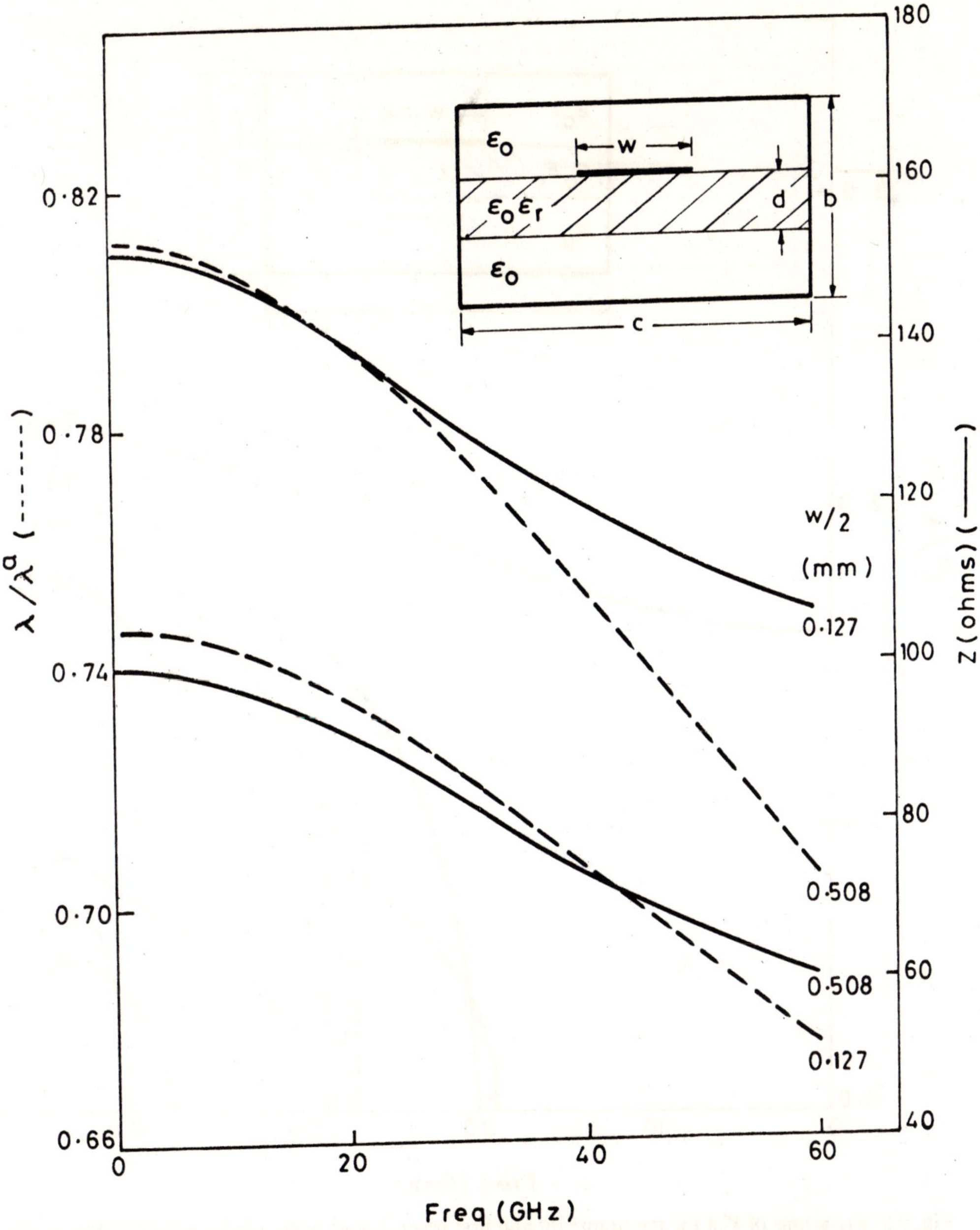

Fig. 9.7 Variations of dominant mode normalized guide wavelength λ/λ^a and characteristic impedance Z of suspended stripline as a function of frequency; $\epsilon_r = 3.78$, $d = 0.508$ mm, $b = 3.048$ mm, $c/w = 20$.

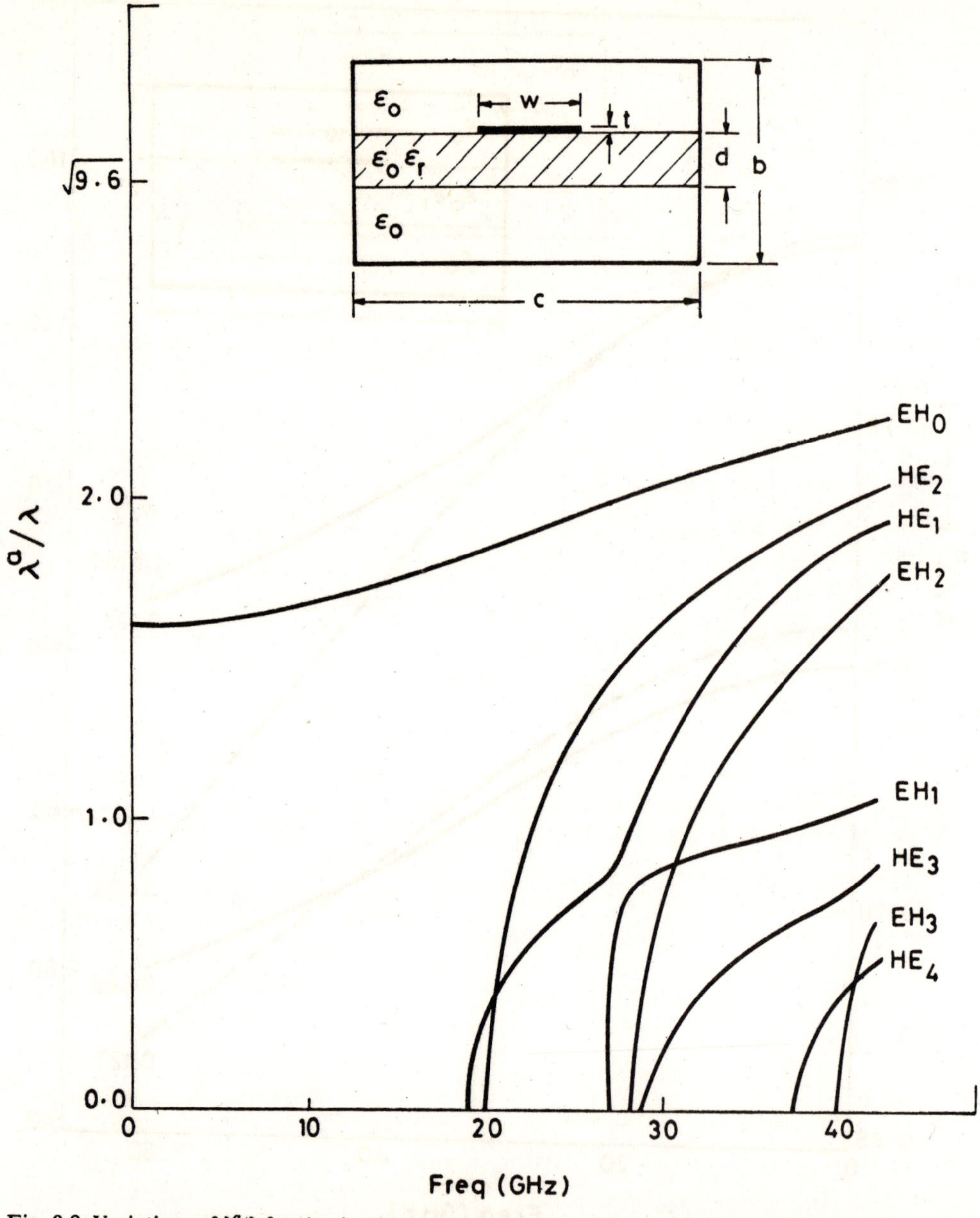

Fig. 9.8 Variations of λ^a/λ for the dominant and first seven higher order modes of suspended stripline as a function of frequency; $c = 2b = 7.112$ mm, $d = 635$ μm, $t = 5$ μm, $w = 1$ mm, $\epsilon_r = 9.6$. (From Bornemann [10], IEE Proc., Pt.H, 1985, Vol. 132, pp. 1-6, reprinted with permission of IEE).

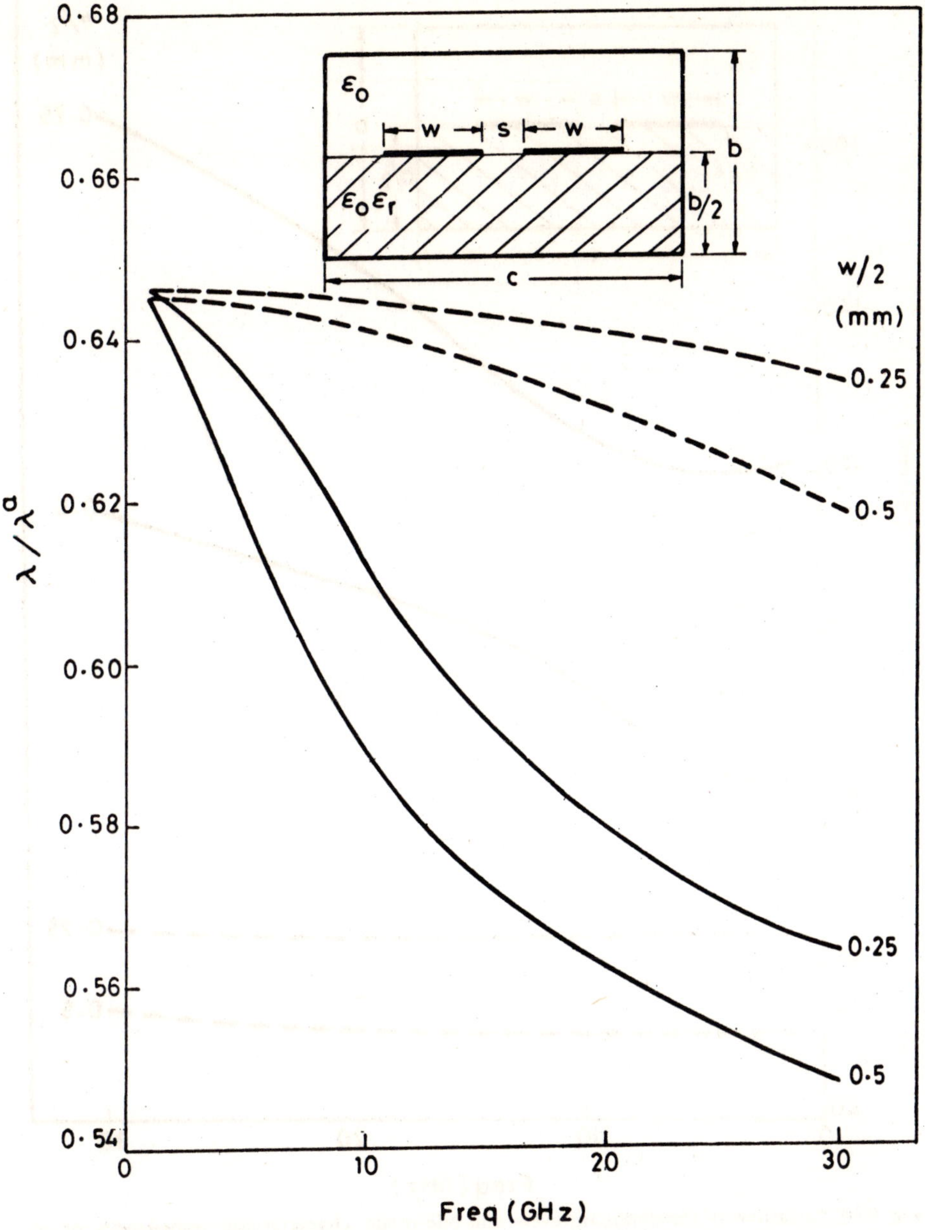

Fig. 9.9 Variations of the dominant even- and odd-mode guide wavelengths of an edge-coupled microstrip-like stripline as a function of frequency; $\epsilon_r = 3.78$, $b = 2$ mm, $s = 0.2$ mm, $c/w = 20$, ——— even mode, – – – odd mode.

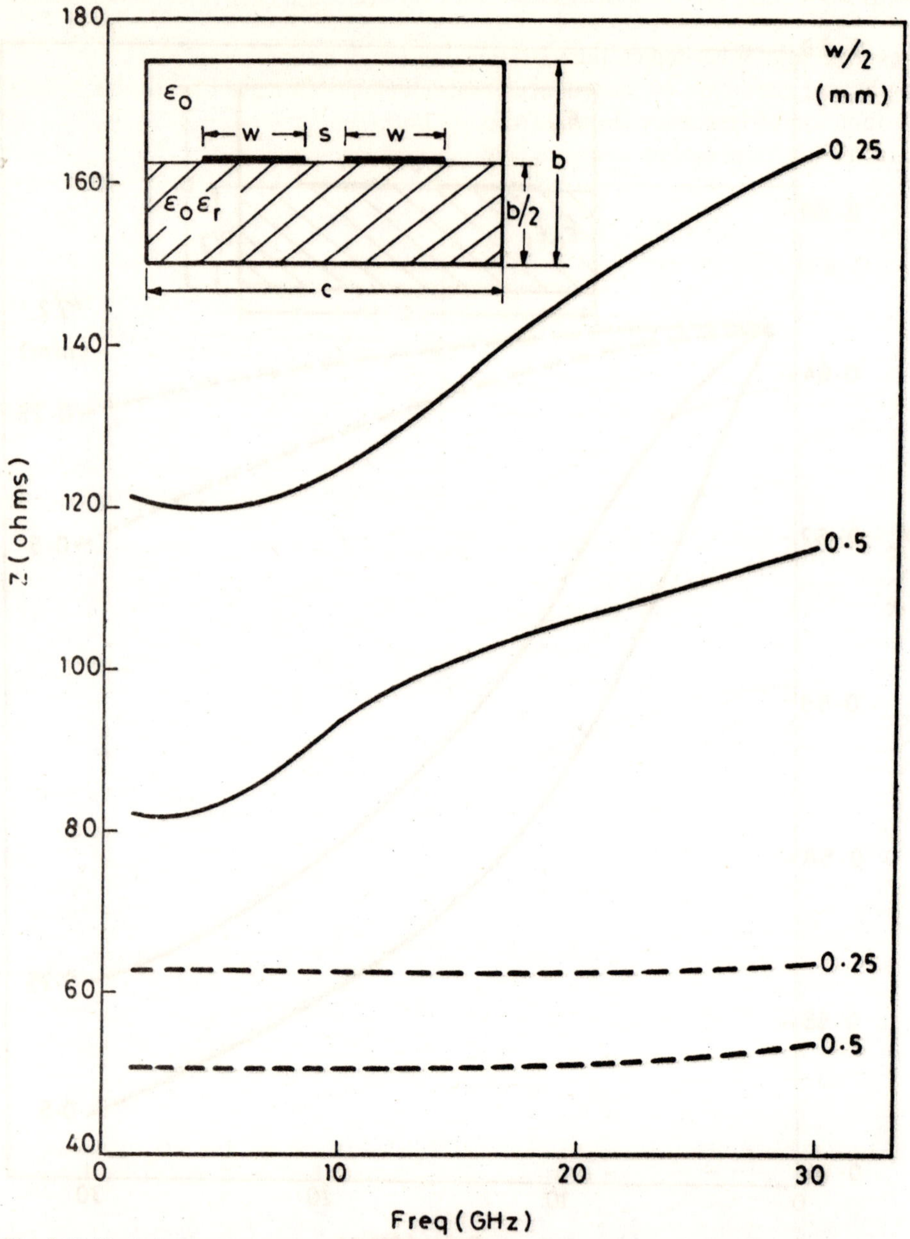

Fig. 9.10 Variations of the dominant even- and odd-mode characteristic impedances of an edge-coupled microstrip-like stripline as a function of frequency; $\epsilon_r = 3.78$, $b = 2$ mm, $s = 0.2$ mm, $c/w = 20$, ——— even mode, – – – odd mode.

increases as the frequency is increased. On the other hand, even-mode characteristic impedance increases rather rapidly as frequency increases, irrespective of the value of w. However, in the lower frequency range, the change in even-mode characteristic impedance is small. Smaller the value of w, wider is the frequency band over which the change in even-mode characteristic impedance is small.

Figures 9.11 and 9.12 depict the variations of the dominant even- and odd-mode guide wavelengths and characteristic impedances, respectively, of an edge-coupled suspended stripline as a function of frequency. In this structure both even- and odd-mode guide wavelengths and characteristic impedances decrease with an increase in frequency. The even-mode pro-

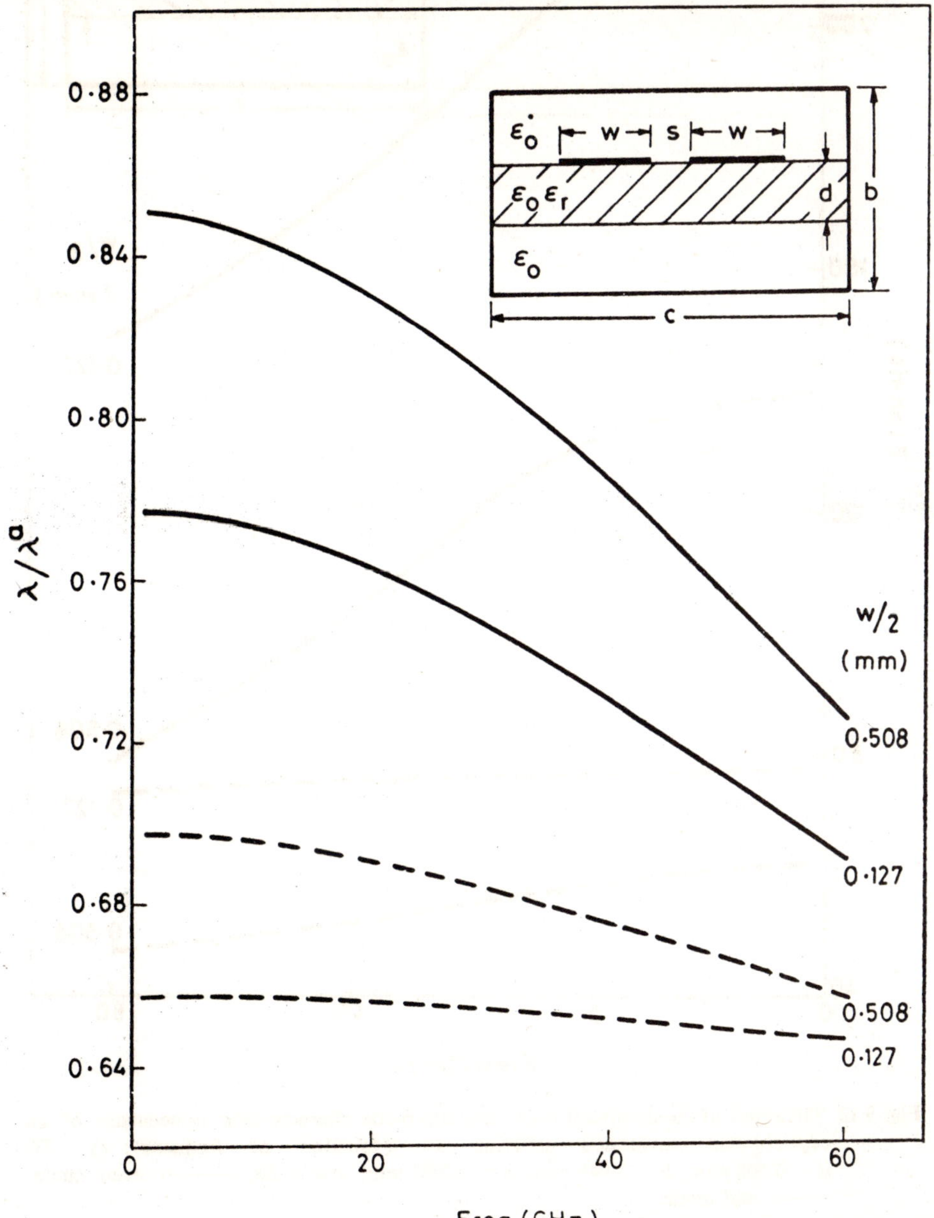

Fig. 9.11 Variations of the dominant even- and odd-mode normalized guide wavelengths of an edge-coupled suspended stripline as a function of frequency; $\epsilon_r = 3.78$, $d = 0.508$ mm, $b = 3.048$ mm, $s = 0.2032$ mm, $c/w = 20$, ——— even mode, – – – odd mode.

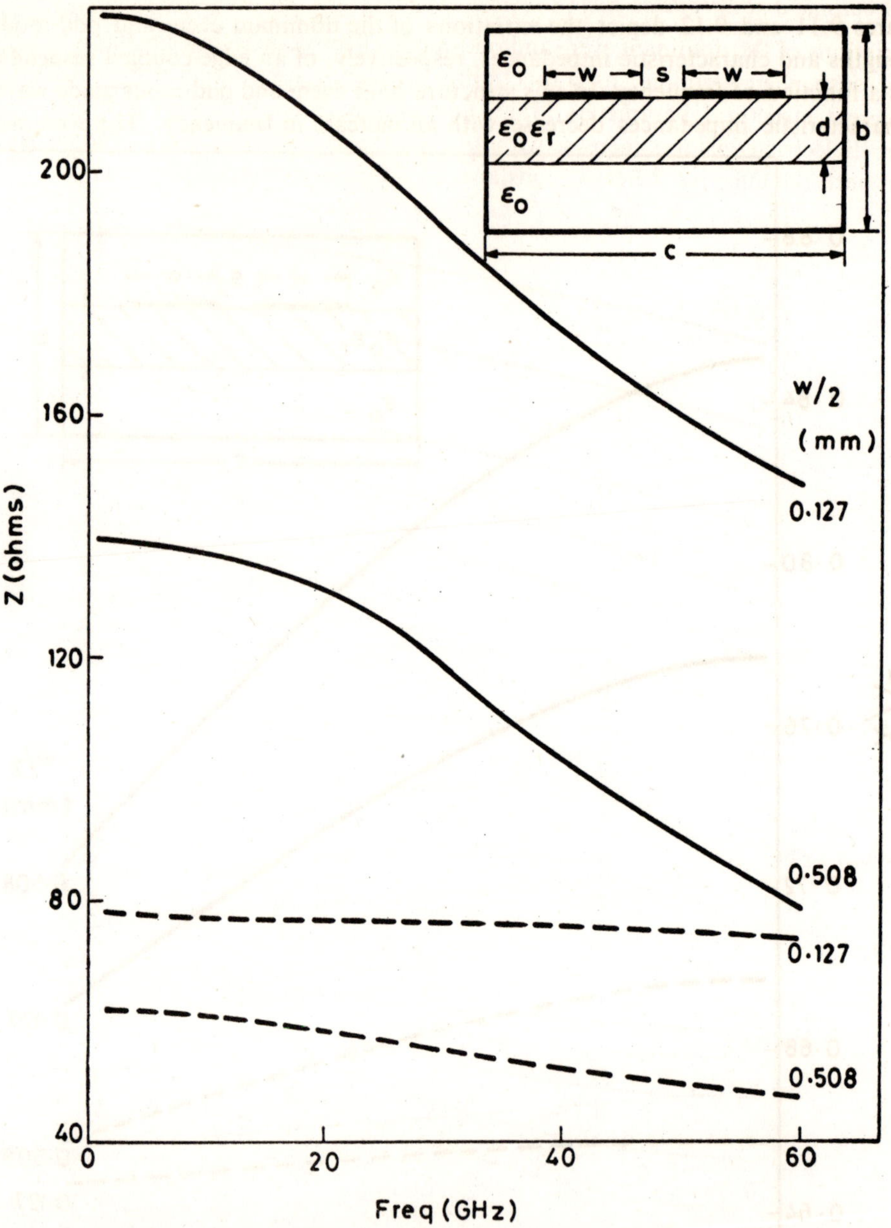

Fig. 9.12 Variations of the dominant even- and odd-mode characteristic impedances of an edge-coupled suspended stripline as a function of frequency; ϵ_r=3.78, d = 0.508 mm, b = 3.048 mm, s = 0.2032 mm, c/w = 20, ——— even mode, – – – odd mode.

pagation parameters decrease at a faster rate than the corresponding odd-mode propagation parameters. Furthermore, for smaller values of w, the odd-mode propagation parameters are rather insensitive to change in frequency.

9.5.3 *Broadside-and Broadside-Edge Coupled Stripline-Like Transmission Lines*

The variations of the dominant even- and odd-mode effective dielectric constants of broadside-coupled suspended stripline, as a function of frequency are plotted in Figs. 9.13 and 9.14 for different structural parameters [18]. These graphs reveal that both even- and odd-mode effective dielectric constants increase with an increase in frequency. Furthermore, the variations in both the effective dielectric constants as a function of frequency is large.

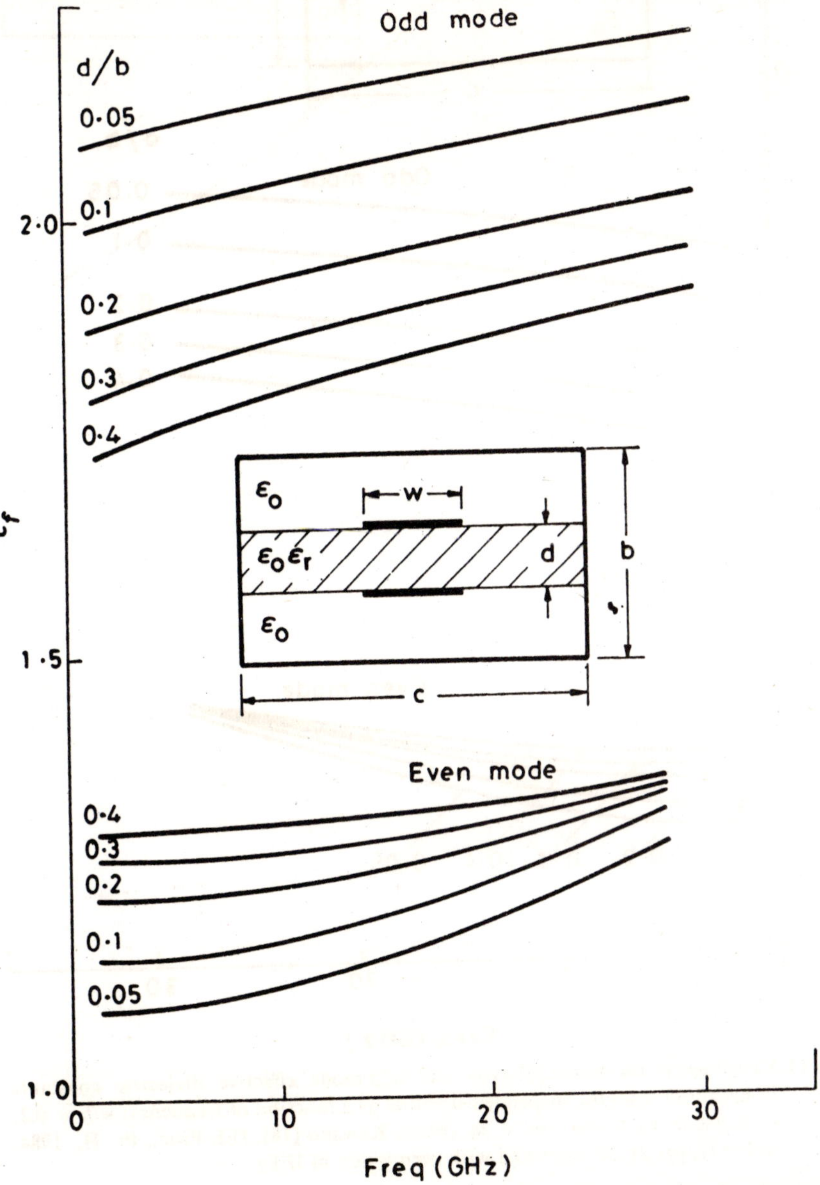

Fig. 9.13 Variations of the dominant even- and odd-mode effective dielectric constants of broadside-coupled suspended stripline as a function of frequency; $w/b = 0.2, \epsilon_r = 2.32$, $d = 1.52$ mm, $c/w = \infty$. (From Kawano [18], IEE Proc., Pt. H, 1984, Vol. 131, pp. 21-24, reprinted with permission of IEE).

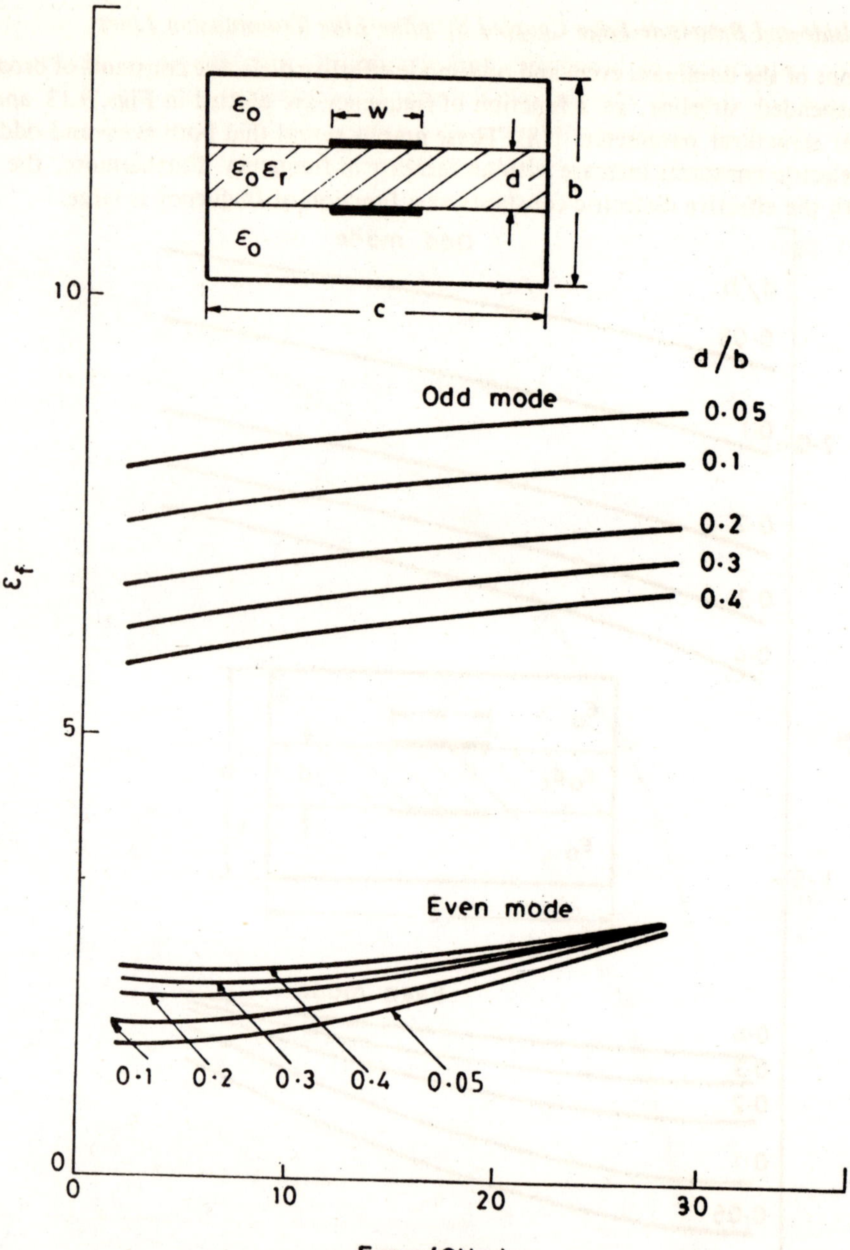

Fig. 9.14 Variations of the dominant even- and odd-mode effective dielectric constants of broadside-coupled suspended stripline as a function of frequency; $w/b = 0.2$, $\epsilon_r = 9.6$, $d = 0.6$ mm, $c/w = \infty$. (From Kawano [18], IEE Proc., Pt. H, 1984, Vol. 131, pp. 21-24, reprinted with permission of IEE).

The variations of the dominant even- and odd-mode normalized guide wavelengths and characteristic impedances of broadside-coupled stripline with double suspended inverted substrate, as a function of w/b for two different frequencies are depicted in Figs 9.15 and 9.16, respectively. Superimposed on these graphs are the results obtained using quasi-TEM

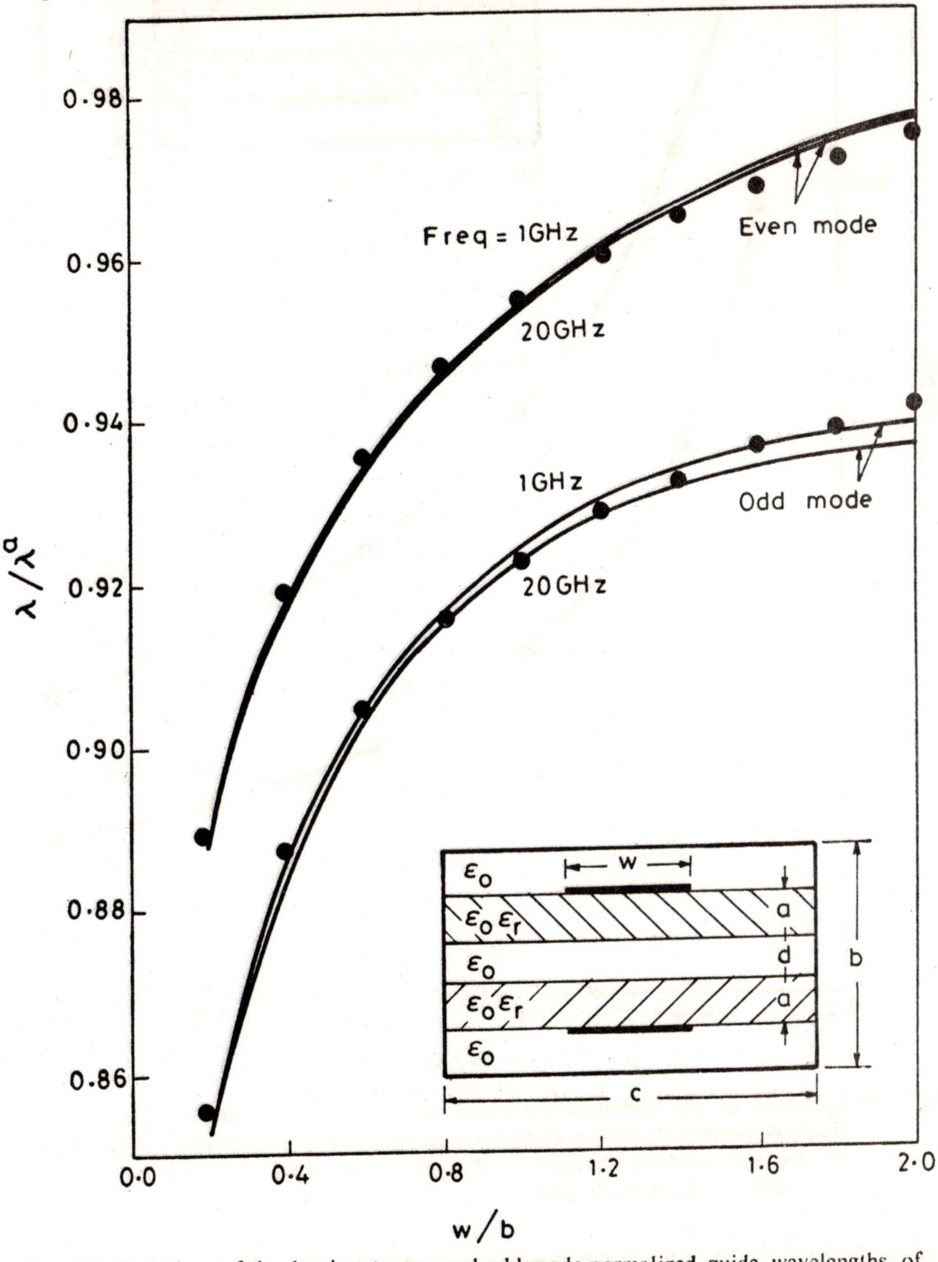

Fig. 9.15 Variations of the dominant even- and odd-mode normalized guide wavelengths of broadside-coupled stripline with double suspended inverted substrates as a function of w/b for two different frequencies; $d/b = 0.4$, $a/b = 0.05$, $c/b = 10$. $\epsilon_r = 3.78$, $b = 1$ mm, . . . TEM results [20].

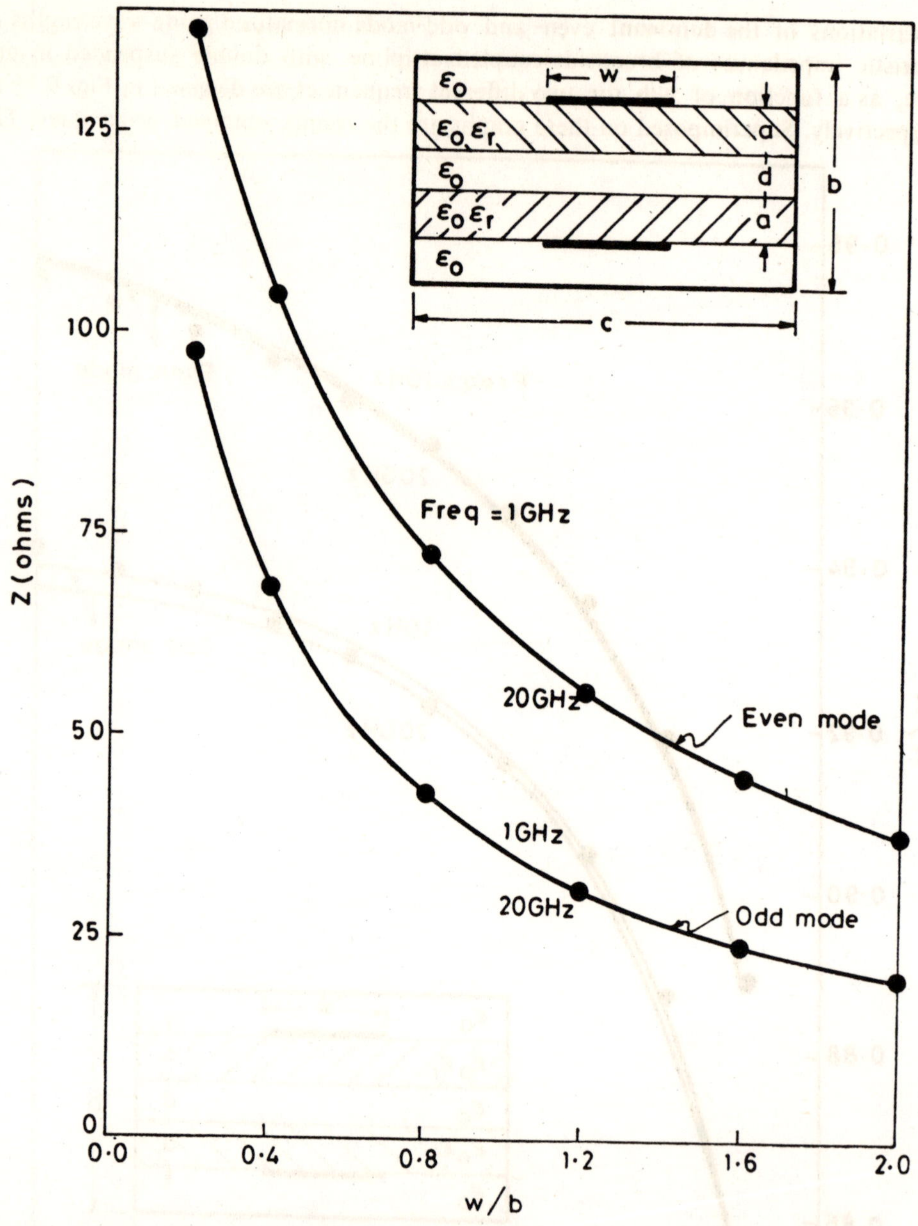

Fig. 9.16 Variations of the dominant even- and odd-mode characteristic impedances of broadside-coupled stripline with double suspended inverted substrates as a function w/b for two different frequencies; $d/b = 0.4$, $a/b = 0.05$, $c/b = 10$, $\epsilon_r = 3.78$, $b = 1$ mm, . . . TEM results [20].

approximation. For the structural parameters considered in Figs 9.15 and 9.16, the quasi-TEM results should suffice upto a frequency of at least 20 GHz. For the same structure, the variations of the dominant even- and odd-mode normalized guide wavelengths and characteristic impedances as a function of frequency are plotted in Figs. 9.17 and 9.18, respectively. The results show that the even-mode propagation parameters exhibit a larger variation with frequency than the corresponding odd-mode propagation parameters.

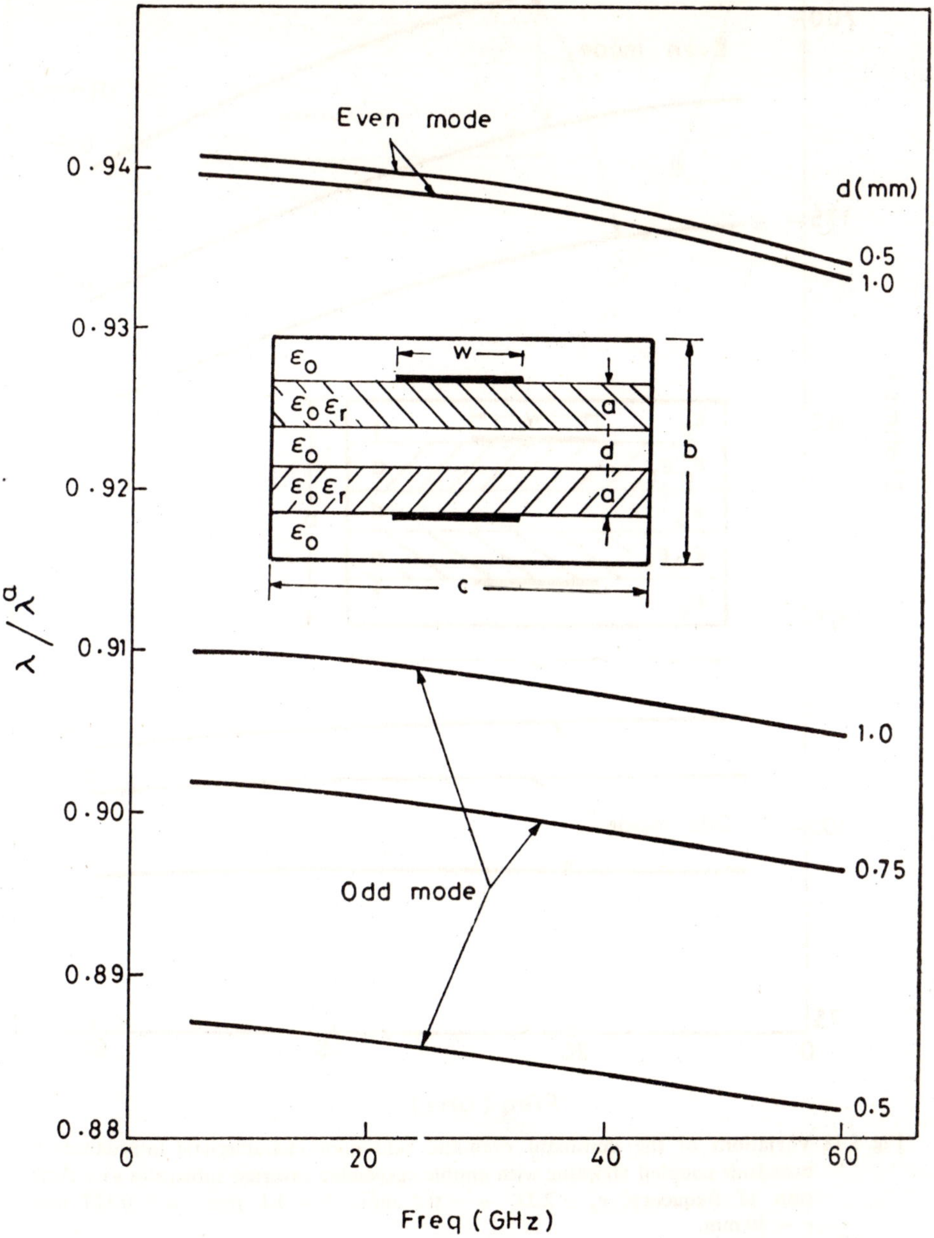

Fig. 9.17 Variations of the dominant even- and odd-mode normalized guide wavelengths of broadside-coupled stripline with double suspended inverted substrates as a function of frequency; $\epsilon_r = 2.22$, $w = 0.5$ mm, $b = 3.1$ mm, $a = 0.127$ mm, $c = 10$ mm.

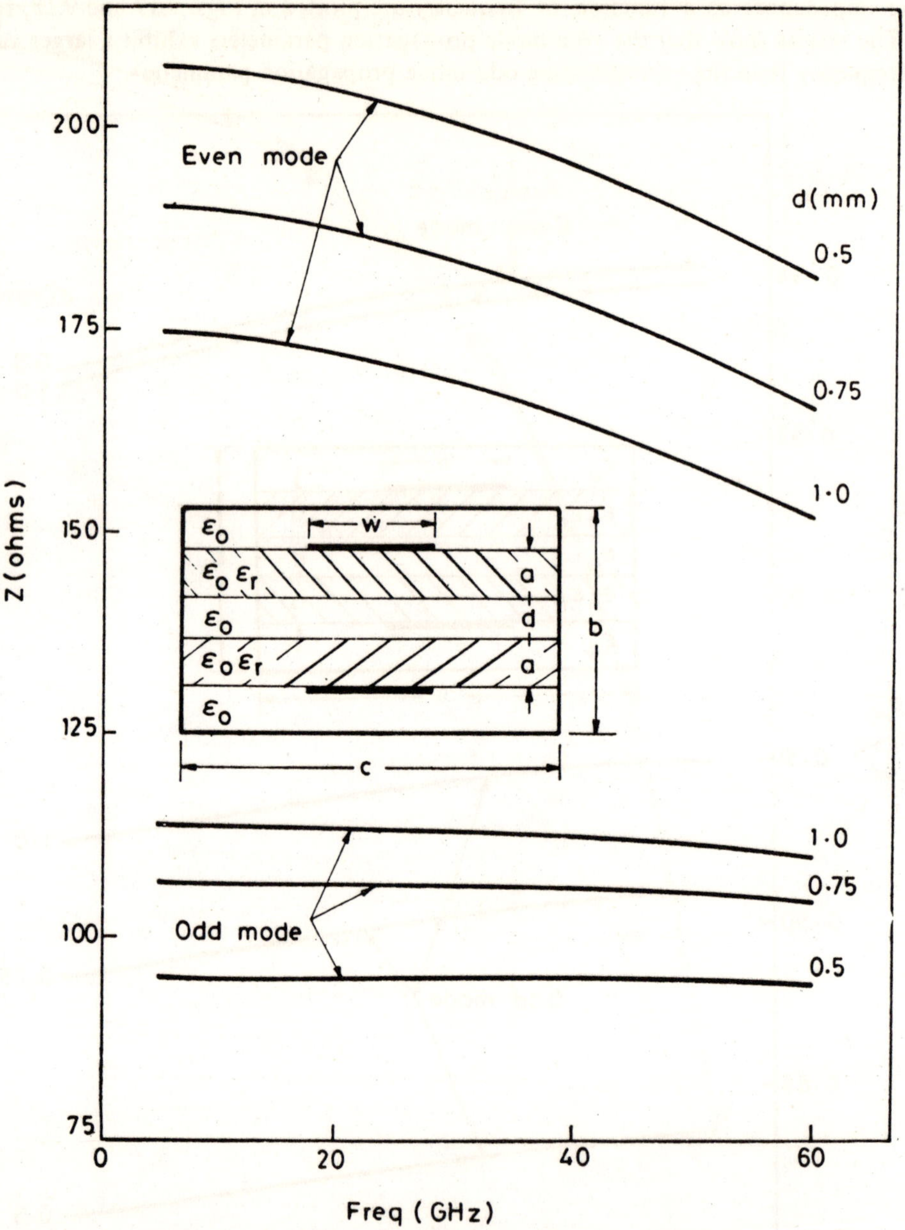

Fig. 9.18 Variations of the dominant even-and odd-mode characteristic impedances of broadside-coupled stripline with double suspended inverted substrates as a function of frequency; $\epsilon_r = 2.22$, $w = 0.5$ mm, $b = 3.1$ mm, $a = 0.127$ mm, $c = 10$ mm.

Figures 9.19 and 9.20 depict the variations of the dominant even- and odd-mode normalized guidewavelengths and characteristic impedances, respectively, of broadside-coupled stripline with double suspended substrate, as a function of w/b for two different

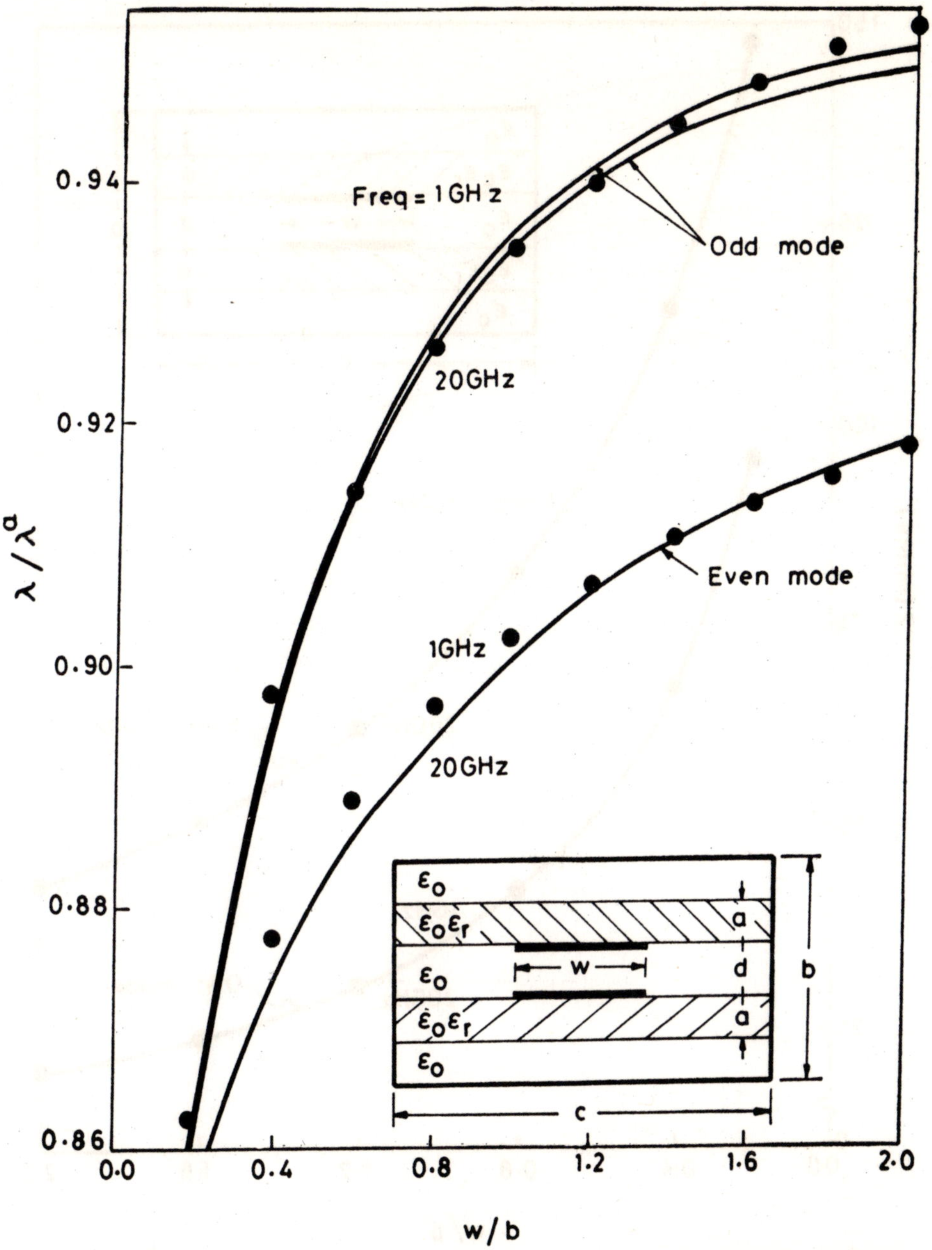

Fig. 9.19 Variations of the dominant even- and odd-mode normalized guide wavelengths of broadside-coupled stripline with double suspended substrates as a function of w/b for two different frequencies; $d/b = 0.4$, $a/b = 0.05$, $c/b = 10$, $\epsilon_r = 3.78$, $b = 1$ mm, . . . TEM results [20].

frequencies. Superimposed on these graphs are the results obtained using quasi-TEM approximation. For the same structure, the variations of the dominant even- and odd-mode normalized guide wavelengths and characteristic impedances as a function of frequency are

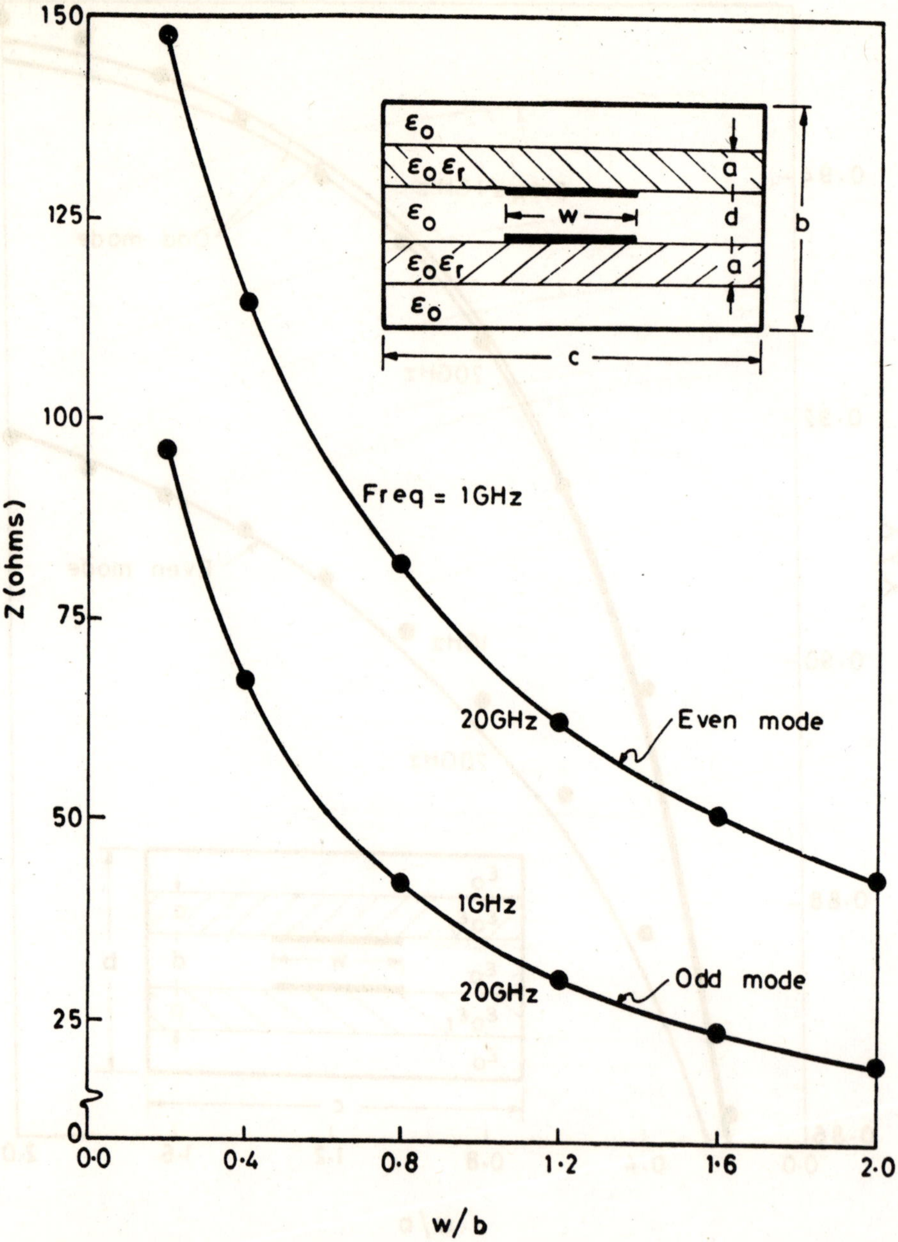

Fig. 9.20 Variations of the dominant even- and odd-mode characteristic impedances of broadside-coupled stripline with double suspended substrates as a function of w/b for two different frequencies; $d/b = 0.4$, $a/b = 0.05$, $c/b = 10$, $\epsilon_r = 3.78$, $b = 1$ mm, . . . TEM results [20].

plotted in Figs. 9.21 and 9.22, respectively. The variations are similar to those observed in the case of broadside-coupled stripline with double suspended inverted substrate.

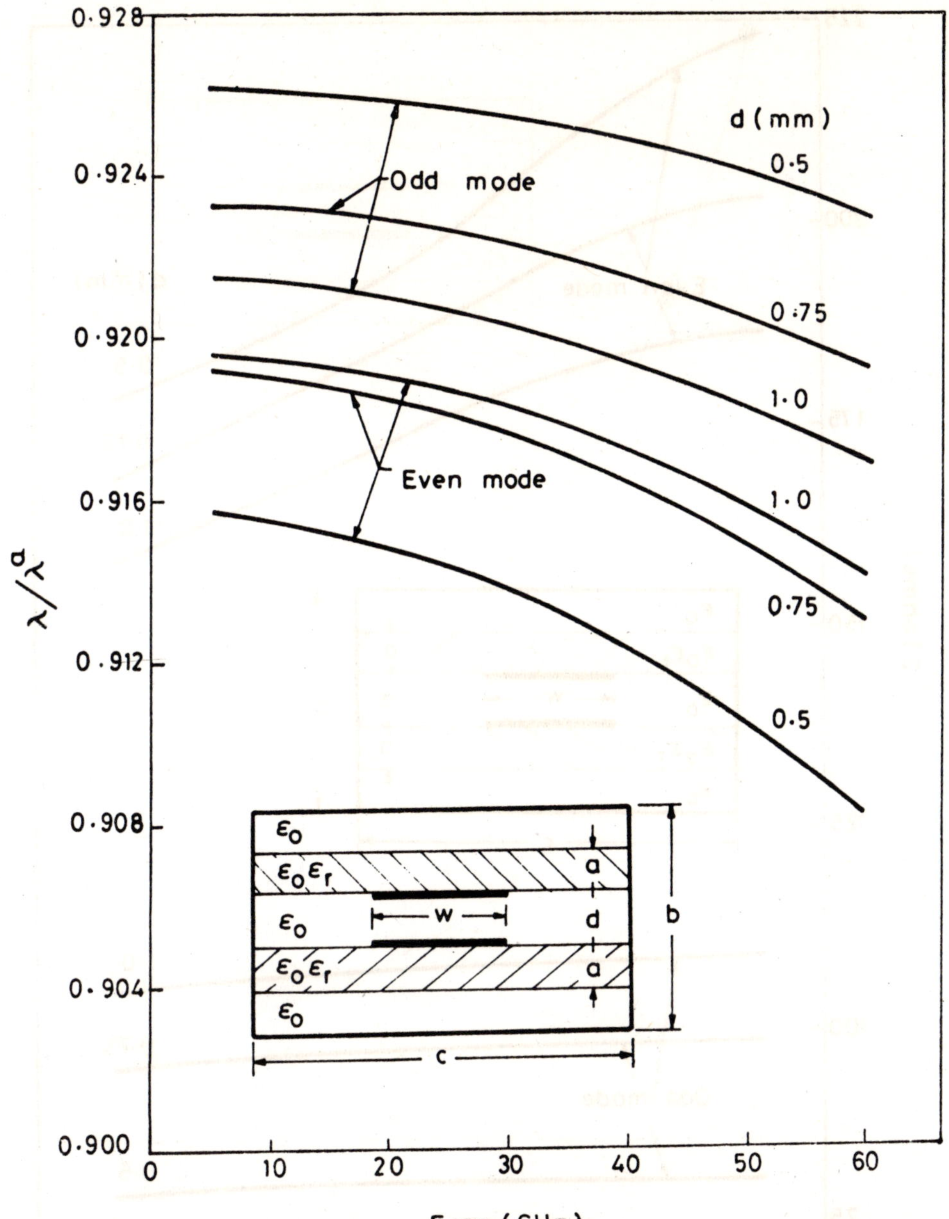

Fig. 9.21 Variations of the dominant even-and odd-mode normalized guide wavelengths of broadside-coupled stripline with double suspended substrates as a function of frequency; $\epsilon_r = 2.22$, $w = 0.5$ mm, $b = 3.1$ mm, $a = 0.127$ mm, $c = 10$ mm.

The modal spectrum of broadside-coupled stripline with double suspended inverted substrates is plotted in Fig. 9.23 [10]. We may observe that the structure supports both the dominant even- and odd-modes over the entire frequency range. However, the dominant mode bandwidth is restricted by the onset of first higher order mode HE_1 at a frequency

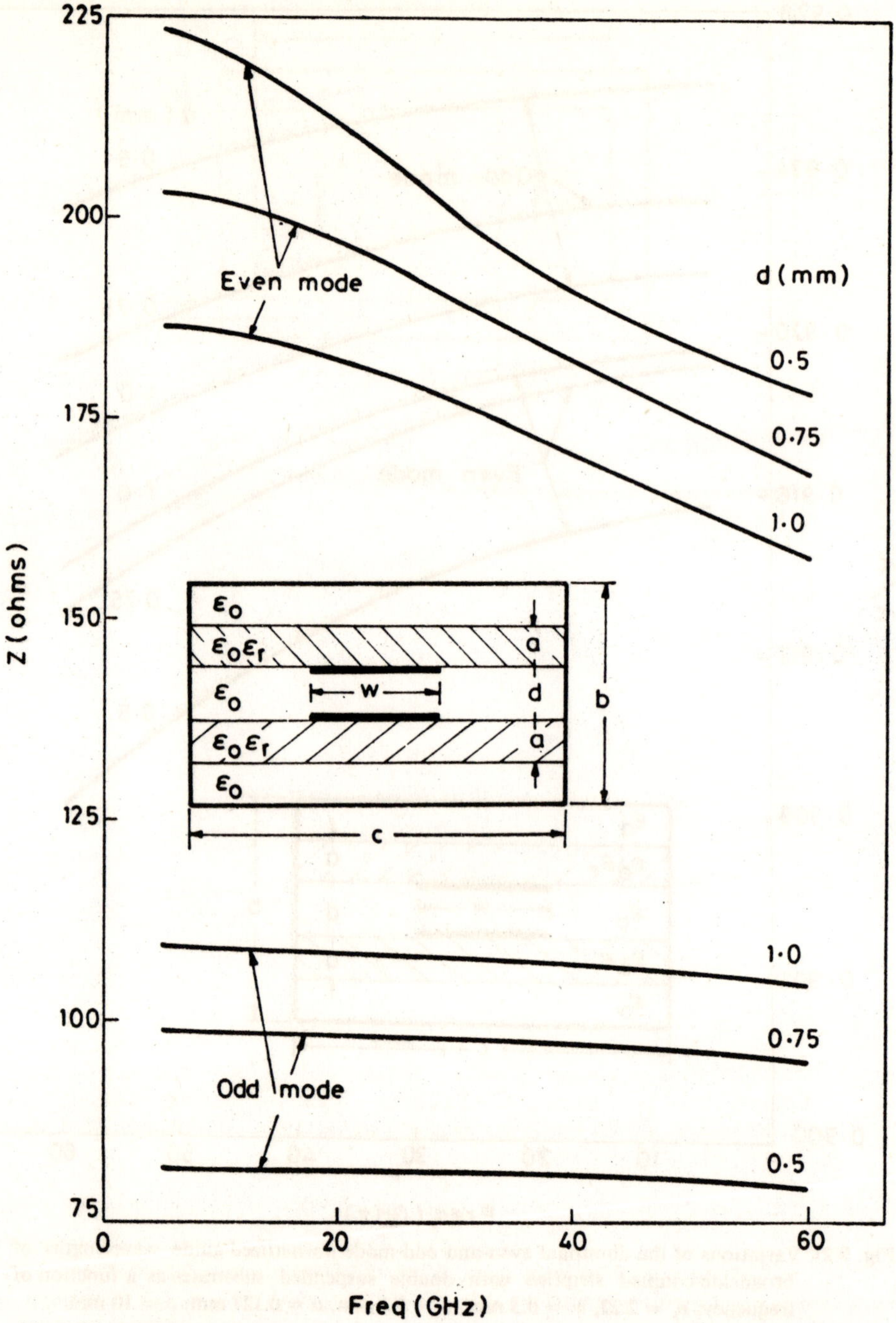

Fig. 9.22 Variations of the dominant even- and odd-mode characteristic impedances of broadside-coupled stripline with double suspended substrates as a function of frequency; $\epsilon_r = 2.22$, $w = 0.5$ mm, $b = 3.1$ mm, $a = 0.127$ mm, $c = 10$ mm.

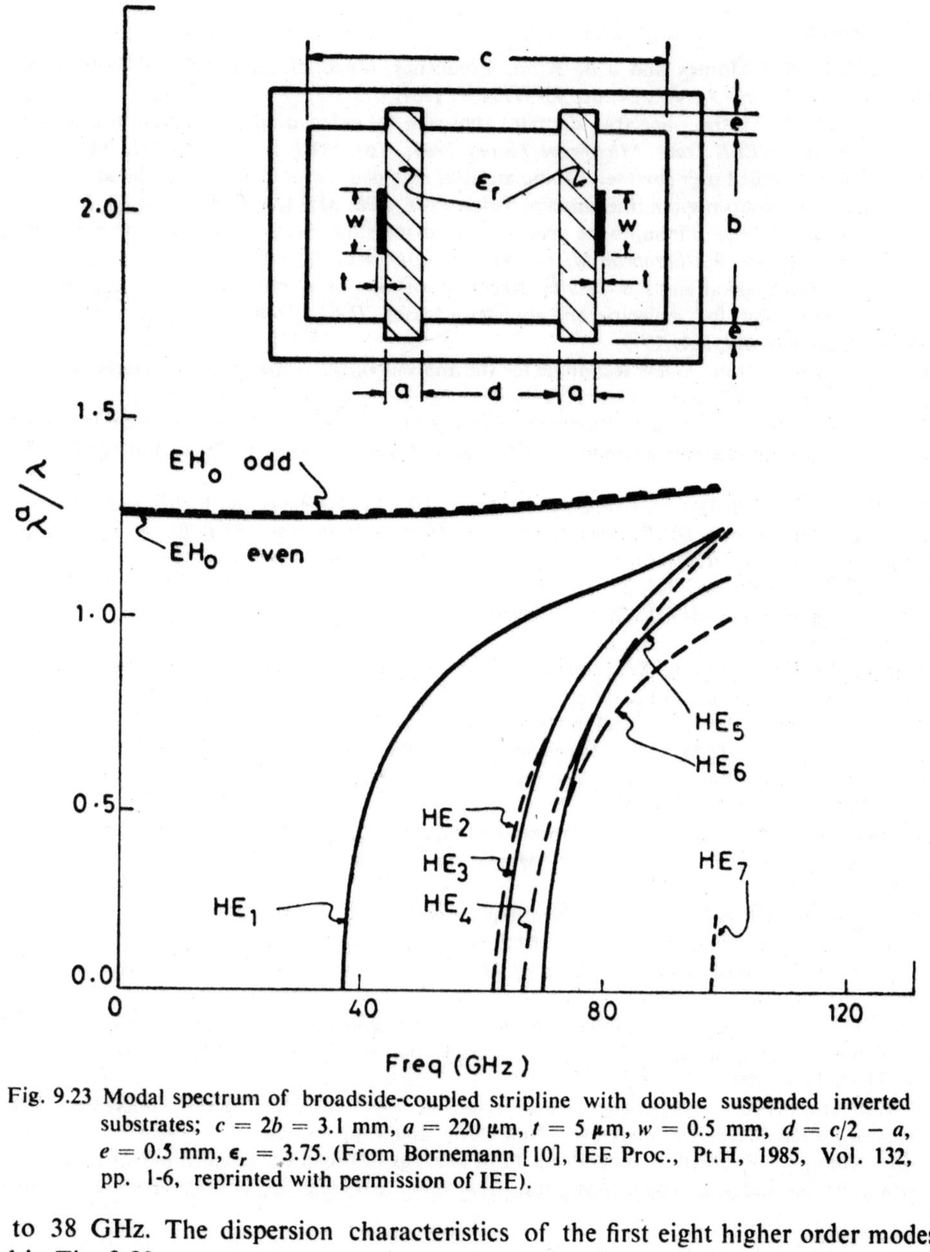

Fig. 9.23 Modal spectrum of broadside-coupled stripline with double suspended inverted substrates; $c = 2b = 3.1$ mm, $a = 220$ µm, $t = 5$ µm, $w = 0.5$ mm, $d = c/2 - a$, $e = 0.5$ mm, $\epsilon_r = 3.75$. (From Bornemann [10], IEE Proc., Pt.H, 1985, Vol. 132, pp. 1-6, reprinted with permission of IEE).

equal to 38 GHz. The dispersion characteristics of the first eight higher order modes are plotted in Fig. 9.23.

The frequency dependent propagation characteristics of broadside-edge coupled stripline—like transmission lines are reported by Wenbiao [19]. Numerical results on even-even, even-odd, odd-even and odd-odd mode characteristic impedances of broadside-edge coupled inverted stripline (structure A3 of Table 4.7) as a function of frequency are available in this paper. It has been shown that the even-even, odd-even and odd-odd mode characteristic impedances change slightly as frequency is increased. However, the even-odd mode impedance exhibits a rapid change with frequency.

REFERENCES

1. C.M. Weil, W.T. Joines and J.B. Kinn, Frequency range of large-scale TEM mode rectangular striplines, *Microwave J.*, Vol. 24, pp. 93–97, Nov. 1981.
2. T. Itoh and R. Mittra., Spectral domain approach for calculating the dispersion characteristics of microstrip lines, *IEEE Trans. Microwave Theory Tech.*, Vol. MTT–21, pp. 496–499, July 1973.
3. R.H. Jansen, Unified user oriented computation of shielded, covered and open planar microwave and millimeter wave transmission line characteristics. *Proc. IEE, MOA*, Vol. 3, pp. 14–22, Jan. 1979.
4. T. Itoh, Spectral domain immittance approach for dispersion characteristics of shielded microstrips with tuning septums, *9th European Microwave Conf. Digest* pp. 435–439, 1979.
5. D. Mirshekar—Syahkal and J.B. Davies, Accurate solution of a microstrip and coplanar structures for dispersion and for dielectric and conductor losses, *IEEE Trans. Microwave Theory Tech* Vol. MTT–27, pp. 694–699, July 1979.
6. R. Mittra and T. Itoh, A new technique for the analysis of the dispersion characteristics of microstrip lines, *IEEE Trans. Microwave Theory Tech.*, Vol. MTT–19, pp. 47–56, Jan. 1971.
7. J.S. Hornsby and A. Gopinath, Numerical analysis of a dielectric loaded waveguide with a microstrip line—Finite difference methods. *IEEE. Trans. Microwave Theory Tech.*, Vol. MTT–17, pp. 684–690, June 1969.
8. D.G. Corr and J.B. Davies, Computer analysis of the fundamental and higher order modes in single and coupled microstrip, *IEEE Trans. Microwave Theory Tech.*, Vol. MTT–20, pp. 669–678, Oct. 1972.
9. G. Kowalski and R. Pregla, Dispersion characteristics of shielded microstrips with finite thickness, *Arch. Elek. Ubertragung*, Vol. 25. pp. 193-196, April, 1971.
10. J. Bornemann, Rigorous field theory analysis of quasi planar waveguides, *Proc. IEE, Pt. H*, Vol. 132, pp. 1–6, Feb. 1985.
11. H. Diestel and S.B. Worm, Analysis of hybrid field problems by the method of lines with nonequidistant discretization, *IEEE Trans. Microwave Theory Tech.*, Vol. MTT–32, pp. 633–638, 1984.
12. S.B. Worm and R. Pregla, Hybrid mode analysis of arbitrarily shaped planar microwave structures by the method of lines, *IEEE Trans. Microwave Theory Tech.*, Vol. MTT–32, pp. 191–196, Feb. 1984.
13. G. Kompa and R. Mehran, Planar Waveguide model for calculating microstrip components. Electron. Lett., Vol. 11, pp. 459-460, 1975.
14. B. Bianco et al, Some consideration about the frequency dependence of the characteristic impedance of uniform microstrip, *IEEE Trans. Microwave Theory Tech.*, Vol. MTT-26, pp. 182-185, March 1978.
15. R.H. Jansen and N.H.L. Koster, New aspects concerning the definition of microstrip characteristic impedance as a function of frequency, *IEEE MTT-S Digest*, pp. 305-307, 1982.
16. I. Stakgold, *Boundary value problems of Mathematical Physics* Vol. 2, McGraw Hill, New York, 1968.
17. J.W. Dettman, *Mathematical Methods in Physics and Engineering*, McGraw Hill, New York, 1968.
18. K. Kawano, Hybrid mode analysis of a broadside coupled microstrip line, *Proc. IEE, Pt. H*, Vol. 131, pp. 21-24, Feb. 1984.
19. Z. Wenbiao, Frequency dependent characteristics of broadside-edge coupled microstrip lines, *Proc. Ist Asia Pacific Microwave Conference*, New Delhi, India, pp. 201-207, Feb. 1986.
20. S.K. Koul and B. Bhat, A generalized TEM analysis of broadside-coupled planar transmission lines with isotropic and anisotropic snbstrates, *Arch. Elek. Ubertragung*, Vol. 38, pp. 37-45, Jan-Feb. 1984.

CHAPTER 10

Discontinuities in Stripline-like Structures

10.1 Introduction

Microwave Integrated circuits based on stripline and stripline-like transmission lines invariably involve discontinuities of one type or another in the strip conductor. These discontinuities result from abrupt changes in the geometry of the strip conductors. Characterization of these discontinuities is therefore essential for accurate design of microwave integrated circuits. The discontinuities generally encountered in practice are of the following types: (a) abruptly ended strip conductor [1, 2], (b) rectangular conductor patch [3], (c) series gap in the strip conductor [4], (d) step change in width [5, 6], (e) right angle bend [7], (f) bends of arbitrary angle [8], (g) Tee-junction [7, 9], (h) cross-junction [10], (i) slot in the strip conductor [2], and (j) transverse slit in the strip conductor [11]. In general, the dimensions of the discontinuity are much smaller than the wavelength, and hence, they can be approximated by equivalent circuits consisting of various lumped elements. The various discontinuities listed above and their equivalent circuits are shown in Table 10.1. Besides these, the characterization of abruptly ended strip conductor in edge-coupled and broadside-coupled stripline and stripline-like transmission lines would aid the design of parallel-coupled and end-coupled filters [12-14].

Most of the aforementioned discontinuity structures in symmetric stripline configuration have been studied both theoretically and experimentally by several authors [2, 5, 10, 13]. In suspended stripline configuration, some results on the end effects of abruptly ended strip conductor have been reported by Jansen [15] and the equivalent circuit of a series gap has been studied by Koster and Jansen [16, 17] and Koul and Raghu Kumar [18].

In this chapter, we present a generalized variational approach towards the determination of lumped capacitance and edge-capacitances of a rectangular conductor patch embedded in a multilayer dielectric configuration. The analysis makes use of the spectral domain formulation under quasi-static approximation combined with the 'Transverse Transmission Line' technique [19]. This method also enables determination of the discontinuity capacitance due to a step change in strip width, equivalent circuit of a series gap in the strip conductor [18], and end effects in broadside-coupled strip conductors [20]. The same method is then extended to include anisotropy in the dielectric substrates [21]. Design data generated using these general formulas as well as the data reported in the literature on the various discontinuity aspects encountered in stripline-like configurations are presented in graphical form.

For the special case of homogeneous symmetric stripline, approximate closed form expressions for the equivalent circuit parameters of most of the discontinuities are available in the

Table 10.1 Discontinuity Structures and their Equivalent Circuits

Discontinuity Structure	Equivalent Circuit

a. Abruptly ended stripconductor

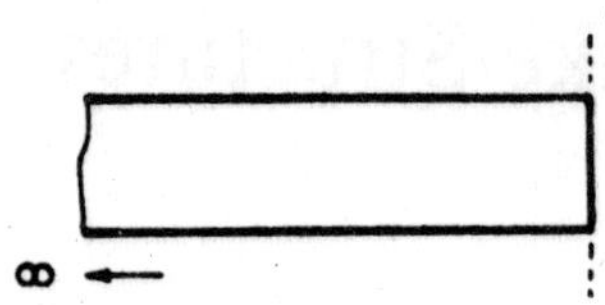

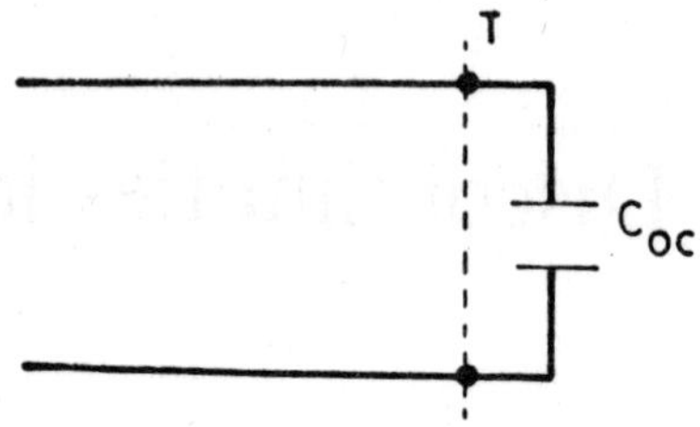

b. Rectangular patch of stripconductor

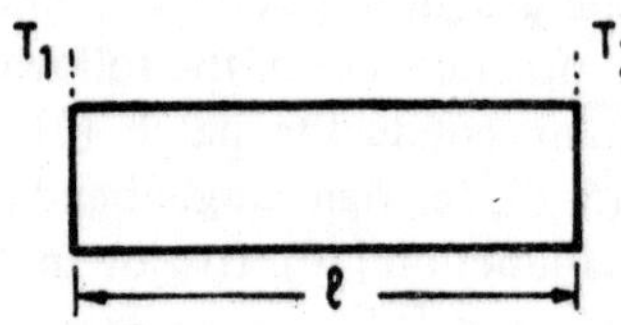

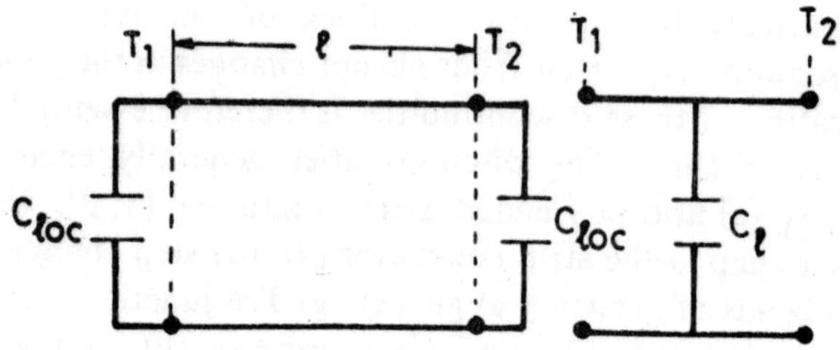

c. Series gap in the stripconductor

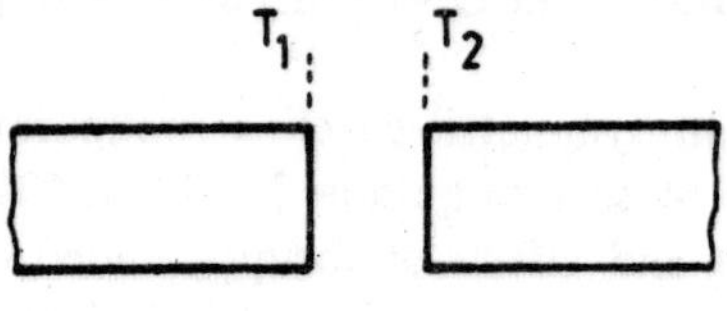

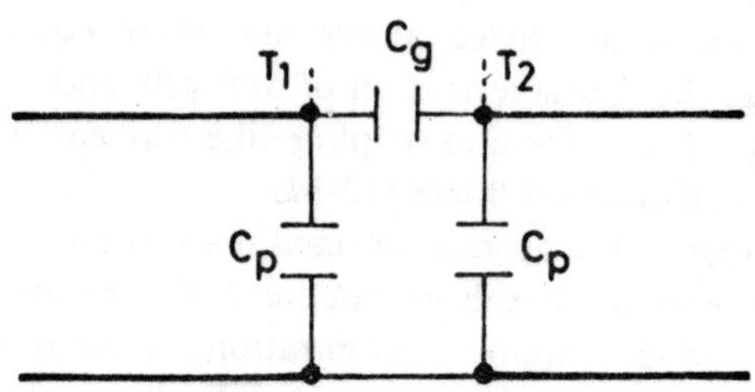

d. Step change in width

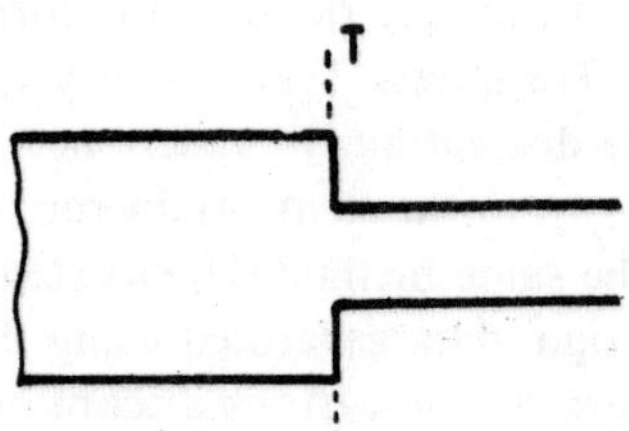

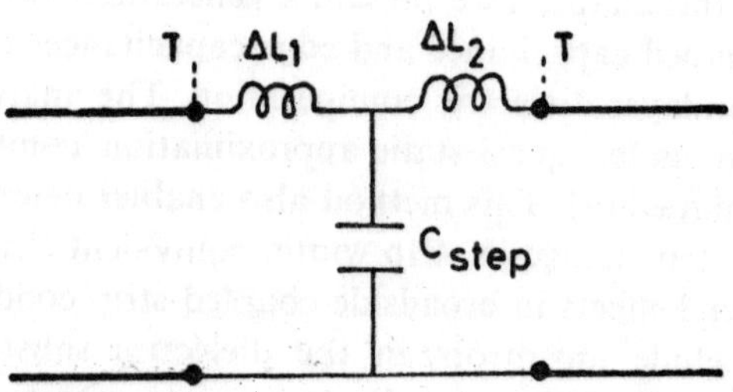

(Contd.)

Table 10.1 (*Contd.*)

Discontinuity Structure	Equivalent Circuit

e. Right angle bend

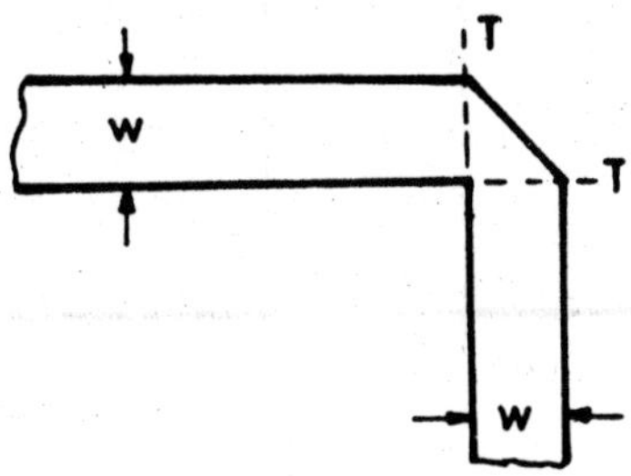

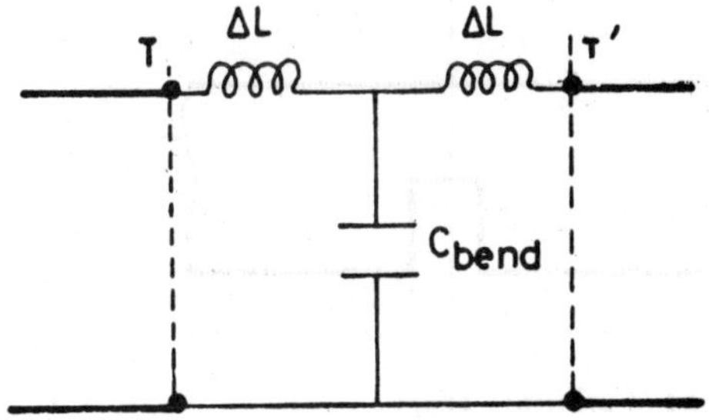

f. Slot in the stripconductor

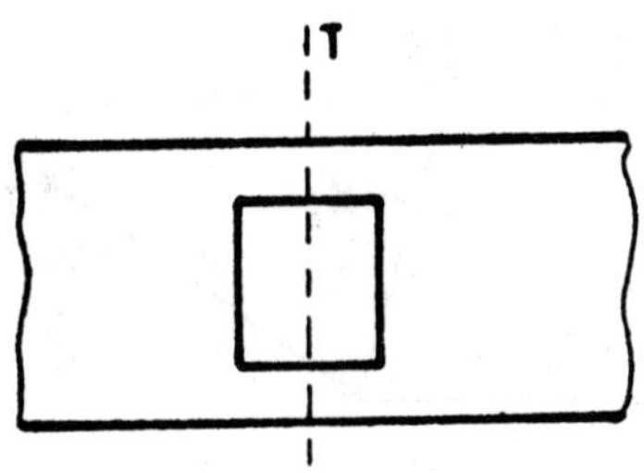

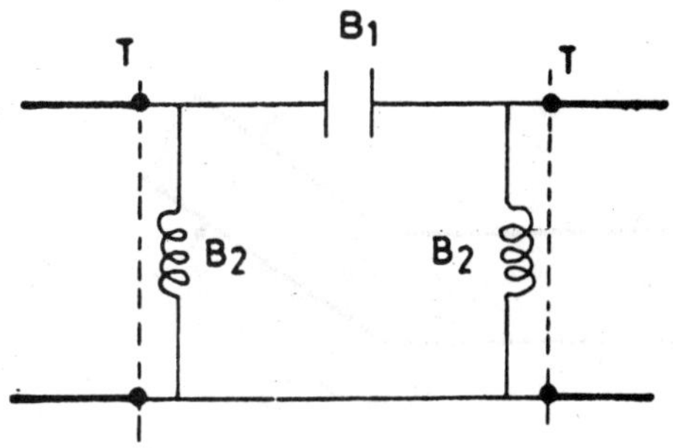

g. Tee-Junction

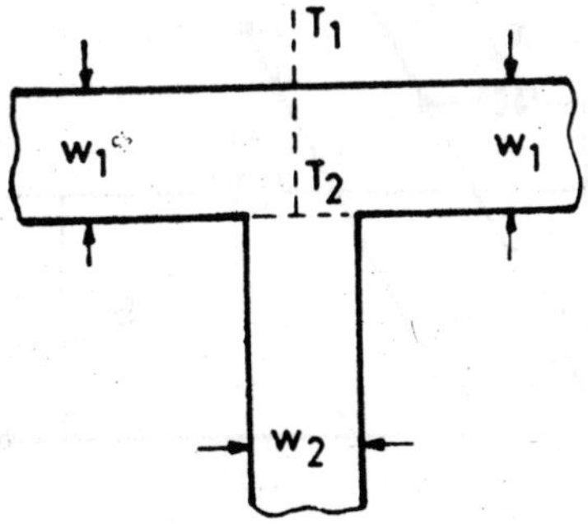

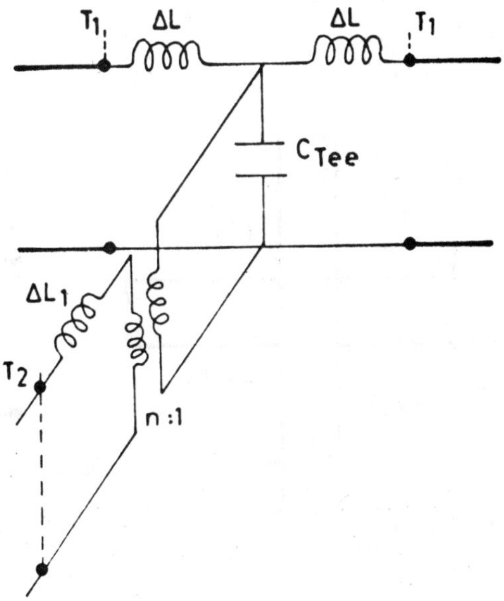

(*Contd.*)

Table 10.1 (*Contd.*)

Discontinuity Structure	Equivalent Circuit
h. Transverse slit in the stripconductor	
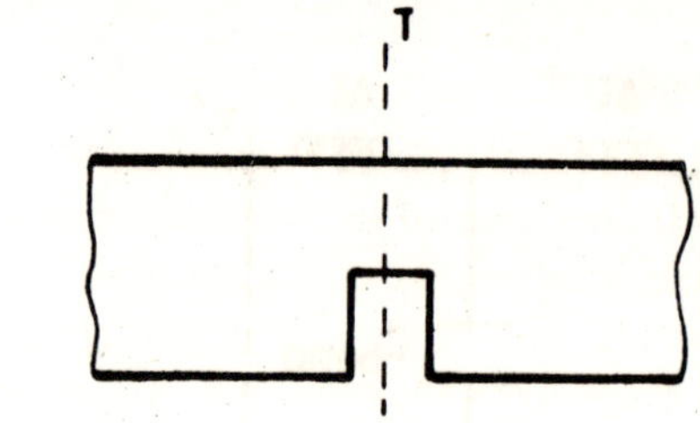	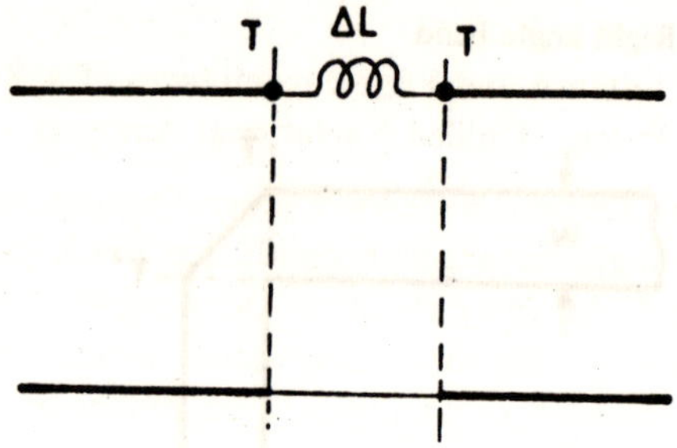
i. Bend of arbitrary angle	
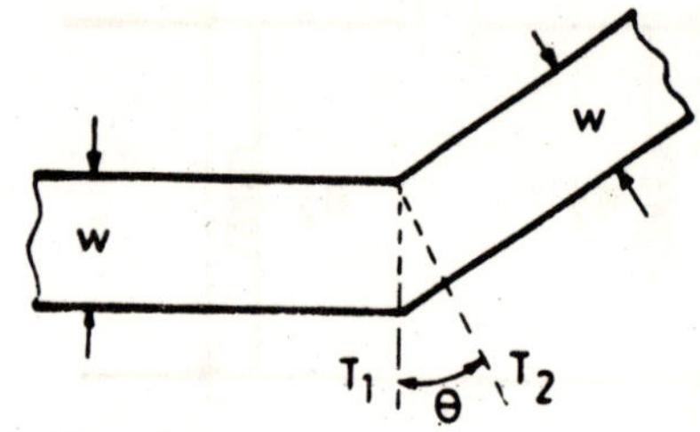	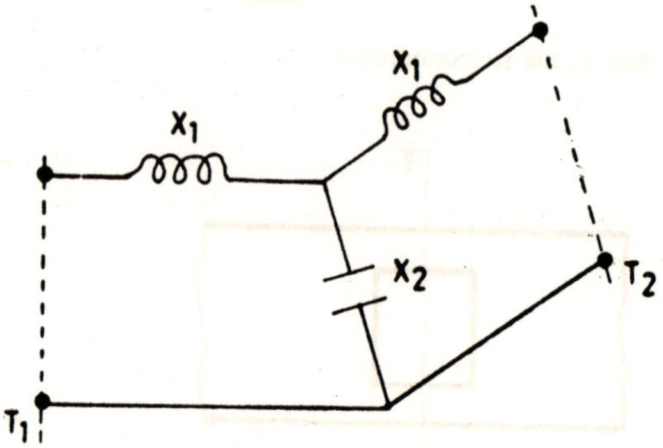
j. Cross-Junction	
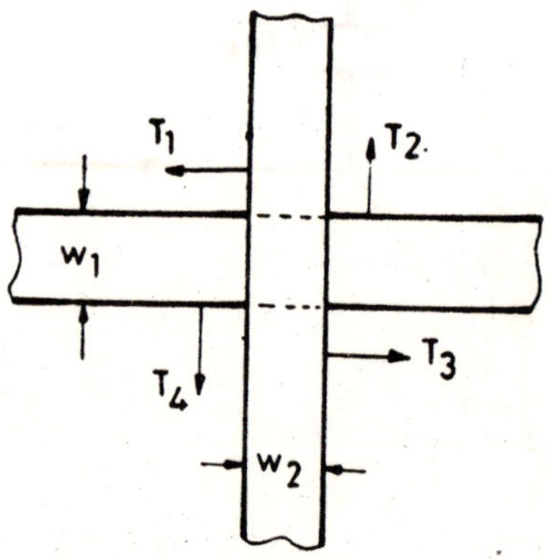	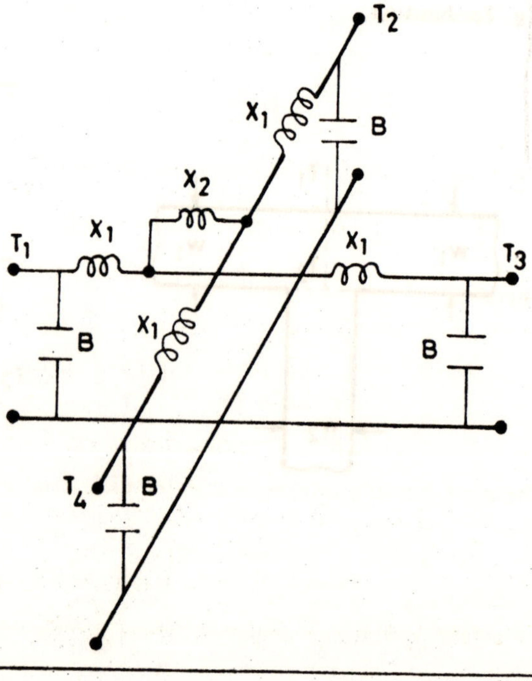

literature [7]. The formulas have been obtained by employing the Babinet equivalence or a small aperture approximation in conjunction with an approximate model of the line. These closed form expressions are listed in Appendix A.

10.2 Lumped and Edge-capacitances of a Rectangular Conductor Patch—Unified Variational Analysis

In this section, we evolve general variational expressions for the lumped and edge-capacitances of a single rectangular conductor patch in iso/anisotropic stripline-like configurations under quasi-TEM approximation. Any one of a variety of structures can be chosen for the purpose of analysis. The final capacitance formula can be easily generalized to multilayer structures. The method of analysis is an extension of the proceduce presented in Chapters 2 and 3 in deriving unified capacitance formula for stripline-like transmission lines. The key feature lies in the adoption of the 'Transverse Transmission Line' technique in obtaining the potential function or the Green's function at the plane of the conductor patch.

10.2.1 Analysis of a Rectangular Conductor Patch in a Three-Layer Isotropic Structure

Consider a three layer isotropic structure with infinitely wide ground planes and with a rectangular conductor patch assumed to be infinitesimally thin (Fig. 10.1a). Since the strip conductor is of finite extent in the z-direction, we begin with the three dimensional Poisson's equation for the potential function.

$$\nabla^2 \phi(x, y, z) = -\frac{1}{\epsilon} f(x, z)\delta(y - y_0) \tag{10.1}$$

where $f(x, z)$ is the charge distribution on the strip conductor situated at $y = y_0$ and ϵ is the absolute permittivity. For the ease of analysis, convert (10.1) into Fourier domain by using the following definition of the two-dimensional Fourier transform.

$$\hat{\phi}(\alpha, y, \beta) = \int_{-\infty}^{\infty}\int_{-\infty}^{\infty} \phi(x, y, z) \exp(j(\alpha x + \beta z))\, dx\, dz \tag{10.2}$$

The Fourier transformed Poisson's equation is then given by

$$\left(\frac{d^2}{dy^2} - \zeta^2\right)\hat{\phi}(\alpha, y, \beta) = -\frac{1}{\epsilon}\hat{f}(\alpha, \beta)\delta(y - y_0) \tag{10.3}$$

where

$$\hat{f}(\alpha, \beta) = \int_{-l/2}^{l/2}\int_{-w/2}^{w/2} f(x, z) \exp(j(\alpha x + \beta z))\, dx\, dz \tag{10.4a}$$

$$\zeta^2 = (\alpha^2 + \beta^2) \tag{10.4b}$$

The boundary conditions at the two dielectric interfaces are,

$$\hat{\phi}(\alpha, y, \beta)|_{y=s_j-0} = \hat{\phi}(\alpha, y, \beta)|_{y=s_j+0}; \quad j = 1, 2 \tag{10.5a}$$

$$\epsilon_0\epsilon_{rj}\frac{d}{dy}\hat{\phi}(\alpha, y, \beta)|_{y=s_j-0} = \epsilon_0\epsilon_{rj+1}\frac{d}{dy}\hat{\phi}(\alpha, y, \beta)|_{y=s_j+0}; \quad j = 1, 2 \tag{10.5b}$$

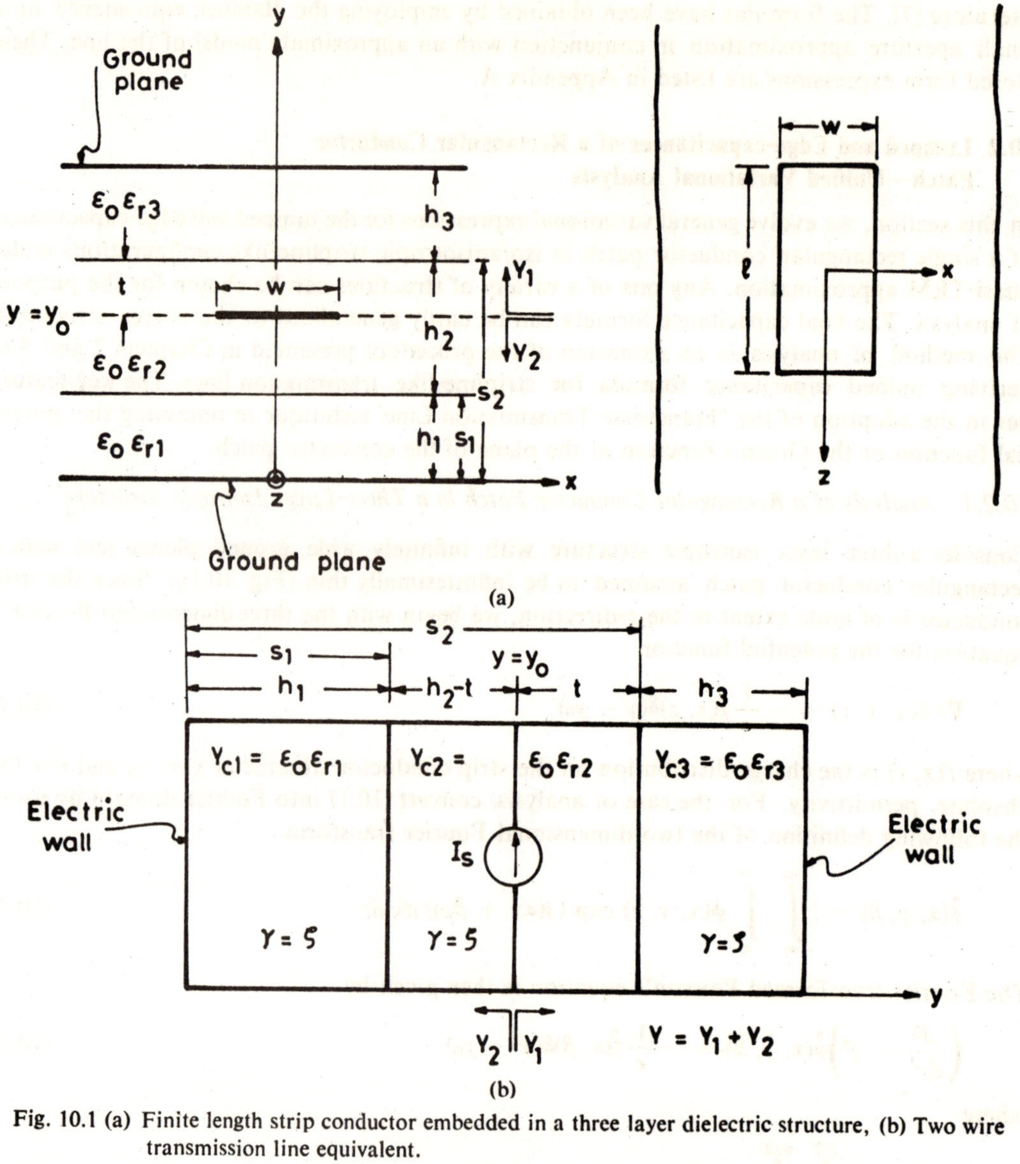

Fig. 10.1 (a) Finite length strip conductor embedded in a three layer dielectric structure, (b) Two wire transmission line equivalent.

In order to solve for $\hat{\phi}(\alpha, y, \beta)$, we now apply the 'transverse transmission line' method. The differential equation governing the voltage V along a transmission line extending along the y-direction with a current source of intensity I_s situated at $y = y_0$ is given by

$$\left(\frac{d^2}{dy^2} - \gamma^2\right) V = -\gamma Z_c I_s \delta(y - y_0) \tag{10.6}$$

where γ is the propagation constant and Z_c is the characteristic impedance of the line. Referring to the two wire transmission line equivalent shown in Fig. 10.1b, the continuity conditions at the two interfaces between sections of different characteristic admittances are given by

$$V|_{y=s_j-0} = V|_{y=s_j+0}; \qquad j = 1, 2 \tag{10.7a}$$

$$Y_{cj} \frac{dV}{dy}\bigg|_{y=s_j-0} = Y_{cj+1} \frac{dV}{dy}\bigg|_{y=s_j+0}; \qquad j = 1, 2 \tag{10.7b}$$

where Y_{cj} is the characteristic admittance of the jth section. Comparing (10.3) and (10.5) with (10.6) and (10.7) respectively, we obtain the following analogous relations

$$V = \hat{\phi}(\alpha, y, \beta), \quad \gamma = \zeta = \sqrt{\alpha^2 + \beta^2}, \quad Z_c = 1/\epsilon, \quad Y_{cj} = \epsilon_0\epsilon_{rj},$$

$$I_s = \hat{f}(\alpha, \beta)/\zeta \tag{10.8}$$

Since $V = I_s/Y$ at $y = y_0$, the transformed potential function at the charge plane can be expressed in the form,

$$\hat{\phi}(\alpha, y_0, \beta) = \frac{\hat{f}(\alpha, \beta)}{\zeta Y} \tag{10.9}$$

where Y is the admittance at the charge plane $y = y_0$, the expression for which can be easily obtained from the transmission line equivalent shown in Fig. 10.1b.

In order to facilitate the remaining part of the analysis, we rewrite (10.9) in the form,

$$\hat{\phi}(\alpha, y_0, \beta) = \hat{G}(\alpha, y_0, \beta)\hat{f}(\alpha, \beta) \tag{10.10}$$

where

$$\hat{G}(\alpha, y_0, \beta) = \frac{1}{\zeta Y} \tag{10.11}$$

The potential $\hat{\phi}(\alpha, y_0, \beta)$ is constituted by $\hat{\phi}_c(\alpha, y_0, \beta)$, the transform of the potential on the strip conductor and $\hat{\phi}_d(\alpha, y_0, \beta)$, the transform of the potential on the complementary region at the interface $y = y_0$. Thus

$$\begin{aligned}\hat{\phi}(\alpha, y_0, \beta) &= \hat{G}(\alpha, y_0, \beta)\hat{f}(\alpha, \beta) \\ &= \hat{\phi}_c(\alpha, y_0, \beta) + \hat{\phi}_d(\alpha, y_0, \beta)\end{aligned} \tag{10.12}$$

Here $\hat{\phi}_d(\alpha, y_0, \beta)$ and $\hat{f}(\alpha, \beta)$ are unknown quantities. We now specifv the charge distribution in terms of a set of basis functions as

$$f(x, z) = \sum_{n=1}^{N} a_n\psi_n(x, z) \tag{10.13}$$

where a_n's are the unknown coefficients. Taking Fourier transform of (10.13) we get

$$\hat{f}(\alpha, \beta) = \sum_{n=1}^{N} a_n\hat{\psi}_n(\alpha, \beta) \tag{10.14}$$

where

$$\hat{\psi}_n(\alpha, \beta) = \int_{-\infty}^{\infty}\int_{-\infty}^{\infty} \psi_n(x, z)e^{j(\alpha x+\beta z)}\, dx\, dz \tag{10.15}$$

Substituting (10.14) in (10.12), we get

$$\hat{\phi}_c(\alpha, y_0, \beta) + \hat{\phi}_d(\alpha, y_0, \beta) = \hat{G}(\alpha, y_0, \beta) \sum_{n=1}^{N} a_n \hat{\psi}_n(\alpha, \beta) \tag{10.16}$$

In this equation $\hat{\phi}_d(\alpha, y_0, \beta)$ can be eliminated through the use of Parseval's theorem. We first define the inner product

$$\langle \hat{\psi}_1(\alpha, \beta), \hat{\psi}_2(\alpha, \beta) \rangle = \int_{-\infty}^{\infty}\int_{-\infty}^{\infty} \hat{\psi}_1(\alpha, \beta)\hat{\psi}_2^*(\alpha, \beta)\, d\alpha\, d\beta \tag{10.17}$$

By virtue of the Parseval's theorem

$$\int_{-\infty}^{\infty}\int_{-\infty}^{\infty} \hat{\psi}_1(\alpha, \beta)\hat{\psi}_2^*(\alpha, \beta)\, d\alpha\, d\beta = (2\pi)^2 \int_{-\infty}^{\infty}\int_{-\infty}^{\infty} \psi_1(x, z)\psi_2(x, z)\, dx\, dz \tag{10.18}$$

Taking the inner product of (10.16) with $\hat{\psi}_m(\alpha, \beta)$, we get

$$\langle \hat{\phi}_c(\alpha, y_0, \beta), \hat{\psi}_m(\alpha, \beta) \rangle = \sum_{n=1}^{N} a_n \langle \hat{G}(\alpha, y_0, \beta)\hat{\psi}_n(\alpha, \beta), \hat{\psi}_m(\alpha, \beta) \rangle,$$

$$m = 1, 2, \ldots N \tag{10.19}$$

The inner product of $\hat{\phi}_d(\alpha, y_0, \beta)$ with $\hat{\psi}_m(\alpha, \beta)$ vanishes since they exist on complementary regions. Applying the relations given in (10.17) and (10.18) to (10.19) and setting the potential of the strip conductor $\phi_c(x, y_0, z)$ equal to 1, we obtain

$$\sum_{n=1}^{N} a_n P_{mn} = F_m \tag{10.20}$$

where

$$P_{mn} = \int_{-\infty}^{\infty}\int_{-\infty}^{\infty} \hat{G}(\alpha, y_0, \beta)\hat{\psi}_n(\alpha, \beta)\hat{\psi}_m^*(\alpha, \beta)\, d\alpha\, d\beta \tag{10.21a}$$

$$F_m = \int_{-\infty}^{\infty}\int_{-\infty}^{\infty} \hat{\phi}_c(\alpha, y_0, \beta)\hat{\psi}_m^*(\alpha, \beta)\, d\alpha\, d\beta \tag{10.21b}$$

$$= (2\pi)^2 \int_{-l/2}^{l/2}\int_{-w/2}^{w/2} \psi_m(x, z)\, dx\, dz \tag{10.21c}$$

Once (10.20) is solved for a_n, the lumped capacitance C_l of the strip conductor may be computed from the expression

$$C_l = \frac{Q}{V} = \int_{-l/2}^{l/2}\int_{-w/2}^{w/2} \sum_{n=1}^{N} a_n\psi_n(x, z)\, dx\, dz \tag{10.22a}$$

$$= \left(\frac{1}{2\pi}\right)^2 \sum_{n=1}^{N} a_n F_n \tag{10.22b}$$

where Q is the total charge and $V = \phi_c(x, y_0, z) = 1$ on the strip conductor.

For specified functions $\psi_n(x, z)$ in the charge distribution, P_{mn} and F_m are evaluated from (10.21a) and (10.21c), respectively. Using these in (10.20), a_n's can be determined, the substitution of which in (10.22) yields the lumped capacitance of the rectangular patch conductor.

10.2.2 Choice of Basis Functions and Expression for Lumped Capacitance

In the case of two-dimensional problems such as the uniform transmission lines, it is known that basis functions in the form of polynomials in $|x|$ represent the charge distribution fairly accurately [22, 23]. For the three dimensional problem, as in the present case, we may therefore choose the following function.

$$\psi_n(x, z) = \begin{cases} |x|^{k-1}\,|z|^{j-1}; & \text{on the strip} \\ & n = 1(k = 1, j = 1),\ n = 2(k = 2, j = 1), \ldots \\ 0 & ;\ \text{Otherwise} \end{cases} \tag{10.23}$$

Using a large number of terms improves the accuracy of the results, however at the expense of increase in the computational time. As a compromise between the accuracy and the computational time, we choose the following one term approximation to the basis function. This choice of the basic function has been found to yield quite accurate results adequate for most practical applications [19].

$$\psi_1(x, z) = \psi_1(x)\psi_1(z) \tag{10.24}$$

where

$$\psi_1(x) = \begin{cases} 1 + \left|\dfrac{2x}{w}\right|^3; & -w/2 \leqslant x \leqslant w/2 \\ 0 & ;\ \text{Otherwise} \end{cases} \tag{10.25a}$$

and

$$\psi_1(z) = \begin{cases} 1 & ;\ -(l/2 - k) \leqslant z \leqslant (l/2 - k) \\ 1 + \left(\dfrac{A}{k}\right)\left[\,|z| - \left(\dfrac{l}{2} - k\right)\right]; & -l/2 \leqslant z \leqslant -(l/2 - k) \\ & (l/2 - k) \leqslant z \leqslant l/2 \\ 0 & ;\ \text{Otherwise} \end{cases} \tag{10.25b}$$

A and k are constants determined by numerically maximizing the capacitance C_l. In (10.25b), the value of k is always less than $l/2$.

With one term approximation of the basis function, the expression for the lumped capacitance C_l can be rewritten as

$$C_l = \left(\frac{1}{2\pi}\right)^2 a_1 F_1 = \left(\frac{1}{2\pi}\right)^2 \frac{F_1^2}{P_{11}} \tag{10.26}$$

Substituting for F_1 and P_{11}, we obtain

$$C_l = \frac{1}{(2\pi)^2} \cdot \frac{(2\pi)^4 \left[\displaystyle\int_{-l/2}^{l/2} \int_{-w/2}^{w/2} \psi_1(x, z)\, dx\, dz\right]^2}{\displaystyle\int_{-\infty}^{\infty} \int_{-\infty}^{\infty} \hat{G}(\alpha, y_0, \beta)\hat{\psi}_1(\alpha, \beta)\hat{\psi}_1^*(\alpha, \beta)\, d\alpha\, d\beta} \tag{10.27}$$

Incorporating the expression for $\hat{G}(\alpha, y_0, \beta)$ from (10.11), the final formula becomes

$$C_l = \frac{(2\pi)^2}{\left[\int_{-\infty}^{\infty}\int_{-\infty}^{\infty}\left|\frac{\hat{\psi}_1(\alpha, \beta)}{Q}\right|^2 \frac{1}{\zeta Y}\, d\alpha\, d\beta\right]} \tag{10.28}$$

For the structure shown in Fig. 10.1, the admittance Y at the charge plane is obtained by applying the standard transmission line formula given in (2.131) of Chapter 2.

$$Y = \epsilon_0\epsilon_{r2}\left[\left\{\frac{\epsilon_{r3}\coth(\zeta h_3) + \epsilon_{r2}\tanh(\zeta t)}{\epsilon_{r2} + \epsilon_{r3}\coth(\zeta h_3)\tanh(\zeta t)}\right\} + \left\{\frac{\epsilon_{r1}\coth(\zeta h_1) + \epsilon_{r2}\tanh(\zeta(h_2 - t))}{\epsilon_{r2} + \epsilon_{r1}\coth(\zeta h_1)\tanh(\zeta(h_2 - t))}\right\}\right] \tag{10.29a}$$

The total charge Q on the conductor patch is given by

$$Q = \int_{-l/2}^{l/2}\int_{-w/2}^{w/2} \psi_1(x, z)\, dx\, dz = \frac{5w}{4}[Ak + l] \tag{10.29b}$$

$\hat{\psi}_1(\alpha, \beta)$, the Fourier transform of $\psi_1(x, z)$ is obtained as

$$\hat{\psi}_1(\alpha, \beta) = \hat{\psi}_1(\alpha)\hat{\psi}_1(\beta) \tag{10.29c}$$

where

$$\hat{\psi}_1(\alpha) = \frac{5w}{4}\left\{1.6\left[\frac{\sin\left(\frac{\alpha w}{2}\right)}{\left(\frac{\alpha w}{2}\right)}\right] + \frac{2.4}{\left(\frac{\alpha w}{2}\right)^2}\left[\cos\left(\frac{\alpha w}{2}\right) - 2\left(\frac{\sin\left(\frac{\alpha w}{2}\right)}{\left(\frac{\alpha w}{2}\right)}\right) + \left(\frac{\sin\left(\frac{\alpha w}{4}\right)}{\left(\frac{\alpha w}{4}\right)}\right)^2\right]\right\} \tag{10.29d}$$

$$\hat{\psi}_1(\beta) = \frac{2}{\beta}(1 + A)\sin\left(\frac{\beta l}{2}\right) - \left[\frac{4A}{k\beta^2}\left\{\sin\left(\frac{\beta l - \beta k}{2}\right)\sin\left(\frac{\beta k}{2}\right)\right\}\right] \tag{10.29e}$$

10.2.3 Open Circuit Edge-Capacitance and End Correction

Once the lumped capacitance C_l is determined from (10.28), the edge-capacitance of finite length strip conductor (C_{loc}) (Fig. 10.2) can be obtained as

$$C_{loc} = 0.5[C_l - lC] \tag{10.30}$$

where C is the capacitance per unit length of a uniform line of the same width w. The expression for C (Eq. (2.137) in Chapter 2) is given by

$$C = \frac{\pi}{\left[\int_0^{\infty}\left|\frac{\hat{\psi}_1(\alpha)}{Q}\right|^2 \frac{1}{\alpha Y}\, d\alpha\right]} \tag{10.31}$$

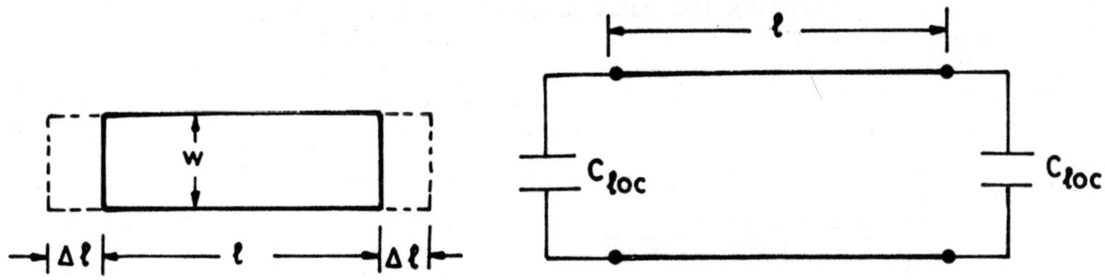

Fig. 10.2 Rectangular conductor showing hypothetical extension and equivalent circuit.

In (10.31), $\hat{\psi}_1(\alpha)$ the Fourier transform of $\psi_1(x)$ is given by (10.29d), the expression for Y is the same as (10.29a) with ζ replaced by α and Q, the total charge per unit length on the stripconductor is given by

$$Q = \int_{-w/2}^{w/2} \psi_1(x)\, dx \tag{10.32}$$

With $\psi_1(x)$ given in (10.25a), $Q = 5w/4$. The edge capacitance can also be expressed in terms of a hypothetical extension of the stripconductor by a small amount Δl (Fig. 10.2) which is given by

$$\Delta l = C_{loc}/C \tag{10.33}$$

In order to determine the open circuit edge-capacitance (C_{oc}) and end correction (Δl) due to an abruptly ended long strip conductor (Fig. a of Table 10.1), we use the following formulas

$$C_{oc} = \lim_{l\to\infty} 0.5[C_l - lC] \tag{10.34}$$

$$\Delta l = C_{oc}/C \tag{10.35}$$

In evaluating (10.34), l is nót infinite but some finitely large value beyond which the change in $(C_l - lC)$ is negligible.

10.2.4 Generalization to Multilayer Iso/Anisotropic Structures

In order to recognize the generality implicit in the expressions derived in Sec. 10.2.1, we first consolidate the relevant formulas for the computation of lumped capacitance C_l of a rectangular patch. For a charge distribution $f(x, z)$ expressed in a general form as

$$f(x, z) = \sum_{n=1}^{N} a_n \psi_n(x, z) \tag{10.36}$$

the lumped capacitance C_l of the rectangular conductor patch can be computed from

$$C_l = \frac{1}{(2\pi)^2} \sum_{n=1}^{N} a_n F_n \tag{10.37}$$

where

$$F_n = (2\pi)^2 \int_{-l/2}^{l/2} \int_{-w/2}^{w/2} \psi_n(x, z)\, dx\, dz \tag{10.38a}$$

F_n's are calculated by specifying the basis functions $\psi_n(x, z)$ and the coefficients a_n's are determined from

$$\sum_{n=1}^{N} a_n P_{mn} = F_m \tag{10.38b}$$

where

$$P_{mn} = \int_{-\infty}^{\infty}\int_{-\infty}^{\infty} \frac{\hat{\psi}_n(\alpha, \beta)\hat{\psi}^*_m(\alpha, \beta)}{\zeta Y}\, d\alpha\, d\beta \tag{10.38c}$$

$$\zeta = \sqrt{\alpha^2 + \beta^2} \tag{10.38d}$$

With the basis function specified, the only parameter which depends on the structural parameters in the computation of C_l is the admittance Y at the charge plane. The admittance parameter Y for the structure can be determined independently from its two wire transmission line equivalent. The lumped capacitance formula (10.27) with (10.38) is therefore general and can be applied to any multilayer structure with a single conductor patch. For one term approximation of the basis function specified by (10.24), (10.28) gives the general formula for C_l. Once the lumped capacitance is determined, the other parameters, namely, the open circuit edge-capacitance and end correction can be computed using the formulas given in Sec. 10.2.3.

It is worth nothing that the two wire transmission line equivalent of a particular structure is the same whether the strip conductor is of finite length as in the present case or infinitely long. Hence the admittance parameters listed in Table 3.3 (Chapter 3) for stripline-like transmission lines are applicable to the corresponding structures with a finite length strip conductor when β in the expressions is replaced by ζ. In Table 3.4 are listed admittance parameters of transmission line structures using anisotropic substrates with optical axis aligned along the vertical y-axis. The same admittance parameters with β replaced by ζ can be used in (10.37)-(10.38) for computing the lumped capacitance of a conductor patch in the corresponding layered structure.

10.3 Broadside-Coupled Rectangular Conductor Patches

10.3.1 Even- and Odd-Mode Lumped Capacitances—Unified Variational Formulas

Figure 10.3 shows the configuration of broadside-coupled rectangular conductor patches embedded in a three layer structure with xz-plane at $y = 0$ representing the plane of symmetry. The even-and odd-mode excitations of the structure would cause a magnetic wall and an electric wall at the plane of symmetry (Fig. 10.4). Following the procedure outlined in Sec. 10.2 for a single conductor patch embedded in a layered dielectric structure, the expressions for the even- and odd-mode lumped capacitances denoted as C_{le} and C_{lo}, turn out to be

$$C_{l\substack{e\\o}} = \frac{(2\pi)^2}{\displaystyle\int_{-\infty}^{\infty}\int_{-\infty}^{\infty} \left|\frac{\hat{\psi}_1(\alpha, \beta)_{\substack{e\\o}}}{Q_{\substack{e\\o}}}\right|^2 \frac{1}{\zeta Y_{\substack{e\\o}}}\, d\alpha\, d\beta} \tag{10.39}$$

where Q_e and $\hat{\psi}_1(\alpha, \beta)_e$ are given by (10.29b)-(10.29e) with A and k replaced by A_e and k_e respectively. Similarly Q_o and $\hat{\psi}_1(\alpha, \beta)_o$ are given by (10.29b)-(10.29e) with A and k replaced

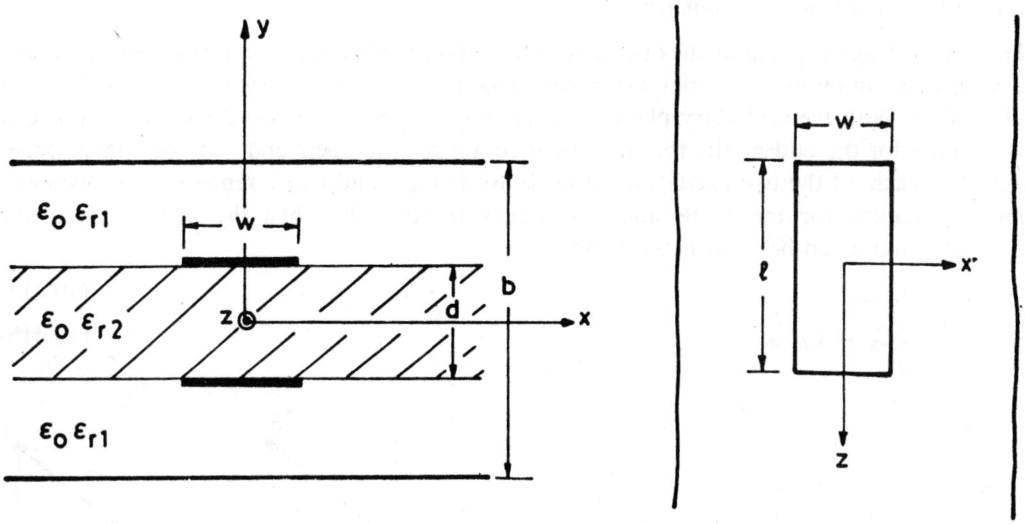

Fig. 10.3 Broadside-coupled rectangular conductor patches in a three layer dielectric structure.

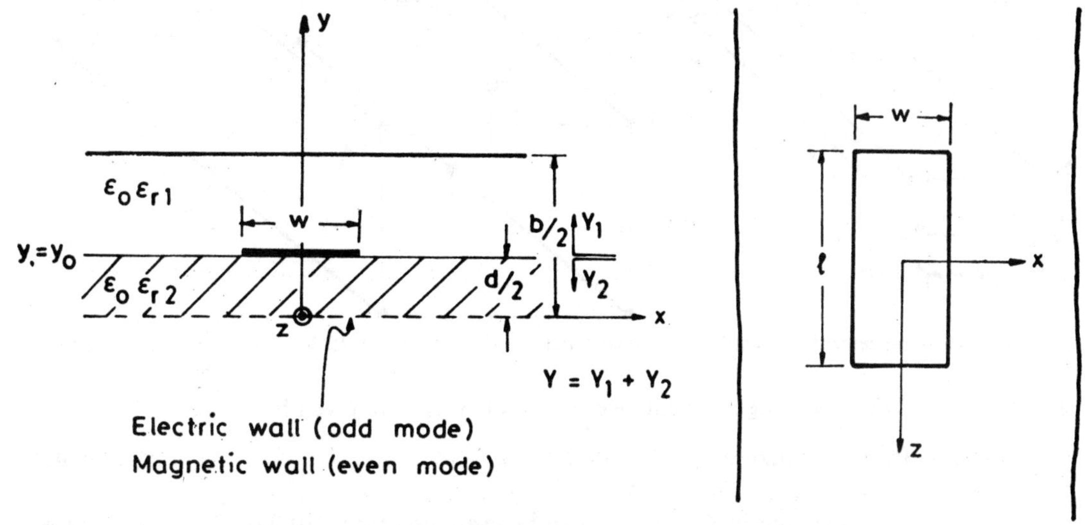

Fig. 10.4 Symmetric half of the structure in Fig. 10.3 for even-and odd-mode analysis.

by A_o and k_o respectively. These correspond to the charge distribution specified in (10.25) with A and k replaced by A_e and k_e for the even-mode and A_o and k_o for the odd-mode. Y_e and Y_o are the even- and odd-mode admittances at the charge plane $y = y_0$. Referring to Fig. 10.4, the expression can be written as

$$Y_{\substack{e\\o}} = \epsilon_o\left[\epsilon_{r1}\coth\zeta\left(\frac{b-d}{2}\right) + \epsilon_{r2}\left(\tanh\left(\frac{\zeta d}{2}\right)\right)^p\right],$$

$$p = \begin{cases} 1, & \text{even mode} \\ -1, & \text{odd mode} \end{cases} \tag{10.40}$$

10.3.2 Equivalent Circuit Parameters

The fringing fields existing at the open-circuited ends of the broadside coupled conductors can be represented in terms of a series gap capacitance C_g and shunt capacitance C_p as depicted in Fig. 10.5. With the symmetry plane representing an open circuit for the even-mode and a short circuit for the odd-mode, the equivalent circuit for the open end can be drawn separately for each of the two excitations (Fig. 10.6). If $C_{loc\,e}$ and $C_{loc\,o}$ represent the open-end edge-capacitances for the even- and odd-modes respectively, then the parameters of the equivalent circuit can be determined from,

$$C_p = C_{loc\,e} \tag{10.41a}$$

$$C_p + 2C_g = C_{loc\,o} \tag{10.41b}$$

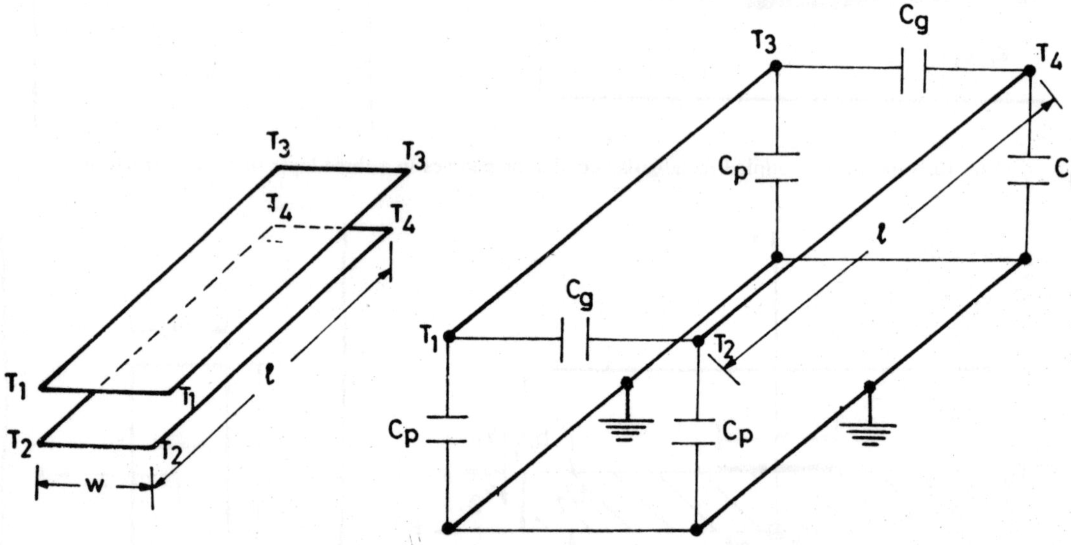

Fig. 10.5 Arrangement of broadside-coupled patch conductors in Fig. 10.3 and the equivalent circuit.

The even- and odd-mode edge capacitances $C_{loc\,e}$ and $C_{loc\,o}$ are given by

$$C_{loc\,\substack{e\\o}} = 0.5[C_{l\substack{e\\o}} - lC_{\substack{e\\o}}] \tag{10.42}$$

where the lumped capacitances C_{le} and C_{lo} can be evaluated from (10.39). C_e and C_o denote the even- and odd-mode capacitances per unit length of the broadside-coupled structure with infinitely long strip conductor. The expressions for these are given by (Eq. (4.93) of Chapter 4)

$$C_{\substack{e\\o}} = \frac{\pi}{\displaystyle\int_0^\infty \left|\frac{\hat{\psi}_1(\alpha)}{Q}\right|^2 \frac{1}{\alpha Y_{\substack{e\\o}}}\, d\alpha} \tag{10.43}$$

where $\hat{\psi}_1(\alpha)$ and Q are given by (10.29d) and (10.32) respectively and Y_e and Y_o are given by (10.40) with ζ replaced by α.

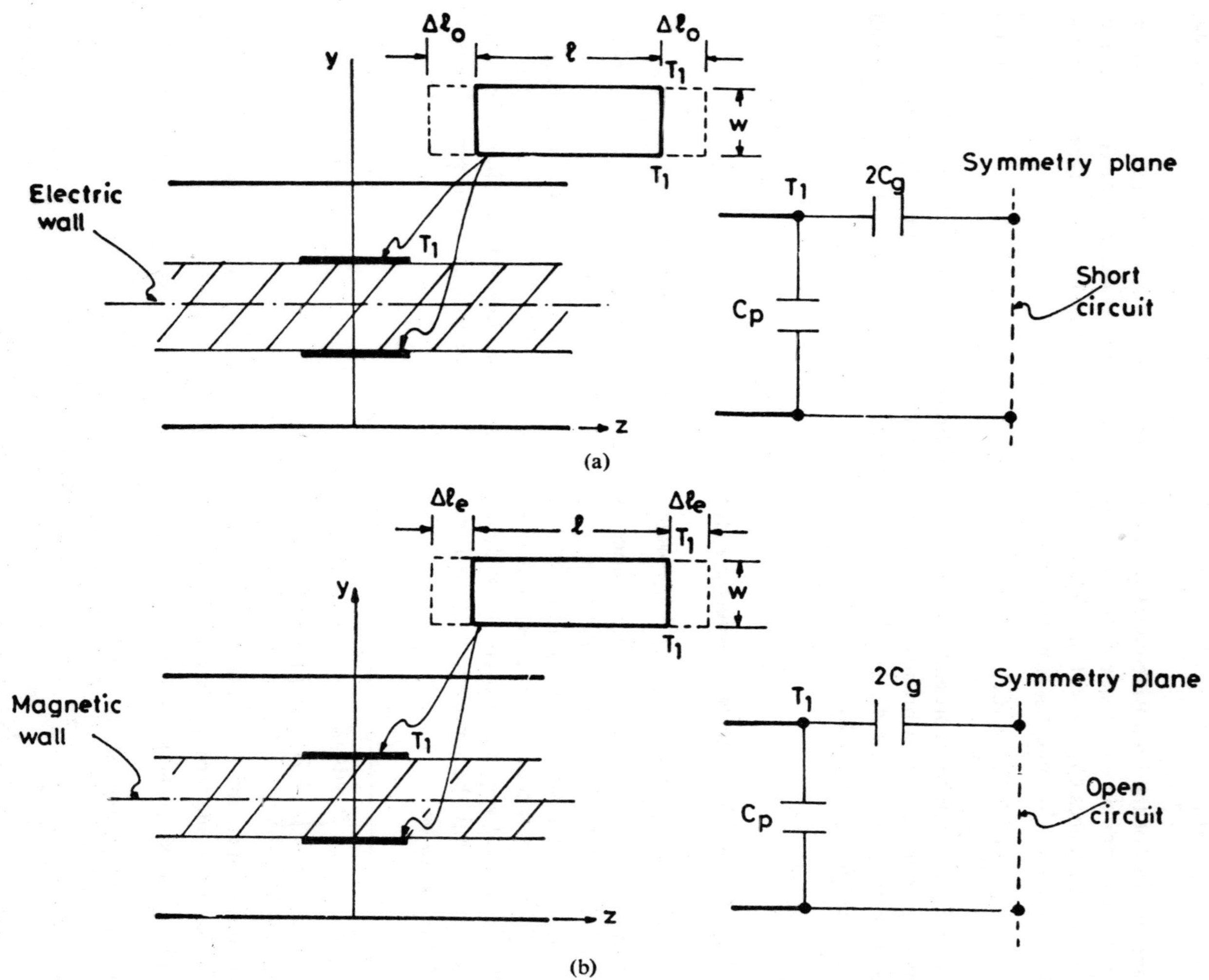

Fig. 10.6 Broadside-coupled patch conductor structure showing hypothetical extension Δl and equivalent circuit at the open end for (a) odd mode and (b) even mode.

The even- and odd-mode edge capacitances $C_{loc\,e}$ and $C_{loc\,o}$ given in (10.42) can be expressed in terms of a hypothetical extension of the strip conductor beyond the open end by a small amount Δl_e and Δl_o respectively (Fig. 10.6).

$$\Delta l_{\substack{e\\o}} = C_{loc\,\substack{e\\o}}/C_{\substack{e\\o}} \tag{10.44}$$

If the length l of the broadside-coupled strip conductor patches is taken sufficiently long, then the open ends on either side of the strip conductor length get decoupled. Alternatively, the edge capacitances $C_{loc\,e}$ and $C_{loc\,o}$ become independent of the length l. Denoting these limiting values as $C_{oc\,e}$ and $C_{oc\,o}$, we can write

$$C_{oc\,\substack{e\\o}} = \lim_{l\to\infty} 0.5[C_{l\substack{e\\o}} - lC_{\substack{e\\o}}] \tag{10.45}$$

$C_{oc\,e}$ and $C_{oc\,o}$ therefore represent the even- and odd-mode edge capacitances of abruptly ended strip conductors in a broadside-coupled strip transmission structure. The corresponding hypothetical extensions Δl_e and Δl_o beyond the abruptly terminated end of the coupled strip conductors are given by

$$\Delta l_{\substack{e\\o}} = C_{oc\,\substack{e\\o}}/C_{\substack{e\\o}} \tag{10.46}$$

10.4 Edge-Coupled Rectangular Conductor Patches

10.4.1 Even- and Odd-Mode Lumped Capacitances—Unified Variational Formulas

The even- and odd-mode lumped capacitance expressions for symmetrically arranged edge-coupled rectangular conductor patches in a stripline-like configuration can be obtained by considering a typical structure shown in Fig. 10.7. The symmetry is with respect to the yz plane at $x = 0$. With magnetic and electric walls placed at the symmetry plane for the even- and odd-modes respectively, the procedure for determining the two lumped capacitances denoted as C_{le} and C_{lo} would be similar to that used for a single conductor patch. The general expression turns out to be

$$C_{l\substack{e\\o}} = \frac{2(2\pi)^2}{\displaystyle\int_{-\infty}^{\infty}\int_{-\infty}^{\infty} \left|\frac{\hat{\psi}_1(\alpha, \beta)_{\substack{e\\o}}}{Q_{\substack{e\\o}}}\right|^2 \frac{1}{\zeta Y}\, d\alpha\, d\beta} \tag{10.47}$$

As usual, the subscript 'e' denotes the even-mode and 'o' denotes the odd-mode. The above expression is for one term approximation to the charge distribution specified by $\psi_1(x, z)_{\substack{e\\o}}$ on one strip. $\hat{\psi}_1(\alpha, \beta)_{\substack{e\\o}}$ is the Fourier transform of $\psi_1(x, z)_{\substack{e\\o}}$ and $Q_{\substack{e\\o}}$ denote the total charge on a strip. The various expressions are

$$\hat{\psi}_1(\alpha, \beta)_e = 2 \int_{s/2}^{s/2+w} \int_{-l/2}^{l/2} \psi_1(x, z)_e \cdot \cos \alpha x \cdot e^{j\beta z}\, dx\, dz \tag{10.48a}$$

$$\hat{\psi}_1(\alpha, \beta)_o = -2j \int_{s/2}^{s/2+w} \int_{-l/2}^{l/2} \psi_1(x, z)_o \cdot \sin \alpha x \cdot e^{j\beta z}\, dx\, dz \tag{10.48b}$$

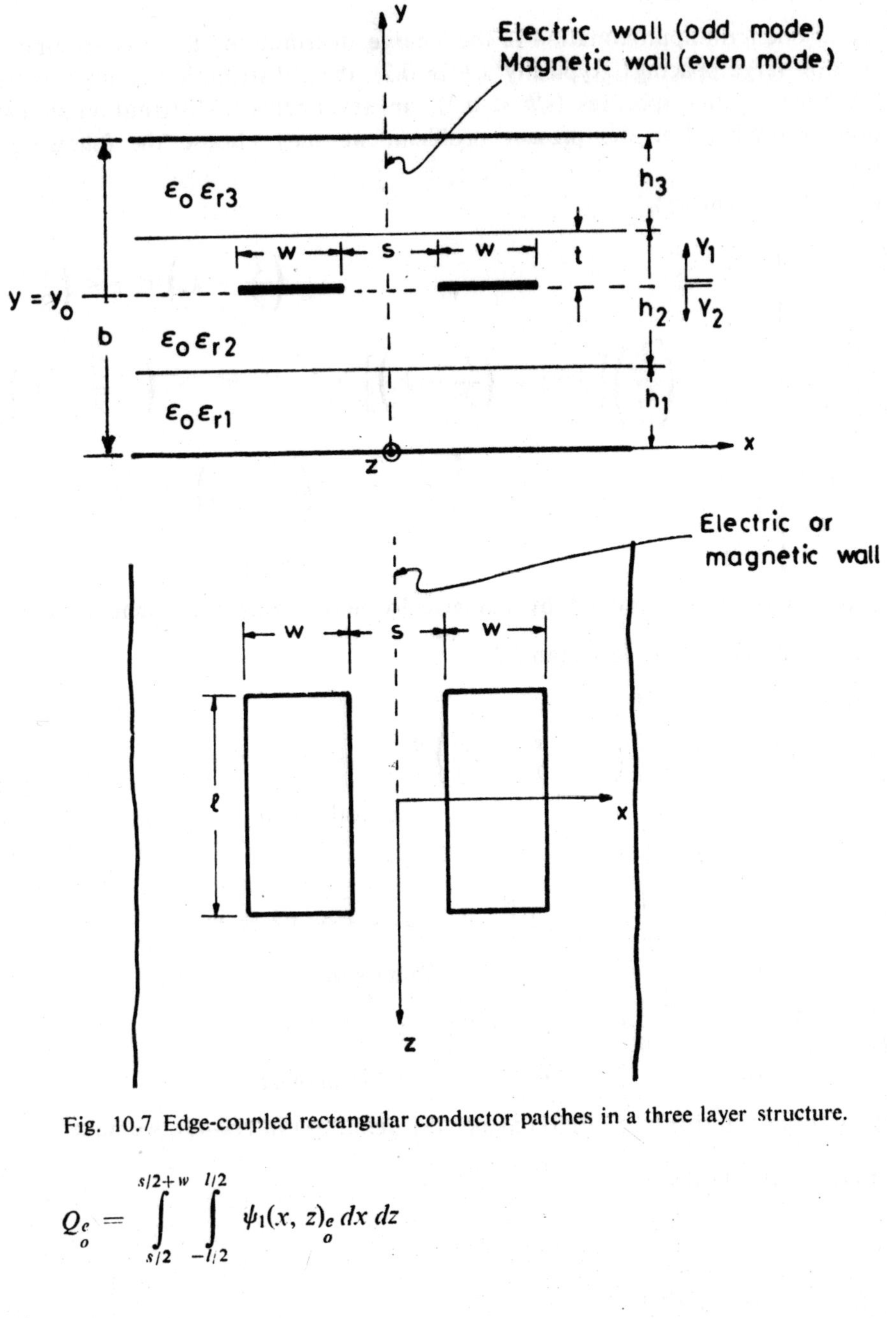

Fig. 10.7 Edge-coupled rectangular conductor patches in a three layer structure.

$$Q_{\substack{e\\o}} = \int_{s/2}^{s/2+w} \int_{-l/2}^{l/2} \psi_1(x, z)_{\substack{e\\o}}\, dx\, dz \tag{10.49}$$

and

$$\zeta = \sqrt{\alpha^2 + \beta^2} \tag{10.50}$$

The parameter Y in (10.47) is the admittance at the charge plane and is the same for even- and odd-modes. For the structure shown in Fig. 10.7, the expression for Y is given in (10.29a).

10.4.2 Choice of Basis Functions

The choice of one term approximation to the charge distribution depends strongly on the spacing s. For large spacings (typically $s/b \geqslant 0.2$), the x-distribution is more or less symmetrical. But for smaller spacings, $(s/b < 0.2)$, an asymmetrical distribution should yield more accurate results. For the present problem, we may choose the following charge distribution.

$$\psi_1(x, z)_{\substack{e\\o}} = \psi_1(x)_{\substack{e\\o}}\psi_1(z)_{\substack{e\\o}} \tag{10.51}$$

$$\psi_1(z)_{\substack{e\\o}} = \begin{cases} 1 & ; \; -\left(\frac{l}{2} - k_{\substack{e\\o}}\right) \leqslant z \leqslant \left(\frac{l}{2} - k_{\substack{e\\o}}\right) \\ 1 + \left(\frac{A_{\substack{e\\o}}}{k_{\substack{e\\o}}}\right)\left[\,|z| - \left(\frac{l}{2} - k_{\substack{e\\o}}\right)\right] & ; \; -\frac{l}{2} \leqslant z \leqslant \left(-\frac{l}{2} - k_{\substack{e\\o}}\right) \\ & \quad \left(\frac{l}{2} - k_{\substack{e\\o}}\right) \leqslant z \leqslant \frac{l}{2} \\ 0 & ; \; \text{Otherwise} \end{cases} \tag{10.52}$$

$A_{\substack{e\\o}}$ and $k_{\substack{e\\o}}$ are constants determined by numerically maximizing the capacitance $C_{l\substack{e\\o}}$. The values of k_e and k_o are always less than $l/2$.

For large spacings $(s/b \geqslant 0.2)$

$$\psi_1(x)_{\substack{e\\o}} = \begin{cases} 1 + B_{\substack{e\\o}}\left|\frac{2}{w}\left(x - \frac{s}{2} - \frac{w}{2}\right)\right|^3 & ; \; s/2 \leqslant x \leqslant s/2 + w \\ 0 & ; \; \text{otherwise} \end{cases} \tag{10.53a}$$

For small spacings $(s/b < 0.2)$

$$\psi_1(x)_e = \begin{cases} 1 + B_{\substack{e\\o}}\left|\frac{1}{w}(x - s/2)\right|^3 & ; \; s/2 \leqslant x \leqslant s/2 + w \\ 0 & ; \; \text{Otherwise} \end{cases} \tag{10.53b}$$

$$\psi_1(x)_o = \begin{cases} 1 + B_{\substack{e\\o}}\left|\frac{1}{w}(w - x + s/2)\right|^3 & ; \; s/2 \leqslant x \leqslant s/2 + w \\ 0 & ; \; \text{Otherwise} \end{cases} \tag{10.53c}$$

Taking the Fourier transform of (10.51) as per the definition in (10.48), we have

$$\hat{\psi}_1(\alpha, \beta)_{\substack{e\\o}} = \hat{\psi}_1(\alpha)_{\substack{e\\o}}\hat{\psi}_1(\beta)_{\substack{e\\o}} \tag{10.54}$$

where

$$\hat{\psi}_1(\beta)_{\substack{e\\o}} = \int_{-l/2}^{l/2} \psi_1(z)_{\substack{e\\o}} e^{j\beta z}\, dz \tag{10.55a}$$

$$\hat{\psi}_1(\alpha)_e = 2\int_{s/2}^{s/2+w} \psi_1(x)_e \cos(\alpha x)\, dx \tag{10.55b}$$

$$\hat{\psi}_1(\alpha)_o = -2j \int_{s/2}^{s/2+w} \psi_1(x)_o \sin(\alpha x)\, dx \tag{10.55c}$$

Substituting for $\psi_1(z)_{\substack{e\\o}}$ from (10.52) and $\psi_1(x)_{\substack{e\\o}}$ from (10.53) and evaluating the above integrals , we obtain

$$\hat{\psi}_1(\beta)_{\substack{e\\o}} = \frac{2}{\beta}(1 + A_{\substack{e\\o}}) \sin\left(\frac{\beta l}{2}\right) - \left[\frac{4A_{\substack{e\\o}}}{\beta^2 k_{\substack{e\\o}}} \cdot \sin\left\{\frac{\beta(l - k_{\substack{e\\o}})}{2}\right\} \sin\left(\frac{\beta k_{\substack{e\\o}}}{2}\right)\right] \tag{10.56a}$$

For large spacings; ($s/b \geqslant 0.2$)

$$\hat{\psi}_1(\alpha)_e = 2w \cos\left\{\alpha\left(\frac{s+w}{2}\right)\right\}\left[(1 + B_e)\left(\frac{\sin\left(\frac{\alpha w}{2}\right)}{\left(\frac{\alpha w}{2}\right)}\right) + \frac{3B_e}{\left(\frac{\alpha w}{2}\right)^2}\left\{\cos\left(\frac{\alpha w}{2}\right) - 2\cdot\left(\frac{\sin\left(\frac{\alpha w}{2}\right)}{\left(\frac{\alpha w}{2}\right)}\right) + \left(\frac{\sin\left(\frac{\alpha w}{4}\right)}{\left(\frac{\alpha w}{4}\right)}\right)^2\right\}\right] \tag{10.56b}$$

$$\hat{\psi}_1(\alpha)_o = -2jw \sin\left\{\alpha\left(\frac{s+w}{2}\right)\right\}\left[(1 + B_o)\left(\frac{\sin\left(\frac{\alpha w}{2}\right)}{\left(\frac{\alpha w}{2}\right)}\right) + \frac{3B_o}{\left(\frac{\alpha w}{2}\right)^2}\left\{\cos\left(\frac{\alpha w}{2}\right) - 2\cdot\left(\frac{\sin\left(\frac{\alpha w}{2}\right)}{\left(\frac{\alpha w}{2}\right)}\right) + \left(\frac{\sin\left(\frac{\alpha w}{4}\right)}{\left(\frac{\alpha w}{4}\right)}\right)^2\right\}\right] \tag{10.56c}$$

For small spacings; ($s/b < 0.2$)

$$\begin{aligned}\hat{\psi}_1(\alpha)_e = \frac{2w}{(\alpha w)^3}\Bigg[&\left\{((2 + B_e)(\alpha w)^2 - 6) \sin\left(\frac{\alpha w}{2}\right)\right. \\ &\left. + 3B_e(\alpha w)\cdot\cos\left(\frac{\alpha w}{2}\right)\right\}\cos\left(\alpha\left(\frac{s+w}{2}\right)\right) + B_e\left\{((\alpha w)^2 - 6)\cos\left(\frac{\alpha w}{2}\right)\right. \\ &\left. - \frac{3}{(\alpha w)}((\alpha w)^2 - 4)\sin\left(\frac{\alpha w}{2}\right)\right\}\cdot\sin\left(\alpha\left(\frac{s+w}{2}\right)\right)\Bigg]\end{aligned} \tag{10.56d}$$

$$\begin{aligned}\hat{\psi}_1(\alpha)_o = -\frac{2jw}{(\alpha w)^3}\Bigg[&\left\{((2 + B_o)(\alpha w)^2 - 6)\sin\left(\frac{\alpha w}{2}\right) + 3B_o(\alpha w)\cdot\cos\left(\frac{\alpha w}{2}\right)\right\} \\ &\sin\left(\alpha\left(\frac{s+w}{2}\right)\right) + B_o\left\{((\alpha w)^2 - 6)\cos\left(\frac{\alpha w}{2}\right)\right. \\ &\left. - \frac{3}{\alpha w}((\alpha w)^2 - 4)\sin\left(\frac{\alpha w}{2}\right)\right\}\cos\left(\alpha\left(\frac{s+w}{2}\right)\right)\Bigg]\end{aligned} \tag{10.56e}$$

Evaluation of $Q_{\substack{e\\o}}$ according to (10.49) gives

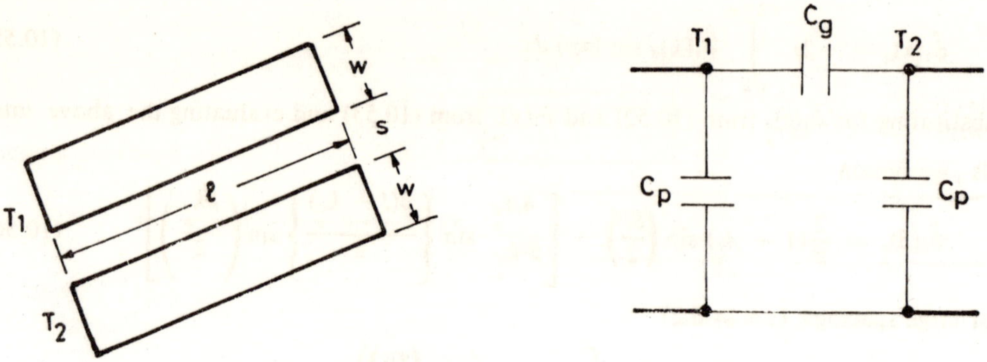

Fig. 10.8 Edge-coupled conductor arrangement in Fig. 10.7 and equivalent circuit at the open end terminal.

$$Q_{\substack{e\\o}} = (1 + 0.25B_{\substack{e\\o}})w \tag{10.57}$$

10.4.3 *Equivalent Circuit Parameters*

The equivalent circuit representing the fringe capacitances at the open end terminal T_1T_2 is shown in Fig. 10.8. Using the equivalent circuit representations for the even- and odd-mode excitations (Fig. 10.9) and the relations given in (10.41) and (10.42), the capacitance C_p and C_g can be determined. In (10.42), lumped capacitances C_{le} and C_{lo} can be evaluated using (10.47) with (10.48)-(10.57). The expressions for C_e and C_o which correspond to the infinitely long strip conductors are given by (refer Chapter 4, eq. (4.60)),

$$C_{\substack{e\\o}} = \frac{2\pi}{\displaystyle\int_0^\infty \left| \frac{\hat{\psi}_1(\alpha)_{\substack{e\\o}}}{Q_{\substack{e\\o}}} \right|^2 \frac{1}{\alpha Y}\, d\alpha} \tag{10.58}$$

where

$$\hat{\psi}_1(\alpha)_e = 2 \int_{s/2}^{s/2+w} \psi_1(x)_e \cos(\alpha x)\, dx \tag{10.59a}$$

$$\hat{\psi}_1(\alpha)_o = -2j \int_{s/2}^{s/2+w} \psi_1(x)_o \sin(\alpha x)\, dx \tag{10.59b}$$

$$Q_{\substack{e\\o}} = \int_{s/2}^{s/2+w} \psi_1(x)_{\substack{e\\o}}\, dx \tag{10.59c}$$

The expressions for $\psi_1(x)_e$ and $\psi_1(x)_o$ are given in (10.53) and the expression for Y is the same as (10.29a) with ζ replaced by α.

With the knowledge of $C_{loc\ e}$, $C_{loc\ o}$, C_e and C_o, the end correction Δl_e and Δl_o of the edge-coupled conductor patches (Fig. 10.9) can be calculated using (10.44). The edge-capacitances

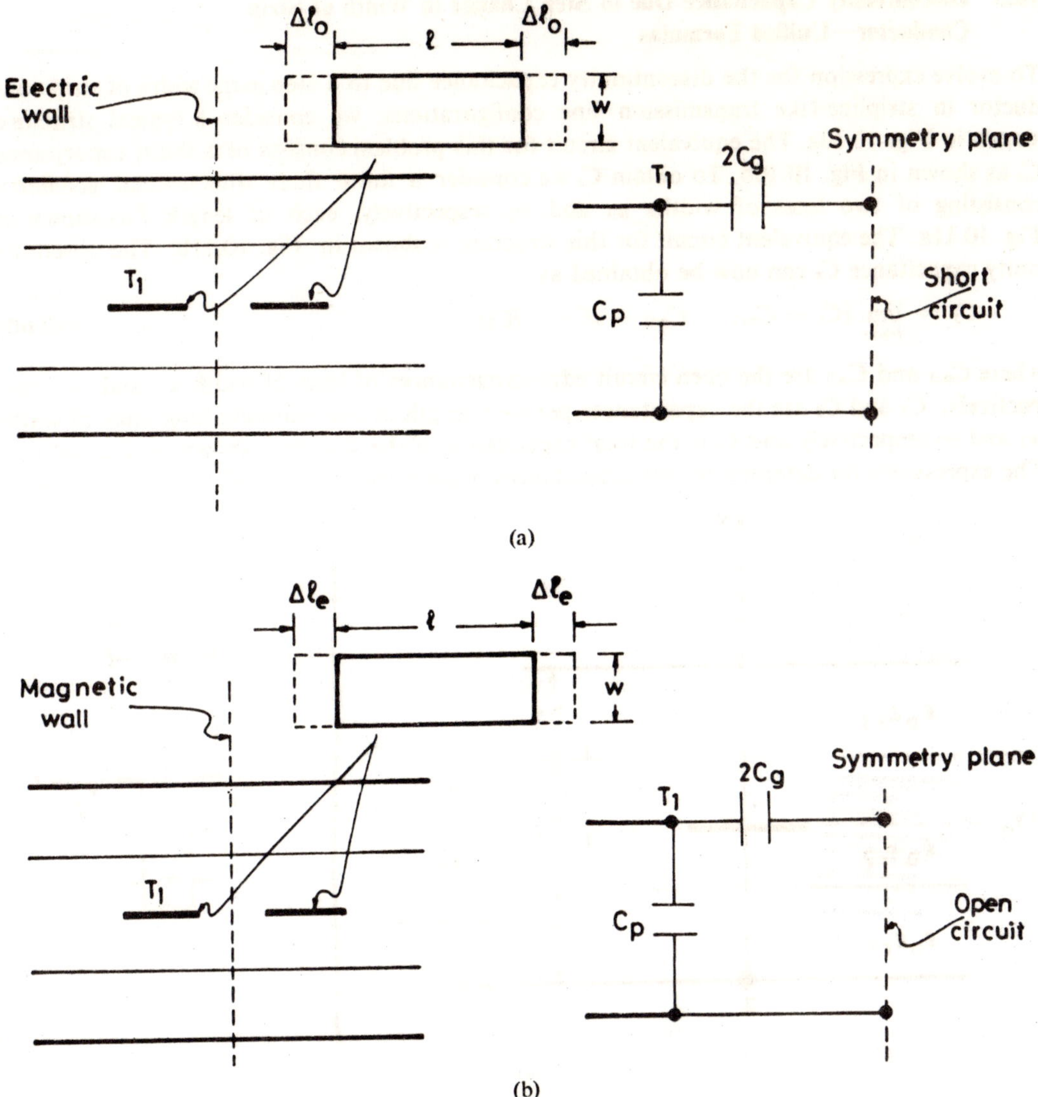

Fig. 10.9 Edge-coupled patch conductor structure showing hypothetical extension Δl and equivalent circuit at the open end for (a) odd mode and (b) even mode.

and end corrections at the open ends of abruptly terminated edge-coupled strip conductors of long length can also be evaluated using (10.45) and (10.46).

It may be noted that the entire set of formulas (10.47)—(10.59) given in this section are applicable to edge-coupled rectangular conductor patches in multilayer iso/anisotropic substrates so long as there is a plane of symmetry between the conductor patches as shown in Fig. 10.7. Only the admittance parameter Y needs to be changed depending on the structure. For structures employing isotropic substrates, the admittance expressions listed in Table 4.1 and for those employing anisotropic substrates, the admittance expressions listed in Table 4.2 can be directly used with β replaced by ζ for calculating C_{le} and C_{lo}.

10.5 Discontinuity Capacitance Due to Step Change in Width of Strip Conductor—Unified Formulas

To evolve expression for the discontinuity capacitance due to a step in the width of strip conductor in stripline-like transmission line configurations, we consider a typical structure shown in Fig. 10.10a. The equivalent circuit for this problem consists of a shunt capacitance C_s as shown in Fig. 10.10b. To obtain C_s we consider a finite three dimensional geometry consisting of two lines of widths w_1 and w_2 respectively, each of length l as shown in Fig. 10.11a. The equivalent circuit for this structure is shown in Fig. 10.11b. The discontinuity capacitance C_s can now be obtained as

$$C_s = \lim_{l\to\infty} [C_t - C_{oc1} - C_{oc2} - lC_1 - lC_2] \tag{10.60}$$

where C_{oc1} and C_{oc2} are the open circuit edge-capacitances of lines of width w_1 and w_2, respectively, C_1 and C_2 are the capacitances per unit length of the infinitely long lines of width w_1 and w_2 respectively and C_t is the total capacitance of the structure shown in Fig. 10.11a. The expressions for determining the capacitances of uniform lines (C_1, C_2) and open-circuit

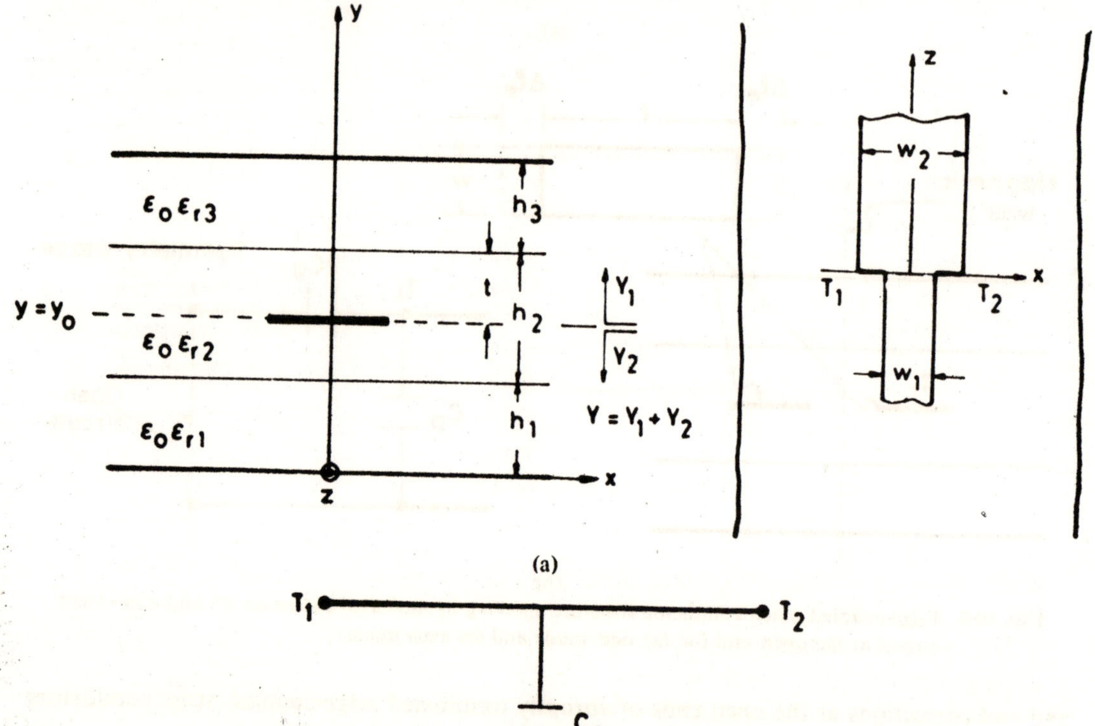

Fig. 10.10 (a) Three layer transmission line showing a step discontinuity in the strip conductor, (b) Equivalent circuit of step discontinuity.

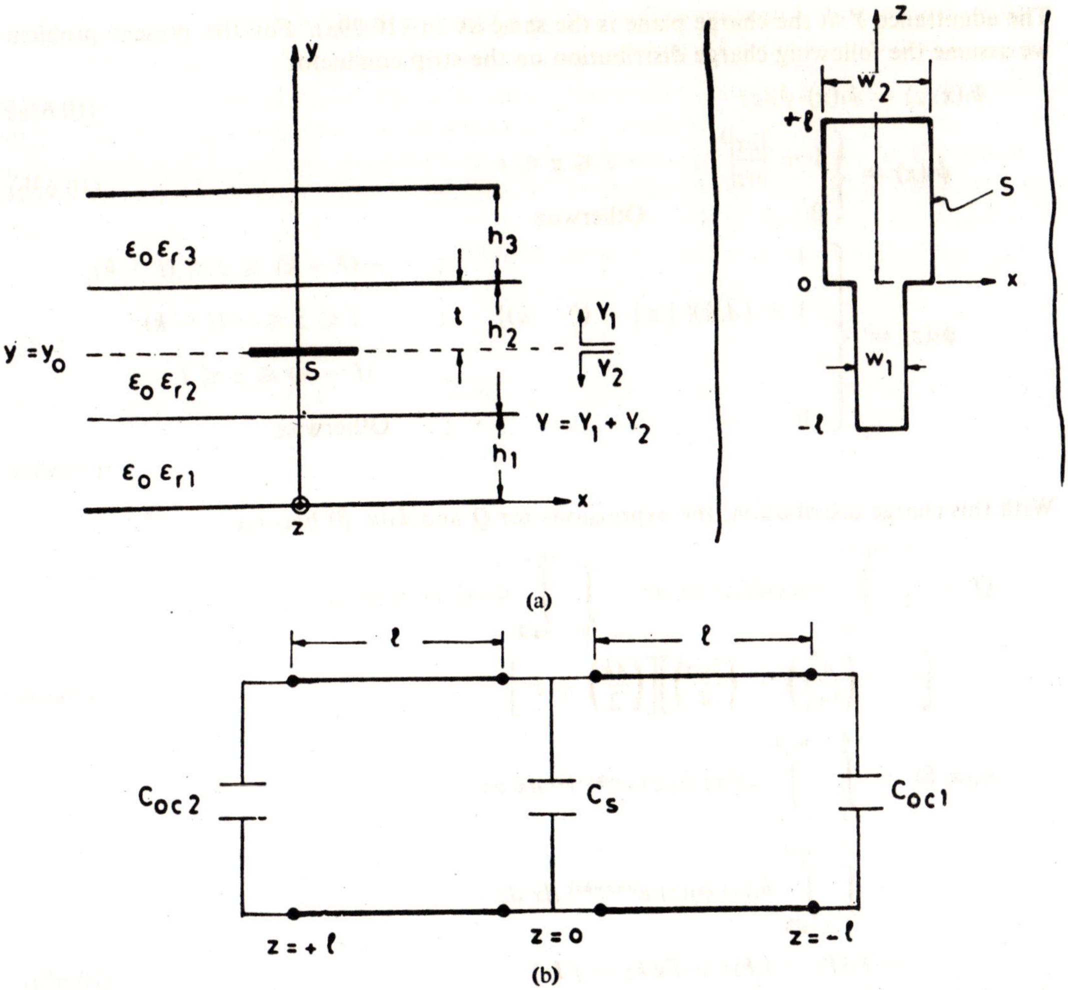

Fig. 10.11 (a) Configuration for calculating capacitance of a step discontinuity and (b) its equivalvent circuit.

edge-capacitances (C_{oc1}, C_{oc2}) are already available in earlier sections (Eqs. (10.31) and (10.34), respectively).

The total lumped capacitance C_t of the structure shown in Fig. 10.11a can be determined using the expression given by (10.28)

$$C_t = \frac{(2\pi)^2}{\left[\int\limits_{-\infty}^{\infty} \int\limits_{-\infty}^{\infty} \left| \frac{\hat{\psi}_1(\alpha, \beta)}{Q} \right|^2 \frac{1}{\zeta Y} \, d\alpha \, d\beta \right]} \tag{10.61}$$

where $\hat{\psi}_1(\alpha, \beta)$ is the fourier transform of the charge distribution $\psi_1(x, z)$ on the stripconductor as defined in (10.15) and Q is total charge on the stripconductor given by

$$Q = \int\limits_S \psi_1(x, z) \, ds \tag{10.62}$$

The admittance Y at the charge plane is the same as in (10.29a). For the present problem we assume the following charge distribution on the strip conductor

$$\psi_1(x, z) = \psi_1(x)\cdot\psi_1(z) \tag{10.63a}$$

$$\psi_1(x) = \begin{cases} 1 + \left|\dfrac{2x}{w_2}\right|^3, & -l \leqslant z \leqslant l \\ 0 ; & \text{Otherwise} \end{cases} \tag{10.63b}$$

$$\psi_1(z) = \begin{cases} 1 ; & -(l-k) \leqslant z \leqslant (l-k) \\ 1 + (A/k)[\,|z| - (l-k)] ; & -l \leqslant z \leqslant -(l-k) \\ & (l-k) \leqslant z \leqslant l \\ 0 ; & \text{Otherwise} \end{cases} \tag{10.63c}$$

With this charge distribution, the expressions for Q and $\hat{\psi}_1(\alpha, \beta)$ become

$$Q = \int_{-l}^{0}\int_{-w_1/2}^{w_1/2} \psi_1(x)\psi_1(z)\, dx\, dz + \int_{0}^{l}\int_{-w_2/2}^{w_2/2} \psi_1(x)\,\psi_1(z)\, dx\, dz$$

$$= \left[w_1 + \left(\frac{w_1^4}{4w_2^3}\right) + \left(\frac{5w_2}{4}\right)\right]\left[\left(\frac{Ak}{2}\right) + l\,\right] \tag{10.64}$$

$$\hat{\psi}_1(\alpha, \beta) = \int_{-l}^{0}\int_{-w_1/2}^{w_1/2} \psi_1(x)\,\psi_1(z)\, e^{j(\alpha x + \beta z)}\, dx\, dz$$

$$+ \int_{0}^{l}\int_{-w_2/2}^{w_2/2} \psi_1(x)\,\psi_1(z)\, e^{j(\alpha x + \beta z)}\, dx\, dz$$

$$= F_1(F_2 - j\,F_3) + F_4(F_2 + j\,F_3) \tag{10.65a}$$

where

$$F_1 = \left(w_1 + \frac{w_1^4}{w_2^3}\right)\left\{\frac{\sin\left(\frac{\alpha w_1}{2}\right)}{\left(\frac{\alpha w_1}{2}\right)}\right\} + \frac{12 w_1^2}{w_2^3 \alpha^2}\left[\cos\left(\frac{\alpha w_1}{2}\right) - 2\left\{\frac{\sin\left(\frac{\alpha w_1}{2}\right)}{\left(\frac{\alpha w_1}{2}\right)}\right\} + \left\{\frac{\sin\left(\frac{\alpha w_1}{4}\right)}{\left(\frac{\alpha w_1}{4}\right)}\right\}^2\right] \tag{10.65b}$$

$$F_4 = 2w_2\left\{\frac{\sin\left(\frac{\alpha w_2}{2}\right)}{\left(\frac{\alpha w_2}{2}\right)}\right\} + \frac{12}{w_2 \alpha^2}\left[\cos\left(\frac{\alpha w_2}{2}\right) - 2\left\{\frac{\sin\left(\frac{\alpha w_2}{2}\right)}{\left(\frac{\alpha w_2}{2}\right)}\right\} + \left\{\frac{\sin\left(\frac{\alpha w_2}{4}\right)}{\left(\frac{\alpha w_2}{4}\right)}\right\}^2\right] \tag{10.65c}$$

$$F_2 = \left[\left(\frac{1+A}{\beta}\right) \sin (\beta l) - \frac{2A}{k\beta^2} \sin \beta(l - k/2) \cdot \sin (\beta k/2)\right] \tag{10.65d}$$

$$F_3 = \left[\frac{2}{\beta} \sin \left(\frac{\beta l}{2}\right) - \frac{A}{\beta} \cos (\beta l) + \frac{2A}{k\beta^2} \cos \beta(l - k/2) \sin (\beta k/2)\right] \tag{10.65e}$$

Substituting the results of (10.64) and (10.65) in (10.61), the total lumped capacitance C_t of the structure shown in Fig. 10.11a can be determined. Once C_t is obtained, the step discontinuity capacitance C_s can be calculated using (10.60).

10.6 Characterization of Gap in Strip Conductor—Unified Variational Analysis

Discontinuity due to a gap in the strip conductor in any stripline-like transmission line can be represented in terms of a pi-equivalent network as shown in Fig. 10.12. The series capacitance C_g is due to the coupling between the adjacent ends of the strip conductors and the shunt capacitance C_p arises due to the fringing field from the open end of the strip with respect to ground. In order to derive general expressions for these capacitances, we consider a three layer transmission structure with shielding side walls and with a strip conductor having two series gaps as shown in Fig. 10.13a. Two specific cases are considered; (a) odd mode designated as '*o*' when the planes at $z = 0$ and a represent electric walls and (b) even mode designated as '*e*' when the planes at $z = 0$ and a represent magnetic walls. The equivalent circuit for both these cases is shown in Fig. 10.13b. We can now isolate the part of the structure between $z = 0$ and a, with boundary walls defined as above. The resulting configuration is a rectangular box type resonating structure as shown in Fig. 10.14.

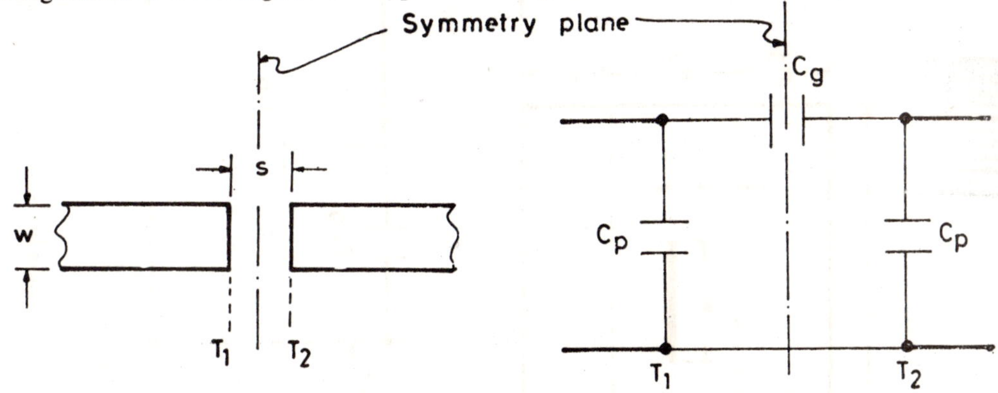

Fig. 10.12 Representation of a gap in strip conductor and its equivalent circuit.

If C_{le} and C_{lo} represent the lumped capacitances of the conductor patch in Fig. 10.14 for even-mode ($z = 0$ and a magnetic walls) and odd-mode ($z = 0$ and a electric walls) respectively, then the capacitances C_g and C_p of the series gap in Fig. 10.12 can be obtained from

$$2C_g + C_p = 0.5 \lim_{a\to\infty} [C_{lo} - (a - s)C] \tag{10.66a}$$

$$C_p = 0.5 \lim_{a\to\infty} [C_{le} - (a - s)\, C] \tag{10.66b}$$

where C is the capacitance per unit length of an infinitely long line of the same width w. In evaluating (10.66), the gap dimension s is kept fixed. The dimension a is not infinite, but a

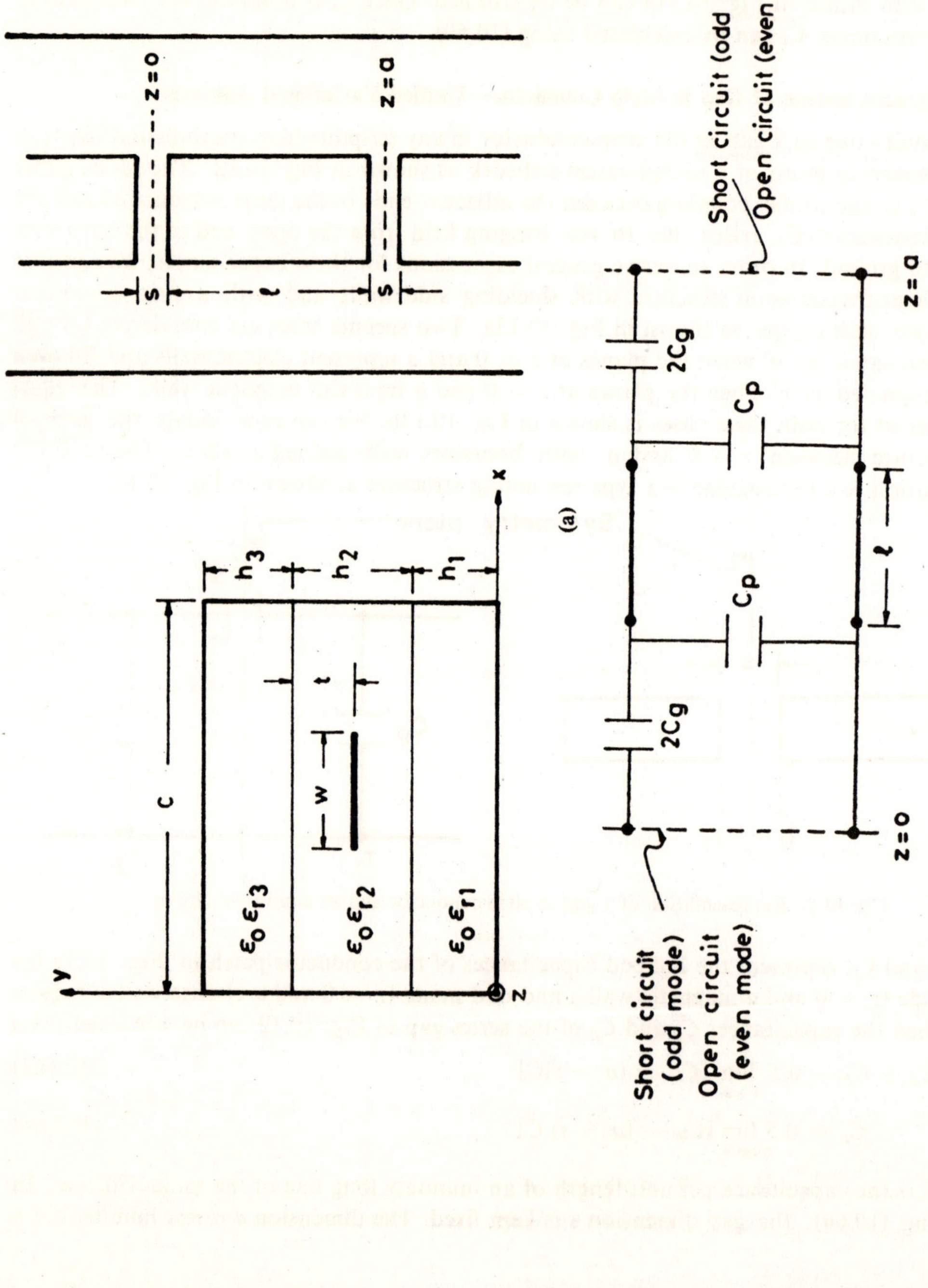

Fig. 10.13 (a) Three layer transmission line showing a strip conductor having two series gaps and (b) its equivalent circuit.

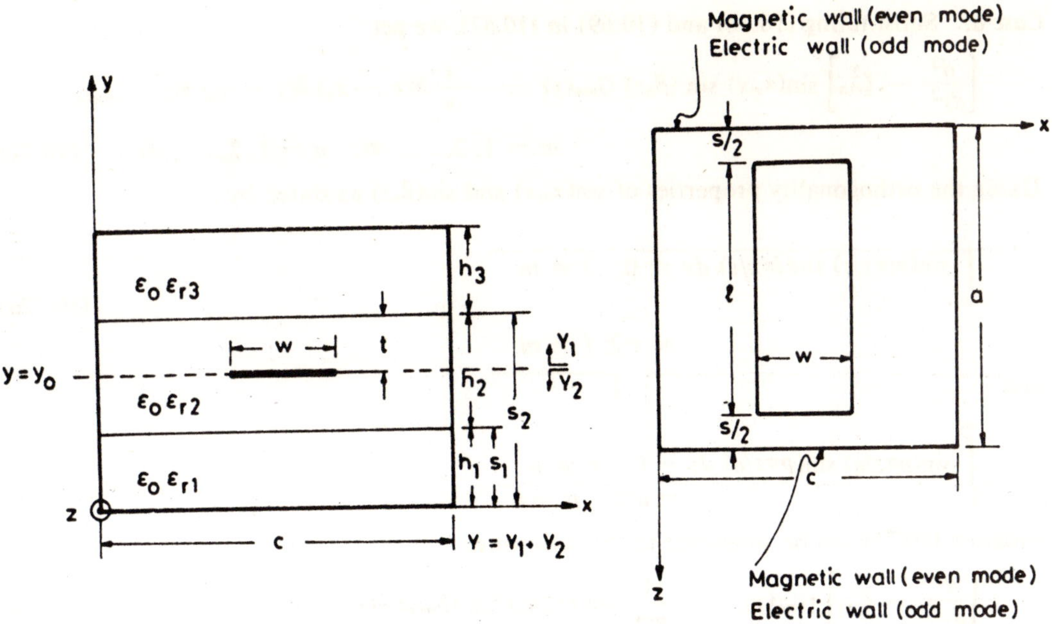

Fig. 10.14 Three layer transmission line structure for calculating even- and odd-mode lumped capacitances of a strip conductor having two series gaps.

large finite value (hence l large) beyond which the change in $[C_{lo} - (a - s)C]$ and also in $[C_{le} - (a - s)C]$ is negligible. In the following, we give the procedure for determining the variational expressions for the lumped capacitances C_{le} and C_{lo}.

The first step in the analysis is to determine the three dimensional Green's function for the structure shown in Fig. 10.14. For a point charge located at (x_o, y_o, z_o), the Green's function $G(x, y, z/x_o, y_o, z_o)$ should satisfy the Poisson's equation

$$\nabla^2 G(x, y, z,/x_o, y_o, z_o) = -\frac{1}{\epsilon}\,\delta(x - x_o)\,\delta(y - y_o)\,\delta(z - z_o) \tag{10.67}$$

The boundary conditions at the side walls $x = 0$ and c are of Dirichlet type ($G = 0$). At the walls $z = 0$ and a, they are of Neumann-type ($\partial G/\partial z = 0$) for even-mode and of Dirichlet type ($G = 0$) for odd-mode. The solution of (10.67) can be expressed in the form

$$G(x, y, z/x_o, y_o, z_o) = \sum_m \sum_n G_m(x) G_n(z)\, G_{mn}(y) \tag{10.68}$$

In order to satisfy the boundary conditions on the walls at $z = 0, a$ and at $x = 0, c$, the following expressions are found for $G_m(x)$ and $G_n(x)$ for the two different cases.

Case a: Electric walls at $x = 0$, c and $z = 0$, a (odd mode)

$$\begin{aligned} G_m(x) &= \sin(\alpha_m x); \quad \alpha_m = m\pi/c, \quad m = 1, 2, \ldots \infty \\ G_n(z) &= \sin(\beta_n z); \quad \beta_n = n\pi/a, \quad n = 1, 2, \ldots \infty \end{aligned} \tag{10.69}$$

Case b: Electric walls at $x = 0$, c and magnetic walls at $z = 0$, a (even mode)

$$\begin{aligned} G_m(x) &= \sin(\alpha_m x); \quad \alpha_m = m\pi/c, \quad m = 1, 2, \ldots \infty \\ G_n(z) &= \cos(\beta_n z); \quad \beta_n = n\pi/a, \quad n = 0, 1, \ldots \infty \end{aligned} \tag{10.70}$$

To determine $G_{mn}(y)$ we use the well known transverse transmission line method.

Case a: Substituting (10.68) and (10.69) in (10.67), we get

$$\left[\frac{d^2}{dy^2} - \zeta_{mn}^2\right] \sin(\alpha_m x) \sin(\beta_n z)\, G_{mn}(y) = -\frac{1}{\epsilon}\,\delta(x - x_o)\,\delta(y - y_o)\,\delta(z - z_o),$$

$$m = 1, 2, \ldots \infty, \quad n = 1, 2, \ldots \infty \qquad (10.71)$$

Using the orthogonality properties of $\sin(\alpha_m x)$ and $\sin(\beta_n z)$ as stated by

$$\int_0^c \sin(m\pi x/c)\sin(l\pi x/c)\,dx = 0; \quad l \neq m$$

$$= c/2; \; l = m \qquad (10.72a)$$

and

$$\int_0^a \sin(n\pi z/a)\sin(p\pi z/a)\,dz = 0; \quad p \neq n$$
$$= a/2; \quad p = n \qquad (10.72b)$$

equation (10.71) can be simplified to the following:

$$\left[\frac{d^2}{dy^2} - \zeta_{mn}^2\right] G_{mn}(y) = -\frac{4}{ac\epsilon}\sin(\alpha_m x_o)\sin(\beta_n z_o)\,\delta(y - y_o) \qquad (10.73)$$

where

$$\zeta_{mn}^2 = \alpha_m^2 + \beta_n^2 = \left[\left(\frac{m\pi}{c}\right)^2 + \left(\frac{n\pi}{a}\right)^2\right] \qquad (10.74)$$

The boundary conditions at the two dielectric interfaces are given by

$$G_{mn}(y)\Big|_{y = s_j - 0} = G_{mn}(y)\Big|_{y = s_j + 0}, \quad j = 1, 2 \qquad (10.75a)$$

$$\epsilon_o \epsilon_{rj}\frac{\partial}{\partial y} G_{mn}(y)\Big|_{y = s_j - 0} = \epsilon_o \epsilon_{rj+1}\frac{\partial}{\partial y} G_{mn}(y)\Big|_{y = s_j + 0}, \quad j = 1, 2 \qquad (10.75b)$$

The transmission line equivalent of the structure shown in Fig. 10.14 is the same as the one shown in Fig. 10.1 with ζ replaced by ζ_{mn}. The voltage V along this transmission line satisfies the differential equation given by

$$\left[\frac{d^2}{dy^2} - \gamma^2\right] V = -\gamma Z_c I_s\,\delta(y - y_o) \qquad (10.76)$$

The continuity conditions at the interfaces are given by

$$V|_{y=s_j-0} = V|_{y=s_j+0}, \quad j = 1, 2 \qquad (10.77a)$$

$$Y_{cj}\frac{dV}{dy}\Big|_{y=s_j-0} = Y_{cj+1}\frac{dV}{dy}\Big|_{y=s_j+0}, \quad j = 1, 2 \qquad (10.77b)$$

Comparing (10.73) with (10.76) and also (10.75) with (10.77), we can draw the following mathematical equivalences:

$$V = G_{mn}(y),\; \gamma = \zeta_{mn},\; Z_c = 1/\epsilon,\; I_s = \frac{4}{ac\,\zeta_{mn}}\sin(\alpha_m x_o)\sin(\beta_n z_o) \qquad (10.78)$$

since $V = I_s/Y$ at $y = y_o$, we can write

$$G_{mn}(y)\Big|_{y=y_o} = \frac{4}{ac\,\zeta_{mn}\,Y} \sin(\alpha_m x_o) \sin(\beta_n z_o) \tag{10.79}$$

Substitution of (10.79) and (10.69) in (10.68) gives the final expression for the Green's function for the odd-mode case.

$$G(x, y_o, z/x_o, y_o, z_o)\Big|_{\text{odd}} = \sum_{m=1}^{\infty} \sum_{n=1}^{\infty} \frac{4}{ac\,\zeta_{mn}\,Y} \sin(\alpha_m x) \sin(\alpha_m x_o) \sin(\beta_n z) \sin(\beta_n z_o) \tag{10.80}$$

Case b: Substituting (10.68) and (10.70) in (10.67), we get

$$\left[\frac{d^2}{dy^2} - \zeta_{mn}^2\right] \sin(\alpha_m x) \cos(\beta_n z)\; G_{mn}(y) = -\frac{1}{\epsilon}\, \delta(x - x_o)\, \delta(y - y_o)\, \delta(z - z_o) \tag{10.81}$$

Using the orthogonality property of $\sin(\alpha_m x)$ defined in (10.72a) and that of $\cos(\beta_n z)$ given by

$$\begin{aligned} \int_0^a \cos\left(\frac{n\pi z}{a}\right) \cos\left(\frac{p\pi z}{a}\right) dz &= a/2, \quad p = n \neq 0 \\ &= 0, \quad p \neq n \\ &= a, \quad p = n = 0 \end{aligned} \tag{10.82}$$

Equation (10.81) can be simplified to

$$\left[\frac{d^2}{dy^2} - \zeta_{mn}^2\right] G_{mn}(y) = \frac{-4\gamma_n}{ac\epsilon} \sin(\alpha_m x_o) \cos(\beta_n z_o)\, \delta(y - y_o) \tag{10.83}$$

where,

$$\gamma_n = \begin{cases} 1/2, & n = 0 \\ 1, & n \neq 0 \end{cases} \tag{10.84}$$

Comparing (10.84) with (10.76) and also (10.75) with (10.77), we get

$$V = G_{mn}(y),\ \gamma = \zeta_{mn},\ Z_c = 1/\epsilon,\ I_s = \frac{4\gamma_n}{ac\,\zeta_{mn}} \sin(\alpha_m x_o) \cos(\beta_n z_o) \tag{10.85}$$

Since $V = I_s/Y$ at $y = y_o$, we obtain

$$G_{mn}(y)\Big|_{y=y_o} = \frac{4\gamma_n}{ac\,\zeta_{mn}\,Y} \sin(\alpha_m x_o) \cos(\beta_n z_o) \tag{10.86}$$

Substituting (10.86) and (10.70) in (10.68), we get the final expression for the Green's function for the even-mode case

$$G(x, y_o, z/x_o, y_o, z_o)\Big|_{\text{even}} = \sum_{m=1}^{\infty} \sum_{n=0}^{\infty} \frac{4\gamma_n}{ac\,\zeta_{mn}\,Y} \sin(\alpha_m x) \sin(\alpha_m x_o) \cos(\beta_n z) \cos(\beta_n z_o),$$

$$\gamma_n = \begin{cases} 1/2, & n = 0 \\ 1, & n \neq 0 \end{cases} \tag{10.87}$$

In equations (10.80) and (10.87), Y is the admittance at the charge plane obtained from the two-wire transmission line equivalent circuit (Fig. 10.1 with ζ replaced by ζ_{mn}). For the present structure the expression for Y is given by (10.29a) with ζ replaced by ζ_{mn}.

Having determined the even- and odd-mode Green's functions, we use the well known variational technique to evaluate the even-and odd-mode lumped capacitances C_{le} and C_{lo}. We assume the strip conductor to be of zero thickness so that the charge distribution can be expressed as

$$\rho(x, y, z)_{\substack{e\\o}} = f(x, z)_{\substack{e\\o}}\, \delta(y - y_o) \tag{10.88}$$

The variational expression for the capacitance is then given by

$$\frac{1}{C_{l\substack{e\\o}}} = \frac{\iint G(x, y_o, z/x_o, y_o, z_o)_{\substack{e\\o}}\, f(x, z)_{\substack{e\\o}}\, f(x_o, z_o)_{\substack{e\\o}}\, ds\, ds_o}{\left[\iint f(x, z)\, ds\right]^2} \tag{10.89}$$

where the integration is to be performed on the surface of the conductor patch. The charge distribution $f(x, z)_{\substack{e\\o}}$ on the strip conductor can be rewritten as

$$f(x, z)_{\substack{e\\o}} = f(x)_{\substack{e\\o}}\, g(z)_{\substack{e\\o}} \tag{10.90}$$

where $f(x)_{\substack{e\\o}}$ and $g(z)_{\substack{e\\o}}$ specify the variations in the x- and z-directions, respectively. For the x-direction we assume a charge distribution similar to the one used for a uniform line (Chapter 3, Eq. (3.38) with $A = 1$)

$$f(x)_{\substack{e\\o}} = \begin{cases} \left[1 + \left|\dfrac{2}{w}\left(x - \dfrac{c}{2}\right)\right|^3\right]; & \dfrac{c - w}{2} \leqslant x \leqslant \dfrac{c + w}{2} \\ 0; & \text{otherwise} \end{cases} \tag{10.91}$$

In the z-direction, the charge distribution will be more or less flat except near the edges. This prompts us to choose the following distribulion for $g(z)$.

$$g(z)_{\substack{e\\o}} = \begin{cases} 1 & ;\ s/2 + y_o \leqslant z \leqslant a - s/2 - y_o \\ \left[1 + \dfrac{k_{\substack{e\\o}}}{y_o}\left\{\left|z - \dfrac{a}{2}\right| - \left(\dfrac{a}{2} - \dfrac{s}{2} - y_o\right)\right\}\right]; & \begin{array}{l} s/2 \leqslant z \leqslant s/2 + y_o \\ a - s/2 - y_o \leqslant z \leqslant a - s/2 \end{array} \\ 0 & ;\ 0 \leqslant z \leqslant s/2 \\ & \ \ a - s/2 \leqslant z \leqslant a \end{cases} \tag{10.92}$$

k_e and k_o are constants determined by numerically maximizing the lumped capacitances C_{le} and C_{lo}, respectively. Using (10.91) and (10.92) in (10.89) and evaluating the integrals, we get

$$C_{le} = \frac{\left(\dfrac{5}{8}\right)^2\left[1 - \dfrac{s}{a} + k_e\dfrac{y_o}{a}\right]^2}{\sum\limits_{m\ \text{odd}}\sum\limits_{n\ \text{even}} \dfrac{4\gamma_n}{ac}\left(\dfrac{s}{a}\right)^2 \dfrac{P_m^2\, R_n^2}{\zeta_{mn} Y}}; \qquad \gamma_n = \begin{cases} 1/2, & n = 0 \\ 1, & n = 2, 4, 6 \ldots \end{cases} \tag{10.93}$$

and

$$C_{lo} = \frac{\left(\frac{5}{8}\right)^2\left[1 - \frac{s}{a} + k_o \frac{y_o}{a}\right]^2}{\sum\limits_{m \text{ odd}} \sum\limits_{n \text{ odd}} \frac{1}{ac}\left(\frac{s}{a}\right)^2 \frac{P_m^2 Q_n^2}{\zeta_{mn} Y}} \tag{10.94}$$

The various parameters appearing in (10.93) and (10.94) are

$$P_m = \left(\frac{2c}{m\pi w}\right)\left[2\left\{1 - 3\left(\frac{2c}{m\pi w}\right)^2\right\} \sin\left(\frac{m\pi w}{2c}\right) + 6\left(\frac{2c}{m\pi w}\right)^3 + 3\left(\frac{2c}{m\pi w}\right)\left\{1 - 2\left(\frac{2c}{m\pi w}\right)^2\right\}\cos\left(\frac{m\pi w}{2c}\right)\right] \tag{10.95a}$$

$$R_n = U_{sn} + k_e V_{sn} \tag{10.95b}$$

$$Q_n = U_{cn} + k_o V_{cn} \tag{10.95c}$$

$$\zeta_{mn} = \left[\left(\frac{m\pi}{c}\right)^2 + \left(\frac{n\pi}{a}\right)^2\right]^{1/2} \tag{10.95d}$$

$$U_{sn} = \left(\frac{2a}{n\pi s}\right) \sin\left(\frac{n\pi s}{2a}\right) \tag{10.95e}$$

$$V_{sn} = \left(\frac{2a}{n\pi s}\right)\left[\sin\left(\frac{n\pi s}{2a}\right) - \frac{2a}{n\pi y_o} \sin\left\{\frac{n\pi}{2a}(s + y_o)\right\} \sin\left(\frac{n\pi y_o}{2a}\right)\right] \tag{10.95f}$$

$$U_{cn} = \left(\frac{2a}{n\pi s}\right) \cos\left(\frac{n\pi s}{2a}\right) \tag{10.95g}$$

$$V_{cn} = \left(\frac{2a}{n\pi s}\right)\left[\cos\left(\frac{n\pi s}{2a}\right) - \frac{2a}{n\pi y_o} \cos\left\{\frac{n\pi}{2a}(s + y_o)\right\} \sin\left(\frac{n\pi y_o}{2a}\right)\right] \tag{10.95h}$$

The constants k_e and k_o are obtained by maximising the lumped capacitances C_{le} and C_{lo}, respectively. For this we set

$$\frac{\partial C_{le}}{\partial k_e} = 0 \tag{10.96a}$$

and

$$\frac{\partial C_{lo}}{\partial k_o} = 0 \tag{10.96b}$$

Evaluating (10.96) and with some algebraic manipulations, we get the following expressions for k_e and k_o.

$$k_e = -\frac{\sum\limits_{m \text{ odd}} \sum\limits_{n \text{ even}} \frac{\gamma_n P_m^2 U_{sn}}{\zeta_{mn} Y}\left[\left(1 - \frac{s}{a}\right) V_{sn} - \frac{y_o}{a} U_{sn}\right]}{\sum\limits_{m \text{ odd}} \sum\limits_{n \text{ even}} \frac{\gamma_n P_m^2 V_{sn}}{\zeta_{mn} Y}\left[\left(1 - \frac{s}{a}\right) V_{sn} - \frac{y_o}{a} U_{sn}\right]} \tag{10.97a}$$

$$k_o = -\frac{\sum\limits_{m \text{ odd}} \sum\limits_{n \text{ odd}} \frac{P_m^2 U_{cn}}{\zeta_{mn} Y}\left[\left(1 - \frac{s}{a}\right) V_{cn} - \frac{y_o}{a} U_{cn}\right]}{\sum\limits_{m \text{ odd}} \sum\limits_{n \text{ odd}} \frac{P_m^2 V_{cn}}{\zeta_{mn} Y}\left[\left(1 - \frac{s}{a}\right) V_{cn} - \frac{y_o}{a} U_{cn}\right]} \tag{10.97b}$$

With the determination of the even-and odd-mode lumped capacitances C_{le} and C_{lo} from (10.93) and (10.94) respectively, the parameters of the equivalent circuit shown in Fig. 10.12 can be obtained using (10.66). In (10.66) the only other parameter required is C, the line capacitance of a uniform transmission line. The expression for C is given by (Chapter 3, Eq. (3.41) with $A = 1$)

$$C = \frac{1.5625}{\sum_{m\ \text{odd}} (L_m + M_m)^2 P_m/Y} \tag{10.98}$$

where

$$L_m = \sin(\alpha_m w/2) \tag{10.99a}$$

$$M_m = \left(\frac{2}{\alpha_m w}\right)^3 \left[\left\{\left(\frac{\alpha_m w}{2}\right)^2 - 2\right\}\cos\left(\frac{\alpha_m w}{2}\right) + \left(\frac{\alpha_m w}{2}\right)\cdot\left\{\left(\frac{\alpha_m w}{2}\right)^2 - 6\right\}\sin\left(\frac{\alpha_m w}{2}\right) + 6\right] \tag{10.99b}$$

$$P_m = \frac{2}{m\pi}\left(\frac{2}{\alpha_m w}\right)^2 \tag{10.99c}$$

$$\alpha_m = m\pi/c \tag{10.99d}$$

Y is the admittance at the charge plane given by (10.29a) with ζ replaced by α_m. In equations (10.66a) and (10.66b), a is not infinite but some finitely large value beyond which the changes in $[C_{lo} - (a - s)C]$ and $[C_{le} - (a - s)C]$ are negligible, respectively.

10.7 Numerical Results

In the following, design data generated using the general formulas reported in this chapter on various discontinuities encountered in stripline-like transmission line are presented in graphical form. In addition, design data already reported in the literature using other numerical methods are also presented.

Figure 10.15 shows the variations of normalized edge-capacitance (C_{oc}/w) and the normalized extension in length due to end effects ($\Delta l/h_1$) of suspended stripline as a function of w/h_1, with d/h_1 as a parameter. For a fixed value of d/h_1, $\Delta l/h_1$ increases and C_{oc}/w decreases with an increase in w/h_1. On the other hand, for a fixed value of w/h_1, both $\Delta l/h_1$ and C_{oc}/w increase as d/h_1 is increased. For the structural parameters chosen, the extension in length due to end effects is of the order of 0.6 to 2.4 times the bottom air gap height.

The variations of normalized extension in length due to end effects ($\Delta l/d$) of suspended stripline as a function of normalized substrate thickness (d/λ^a) is depicted in Fig. 10.16. For a fixed value of d/λ^a, as w/d is increased, $\Delta l/d$ increases. Further, for a fixed value of w/d, $\Delta l/d$ first decreases slightly and then increases as d/λ^a increases. We may observe that values of Δl are considerably higher, thereby meaning that a considerable fraction of electromagnetic field energy extends into the air regions around the strip conductor open end.

The variations of normalized capacitance C_N for the even- and odd-modes of square conductor patches in broadside-coupled suspended stripline are shown in Fig. 10.17. For a fixed value of w/b and ϵ_r, as d/b increases, the normalized even-mode lumped capacitance increases and the normalized odd-mode lumped capacitance decreases.

Figures 10.18 and 10.19 depict the variations of $\Delta le/d$ and $\Delta lo/d$, respectively as a function of d/λ^a for edge-coupled suspended stripline. For a fixed value of d/λ^a, both $\Delta le/d$ and $\Delta lo/d$

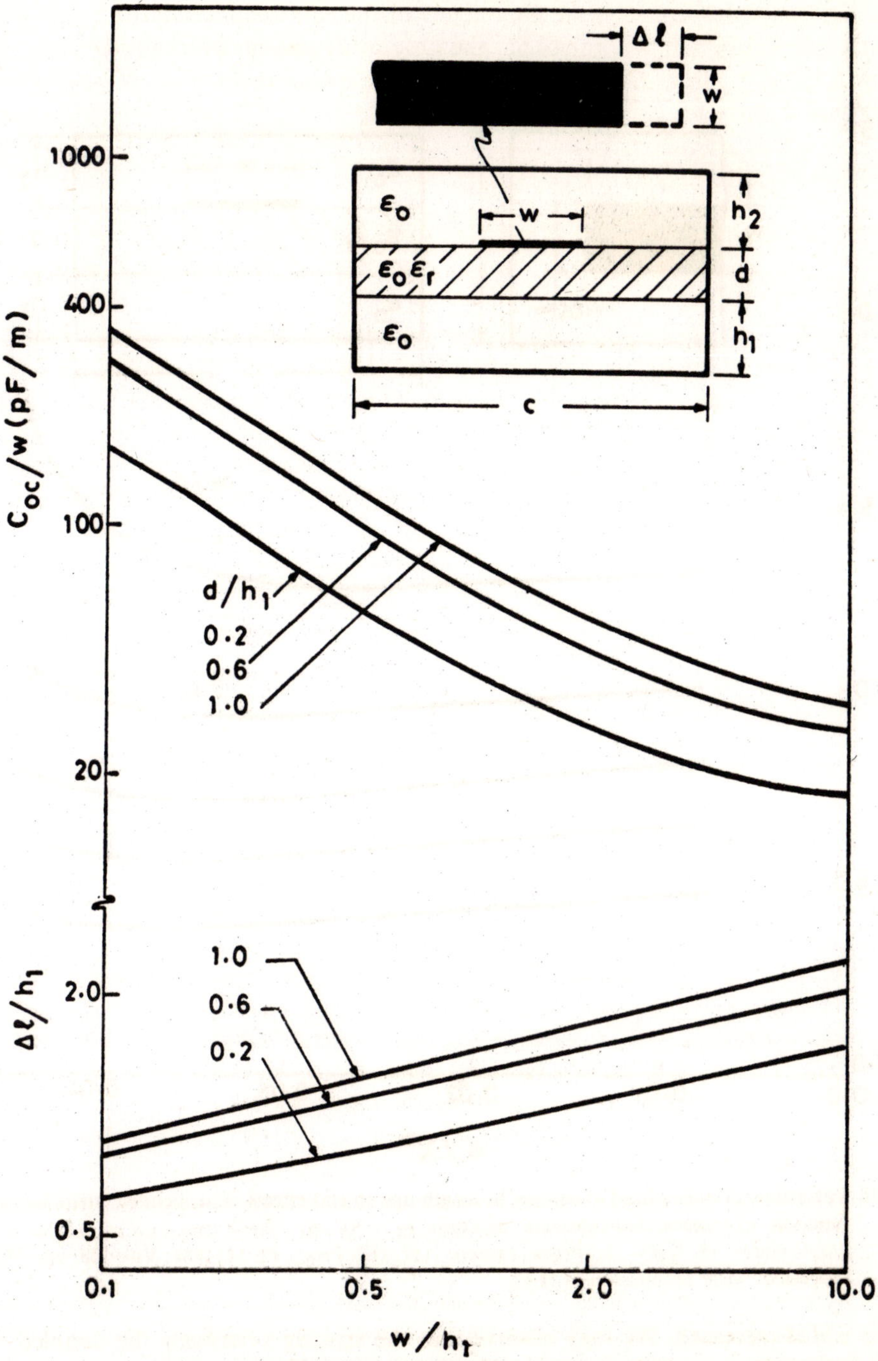

Fig. 10.15 Variations of normalized edge-capacitance and normalized extension in length of suspended stripline versus stripwidth with the thickness of the dielectric substrate as a parameter; $c/w = 40$, $h_2/h_1 = 10$, $\epsilon_r = 9.6$

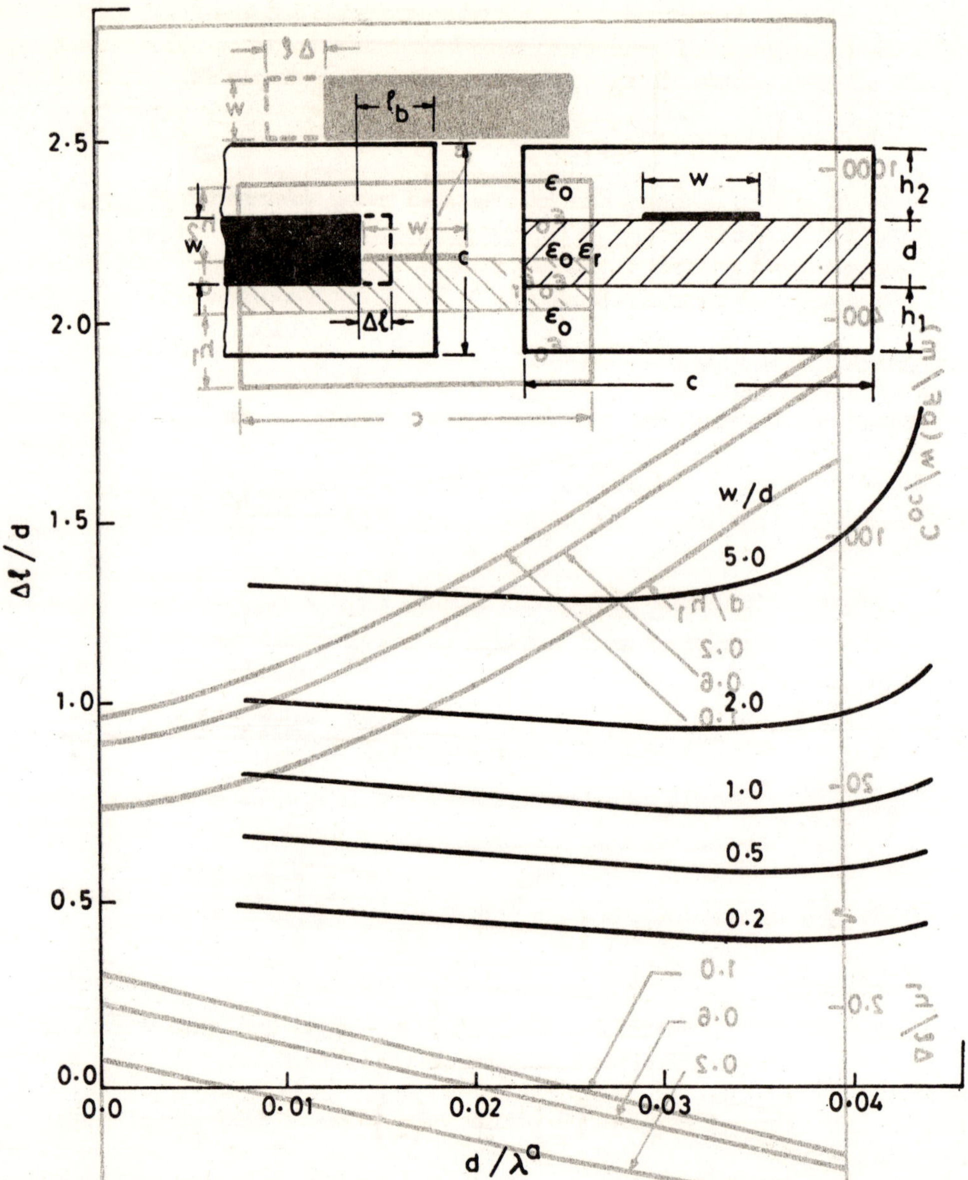

Fig. 10.16 Variations of normalized extension in length due to end effects of suspended stripline as a function of normalized substrate thickness; $\epsilon_r = 9.7$, $d = 0.635$ mm, $h_1/d = 1$, $h_2/d = 10$, $c/d = w/d + 10$, $l_b/d = 5$ (From Jansen [15], IEE Proc., Pt. H. 1981, Vol. 128, pp. 77-86, reprinted with permission of IEE).

increase as w/d is increased. We may observe that the spacing s between the coupled strips has only a little influence on the reference plane extension Δle. Furthermore, Δlo exhibits a tendency to increase with an increase in the gap width s.

The variations of the equivalent circuit parameters of a gap in the centre conductor of a microstrip-like stripline as a function of s/b are plotted in Fig. 10.20. As s/b increases, C_p

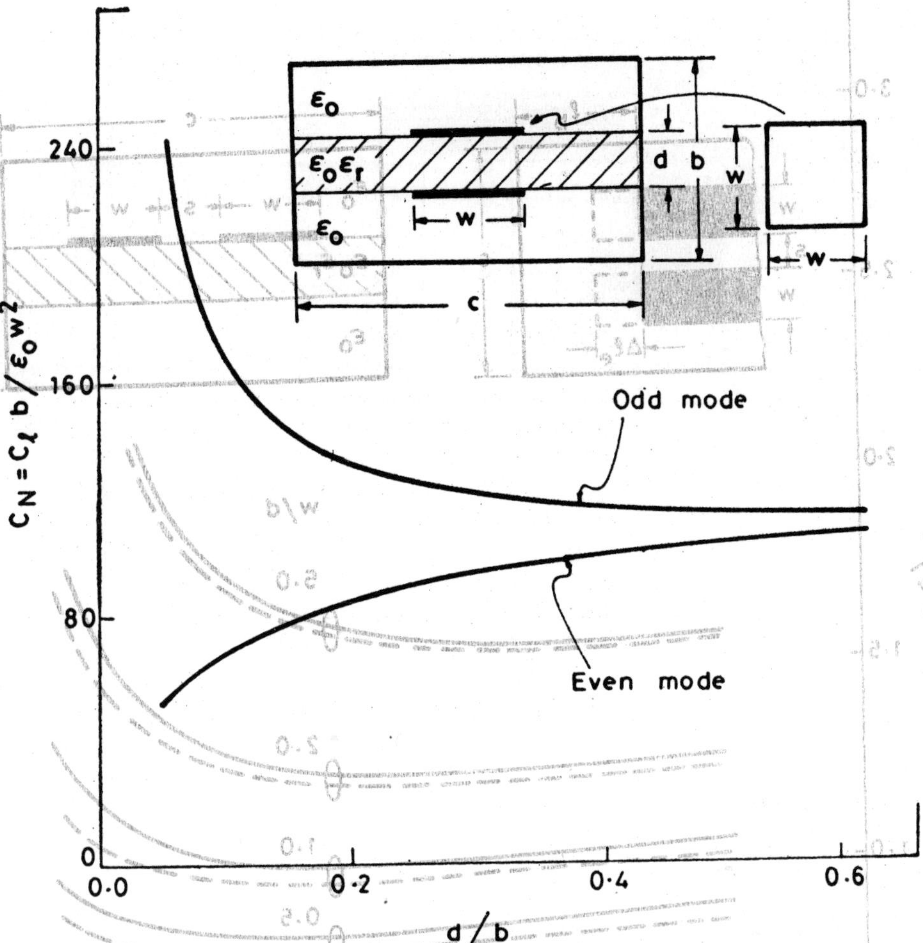

Fig. 10.17 Variations of the normalized capacitances for the even- and odd-modes of square conductor patches in broadside-coupled suspended stripline; $c/w = 40$, $w/b = 0.1$, $\epsilon_r = 3.78$

increases, and C_g decreases. In the limiting case when the two strips are decoupled ($s \to \infty$), C_g tends to zero whereas C_p saturates to a value equal to the edge-capacitance of the open-ended strip conductor (C_{oc}). Figure 10.21 depicts the variations of the equivalent circuit parameters of a gap in strip conductor of a suspended stripline as a function of w/b for two different values of s/b. For a fixed value of s/b, as w/b is increased, both C_p and C_g increase. Furthermore, for a fixed value of w/b, as s/b is increased, C_p increases and C_g decreases. The data presented in Figs 10.20 and 10.21 can be directly used to design end-coupled filters in the microstrip-like stripline and suspended stripline configurations, respectively.

The gap capacitance C_g and the stray capacitances C_{p1} and C_{p2} of the suspended stripline are depicted in Fig. 10.22 as a function of the normalized width w_2/w_1 of the second strip conductor. It is observed that C_g increases as w_2/w_1 increases. Furthermore, the capacitance

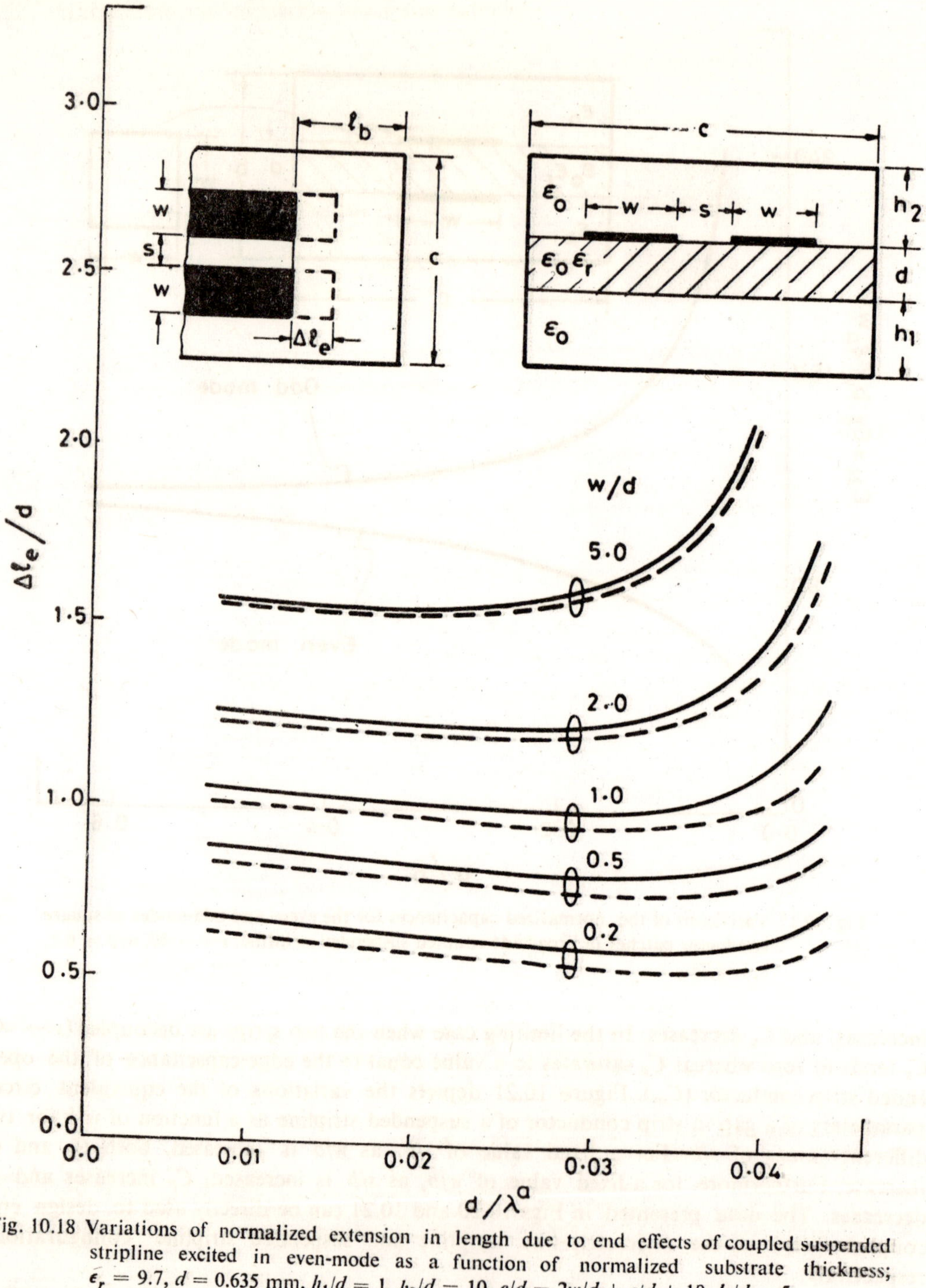

Fig. 10.18 Variations of normalized extension in length due to end effects of coupled suspended stripline excited in even-mode as a function of normalized substrate thickness; $\epsilon_r = 9.7$, $d = 0.635$ mm, $h_1/d = 1$, $h_2/d = 10$, $c/d = 2w/d + s/d + 10$, $l_b/d = 5$, — — — $s/d = 0.2$, ———— $s/d = 1.0$. (From Jansen [15], IEE Proc., Pt. H, 1981, Vol. 128, pp. 77-86, reprinted with permission of IEE)

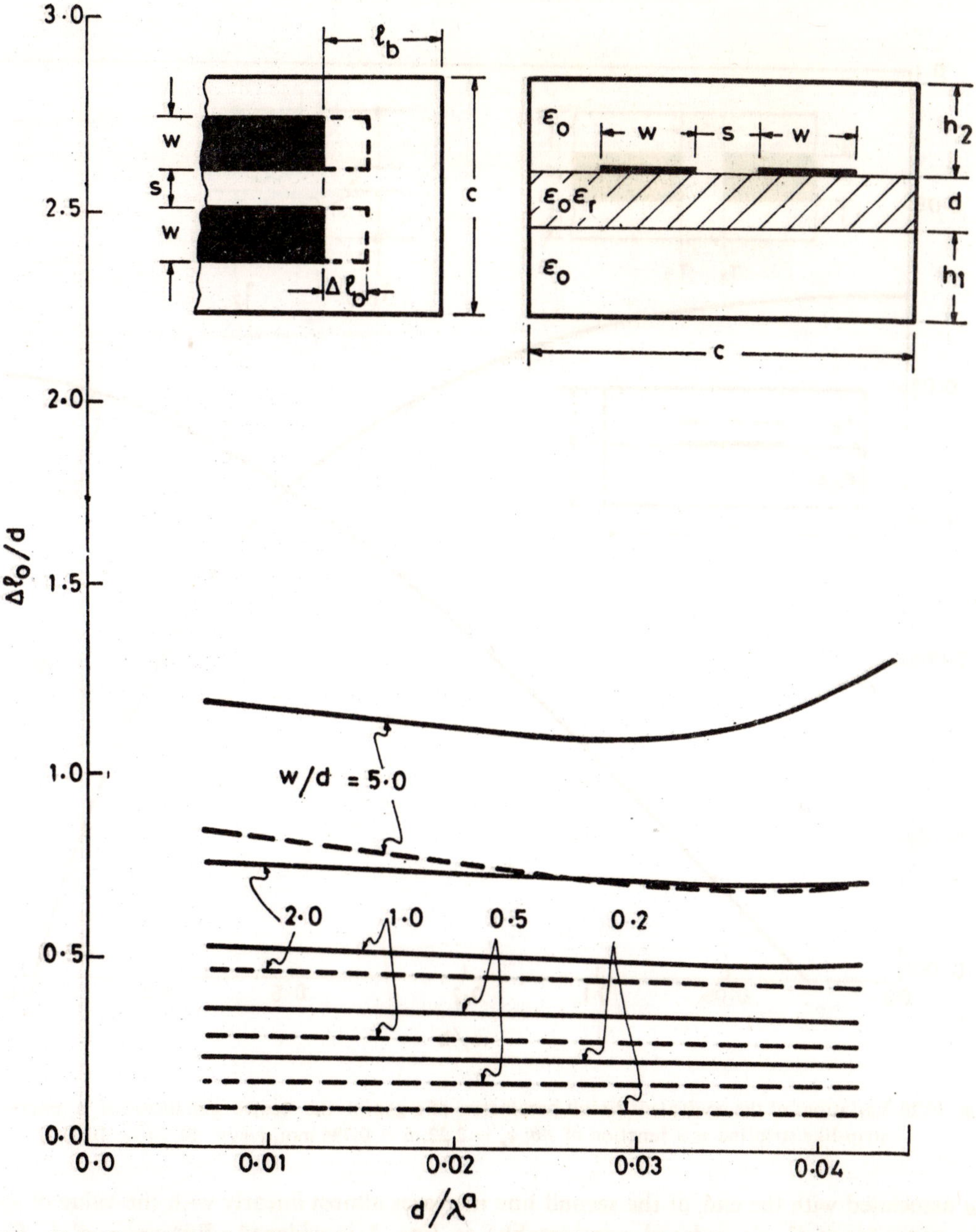

Fig. 10.19 Variations of normalized extension in length due to end effects of coupled suspended stripline excited in odd-mode as a function of normalized substrate thickness, $\epsilon_r = 9.7$, $d = 0.635$ mm, $h_1/d = 1$, $h_2/d = 10$, $c/d = 2w/d + s/d + 10$, $l_b/d = 5$, – – – $s/d = 0.2$, ——— $s/d = 1.0$. (From Jansen [15], IEE Proc. Pt. H, 1981, Vol. 128, pp. 77-86, reprinted with permission of IEE).

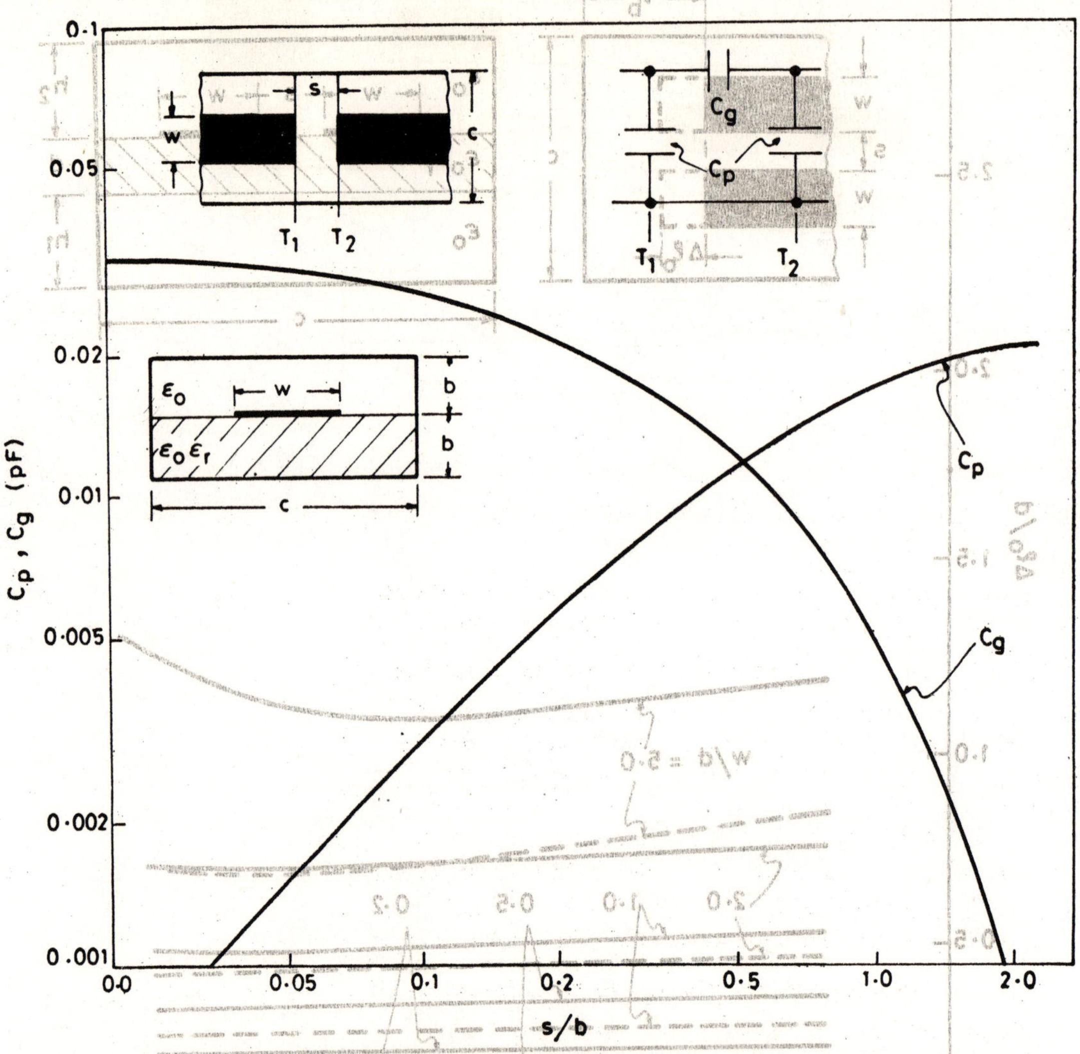

Fig. 10.20 Variations of the equivalent circuit parameters of a gap in the centre conductor of a microstrip-like stripline as a function of s/b; $\epsilon_r = 2.22$, $b = 0.794$ mm, $c/w = 20$, $l/b = 10$.

C_{p2} associated with the end of the second line increases almost linearly with the value of w_2. The capacitance C_{p1} is reduced considerably as line 2 is widened. For $w_2/w_1 \ll 1$, C_{p1} reduces nearly to the edge-capacitance (C_{oc}) of the open ended line 1. For nearly equal line widths, its relative change reaches a maximum. In the limit $w_2/w_1 \gg 1$, the value of C_{p1} approaches a constant lower capacitance.

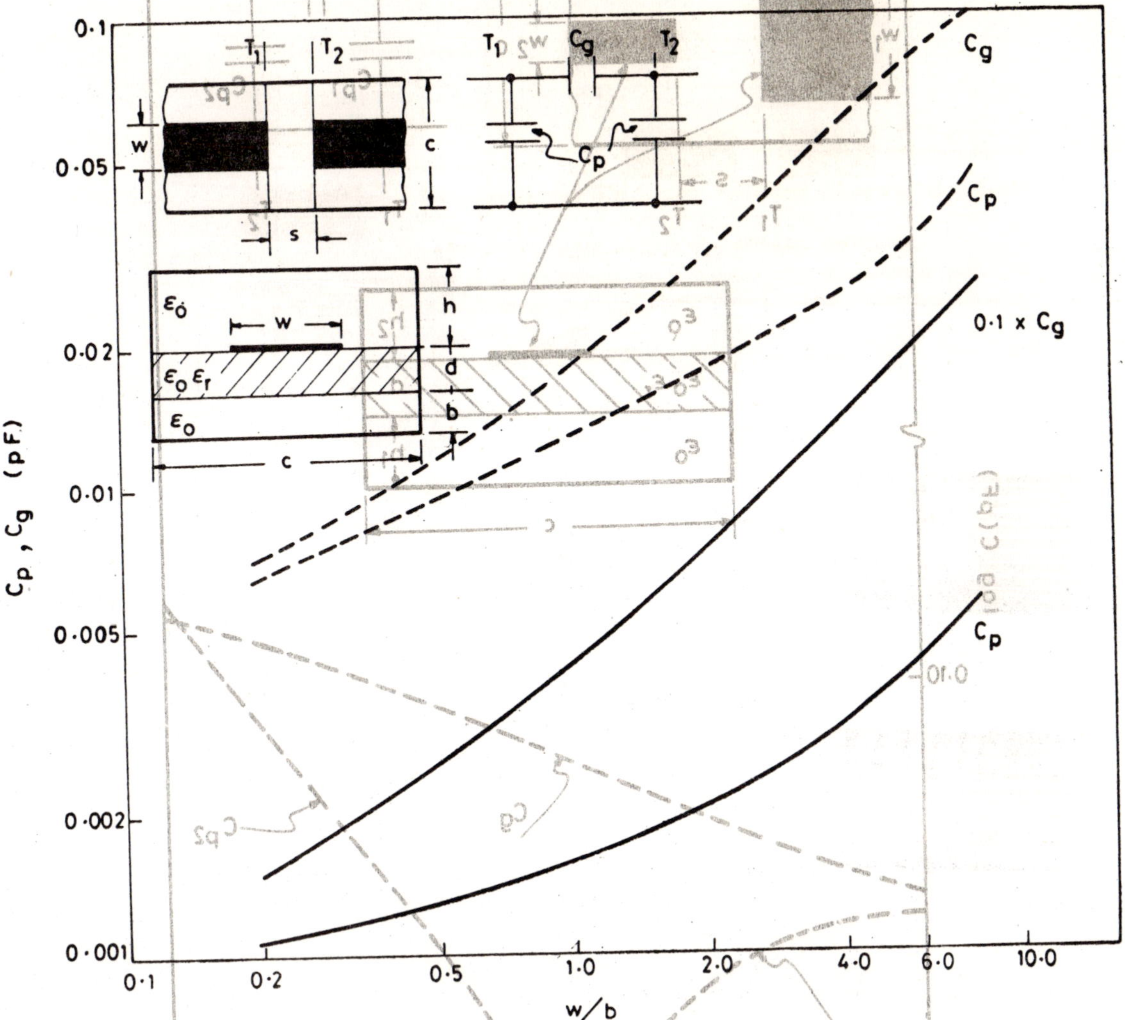

Fig. 10.21 Variations of the equivalent circuit parameters of a gap in the centre conductor of a suspended stripline as a function of w/b; $\epsilon_r = 3.78$, $d = 0.508$ mm, $b = 1.0$ mm, $c/w = 20$, $l/b = h/b = 10$, ——— $s/b = 0.1$, – – – $s/b = 1.0$.

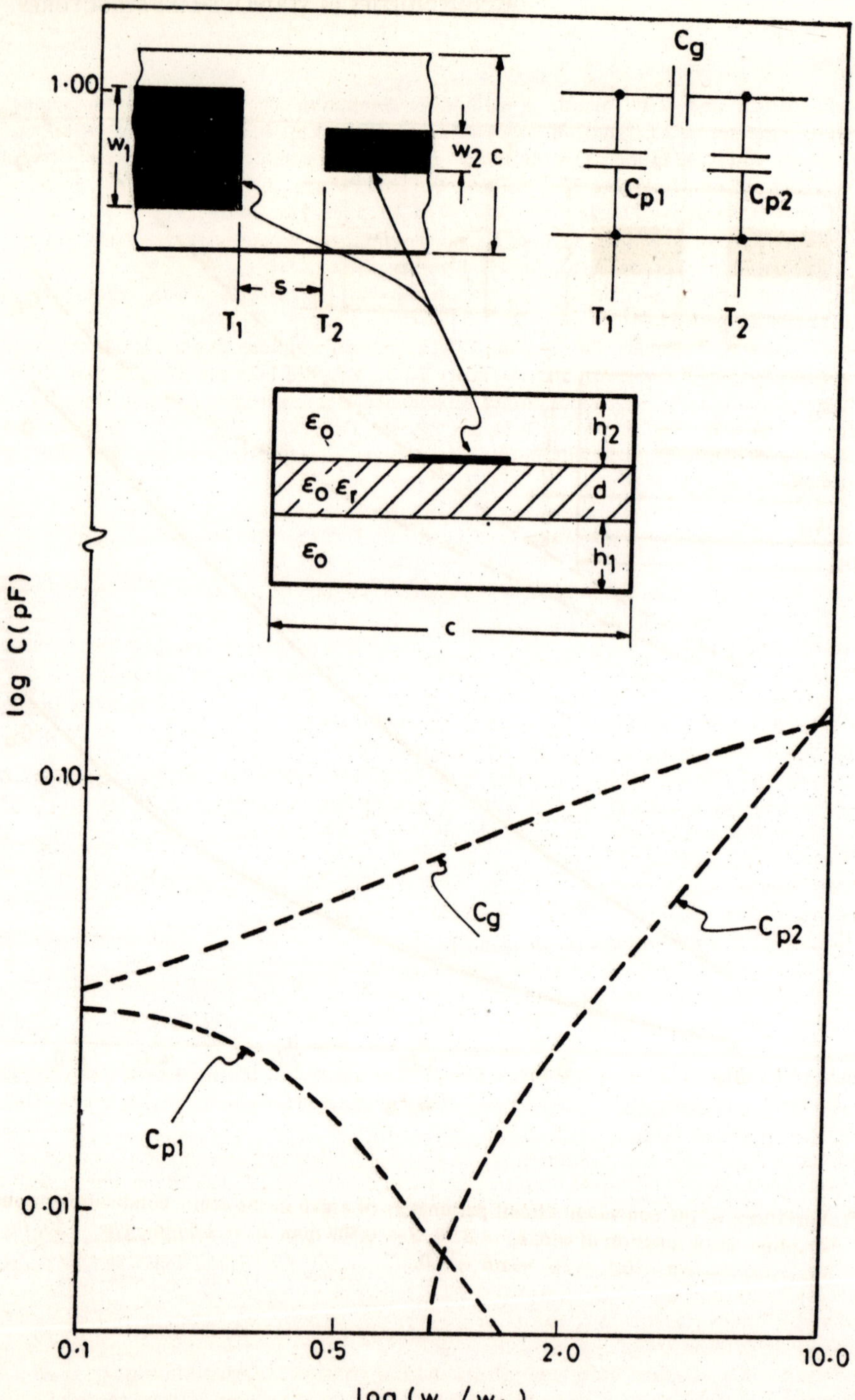

Fig. 10.22 Variations of the equivalent circuit parameters of an asymmetric gap in suspended stripline as a function of w_2/w_1; $\epsilon_r = 9.7$, $d = 0.635$ mm, $h_1/d = 1$, $h_2/d = 10$, $s/d = 0.5$, $w_1/d = 1$, frequency = 4 GHz, $c/d = w_2/d + 10$. (From Jansen et al. [17], first presented at the 11th European Microwave Conference, reprinted with permission).

REFERENCES

1. T. Itoh, R. Mittra and R.D. Ward, A method for computing edge-capacitance of finite and semi-infinite microstrip lines, *IEEE Trans. Microwave Theory Tech.*, Vol. MTT-20, pp. 847-849, Dec. 1972.
2. H.M. Altschuler and A.A. Oliner, Discontinuities in the centre conductor of symmetric strip transmission line, *IRE Trans. Microwave Theory Tech.*, Vol. MTT-8, pp. 328-339, May 1960. Also correction in *IRE Trans. Microwave Theory Tech.*, Vol. MTT-10, p. 43, Feb. 1962.
3. A. Farrar and A.T. Adams, Matrix methods for microstrip three dimensional problems, *IEEE Trans. Microwave Theory Tech.*, Vol. MTT-20, pp. 497-504, Aug. 1972.
4. M. Maeda, An analysis of gap in microstrip transmission lines, *IEEE Trans. Microwave Theory Tech.*, Vol. MTT-20, pp. 390-396, June 1972.
5. V. Nalbandian and W. Steenaart, Discontinuities in symmetric striplines due to impedance steps and their compensation, *IEEE Trans. Microwave Theory Tech.*, Vol. MTT-20, pp. 573-578, Sept. 1972.
6. C. Gupta and A. Gopinath, Equivalent circuit capacitance of microstrip step change in width, *IEEE Trans. Microwave Theory Tech.*, Vol, MTT-25, pp. 819-822, Oct. 1977.
7. A.A. Oliner, Equivalent circuits for discontinuities in balanced strip transmission line, *IRE Trans. Microwave Theory Tech.*, Vol. MTT-3, pp. 134-143, March 1955.
8. B. Easter, A. Gopinath and I.M. Stephenson, Theoretical and experimental methods for evaluating discontinuities in microstrip, *The Radio and Electronic Engineer*, Vol. 48, No. 112, pp. 73-84, Jan./Feb. 1970.
9. A.G. Franco and A.A. Oliner, Symmetric strip transmission line Tee junction, *IRE Trans. Microwave Theory Tech.*, Vol. MTT-10, pp. 118-124, Feb. 1962.
10. J.J. Campbell, Application of the solution of certain boundary value problems to the symmetrical four-port junction and specially truncated bends in parallel-plate waveguides and balanced strip transmission lines, *IEEE Trans. Microwave Theory Tech.*, Vol. MTT-16, pp. 165-176, March 1968.
11. W.J.R. Hoefer, Equivalent series inductivity of a narrow transverse slit in microstrip, *IEEE Trans. Microwave Theory Tech.*, Vol. MTT-25, pp. 822-824, Oct. 1977.
12. C.I. Mobbs and J.D. Rhodes, A generalized chebyshev suspended substrate stripline bandpass filter, *IEEE Trans. Microwave Theory Tech.*, Vol. MTT-31, pp. 397-402, May 1983.
13. J.K. Richardson, Gap spacing for narrow-bandwidth end-coupled symmetric stripline filters, *IEEE Trans. Microwave Theory Tech.*, Vol. MTT-16, pp. 559-560, Aug. 1968.
14. E.H. Bradley, Design and development of stripline filters, *IRE Trans. Microwave Theory Tech.*, Vol. MTT-4, pp. 86-93, Feb. 1956.
15. R.H. Jansen, Hybrid mode analysis of end effects of planar microwave and millimeter wave transmission line, *Proc. IEE, Pt. H.*, Vol. 128, pp. 77-86, April 1981.
16. N.L. Koster and R.H. Jansen, The equivalent circuit of the asymmetrical series gap in microstrip and suspended substrate line, *IEEE Trans. Microwave Theory Tech.*, Vol. MTT-30, pp. 1273-1279, Aug. 1982.
17. R.H. Jansen and N.H.L. Koster, A unified CAD basis for the frequency dependent characterization of strip, slot and coplanar MIC components, *11th European Microwave Conf. Digest*, pp. 682-687, 1981.
18. S.K. Koul and S. Raghu Kumar, Characterization of a gap in the centre conductor of microstrip and microstrip-variant structures with isotropic and anisotropic substrates, *Presented at the second Asia Pacific Microwave Conference*, Beijing, China, pp. 65-66, Oct. 26-28, 1988.
19. B. Bhat and S.K. Koul, Lumped capacitance, open-circuit end effects and edge-capacitance of microstrip-like transmission lines for microwave and millimeter wave applications, *IEEE Trans. on Microwave Theory Tech.*, Vol. MTT-32, pp. 433-439, April 1984.
20. S.K. Koul and B. Bhat, Equivalent circuit of broadside-coupled microstrip open ends, *IEEE MTT-S, Int. Microwave Symposium Digest*, pp. 497-498, June 1985.
21. S.K. Koul and B. Bhat, Evaluation of end effects and lumped capacitance of microstrip with anisotropic substrates—Transverse Transmission Line Technique, *14th European Microwave Conf. Digest*, pp. 148-153, Sept. 1984.
22. E. Yamashita and R. Mittra, Variational method for the analysis of microstrip lines, *IEEE Trans. Microwave Theory Tech.*, Vol. MTT-16, pp. 251-256, April 1968.
23. E. Yamashita, Variational method for the analysis of microstrip-like transmission lines, *IEEE Trans. Microwave Theory Tech.*, Vol. MTT-16, pp. 529-535, Aug. 1968.

APPENDIX A

Closed Form Expressions for the Equivalent Circuit Parameters of Discontinuities in Homogeneous Balanced Strip Transmission Line

In the following approximate closed form expressions reported by Oliner [7] for the equivalent circuit parameters of various discontinuities in the centre conductor of a homogeneous balanced strip transmission line (first structure in Table 3.3) are listed. The thickness of the strip conductor is assumed to be zero.

1. Round hole in the Inner Conductor

The physical structure of a round hole in the inner conductor of the balanced stripline is shown in Fig. A.1a. The equivalent circuits for centre line and edge-representations are shown in Figs. A.1b and c, respectively. The closed form expressions for equivalent circuit parameters are given by:

(a) Centre line representation

$$\frac{B_1}{Y_0} = \frac{-3}{4\pi}\frac{\lambda bD}{d^3} \tag{A.1}$$

$$\frac{B_2}{Y_0} = \frac{1}{4(B_1/Y_0)} \tag{A.2}$$

where b is the ground plane spacing, λ is the wavelength of the TEM mode, Y_0 is the characteristic admittance of the line and D is given by

$$D = \frac{bK(k)}{K(k')} \tag{A.3}$$

where $k = \tanh(\pi w/2b)$, $k' = \sqrt{1-k^2}$ and $K(k)$ is the complete elliptic integral of the first kind of modulus k.

(b) Edge-Representation

$$\frac{B_2'}{Y_0} = \frac{1 + (B_2/Y_0)\cot(\pi d/\lambda)}{\cot(\pi d/\lambda) - (B_2/Y_0)} \tag{A.4}$$

$$\frac{B_1'}{Y_0} = \frac{1}{2}\left[\frac{1 + 2(B_1/Y_0)\cot(\pi d/\lambda)}{\cot(\pi d/\lambda) - 2(B_1/Y_0)}\right] - \frac{1}{2}\frac{B_2'}{Y_0} \tag{A.5}$$

2. Gap in the Inner Conductor

The physical structure of a gap in the inner conductor of the balanced stripline is shown in Fig. A.2a. The equivalent circuits for centre line and edge representation are shown in

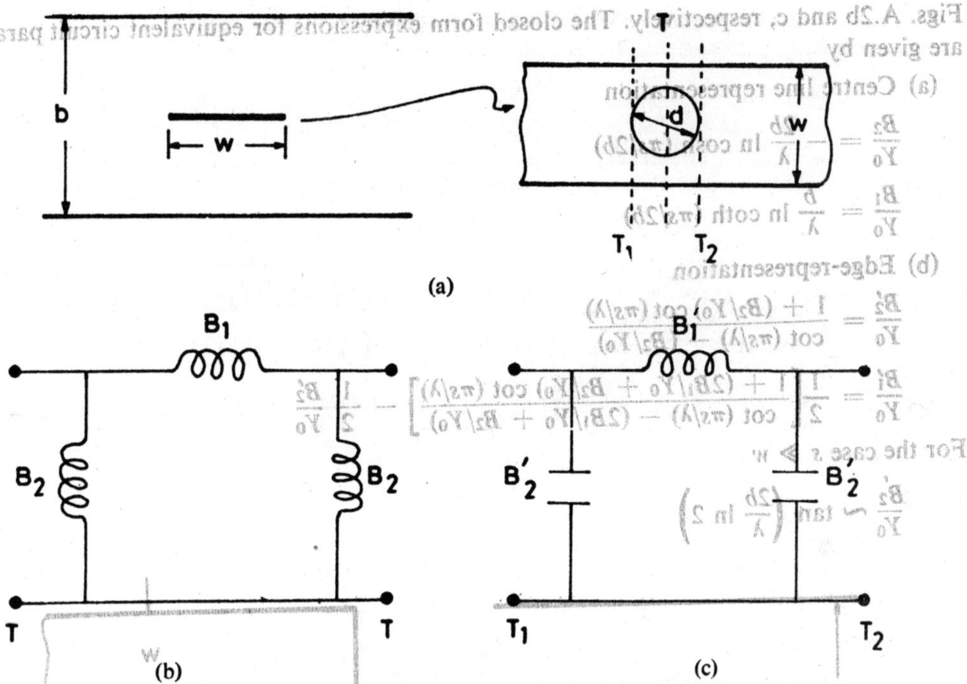

Fig. A.1 (a) Balanced stripline with a round hole in the inner conductor, (b) The equivalent circuit for centre line representation, (c) The equivalent circuit for edge-representation.

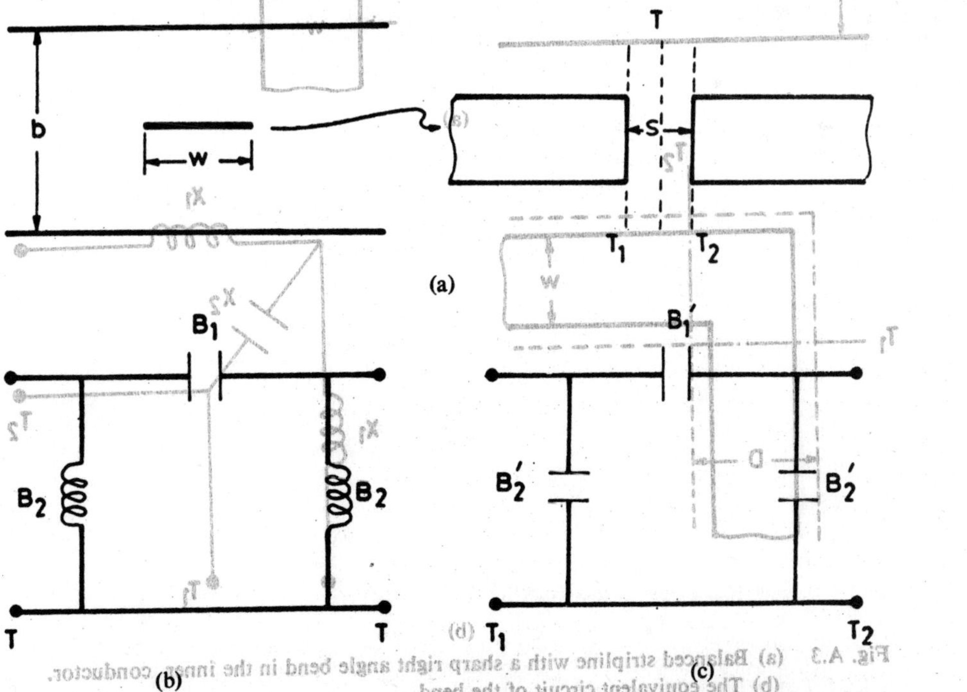

Fig. A.2 (a) Balanced stripline with a gap in the inner conductor, (b) The equivalent circuit for centre line representation, (c) The equivalent circuit for edge representation.

Figs. A.2b and c, respectively. The closed form expressions for equivalent circuit parameters are given by

(a) Centre line representation

$$\frac{B_2}{Y_0} = -\frac{2b}{\lambda} \ln \cosh (\pi s/2b) \tag{A.6}$$

$$\frac{B_1}{Y_0} = \frac{b}{\lambda} \ln \coth (\pi s/2b) \tag{A.7}$$

(b) Edge-representation

$$\frac{B_2'}{Y_0} = \frac{1 + (B_2/Y_0) \cot (\pi s/\lambda)}{\cot (\pi s/\lambda) - (B_2/Y_0)} \tag{A.8}$$

$$\frac{B_1'}{Y_0} = \frac{1}{2}\left[\frac{1 + (2B_1/Y_0 + B_2/Y_0) \cot (\pi s/\lambda)}{\cot (\pi s/\lambda) - (2B_1/Y_0 + B_2/Y_0)}\right] - \frac{1}{2}\frac{B_2'}{Y_0} \tag{A.9}$$

For the case $s \gg w$

$$\frac{B_2'}{Y_0} \sim \tan\left(\frac{2b}{\lambda} \ln 2\right) \tag{A.10}$$

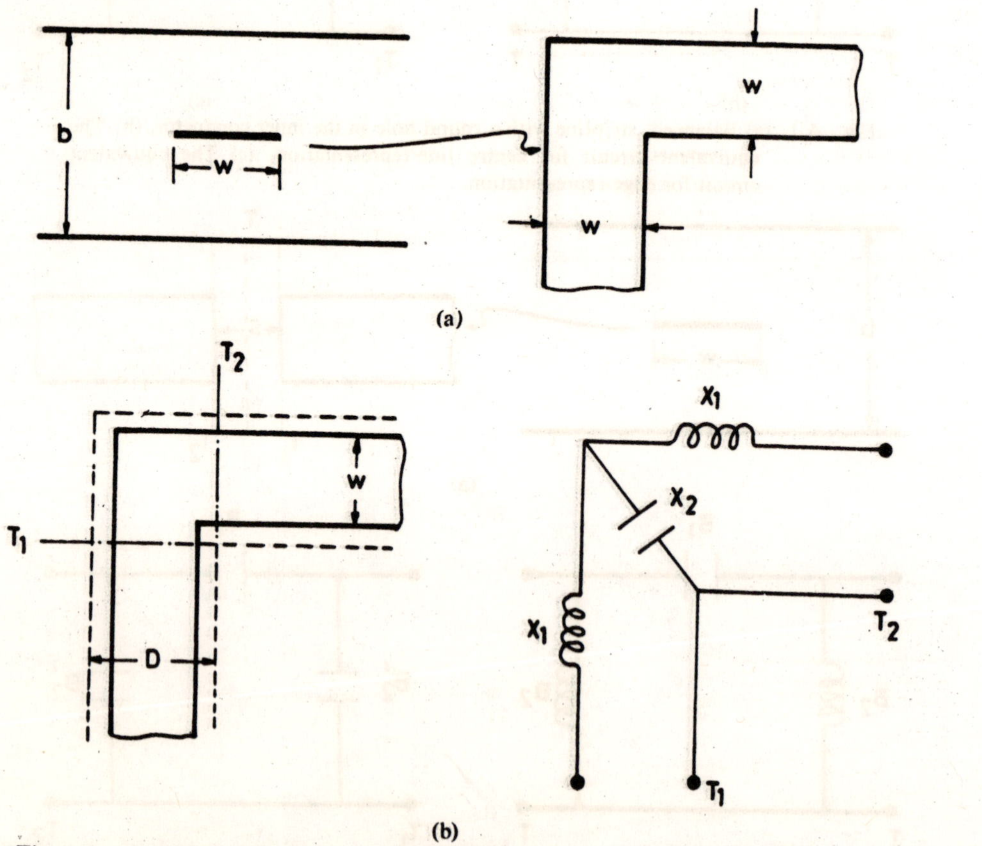

Fig. A.3 (a) Balanced stripline with a sharp right angle bend in the inner conductor. (b) The equivalent circuit of the bend.

3. Sharp Right Angle Bend

The physical structure of a sharp right angle bend in the inner conductor of a balanced stripline is shown in Fig. A.3a. The equivalent circuit of this bend is shown in Fig. A.3b. The closed form expressions for the equivalent circuit parameters are given by

$$\frac{X_1}{Z_0} = \frac{2D}{\lambda}(0.878 + 2(D/\lambda)^2) \tag{A.11}$$

$$\frac{X_2}{Z_0} = \frac{-\lambda}{2\pi D}(1 - 0.114(2D/\lambda)^2) \tag{A.12}$$

where Z_0 is the characteristic impedance of the line and D is given by A.3.

4. Sharp Bend of any Angle

The physical structure of a sharp bend of any angle in the inner conductor of a balanced stripline is shown in Fig. A.4a. The equivalent circuit of this bend is shown in Fig. A.4b. The closed form expressions for equivalent circuit parameters are given by

$$\frac{X_2}{Z_0} = \frac{-\lambda}{2\pi D}\cot(\theta/2) \tag{A.13}$$

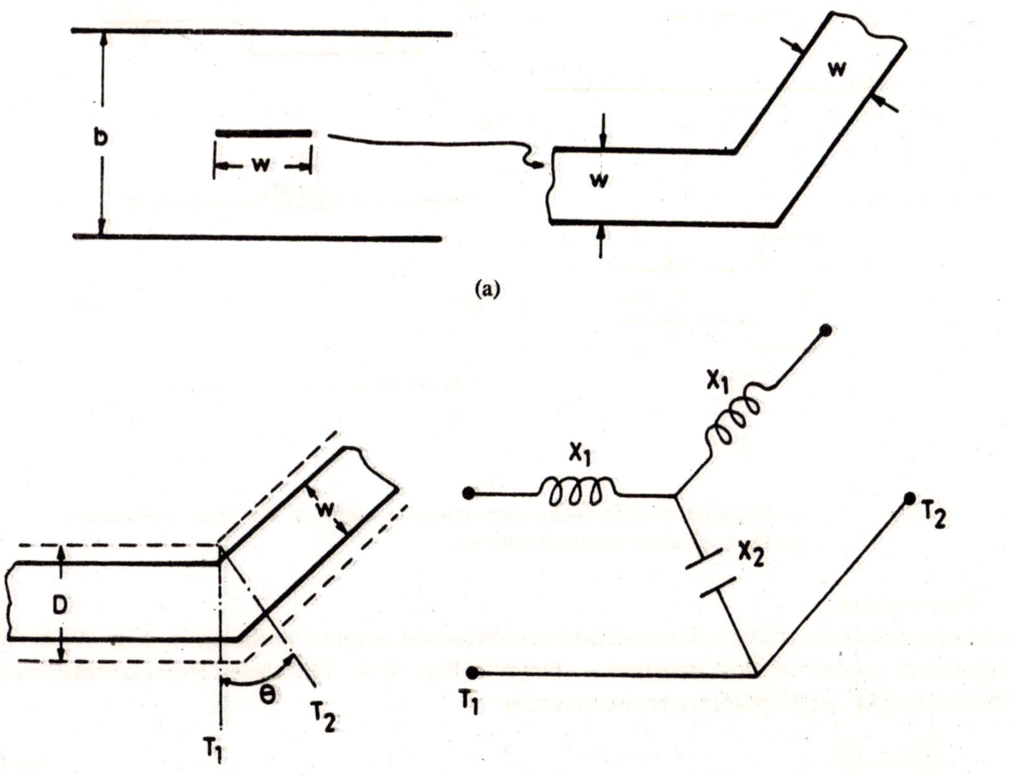

(a)

(b)

Fig. A.4 (a) Balanced stripline with a sharp bend of any angle in the inner conductor, (b) The equivalent circuit of the bend.

$$\frac{X_1}{Z_0} = \frac{2D}{\lambda}\{\Psi(-0.5(1-\theta/\pi)) - \Psi(-0.5)\} \tag{A.14}$$

where $\Psi(x)$ is the logarithmic derivative of x and D is given by A.3.

5. Step Change in Width

The physical structure of a step in the inner conductor of a balanced stripline is shown in Fig. A.5a. The approximate equivalent circuit of this structure is shown in Fig. A.5b. The closed form expressions for the equivalent circuit parameters are given by

$$\frac{X}{Z_{01}} = \frac{2D_1}{\lambda} \ln \operatorname{cosec}\left(\frac{\pi D_2}{2D_1}\right) \tag{A.15}$$

where D_1 is given by A.3 with w replaced by w_1 and

$$\frac{Z_{02}}{Z_{01}} = \frac{D_1}{D_2} \tag{A.16}$$

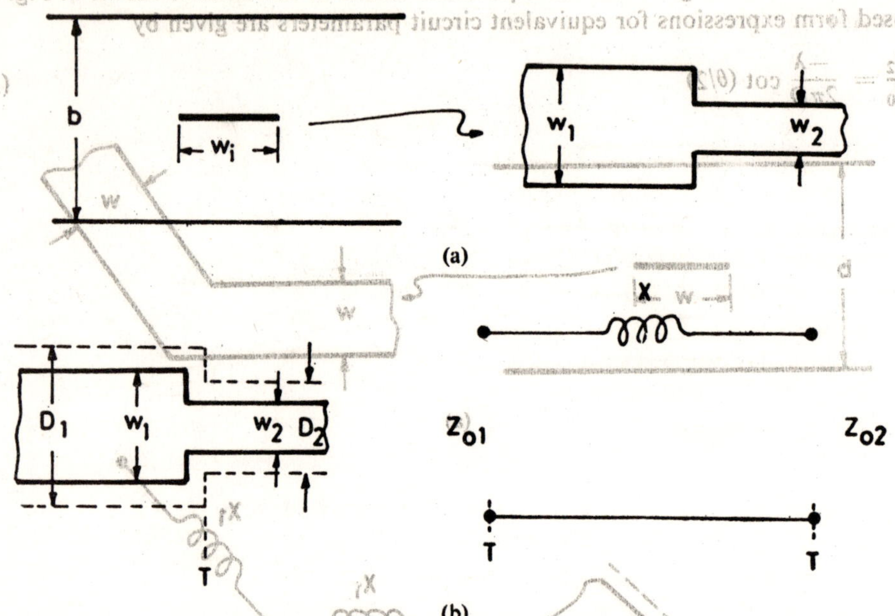

Fig. A.5 (a) Balanced stripline with a step change in width in the inner conductor, (b) The equivalent circuit of the step.

6. Tee-Junction

The physical structure of a Tee-junction in a balanced stripline is shown in Fig. A.6a. The equivalent circuit of this structure is shown in Fig. A.6b. The closed form expressions for the equivalent circuit parameters are given by

$$\frac{Z_{02}}{Z_{01}} = \frac{D_1}{D_2} \tag{A.17}$$

$$n = \frac{\sin(\pi D_2/\lambda)}{(\pi D_2/\lambda)} \tag{A.18}$$

$$\frac{X_1}{Z_{01}} = -\frac{D_2}{\lambda}\frac{D_2}{D_1}\frac{\pi^2}{16}n^2 \tag{A.19}$$

$$\frac{X_2}{Z_{01}} = \frac{1}{n^2}\frac{2D}{\lambda}\left[\ln\operatorname{cosec}\left(\frac{\pi D_2}{2D_1}\right) + \ln 2 + \frac{\pi D_2}{6D_1} + \frac{3}{2}\left(\frac{D_1}{\lambda}\right)^2\right] \tag{A.20}$$

where D_1 is given by A.3 with w replaced by w_1.

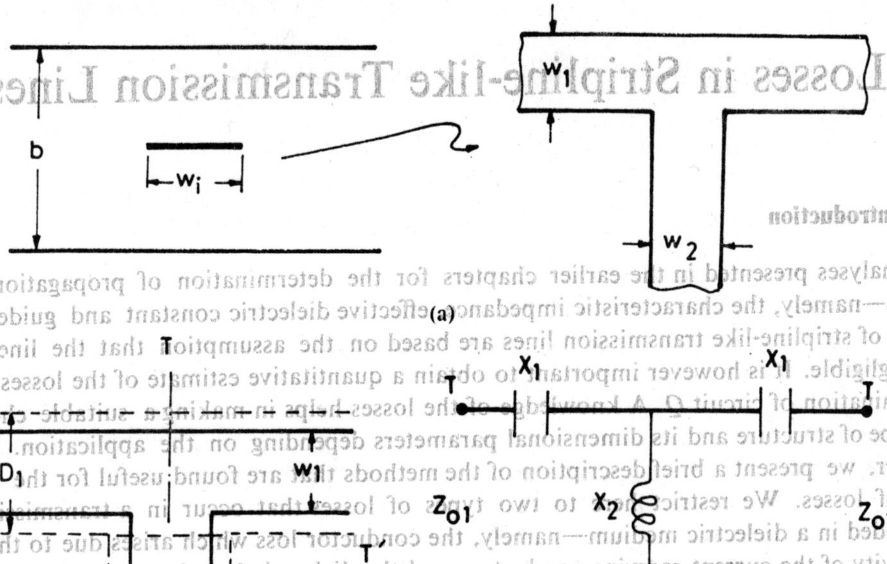

Fig. A.6 (a) Physical structure of a Tee-junction in a balanced stripline, (b) The equivalent circuit of the Tee-junction.

CHAPTER 11

Losses in Stripline-like Transmission Lines

11.1 Introduction

The analyses presented in the earlier chapters for the determination of propagation parameters—namely, the characteristic impedance, effective dielectric constant and guide wavelength of stripline-like transmission lines are based on the assumption that the line losses are negligible. It is however important to obtain a quantitative estimate of the losses for the determination of circuit Q. A knowledge of the losses helps in making a suitable choice of the type of structure and its dimensional parameters depending on the application. In this chapter, we present a brief description of the methods that are found useful for the evaluation of losses. We restrict here to two types of losses that occur in a transmission line embedded in a dielectric medium—namely, the conductor loss which arises due to the finite resistivity of the current carrying conductors and the dielectric loss due to the dissipation in the dielectric medium.

For a TEM transmission line, if $(R + j\omega L)$ and $(G + j\omega C)$ denote the series impedance and shunt admittance respectively per unit length, then its propagation constant is given by [1]

$$\gamma = (\alpha + j\beta) = [(R + j\omega L)(G + j\omega C)]^{1/2} \tag{11.1}$$

In (11.1), α and β represent the attenuation constant and phase constant, respectively, per unit length of the line. In practical lines, the losses are sufficiently small to satisfy the relations, $\omega L \gg R$ and $\omega C \gg G$. With this approximation, the attenuation constant α reduces to

$$\alpha = \frac{1}{2}\left(R\sqrt{\frac{C}{L}} + G\sqrt{\frac{L}{C}}\right) \tag{11.2}$$

The first term represents the conductor loss and the second term, the dielectric loss. For low loss lines, therefore, the total attenuation can be expressed as a simple sum of the attenuation in the conductor and the dielectric computed individually.

Practical conductors have a large conductivity σ such that the condition $\sigma \gg \omega\epsilon$ (ϵ is the dielectric constant of the conductor) can be satisfied in most microwave and millimeter wave applications. At these frequencies, the currents and magnetic fields get confined to a small depth near the surface of the conductor. Owing to this skin effect phenomenon, two parameters—namely, the 'skin depth' and the 'surface impedance' are defined which are convenient to use in the determination of the loss. The skin depth δ_s is defined as the depth into the conductor at which the current density or magnetic flux gets attenuated to $1/e$ of its value at the surface. It is given by [1],

$$\delta_s = \sqrt{\frac{2}{\omega\mu_0\sigma}} \tag{11.3}$$

where μ_0 is the permeability, which, for a good conductor is the same as that of free space. The surface impedance, denoted as Z_s, is defined as the ratio of the tangential electric field strength at the surface of the conductor to the linear current density that flows as a result of this. This linear current density represents the total conduction current per unit width of the conductor. The expression for the surface impedance is given by [1],

$$Z_s = R_s + jX_s = (1 + j)\sqrt{\frac{\omega\mu_0}{2\sigma}} \tag{11.4}$$

The surface resistance R_s has the same magnitude as the surface reactance X_s at all frequencies. Using (11.3), these parameters can be expressed in terms of the skin depth δ_s as

$$R_s = X_s = \omega\left(\frac{\mu_0\delta_s}{2}\right) \tag{11.5}$$

11.2 General Formulas for Attenuation in a Transmission Line

The analysis presented here is based on a perturbational approach, which is valid when the loss per unit length of the line is small. Consider a uniform transmission line extending to infinity in the z-direction. For a TEM wave propagating along the line, the voltage and current along the line can be expressed as

$$V(z) = V_0 e^{-(\alpha+j\beta)z} \tag{11.6a}$$

$$I(z) = I_0 e^{-(\alpha+j\beta)z} \tag{11.6b}$$

The average power transmitted P_T is given by

$$P_T = \frac{1}{2} R_e[V(z)I^*(z)] = \frac{1}{2} V_0 I_0 e^{-2\alpha z} \tag{11.7}$$

The power lost per unit length, denoted as P_l, is the same as the decrease of transmitted power per unit length. It is given by

$$P_l = -\frac{\partial P_T}{\partial z} = 2\alpha P_T \tag{11.8}$$

The total attenuation α per unit length is given by

$$\alpha = \frac{P_l}{2P_T} \tag{11.9}$$

The total power lost P_l can be considered as the sum of the power dissipated in the conductors (P_c) and that in the dielectric material (P_d), Identifying the total attenuation constant α as the sum of the conductor attenuation constant α_c and the dielectric attenuation constant α_d, we can write

$$\alpha = \alpha_c + \alpha_d \tag{11.10}$$

where

$$\alpha_c = \frac{P_c}{2P_T} \tag{11.11a}$$

$$\alpha_d = \frac{P_d}{2P_T} \tag{11.11b}$$

The perturbational analysis for loss calculation is based on the assumption that the electric and magnetic field distributions in the structure are not altered due to the presence of losses. Further, we assume quasi-TEM mode approximation for all inhomogeneous transmission structures so that the electric field $\vec{E}$ can be expressed as the gradient of the electric scalar potential ϕ.

The power transmitted P_T can be determined by integrating $\frac{1}{2}\vec{E}\times\vec{H}^* = \frac{E^2}{2\eta} = \frac{1}{2}v\epsilon(\nabla\phi)^2$ over the cross-sectional area of the transmission line perpendicular to the direction of propagation. Here $\vec{E}$ and $\vec{H}$ are the electric and magnetic field vectors in the cross-sectional $(x\text{-}y)$ plane, η is the wave impedance, ϵ is the absolute dielectric constant of the dielectric medium, and v is the phase velocity. Thus

$$P_T = \frac{1}{2}\iint v\epsilon(\nabla\phi)^2\, dx\, dy \tag{11.12}$$

The power lost in the conductor P_c can be calculated from the knowledge of the surface resistance R_s and the surface current i_s in the conductor.

$$P_c = \frac{1}{2}\oint R_s i_s^2\, dl \tag{11.13}$$

where the line integral is taken around the strip conductor along the ground conductor surface. The surface current can be expressed as, $i_s = \rho_s v = (\hat{n}\cdot\epsilon\nabla\phi)v$, where ρ_s is the surface charge density and $\hat{n}$ is the unit vector normal to the surface of the conductor. Using this relation, (11.13) can be expressed in terms of the potential function ϕ, as

$$P_c = \frac{1}{2}R_s v^2\oint(\hat{n}\cdot\epsilon\nabla\phi)^2\, dl \tag{11.14}$$

The expression for the power dissipated in the dielectric material is given by

$$P_d = \frac{1}{2}\iint \sigma_d(\nabla\phi)^2\, dx\, dy \tag{11.15}$$

where σ_d is the conductivity of the dielectric material. Using (11.12), (11.14) and (11.15) in (11.11), the conductor and dielectric attenuation constants can be written as

$$\alpha_c = \frac{\oint R_s v^2(\hat{n}\cdot\epsilon\nabla\phi)^2\, dl}{2\iint v\epsilon(\nabla\phi)^2\, dx\, dy} \text{ nepers/unit length} \tag{11.16}$$

$$\alpha_d = \frac{\iint \sigma_d(\nabla\phi)^2\, dx\, dy}{2\iint v\epsilon(\nabla\phi)^2\, dx\, dy} \text{ nepers/unit length} \tag{11.17}$$

α_c and α_d given by (11.16) and (11.17) can be converted to *dB*/unit length by multiplying the equations by a factor 20/ln (10).

11.3 Approximate Formulas

11.3.1 Conductor Attenuation Using Incremental Inductance Rule [2]

In the 'incremental inductance rule' proposed by Wheeler [2], the series surface resistance per unit length of the conductor is expressed in terms of that part of the total inductance per unit length which is attributed to the skin effect, that is, the inductance caused by the magnetic flux inside the conductor. This inductance, denoted as L_s, can be obtained from (11.5) as

$$L_s = \frac{X_s}{\omega} = (\mu_o\ \delta_s/2) \tag{11.18}$$

The internal inductance L_s can therefore be interpreted as the inductance of a layer of conductive material having a thickness equal to half the skin depth. This inductance is the increment of the total inductance caused by the penetration of the magnetic flux into the conductor and is the same as would be caused by the surface receding to a depth $\delta_s/2$. The 'incremental inductance rule' [2] states that, the effective resistance in a circuit is equal to the change of reactance caused by the penetration of magnetic flux into the conductor constituting the circuit. It is assumed that the conductor thickness is much greater than the depth of penetration. The effective resistance, and hence the conductor attenuation in any circuit can thus be determined if its inductance as a function of the geometrical conductor parameters is calculated. The process of determining the attenuation constant α_c involves, therefore, computing first the rate of change of inductance of the circuit with recession into each of the metal surfaces constituting the circuit and then finding the effective resistance of each surface as governed by the skin effect. The increment of inductance L' caused by penetration into each surface is

$$L' = \frac{\delta_s}{2}\left(\frac{\delta L}{\delta n}\right) \tag{11.19}$$

where δn is a small depth normal to and directed into the surface of the conductor and L is the external inductance caused by the magnetic flux existing in the wave propagating space around the conductor. The effective resistance R contributed by the surface is given by

$$R = \omega L' = \frac{\omega\delta_s}{2}\cdot\left(\frac{\delta L}{\delta n}\right) \tag{11.20}$$

Using the expression (11.5) for the skin effect resistance R_s in (11.20), we can write

$$R = \frac{R_s}{\mu_o}\cdot\frac{\partial L}{\partial n} \tag{11.21}$$

For a stripline-like transmission structure operating in quasi-TEM mode under low loss approximation, the characteristic impedance Z can be expressed as

$$Z = \sqrt{\frac{L}{C}} \tag{11.22}$$

where L and C are the inductance and capacitance per unit length of the line. The conductor attenuation constant α_c is given by the first term in (11.2)

$$\alpha_c = \frac{R}{2}\sqrt{\frac{C}{L}} \tag{11.23}$$

Using (11.21) and (11.22) in (11.23) gives

$$\alpha_c = \frac{R_s}{2Z\mu_o}\cdot\frac{\partial L}{\partial n} \tag{11.24}$$

The phase velocity v in a low-loss line is given by

$$v = \frac{1}{\sqrt{LC}} = \frac{1}{\sqrt{\mu_o\epsilon_o\epsilon_f}} \tag{11.25}$$

where ϵ_f is the effective dielectric constant of the medium. From (11.22) and (11.25), we obtain

$$L = \frac{Z}{v} = Z\sqrt{\mu_o\epsilon_o\epsilon_f}$$

so that

$$\frac{\partial L}{\partial n} = \frac{\partial Z}{\partial n}\cdot\sqrt{\mu_o\epsilon_o\epsilon_f} \tag{11.26}$$

Substituting for $\partial L/\partial n$ from (11.26) in (11.24), gives the following expression for the conductor attenuation constant:

$$\alpha_c = \frac{R_s\sqrt{\epsilon_f}}{240\pi Z}\cdot\frac{\partial Z}{\partial n} \tag{11.27}$$

11.3.2 Dielectric Attenuation [3]

Spielman [3] has derived an approximate closed form expression for the dielectric attenuation constant in a layered transmission line. If $\bar{\epsilon_j} = (\epsilon_j' - j\epsilon_j'')$ represents the complex permittivity of the jth dielectric layer, its conductivity σ_{dj} is given by $\omega\epsilon_j''$. The expression (11.15) for the time average power dissipated in the dielectric, applied to a N-layered structure becomes

$$P_d = \frac{1}{2}\sum_{j=1}^{N}\iint_{S_j}\omega\epsilon_j''\,|E_j|^2\,dx\,dy \tag{11.28}$$

where S_j represents the cross-sectional area of the jth dielectric layer. Denoting $\tan\delta_{dj} = \epsilon_j''/\epsilon_j'$ as the dielectric loss tangent of the jth layer, (11.28) can be rewritten as

$$P_d = \sum_{j=1}^{N}\omega W_{ej}\tan\delta_{dj} \tag{11.29}$$

where

$$W_{ej} = \frac{1}{2}\iint_{S_j}\epsilon_j'\,|E_j|^2\,dx\,dy \tag{11.30}$$

is the time averaged electric energy stored in the jth dielectric layer. We now relate W_{ej} to W_e, the time average of the total energy stored in the electric field per unit length of the transmission line.

$$W_e = \frac{1}{2}\int_V \frac{D^2}{\epsilon}\, dV \tag{11.31}$$

If the dielectric constant ϵ is perturbed by $\delta\epsilon$, then

$$\delta W_e = \int_V \frac{\vec{D}.\delta\vec{D}}{\epsilon}\, dV - \int_V \frac{D^2\,\delta\epsilon}{2\epsilon^2}\, dV \tag{11.32}$$

Consider the relation

$$\nabla\cdot(\phi\delta\vec{D}) = -\vec{E}\cdot\delta\vec{D} + \phi\nabla\cdot\delta\vec{D} \tag{11.33a}$$

where

$$\vec{E} = -\nabla\phi \tag{11.33b}$$

$$\nabla\cdot\vec{D} = \rho \tag{11.33c}$$

Taking the volume integral of (11.33a),

$$\int_V \vec{E}\cdot\delta\vec{D}\, dV = -\int_V \nabla\cdot(\phi\delta\vec{D})\, dV + \int_V \phi\delta\rho\, dV \tag{11.34}$$

Substituting (11.34) in (11.32), we obtain

$$\delta W_e = \int_V \phi\delta\rho\, dV - \int_V \nabla\cdot(\phi\delta\vec{D})\, dV - \frac{1}{2}\int_V E^2\,\delta\epsilon\, dV \tag{11.35}$$

The first integral on the right hand side of (11.35) is zero since there is no space charge in the dielectric ($\delta\rho = 0$). The second integral can be expressed as

$$\int_V \nabla\cdot(\phi\delta\vec{D})\, dV = \sum_{m=1}^{N_c} \phi_m \oint_{C_m} (\delta\vec{D}\cdot\hat{n})\, dl = -\sum_{m=1}^{N_c} \phi_m\,\delta Q_m \tag{11.36}$$

where C_m is the contour encircling the mth conductor and $\hat{n}$ is the unit normal perpendicular to and pointing into the conductor. Expression (11.35) can now be written as

$$\delta W_e = \sum_{m=1}^{N_c} \phi_m\,\delta Q_m - \frac{1}{2}\int_{V_j} E_j^2\,\delta\epsilon_j'\, dV \tag{11.37}$$

where V_j is the volume associated with the jth dielectric layer. The summation $\sum_{m=1}^{N_c} \phi_m\,\delta Q_m$ can be written in terms of coefficient of capacitance as

$$\sum_{m=1}^{N_c} \phi_m\,\delta Q_m = \sum_{m=1}^{N_c}\sum_{n=1}^{N_c} \phi_m\,\phi_n\,\delta C_{mn} = 2\delta W_e \tag{11.38}$$

Using (11.38) in (11.37), we obtain

$$\delta W_e = \frac{1}{2}\int_{V_j} E_j^2\,\delta\epsilon_j'\, dV = \frac{\delta\epsilon_j'}{\epsilon_j'}\, W_{ej} \tag{11.39}$$

Substituting for W_{ej} from (11.39) in (11.29) gives

$$P_d = \sum_{j=1}^{N} \omega\epsilon_j' \frac{\partial W_e}{\partial \epsilon_j'} \tan \delta_{dj} \tag{11.40}$$

The total electric energy stored in the structure can also be expressed in terms of the line capacitance C per unit length of the structure as

$$W_e = \frac{1}{2} C \,|\, V \,|^2 = \frac{\epsilon_f}{2} C^a \,|\, V \,|^2 \tag{11.41}$$

where C^a is the line capacitance with all dielectrics replaced by air. ϵ_f is the relative effective dielectric constant and is given by

$$\epsilon_f = C/C^a \tag{11.42}$$

With the use of (11.41) in (11.40), we obtain

$$P_d = \frac{\omega}{2} C^a \,|\, V \,|^2 \sum_{j=1}^{N} \epsilon_j' (\tan \delta_{dj}). \frac{\partial \epsilon_f}{\partial \epsilon_j'} \tag{11.43}$$

Using the relations, $V_o/I_o = Z = \frac{1}{vC}$, and $|\, V \,| = V_o \, e^{-\alpha z}$, we can rewrite (11.7) as

$$P_T = \frac{1}{2} vC \,|\, V \,|^2 \tag{11.44}$$

Using (11.42)–(11.44) in (11.11b), the expression for α_d becomes

$$\alpha_d = \frac{\omega}{2v\epsilon_f} \sum_{j=1}^{N} \epsilon_j' (\tan \delta_{dj}) \frac{\partial \epsilon_f}{\partial \epsilon_j'}$$

$$= \frac{\pi}{\lambda^a \sqrt{\epsilon_f}} \sum_{j=1}^{N} \epsilon_j' (\tan \delta_{dj}) \cdot \frac{\partial \epsilon_f}{\partial \epsilon_j'} \text{ nepers/unit length} \tag{11.45}$$

where λ^a is the free space wavelength. For computational ease, the partial derivative in (11.45) is expressed in terms of a 'forward difference quotient' as

$$\frac{\partial \epsilon_f}{\partial \epsilon_j'} \simeq \frac{(\hat{\epsilon}_f - \epsilon_f)}{(\hat{\epsilon}_j' - \epsilon_j')} \tag{11.46}$$

where $\hat{\epsilon}_f$ is the value of ϵ_f of the structure when ϵ_j' of the jth layer is perturbed to a slightly higher value $\hat{\epsilon}_j'$. Thus (11.45) can be rewritten as

$$\alpha_d = \frac{\pi}{\lambda^a \sqrt{\epsilon_f}} \cdot \sum_{j=1}^{N} \epsilon_j' (\tan \delta_{dj}) \cdot \frac{(\hat{\epsilon}_f - \epsilon_f)}{(\hat{\epsilon}_j' - \epsilon_j')} \text{ nepers/unit length} \tag{11.47}$$

11.4 Homogeneous Symmetric Stripline

11.4.1 Conductor Loss Coefficient

A comprehensive analysis of conductor loss in a shielded stripline based on the 'incremental inductance rule' has been reported by Cohn [4]. Salient steps of this analysis and the resulting approximate closed form formulas are presented here.

For a homogeneous stripline filled with a dielectric of relative dielectric constant ϵ_r, the expression for conductor attenuation constant α_c is the same as (11.27) with ϵ_f replaced by ϵ_r. Thus,

$$\alpha_c = \frac{R_s\sqrt{\epsilon_r}}{240\pi Z}\cdot\frac{\partial Z}{\partial n} \tag{11.48}$$

In order to apply this formula to a shielded stripline, δ_n must be considered on the inner surfaces of the two ground planes and on the four surfaces of the strip conductor as shown in Fig. 11.1. Thus a change δb pointing into and normal to the two ground planes requires a change equal to $2\delta n$ in the ground plane spacing. Similarly, the change required in the width of the strip is $\delta W = -2\,\delta n$, and that in the thickness of the strip is $\delta t = -2\delta n$. The expression for $\partial Z/\partial n$ in (11.48) can be written as

$$\frac{\partial Z}{\partial n} = \frac{\frac{\partial Z}{\partial b}\cdot\delta b + \frac{\partial Z}{\partial w}\cdot\delta w + \frac{\partial Z}{\partial t}\cdot\delta t}{\delta n} \tag{11.49}$$

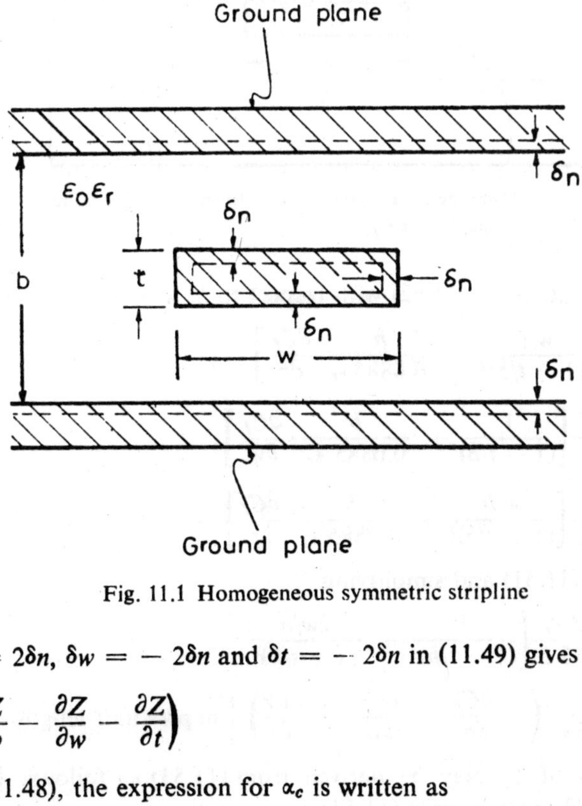

Fig. 11.1 Homogeneous symmetric stripline

Substituting, $\delta b = 2\delta n$, $\delta w = -2\delta n$ and $\delta t = -2\delta n$ in (11.49) gives

$$\frac{\partial Z}{\partial n} = 2\left(\frac{\partial Z}{\partial b} - \frac{\partial Z}{\partial w} - \frac{\partial Z}{\partial t}\right) \tag{11.50}$$

Using (11.50) in (11.48), the expression for α_c is written as

$$\alpha_c = \frac{R_s\sqrt{\epsilon_r}}{120\pi Z}\left(\frac{\partial Z}{\partial b} - \frac{\partial Z}{\partial w} - \frac{\partial Z}{\partial t}\right) \tag{11.51}$$

(i) *Wide strip case,* $w/(b-t) \geqslant 0.35$: For a wide strip approximation, that is, $w/(b-t) \geqslant 0.35$, the expression for the characteristic impedance of shielded stripline is given by [4]

$$Z = \frac{94.15}{\sqrt{\epsilon_r}\left[\dfrac{w/b}{(1-t/b)} + \dfrac{C_f}{0.0885\,\epsilon_r}\right]} \tag{11.52}$$

where C_f is the fringing capacitance in pF/cm from one corner of the strip to the adjacent ground as shown in Fig. 11.2. In the wide strip case, the fringe field C_f is not significant. An approximate formula for C_f is given by

$$C_f = \frac{0.0885\,\epsilon_r}{\pi}\left[\frac{2}{(1-t/b)}\ln\left\{\frac{1}{(1-t/b)}+1\right\} - \left\{\frac{1}{(1-t/b)}-1\right\}\ln\left\{\frac{1}{(1-t/b)^2}-1\right\}\right] pF/cm \tag{11.53}$$

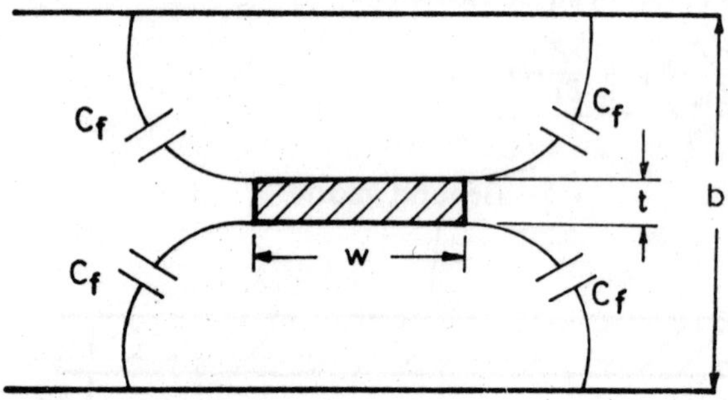

Fig. 11.2 Homogeneous symmetric stripline showing fringe capacitance C_f

Taking the partial derivatives of (11.52) with respect to b, w and t, we get

$$\frac{\partial Z}{\partial b} = \frac{\sqrt{\epsilon_r}\,Z^2}{94.15\,b}\left[\frac{w/b}{(1-t/b)^2} - \frac{b}{0.0885\,\epsilon_r}\cdot\frac{\partial C_f}{\partial b}\right] \tag{11.54a}$$

$$\frac{\partial Z}{\partial w} = -\frac{\sqrt{\epsilon_r}\,Z^2}{94.15\,b}\left[\frac{1}{(1-t/b)} + \frac{b}{0.0885\,\epsilon_r}\cdot\frac{\partial C_f}{\partial w}\right] \tag{11.54b}$$

$$\frac{\partial Z}{\partial t} = -\frac{\sqrt{\epsilon_r}\,Z^2}{94.15\,b}\left[\frac{w/b}{(1-t/b)^2} + \frac{b}{0.0885\,\epsilon_r}\cdot\frac{\partial C_f}{\partial t}\right] \tag{11.54c}$$

Substituting (11.54) in (11.51) and simplifying

$$\alpha_c = \frac{1}{120\pi}\cdot\frac{R_s\,Z\,\epsilon_r}{94.15\,b}\left[\frac{1}{(1-t/b)} + \frac{2w/b}{(1-t/b)^2} + \frac{b}{0.0885\,\epsilon_r}\left(-\frac{\partial C_f}{\partial b} + \frac{\partial C_f}{\partial w} + \frac{\partial C_f}{\partial t}\right)\right] \text{ nepers/unit length} \tag{11.55}$$

The partial derivatives of C_f may be derived from (11.53) as follows. Using a change of variable $x = 1/(1-t/b)$, we can write (11.53) as

$$C_f = \frac{0.0885\,\epsilon_r}{\pi}\left[2x\ln(x+1) - (x-1)\ln(x^2-1)\right] \tag{11.56a}$$

The various partial derivatives are given by

$$\frac{\partial C_f}{\partial b} = \frac{\partial C_f}{\partial x}\cdot\frac{\partial x}{\partial b}, \quad \frac{\partial C_f}{\partial w} = \frac{\partial C_f}{\partial x}\cdot\frac{\partial x}{\partial w}, \quad \frac{\partial C_f}{\partial t} = \frac{\partial C_f}{\partial x}\cdot\frac{\partial x}{\partial t} \tag{11.56b}$$

$$\frac{\partial C_f}{\partial x} = \frac{0.0885\epsilon_r}{\pi}\left[\ln\left(\frac{x+1}{x-1}\right)\right] \tag{11.56c}$$

$$\frac{\partial x}{\partial b} = -\frac{t/b}{b(1-t/b)^2}, \quad \frac{\partial x}{\partial w} = 0, \quad \frac{\partial x}{\partial t} = \frac{1}{b(1-t/b)^2} \tag{11.56d}$$

With (11.56), the expression for α_c in (11.55) reduces to

$$\alpha_c = \frac{R_s}{120\pi}\cdot\frac{Z\epsilon_r}{94.15b}\left[\frac{1}{(1-t/b)} + \frac{2w/b}{(1-t/b)^2} + \frac{(1+t/b)}{\pi(1-t/b)^2}\ln\left(\frac{2-t/b}{t/b}\right)\right] \text{ nepers/unit length} \tag{11.57}$$

For Copper, $R_s = 8.2535\times 10^{-3}\sqrt{f}$, where f is in GHz. The expression for α_c is then given by

$$\alpha_c = \frac{2.02\times 10^{-6}\,\epsilon_r Z\sqrt{f(\text{GHz})}}{b}\left[\frac{1}{(1-t/b)} + \frac{2w/b}{(1-t/b)^2} + \frac{(1+t/b)}{\pi(1-t/b)^2}\cdot\ln\left(\frac{2-t/b}{t/b}\right)\right] \text{ dB/unit length} \tag{11.58}$$

(ii) *Narrow strip case, $w/(b-t) \leqslant 0.35$, $t/b \leqslant 0.25$; $t/w \leqslant 0.11$*: The characteristic impedance of a stripline for the narrow strip case can be expressed in the form

$$Z = \frac{60}{\sqrt{\epsilon_r}}\ln\left(\frac{4b}{\pi d_0}\right) \tag{11.59}$$

where d_0 is the diameter of a circular cross-section conductor equivalent to the rectangular strip. In terms of the width w and thickness t of the strip conductor, the expression for d_0 is

$$d_0 = \frac{w}{2}\left[1 + \frac{t}{\pi w}\left(1 + \ln\frac{4\pi w}{t}\right) + 0.51\left(\frac{t}{w}\right)^2\right] \tag{11.60}$$

Differentiating Z with respect to b, w and t gives

$$\frac{\partial Z}{\partial b} = \frac{60}{\sqrt{\epsilon_r}\,b}, \quad \frac{\partial Z}{\partial w} = -\frac{60}{\sqrt{\epsilon_r}\,d_0}\cdot\frac{\partial d_0}{\partial w}, \quad \frac{\partial Z}{\partial t} = -\frac{60}{\sqrt{\epsilon_r}\,d_0}\cdot\frac{\partial d_0}{\partial t} \tag{11.61}$$

Substituting (11.61) in (11.51) gives

$$\alpha_c = \frac{R_s}{2\pi Zb}\left[1 + \frac{b}{d_0}\left(\frac{\partial d_0}{\partial w} + \frac{\partial d_0}{\partial t}\right)\right] \tag{11.62}$$

Differentiating (11.60) with respect to w and t and substituting in (11.62), we obtain

$$\alpha_c = \frac{R_s}{2\pi Zb}\left[1 + \frac{b}{d_0}\left\{0.5 + 0.669\left(\frac{t}{w}\right) - 0.255\left(\frac{t}{w}\right)^2 + \left(\frac{1}{2\pi}\right)\ln\left(\frac{4\pi w}{t}\right)\right\}\right] \text{ nepers/unit lengtn} \tag{11.63}$$

Considering copper as the conductor material, the expression for α_c becomes

$$\alpha_c = \frac{0.0114\sqrt{f(\text{GHz})}}{Zb}\left[1 + \frac{b}{d_0}\left\{0.5 + 0.669\left(\frac{t}{w}\right) - 0.255\left(\frac{t}{w}\right)^2 + \left(\frac{1}{2\pi}\right)\ln\left(\frac{4\pi w}{t}\right)\right\}\right] \quad dB/\text{unit length} \tag{11.64}$$

If the expression for Z is used in (11.58) and (11.64), it becomes apparent that the conductor attenuation constant for a homogeneous stripline is proportional to ϵ_r and also $\sqrt{f}$. In Fig. 11.3, the attenuation constant α_c computed using (11.58) and (11.64) is plotted as a

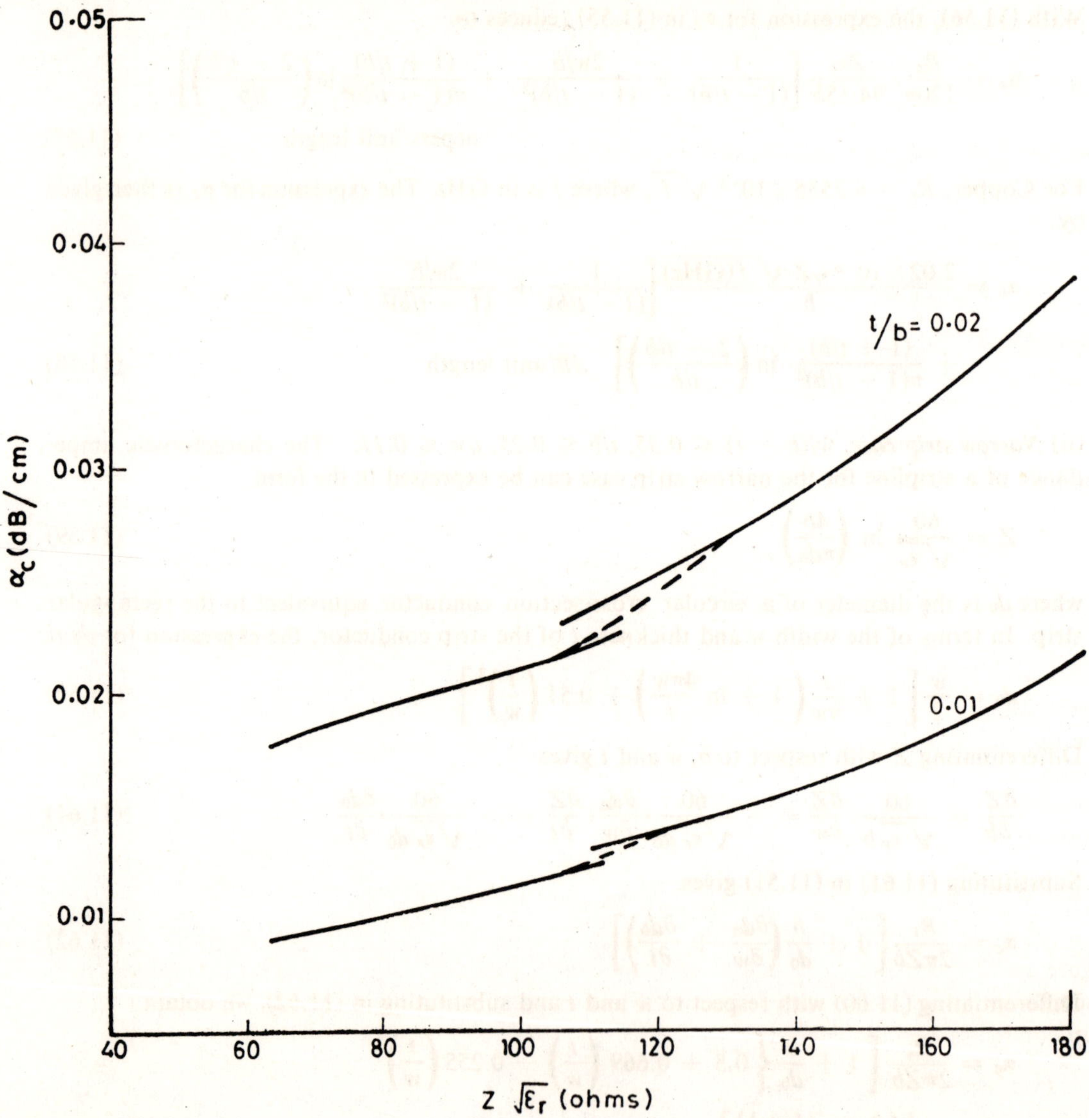

Fig. 11.3 Conductor attenuation constant as a function of $Z\sqrt{\epsilon_r}$ for homogeneous symmetric stripline; $\epsilon_r = 2.22$

function of $Z\sqrt{\epsilon_r}$ for two values of ground plane spacing (b = 1/8″ and 1/16″). The thickness of the strip conductor is kept constant so that $t/b = 0.01$ and 0.02 for b = 1/8″ and 1/16″, respectively. The dielectric is *RT*-duroid having $\epsilon_r = 2.22$ and the frequency is 10 GHz. It can be seen that in the vicinity of $w/b = 0.35$, the values of attenuation constant computed using (11.58) and (11.64) differ by about 8%. The dotted line between the two curves shows the average and should give reasonably accurate values of α_c in the transition region. The plots for $t/b = 0.01$ and 0.02 show that, for a given impedance level, the conductor attenuation can be reduced by increasing the ground plane spacing. The width of the conductor also increases accordingly. The largest ground plane spacing that can be employed in practice is limited by the onset of higher order modes.

11.4.2 Dielectric Loss Coefficient and Q-factor

The dielectric loss coefficient α_d for the homogeneous stripline can be written from (11.45) by taking one term in the summation and setting $\epsilon_f = \epsilon_r$ and $\epsilon_j' = \epsilon_0\epsilon_r$. Further, replacing $\tan\delta_{dj}$ by $\tan\delta_d$, we can write

$$\alpha_d = \left(\frac{\pi\sqrt{\epsilon_r}}{\lambda^a}\right)\tan\delta_d \qquad \text{nepers/unit length} \tag{11.65}$$

The Q-factor of a homogeneous stripline is given by

$$Q = \frac{\beta}{2\alpha} = \frac{8.68\pi\sqrt{\epsilon_r}}{\lambda^a\alpha} \tag{11.66}$$

where α is the total loss coefficient expressed in dB/unit length.

11.5 Three Layer Stripline

The conductor and dielectric loss coefficients in any stripline-like transmission line employing layered dielectrics can be computed from the general formulas (11.16) and (11.17). The evaluation, however, requires the determination of the potential function in each of the dielectric layers. In this section, we evaluate (11.16) and (11.17) for a three layer stripline. The loss coefficients of the important practical stripline-like transmission lines—namely, the homogeneous stripline with centered and off-centered strips, microstrip-like stripline and the suspended stripline can then be obtained as special cases of the expression for the three-layer stripline.

11.5.1 Potential Distribution Functions in a Three Layer Stripline

Figure 11.4a shows the cross-section of a three layer shielded stripline. The strip conductor is assumed to be infinitesimally thin and is located at $y = y_0$. The charge distribution on the strip conductor is represented as

$$\rho(x, y) = f(x)\delta(y - y_0) \tag{11.67}$$

If $G(x, y/x_0, y_0)$ represents the Green's function for the structure due to a point charge at (x_0, y_0), the potential distribution function $\phi(x, y)$ at any point within the structure can be

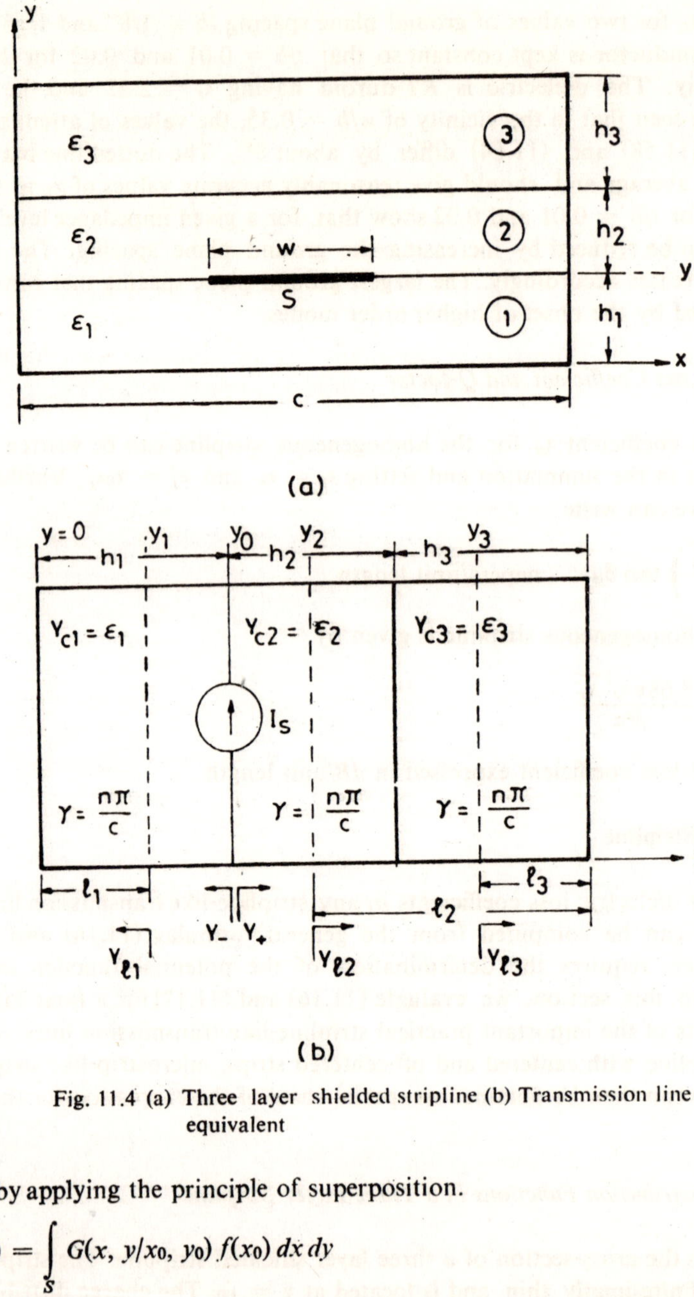

Fig. 11.4 (a) Three layer shielded stripline (b) Transmission line equivalent

determined by applying the principle of superposition.

$$\phi(x, y) = \int_S G(x, y/x_0, y_0)\, f(x_0)\, dx\, dy \tag{11.68}$$

where the integration is carried over the strip conductor S. The Green's function $G(x, y_0/x_0, y_0)$ at the charge plane $y = y_0$ is given by (see (2.77) in Chapter 2).

$$G(x, y_0/x_0, y_0) = \sum_{n=1}^{\infty} G_n(y_0) \sin \frac{n\pi x}{c} \tag{11.69a}$$

where

$$G_n(y_0) = \frac{2}{n\pi Y} \sin \frac{n\pi x_0}{c} \tag{11.69b}$$

In (11.69b), Y is the admittance at the charge plane which can be easily obtained from the transmission line equivalent shown in Fig. 11.4b.

$$Y = \epsilon_1 \coth \left(\frac{n\pi h_1}{c}\right) + \epsilon_2 \left[\frac{\epsilon_3 \coth\left(\frac{n\pi h_3}{c}\right) + \epsilon_2 \tanh\left(\frac{n\pi h_2}{c}\right)}{\epsilon_2 + \epsilon_3 \coth\left(\frac{n\pi h_3}{c}\right) \cdot \tanh\left(\frac{n\pi h_2}{c}\right)}\right] \tag{11.70}$$

In order to find the Green's function at any point in the structure, we make use of the following formula for the voltage along the transmission line shown in Fig. 11.4b.

$$V(y_0) = V(y_j)\left[\cosh | \gamma(y_j - h_1) | + \frac{Y_{lj}}{Y_{cj}} \sinh | \gamma(y_j - h_1) | \right], \quad j = 1, 2, 3 \tag{11.71}$$

where $V(y_0)$ is the voltage at the charge plane $y = y_0$, y_j is the distance from the origin at any point in the jth layer, Y_{cj} is the characteristic impedance of the jth layer, and Y_{lj} is the input admittance of a short circuited line of length l_j where $l_j = y_1$, $(h_1 + h_2 + h_3 - y_2)$, and $(h_1 + h_2 + h_3 - y_3)$ for $j = 1, 2$ and 3, respectively (see Fig. 11.4b). Since $G_n(y)$ is analogous to V, we can write

$$G_n(y_0) = G_n(y_j)\left[\cosh\left|\frac{n\pi}{c}(y_j - h_1)\right| + \frac{Y_{lj}}{Y_{cj}} \sinh\left|\frac{n\pi}{c}(y_j - h_1)\right|\right], j = 1, 2, 3 \tag{11.72}$$

Using (11.69) in (11.72), the Green's functions in the three dielectric layers can be expressed as follows:

Region 1; $0 \leqslant y \leqslant h_1$

$$G(x, y/x_0, y_0) = \sum_{n=1}^{\infty} A_n \sinh\left(\frac{n\pi y}{c}\right) \cdot \sin\left(\frac{n\pi x}{c}\right) \sin\left(\frac{n\pi x_0}{c}\right) \tag{11.73a}$$

Region 2; $h_1 \leqslant y \leqslant h_1 + h_2$

$$G(x, y/x_0, y_0) = \sum_{n=1}^{\infty} \left[B_n + C_n \coth\left(\frac{n\pi y}{c}\right)\right] \sinh\left(\frac{n\pi y}{c}\right) \sin\left(\frac{n\pi x}{c}\right) \sin\left(\frac{n\pi x_0}{c}\right) \tag{11.73b}$$

Region 3 ; $h_1 + h_2 \leqslant y \leqslant h_1 + h_2 + h_3$

$$G(x, y/x_0, y_0) = \sum_{n=1}^{\infty} D_n \sinh\left\{\frac{n\pi}{c}(h_1 + h_2 + h_3 - y)\right\} \sin\left(\frac{n\pi x}{c}\right) \sin\left(\frac{n\pi x_0}{c}\right) \tag{11.73c}$$

where

$$A_n = \frac{2\eta_n}{n\pi P_n \sinh\left(\frac{n\pi h_1}{c}\right)} \tag{11.74a}$$

$$B_n = \frac{2\left[\epsilon_3\left(\coth\left(\frac{n\pi h_3}{c}\right)\right)^2 - \epsilon_2\right] \sinh\left(\frac{n\pi h_3}{c}\right)}{n\pi P_n \sinh\left(\frac{n\pi h_1}{c}\right)} \tag{11.74b}$$

$$C_n = \frac{2(\epsilon_2 - \epsilon_3)\cosh\left(\frac{n\pi h_3}{c}\right)}{n\pi p_n \sinh\left(\frac{n\pi h_2}{c}\right)} \tag{11.74c}$$

$$D_n = \frac{2\epsilon_2}{n\pi p_n} \tag{11.74d}$$

$$p_n = \epsilon_2 \xi_n + \epsilon_1 \eta_n \coth\left(\frac{n\pi h_1}{c}\right) \tag{11.74e}$$

$$\xi_n = \epsilon_3 \coth\left(\frac{n\pi h_3}{c}\right)\coth\left(\frac{n\pi h_2}{c}\right) + \epsilon_2 \tag{11.74f}$$

$$\eta_n = \epsilon_3 \coth\left(\frac{n\pi h_3}{c}\right) + \epsilon_2 \coth\left(\frac{n\pi h_2}{c}\right) \tag{11.74g}$$

we now assume a charge distribution on the strip conductor S to be of the form,

$$f(x) = \frac{1}{w}\left[1 + A\left|\left(\frac{2}{w}\right)\left(x - \frac{c}{2}\right)\right|^3\right]; \qquad \frac{c-w}{2} \leqslant x \leqslant \frac{c+w}{2}$$
$$= 0 \qquad ; \qquad \text{otherwise} \tag{11.75}$$

Substituting (11.73) and (11.75) in (11.68) and evaluating the integral, we obtain the following expressions for the potential distribution functions in each of the three regions.

$$\phi_1(x, y) = \sum_{n=1}^{\infty} A_n T_n \sinh\left(\frac{n\pi y}{c}\right)\sin\left(\frac{n\pi x}{c}\right) \tag{11.76a}$$

$$\phi_2(x, y) = \sum_{n=1}^{\infty}\left[B_n + C_n \coth\left(\frac{n\pi y}{c}\right)\right] T_n \sinh\left(\frac{n\pi y}{c}\right)\cdot\sin\left(\frac{n\pi x}{c}\right) \tag{11.76b}$$

$$\phi_3(x, y) = \sum_{n=1}^{\infty} D_n T_n \sinh\left(\frac{n\pi}{c}(h_1 + h_2 + h_3 - y)\right)\cdot\sin\left(\frac{n\pi x}{c}\right) \tag{11.76c}$$

where

$$T_n = \left(\frac{2c}{n\pi w}\right)(L_n + AM_n) \tag{11.76d}$$

$$L_n = \sin\left(\frac{n\pi w}{2c}\right) \tag{11.76e}$$

$$M_n = \left(\frac{2c}{n\pi w}\right)^3\left[3\left\{\left(\frac{n\pi w}{2c}\right)^2 - 2\right\}\cos\left(\frac{n\pi w}{2c}\right) + \left(\frac{n\pi w}{2c}\right)\left\{\left(\frac{n\pi w}{2c}\right)^2 - 6\right\}\sin\left(\frac{n\pi w}{2c}\right) + 6\right] \tag{11.76f}$$

$$A = -\frac{\sum_{n\ \text{odd}} (L_n - 4M_n)\, L_n P_n / Y}{\sum_{n\ \text{odd}} (L_n - 4M_n)\, M_n P_n / Y} \tag{11.76g}$$

The expression for constant A has been obtained by maximizing the capacitance of the structure as described in Chapter 3.

11.5.2 Conductor Loss Coefficient

The conductor loss coefficient α_c in the three layers of stripline can now be determined from (11.16) using the potential functions given by (11.76). We have

$$\alpha_c = \frac{P_c}{2P_T} \tag{11.77}$$

where

$$P_c = R_s v^2 \oint (\hat{n} \cdot \epsilon \nabla \phi)^2 \, dl$$

$$= \sqrt{\frac{\omega \mu_0}{2\sigma_c}} \, v^2 \oint \hat{n} \cdot (\epsilon \nabla \phi)^2 \, dl \tag{11.78}$$

In (11.78), σ_c is the conductivity of the strip conductor as well as the shielding enclosure. The integration is to be carried out around the strip conductor and along the inner surface of the shielding enclosure. Thus

$$P_c = \sqrt{\frac{\omega\mu_0}{2\sigma_c}} \, v^2 \Bigg[\int_0^c \left\{ \epsilon_1^2 \left(\frac{d\phi_1}{dy}\right)^2 + \epsilon_3^2 \left(\frac{d\phi_3}{dy}\right)^2 \right\} dx + 2 \int_0^{h_1} \epsilon_1^2 \left(\frac{d\phi_1}{dx}\right)^2 dy$$

$$+ 2 \int_{h_1}^{h_1+h_2} \epsilon_2^2 \left(\frac{d\phi_2}{dx}\right)^2 dy + 2 \int_{h_1+h_2}^{h_1+h_2+h_3} \epsilon_3^2 \left(\frac{d\phi_3}{dx}\right)^2 dy$$

$$+ \int_{(c-w)/2}^{(c+w)/2} \left[\epsilon_1^2 \left(\frac{d\phi_1}{dy}\right)^2 + \epsilon_2^2 \left(\frac{d\phi_2}{dy}\right)^2 \right] dx \Bigg] \tag{11.79}$$

The expression for transmitted power is given by

$$P_T = v \Bigg[\int_0^c \int_0^{h_1} \epsilon_1 (\nabla \phi_1)^2 \, dx \, dy + \int_0^c \int_{h_1}^{h_1+h_2} \epsilon_2 (\nabla \phi_2)^2 \, dx \, dy$$

$$+ \int_0^c \int_{h_1+h_2}^{h_1+h_2+h_3} \epsilon_3 (\nabla \phi_3)^2 \, dx \, dy \Bigg] \tag{11.80}$$

Substituting for ϕ_1, ϕ_2 and ϕ_3 from (11.76) and evaluating the various integrals, we obtain

$$\alpha_c = \sqrt{\frac{\omega\mu_0}{2\sigma_c}} \, v \left[\frac{I_{x1} + I_{x2} + I_{x3} + I_{x4} + 2(I_{y1} + I_{y2} + I_{y3})}{\epsilon_1 I_1 + \epsilon_2 I_2 + \epsilon_3 I_3} \right]$$

nepers/unit length (11.81)

where

$$I_{x1} = \sum_{n \text{ odd}} \epsilon_1^2 \left(\frac{T_n}{\rho_n}\right)^2 \eta_n \left[\frac{1}{\sinh^2 \left(\frac{n\pi h_1}{c}\right)} \right] \tag{11.82a}$$

$$I_{x2} = \sum_{n \text{ odd}} \sum_{m \text{ odd}} \epsilon_1^2 \frac{T_n T_m}{\rho_n \rho_m} \xi_{nm} \coth\left(\frac{n\pi h_1}{c}\right) \cdot \coth\left(\frac{m\pi h_1}{c}\right) \tag{11.82b}$$

$$I_{x3} = \sum_{n\text{ odd}} \sum_{m\text{ odd}} \epsilon_2^2 \frac{T_n T_m}{p_n p_m} \xi_n \xi_m \cdot \xi_{nm} \tag{11.82c}$$

$$I_{x4} = \sum_{n\text{ odd}} \epsilon_2^2 \epsilon_3^2 \left(\frac{T_n}{p_n}\right)^2 \cdot \frac{1}{\left[\sinh^2\left(\frac{n\pi h_3}{c}\right) \cdot \sinh^2\left(\frac{n\pi h_1}{c}\right)\right]} \tag{11.82d}$$

$$I_{y1} = \sum_{n\text{ odd}} \sum_{m\text{ odd}} \epsilon_2 \epsilon_3 \frac{2 T_n T_m}{p_n p_m} \eta_m \eta_n K_{nm} \tag{11.82e}$$

$$I_{y2} = \sum_{n\text{ odd}} \sum_{m\text{ odd}} \frac{2\epsilon_2^2 T_n T_m}{\pi p_n p_m (n^2 - m^2)} [n\xi_n \eta_m - m\xi_m \eta_n] - \frac{\epsilon_2}{\epsilon_3} I_{y3} \tag{11.82f}$$

$$I_{y3} = \sum_{n\text{ odd}} \sum_{m\text{ odd}} \frac{2\epsilon_1^2 T_n T_m}{p_n p_m} \cdot \frac{P_{nm}}{\left[\sinh\left(\frac{n\pi h_2}{c}\right) \sinh\left(\frac{m\pi h_2}{c}\right)\right]} \tag{11.82g}$$

$$K_{nm} = \frac{1}{\pi(n^2 - m^2)} \left[n \coth\left(\frac{n\pi h_1}{c}\right) - m \coth\left(\frac{m\pi h_1}{c}\right)\right] \tag{11.82h}$$

$$\xi_{nm} = \frac{4}{\pi(n^2 - m^2)} \left[n \sin\left(\frac{n\pi w}{2c}\right) \cos\left(\frac{m\pi w}{2c}\right) - m \cos\left(\frac{n\pi w}{2c}\right) \sin\left(\frac{m\pi w}{2c}\right)\right] \tag{11.82i}$$

$$I_1 = \sum_{n\text{ odd}} \frac{1}{n\pi} \left(\frac{T_n}{p_n}\right)^2 \eta_n^2 \coth\left(\frac{n\pi h_1}{c}\right) \tag{11.82j}$$

$$I_2 = \sum_{n\text{ odd}} \frac{1}{n\pi} \left(\frac{T_n}{p_n}\right)^2 \left[\epsilon_3 \xi_n \coth\left(\frac{n\pi h_3}{c}\right) + \epsilon_2 \eta_n\right] \tag{11.82k}$$

$$I_3 = \sum_{n\text{ odd}} \frac{1}{n\pi} \left(\frac{T_n}{p_n}\right)^2 \epsilon_2^2 \coth\left(\frac{n\pi h_3}{c}\right) \frac{1}{\sinh^2\left(\frac{n\pi h_1}{c}\right)} \tag{11.82l}$$

$$P_{nm} = \frac{1}{\pi(n^2 - m^2)} \left[n \coth\left(\frac{n\pi h_3}{c}\right) - m \coth\left(\frac{m\pi h_3}{c}\right)\right] \tag{11.82m}$$

p_n, ξ_n, η_n and T_n are defined as in (11.74e), (11.74f), (11.74g) and (11.76d), respectively.

11.5.3 Dielectric Loss Coefficient

The dielectric loss coefficient α_d for the three layer stripline can be obtained from (11.17) using the potential functions given by (11.76). If σ_{d1}, σ_{d2} and σ_{d3} represent the conductivities of the dielectric regions 1, 2 and 3, respectively, we can express (11.17) as

$$\alpha_d = \frac{1}{2v} \left[\frac{\sigma_{d1} I_1 + \sigma_{d2} I_2 + \sigma_{d3} I_3}{\epsilon_1 I_1 + \epsilon_2 I_2 + \epsilon_3 I_3}\right] \quad \text{nepers/unit length} \tag{11.83}$$

where

$$I_1 = \int_0^c \int_0^{h_1} (\nabla \phi_1)^2 \, dx \, dy \tag{11.84a}$$

$$I_2 = \int_0^c \int_{h_1}^{h_1+h_2} (\nabla \phi_2)^2 \, dx \, dy \tag{11.84b}$$

$$I_3 = \int_0^c \int_{h_1+h_2}^{h_1+h_2+h_3} (\nabla\phi_3)^2 \, dx \, dy \tag{11.84c}$$

The expressions for the integrals I_1, I_2 and I_3 are given in (11.82j), (11.82k) and (11.82l), respectively.

11.6 Power Handling Capability

11.6.1 Peak Power Handling Capability

The peak power handling capability of any stripline-like transmission line is limited by the breakdown between the strip conductor and the ground plane. If V_0 is the peak voltage the structure can withstand and Z is its characteristic impedance, the peak power handling limit P_p is given by

$$P_p = \frac{V_0^2}{2Z} \tag{11.85}$$

It is apparent that low impedance lines have higher peak power capability.

The dielectric strength of the substrate material and that of air region are primary factors in determining the breakdown voltage. In practice, sharp edges of the strip conductor serve as field concentrators. Thick conductors with rounded edges can withstand larger voltage as compared with those with sharp edges. Breakdown can also occur in connectors and transitions much before that in the line itself.

11.6.2 Average Power Handling Capability of a Homogeneous Stripline [5, 6]

The average power handling capability of a stripline is limited by the heating caused due to the dissipative losses in the inner conductors and the dielectric. These dissipative losses are a function of the cross-sectional dimensions of the stripline, resistivity of the conductors, the dielectric loss tangent and thermal conductivity of the dielectric material.

(i) *Choice of Stripline Dimensions*: For a given characteristic impedance, the conductor loss in a homogeneous symmetric stripline decreases as the ground plane spacing is increased (see Fig. 11.3). The maximum permissible ground plane spacing is governed by the onset of TE and TM modes. Figure 11.5 shows the electric field configuration for these modes. The lowest order TE mode can propagate when the width w and spacing b are such that the average circumferential distance (as indicated by the dotted line in Fig. 11.5a) exceeds about one wavelength. The cut-off wavelengths of TE and TM modes, denoted as λ_{ce} and λ_{cm} respectively, can be expressed in terms of stripline parameters as

$$\lambda_{ce} = \sqrt{\epsilon_r}\,(2w + 0.5\pi b) \tag{11.86}$$

$$\lambda_{cm} = 2b\sqrt{\epsilon_r} \tag{11.87}$$

The ratio of the cut-off frequency f_{cm} of the TM mode to the cut-off frequency f_{ce} of the TE mode is given by

$$\frac{f_{cm}}{f_{ce}} = \frac{\lambda_{ce}}{\lambda_{cm}} = \left(\frac{w}{b} + 0.785\right) \tag{11.88}$$

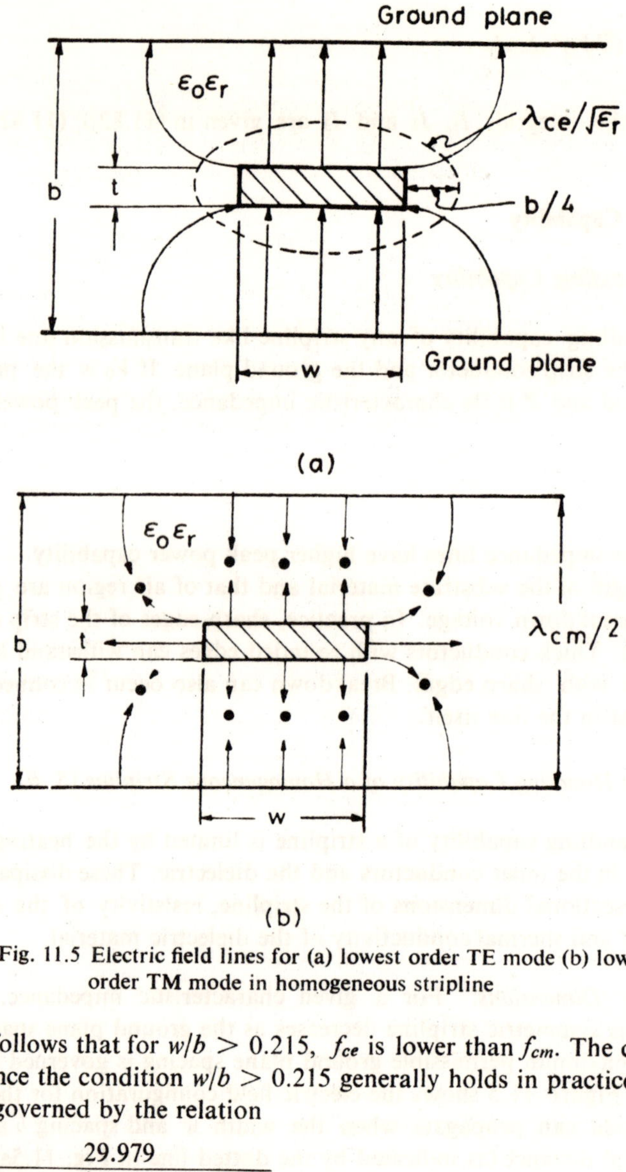

Fig. 11.5 Electric field lines for (a) lowest order TE mode (b) lowest order TM mode in homogeneous stripline

From (11.88) it follows that for $w/b > 0.215$, f_{ce} is lower than f_{cm}. The converse is true for $w/b < 0.215$. Since the condition $w/b > 0.215$ generally holds in practice, maximum ground plane spacing is governed by the relation

$$b_{\max} = \frac{29.979}{\left(\frac{2w}{b} + \frac{\pi}{2}\right)\sqrt{\epsilon_r}\, f_{ce}(\text{GHz})} \tag{11.89}$$

(ii) *Thermal Conductance of Stripline:* The thermal conductance of a stripline can be calculated by using the analogy between the heat flow field and the electric field. Considering the example of two parallel plates, the expressions for its electrical capacity C_p and thermal conductance G_p can be written as

$$C_p = \frac{\epsilon A}{d} \tag{11.90}$$

$$G_p = g\frac{A}{d} \tag{11.91}$$

where ϵ is the real part of the complex dielectric constant $\tilde{\epsilon}$ of the medium between the plates, A is the area of each plate, d is the plate separation and g is the thermal conductivity of the dielectric. The ratio G_p/C_p is independent of the geometrical parameters and is given by

$$\frac{G_p}{C_p} = \frac{g}{\epsilon} \tag{11.92}$$

Because of the analogy between the solutions of heat flow and electrostatic field problems, (11.92) can be extended to homogeneous dielectric cross-sections of arbitrary shape [6]. Figure 11.6 shows thermal field pattern in a homogeneous stripline. The heat flow lines are identical to the electric field lines and the isotherms are identical to the equipotential lines or magnetic field lines. Using the relation (11.92), the thermal conductance per unit length of the stripline can be expressed as

$$G_t = \frac{Cg}{\epsilon_0\epsilon_r} \tag{11.93}$$

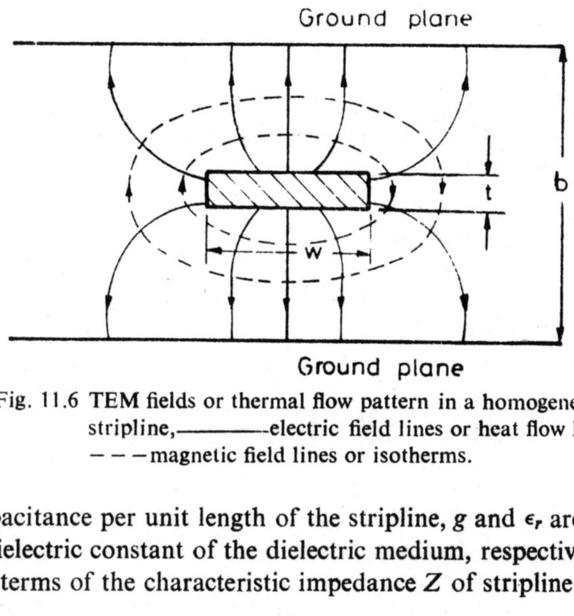

Fig. 11.6 TEM fields or thermal flow pattern in a homogeneous stripline, ——— electric field lines or heat flow lines - - - magnetic field lines or isotherms.

where C is the capacitance per unit length of the stripline, g and ϵ_r are the thermal conductivity and relative dielectric constant of the dielectric medium, respectively. The capacitance can be expressed in terms of the characteristic impedance Z of stripline as

$$C = \frac{\sqrt{\epsilon_r}}{Zv^a} \tag{11.94}$$

where v^a is the velocity in air. Substituting for C from (11.94) in (11.93), we obtain

$$G_t = \frac{120\pi g}{Z\sqrt{\epsilon_r}} = \frac{120\pi g}{Z^a} \tag{11.95}$$

where Z^a is the characteristic impedance of the stripline with dielectric replaced by air.

(iii) *Temperature Rise in Homogeneous Stripline*: From the thermal equivalent circuit of the stripline, we may consider that all the heat generated by the inner strip conductor and dielectric losses flows towards the outer conductors, which in turn act as heat sinks. Since the heat flow is in the transverse plane perpendicular to the direction of propagation, a temperature gradient exists between the inner and outer conductors. The temperature rise ΔT of the inner conductor with respect to the outer conductor can be determined from the relation

$$\Delta T = \frac{(P_{ic} + P_d)}{G_t} \tag{11.96}$$

where P_{ic} and P_d represent the power dissipated per unit length of the line due to the losses in the inner conductor and dielectric, respectively. If P_T denotes the incident power in watts, α_{ic} and α_d denote the loss coefficients in nepers/cm due to the inner conductor and dielectric respectively, and G_t is expressed in W/(°C − cm), the temperature rise is given by

$$\Delta T = \frac{2(\alpha_{ic} + \alpha_d)P_T}{G_t}\ ^\circ\text{C} \tag{11.97}$$

Substituting for G_t from (11.95), the temperature rise per one watt of incident power becomes,

$$\Delta T = \frac{(\alpha_{ic} + \alpha_d)\, Z\sqrt{\epsilon_r}}{60\pi g}\ ^\circ\text{C/W} \tag{11.98}$$

where g is expressed in W/(°C − cm)
The dielectric loss coefficient α_d for a stripline can be calculated from (11.65). The loss coefficient α_{ic} due to the inner conductor can be separated from the total conductor loss coefficient given in (11.51). It is given by

$$\alpha_{ic} = -\frac{R_s\sqrt{\epsilon_r}}{120\pi Z}\left(\frac{\partial Z}{\partial w} + \frac{\partial Z}{\partial t}\right) \tag{11.99}$$

For the wide strip case ($w/(b-t) \geqslant 0.35$), we substitute for $\partial Z/\partial w$ and $\partial Z/\partial t$ from (11.54b) and (11.54c) in (11.99) to obtain

$$\alpha_{ic} = \frac{R_s}{120\pi}\cdot\frac{Z\sqrt{\epsilon_r}}{94.15b}\left[\frac{1}{\left(1-\frac{t}{b}\right)} + \frac{w/b}{\left(1-\frac{t}{b}\right)} + \frac{1}{\pi\left(1-\frac{t}{b}\right)^2}\ln\left(\frac{2-t/b}{t/b}\right)\right] \quad \text{nepers/cm} \tag{11.100}$$

For the narrow strip case ($w/(b-t) \leqslant 0.35$, $t/b \leqslant 0.35$, $t/w \leqslant 0.11$), the expressions for $\partial Z/\partial w$ and $\partial Z/\partial t$ obtained from (11.61) are substituted in (11.99) to yield

$$\alpha_{ic} = \frac{R_s}{2\pi Z d_0}\left[0.5 + 0.669\left(\frac{t}{w}\right) - 0.255\left(\frac{t}{w}\right)^2 + \frac{1}{2\pi}\ln\left(\frac{4\pi w}{t}\right)\right] \quad \text{nepers/unit length} \tag{11.101}$$

(iv) *Maximum Average Power Handling Capability:* The maximum average power that the line can handle can be calculated from

$$P_{av} = (T_{max} - T_s)/\Delta T \text{ Watts} \tag{11.102}$$

where ΔT is given by (11.98). T_s is the temperature at the outer surface of stripline. T_{max} is the maximum allowable temperature of the dielectric. The properties of dielectric substrates used in loss calculations are listed in [7]. The maximum operating temperature for polystyrene is known to be 100°C and more than 100°C for other dielectrics used in practice. Hence, the maximum allowable temperature can be safely taken as 100°C for average power handling calculations.

REFERENCES

1. E.C. Jordan and K.G. Balmain, Electromagnetic Waves and Radiating Systems, Prentice-Hall, New Delhi, 1971.
2. H.A. Wheeler, Formulas for the skin effect, *Proc. IRE*, Vol. 30, pp. 412-424, Sept. 1942.
3. B.E. Spielman, Computer-aided analysis of dissipation losses in isolated and coupled transmission lines for microwave and millimeter-wave integrated circuit applications, *Naval Research Laboratory*, Washington, D.C., NRL Rep. 8009, July 1976.
4. S.B. Cohn, Problems in strip transmission lines, *IRE Trans. Microwave Theory Tech.*, Vol. MTT-3, pp. 119-126, March 1955.
5. P. Schiffres, How much CW Power can stripline handle, *Microwaves*, pp. 25-34, June 1966.
6. R.C. Levine, Determination of thermal conductance of dielectric filled strip transmission line from characteristic impedance, *IEEE Trans. Microwave Theory Tech.*, Vol. MTT-15, pp. 645-646, Nov. 1967.
7. I.J. Bahl and K.C. Gupta, Average power-handling capability of microstrip lines, *Microwaves, Optics and Acoustics*, Vol. 3, pp. 1-4, Jan. 1979.

CHAPTER 12

Circuit Design and Applications

12.1 Introduction

This chapter is devoted to the circuit design and applications of stripline-like transmission structures. First, a brief review of the general network representation of microwave circuits is provided. This is followed by a discussion on transitions from stripline-like transmission lines to coaxial lines and waveguides. Next, the design aspects of the commonly used basic circuit elements in the form of transmission line sections and stubs, coupled line sections, lumped elements etc. are considered. Bulk of the chapter contains design techniques for the realization of various passive components—namely, directional couplers, hybrids, power dividers and filters. In addition, design aspects of components using active devices—namely P-I-N diode switches, attenuators, phase shifters and mixers are included. A variety of practical configurations, their relative merits and certain broadbanding techniques are presented. Design formulas are furnished in consolidated form to aid the component designer. No attempt is made to present detailed analyses leading to design formulas although a brief account of the analytical steps involved is provided in some specific cases. For detailed analysis, the reader is referred to the appropriate references cited in the chapter.

12.2 Microwave Network Parameters

12.2.1 Matrix Representations

The circuit performance of a microwave network is generally evaluated with the aid of one of the following matrix representations: (i) impedance/admittance matrix, (ii) transmission or '*ABCD*' matrix, and (iii) scattering matrix. Detailed description of these network parameters is available in several books [1–4]. Only the salient features are summarized here for the purpose of easy reference to the rest of the chapter.

Impedance Matrix

Consider a two-port network representation as shown in Fig. 12.1. The terminal voltages V_1 and V_2 are related to the terminal currents I_1 and I_2 via the impedance matrix of the network as per the following:

$$\begin{bmatrix} V_1 \\ V_2 \end{bmatrix} = \begin{bmatrix} Z_{11} & Z_{12} \\ Z_{21} & Z_{22} \end{bmatrix} \begin{bmatrix} I_1 \\ I_2 \end{bmatrix} \tag{12.1}$$

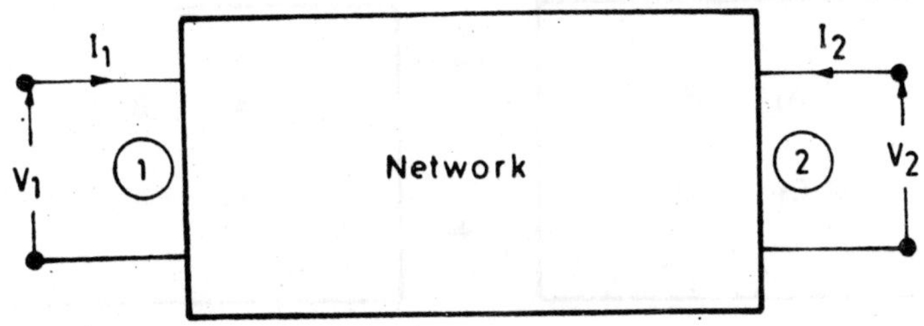

Fig. 12.1 Two-port network representation in terms of voltage and currents.

Symbolically,

$$[V] = [Z][I] \tag{12.2}$$

For a n-port network, $[Z]$ is given by a $n \times n$ matrix. If the network is reciprocal, $Z_{ij} = Z_{ji}$, where Z_{ij} is the element in the ith row j^{th} column. If the network is lossless, all the elements of the impedance matrix are imaginary. The elements of the admittance matrix Y are obtained by inverting the impedance matrix. Thus

$$[Y] = [Z]^{-1} \tag{12.3}$$

Transmission (ABCD) Matrix

The transmission matrix is particularly useful in relating the overall performance of cascaded two-port networks to the performance of an individual network. It relates the voltage and current at the input terminal to those at the output terminal. Referring to the general two-port network shown in Fig. 12.1, we can write

$$\begin{bmatrix} V_1 \\ I_1 \end{bmatrix} = \begin{bmatrix} A & B \\ C & D \end{bmatrix} \begin{bmatrix} V_2 \\ -I_2 \end{bmatrix} \tag{12.4}$$

When two networks are cascaded together as shown in Fig. 12.2, the overall transmission matrix is obtained by multiplying the matrices of the individual networks. We can therefore write

$$\begin{bmatrix} A & B \\ C & D \end{bmatrix} = \begin{bmatrix} A_1 & B_1 \\ C_1 & D_1 \end{bmatrix} \begin{bmatrix} A_2 & B_2 \\ C_2 & D_2 \end{bmatrix} \tag{12.5}$$

For a reciprocal network,

$$AD - BC = 1 \tag{12.6}$$

If the network is symmetrical,

$$A = D \tag{12.7}$$

For a lossless network, A and D are real, and, B and C are imaginary. The elements of the impedance matrix of a network can be obtained directly from the elements of its transmission matrix using the following relation:

$$\begin{bmatrix} Z_{11} & Z_{12} \\ Z_{21} & Z_{22} \end{bmatrix} = \frac{1}{C} \begin{bmatrix} A & AD - BC \\ 1 & D \end{bmatrix} \tag{12.8}$$

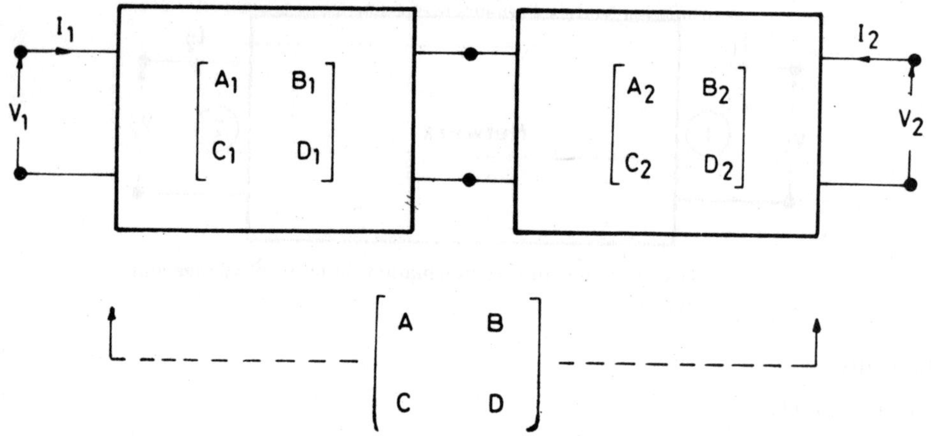

Fig. 12.2 Cascade of two-port networks.

Alternatively,

$$\begin{bmatrix} A & B \\ C & D \end{bmatrix} = \frac{1}{Z_{21}} \begin{bmatrix} Z_{11} & (Z_{11}Z_{22} - Z_{12}Z_{21}) \\ 1 & Z_{22} \end{bmatrix} \tag{12.9}$$

Table 12.1 lists the *ABCD* parameters of some of the basic circuit elements.

Scattering Matrix

The main advantage of the scattering matrix over the impedance or transmission matrix is that its parameters for any microwave network can be easily measured. The scattering parameters are defined from the wave point of view in terms of the normalized incident and reflected voltage waves at the terminals of the network. Referring to Fig. 12.3, we can write

$$\begin{bmatrix} V_1^r \\ V_2^r \end{bmatrix} = \begin{bmatrix} S_{11} & S_{12} \\ S_{21} & S_{22} \end{bmatrix} \begin{bmatrix} V_1^i \\ V_2^i \end{bmatrix} \tag{12.10}$$

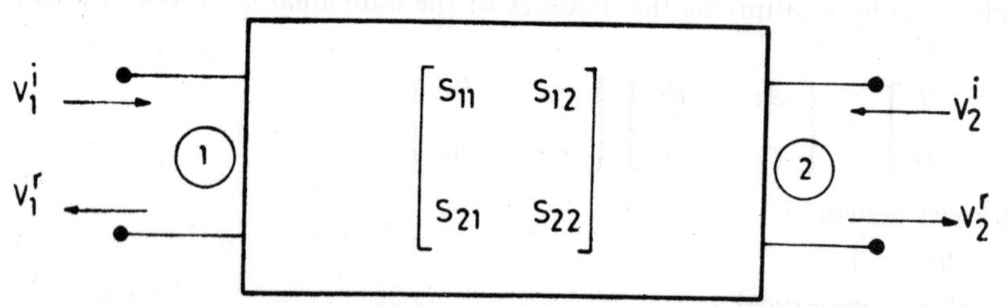

Fig. 12.3 Two-port network representation in terms of incident and reflected voltage waves.

In symbolic form,

$$[V^r] = [S][V^i] \tag{12.11}$$

For a *n*-port network, $[S]$ is given by a $n \times n$ matrix, in which S_{ii} gives the reflection coefficient at port *i*, and S_{ij} gives the transmission coefficient from port *j* to port *i*. An important feature of the scattering matrix is that, while the amplitude of its elements S_{ii} and S_{ij} are

Table 12.1 Transmission Matrix Parameters of Some Basic Circuit Elements

Network	$\begin{bmatrix} A & B \\ C & D \end{bmatrix}$
● Transmission Line Section 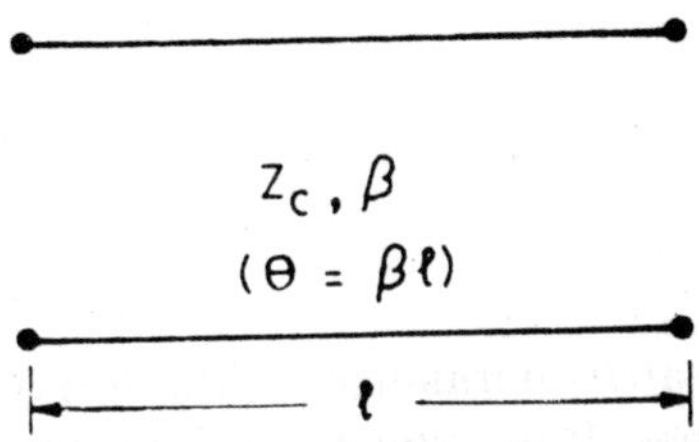 	$\begin{bmatrix} \cos\theta & jZ_c \sin\theta \\ \dfrac{j \sin\theta}{Z_c} & \cos\theta \end{bmatrix}$
● Series Impedance 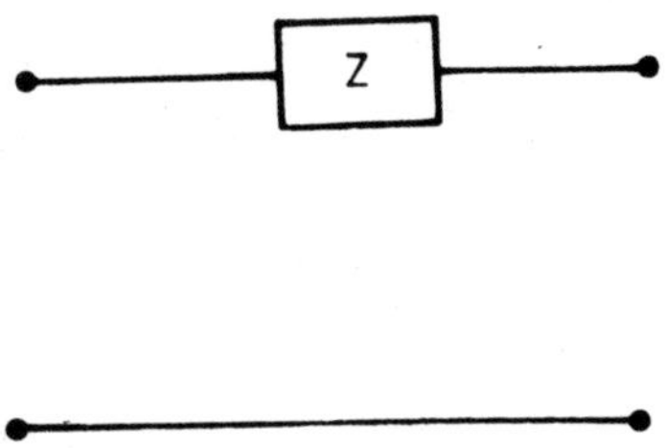 	$\begin{bmatrix} 1 & Z \\ 0 & 1 \end{bmatrix}$
● Shunt Admittance 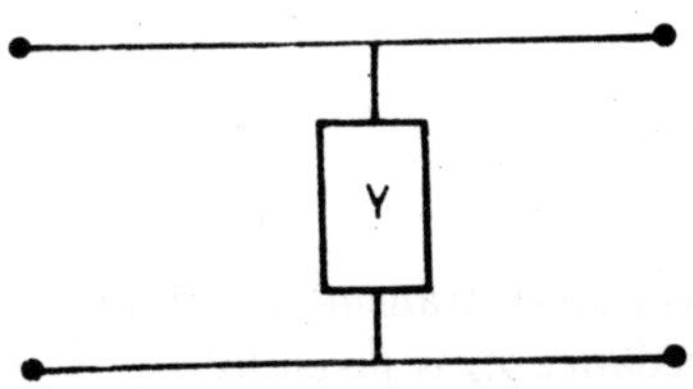 	$\begin{bmatrix} 1 & 0 \\ Y & 1 \end{bmatrix}$
● Transformer 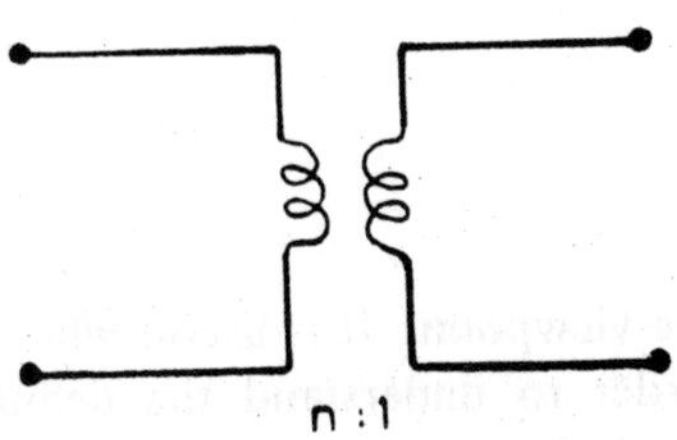 	$\begin{bmatrix} n & 0 \\ 0 & \dfrac{1}{n} \end{bmatrix}$

fixed, their phase values may be altered depending on the choice of the reference planes at the port terminals. For a reciprocal network

$$S_{ij} = S_{ji} \tag{12.12}$$

If the network is lossless,

$$[S][S]^* = [U] \tag{12.13}$$

where U is a $n \times n$ unit matrix. The scattering parameters of a network can be derived from its impedance parameters using the relation,

$$[S] = ([\bar{Z}] - [U])([\bar{Z}] + [U])^{-1} \tag{12.14a}$$

or

$$[S] = ([\bar{Z}] + [U])^{-1}([\bar{Z}] - [U]) \tag{12.14b}$$

where $[\bar{Z}]$ gives the elements of the impedance matrix normalized with respect to the characteristic impedance Z_c of the input and output lines. It is assumed that all the ports of the network are terminated in lines having the same characteristic impedance Z_c.

For a two-port network, the scattering parameters can be obtained from its transmission matrix parameters using the following relations:

$$S_{11} = \frac{\left(A + \dfrac{B}{Z_c} - CZ_c - D\right)}{\Delta} \tag{12.15a}$$

$$S_{12} = \frac{2(AD - BC)}{\Delta} \tag{12.15b}$$

$$S_{21} = \frac{2}{\Delta} \tag{12.15c}$$

$$S_{22} = \frac{\left(-A + \dfrac{B}{Z_c} - CZ_c + D\right)}{\Delta} \tag{12.15d}$$

where

$$\Delta = \left(A + \frac{B}{Z_c} + CZ_c + D\right) \tag{12.15e}$$

As an example, the scattering matrix of a uniform section of transmission line of length l having a propagation constant β and characteristic impedance Z_c is given by

$$[S] = \begin{bmatrix} 0 & e^{-j\beta l} \\ e^{-j\beta l} & 0 \end{bmatrix} \tag{12.16}$$

12.2.2 Image Parameters

Image Impedance

The concept of image matching is based on the wave viewpoint. It is useful while dealing with a chain of identical networks as in a filter. In order to understand the definition of image parameters, we consider a reciprocal unsymmetrical two-port network (Fig. 12.4).

Let Z_g be the generator impedance and Z_l the load impedance. Furthermore, let Z_{I1} be the impedance at terminal 1 with Z_l connected and Z_g disconnected, and Z_{I2} be the impedance at terminal 2 with Z_g connected and Z_l disconnected. When impedances Z_g and Z_l are adjusted such that $Z_g = Z_{I1}$ and $Z_l = Z_{I2}$, an image match is said to exist at terminals 1 and 2. Then Z_{I1} and Z_{I2} are called image impedances at ports 1 and 2. If an infinite number of identical networks of this type are cascaded such that at each junction, either terminals marked 2 are connected together or terminals marked 1 are connected, then, from the wave point of view there is no discontinuity at any of the junctions. The wave passes from one network to the next without any reflection.

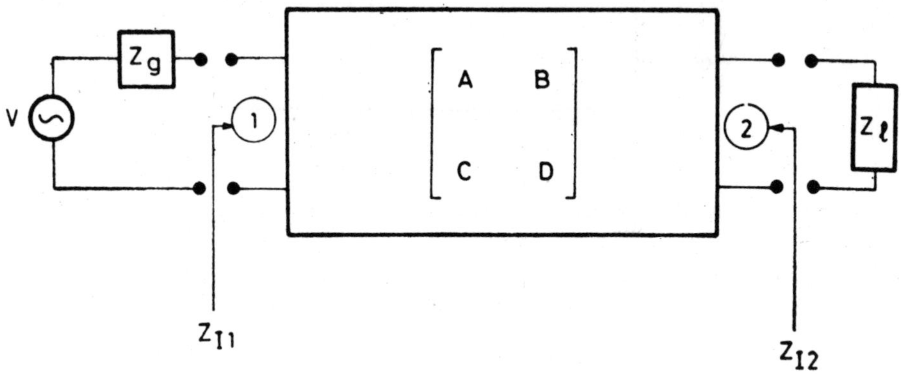

Fig. 12.4 Network for defining image impedance.

The image impedances of a two-port network can be obtained from its $ABCD$ parameters using the following relations:

$$Z_{I1} = \left(\frac{AB}{CD}\right)^{1/2} \tag{12.17a}$$

$$Z_{I2} = \left(\frac{BD}{AC}\right)^{1/2} \tag{12.17b}$$

In terms of the $[Z]$ and $[Y]$ parameters of the network, they are expressed as

$$Z_{I1} = \left[\frac{Z_{11}\Delta Z}{Z_{22}}\right]^{1/2} = \left[\frac{Y_{22}}{Y_{11}\Delta Y}\right]^{1/2} \tag{12.18a}$$

$$Z_{I2} = \left[\frac{Z_{22}\Delta Z}{Z_{11}}\right]^{1/2} = \left[\frac{Y_{11}}{Y_{22}\Delta Y}\right]^{1/2} \tag{12.18b}$$

where

$$\Delta Z = Z_{11}Z_{22} - Z_{12}^2; \qquad Z_{12} = Z_{21} \tag{12.18c}$$

$$\Delta Y = Y_{11}Y_{22} - Y_{12}^2; \qquad Y_{12} = Y_{21} \tag{12.18d}$$

Image Transfer Constant

Image transfer constant (or image propagation constant) defines transmission through the network when the terminations at the two ports match the image impedances. This situation is illustrated in Fig. 12.5. If γ denotes the image transfer constant through the network, then the voltage and current at the output terminal can be expressed in terms of the corresponding quantities at the input terminal via the relations; $V_2 = V_1 e^{-\gamma}$ and $I_2 = I_1 e^{-\gamma}$. We also have $I_1 = V_1/Z_{I1}$ and $I_2 = V_2/Z_{I2}$. Using these relations, we obtain

$$\gamma = \ln\left[\frac{V_1}{V_2}\left(\frac{Z_{I2}}{Z_{I1}}\right)^{1/2}\right] = \alpha_I + j\beta_I \tag{12.19}$$

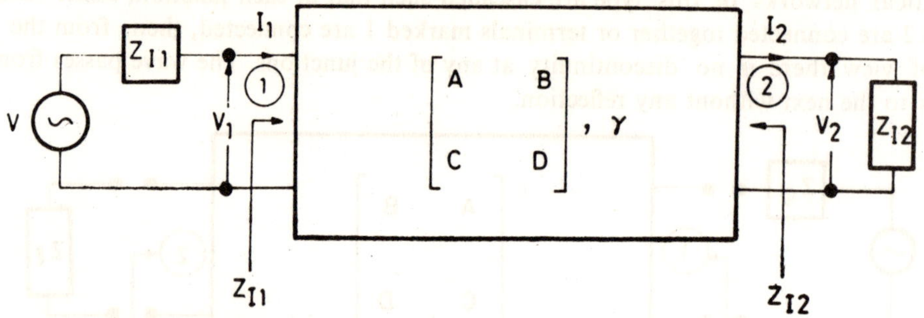

Fig. 12.5 Network for defining image transfer constant

In (12.19), α_I is the image attenuation in nepers and β_I is the image phase in radians for the two-port network. The ratio V_1/V_2 can be expressed in terms of the *ABCD* parameters as

$$\frac{V_1}{V_2} = A + \left(\frac{ABC}{D}\right)^{1/2} \tag{12.20}$$

Using (12.17) and (12.20) in (12.19), we obtain,

$$\gamma = \ln\,[(AD)^{1/2} + (BC)^{1/2}] \tag{12.21}$$

For a reciprocal network, we have $AD - BC = 1$. Using this condition, (12.21) can be expressed in the form

$$\gamma = \cosh^{-1}(AD)^{1/2} = \sinh^{-1}(BC)^{1/2} \tag{12.22a}$$

In terms of the [Z] and [Y] parameters, we can write

$$\gamma = \cosh^{-1}\left[\frac{(Z_{11}Z_{22})^{1/2}}{Z_{12}}\right] = \sinh^{-1}\left[\frac{(\Delta Z)^{1/2}}{Z_{12}}\right] \tag{12.22b}$$

$$\gamma = \cosh^{-1}\left[\frac{(Y_{11}Y_{22})^{1/2}}{Y_{12}}\right] = \sinh^{-1}\left[\frac{(\Delta Y)^{1/2}}{Y_{12}}\right] \tag{12.22c}$$

It may be noted that in the case of a uniform transmission line of length l, the image impedance is the same as its characteristic impedance, and the image transfer constant γ is equal to the propagation constant per unit length multiplied by the length l of the section.

12.2.3 General Three-port and Four-port Networks

Three-port Network

Three-port networks normally encountered in practice are *T*-junctions, circulators and two-way power dividers. From the properties of the scattering matrix, it can be shown that it is not possible to construct a perfectly matched lossless three-port reciprocal junction. The reflection coefficient of at least one of the ports must be nonzero. On the other hand, a perfectly matched, lossless, nonreciprocal three-port junction behaves like a perfect three-port circulator. Its scattering matrix has the form

$$[S] = \begin{bmatrix} 0 & S_{12} & 0 \\ 0 & 0 & S_{23} \\ S_{31} & 0 & 0 \end{bmatrix} \tag{12.23a}$$

or

$$[S] = \begin{bmatrix} 0 & 0 & S_{13} \\ S_{21} & 0 & 0 \\ 0 & S_{32} & 0 \end{bmatrix} \tag{12.23b}$$

Four-port Network

The commonly used microwave components possessing four ports are directional couplers and rat-race hybrids. Both these components possess a plane of symmetry. The scattering parameters of such symmetric four port networks can be determined by appropriate superposition of the two-port scattering parameters of the symmetric one half of the network for even- and odd-mode excitations. In order to illustrate the procedure, we consider a symmetric four-port network as shown in Fig. 12.6a. Let a voltage signal V be applied to port 1 of the network. In order to find the signals emerging out of the various ports, we consider two independent excitations (even-mode and odd-mode) of the ports 1 and 4. In the even-mode, ports 1 and 4 are excited with signals which are of equal amplitude $V/2$ and in phase, and in the odd-mode, the two signals are equal in amplitude but out of phase (Fig. 12.6b). At the plane of symmetry pp', a voltage maximum and minimum occur for the even-mode and odd-mode, respectively. For the purpose of analysis, only one half of the structure needs to be considered. For each case, the problem reduces to that of a two-port network with the line of symmetry terminated in a magnetic wall (open circuit) for the even-mode, and in an electric wall (short circuit) for the odd-mode (Fig. 12.6c). Let S_{11e} and S_{12e} represent the reflection coefficient and transmission coefficient, respectively, for the even-mode, and S_{11o} and S_{12o} represent the corresponding values for the odd-mode. By superposition, the sum of the two excitations is equal to the signal voltage V being applied to port 1. The output signals at the four ports are then the sum of the signals emerging in the even- and odd-mode excitations. They are given by

$$V_1^r = S_{11}V = \frac{1}{2}(S_{11e} + S_{11o})V \tag{12.24a}$$

$$V_2^r = S_{12}V = \frac{1}{2}(S_{12e} + S_{12o})V \tag{12.24b}$$

$$V_3^r = S_{13}V = \frac{1}{2}(S_{12e} - S_{12o})V \tag{12.24c}$$

$$V_4^r = S_{14}V = \frac{1}{2}(S_{11e} - S_{11o})V \tag{12.24d}$$

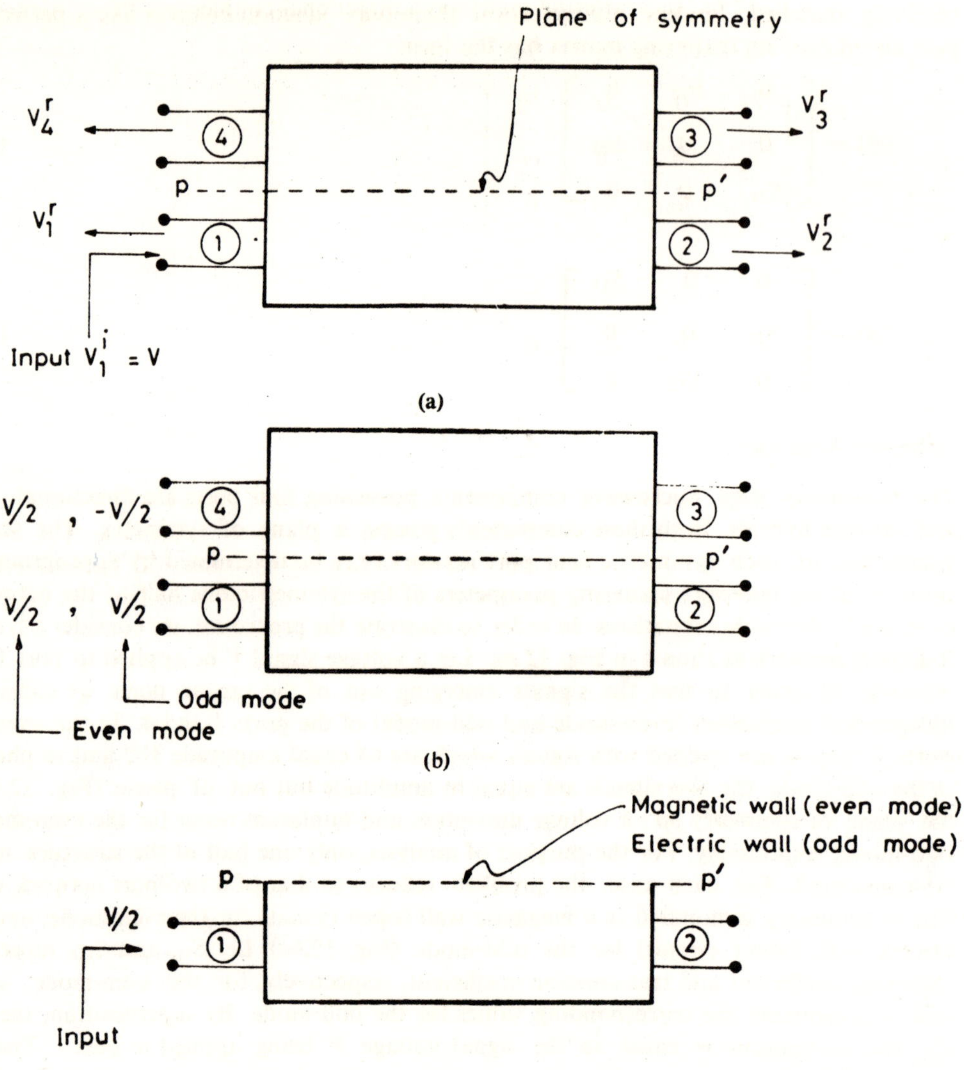

(c)

Fig. 12.6 (a) General symmetrical four-port network (b) Even- and odd-mode excitations (c) One half of the structure for the purpose of analysis.

In view of the symmetry in the structure,

$$S_{11} = S_{22} = S_{33} = S_{44} \tag{12.25a}$$

$$S_{12} = S_{21} = S_{34} = S_{43} \tag{12.25b}$$

$$S_{13} = S_{31} = S_{24} = S_{42} \tag{12.25c}$$

$$S_{14} = S_{41} = S_{23} = S_{32} \tag{12.25d}$$

A basic property of a lossless reciprocal four-port network is that if all the four ports are matched, the device behaves like a directional coupler. Referring to the four-port network shown in Fig. 12.6a, its scattering matrix as a directional coupler can be written as

$$[S] = \begin{bmatrix} 0 & c_1 & jc_2 & 0 \\ c_1 & 0 & 0 & jc_2 \\ jc_2 & 0 & 0 & c_1 \\ 0 & jc_2 & c_1 & 0 \end{bmatrix} \tag{12.26a}$$

or

$$[S] = \begin{bmatrix} 0 & c_1 & 0 & jc_2 \\ c_1 & 0 & jc_2 & 0 \\ 0 & jc_2 & 0 & c_1 \\ jc_2 & 0 & c_1 & 0 \end{bmatrix} \tag{12.26b}$$

so that $c_1^2 + c_2^2 = 1$. The scattering matrix (12.26a) is applicable for forward wave couplers such as the branch line coupler, in which case no coupling exists between ports 1 and 4, and, between ports 2 and 3. The scattering matrix (12.26b) is applicable to backward wave couplers in which no coupling exists between ports 1 and 3, and, between ports 2 and 4.

12.3 Launching Methods and Transitions

While stripline and its derivatives form the base for integrated component design, they are generally not used for transport of power or for system interconnections. The latter functions are invariably performed by coaxial lines or waveguides. Suitable transitions, therefore, are to be incorporated in any system integration. Among the class of stripline-like transmission lines dealt with in the earlier chapters, the homogeneous stripline and the suspended stripline are the commonly used media in integrated components. Transitions have been reported from these transmission lines to coaxial lines [5-10] and waveguides [11-14]. A brief description of these transitions is provided below.

12.3.1 Coaxial to Stripline Launcher

Coaxial feeds are the most commonly used in all stripline components [5-8]. Since the dominant mode in a coaxial line is TEM, it offers the smoothest transition to the stripline TEM mode. Commercially available end-launch connectors with the center conductor protruding in the form of a tab are generally used at the input and output ports of the stripline components. Figure 12.7 shows typical constructional features of two types of coaxial end-launchers feeding a stripline. The stripline is a sandwiched structure, consisting of the circuit pattern in between two dielectric sheets each of thickness equal to half the ground plane spacing. Copper-clad RT-duroid sheets are generally used. The inner surface of one of the substrates carries the circuit pattern and that of the other is bare. The conductive layers on the two outer surfaces serve as ground planes. The center tab of the coaxial connector is sandwiched between the two substrates and launches a wave onto the strip conductor. Care should be taken to ensure that there is no air gap between the two substrates near the tab. Best match over the largest bandwidth is obtained when the outer diameter of the coaxial

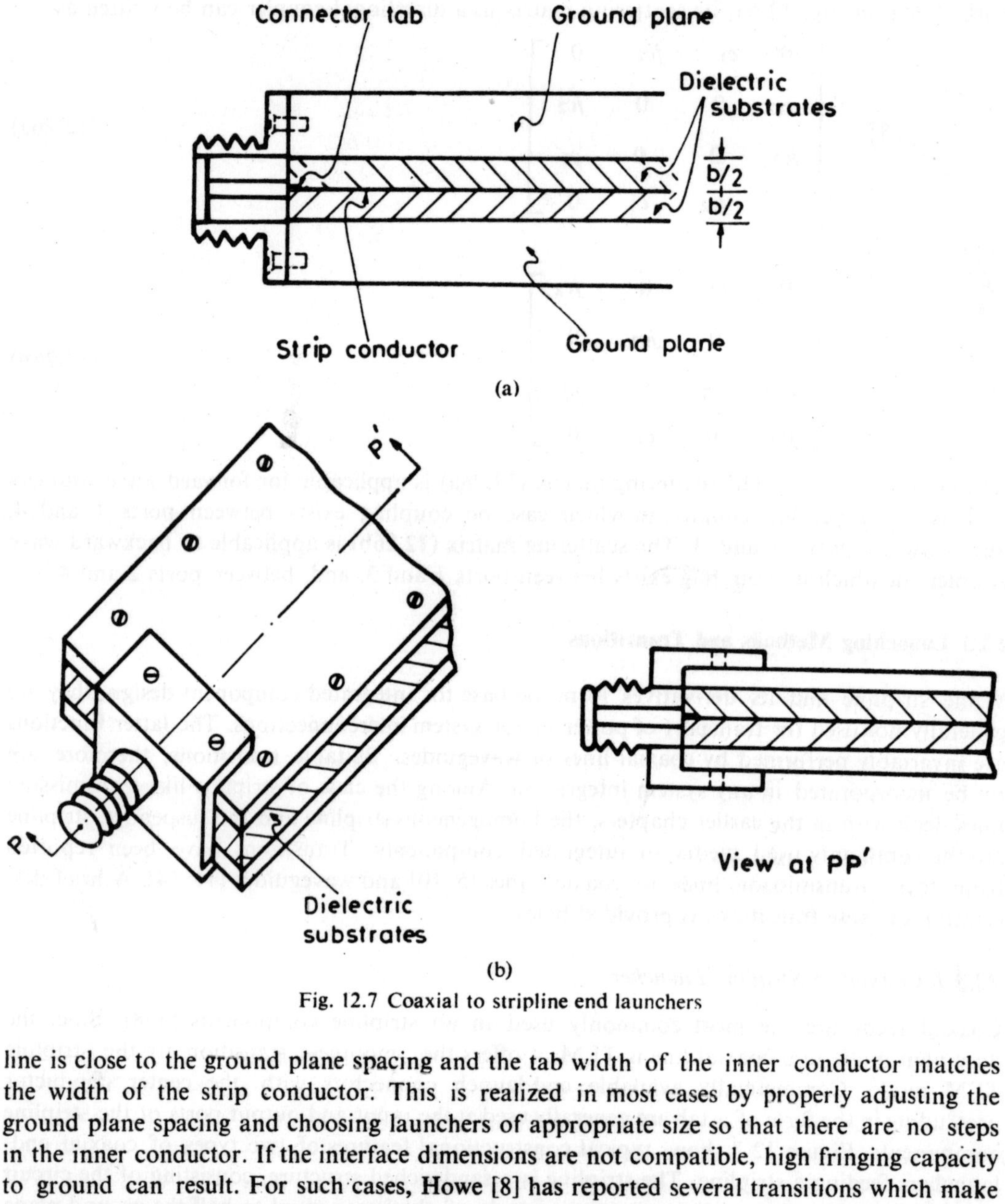

Fig. 12.7 Coaxial to stripline end launchers

line is close to the ground plane spacing and the tab width of the inner conductor matches the width of the strip conductor. This is realized in most cases by properly adjusting the ground plane spacing and choosing launchers of appropriate size so that there are no steps in the inner conductor. If the interface dimensions are not compatible, high fringing capacity to ground can result. For such cases, Howe [8] has reported several transitions which make use of an intermediate section of either a coaxial line, stripline, or slab line. For a broadband match with a minimum VSWR, the diameters of the inner and outer conductors of the launcher, and the ground plane spacing should be adjusted such that there are no steps in the inner conductor.

Levy [9] has reported a broadband coaxial to stripline transition which uses a tapered length of rectangular coaxial line. The VSWR of this transition is reported to be less than 1.02 upto 11 GHz. Pyle [10] has described a compensated butt transition between a coaxial line and an air filled shielded stripline. In this, the inner conductor of the coaxial line projects into the stripline by a short distance, which acts as a compensating section. With a compensating section of length 0.033 inches, an operating bandwidth of 3 to 12 GHz with a maximum VSWR of 1.035 has been reported by Pyle.

There are practical situations in which it is convenient to feed the stripline at right angles from the broad wall rather than from the edge of the package. The method of feeding by means of surface-launch connectors in such cases is illustrated in Fig. 12.8. In this, the center pin of the connector has a flat tip. The pin pierces through the top substrate and makes pressure contact with the strip conductor at the feed point. The connector flange is mounted in flush with the top ground plane and is held in position by means of a set of screws passing through the stripline package. These screws also serve as mode suppressors.

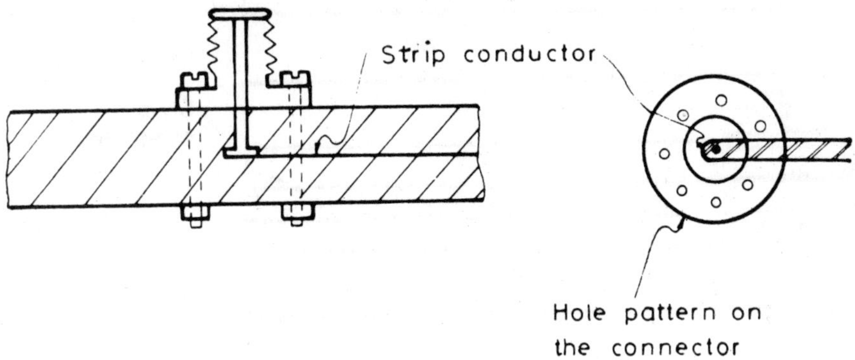

Fig. 12.8 Coaxial to stripline surface launcher

12.3.2 Coaxial to Suspended Stripline Launcher

Figure 12.9a shows the basic constructional features of a suspended stripline. The shielding box consists of two parts with grooves on the side walls to support the dielectric substrate. The substrate carrying the circuit pattern is lowered into the grooves of the lower part of the housing so as to make a firm fit. The upper part is then placed in position and the two sections are screwed together. The groove height is equal to the substrate thickness and the groove depth g is large enough ($g/b > 0.1$) to provide adequate mechanical rigidity. For small aspect ratios of the box ($c/b \leqslant 5$) and wider strip conductors ($w/b \geqslant 0.4$), the effect of grooves on the characteristic impedance Z and the guide wavelength ratio λ/λ^a, should be taken into account while designing the circuit. This aspect has been discussed in Chapter 5. As can be seen from a representative graph for $c/b = 1$ (Fig. 5.9), both Z and λ/λ^a increase with an increase in the groove depth initially and for g/b greater than about 0.1, both Z and λ/λ^a are rather insensitive to further increase in groove depth.

As in the case of homogeneous striplines, the suspended striplines can be easily fed by standard in-line coaxial connectors at microwave frequencies (Fig. 12.9b). In this case, the ground plane spacing is generally larger than the inner diameter of the outer conductor of the coaxial connector with the result the fringing field of the suspended stripline gets intercepted by the outer conductor of the coaxial line. The extra capacitance thus introduced causes mismatch of the junction and reduces the bandwidth of operation. Levy [9] has described a broadband coaxial to suspended stripline transition which incorporates a tapered

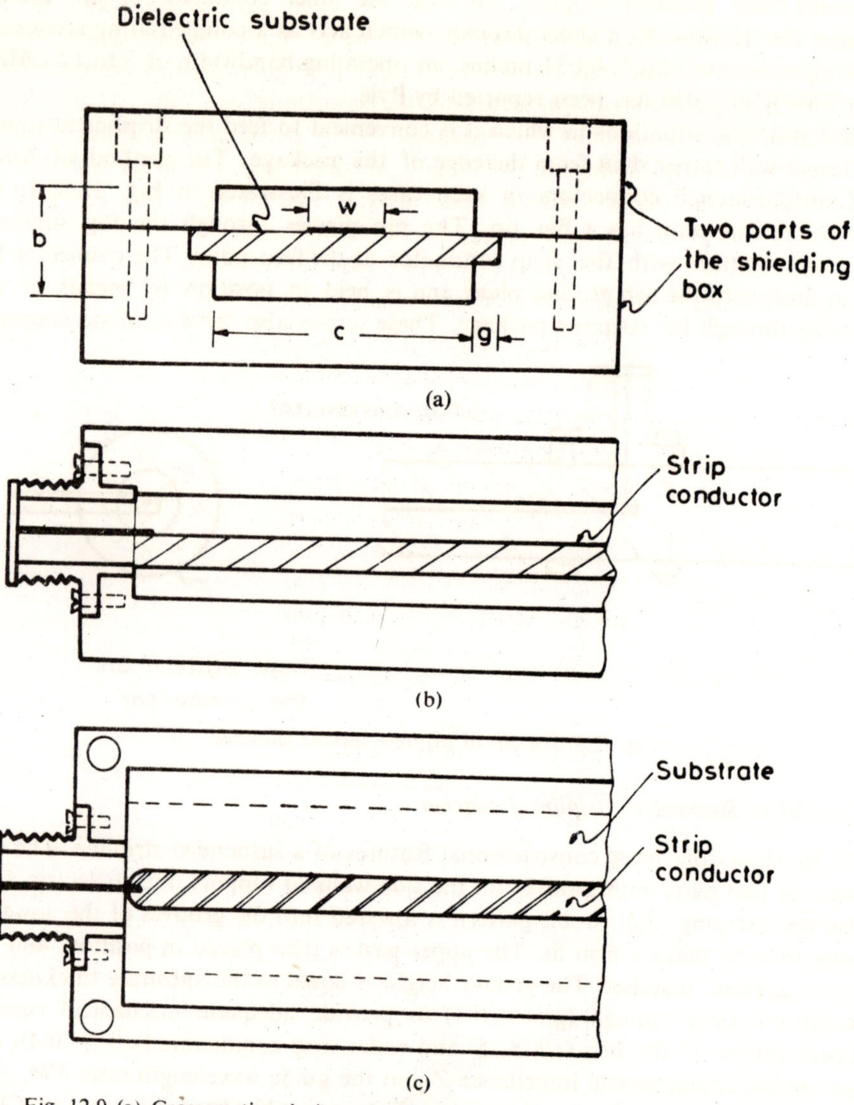

Fig. 12.9 (a) Cross-sectional view of suspended stripline (b) Coaxial to suspended stripline launcher—side cross-sectional view (c) View of the launching end with top portion of the housing removed.

length of rectangular coaxial air line. This transition is reported to have a VSWR of less than 1.02 upto 11 GHz. Another broadband transition which is much simpler in construction has been reported by Pyle [10]. It is a compensated butt transition in which the inner conductor of the coaxial line projects into the suspended stripline by a small amount. A portion of the dielectric substrate is removed to reduce the capacitance near the launching region. With this compensating section, an operating bandwidth from 4 to 11 GHz with a maximum VSWR of 1.05 has been reported by Pyle. Reduction of capacitance at the junction by slicing a portion of the dielectric is easy when soft substrates such as rexolite and RT-duroid are employed. For hard substrates made of alumina or quartz, this is a difficult proposition. Junction capacitance at the launching end can be reduced to some extent by simply tapering the strip conductor at the launching end (Fig. 12.9c). Either a curved taper or a linear taper may be provided. When miniature coaxial connectors are adapted, the center pin of the connector is generally narrower than the strip conductor width. By tapering the end of the strip conductor, better dimensional match is also achieved at the junction.

12.3.3 Waveguide to Suspended Stripline Transition

The need for a direct transition from waveguide to suspended stripline arises in integrated systems where the source is generally constructed in waveguide. Probe-type transitions have been reported [11–13], in which the dielectric substrate carrying the strip conductor of the suspended stripline is lowered into the waveguide through a slit in its broadwall. The dielectric substrate with the printed strip can be oriented perpendicular to the side walls of the rectangular waveguide as shown in Fig. 12.10 or can be placed parallel to the side walls. The probe picks up energy from the TE_{10} mode of the waveguide and launches a nearly TEM mode into the suspended stripline. Best pick up is achieved when the probe depth is nearly 0.6 times the

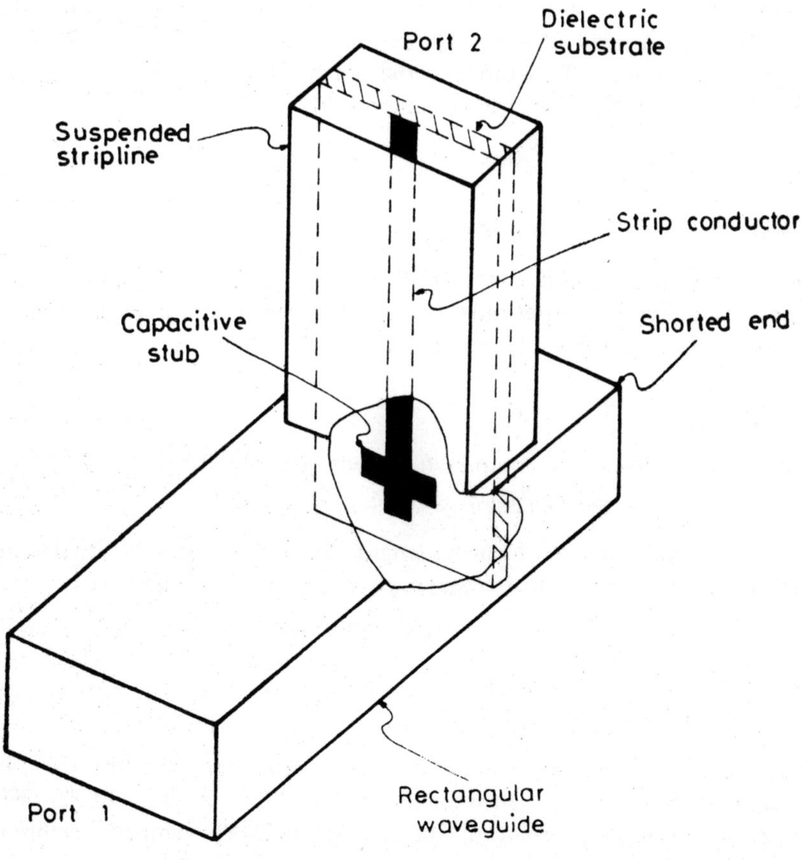

Fig. 12.10 Waveguide to suspended stripline transition

waveguide height and when the waveguide is shorted at the farther end at a distance of about one quarter guide wavelength from the strip conductor. A capacitive stub (see Fig. 12.10) serves to tune out the reactance at the feed end of the suspended stripline. In practice, the probe depth, the exact position of the waveguide short and the capacitive stub dimension are adjusted experimentally to provide the best broadband match centered around the desired operating frequency. For a transition from waveguide to suspended stripline using alumina substrate, Knerr [11] has reported a return loss greater than 30dB over a 15% bandwidth centered around 4 GHz.

A different type of fully integrated transition which is particularly attractive for millimeter wave applications is illustrated in Fig. 12.11 [14]. This transition incorporates two identical, gradually tapered ridges built on the inside of the narrow walls of the waveguide. A dielectric substrate with a printed ridge on one of the two sides is positioned in the symmetric E-plane bridging the broad walls of the guide. The printed ridge on the substrate is slowly tapered so as to complement the two ridges built on the narrow walls. Where the width of narrow wall ridges occupy half the waveguide height, the two ridges are abruptly widened to bridge the opposite broadside wall of the waveguide. At this plane, the printed ridge on the substrate is cascaded with the printed suspended substrate stripline. With this arrangement, the incident electric field of the TE_{10} mode is symmetrically split and each portion as it travels along the tapered ridge section is rotated by 90° so as to transform the field into the nearly TEM mode of the suspended stripline. The minimum length of the tapered ridge section for a smooth transition from the TE_{10} mode to the nearly TEM mode is estimated to be approximately 1.5 times the wavelength in the waveguide [14].

12.4 Basic Circuit Elements

12.4.1 Transmission Line Sections and stubs

Consider a lossless transmission line section of electrical length θ and characteristic impedance Z_c. If Z_{oc} and Z_{sc} denote the input impedance of the line when the output port is terminated in an open and a short, respectively, we can write

$$Z_{oc} = -jZ_c \cot \theta \quad (12.27a)$$

$$Z_{sc} = jZ_c \tan \theta \quad (12.27b)$$

We now introduce the Richards' complex frequency variable [15]

$$S = j \tan \theta = j \tan (\pi f/2f_0) \quad (12.28)$$

where f_0 is the real frequency at which the length of the line is exactly equal to quarter wavelength of the medium and f is a real frequency variable. Using (12.28), we can express (12.27) as

$$Z_{oc} = \frac{Z_c}{S} \quad (12.29a)$$

$$Z_{sc} = Z_c S \quad (12.29b)$$

In the S-plane, an open circuited line behaves like a capacitor of value $C = 1/Z_c$, and a short circuited line behaves like an inductor of value $L = Z_c$ (Fig. 12.12). Richards' [15] has shown that with this transformation, the lumped network synthesis techniques can be directly applied to microwave circuits.

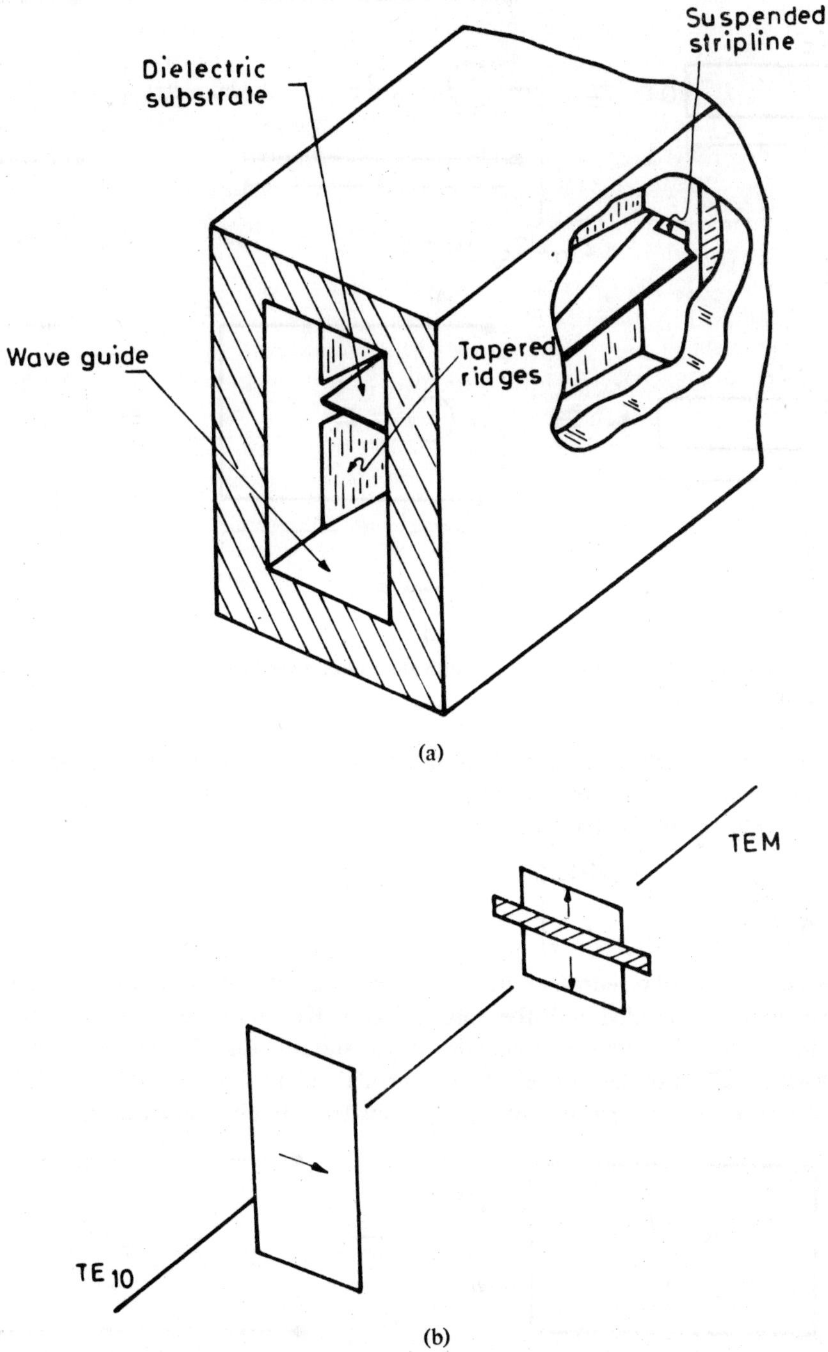

Fig. 12.11 (a) Waveguide to suspended stripline transition using tapered ridges (b) Concept of the transition—splits E field and rotates the components by 90° (From Galin [14], reprinted with permission of Microwaves and *RF*, copyright © 1983 Hayden Publishing Co.)

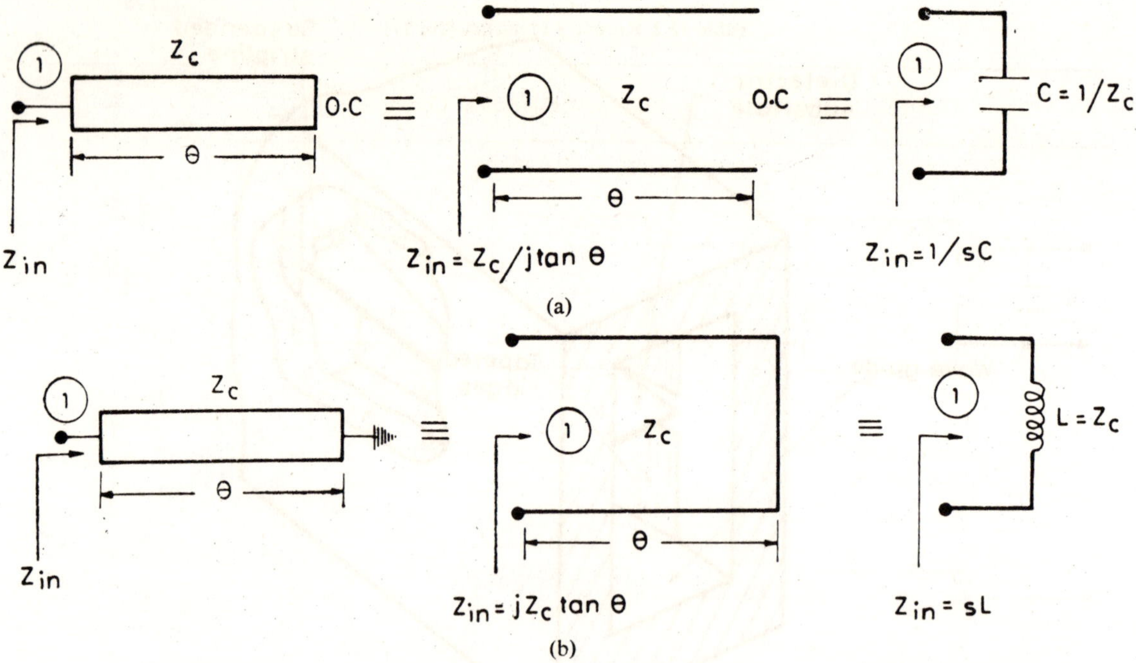

Fig. 12.12 (a) Open-circuited stripline and its equivalent circuit (b) Short-circuited stripline and its equivalent circuit

An important circuit element used in the synthesis of filter networks is the 'unit element'. The unit element with a characteristic impedance Z_c is a distributed element having the following transmisson (*ABCD*) matrix:

$$[T] = \frac{1}{\sqrt{1-S^2}} \begin{bmatrix} 1 & Z_c S \\ \frac{S}{Z_c} & 1 \end{bmatrix} \tag{12.30}$$

Figure 12.13 shows symbolic representation of a unit element. Introduction of unit elements in the lumped network coupled with the application of Kuroda's identities [16] offers considerable flexibility in translating the network to a suitable distributed circuit that can be realized in practice [17]. For ease of reference, Kuroda's identities are listed in Table 12.2. Transformation from the original circuit to the equivalent circuit is achieved by equating the

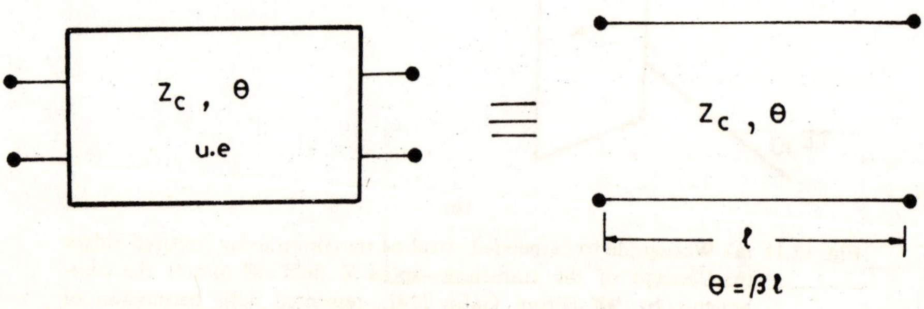

Fig. 12.13 Representation of 'unit element'

Table 12.2 Kuroda's Identities [16, 17]

Original Circuit	Equivalent Circuit	
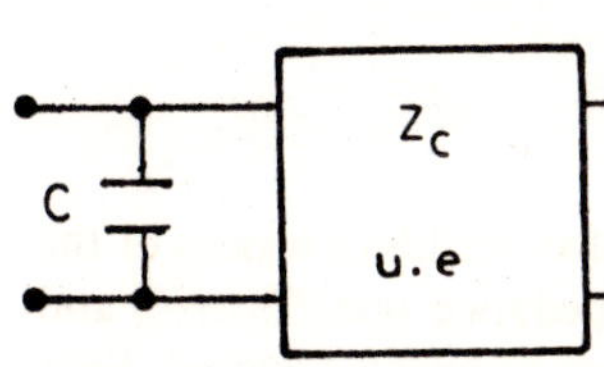	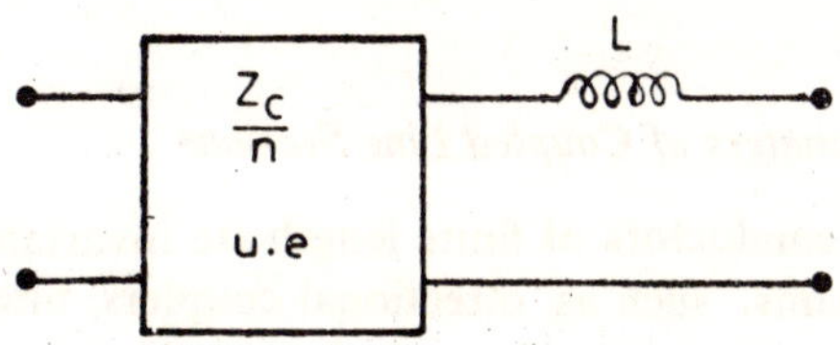	$L = \left(\frac{n-1}{n}\right) Z_c$ $n = 1 + Z_c C$
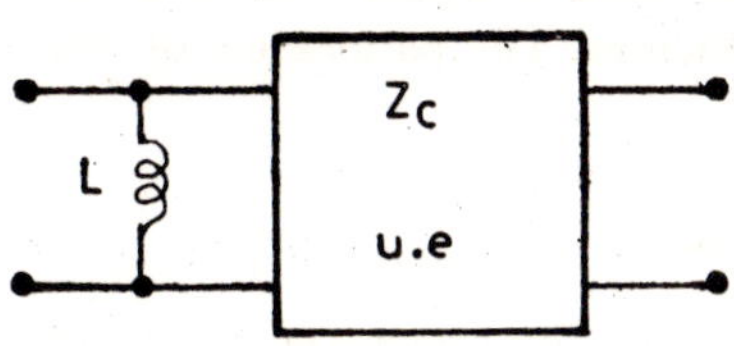	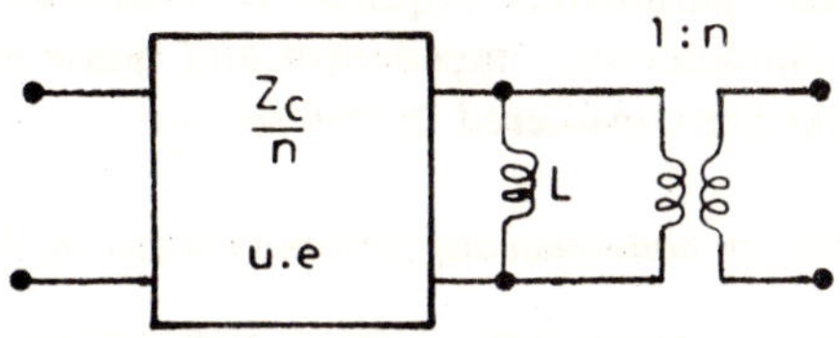	$L = \frac{Z_c}{n(n-1)}$ $n = 1 + \frac{Z_c}{L}$
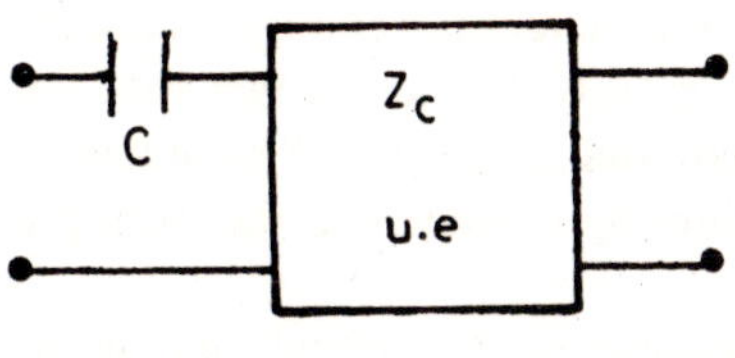	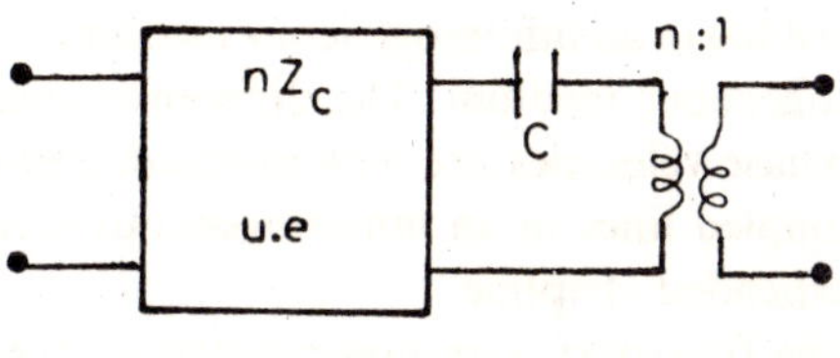	$C = \frac{1}{n(n-1)Z_c}$ $n = 1 + \frac{1}{Z_c C}$
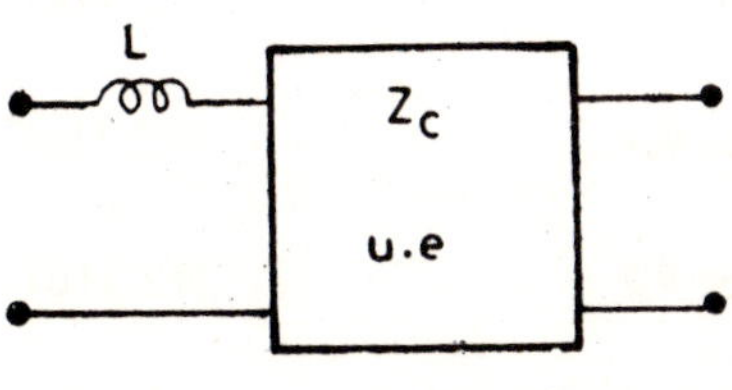	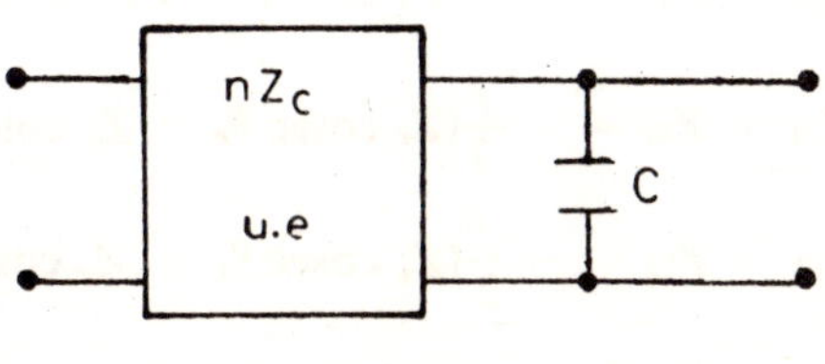	$C = \frac{(n-1)}{nZ_c}$ $n = 1 + \frac{L}{Z_c}$

overall *ABCD* matrices of the two citcuits. All the identities remain valid if the two equivalent networks are turned around by interchanging their input and output ports.

Table 12.3 illustrates some of the typical stripline circuit elements with open or short circuited stubs along with their equivalent *S*-plane circuits and also the two-wire transmission line equivalents [17]. These equivalent circuit representations have utility in the design of filters.

12.4.2 Four-Port Parameters of Coupled Line Sections

Parallel-coupled strip conductors of finite length are invariably used as building blocks in the design of several circuits, such as, directional couplers, filters, impedance transformers and phase shifters. These coupled structures may be classified as; (1) symmetrical coupled lines embedded in homogeneous or inhomogeneous medium (Fig. 12.14), and (2) asymmetrical coupled lines embedded in homogeneous or inhomogeneous medium (Fig. 12.15).

From these four-port networks, a variety of useful two-port networks can be derived by placing open- and/or short-circuits on various terminal pairs or by connecting ends of the lines together. The basic parameters required to characterize these coupled lines are the even- and odd-mode characteristic impedances and phase velocities. The parameters of the basic four-port networks are considered in this section.

Symmetrical coupled lines in homogeneous/inhomogeneous medium

The network proporties of symmetrical coupled lines embedded in a homogeneous dielectric medium have been investigated by Jones and Bolljahn [18]. The analysis makes use of the conventional two-wire transmission line voltage-current relations when the coupled lines are excited in even- and odd-modes. An inherent property of a homogeneous structure is that the even- and odd-modes propagate with the same phase velocity. This property leads to coupler designs having infinite directivity and perfect match at all frequencies. When symmetrical coupled lines are embedded in an inhomogeneous medium, the characteristics change markedly from those for a homogeneous medium. The difference arises mainly due to the fact that the even- and odd-mode phase velocities are now unequal. Zysman and Johnson [19] have analysed the symmetrical coupled lines in an inhomogeneous medium by considering the example of an edge-coupled suspended stripline.

The parameters of the four-port impedance matrix of the symmetrical coupled line in an inhomogeneous medium (Fig. 12.14a and c) are given in [19]. They are:

$$Z_{11} = Z_{22} = Z_{33} = Z_{44} = -\frac{j}{2}(Z_e \cot\theta_e + Z_o \cot\theta_o) \tag{12.31a}$$

$$Z_{12} = Z_{21} = Z_{34} = Z_{43} = -\frac{j}{2}(Z_e \cot\theta_e - Z_o \cot\theta_o) \tag{12.31b}$$

$$Z_{13} = Z_{31} = Z_{24} = Z_{42} = -\frac{j}{2}(Z_e \operatorname{cosec}\theta_e - Z_o \operatorname{cosec}\theta_o) \tag{12.31c}$$

$$Z_{14} = Z_{41} = Z_{23} = Z_{32} = -\frac{j}{2}(Z_e \operatorname{cosec}\theta_e + Z_o \operatorname{cosec}\theta_o) \tag{12.31d}$$

where Z_e and θ_e represent the characteristic impedance and electrical length, respectively, for the even-mode, and Z_o and θ_o are corresponding quantities for the odd-mode. The impe-

Table 12.3 Stripline Circuit Elements and Equivalent Circuits [17]

Stripline Circuit	Equivalent *S*-Plane Circuit	Two-Wire Transmission Line Equivalent
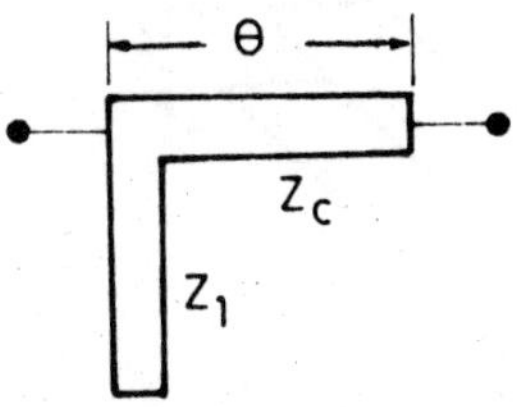	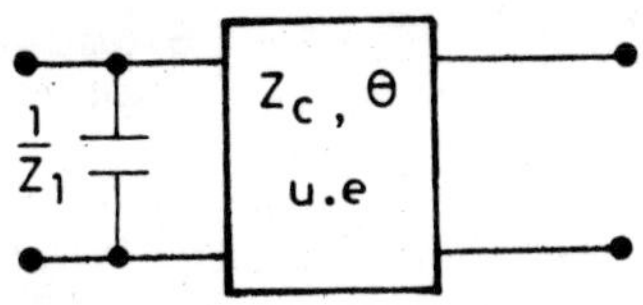	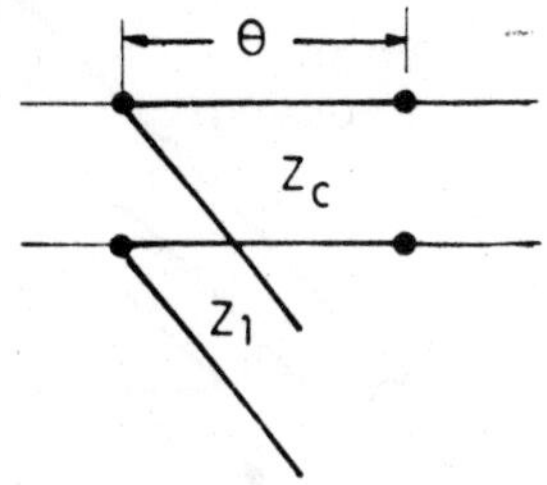
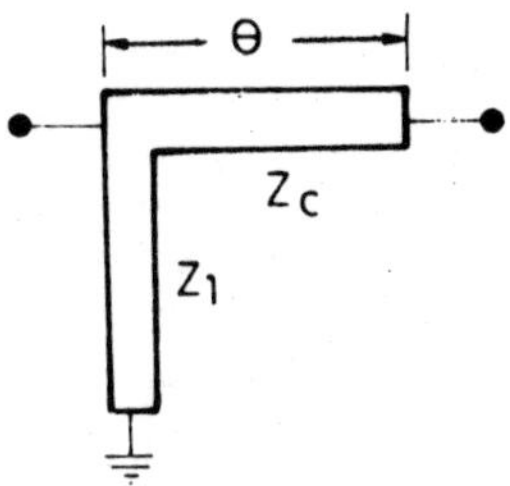	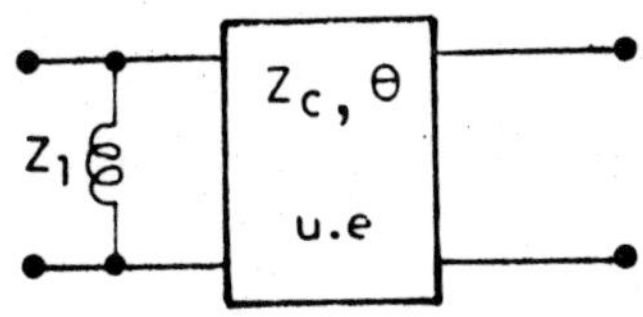	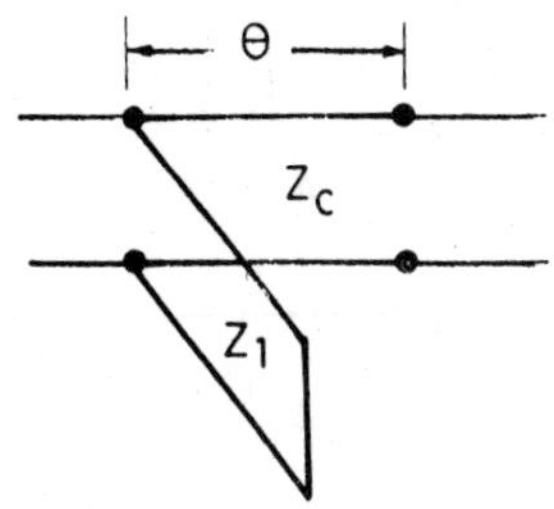
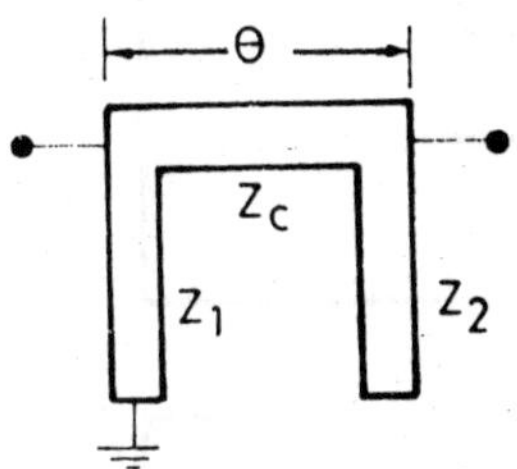	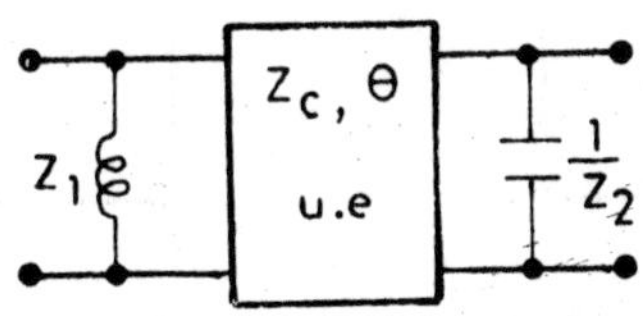	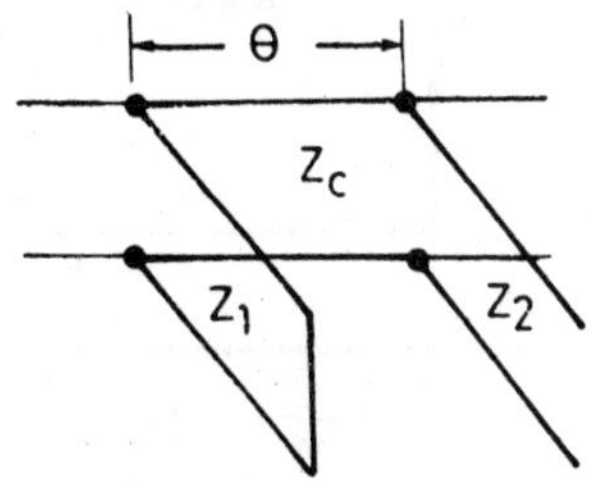
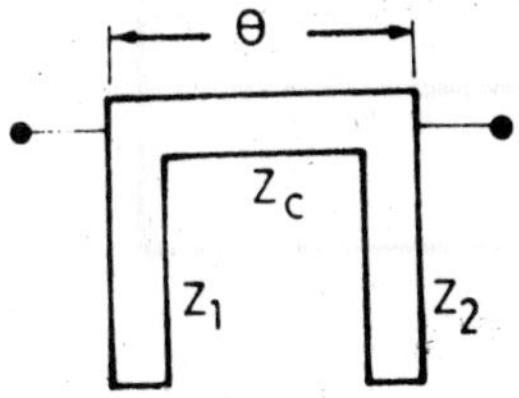	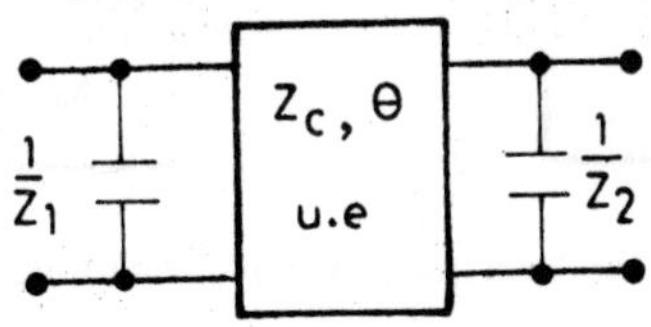	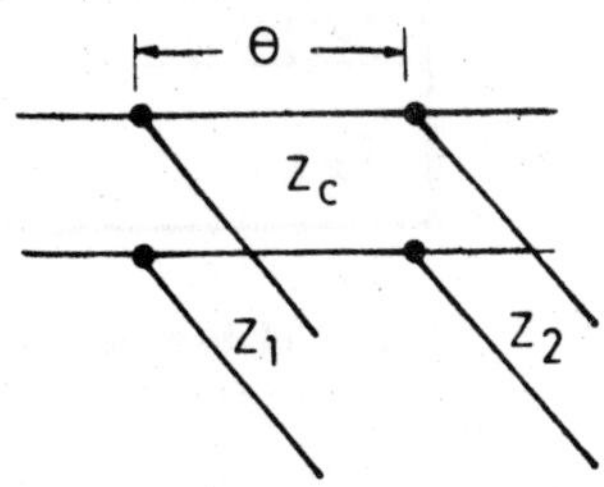
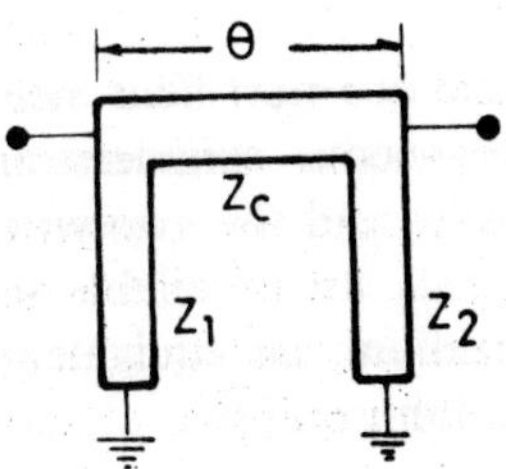	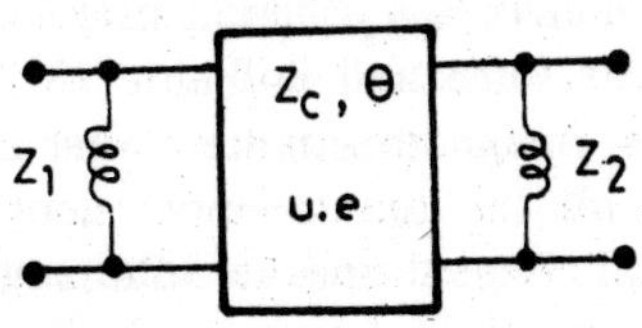	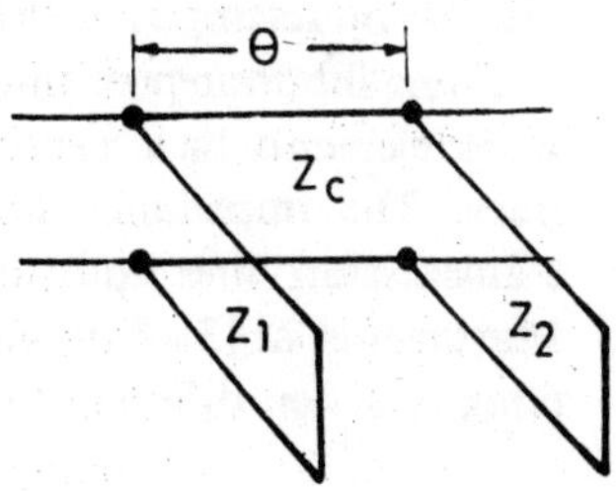

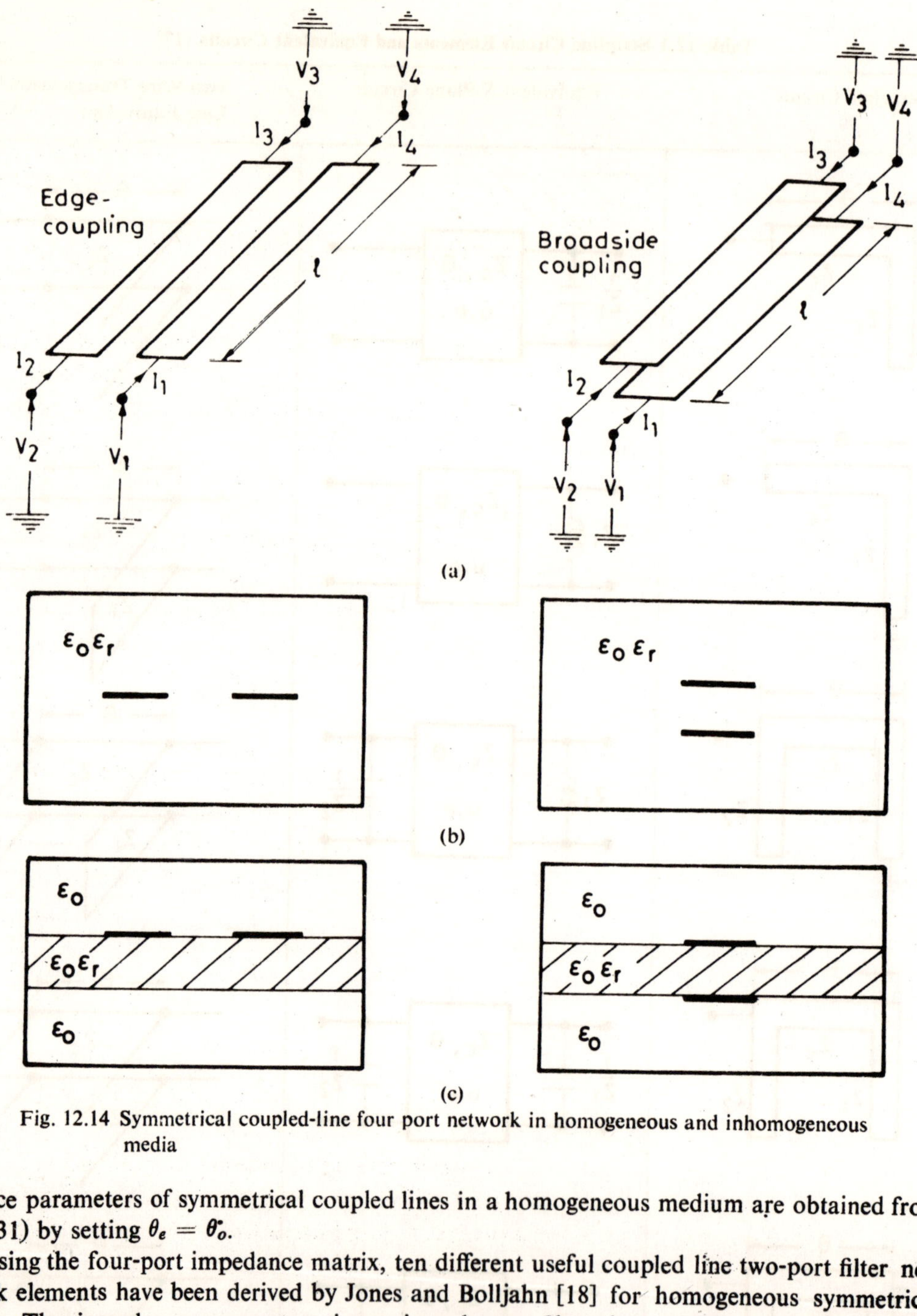

Fig. 12.14 Symmetrical coupled-line four port network in homogeneous and inhomogeneous media

dance parameters of symmetrical coupled lines in a homogeneous medium are obtained from (12.31) by setting $\theta_e = \theta_o$.

Using the four-port impedance matrix, ten different useful coupled line two-port filter network elements have been derived by Jones and Bolljahn [18] for homogeneous symmetrical lines. The impedance parameters, image impedances, filter characteristics and the two-wire transmission line equivalents of all the ten two-port network elements are available in Matthaei et al [3]. For symmetrical coupled lines in inhomogeneous medium, the equivalent circuits of ten different two-port networks are reported in Zysman and Johnson [19].

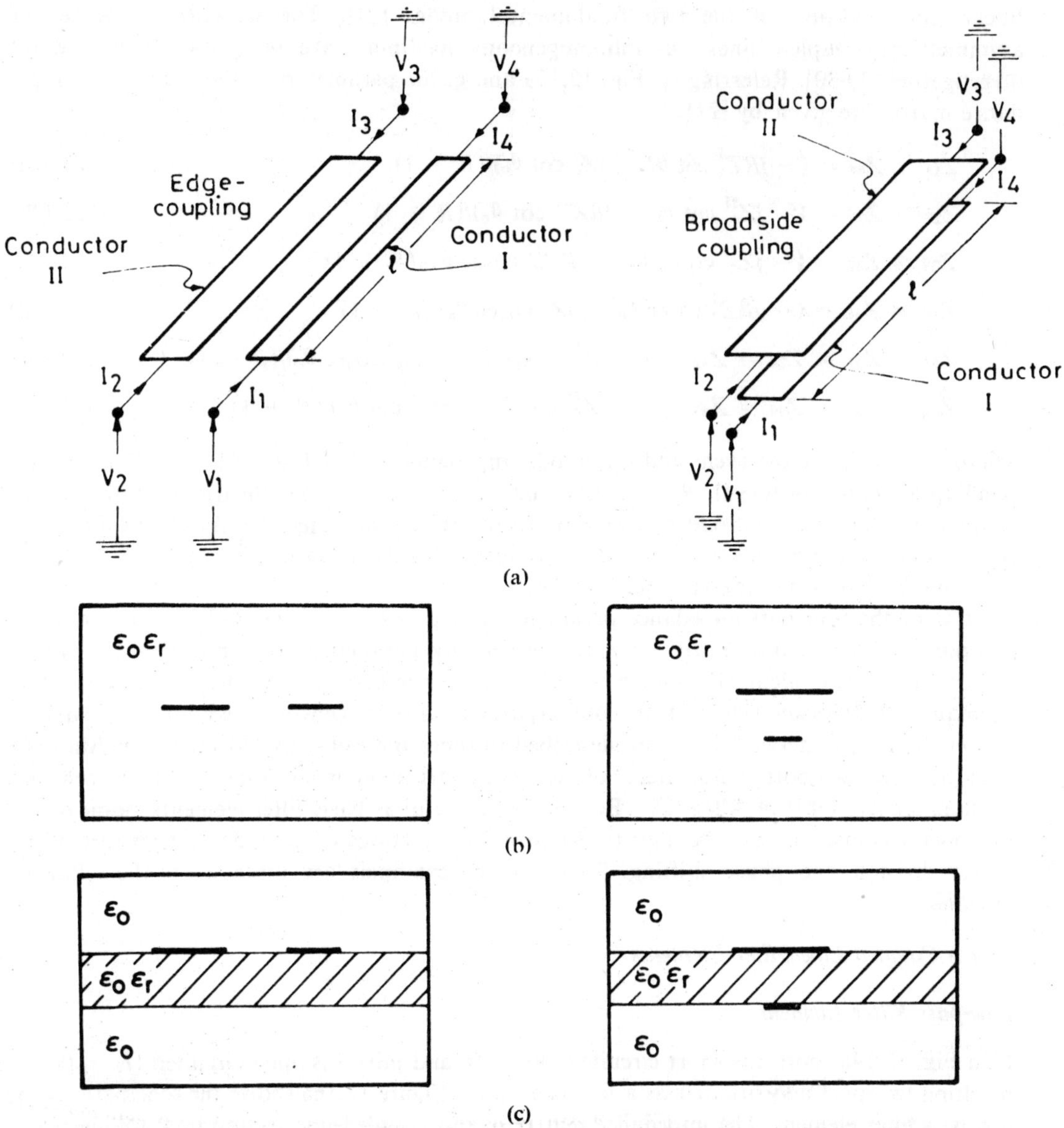

Fig. 12.15 Asymmetrical coupled-line four-port network in homogeneous and inhomogeneous media

Asymmetrical coupled lines in homogeneous/inhomogeneous medium

Asymmetrical coupled lines offer inherent impedance transforming capability. Furthermore, considerable design flexibility is introduced due to the unequal widths of the two strip conductors.

The fundamental modes propagating in asymmetrical coupled lines embedded in homogeneous or inhomogeneous medium are the equal-voltage even-mode and the equal-current odd-mode [20]. The equal voltage even- and odd-modes of symmetrical lines form simple linear combinations of the two fundamental modes [21]. The network properties of asymmetrical coupled lines in inhomogeneous medium have been studied by several investigators [20-30]. Referring to Fig. 12.15a and c, the parameters of the four port impedance matrix are given by [21].

$$Z_{11} = Z_{44} = (-jRZ_e^{\mathrm{I}} \cot \theta_e - jZ_o^{\mathrm{I}} \cot \theta_o)/(R + 1) \tag{12.32a}$$

$$Z_{22} = Z_{33} = (-jZ_e^{\mathrm{II}} \cot \theta_e - jRZ_o^{\mathrm{II}} \cot \theta_o)/(R + 1) \tag{12.32b}$$

$$Z_{23} = Z_{32} = (-jZ_e^{\mathrm{II}} \operatorname{cosec} \theta_e - jRZ_o^{\mathrm{II}} \operatorname{cosec} \theta_o)/(R + 1) \tag{12.32c}$$

$$Z_{14} = Z_{41} = (-jRZ_e^{\mathrm{I}} \operatorname{cosec} \theta_e - jZ_o^{\mathrm{I}} \operatorname{cosec} \theta_o)/(R + 1) \tag{12.32d}$$

$$Z_{13} = Z_{31} = Z_{24} = Z_{42} = (-jZ_e^{\mathrm{II}} \operatorname{cosec} \theta_e + jZ_o^{\mathrm{II}} \operatorname{cosec} \theta_o)/(R + 1) \tag{12.32e}$$

$$Z_{12} = Z_{21} = Z_{34} = Z_{43} = (-jZ_e^{\mathrm{II}} \cot \theta_e + jZ_o^{\mathrm{II}} \cot \theta_o)/(R + 1) \tag{12.32f}$$

where Z_e^{I} and Z_o^{I} are the even- and odd-mode impedances of line I and Z_e^{II} and Z_o^{II} are corresponding quantities of line II. $R = Z_e^{\mathrm{II}}/Z_e^{\mathrm{I}}$, and θ_e and θ_o are the electrical lengths of the coupled line for even- and odd-mode, respectively. It may be noted that for identical coupled conductors, the expessions given in (12.32) reduce to (12.31) if we set $Z_e^{\mathrm{I}} = Z_e^{\mathrm{II}}$ and $Z_o^{\mathrm{I}} = Z_o^{\mathrm{II}}$. If the medium is homogeneous, then $\theta_e = \theta_o$.

Utilizing the four-port impedance parameters given in (12.31) and (12.32), a variety of two-port networks can be easily derived. The network parameters of ten different two-port circuits made of identical coupled lines in an inhomogeneous medium are derived in Zysman and Johnson [19]. Their counterparts in a homogeneous medium correspond to various basic filter networks and are described in Jones and Bolljahn [18] and also in Matthaei et al [3]. The two-port parameters of non-symmetrical coupled lines in inhomogeneous, medium are available in Allen [21]. Besides being useful as basic filter elements, some of the two-port networks can also be used to perform the functions of impedance transformation, power division and phase shifting. Some of the examples are discussed in the following sections.

12.4.3 Coupled Line Filter Elements

Low-pass Filter Element

If in Fig. 12.14a, port 2 is short circuited ($V_2 = 0$) and port 3 is open circuited ($I_3 = 0$), the resulting two-port network acts as a low-pass filter. Figure 12.16a shows the schematic of the low-pass filter element. The impedance matrix of this coupled-line section in a homogeneous medium can be easily obtained as

$$[Z] = \begin{bmatrix} \dfrac{2Z_eZ_o}{(Z_e + Z_o)S} & \dfrac{2Z_eZ_o\sqrt{1 - S^2}}{(Z_e + Z_o)S} \\[2ex] \dfrac{2Z_eZ_o\sqrt{1 - S^2}}{(Z_e + Z_o)S} & \dfrac{S^2(Z_e - Z_o)^2 + 4Z_eZ_o}{2S(Z_e + Z_o)} \end{bmatrix} \tag{12.33}$$

where

$$S = j \tan \theta \tag{12.34}$$

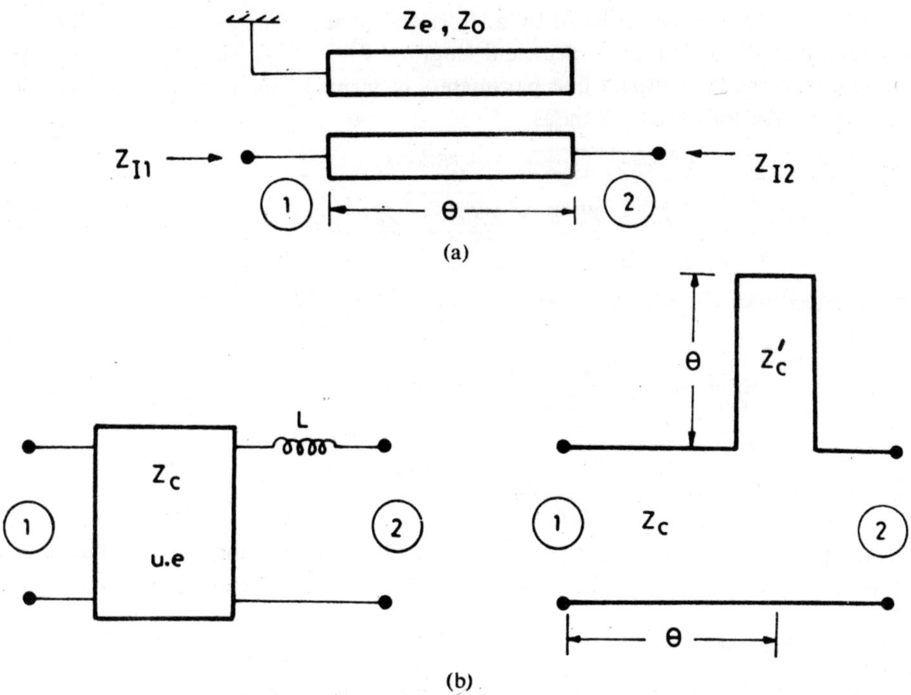

Fig. 12.16 (a) Low-pass filter element (b) Equivalent circuits
$Z_c = 2Z_eZ_o/(Z_e + Z_o)$, $L = Z_c' = (Z_e - Z_o)^2/2(Z_e + Z_o)$

Now consider a unit element cascaded to a S-plane inductor as shown in Fig. 12.16b. Its overall $ABCD$ matrix (denoted as $[M]$) is given by

$$[M] = \frac{1}{\sqrt{1 - S^2}} \begin{bmatrix} 1 & Z_cS \\ \dfrac{S}{Z_c} & 1 \end{bmatrix} \begin{bmatrix} 1 & LS \\ 0 & 1 \end{bmatrix}$$

$$= \frac{1}{\sqrt{1 - S^2}} \begin{bmatrix} 1 & (L + Z_c)S \\ \dfrac{S}{Z_c} & \left(\dfrac{S^2L}{Z_c} + 1\right) \end{bmatrix} \tag{12.35}$$

The corresponding Z-matrix is obtained as

$$[Z] = \begin{bmatrix} \dfrac{Z_c}{S} & \dfrac{Z_c\sqrt{1 - S^2}}{S} \\ \dfrac{Z_c\sqrt{1-S^2}}{S} & \dfrac{S^2L + Z_c}{S} \end{bmatrix} \tag{12.36}$$

Comparing the elements of the Z-matrices in (12.33) and (12.36) we obtain

$$Z_c = \frac{2Z_eZ_o}{Z_e + Z_o} \tag{12.37a}$$

$$L = \frac{(Z_e - Z_o)^2}{2(Z_e + Z_o)} \tag{12.37b}$$

The S-plane inductor can be replaced by a short-circuited transmission line section of characteristic impedance $Z_c' = L$ and its electrical length is $\theta = \pi f/2f_o$ (see Fig. 12.16b). We can, alternatively, express the coupled line parameters in terms of its two-wire transmission line equivalent using the following formulas:

$$Z_e = Z_c + Z_c' + \sqrt{Z_c'(Z_c + Z_c')}, \qquad Z_c' = L \tag{12.38a}$$

$$Z_o = \frac{Z_c Z_e}{2Z_e - Z_c} = \frac{Z_c + Z_c' + \sqrt{Z_c'(Z_c + Z_c')}}{1 + 2(Z_c' + \sqrt{Z_c'(Z_c + Z_c')})/Z_c} \tag{12.38b}$$

The image impedance Z_{I1} and Z_{I2}, and the image phase shift ϕ through the coupled line are given by [3].

$$Z_{I1} = \frac{2Z_e Z_o \cos\theta}{[-(Z_e - Z_o)^2 + (Z_e + Z_o)^2 \cos^2\theta]^{1/2}} \tag{12.39}$$

$$Z_{I2} = Z_e Z_o / Z_{I1} \tag{12.40}$$

$$\cos\phi = \frac{[(Z_e + Z_o)^2 \cos^2\theta - (Z_e - Z_e)^2]^{1/2}}{2\sqrt{Z_e Z_o}} \tag{12.41}$$

The variation of Z_{I1} as a function of θ shows that the coupled line section has a low-pass characteristic with its cut-off (at $Z_{I1} = \infty$) located at $\theta = \cos^{-1}((Z_e - Z_o)/(Z_e + Z_o))$.

Bandpass Filter Elements

Figures 12.17 and 12.18 show schematic diagrams of two commonly used coupled line bandpass filter elements along with their equivalent circuits. Both are symmetrical networks. The

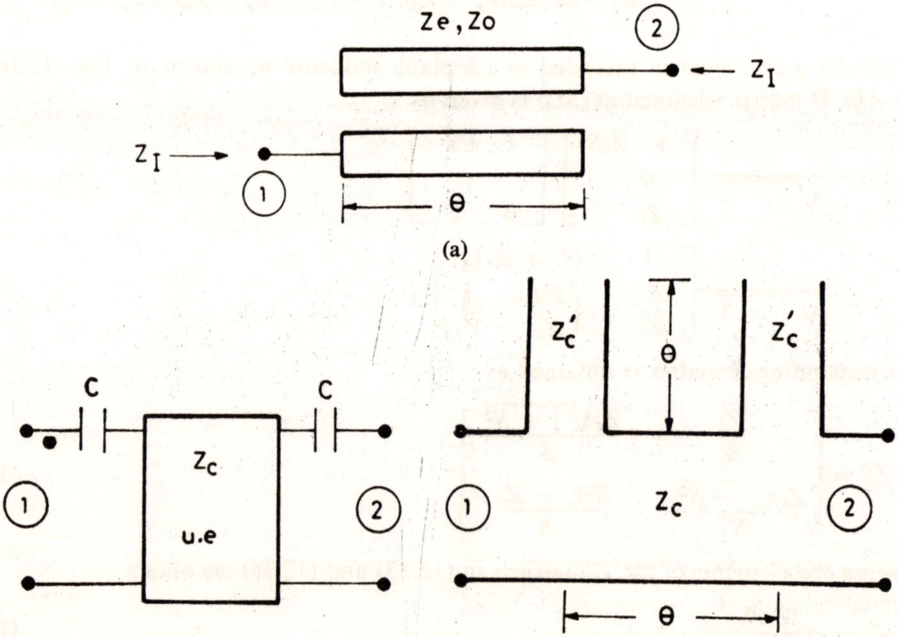

Fig. 12.17 (a) Bandpass filter element (b) Equivalent circuits

$Z_c = (Z_e - Z_o)/2$, $1/C = Z_c' = Z_o$

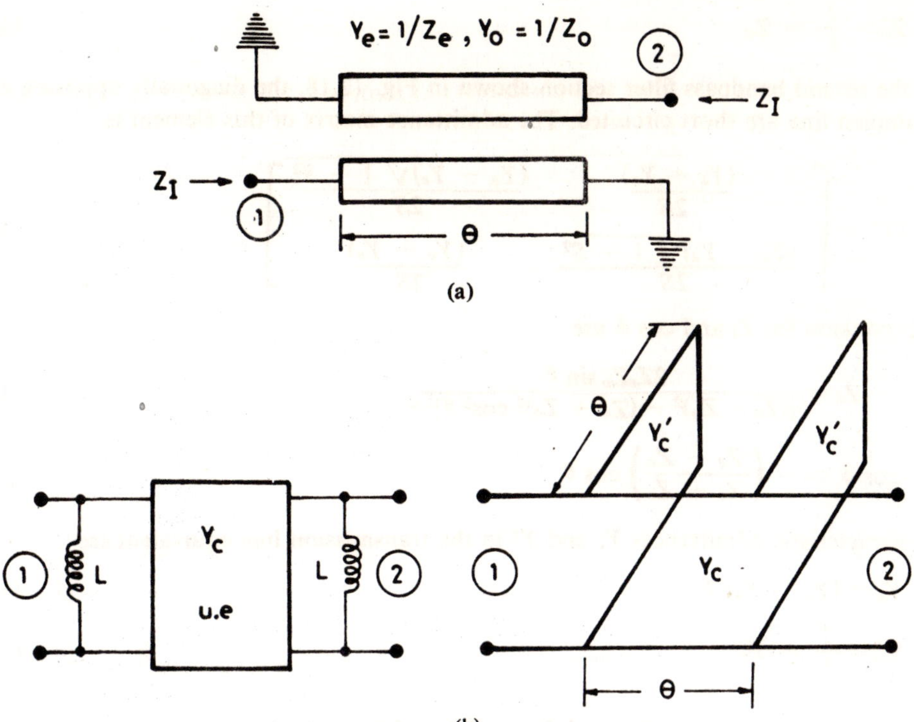

Fig. 12.18 (a) Bandpass filter element (b) Equivalent circuits
$Y_c = (Y_o - Y_e)/2,\ Y_c' = 1/L = Y_e$

procedure for deriving the equivalence between the coupled line section and the two-wire transmission line network is the same as outlined above for the low-pass filter element.

For the coupled-line section with diagonally opposing ends open circiuted (Fig. 12.17), the impedance matrix is

$$[Z] = \begin{bmatrix} \dfrac{Z_e + Z_o}{2S} & \dfrac{(Z_e - Z_o)\sqrt{1 - S^2}}{2S} \\ \dfrac{(Z_e - Z_o)\sqrt{1 - S^2}}{2S} & \dfrac{Z_e + Z_o}{2S} \end{bmatrix} \tag{12.42}$$

The image impedance Z_I and the image phase shift ϕ can be obtained from

$$Z_I = \frac{[(Z_e - Z_o)^2 - (Z_e + Z_o)^2 \cos^2 \theta]^{1/2}}{2 \sin \theta} \tag{12.43}$$

$$\cos \phi = \left(\frac{Z_e + Z_o}{Z_e - Z_o}\right) \cos \theta \tag{12.44}$$

The equivalent network parameters are related to the coupled line even- and odd-mode impedances by the following expressions:

$$Z_c = \frac{Z_e - Z_o}{2} \tag{12.45a}$$

$$Z_c' = \frac{1}{C} = Z_o \tag{12.45b}$$

In the second bandpass filter section shown in Fig. 12.18, the diagonally opposing ends of the coupled line are short circuited. The admittance matrix of this element is

$$[Y] = \begin{bmatrix} \dfrac{(Y_e + Y_o)}{2S} & \dfrac{(Y_o - Y_e)\sqrt{1 - S^2}}{2S} \\ \dfrac{(Y_o - Y_e)\sqrt{1 - S^2}}{2S} & \dfrac{(Y_e + Y_o)}{2S} \end{bmatrix} \tag{12.46}$$

The expression for Z_I and $\cos\phi$ are

$$Z_I = \frac{2Z_eZ_o \sin\theta}{[(Z_e - Z_o)^2 - (Z_e + Z_o)^2 \cos^2\theta]^{1/2}} \tag{12.47}$$

$$\cos\phi = -\left(\frac{Z_e - Z_o}{Z_e + Z_o}\right)\cos\theta \tag{12.48}$$

The characteristic admittances Y_c and Y_c' in the transmission line equivalent are

$$Y_c = (Y_o - Y_e)/2 \tag{12.49a}$$

$$Y_c' = \frac{1}{L} = Y_e \tag{12.49b}$$

12.4.4 Impedance Matching Network Using Coupled Line Section

A section of parallel-coupled transmission line can be designed to serve as a DC block [24-27], and also as an impedance transformer [28]. Impedance matching network made of coupled-line section is useful in MIC amplifiers and has the advantage that it eliminates the need for a separate DC blocking capacitor in the interstage coupling network. Figure 12.19a shows the configuration of a symmetrical coupled line section with ports 2 and 4 open-circuited, and ports 1 and 3 terminated in impedances Z_1 and Z_3 respectively. For an inhomogeneous medium, the elements of the impedance matrix Z_{11} $(= Z_{33})$ and Z_{13} $(= Z_{31})$ are given by (12.31a) and (12.31c) respectively, and equivalent circuit is shown in Fig. 12.17b. We now normalize the impedance matrix with respect to the terminating impedances Z_1 and Z_3. Denoting the normalized impedance matrix as $[\bar{Z}]$, we have

$$[\bar{Z}] = \begin{bmatrix} \sqrt{Z_1} & 0 \\ 0 & \sqrt{Z_3} \end{bmatrix}^{-1} \begin{bmatrix} Z_{11} & Z_{13} \\ Z_{13} & Z_{11} \end{bmatrix} \begin{bmatrix} 1/\sqrt{Z_1} & 0 \\ 0 & 1/\sqrt{Z_3} \end{bmatrix}$$

$$= \frac{1}{\sqrt{Z_1Z_3}} \begin{bmatrix} Z_{11}\sqrt{Z_3/Z_1} & Z_{13} \\ Z_{13} & Z_{11}\sqrt{Z_1/Z_3} \end{bmatrix} \tag{12.50}$$

The scattering matrix $[S]$ of the network can now be obtained by substituting (12.50) in (12.14). The transmission coefficient S_{31} is then given by

$$S_{31} = \frac{2Z_{13}\sqrt{Z_1Z_3}}{[(Z_{11} + Z_1)(Z_{11} + Z_3) - Z_{13}^2]} \tag{12.51}$$

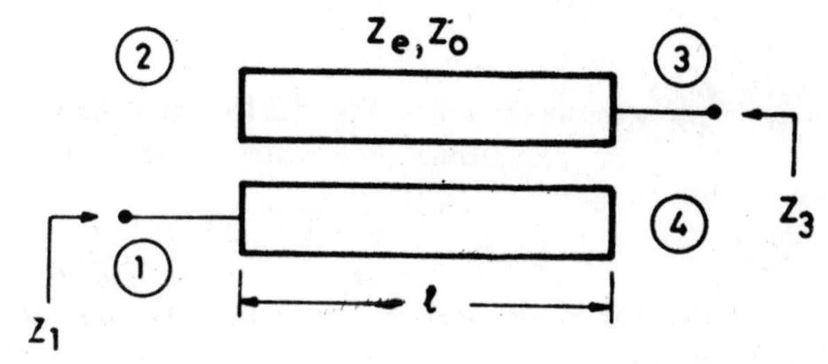

(a)

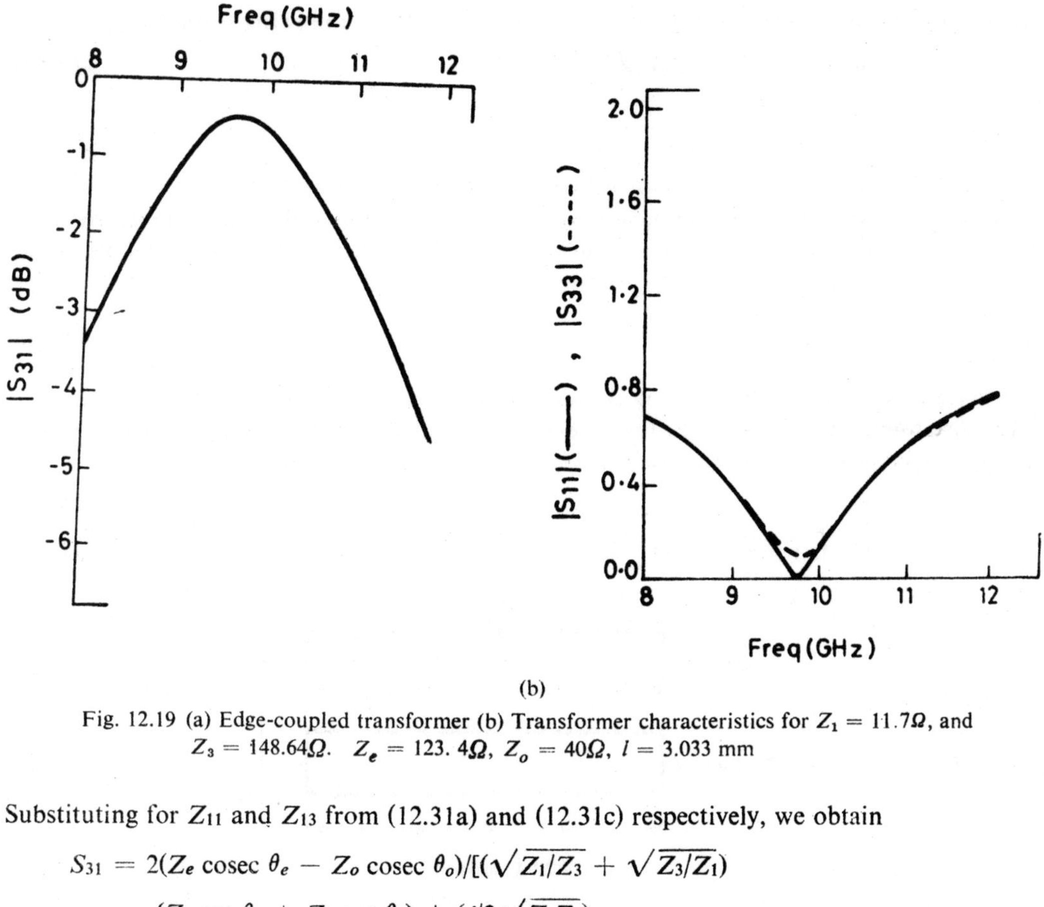

(b)

Fig. 12.19 (a) Edge-coupled transformer (b) Transformer characteristics for $Z_1 = 11.7\Omega$, and $Z_3 = 148.64\Omega$. $Z_e = 123.4\Omega$, $Z_o = 40\Omega$, $l = 3.033$ mm

Substituting for Z_{11} and Z_{13} from (12.31a) and (12.31c) respectively, we obtain

$$\begin{aligned} S_{31} = \; & 2(Z_e \operatorname{cosec} \theta_e - Z_o \operatorname{cosec} \theta_o)/[(\sqrt{Z_1/Z_3} + \sqrt{Z_3/Z_1}) \\ & \cdot (Z_e \cot \theta_e + Z_o \cot \theta_o) + (j/2\sqrt{Z_1 Z_3}) \\ & \cdot \{(Z_e \operatorname{cosec} \theta_e - Z_o \operatorname{cosec} \theta_o)^2 - (Z_e \cot \theta_e + Z_o \cot \theta_o)^2 + 4Z_1Z_3\}] \end{aligned} \tag{12.52}$$

In the design of the DC block or impedance transformers, the quantity of interest is $|S_{31}|^2$. The parameters of the network are chosen so as to give the desired transmission response about the operating frequency.

As an example, consider the design of an impedance transformer using a coupled-line section in a homogeneous medium. Let the length of the coupled line be quarter wavelength at the operating frequency. We then have $\theta_e = \theta_o = \theta = \pi/2$. From (12.52), S_{31} at $\theta = \pi/2$

reduces to

$$S_{31} = \frac{4(Z_e - Z_o)\sqrt{Z_1 Z_3}}{j[(Z_e - Z_o)^2 + 4Z_1 Z_3]} \tag{12.53}$$

Setting $|S_{31}| = 1$, we obtain

$$Z_e = Z_o + 2\sqrt{Z_1 Z_3} \tag{12.54}$$

The selection of a suitable value for Z_o depends on three factors—namely, bandwidth, insertion loss and physical realizability.

If the medium is inhomogeneous, than $\theta_e \neq \theta_o$. An approximate value of the physical length l of the coupled line section can be calculated using the following formula:

$$l = \pi/2[K_+ + K_-(Z_e - Z_o)/(Z_e + Z_o)] \tag{12.55}$$

where

$$K_\pm = (\beta_e \pm \beta_o)/2 \tag{12.56}$$

and β_e and β_o are even- and odd-mode propagation constants, respectively.

Figure 12.19b shows the typical measured characteristics of a coupled-line transformer in microstrip configuration employing alumina substrate ($\epsilon_r = 9.6$) of thickness 0.64 mm. The transformer was designed using the approximate formulas (12.54) and (12.55).

12.4.5 *C*-Section

Microwave *C*-sections are obtained by connecting together ports 3 and 4 of Figs. 12.14a and 12.15a. In a homogeneous medium, the *C*-section (Fig. 12.20) behaves as an all-pass network. Its input impedance (Z_{in}) and the transmission phase shift (ϕ) are determined from

$$Z_{in} = \sqrt{Z_e Z_o} \tag{12.57}$$

$$\cos\phi = (R - \tan^2\theta)/(R + \tan^2\theta), \qquad R = Z_e/Z_o \tag{12.58}$$

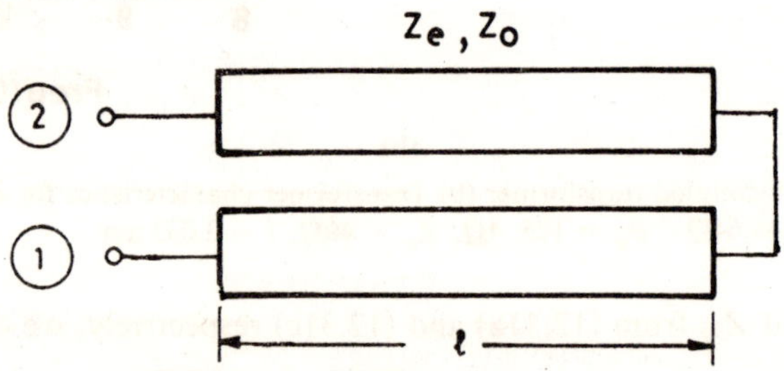

Fig. 12.20 *C*-section

Typical dispersive phase response of a *C*-section in a homogeneous medium is displayed in Fig. 12.21. Schiffman [29] observed that by switching between a *C*-section and a uniform non-dispersive line of suitably chosen length, a nearly constant phase shift can be achieved over a wide bandwidth. For example, an octave-wide 90° phase shifter with a maximum phase error of $\pm$ 2.8° can be designed using a coupled section with a ratio of $R = 2.73$ and a reference line of length $\frac{3\lambda}{4}$ at the center frequency [29]. Schiffman phase shift sections of this type are used in switched line PIN-diode phase shifters to achieve broadband performance.

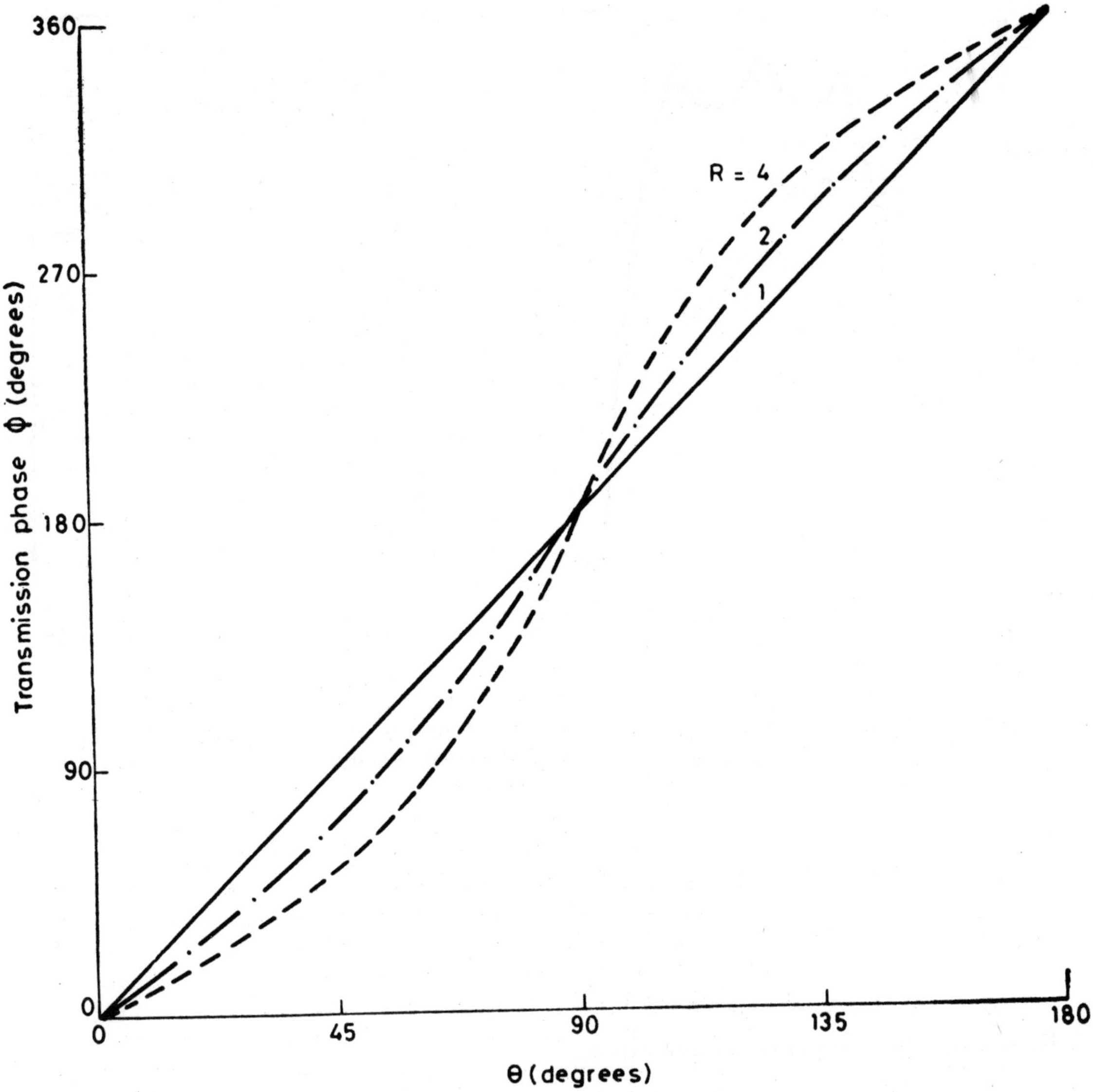

Fig. 12.21 Transmission phase response of homogeneous C-section

The inhomogeneous counterpart of the C-section resonates at some frequencies creating a stopband. When the even- and odd-mode phase velocity ratio ($k = v_e/v_o$) is larger, the C-section behaves like an excellent bandpass network with equiripple passband and stop-band response. Figure 12.22 shows the typical response of a single C-section designed in inhomogeneous broadside-coupled suspended stripline configuration having a velocity ratio $k = 2$ [30]. The design formulas for C-section with $k = 2$ used in [30] are outlined below:

The zeros of attenuation are located at $\theta_e = 0°, 180°, 360°, \ldots$ for $M = (Z_c^2/Z_e Z_o) < 2$, where Z_c is the characteristic impedance of input and output lines. For $M \geqslant 2$, additional

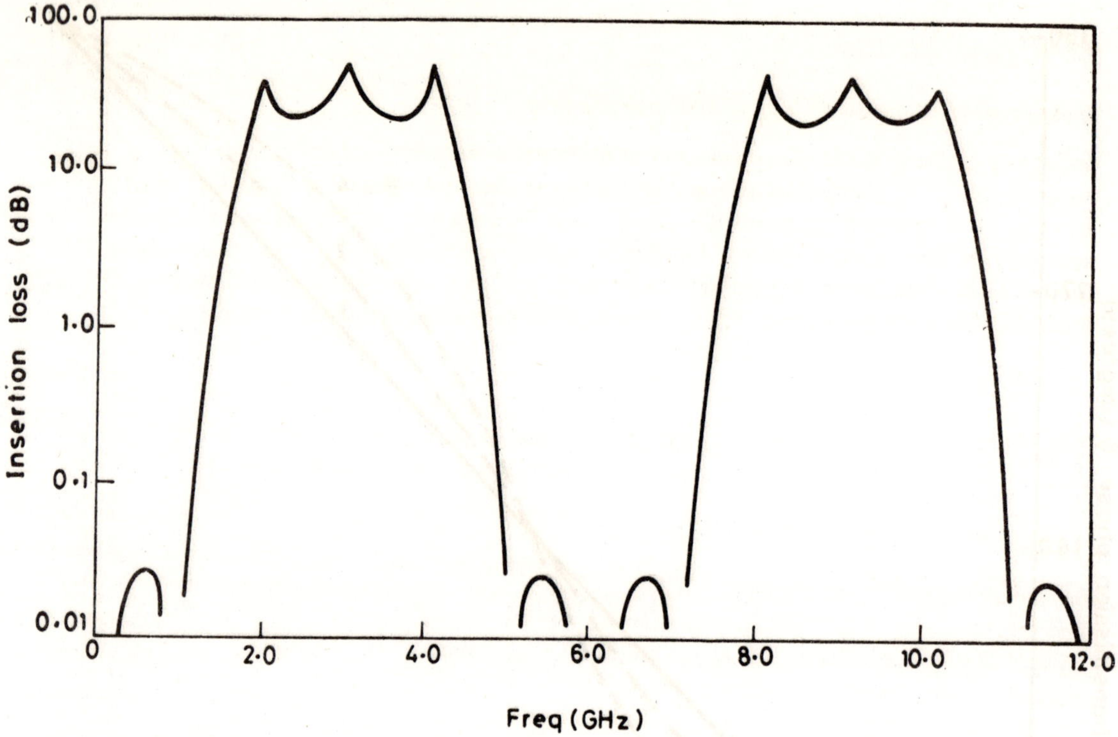

Fig. 12.22 Response of an inhomogeneous C-section $k = 2$, $M = 3$, $R = 3$, $v_e = 2.28\times10^{10}$ cm/sec, $l = 1.875$ cm, $Z = 50\,\Omega$ (From Allen [30], copyright © 1975 IEEE, reprinted with permission)

zeros occur at

$$\theta_e = \sin^{-1}(\pm\sqrt{(M-2)/(2M-2)}) \tag{12.59}$$

The poles of attenuation are located at $\theta_e = 90°, 270°, \ldots$ for $R = Z_e/Z_o < 2$. For $R \geqslant 2$, additional poles occur at

$$\theta_e = \cos^{-1}(\pm\sqrt{(R-2)/(2R-2)}) \tag{12.60}$$

Relative maxima and minima are located at

$$\theta_e = \sin^{-1}(\pm\sqrt{x}) \tag{12.61}$$

where

$$x = (-B \pm \sqrt{B^2-4AC})/2A, \; x \text{ positive and} \leqslant 1 \tag{12.62a}$$

$$A = 4(RM-1) \tag{12.62b}$$

$$B = 2(R-M) + 4(1-RM) \tag{12.62c}$$

$$C = R(M-2) \tag{12.62d}$$

The value of attenuation at each relative maximum or minimum is given by

$$L(dB) = 10\log_{10}\left[1 + \frac{R}{M}\left\{\left(\frac{x}{1-x}\right)\left(\frac{2x-2Mx+M-2}{2x-2xR+R}\right)\right\}^2\right] \tag{12.63}$$

12.4.6 Lumped Elements [31—37]

Inductors and Capacitors Using Distributed Elements

Inductors and capacitors for operation at microwave frequencies can be realized using short sections of TEM transmission line, typically less than $\lambda/4$ where λ is the guide wavelength. In order to illustrate this we consider the π—and T-equivalent networks of a transmission line section (see Fig. 12.23a and b). Inspection of the π—network shows that a short length of line of high characteristic impedance Z_c behaves predominantly like a series inductance. The length l_L of the inductive line (Fig. 12.23c) is related to the inductance L by the formula

$$l_L = \frac{\lambda}{2\pi} \sin^{-1}\left(\frac{\omega L}{Z_c}\right) \tag{12.64}$$

The value of the shunt susceptance ωC_p is very small. C_p can be considered as the capacitive end correction of the inductive line and can be obtained using the expression

$$C_p = \frac{1}{\omega Z_c} \tan\left(\frac{\pi l_L}{\lambda}\right) \tag{12.65}$$

Similarly from the T-equivalent network, it can be seen that a short length of low impedance line terminated at either end by a relatively high impedance has the effect of shunt capacitance. This is realized by having a wide rectangular patch as shown in Fig. 12.23d. The length l_C of the rectangular patch is related to the capacitance C by the formula

$$l_C = \frac{\lambda}{2\pi} \sin^{-1}(\omega C Z_c) \tag{12.66}$$

The end inductances L_s can be obtained using the formula

$$L_s = \frac{Z_c}{\omega} \tan\left(\frac{\pi l_C}{\lambda}\right) \tag{12.67}$$

The value of this inductance is generally so small that it can be neglected. Such short sections of high impedance and low impedance strip conductors are alternatively cascaded to realize series inductances and shunt capacitances, respectively, in stripline low-pass filters.

There are several filter configurations which require the use of series capacitance. As discussed in Chapter 10, a narrow gap in a strip conductor acts as a series capacitance. Higher values of capacitances can be achieved in suspended stripline by overlapping broad-side-coupled strips as shown in Fig. 12.24. The value of the capacitance depends on the degree of overlap.

Interdigital Capacitors

Lumped capacitors with capacitance value in the approximate range 0.1 to 10 pF may be realized using planar interdigital geometry as shown in Fig. 12.25. The capacitor is formed by the fringing electric field between the interdigital gap. In order for the interdigital capacitor to remain lumped at microwave frequencies, its overall length should be small and its gap width g should be about 5 microns or less. The Q-factor is dependent on the width to gap ratio of the fingers. By maximizing the width w, Q-factors in excess of 400 can be achieved. Care should be taken to ensure that there is no transverse resonance in the structure when a large number of fingers are used.

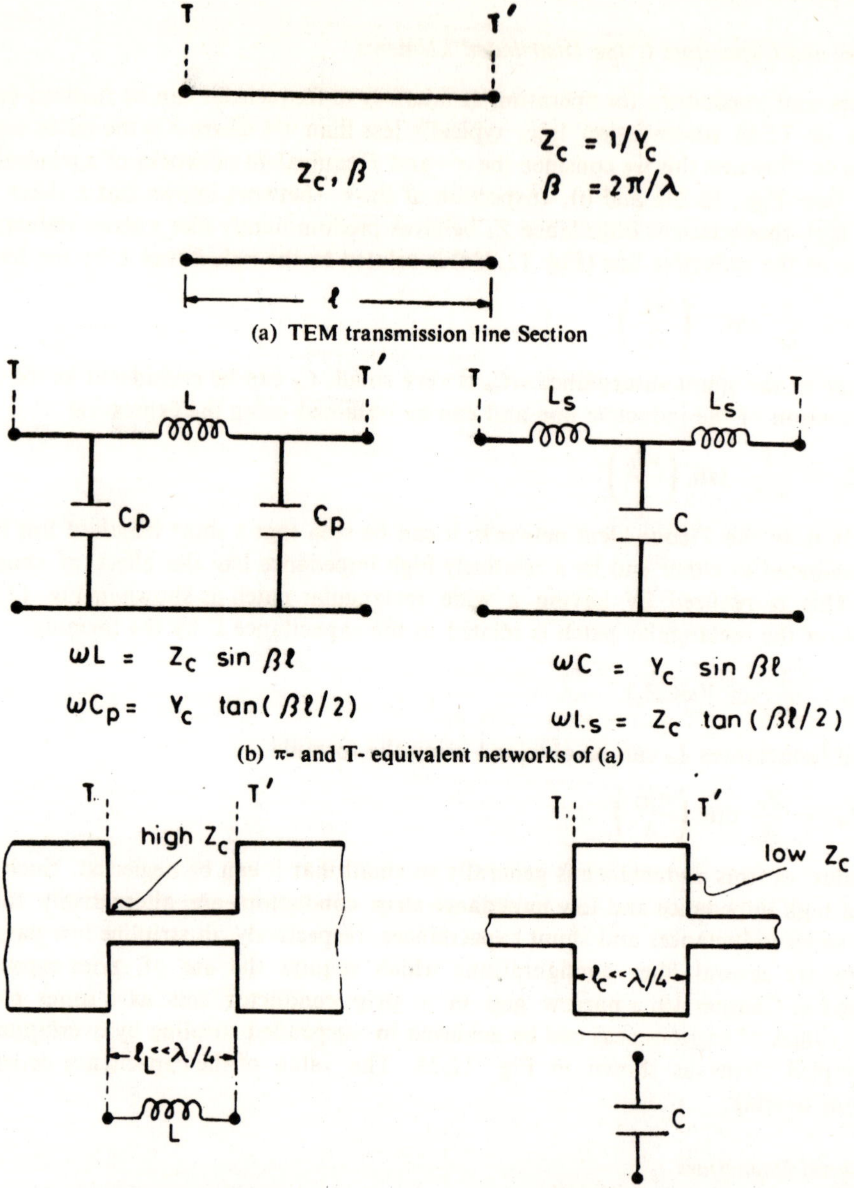

(a) TEM transmission line Section

(b) π- and T- equivalent networks of (a)

(c) Series inductance

(d) Shunt capacitance

Fig. 12.23 Realization of lumped elements in stripline

The analysis of interdigital capacitor has been reported by Alley [35]. If the ground plane effects are negligible, then the capacitance between two sets of digits can be obtained from the odd-mode capacitance formula of coupled lines. The expression for the capacitance for the special case of $w = g$ is given by [35].

$$C = l(\epsilon_r + 1)[(n-3)A_1 + A_2]pF \tag{12.68}$$

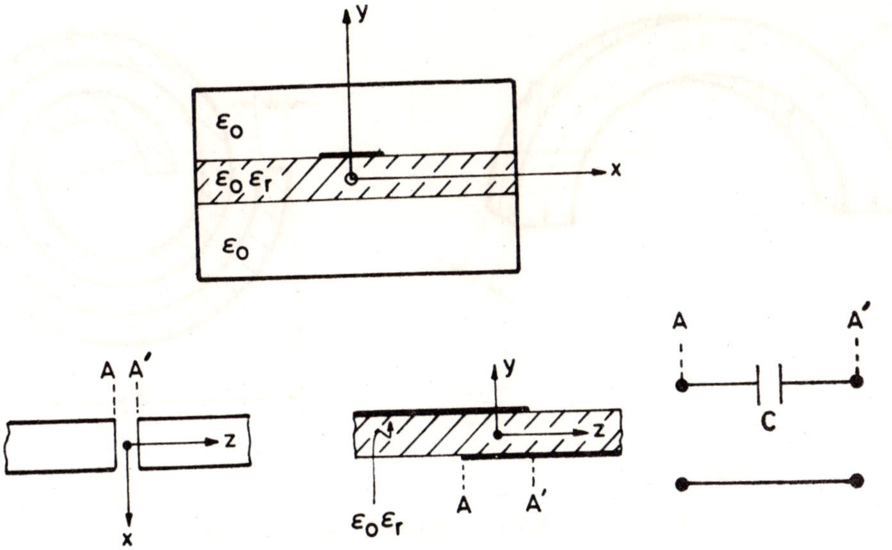

Fig. 12.24 Realization of series capacitance using end-coupled and broadside-coupled strip conductors

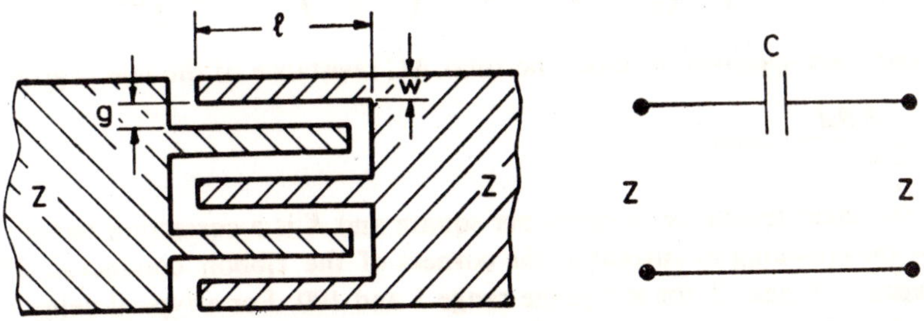

Fig. 12.25 Interdigital capacitor

where n is the number of fingers, l is the length of finger in inches, ϵ_r is the relative dielectric constant of the substrate considered to be of infinite thickness or with no ground plane, $A_1 = 0.225\ pF/\text{inch}$ and $A_2 = 0.252\ pF/\text{inch}$. The Q-factor of the Capacitor is given by

$$Q = \frac{1}{\omega CR} \tag{12.69}$$

where

$$R = \frac{4lR_s}{3wn} \tag{12.70}$$

Ribbon Inductor

A thin strip conductor on a dielectric substrate without the ground plane acts as an inductor (Fig. 12.26a). If l, w and t denote the length, width and thickness of the strip conductor,

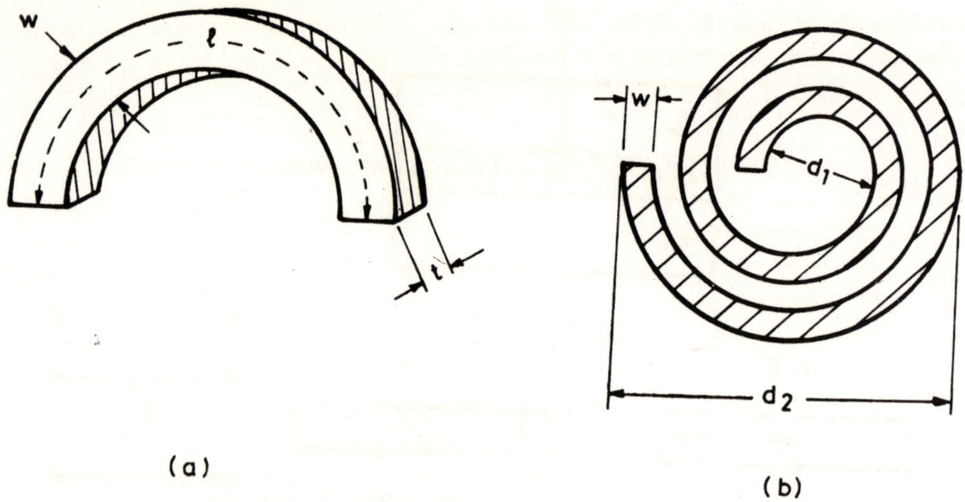

Fig. 12.26 (a) Ribbon inductor (b) Spiral inductor.

then the total inductance of the strip can be calculated using the formula [32].

$$Ll = 5.08\times 10^{-3} l\left[\ln\left(\frac{l}{w+t}\right) + 0.2235\left(\frac{w+t}{l}\right) + 1.193\right] nH \tag{12.71}$$

where l, w and t are specified in mils. The total AC resistance of the line is given by

$$R = \frac{KR_s l}{2(w+t)} \text{ ohms} \tag{12.72}$$

where R_s is the sheet resistance in ohms per square and K is a correction factor which takes into account the crowding of current at the corners of the ribbon conductor. The value of K lies between 1.3 and 2 for w/t in the range 1 and 100. For a typical value of $w/t = 10$, the value of K is approximately 1.55. Ribbon inductors of this type are useful for realizing low inductance values of the order of 2 to 3 *nH*.

Spiral Inductor

By printing the strip conductor in the form of a spiral, higher inductance values can be achieved. For a *n*-turn spiral inductor (Fig. 12.26b), the inductance expression is given by [32-34]

$$L = 0.01\pi n^2 a\left[\ln\left(\frac{8a}{b}\right) + \frac{1}{24}\left(\frac{b}{a}\right)^2\left\{\ln\left(\frac{8a}{b}\right) + 3.583\right\} - \frac{1}{2}\right] nH \tag{12.73}$$

where

$$a = \frac{1}{4}(d_1 + d_2) \tag{12.74a}$$

$$b = \frac{1}{2}(d_2 - d_1) \tag{12.74b}$$

The parameters d_1 and d_2 are the inner and outer diameter, respectively of the spiral, and are specified in mils. The resistance of the spiral inductor is given by

$$R = \frac{K' n a R_s}{w} \text{ ohms} \tag{12.75}$$

where K' is a correction factor which accounts for current crowding. A typical value for K' is 1.5. The Q-factor for a spiral is typically 100 which is higher than that for a straight ribbon. The expression for Q is

$$Q = \frac{2 \times 10^{-9}\, nafw}{K'w(8a + 11b)} \tag{12.76}$$

where the dimensions a, b and w are in mils.

Resonant Circuits

Resonant circuits are realized by suitably combining inductors and capacitors. For example, Fig. 1.5a and b shows the series- and parallel-resonant circuits obtained by combining single turn inductor with interdigital capacitor. Figure 12.27 shows the layout of series-resonant shunt circuit and parallel-resonant shunt circuit using distributed elements considered in Fig. 12.23. In Fig. 12.27b, the shunt inductor is realized by shorting the end of the thin inductive strip to the ground plane.

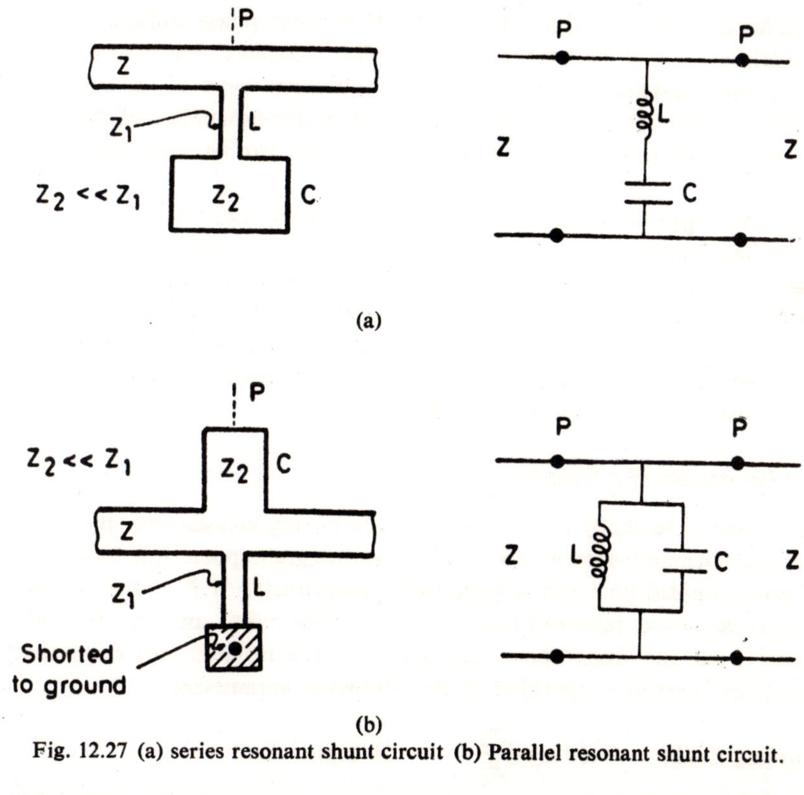

Fig. 12.27 (a) series resonant shunt circuit (b) Parallel resonant shunt circuit.

Figure 12.28 shows the layout of open circuited quarter-wave stub attached to the main line. At the resonant frequency, the stub presents a short circuit at the junction point. The element thus acts as a bandstop filter around the resonance [38].

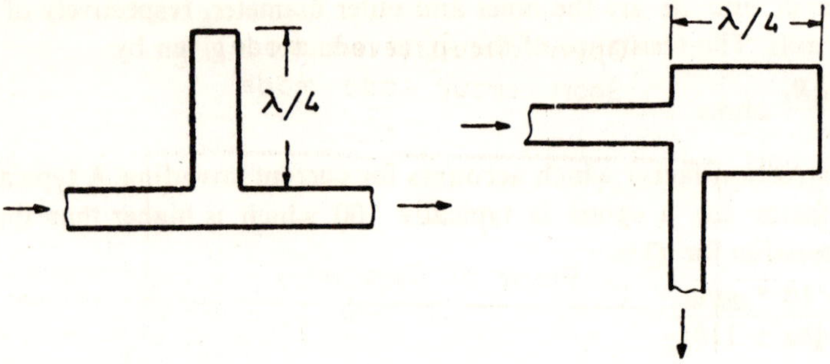

Fig. 12.28 Quarter-wave stubs as bandstop filter element.

12.5 Branch-line Directional Couplers [3, 39-46]

In stripline-like configurations, two basic types of directional couplers find extensive practical applications: (a) branch-line directional coupler and (b) parallel-coupled directional coupler. They are essentially power division networks with two important features—namely, the two output ports are mutually isolated, and the signals at these two ports are out of phase by 90°. These components form the building blocks of several other components, as for example, balanced mixers, variable attenuators, PIN diode phase shifters, directional filters, diplexers, multiplexers and TR switches. Branch-line couplers, are also used extensively in antenna array feed systems in preference to *Y*-junction type of power dividers and as impedance transformers. In active devices they provide the advantage of direct d.c. coupling. Branchline couplers are forward-wave couplers and hence can be cascaded without crossing lines.

The theory and design of branch-line couplers have been reported in several publications [3, 39-41]. In this section, we restrict our discussion to two-branch and three-branch couplers. Couplers having more than three branches are difficult to realize in practice, because, the end sections require increasingly larger impedance levels resulting in impractically narrow line widths. Furthermore, with an increase in the number of sections the insertion loss increases because of the requirement of larger circuit area. Discussion on the parallel-coupled directional couplers is provided in Section 12.6.

12.5.1 Basic Parameters of a Coupler

Directional coupler is basically a four-port network having at least one plane of symmetry. Referring to the schematic shown in Fig. 12.29 we designate ports 1, 2, 3 and 4 as the input port, direct port, coupled port and isolated port, respectively. If P_1 is the power fed to the input port, P_1' is the power reflected back from the input port, and P_2, P_3 and P_4 are the output powers from the direct port, coupled port and isolated port respectively, then the performance of the coupler is specified by the following parameters:

$$\text{Coupling } C(dB) = -10 \log_{10}\frac{P_3}{P_1} = -20 \log_{10} |S_{13}| \qquad (12.77a)$$

$$\text{Directivity } D(dB) = -10 \log_{10}\frac{P_4}{P_3} = -20 \log_{10} |S_{34}| \qquad (12.77b)$$

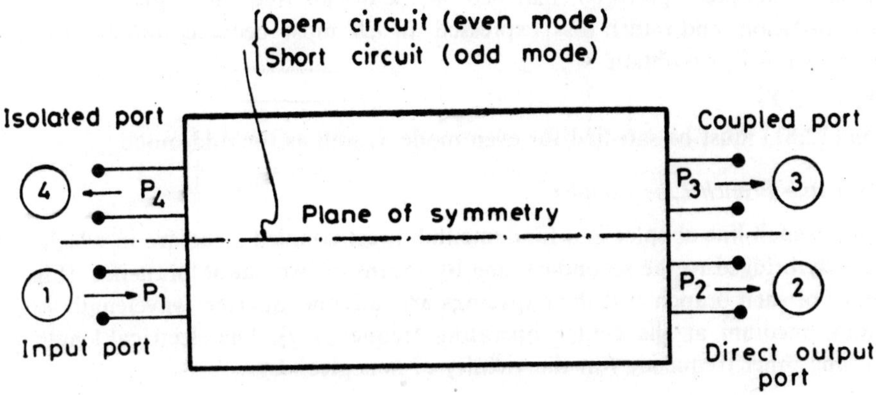

Fig. 12.29 Four-port network representation of a directional coupler

$$\text{Isolation } L(dB) = -10 \log_{10}\frac{P_4}{P_1} = -20 \log_{10} |S_{14}| \tag{12.77c}$$

$$\text{Return loss } R(dB) = -10 \log_{10}\frac{P_1'}{P_1} = -20 \log_{10} |S_{11}| \tag{12.77d}$$

$$\text{Input VSWR} = \frac{1+|S_{11}|}{1-|S_{11}|} \tag{12.78}$$

The scattering parameters of the coupler can be obtained from the two-port parameters of one half of the structure about the symmetry plane for even- and odd-modes of excitation, as described in Sec. 12.2. From (12.24), they can be expressed as

$$S_{11} = \frac{1}{2}(S_{11e} + S_{11o}) \tag{12.79a}$$

$$S_{12} = \frac{1}{2}(S_{12e} + S_{12o}) \tag{12.79b}$$

$$S_{13} = \frac{1}{2}(S_{12e} - S_{12o}) \tag{12.79c}$$

$$S_{14} = \frac{1}{2}(S_{11e} - S_{11o}) \tag{12.79d}$$

where the subscripts e and o refer to the even- and odd-mode excitations respectively. Taking into consideration that the symmetric half two-port network is symmetrical and reciprocal, its scattering parameters for the two modes of excitation can be easily determined from its $ABCD$ parameters.

$$S_{11\substack{e\\o}} = \left[\frac{BY_c - C/Y_c}{2A + BY_c + C/Y_c}\right]_{\substack{e\\o}} \tag{12.80a}$$

$$S_{12\substack{e\\o}} = \left[\frac{2}{2A + BY_c + C/Y_c}\right]_{\substack{e\\o}} \tag{12.80b}$$

Here Y_c is the characteristic admittance of the input and output arms and the subscripts e and o outside the bracket refer to the parameters A, B and C.

In an ideal coupler, perfectly matched at the design frequency, $P_1' = P_4 = 0$. Hence, directivity, isolation, and return loss expressed in dB must become infinite. Furthermore, when we set $S_{11} = 0$, we obtain

$$BY_c = C/Y_c \tag{12.81}$$

Condition (12.81) must be satisfied for even-mode as well as the odd-mode.

12.5.2 Two-stub Branch-Line Coupler

A two-stub branch line coupler is a fundamental direct-coupled structure in which the main line is directly bridged to the secondary line by means of two-shunt branches (Fig. 12.30a). The length l of each branch and their spacings are all one quarter wavelength $\lambda_0/4$ in the transmission medium at the center operating frequency f_0. The electrical length θ of each branch at any other frequency f in the vicinity of f_0 is given by

$$\theta = \frac{2\pi l}{\lambda} = \frac{\pi}{2}\frac{\lambda_0}{\lambda} \tag{12.82}$$

In Fig. 12.30a, Y_a and Y_b denote the characteristic admittance of the series branch and shunt branch respectively. All input and output lines have the same characteristic admittance Y_c. Figure 12.30b shows half-section of the coupler considered for the purpose of analysis. In the even-mode excitation, the shunt stubs are open circuited and in the odd-mode excitation, they are short circuited at a distance of $\theta/2$ from the main line. As shown in the equivalent circuit (Fig. 12.30c), let Y_{se} and Y_{so} denote the input admittances of the stub of length $\theta/2$ for the even- and odd-mode respectively. They are given by

$$Y_{se} = jY_b \tan(\theta/2) \tag{12.83a}$$

$$Y_{so} = -jY_b \cot(\theta/2) \tag{12.83b}$$

Let $[M]_e$ and $[M]_o$ donote the overall $ABCD$ matrix of the equivalent circuit for the even- and odd-mode respectively. Using the $ABCD$ matrix of the shunt element and the section of main line of length θ as given in Table 12.1, we can obtain the elements of $[M]_e$ and $[M]_o$. Thus,

$$[M]_e = \begin{bmatrix} 1 & 0 \\ jY_bT & 1 \end{bmatrix} \begin{bmatrix} \cos\theta & \dfrac{j}{Y_a}\sin\theta \\ jY_a\sin\theta & \cos\theta \end{bmatrix} \begin{bmatrix} 1 & 0 \\ jY_bT & 1 \end{bmatrix}$$

$$= \begin{bmatrix} \left(\cos\theta - \dfrac{Y_bT\sin\theta}{Y_a}\right) & \left(\dfrac{j\sin\theta}{Y_a}\right) \\ j\left(2Y_bT\cos\theta - \dfrac{Y_b^2T^2\sin\theta}{Y_a} + Y_a\sin\theta\right) & \left(\cos\theta - \dfrac{Y_bT\sin\theta}{Y_a}\right) \end{bmatrix} \tag{12.84}$$

where

$$T = \tan(\theta/2) \tag{12.85}$$

The expression for $[M]_o$ is obtained by replacing Y_b by $-Y_b$ and T by $(1/T)$ in (12.84)

$$[M]_o = \begin{bmatrix} \left(\cos\theta + \dfrac{Y_b\sin\theta}{TY_a}\right) & \left(\dfrac{j\sin\theta}{Y_a}\right) \\ j\left(-\dfrac{2Y_b\cos\theta}{T} - \dfrac{Y_b^2\sin\theta}{Y_aT^2} + Y_a\sin\theta\right) & \left(\cos\theta + \dfrac{Y_b\sin\theta}{Y_aT}\right) \end{bmatrix} \tag{12.86}$$

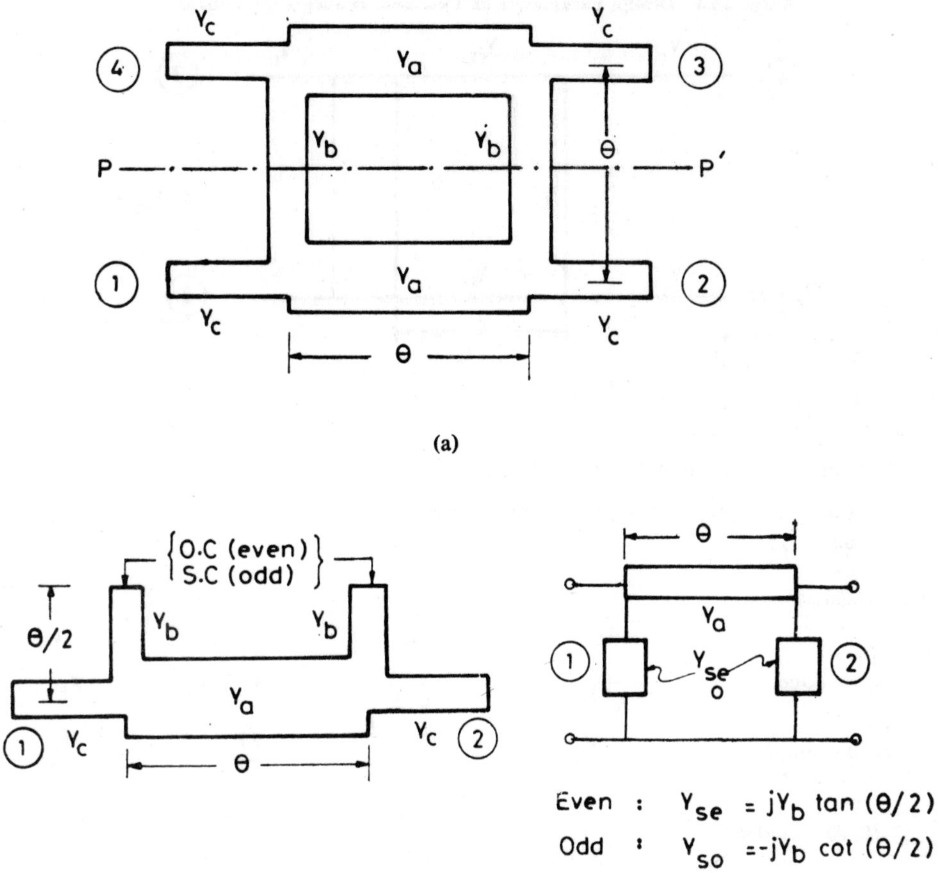

Fig. 12.30 (a) Two-stub branch-line coupler, $\theta = 2\pi l/\lambda$, $l = \lambda_0/4$; at midband ($f = f_0$), $\lambda = \lambda_0$, $\theta = \theta_0 = \pi/2$, (b) Half-section for even- and odd-mode analysis, (c) Equivalent circuit of (b).

The elements of $[M]_e$ give the even-mode parameters; A_e, B_e, C_e and D_e, and, the elements of $[M]_o$ give the odd-mode parameters, A_o, B_o, C_o and D_o. The scattering parameters of the coupler as a function of θ are then determined by using these expressions in (12.79) and (12.80).

Properties of the Coupler at Midband ($f = f_0$)

At midband, $f = f_0$, $\theta = \pi/2$ and $T = 1$. Expressions (12.84) and (12.86) reduce to

$$[M]_{\substack{e\\o}} = \begin{bmatrix} \mp Y_b/Y_a & j/Y_a \\ j(Y_a - Y_b^2/Y_a) & \mp Y_b/Y_a \end{bmatrix} \tag{12.87}$$

We now assume that the coupler is matched at midband, so that $S_{11e} = S_{11o} = 0$. The condition given in (12.81) applies to both the even- and odd-modes. Using the parameters B and C from (12.87) in (12.81) yields

$$Y_c^2 = Y_a^2 - Y_b^2 \tag{12.88}$$

Table 12.4 Design Parameters of Two-Stub Branch-Line Coupler

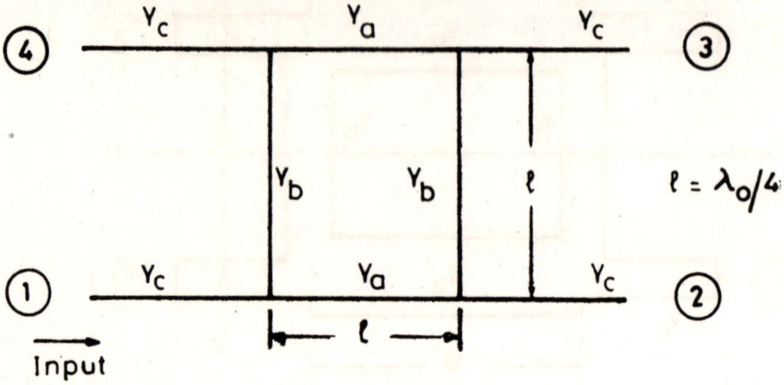

Midband parameters:

f_0 : midband frequency

λ_0 : midband wavelength in the medium

$S_{11} = S_{14} = 0, \quad S_{12} = -jY_c/Y_a, \quad S_{13} = -Y_b/Y_a$

$Y_c^2 = Y_a^2 - Y_b^2$

Branch admittances:

	S_{12}	S_{13}	Y_a	Y_b
3 *dB*, 90° coupler	$-\dfrac{j}{\sqrt{2}}$	$-\dfrac{1}{\sqrt{2}}$	$\sqrt{2}\,Y_c$	Y_c
6 *dB*, 90° coupler	$-j\dfrac{\sqrt{3}}{2}$	$-\dfrac{1}{2}$	$\dfrac{2Y_c}{\sqrt{3}}$	$\dfrac{Y_c}{\sqrt{3}}$
10 *dB*, 90° coupler	$-j\dfrac{3}{\sqrt{10}}$	$-\dfrac{1}{\sqrt{10}}$	$\dfrac{\sqrt{10}\,Y_c}{3}$	$\dfrac{Y_c}{3}$

Parameters for computing frequency response:

λ : guide wavelength in the medium at a frequency f.

$$\theta = \frac{\pi}{2}\cdot\frac{\lambda_0}{\lambda} \quad ; \quad T = \tan\left(\frac{\theta}{2}\right)$$

$$A_e = \left(\cos\theta - \frac{Y_b}{Y_a} T \sin\theta\right) = D_e$$

$$B_e = \frac{j\sin\theta}{Y_a}$$

$$C_e = j\left(2Y_b T\cos\theta - \frac{Y_b^2 T^2 \sin\theta}{Y_a} + Y_a \sin\theta\right)$$

$$A_o = \left(\cos\theta + \frac{Y_b \sin\theta}{Y_a T}\right) = D_o$$

$$B_o = \frac{j\sin\theta}{Y_a}$$

$$C_o = j\left(-\frac{2Y_b}{T}\cos\theta - \frac{Y_b^2}{Y_a}\cdot\frac{\sin\theta}{T^2} + Y_a\sin\theta\right)$$

Substitute these parameters in (12.79) and (12.80) to determine S_{11}, S_{12}, S_{13} and S_{14}.

The various scattering parameters of the coupler at midband can now be obtained by using the even- and odd-mode *ABCD* parameters given by (12.87) in (12.79) and (12.80). They are,

$$S_{11} = S_{14} = 0 \tag{12.89a}$$

$$S_{12} = -jY_c/Y_a \tag{12.89b}$$

$$S_{13} = -Y_b/Y_a \tag{12.89c}$$

The design formulas derived above for the two-stub branch-line coupler are summarized in Table 12.4. The coupling depends on the ratio of the impedances of series and shunt branches. The structure is convenient to fabricate for values of coupling in the range of approximately 3 *dB* to 10 *dB*. For loose couplings greater than 10 *dB*, the shunt arm impedances assume impractically large values. Figures 12.31, 12.32 and 12.33 show the theoretical frequency response characteristics of couplers having a mean coupling of 3 *dB*, 6 *dB*, and 10 *dB*, respectively. The bandwidth of the coupler is limited essentially by its directivity. It can be seen from Figs. 12.31 to 12.33 that the directivity of all the three couplers is better than 15 *dB* over nearly 10% bandwidth.

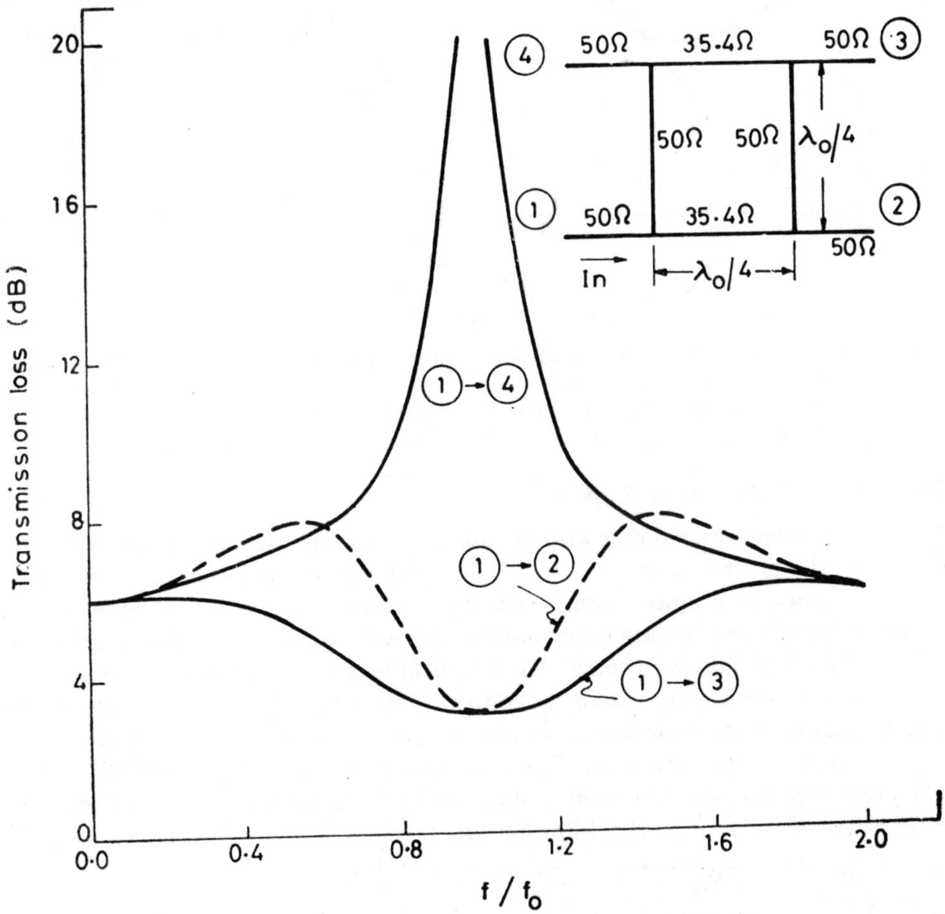

(a)

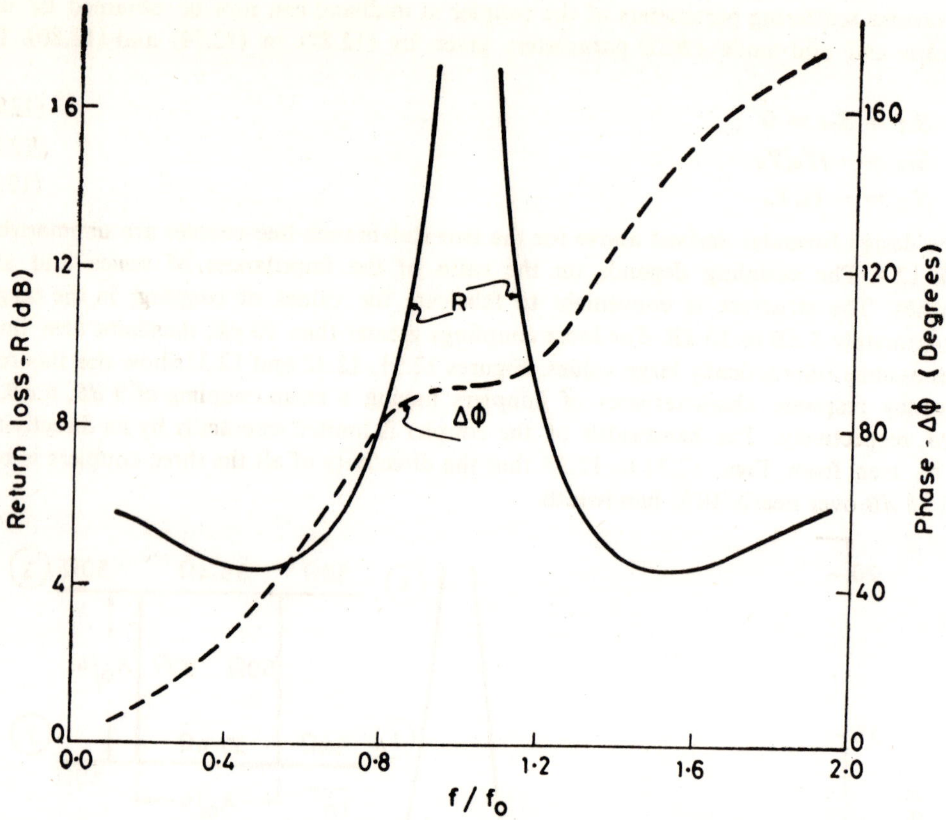

(b)

Fig. 12.31 Frequency response of a 3 *dB*, 90° two-stub branch-line coupler: (a) Transmission loss in *dB* between input port 1 and ports 2, 3 and 4, (b) Return loss $R = -20 \log_{10} | S_{11} |$ and phase difference $\Delta\phi$ between ports 3 and 2, as a function of normalized frequency.

Modifications for Flat Coupling Response

Riblet [42] has demonstrated that with the use of an appropriate matching network at each of the four ports of the coupler, a very flat coupling characteristic can be obtained. Figure 12.34 shows the coupler pattern with the matching network reported by Riblet. All the admittances with bar notation are normalized with respect to the characteristic admittance Y_c of the input and output lines. The electrical length of the various sections is $\pi/2$ at the centre frequency f_0. The matching network consists of a quarter wave transformer of normalized characteristic admittance $\bar{Y}_1$ and a short circuited stub of length θ_0 with a normalized characteristic admittance $\bar{Y}_2$ (or an open circuited stub of length $2\theta_0$ and having a normalized characteristic admittance equal to $\bar{Y}_2/2$). The design of this matched coupler can be easily carried out using the following equations [42]

$$K = \frac{\bar{Y}_a}{\bar{Y}_b} = \sqrt{\left|\frac{S_{12}}{S_{13}}\right|^2 + 1} \tag{12.90}$$

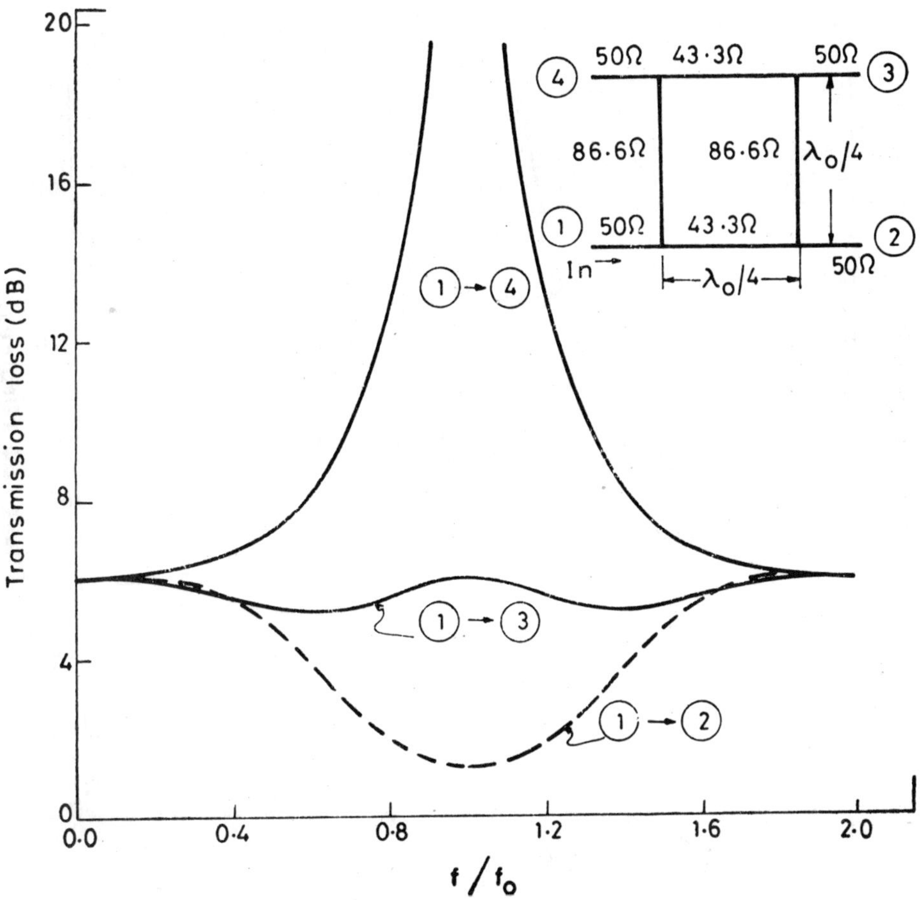

(a)

$$\theta_0 = \cos^{-1}\left[\frac{1}{\sqrt{2}} \cos \left\{\frac{\pi}{2}\left(1 + \frac{\Delta f}{2f_0}\right)\right\}\right] \tag{12.91}$$

$$\frac{\sin^2 \theta_0}{\bar{Y}_1^2} + \left(1 + \frac{\bar{Y}_2}{\bar{Y}_1}\right)^2 \cos^2 \theta_0 = \frac{\sin \theta_0}{\bar{Y}_b\sqrt{K^2 - 1}} \tag{12.92}$$

$$\frac{\sin \theta_0}{\bar{Y}_1} - \left(1 + \frac{\bar{Y}_2}{\bar{Y}_1}\right)\left(\bar{Y}_1 \sin \theta_0 - \bar{Y}_2 \frac{\cos^2 \theta_0}{\sin \theta_0}\right) = -\left(\frac{K+1}{K-1}\right)^{1/2} \tag{12.93}$$

For specified values of coupling and fractional bandwidth $\Delta f/f_0$, (12.90) and (12.91) are used to calculate K and θ_0. $\bar{Y}_b$ is an independent parameter, which in most cases can be chosen close to 1 in order to make the branch impedances easily realizable. $\bar{Y}_1$ and $\bar{Y}_2$ are then calculated from (12.92) and (12.93) by substituting for K and θ_0.

12.5.3 Three-Stub Branch Line Coupler

Compared with the conventional two-stub branch line coupler, the three-stub branch line

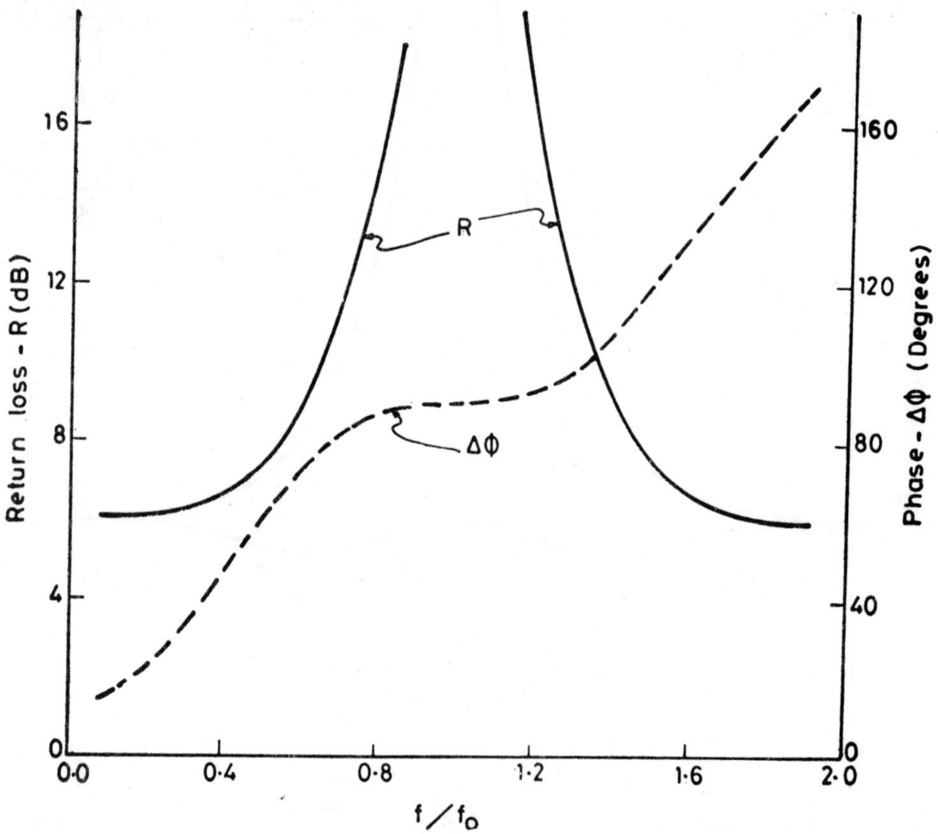

(b)

Fig. 12.32 Frequency response of 6 *dB*, 90° two-stub branch-line coupler: (a) Transmission loss in *dB* between input port 1 and ports 2, 3 and 4, (b) Return loss $R = 20 \log_{10} |S_{11}|$ and phase difference $\Delta\phi$ between ports 3 and 2, as a function of normalized frequency.

coupler gives a larger bandwidth (on the order of 25%). Figure 12.35 shows the typical pattern of such a coupler having end-to-end symmetry. Table 12.5 summarizes the design formulas for any desired coupling at midband and also formulas for computing the frequency response. In this configuration, 0 *dB* coupler is possible, in which all the power fed to input port 1 appears at port 3. Such a coupler can be used as a duplexer. In the 3 *dB*-coupler, with Z_c = 50 ohms, the impedances of the two outer shunt branches are 120.8 ohms, which is about the upper limit for practical realization.

12.5.4 Branch Line Coupler-Cum-Impedance Transformer

Branch line couplers with inherent impedance transformation capability between the input and output ports are useful in several applications. A classic example of their application is in combining outputs of parallel amplifiers [43].

Following the exact synthesis procedure presented by Levy and Lind [44] for the symmetrical branch guide directional coupler, Lind [45] has presented the synthesis of asymmetrical branch guide coupler-cum-impedance transformer including tables of element values for couplers having three and five branches. Design equations for a two-stub stripline directional coupler-cum-impedance transformer have been presented by Lind [45] and also by HO [46]. Table 12.6 summarizes the design equations due to HO for a two-branch coupler with unequal input and output port impedances and equal or unequal power split. The theoretical frequency response curves of couplers with 50 ohm input and various output impedances and a power split ratio of 2 : 1 reported in HO [46] are reproduced in Fig. 12.36. The useful bandwidth is approximately 10% although it deteriorates slightly as the output impedance deviates more from the input impedance.

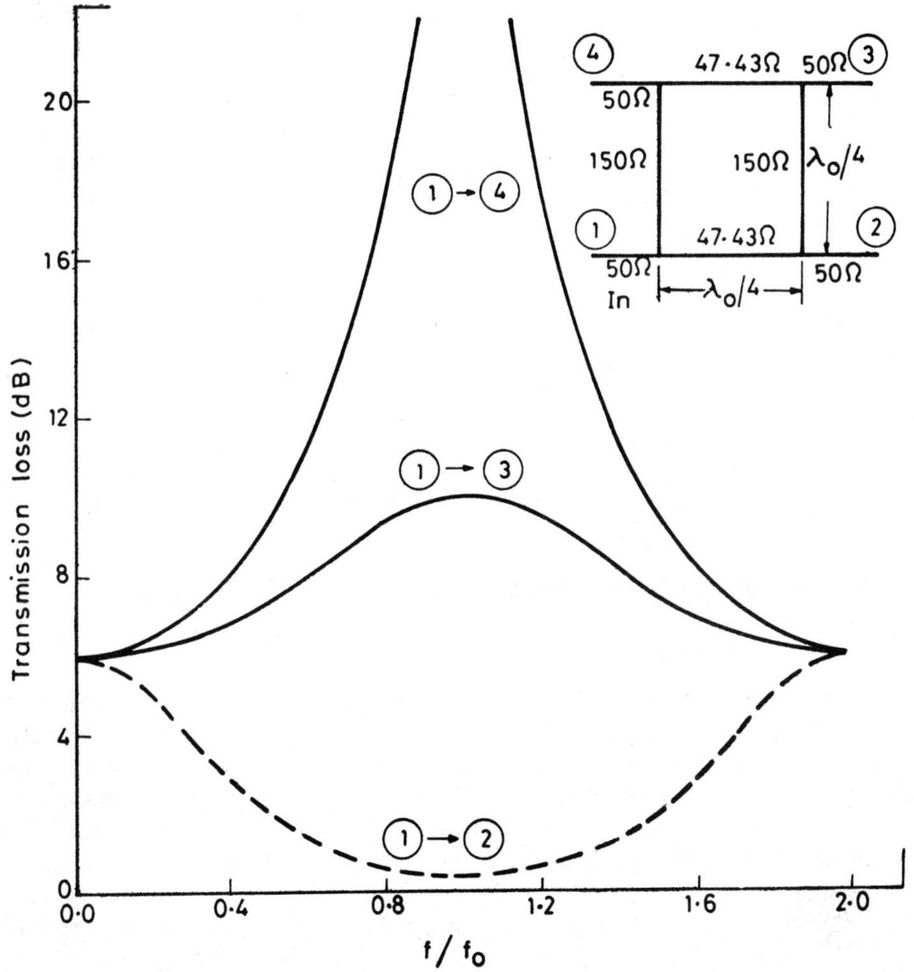

(a)

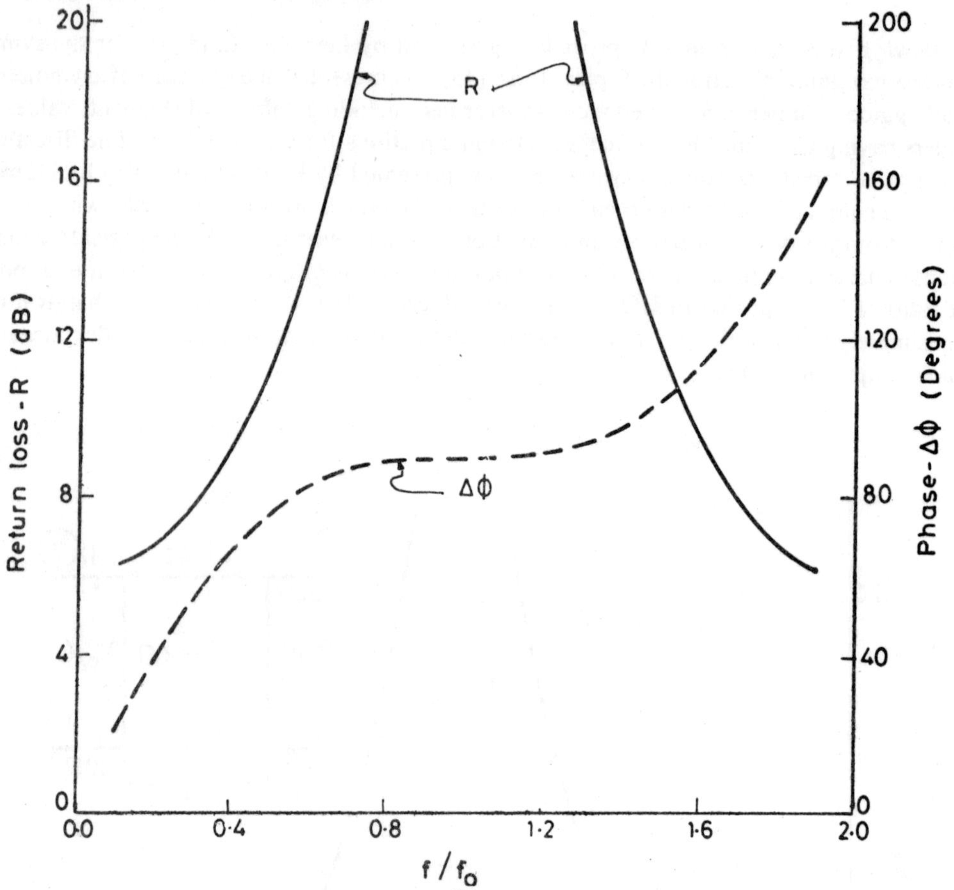

(b)

Fig. 12.33 Frequency response of 10 *dB*, 90° two-stub branch-line coupler, (a) Transmission loss in *dB* between input port 1 and ports 2, 3 and 4, (b) Return loss $R = 20 \log_{10} | S_{11} |$ and phase difference $\Delta\phi$ between ports 3 and 2, as a function of normalized frequency.

12.6 Parallel-Coupled Directional Couplers [47-52]

12.6.1 Single-Section Backward-Wave Coupler

Parallel-coupled directional couplers offer much larger bandwidths as compared with the branch-line couplers. They are mostly backward-wave couplers although forward-wave couplers are also possible with the use of the inhomogeneous medium. The most commonly used parallel-coupled directional coupler is the TEM-mode single-section backward-wave coupler. The schematic of such a coupler is shown in Fig. 12.37. As the term 'backward-wave coupler' implies, the electric and magnetic field interaction between the parallel-coupled conductors causes the coupled signal to travel in the direction opposite to that of the input signal [41, 47]. Maximum coupling occurs when the length of the coupling region is equal to one quarter wavelength (or an odd-multiple of quarter wavelength) in the propagating medium.

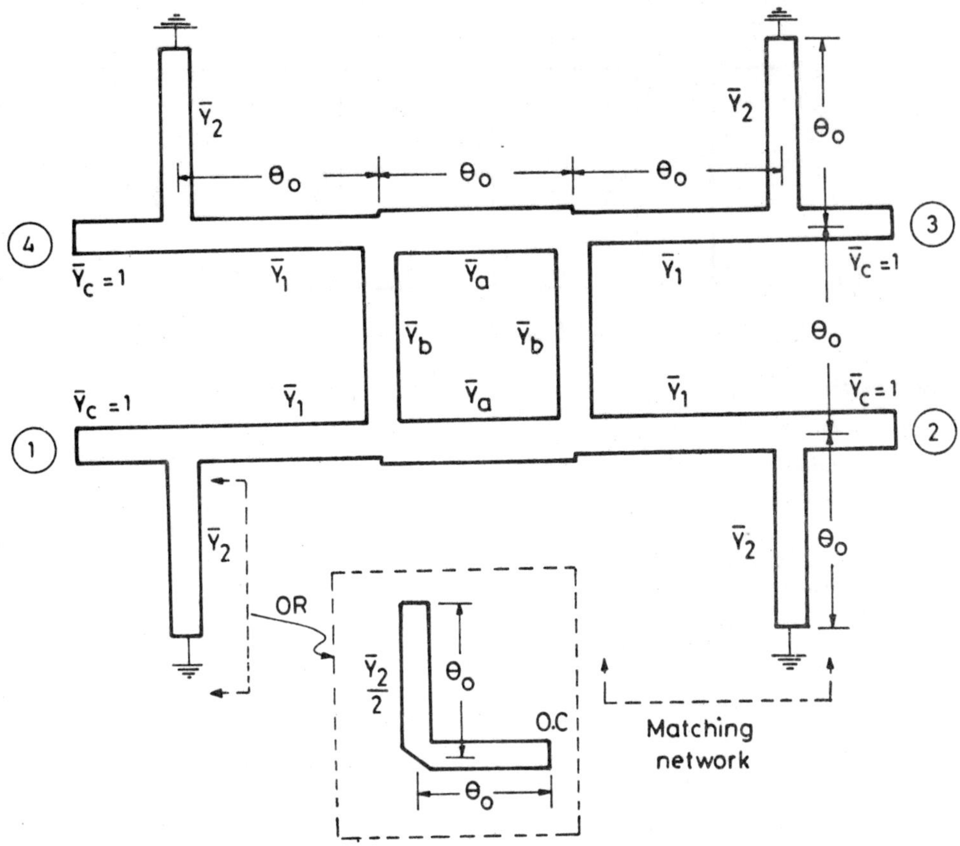

Fig. 12.34 Matched two-stub branch-line coupler with very flat coupling (From Riblet [42], Copyright © 1978 IEEE, reprinted with permission).

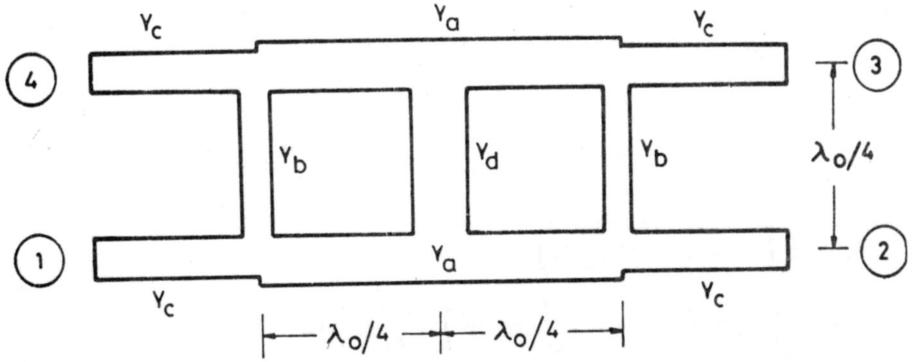

Fig. 12.35 Three-stub branch-line coupler.

The analysis of backward-wave coupler in a TEM-mode symmetric stripline has been reported by Jones and Bolljahn [48]. The basic problem is that of determining the voltage signals appearing at the four ports of the coupler when an arbitrary voltage signal is applied to one of its ports. This is illustrated in Fig. 12.38a in which each port is terminated in a

Table 12.5 Design Parameters of Three-Stub Branch-Line Coupler

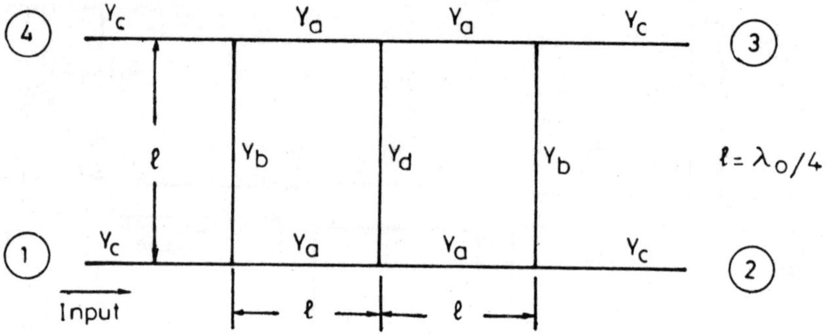

Midband formulas:

f_0 : midband frequency

λ_0 : midband wavelength in the medium

$$S_{11} = S_{14} = 0$$

$$S_{12} = \frac{Y_a^2(Y_bY_d - Y_a^2)}{[(Y_bY_d - Y_a^2)^2 + Y_d^2Y_c^2]}$$

$$S_{13} = \frac{jY_cY_a^2Y_d}{[(Y_bY_d - Y_a^2)^2 + Y_d^2Y_c^2]}$$

$$(Y_b^2 + Y_c^2)\, Y_d = 2Y_bY_a^2$$

Branch admittances:

	S_{12}	S_{13}	Y_a	Y_b	Y_d
0 *dB*, 90° coupler	0	$-j$	Y_c	Y_c	Y_c
3 *dB*, 90° coupler	$\frac{1}{\sqrt{2}}$	$-\frac{j}{\sqrt{2}}$	$\sqrt{2}\,Y_c$	$(\sqrt{2}-1)Y_c$	$\sqrt{2}\,Y_c$

Parameters for computing frequency response:

λ : guide wavelength in the medium at a frequency f.

$$\theta = \frac{\pi}{2} \cdot \frac{\lambda_0}{\lambda}, \qquad T = \tan\frac{\theta}{2}.$$

$$A_e = \left(\cos 2\theta - \frac{Y_b T \sin 2\theta}{Y_a}\right) - \frac{Y_d T}{Y_a}\left(\frac{\sin 2\theta}{2} - \frac{Y_b T \sin^2\theta}{Y_a}\right) = D_e$$

$$B_e = \frac{j}{Y_a}\left(\sin 2\theta - \frac{Y_d T \sin^2\theta}{Y_a}\right)$$

$$C_e = j\left[Y_bT\left\{\left(\cos 2\theta - \frac{Y_bT}{Y_a}\cdot\sin 2\theta\right) - \frac{Y_bT}{Y_a}\left(\sin 2\theta - \frac{Y_bT\sin^2\theta}{Y_a}\right)\right\} + \left(Y_a \sin 2\theta + Y_bT\cos 2\theta\right) + Y_dT\cos^2\theta\right]$$

$$A_o = \left(\cos 2\theta + \frac{Y_b}{Y_aT}\sin 2\theta\right) + \frac{Y_d}{Y_aT}\left(\frac{\sin 2\theta}{2} + \frac{Y_b}{Y_aT}\cdot\sin^2\theta\right) = D_o$$

(Contd.)

$$B_o = \frac{j}{Y_a}\left(\sin 2\theta + \frac{Y_d}{Y_a T} \sin^2 \theta \right)$$

$$C_o = j\left[-\frac{Y_b}{T}\left\{\left(\cos 2\theta + \frac{Y_b \sin 2\theta}{Y_a T}\right) + \frac{Y_d}{Y_a T}\left(\sin 2\theta + \frac{Y_b}{Y_a T} \sin^2 \theta \right)\right\} + \left(Y_a \sin 2\theta - \frac{Y_b}{T} \cos 2\theta \right) - \frac{Y_d \cos^2 \theta}{T} \right]$$

Substitute these parameters in (12.79) and (12.80) to determine S_{11}, S_{12}, S_{13} and S_{14}.

constant resistance Z_c and a signal of voltage $2V_0$ is applied to port 1. An equivalent representation in terms of the even- and odd-mode excitations is illustrated in Fig. 12.38b and c. As shown in the figure, ports 1 and 3 are fed with in-phase voltages in the even-mode and out-of-phase voltages in the odd-mode. By virtue of the symmetry with respect to PP', it suffices to consider a two-port network, between ports 1 and 2, with a magnetic wall through PP' in the even-mode case and an electric wall through PP' in the odd-mode case. The voltage outputs at the four ports of the coupler can now be expressed as a superposition of the terminal voltages in the even- and odd-mode excitations. They are

$$V_1 = V_{1e} + V_{1o} \tag{12.94a}$$

Table 12.6 Design Parameters of Two-Stub Coupler-Cum-Transformer [46]

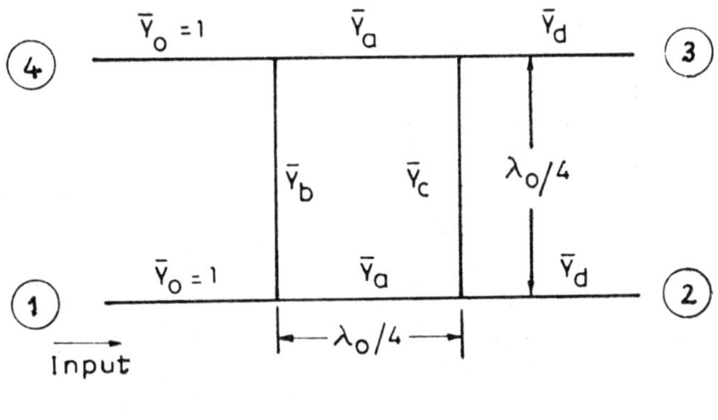

At midband:

λ_0 : midband wavelength in the medium

$$S_{11} = S_{14} = 0$$

$$k = \frac{|S_{13}|^2}{|S_{12}|^2}, \text{ power splitting ratio}$$

$$\bar{Z}_d = \frac{1}{\bar{Y}_d}, \quad \text{impedance ratio between output and input}$$

$$\bar{Y}_a = \left[\frac{(k+1)}{k}\,\bar{Y}_d\right]^{1/2}$$

$$\bar{Y}_b = \frac{1}{\sqrt{k}}$$

$$\bar{Y}_c = \frac{\bar{Y}_d}{\sqrt{k}}$$

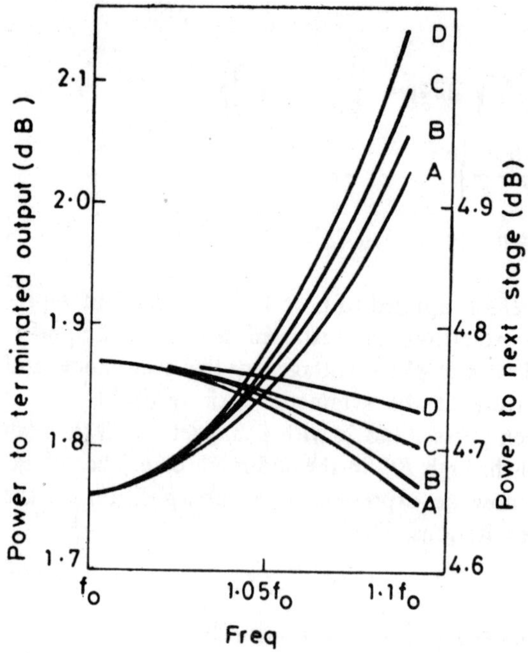

k=2	curve A	curve B	curve C	curve D
$\overline{Y}_b$	0.7071	0.7071	0.7071	0.7071
$\overline{Y}_a$	1.225	1.732	2.121	2.449
$\overline{Y}_c$	0.7071	1.414	2.121	2.828
$\overline{Y}_d$	1.000	2.000	3.000	4.000
output impedance	50	25	16.$\bar{6}$6	12.5

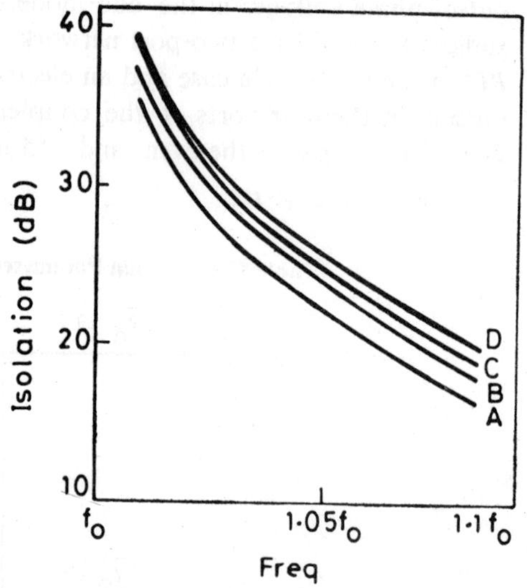

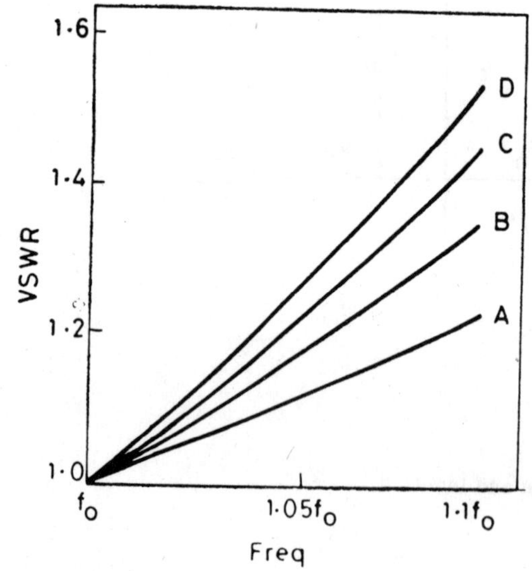

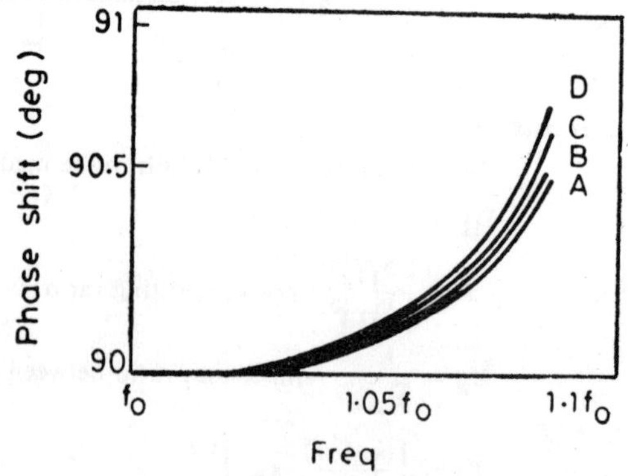

Fig. 12.36 Frequency response of two-stub coupler-cum-impedance transformer. Input impedance = 50 Ω (From HO [46], reprinted with permission of Microwaves, Copyright © 1976 Hayden Publishing Co.)

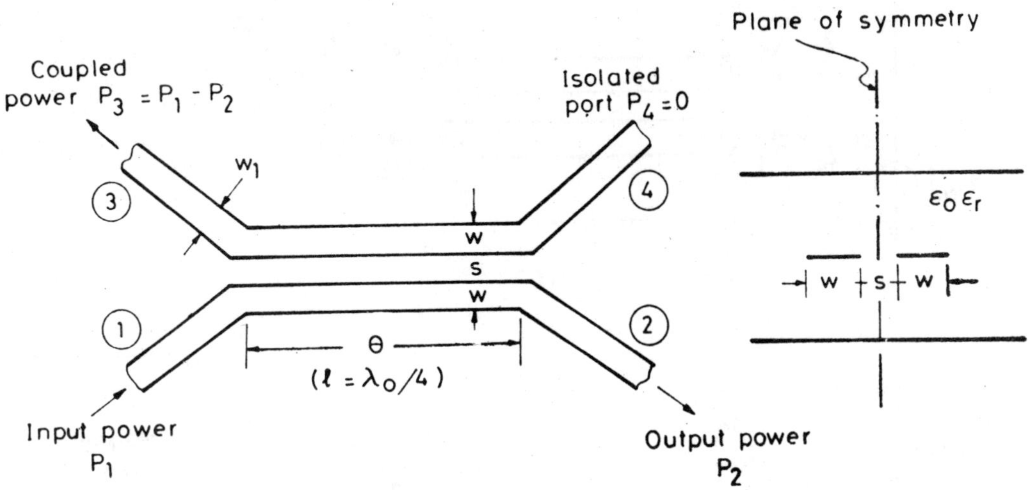

(a)

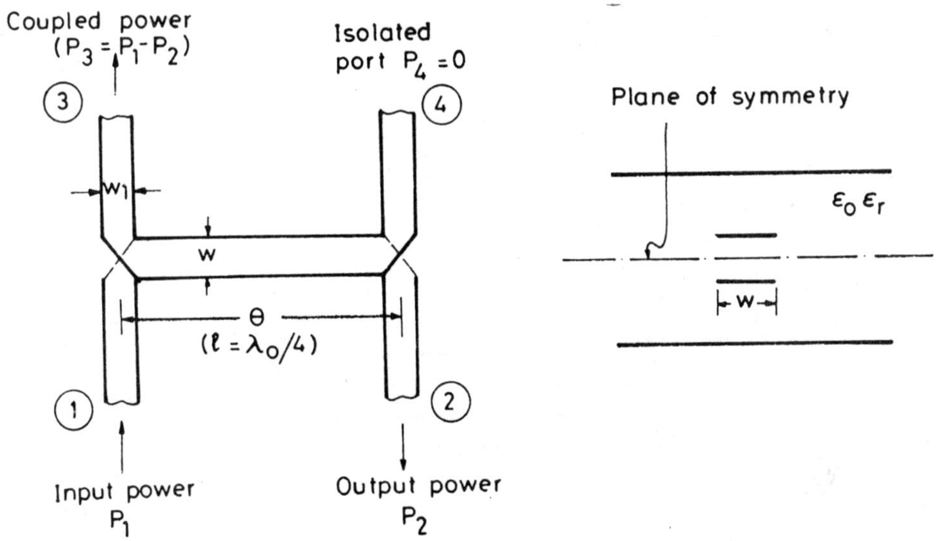

(b)

Fig. 12.37 Single section backward-wave coupler in (a) edge-coupled stripline, (b) broadside-coupled stripline

$$V_2 = V_{2e} + V_{2o} \tag{12.94b}$$

$$V_3 = V_{3e} + V_{3o} = V_{1e} - V_{1o} \tag{12.94c}$$

$$V_4 = V_{4e} + V_{4o} = V_{2e} - V_{2o} \tag{12.94d}$$

where the subscripts 'e' and 'o' refer to the even-mode and odd-mode respectively. The input impedance of the coupler can be obtained from

$$Z_{in} = \frac{V_{1e} + V_{1o}}{I_{1e} + I_{1o}} \tag{12.95}$$

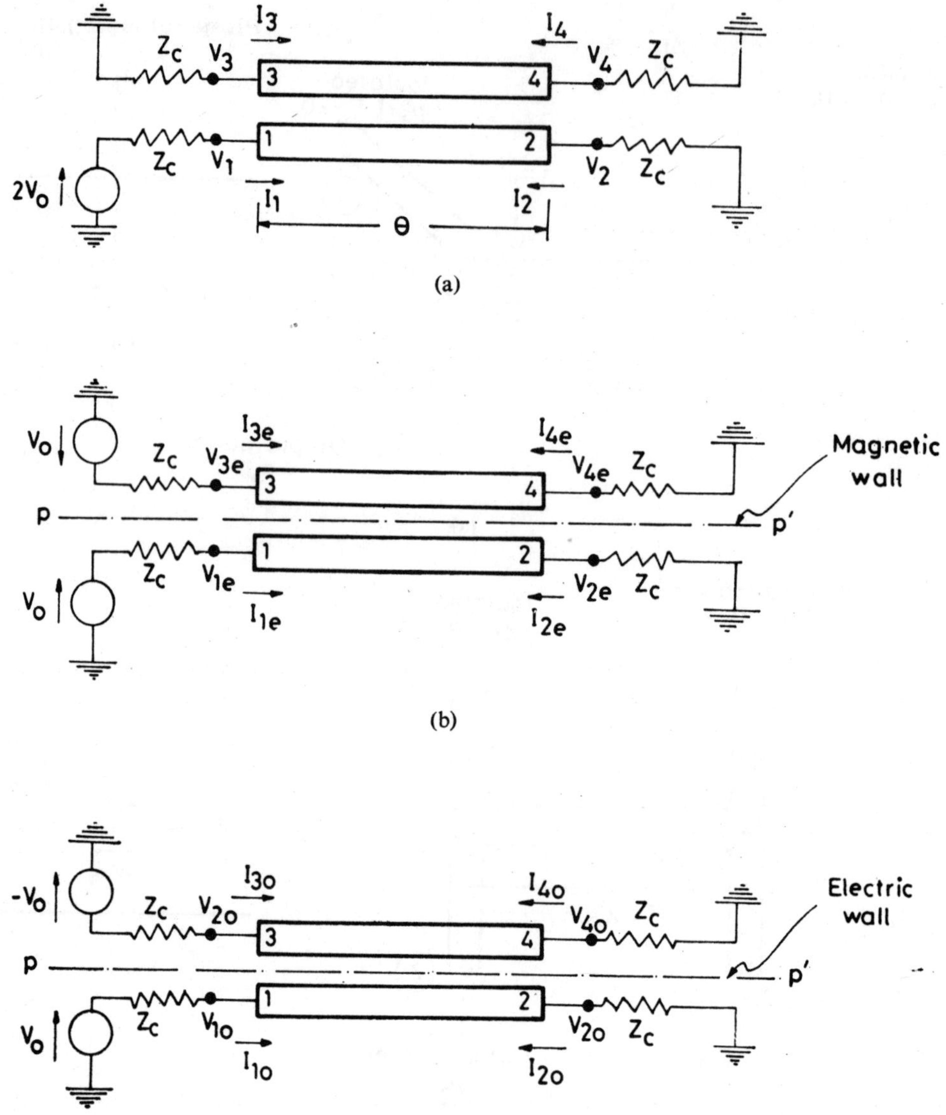

Fig. 12.38 (a) Stripline directional coupler, (b) Schematic for even-mode excitation, (c) Schematic for odd-mode excitation

where

$$I_{1e} = \frac{V_0}{Z_{1e} + Z_c}, \qquad I_{1o} = \frac{V_0}{Z_{1o} + Z_c} \tag{12.96a}$$

$$V_{1e} = I_{1e}Z_{1e}, \qquad V_{1o} = I_{1o}Z_{1o} \tag{12.96b}$$

Z_{1e} and Z_{1o} are the input impedances looking into port 1 when the coupler is excited in even-mode and odd-mode respectively. If Z_e denotes the even-mode characteristic impedance and

Z_o denotes the odd-mode characteristic impedance of the coupled transmission line, the expressions for Z_{1e} and Z_{1o} are given by

$$Z_{1e} = \frac{Z_e(Z_c + jZ_e \tan \theta)}{(Z_e + jZ_c \tan \theta)} \tag{12.97a}$$

$$Z_{1o} = \frac{Z_o(Z_c + jZ_o \tan \theta)}{(Z_o + jZ_c \tan \theta)} \tag{12.97b}$$

In order that the directional coupler is perfectly matched at all frequencies, its input impedance Z_{in} must be equal to the port impedance Z_c. Substituting (12.96) and (12.97) in (12.95), it can be shown that the condition $Z_{in} = Z_c$ is satisfied when

$$Z_c = \sqrt{Z_e Z_o} \tag{12.98}$$

It may be noted that this relation holds for all values of θ. Now, from the straightforward analysis of a section of transmission line of length θ terminated at either end by an impedance equal to $\sqrt{Z_e Z_o}$ and fed by a voltage V_0, the terminal voltages V_{1e} and V_{2e} for the even-mode, and, V_{1o} and V_{2o} for the odd-mode can be determined. These, when substituted in (12.94), give the output voltages at the four ports of the coupler in terms of the input voltage at port 1. The results are

$$\frac{V_1}{V_0} = 1 \tag{12.99a}$$

$$\frac{V_2}{V_0} = \frac{2}{2 \cos \theta + j \sin \theta \left(\sqrt{\dfrac{Z_e}{Z_o}} + \sqrt{\dfrac{Z_o}{Z_e}} \right)} \tag{12.99b}$$

$$\frac{V_3}{V_0} = \frac{j \sin \theta \left(\sqrt{\dfrac{Z_e}{Z_o}} - \sqrt{\dfrac{Z_o}{Z_e}} \right)}{2 \cos \theta + j \sin \theta \left(\sqrt{\dfrac{Z_e}{Z_o}} + \sqrt{\dfrac{Z_o}{Z_e}} \right)} \tag{12.99c}$$

$$\frac{V_4}{V_0} = 0 \tag{12.99d}$$

It is obvious from (12.99b) and (12.99c) that the coupled voltage signals at ports 2 and 3 are out of phase by 90°. Further, the voltage at the coupled port 3 attains a maximum value when the electrical length $\theta = n\pi/2$ for n odd, For minimum size, the physical length l of the coupled section is chosen equal to $\lambda_0/4$ at midband. Setting $\theta = \pi/2$ in (12.99c), the maximum voltage coupling coefficient of the coupler, denoted as C_v is given by

$$C_v = \frac{Z_e - Z_o}{Z_e + Z_o} \tag{12.100}$$

This is the midband coupling factor and is taken as the basic starting parameter in the design of the coupler. The voltage ratios V_2/V_0 and V_3/V_0 can be expressed in terms of C_v as:

$$\frac{V_2}{V_0} = \frac{\sqrt{1 - C_v^2}}{[\sqrt{1 - C_v^2} \cos \theta + j \sin \theta]} \tag{12.101a}$$

$$\frac{V_3}{V_0} = \frac{jC_v \sin \theta}{[\sqrt{1 - C_v^2} \cos \theta + j \sin \theta]} \tag{12.101b}$$

The design equations derived above for an ideal TEM mode coupler are listed in Table 12.7. In the design of the coupler, the midband coupling factor C_v and the port impedances Z_c are specified. Using (12.98) and (12.100) Z_e and Z_o are calculated which in turn are used to determine the dimensions of the coupled line. The frequency response of the coupler can be obtained from (12.101). For small values of C_v the coupling factor V_3/V_0 varies approximately as a sinusoid as a function of θ. Theorètically, the coupler offers perfect match and infinite directivity at all frequencies. This result is a direct consequence of the inherent property that the even- and odd-mode phase velocities are equal when the propagating mode is pure TEM. The bandwidth of the coupler is typically around 40%. Couplers with tightly coupled lines offer slightly higher bandwidth as compared with the weakly coupled lines. For example, for a tolerance on coupling variation within ± 0.2 *dB*, a coupler with 3 *dB* mean coupling offers approximately 55% bandwidth whereas a coupler with 10 *dB* mean coupling gives about 40% (Fig. 12.39).

Table 12.7 Design Equations of a Single Section Backward-Wave Coupler

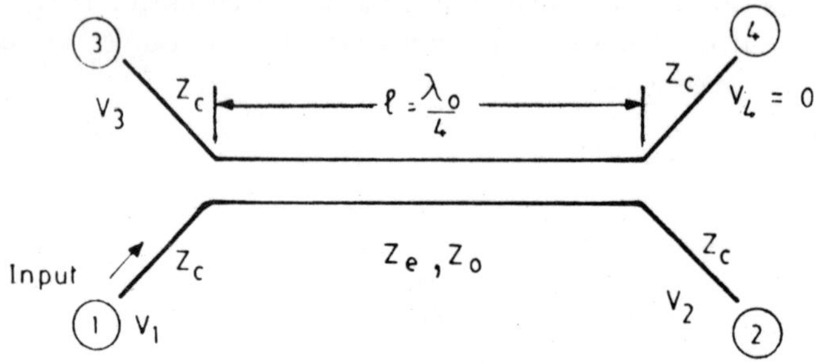

Specify:

Midband operating frequency: f_0

Port impedances: Z_c

Mean coupling in *dB*: $-20 \log_{10} C_v$

Coupler Parameters:

Coupling length $l = \frac{\lambda_0}{4}$ (λ_0 : guide wavelength at f_0).

$$Z_e = Z_c \sqrt{\frac{1 + C_v}{1 - C_v}}, \quad Z_o = Z_c \sqrt{\frac{1 - C_v}{1 + C_v}}$$

Frequency Response:

$$\left|\frac{V_3}{V_1}\right|^2 = \frac{C_v^2 \sin^2 \theta}{(1 - C_v^2 \cos^2 \theta)} \quad ; \quad \theta = \frac{\pi}{2} \cdot \frac{\lambda_0}{\lambda}$$

$$\left|\frac{V_2}{V_1}\right|^2 = \frac{1 - C_v^2}{(1 - C_v^2 \cos^2 \theta)}$$

$$\text{Coupling in } dB = -20 \log_{10} \left|\frac{V_3}{V_1}\right|$$

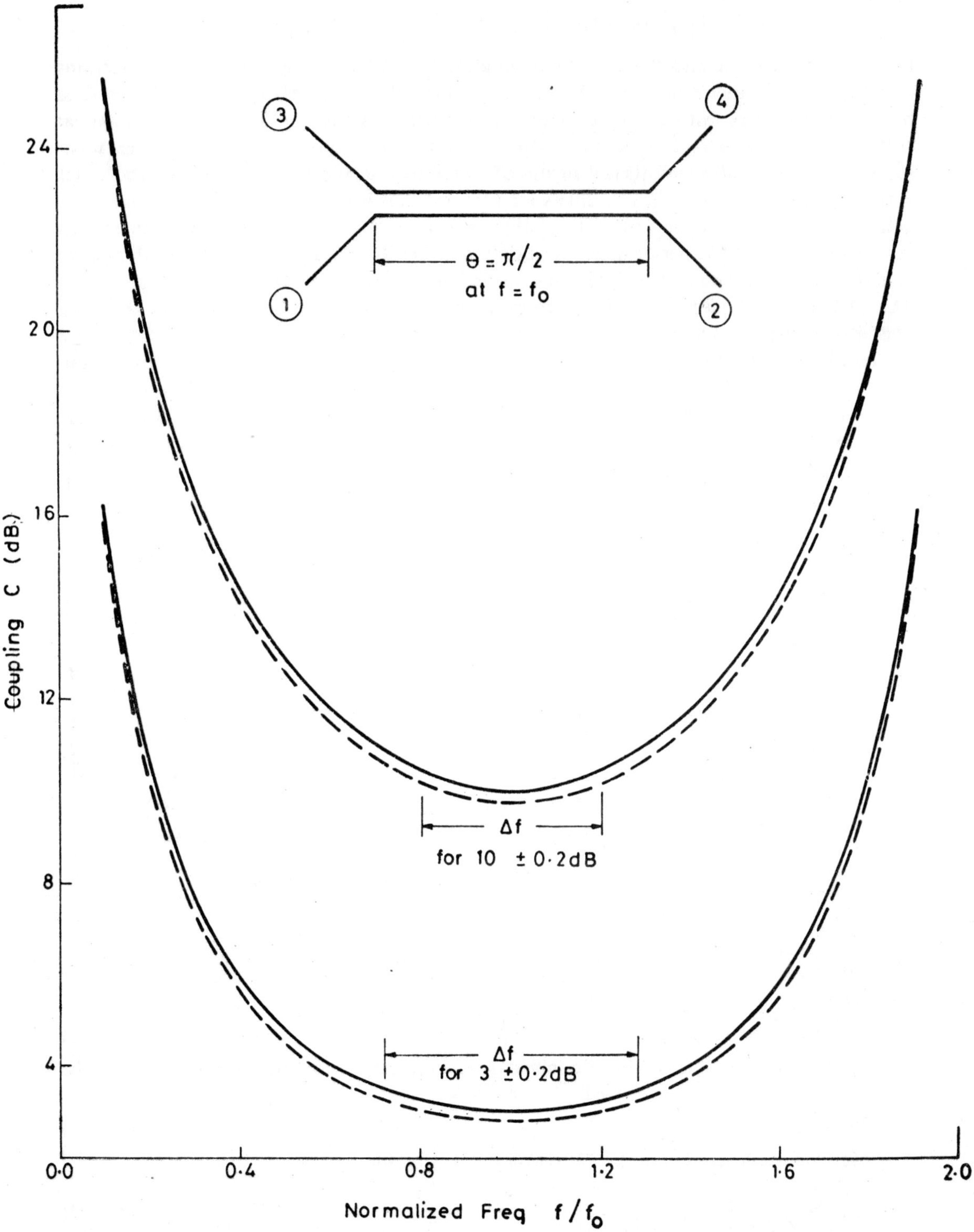

Fig. 12.39 Coupling variation as a function of frequency for a backward-wave coupler, $C(dB) = -20 \log_{10} | S_{13} |$.

12.6.2 Multi-Section Symmetrical Backward-Wave Coupler

Improved bandwidth and flatness of coupling can be achieved by cascading several sections of coupled lines, each section having an electrical length $\theta=\pi/2$ at midband frequency. The bandwidth increases with an increase in the number of sections. TEM mode symmetrical couplers with symmetry around the center section possess odd number of quarter wave sections. Figure 12.40 shows the schematic of a multisection symmetrical coupler. The exact design of optimum symmetrical couplers for three sections has been reported by Shimizu and Jones [49] and also Young [50]. The design incorporates an equal-ripple approximation to the mean coupling. The performance is optimum in the sense that the bandwidth obtainable is maximum when the number of sections, mean coupling, and coupling tolerance are fixed. The design of optimum symmetrical couplers having more than three sections has been presented by Cristal and Young [51]. The theory makes use of the mathematical equivalence between the TEM-mode parallel-coupled directional coupler and sections of transmission lines established by Young [50] and Levy [52]. The equivalence is that the backward coupled wave (at port 3) and the forward coupled wave (at port 2) of the coupler (Fig. 12.40) correspond to the reflected wave and transmitted wave, respectively, of the cascaded transmission line.

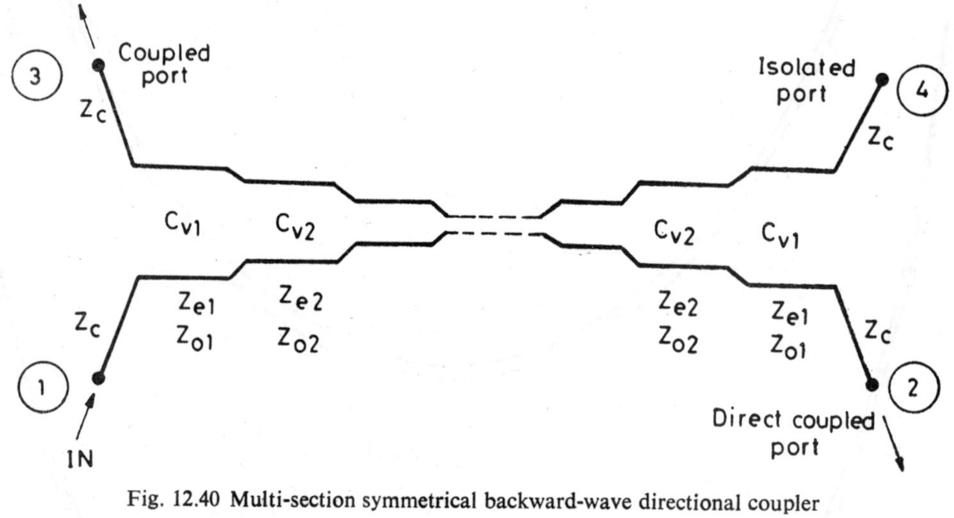

Fig. 12.40 Multi-section symmetrical backward-wave directional coupler

The multi-section symmetrical coupler retains all the basic properties of a single quarter-wave section coupler. For retaining perfect match and perfect directivity, however the even- and odd-mode impedances of various sections must satisfy the following relation:

$$\sqrt{Z_{e1}Z_{o1}} = \sqrt{Z_{e2}Z_{o2}} = \ldots = \sqrt{Z_{ei}Z_{oi}} = \ldots \sqrt{Z_{en}Z_{on}} = Z_c \tag{12.102}$$

Normalizing the even- and odd-mode impedances with respect to the impedance Z_c of the ports, condition (12.102) for a typical i^{th} section is

$$Z_{ei}Z_{oi} = Z_c^2 \tag{12.103}$$

The coupling coefficient C_{vi} of the ith section is given by the formula

$$C_{vi} = \frac{Z_{ei}^2 - 1}{Z_{oi}^2 + 1} \tag{12.104}$$

12.6.3 Asymmetric Backward Wave Coupler

The advantage of the asymmetric backward wave couplers is that they can offer much larger bandwidth as compared with symmetrical backward wave couplers having the same number of sections. The general synthesis and detailed design of asymmetric multielement coupled transmission line directional couplers have been reported by Levy [52, 53]. Figure 12.41 shows the layout of a two-element asymmetric coupler. Each element is a quarter wavelength long at the centre frequency. The procedure for determining the parameters of this coupler for a specified mean coupling and coupling tolerance is as follows [52].

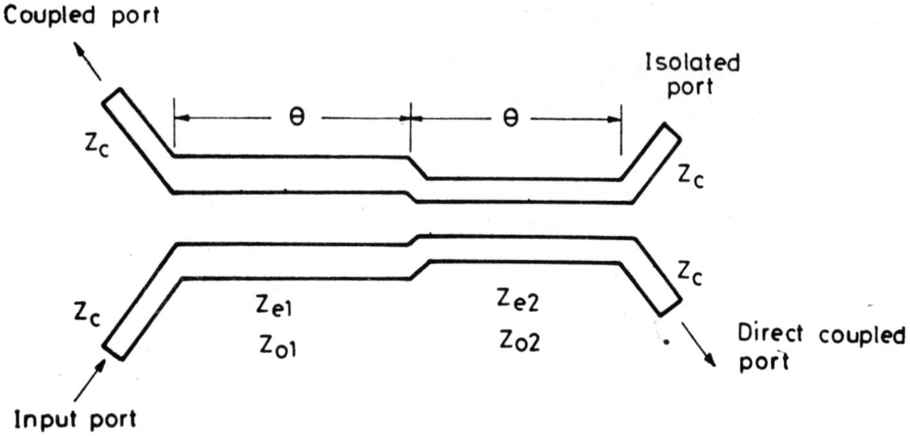

Fig. 12.41 Two-element asymmetric backward-wave coupler.

For a specified mean coupling C_v *dB* and coupling tolerance $\pm R$ *dB*, we first determine the parameters H and J using

$$C_v = 10 \log_{10}\left(\frac{\sinh 2H}{\sinh 2J}\right) \tag{12.105a}$$

$$R = 10 \log_{10}\left(\frac{\tanh H}{\tanh J}\right) \tag{12.105b}$$

If θ is the electrical length of each section, then the bandwidth of the coupler extends from θ to $\pi - \theta$, where θ is obtained from

$$\cos \theta = \frac{1}{\cosh (J/2)} \tag{12.106}$$

Using the values of H and J as obtained from (12.105), we calculate another set of parameters β and h from two of the following three equations:

$$\frac{\cosh^2 (J/2) + \sinh^2 (J/2)}{\sinh^2 (J/2)} = \frac{2\beta}{(\beta - h)} \tag{12.107a}$$

$$\frac{\sinh^2 (H/2) + \cosh^2 (J/2)}{\sinh^2 (H/2)} = \frac{\sqrt{1 + \beta^2} + \beta}{\sqrt{1 + \beta^2} - h} \tag{12.107b}$$

$$\frac{\cosh^2 (H/2) - \cosh^2 (J/2)}{\cosh^2 (H/2)} = \frac{\sqrt{1 + \beta^2} - \beta}{\sqrt{1 + \beta^2} + h} \tag{12.107c}$$

The even-mode impedance Z_{e1} and Z_{e2} of the two sections are calculated using

$$Z_{e1} = Z_c\left(\frac{p}{q+1}\right) \tag{12.108a}$$

$$Z_{e2} = Z_c Z_{e1} q \tag{12.108b}$$

where

$$p = [(1+\beta^2) + (\beta + h)\sqrt{1+\beta^2} + \beta h]^{1/2} + [(1+\beta^2) - (\beta+h)\sqrt{1+\beta^2} + \beta h]^{1/2} + [2\beta(\beta+h)]^{1/2} \tag{12.109a}$$

$$q = (1 + \beta^2 - h^2)^{1/2} - (\beta^2 - h^2)^{1/2} \tag{12.109b}$$

and Z_c is the characteristic impedance of the terminating ports. The odd-mode impedances Z_{o1} and Z_{o2} are then obtained from the relation

$$Z_{e1}Z_{o1} = Z_{e2}Z_{o2} = Z_c^2 \tag{12.110}$$

It has been shown [52] that with a two-element coupler, one can achieve a bandwidth ratio of 5 : 1 for a coupling of 3 ± 0.5 *dB* and a bandwidth ratio of 4 : 1 for a coupling of 10 ± 0.5 *dB*. By a careful choice of reference planes in the output ports, the 3 *dB* couplers may be used as 90° hybrids.

12.6.4 Directional Coupler in Inhomogeneous Medium

Figure 12.42 shows the configuration of a broadside-coupled directional coupler in suspended stripline. The medium is inhomogeneous and the dielectric loading between the two strips

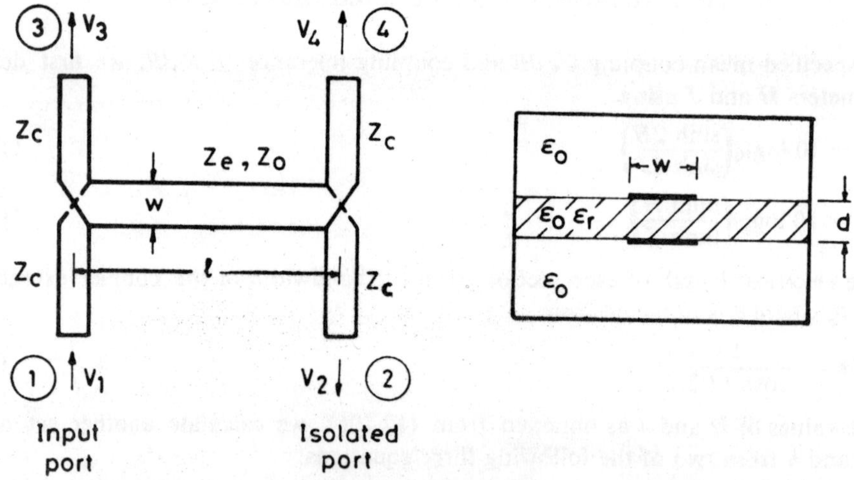

Fig. 12.42 Broadside-coupled directional coupler in suspended stripline

causes the odd-mode to propagate with a smaller phase velocity than the even-mode ($v_o < v_e$). Consequently, the electrical length of the coupled line for the odd-mode is larger than that for the even-mode ($\theta_o > \theta_e$). Dalley [54] has reported the general analysis of this coupler taking into account the unequal phase velocities for the even- and odd-modes. Referring to Fig. 12.42 for the port voltages and impedances, the general coupler equations due to Dalley [54] are summarized below:

$$\frac{Z_{in}}{Z_c} = \frac{1 - P_1 \tan\theta_e \tan\theta_o + j(P_2 \tan\theta_e + P_3 \tan\theta_o)}{1 - P_1 \tan\theta_e \tan\theta_o + j(P_3 \tan\theta_e + P_2 \tan\theta_o)} \tag{12.111}$$

$$\frac{V_2}{V_1} = \frac{1}{Q_e} + \frac{1}{Q_o} \tag{12.112a}$$

$$\frac{V_3}{V_1} = \frac{(\cos\theta_e + j\sqrt{R}\sin\theta_e)}{Q_e} - \frac{\cos\theta_o + j(1/\sqrt{R})\sin\theta_o}{Q_o} \tag{12.112b}$$

$$\frac{V_4}{V_1} = \frac{1}{Q_e} - \frac{1}{Q_o} \tag{12.112c}$$

where

$$P_1 = \left(\frac{Z_e + Z_o}{2Z_c}\right)^2 \tag{12.113a}$$

$$P_2 = \frac{3Z_e + Z_o}{4Z_c} \tag{12.113b}$$

$$P_3 = \frac{3Z_o + Z_e}{4Z_c} \tag{12.113c}$$

$$Q_e = 2\cos\theta_e + j(\sqrt{R} + 1/\sqrt{R})\sin\theta_e \tag{12.113d}$$

$$Q_o = 2\cos\theta_o + j(\sqrt{R} + 1/\sqrt{R})\sin\theta_o \tag{12.113e}$$

$$Z_c = \sqrt{Z_e Z_o} \tag{12.113f}$$

$$R = Z_e/Z_o \tag{12.113g}$$

In (12.111), Z_{in} is the driving point impedance at the input port, and Z_c is the characteristic impedance of the four output ports. Imposing the condition of perfect impedance match at the input ($Z_{in} = Z_c$), yields

$$\tan\theta_e = \tan\theta_o \tag{12.114}$$

Because θ_o is greater than θ_e in this structure, we can write

$$\theta_o = \theta_e + \pi \tag{12.115}$$

For this matched condition, the coupler equations reduce to

$$\frac{V_2}{V_1} = 0 \tag{12.116a}$$

$$\frac{V_3}{V_1} = \frac{j\left(\sqrt{\dfrac{Z_e}{Z_o}} - \sqrt{\dfrac{Z_o}{Z_e}}\right)\sin\theta_e}{\left[2\cos\theta_e + j\left(\sqrt{\dfrac{Z_e}{Z_o}} + \sqrt{\dfrac{Z_o}{Z_e}}\right)\sin\theta_e\right]} \tag{12.116b}$$

$$\frac{V_4}{V_1} = \frac{2}{\left[2\cos\theta_e + j\left(\sqrt{\dfrac{Z_e}{Z_o}} + \sqrt{\dfrac{Z_o}{Z_e}}\right)\sin\theta_e\right]} \tag{12.116c}$$

The length of the coupled section is given by

$$l = \frac{v^a}{2f_0(\sqrt{\epsilon_{fo}} - \sqrt{\epsilon_{fe}})} \tag{12.117}$$

where v^a is the free space velocity, f_0 is the center frequency, and ϵ_{fe} and ϵ_{fo} are the effective dielectric constants for the even-mode and odd-mode respectively.

From (12.116), it can be seen that port 2 of the main input line is isolated and the two output signals appear at the two ports of the auxiliary line, thus providing d.c. isolation between the input port and both the output ports. However, unlike the backward wave coupler, the condition for input match given by (12.114) is valid only at the center frequency. The bandwidth of the coupler is therefore smaller than that of a backward wave coupler and is limited essentially due to the input return loss. Dalley [54] has reported a theoretical bandwidth of 20% for a minimum return loss of 27 *dB*.

12.6.5 Practical Considerations

As discussed in the preceding sections, the bandwidth of parallel-coupled backward wave directional coupler can be increased by increasing the number of quarter-wave sections. The practical realization of such multi-section couplers, however, requires a judicious choice of the transmission structure. In a symmetric coupler, the tightest coupled element forms the centre section, and in an asymmetric coupler, it forms one of the end sections. The coupling of this tightest coupled element is always considerably tighter than the mean coupling level of the coupler. For the same coupling level and tolerance on coupling, as the number of sections is increased, the coupling of the tightest coupled section becomes increasingly tighter. For example, a two-element, 3 *dB* asymmetric TEM-mode coupler designed for a bandwidth ratio of 3 : 1 (coupling tolerance ± 0.128 *dB*) requires the tightest coupled element to have a normalized even-mode impedance of 3.13. For an input impedance of 50 ohms, the even- and odd-mode impedances of this element are 156.5 ohms and 16 ohms respectively. On the other hand, a five-element, 3 *dB* coupler designed for a bandwidth ratio of 8 : 1 (coupling tolerance ± 0.123 *dB*) requires the tightest coupled element to have a normalized even-mode impedance of 4.2. This corresponds to an even-mode impedance of 210 ohms and an odd-mode impedance of 11.9 ohms. These impedances are difficult to realize in practice in edge-coupled transmission structures, because the spacing between the coupled conductors becomes too narrow to fabricate within the required tolerance. Impedances of this order can be realized using broadside-coupled and broadside-offset coupled strip conductors in either sandwiched stripline or suspended stripline configuration (see Chapters 7 and 8).

An important parameter which limits the practical bandwidth of parallel-coupled backward wave directional coupler is its directivity. For perfect directivity, the even- and odd-mode phase velocities of coupled lines must be equal. This is satisfied in an ideal TEM-mode coupler. Couplers fabricated in quasi-TEM lines such as the suspended stripline do not meet this requirement. There are however schemes for equalizing the two phase velocities; (a) by using a dielectric overlay on the conductor pattern and (b) by using anisotropic dielectric substrates instead of isotropic substrates. These are discussed in Chapters 6 and 7. The even- and odd-mode phase velocities can also be equalized by altering the conductor pattern. Examples are, the meander-folded Schiffman section reported by Rehnmark [55], and the wiggly line microstrip coupler reported by Podell [56]. These couplers, however, are reported to be useful for lower microwave frequencies. Furthermore, in order to be able to achieve high directivity, they require stringent tolerances to be met during the preparation of the photomask as well as in the etching process.

The unequal mode velocities inherent in inhomogeneous coupled lines can be utilized to obtain coupled outputs at the two ports of the auxiliary line. As discussed in Sec. 12.6.4, this requires a proper choice of the coupling length. A structure which makes use of unequal mode velocities to achieve only forward wave coupling is the coupled combline directional coupler reported by Gunton and Paige [57]. The coupler makes use of two interleaved planar comb transmission lines. The coupling is predominantly capacitive and the coupling level is dependent on the difference in the mode velocities rather than the mode impedances. This type of coupler is reported to offer a bandwidth of about 35% for a VSWR of less than 1.25 and a minimum isolation of 20 *dB*.

Brenner [58] has carried out the effect of variations of the critical parameters of quarter wave parallel coupled directional couplers on their performance characteristics. If ΔZ_e, ΔZ_o, Δv_e and Δv_o denote small variations from the ideal values of even-mode impedance, odd-mode impedance, even-mode phase velocity and odd-mode phase velocity, respectively, then the first-order approximation of the scattering parameters of the coupler (refer Fig. 12.37) at the centre frequency are given as follows [58]:

$$S_{11} = \frac{1}{2}\left(\frac{\Delta Z_e}{Z_e} + \frac{\Delta Z_o}{Z_o}\right)(1 - C_v^2) + j\frac{\pi}{4}\left(\frac{\Delta v_e}{v_e} - \frac{\Delta v_o}{v_o}\right)C_v\sqrt{1 - C_v^2} \tag{12.118a}$$

$$S_{21} = -j\sqrt{1 - C_v^2} + \frac{\pi}{4}\left(\frac{\Delta v_e}{v_e} + \frac{\Delta v_o}{v_o}\right)(1 - C_v^2) + \frac{j}{2}\left(\frac{\Delta Z_e}{Z_e} - \frac{\Delta Z_o}{Z_o}\right)C_v\sqrt{1 - C_v^2} \tag{12.118b}$$

$$S_{31} = C_v + \frac{1}{2}\left(\frac{\Delta Z_e}{Z_e} - \frac{\Delta Z_o}{Z_o}\right)(1 - C_v^2) + j\frac{\pi}{4}\left(\frac{\Delta v_e}{v_e} + \frac{\Delta v_o}{v_o}\right)C_v\sqrt{1 - C_v^2} \tag{12.118c}$$

$$S_{41} = \frac{\pi}{4}\left(\frac{\Delta v_e}{v_e} - \frac{\Delta v_o}{v_0}\right)(1 - C_v^2) + \frac{j}{2}\left(\frac{\Delta Z_e}{Z_e} + \frac{\Delta Z_o}{Z_o}\right)C_v\sqrt{1 - C_v^2} \tag{12.118d}$$

where

$$C_v = \frac{Z_e - Z_o}{Z_e + Z_o} \tag{12.119}$$

It has been shown [58] that small velocity perturbations have a strong effect on directivity but very little on the input VSWR, especially in loose couplers. For example a 10% change in v_e or v_o results in a return loss of 33 *dB* and a directivity of 13 *dB* in a 10 *dB* coupler, and in a 20 *dB* coupler, it results in a return loss of 42 *dB* but the directivity deteriorates to 2 *dB*. On the other hand, for a 10% change in Z_e or Z_o, the return loss and the directivity are both of the order of 26 *dB* in a 10 *dB* coupler as well as a 20 *dB* coupler.

12.7 Hybrid Rings and Baluns

12.7.1 Hybrid Ring Coupler [59, 60]

The branch-line coupler as well as the parallel-coupled backward wave coupler provide an inherent phase difference of 90° between the voltages at the two output arms. The hybrid ring coupler (Fig. 12.43a) is a four port network with the difference that the two output signals are either in-phase or 180° out-of-phase depending on the choice of the input arm. The mean circumference of the ring is $3\lambda_0/2$, where λ_0 is the wavelength in the transmission medium at midband frequency f_0. If we denote $12\theta_0$ as the mean electrical length at f_0, then $\theta_0 = \pi/4$. The electrical length θ at any other frequency f is given by

$$\theta = \frac{\pi}{4} \frac{\lambda_0}{\lambda} \tag{12.120}$$

where λ is the guide wavelength at f.

For the purpose of analysis, we take advantage of the symmetry of the structure about the plane PP' and consider one symmetric half with an open circuit at PP' for even-mode excitation and a short circuit at PP' for odd-mode excitation. This is shown in Fig. 12.43b along with its equivalent circuit. The parameters Y_{1e} and Y_{1o} denote the input admittance of the stub of length 3θ for the even-mode and odd-mode respectively, and Y_{2e} and Y_{2o} denote the corresponding quantities for the stub of length θ. They are given by

$$Y_{1e} = jY_aT_1 \tag{12.121a}$$

$$Y_{1o} = -jY_a/T_1 \tag{12.121b}$$

$$Y_{2e} = jY_aT_2 \tag{12.121c}$$

$$Y_{2o} = -jY_a/T_2 \tag{12.121d}$$

where

$$T_1 = \tan 3\theta, \qquad T_2 = \tan \theta \tag{12.122}$$

Let $[M]_e$ and $[M]_o$ denote the overall $ABCD$ matrix of the equivalent network for even-mode and odd-mode respectively. They are given by

$$[M]_e = \begin{bmatrix} (\cos 2\theta - Y_aT_2 \sin 2\theta/Y_b) & j \sin 2\theta/Y_b \\ jY_aT_1 (\cos 2\theta - Y_aT_2 \sin 2\theta/Y_b) + j(Y_b \sin 2\theta + Y_aT_2 \cos 2\theta) & (-Y_aT_1 \sin 2\theta/Y_b + \cos 2\theta) \end{bmatrix} \tag{12.123}$$

The elements of $[M]_o$ are obtained by replacing Y_a by $-Y_a$, T_1 by $1/T_1$ and T_2 by $1/T_2$ in (12.123). At the centre frequency f_0, we have $\theta = \theta_0 = \pi/4$. Setting $\theta = \pi/4$ in (12.123), we obtain

$$[M]_e = \begin{bmatrix} -Y_a/Y_b & -j/Y_b \\ j(Y_a^2 + Y_b^2)/Y_b & Y_a/Y_b \end{bmatrix} \tag{12.124}$$

Using the formulas given in (12.15), we now convert the elements of $[M]_e$ and $[M]_o$ into the scattering parameters. Denoting $[S]_e$ and $[S]_o$ as the scattering matrices for the even-mode and odd-mode respectively, we get

$$[S]_e = \begin{bmatrix} \dfrac{1 - Y_a^2 - Y_b^2 + j2Y_a}{\Delta} & \dfrac{-j2Y_b}{\Delta} \\ \dfrac{-j2Y_b}{\Delta} & \dfrac{1 - Y_a^2 - Y_b^2 - j2Y_a}{\Delta} \end{bmatrix} \tag{12.125}$$

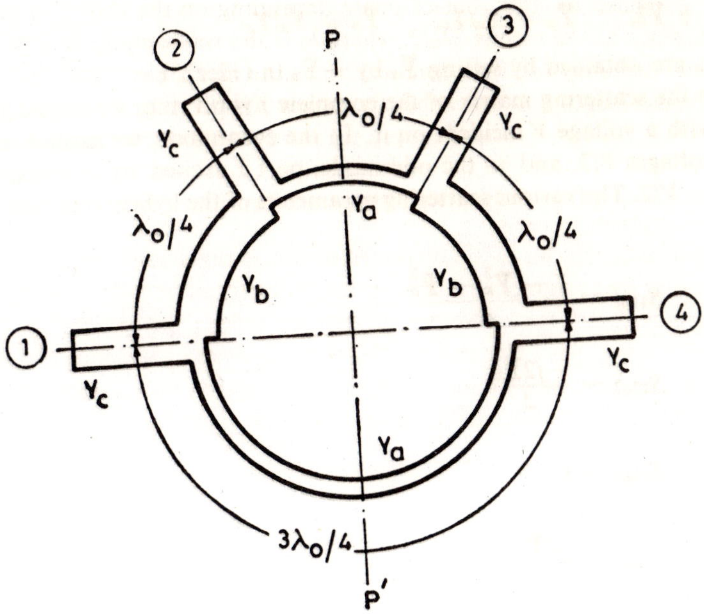

(a)

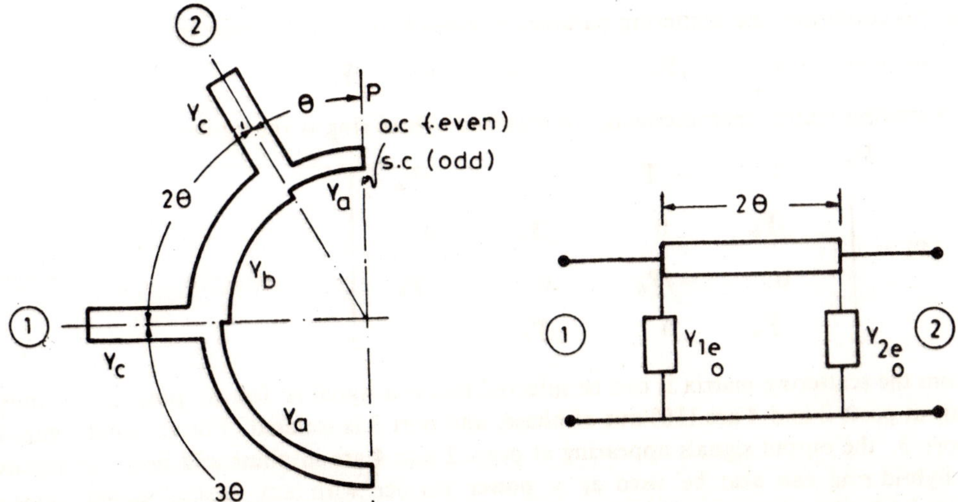

(b)

Fig. 12.43 (a) Hybrid ring coupler, (b) One-half of the structure for the purpose of analysis and its equivalent circuit

where

$$\Delta = 1 + \bar{Y}_a^2 + \bar{Y}_b^2, \qquad \bar{Y}_a = Y_a/Y_c, \qquad \bar{Y}_b = Y_b/Y_c \tag{12.126}$$

The elements of $[S]_o$ are obtained by setting $\bar{Y}_a$ by $-\bar{Y}_a$ in (12.125).

In order to derive the scattering matrix of the complete hybrid ring, we consider port 1 to be the input port with a voltage V incident on it. In the even-mode, we assume ports 1 and 4 to be excited by voltages $V/2$, and in the odd-mode, port 1 excited by a voltage $V/2$ and port 4 by a voltage $-V/2$. The various scattering parameters of the hybrid ring (see Fig. 12.43) are given by

$$S_{11} = \frac{1}{2}(S_{11e} + S_{11o}) = \frac{1 - \bar{Y}_a^2 - \bar{Y}_b^2}{\Delta} \tag{12.127a}$$

$$S_{12} = \frac{1}{2}(S_{12e} + S_{12o}) = \frac{-j2\bar{Y}_b}{\Delta} \tag{12.127b}$$

$$S_{13} = \frac{1}{2}(S_{12e} - S_{12o}) = 0 \tag{12.127c}$$

$$S_{14} = \frac{1}{2}(S_{11e} - S_{11o}) = \frac{j2\bar{Y}_a}{\Delta} \tag{12.127d}$$

For matched condition at the input, $S_{11} = 0$ which yields

$$\bar{Y}_a^2 + \bar{Y}_b^2 = 1 \tag{12.128}$$

With this condition, the scattering parameters given by (12.127) reduce to

$$S_{11} = 0, \qquad S_{12} = -j\bar{Y}_b, \qquad S_{13} = 0, \qquad S_{14} = j\bar{Y}_a \tag{12.129}$$

The scattering matrix characterizing the matched hybrid ring is then given by

$$[S] = \begin{bmatrix} 0 & -j\bar{Y}_b & 0 & j\bar{Y}_a \\ -j\bar{Y}_b & 0 & -j\bar{Y}_a & 0 \\ 0 & -j\bar{Y}_a & 0 & -j\bar{Y}_b \\ j\bar{Y}_a & 0 & -j\bar{Y}_b & 0 \end{bmatrix} \tag{12.130}$$

From the scattering matrix it can be inferred that if a signal is fed to port 1, the output signals at ports 2 and 4 are 180° out of phase, and port 3 is isolated. For an input signal fed to port 3, the output signals appearing at ports 2 and 4 are in-phase and port 1 is isolated. The hybrid ring can also be used as a power divider with any desired power ratio by appropriately choosing the admittances Y_a and Y_b.

12.7.2 Rat-race Hybrid [8, 60, 61]

The rat-race hybrid is a special case of the hybrid ring shown in Fig. 12.43 when $Y_a = Y_b = Y_c/\sqrt{2}$. It is a planar version of the waveguide magic-*T*. The scattering matrix of the rat-race hybrid (see Fig. 12.44) is given by

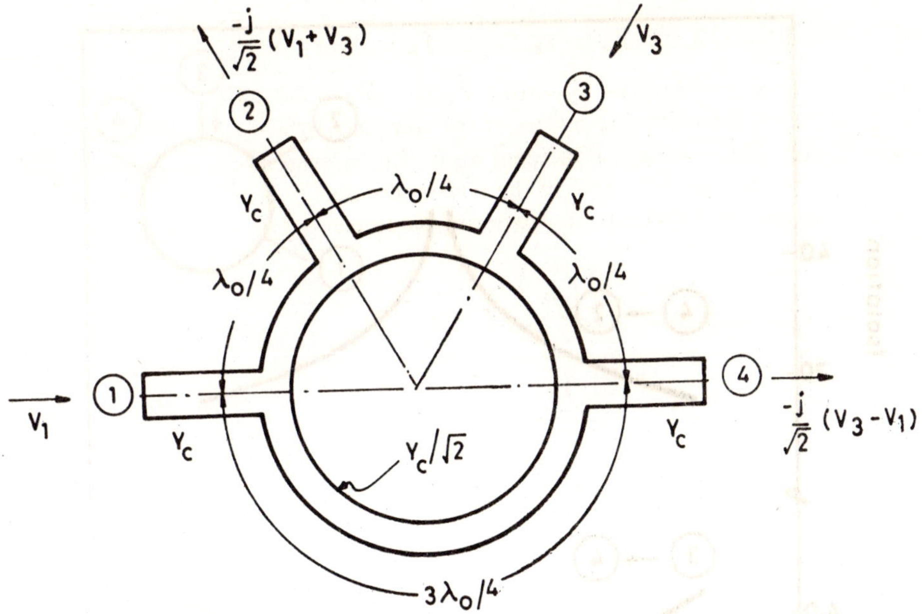

Fig. 12.44 Rat-race hybrid

$$[S] = \begin{bmatrix} 0 & \frac{-j}{\sqrt{2}} & 0 & \frac{j}{\sqrt{2}} \\ \frac{-j}{\sqrt{2}} & 0 & \frac{-j}{\sqrt{2}} & 0 \\ 0 & \frac{-j}{\sqrt{2}} & 0 & \frac{-j}{\sqrt{2}} \\ \frac{j}{\sqrt{2}} & 0 & \frac{-j}{\sqrt{2}} & 0 \end{bmatrix} \tag{12.131}$$

The scattering parameters given by (12.131) are valid at the centre frequency. Setting $Y_a = Y_b = Y_c/\sqrt{2}$ in the general *ABCD* matrix given by (12.123) and following the same procedure as outlined in Sec. 12.7.1, the frequency dependent scattering parameters of the rat-race hybrid can be derived.

Figure 12.45 shows the response curves of a 3 *dB* rat-race hybrid [8]. It can be seen that the useful practical bandwidth in terms of isolation, power split and VSWR is approximately 20%. Figure 12.46 shows the phase response as a function of frequency for both 0° and 180° conditions at midband. It may be observed that for a specified phase deviation, the 0° condition has a much wider bandwidth than the 180° condition. However, the performance is acceptable for most circuit applications operating upto about 20% bandwidth. Rat-race hybrids with equiphase outputs are particularly useful as power division networks in symmetrical parallel-feed antenna arrays, because no phase compensating element is necessary. They are also commonly used in balanced mixers, monopulse comparators and single sideband generators.

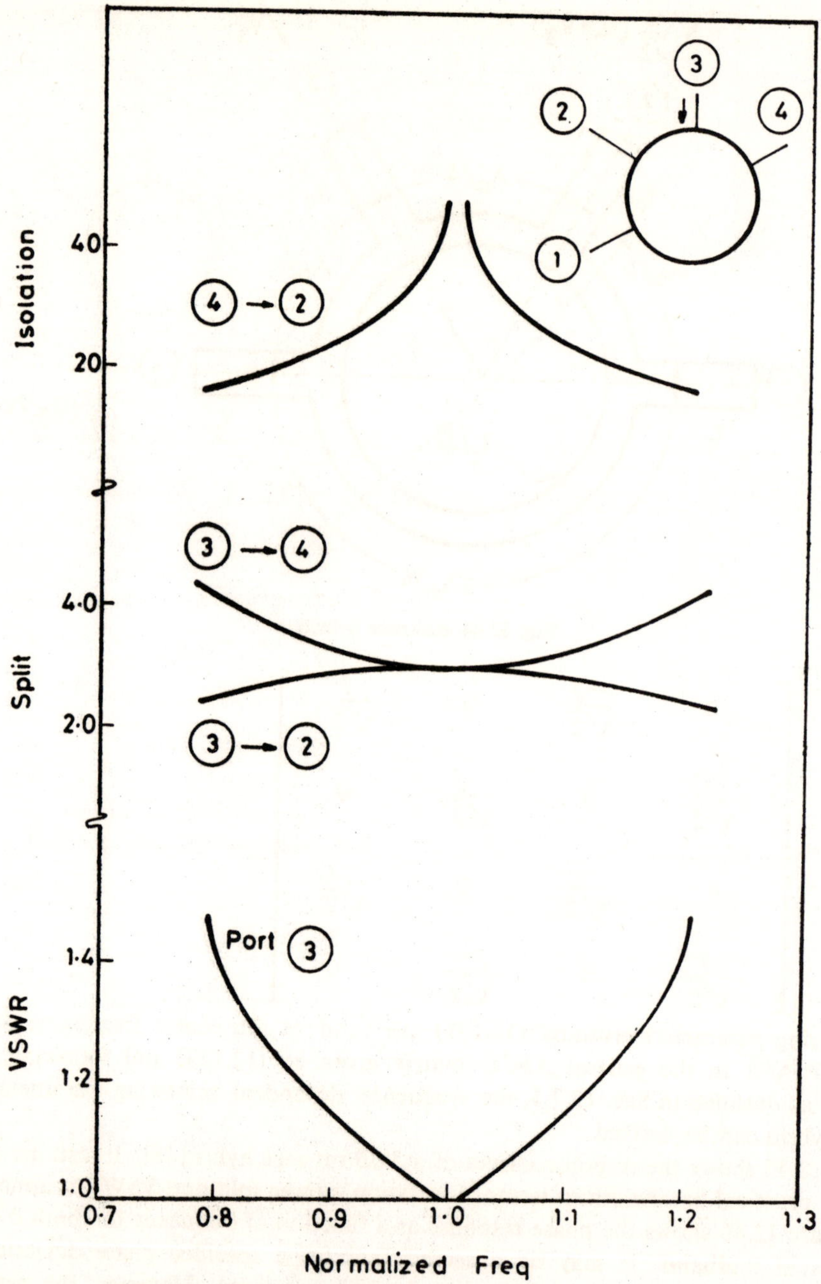

Fig. 12.45 Response curves for a rat-race hybrid (From Howe [8], reprinted with permission of Artech House Inc.)

The bandwidth of the rat-race hybrid shown in Fig. 12.44 is limited mainly due to the three quarter-wavelength section of the ring. Figure 12.47 shows a broadband version of the hybrid reported by March [61]. In this, the three quarter-wavelength section is replaced by a

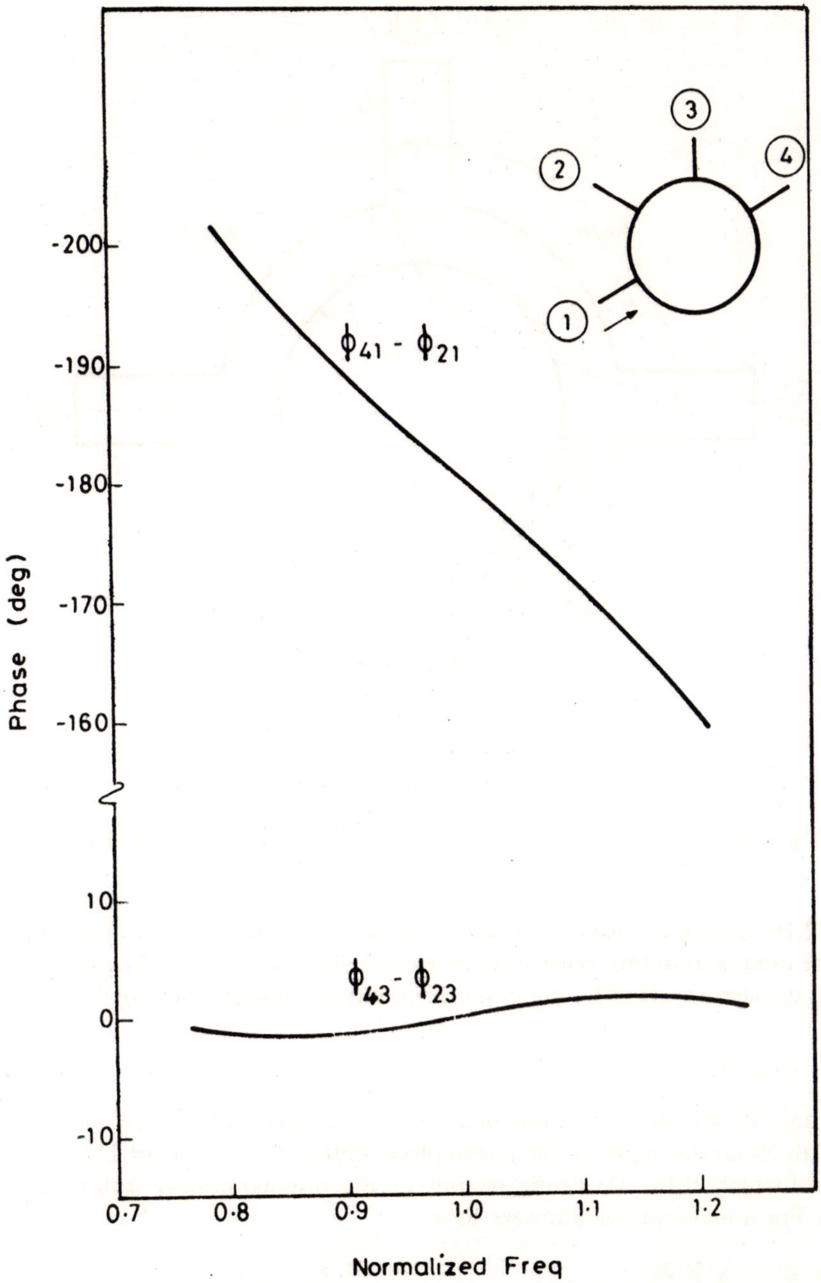

Fig. 12.46 Phase response of a rat-race hybrid (From Howe [8], reprinted with permission of Artech House, Inc.)

quarter-wavelength coupled line section with diametrically opposite ends short circuited. This coupled section is basically a bandpass filter element and approximates a phase-reversing network over a wide frequency range. The image impedance Z_I of such a coupled section

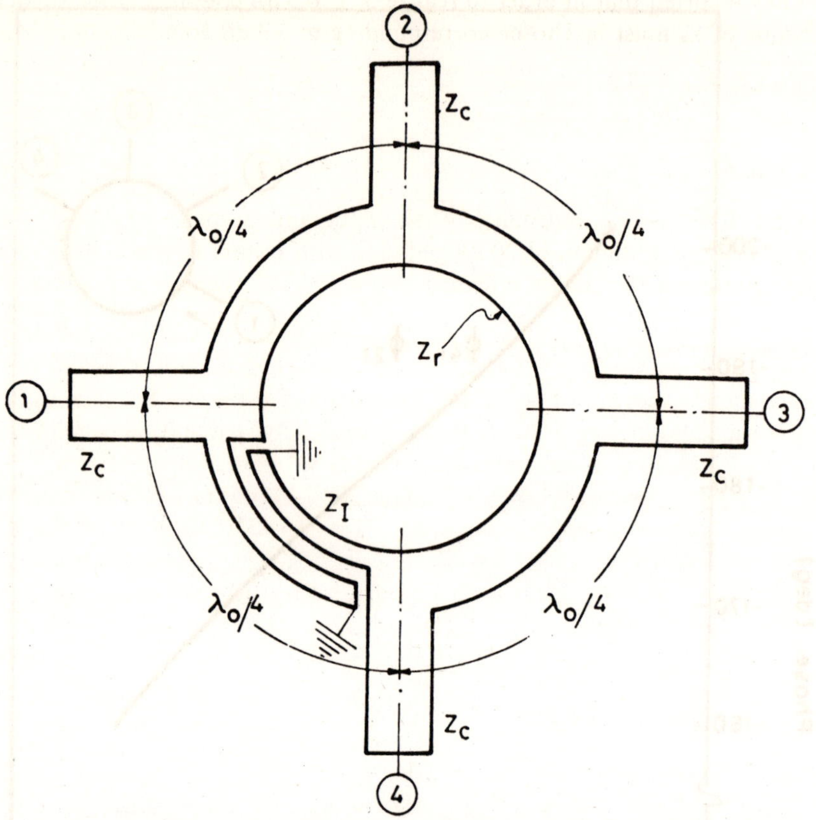

Fig. 12.47 Wideband rat-race hybrid (From March [61], Copyright © 1968 IEEE, reprinted with permission)

(see Fig. 12.18) having an electrical length θ is given by (12.47) and the image phase shift ϕ between the input and output ports is given by (12.48). At midband frequency, the coupled section is a quarter-wavelength long in the transmission medium. Setting $\theta = \pi/2$ in (12.47) gives

$$Z_I = 2Z_eZ_o/(Z_e - Z_o) \tag{12.132}$$

where Z_e and Z_o are the even- and odd-mode impedances respectively of the coupled line section. With Z_e greater than Z_o, the image phase shift at $\theta = \pi/2$ is $3\pi/2$ which is equal to the electrical length of the $3\lambda_0/4$ long section of the original rat-race hybrid considered in Fig. 12.44. For a matched condition we have

$$Z_r = Z_I = \sqrt{Z_eZ_o} \tag{12.133}$$

where Z_r is the characteristic impedance of the remaining section of the ring made of single strip conductor. From (12.132) and (12.133), we can write

$$Z_e = (\sqrt{2} + 1)Z_r \tag{12.134a}$$

$$Z_o = (\sqrt{2} - 1)Z_r \tag{12.134b}$$

March [61] has reported that in order to realize 3 ± 0.3 *dB* coupling over an octave bandwidth, the value of Z_r must be chosen corresponding to 3.3 *dB* coupling at midband. Thus

$$20 \log_{10} (Z_r/Z_c) = 3.3 \tag{12.135}$$

or

$$Z_r = 1.46 Z_c \tag{12.136}$$

where Z_c is the characteristic impedance of the input and output lines. With this value of Z_r, the hybrid offers a minimum of 20 *dB* isolation with respect to the isolated port and the ring outputs maintain a 180° phase difference within ± 15° over the octave bandwidth.

12.7.3 Broadband Baluns [62, 63]

In the rat-race hybrid shown in Fig. 12.44, if a signal is fed to port 1, the output signals at ports 2 and 4 are equal in magnitude and out of phase. By terminating port 3 (isolated port) in an open circuit, this hybrid can be used as a balun. The device offers a bandwidth of about 30% for a phase deviation of ± 10° about 180°. Figure 12.48 shows the configuration of a broadband balun reported by Bex [62]. The bandwidth depends on the line lengths l and l_{opt}, and, to a lesser extent on the characteristic impedances of the ring sections. For a specified value of λ_0, l and Z_c, l_{opt} is to be optimized. Bex [62] has reported a bandwidth of about 50% for a balun designed at 2.5 GHz.

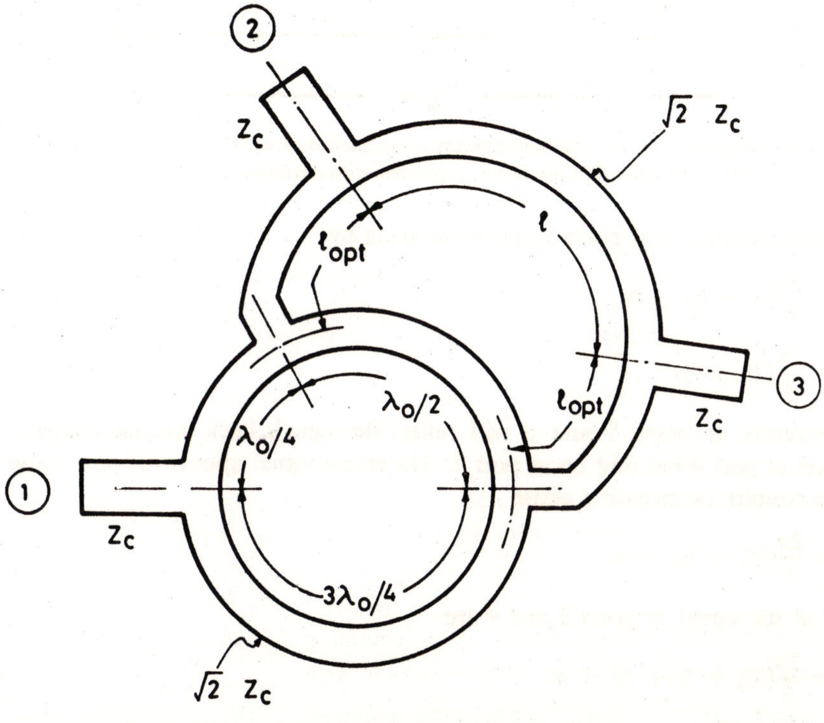

Fig. 12.48 Broadband balun (From Bex [62], Electronics Letters, Jan. 1975, Vol. 11. No. 2, pp. 47-48, reprinted with permission of IEE.)

Figure 12.49 shows the schematic of another broadband balun employing 3 *dB*, 90° backward wave couplers [63]. Broad bandwidths can be achieved in this balun by using multisection backward wave couplers. With signal fed to port 1, the first coupler divides the power equally but with a phase difference of 90° between ports 2 and 3. The phase of the signal at port 4 with reference to the input port 1 is

$$\phi_{14} = \frac{2\pi}{\lambda}(l_1 + l_2) - \frac{\pi}{2} \tag{12.137}$$

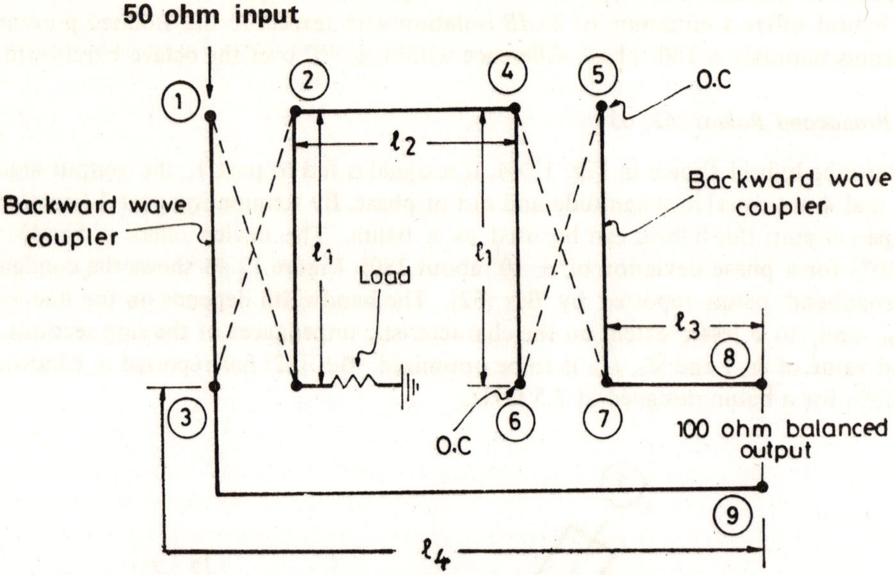

Fig. 12.49 Schematic of a broadband balun using backward wave couplers (From Walker [63], Copyright © 1968 IEEE, reprinted with permission)

The corresponding phases at ports 5 and 6 are given by,

$$\phi_{15} = \frac{2\pi}{\lambda}(l_1 + l_2) - \pi \tag{12.138}$$

$$\phi_{16} = \frac{2\pi}{\lambda}(2l_1 + l_2) - \frac{\pi}{2} \tag{12.139}$$

The open circuits at ports 5 and 6 will reflect the signals back into the coupler. The two signals cancel at port 4 but add up at port 7. The entire signal appears at port 7 after passing through the coupler twice with a phase

$$\phi_{17} = \frac{2\pi}{\lambda}(3l_1 + l_2) - \pi \tag{12.140}$$

The phases of the signal at ports 8 and 9 are

$$\phi_{18} = \frac{2\pi}{\lambda}(3l_1 + l_2 + l_3) - \pi \tag{12.141}$$

$$\phi_{19} = \frac{2\pi}{\lambda}(l_1 + l_4) \tag{12.142}$$

In order that this device functions as a balun, we require $\phi_{19} - \phi_{18} = \pi$. This condition is satisfied when

$$l_4 = 2l_1 + l_2 + l_3 \tag{12.143}$$

If the impedance of the input line is 50 ohms, the output impedance of the balun is 100 ohms, making it quite useful for feeding cavity backed spiral antennas.

12.8 Power Dividers

As discussed in Secs. 12.5 to 12.7, directional couplers and hybrid rings can be used as two-way power dividers. The fourth port which forms the isolated port is terminated in a matched load. The branchline and parallel-coupled directional couplers provide quadrature phase difference between the two output-signals. The hybrid ring provides output signals which are either in-phase or out-of-phase. In several microwave applications, as for example, a parallel feed for a phased array antenna, the input signal is required to be divided into several equi-amplitude, equiphase output signals. A matched, symmetric n-way power divider has the advantage that it gives neither amplitude nor phase imbalance at any frequency. Such a power divider can also be used as a n-way power combiner by simply reversing the input and output ports. Using such combiners, output powers of a number of solid state amplifiers and oscillators can be combined over a wide frequency range.

A n-way hybrid power divider was first described by Wilkinson [64]. Following this publication, several investigators [65–70] have reported further studies on various aspects of n-way power dividers. In particular, the special case of a two-way broadband power divider has received extensive attention [71–74]. Design curves for a number of power divider networks have been reported by Howe [8]. In this section, we shall first discuss the design aspects of three port hybrids as power dividers/combiners and then review the salient aspects of multiport power dividers.

12.8.1 Basic Two-Way Power Divider

It is well known that a reciprocal three-port T-junction cannot be matched at all the three ports simultaneously. For example, consider a three port junction as a power divider. Referring to Fig. 12.50, let port 1 be a matched input port. Power incident at port 1 is equally divided between the output ports 2 and 3. The phase factors ϕ_{21} and ϕ_{31} of S_{21} and S_{31}, respectively, can be equalized to zero by properly selecting the reference planes and

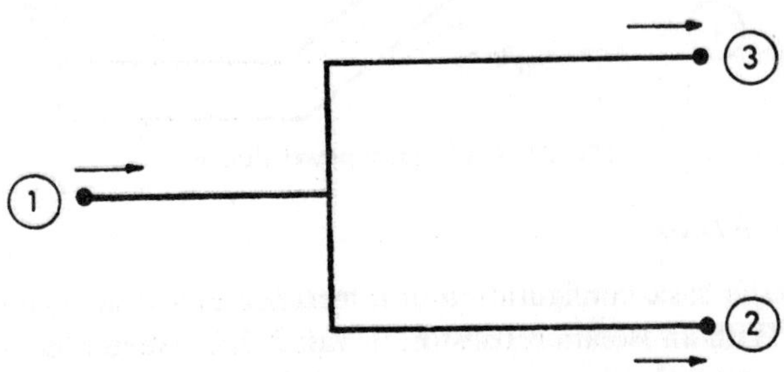

Fig. 12.50 Schematic of a T-junction

the scattering matrix can be written as

$$[S] = \begin{bmatrix} 0 & \frac{1}{\sqrt{2}} & \frac{1}{\sqrt{2}} \\ \frac{1}{\sqrt{2}} & S_{22} & S_{23} \\ \frac{1}{\sqrt{2}} & S_{32} & S_{33} \end{bmatrix}, \quad S_{23} = S_{32} \tag{12.144}$$

Applying the unitary condition given by (12.13), it is easy to show that

$$S_{22} = -S_{23} \tag{12.145a}$$

and

$$| S_{22} | = | S_{33} | = | S_{23} | = \frac{1}{2} \tag{12.145b}$$

Ports 2 and 3 are not matched and the voltage reflection coefficient at each of these is 1/2. Furthermore, the two output ports are not mutually isolated. Figure 12.51 shows two possible layouts of this power divider.

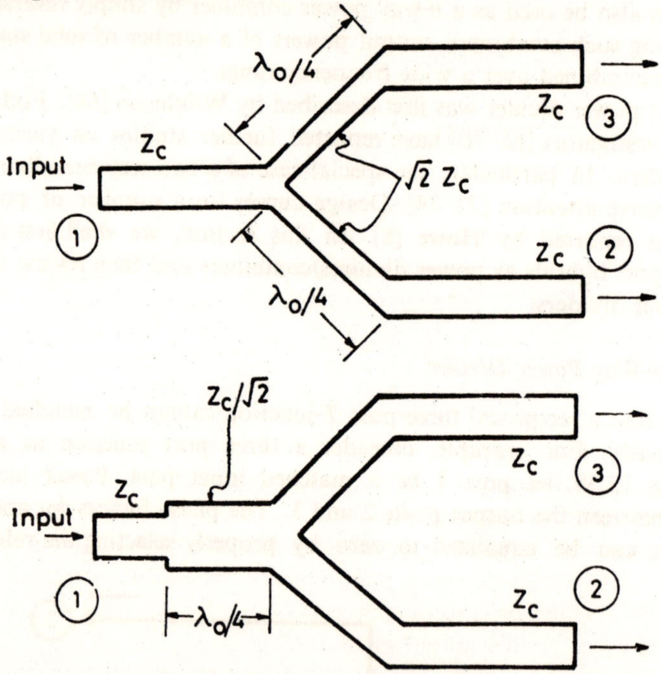

Fig. 12.51 Two-way power dividers

12.8.2 Matched Power Divider

Figure 12.52 shows the basic configuration of a matched power divider/combiner introduced by Wilkinson [64]. With an isolation resistor of value $2Z_c$ between ports 2 and 3, the device is perfectly matched at all the three ports, and ports 2 and 3 are isolated from each other at the centre frequency. The properties of this network can be derived by considering its equivalent circuit when excited by a voltage generator at port 1 (Fig. 12.53a) and at port 2 (Fig. 12.53b) independently. For excitation at port 1, the plane of symmetry *PP'* represents an open circuit. The excitation at port 2 may be considered as a superposition of even- and odd-mode excitations for which the plane of symmetry *PP'* represents an open circuit and a short circuit respectively. The scattering parameters are obtained by applying the conventional transmission line theory to the symmetric one half of the circuit. The overall scattering matrix of the device shown in Fig. 12.52 is obtained as

$$[S] = \begin{bmatrix} 0 & \frac{-j}{\sqrt{2}} & \frac{-j}{\sqrt{2}} \\ \frac{-j}{\sqrt{2}} & 0 & 0 \\ \frac{-j}{\sqrt{2}} & 0 & 0 \end{bmatrix} \tag{12.146}$$

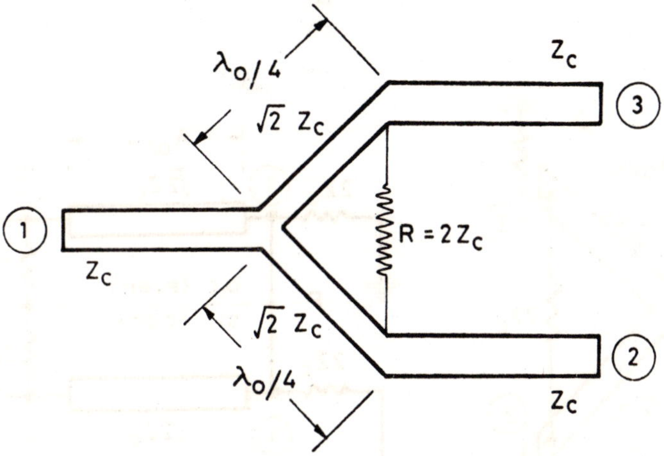

Fig. 12.52 Matched two-way power divider

As a power divider, the input signal is fed to port 1, and the power is equally divided between ports 2 and 3. When equi-amplitude, in-phase signals are fed to ports 2 and 3, the device acts as a power combiner and the entire power appears at port 1. If power is fed only to port 2 or port 3 then half the power appears at port 1 and the remaining half gets dissipated in the isolation resistor. In all cases, ports 2 and 3 are mutually isolated. This device gives a theoretical bandwidth of approximately 1.44 : 1 for an input VSWR of 1.22 and isolation of 20 *dB* between the two output ports.

Cohn [72] has shown that the bandwidth of the three-port hybrid power divider can be increased by cascading several pairs of quarter-wave transmission lines in the output path and bridging the junctions between successive pairs by isolation resistors. Figure 12.54 shows the schematic of a two-section symmetric power divider along with its response curves. The design formulas for obtaining the resistor values for this two-section case are given by [72]

$$R_1 = \frac{2Z_1Z_2}{[(Z_1 + Z_2)(Z_1 - Z_2 \cot^2 \phi)]^{1/2}} \tag{12.147a}$$

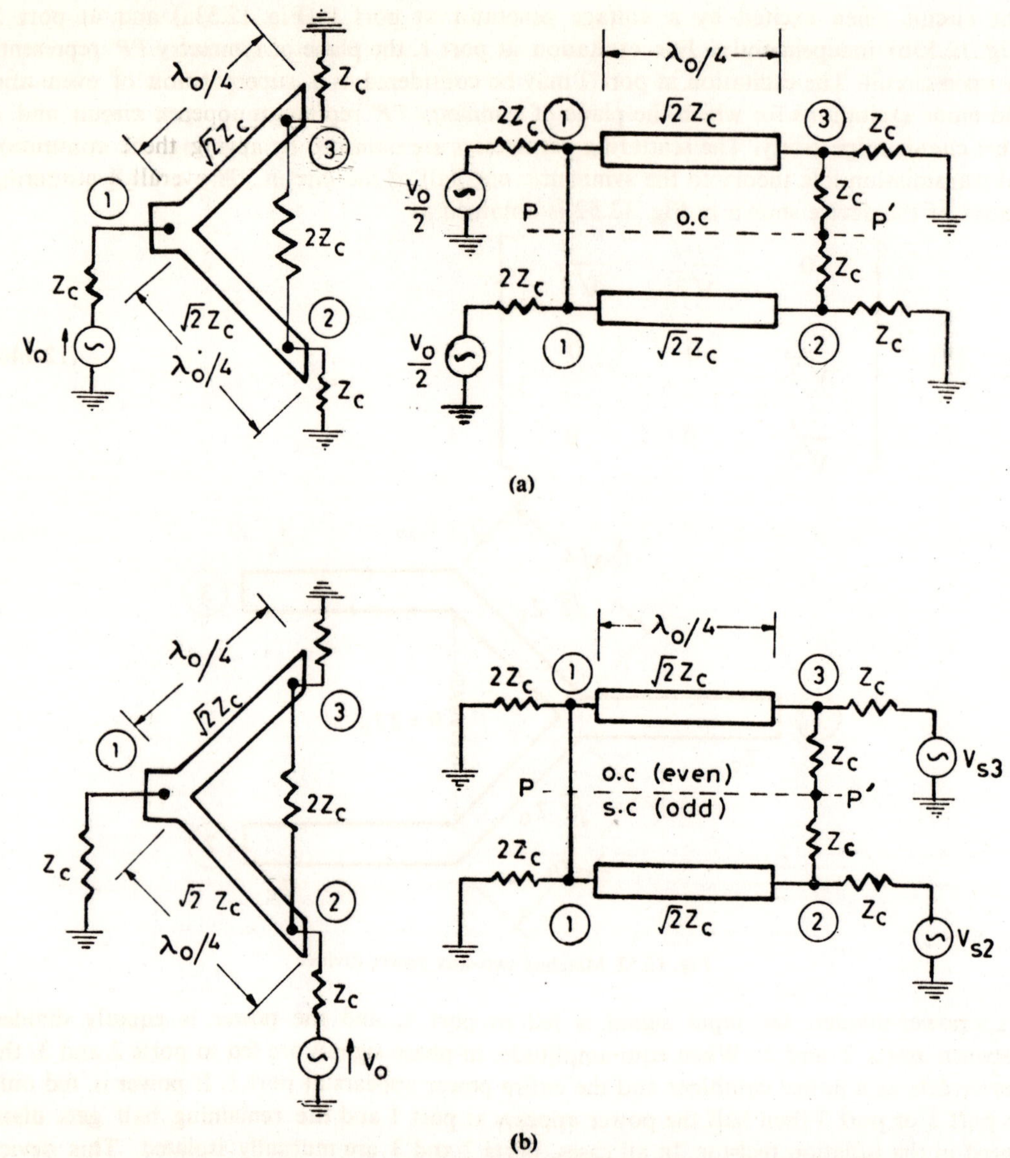

Fig. 12.53 Excitations of the power divider from (a) port 1 and (b) port 3 and their equivalent circuits: $V_{s2} = V_{s3} = V_0/2$ for even-mode, $V_{s2} = -V_{s3} = V_0/2$ for odd-mode.

$$R_2 = \frac{2R_1(Z_1 + Z_2)}{[R_1(Z_1 + Z_2) - 2Z_1]} \tag{12.147b}$$

where

$$\phi = \frac{\pi}{2}\left[1 - \frac{1}{\sqrt{2}}\left(\frac{f_2 - f_1}{f_2 + f_1}\right)\right] \tag{12.147c}$$

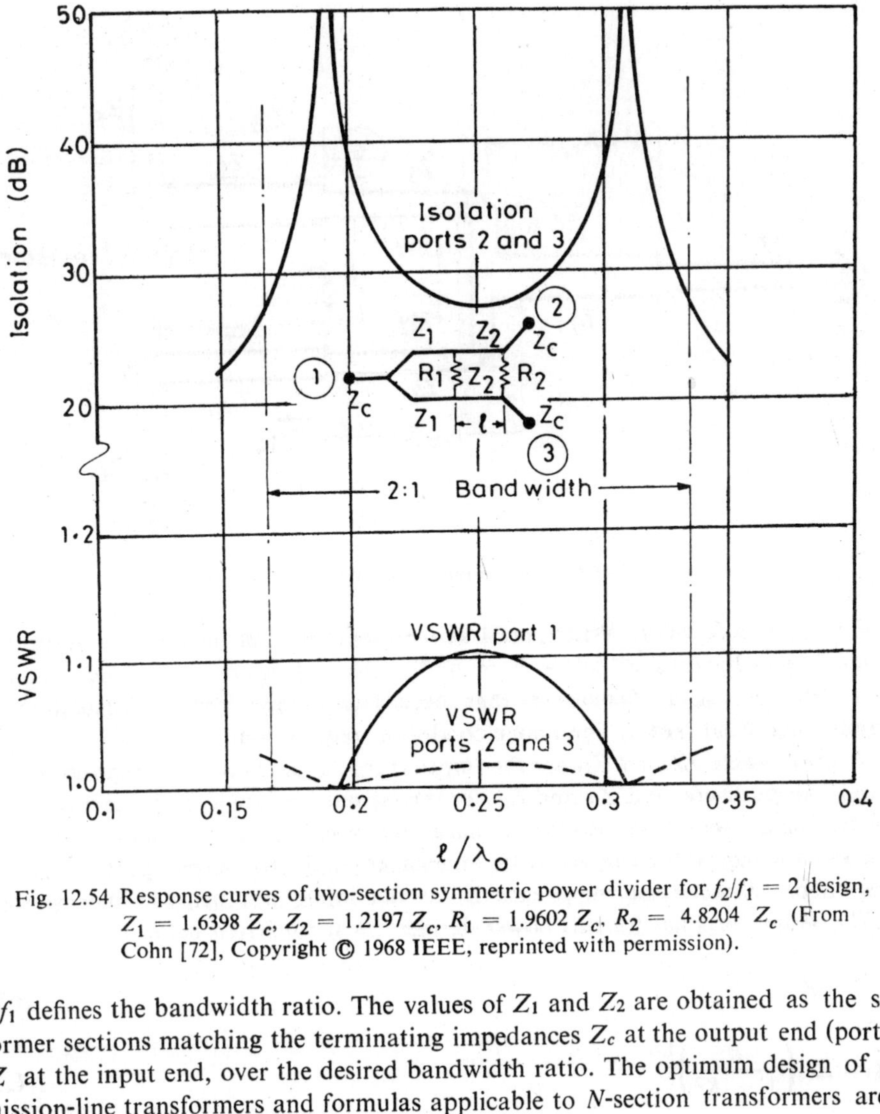

Fig. 12.54. Response curves of two-section symmetric power divider for $f_2/f_1 = 2$ design, $Z_1 = 1.6398\ Z_c$, $Z_2 = 1.2197\ Z_c$, $R_1 = 1.9602\ Z_c$, $R_2 = 4.8204\ Z_c$ (From Cohn [72], Copyright © 1968 IEEE, reprinted with permission).

and f_2/f_1 defines the bandwidth ratio. The values of Z_1 and Z_2 are obtained as the stepped-transformer sections matching the terminating impedances Z_c at the output end (port 2 or 3) and $2Z$ at the input end, over the desired bandwidth ratio. The optimum design of stepped transmission-line transformers and formulas applicable to N-section transformers are available in Cohn [75], and the various impedance values for $N = 2$, 3 and 4 are listed in tabular form in Young [76]. From the response curves depicted in Fig. 12.54, it can be seen that over an octave band, the maximum VSWR is 1.1 at port 1 and 1.02 at ports 2 and 3, and the minimum isolation between ports 2 and 3 is 27.3 *dB*. Cohn [72] has reported that the useful bandwidth increases to about 4 : 1 for three sections, 5.5 : 1 for four sections and 10 : 1 for seven sections. For most practical applications, the bandwidth offered by the two-section power divider is adequate.

12.8.3 Power Divider for Unequal Power Division

Parad and Moynihan [71] have derived design formulas for a matched two-way power divider which provides two isolated equiphase, unequal amplitude outputs. Figure 12.55 shows the

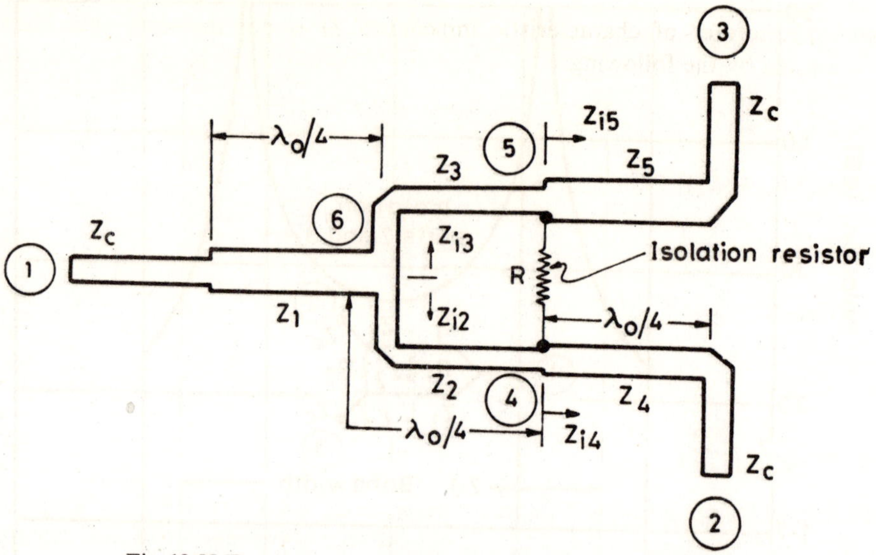

Fig. 12.55 Two-way power divider for unequal power division

layout of the power divider. With port 1 as the input port, let the power division between ports 3 and 2 be in the ratio $K^2 : 1$ or $P_3 = K^2P_2$. The voltages in arms 2 and 3 equidistant from port 1 must be equal. This ensures that there is no voltage drop in the isolation resistor when power is fed from port 1. The above conditions require that $Z_{i2} = K^2Z_{i3}$, $Z_{i4} = K^2Z_{i5}$ and $Z_2 = K^2Z_3$, where Z_{i2} and Z_{i3} are the impedances at junction 6 looking into the output arms 2 and 3 respectively, and Z_{i4} and Z_{i5} are the output impedances at junctions 4 and 5 respectively. The quarter-wave sections of characteristic impedances Z_4 and Z_5 serve as output transformers to match to the characteristic impedance Z_c of the output ports. The quarter-wave section of characteristic impedance Z_1 in the input arm helps to improve the input VSWR. The design formulas for this power divider are summarized below [71]:

$$\frac{P_3}{P_2} = K^2 \tag{12.148a}$$

$$Z_1 = Z_c\left(\frac{K}{1+K^2}\right)^{1/4} \tag{12.148b}$$

$$Z_2 = Z_cK^{3/4}(1+K^2)^{1/4} \tag{12.148c}$$

$$Z_3 = \frac{Z_c(1+K^2)^{1/4}}{K^{5/4}} \tag{12.148d}$$

$$Z_4 = Z_c\sqrt{K} \tag{12.148e}$$

$$Z_5 = \frac{Z_c}{\sqrt{K}} \tag{12.148f}$$

$$R = Z_c\left(\frac{1+K^2}{K}\right) \tag{12.148g}$$

If the input transformer of characteristic impedance Z_1 is not included, then the design equations are given by the following:

$$\frac{P_3}{P_2} = K^2 \tag{12.149a}$$

$$Z_2 = Z_c[K(1 + K^2)]^{1/2} \tag{12.149b}$$

$$Z_3 = Z_c\left(\frac{1 + K^2}{K^3}\right)^{1/2} \tag{12.149c}$$

$$Z_4 = Z_c\sqrt{K} \tag{12.149d}$$

$$Z_5 = \frac{Z_c}{\sqrt{K}} \tag{12.149e}$$

$$R = Z_c\left(\frac{1 + K^2}{K}\right) \tag{12.149f}$$

For equal power division, $K^2 = 1$. For this special case, the element values of the power divider reduce to those given in Fig. 12.52. The bandwidth of the power divider is maximum for equal power split and becomes narrower as the ratio of unequal power split increases [8].

12.8.4 N-Way Power Divider

Figure 12.56 shows the schematic of a N-way power divider for equal power division. The input line of characteristic impedance Z_c is connected to M-sections of N-uncoupled transmission line sections with isolation resistors bridging the ends of neighbouring sections. Nagai et al [69] have analysed such a N-way circuit by obtaining the eigenvalues and corresponding eigenvectors of the characteristic admittance matrices of the M-sections and admittance matrices for the isolation resistors. The design equations are

$$G_B = h_i G_M + \cfrac{Y_M^2}{h_i G_{M-1} + \cfrac{Y_{M-1}^2}{\ddots\; h_i G_3 + \cfrac{Y_3^2}{h_i G_2 + \cfrac{Y_2^2}{h_i G_1}}}}, \quad i = 2, 3, \ldots, N \tag{12.150}$$

where

$$h_i = 2 - 2\cos\left[\frac{\pi(i-1)}{N}\right], \quad i = 1, 2, \ldots, N \tag{12.151}$$

At the centre frequency, the electrical length of each section is $\pi/2$. For input match, it is necessary that

$$Z_A = NZ_c \tag{12.152}$$

where Z_A is the impedance looking into each of the output lines from the input port at junction A. The characteristic impedances $Z_1, Z_2 \ldots Z_M$ of the quarter-wave sections in each

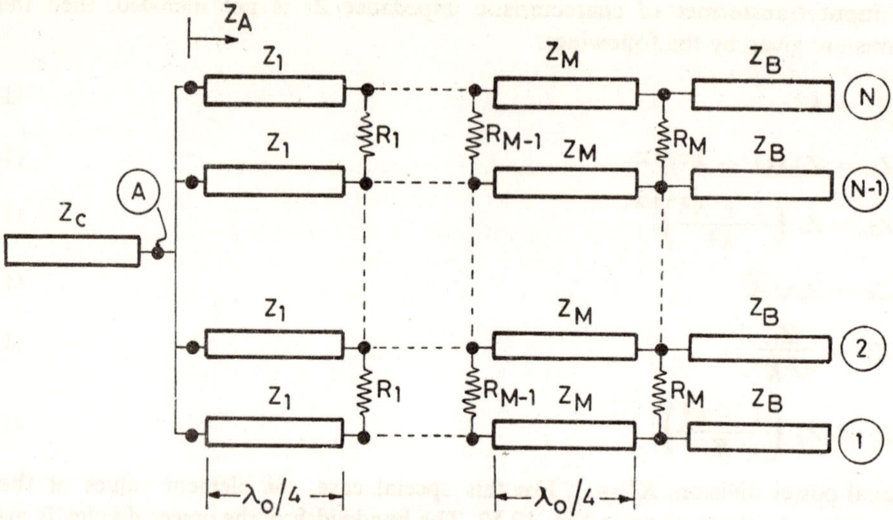

Fig. 12.56 *N*-way power divider

of the output arms are obtained from the transducer-loss characteristic of the maximally flat transformer or Chebyshev transformer [1]. The isolation resistors $R_1, R_2 \ldots R_M$ required for matching the output ports are determined by solving (12.150) (with $M = N - 1$).

As an example, we consider a three-way power divider. It is assumed that the impedance of the input port and the three output ports is 50 ohm ($Z_c = Z_B = 50$ ohms). Each of the three output arms consists of 2-section quarter wave transformers with terminating impedances $Z_A = 150$ ohm and $Z_B = 50$ ohm. For a maximally flat transformer, the characteristic impedances Z_1 and Z_2 are obtained as

$$Z_2 = Z_A^{1/4} Z_B^{3/4} = 65.8 \text{ ohm} \tag{12.153a}$$

$$Z_1 = Z_A^{3/4} Z_B^{1/4} = 114.0 \text{ ohm} \tag{12.153b}$$

With $N = 3$, we have $M = 2$, $h_2 = 1$ and $h_3 = 3$. From (12.150) we can write the following relations for $i = 2$ and 3

$$\frac{1}{Z_B} = \frac{1}{R_2} + \frac{R_1}{Z_2^2} \tag{12.154a}$$

$$\frac{1}{Z_B} = \frac{3}{R_2} + \frac{R_1}{3Z_2^2} \tag{12.154b}$$

Solving for R_1 and R_2, we obtain

$$R_2 = 4Z_B = 200 \text{ ohm} \tag{12.155a}$$

$$R_1 = \frac{3}{4}\frac{Z_2^2}{Z_B} = 64.95 \text{ ohm} \tag{12.155b}$$

A three-way power divider of this design is reported to operate over an octave bandwidth with an isolation greater than 20 *dB* between the three output ports and a VSWR of less than 1.4 at all the four ports [69].

12.9 Filters

MIC filters generally make use of a cascade of distributed circuit elements in the form of coupled resonators, stub sections, etc. A variety of such filters can be designed using the low-frequency prototype filter synthesis techniques. One such technique that is extensively used is called the 'insertion-loss' method. This method begins with a complete specification of the attenuation characteristic of the microwave filter. From this, a basic prototype low-pass filter having the desired passband characteristic is first designed. Using suitable frequency transformations and element realizations, the required type of filter—low-pass, bandpass, high-pass or bandstop filter—is derived. The lumped element values of these filters are then realized in terms of the distributed circuit elements.

12.9.1 Low-Pass Prototype Filter [3, 8, 77–80]

In this section, we summarise the low-frequency prototype filter design technique using the insertion-loss method. The method begins with a specification of the power loss ratio P_{LR} or reflection coefficient $|\Gamma|$ as a function of the angular frequency ω. The power loss ratio is defined as the incident power divided by the actual power delivered to the load. The insertion loss of the network is given by

$$L(dB) = 10 \log_{10} P_{LR} = 10 \log_{10}\left(\frac{1}{1 - |\Gamma|^2}\right) \tag{12.156}$$

The power loss ratio that is commonly used for microwave filter design is that which gives a maximally-flat passband response (Butterworth) or an equi-ripple passband response (Chebyshev). In certain broadband applications having exceptionally sharp selectivity requirements, elliptic-function type filter response is used.

Figure 12.57 shows typical attenuation characteristics of maximally-flat and Chebyshev filters. These characteristics can be mathematically represented as follows:

Maximally flat response ($\omega_1' = 1$):

$$L(dB) = 10 \log_{10} (1 + (\omega')^{2n}) \tag{12.157}$$

Chebyshev response ($\omega_1' = 1$):

$$\begin{aligned} L(dB) &= 10 \log_{10} [1 + \{10^{(L_a/10)} - 1\} \cos^2 (n \cos^{-1} \omega')], \qquad \omega' \leqslant 1 \\ &= 10 \log_{10} [1 + \{10^{(L_a/10)} - 1\} \cosh^2 (n \cosh^{-1} \omega')], \qquad \omega' \geqslant 1 \end{aligned} \tag{12.158}$$

where n is the order of the filter and L_a is the maximum attenuation of the Chebyshev filter in the passband. Both the characteristics can be realized by means of ladder networks made of inductors and capacitors. Figure 12.58 shows a general ladder network for prototype low-pass filter. The element values g_k of the prototype filter can be calculated from the following relations:

Maximally flat response ($\omega_1' = 1$):

$$\begin{aligned} g_0 &= 1 \\ g_k &= 2 \sin\left[\frac{(2k-1)\pi}{2n}\right], \qquad k = 1, 2, \ldots, n \\ g_{n+1} &= 1 \end{aligned} \tag{12.159}$$

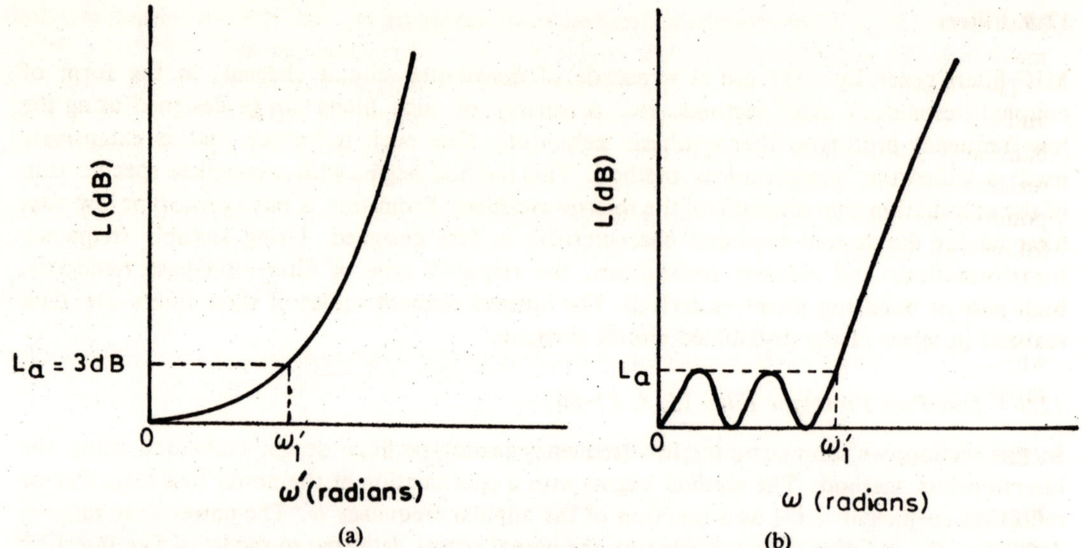

Fig. 12.57 (a) Maximally-flat low-pass characteristic, (b) Chebyshev low-pass characteristic

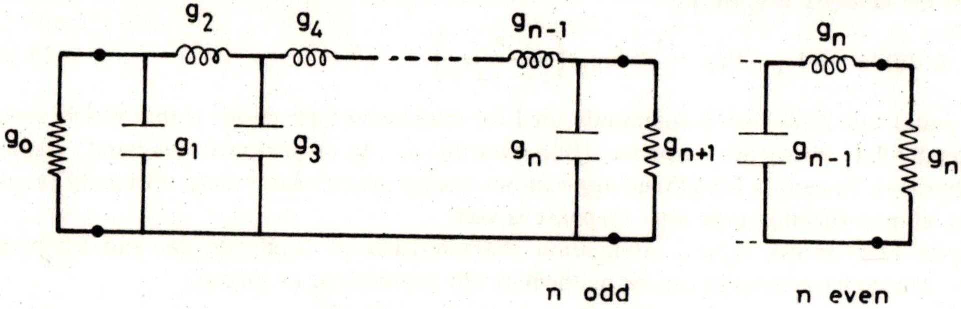

Fig. 12.58 Prototype low-pass filter

Chebyshev response ($\omega_1' = 1$):

$$g_0 = 1$$

$$g_1 = \frac{2a_1}{\gamma}$$

$$g_k = \frac{4a_{k-1}a_k}{b_{k-1}g_{k-1}}, \; k = 2, 3, \ldots, n \tag{12.160}$$

$$g_{n+1} = \coth^2(\beta/4), \; n \text{ even}$$
$$= 1, \; n \text{ odd}$$

where

$$\gamma = \sinh\left(\frac{\beta}{2n}\right)$$

$$a_k = \sin\left[\frac{(2k-1)\pi}{2n}\right], \quad k = 1, 2, \ldots, n$$

$$b_k = \gamma^2 + \sin^2(k\pi/n), \quad k = 1, 2, \ldots n \tag{12.161}$$

$$\beta = \ln[\coth(L_a/17.37)]$$

Matthaei et al [3] have provided extensive design tables listing the element values for both maximally-flat and Chebyshev filters having $n = 1$ to 15 reactive elements.

The element values $g_1, g_2, \ldots, g_{n+1}$ of the prototype filter are normalized to make $g_0 = 1$ and $\omega_1' = 1$. This prototype forms the basis for designing the desired filter with a given band-edge and input line impedance. If g_R, g_L and g_C are the g values corresponding to resistances, inductances and capacitances of the normalized prototype filter, then the corresponding elements R, L and C of the actual scaled filter circuit can be obtained by using the following relations:

$$R = R_0 g_R, \qquad L = R_0 g_L/\omega_1, \qquad C = g_C/\omega_1 R_0 \tag{12.162}$$

where ω_1 is the angular frequency corresponding to the band-edge and R_0 is the generator resistance of the scaled filter.

12.9.2 Low-Pass Filters Using High- and Low-Impedance Lines

Figure 12.59 shows the layout of a three-element low-pass filter using alternating short sections of low impedance and high impedance lines. The low impedance line behave predominantly like a shunt capacitance and the high impedance line acts as a series inductance. Formulas for designing these elements are given in Sec. 12.4. We now outline the procedure

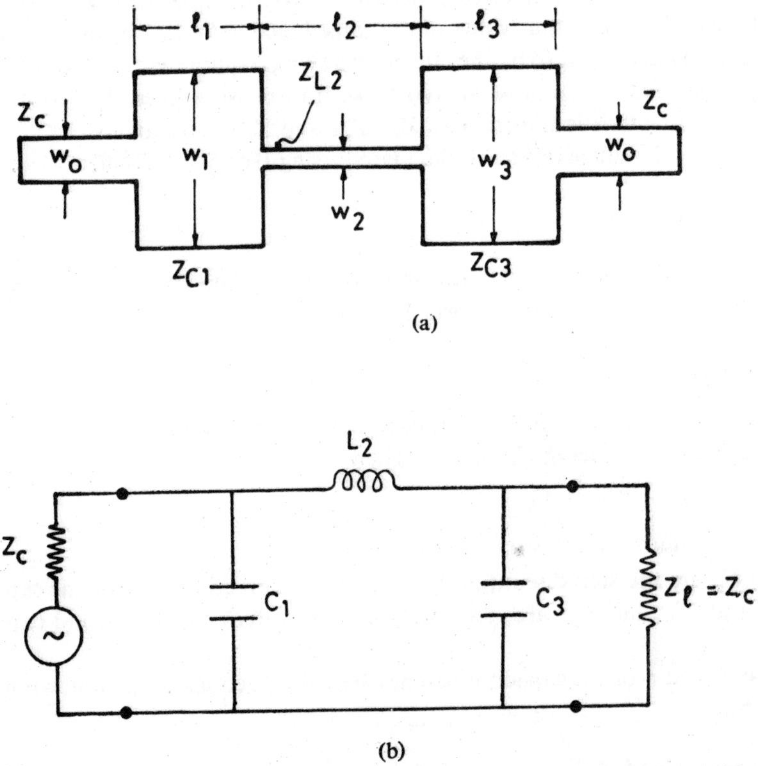

Fig. 12.59 (a) Low-pass filter using high-and low-impedance lines, (b) Equivalent circuit

for designing such filters from the lumped prototype by considering the three-element filter in Fig. 12.59 as an example.

•First obtain the elements g_1, g_2 and g_3 of the prototype filter using the standard formulas (12.159) or (12.160) [3]: For example, for a maximally-flat filter with $g_0 = 1$ and $\omega_1' = 1$ and 3 dB cut-off, we obtain from (12.159)

$$g_1 = 1, \quad g_2 = 2, \quad g_3 = 1 \quad \text{and} \quad g_4 = 1 \tag{12.163}$$

•Determine the values of C_1, L_2 and C_3 of the actual low-pass filter (see. Fig. 12.59b) using the following transformations:

$$C_1 = \frac{g_1}{\omega_1 Z_c}, \quad L_2 = \frac{Z_c g_2}{\omega_1}, \quad C_3 = \frac{g_3}{\omega_1 Z_c} \tag{12.164}$$

The terminating impedance (real) on the output side is given by

$$Z_l = Z_c g_4 = Z_c \tag{12.165}$$

In (12.164), ω_1 is the angular frequency corresponding to the 3 dB band-edge of the actual filter, and Z_c is the terminating impedance on the input side.

•The input and output lines generally are of characteristic impedance 50 ohm. Thus, the width w_o of these lines correspond to $Z_c = 50$ ohm.

•The characteristic impedances Z_{c1} and Z_{c3} of the two capacitor patches must be smaller than $Z_c (= 50\,\Omega)$. But the value should not be so low as to cause transverse resonances across the width of the patch conductors. A convenient value for Z_{c1} and Z_{c3} would be around 20 ohm. For the inductive line, the characteristic impedance Z_{L2} should be much larger than 50 ohm. The maximum value of the impedance is limited by the thinnest possible line that can be fabricated and also by the current carrying requirement. A value of Z_{L2} around 100 ohm is practically feasible in most stripline-like transmission lines. The widths w_1, w_2 and w_3 correspond to characteristic impedances Z_{c1}, Z_{L2} and Z_{c3}, respectively.

•The length l_2 of the inductive line is calculated using (see Sec. 12.4.6)

$$l_2 = \frac{\lambda}{2\pi} \sin^{-1}\left(\frac{\omega_1 L_2}{Z_{L2}}\right) \tag{12.166}$$

where λ is the guide wavelength in the transmission lines. The capacitive end-correction of the inductive line is obtained from (see Sec. 12.4.6)

$$C_{L2} = \frac{\pi l_2}{\omega Z_{L2} \lambda} \tag{12.167}$$

•Incorporate end corrections in the lumped capacitances C_1 and C_3. Let C_1' and C_3' be the new capacitance values replacing C_1 and C_3. Then

$$C_1' = C_1 - C_{L2} - C_{t1} - C_{t2} \tag{12.168a}$$

$$C_3' = C_3 - C_{L2} - C_{t3} - C_{t4} \tag{12.168b}$$

where C_{t1} and C_{t2} are the step discontinuity capacitances on either side of the capacitor patch of length l_1, and C_{t3} and C_{t4} are the corresponding values for the second capacitor patch of length l_3.

•The lengths l_1 and l_3 of the capacitor patches are calculated using the following expressions (see Sec. 12.4.6):

$$l_1 = \frac{\lambda}{2\pi} \sin^{-1}(\omega_1 C_1' Z_{c1}) \tag{12.169a}$$

$$l_3 = \frac{\lambda}{2\pi} \sin^{-1} (\omega_1 C_3' Z_{c3}) \tag{12.169b}$$

Another type of low-pass filter which can be realized using high- and low- impedance lines is the elliptic function filter (Fig. 12.60a). In this circuit, the shunt element uses a series combination of inductance and capacitance. Elliptic function filters have equi-ripple response in the passband as well as in the stopband (Fig. 12.60c). These filters have the advantage that the poles of attenuation lie near the cut-off frequencies thus offering higher rates of cut-off than Chebyshev filters of equal degree and equal passband insertion loss. Comprehensive tables of data for the element values of lumped constant prototype are available in the literature [8, 77–79]. The normalized values of inductances and capacitances can be translated into the actual values for the filter under consideration by using (12.162). The various dimensions of the filter layout can then be calculated using the procedure similar to the one outlined above for the maximally flat low-pass filter.

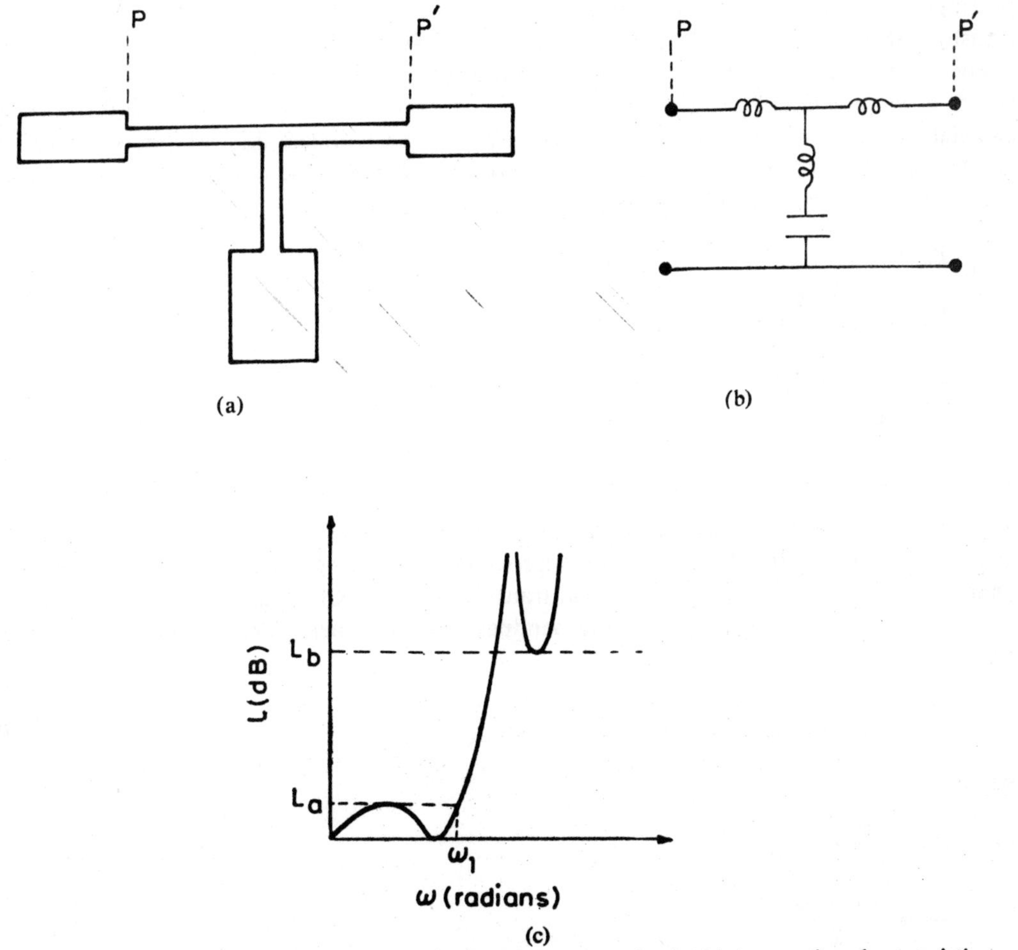

Fig. 12.60 (a) Layout of elliptic-function filter, (b) Equivalent circuit, (c) Attenuation characteristic

12.9.3 Coupled-Line Low-Pass Filter [81, 82]

Transformation of lumped element prototype into a physically realizable configuration consisting of only quarter-wave parallel coupled lines has been reported by Ozaki and Ishii [81] and Wenzel [82]. This transformation involves inserting a series of unit elements with 1 ohm characteristic impedance between the lumped element prototype network and its load (also of impedance 1 ohm). With this, the input impedance of the network as well as its insertion loss characteristic remains unaltered. Unit elements of characteristic impedance 1 ohm can also be introduced in front of the network. Using Kuroda's identities, the network is transformed to another equivalent network in which the lumped elements of appropriate value and unit elements are cascaded alternatively. Network equivalences [16, 17] which enable such transformations are listed in Tables 12.2 and 12.3. The transformed circuit is then directly realized in terms of a suitable distributed element configuration.

As an example, the transformation from a third order maximally flat low-pass prototype into a coupled-line low-pass filter is illustrated in Fig. 12.61 [82]. Figure 12.61b shows the prototype filter (Fig. 12.61a) with unit elements added on either side of the network. The transformation to the network shown in Fig. 12.61c can be easily made using the first network equivalence presented in Table 12.2. Each unit element with a series inductor is then replaced by a parallel-coupled quarter-wave section as shown in Fig. 12.61d. The formulas for calculating the even- and odd-mode impedances of the coupled line section are given by (12.38) (see Fig. 12.16). For example, the normalized even- and odd-mode impedance (Z_{e1} and Z_{o1}) of the first coupled section in Fig. 12.61d are given by,

$$\bar{Z}_{e1} = \bar{Z}_{c1} + L_1 + \sqrt{L_1(\bar{Z}_{c1} + L_1)} = 2.366 \tag{12.170a}$$

$$\bar{Z}_{o1} = \frac{\bar{Z}_{c1}\bar{Z}_{e1}}{2\bar{Z}_{e1} - \bar{Z}_{c1}} = 0.634 \tag{12.170b}$$

For the second section, $\bar{Z}_{e2} = 4.736$ and $\bar{Z}_{o2} = 0.264$, and for the third section, $\bar{Z}_{e3} = 1.707$ and $\bar{Z}_{o3} = 0.293$.

12.9.4 Direct-Coupled Bandpass Filter

The lumped element low-pass prototype filter design discussed in Sec. 12.9.1 can be used to achieve bandpass filter designs having approximately the same maximally-flat or Chebyshev characteristic. Figure 12.62 shows typical transformation of Chebyshev response of a lumped element prototype to the corresponding bandpass filter response. A commonly used mapping function is

$$\frac{\omega'}{\omega_1'} = \frac{2}{\Delta\omega}\left(\frac{\omega - \omega_0}{\omega_0}\right) \tag{12.171a}$$

where

$$\Delta\omega = \frac{\omega_2 - \omega_1}{\omega_0} \tag{12.171b}$$

$$\omega_0 = \frac{\omega_1 + \omega_2}{2} \tag{12.171c}$$

The symbols used in (12.171) are marked in Fig. 12.62.

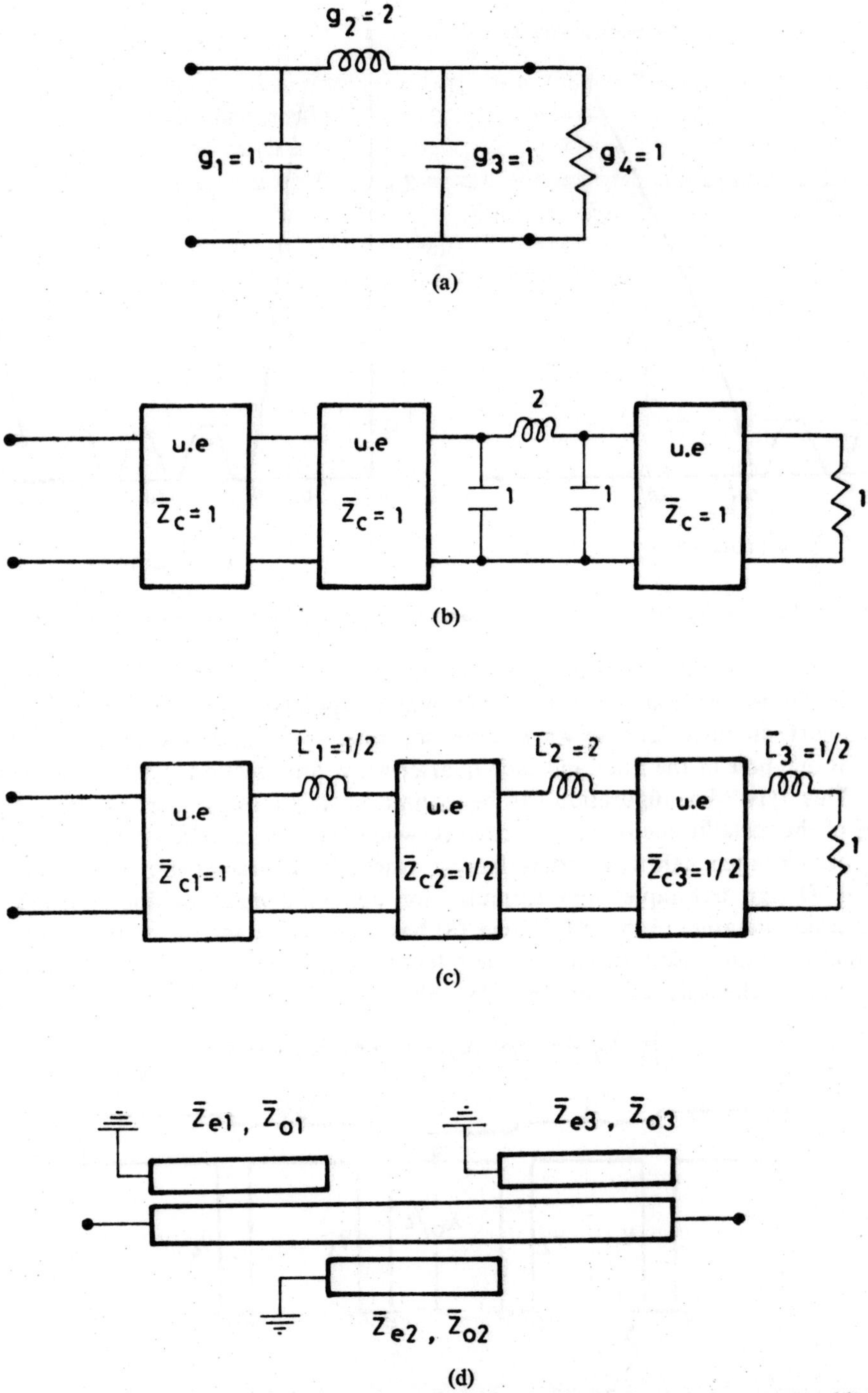

Fig. 12.61 (a) Third-order Butterworth filter, (b) Butterworth filter with unit elements added, (c) Transformed circuit using Kuroda's identities, (d) Realization in parallel-coupled configuration (From Wenzel [82], Copyright © 1964 IEEE, reprinted with permission)

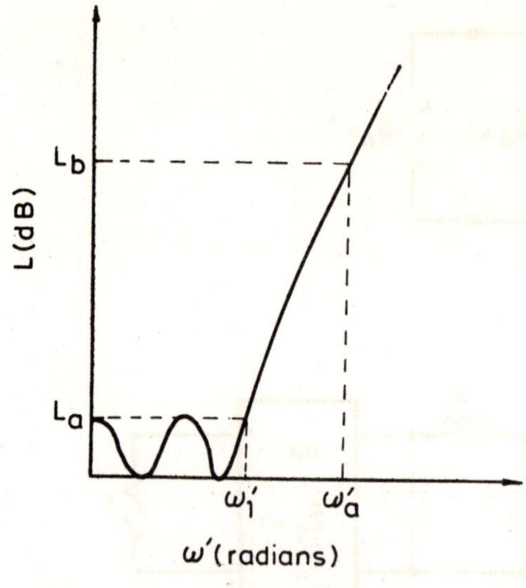

(a)

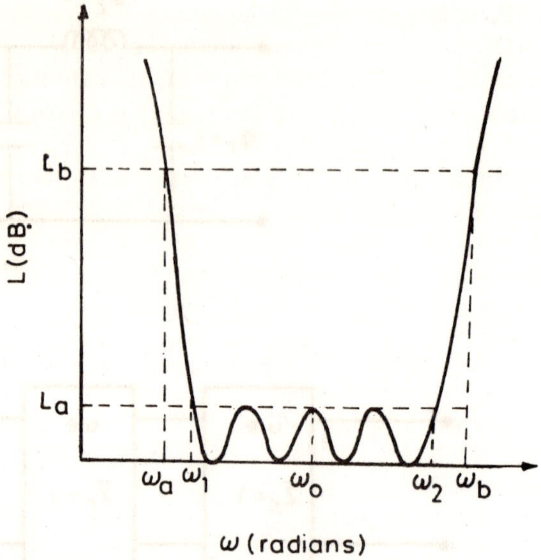

(b)

Fig. 12.62 Transformation from (a) low-pass prototype response to (b) bandpass response

Figure 12.63 shows the layout of a typical direct-coupled bandpass filter. Filter of this type are useful for wideband (> 25% bandwidth) applications. The filter comprises of a series of short -circuited quarter-wave stubs separated by quarter-wave line sections. Each stub acts as one pole of the filter and each quarter-wave line section. acts as an impedance inverter. This type of configuration can be realized in striplines by shorting the stubs to the end-walls of the metallic enclosure. In circuits where short-circuited stubs are difficult to realize, open-circuited half-wave stubs having twice the characteristic impedance may be substituted. Design techniques and formulas for designing *n*-section filters of this type are available in the literature [3, 83, 84]. Howe [8] has provided a set of useful design graphs for maximally flat and also 0.01 equi-ripple filters having 2 to 9 sections. For ready use, we summarize below the salient design formulas [3]:

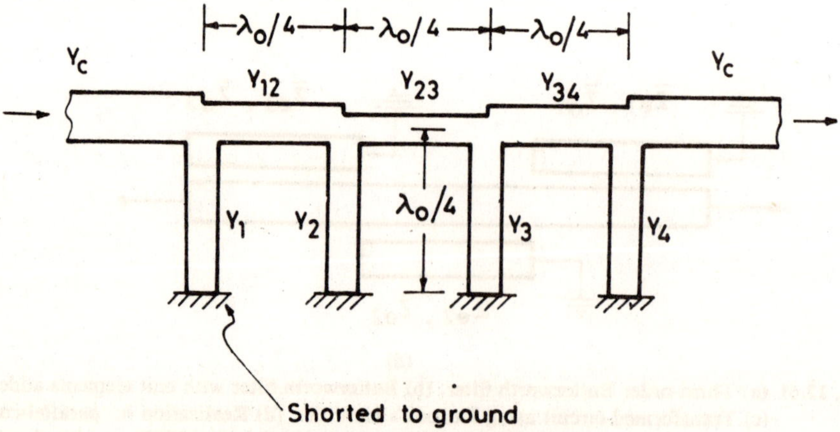

Fig. 12.63 Direct-coupled bandpass filter

- First, use the low-pass to bandpass mapping as per (12.171)
- Obtain the number (n) of reactive elements in the prototype from the specification of the passband insertion loss and the rate of cut-off of the filter (refer Fig. 4.04 of [3]). The element values $g_0, g_1, \ldots g_{n+1}$ of the prototype can be obtained from (12.159) or (12.160)
- Specify the fractional bandwidth $\Delta\omega$ and the terminating admittances Y_c.
- The number of short-circuited stubs in the bandpass filter is equal to the number of reactive elements in the prototype. For the filter shown in Fig. 12.63, $n = 4$. The lengths of all stubs and connecting lines are chosen equal to $\lambda_0/4$, where λ_0 is the guide wavelength in the stripline structure at midband frequency ($\omega = \omega_0$).
- Compute the following parameters:

$$\theta_1 = \frac{\pi}{2}\left(1 - \frac{\Delta\omega}{2}\right) \tag{12.172}$$

$$\frac{J_{12}}{Y_c} = g_0\left[\frac{2Kg_1}{g_2}\right]^{1/2}$$

$$\frac{J_{k,k+1}}{Y_c} = \frac{2Kg_0g_1}{(g_kg_{k+1})^{1/2}}, \qquad k = 2 \text{ to } n-2$$

$$\frac{J_{n-1,n}}{Y_c} = g_0\left[\frac{2Kg_1g_{n+1}}{g_0g_{n-1}}\right]^{1/2} \tag{12.173}$$

$$N_{k,k+1} = \left[\left(\frac{J_{k,k+1}}{Y_c}\right)^2 + \left(\frac{2Kg_0g_1\omega_1' \tan\theta_1}{2}\right)^2\right]^{1/2}, \qquad k = 1 \text{ to } n-1$$

where K is a constant, the value of which may be chosen to obtain realizable values of admittances for the filter sections. Choice of K between 0.5 and 1.0 is reported to have negligible effect on the transmission characteristics of the filter [3]. A typical practical value of K is 1.0.

- Calculate the characteristic admittances of the stubs:

$$Y_1 = Y_c\omega_1'(1 - K)g_0g_1 \tan\theta_1 + Y_c\left(N_{12} - \frac{J_{12}}{Y_c}\right)$$

$$Y_k = Y_c\left[N_{k-1,k} + N_{k,k+1} - \frac{J_{k-1,k}}{Y_c} - \frac{J_{k,k+1}}{Y_c}\right], \qquad k = 2 \text{ to } n-1 \tag{12.174}$$

$$Y_n = \left[Y_c\omega_1'(g_ng_{n+1} - Kg_0g_1)\tan\theta_1 + Y_c\left(N_{n-1,n} - \frac{J_{n-1,n}}{Y_c}\right)\right]$$

- The characteristic admittances of the connecting quarter-wave lines are given by

$$Y_{k,k+1} = J_{k,k+1}, \qquad k = 1 \text{ to } n-1 \tag{12.175}$$

12.9.5 Coupled Line Bandpass Filter

Parallel-Coupled Resonator Filter

Coupled line bandpass filters make use of a series of half-wavelength long resonant conducting strips. The parallel-coupled resonator filter shown in Fig. 12.64a is a popularly used configuration in applications requiring narrow to moderate bandwidths (upto about 20%). The adjacent resonators are parallel coupled along half of their length.

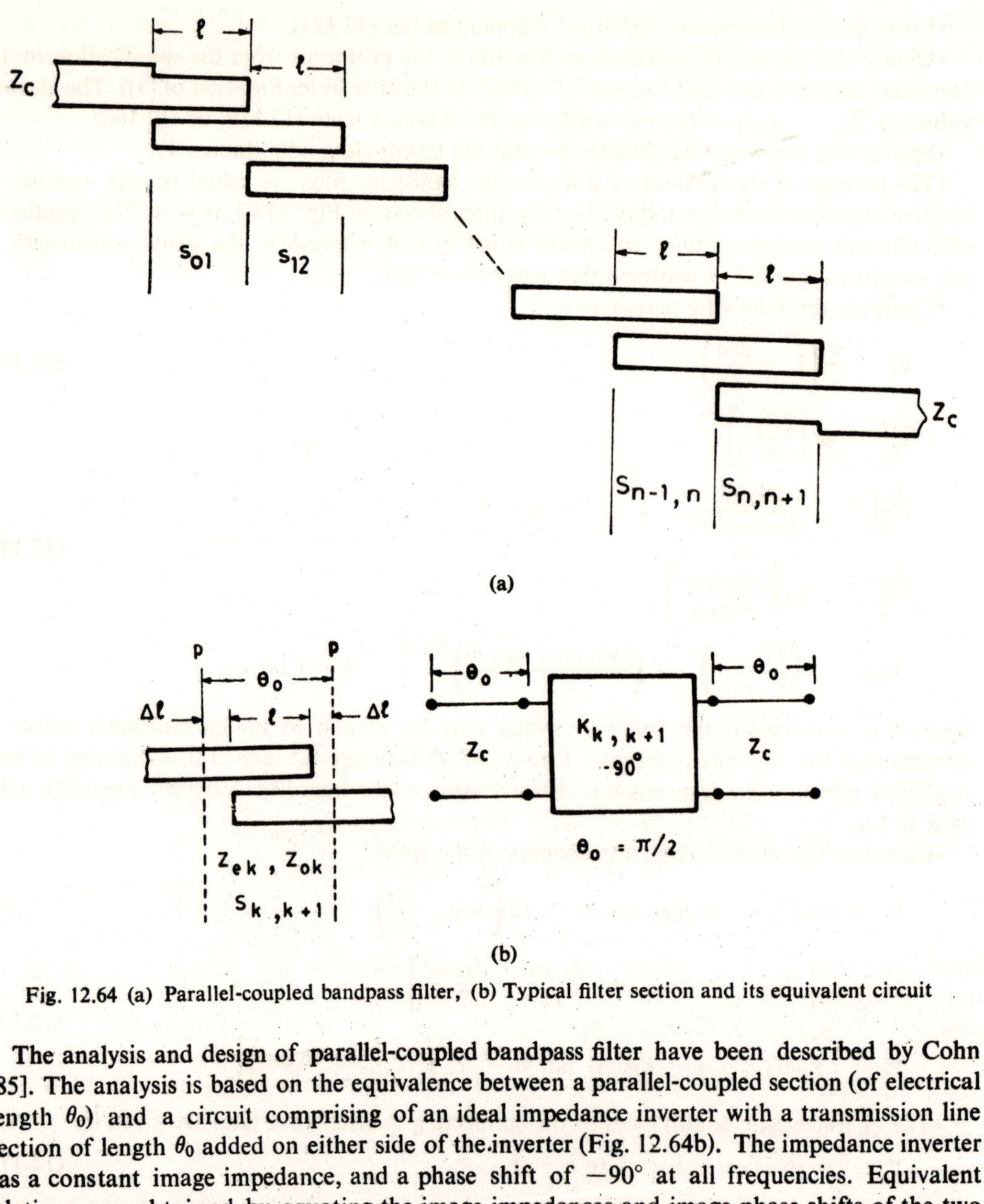

Fig. 12.64 (a) Parallel-coupled bandpass filter, (b) Typical filter section and its equivalent circuit

The analysis and design of parallel-coupled bandpass filter have been described by Cohn [85]. The analysis is based on the equivalence between a parallel-coupled section (of electrical length θ_0) and a circuit comprising of an ideal impedance inverter with a transmission line section of length θ_0 added on either side of the inverter (Fig. 12.64b). The impedance inverter has a constant image impedance, and a phase shift of $-90°$ at all frequencies. Equivalent relations are obtained by equating the image impedances and image phase shifts of the two circuits. For $\theta_0 = \pi/2$ and $Z_c/K_{k,k+1} \ll 1$, the relations are

$$\frac{Z_{ek}}{Z_c} = 1 + \frac{Z_c}{K_{k,k+1}} + \frac{Z_c^2}{K_{k,k+1}^2} \tag{12.176a}$$

$$\frac{Z_{ok}}{Z_c} = 1 - \frac{Z_c}{K_{k,k+1}} + \frac{Z_c^2}{K_{k,k+1}^2} \tag{12.176b}$$

where Z_{ek} and Z_{ok} are the even- and odd-mode impedances of a coupled section, $K_{k,k+1}$ is the impedance of the inverter and Z_c is the impedance of the connecting lines. The relations given by (12.176) are valid over narrow and moderate bandwidths. The parallel-coupled bandpass filter shown in Fig. 12.64a can now be replaced by a chain of equivalent networks of the type shown in Fig. 12.64b. In the following, we summarize the salient design formulas for the filter [85]:

- Use the mapping function given by (12.171) to determine ω'/ω_1'. Select the number of resonator sections that will yield the desired insertion loss response (use Fig. 4.04 of [3])
- Obtain the low-pass prototype element values along with the cut-off frequency ω_1' using (12.159) or (12.160)
- Specify the terminating impedances Z_c
- Compute the inverter impedances

$$\frac{Z_c}{K_{01}} = \left[\frac{\pi\Delta\omega}{2\omega_1' g_0 g_1}\right]^{1/2}$$

$$\frac{Z_c}{K_{k,k+1}} = \left[\frac{\pi\Delta\omega}{2\omega_1'\sqrt{g_k g_{k+1}}}\right], \qquad k = 1 \text{ to } n-1 \tag{12.177}$$

$$\frac{Z_c}{K_{n,n+1}} = \left[\frac{\pi\Delta\omega}{2\omega_1' g_n g_{n+1}}\right]^{1/2}$$

- Determine the even- and odd-mode impedances of each coupled line section

$$\frac{Z_{ek+1}}{Z_c} = \left[1 + \frac{Z_c}{K_{k,k+1}} + \frac{Z_c^2}{K_{k,k+1}^2}\right], \qquad k = 0 \text{ to } n \tag{12.178a}$$

$$\frac{Z_{ok+1}}{Z_c} = \left[1 - \frac{Z_c}{K_{k,k+1}} + \frac{Z_c^2}{K_{k,k+1}^2}\right], \qquad k = 0 \text{ to } n \tag{12.178b}$$

- The electrical length θ_0 of each coupled line section is given by

$$\theta_0 = \frac{2\pi}{\lambda_0}(l + 2\Delta l) = \frac{\pi}{2} \tag{12.179}$$

where l is the physical length, and Δl is the end-correction due to the fringing capacitance in the coupled line (see [86–90]). The guide wavelength λ_0 in the structure can be calculated using the following approximate relation:

$$\lambda_0 = \frac{2\lambda^a}{(\sqrt{\epsilon_{fe}} + \sqrt{\epsilon_{fo}})} \tag{12.180}$$

where ϵ_{fe} and ϵ_{fo} are the effective dielectric constants for the even-mode and odd-mode respectively, and λ^a is the free space wavelength at the centre frequency ω_0.

In the filter configuration shown in Fig. 12.64, the input and output lines are connected directly to the first and the last quarter-wave coupled section respectively. An alternate technique is to make use of a tapped line at the input as well as at the output as shown in Fig. 12.65. Design techniques for using input/output tapped lines in parallel-coupled bandpass filters have been described by Wong [91]. The lengths l_1 and l_2 (see Fig. 12.65) are suitably chosen so as to find an optimum location for the tapped point. This type of arrangement is very useful in situations where the parallel coupling at the end sections becomes too tight to realize in practice.

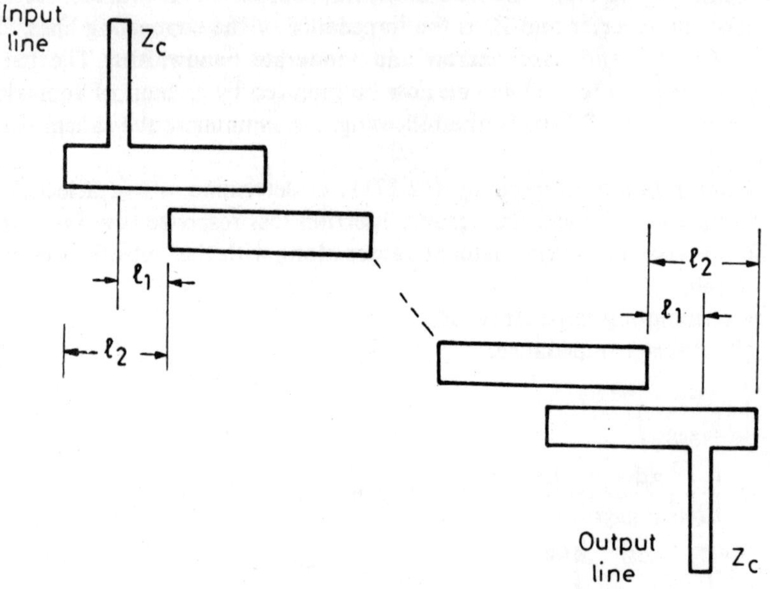

Fig. 12.65 Bandpass filter with tapped input/output lines

End-coupled Resonator Filter

Another useful type of coupled line filter is the end-coupled resonator filter shown in Fig. 12.66a. In this configuration, half-wavelength long resonators are coupled end-to-end by small capacitive gaps. As compared with the parallel-coupled bandpass filter, the end-coupled filter is twice as long and the gap dimensions are much smaller. This type of filter is suitable for narrow bandwidth of approximately 5%. For wider bandwidths, the gap width becomes so narrow that it poses fabricational difficulty. This problem can be circumvented in suspended stripline by placing the adjacent resonators alternately on the top and bottom sides of the substrate. As illustrated in Fig. 12.66b, the ends of the adjacent resonators can even be overlapped to achieve the desired capacitance value. In both the filter layouts, a gap between adjacent conductors can be represented by a π-equivalent network. The resulting equivalent circuit for the n-element filter is shown in Fig. 12.66c. For small gap widths, we can represent the gap as just a series capacitance dsicontinuity ($B'_{k,k+1} = 0$). Figure 12.66d shows another equivalent representation of the filter in terms of admittance inverters. The design technique and formulas for end-coupled filters are available in the literature [3, 92-94]. In the following, we summarize the design steps which can be used for designing narrow-band filters with Chebyshev response in the passband:

• Specify the passband insertion loss L_a (in dB) and band-edge frequencies ω_1 and ω_2. Also, specify the skirt selectivity, that is, the attenuation L_b (in dB) at a frequency ω just outside the passband (Refer Fig. 12.62 for definition of parameters).

• Determine the ratio $|\omega'/\omega'_1|$ using the following mapping function:

$$\frac{\omega'}{\omega'_1} = \frac{2}{\Delta\omega}\left(\frac{\omega - \omega_0}{\omega}\right) \tag{12.181}$$

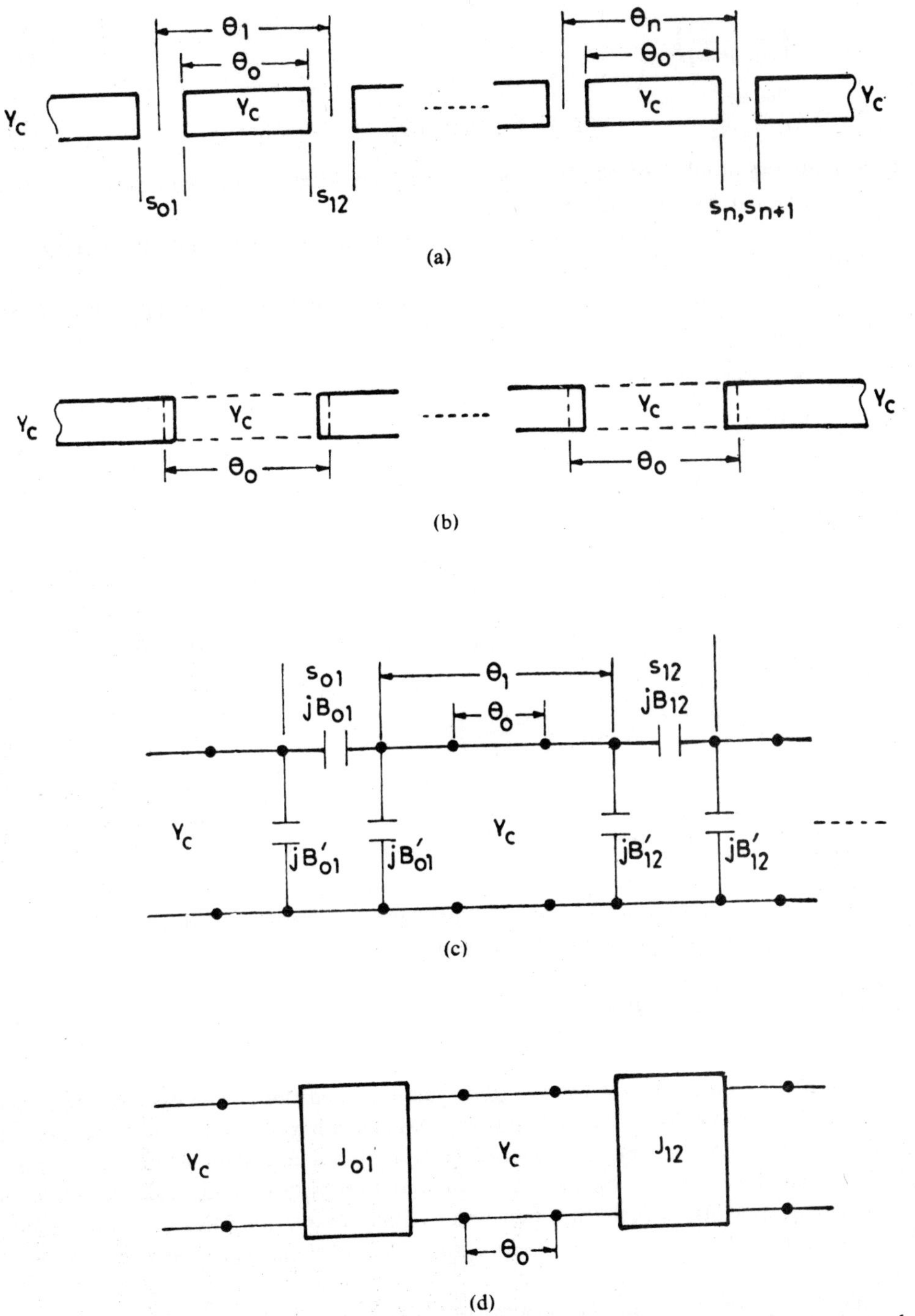

Fig. 12.66 (a) Layout of end-coupled resonator filter in stripline ($\theta_0 = \pi$), (b) Filter in suspended stripline with resonators alternating on the two sides of substrate, (c) and (d) Equivalent circuits

where

$$\Delta\omega = 2\left(\frac{\omega_2 - \omega_1}{\omega_2 + \omega_1}\right) \tag{12.182a}$$

$$\omega_0 = \frac{2\omega_1\omega_2}{\omega_1 + \omega_2} \tag{12.182b}$$

• Determine the number of elements (n) for the filter which meets the desired values of L_a, L_b and ω'/ω_1' (Refer Fig. 4.04 of [3]).

• Determine the element values $g_0, g_1, \ldots g_{n+1}$ of the low-pass prototype using (12.160) and (12.161).

• Calculate the admittance inverter parameters $J_{k,k+1}$ using the following formulas [3]:

$$\frac{J_{01}}{Y_c} = \left[\frac{\pi}{2}\frac{\Delta\omega}{g_0 g_1 \omega_1'}\right]^{1/2}$$

$$\frac{J_{k,k+1}}{Y_c} = \frac{\pi\Delta\omega}{2\omega_1'}\frac{1}{(g_k g_{k+1})^{1/2}}, \qquad k = 1 \text{ to } n-1 \tag{12.183}$$

$$\frac{J_{n,n+1}}{Y_c} = \left[\frac{\pi\Delta\omega}{2g_n g_{n+1}\omega_1'}\right]^{1/2}$$

where Y_c is the characteristic admittance of the resonators and also of the input and output lines.

• The admittance inverter parameter $J_{k,k+1}$ is related to the gap susceptances by the following expression [3]:

$$\frac{J_{k,k+1}}{Y_c} = \left|\tan\left\{\frac{\phi_{k,k+1}}{2} + \tan^{-1}\left(\frac{B'_{k,k+1}}{Y_c}\right)\right\}\right| \tag{12.184a}$$

where

$$\phi_{k,k+1} = -\tan^{-1}\left(\frac{2B_{k,k+1} + B'_{k,k+1}}{Y_c}\right) - \tan^{-1}\left(\frac{B'_{k,k+1}}{Y_c}\right) \tag{12.184b}$$

If we assume the gap to be a perfect series capacitance discontinuity ($B'_{k,k+1} = 0$), then the series susceptance can be directly calculated using

$$\frac{B_{k,k+1}}{Y_c} = \frac{J_{k,k+1}}{\left[1 - \left(\frac{J_{k,k+1}}{Y_c}\right)^2\right]} \tag{12.185}$$

In the case of homogeneous stripline, approximate closed form expressions are available which relate the gap widths to the gap susceptances (see Chapter 10). For suspended stripline and other structures, the gaps are theoretically analyzed and data which relate the gap widths to the susceptances is presented in Chapter 10. The series susceptance calculated from (12.185) yields gap width directly to a first order of approximation. For synthesizing the gap width more accurately, it is necessary to program (12.184) and use computer optimization technique.

• Calculate the electrical length θ_k between the gaps $s_{k-1,k}$ and $s_{k,k+1}$ using

$$\theta_k = \pi + \frac{1}{2}[\phi_{k-1,k} + \phi_{k,k-1}] \tag{12.186}$$

The length l of each resonant strip is equal to $\lambda_0/2$ where λ_0 is the wavelength in the structure at ω_0.

Nguyen and Chang [93] have applied an alternate design technique for designing end-coupled filters in suspended stripline for millimeter-wave applications. The design steps are summarized below:

• Specify the passband ripple level L_a (in dB). Determine the parameter ϵ using the formula

$$-10\log_{10}\left(\frac{1}{1+\epsilon^2}\right) = L_a(dB) \tag{12.187}$$

• Determine the guide wavelengths λ_1 and λ_2 corresponding to the band-edge frequencies ω_1 and ω_2.

• Obtain the bandwidth scaling factor α and the guide wavelength λ_0 (corresponding to the design frequency) using the following formulas:

$$\lambda_1 \sin\left(\frac{\pi\lambda_0}{\lambda_1}\right) + \lambda_2 \sin\left(\frac{\pi\lambda_0}{\lambda_2}\right) = 0 \tag{12.188}$$

$$\alpha = \frac{\lambda_1}{\lambda_0}\sin\left(\frac{\pi\lambda_0}{\lambda_1}\right) \tag{12.189}$$

• At the design frequency, $\theta_0 = \pi$. The length l of the resonant strips is equal to $\lambda_0/2$. Determine the number (n) of filter elements using the following formula for transducer loss. Obtain the stop band specifications.

$$|\,s_{12}\,|^2 = \left\{1 + \epsilon^2 T_n^2\left[\frac{\lambda}{\alpha\lambda_0}\sin\left(\frac{\pi\lambda_0}{\lambda}\right)\right]^{-1}\right\} \tag{12.190}$$

where $T_n(x)$ is the nth order Chebyshev polynomial of the first kind and is given by

$$T_n(x) = \cosh\,(n\cosh^{-1} x) \tag{12.191}$$

The transducer loss at the band-edges is equal to $1/(1+\epsilon^2)$.

• Determine the admittance inverter parameters using

$$J_{k,k+1} = \frac{1}{y}\left[y^2 + \sin^2\left(\frac{k\pi}{n}\right)\right], \qquad k = 0 \text{ to } n \tag{12.192}$$

where

$$y = \sinh\left[\frac{1}{n}\sinh^{-1}\left(\frac{1}{\epsilon}\right)\right] \tag{12.193}$$

The remaining design steps for calculating the gap susceptances, gap widths and electrical lengths θ_k between the gaps are the same as enumerated above from (12.184) through (12.186).

12.9.6 High-Pass Filters

Figure 12.67 shows transformation of a typical low-pass prototype response to a corresponding high-pass filter response. The frequency transformation is effected by means of the formula

$$\omega' = -\frac{\omega_1\omega_1'}{\omega} \tag{12.194}$$

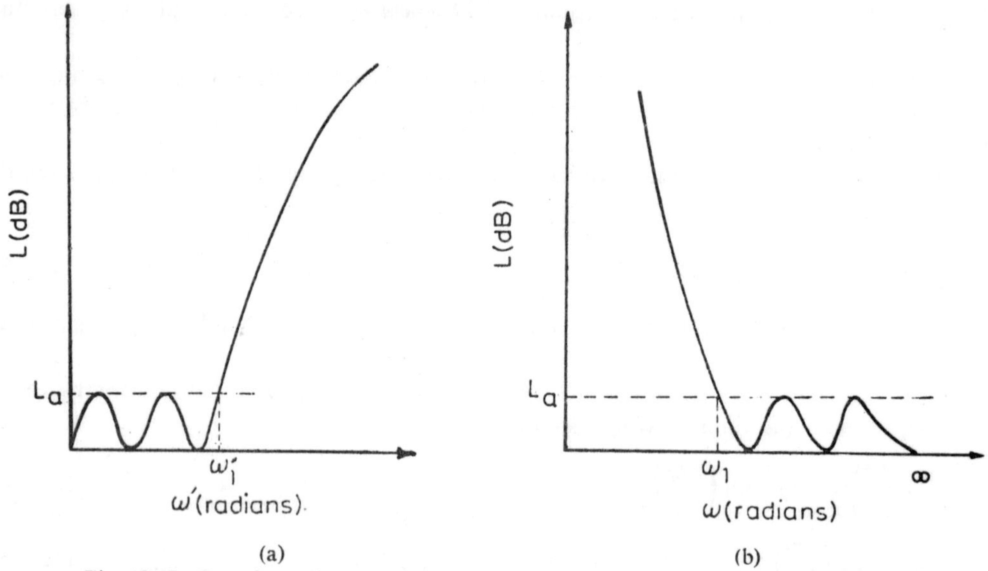

Fig. 12.67 Transformation from (a) low-pass prototype response to (b) high-pass response

A lumped element high-pass filter can be represented as a ladder network consisting of shunt inductors and series capacitors, and hence forms a complement of the low-pass filter. If g_L represents the g value of any series inductance in the low-pass prototype, the corresponding element in the high-pass filter is a series capacitance of value

$$C = \frac{1}{\omega_1 \omega_1' g_L} \tag{12.195}$$

Similarly, if g_C represents the g value of any shunt capacitance in the low-pass prototype, the corresponding element in the high-pass filter is a shunt inductance of value

$$L = \frac{1}{\omega_1 \omega_1' g_C} \tag{12.196}$$

Figure 12.68a shows a high-pass network consisting of three elements. Using Kuroda's identities and TEM equivalents the network can be transformed to quarter-wave coupled-line filters as shown in Fig. 12.68b and c. The design technique for realizing filters in these configurations has been reported by Wenzel [82]. In the third layout shown in Fig. 12.68c, the shunt inductance is realized by a short thin high impedance line connected directly to the main line with the other end connected to ground. The series capacitance has rather a large value that it becomes difficult to realize it by a gap capacitance. Series capacitances of sufficiently large value can be realized in suspended stripline by using broadside-coupled strips as shown in Fig. 12.68d. Alternatively, lumped element chip-type capacitors may be used.

12.9.7 Bandstop Filters [95-99]

Bandstop filters with high selectivity find important applications in communication receivers and ESM systems for rejecting strong interfering frequencies and unwanted harmonics. Broad stopband requirements are generally met by the use of low-pass, bandpass or high-pass filters.

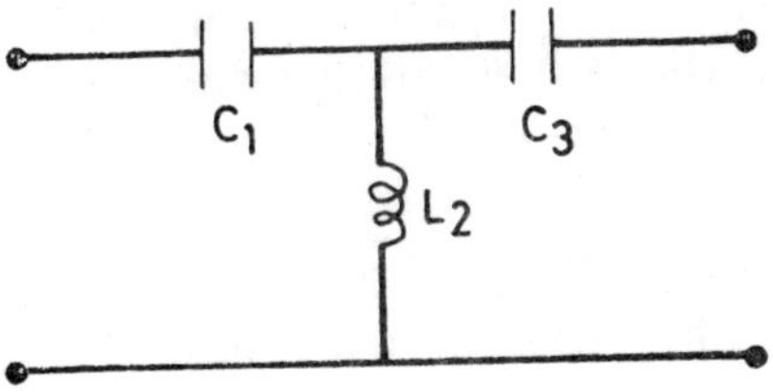

(a)

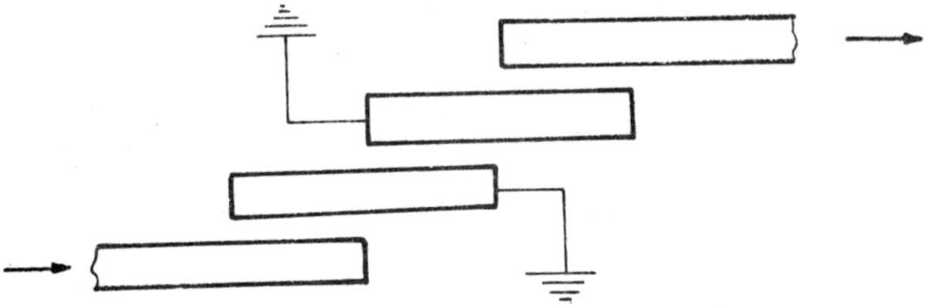

(b)

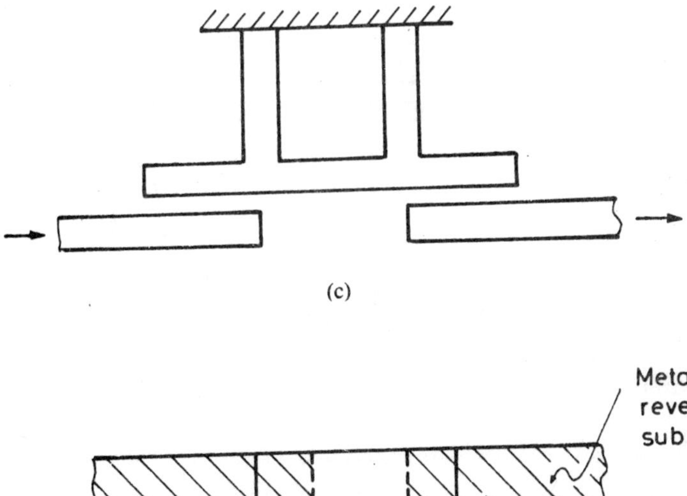

(c)

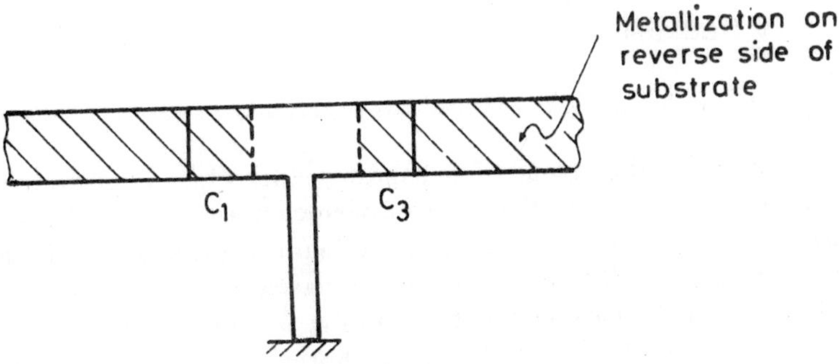

(d)

Fig. 12.68 (a) Lumped element high-pass filter, (b) and (c) Filter layout using quarter-wave coupled lines, (d) Realization in suspended stripline with broad side-coupled strips as series capacitor

There are however specific requirements for bandstop filters with narrow stopbands. Single section narrow stopband filters are particularly useful as image rejection filters in balanced mixers.

Bandstop Filter with Gap-Coupled Shunt Stubs

As in the case of bandpass filters, the design of the bandstop filter can be based on the transformation from a low-pass prototype. Figure 12.69 shows transformation from a Chebyshev low-pass characteristic to bandstop characteristic. The frequency transformation is effected by means of the formula

$$\frac{1}{\omega'} = \frac{1}{\Delta\omega\, \omega_1'}\left(\frac{\omega}{\omega_0} - \frac{\omega_0}{\omega}\right) \tag{12.197}$$

where

$$\Delta\omega = \frac{\omega_2 - \omega_1}{\omega_0} \tag{12.198a}$$

$$\omega_0 = (\omega_1\omega_2)^{1/2} \tag{12.198b}$$

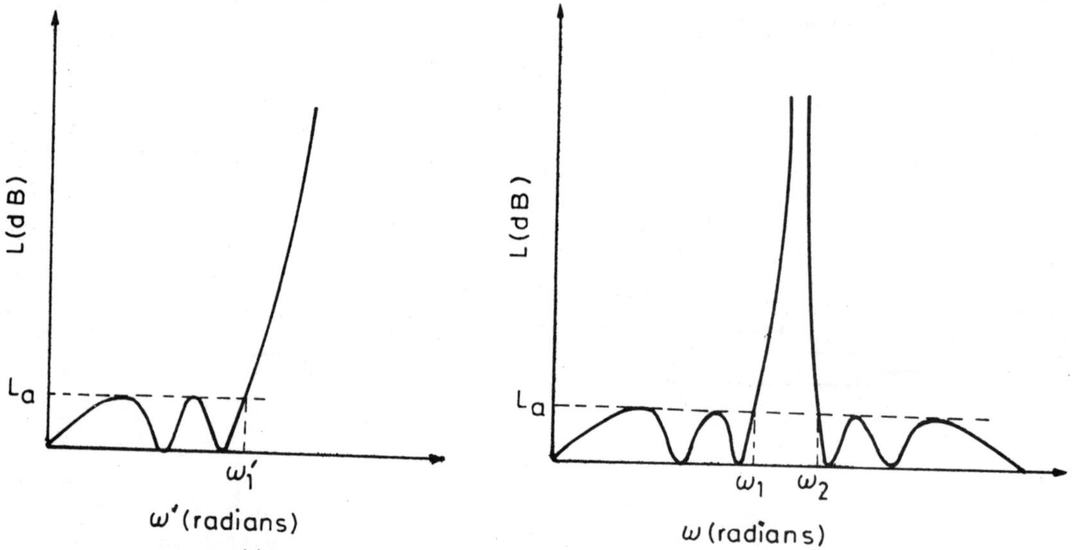

(a) (b)

Fig. 12.69 Transformation from (a) low-pass prototype response to (b) bandstop response

Figure 12.70 shows the layout of a bandstop filter which is useful for narrow stopband applications. The filter consists of a series of open-circuited shunt stubs capacitively coupled to the main line at intervals of quarter-wavelength at the stopband center frequency. The shunt stubs, which are slightly shorter than half-wavelength, approximate inductances, and the gaps approximate capacitances as shown in Fig. 12.70b. The use of open-circuited stubs instead of the nearly quarter-wavelength long short-circuited stubs eliminate any problem associated with the connection of stub ends to ground. Furthermore, the open-circuited stubs permit a wider tuning range to be achieved by capacitively loading the open ends. An approximate but general design technique for a n-section filter of the type shown in Fig. 12.70a is available in Matthaei et al [3]. We summarize below their design procedure for a bandstop filter with Chebyshev response outside the stopband:

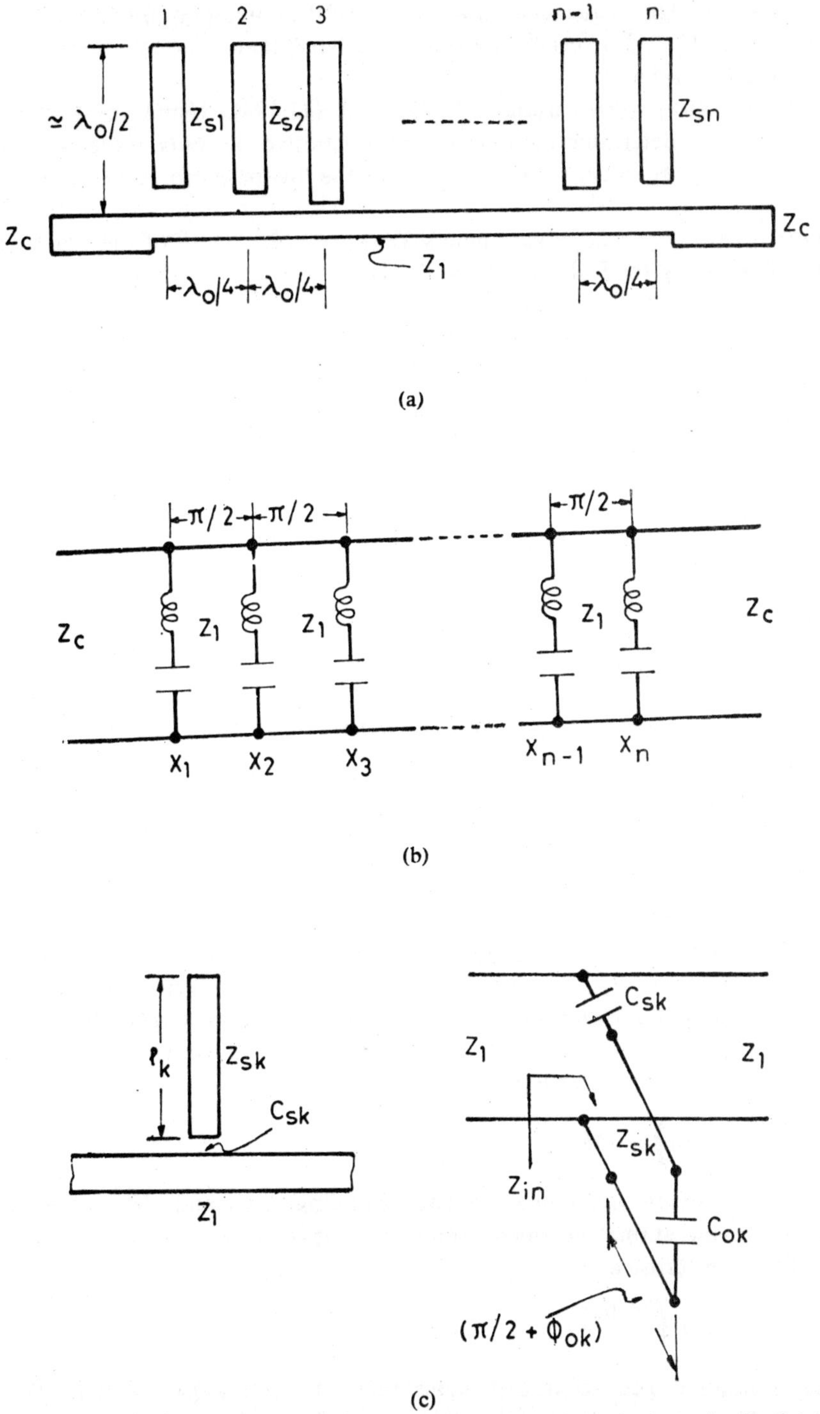

Fig. 12.70 (a) Bandstop filter with gap coupled stubs, (b) Equivalent circuit of (a), (c) Single stub coupled to the main line and its equivalent representation

•Specify the band-edge frequencies ω_1 and ω_2, and the corresponding insertion loss L_a (in dB). (Refer Fig. 12.69). Also specify the skirt selectivity, that is, attenuation L_b (in dB) at a frequency ω in the stop-band.

•Calculate the ratio $|\omega'/\omega_1'|$ using (12.197) and (12.198). Using the known values of $|\omega'/\omega_1'|$, L_a and L_b, determine the number (n) of reactive elements in the prototype.

•Calculate the element values $g_0, g_1, \ldots, g_{n+1}$ of the low-pass prototype. (Use (12.160) and (12.161)).

•Determine the reactance slope parameters X_k ($k = 1, 2, \ldots, n$) of the shunt elements of the bandstop filter using the following formulas [3].

$$\frac{X_1}{Z_c} = \frac{1}{\omega_1' g_0 g_1 \Delta\omega}$$

$$\frac{X_k}{Z_c} = \left(\frac{Z_1}{Z_c}\right)^2 \frac{g_0}{\omega_1' g_k \Delta\omega}, \qquad k \text{ even} \tag{12.199}$$

$$\frac{X_k}{Z_c} = \frac{1}{\omega_1' g_0 g_k \Delta\omega}, \qquad k \text{ odd}$$

where

$$Z_1 = Z_c, \qquad n \text{ even} \tag{12.200a}$$

$$\left(\frac{Z_1}{Z_c}\right)^2 = \frac{1}{g_0 g_{n+1}}, \qquad n \text{ odd} \tag{12.200b}$$

•The reactance slope parameter X_k, the characteristic impedance Z_{sk} of the kth stub, and the gap capacitance C_{sk} (see Fig. 12.70c) are related by the following expressions:

$$2\left(\frac{X_k}{Z_{sk}}\right) = \left(\frac{\pi}{2} + \phi_{ok}\right) \sec^2 \phi_{ok} + \tan^2 \phi_{ok} \tag{12.201a}$$

$$\omega_0 C_{sk} = \frac{1}{Z_{sk} \tan \phi_{ok}} \tag{12.201b}$$

where ω_0 is given by (12.198b). In a practical design, Z_{sk} is chosen so as to give a convenient and realizable width for the stub section, and (12.201a) is then solved to obtain ϕ_{ok}. Substituting the values of Z_{sk} and ϕ_{ok} in (12.201b) gives the gap capacitance C_{sk}.

•The physical length l_k of the stub is obtained from its electrical length by properly accounting for the open-circuit end-effect.

$$l_k = \frac{\lambda_0}{2\pi}\left(\frac{\pi}{2} + \phi_{ok}\right) - \Delta l_k \tag{12.202}$$

where λ_0 is the wavelength in the structure corresponding to ω_0 and Δl_k is the open-circuit end correction of the stub. The open-circuit end capacitance C_{ok} is related to the end-correction Δl_k by the formula

$$\omega_0 C_{ok} = \frac{1}{Z_{sk}} \tan\left(\frac{2\pi \Delta l_k}{\lambda_0}\right) \tag{12.203}$$

Methods for evaluating open-circuit end capacitances are discussed in Chapter 10.

•The gap width corresponding to the capacitance C_{sk} can be experimentally evaluated by testing a single stub coupled to the main line. Referring to the equivalent circuit shown in

Fig. 12.70c, the expression for the input admittance $Y_{in,k}$ presented by the stub at the main line is given by

$$Y_{in,k} = \frac{j\omega C_{sk}\left[Z_{sk}\omega C_{sk} + \tan\left(\frac{2\pi}{\lambda}l_k\right)\right]}{\omega Z_{sk}(C_{sk} + C_{ok}) - \tan\left(\frac{2\pi}{\lambda}l_k\right)(Z_{sk}^2\omega^2 C_{sk}C_{ok} - 1)} \tag{12.204}$$

Using this expression, the 3 dB bandwidth of a single resonator can be computed for various values of C_{sk}. The measured value of 3 dB bandwidth is then compared with the computed value to characterize the gap spacing as a function of C_{sk}.

Single Section Coupled-line Bandstop Filter

A second type of bandstop filter suitable for narrow stopbands is shown in Fig. 12.71a. As discussed in sec. 12.4.3, this is basically a low-pass filter element. From the image impedance expressions given by (12.39) and (12.40), it can be seen that when the electrical length θ of the coupled section equals $\pi/2$, $Z_{I1} = 0$ and $Z_{I2} = \infty$. Thus, a quarter-wavelength long coupled line section behaves like a bandstop filter. Figure 12.71b shows a practical layout of the bandstop filter where the short circuit is realized by a quarter-wavelength long open-circuited stub. This type of filter is commonly used in balanced mixers for reflecting image frequency.

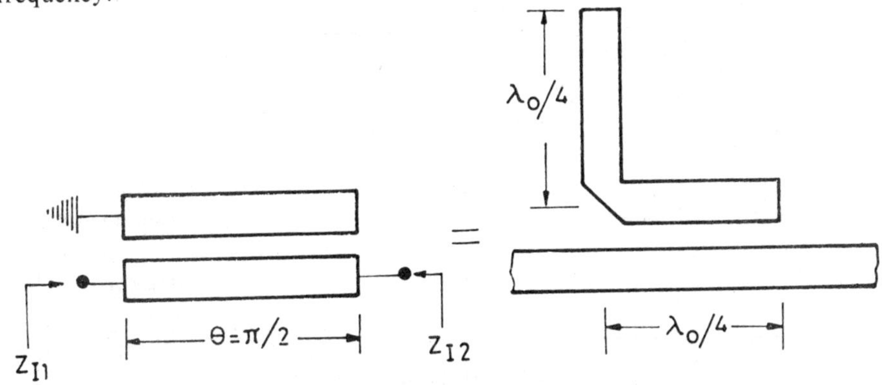

(a) (b)

Fig. 12.71 Coupled line bandstop filter

Bandstop Filters with Wide Stopband

Figure 12.72 shows two configurations of the bandstop filter suitable for wide stopband. The stub-coupled filter shown in Fig. 12.72a consists of $\lambda_0/4$ open circuited stubs separated by connecting lines which are either $\lambda_0/4$ or $3\lambda_0/4$ long, where λ_0 is the guide wavelength at the stopband centre frequency. Exact design equations for designing this type of filter upto five stubs have been reported by Schiffman and Matthaei [96].

In the spur-line filter shown in Fig. 12.72b, the shunt stub is folded parallel to the main line so that coupling is both by direct connection and by fringing fields. The coupling length is quarter wavelength at the stopband centre frequency. The energy stored in the resonant structure is determined predominantly by the odd-mode fields. The stopband bandwidth of the filter can be narrowed by reducing the odd-mode impedance, that is, by reducing the

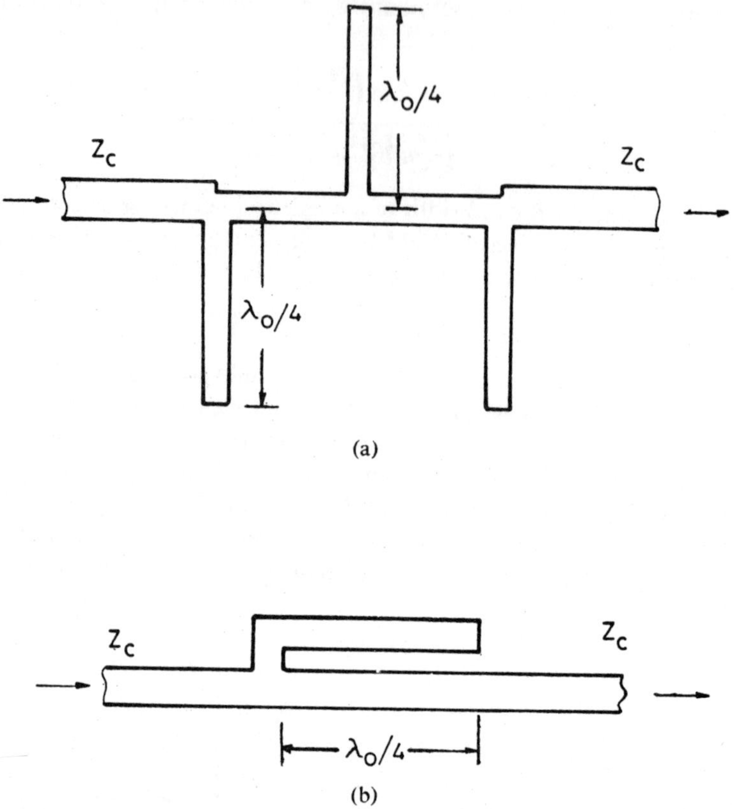

Fig. 12.72 (a) Direct stub-coupled bandstop filter, (b) Spur-line bandstop filter

spacing between the coupling lines. The analysis and design of spur-line filters using symmetrical and asymmetrical lines have been reported by several investigators [96-99]. Spur-line filters with asymmetrical coupled lines have the advantage of being able to act as impedance transformers and can be designed to offer wider stopband [99].

12.10 P-I-N Diode Switches and Attenuators [100-105]

The p-i-n diode acts as an electronic switch when operated between fixed forward and reverse bias states. As a basic switch, it is extensively used in the realization of multiple-throw switches, phase shifters, modulators, limiters and duplexers. With a variable forward bias the forward bias resistance of the p-i-n diode can be varied over a wide range. This property is used in realizing electronically variable attenuators. An exhaustive treatment of the p-i-n diode theory and design techniques of p-i-n diode control components is available in the books by White [100] and Garver [101].

12.10.1 P-I-N Diode Equivalent Circuit

The p-i-n diode consists of an intrinsic semiconductor (silicon) layer sandwiched between heavily doped p- and n-regions (Fig. 12.73a). The general equivalent circuit of a p-i-n diode is shown in Fig. 12.73b where R_j and C_j represent the junction resistance and capacitance respectively, R_s is the bulk semiconductor resistance, L_p is the package or lead inductance and C_p is the package capacitance. Typical values for L_p and C_p are in the range 0.1 to 2 nH and 0.1 to 0.5 pF respectively. For a chip diode, L_p and C_p may be omitted. The other parameters of the diode take different values under different bias conditions. Under forward bias, the majority charge carriers from the p- and n-regions flood into the I-layer setting it into conducting state. The stored charge in the I-layer permits the diode to be conductivity modulated. The diode behaves essentially like a short circuit thus allowing microwave signals to pass through. For a practical p-i-n diode in chip form, the equivalent circuit can be represented by a small resistance R_f (Fig. 12.73c). It is the sum of the resistances of the p- and n-layers and the contact resistance and is typically in the range 0.2 to 5 ohms. Under

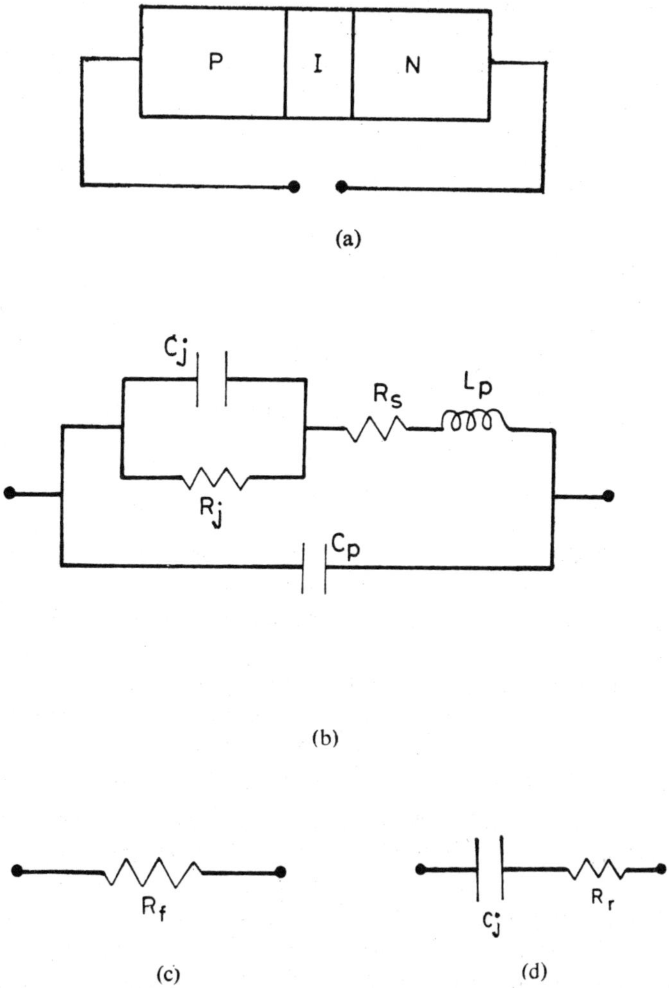

Fig. 12.73 (a) P-I-N diode, (b) General equivalent circuit, (c) Equivalent circuit of p-i-n diode chip under forward bias, (d) Equivalent circuit of p-i-n diode chip under reverse bias

reverse bias, the charge carriers are completely swept out of the I-layer. The equivalent circuit can be represented by a capacitance C_j in series with a small resistance R_r as shown in Fig. 12.73d. Typical values of C_j and R_r are in range 0.01 to 2 pF and 0.2 to 5 ohms respectively. The diode presents a high impedance and is predominantly governed by the capacitive reactance.

12.10.2 Basic Series and Shunt Switches

Figure 12.74a shows a p-i-n diode mounted in series with a transmission line of characteristic impedance Z_c. The transmission coefficient S_{21} of the network can be easily obtained from its $ABCD$ parameters. The transmission loss α is given by

$$\alpha\,(dB) = 20 \log_{10} \frac{1}{|S_{21}|}$$

$$= 20 \log_{10} \left| \frac{A + BY_c + CZ_c + D}{2} \right|, \qquad Y_c = \frac{1}{Z_c} \tag{12.205}$$

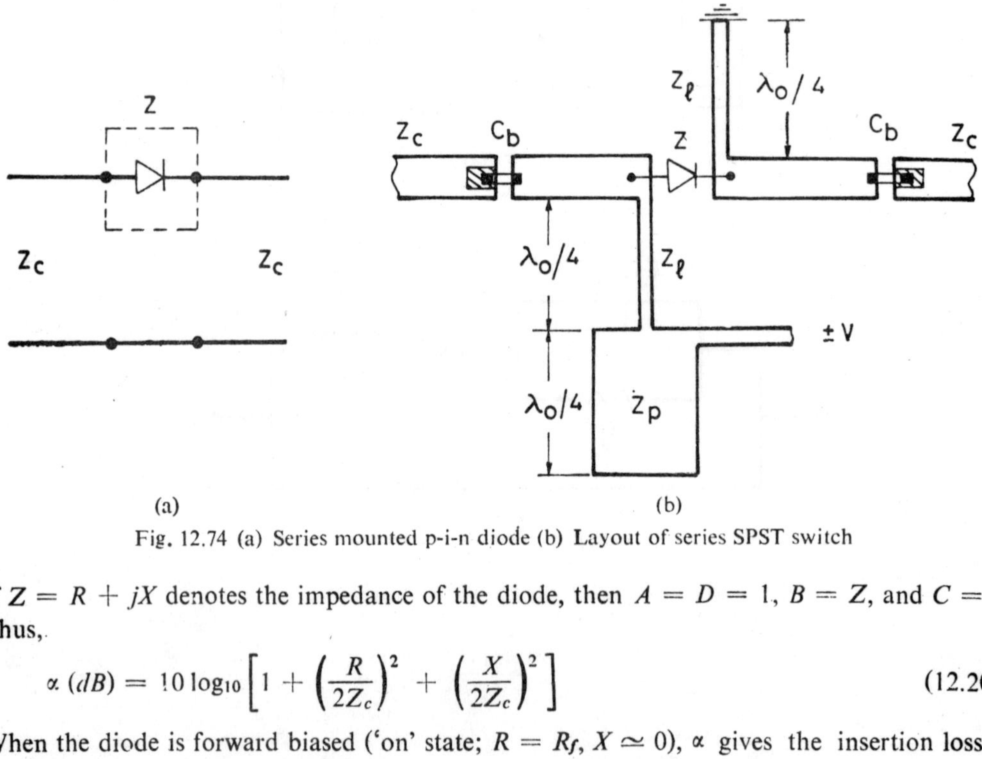

(a) (b)

Fig. 12.74 (a) Series mounted p-i-n diode (b) Layout of series SPST switch

If $Z = R + jX$ denotes the impedance of the diode, then $A = D = 1$, $B = Z$, and $C = 0$. Thus,

$$\alpha\,(dB) = 10 \log_{10} \left[1 + \left(\frac{R}{2Z_c} \right)^2 + \left(\frac{X}{2Z_c} \right)^2 \right] \tag{12.206}$$

When the diode is forward biased ('on' state; $R = R_f$, $X \simeq 0$), α gives the insertion loss of the switch, and when the diode is reverse biased ('off' state; $R = R_r$, $X = 1/\omega C_j$), α gives the isolation.

Figure 12.74b shows the layout of the series switch with the biasing circuit. The d.c. blocking capacitor C_b connected in series with the main line is normally a ceramic chip capacitor. It is soldered directly on the strip conductor on one-side of the gap with its strap soldered to the adjacent strip conductor on the other side. D.C. bias is introduced via a bypassed bias inductor line with a D.C. return inductor to complete the bias current circuit.

These inductors are quarter-wavelength long, high impedance ($Z_l > 100$ ohms) strip conductors. The RF by pass capacitor consists of a quarter-wavelength long conductor patch of low characteristic impedance ($Z_p < 25$ ohms).

Figure 12.75 shows the basic configuration of a shunt switch and the circuit layout. The transmission loss in a shunt switch is given by

$$\alpha\,(dB) = 10 \log_{10}\left[\left(1 + \frac{GZ_c}{2}\right)^2 + \left(\frac{BZ_c}{2}\right)^2\right] \tag{12.207}$$

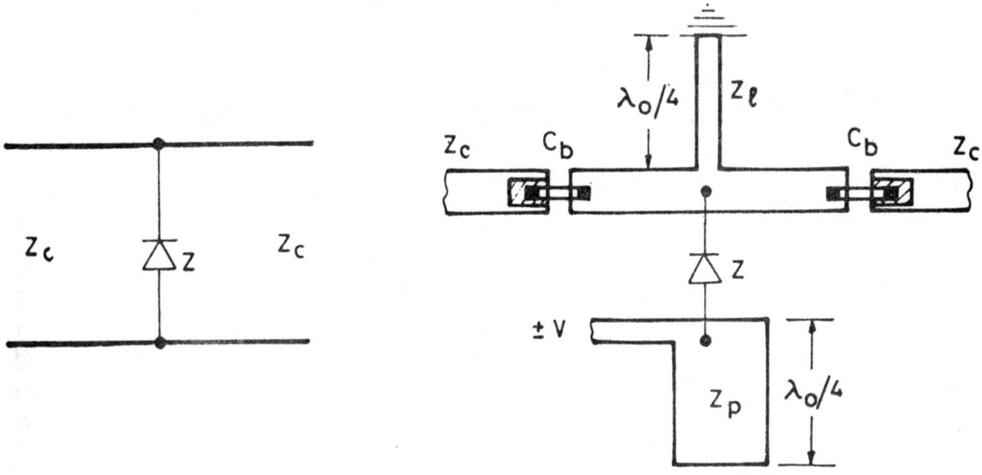

(a) (b)

Fig. 12.75 (a) Shunt mounted, p-i-n diode (b) Layout of shunt SPST Switch

where $(G + jB) = 1/Z$ is admittance of the p-i-n diode. In this case, α corresponds to insertion loss when the diode is 'off', and isolation when the diode is 'on'.

Switches employing series mounted diodes are suitable for low power designs (typically less than 1 watt) whereas those using diodes in shunt can handle higher power levels because of superior thermal mounting. The peak power handling capability of a switch is determined by the reverse bias breakdown voltage V_b, and the characteristic impedance Z_c of the transmission line in which the diode is mounted. For a series diode, the peak power limit is given by $V_b^2/32Z_c$, and for the shunt diode it is $V_b^2/8Z_c$. The bandwidth of these switches is limited primarily by the parasitic inductance and capacitance of the p-i-n diode and to a lesser extent by the bias circuit. Broadband performance (over several octaves) with very low VSWR can be achieved by designing the switch based on filter theory [103]. The reactive elements of the diode and the bias circuit are made part of the low-pass filter elements so that the response of the switch is governed by the filter response.

12.10.3 SPDT Switches

Figure 12.76 shows the layout of a series-shunt SPDT (single-pole, double-throw) switch. D_1, D_2, D_3 and D_4 are identical p-i-n diodes. When a positive bias is applied to bias point B_1 and negative bias to bias point B_2, diodes D_1 and D_4 are in the 'on' state (conducting), and D_2 and D_3 are in the 'off' state. Microwave signal fed to the input port appears at output port 1 and port 2 is isolated. When the bias at B_1 and B_2 are reversed, signal flows to port 2 and no output appears at port 1.

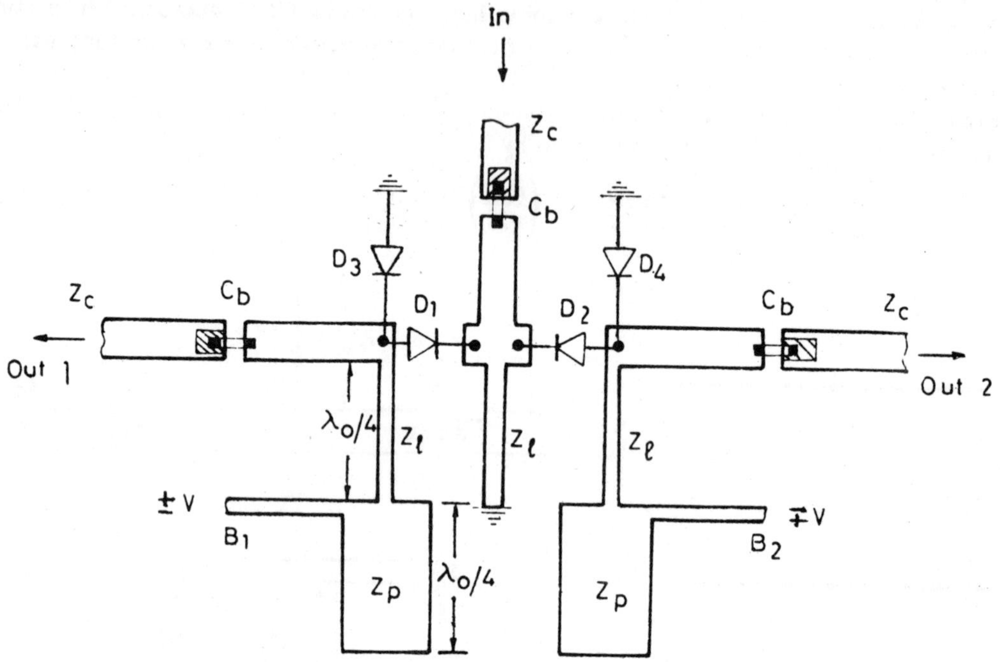

Fig. 12.76 Layout of series-shunt SPDT switch

When the signal flow is from input port to port 1, the insertion loss α_I in the transmission path can be approximated by the sum of the losses due to individual elements. To a first order approximation, it is given by [104]

$$\alpha_I(dB) = 20 \log_{10}\left(1 + \frac{R_{f1}}{2Z_c}\right) + 10 \log_{10}\left[1 + \left(\frac{Z_c}{2X_2}\right)^2\right] + 10 \log_{10}\left[1 + \left(\frac{Z_c}{2X_3}\right)^2\right] \quad (12.208)$$

where R_{f1} is the forward bias resistance of D_1, and X_2 and X_3 are the reverse bias capacitive reactances of D_2 and D_3 respectively.

The isolation in a series SPDT switch (Fig. 12.76 with D_3 and D_4 omitted) is 6 dB higher than in a corresponding SPST switch because the voltage at the 'off' series diode is only half of the voltage in a SPST switch. An additional 6 dB isolation is obtained by cascading series and shunt diodes. Thus, for the series-shunt SPDT switch of Fig. 12.76, the isolation α_s, to a first order approximation, is given by [104]

$$\alpha_s(dB) = 10 \log_{10}\left[1 + \left(\frac{X_2}{2Z_c}\right)^2\right] + 20 \log_{10}\left(1 + \frac{Z_c}{2R_{f4}}\right) + 12 \quad (12.209)$$

where R_{f4} is the forward bias resistance of D_4. With identical diodes, the insertion loss and isolation calculated from (12.208) and (12.209), respectively, also apply when the transmission is from the input port to port 2.

Although the addition of the two shunt diodes improves the isolation of the series SPDT switch by 6 dB, it also results in a marginal increase in the insertion loss as indicated by the third term in (12.208). The series-shunt configuration offers high isolation as well as broad bandwidth of several octaves.

From the point of view of isolation as well as power handling capability, the shunt SPDT switch is preferred over the series SPDT switch. Waugh [105] has reported a simple matching scheme for broadbanding a shunt SPDT switch while preserving its high isolation property. The schematic of the switch is shown in Fig. 12.77. The circuit employs a matching *T*-network at the common junction between the input port and the two output ports. Quarter-wavelength long lines of characteristic impedance Z (less than Z_c) form the three arms of the *T*-network. This matching network offers an equiripple VSWR response within the operating frequency range. The value of Z must be chosen so as to achieve the best compromise between bandwidth and VSWR. Waugh [105] has reported a VSWR of 1.43 : 1 over a bandwidth of 105% (3.2 : 1) for a value of Z equal to 32 ohms (Z_c = 50 ohms). Without this compensation, that is, with Z = Z_c = 50 ohms, the bandwidth of the same switch is only 40%.

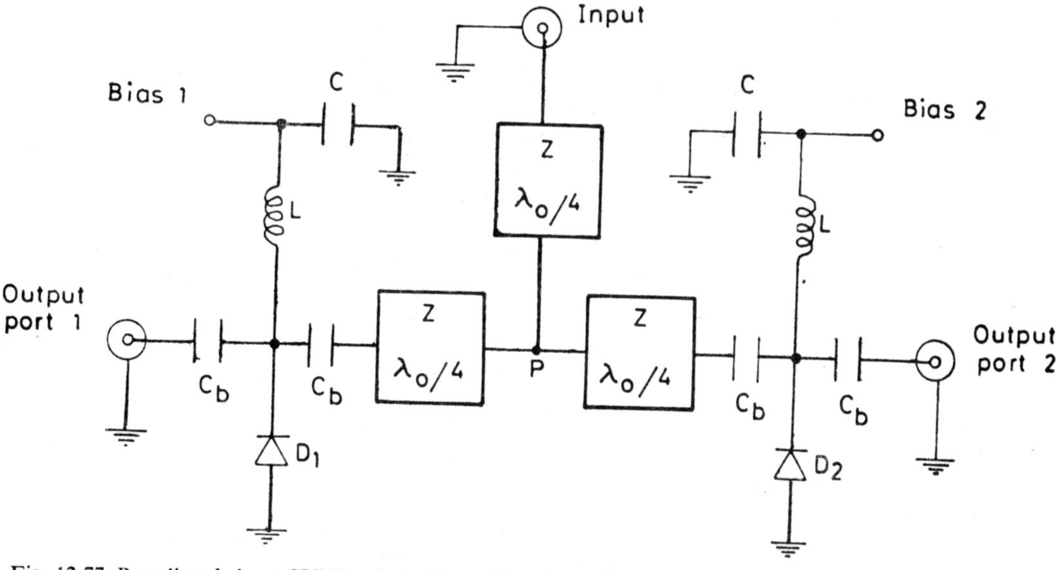

Fig. 12.77 Broadband shunt SPDT switch (From Waugh [105], reprinted with permission of Microwave Journal, Copyright © 1975 Horizon House, Inc.)

The above broadbanding technique has also been applied to shunt multi-throw switches [105]. For example, for realizing a shunt SP3T switch, another section identical to the one that appears between the common junction point *P* and port 1 (or port 2) is attached to point *P* in Fig. 12.77. This additional section allows signal to be switched to the third output port.

12.10.4 Hybrid-Coupled Attenuator

The switching circuits described in the preceding section can also be used as attenuators by varying the forward bias current of the diodes. They form reflective-type attenuators in which the input VSWR varies with the bias. In many microwave systems, such high VSWR is not desirable. P–I–N diode attenuators with constant input impedance characteristic can be built by incorporating a circulator or a hybrid coupler in the circuit. Figure 12.78 shows the schematic of a single hybrid coupled attenuator. Two identical p–i–n diodes (of impedance Z) are mounted symmetrically in the two output arms of the 3 dB, 90° hybrid coupler and are shunted by matched loads. The bias circuitry of the diodes is not shown. The two p–i–n diodes are forward biased at the same level. Power fed to port 1 is divided equally between ports 3 and 4. Depending on the diode impedance, part of the signal gets reflected back from each diode. The two reflected signals combine and emerge out of port 2. Port 1 is isolated and appears matched to the input signal irrespective of the diode impedance. The maximum attenuation that can be achieved in this type of attenuator depends on the directivity of the coupler and the quality of terminations when the diodes are unbiased.

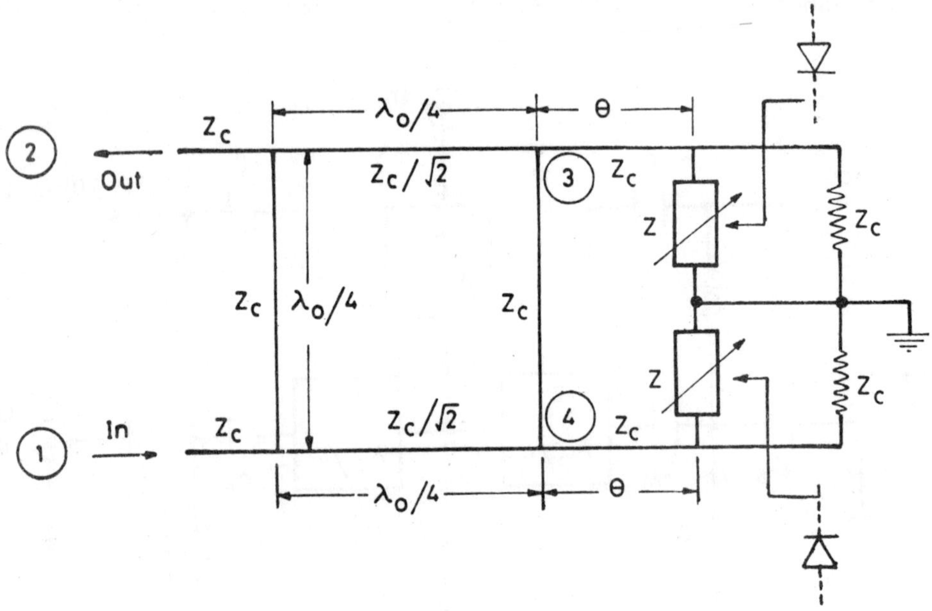

Fig. 12.78 Hybrid-coupled p-i-n diode attenuator

12.10.5 Duplexer

P–I–N diode duplexers are used in radars. A duplexer is a device which permits the same antenna to be used for transmission and reception. The circuit consists of two high directivity 3 dB, 90° hybrid couplers and shunt mounted diodes. The functions of the duplexer in the transmit mode and receive mode are illustrated in Fig. 12.79. The duplexer is self actuated by the p–i–n diode limiters. In the transmit mode, the high power of the transmitter sets the diodes into conducting state. The power reflected from the diodes adds up at the antenna port. Any power which passes the limiter is mostly absorbed by the resistive termination, thus protecting the receiver. In the receive mode, the low power signal passes through the p–i–n diodes. Because of the 180° phase difference introduced by the two couplers, the received signal adds up in phase at the receiver but cancels at port 7.

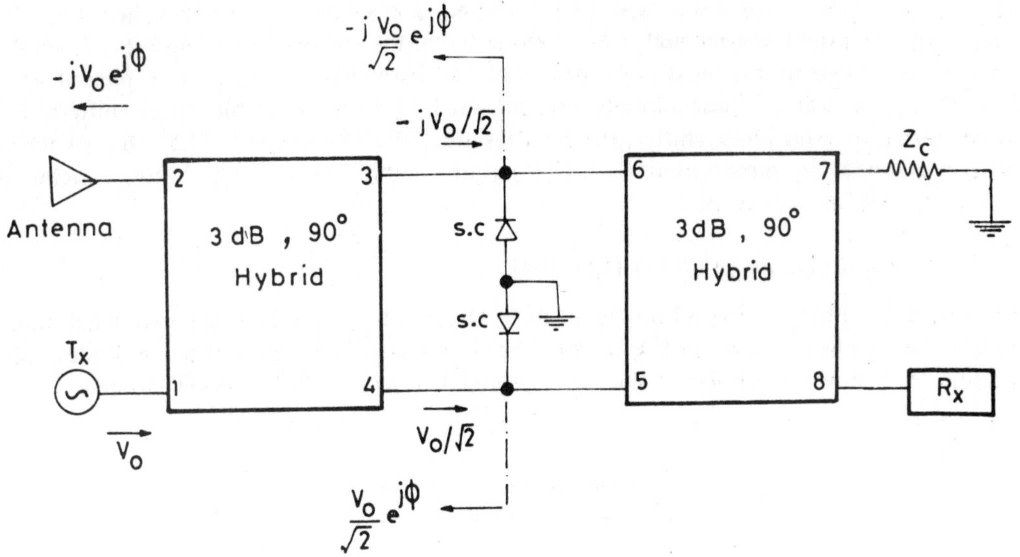

(a)

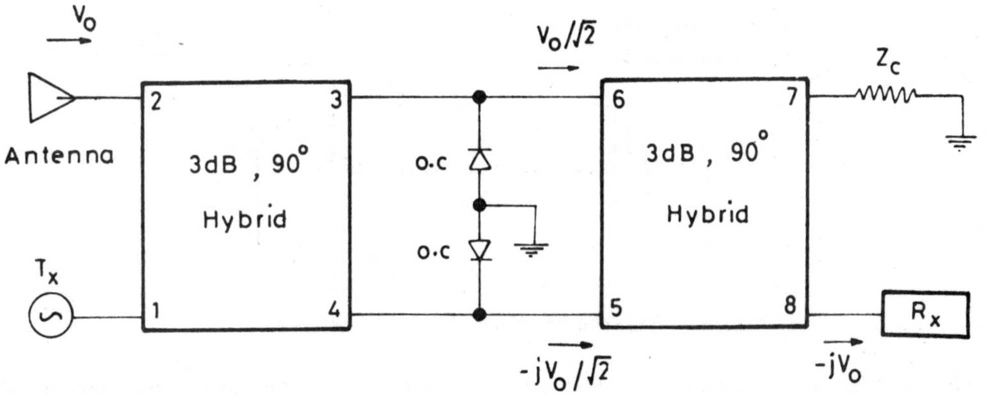

(b)

Fig. 12.79 Schematic of a duplexer (a) transmit mode and (b) receive mode

12.11 Phase Shifters [100, 101, 106-121]

MIC phase shifters can be built using p–i–n diodes, varactors or GaAs FETs. Of the three semiconductor devices, p–i–n diodes are the most commonly used because of their reproducibility of characteristics and high power handling capability. Digital p–i–n diode phase shifters are extensively employed in phased arrays to electronically switch the radiated beam. These phase shifters can be broadly classified as either the reflection type or the transmission type. The reflection type phase shifters can be realized using a circulator or a hybrid coupler.

Some of the useful transmission type phase shifters are known as the switched line, the loaded line and the low-pass high-pass. Among these, the switched line, the hybrid coupled and the loaded line are the most popularly employed basic circuit forms. For phased array applications, several of these circuits are cascaded to form multi-bit phase shifters. For example, in a four-bit phase shifter, the bit sizes are 180°, 90°, 45° and 22.5°. By switching the p-i-n diodes in these circuits in appropriate combinations, phase shifting is achieved from 0° to 360° in steps of 22.5°.

12.11.1 Switched-Line Phase Shifter [106–109]

A switched line phase shifter adopts two SPDT switches to switch between two fixed transmission line paths as shown in Fig. 12.80. The circuit uses four identical p-i-n diodes. Bias circuitry (not shown) is similar to the one described in sec. 12.10 for SPDT switches.

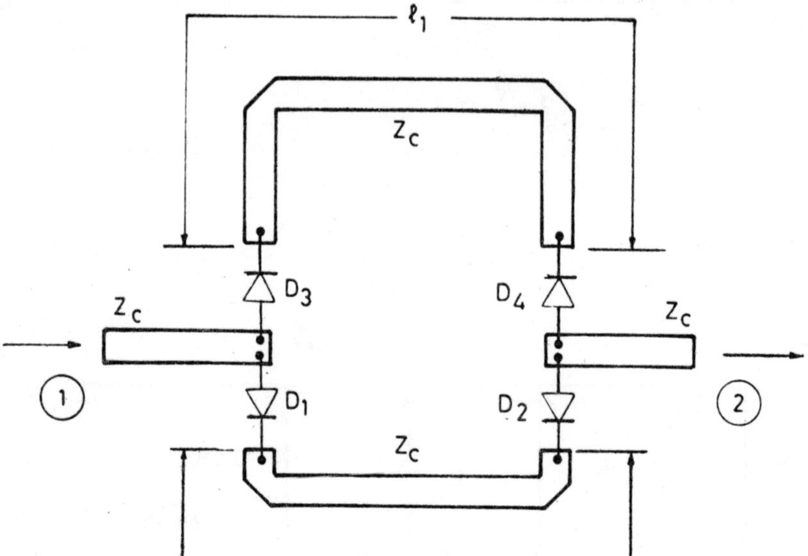

Fig. 12.80 Switched line phase shifter with series diodes

The transmission coefficient of this circuit can be easily determined by analysing it in terms of the even- and odd-mode excitations. Let S_{21} denote the transmission coefficient of the network when the diodes D_1 and D_2 are forward biased and D_3 and D_4 are reverse biased. If Z_f and Z_r are the forward and reverse bias impedances of the diode normalized with respect to Z_c, the expression for S_{21} can be written as

$$S_{21} = |S_{21}|\, e^{j\phi_0} = \left[1 + \frac{1}{(\bar{Z}_f - j\cot(\beta l_0/2))} + \frac{1}{(\bar{Z}_r - j\cot(\beta l_1/2))}\right]^{-1}$$

$$- \left[1 + \frac{1}{(\bar{Z}_f + j\tan(\beta l_0/2))} + \frac{1}{(\bar{Z}_r + j\tan(\beta l_1/2))}\right]^{-1} \qquad (12.210)$$

where $\beta = 2\pi/\lambda$ is the propagation constant in the structure. In the second switching state, when D_3 and D_4 are forward biased, and D_1 and D_2 are reverse biased, the transmission coefficient S'_{21} is given by

$$S'_{21} = |S'_{21}| e^{j\phi_1} = \left[1 + \frac{1}{(\overline{Z}_r - j\cot(\beta l_0/2))} + \frac{1}{(\overline{Z}_f - j\cot(\beta l_1/2))}\right]^{-1} - \left[1 + \frac{1}{(\overline{Z}_r + j\tan(\beta l_0/2))} + \frac{1}{(\overline{Z}_f + j\tan(\beta l_1/2))}\right]^{-1} \quad (12.211)$$

The differential phase shift is then given by

$$\Delta\phi = (\phi_1 - \phi_0) \quad (12.212)$$

and the insertion loss (in *dB*) in the two switching states can be obtained as 20 $\log_{10} |S_{21}|$ and 20 $\log_{10} |S'_{21}|$. If the p-i-n diodes are ideal, the phase shift is simply a linear function of frequency and is given by $\Delta\phi = \beta(l_1 - l_0)$. With nonideal diodes also, the phase-frequency characteristic is nearly linear except at some resonant frequencies where the effective length of the 'off' path (electrical length of the line plus the equivalent length of the capacitive 'off' diode) is equal to $\lambda/2$ (or integral multiples of $\lambda/2$). At resonance, the reflected signal adds up in phase with the incident signal so as to give large phase errors and also insertion loss spikes. However, for a 180° phase bit, with l_1 longer than l_0 by $\lambda/2$, the phase error would be negligible for all values of l_0 less than $\lambda/2$. In general, in order to ensure minimum phase error and equal insertion loss in both the switching states, the value of $2\pi l_0/\lambda$ should be chosen between 20° and 50° [107].

The switched line phase shifter considered above is essentially a time-delay device. Wide bandwidth of the order of an octave can be achieved by using a Schiffman coupled section [29] in one of the transmission paths as shown in Fig. 12.81a. The phase-frequency response of Schiffman section has been discussed in Sec. 12.4.5. The phase shift ϕ through the section

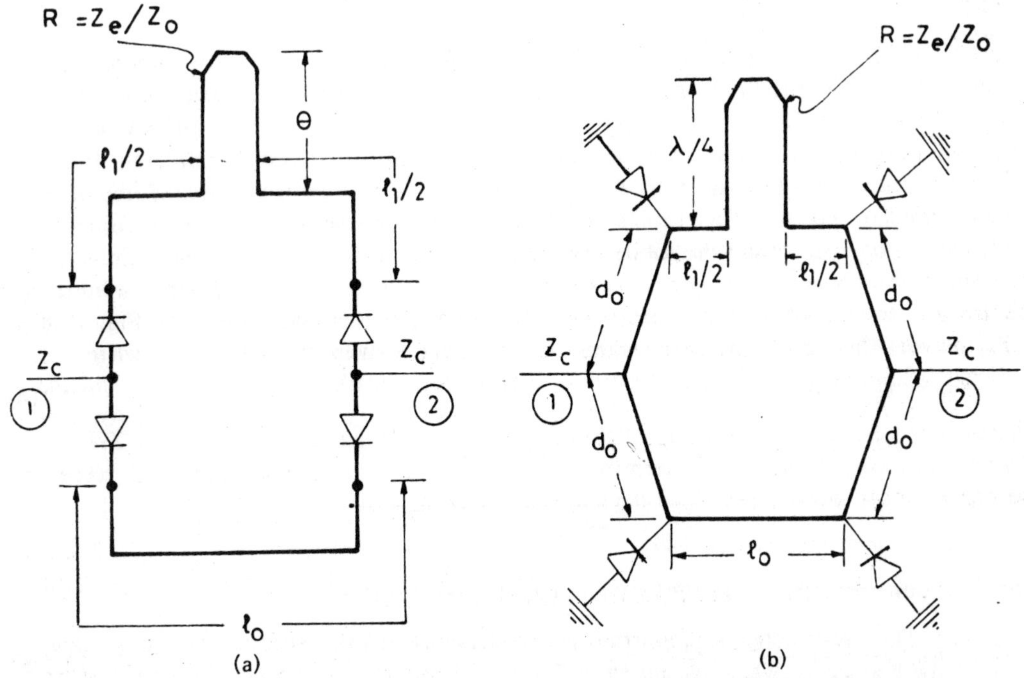

Fig. 12.81 Switched line Schiffman phase shifter with (a) series diodes and (b) shunt diodes

as a function of frequency consists of a linear component and a sinusoidal component, and is given by (12.58). Referring to the notations used in Fig. 12.81a, the differential phase shift of the phase shifter is given by

$$\Delta\phi = \beta(l_0 - l_1) - \cos^{-1}\left(\frac{R - \tan^2\theta}{R + \tan^2\theta}\right) \tag{12.213}$$

At the centre frequency, the length of the coupled line section is $\lambda/4\,(\theta = \pi/2)$. The lengths l_0 and l_1 are chosen such that the slope of $\beta(l_0 - l_1)$ is parallel to the slope of the Schiffman curve at $\theta = \pi/2$. This ensures relatively constant phase shift over a wide bandwidth.

Figure 12.81b shows the configuration of the switched line Schiffman phase shifter with shunt mounted diodes. The diodes are mounted at a distance of d_0 from the main junction. If the p–i–n diodes are purely resistive in the "on" condition, then choosing $d_0 = \lambda/4$ ensures high-impedance at the main junction and hence a high isolation of the "on" diode path. In practice, when the lead inductance of the diode needs to be taken into account, the distance d_0 will have to be accordingly adjusted so as to achieve maximum impedance at the main junction due to the "on" diode. The design of the phase shifter therefore involves the correct choice of d_0, l_0 and l_1 so as to obtain the desired phase shift and minimum insertion loss. This can be easily done by first obtaining the expressions for the transmission and reflection coefficients of the network using the even- and odd-mode technique and then computationally optimizing the dimensional parameters.

Another variation of the switched line phase shifter is to use Schiffman sections of different lengths and coupling factors in both the transmission paths. This design enables elimination of off-path resonance thus offering resonant free wideband phase shifter.

12.11.2 Hybrid-Coupled Phase Shifter [109-115]

Figure 12.82a shows the schematic of a hybrid-coupled phase shifter with identical p–i–n diodes in the two output arms and terminated in short circuited transmission line sections. Other hybrids—namely, the 3 *dB* backward wave coupler and the rat race hybrid have also been used to realize such reflective-type phase shifters [109], There are also different forms of switching circuits by which the output arms of the hybrids may be terminated. Some of these circuit forms are shown in Fig. 12.82b–d. The diodes can be series or shunt mounted and the transmission line behind the diode can be terminated in a short or an open circuit. The analysis of the phase shifter can be carried out by first determining the input admittance looking into the reflective load at plane *PP'*. Considering the circuit shown in Fig. 12.82a, if Y_{inf} denotes the input admittance when the diodes are forward biased, we can write

$$Y_{inf} = G_f + jB_f - jY_1 \cot\theta \tag{12.214}$$

where $(G_f + jB_f)$ is the admittance of the diode under forward bias and $Y_1(=1/Z_1)$ is the characteristic admittance of the terminating transmission line. Normalizing with respect to the characteristic admittance Y_c of the main line, we can write

$$y_{inf} = (g_f + jb_f - j\,y_1 \cot\theta) \tag{12.215}$$

The reflection coefficient Γ_f at *PP'* is then given by,

$$\Gamma_f = \frac{(1 - g_f) - j(b_f - y_1 \cot\theta)}{(1 + g_f) + j(b_f - y_1 \cot\theta)} \tag{12.216}$$

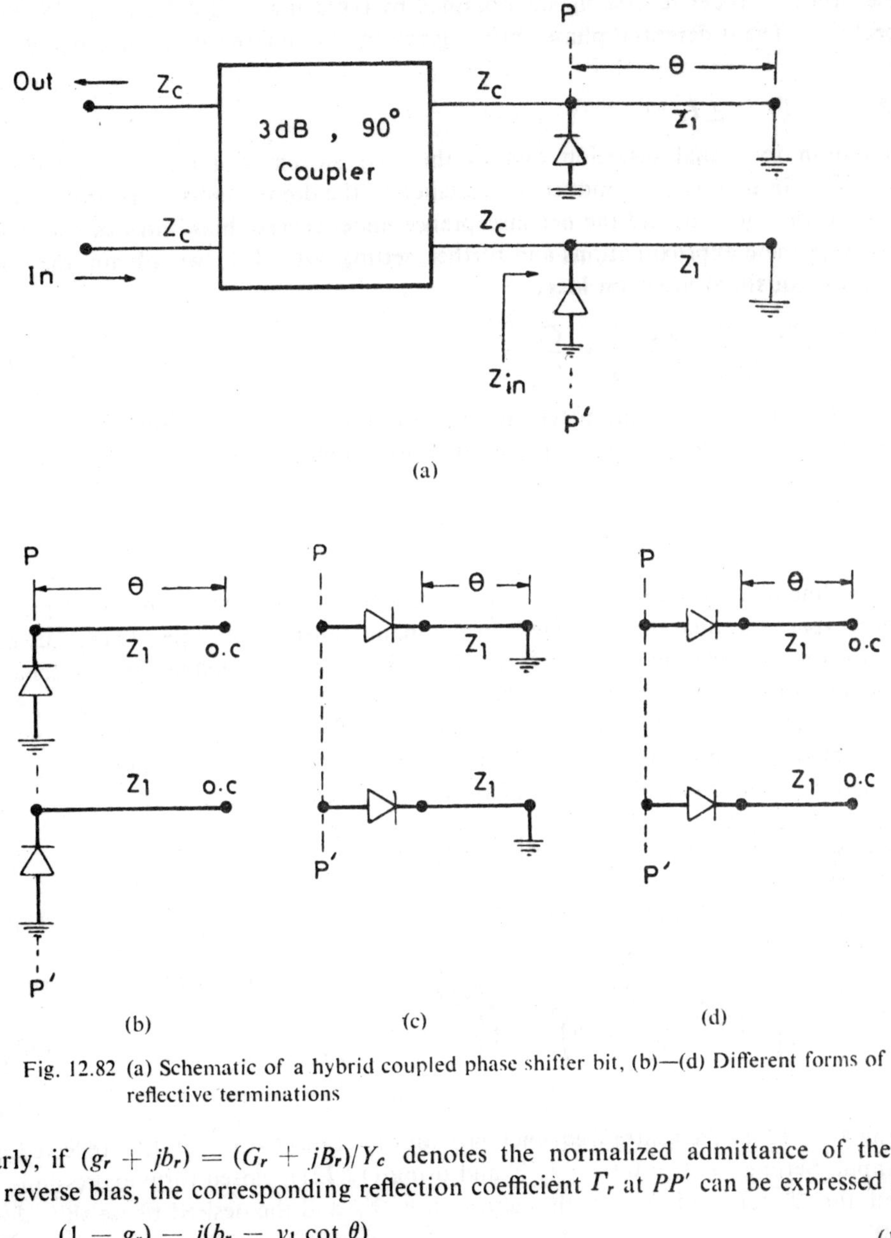

Fig. 12.82 (a) Schematic of a hybrid coupled phase shifter bit, (b)—(d) Different forms of reflective terminations

Similarly, if $(g_r + jb_r) = (G_r + jB_r)/Y_c$ denotes the normalized admittance of the diode under reverse bias, the corresponding reflection coefficient Γ_r at PP' can be expressed as

$$\Gamma_r = \frac{(1 - g_r) - j(b_r - y_1 \cot \theta)}{(1 + g_r) + j(b_r - y_1 \cot \theta)} \tag{12.217}$$

From the reflection coefficient, we can now obtain the return loss. Assuming the coupler to be ideal, the overall insertion loss of the phase shifter is the same as the return loss. The expression for the insertion loss under the forward bias condition is given by

$$\alpha_{If} = 10 \log_{10} \left[\frac{(1 + g_f)^2 + (b_f - y_1 \cot \theta)^2}{(1 - g_f)^2 + (b_f - y_1 \cot \theta)^2}\right] dB \tag{12.218}$$

The insertion loss under reverse bias is obtained by replacing g_f and b_f in (12.218) by g_r and b_r respectively. The differential phase shift is given by the difference in the arguments of Γ_r and Γ_f.

$$\Delta\phi = \angle\Gamma_r - \angle\Gamma_f \tag{12.219}$$

The condition for equal insertion loss in the two switching states is obtained by setting $|\Gamma_r| = |\Gamma_f|$. In practice, the inductive reactance of the diode under forward bias can be tuned out so that $b_f \simeq 0$, and the net susceptance under reverse bias satisfies the condition, $b_r \gg g_r$. With these approximations and further setting $g_f g_r \ll 1$, we obtain the following condition for equalized insertion loss:

$$\left[\frac{Y_c^2 + (B_r - Y_1 \cot\theta)^2}{Y_c^2 - Y_1^2 \cot^2\theta}\right] = \frac{G_r}{G_f} \tag{12.220}$$

For a specified set of diode parameters and main line admittance Y_c, the parameters θ and Y_1 are optimized to give the desired phase shift $\Delta\phi$ commensurate with minimum insertion loss.

12.11.3 Loaded Line Phase Shifter [116–121]

The loaded line phase shifter makes use of a symmetric pair of switchable susceptances (or reactances) spaced at quarter-wavelength in the main transmission line. These susceptances can be realized in different ways. Figure 12.83 shows the stub loaded and switchable stub loaded configurations and also their equivalent circuit representation. Let jB_1 and jB_2 denote the susceptances presented by the shunt stub at the main line in the forward and reverse bias states of the p–i–n diode. The analysis of the circuit can be easily carried out using the *ABCD* matrix method. The differential phase shift is given by [117]

$$\Delta\phi = \cos^{-1}\left(\cos\theta - \frac{B_1}{Y_{c1}}\sin\theta\right) - \cos^{-1}\left(\cos\theta - \frac{B_2}{Y_{c1}}\sin\theta\right) \tag{12.221}$$

The magnitude of the input reflection coefficient in the two bias states is given by [118].

$$|\rho_i| = \frac{\left|\dfrac{B_i}{Y_c}\cos\theta + \dfrac{1}{2Y_cY_{c1}}(Y_{c1}^2 - Y_c^2 - B_i^2)\sin\theta\right|}{\left[\left(\cos\theta - \dfrac{B_i}{Y_{c1}}\sin\theta\right)^2 + \left\{\dfrac{B_i}{Y_c}\cos\theta + \dfrac{1}{2Y_cY_{c1}}(Y_c^2 + Y_{c1}^2 - B_i^2)\sin\theta\right\}^2\right]^{1/2}}$$

$$\text{for } i = 1, 2 \tag{12.222}$$

Choosing $\theta = \pi/2$ at the centre frequency provides for cancellation of the reflected signals at the input. Setting $|\rho_i| = 0$ $(i = 1, 2)$ and using (12.221), closed form expressions can be obtained for B_1, B_2 and Y_{c1} in the terms of θ, Y_c and the desired phase shift $\Delta\phi$. They are [120]

$$B_1 = Y_c\,[\sec(\Delta\phi/2)\cos\theta + \tan(\Delta\phi/2)] \tag{12.223a}$$

$$B_2 = Y_c\,[\sec(\Delta\phi/2)\cos\theta - \tan(\Delta\phi/2)] \tag{12.223b}$$

$$Y_{c1} = Y_c \sec(\Delta\phi/2)\sin\theta \tag{12.223c}$$

We can now relate B_1 and B_2 to the stub parameters and the diode impedances depending on the type of loading circuit. Bahl and Gupta [118] have reported the design parameters

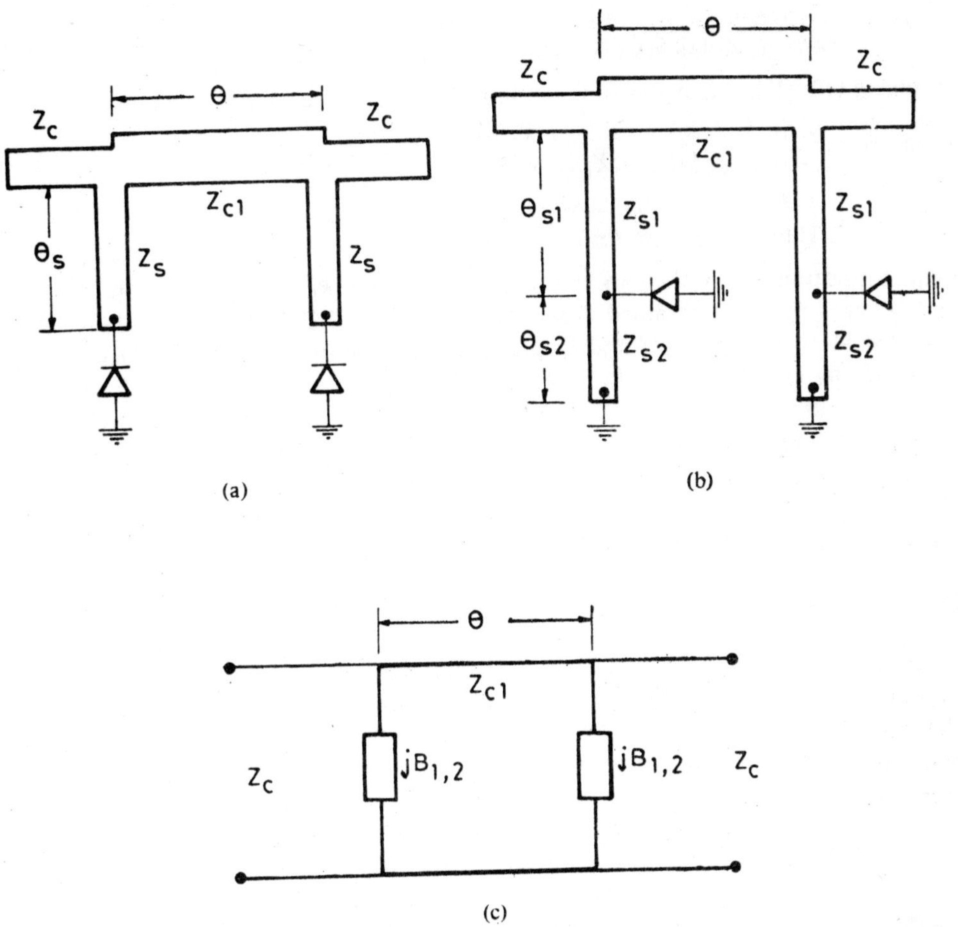

Fig. 12.83 Loaded line phase shifter bits, (a) Stub-mounted type, (b) Switchable stub length type, (c) Equivalent circuit of (a) and (b)

for stub-mounted loaded line phase shifters and have pointed out that the stub-loaded configuration (Fig. 12.83a) is suitable for nonideal diodes whereas the switchable stub loaded configuration (Fig. 12.83b) is suitable for nearly ideal diodes. The design formulas for the stub loaded circuit (Fig. 12.83a) are given below [118].

If Z_{d1} $(= R_f + jX_f)$ and Z_{d2} $(= R_r + jX_r)$ denote the forward and reverse bias impedances of the diode, the admittance Y_i of the stub at the main line is given by

$$Y_i = \frac{(Z_s + jZ_{di} \tan \theta_s}{Z_s(Z_{di} + jZ_s \tan \theta_s)} \quad \text{for} \quad i = 1, 2 \tag{12.224}$$

If we assume low-loss diodes so that $R_f \simeq 0$ and $R_r \simeq 0$, then the real part of Y_i can be neglected in comparison with its imaginary part. Setting $Y_i \simeq jB_i$ in (12.224), we get

$$B_1 = \frac{X_f \tan \theta_s - Z_s}{Z_s(X_f + Z_s \tan \theta_s)} \tag{12.225a}$$

$$B_2 = \frac{X_r \tan \theta_s - Z_s}{Z_s(X_r + Z_s \tan \theta_s)} \tag{12.225b}$$

Rearranging (12.225), the expressions for $\tan \theta_s$ and Z_s of the stub are obtained as

$$\tan \theta_s = \frac{Z_s(1 + X_f B_1)}{(X_f - Z_s^2 B_1)} = \frac{Z_s(1 + X_r B_2)}{(X_r - Z_s^2 B_2)} \tag{12.226}$$

$$Z_s = \left[\frac{(X_f - X_r) - X_f X_r(B_1 - B_2)}{(B_1 - B_2) - B_1 B_2(X_f - X_r)}\right]^{1/2} \tag{12.227}$$

For equal insertion loss in the two bias states, we choose $B_1 = -B_2$. The stub parameters θ_s and Z_s can then be obtained from (12.226) and (12.227). The above design formulas assume that $R_f = R_r = 0$. Bahl and Gupta [118] have reported that the values of $\Delta\phi$ and P_i calculated by assuming $R_f = R_r = 2$ ohms differs from the corresponding values for the lossless diode case by less than 2%. The conditions $\theta = 90°$ and $B_1 = -B_2$ provide for maximum bandwidth of the phase shifter. For effective cancellation of the reflected signals at the input over the desired bandwidth, the normalized susceptance must be small compared to unity. The loaded line phase shifter is therefore useful for small phase bits of size upto 45°. Typical bandwidth is about 20% for a 22.5° phase bit and 10% for a 45° bit.

12.12 Mixer

Mixer is an essential component of almost all receivers used in communication, radar and radio astronomy applications. Its primary function is to convert the frequency of an incoming RF signal to a lower frequency known as the IF so that the subsequent operations such as amplification, detection etc. can be performed more effectively. Microwave mixers make use of nonlinear semiconductor devices, usually Schottky barrier diodes for mixing operation. A typical mixer consists of a nonlinear mixer diode together with coupling networks for feeding the signal (RF) and local oscillator (LO) power and for extracting the IF signal. Practical mixer configurations can be broadly divided into three categories. They are, single ended mixers [122, 123], balanced mixers [124–129], and double-balanced mixers [130, 131]. A detailed account of various types of mixer diodes and mixer configurations is available in the book by Maas [132]. Of the various types of mixers, the balanced mixer employing Schottky barrier diodes is the most commonly used configuration in practical application. In the following, we first discuss the principle of mixer action and then describe typical balanced mixer circuits.

12.12.1 Principle of Mixer Action

The most commonly used nonlinear device for mixing operation is the Schottky barrier diode. The Schottky diode makes use of a rectifying metal semiconductor junction. In the forward biased state, the current flow is due to the injection of majority carriers from the semiconductor into the metal, and minority carrier effects are virtually eliminated. Being a majority carrier device, it is free from recombination time limitations and of charge storage capacitance. This is a distinct advantage over the PN-junction diode in which the switching time is limited by the charge storage effects. Schottky barrier diodes make use of either Si or GaAs as the basic semiconductor material. GaAs has higher mobility and saturation velocity as compared with Si. GaAs is therefore preferred over Si for mixer diodes required to operate at millimeter wave frequencies. Besides Schottky barrier diodes, GaAs MESFETs have also been used in millimeter wave mixers. In a MESFET, the gate forms a Schottky barrier with the semiconductor, whereas the drain and source are ohmic contacts. GaAs MESFETs are ideally suited for monolithic microwave integrated circuits.

Figure 12.84 shows a general equivalent circuit of a mixer diode. The diode is represented by a time varying non-linear resistance $R(t)$ in parallel with a junction capacitance C_j. Although C_j also varies with the time varying applied voltage, the current through it contributes negligibly to mixing action. Hence, for practical design of mixer circuits, C_j can be treated as a fixed capacitance. The other parameters R_s, L_s and C_p in the equivalent circuit represent the series resistance, series inductance and package capacitance respectively.

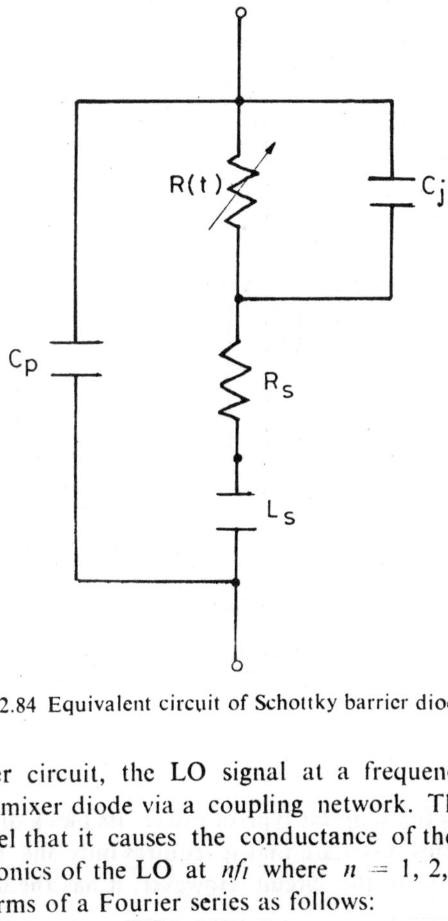

Fig. 12.84 Equivalent circuit of Schottky barrier diode

In a conventional mixer circuit, the LO signal at a frequency f_l and a RF signal at a frequency f_s are fed to the mixer diode via a coupling network. The LO signal is fed at a sufficiently high power level that it causes the conductance of the diode to vary with time. The device generates harmonics of the LO at nf_l where $n = 1, 2, 3, \ldots$ The conductance $G(t)$ can be expanded in terms of a Fourier series as follows:

$$G(t) = g_0 + g_1 \cos \omega_l t + g_2 \cos 2\omega_l t + \ldots \tag{12.228}$$

where $\omega_l = 2\pi f_l$. If a RF signal which is at a low level as compared with the LO signal is also applied to the diode, currents are generated at various mixed product frequencies. The diode current can be represented as

$$i(t) = G(t)\, V_0 \cos(\omega_s t) \tag{12.229}$$

where $\omega_s = 2\pi f_s$ and V_0 is a constant. Substituting for $G(t)$ from (12.228) in (12.229), wc can write

$$i(t) = V_0 \Big[g_0 \cos \omega_s t + \frac{g_1}{2} \{\cos (\omega_s + \omega_l)t + \cos (\omega_s - \omega_l)t\} + \frac{g_2}{2} \{\cos (2\omega_s + \omega_l)t + \cos (2\omega_s - \omega_l)t\} + \ldots \Big] \quad (12.230)$$

Figure 12.85 shows the spectrum of frequencies generated at the diode. The frequency of the desired IF is $f_{IF} = (f_s - f_l)$. This frequency is extracted from the mixer circuit via a low-pass filter.

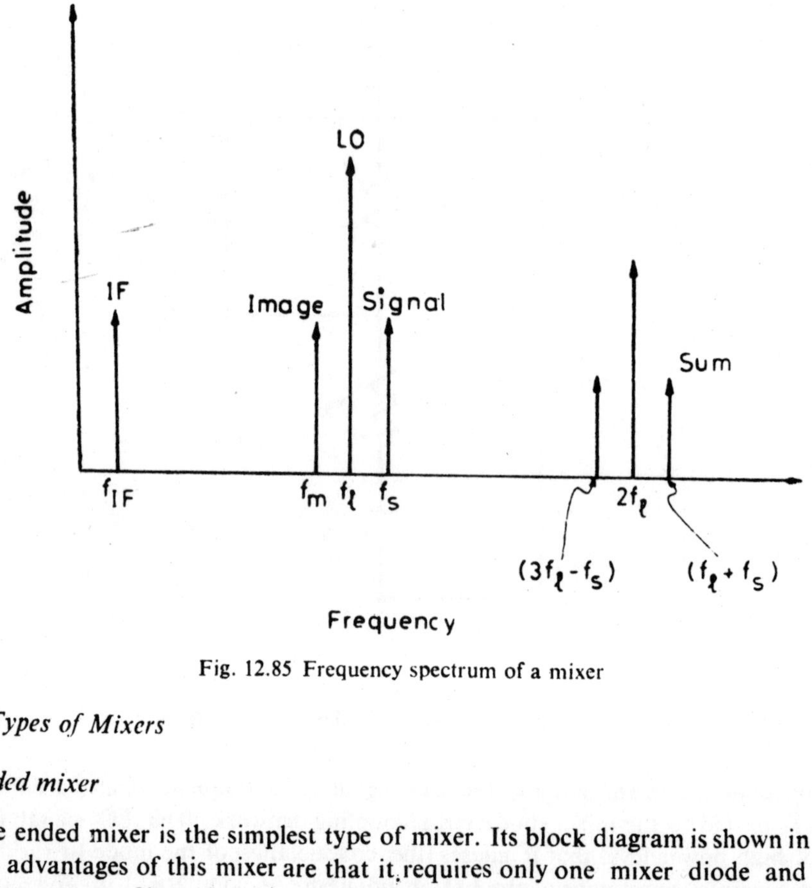

Fig. 12.85 Frequency spectrum of a mixer

12.12.2 *Types of Mixers*

Single-ended mixer

The single ended mixer is the simplest type of mixer. Its block diagram is shown in Fig. 12.86. The main advantages of this mixer are that it requires only one mixer diode and it is very easy to incorporate filters in the circuit. However, it has the disadvantage that there is no

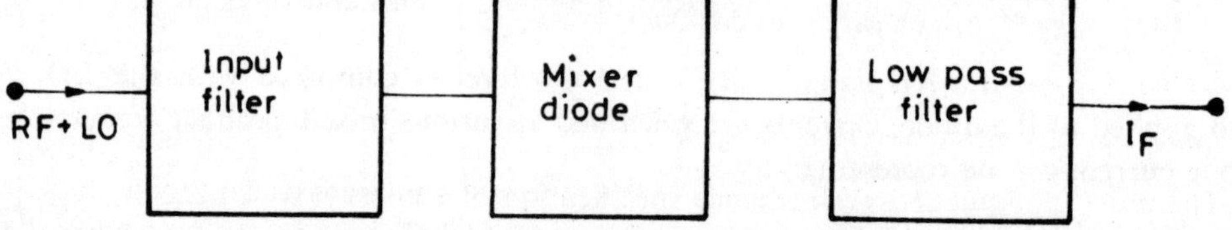

Fig. 12.86 Block diagram of single ended mixer

isolation between the LO port and the RF port. Secondly, any amplitude modulation noise side bands of the LO can mix with the IF thereby degrading the noise figure of the mixer.

Balanced mixer

The balanced mixer makes use of two identical mixer diodes in a balanced configuration as shown in Fig. 12.87. The RF and LO signals are fed to two different input ports of the coupling network and are isolated from each other even without the use of input filters. The balanced circuit arrangement enables suppression of LO noise and rejection of certain spurious responses.

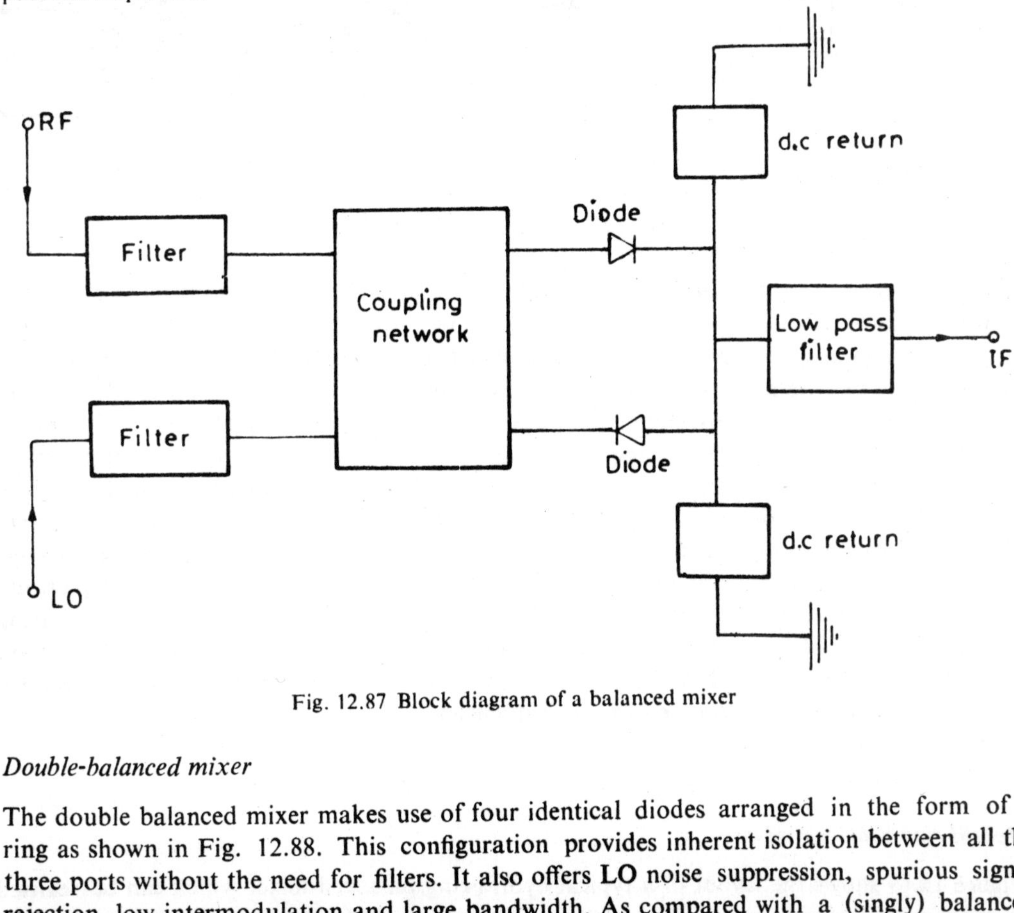

Fig. 12.87 **Block diagram of a balanced mixer**

Double-balanced mixer

The double balanced mixer makes use of four identical diodes arranged in the form of a ring as shown in Fig. 12.88. This configuration provides inherent isolation between all the three ports without the need for filters. It also offers LO noise suppression, spurious signal rejection, low intermodulation and large bandwidth. As compared with a (singly) balanced mixer, the double balanced mixer requires twice the number of diodes and higher LO power. In a (singly) balanced mixer, it is easier to apply d.c. bias and tune the diodes to minimize the conversion loss.

Image recovery in a mixer

The most fundamental performance specification of a mixer is its conversion loss. It is defined as the ratio of the signal power to the available IF output power. A typical conversion

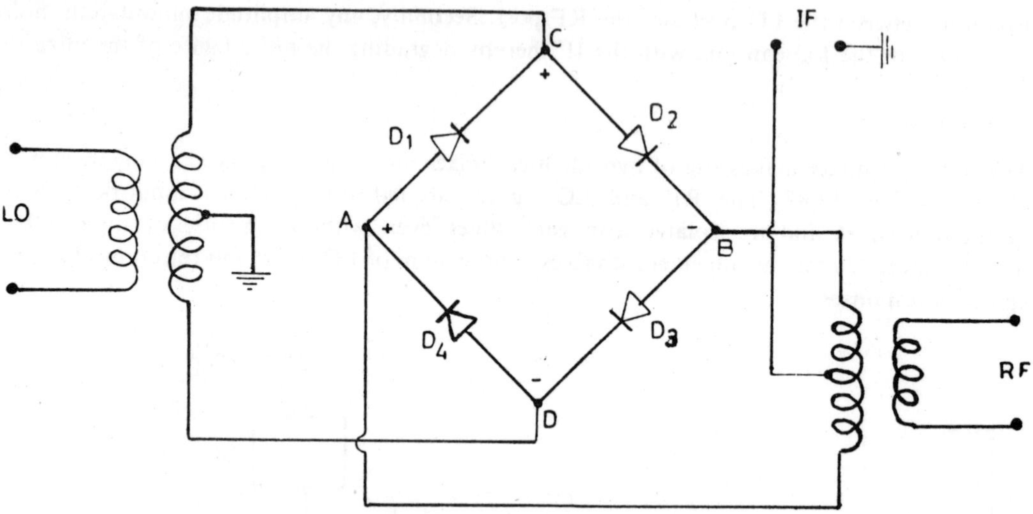

Fig. 12.88 Schematic diagram of a double balanced mixer

loss for a mixer lies between 5 and 7 *dB*. The conversion loss is influenced by several factors—namely, the properties of the mixer diode, parasitic mixer products, impedance mismatches and circuit losses. The conversion efficiency of a practical mixer can be improved by recovering the energy that is converted to unwanted parasitic frequencies. Of the various parasitic frequencies, the most important ones are, the image frequency ($2f_l - f_s$) and the sum frequency ($f_s + f_l$). These frequencies can be reactively terminated with appropriate phase by the use of band-stop filters so that remixing takes place at the diode to recover additional IF power. Practical mixers have been reported which use bandstop filters to reactively terminate only the image frequency [125]. These are called image recovery mixers and are known to offer conversion loss of the order of 4 to 5 *dB* at microwave frequencies. Neuf [133] has reported another type of image recovery mixer which uses a phasing technique instead of filters. Low noise mixers with both image and sum enhancement have also been reported [122-124]. Such mixers generally use separate bandstop filters for reactively terminating the sum and image frequencies.

12.12.3 Balanced Mixers

Hybrid-ring balanced mixers

Figure 12.89 shows the layout of a typical hybrid-ring balanced mixer which can be realised in stripline-like configuration. The circuit consists of a 3 *dB*,180° rat-race hybrid, a matched pair of mixer diodes, two matching transformers, two band-stop filters, two low-pass filters and a d.c. return.

The RF power fed to port 1 and the LO power fed to port 3 are split equally between the output ports 2 and 4 of the hybrid ring. At the two mixer diodes situated in arms 2 and 4, the RF signals appear 180° out-of-phase with each other whereas the LO signals appear in phase. As described in Sec. 12.7.2, for a 3 *dB*, 180° hybrid, the ring is of uniform characteristic impedance equal to $\sqrt{2}\,Z_c$ where Z_c is the characteristic impedance of the connect-

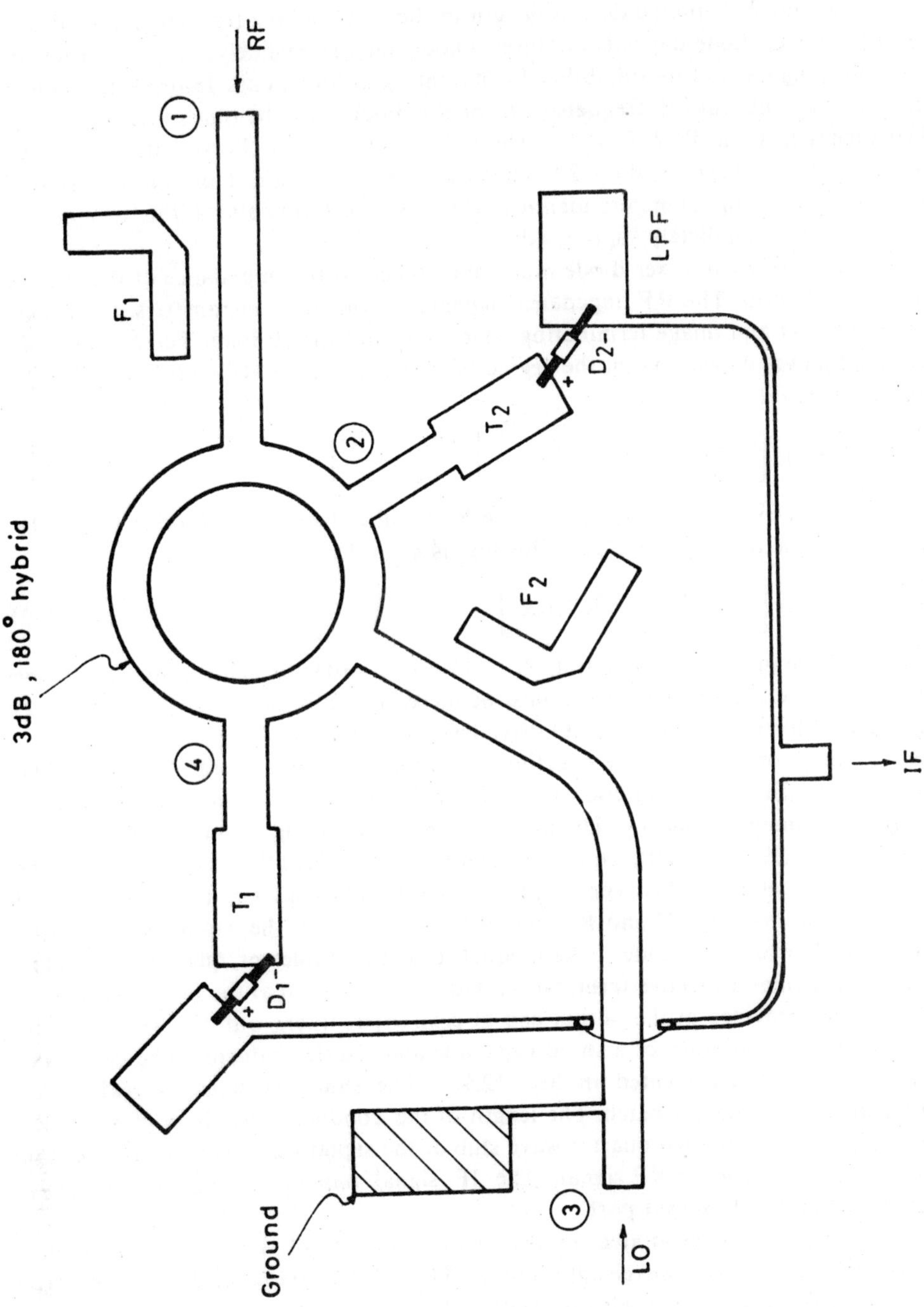

Fig. 12.89 **Hybrid ring balanced mixer,** D_1, D_2: mixer diodes, F_1, F_2: image reject filters, T_1, T_2: matching transformers

ing arms. The mean circumference of the ring is $3\lambda_0/2$ at the centre frequency f_0, where $f_0 = (f_s + f_i)/2$ and λ_0 is the guide wavelength in the stripline-like transmission medium. The choice of the mixer diode depends on three major factors—namely, the frequency of operation, its noise figure and its suitability for mounting in the circuit. In order to achieve low conversion loss, the cut-off frequency, f_c, of the diode must be much higher than the RF signal frequency f_s (typically $f_c/f_s > 10$, where $f_c = 1/(2\pi R_s C_j)$). Furthermore, the two diodes (shown as D_1 and D_2 in Fig. 12.89) must be a matched pair. This is to ensure that the converted IF noise components are identical at the output terminals of the two diodes so that the LO noise is completely suppressed.

The RF impedance of each mixer diode must be matched to the impedance of the output arm of the rat-race hybrid. The RF impedance depends on the diode parameters, frequency, LO drive level, the LO and image terminating conditions and the IF load. There is a certain loss of RF signal power depending on the degree of mismatch at the RF and IF ports. This loss is expressed as [134]

$$\alpha_1(dB) = 10 \log_{10}\left[\frac{(S_1 + 1)^2}{4S_1}\right] + 10 \log_{10}\left[\frac{(S_2 + 1)^2}{4S_2}\right] \tag{12.231}$$

where S_1 is the RF VSWR and S_2 is the IF VSWR. Another factor contributing to the RF signal loss is due to the diode parameters. This loss is given by [134]

$$\alpha_2(dB) = 10 \log_{10}\left[1 + \frac{R_s}{R_j} + \omega^2 C_j^2 R_s R_j\right] \tag{12.232}$$

The value of α_2 can be minimized by letting $R_j = 1/\omega C_j$. In order to achieve this condition at a specified frequency, the LO drive level must be increased until the value of R_j reaches the required value. This is the optimum LO drive level for a given RF signal.

The mixer circuit incorporates image rejection filters (marked as F_1 and F_2) in the RF and LO input arms. The image rejection filter is essentially a bandstop filter with the stopband centered at the image frequency. The filter is formed by having a coupled line section (as shown in Fig. 12.89) of length equal to one-quarter wavelength in the medium at the image frequency. The details of this type of filter are given in Sec. 12.9.1. The bandwidth of this filter is narrow so that the LO and RF signals lie well outside the stopband. The distance between the filter and the diodes is kept equal to odd-multiples of quarter-wavelength so that the diode provides a reactive termination with a short circuit at the image frequency.

The low-pass filter provided at the output end of each diode filters out the IF signal. It is a two-section filter consisting of a shunt capacitor and a series inductor. The design details of this type of filter are presented in Sec. 12.9.7. The shunt capacitor is realized by means of a rectangular conductor patch. The length of the conductor patch is chosen such that it acts as an open circuited quarter-wave stub at the signal frequency. Thus, the stub presents an effective short for the RF signal. The IF signal passing through the low-pass filters are combined at the IF output port.

The d.c. bypass circuit incorporated in the circuit consists of a short-circuited high impedance stub which is quarter-wavelength long at RF and LO frequencies. Besides providing a d.c. return path, this circuit also presents a short circuit at the sum frequency and the second harmonic of RF and LO frequencies. The characteristic impedance of the stub should be about 100 ohms or even more so as to provide a low frequency isolation of the order of 40 *dB* at the diode terminals.

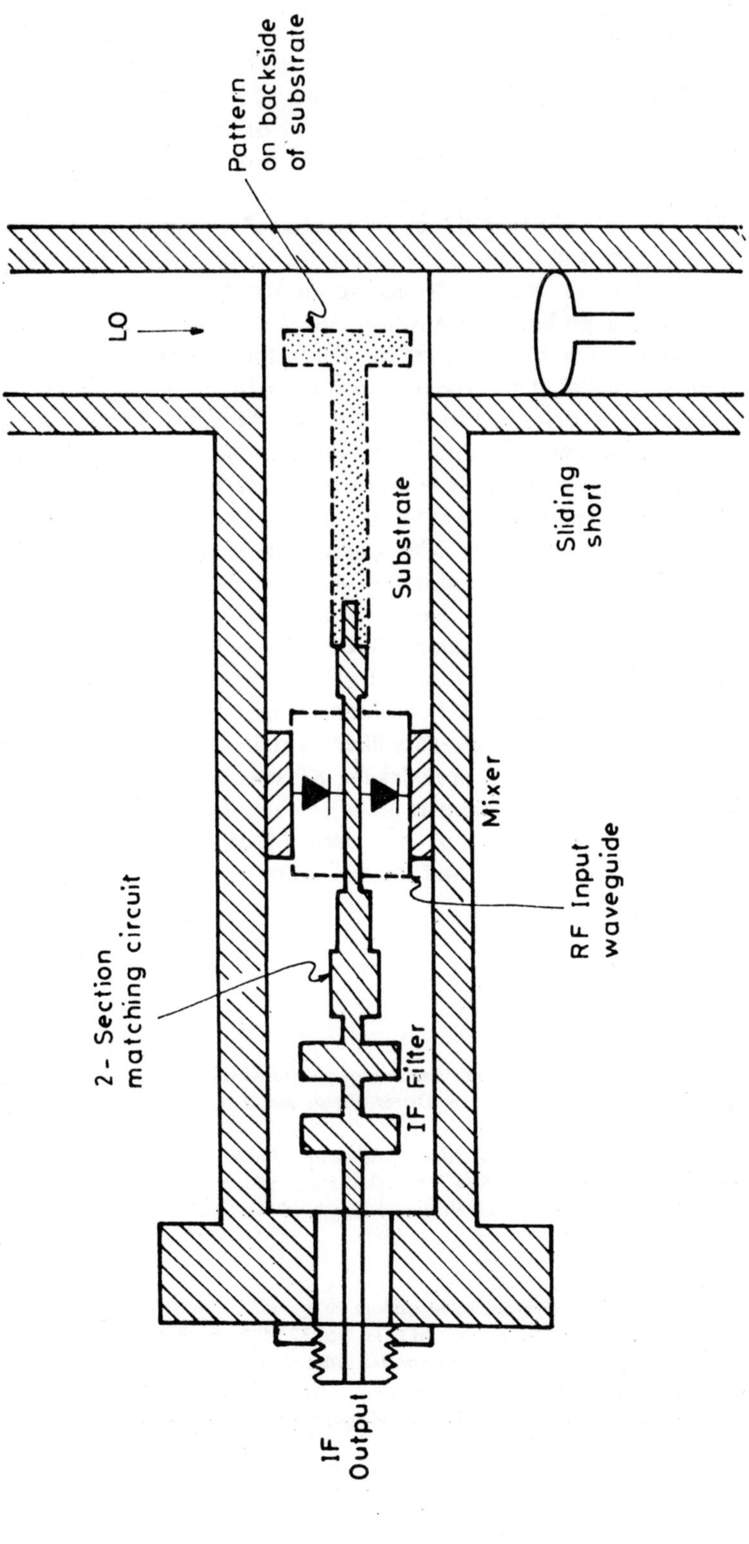

Fig. 12.90 Crossbar Suspended stripline Mixer, (From Tahim et al [128], copyright © 1983 IEEE, reprinted with permission)

Cross-bar suspended stripline mixer

The cross-bar type of mixer in suspended stripline is reported to be particularly useful at millimeter-wave frequencies [127-129, 135]. Figure 12.90 shows the view of a mixer circuit as reported by Tahim et al [128]. It consists of a pair of beam lead diodes, a low-pass filter, matching network, broadside coupler and an electric probe type transition. The RF signal is fed to the mixer diodes by means of a rectangular waveguide which is oriented perpendicular to the circuit board. The two diodes are connected in opposite polarity between the center conductor and the side walls of the suspended stripline such that they are in series across the broadwall of the waveguide feeding the RF signal. The LO signal is coupled to the diodes via an electric probe type transition and a broadside coupler. The transition is formed by extending the substrate carrying the printed probe into the rectangular waveguide carrying the LO signal. This type of transition between a suspended stripline and a rectangular waveguide is described in Sec. 12.3. The RF and LO input waveguides are oriented orthogonal to each other, thus providing inherent isolation between the two signals. The broadside coupled section situated between the diode junction and the probe presents an open circuit to the IF and also serves as a d.c. block. The design of parallel-coupled directional couplers is covered in Sec. 12.6. The coupled line section is quarter wavelength long including correction due to the discontinuities. As an alternative to the broadside coupler, Low [135] has used an end-coupled bandpass filter, thus enabling the entire circuit of the cross-bar mixer to be printed on the same side of the substrate. The IF signal is extracted via a low-pass filter consisting of alternating inductive and capacitive sections, the design details of which are covered in Sec. 12.9. For IF matching, a two-section matching transformer is used between the diodes and the low-pass filter. A reduced height waveguide transformer and a backshort behind the diodes provides for RF matching. LO matching is achieved by means of a sliding short in the rectangular guide carrying LO signal. This type of mixer is reported to offer a conversion loss of less than 7.2 *dB* over a 20 GHz instantaneous IF bandwidth with LO fixed at 85 GHz. and the RF swept from 86 to 106 GHz [128].

REFERENCES

Microwave Networks:

1. R.E. Collin, *Foundations for Microwave Engineering*, McGraw-Hill, New York, 1966.
2. S. Ramo, J.R. Whinnery and T. Van Duzer, *Fields and Waves in Communication Electronics*, John Wiley & Sons, 1965.
3. G.L. Matthaei, L. Young and E.M.T. Jones, *Microwave Filters, Impedance Matching Networks and Coupling Structures*, McGraw-Hill, New York, 2nd Ed., 1980.
4. J.L. Altman, *Microwave Circuits*, Van-Nostrand Reinhold New York, 1964.

Transitions:

5. E.N. Torgow and J.W.E. Griemsmann, Miniature strip transmission line for microwave applications, *IRE Trans. Microwave Theory Tech.*, Vol. MTT-3, pp. 57-64, Feb. 1955.
6. R.N. Wild, Photoetched microwave transmission lines, *IRE Trans Microwave Theory Tech.*, Vol. MTT-3, pp. 21-30, Feb. 1955.
7. J.H. Craven, Coaxial to strip transmission line adaptor, *IRE Trans. Microwave Theory Tech.*, Vol. MTT-9, pp. 200-201, Feb. 1961.

8. H. Howe, *Stripline Circuit Design*, Artech House, Dedham, Mass. (USA), 1974.
9. R. Levy, New coaxial-to-stripline transformers using rectangular lines, *IRE Trans. Microwave Theory Tech.*, Vol. MTT-9, pp. 273-274, March 1961.
10. J.R. Pyle, Broad-band coaxial-to-stripline transitions, *IRE Trans. Microwave Theory Tech.*, Vol. MTT-12, pp. 364-365, May 1964.
11. R.H. Knerr, A new type of waveguide-to-stripline transition, *IRE Trans. Microwave Theory Tech.*, Vol. MTT-16, pp. 192-194, March 1968.
12. G. Glance and R. Trambarulo, A waveguide to suspended stripline transition, *IEEE Trans. Microwave Theory Tech.*, Vol. MTT-21, pp. 117-118, Feb. 1973.
13. D. Rubin and A.R. Hislop, Millimeter Wave coupled line filters—Design techniques for suspended substrate and microstrip, *Microwave J.*, Vol., 23, pp. 67, 69-71, 78, Oct. 1980.
14. I. Galin, New transition expands options for SSS users, *Microwaves and RF*, Vol. 22, pp. 72-73, Jan. 1983.

Basic Circuit Elements:

15. P.I. Richards, Resistor transmission-line circuits, *Proc. IRE*, Vol. 36, pp. 217-220, Feb. 1948.
16. K. Kuroda, Derivation methods of distributed constant filters from lumped constant filters, *Text for lectures at joint meeting of Kansai Branch of Institute of Elec. Comm., and of Illumin. Engrs. of Japan*, p. 32, Oct. 1952 (In Japanese).
17. R.J. Wenzel, Exact design of TEM microwave networks using quarter-wave lines, *IEEE Trans. Microwave Theory Tech.*, Vol. MTT-12, pp. 94-111, Jan. 1964.
18. E.M.T. Jones and J.T. Bolljahn, Coupled strip-transmission line filters and directional couplers, *IRE Trans. Micrawove Theory Tech.*, Vol. MTT-4, pp. 75-81, April 1956.
19. G.I. Zysman and A.K. Johnson, Coupled transmission line networks in an inhomogeneous dielectric medium, *IEEE Trans. Microwave Theory Tech.*, Vol. MTT-17, pp. 753-759, Oct. 1969.
20. R.A. Speciale, Even- and odd-mode waves for nonsymmetrical coupled lines in nonhomogeneous media, *IEEE Trans. Microwave Theory Tech.*, Vol. MTT-23, pp. 897-908, Nov. 1975.
21. J.L. Allen, Non-symmetrical coupled lines in an inhomogeneous dielectric medium, *Int. J. Electronics*, Vol. 38, No. 3, pp. 337-347, 1975.
22. V.K. Tripathi, Asymmetric coupled transmission lines in an inhomogeneous medium, *IEEE Trans. Microwave Theory Tech.*, Vol. MTT-23, pp. 734-739, Sept. 1975.
23. V.K. Tripathi, Equivalent circuits and characteristics of inhomogeneous nonsymmetrical coupled-line two-port circuits, *IEEE Trans. Microwave Theory Tech.*, Vol. MTT-25, pp. 140-142, Feb. 1977.
24. D. Lacombe and J. Cohen, Octave-band microstrip DC blocks, *IEEE Trans. Microwave Theory Tech.*, Vol. MTT-20, pp. 555-556, Aug. 1972.
25. D. Kajfez and B.S. Vidula, Design equations for symmetric microstrip DC blocks, *IEEE Trans. Microwave Theory Tech.*, Vol. MTT-28, pp. 974-981, Sept. 1980.
26. C.Y. Ho, Analysis of d.c. blocks using coupled lines, *IEEE Trans. Microwave Theory Tech.*, Vol. MTT-23, pp. 773-777, 1975.
27. V. Rizzoli, Analysis and design of *DC* blocks, *Microwave J.*, Vol., 20, pp. 109-110, June 1977.
28. B. Easter and B.S. Shivashankaran, Some results on the edge-coupled microstrip section as an impedance transformer, *IEE J. Microwaves, Opt. and Acoustics*, Vol. 3, pp. 63-66, March 1979.
29. B.M. Schiffman, A new class of broadband microwave 90° phase shifters, *IRE Trans. Microwave Theory Tech.*, Vol. MTT-6, pp. 232-237, April 1958.
30. J.L. Allen, Inhomogeneous coupled-line filters, *IEEE MTT-S Int. Microwave Symp., Digest*, pp. 159-161, May 1975.
31. M. Caulton, Lumped elements in microwave integrated circuits, *in Advances in Microwaves*, Vol. 8, by L. Young and H. Sobol (Eds.), Academic Press, New York, 1974.
32. F.E. Terman, *Radio Engineer Handbook*, McGraw-Hill, New York, 1943.
33. F.W. Grover, *Inductance Calculations*, Van Nostrand, New Jersey, 1946.
34. J.M.C. Dukes, *Printed Circuits : Their design and Application*, Macdonald, London, 1961.
35. G.D. Alley, Interdigital capacitors and their applications in lumped element microwave integrated circuits, *IEEE Trans. Microwave Theory Tech.*, Vol. MTT-18, pp. 1028-1033, Dec. 1970.

36. C.S. Aitchison, R. Davies, I.D. Higgins, S.R. Longley, B.H. Newton, J.F. Wells and J.C. Williams, Lumped circuit elements at microwave frequencies, *IEEE Trans. Microwave Theory Tech.*, Vol. MTT-19, pp. 928-937, Dec. 1971.
37. K.C. Gupta, R. Garg and R. Chadha, *Computer-aided Design of Microwave Circuits*, Artech House, Inc., Dedham, Mass, 1981.
38. J.H.C. Van Heuven and A.G. Van Nie, *Microwave integrated circuits*, Philiph Tech. Review, Vol. 32, No. 9/10/11/12, pp. 292-304, 1971.

Branch Line Directional Couplers:

39. J. Reed and G.J. Wheeler, A method of analysis of symmetrical four port networks, *IRE Trans. Microwave Theory Tech.*, Vol. MTT-4, pp. 246-252, Oct. 1956.
40. L. Young, Synchronous branch guide directional couplers for low and high power applications, *IRE Trans. Microwave Theory Tech.*, Vol. MTT-10, pp. 459-475, Nov. 1962.
41. L. Young (ed.), *Advances in Microwaves*, Vol. 1, Academic Press, New York, 1966.
42. G.P. Riblet, A directional coupler with very flat coupling, *IEEE Trans. Microwave Theory Tech.*, Vol. MTT-26, pp. 70-74, Feb. 1978.
43. L.F. Lind, Synthesis of asymmetrical branch guide directional coupler-impedance transformers, *IEEE Trans. Microwave Theory Tech.*, Vol. MTT-17, pp. 45-48, Jan. 1969.
44. R. Levy and L.F. Lind, Synthesis of symmetrical branch guide directional couplers, *IEEE Trans. Microwave Theory Tech.*, Vol. MTT-16, pp. 80-89, Feb. 1968.
45. L.F. Lind, Design equations for a 2-branch stripline directional coupler/impedance tranformer, *Electron. Lett.*, Vol. 5, No. 20, pp. 495-496, Oct. 1969.
46. C.Y. Ho, Transform impedance with a branch-line coupler, *Microwaves*, Vol. 15, p. 47, 50, 52, May 1976.

Parallel Coupled Directional Couplers:

47. B.M. Oliver, Directional electromagnetic couplers, *Proc. IRE*, Vol. 42, No. 11, pp. 1686-1692, Nov. 1954.
48. E.M.T. Jones and J.T. Bolljahn, Coupled strip transmission line filters and directional couplers, *IRE Trans. Microwave Theory Tech.*, Vol. MTT-4, pp. 75-81, April 1956.
49. J.K. Shimizu and E.M.T. Jones, Coupled transmission line directional couplers, *IRE Trans. Microwave Theory Tech.*, Vol. MTT-6, pp. 403-410, Oct. 1958.
50. L. Young, The analytical equivalence of TEM-mode directional couplers and transmission line stepped impedance filters, *Proc. IEE (Lond)*, Vol. 110, pp. 275-281, Feb. 1963.
51. E.G. Cristal and L. Young, Theory and tables of optimum symmetrical TEM mode coupled transmission line directional couplers, *IEEE Trans. Microwave Theory Tech.*, Vol. MTT-13, pp. 544-548, Sept. 1965.
52. R. Levy, General synthesis of asymmetric multi-element coupled-transmission line directional couplers, *IRE Trans. Microwave Theory Tech.*, Vol. MTT-11, pp. 226-237, July 1963.
53. R. Levy, Tables for asymmetric multi-element coupled transmission-line directional couplers, *IEEE Trans. Microwave Theory Tech.* Vol. MTT-12, pp. 275-279, May 1964.
54. J.E. Dalley, A strip-line directional coupler utilizing a non-homogeneous dielectric medium, *IEEE Trans. Microwave Theory Tech.*, Vol. MTT-17, pp. 706-711, Sept. 1969.
55. S. Rehnmark, Meander folded coupled lines, *IEEE Trans. Microwave Theory Tech.*, Vol. MTT-26, pp. 225-231, April 1978.
56. A. Podell, A high directivity microstrip coupler technique, *IEEE G-MTT Int. Microwave Symp.*, Digest pp. 33-36, May 1970.
57. D.G. Gunton and E.G.S. Paige, Directional coupler for gigahertz frequencies, based on the coupling properties of two planar comb transmission lines, *Electron. Lett.*, Vol. 11, No. 17. pp., Aug. 1975.
58. H.E. Brenner, Perturbations of the critical parameters of quarter-wave directional couplers, *IEEE Trans. Microwave Theory Tech.*, Vol. MTT-15, pp. 384-385, June 1967.

Hybrid Rings and Baluns:

59. C.Y. Pon, Hybrid ring directional coupler for arbitrary power divisions. *IRE Trans. Microwave Theory Tech.*; Vol. MTT-9, pp. 529-535, Nov. 1961.
60. C.Y. Pon, Control your power split using a hybrid rat race, *Microwaves*, Vol. 9, pp. 34-37, April 1970.
61. S. March, A wideband stripline hybrid ring, *IEEE Trans. Microwave Theory Tech.*, Vol. MTT-16, pp. 361, June 1968.
62. H. Bex, New broadband balun, Electron. Lett., Vol. 11, No. 2, pp. 47-48, Jan. 1975.
63. S. Walker, Broadband stripline balun using quadrature couplers, *IEEE Trans. Microwave Theory Tech.*, Vol. MTT-16, pp. 132-133, Feb. 1968.

Power Dividers:

64. E.J. Wilkinson, An *N*-way hybrid power divider, *IRE Trans. Microwave Theory Tech.*, Vol. MTT-8, pp. 116-118, Jan. 1960.
65. J.J. Taub. and G.P. Kurpis, A more general *N*-way hybrid power divider, *IEEE Trans. Microwave Theory Tech.*, Vol. MTT-17, pp. 406-408, July 1969.
66. H.Y. Yee, F.C. Chang and N.F. Audeh, *N*-way TEM-mode broadband power dividers, *IEEE Trans. Microwave Theory Tech.*, Vol. MTT-18, pp. 682-688, Oct. 1970.
67. A.A.M. Saleh, Planar, electrically symmetric, *N*-way, hybrid power dividers/combiners, *IEEE Trans. Microwave Theory Tech.*, Vol. MTT-28, pp. 555-563, June 1980.
68. N. Yau, J. M. Schellenberg and Y.C. Shih, A new *N*-way broadband planar power combiner/divider, *IEEE MTT-S Int. Microwave Symp.*, Digest pp. 147-149, June, 1986.
69. N. Nagai, E. Mackawa and K. Ono, New *N*-way hybrid power dividers, *IEEE Trans. Microwave Theory Tech.*, Vol. MTT-25, pp. 1008-1012, Dec. 1977.
70. A.A.M. Saleh, Computation of the frequency response of a class of symmetric *N*-way power dividers, *Bell Syst. Tech. Journal*, Vol. 59, No. 8. pp. 1493—1511, Oct. 1980.
71. L.I. Parad and R.L. Moynihan, Split-tee power divider, *IEEE Trans. Microwave Theory Tech*,, Vol. MTT-13, pp. 91-95, Jan. 1965.
72. S.B. Cohn, A class of broadband three-port TEM-mode hybrids, *IEEE Trans. Microwave Theory Tech.*, Vol. MTT-16, pp. 110-116, Feb. 1968.
73. R.B. Ekinge, A new method of synthesizing matched broadband TEM-mode three-ports, *IEEE Trans. Microwave Theory Tech.*, Vol. MTT-29, pp. 81-88, Jan. 1981.
74. R.P. Tetarenko and P.A. Goud, Broadband properties of a class of TEM-mode hybrids, *IEEE Trans. Microwave Theory Tech.*, Vol. MTT-19, pp. 887-889, Nov. 1971.
75. S.B. Cohn, Optimum design of stepped transmission line transformers, *IRE Trans. Microwave Theory and Tech.*, Vol.MTT-3, pp. 16-21, April 1955.
76. L. Young, Tables for cascaded homogeneous quarter-wave transformers, *IRE Trans. Microwave Theory Tech.*, Vol. MTT-7, pp. 233-238, April 1959 (also correction in *IRE Trans. Microwave Theory Tech.*, Vol. MTT-8, pp. 243-244, March 1960.

Filters:

77. R. Saal and E. Ulbrich, On the design of filters by synthesis, *IRE Trans. Circuit Theory*, Vol. CT-5, pp. 284-327, Dec. 1958.
78. A.I. Zverev, Handbook of Filter Synthesis, John Wiley and Sons, New York, 1967.
79. R. Saal, Der Entwurf von Filtern mit Hilfe des Kataloges Normierter Tiefpasse, Telefunken GMBH, Backnang, W. Germany, 1963.
80. R. Levy and I. Whiteley, Synthesis of distributed elliptic-function filters from lumped-constant prototypes, *IEEE Trans. Microwave Theory Tech.*, Vol. MTT-14, pp. 506-517, Nov. 1966.
81. H. Ozaki and J. Ishii, Synthesis of a class of stripline filters, *IRE Trans. on Circuit Theory*, Vol. CT-5, pp. 104-109, June 1958.
82. R.J. Wenzel, Exact design of TEM microwave networks using quarter-wave lines, *IEEE Trans. Microwave Theory Tech.*, Vol. MTT-12, pp. 94-111, Jan. 1964.

83. G.L. Matthaei, Design of wide-band (and narrow-band) bandpass microwave filters on insertion loss basis, *IRE Trans. Microwave Theory Tech.*, Vol. MTT-8, pp. 580-593, Nov. 1963.
84. E.G. Cristal, New design equations for a class of microwave filters, *IEEE Trans. Microwave Theory Tech.*, Vol. MTT-19, pp. 486-490, May 1971.
85. S.B. Cohn, Parallel-coupled transmission line resonator filters, *IRE Trans. Microwave Theory Tech.*, Vol. MTT-6, pp. 223-231, April 1958.
86. A.K. Sharma and B. Bhat, Spectral domain analysis of interacting microstrip resonant structures, *IEEE Trans. Microwave Theory Tech.*, Vol. MTT-6, pp. 681-685, Aug. 1983.
87. R. Khanna, S.K. Koul and B. Bhat, Characteristics of single and coupled rectangular resonators in suspended substrate stripline, *Electron. Lett.*, Vol. 22, No. 7, pp. 376-378, March 1986.
88. S.K. Koul and B. Bhat, Equivalent circuit of broadside-coupled microstrip open ends, *IEEE MTT-S Int. Microwave Symp.*, Digest pp. 497-498, June 1985.
89. B. Easter and C. Gupta, More accurate model of the coupled microstripline section, *IEE. J. Microwaves Optics and Acoustics*, Vol. 3, No. 3, pp. 99-103, May 1979.
90. R.H. Jansen and N.H.L. Koster, Accurate results on the end effect of single and coupled microstrip lines for use in microwave circuit design, *Arch. Elek. Ubertragung*, Vol. 34, pp. 453-459, 1980.
91. J.S. Wong, Microstrip tapped line filter design, *IEEE Trans. Microwave Theory Tech.*, Vol. MTT-27, pp. 44-50, Jan. 1979.
92. J.K. Richardson, Semiempirical formula for the end-coupled bandpass symmetric stripline filter, *Electron. Lett.*, Vol. 6, No. 19, pp. 616-618, Sept. 1970.
93. C. Nguyen and K. Chang, Design and performance of millimeter-wave end-coupled bandpass filters *Int. J. of Infrared and Millimeter Waves*, Vol. 6, No. 7, pp. 497-509, 1985.
94. J.D. Rhodes, *Theory of Electrical Filters*. John Wiley and Sons, New York, 1976.
95. L. Young, G.L. Matthaei and E.M.T. Jones, Microwave bandstop filters with narrow stopbands, *IRE Trans. Microwave Theory Tech.*, Vol. MTT-10, pp. 416-427, Nov. 1962.
96. B.M. Schiffman and G.L. Matthaei, Exact design of bandstop microwave filters, *IEEE Trans. Microwave Theory Tech.*, Vol. MTT-12, pp. 6-15, Jan. 1964.
97. R.N. Bates, Design of microstrip spur-line bandstop filters, *IEE. J. Microwaves Optics and Acoustics*, Vol. 1, No. 6, pp. 209-214, Nov. 1977.
98. R.N. Bates and R.E. Pearson, Designing bandstop filters for microwave frequencies, *Electronic Engineering*, pp. 39-41, April 1978.
99. C. Nguyen and K. Chang, Analysis and design of spurline bandstop filters, *IEEE MTT-S Int. Microwave Symp.*, Diyest pp. 445-447, June 1985.

P-I-N Diode Switches and Attenuators:

100. J.F. White *Microwave Semiconductor Engineering*, Van-Nostrand Reinhold, New York 1982.
101. R.V. Garver, *Microwave Diode Control Devices*, Artech House, Dedham, Mass (USA), 1976.
102. J.F. White and K.E. Mortenson, Diode SPDT switching at high power with octave microwave band-width, *IEEE Trans. Microwave Theory Tech.*, Vol. MTT-16, pp. 30-36, Jan. 1968.
103. R.V. Garver, Diode switching, *Microwave Syst. News*. Vol. 13, No. 12, pp. 194-204, Nov. 1983.
104. P. Chorney, Multi-octave, multi-throw PIN-diode switches, *Microwave J.*, Vol. 17, pp. 39-42, 48, Sept. 1974.
105. R.W. Waugh, Broadbanding the shunt PIN-diode SPDT switch, *Microwave J.*, Vol. 18, pp. 55-60, Feb. 1975.

Phase Shifters:

106. J.F. White, Review of semiconductor microwave phase shifters, *Proc. IEEE*, Vol. 56, pp. 1924-1931, Nov. 1968.
107. R.V. Garver, Broadband diode phase shifters, *IEEE Trans. Microwave Theory Tech.*, Vol. MTT-20, pp. 314-323, May 1972.
108. R.P. Coats, An octave-band switched-line microstrip 3-*db* diode phase shifter, *IEEE Trans. Microwave Theory Tech.*, Vol. MTT-21, pp. 444-449, July 1973.
109. J.F. White, Diode phase shifters for array antennas, *IEEE Trans. Microwave Theory Tech.*, Vol. MTT-22, pp. 658-674, June 1974.

110. R.W. Burns and L. Stark, PIN diodes advance high power shifting, *Microwaves*, Vol. 4, No. 11, pp. 38-48, Nov. 1968.
111. P. Wahi and K.C. Gupta, Effect of diode parameters on reflection type phase shifters, *IEEE Trans. Microwave Theory Tech.*, Vol. MTT-24, pp. 619-621, Sept. 1976.
112. P.J. Starski, Optimization of matching network for a hybrid coupled phase shifter, *IEEE Trans. Microwave Theory Tech.*, Vol. MTT-25, pp. 662-666, Aug. 1977.
113. S. Mahapatra and Q.H. Bakir, Computer-aided design and evaluation of MIC hybrid coupled PIN diode phase shifter, *IEE-IERE Proc.* (India), Vol. 16, pp. 19-28, Jan. 1978.
114. A.H. Atwater, Reflection coefficient transformations for phase shifting circuits, *IEEE Trans. Microwave Theory Tech.*, Vol. MTT-28, pp. 563-568, June 1980.
115. K. Watanabe, M. Arima and T. Yamamoto, Graph design of p-i-n diode phase shifter, *IEEE Trans. Microwave Theory Tech*,. Vol. MTT-29, 829-831, Aug. 1981.
116. F.L. Opp and W.F. Hoffman, Design of digital loaded line phase shift networks for microwave thin film applications, *IEEE Trans. Microwave Theory Tech.*, Vol. MTT-16, pp. 462-468, July 1986.
117. J.F. White, High power, p-i-n diode controlled, microwave transmission phase shifters, *IEEE Trans. Microwave Theory Tech.*, Vol. MTT-13, pp. 233-242, March 1965.
118. I.J. Bahl and K.C. Gupta, Design of loaded line PIN diode phase shifter circuits, *IEEE Trans. Microwave Theory Tech.*, Vol. MTT-28, pp. 219-224, March 1980.
119. T. Yahara, A note on designing digital loaded-line phase shifters, *IEEE Trans. Microwave Theory Tech.*, Vol. MTT-20, pp. 703-704, Oct. 1972.
120. W.A. Davis, Design equations and bandwidth of loaded line phase shifters, *IEEE Trans. Microwave Theory Tech.*, Vol. MTT-22, pp. 561-563, May 1974.
121. T. Yahara, Y. Kadowaki and K. Shirahata, Optimum design of digital diode-loaded line phase shifters, *Electron. Commun.* (Japan), Vol. 57B, pp. 43-52, 1974.

Mixers:

122. J.E. Degenford and B.A. Newman, Analysis and design of a low noise *X*-band mixer with 1 GHz *IF* amplifier, *IEEE Trans. Microwave Theory Tech.*, Vol. MTT-23, pp. 375-376. April 1975.
123. R.N. Bates, Microwave low noise mixers with image and sum frequency enhancement, *IERE Conf. on Radio Receivers and Associated Systems*, Conf. Digest pp. 317-322, July 1978.
124. L.E. Dickens and D.W. Maki, An integrated-circuit balanced mixer, image and sum enhanced, *IEEE Trans. Microwave Theory Tech.*, Vol. MTT-23, pp. 276-281, March 1975.
125. K.M. Johnson, *X*-band integrated circuit mixer with reactively terminated image, *IEEE Trans. Microwave Theory Tech.*, Vol. MTT-16, pp. 388-397, July 1968.
126. T.H. Oxley, Review of some microwave integrated circuit components utilizing microstrip techniques *The Radio and Electronics Engineer*, Vol. 48, No. 1/2, pp. 3-12, Jan./Feb. 1978.
127. R.S. Tahim, G.M. Hayashibara and K. Chang, Design and performance of *W*-band broadband integrated circuit mixers, *IEEE Trans. Microwave Theory and Tech.*, Vol. MTT-31, pp. 277-283, March 1983.
128. R.S. Tahim, T. Pham and K. Chang, *W*-band low-noise integrated circuit crossbar and finline mixers with over 20 GHz instantaneous *RF* bandwidth, *IEEE MTT-S Int. Microwave Symp.*, Digest pp. 290-292, 1983.
129. C. Nguyen and K. Chang, 140 GHz broadband crossbar stripline mixers with over 20 GHz instantaneous *RF* bandwidh, *Electron. Lett.*, Vol. 20, No. 11, p. 441, May 1986.
130. R.B. Mouw and S.M. Fukuchi, Broadband double balanced mixer/modulators, *Microwave J.*, Vol. 12, pp. 131-134, March 1969; also Vol. 12, pp. 71-76, May 1969.
131. R.M. Narayanan, Multi-octave, double-balanced mixers designed using MIC technology, *Microwave Syst. News*, pp. 108-115, Jan. 1987.
132. S.A. Maas, *Microwave Mixers*, Artech House Inc., 1986.
133. D. Neuf, A quiet mixer, *Microwave J.*, Vol. 16, No. 5, pp. 29, May 1973.
134. Hewlett Packard Application Note. 907, *The hot-carrier diode theory, design and application*, Sept. 1970.
135. S. Low, *W*-band Crossbar mixers integrated entirely on a single-sided substrate yielding 15 GHz instantaneous bandwidth, *IEEE MTT-S Int. Microwave Symp.*, Digest pp. 595-596, 1987.

Index